TOP 100 FOOD PLANTS

TOP 100 FOOD PLANTS

ERNEST SMALL

Principal Research Scientist
Research Branch
Agriculture and Agri-Food Canada
Ottawa, Ontario, Canada

An Imprint of NRC Research Press
Ottawa 2009

ISBN 978-0-660-19858-3
NRC No. 49730

Library and Archives Canada Cataloguing in Publication

Small, Ernest, 1940-

Top 100 food plants / Ernest Small.

Issued by: National Research Council Canada.
Includes bibliographical references and index.
ISBN 978-0-660-19858-3

1. Plants, Edible—Identification. 2. Plants, Edible—Nutrition. 3. Food crops.
I. National Research Council Canada II. Title.

QK98.5.A1 S63 2009 581.6'32 C2008-980341-8

Inquiries: Monograph Publishing Program, NRC Research Press, National Research Council of Canada, Ottawa, Ontario K1A 0R6, Canada. Web site: http://pubs.nrc-cnrc.gc.ca

Correct citation for this publication: Small, Ernest. 2009. *Top 100 food plants.* NRC Press, Ottawa, Ontario, Canada. 636 p.

CONTENTS

Common Name Guide to Culinary Plants Discussed in Detail

The 100 common names or name combinations used as chapter headings are given in boldface. The most frequently encountered synonyms are also given here. For additional synonyms, as well as names of plants mentioned incidentally in the book, see Complete Index of Common Names at the end of the book.

Scientific Name Guide to Culinary Plants Discussed in Detail

This index presents the correct scientific names for the plants discussed in detail. For scientific synonyms of these as well as for scientific names of plants mentioned incidentally in the book, see Complete Index of Scientific Names at the end of the book.

Scientific name	Chapter
Agave americana (century plant, maguey)	Tequila and Mezcal (p. 526)
Agave angustifolia (maguey lechugilla)	Tequila and Mezcal (p. 526)
Agave salmiana (pulque agave)	Tequila and Mezcal (p. 526)
Agave tequilana (blue agave, tequila)	Tequila and Mezcal (p. 526)
Allium canadense (Canada garlic)	Garlic (p. 253)
Allium cepa (common onion)	Onion (p. 363)
Allium fistulosum (Welsh onion)	Onion (p. 363)
Allium ×proliferum (Egyptian onion)	Onion (p. 363)
Allium sativum var. *ophioscorodon* (rocambole)	Garlic (p. 253)
Allium sativum var. *sativum* (common garlic)	Garlic (p. 253)
Amaranthus blitum (livid amaranth)	Amaranth (p. 49)
Amaranthus caudatus (pendant amaranth)	Amaranth (p. 49)
Amaranthus cruentus (purple amaranth)	Amaranth (p. 49)
Amaranthus dubius (spleen amaranth)	Amaranth (p. 49)
Amaranthus hypochondriacus (prince's feather)	Amaranth (p. 49)
Amaranthus tricolor (Chinese amaranth)	Amaranth (p. 49)
Ananas comosus	Pineapple (p. 407)
Arachis hypogaea	Peanut (p. 389)
Avena sativa	Oat (p. 350)
Bertholletia excelsa	Brazil nut (p. 111)
Beta vulgaris subsp. *cicla* (Swiss chard)	Beet, Sugar beet, and Swiss chard (p. 97)
Beta vulgaris subsp. *vulgaris* (beet and sugar beet)	Beet, Sugar beet, and Swiss chard (p. 97)
Brassica alboglabra (Chinese kale)	Kale and Collard (p. 284)
Brassica juncea (brown mustard, yellow mustard, mustard greens, canola)	Mustard, Mustard greens, and Spinach mustard (p. 342)
Brassica napus subsp. *napus* (Canola, Argentine rape, Swedish rape)	Rapeseed and Canola (p. 440)
Brassica napus subsp. *napus* (Siberian kale)	Kale and Collard (p. 284)
Brassica oleracea var. *botrytis*	Cauliflower (p. 146)
Brassica oleracea var. *capitata* (cabbage)	Cabbage (p. 126)
Brassica oleracea var. *costata* (Portuguese kale)	Kale and Collard (p. 284)
Brassica oleracea var. *fruticosa* (thousand-headed kale)	Kale and Collard (p. 284)

Scientific name	Chapter
Brassica oleracea var. *gemmifera*	Brussels sprouts (p. 123)
Brassica oleracea var. *italica* (broccoli)	Broccoli and Broccoli raab (p. 120)
Brassica oleracea var. *medullosa* (marrow-stem kale)	Kale and Collard (p. 284)
Brassica oleracea var. *sabauda* (Savoy cabbage)	Cabbage (p. 126)
Brassica oleracea var. *sabellica* (Scotch kale)	Kale and Collard (p. 284)
Brassica oleracea var. *viridis* (kale, collard, borecole)	Kale and Collard (p. 284)
Brassica perviridis (spinach mustard)	Mustard, Mustard greens, and Spinach mustard (p. 342)
Brassica rapa subsp. *chinensis* (pak-choi)	Cabbage (p. 126)
Brassica rapa subsp. *dichotoma* (toria, Indian rape, brown sarson)	Rapeseed and Canola (p. 440)
Brassica rapa subsp. *narinosa* (Chinese savoy)	Cabbage (p. 126)
Brassica rapa subsp. *nipposinica* (mizuna)	Cabbage (p. 126)
Brassica rapa subsp. *oleifera* (Canola, turnip rape, Polish rape)	Rapeseed and Canola (p. 440)
Brassica rapa subsp. *pekinensis* (pe-tsai)	Cabbage (p. 126)
Brassica rapa subsp. *trilocularis* (yellow sarson, Indian colza)	Rapeseed and Canola (p. 440)
Brassica ruvo (broccoli raab)	Broccoli and Broccoli raab (p. 120)
Brassica ruvo (ruvo kale)	Kale and Collard (p. 284)
Cajanus cajan	Pea, pigeon (p. 404)
Camellia sinensis	Tea (p. 514)
Capsicum annuum (chile pepper and sweet pepper)	Chile pepper and Sweet pepper (p. 157)
Capsicum baccatum var. *pendulum* (Peruvian pepper)	Chile pepper and Sweet pepper (p. 157)
Capsicum chinense (habañero pepper)	Chile pepper and Sweet pepper (p. 157)
Capsicum frutescens (Tabasco pepper)	Chile pepper and Sweet pepper (p. 157)
Capsicum pubescens (apple pepper)	Chile pepper and Sweet pepper (p. 157)
Carica papaya	Papaya (p. 377)
Carthamus tinctorius	Safflower (p. 459)
Chenopodium quinoa	Quinoa (p. 436)
Chondrus crispus (Irish moss)	Seaweeds (p. 462)
Cicer arietinum	Chickpea (Garbanzo bean) (p. 154)
Cinnamomum aromaticum (cassia)	Cinnamon and Cassia (p. 177)
Cinnamomum verum (cinnamon)	Cinnamon and Cassia (p. 177)
Citrullus lanatus	Watermelon (p. 552)
Citrus aurantifolia	Lime (p. 307)
Citrus aurantium (Seville orange)	Orange (p. 370)
Citrus bergamia (bergamot orange)	Orange (p. 370)
Citrus limon	Lemon (p. 289)
Citrus maxima (pomelo)	Grapefruit and Pomelo (p. 276)
Citrus ×paradisi (grapefruit)	Grapefruit and Pomelo (p. 276)
Citrus reticulata (mandarin orange, tangerine, Satsuma orange)	Orange (p. 370)
Citrus sinensis (sweet orange)	Orange (p. 370)
Cocos nucifera	Coconut (p. 181)
Coffea arabica (Arabian coffee)	Coffee (p. 187)

Scientific name	Chapter
Coffea canephora (robusta coffee)	Coffee (p. 187)
Coffea liberica (Liberian coffee)	Coffee (p. 187)
Cola acuminata (abata cola)	Cola (p. 199)
Cola nitida (bitter cola)	Cola (p. 199)
Colocasia esculenta	Taro (p. 510)
Corylus americana (American hazelnut)	Filbert and Hazelnut (p. 244)
Corylus avellana (European filbert)	Filbert and Hazelnut (p. 244)
Corylus colurna (Turkish filbert)	Filbert and Hazelnut (p. 244)
Corylus cornuta (beaked hazelnut)	Filbert and Hazelnut (p. 244)
Corylus maxima (giant filbert)	Filbert and Hazelnut (p. 244)
Cucumis anguria (West Indian gherkin)	Cucumber (p. 219)
Cucumis melo	Melon (p. 327)
Cucumis melo var. *conomon* (Chinese cucumber)	Cucumber (p. 219)
Cucumis sativus (cucumber)	Cucumber (p. 219)
Cucurbita argyrosperma (pumpkin, winter squash, cushaw, cymlin)	Pumpkin and Squash (p. 429)
Cucurbita ficifolia (Malabar gourd)	Pumpkin and Squash (p. 429)
Cucurbita foetidissima (buffalo gourd)	Pumpkin and Squash (p. 429)
Cucurbita maxima (pumpkin, squash, marrow)	Pumpkin and Squash (p. 429)
Cucurbita moschata (winter squash, pumpkin)	Pumpkin and Squash (p. 429)
Cucurbita pepo (pumpkin, squash, marrow)	Pumpkin and Squash (p. 429)
Cynara scolymus	Artichoke (Globe artichoke) (p. 72)
Daucus carota	Carrot (p. 137)
Digitaria exilis (white fonio)	Fonio (p. 250)
Digitaria iburua (black fonio)	Fonio (p. 250)
Dioscorea alata (winged yam)	Yam (p. 564)
Dioscorea batatas (Chinese potato)	Yam (p. 564)
Dioscorea bulbifera (air potato)	Yam (p. 564)
Dioscorea ×cayenensis (yellow Guinea yam)	Yam (p. 564)
Dioscorea elephantipes (elephant's foot)	Yam (p. 564)
Dioscorea esculenta (Asiatic yam)	Yam (p. 564)
Dioscorea rotundata (white Guinea yam)	Yam (p. 564)
Dioscorea trifida (cush-cush)	Yam (p. 564)
Echinochloa frumentacea (Japanese millet)	Millet (p. 332)
Elaeis guineensis	Oil palm (p. 354)
Elettaria cardamomum	Cardamom (p. 133)
Eleusine coracana (finger millet)	Millet (p. 332)
Eragrostis tef	Teff (p. 522)
Ficus carica	Fig (p. 239)
Fragaria ×ananassa	Strawberry (p. 491)
Glycine max	Soybean (p. 481)
Gossypium arboreum (tree cotton)	Cotton (p. 214)

Scientific name	Chapter
Gossypium barbadense (sea island cotton)	Cotton (p. 214)
Gossypium herbaceum (Levant cotton)	Cotton (p. 214)
Gossypium hirsutum (American upland cotton)	Cotton (p. 214)
Helianthus annuus	Sunflower (p. 502)
Hordeum vulgare	Barley (p. 87)
Ilex paraguariensis	Maté (p. 323)
Illicium verum (star anise)	Anise and Star anise (p. 57)
Ipomoea batatas	Sweet potato (p. 506)
Juglans ailantifolia (Japanese walnut)	Walnut (p. 545)
Juglans ailantifolia var. *cordiformis* (heartnut)	Walnut (p. 545)
Juglans nigra (black walnut)	Walnut (p. 545)
Juglans regia (Persian walnut)	Walnut (p. 545)
Lablab purpureus	Hyacinth bean (p. 281)
Lactuca sativa	Lettuce (p. 298)
Lagenaria siceraria	Bottle gourd (p. 107)
Laminaria (kelp)	Seaweeds (p. 462)
Lens culinaris	Lentil (p. 294)
Lupinus albus (white lupin)	Lupin (p. 311)
Lupinus angustifolius (blue lupin)	Lupin (p. 311)
Lupinus luteus (yellow lupin)	Lupin (p. 311)
Lupinus mutabilis (Andean lupin)	Lupin (p. 311)
Lycopersicon esculentum (tomato)	Tomato (p. 533)
Lycopersicon pimpinellifolium (currant tomato)	Tomato (p. 533)
Malus domestica	Apple (p. 61)
Mangifera indica	Mango (p. 319)
Manihot esculenta	Cassava (p. 142)
Musa ×paradisiaca	Banana and Plantain (p. 81)
Olea europaea	Olive (p. 357)
Oryza glaberrima (African rice, red rice)	Rice (p. 444)
Oryza sativa (Asian rice)	Rice (p. 444)
Palmaria palmata (dulse)	Seaweeds (p. 462)
Panicum miliaceum (proso millet)	Millet (p. 332)
Pennisetum glaucum (pearl millet)	Millet (p. 332)
Persea americana var. *americana* (West Indian avocado)	Avocado (p. 76)
Persea americana var. *drymifolia* (Mexican avocado)	Avocado (p. 76)
Persea americana var. *nubigena* (Guatemalan avocado)	Avocado (p. 76)
Phaseolus lunatus	Lima bean (p. 303)
Phaseolus vulgaris	Bean (Common bean) (p. 91)
Phoenix dactylifera	Date (p. 230)
Pimpinella anisum (anise)	Anise and Star anise (p. 54)
Piper nigrum	Pepper (p. 400)
Pistacia vera	Pistachio (p. 412)

Scientific name	Chapter
Pisum sativum	Pea (p. 381)
Porphyra (laver, nori)	Seaweeds (p. 462)
Prunus americana (American plum)	Plum (p. 415)
Prunus angustifolia (Chicasaw plum)	Plum (p. 415)
Prunus armeniaca	Apricot (p. 68)
Prunus avium (sweet cherry)	Cherry (p. 149)
Prunus cerasifera (cherry plum, myrobalan plum)	Plum (p. 415)
Prunus cerasus (sour cherry)	Cherry (p. 149)
Prunus domestica (common plum)	Plum (p. 415)
Prunus dulcis	Almond (p. 44)
Prunus hortulana (wild-goose plum)	Plum (p. 415)
Prunus insititia (damson plum)	Plum (p. 415)
Prunus maritima (beach plum)	Plum (p. 415)
Prunus munsoniana (munson plum)	Plum (p. 415)
Prunus nigra (Canada plum)	Plum (p. 415)
Prunus persica	Peach and Nectarine (p. 385)
Prunus salicinia (Japanese plum)	Plum (p. 415)
Prunus simonii (apricot plum)	Plum (p. 415)
Prunus subcordata (Pacific plum)	Plum (p. 415)
Pyrus communis (common pear)	Pear (p. 395)
Pyrus pyrifolia (Asian pear)	Pear (p. 395)
Ribes nigrum (black currant)	Currant (p. 224)
Ribes odoratum (golden currant)	Currant (p. 224)
Ribes petraeum (rock currant)	Currant (p. 224)
Ribes rubrum (red currant)	Currant (p. 224)
Ribes rubrum × *R. spicatum* (white currant)	Currant (p. 224)
Ribes spicatum (northern red currant)	Currant (p. 224)
Saccharum officinarum	Sugar cane (p. 496)
Secale cereale (rye)	Rye and Triticale (p. 454)
Sesamum indicum	Sesame (p. 471)
Setaria italica (foxtail millet)	Millet (p. 332)
Sinapis alba (white mustard, yellow mustard)	Mustard, Mustard greens, and Spinach mustard (p. 342)
Solanum melongena	Eggplant (p. 234)
Solanum tuberosum	Potato (p. 419)
Sorghum bicolor	Sorghum (p. 477)
Spinacia oleracea	Spinach (p. 487)
Theobroma cacao	Chocolate and Cacao (p. 167)
×*Triticosecale* (triticale)	Rye and Triticale (p. 454)
Triticum aestivum subsp. *aestivum* (common wheat, bread wheat)	Wheat (p. 556)
Triticum aestivum subsp. *compactum* (club wheat)	Wheat (p. 556)
Triticum dicoccon (emmer)	Wheat (p. 556)

Scientific name	Chapter
Triticum durum (durum wheat, macaroni wheat)	Wheat (p. 556)
Triticum monococcum (einkorn)	Wheat (p. 556)
Triticum polonicum (Polish wheat)	Wheat (p. 556)
Triticum spelta (spelt)	Wheat (p. 556)
Triticum turgidum (poulard wheat)	Wheat (p. 556)
Ulva (laver)	Seaweeds (p. 462)
Ulva lactuca (sea lettuce)	Seaweeds (p. 462)
Undaria (kelp)	Seaweeds (p. 462)
Vanilla planifolia	Vanilla (p. 541)
Vicia faba	Broad bean (p. 116)
Vigna angularis	Adzuki bean (p. 41)
Vigna radiata	Mung bean (p. 338)
Vigna unguiculata subsp. *unguiculata*	Black-eyed pea (Cowpea) (p. 104)
Vitellaria paradoxa	Shea nut (p. 474)
Vitis labrusca (fox grape)	Grape (p. 259)
Vitis rotundifolia (muscadine grape)	Grape (p. 259)
Vitis vinifera (wine grape)	Grape (p. 259)
Xanthosoma sagittifolium	Malanga (p. 315)
Zea mays	Corn (maize) (p. 204)

ABSTRACT

This monograph reviews scientific and technological information about the world's major food plants and their culinary uses. An introductory chapter discusses nutritional and other fundamental scientific aspects of plant foods. Most of the book is made up of 100 chapters, each dealing with a particular species or group of species. Detailed information is presented on the most important food crops, particularly those commonly encountered in the Western World. All categories of food plants are covered, including cereals, oilseeds, fruits, nuts, vegetables, legumes, herbs, spices, beverage plants, and sources of industrial food extracts. A user-friendly standard format is employed. The initial section entitled "Names" provides extensive information on scientific and English common names of the plants. Next is a section called "Plant Portrait," which comprises a description of the plant, its history, and its economic and social importance. This is followed by "Culinary Portrait," which details food uses and gives, particularly for the more obscure food plants, practical information on storage and preparation, and potential toxicity. In a subsection titled "Culinary Vocabulary," information is given on the names of especially important foods prepared from the plants, and on a variety of related culinary words, phrases, and terms. This is followed by "Curiosities of Science and Technology," which contains notable and interesting scientific or technological observations and accomplishments that complement the main textual material. Finally, a section called "Key Information Sources" provides selected references to books and articles on the subject of each chapter. A subsection of this entitled "Specialty Cookbooks" presents references to food preparation using the particular plant in question. There are more than 3000 literature citations in the book, and the text is complemented by over 250 drawings of very high quality.

RÉSUMÉ

Cette monographie passe en revue l'information scientifique et technique sur les principales plantes comestibles du monde et leur usage culinaire. L'introduction aborde les aspects nutritionnels et d'autres aspects scientifiques fondamentaux des plantes comestibles. Chacun des 100 chapitres qui composent l'essentiel de l'ouvrage traite d'une espèce particulière ou d'un groupe d'espèces en fournissant de l'information détaillée sur les principales cultures vivrières, et plus particulièrement sur celles qui sont présentes en Occident. Toutes les catégories de plantes comestibles y trouvent leur place, y compris les céréales, les graines oléagineuses, les fruits, les noix, les légumes, les légumineuses, les herbes, les épices, les plantes à boire et les sources d'extraits alimentaires industriels. Le tout est présenté de façon conviviale. La première section, intitulée « Names », dresse une liste exhaustive des noms scientifiques et des noms communs donnés aux plantes en anglais. La section suivante, intitulée « Plant Portrait », décrit la plante, retrace son histoire et expose son importance économique et sociale. Ensuite, la section « Culinary Portrait » précise son usage culinaire et fournit — surtout dans le cas des plantes méconnues — des renseignements pratiques sur l'entreposage, la préparation et la toxicité potentielle. Une sous-section intitulée « Culinary Vocabulary » indique le nom des préparations culinaires particulièrement importantes qui sont tirées de la plante et renseigne sur une variété de termes ou d'expressions culinaires qui y sont reliés. Après vient la section « Curiosities of Science and Technology », qui fait état d'observations et de réalisations scientifiques ou technologiques d'importance ou dignes d'intérêt, lesquelles enrichissent le texte principal. Enfin, la section intitulée « Key Information Sources » propose des références à des livres et à des articles portant sur le sujet couvert par chaque chapitre. Une sous-section intitulée « Specialty Cookbooks » renvoie aux préparations alimentaires basées sur la plante en question. Le livre comprend plus de 3 000 citations, et le texte est appuyé par plus de 250 illustrations d'une très grande qualité.

Acknowledgments and Dedication

This book is dedicated to Brenda Brookes, my principal cooperative in research for more than a quarter century, who is responsible for selecting most of the illustrations presented in this book, personally drawing some of them, and expertly electronically enhancing all of them. Her skill and patience in carrying out this very large and challenging task is greatly appreciated.

The original, copyright-free drawings on which the majority of drawings in this book are based are from publications produced from the late 17th to the very early 20th century (all sources are individually acknowledged in the captions, and original publications from which illustrations were taken are cited in Appendix 1). Beginning in the late 19th century, photography quickly and very substantially reduced the drawing of plants for scientific presentation. Despite the obvious accuracy of photography, biologists are well acquainted with the fact that drawings are usually vastly superior to photographs for showing details. Fortunately, during the golden age of plant illustration referred to above, masterpieces of botanical art were prepared of most food plants, and enhanced versions of many of these are presented here. The contribution of the original artists, albeit involuntary, is appreciated.

Some of the illustrations are by artists associated with the research facilities of Agriculture and Agri-Food Canada, Ottawa. These include six figures by Barry Flahey and five by Marcel Jomphe, as acknowledged in the figure captions.

Additional illustrations obtained through the courtesy of various individuals and institutions are acknowledged in the corresponding captions. The frontispiece was provided by the United States Department of Agriculture. Information obtained from experts that was used in particular chapters is acknowledged where used in the text.

My professional colleagues have been a constant source of information for many years, and no doubt their wisdom is reflected in some of the information I have repeated here. Certainly, in numerous cases, I have adopted ideas, interpretations, and kinds of presentations that originated with others—too many to acknowledge individually. I can only point out that imitation is the sincerest form of flattery.

Preface

Despite food constituting the *raison d'être* for food plants, the botanical and agricultural aspects of most food crops have been remarkably separated in literature from the culinary usages of the same plants. Except for the most common species, specialists in culinary usage are usually unfamiliar with the majority of food plant possibilities, and indeed even with the correct names of the plants from which most foods originate. This monograph is unique in combining an extensive presentation of botanical and agricultural aspects of food plants with their related culinary technology, and should therefore serve to expand the information base of specialists in these fields. This work is also intended to serve as a vehicle for public education in the realm of science and technology. The general public should find this book attractive, because of the huge interest today in the subjects of gardening, cooking, and human health in relation to diet. Indeed, with a large audience in mind, the book has been prepared in a style and level of language that most people will find user-friendly.

Warning

Some food plants contain chemicals that are potentially dangerous for some people, particularly pregnant women, infants, the elderly, and those with medical conditions. Some contain chemicals that are hazardous to all people, but can be made edible by expert preparation. Despite extensive research to date, the toxicity of most plant constituents to date has not been established. Virtually everyone is potentially liable to experience allergic reactions to some foods not previously consumed, and so cautious sampling of new foods over a period of time is recommended in order to establish personal tolerances. Some of the plants described in this book are meant to be used in small amounts at any meal, and sparingly in the total diet. Extensive information on the toxic potential of most of the plants discussed in this book is given in the Introduction and in the individual chapters on given food plants, but it should not be assumed that the safety information is complete. The impression should not be left that the world of previously uneaten food plants is a mine field; as discussed in detail in the Introduction, exploring new plant foods in order to increase variety in the diet is one of the best ways of improving health. It is abundantly clear that plants are gold mines of nutrients, many of these imperfectly understood or not yet even discovered. The only way of guaranteeing adequate intake of all of these health-promoting, disease-fighting phytochemicals is to consume a wide variety of plant foods.

INTRODUCTION

Food, clothing, and shelter are generally cited as the fundamental "necessities of life." The raw materials for these come mostly from plants. Photosynthetic plants are not only the major source of direct nutrition for humans, but also the ultimate source of all food, since animals and fungi used as food are nourished by eating plants. Food is the first priority for survival, and of the millions of species on Earth those discussed in this book dominate the human diet. Notwithstanding the importance of numerous plant species as sources of building and industrial materials, clothing, medicines, and other important staples of life, food plants are our most indispensable companions on the planet. This book is basically a presentation of key information about the world's most important plants.

How the Choice of the Leading 100 Food Plants Was Made

Today, a small number of plant species is responsible for most of human nutrition. Only about 150 food plants have entered world commerce, and just 12 species provide 75% of our food. Authorities have differed in their recognition of the leading food plants because different sources of information have been used and different criteria have been adopted (including reported market value, dry weight, and nutritional values). It should be emphasized that given crops differ in their global and local values and numbers of people who depend on them. The choice of the leading 100 food crops presented in this book is based mostly on the 103 species that according to Christine and Robert Prescott-Allen (1990) contribute at least 90% of the plant food supply within every country of the world (based on 146 countries). In some cases, a given food plant is important only in one or a few countries, and so is not significant in world commerce. This way of defining the leading food plants places importance on the plants that keep large numbers of people from starving, but except for inclusion of a small number of species of critical importance in some developing regions of the world, most of the leading food plants are also the ones that are of chief commercial importance on a global scale.

In this book a "food plant" usually includes just one species, but in some cases it includes several species, and in a few cases some species provide more than one "food plant" (e.g., the species *Brassica oleracea* includes broccoli and cauliflower).

The Prescott-Allens grouped certain related species together for their analysis. The groupings of the Prescott-Allens have been followed in this book with respect to the following:

Coffee (i.e., all of the coffee species are treated in one chapter, entitled Coffee), Cottonseed, Currant, Filbert and Hazelnut, Millet, Rapeseed and Canola, Rice, Onion, Orange (however, Tangarine has been added to this), Pumpkin and Squash, Wheat, and Yam. Several additional entries in this monograph also include several species. In most cases where several species are included in a chapter, one species is of chief importance and the others have relatively limited value. However, it should be noted that frequently the plants that are of minor importance on a global scale are of great importance in particular countries.

The contents of the groupings of the Prescott-Allens have been separated with respect to the following: Sugar Cane and Sugar Beet (grouped together by the Prescott-Allens, these are treated as separate entries, Sugar Cane and Beet, in this book). Most of the "Beans" were grouped together by the Prescott-Allens; in this monograph, they constitute the following entries: Adzuki bean, Bean (Common bean), Hyacinth bean, Lima bean, and Mung bean. The "Cabbages" of the Prescott-Allens included Broccoli, Brussels Sprouts, Cauliflower, and Kale. Each of these is treated in this monograph as a separate entry. Lemon and Lime were grouped together by the Prescott-Allens; they are treated in this monograph as separate entries. Whether all of these crops really constitute a "top 100 food crop" is not certain, since they were not analyzed separately, but they are so different that treating them in collective groups did not seem appropriate.

The above analysis resulted in 93 food plants (some of these with several species). To bring the total to 100, the following were added: Seaweeds (collectively, these furnish considerable food); Teff (the most important food plant of Ethiopia); Amaranth (an extraordinarily important food plant that appears to have been omitted from the analysis of the Prescott-Allens because it is mostly grown in home gardens in Developing Nations, and so not reported in national statistics);

and the plants that furnish Cinnamon, Cola, Tequila, and Vanilla (which are so popular that their inclusion seemed essential). After some soul-searching, grass pea (*Lathyrus sativus*) has not been included as a "top food plant," for although this poverty food sustains huge numbers of people in southern Asia, the presence of a neurotoxin has resulted in permanent crippling of millions (Small and Catling 2004).

Format of Presentation

Each of the 100 chapters dealing with food plants dedicated to a given plant or group of plants begins with a name in bold font. This is the English common name that is most frequently encountered. Occasionally, plants are known by several names with about equal frequency, in which case one or two of the names were chosen and presented in bold. Immediately below this in regular font is the Latin name of the plant family, followed by the most frequent English name of the family. Traditional family names end in *aceae*. For eight of the first plant families recognized by botanists, there are alternative names that do not end in this suffix. In these cases, the modern family name is given first, and the traditional name later in parentheses. The following example is illustrative:

CORN (MAIZE)

Family: Poaceae (Gramineae; grass family)

The first section is entitled "Names." This begins with the correct scientific (Latin) name, and is followed, where it might lead to confusion, by some commonly encountered scientific synonyms. (Some species have dozens of old scientific names.) Scientific nomenclature is somewhat complicated, and some of the complications that are encountered in this book are explained in Appendix 1, Technical Aspects of Plant Names. Following the initial presentation of the scientific name, details are presented (in bullet form) about additional English common names, and about etymological details of the English and Latin names. Concealed in these names is a wealth of information about the nature and utility of the plants in question.

The next section, called "Plant Portrait," provides a basic botanical and agricultural description of the plant, as well as of its history, and its economic and social importance.

The following section, "Culinary Portrait," is concerned with food uses, particularly industrial and technological aspects. The intent is not to provide details appropriate for popular cookbooks, but rather to indicate basic characteristics of the plant that make it useful for some applications, less so for others. Under a subsection titled "Culinary Vocabulary," information is given (in bulleted format) on the names of especially important foods prepared from the plants, as well as on a variety of related culinary words, phrases, and terms. The selection of such foods and phrases is limited, and is intended to illustrate some of the more important, interesting, and popular applications.

The section entitled "Curiosities of Science and Technology," again in bulleted format, contains notable and interesting scientific or technological observations and accomplishments that complement the main text.

The final section called "Key Information Sources" provides selected references to books and articles on the subject of each chapter. A subsection of this entitled "Specialty Cookbooks" presents references to food preparation using the particular plant in question. These references are intended to provide the interested reader with extensive background material. In compiling these, the following databases were examined: AGRIS (1975–2004), AGRICOLA (1970–2004), Biological Abstracts (1969–2004), FoodLine/ScienceSight (1972–2004), FSTA (1969–2004), and WorldCat (catalogue holdings of 50 540 libraries of the world). As well, many of the cited references were obtained by checking the reference lists of hundreds of review articles and books. Most of the references are in English, and non-English literature is cited basically when articles and books were simply not available in English, or when particularly important foreign-language reviews are available. (For some species that are important primarily in tropical regions, or in temperate regions of the world where English is not the dominant language, the non-English literature is considerably more extensive than information available in English.) Material on the Web has also been extensively examined, but URLs are rarely provided in this monograph, since experience has shown that these are mostly short-lived.

Literature cited

Prescott-Allen, C., and Prescott-Allen, R. 1990. How many plants feed the world? Conserv. Biol. **4**: 365–374.

Small, E., and Catling, P.M. 2004. Blossoming treasures of biodiversity 14. Grass pea (*Lathyrus sativus*). Can a last resort food become a first choice? Biodiversity, **5**(**4**): 29–32.

HUMAN NUTRITION IN RELATION TO PLANT FOODS

Top 100 Food Plants is mainly a detailed review of the world's most important plant species used as human food. Food and food plants are vital to our existence, and science has produced an enormous literature on these topics (for key sources of information, see Appendix 3). Every edible plant has particularly desirable and occasionally undesirable characteristics, but since humans consume a great variety of foods, it is critical to evaluate the contributions of individual food plants in the light of key accumulated information from several sciences. This introductory chapter summarizes knowledge of food plants from the points of view of biology, evolution, history, geography, medicine, and most particularly, nutrition.

What People Eat

Humans are naturally omnivorous, as evidenced by thousands of years of recorded observations, our natural appetites, and our anatomy and physiology. Our teeth and digestive systems are clearly designed to exploit a wide variety of both plant and animal foods. Human teeth are constructed for generalized eating, and are much less robust than those of animals adapted to live exclusively as either carnivores (animal eaters) or herbivores (plant eaters). Our intestines contain bacteria capable of digesting hemicelluloses and hydrated celluloses, which come only from plants, but we lack the specialized guts of animals that live exclusively on plants. Such plant eaters house special bacteria or fungi in their stomachs which are capable of extracting extraordinary amounts of energy from plant tissues. Our taste buds are naturally attracted to sweetness, much more than any other taste provided by food, and overwhelmingly, nature provides sweetness in the form of fruit, clearly indicating a heritage of at least partial dependency on plants. As noted below, a diet restricted to very fresh, raw animals virtually consumed whole (the life style of carnivores) seems capable of sustaining humans for prolonged periods in good health. However, dietary diseases associated with people who avoid plant foods suggest that humans are not adapted like carnivores to a diet high in both fat and protein. Most animals can synthesize their own vitamin C, indispensable for life, but humans must get it from plants. On the other hand, vitamin B_{12}, also indispensable for life, is normally obtained by humans only by eating animals. Plants are superior sources of some of the essential nutrients described below, while animals are highly superior sources of others. These facts indicate unambiguously that we are naturally adapted to consuming

Table 1. Estimated numbers of kinds of species that humans regularly eat.

Group[1]	Estimated number of species in the world	Number of species regularly consumed by humans	Number of species that are commercially[2] raised for food for humans
Animals:			
Vertebrates	41 000	Thousands	Hundreds
Invertebrates	5 million–50 million	Hundreds	Tens
Fungi[3]	1 000 000	200	20
Seaweeds	5000–10 000	Hundreds	Tens
Higher plants	400 000	Thousands	Hundreds

[1] The traditional "kingdoms" of organisms (animals, fungi, algae, etc.) are now known to be artificial.
[2] "Commerce" in this sense means marketed, at least locally. The numbers of species providing internationally traded edible goods are smaller.
[3] Estimates provided by Dr. S. Redhead.

both plants and animals. As noted below, however, a predominantly plant-based diet is almost universally recommended.

We humans are not only omnivorous, we are also generalist feeders; that is, we do not just eat some plants and some animals, we eat a wide range of both. Generalist feeders are among the world's most successful organisms (rats and cockroaches are other examples), because being able to consume a wide range of foods enables species to exploit a wide range of habitats. As a species, we consume a greater variety of the other inhabitants of the world than all other creatures. Table 1 indicates what kinds of our fellow species we currently eat as regular components of the human diet (during times of famine, desperate people have eaten a far greater range of living things).

Relative Importance of Animal Foods in the Human Diet

Most of the world's millions of species are insects (the largest group of invertebrate animals), but these represent a very minor source of human food. Although they are high in protein, and immature stages ("grubs") are not that different nutritionally from meat, insects have a chitinous exoskeleton that is indigestible and seems to have discouraged people from eating them. Most of the more than 40 000 vertebrates (fish, mammals, birds, amphibians, and reptiles) are edible. Vertebrates are often large, providing considerable complete protein (as discussed below), and wherever available have been a first choice of humans. Many invertebrate animals (especially those that constitute "sea food") also are an invaluable source of protein. Edible animal products have great appeal in almost all cultures, but for most of human existence, gathering plant foods has been more efficient than hunting, and most food consumed has been from plants. Today, meat remains the first preference of most people, but its expense keeps consumption down. The pie chart below illustrates the overall composition of categories of food consumed in the world, and indicates that plant sources are still considerably more important than animal sources.

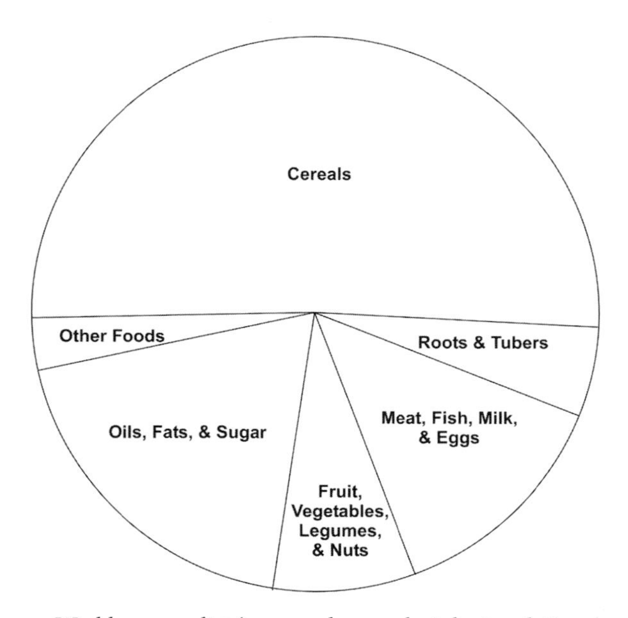

World average diet (expressed on a caloric basis; relative amounts based on 1988–1990 data). After a Food and Agriculture Organization online report: http://www.fao.org/documents/show_cdr.asp?url_file=/docrep/U8480E/U8480E07.htm

More than half of the species of vertebrates are fish, which have been the greatest source of wild animal food. Unfortunately, the natural supply from the seas of the world is rapidly being depleted, and increasingly, people everywhere are becoming dependent on animals and plants that are produced outside of nature. An extremely small number of animal species (predominantly pigs, cattle, chickens, and a lesser number of sheep) now dominates livestock production, and this is a cause of great concern for agriculture. The same thing has happened with respect to plants, as discussed in this chapter.

Below, information is presented that argues for the advisability of a diet predominant in plant foods. The traditional, mostly meat and fish diet of the Inuit of Arctic regions has long been a curiosity that has intrigued nutritionists, because it seemed to necessarily be deficient in nutrients normally supplied by plants. Nevertheless, Inuit people living on meat and fish appeared to be in excellent health, despite the predominance of protein and fat, with almost no carbohydrate consumption. The kinds of animal fats eaten were found to be good (see below for the importance of consuming unsaturated fats). The extraordinary high protein consumption resulted in development of large livers and frequent urination (to get rid of nitrogen from the protein), but this did not seem to affect health. Vitamin C, as noted below, is obtained from plants, not animals. However, liver from just-killed animals (while still warm) provided the Inuit with sufficient amounts (and in the summer, wild plants like scurvy grass (*Cochlearia officinalis* L.) were sources of the vitamin). Meat has a low calcium content, but the Inuit practice of bone-chewing, especially of soft fish bones, provided sufficient calcium. Eating of virtually the entire animals that were caught provided other nutrients not normally obtained from simply eating choice cuts of meat. Several other dietary practices seem to have also helped the Inuit survive on a predominantly meat diet (Draper 2000). Unfortunately, modern Inuits have turned to junk food and other poor offerings of the modern prepared-food industry, with disastrous consequences for their health. The traditional meat-centered diet of the Inuit does illustrate the adaptability of humans from a nutritional viewpoint, but since very few today would be willing to consume fresh, raw meat and follow other difficult dietary practices, plant-centered diets remain the optimum style of nutrition.

Vegetarianism

The term "vegetarianism" usually refers to the voluntary abstention from consuming animal foods. This is commonly based on one or more of religious, spiritual, ethical, dietary, or environmental concerns. Most vegetarians are "lacto-ovo vegetarians," who consume dairy products and eggs. "Vegans" abstain from all animal foods, including dairy products and milk. Several Asian religions, including Hinduism, Buddhism, and Jainism, either recommend or insist on some form of vegetarianism. Several religious movements in Western countries have also been founded on vegetarianism. Since the 1960s, vegetarianism has also been taken up by many in Western countries for nonreligious reasons (cited above). It is now clear that humans can live at least predominantly without animal products (excepting, of course, human milk for infants), in excellent health, although care is necessary to insure sufficient intake of protein, because of the inadequate balance of protein constituents (amino acids) in plants. What is not altogether clear, and what provokes fierce debate, is whether a 100% plant-based diet is sufficient to promote perfect health. Vitamin B_{12} is not present in plants, but can be provided in plant foods contaminated by either bacteria or insects, which seems to explain why some strict vegetarians do not develop dietary diseases. Another vitamin, riboflavin, as well as the minerals calcium, zinc, phosphorus, and iron, are relatively difficult to obtain from plants compared to animals. Infants, children, and pregnant women particularly can suffer from nutrient deficiencies that can develop if a strictly vegetarian diet is not very carefully formulated. Vegetarians tend to be extremely diet conscious and health-oriented, but malnutrition from vegetarianism does sometimes occur and is a risk that needs to be kept in mind. Numerous nutritionists recommend taking a vitamin/mineral supplement daily, and this advice is particularly useful for vegetarians.

The Evolution of the Human Diet with Respect to Plant Foods

"NOTHING WILL BENEFIT HUMAN HEALTH AND INCREASE CHANCES FOR SURVIVAL OF LIFE ON EARTH AS MUCH AS THE EVOLUTION TO A VEGETARIAN DIET."

—Albert Einstein (1879–1955)

Most plant-eating animals are food specialists, consuming a very narrow range of plant species (technically, they are *oligophagous*). Humans, by contrast, are generalists, consuming a wide variety of plants (technically, we are *polyphagous*). Anthropological studies have shown that tribal societies dependent on wild food typically eat hundreds of plants, sometimes well over 1000 species. This wide range of consumed plants is indicative of natural consumption of a wide range of chemical constituents, and also of biochemical adaptation to metabolizing these compounds. Humans have a set of enzymes that detoxify numerous natural toxins. However, the system appears designed to deal with small amounts of numerous toxins, rather than a large amount of any given toxin. This is clearly indicative of an evolutionary history of eating an extensive variety of plant materials, and suggests the wisdom of consuming a wide variety of plant foods.

Two general classes of chemical constituents in plants have been recognized. "Primary metabolites" are chemicals that are obviously critical to key functions of life. They include proteins (and their building blocks, amino acids), simple sugars, carbohydrates, fatty acids, and vitamins. Most primary metabolites occur in all plants, often in considerable quantity, and they provide all of the essential nutritional constituents that animals must consume to live. (Chlorophyll and plant hormones are vital primary metabolites for the plants, but not subsequently for animals.) Additionally, as discussed below, "dietary fiber" is increasingly viewed as essential, although not contributing directly to body constituents and (or) functions in the manner of primary metabolites. "Secondary metabolites" include a very wide array of chemicals, which plants manufacture. Plant species differ widely in their ability to produce given secondary chemicals. The function of these is usually unclear, and various hypotheses have been advanced (for example, some may be metabolic waste products, and some may serve to attract pollinators). In numerous species— probably in the majority of cases— secondary chemicals are protective against other organisms. Many secondary chemicals are poisonous to humans, but many others are responsible for the taste and smell that are greatly valued in edible plants. It has been hypothesized (Johns 1990, Johns and Chapman 1995) that in consuming plants for millions of years, since the time that humans evolved from other primates, people have come to depend on certain plant constituents as therapeutic agents (as tonics and medicine). Indeed, as outlined below, it is clear that there are many plants constituents that are health-promoting and (or) curative.

Plants concentrate energy reserves mostly in reproductive organs (fruits, nuts, seeds) and in underground storage organs such as roots and rhizomes (underground stems), and it is primarily these that nourished humankind in the past, and continue to sustain us. Leaves and aboveground stems are much less "nutrient-dense," but undoubtedly have always been consumed by humans (at least when young and tender), because they are available throughout the growing season, unlike the other parts mentioned, which are most developed at a particular time.

In theory, all plant tissues can be converted to energy, but humans lack the digestive enzymes to process all plant constituents. Much more serious is the fact that many plant constituents are toxic. Poisonous chemicals tend to be absent in mobile creatures that can escape predators simply by moving away, which explains why there are very few vertebrate animals that are poisonous when consumed. By contrast, sedentary species generally have developed chemicals that discourage other organisms from attacking them. Illustrative of this, fungi and higher plants, which are of course immobile, are rich in toxins. Nevertheless, there are thousands of higher plants as well as numerous seaweeds and fungi that are edible. Fruit in particular tend to be edible, in order to attract animals to disperse the seeds.

Before the dawn of agriculture some 10 000 years ago, food was acquired entirely by foraging and hunting, and nature still provides substantial amounts of food. By trial and error, people throughout the world learned what plants were edible, and if poisonous, they often found ways of processing the food to denature or remove the toxic constituents. With the progress of civilization and agriculture, humankind first learned how to produce higher yields by cultivating certain especially desirable plants. Equally significant, people progressively altered the genetics of a small number of species to increase yield and decrease natural toxins. Hundreds of plant species have been changed genetically to serve as sources of food (by comparison, only about 50 animal species have been genetically altered for food purposes). As detailed below, a remarkably small number of such "domesticated" plant species now is responsible for most of the world's food. Nevertheless, numerous wild plants are still consumed. The above pie diagram, based on over a thousand species, shows that some categories of food plant are represented by numerous species. As is show, there are numerous species of fruits and vegetables, far exceeding the number of species of cereals and "pulses."

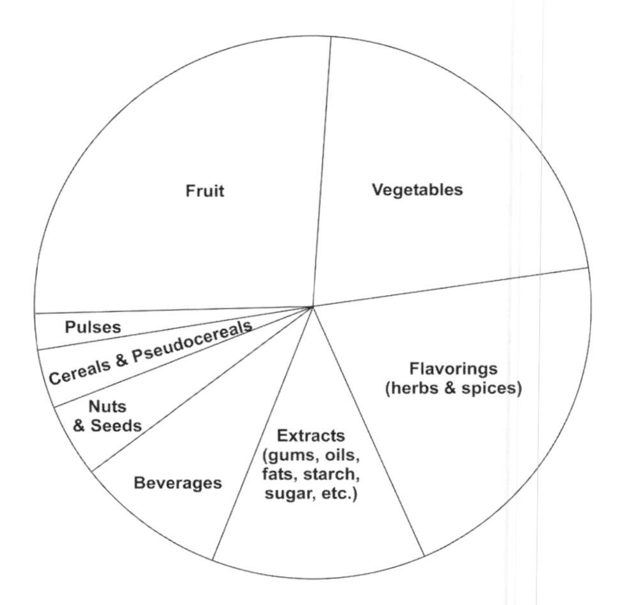

Relative numbers of species in categories of food plants, based on 1431 species listed in Wiersema and León 1999, of which about 200 are represented in more than one category (for example, a given species may be used as both a source of a fruit and a vegetable).

Nutritional Success by Combining Cereals and Pulses

Despite their relatively small numbers, cereals and pulses are of dominating importance, and in combination have been one of the foundations upon which civilizations have been built. "Cereals," by definition, are simply members of the grass family that furnish edible seeds for people. ("Pseudocereals" are plants of other plant families, such as buckwheat, which develop small, cereal-like seeds.) "Pulses" are species of the pea family that produce edible pods, seeds, or "beans." Technically, cereal grains and the pods of the pea family are fruits, and represent the plants' means of reproduction. Because of this central importance to the plants, they invest a huge proportion of their energy into packing nutrients into the seeds, to nourish the young seedlings. Just like the eggs of birds, they are superb, concentrated food packages from the point of view of humans. But there is a nutritional problem. Pulses are a rich source of protein, particularly the essential amino acid lysine, but are usually deficient in the sulfur-containing amino acids methionine and cystine. By contrast, cereals contain lower amount of proteins, and these are deficient in lysine but adequate in sulfur-containing amino acids. Together, cereals and pulses provide a balance of essential protein components, which is necessary for survival, especially when adequate meat supplies are unavailable. Anthropologists have observed that wild-food-gathering societies often make particular efforts to acquire both cereals and pulses. The same is true of more advanced agricultural societies, which concentrate on growing both cereals and pulses. In the Mediterranean area, the cereals wheat and barley have commonly been combined with the pulse lentil. In Asia, the cereal rice has been combined with several Asian legumes, such as soybean and the mung bean. In Africa, sorghum is the traditional main cereal, and is consumed with several legume vegetables. In the New World, the cereals corn (maize), wild rice, and the pseudo-cereals amaranth and quinoa (which played the role of cereals) were paired with the common bean (today, Old World cereals have become important). Throughout this book, there are numerous examples of ethnic dishes that combine cereals and pulses, a tradition borne out of the natural nutritional advantages of this association.

The Principal Food Plants

More than 20 000 plant species have been used as food by humans (Tanaka's 1976 encyclopedia lists over 10 000 edible plants). At least 5000 of these are eaten regularly, and cultivated forms have been selected from about 3000. But today, very few plant species are responsible for most of human nutrition. Only about 150 food plants have entered world commerce and just 12 species provide 75% of our food. The grass family (which includes wheat, rice, corn (maize), barley, sorghum, millets, oat, rye, and sugar cane) furnishes 80% of calories consumed by humans. Only three grass family crops—wheat, rice, and corn—account for about 60% of the calories and 56% of the protein that humans get directly from plants.

Cornucopia, from Corel Corporation (1998).

Rank-ordering the importance of the world's food crops is somewhat problematical, since importance changes from year to year, and there are different ways of expressing importance: as done for this book (as described earlier, on the basis of how important the crop is to prevent hunger in more than 100 countries), on a dry-weight basis, on an calorie (energy) basis, and on a monetary basis. Harlan's (1992) widely quoted, if somewhat dated analysis of what are interpreted as the top 30 food crops is shown in Table 2, based on calculated edible dry matter. For this chapter, an analysis of recent fresh-weight data of all of the crops reported by the Food and Agriculture Organization was conducted, and is given in Table 3, with the data somewhat arbitrarily reported in nine food categories. Harlan's list of the top 30 food crops corresponds with the following in Table 3: the top seven cereal species; the top seven vegetable species; the top four legumes ("beans" means *Phaseolus communis*, and is reported three times in the legume list); five of the top six fruits (watermelon is not in Harlan's list, and fresh weight is misleading for this fruit); five of the top six "oil sources" (Harlan did not include oil palm, which certainly is one of the world's leading food plants); and the two sugar sources (sugar cane and sugar beet).

Table 2. Professor Jack Harlan's 1992 analysis of the world's 30 leading food crops measured by annual edible dry matter production (based on averages of Food and Agriculture Organization Yearbooks for 1985, 1986, and 1987; cf. (Janick 1999)).

Crop	Million metric tons
Wheat	468
Maize	429
Rice	330
Barley	160
Soybean	88
Sugar cane	67
Sorghum	60
Potato	54
Oat	43
Cassava	41
Sweet potato	35
Sugar beet	34
Rye	29
Millets	26
Rapeseed	19
Bean	14
Peanut	13
Pea	12
Banana	11
Grape	11
Sunflower	9.7
Yams	6.3
Apple	5.5
Coconut	5.3
Cottonseed oil	4.8
Orange	4.4
Tomato	3.3
Cabbage	3.0
Onion	2.6
Mango	1.8

Table 3. World Production Statistics, organized into nine categories by principal food use, and ranked by tonnage (means of 5 years (2000–2004), in thousands of tonnes). Source of raw data: Food and Agriculture Organization of the United Nations; http://faostat.fao.org/site/408/DesktopDefault.aspx?PageID=408

Note that some commodities such as karite nut, linseed, and tallow tree are used only partly for food; "nes" refers to categories in which specific species were not identified; see website for additional details. Also note that fresh weight is only a crude indicator of dollar and nutritional values (e.g., three times as much raw potato would have to be eaten to achieve equivalent energy to that of cereals), and water content of fruits, leaves, and most roots considerably exceeds that of seeds.

Cereals	Tonnage
Rice	591 000
Wheat	589 000
Barley	143 000
Sorghum	57 646
Millets	28 428
Oat	26 380
Rye	19 332
Triticale	11 244
Cereals nes	2275
Fonio	259

Pseudocereals	Tonnage
Buckwheat	2605
Quinoa	52

Vegetables	Tonnage
Potato	321 000
Vegetables, fresh nes	236 000
Cassava	189 000
Potato, sweet	133 000
Tomato	114 000
Cabbage	62 594
Onions, dry	52 517
Yams	38 899
Cucumbers and gherkins	38 223
Plantain	31 780
Eggplant	28 433

Table 3 (*continued*).

Vegetables (*continued*)	**Tonnage**
Chillies and peppers	22 684
Carrot	22 447
Lettuce	19 884
Pumpkins, squash, gourds	18 083
Cauliflower	15 504
Spinach	10 816
Taro	10 355
Corn (maize), green	8822
Roots and tubers nes	6740
Asparagus	5608
Okra	4889
Onions + shallots, green	4267
Mushrooms	2963
Avocado	2917
Artichoke	1300
Yautia	372

Legumes	**Tonnage**
Soybean	184 000
Peanuts in shell	35 434
Beans, dry	18 009
Pea, dry	10 652
Pea, green	8392
Chickpea	7764
Beans, green	5923
Broad bean, dry	4134
Pulses nes	3784
Cowpea, dry	3628
Lentil	3282
Pigeon pea	3113
Lupins	1194
Broad bean, green	1044
Beans, string	1660
Bambara bean	58

Fruits	**Tonnage**
Watermelon	87 232
Banana	68 609
Grape	63 519
Orange	62 459
Apple	59 119
Mangoes	26 065
Cantaloupe and other melons	24 669

Fruits (*continued*)	**Tonnage**
Fruit, fresh new	24 297
Tangarines	21 200
Pear	17 489
Pineapple	15 199
Fruit, tropical, fresh nes	14 882
Peach and nectarine	14 484
Lemon and lime	12 143
Plum	9450
Date	6483
Papaya	6420
Citrus fruit nes	5862
Grapefruit and pomelo	4900
Strawberry	3321
Apricot	2674
Persimmon	2460
Cherry	1826
Cashew apple	1641
Cherry, sour	1142
Kiwi	1031
Fig	1020
Currants	778
Berries nes	634
Raspberries	451
Quince	380
Stone fruits, fresh, nes	378
Cranberry	334
Blueberries	236
Gooseberries	179

Oil sources	**Tonnage**
Corn (maize)	636 000
Oil palm fruit	140 000
Cotton seed	58 200
Coconut	53 029
Rapeseed	38 494
Sunflower	25 091
Olive	16 482
Linseed	2006
Oilseeds nes	1809
Tallow tree seeds	848
Karite nut (shea nut)	665
Safflower seed	645
Hempseed	33

Table 3 (*concluded*).

Spices and condiments	Tonnage
Garlic	12 189
Coffee, green	7531
Cocoa beans	3467
Tea	3150
Sesame seed	3072
Pimento, allspice	2379
Spices nes	1890
Ginger	963
Chicory root	919
Maté	894
Areca nut	703
Mustard seed	555
Anise, badian, fennel	399
Pepper, white, long, and black	369
Kolanut	226
Carob	192
Clove, whole + stems	128
Cinnamon	123
Hop	102
Peppermint	77
Nutmeg, mace, cardamom	76
Poppy seed	65
Vanilla	8

Sugar sources	Tonnage
Sugar cane	1 310 000
Beet, sugar	242 000
Sugar crops nes	795

Nuts and edible seeds	Tonnage
Cashew	1906
Almond	1665
Walnuts	1360
Chestnut	1035
Hazelnuts (filberts)	739
Nuts nes	690
Melonseed	577
Pistachio	438
Brazil nut	73

Economic and Environmental Reasons for Diversifying Food Plant Production

The overdependence on the leading crops is associated with severe economic problems because of surpluses and unwise subsidies. As well, it is ancient wisdom that growing only a few crops is like putting all your eggs in one basket. All crops are subject to disasters because of pests, diseases, and the climate, so growing several crops ensures that at least some will produce a reasonable harvest. Moreover, growing huge, uniform expanses of given crops compromises biodiversity and the environment. Most people have little conception of how much of the planet is devoted to food. In 1991, food production occupied about 37% of the world's land area (World Resources Institute 1994). Most people also do not appreciate the extent of the monopoly that cereals hold over the planet. Most of the world's cropland is used to produce grain, of which humans consume about 50% directly and the other 50% indirectly as livestock products (Brown et al. 1998). Still another consideration is that the efficiency of growing huge monocultures is at the price of decreasing employment and rural diversity. Socially, there are advantages to having more small farms specializing in crops that require great skill and knowledge.

The Geographical Origin of Food Plants

The Russian geneticist and agronomist N.I. Vavilov (1887–1943), widely considered to have been the foremost plant geographer of the world, developed concepts regarding the origin of cultivated plants (Vavilov 1926) that are still widely respected. He proposed that there are eight geographical centers of origin of crops, with some sub-centers (see map on page 11), identified by their having the greatest genetic diversity of sets of species (Table 4). In the 1930s, the bureaucrat T.D. Lysenko (1898–1976), a pseudo-scientist whose ideology suited the totalitarian, communist Soviet Union, challenged Vavilov's ideas. At great personal risk and demonstrating extraordinary intellectual honesty, Vavilov tried to defend his scientific achievements, but this resulted in imprisonment and death. Lysenko's ideas led to the discredit of Soviet genetics, the failure of Soviet agriculture, and in no small way to the end of the Soviet Union. Vavilov's centers reflect to a considerable extent ancient centers of civilization, where crops were developed (often explaining why there are so many different kinds there), and not necessarily where the ancestral wild species are located. Moreover, for many crops, the

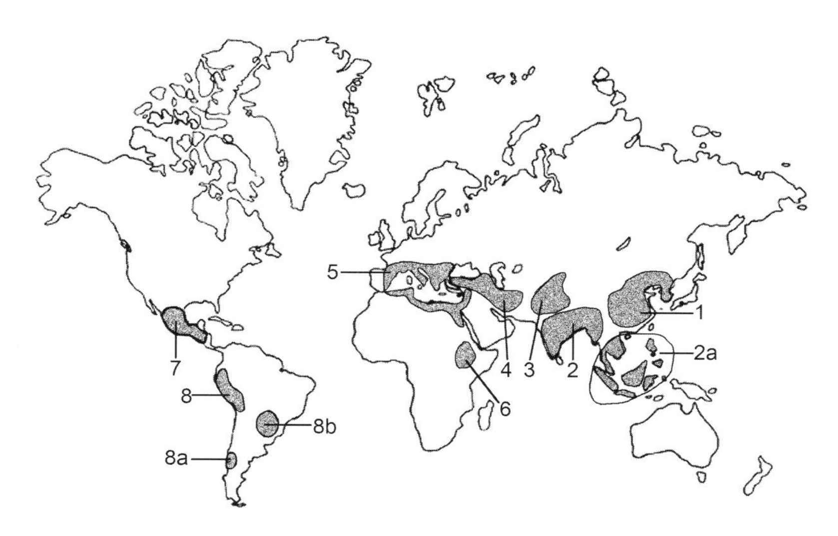

Vavilov's primary centers of diversity (cf. Table 4).

place of origin appears to have been much more diffuse than indicated by the classical Vavilovian centers, and modern genetics is now relating the geographical relationships of modern crops with their wild ancestors with much more precision than was possible in the past. Nevertheless, Vavilov's work still dramatically illustrates the debt the modern world owes to different areas of the world in providing the wealth of plant foods that are available today.

Nikolai Ivanovich Vavilov. Photograph courtesy of the Vavilov Institute of Plant Industry, St. Petersburg, Russia.

Table 4. Sites of origin of food plants according to Vavilov's primary centers of diversity[1] (cf. above map). Crops with multiple sites of variation or with species or varieties with notable diversity in different sites are indicated in italics. Vavilov's original numbering system (which has been widely reproduced) is followed. Vavilov's use of crop diversity as a guide to where the respective plants originated is now known to be quite incorrect in some cases (i.e., the plants originated elsewhere and were transported to the center of diversity where numerous kinds were selected). For more up to date information, consult the chapter in which each species or group of species is discussed.

1. Chinese Center
Adzuki bean
Apricot
Arrowhead
Asian ginseng
Asian pear
Bamboo
Barley
Buckwheat
Cassia
Chinese artichoke
Chinese cabbage
Chinese kale
Chinese water chestnut
Cucumber
Date plum
Eggplant
European water chestnut
Garland chrysanthemum
Ginkgo

Table 4 (*continued*).

1. Chinese Center (*continued*)
Gobo
Hawthorn
Hemp
Japanese chestnut
Japanese walnut
Jujube
Korean pine
Kumquat
Lettuce
Litchi
Loquat
Lotus
Malabar spinach
Manchurian wild rice
Millet
Oat
Opium poppy
Oriental persimmon
Peach
Perilla
Radish
Sago palm (*Cycas revoluta*)
Smooth loofah
Soybean
Star anise
Taro
Tea
Turkish filbert
Turnip
Wasabi
Water spinach
Welsh onion
White mulberry
Yam
Yard-long bean

2. Indian Center
Amaranth
Angled loofah
Bamboo
Betelnut
Betel pepper
Bitter gourd
Black mustard
Black pepper

2. Indian Center (*continued*)
Bottle gourd
Brown mustard
Carambola
Cardamom
Chickpea
Citron
Coconut
Cotton
Cowpea
Cucumber
Cumin
Date palm
Dill
Eggplant
Fenugreek
Galangal
Hemp
Hyacinth bean
Jicama
Kaempferia galangal
Lemon
Lime
Malabar spinach
Mango
Millet
Mung bean
Noni
Orange
Palmyra palm
Pigeon pea
Radish
Rice
Rose apple
Roselle
Safflower
Sesame
Seville orange
Snake gourd
Sorghum
Sugar cane
Tamarind
Taro
Turmeric
Yam
Zedoary

Table 4 (*continued*).

2a. Indo-Malayan Center
Banana
Betelnut
Black pepper
Breadfruit
Candlenut
Cardamom
Clove
Coconut
Durian
Ginger
Jackfruit
Job's tears
Kaempferia galangal
Mangosteen
Nutmeg
Pomelo
Sago palm
Sugar cane
Turmeric
Wax gourd
Yam

3. Inner Asiatic Center
Almond
Apple
Apricot
Basil
Broad bean
Brown mustard
Carrot
Chickpea
Coriander
Cotton
Flax
Garden cress
Garlic
Grass pea
Hemp
Jujube
Lentil
Mung bean
Onion
Pea
Pear
Persian walnut

3. Inner Asiatic Center (*continued*)
Pistachio
Polish rape
Radicchio
Radish
Rye
Safflower
Sesame
Spinach
Summer purslane
Turkish filbert
Turnip
Wheat
Wine grape

4. Asia Minor Center
Alfalfa
Almond
Anise
Apple
Barley
Black mustard
Cabbage
Carrot
Coriander
Cucumber
Date plum
European chestnut
European filbert
Fig
Flax
Giant filbert
Hawthorn
Leek
Lentil
Lettuce
Lupin
Medlar
Melon
Oat
Opium poppy
Pear
Persian walnut
Pistachio
Polish rape
Pomegranate

Table 4 (*continued*).

4. Asia Minor Center (*continued*)
Pumpkin/squash
Quince
Radicchio
Rye
Saffron
Sesame
Sour cherry
Summer purslane
Turkish filbert
Wheat
Wine grape

5. Mediterranean Center
Alexanders
Anise
Artichoke
Asparagus
Bay
Beet
Black cumin
Black mustard
Broad bean
Cabbage
Caraway
Carob
Celery
Chervil
Chickpea
Chicory
Chufa
Dill
Endive
Fennel
Flax
Grass pea
Hop
Hyssop
Lavender
Leek
Lentil
Lettuce
Lupin
Oat
Olive
Parsley

5. Mediterranean Center (*continued*)
Parsnip
Pea
Peppermint
Polish rape
Radicchio
Rhubarb
Rosemary
Rue
Rutabaga
Sage
Salsify
Scorzonera
Sea kale
Sorrel
Spanish oyster
Summer purslane
Summer savory
Swedish rape
Thyme
Turnip
Wheat
White mustard

6. Abyssinian (Ethiopian) Center
Barley
Black cumin
Black-eyed pea
Chickpea
Coffee
Coriander
Fenugreek
Flax
Garden cress
Grass pea
Hyacinth bean
Khat
Lentil
Millet
Myrrh
Okra
Pea
Safflower
Sesame
Sorghum
Teff
Wheat

Table 4 (*concluded*).

7. South Mexican and Central American Center
Amaranth
Arrowroot
Avocado
Cactus pear
Cashew
Cassabanana
Chayote
Cherry tomato
Chocolate
Common bean
Corn (maize)
Cotton
Epazote
Green sapote
Guava
Jicama
Lima bean
Malabar gourd
Mamey sapote
Nopal cactus
Papaya
Pumpkin/squash
(Red) Pepper
Scarlet runner bean
Sweet potato
Sweet sop
Tepary bean
Tequila/Mezcal
Tobacco
Tomatillo
White sapote
Yellow sapote

8. South American (Peru, Ecuador, Bolivia) Center
Amaranth
Cape gooseberry
Cherimoya
Coca
Common bean
Corn (maize)
Cotton
Guava
Lima bean
Lupin
Malanga
Mashua
Muster-John-Henry (marigold)
Oca
Passionfruit
Peach palm
Pepino
Potato
Pumpkin/squash
Quinine
Quinoa
(Red) Pepper
Tobacco
Tomato
Tree tomato
Ullucu

8a. Chilean Center
Potato
Strawberry

8b. Brazilian–Paraguayan Center
Brazil nut
Cashew
Cassava
Feijoa
Maté
Passionfruit
Peanut
Pineapple

[1] As presented in Vavilov, N.I. 1935. The phyto-geographical basis for plant breeding. Teoreticheskie osnovy selecktsii (Theoretical bases for plant breeding), vol. 1, Moscow-Leningrad. Available translated from the original Russian in Vavilov (1992); see Literature Cited.

Natural Toxins in Food Plants and their Potential for Harm

Natural chemicals that plants manufacture to protect themselves against microorganisms, insects, and animal predators make up 5–10% of the dry weight of plants that we consume (Ames et al. 1987; Beier 1990). While this suggests that people are consuming an enormous amount of poison, these chemicals are not necessarily poisonous to humans. Ironically, many of the chemicals that produce flavor in plants, especially in herbs and spices, are likely there to protect those species against being eaten, although in the case of humans these same chemicals make the plants more likely to be eaten.

Nevertheless, there are numerous compounds in our major food plants and food products that in theory at least are poisonous for humans. Examples of such toxins include the following: carotatoxin (a poison) in carrot, goitrogens (which block iodine absorption) in cabbage, oxalic acid (which can cause kidney problems) in spinach, solanine (a poison) in potatoes, a cyanogenic glycoside (which yields cyanide) in lima beans, pulegone (a chemical that is toxic to the liver and lungs) in pennyroyal and in the essential oil of many culinary herbs, thujone (a highly neurotoxic chemical) in wormwood (and many other herbs), canavanine (a toxic, nonprotein amino acid) in the seeds of many leguminous plants such as alfalfa, and diacetyl (a mutagen) in wine. Ames and Gold (1990) noted that of over 800 chemicals present in roasted coffee, the safety of only 21 had been tested on animals. Moreover, 16 of the tested chemicals were capable of causing cancer in rats. For example, dicarbonyl aldehyde methylglyoxal is a potent mutagen produced by fermentation in prepared coffee. While many of the thousands of chemicals in plants can be poisonous, their concentrations are for the most part too low to be of concern. However, even if many natural chemicals are safe to consume in normal quantities, too much of a good thing can be bad (probably the most familiar example is coffee). The liver serves the body by sieving out toxins, but when overloaded with a particular chemical, toxic reactions may occur.

The presence of cancer-causing agents in plants often causes great fear. Examples include the pyrrolizidine alkaloids in comfrey (which should not be consumed), and in borage, although in much lower concentrations in the latter; furocoumarins in angelica (these have been shown to be carcinogenic in laboratory animals); estragole in tarragon and basil; allyl isothiocyanate in brown mustard; psoralens in celery; and hydrazine compounds (some of which cause tumors in mice) in the common mushroom of commerce. For the most part, like the toxins mentioned above, carcinogens occur in sufficiently limited concentrations in most plants used as food that they are not alarming.

The presence of abortifacients in plants is also a cause of great concern. Rue, which should not be consumed, is a culinary herb that has been used historically as an abortifacient. Parsley seeds and juniper berries, and more particularly the oil obtained from them, should also be avoided by expectant mothers. The sweet herb stevia allegedly has been used to control fertility, but as a contraceptive rather than an abortifacient.

Another class of chemicals that can cause problems for pregnant women is teratogens, which cause birth defects. Alkaloids, a widespread class of chemicals in plants, are of particular concern. Examples include nicotine in tobacco, solanine produced by green potato sprouts, and caffeine in coffee, tea, cola, and other plants. Women who are or may become pregnant are urged to avoid or limit consumption of foods and drugs containing caffeine.

Some plants contain compounds that act like hormones. Species of sarsaparilla, in fact, are the source of chemicals used to prepare the birth-control pill. Licorice also contains potent hormone-like chemicals, and eating too much can produce pseudo-aldosteronism, characterized by sodium and water retention, potassium depletion, hypertension, and even heart failure. It has been alleged that sufficient estrogen is present in fenugreek that overconsumption could result in problems.

Numerous plants are capable of producing hydrocyanic acid (prussic acid), and plants with this capability are said to be cyanogenetic (= cyanogenic). This gas is so toxic that inhalation of a small amount may produce death in a few minutes. Fortunately, the gas is not produced normally by such plants, but is chemically combined in the form of a cyanogenetic glucoside. However, this glucoside is often associated with an enzyme capable of decomposing it and freeing the hydrocyanic acid, and this can occur in a bruised or wilting plant, or much more dangerously in the alimentary tract of an animal (even without the enzyme). Cases of poisoning of stock animals and humans from hydrocyanic plants are occasionally reported. In some plants, this potential for poisoning animals protects the plants against being consumed. Commonly consumed cyanogenic plants include rice, oats, sugar cane, corn, lima bean, cherries, apples, and (as note below) cassava.

Several plants used as food sources are classified as "narcotics." These include hemp, opium poppy, khat, betel, kava, and coca leaf. Once again, the key issue is whether or not the amounts of the narcotic chemical are high enough to be harmful. Seeds of several plants of the carrot family contain minute quantities of a hallucinogen. Fennel oil is a narcotic that can cause epileptiform madness and hallucination. The oils of dill, anise, and parsley have similar properties, apparently owing to a chemical transformation in the living body after ingestion, which results in the production of hallucinogenic amphetamines.

As noted above, natural toxins that protect wild plants against herbivorous animals, bacteria, and fungi have often been greatly reduced in their domesticated food plant relatives. Many food plants, however, have not been altered very much from their wild ancestors, and indeed many are simply unaltered wild plants. In these cases, there remains a definite possibility of poisoning should a considerable amount of material be consumed. There are numerous minor food plants that need to be consumed sparingly because of the danger of toxicity (e.g., buffaloberry, epazote, juniper, nutmeg, and saffron). For some plants, the toxic potential is so significant that it is quite inadvisable to consume them (e.g., absinthe, comfrey, kava, pennyroyal, pokeweed, rue, and tansy), despite the fact that they continue to be used as food.

Some plants simply require knowledge and expertise to avoid being poisoned. For example, ginkgo fruits are widely gathered for consumption by Asians when they discover fruiting trees growing as ornamentals along roadsides, as ginkgo is a familiar culinary delicacy in China and Japan. However, collecting and preparing the seeds for eating by those who aren't familiar with this food has a high probability of producing dermatitis first, then food poisoning. Crude extracted cycad starch is also very toxic, and if not properly detoxified, extreme illness results.

Often, some parts of a plant are edible, but other parts can be extremely poisonous. While the leaves and flowers of sweet violet are edible, the seeds, rhizomes (underground stems), and roots can be poisonous in large doses, causing severe gastroenteritis, nervousness, and respiratory and circulatory depression. The roots of the horseradish tree are pungent like horseradish and are much used in India as a substitute for horseradish. However, the root bark must be completely removed, as it contains potent, toxic constituents. The fully ripe fruits of May-apple make excellent jellies and marmalades, but the unripe fruits and other parts of the plant are poisonous. Many who have not exercised care in collecting the edible part of akee have died from consuming the poisonous parts. The fruits and foliage of the common potato can be quite poisonous (pet rabbits have been poisoned by inadvertent feeding).

For some species, there are varieties that are very toxic, but others are much less so (e.g., grass pea, Labrador tea, neem, taro). It might seem obvious that the less toxic varieties should be adopted to reduce risk, but there are other considerations. In the case of grass pea, consumption of which paralyses and cripples thousands in Asia and Africa every year, cultural resistance against adopting the safer varieties has been a critical factor. Labrador tea and neem are extremely minor as food, and there has as yet not been much motivation to develop the safer types. In the case of taro, which is a staple crop in many developing countries, the relatively poisonous varieties are naturally much more resistant to microorganisms and insects, and are grown because the farmers cannot afford to buy insecticides and other protective chemicals.

Some plants can accumulate toxic quantities of elements, depending on growth conditions. Brazil nuts are extraordinarily rich in selenium. The nuts can contain 250 times as much selenium as in most foods, depending on the soil where they are grown, and a single Brazil nut can exceed the US *Recommended Daily Allowance* of selenium. Too much can be toxic and cause balding.

Toxic levels of nitrates in some plants, such as lettuce, can result from heavy fertilization with nitrogen. Like many other leafy, green vegetables, spinach tends to accumulate nitrogen in the nitrate form (NO_3), particularly when nitrate fertilizer is added to increase green color and succulence. The result can be a poisoning in the digestive tract of mammals, called methemoglobinemia. This can be serious for ruminant (cud-chewing) animals such as cows, but also for human infants who consume large amounts of high-nitrate leafy vegetables. The disease has been dubbed "blue baby" (blue baby syndrome is a condition generally associated with low-oxygen availability in babies; in fact, ingestion of high levels of nitrates can destroy haemoglobin in humans, the molecule responsible for carrying oxygen). Different varieties, soils, and methods of growing spinach and other plants have been found to influence nitrate accumulation, and growers are well aware of the problem. In Queensland, Australia, green papayas are often canned. Because of local conditions, the plants may take up high levels of nitrates from the soil, and this causes detinning of ordinary cans. All papayas with over 30 parts per million of nitrates must be packed in cans lacquered on the inside. In situations where toxic levels of chemicals can develop in food plants, it is the responsibility of producers and marketers to ensure that their goods can be safely consumed.

The edibility of some species is controversial; there is suspicion that they contain toxic ingredients, but not a universally accepted recommendation that they not be consumed. Stevia is widely consumed in Asia, but illegal in food in the United States and Canada because of suspicion that the steviosides present are toxic. Mexican oregano has been shown to have the ability to reduce reproductive capacity. The chemical that might contribute to the antifertility, lapachenole, is carcinogenic. Nevertheless, little concern has been expressed to date about the use of Mexican oregano as food.

In many cases, a food plant is simply highly inadvisable for some people because they are young, old, or suffering from a medical condition. Peppermint tea should not be given to infants or very young children, since they may experience a choking sensation from the menthol. Consumption of fiddleheads (*Matteuccia struthiopteris* (L.) Tod.) has produced some disturbing cases of diarrhea, nausea, vomiting, abdominal cramps, headaches, and dehydration, particularly among the elderly and infants, but whether constituents in this vegetable are responsible remains unclear. Because of its high potassium content, there has been a recommendation that carambola (star fruit) be avoided by uremic patients. High-oxalate, high-calcium plants like rhubarb, sorrel, and spinach are inadvisable for people susceptible to calcium-oxalate kidney stones, and may also be imprudent for those with arthritis or gout. People with gallstones should avoid eating large amounts of radish, as this could cause painful spasms in the biliary (bile) tract. Goitrogenic plants (especially the cabbage family) can cause problems for those suffering from goiter; these sulfur compounds slow the thyroid gland's production of hormones, and might be troublesome for people with thyroid disorders. The high iodine content in seaweeds can also exacerbate goiter. The coumarins of angelica are known to thin blood, and consumption of large quantities can result in dangerous thinning, resulting in internal bleeding, the blood leaking from capillaries and vessels. Individuals on anticoagulant therapy should avoid this herb. The same is true for garlic, which can retard clotting time, and individuals regularly taking aspirin or other anticoagulant drugs, or susceptible to low blood clotting conditions, should avoid consuming large amounts. A toxic substance (antipyridoxine) occurs in the fresh seeds of ginkgo, and has poisoned people consuming the nuts both for medicinal and food purposes. The toxin has an anti-vitamin effect, and it is dangerous for people suffering from vitamin deficiency to eat large quantities of fresh nuts. Grapefruit can amplify the effect of many medications, and minimize the effect of a few, and in recent years, most medical professionals in the United States have instructed their patients to avoid drinking grapefruit juice at any time while taking any oral medications. Warnings are given throughout this book of hazards associated with particular plants.

Most people can tolerate up to a given amount of pungent or spicy foods, but if this level of consumption is exceeded, discomfort results. The mustards discussed in this book, contain compounds called glucosinolates. These release mustard oils (thiocyanates or isothiocyanates) during digestion. Sulfur-containing mustard oils contribute to the taste of the mustards and horseradish, but can be severe irritants of the lining of the digestive tract if consumed in large quantities. Chile peppers contain the irritant capsaicin, and once again, if too much is consumed problems arise. Overconsumption of watercress can lead to kidney problems, and the undiluted juice can produce throat and stomach inflammations.

In the category of "just plain weird" is a problem associated with eating persimmons. "Bezoars" are rounded, layered stones found in internal organs such as the stomach, gall bladder, and kidneys of large herbivorous mammals, primarily sheep and goats. A bezoar is most often formed from an undigested mass of ingested material that remains in the stomach. Such stones can obstruct the passage of food into the intestine, and surgery is usually necessary to remove them. Bezoars of vegetable origin are most common in areas where persimmons are popular. So common is the persimmon as a cause of the formation of bezoars that the term "diospryobezoar" was once proposed for persimmon bezoars. Persimmons seem extraordinarily able to cause bezoars because their tannins, when liberated in the stomach, can consolidate vegetable material into a mass, especially when the stomach is empty. Fortunately, bezoars are relatively rare, although people who have had stomach surgery or suffer from slow stomach action seem prone to developing them. Because of the possibility of developing bezoars, it has been suggested that gorging on persimmons is not a good idea, especially on an empty stomach.

People who work in occupations that bring them into regular contact with food plants, particularly skin contact and exposure to generated dust, are in heightened danger of experiencing severe toxic reactions. For example, workers who cut up pineapples for long periods often have their fingerprints almost completely obliterated from the effects of the enzyme bromelain. Celery is stimulated by infection with fungal pathogens to produce sensitizing chemicals, and this has led to outbreaks of dermatitis among commercial celery handlers. Hand-harvesting of chickpea must be done with gloves for protection from the malic and oxalic acids that are exuded by the plants. The wood of lemon trees and its sawdust may induce skin reactions in sensitive woodworkers.

Tea is illustrative of the ambivalence of scientific evidence about the advisability of consuming some plants. Chemicals in tea (condensed catechin tannins) have been linked to high rates of esophageal cancer in some areas where tea is heavily consumed. (Tannins are also undesirable in that they hinder the absorption of iron from foods. Adding milk binds the tannins preventing harmful effects.) However, the relation of tea and cancer is mixed, some reports suggesting both black and green tea reduce some cancers. In particular, green tea is not fermented, so many of its active constituents remain unaltered. Chemicals called polyphenols are believed to be responsible for most of green tea's claimed roles in promoting good health, including protection against cardiovascular disease, some cancers, and even dental plaque. Because several cupfuls daily seem necessary to produce significant benefits, decaffeinated tea is helpful in reducing insomnia, anxiety, and other symptoms that may be caused by the herb's caffeine content.

Plant Foods as Medicine

"LET FOOD BE THY MEDICINE."

—Hippocrates (460?–?377 BC), Greek physician, the Father of Western Medicine

The distinction between medicine and food is somewhat arbitrary, because a healthy diet promotes health and is often a basis for curing disease. "Food" provides needed chemicals that the body uses to produce energy, build or repair tissues, and conduct the complex biochemical transformations that sustain life. Deficiencies or imbalances of food constituents may lead directly to illnesses or physiological disorders, or may predispose the body to disease or injury. Most so-called "medicinal plants" contain drugs that are medicinally useful because they alter metabolism in a desirable way (for example, reducing fever). However, common fruits and vegetables are also medicinal plants, although rarely thought of as such. Virtually every food plant has been used historically as a medicine. No matter how one chooses to make distinctions between food and medicine, it is clear that there is a demand today for foods that are *both* more nutritional and health-promoting than in the past.

Numerous extracts from edible plants are used directly as prescription medicines. For example, rutin, a chemical harvested from buckwheat leaves, is used to evaluate and treat blood vessels for hemorrhagic diseases such as retinal bleeding and stroke, and also in the treatment of frostbite. Rutin is abundant in buckwheat bran, and in the diet is said to reduce cholesterol, lower blood pressure, and strengthen diseased, weakened veins and arteries. Fagopyritol, another compound derived from buckwheat, has been used to help manage diabetes.

Food plants are now widely used as sources of plant extracts, marketed, without prescriptions, as "nutritional supplements" in pills, powders, and beverages. For example, health-food stores today might sell extracts of garlic, bilberry, and cranberry, bran extract from cereals, and edible oils high in gamma-linolenic acid from hempseed. Because they are much less expensive than drugs, such herbal preparations or extracts used as additions to diet have been advanced as a new, cost-effective health-care system. The marketing phrase that has become most popular for extracts with healthful properties is "nutraceutical" (often spelled nutriceutical in the past), a term coined by Dr. Stephen DeFelice of the Foundation for Innovative Medicine, a New Jersey based industry group. His original definition was "a food derived from naturally occurring substances which can and should be consumed as part of the daily diet, and which serves to regulate or otherwise affect a particular body process when ingested." However, the term "nutraceutical" is now commonly applied to an extremely wide variety of preparations with perceived medicinal value but not necessarily with apparent food value (such as amino acids, essential fats, dietary fibers and fiber-enriched foods, plant and animal pigments, antioxidants, vitamins, minerals, sugar and fat substitutes, fatless meat, skim milk, genetically engineered designer foods, herbal products, and even processed foods such as cereals, soups, and beverages).

In addition to the direct consumption of food extracts, constituents from plants are now used to enrich other foods. Among the marketing terms that have developed for foods fortified with food extracts are "medical foods," "pharma foods," "phytofoods," and (most popular) "functional foods." The first commercial functional food seems to have been calcium-fortified orange juice, and today, there are thousands of fortified food preparations offered in the marketplace.

Some food plants, especially herbs and spices, produce well-known physiological changes, which may be viewed as problems rather than health benefits, depending on individual perceptions. For examples, consumption of dandelion and asparagus increases urination, and many plants act as laxatives (e.g., elder flowers, violet flowers).

Natural components of plants, when consumed in normal amounts, almost certainly act as health promoters, although the mechanisms are not well understood. It appears, for example, that such natural food components as vitamin C, vitamin A and its precursors, vitamin E, selenium, fiber, purines, and glutathione act as "anticarcinogens," counteracting the effect of some cancer-promoting chemicals. There are hundreds of natural plant chemicals, such as carotenoids, flavonoids, isoflavones, and protease inhibitors, which are thought to protect against chronic health conditions.

There is a general consensus that while extracts, vitamins, and minerals may be useful as *supplements* to a diet, they should never be used as *substitutes* for a healthy, well-balanced diet. Taken in isolation from a normal diet, supplements often seem to be less effective.

There is considerable justification for the old saying that "good medicine has to taste bad," and this is illustrated by the presence of bitter ingredients in food that are therapeutic. Broccoli is bitter because of the presence of glucosinolates, notably sinigrin, which has anticancer properties. The bitterness of green tea is partly due to the presence of flavans, particularly catechin and its derivatives, which also have anticancer properties. Soybean isoflavones, notably genistein and daidzein, are responsible for bitterness but again are thought to help prevent cancer. Grapefruit is bitter because of the presence of flavanones, especially naringin, which counteracts cholesterol accumulation. Bitter flavonoids (notably quercitin and apeginin) in chamomile are anti-inflammatory. Recently, broccoli cultivars have been bred with highly-reduced bitterness. Unfortunately, with reduced glucosinolates, they are also less health-promoting. Researchers are making progress developing "bitter blockers"—additives that could fool the taste buds into ignoring bitter elements in food, beverages, and medicine. In the meantime, for many, the only way to consume foods that taste bad is to combine them with foods that taste good, in the manner of serving broccoli with melted cheese.

Culinary Purposes: Why Most Food Is Not Consumed Raw

Food preparation has reached such remarkable levels of sophistication—both scientifically and artistically—that the original functions of altering food from its raw state before it is consumed have become obscure. Historically, the most important reason for processing raw foods was to detoxify them (Johns and Kubo 1988; cf. Katch 1990, Saad 2007). This is most commonly accomplished by heating, which degrades many toxins. Another method is fermentation, which is widely practiced by primitive societies to get rid of toxins (most often, a container of food is simply buried in the ground for a period). Leaching—simply soaking in changes of water, is an effective way of getting rid of certain toxins. Pickling sometimes also denatures toxins. In the same way that charcoal adsorbs (i.e., surface-binds) tiny particles, components of dietary fiber and clay particles have been used as a way to bind toxins in foods. Clays are either added to foods prior to eating, or clay is simply consumed along with food. Such *geophagy*, the consumption of soil, is remarkably widespread, often found in animals other than humans, and undoubtedly is a key contributor to nutritional health. Simply drying sometimes is effective by removing volatile toxins. Drying stinging nettle completely inactivates the stinging hairs, and the herb can be safely handled and eaten. Simply grinding up toxic varieties of cassava is a widely practiced way of getting rid of the poisonous hydrogen cyanide that is generated (and would be generated in the stomach if the cassava had been chewed directly and swallowed).

The second most important function of culinary transformation of raw foods has been to make them more palatable. Peeling or cutting away harder or less palatable parts is the simplest way of making many foods more pleasant to eat. Cooking, soaking, grating, and other techniques make food softer, and therefore easier to chew. Grinding food into smaller particles before ingestion (a way of reducing the need to chew) increases the surface area on which digestive enzymes act, therefore making the material more digestible.

Still another significant historical function of food processing is preservation, so that the food will not spoil, and can be stored. This is commonly accomplished by sun-drying (occasionally by freeze-drying), frequently also by adding salt (sugar is also a preservative, but this function of sugar is much more recent than for salt). In cold areas, food could simply be frozen. Pickling and fermenting are additional old preservation techniques. Before refrigeration, the very widespread use of spices, almost all of which have very strong antimicrobial properties, was very important in food preservation.

Some culinary procedures make food more nutritional. Adding alkaline substances (such as plant ashes) to some foods improves the availability of certain nutrients. Corn is low in niacin (one of the B vitamins), low in calcium, and has a relatively poor balance of essential amino acids. Native peoples in the Americas learned to add lye to corn (maize). This frees bound niacin and adds calcium. Moreover, corn has long been served with beans, which improves the amino acid pattern (as outlined above in NUTRITIONAL SUCCESS BY COMBINING CEREALS AND PULSES).

The culinary world has progressed from the times that food preparation was mainly concerned with getting enough good food. Since at least the Renaissance, the natural artistic sense of humans has witnessed food preparation elevated to a high art. Much of this was carried out at the expense of nutrition, but modern cooks are increasingly sophisticated in choice and preparation of nutritious foods.

Food Preferences

> "IT'S BROCCOLI, DEAR."
> "I SAY IT'S SPINACH, AND I SAY THE HELL WITH IT."
>
> —Caption authored by E.B. White for a cartoon by Carl Rose in *The New Yorker*, 1928

People as a whole are extremely varied in their food preferences. One need only observe the choices made by others in a restaurant or in the checkout line of a cafeteria. However, individuals are generally narrow in their personal food preferences. Of the thousands of foods that can be purchased in a supermarket, most people restrict their choices to a set of no more than a few dozen. Food likes and dislikes are a powerful force in nutrition. Dozens of books and thousands of articles are available, detailing how cultural influences, psychology, and physiology determine relative likes and dislikes for food (e.g., Anderson et al. 2002, Barker 1982, Capaldi 1996, Macbeth 1997, Moskowitz et al. 2002, Rozin 2000, Rozin and Vollmecke 1986, Sokolov 1991).

Genetics plays a significant role in human food preferences. Humans are hard-wired to like certain tastes, and detest others. There is an innate preference for sweet tastes, which is obviously adaptive, since in nature most sweet things have sugars, which are a source of calories. An aversion to bitter tastes also appears to be universal in humans, and this is adaptive because bitter tastes are typically correlated in nature with poisonous substances. The adaptive importance of avoiding bitter substances is indicated by the fact that there are dozens of kinds of bitter receptors in the taste buds of humans, each sensitive to slightly different substances. Acid (sour) tastes seem to be avoided by humans, presumably because they indicate that foods have spoiled (or, in the case of fruit, that it is immature). Bitter foods (such as broccoli, coffee, and olives) and sour foods (including fermented foods such as sauerkraut) often require habituation by eating over a lengthy period before a preference for them is developed. Irritating substances, such as calcium oxalate needles in a number of species, are avoided because of the universal biological adaptation to avoid pain and associated physical damage. As with bitter and sour foods, irritating foods (like hot chile pepper) require lengthy conditioning by repeated consumption. As with taste, humans naturally dislike the bad smell of spoiled or rotting food, and find the natural scents of numerous fresh foods to be quite attractive.

It is interesting that we have specific taste receptors to attract us to sweet foods, but we lack taste organs to attract us to fat, starch, and protein. Clearly, sweetness was of particular importance to humans during our evolution, and this must mean that we have a "frugivorous" genetic heritage (we are natural fruit-eaters). This makes sense because the number of species with palatable fleshy fruits greatly exceeds the numbers with edible leaves, stems, roots, or nuts.

Genetic differences among humans explain some of the differences in food preferences. Asparagus provides an interesting demonstration of how genetic variation in humans can lead to a different appreciation of the same food. Asparagus has sulfur-containing amino acids that break down during digestion into sulfur-containing compounds, particularly methyl mercaptan. These sulfurous compounds are similar to those that make skunks smell. In the past, it was thought that about one in two people has a gene enabling the sulfurous amino acids in asparagus to be broken down into their smellier components (these people were called "stinkers"). The prevailing opinion now is that everyone digests asparagus the same way, but only about half of people have a gene that enables them to smell the compounds. Those who find that asparagus makes their urine stink are likely to avoid this vegetable in the future.

Clearly, culture plays a predominating role in determining life-long patterns of food preferences. The particular foods served, including how they were prepared, produce very strong acquired preferences, as well as aversions. Determining a person's culture or ethnic group is the simplest and by far the most reliable way of predicting his or her food preferences. (Note C.W. Schwabe's *Unmentionable Cuisine* (Schwabe 1979), which details numerous foods that are eaten only by people of certain cultures.) Even the odors of many ethnic foods evoke strong positive or negative reactions, depending on how accustomed one has become to them.

Humans generally are ambivalent with respect to unfamiliar foods, and this is likely adaptive. Before people learned to distinguish all of the species that are poisonous from those that are edible, trying new foods represented a way of discovering new sources of nutrition, but it was at the risk of being poisoned. Eating something that subsequently produces (or at least is thought to produce) sickness is likely to result in a permanent dislike of that food, and this innate tendency has kept people from repeated exposure to toxins. Today, the chances of being poisoned are slight, and with the understanding that it takes time to develop a taste for unfamiliar foods, many people try new ethnic foods as an adventure. Nevertheless, for most adults, a single exposure to a new food that proves to be strange or of an unfamiliar taste is often sufficient to reject it forever.

Even when the strong biological and cultural determinants of what foods people like and dislike are taken into account, it has been found that there are very large differences among individuals, even within the same family. Moreover, just why the majority of these differences develop is not clear.

Some people are motivated to consume certain foods that are perceived as having prestige. Probably because the name "arugula" seems exotic, this vegetable became a favorite salad green of yuppies, who have willingly paid fancy specialty store prices for it. Conversely, many have avoided garlic and cabbage, associating these foods with poverty. One of the saddest manifestations of this phenomenon occurs in some poor countries, where people have been observed to avoid extremely healthy wild leafy vegetables while choosing less nutritious but much more costly imported lettuce.

Food Allergies

"WHAT'S ONE MAN'S POISON, SIGNIOR, IS ANOTHER'S MEAT OR DRINK."

—Francis Beaumont (1584–1616) and John Fletcher (1579–1625), English playwrights of Shakespeare's time.

As noted above, tolerance to certain foods is often determined by whether or not one has become accustomed to them. Garlic and hot chile peppers are good examples. Some people are able to eat them in very large quantities, while others quickly get heartburn from very small amounts. Needle-like crystals of calcium oxalate in taro produce a stinging sensation on the skin when the plant is handled, but many taro farmers seem to gradually lose sensitivity to the irritant.

Much more serious for many, however, is an inborn or acquired intolerance of certain foods, which usually can not be overcome by becoming habituated to them. Indeed, frequently foods that some people can enjoy with immunity are potentially deadly for others. For virtually every plant, there are individuals who will develop allergic reactions from consumption and even from just physical contact. The most commonly reported allergenic plant foods are legumes (such as soybeans), tree nuts, and wheat, and the substances responsible are usually proteins. Peanut (which see) is the most common cause of life-threatening reactions. People who are susceptible to hay fever caused by ragweed pollen may well also have allergic reactions to the flowering tops of related herbs, such as Roman or German chamomile (also in the sunflower family), which contain pollen. ("Cross-reaction" is a phenomenon whereby people who acquire allergies to one plant also become sensitized to react the same way to other plants that contain the same or similar substances.)

Dermatitis (skin rash) is a very common reaction to both handling and eating plants. Ceriman is an example of a plant with needle-like calcium oxalate crystals, and handling and eating the unripe fruit can cause skin and oral reactions. The ripe fruit is edible for most people, but some have reported throat irritation, hives, diarrhea, intestinal gas, and even anaphylaxis after eating ceriman. Some individuals are allergic to the fruits of pawpaw, getting a skin rash from handling them, or serious nausea, vomiting, or diarrhea from eating them. "Photophytodermatitis" is a rash produced in the presence of sunlight in some individuals after consuming some food, or even just touching it. Several species (including angelica, buckwheat, Japanese knotweed, lemon, parsnip, rau ram, and rue) can produce this condition in susceptible people.

Some people are susceptible to products that form only after foods are processed. The chemical tyramine provides a good example. Tyramine is formed when bacteria break down protein, and it causes migraines in some people by constricting blood vessels. It is present in aged wine (as well as other aged foods such as cheese, salami, and bacon). Studies have shown that many people susceptible to migraine excrete tyramine relatively slowly, and chemical analysis of a urine sample taken during a migraine attack can be used to determine whether a person is having a tyramine migraine. It has been estimated that perhaps 10% of migraine sufferers get migraines if they eat more than a small amount (6 mg) of tyramine.

For most people, common food plants can be handled and eaten safely. As a matter of prudence, unfamiliar foods should be sampled in small quantities at several meals to be certain that they can be consumed safely.

Ethnic/Racial/Geographical Genetically Determined Nutritional Susceptibilities

Numerous widespread species exhibit genetic differences in different parts of their natural distribution, and humans are no exception. Of course, peoples have migrated and intermarried, obscuring the differences. Nevertheless, there remain significant differences, and knowledge of different racial susceptibility to certain diseases is medically invaluable. In particular, there is relatively high occurrence of illness in certain populations of ethnic, racial, or geographical groups in response to consumption of certain foods.

For animal foods, milk drinking is the best-known example of geographically linked genetic susceptibility to illness. For most mammals, lactose (milk sugar) is not digestible after the period of weaning. The gut enzyme lactase splits lactose into digestible components, and when production of the enzyme drops after weaning, lactose remain in the digestive system, causing cramps, gas, and diarrhea. Some human groups in the past took up the use of livestock as sources of milk for adults, and were selected for tolerance to lactose. Lactose tolerance is associated with regions where milk has been produced historically (Newcomer 1978), most notably with northern Europeans and some pastoral groups in Asia and Africa. The best-known examples of geographical/racial differences in dietary susceptibilities associated with the plant world follow.

Celiac disease is a condition in some people resulting from their consumption of gluten. Intestinal damage, malabsorption of nutrients, and wasting are typical symptoms. The disease is controlled by removing gluten sources from the diet. Gluten occurs mainly in wheat (which see), and in lesser amounts in related cereals (particularly barley). Gluten, formed by the proteins in wheat flour, is responsible for the texture, appearance, and volume of dough. Gluten gives dough the ability to rise by allowing it to retain gas released by a leavening agent such as baking powder or yeast. A high prevalence of celiac disease has been observed in northwestern Europe and northwest India, both regions at the outer limits of wheat cultivation in Eurasia. It is suspected that in historical times in wheat-growing regions, people with a genetic predisposition to celiac disease were eliminated (before modern medicine, about 15% of celiacs died from the condition).

Favism is a severe hemolytic anemia resulting from an inherited susceptibility to the broad bean (which see). This disease varies considerably, depending on racial group and geography, occurring in Oriental Jews, Mediterranean Europeans, Arabs, Asians, and Blacks (5–50% of the population), but virtually absent in Northern Europeans, European Jews, American Indians, and Eskimos (Cheeke and Shull 1985).

There is some evidence of different tolerances to alcohol among the races of humans. Asian and Amerindian populations are deficient in an enzyme (aldehyde dehydrogenase) that metabolizes the acetaldehyde formed from digesting ethanol, and accordingly are less tolerant of alcohol.

Inuit and Amerindian populations of northern Canada have an unusually high intolerance (3–10%) of sucrose. It has been suggested that the traditional Inuit diet, high in meat and low in sugar, led to the evolutionary loss of the normal insulin response to sugar.

Achieving Nutritional Balance in the Modern Diet

"YOU ARE WHAT YOU EAT."

—Ludwig Andreas Feuerbach (1804–1872), German philosopher
This famous quotation is often incorrectly attributed to Brillat-Savarin (cited below), and dozens of others.

"TELL ME WHAT YOU EAT AND I WILL TELL YOU WHO YOU ARE."

—Anthelme Brillat-Savarin (1755–1826), French politician, best known for his book *The Physiology of Taste*

It has been estimated that humans require 40–50 food constituents for good health. Six classes of these nutrients are commonly recognized: carbohydrates, fats, proteins, vitamins, minerals, and dietary fiber (nutritionists sometimes consider water, which makes up two-thirds of the body, to represent a seventh class). Carbohydrates and fats are the primary energy sources. Proteins also act as energy sources, but their primary functions are related to tissue growth, maintenance, and repair, and they play many physiological roles. Fats are the principal form of food energy storage of the body, and constitute 15–30% of body weight. By contrast, carbohydrates make up only about 0.5% of the body.

Carbohydrates

Most nutritionists recommend that carbohydrates should be the main source of energy or calories. Fats and proteins also supply energy, but have other contributions. Carbohydrates are classified into simple and complex. Simple carbohydrates are sugars, such as glucose (= dextrose, found naturally in sweet fruits), fructose ("fruit sugar," found naturally in ripe fruits, some vegetables, and honey), sucrose ("table sugar," found in sugar cane and sugar beet), and lactose ("milk sugar," found in the milk of mammals), which basically supply energy. Complex carbohydrates are long chains of linked sugars. The main edible complex carbohydrate is starch, which is a chain of glucose molecules. Digestible carbohydrates are broken down to glucose, the common fuel for cells, regardless of whether simple or complex. A chain of events is set off whenever substantial amounts of highly digestible carbohydrates are consumed: they break down very quickly to glucose (blood sugar), a rush of glucose floods the blood, this raises the level of insulin which in turn results in the sugar quickly being consumed by the cells of the body, and then the brain sends out hunger signals to eat again. Such very rapid digestion is considered undesirable, and nutritionists have long recommended limiting intake of simple carbohydrates, because they have been linked with heart disease, diabetes, and obesity. Recently, similar concern has been expressed about the too-rapid digestion of some complex carbohydrates, particularly from such staples as white bread, potatoes, and white rice. Much more desirable is the slow digestion of certain carbohydrates from whole grains (such as whole oats, popcorn, brown rice, and "whole wheat" products, including pasta and bread) and beans. This regularizes blood sugar and insulin levels, and also produces a sense of fullness, lowering the tendency to overeat. Moreover, whole grains and beans contribute fiber, vitamins, and minerals, discussed below.

Grains are the predominant source of carbohydrates, and are therefore of great dietary significance. A cereal grain is composed of three parts—the germ (the embryo, i.e., the tiny plantlet that is intended to produce a new plant), the endosperm (a food supply for the embryo, like the yolk of an egg), and the bran (tough protective tissues covering the germ and endosperm, derived from the seed coat and fruit walls). The germ is often removed from wheat and rice because this improves keeping qualities (unsaturated fats present tend to go rancid). Much of the bran is also removed because this improves taste. Unfortunately, removing the germ and bran also removes essential nutrients. Many specialty cereals (such as fonio, teff, the millets, sorghum, and wild rice), as well as such pseudo-cereals as amaranth, buckwheat, and quinoa are mostly sold as whole grains. However, what are marketed as "whole grain products," particularly breads with such names as "multi-grain" and "12-grain," are frequently actually made predominantly from refined, white flour, with caramel or molasses added to give the dark color suggestive of wholesomeness.

Fats

Fats (which are solid) and oils (which are liquid fats) contain more than twice as many calories on a volume basis compared to carbohydrates and proteins. Chemically, fats are chains of carbon atoms bonded to hydrogen atoms, and the number of hydrogen atoms present relative to the carbon atoms defines the degree of "saturation" (more hydrogen atoms for each carbon atom means greater saturation). The degree of saturation is a principal factor determining whether *consumed* fat is healthy or unhealthy (the body manufactures saturated fats for particular purposes, so it can't be said that all saturated fat is bad). Polyunsaturated fat is less saturated than monounsaturated fat (both are "good"). Polyunsatured fat and monounsatured fat are required to form the membranes covering the body's cells. Good sources of polyunsaturated fat are oils made with corn, soybean, safflower, and cottonseed, and fish. Good sources of monounsaturated fat are olive oil, Canola oil, peanut oil (also peanuts and peanut butter), most nuts, and avocados. By definition, "saturated fats" are carbon chains that have the maximum number of hydrogen atoms attached. There are about two dozen kinds of saturated fat in nature. The undesirable saturated fats are found in red meats, most margarines, vegetable shortening, many fast foods (notably French fries), most commercial baked goods, whole milk, butter, cheese, ice cream, chocolate, and some vegetable oils (particularly coconut and palm oils). They raise blood cholesterol, thereby clogging arteries, leading to heart disease, stroke, and other problems when consumed in large amounts. However, some saturated fats tend to increase "bad cholesterol" (LDL) more than others: those in dairy products are worse than those in beef fat, which is worse than those in chocolate.

Trans unsaturated fat is a special kind of unsaturated fat that is manufactured from polyunsaturated vegetable oils using a process called partial hydrogenation. This changes liquid oils to solid oils that are more easily stored and transported, become rancid less quickly, and can be used to make solid products like margarine and Crisco. Trans fats are considered very harmful, and efforts are underway in some countries to ban the use of trans fats in commercial food products (currently widely used in fast-food French fries and commercial cookies that list "partially hydrogenated vegetable oil" or "vegetable shortening").

Omega-3 fatty acids (also known as n-3 fatty acids) make up a special class of polyunsaturated fats. Although consumed in relatively small amounts, they are of exceptional importance. One of these, alpha-linolenic acid, must be obtained from food (most fatty acids can be synthesized from other components by the body). A lack of it results in neurological abnormalities and poor growth. The omega-3 fatty acids are of special benefit in preventing or treating heart disease and stroke, possibly also such autoimmune diseases as lupus, eczema, and rheumatoid arthritis, and other problems. Good sources of omega-3 fatty acids are fish, shellfish, flaxseed, pumpkin seeds, walnuts, Canola, and unhydrogenated soybean oil.

Omega-6 fatty acids (also known as n-6 fatty acids) constitute another class of polyunsaturated fatty acids of special dietary concern. One of these, linoleic acid, must be obtained from the diet; a lack of it results in a scaly rash and reduced growth. Good sources are oils from sunflower, safflower, sesame, and corn. Normally, linoleic is converted in the body to gamma-linolenic acid (GLA for short), but as many as 15% of people lack the ability to synthesize sufficient GLA, which is vital to good health and accordingly should be supplemented in the diet if it is deficient. Good sources of GLA are oils from evening primrose, borage, black currant, grape, and hempseed.

Alpha-linolenic acid (an omega-3 fatty acid) and linoleic acid (an omega-6 fatty acid) are the only two fatty acids that *must* be eaten, and so are termed "essential fatty acids." At one time, these were called "vitamin F" (F for fat). Omega-3 and omega-6 fatty acids are important because they are converted into a family of compounds called prostaglandins, which are hormone-like chemicals with numerous important functions, including the enhancement of immune system function and the control of inflammation. Much of the Western World's diet is serious nutritionally unbalanced in its *ratio* of consumption of omega-3 and omega-6 fatty acids—too much omega-3 is consumed in relation to the amount of omega-6, and this is deleterious to health. A 3:1 balance of omega-3 to omega-6 is considered ideal. Practically, this is achieved by increasing consumption of foods rich in linoleic acid.

Proteins

Proteins are long, intricate chains of basic units called amino acids. About 22 amino acids occur in human proteins. Protein makes up much of the body, including hair, skin, muscles, blood hemoglobin, and enzymes. There are over 10 000 different proteins in the body. Amino acids are not stored in the body like fat, and need to be ingested regularly. Eight amino acids are generally regarded as "essential," and must be obtained by humans from their diet. (The body synthesizes nonessential amino acids.) All essential amino acids must be consumed simultaneously for protein synthesis, and if some are absent, required proteins can not be manufactured by the body. Relative amounts of the essential amino acids that must be consumed by humans is a quite controversial subject. The following data are from the U.S. National Academies Press Dietary Reference Intakes for Energy, Carbohydrate, Fiber, Fat, Fatty Acids, Cholesterol, Protein, and Amino Acids (Macronutrient) (2002) website (http://books.nap.edu/books/0309085373/html/524.html#pagetop). Recommended relative estimates of daily requirements for individual amino acids, expressed as mg amino acid/kg body weight/day, for the eight essential amino acids, follow (for adults): isoleucine (23), leucine (49), lysine (48), methionine (methionine + cysteine = 23), phenylalanine (phenylalanine + tyrosine = 48), threonine (28), tryptophan (8), and valine (32). In addition, arginine and histidine are considered essential for children. "Complete protein," containing all of the essential amino acids, is found mostly in meat, poultry, fish, eggs, and dairy products. Unfortunately, many protein-rich animal foods contain high levels of saturated fat. Beans, seeds, and nuts often are rich in proteins. However, most plant foods are deficient in certain amino acids, notably tryptophan, methionine, lysine, and (or) threonine. Vegetarians require a carefully balanced diet in order to achieve adequate protein nutrition. Most people eat more protein that the body needs. Adult women only require about 50 g (less than 2 ounces) daily while adult men need about 65 g. This is supplied by any one of the following: about 250 g (9 ounces) of beef, chicken, or fish; 9 large eggs; 3 cups of cooked lentils or peas; or 120 g (4 ounces) of almonds (but keep in mind that plants tend not to have complete protein).

Vitamins

Vitamins have been defined simply as "carbon-containing compounds essential in small quantities for normal functioning of the body." (Strictly, vitamin D is not a vitamin, since it can be synthesized by the skin under the influence of ultraviolet light. However, in northern regions for much of the year when exposure to sunlight is limited, vitamin D is required in the diet.) If any vitamin is not present in sufficient amounts, a deficiency disease results. Vitamins facilitate chemical reactions in body cells, particularly in processing food. There are 13 essential vitamins (at least 16 compounds have been called vitamins), classified into those that are soluble in fat and those that are soluble in water (Table 5). Fat-soluble vitamins tend to accumulate in the body, water-soluble vitamins do not, and because they are readily excreted, they must be provided by the daily diet. Vitamins are obtained by eating a combination of animal and plant foods. (Resident intestinal bacteria, if supplied with the appropriate precursors, may synthesize vitamin K and most of the B vitamins.) Vitamin B_{12} is not found in plant foods (unless fermented). Vitamin C is largely obtained from fruits and vegetables; there is a small amount in meat, but this is destroyed when cooked. Choline is not strictly a vitamin, since the body can synthesize small amounts, but not enough to meet all metabolic needs. In the future, additional compounds may be added to the list of vitamins.

Table 5. Major vitamins. (Vitamins A, D, E, and F are fat-soluble, the remainder are water-soluble. There are eight B vitamins, including B_1 (thiamine), B_2 (riboflavin), B_3 (niacin), B_6 (pyridoxine), B_{12} (cobalamin), pantothenic acid, biotin, and folic acid.)

Vitamin	Common non-plant sources	Common plant sources	Importance
A_1 (retinol) A_2 (dehydroretinol)	Fortified milk, butter, cheese, eggs, liver, fish oils	Vitamin A does not occur in plants, but plants (mostly the green parts) supply animals with precursors ("provitamins"), including beta-carotene and other carotenes	Essential for eye health, night vision, good bones and teeth, growth, proper digestion, production of red and white corpuscles, healthy immune system function, and lactation
B_1 (thiamine = thiamin)	Yeast, meat, eggs, milk	Green vegetables, legumes, seeds, whole grains, cereals	Essential to production of energy through carbohydrate breakdown; necessary for nerve function, muscle function; prevents beriberi
B_2 (riboflavin)	Yeast, liver, kidney, eggs, ham, milk, cheese, yogurt, beef	Leafy green vegetables, whole-grain breads, cereals	Essential for cell growth (regulates hormones); involved in enzymatic reactions for metabolizing proteins, fats, and carbohydrates; helps maintain healthy skin, eyes, mucous membranes, nerve function
B_3 (niacin = nicotinamide = nicotinic acid)	Yeast, meats, fish	Peanuts and other nuts, legumes, whole grains, spinach, enriched bread and cereals	An essential component of some enzymes producing energy and tissue respiration; aids in formation of red blood cells; prevents pellagra
B_6 (A group of three closely related compounds: pyridoxine, pyridoxal, pyridoxamine)	Liver, yeast, chicken, fish, eggs	Brown rice, whole-wheat products, bananas, potatoes, watermelon, avocado, etc.	Associated with enzymes in various metabolic pathways, particularly for brain and nervous system functions; protects against cardiovascular disease; assists in fighting infection
B_{12} (cobolamin)	Meats, liver, kidney, fish, milk products, eggs	Usually not considered available in sufficient amounts from plants	Participates in nucleic acid synthesis; necessary for healthy red blood cells, nerve function, and protection against cardiovascular disease; deficiency produces anemia, neurological disorders
Biotin	Widely present in animal foods, in kidney, yeast, egg yolk	Uncommon or in low amounts in plants	Involved in several metabolic processes, including synthesis of fatty acids (rarely observed to be lacking in humans)
C (ascorbic acid)	Absent (except in some raw meat, as explained above)	Citrus fruits, berries, green vegetables, peppers, fortified cereals	Needed for normal cell functioning, for formation of healthy collagen, bones, teeth, cartilage, skin, capillary walls; promotes effective use of other nutrients; an antioxidant
D (calciferol)	Egg yolks, fortified milk, liver, fish oils	Plants are not a significant source	Promotes absorption and use of calcium and phosphorus; necessary for healthy bones, teeth; assists functioning of nervous system and muscles; may play a role in decreasing the risk of some autoimmune diseases

Table 5 (*concluded*).

Vitamin	Common non-plant sources	Common plant sources	Importance
E (two chemical families: tocopherols and tocotrienols, with various forms, any one of which is sufficient; α-tocopherol is the chief kind used by humans, and many equate it with vitamin E)	Poorer sources than plants, but present in animal fats and fish oils	In many foods, including leafy green vegetables, vegetable oils, nuts, seeds, wheat germ	An antioxidant that protects red blood cells, vitamin A, and unsaturated fatty acids from oxidative destruction; maintains healthy membranes, fertility; protects against a range of diseases including heart disease and some cancers
Folic acid (= folate = folic acid = folacin)	Liver, kidney, yeast	Fortified cereals, dark green leafy vegetables, fruits, legumes, wheat germ	Nucleoprotein synthesis; necessary in mitosis; protects against neural tube defects in babies, cardiovascular disease, and colon cancer; helps in red blood cell formation; deficiencies produce anemia, digestive problems, fetal abnormalities
K (menadione)	Milk, eggs	Green vegetables, cereals, tomatoes, orange peel	Participates in synthesis of prothrombin in liver for normal blood clotting; maintains bone metabolism (widely present in both plant and animal foods)
Pantothenic acid	Liver, kidney, yeast	Wheat, bran, peas	Necessary for cellular metabolism (widely present in both plant and animal foods, never observed to be lacking in humans)
Choline	Milk, eggs, liver	Peanuts; widely distributed in foods	Important in nerve function; natural deficiency is very rare; it has been speculated that strict vegetarians may be liable to insufficient intake

Minerals

Minerals required in large amounts are called "major minerals" (Table 6). These include sodium potassium, chlorine, magnesium, sulfur, calcium, and phosphorus. Calcium and phosphorus occur in the body in particularly large amounts, contributing especially to bone structure. The remaining minerals are "trace elements" present in very small amounts, and functioning to regulate numerous body functions. At least 20 elements are candidates as essential for human nutrition in extremely small trace amounts. Cobalt, iodine, manganese, and selenium are known to be involved in biochemical functions, and are therefore definitely necessary. Molybdenum also has a defined physiological function, but is considered relatively unimportant. There is relatively strong evidence for nutritional significance of boron and chromium, less evidence for the essentiality of arsenic, nickel, silicon, and vanadium, and quite limited evidence of the nutritional relevance of aluminum, bromine, cadmium, fluorine, germanium, lead, lithium, rubidium, and tin. Minute amounts of arsenic have often been found in turnips and cabbages, perhaps the most common vegetables known to accumulate this element. Although the presence of arsenic is considered a danger sign, indicating contamination, it is of interest that this deadly poison may in fact be necessary for human life. Essential minerals are obtained from both plant and animal tissues. Plants are not a good source of sodium, which was a problem in past times, but hardly at present, with the high amounts of salt routinely added to food products actually posing a hazard to health.

Table 6. Major nutritional minerals.

Mineral	Common non-plant sources	Common plant sources	Importance
Calcium	Dairy products, seafood	Green leafy vegetables, tofu	The body contains about 1 kg (2 pounds) of calcium, almost all in bones; necessary for healthy bones, teeth, for clotting of blood, functioning of nerve tissue and muscles, for enzymatic processes; controls movement of fluids through cell walls
Chromium	Yeast, calf's liver, American cheese	Whole-grain products, bran cereals, wheat germ	Assists insulin to convert carbohydrates and fat into energy
Copper	Shellfish, liver	Nuts, seeds, legumes, whole grains	Necessary for formation of skin and connective tissues; required for many chemical reactions producing energy; necessary for heart function
Iodine	Iodized salt, seafood, animals and plants raised near seacoast (inadequate soil levels decrease amounts in both plants and animals)		A component of thyroid hormone; deficiency leads to goiter; in pregnancy, deficiency can cause mental retardation, muscular and skeletal malfunctions in child
Iron	Meats, poultry, fish	Cereals, fruits, green vegetables, whole-grain products	Essential to form oxygen-carrying hemoglobin in red blood cells; involved in muscle functioning and enzyme reactions for energy production
Magnesium	Generally present in animal tissues	Nuts, legumes, whole grains, green vegetables, bananas	Essential enzyme activator, probably involved in formation and maintenance of body protein; important for cardiac function
Phosphorus	Milk, meats, poultry, fish	Cereals, legumes, fruits	Involved in production of energy for body, a constituent of bones and teeth, necessary for functions of brain, nerves, muscles, enzyme formation
Potassium	Abundant in both plant and animal tissue		Promotes certain enzyme reactions, acts with sodium to maintain normal pH levels and balance between fluids inside and outside cells; assists in nerve transmission, contraction of muscles including heart muscle
Selenium	Seafood, kidney, liver	Cereals, grains	Functions as an antioxidant, protecting cells from damage, essential for heart
Sodium	Table salt, many prepared foods, common in both plant and animal tissues		Regulates volume of body fluids; in balance with potassium, maintains balance between fluids inside and outside cells; necessary for nerve and muscle functioning, functioning of heart muscle
Zinc	Meats, poultry, oysters, eggs, milk, yogurt	Legumes, nuts, whole-grain cereals	Component of enzymes involved in energy production and digestion of protein; required for normal function of several metal-protein complexes; important roles in functioning of immune system, cell division, cell growth, blood-clotting and wound healing

Fiber

Fiber comes from plant foods. It is not strictly a nutrient (because it is not digested and absorbed into the body), but is important for the normal function of the gastrointestinal tract. Insoluble fiber (which does not dissolve in water) originates from the cell walls of plants, and is mainly cellulose. It occurs in cereal bran, whole grains, and vegetables, and provides bulk, which helps the movement of waste through the large intestine. Diets high in insoluble fiber are desirable to prevent constipation, type 2 diabetes, and diseases of the intestines. Insoluble fiber also seems to slow digestion and thereby keep hunger pangs at bay. Soluble fiber (which dissolves in water) occurs in legumes, oats, barley, and some fruits and vegetables. Soluble fiber tends to lower blood cholesterol, by forming a jelly-like material that traps cholesterol in the intestine, increasing the excretion of cholesterol.

Antioxidants

Antioxidants are a group of substances that protect cells and certain of their constituents against the destructive effects of oxygen. These include vitamins C and E, beta-carotene and other carotenoids, the minerals selenium and manganese, and probably hundreds of other compounds that are present in food. Oxygen-using reactions required to burn fats and carbohydrates and other oxygen-using processes generate oxygen-based by-products called free radicals. Free radicals are also generated in the environment, occurring in the air (in cigarette smoke, for example), and even as a result of sunlight contacting skin. These damage proteins, membranes, nucleic acids, and other critical components of cells. Free radicals may contribute to illnesses such as cancer, arthritis, heart diseases, cataracts, memory loss, and aging. Antioxidants neutralize free radicals.

Eating Patterns

Nutritionists are concerned with three distinct aspects of eating: getting enough food (i.e., a sufficient amount to supply the body's energy needs), getting enough of key food constituents (for example, protein, vitamins, and minerals), and avoiding overeating. Getting enough food is a problem associated with poverty, periods of growth (in the young and during pregnancy), for invalids, and with certain groups prone to health problems (for example, the elderly and alcoholics). The next two areas (getting enough of key food constituents and avoiding overeating) need to be considered together. In the Western World, most people naturally select a set of foods with an overall balance of nutrients appropriate to maintain health (this is evidenced by the comparative rarity of deficiency diseases). However, most people also tend to eat too much of some food constituents (notably sugar and unhealthy fats). Moreover, it is becoming increasingly clear that plant foods contain numerous ingredients that are beneficial to health (as mentioned in the section PLANT FOODS AS MEDICINE), and many people do not eat enough plant foods. The following information deals with the question of food consumption patterns that best meet overall nutritional goals.

The "food pyramid" is an old nutritional tool formed on a base of those foods that are recommended as staples to be eaten in greater quantities, with progressive layers toward the apex formed by foods to be consumed in smaller amounts. Variety, balance, and moderation are the key considerations. The United States Department of Agriculture Food Guide Pyramid was first made public in 1992. More than any other presentation in the world of nutrition, the food pyramid has been considered to represent the key to healthy eating, and has been very widely used as a teaching tool, pointing the way to those foods that are good and should be eaten often, in contrast to those that are less desirable, and should be eaten sparingly. However, many nutritionists have argued that the classic US food guide is flawed. First, the base (6–11 servings of bread, cereal, rice, and pasta), primarily of carbohydrates, suggests that somehow carbohydrates are the key to health, and certainly putting fats at the apex suggests that all fats are unhealthy. However, it is now clear that not all fats are unhealthy, and not all carbohydrate sources are equally healthy. Whole grains (like whole wheat and brown rice), which retain the outer bran and inner germ layers, along with the energy-rich starch that grains deliver, are notably healthier than highly milled cereals such as used to produce refined white flour and white bread (refining removes most of the bran and some of the germ). Whole grains are digested relatively slowly, keeping blood sugar and insulin levels stable, and perhaps preventing the development of type 2 (adult onset) diabetes. Second, the next layer, including vegetables (3–5 servings) and fruit (2–4 servings) underestimates their importance. There is good reason to consider vegetables and fruits as the base of a food pyramid, and to increase the relative quantities recommended in relation to carbohydrates. It is clear that a diet high in vegetables and fruits is a key

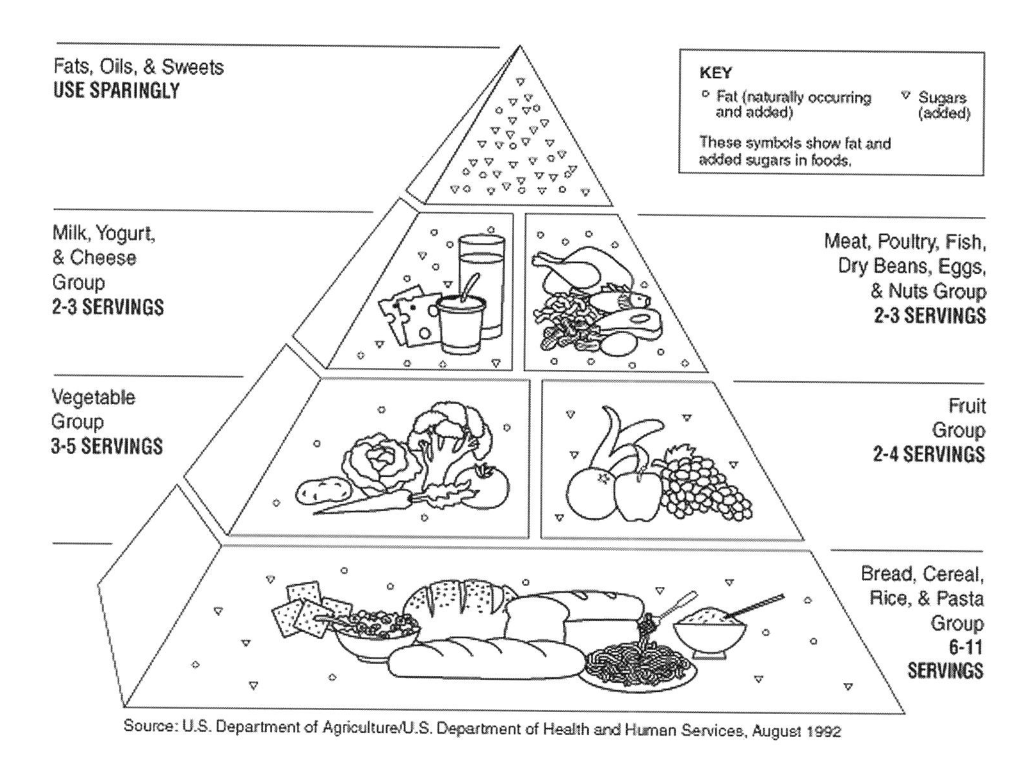

The old Food Guide Pyramid of the U.S. Department of Agriculture and the U.S. Department of Health and Human Services.

to reducing chronic diseases, heart disease, and some cancers. The third layer of the US food pyramid presents two groups. On the right is the protein group, with the recommendation to consume 2–3 servings from meat, poultry, fish, beans, and nuts. On the left is the dairy group (milk, yogurt, cheese), which is the recommended source of calcium. For both of these groups, nutritionists have debated the relative merits of exactly what sources should be used to supply protein and other nutrients. The top of the pyramid has fats, oils, and sweets. While sugar (responsible for a startling 30% of the average American's intake of calories) remains a dietary villain, some forms of fats and oils have regained respectability. Unsaturated oil, notably from olive, Canola, soy, corn, sunflower, and peanut, are healthy, and tend to improve cholesterol levels. Most nuts and avocados also contain unsaturated oil that is healthy, lowering cholesterol levels. The widespread use of (very unhealthy) hydrogenated oils producing trans fats obviously requires legislative correction.

One of the difficulties with governmental recommendations for food consumption is lobbying from food industries. In the case of the US food pyramid, organizations such as the American Meat Institute, the National Cattlemen's Beef Association, the Wheat Foods Council, the National Dairy Council, the Soft Drink Association, and the United Fresh Fruit and Vegetable Association, have naturally argued that their products deserve to be purchased by consumers in large quantities. [Nestle (2002) provides an analysis of how business influences the politics of eating in the United States.]

Willett et al. (2001) published an alternative "Healthy Eating Pyramid," which takes into account many of the criticisms of the classical US food pyramid. Their pyramid, however, is not a pure food pyramid, as its base is "daily exercise and weight control," and it advocates multiple vitamin/mineral supplements for most people.

Anatomy of MyPyramid

One size doesn't fit all

USDA's new MyPyramid symbolizes a personalized approach to healthy eating and physical activity. The symbol has been designed to be simple. It has been developed to remind consumers to make healthy food choices and to be active every day. The different parts of the symbol are described below.

Activity
Activity is represented by the steps and the person climbing them, as a reminder of the importance of daily physical activity.

Moderation
Moderation is represented by the narrowing of each food group from bottom to top. The wider base stands for foods with little or no solid fats or added sugars. These should be selected more often. The narrower top area stands for foods containing more added sugars and solid fats. The more active you are, the more of these foods can fit into your diet.

Personalization
Personalization is shown by the person on the steps, the slogan, and the URL. Find the kinds and amounts of food to eat each day at MyPyramid.gov.

Proportionality
Proportionality is shown by the different widths of the food group bands. The widths suggest how much food a person should choose from each group. The widths are just a general guide, not exact proportions. Check the Web site for how much is right for you.

Variety
Variety is symbolized by the 6 color bands representing the 5 food groups of the Pyramid and oils. This illustrates that foods from all groups are needed each day for good health.

Gradual Improvement
Gradual improvement is encouraged by the slogan. It suggests that individuals can benefit from taking small steps to improve their diet and lifestyle each day.

The new Food Guide Pyramid of the U.S. Department of Agriculture.

In 2005, the United States government issued "New U.S. Food Guidelines" and a series of guides geared to particular nutritional needs and lifestyles (see following URL). The new guidelines recommended eating (on average) nine servings a day (about 4½ cups) of fruits and vegetables, almost double the five servings previously recommended. The traditional US food pyramid was replaced with a pyramid in which six basic food groups are depicted as vertical bands running from the base to the tip (with width proportional to the amounts that should be consumed), and the importance of exercise symbolized by a person climbing one side of the pyramid. The rather elaborate symbolism associated with the new pyramid is explained in the above illustration.

This book refers the reader to very few websites, because most go extinct or quickly mutate into different forms. However, the United States government's online health-related resources are particularly extensive and informative, and can be depended on to be accessible. From the point of view of food nutrition, a useful URL for getting started is http://www.nal.usda.gov/fnic/dga/. This provides access to the 2005 edition of the new U.S. Food Guidelines. Another very useful subsite deals with "Fruits and Vegetables of the Month," and presents, for each vegetable or fruit, extensive background, detailed nutritional information, and recipes. As of 2009, information was posted on the following fruits and vegetables:

Fruits
Apple
Apricot
Avocado
Banana
Berries (blackberry, blueberry, currant, raspberry, strawberry)
Carambola
Cherry
Cranberry
Exotic fruit (cherimoya, kumquat, melon, pepino, sapote, ugli fruit)
Exotic winter fruit [tamarillo (a vegetable), feijoa, red banana, horned melon, guava]
Fig
Gooseberry
Grape
Grapefruit
Kiwi
Lemon
Lime
Mango
Melon
Nectarine
Orange
Papaya
Passion fruit
Peach
Pear
Pear, Asian
Persimmon
Pineapple
Plum
Pluot and aprium (plum–apricot hybrids)
Pomegranate
Quince
Watermelon

Vegetables
Artichoke
Asparagus
Beans— dried
Beans—fresh
Bell pepper
Broccoli
Brussels sprouts
Cabbage
Cactus
Carrot
Cauliflower
Celery and "fennel" (Florence fennel)
Chile pepper
Corn
Cucumber
Eggplant
Exotic vegetables (bitter melon, calabaza squash, chayote)
Garlic
Greens (broccoli raab, collard, kale, mustard greens, Swiss chard)
Leek and green onion
Lettuce
Mushrooms
Okra
Onion
Pea
Plantain
Potato
Radish
Rhizomes (ginger, galangal, and turmeric)
Rhubarb
Root vegetables (beet, celeriac, parsnip, rutabaga, salsify, scorzonera, turnip)
Spinach
Sprouts (alfalfa, mung bean, etc.)
Squash
Sweet potato
Tomatillo
Tomato
Tubers (cassava, Jerusalem artichoke, jicama, taro, water chestnut)

The Addictive Nature of Food and the Resulting Fat Pandemic

This section emphasizes that it is the warping of the original purpose of food, the sustenance of life, to a means of sensory gratification that has been responsible for the general misuse of food to produce illness rather than health.

In order to explain why people today in much of the world are eating too much and too unwisely, Rozin (2002) contrasted the present food situation with the environment in which humans used to exist:

- In the past, food was scarce, now it is abundant.
- In the past, the variety of foods available to a population of people was limited, now it is abundant.
- In the past, people had to work, forage, or hunt to obtain food. Now it is obtainable with very little effort.
- In the past, people usually died in middle adulthood, often from injuries and infection. Medical science now prolongs life, and degenerative diseases resulting from diet have become more apparent.
- In the past, people were primarily concerned with short-term consequences of consuming unhealthy foods, and exercised care in choosing foods that might produce sickness or death. Today, food laws make short-term poisoning insignificant, and people are much less reluctant to eat large amounts.
- Addictive foods that humans crave were rarely obtainable in the past (honey is a good example), while today addictive foods are widely available.

Extremely sweet food was such a rarity in the past that it is difficult to understand now just how intense the human craving for sweetness can be. Consider the following historical information about sweet peas, which were once extremely rare, so much so that they were reserved for royalty, amongst whom gluttony was common. English King John (1167–1216) is thought to have died from overeating green peas. English King William III (1650–1702) was well known for boorishly hogging all of the green peas that appeared on a dining table. In France in the late 17th century, green peas were still uncommon and were in remarkable demand. A letter of Madame de Maintenon (last mistress of French King Louis XIV and eventually his second wife) in 1696 documents the kind of excess that prevailed at the French Court: "The subject of Peas, continues to absorb all others...the anxiety to eat them, the pleasures of having eaten them, and the desire to eat them again, are the three great matters which have been discussed by our Princes for four days past. Some ladies, even after having supped at the royal table, and well supped too, returning to their own homes, at the risk of suffering from indigestion, will again eat Peas before going to bed. It is both a fashion and a madness."

Food clearly is addictive [see Wadley and Martin (1993) for a theory that agriculture evolved because people became progressively addicted to increasingly available foods]. English economist Thomas Malthus (1766–1834) and other historical figures have noted how nature controls populations of animals largely by starvation, so it is not surprising that when food is available, humans are highly motivated to eat. Alcohol is an example of a particular food that is so addictive for some that it literally destroys their health.

There is clinical evidence that the consumption of foods rich in fat and (or) sugar is associated with endorphin release. Endorphins are chemicals, secreted in the brain and nervous system generally, that result in pain relief or pleasure (in effect, they are painkillers). Opiates such as morphine are chemically very similar to endorphins (indeed, the word "endorphin" was based in part on the word "morphine"), and "narcotics" often mimic the effect of natural painkillers (but on a larger scale). Thus there is good reason to believe that food consumption shares, albeit on a smaller scale, the addictive qualities of narcotics.

For most people, the attractive taste of certain classes of foods leads to overeating. Nature designed humans to be attracted to the sweetness of fruits, but unfortunately, this has led to addiction to sugar and the oversupply of sugar in prepared foods. Nature also designed humans to be attracted to the thick or oily texture of fats (like those in chocolate), because foods with this quality are nutrient-rich. Unfortunately, some fats, as noted above, are very unhealthy. Humans are generally attracted to nutrient-rich foods, particularly in the form of animal foods, nuts, and seeds, because such foods in the past promoted survival, but today, it is too easy to eat too much.

There is no better example than chocolate to illustrate the contrast of how many modern foods differs from their nearest counterparts of past times. Chocolate is naturally bitter (a mechanism, as noted above, to deter animals from eating the seeds), limiting its attractiveness to humans in the past. As documented in the chapter on chocolate, a series of technological developments turned it into the world's most craved food. Modern chocolate preparations are simultaneously rich in the two food sensations that humans innately crave: sweet taste and fatty texture (to say nothing of the aroma, one of the most attractive in the world).

Studies have demonstrated (both in laboratory animals and in humans) that a monotonous diet is strongly correlated with weigh control. Quite possibly in the past, when fewer foods were available in a given area, this factor was important in avoiding obesity. However, this is now of no value to avoiding gain of weight in humans, because a huge variety of foods is now available, at least in most countries. Constant consumption of a given food tends to decrease interest in it, but a monotonous diet of a food that is strongly liked tends to remain fairly satisfying, while repetitive serving of a food that is not attractive tends to result in that food becoming very strongly disliked. One need only recall the pet name for the much-served army ration, chipped beef on toast. There is a lesson here for serving vegetables, which while valuable for weight control, do not have the appeal of sweet and fatty foods. It may be useful to serve many different vegetables, to prevent a dislike of "rabbit food" from developing or becoming stronger.

Too much rather than too little food is the problem in most of the Western World today, as evidenced by widespread obesity. Concern with weight reduction has spawned a huge industry, chiefly concerned with dieting. Fad diets are often dangerous, and have not worked for most people. The best diet is one that involves a wide variety of nutritious foods in moderate amounts, coupled with exercise, but how can this be accomplished in an environment saturated with irresistible food and constant advertising? Education is obviously a key to solving the problem, and shielding children from becoming addicted to junk food should be a first priority. Excessive portion size has been demonstrated to be a key factor in stimulating overeating. Fast-food chains are now adding healthy items to their menus, but remain a major source of not only too much eating, but unhealthy eating. In a free society, people should be allowed free choices, but given the genuinely addictive nature of food, just as limits have been placed on the purveying of tobacco, it is reasonable to consider limits on the marketing of food.

The cost of food is an important factor in determining food choices. Unfortunately, on an energy basis, sugar and fat are the cheapest way of delivering calories: one dollar's worth of sugar can provide 19 000 kilocalories (enough to sustain an adult for five days); one dollar's worth of vegetable fats can provide 5000 kilocalories. In the short-term, whole grains, vegetables, and fruits, the basis of a healthy diet, can not compete with the low price of sugar and fat, and inevitably, those in the lower socioeconomic levels of society have turned to foods with excessive sugar and fat. In the long run, the costs of dietary-induced illnesses and the associated costs in human suffering make the choice of sweet, fatty foods extremely unwise.

The reason that no one ever recommends against eating too much fruit or most vegetables is that humans naturally self-regulate their intake (although it is recommended that drinking too much fruit juice be avoided). These plant foods are moderately pleasant in mouth feel, and produce a satisfying sense of fullness. Fleshy fruits (the kind used as dessert) are mostly water, so they are particularly useful for filling up without excess calories. The much more nutrient-dense plant foods (nuts, cereals, some tubers and roots), and meat, by contrast, require that people control their cravings, and for the most part, these foods are so familiar that people have learned when enough is enough (a psychological feeling of fullness is produced much more quickly than when eating fruits and vegetables). Up to this point, the world of culinary preparation has contributed nobly to society. But now we come to the problem of substituting natural foods with mass-marketed commercially prepared foods, most of which are visibly unrecognizable, and provide reduced cues about how much may be safely consumed. Much worse, sensory cues have been engineered to urge the consumer to eat as much as possible. By deliberately maximizing sweetness, fat, and salt, the industry has perverted and exploited natural human cravings, with inevitable overeating.

"Body mass index" (BMI), expressed as weight/height^2 (kg/m^2; or weight in pounds divided by height in inches squared and then multiplied by 703), is commonly used to classify adults as "overweight" (BMI 25.0–29.9) or "obese" (BMI greater than or equal to 30.0). As of 2009, in the US almost two-thirds of adults are either overweight or obese, and one-third are obese. By comparison, in France, about 30% of adults are either overweight or obese, and about 10% are obese.

No people on earth have advanced the art of food preparation and consumption more then the French. Curiously, although many French foods appear to be decadently and spectacularly rich in calories, the pandemic of obesity and diet-related diseases that is plaguing much of the Western World has only in recent years begun to affect the French. This is in part because the French, at least until recently, have followed the "Mediterranean diet" in which there is a predominance of healthy plant foods (including fruits, vegetables, coarsely ground grains, beans, nuts, seeds, olive oil, wine with meals, and small amounts of animal products). Also important is the fact that the French have a tradition of eating moderately, eating substantially less than, for example, Americans. Clearly, it is possible to eat wonderful food without becoming overweight and (or) unhealthy, but it needs to be done wisely.

Modern Nutritional Crimes, Culinology, and the Mediocrity of Mass-Marketed Prepared Foods

> "FOOD IS NO LONGER VIEWED FIRST AND FOREMOST AS A SUSTAINER OF LIFE. RATHER, TO THOSE WHO SEEK TO COMMAND OUR FOOD SUPPLY IT HAS BECOME INSTEAD A MAJOR SOURCE OF CORPORATE CASH FLOW."
>
> —A.V. Krebs, 1992, *The Corporate Reapers: The Book of Agribusiness*

In the past, nutritionally related diseases were mostly a problem of the poor. In today's affluent societies, food consumption continues to be associated with nutritionally related diseases, albeit of a different nature. In the past, dietary diseases were largely the result of deficiencies in essential nutrients (i.e., malnutrition). Today, at least in most industrialized nations, dietary diseases are largely the result of excesses of essential nutrients. Indeed, almost all nations are increasingly adopting dietary patterns that generate obesity, heart disease, hypertension, adult-onset diabetes, dental caries, and some kinds of cancer.

Indigenous Peoples of the world, including many North American Indians and Mexican-Americans, are thought to have descended from hunter–gatherer societies in which people evolved to store fat more easily for times of famine. Overconsumption for these peoples has led directly to an epidemic of diabetes.

Although not yet found in dictionaries, a web search for the term "culinology" will turn up thousands of hits. The term was coined by Winston Riley, former president and a founder of the Research Chefs Association (RCA, based in Atlanta, Georgia), to describe a fusion of two disciplines—culinary art and food technology. The RCA (which describes itself as the leading professional community for food research and development) employs "Culinology" as its trademarked term, described as "a blending of culinary arts and the science of food." The word "culinology" was featured by the Institute of Food Technologists for their 2004 annual meeting and exposition (in Las Vegas) to describe "the juxtaposition of culinary art and science underway in research kitchens worldwide." Culinology, in the context of the meeting, was concerned with "the development of innovative food products which can be produced and distributed at a large-scale, while ensuring food safety, nutritional benefits, and the likelihood of commercial acceptance." Courses in "Culinology™" are taught at several American universities.

Increasingly, food is consumed in a prepackaged, preprepared form, whether bought in supermarkets or served in restaurants. Food safety and the healthfulness of foods are matters of concern, but since large corporations monopolize the prepared food industry, wealth, not health, is the principal motive. Tastes, smells, and textures are the result of carefully contrived addition of often artificial flavors and other chemically synthesized substances. The food industry now uses almost 5000 additives in prepared foods to change their color, flavor, surface appearance, and texture (Millstone and Lange 2003). Certainly, there is a need for combining the culinary arts with culinary sciences, as claimed by *Culinology*, but there is also a need to do this *principally* for the delivery of healthy, varied, and tasty foods to society, not primarily for commercial motives. The present mass-marketing of contrived and often artificial "convenience" foods, prepared at the lowest cost possible from the cheapest ingredients that can be obtained, is an extremely unwelcome development. [For a review of dozens of leading commercial successes, like Minute Rice, Wonder Bread, and Carnation Instant Breakfast, see Wyman (2004).] The prepared food industry bears considerable responsibility for widespread obesity, and for the production of bland, precooked concoctions that insult the taste buds of true food lovers. The cost of foods has been shown to be extremely important—often the determining factor in consumer choice—and the relatively cheap prepared foods now dominating the marketplace inevitably worsen the obesity problem. The problem has been compounded by saturation advertising, and the use of labor-saving devices that have greatly reduced the amount of exercise. These developments reflect badly on science, on the food professions, and on many who have either directly guided the food industry, or who have allowed it to develop in this way. There are now vociferous consumer movements urging the marketing of prepared foods that are healthier and less likely to lead to overeating. The wise consumer would do well to minimize intake of many of the prepared food products sold today, and instead return to the basic foods provided by nature. Toward this goal, this book offers choices from the plant world.

Slow Food

The phrase "slow food" was coined in an attempt to counter the negative aspects associated with the development of fast food. Slow Food®, founded by Carlo Petrini in Italy in 1986, is an international association that promotes food and wine culture, and defends food and agricultural biodiversity. According to its mission statement, "it opposes the standardization of taste, defends the need for consumer information, protects cultural identities tied to food and gastronomic traditions, safeguards foods and cultivation and processing techniques inherited from tradition, and defend domestic and wild animal and vegetable species... Slow Food believes the enjoyment of excellent foods and wines should be combined with efforts to save the countless traditional cheeses, grains, vegetables, fruits, and animal breeds that are disappearing due to the prevalence of convenience food and agribusiness."

The Future of Food Plants

> "YOU'VE GOTTA TELL 'EM! SOYLENT GREEN IS PEOPLE! WE'VE GOT TO STOP THEM! SOMEHOW!"
>
> —Stanley Greenberg (US screenwriter) and Richard Fleischer (US motion picture director) (uttered by the character Thorn, played by Charlton Heston in the 1973 movie *Soylent Green*, on learning that the synthetic foodstuff of this name, fed to an overpopulated planet, is made from dead bodies, not soybean and lentil as stated)

Above, the viewpoint is expressed that humankind is naturally adapted to the balance of dietary constituents supplied naturally by a suitable variety of plant foods. This is not to disagree that there is a continuing need to improve the quality, nutritional value, and variety of foods, as well as the efficiency of food production. But it is to state, what is surely common knowledge, that the world already possesses a veritable cornucopia of culinary delights, and people who choose to consume a suitable balance and quantity of these are likely to live a life of health. The movement to produce foods organically is representative of one basic attitude about the future of food. Organic agriculture amounts to respecting the natural world by utilizing production techniques that thousands of years have validated. Supplementing this view of natural cultivated food sources is a similar attitude about wild food sources. Admittedly, so much of the world's surface has been converted to urban landscapes and agriculture that it is no longer feasible for wild nature to furnish more than a fraction of food needs, but wildlands still supply appreciable quantities of some fruits, nuts, and vegetables, as well as animal foods, and the seas still supply very large amounts of plant and animal foods. These wild resources need to be conserved.

Above, the viewpoint is expressed that the pervasive substitution of artificial foods for natural ones has had disastrous consequences for health. This development is representative of an alternative attitude about the future of food, which in essence holds that new, high-tech foods as healthy as those supplied by nature can be synthesized and engineered. Essential to this viewpoint is the supposition that nutrients can be synthesized or at least isolated and extracted from their original food sources, and then combined in novel foods to reproduce the nutritional value of natural foods. However, it is abundantly clear that the benefits of many natural food components are very imperfectly understood at present, and that extracts often do not have the same benefits as food components still in their original food sources. Moreover, almost certainly there are hundreds of compounds ("phytochemicals") in plants that fight illness and promote health in ways that are simply not understood at present. It is also likely that humans are dependent on at least dozens of such plant compounds that have not yet even been identified. It follows that the artificial foods now marketed are not an adequate substitute for natural foods.

In the future, science is very likely capable of producing artificial foods with an ideal balance of all of the nutrients and health-promoting compounds in natural foods, and one day people may simply consume pills. The problem to date has been, as so often is the case, that technological innovations have been adopted before the science has been adequately established, and before social, environmental, and consumer concerns could be adequately considered. When it comes to something as basic as food, most people have an intuitive reluctance to accept new technologies, some of which have obviously been forced on the public without proper analysis and oversight. The opposition to genetically engineered plants and animals for food use, to the use of irradiation as a means of food preservation, and to the use of hormones to stimulate milk production in cows, all stem from a distrust of the influential business interests that seem hell-bent on profit, and of governments that seem unconcerned with adequate regulation.

Although living plants have been the foundation of human nutrition, it is by no means certain that this will always be the case. It is conceivable that artificial biochemical systems can be created that will capture light or other forms of energy and produce edible foodstuffs, designed to meet whatever consumer demands exist in the future. One wonders, however, if those future generations will look back, with nostalgia, to the times when nature provided such a vast range of attractive choices that there was more reason to ask for less than for more.

Literature cited

Ames, B.N., and Gold. L.S. 1990. Too many rodent carcinogens: mitogenesis increases mutagenesis. Science, **249**(4972): 970–971. [Mitogenesis is cell division; the title indicates that dividing cells are more likely to mutate than non-dividing cells.]

Ames, B.N., Magaw, R., and Gold, L.S. 1987. Ranking possible carcinogenic hazards. Science, **236**: 271–280.

Anderson, H., Blundell, J., and Chiva, M. (*Editors*). 2002. Food selection from genes to culture. Proceedings international symposium, Paris, France, December 1–2, 2000. Nature Publishing Group, Houndmills, Basingstoke, Hampshire, UK. 202 pp.

Barker, L.M. (*Editor*). 1982. The psychobiology of human food selection. AVI Publishing Co., Westport, CT. 262 pp.

Beier, R.C. 1990. Natural pesticides and bioactive components in foods. Rev. Environ. Contam. Toxicol. **113**: 47–137.

Brown, L.R., Flavin, C., French, H., and 10 others. 1998. State of the world, 1998, Worldwatch Institute report on progress toward a sustainable society. 15th edition. W.W. Norton & Co., New York, NY. 251 pp.

Capaldi, E.D. (*Editor*). 1996. Why we eat what we eat: the psychology of eating. American Psychological Association, Washington, DC. 339 pp.

Cheeke, P.R. 1998. Natural toxicants in feeds, forages and poisonous plants. 2nd edition. Interstate Publishers, Danville, IL. 479 pp.

Cheeke, P.R., and Shull, L.R. 1985. Natural toxicants in feeds and poisonous plants. Avi Publishing Company, Westport, CT. 492 pp.

Draper, H.H. 2000. Human nutritional adaptation: biological and cultural aspects. *In* The Cambridge world history of food, vol. 2. *Edited by* K.F. Kiple and K.C. Ornelas. Cambridge University Press, Cambridge, UK. pp. 1466–1476.

Harlan, J.R. 1992. Crops & man. 2nd edition. American Society of Agronomy, Crop Science Society of America, Madison, WI. 284 pp.

Janick, J. 1999. New crops and the search for new food resources. *In* Perspectives on new crops and new uses. *Edited by* J. Janick. ASHS Press, Alexandria, VA. pp. 104–110.

Johns, T. 1990. With bitter herbs they shall eat it: chemical ecology and the origins of human diet and medicine. University of Arizona Press, Tucson, AZ. 356 pp.

Johns, T., and Chapman, L. 1995. Phytochemicals ingested in traditional diets and medicines as modulators of energy metabolism. *In* Phytochemistry of medicinal plants. *Edited by* T. Arnason. Plenum Press, New York, NY. pp. 161–187.

Johns, T., and Kubo, I. 1988. A survey of traditional methods employed for the detoxification of plant foods. J. Ethnobiol. **8**: 81–129.

Katz, S.H. 1990. An evolutionary theory of cuisine. Human Nature **1**: 233–259.

Macbeth, H. (*Editor*). 1997. Food preferences and taste. Continuity and change. Berghahn Books, Providence, RI. 218 pp.

Millstone, E., and Lang, T. 2003. The atlas of food. Who eats what, where and why. Earthscan, London, UK. 128 pp.

Moskowitz, H., Beckley, J., and Adams, J. 2002. What makes people crave fast foods? Nutrition Today, **37**: 237–242.

Nestle, M. 2002. Food politics: how the food industry influences nutrition and health. University of California Press, Berkeley, CA. 457 pp.

Newcomer, A.D. 1978. Milk, lactase, and gene distribution. Dig. Dis. **23**: 961–962.

Rozin P. 2000. The psychology of food choice. *In* The Cambridge World History of Food. Vol. 2. *Edited by* K.F. Kiple and K.C. Ornelas. Cambridge University Press, Cambridge, UK. pp. 1476–1486.

Rozin, P. 2002. Human food intake and choice: biological, psychological and cultural perspectives. *In* Anderson et al. 2002 (cited above). pp. 8–25.

Rozin, P., and Vollmecke, T. 1986. Food likes and dislikes. Ann. Rev. Nutr. **6**: 433–456.

Saad, G. 2007. The evolutionary bases of consumption. Lawrence Erlbaum Associates, Mahwah, NJ. 339 pp.

Schwabe, C.W. 1979. Unmentionable cuisine. University Press of Virginia, Charlottesville, VA. 476 pp.

Sokolov, R.A. (*Editor*). 1991. Why we eat what we eat. How the encounter between the New World and the Old changed the way everyone on the planet eats. Summit Books, New York, NY. 254 pp.

Tanaka, T. 1976. Tanaka's cyclopedia of edible plants of the world. Keigaku Publishing Co., Tokyo, Japan. 924 pp.

Vavilov, N.I. 1926. Studies on the origin of cultivated plants. Institute of Applied Botany and Plant Breeding, Leningrad, USSR. 248 pp. (In Russian and English)

Vavilov, N.I. 1992. Origin and geography of cultivated plants. Cambridge University Press, Cambridge, UK. 498 pp. [Based on a 1987 collection of Vavilov's publications in Russian, translated into English by D. Löve.]

Wadley, G., and Martin, A. 1993. The origins of agriculture—a biological perspective and a new hypothesis. Austral. Biol. **6**: 96–105.

Wiersema, J.H.. and León, B. 1999. World economic plants. A standard reference. CRC Press, New York, NY. 749 pp.

Willett, W.C., Skerrett, P.J., Giovannucci, E.L., and Callahan, M. 2001. Eat, drink, and be healthy. The Harvard Medical School guide to healthy eating. Simon & Schuster, New York, NY. 299 pp.

World Resources Institute. 1994. World resources 1994–95. Oxford University Press, New York, NY. 400 pp.

Wyman, C. 2004. Better than homemade. Amazing foods that changed the way we eat. Quirk Books, Philadelphia, PA. 144 pp.

Detailed Information on Food Plants

(100 chapters, alphabetized by common name, from Almond to Yam; see Common Name Guide)

ADZUKI BEAN

Family: Fabaceae (Leguminosae; pea family)

Names

Scientific name: *Vigna angularis* (Willd.) Ohwi & H. Ohashi [*Azukia angularis* (Willd.) Ohwi; *Dolichos angularis* Willd.; *Phaseolus angularis* (Willd.) W. Wight]

- "Adzuki" is based on the Japanese word for the plant, azuki, which in turn is derived from the Chinese xiâo dòu, small bean.
- The spellings "adzuki" and "adsuki" are common, and "azuki" is also found. "Aduki" is sometimes encountered among North America food faddists. Regardless of spelling, the pronunciation is most commonly either ah-ZOO-kee or AH-zoo-kee.
- The name "red bean" reflects the typically reddish color of the seeds, popular in Japan.
- The name "adanka bean" is occasionally found.
- See MUNG BEAN for information on the genus name *Vigna*.
- *Angularis* in the scientific name is Latin for having corners or angles, descriptive of the seeds.

Plant Portrait

Adzuki bean is an erect, usually bushy, or sometimes slightly vining or prostrate yellow-flowered herb 25–90 cm (about 1–3 feet) tall. The pods are cylindrical, pendent, quite narrow, 6–13 cm (about 2–5 inches) long, straw colored to blackish or brown, and somewhat constricted between the 4–12 seeds. The seeds are small, subcylindrical, and smooth. Seed colors range from solid maroon or wine-red to solid black, blue-black, gray, brown, buff, white, and various mottled combinations of these. The plant is probably native to India or Japan, has long been established in China and Sarawak, and was more recently introduced for cultivation in warm parts of the Americas and in some African and Pacific countries. The crop requires about the same climatic conditions as soybean. The adzuki bean is one of the 12 most important grain legumes in the world (grain legumes are plants of the pea family, the seeds of which are used as human food). This bean has been consumed in East Asia for over 2000 years. Today, it is raised mainly in Japan, China, Taiwan, and South Korea. In Japan, adzuki bean is the second most important dry bean, and indeed is known as the "king of beans."

Culinary Portrait

The principal use of adzuki throughout the Far East is in confectionery, particularly as fillings for breads, dumplings, and sweet cakes. The adzuki bean is prized for these uses, in large part owing to its desirable red color, but also because of its delicate sweet flavor and the characteristic grainy texture of the pastes made from it. These pastes are used in numerous dishes. Red adzuki dishes are traditionally served in Asia on festive days such as weddings, birthdays, or New Year parties. Adzuki beans are also made into a meal or flour employed in soups, cakes, milk substitutes, and confections. The beans are frequently dried and cooked whole, and in Asia, they are typically served with rice. Dried adzuki beans require 1–3 hours of soaking in cold water before boiling (simmering for 1½–2 hours is recommended; alternatively, 20 minutes in a pressure cooker for soaked beans, 25 minutes for unsoaked beans). The beans may also be popped like corn or roasted and used as a coffee substitute. Young tender pods can be harvested as snap beans and cooked like common green beans. Adzuki bean sprouts are the major product consumed in the United States. The beans are high in protein—20–25%—and therefore quite nutritious. Like other species of *Vigna*, anti-nutritional (toxic) constituents have been found in the seeds, but are usually insignificant after cooking.

Culinary Vocabulary

- *Chamanju* (prononced cha-mahn-joo) is a Japanese confection made of sweet bean paste formed into balls, encased with wheat flour dough, and steamed. This is served traditionally at afternoon tea.

Curiosities of Science and Technology

- Commodore Matthew C. Perry (1794–1858) led a US expedition to Japan and the China seas during 1852–1854. The adzuki bean was brought to the United States by Perry in 1854.
- In Chinese traditional medicine, yin and yang are the two complementary, antagonistic qualities, imbalance of which causes disease. Foods are characterized as to their content of yin and yang, and are prescribed to cure diseases believed to

Adzuki bean (*Vigna angularis*), by B. Flahey.

result from an imbalance. Adzuki beans are considered to be the most yang of beans, and have been recommended for such yin sickness as kidney trouble, constipation, and difficult labor.

- Adzuki flour has been used for shampoos and facial creams.

Key Information Sources

Engquist, A., and Swanson, B.G. 1992. Microstructural differences among adzuki bean (*Vigna angularis*) cultivars. Food Struct. **11**: 171–179.

Falconer, G.W., and Desborough, P.J. 1994. Growing adzuki beans. 2nd edition. New South Wales Agriculture, Orange, NWS, Australia. 5 pp.

Gupta, V.P., Sharma, V.K., Sharma, J.D., and Rathore, P.K. 1994. Genetic divergence in adzuki bean. Crop Improv. (India), **21(1–2)**: 49–53.

Hsieh, H.M., Pomeranz, Y., and Swanson, B.G. 1992. Composition, cooking time, and maturation of azuki (*Vigna angularis*) and common beans (*Phaseolus vulgaris*). Cereal Chem. **69**: 244–248.

ICON Health Publications. 2003. Adzuki beans: a medical dictionary, bibliography, and annotated research guide to internet references. ICON Health Publications, San Diego, CA. 56 pp.

Lumpkin, T.A., and McClary, D.C. 1994. Azuki bean: botany, production and uses. CABI, Wallingford, Oxon, UK. 268 pp.

Lumpkin, T.A., Konovsky, J.C., Larson, K.J., and McClary, D.C. 1993. Potential new specialty crops from Asia: azuki bean, edamame soybean, and astragalus. *In* New crops. *Edited by* J. Janick and J.E. Simon. Wiley, New York, NY. pp. 45–51.

McGill, J.A., Jr. 1995. Michigan–Japan and azuki beans. Michigan Dry Bean Dig. **19**(**3**): 4–7.

Mimura, M., Yasuda, K., and Yamaguchi, H. 2000. RAPD variation in wild, weedy and cultivated azuki beans in Asia. Genet. Resour. Crop Evol. **47**: 603–610.

Oers, C.C.C.M. van. 1989. *Vigna angularis* (Willd.) Ohwi & Ohashi. *In* Plant resources of South-East Asia, 1, pulses. *Edited by* L.J.G. van der Maesen and S. Somaatmadja. Pudoc, Leiden, Netherlands. pp. 67–69.

Redden, R., Desborough, P., Tompkins, W., Usher, T., and Kelly, A. 2001. Responses of adzuki bean as affected by row spacing, plant density and variety. Austral. J. Exp. Agric. **41**: 235–243.

Sacks, F.M. 1977. A literature review of *Phaseolus angularis*—the adzuki bean. Econ. Bot. **31**: 9–15.

Takehisa, I., Noda, C., Mori, S., Yamashita, M., Nakanishi, H., Inoue, M., and Kamijima, O. 1996. Genetic variation and geographical distribution of azuki bean (*Vigna angularis)* landraces based on the electrophoregram of seed storage proteins. Breeding Sci. (Japan), **51**: 225–230.

Uwaegbute, A.C. 1996. Adzuki bean (*Vigna angularis* (Willd.) Ohwi et Ohashi). *In* Food and feed from legumes and oilseeds. *Edited by* J. Smartt and E. Nwokolo. Chapman and Hall, London, UK. pp. 206–208.

Wang, S.M., Redden, R.J., Jiapeng, J.P.H., Desborough, P.J., Lawrence, P.L., and Usher, T. 2001. Chinese adzuki bean germplasm: 1. Evaluation of agronomic traits. Austral. J. Agric. Res. **52**: 671–681.

Xu, R.Q., Tomooka, N., and Vaughan, D.A. 2000. AFLP markers for characterizing the azuki bean complex. Crop Sci. **49**: 808–815.

Yamaguchi, H. 1992. Wild and weed azuki beans in Japan. Econ. Bot. **46**: 384–394.

Yamaguchi, H., and Nikuma, Y. 1996. Biometric analysis of classification of weed, wild and cultivated azuki beans. Weed Res. (Japan), **41:** 55–62.

Yousif, A.M., Deeth, H.C., Caffin, N.A., and Lisle, A.T. 2002. Effect of storage time and conditions on the hardness and cooking quality of adzuki (*Vigna angularis*). Lebens. Wissenschaft Technol. **35**: 338–343.

Zong, X.X., Kaga, A., Tomooka, N., Wang, X.W., Han, O.K., and Vaughan, D. 2003. The genetic diversity of the *Vigna angularis* complex in Asia. Genome, **46**: 647–658.

ALMOND

Family: Rosaceae (rose family)

Names

Scientific names: *Prunus dulcis* (Mill.) D.A. Webb [*P. amygdalus* Batsch, *P. communis* (L.) Arcang., *Amygdalus communis* L., *A. dulcis* Mill.]

- The English word "almond" traces to the Latin words for the almond, *amandula* and *amygdala*, and the Greek amygdalē. (Early human anatomists applied the word amygdala to a small almond-shaped mass of gray matter inside the brain's temporal lobe. "Amygdaloid" rocks have almond-shaped cavities.)
- The "l" in "almond" is often not pronounced.
- The classical Romans called almonds "Greek nuts," probably because they were introduced to Italy from Greece.
- "Chinese almonds" are the seed kernels of several sweet varieties of apricots (see APRICOT). The use of certain apricot seeds as almonds is frequent in Asia. However, one should never consume "apricot seeds" unless there is reason to believe that they are of edible varieties; two kernels of conventional apricots are said to be sufficient to kill a man.
- The "Jordan almond" is a long, flat, slender nut considered to be the best cocktail almond. It is actually from Spain, not Jordan, the word Jordan having arisen from the Spanish *jardín*, garden.
- "Philippine almonds" are not almonds from the Philippines. They are almonds that occur in pairs in the shell, rather than singly as is normal. "Philippine" in this case relates to a Greek myth about a beautiful princess, Phyllis, who was turned into an almond tree after she committed suicide in frustration at having been abandoned at the altar on her wedding day. Her lover did return, seeking Phyllis after her death, so presumably the pair of almonds symbolizes their eventual union.
- For information on the genus name *Prunus*, see PLUM.
- *Dulcis* in the scientific name *P. dulcis* is Latin for sweet, a reference to the kernels (although as noted below, some almonds are sweet, others are bitter).

Plant Portrait

The almond is a small tree, normally 3–7 m (9–22 feet) tall, but sometimes reaching 30 m (about 100 feet). It produces pink or white flowers in the early spring. The nuts are enclosed in a fleshy husk that splits open when mature, allowing them to drop out or be easily separated at maturity. The shell encloses an oblong, flattened kernel, which may be longer than 2.5 cm (1 inch) and half as wide.

One of the earliest nut trees to be cultivated, the almond was probably domesticated in the eastern part of the Mediterranean Basin before 3000 BC. The species was cultivated in China by 1000 BC, and in Greece by 500 BC. Almonds are mentioned in the Old Testament. The almond has become the most important tree-nut crop in the world. It is now grown in several Mediterranean countries, notably Spain, Italy, France, and Portugal, and also in South Africa, China, and Australia. California is the world's major producer, growing some 100 different cultivars. The leading variety of almonds in California is the Nonpareil, which makes up over 50% of total almond production. The next most important variety is the Mission. Ninety-nine percent of the almonds grown in the United States come from a 644 km (400 mile) stretch of land in California, extending north from Bakersfield to Red Bluff. Almond orchards occupy more farming land in California than any other crop, except grapes. In addition to numerous cultivars grown for their nuts, there are ornamental cultivars. However, most "flowering almonds" are not *P. dulcis*, but other species of the genus *Prunus*.

There are two classes of almond—bitter and sweet. The bitterness is due to a substance known as amygdalin, which is present in large amounts in the kernels of bitter almonds, but only in trace amounts in sweet almonds. Under certain conditions, amygdalin is converted into toxic prussic acid (hydrogen cyanide). Hydrocyanic acid is dangerous—one twentieth of a gram (less than 0.002 ounce) can kill an adult, but it is very volatile and is denatured at higher temperatures, so cooked dishes prepared with bitter almonds are unlikely to be poisonous. However, eating whole raw bitter almonds is dangerous because hydrocyanic acid is formed in the stomach. Death from consuming bitter almonds is almost unknown in adults (one would have to consume about 50 of the very bitter kernels), but children may die by consuming just a few. Peach, apricot, and to a lesser extent, cherry and plum kernels are also similarly toxic. Sweet and bitter almonds should never be substituted for each other. In North America, bitter almonds are difficult to obtain (the sale of unrefined bitter almonds is prohibited in the United States). The sweet almond is grown for its edible nuts, while detoxified bitter almonds provide oil for culinary purposes and cosmetic skin preparations. Bitter almond oil furnishes the chemical benzaldehyde, which is produced from amygdalin, and is used mostly in pharmaceuticals.

Culinary Portrait

Almonds are important dessert nuts, and are used in confectionery, nut butters and pastes, cereals, cakes, macaroons, ice cream, and other sweet snacks. Cakes made with almond powder are especially rich in texture and taste. Dry, ground almonds are used to make stuffings and desserts. The nuts go very well with dates, coffee, and chocolate, and are often eaten alone or in combination as appetizers and snacks. The most famous almond product of Western cuisine containing almonds is marzipan (traditionally known as marchpane), a mixture of ground almonds, sugar, egg whites, and aromatic essences, frequently rose water. Marzipan probably originated in ancient Persia, and was brought to Europe, where Italians raised the art of making objects out of marzipan into an art form. "Almond paste," a pliable paste made from ground blanched almonds, sugar and glucose or egg whites, is similar to marzipan, but coarser, darker, and less sweet. Almond paste is used in pastries as a flavoring agent, and is also employed for decorative work. Marzipan is usually stiffer and lighter in color than almond paste, contains less sugar, and is more expensive. Almond butter has a much milder flavor than peanut butter, and this creamy spread can be used as a sandwich spread, and as a flavoring for sauces, soups, and stews. Almond milk is made from almonds that have been soaked, crushed, and strained of the pulp. Containers of almond milk are available commercially, and can be substituted for cow's milk for those avoiding dairy products. Detoxified bitter almond oil is used as a flavoring in baked goods and confectionery. Almonds are also used to flavor some fish, meat, and poultry dishes, as well as stuffings, sauces, and some vegetable dishes. The nuts complement fish (especially trout) and chicken.

Almonds are marketed both in the shell and as shelled kernels. They are also available in various processed forms: sliced into slivers or small sticks, ground or minced, plain or roasted, unpeeled or peeled, salted, sweetened with sugar or chocolate, smoked, and as paste, butter, or oil. Almonds in the shell have good keeping qualities, but should be purchased (like all nuts) from stores with good turnover. To peel the thin layer of brown skin from whole almonds, they can be blanched (placed in boiling water for 2–3 minutes, which swells the skin), then rinsed under cold water, followed by simply picking off the skin. There are several methods of cooking whole almonds: (1) roast them in a little oil in a dry, nonstick pan, over medium heat, stirring constantly; (2) cook in an oven at 175 °C (350 °F), placing the nuts in a single layer on a baking sheet, cooking until golden brown, stirring occasionally to ensure even roasting; (3) follow the instruction for dry-roasting given in (2), but coat the nuts with a little oil and reduce the temperature to 100–140 °C (200–275 °F). Unshelled almonds can be stored for up to 1 year in a cool dry place. Shelled almonds are best packed in hermetically sealed jars, bags, or cans to ensure maximum freshness. Shelled nuts should be refrigerated and stored for no longer than 6 months (frozen nuts will keep for up to 1 year).

Culinary Vocabulary

- Spain produces several specialty almond sweets. *Almendras garrapiñadas* is toasted almonds cooked in a crunchy sugar or caramelized with sugar and honey. *Guirlache* contains toasted almonds. There are two main kinds of *turrón*, one white, hard, and studded with whole nuts, the other made from a soft paste of ground almonds.
- *Almendrado* is a Mexican almond-flavored dessert made with beaten egg whites bound with gelatin and served with a creamy custard sauce. It is frequently tinted white, green, and red to resemble the Mexican flag.
- In France, a "praline" (from *pralin*) is a sugar-coated almond. In English, the word is used more generally for a confection of nuts (usually almonds or pecans) and caramelized sugar, often as the hard centers of chocolates. In Louisiana, praline means a cookie of almonds and brown sugar. "Praline" is based on the name of French diplomat Compte César du Plessis-Praslin (1598–1675), whose chef is thought to have invented the method of sugar-coating.

Almond (*Prunus dulcis*), from Köhler (1883–1914).

- *Linzer Torte* (torte is German for cake) is claimed to be the oldest cake in the world. It is made in Linz, Austria, from almond short-crust pastry with a characteristic lattice pattern on top, the cake kept moist by red currant jelly.
- "Orgeat" (pronounce OHR-zhat) refers to (1) a sweet syrup made from almonds, sugar, and rose-water or orange-flower water, and used to flavor cocktails and baked goods or (2) a fluid mixture of barley and almonds used for flavoring purposes.
- "Frangipane" (pronounced fran-juh-pahn) is a custard usually flavored with almonds. The name is widely said to be based on a 16th century Italian aristocrat, Muzio Frangipani, famous as the inventor of a perfume based on bitter almonds for scenting gloves called frangipani (King Louis XIII of France was well known for this use). Frangipani's perfume was so similar to the red jasmine, *Plumeria rubra* L., that the plant is now commonly called frangipani. The word frangipane has different old culinary meanings in Italian, including a vanilla-flavored cream, and the modern frangipane custard may in fact have no relationship with the Italian nobleman. In French cookery, frangipane also refers to a particular kind of pastry dough.
- "Tortoni" (pronounced tohr-TOH-nee) is a rich frozen Italian dessert made with eggs, heavy cream or

ice cream (and perhaps rum and cherries), and a coating of crushed almonds or macaroon crumbs. It is said to have originated in the 18th century at the Café Napolitain in Paris, and to be named after its Italian owner. The dish first became popular in North America when it was featured in Italian-American restaurants (often called "biscuit tortoni"), and the word tortoni started circulating in English in 1944. Tortoni is sometimes confused with spumoni, another Italian-type ice cream with fruit and nuts.

- *Amandine* (pronounced AH-mahn-deen and ah-mahn-DEEN, incorrectly "almondine") is a French word for a dish garnished with almonds, or cooked with chopped or grated almonds.
- *Quatre mendiants* (also known as mendiants) is an old-fashioned French dessert made up of a plate of peeled almonds, dried figs, filberts, and raisins. The name means "four beggars," and commemorates the begging life style of the four chief orders of mendicant friars: Carmelites (almonds), Augustinians (filberts), Franciscans (figs), and Dominicans (raisins). The color of an order's fruit or nut was the same as the color associated with that order.
- A "Tipsy Parson" is a type of sherry-soaked almond-topped sponge cake served with custard, which was popular in the late 19th century in the southern United States. The "tipsy" (i.e., slightly drunk) in the name reflects the alcohol content. The dish is also known as a "tipsy pudding," and originated in Britain as a trifle made with layers of custard and wine-soaked sponge cake.
- A "Maid of Honor" is a small tart filled with almond custard. It is often served at afternoon tea in Britain.
- "Mandelbrot" or "mandelbroit" (pronounced MAHN-duhl-broht), based on the German *mandel*, almond, + *brot*, bread, is a crisp, twice-baked almond bread traditionally eaten as a cookie by Jews.
- "Dundee Cake" is a Scottish fruitcake topped with almonds, added during the middle of the baking cycle so that they will become brown but not burnt.
- "Twelfth-Night Cake" is a rich British spice cake prepared with almonds, almonds paste, and candied fruit, and traditionally served on January 6, the 12th day after Christmas.
- "Mock almonds" were prepared, according to the Fannie Farmer classic 1918 *The Boston Cooking-School Cook Book*, as follows: "Cut stale bread into one-eighth inch slices, shape with a round cutter one and one-half inches in diameter, then shape in almond-shaped pieces. Brush over with melted butter, put in a pan, and bake until delicately browned."
- "Amaretto" (pronounced am-ah-REHT-toh) is an Italian liqueur, now made in other countries, with the flavor of almond (although it may contain no almonds, and be flavored with apricot pits). The word amaretto (literally "a little bitter") is based on the Italian *amaro*, bitter, an allusion to the bitter almonds that are its principal flavoring ingredient. According to legend, amaretto was first made in 1525 by the landlady of an inn in the town of Saronno as a present for an artist who had painted her portrait. As well as being used as a cordial, amaretto is employed as a topping for ice cream desserts, and to flavor cakes, pies, and coffee.
- "Amaretti" (pronounced am-ah-REHT-tee) are sweet Italian almond biscuits (made with either bitter almond or apricot kernel paste), similar to small macaroons, and typically provided in individual paper wrappers. Amarettini (pronounced am-ah-REHT-teen-ee) are miniature amaretti.
- "Burnt almonds" refers to sweet, blanched almonds, roasted to a golden brown. The phrase probably originated in the past when oven temperatures were difficult to control, and more often than not, the almonds were actually burned.
- "Green almonds" are young, sweet almonds marketed while their shells are soft and their outer skins are green and tender. The kernels are white, creamy, and delicious, and are peeled and consumed fresh.
- Salicyaldehyde is a food additive that has an almond-like taste and smell, and is used as a flavoring agent.

Curiosities of Science and Technology

- The ancient Persians perfumed almonds before eating by placing them in jars with flower petals.
- The classical Greeks and Romans believed that almonds could cure drunkenness. In Medieval times, sufferers reportedly consumed a mixture of bitter almonds and raw eel. To this day, the superstition persists that if you eat almonds before taking a drink, you will reduce your chances of getting drunk and avoid having a hangover. Almonds are, accordingly, a traditional accompaniment to alcoholic drinks.

- Forty percent of the world's almonds go into chocolates.
- Unlike its relative the peach, where the flesh of the peach is eaten and the pit thrown away, the almond pit containing the edible seed or nut is kept and the thin fibrous outer flesh (the hull) is thrown away.
- People sometimes become alarmed when they find that unblanched, whole almonds have scars on the skin that look like teeth marks from rodents. This is often the case, not from mice nibbling the nuts, but because whole almonds with skins are packed loosely in large containers, which may result in skin damage during transport. Almonds may also be scratched as they are scooped out of bins.
- Because almonds have a high ratio of the amino acid arginine to another amino acid, lysine, it has been recommended that they should be avoided by individuals susceptible to cold sores or herpes infections. Arginine has been found to promote, and lysine to prevent, the activation of the virus.

Key Information Sources

Campbell, R.D. 1988. Hardy almonds. North. Nut Grow. Assoc. Ann. Rep. **79**: 59–60.

Ferguson, L., and Kester, D.E. (*Editors*). 1998. Second international symposium on pistachios and almonds, Davis, California, August 24–29, 1997. International Society for Horticultural Science, Leiden, Netherlands. 611 pp.

Horoschak, T., and Bryan, H.C. 1971. The almond industries of Italy and Spain. U.S. Foreign Agricultural Service, Washington, DC. 19 pp.

Kester, D.E. 1979. Almonds. *In* Nut tree culture in North America. *Edited* by R.A. Jaynes. The Northern Nut Growers Association, Hamden, CT. pp. 148–162.

Kester, D.E., and Asay, R.A. 1975. Almonds. *In* Advances in fruit breeding. *Edited by* J. Janick and J.N. Moore. Purdue University Press, Lafayette, IN. pp. 382–419.

Kester, D.E., Gradziel, T.M., and Grasselly, C. 1990. Almonds (*Prunus*). *In* Genetic resources of temperate fruit and nut crops. *Edited by* J.N. Moore and J.R. Ballington, Jr. International Society for Horticultural Science, Wageningen, Netherlands. pp. 701–758.

Micke, W. 1996. Almond production manual. Division of Agriculture & Natural Resources, University of California, Davis, CA. 294 pp.

Quin, A. 2001. Almonds. Western Australia Dept. of Agriculture, South Perth, WA, Australia. 36 pp.

Rowell, J.D., and Portello, L. 1965. Almonds: a review of the past and present and some alternatives for the future. Bureau of Marketing, California Dept. of Agriculture, Sacramento, CA. 38 pp.

Socias, R. (*Editor*). 2002. Proceedings of the 3rd international symposium on pistachios and almonds, 12th colloquium of Grempa, Zaragoza, Spain, 20–24 May, 2001. International Society for Horticultural Science, Leuven, Belgium. 596 pp.

Summers, F.M. 1962. Insect and mite pests of almonds. Division of Agricultural Sciences, University of California, Berkeley, CA. 15 pp.

Traynor, J. 1993. Almond pollination handbook for almond growers and for beekeepers. Kovak Books, Bakersfield, CA. 86 pp.

University of California. 2002. Integrated pest management for almonds. University of California Integrated Pest Management Program. 2nd edition. University of California, Statewide Integrated Pest Management Project, Division of Agriculture and Natural Resources, Oakland, CA. 199 pp.

Specialty Cookbooks

Almond Board of California. 1970. The crazy about almonds and chocolate cookbooklet. Almond Board of California, Sacramento, CA. 49 pp.

Almond Board of California. 2001. Almonds are in the kitchen. Almond Board of California, Modesto, CA. 104 pp.

American Heart Association. 2001. California almonds and a healthy heart: featuring recipes adapted from American Heart Association cookbooks. Clarkson/Potter, New York, NY. 61 pp.

California Almond Growers Exchange. 1979–1998. The new treasury of almond recipes. California Almond Growers Exchange, Sacramento, CA. 49 pp.

Hood, K.J.M.H., et al. 2004. Almond delights cookbook: a collection of almond recipes. Whispering Pine Press, Spokane Valley, WA. 320 pp.

Schmidt, M. 1984. New almond cookery. Simon & Schuster, New York, NY. 193 pp.

AMARANTH

Family: Amaranthaceae (pigweed family)

Names

Scientific names: *Amaranthus* species

Principal grain species

- Prince's feather—*A. hypochondriacus* L. (*A. hybridus* L. var. *erythrostachys* Moq.)
- Pendant amaranth, blue amaranth, cat-tail, Inca wheat, love-lies-bleeding, slender cockscomb, tassel flower, tumbleweed, velvet flower—*A. caudatus* L. (*A. edulis* Speg., *A. leucocarpus* S. Watson, *A. mantegazzianus* Pass.)

Principal vegetable species

- Chinese amaranth, Chinese spinach, Tampala, Joseph's coat—*A. tricolor* L. (*A. gangeticus* L.)
- Spleen amaranth—*A. dubius* C. Mart. ex Thell.
- Livid amaranth, slender amaranth—*A. blitum* L. (*A. lividus* L.)

Used Both as Grain and Vegetable

- Purple amaranth, Mexican grain amaranth, red amaranth, red shank—*A. cruentus* L. [*A. hybridus* L. subsp. *cruentus* (L.) Thell., *A. paniculatus* L.]
- The name "amaranth" and the genus name *Amaranthus* are based on the classical Latin *amarantus*, from the Greek *amarantos* (*a*, not, + *maraino*, to fade or wither), unfading. Some amaranth flowers indeed retain their color indefinitely. One of the dictionary meanings of amaranth is "an imaginary flower that never fades," while amaranthine means "eternally beautiful and unfading." Amaranths have been a poetic symbol of immortality from the time of ancient Greece.
- Reddish (and some green) varieties of Chinese amaranth are called "red-in-snow" (also red-in-snow cabbage and snow cabbage) because they grow early in the spring in China when the ground is often covered with snow. The "cabbage" in the name is due to the broccoli-like taste (broccoli is a member of the cabbage clan). Cans of pickled red-in-snow cabbage are sometimes available in Chinese food stores.
- The name "prince's feather" (for *A. hypochondriacus*) is thought to reflect resemblance to the headdress worn by a Prince of Wales.
- The name "velvet flower" (for *A. caudatus*) reflects the velvety floral tassels of the plant. The name "love-lies-bleeding" is said to be based on the crimson color of the flowers. However, it has been suggested that the Latin name *Amaranthus* was misinterpreted as based on the Latin *amor*, love, + the Greek *anthos*, flower, literally "love flower"; hence the "love" in the name.
- The name "Joseph's coat" is applied to *A. tricolor* L. cultivars with leaves that are blotched and colored.
- When used for greens, the name "bush greens" is applied to *A. cruentus* and other species of *Amaranthus* used as a vegetable in Africa; "calaloo" is often the name in the Caribbean; and "Chinese spinach" and "een choy" are other names that are encountered.

Plant Portrait

Amaranths are members of the genus *Amaranthus*, which includes about 60 species. These coarse annual herbs are native to mild and tropical climates throughout the world. The plants have an enchanting, beautiful appearance, and several species are grown as ornamentals. Perhaps the best-known ornamental amaranth in North America is Joseph's coat, used as a bedding plant. Grain-type amaranth plants have a main stem axis on top of which there develops a large branched flower head. The flowers are purple, orange, red, or gold in color. By contrast, the flowers of vegetable-type amaranth plants develop not just at the top of the plant but also on side branches. Amaranth seed is quite small, 0.9–1.7 mm (0.035–0.067 inch) in diameter. Seed color can vary from cream to gold and pink to black. The seed of most commercially produced grain amaranth is white or light tan. Plant height varies greatly, depending on species and environment. Some plants are less than a meter (about a yard), others almost 3 m (about 10 feet) in height. Grain amaranth plants grown in the United States are about 2 m (6½ feet) tall, with thick, tough stems similar to sunflower. Amaranth species utilized as a vegetable generally are short, with wide leaves.

Many crops were first aggressive weeds, adapted to grow well in the disturbed conditions around human campsites. People would sooner or later notice that some of these weeds were edible, and began to rely on them. Many amaranths are very weedy, and likely began to be used as crops when people became aware that they were useful as a grain crop, for greens for human consumption, for forage for livestock, and even as dye plants.

Grain amaranth

Grain amaranth species are "pseudocereals." Cereals are grasses that yield abundant, small, edible seeds. Pseudocereals also yield abundant grain, but simply don't belong to the grass family. Well-known grain amaranths include *A. cruentus* and *A. hypochondriacus*, natives of Mexico and Guatemala, and *A. caudatus*, native of the Andean regions of Ecuador, Peru, and Bolivia.

Amaranth was eaten by hunter-gatherers in both North and South America before the domestication of crops. In Mexico, the crop was domesticated about 5000 BC, along with maize, beans, and gourds. The plants were domesticated independently in the Andes. Amaranths were chiefly grown for grain, not vegetables, on the South American continent. Pre-Columbian New World civilizations, particularly the ancient Aztec and Incas, grew large amounts of grain amaranth. Amaranth was the principal grain crop grown by the Aztecs for thousands of years in central Mexico prior to the destruction of South American civilization by the Spanish Conquistadors. The esteem in which the crop was held is indicated by the ancient phrases "mystical grains of the Aztecs," "super grain of the Aztecs," and "golden grain of the Gods." After the arrival of the Spanish, the cultivation of amaranth declined, and by the middle of the 20th century, grain amaranth was grown only in small plots in Mexico, the Andean highlands, and in the Himalayan foothills of India and Nepal. Since the mid 1970s, amaranth has received attention as a new crop for North America. Consumer interest in amaranth as a food ingredient has been mostly due to the positive nutritional characteristics of the grain, although some of the attraction is undoubtedly due to a colorful history as one of the "lost crops of the Incas" (see Quinoa for additional information). The United States has been the leading commercial producer of grain amaranth in recent years. The US production of grain amaranth is concentrated in the Great Plains area, particularly Nebraska, with numerous smaller production areas throughout the Midwest. Amaranth is particularly well suited to the dry areas of the western United States.

Vegetable Amaranth

Amaranths have been cultivated as a vegetable crop by early civilizations dating back at least 2000 years. Vegetable amaranth is grown today in many regions, including China, India, Africa, Nepal, South Pacific Islands, the Caribbean, Greece, Italy, and Russia. While amaranths are probably the most important leaf vegetables of the lowland tropics of Africa and Asia, they are scarcely known in the Americas. The species grown as vegetables are represented primarily by *A. tricolor*, *A. dubius*, *A. lividus*, and *A. cruentus*. *Amaranthus palmeri* S. Watson (carelessweed) and *A. hybridus* L. (green amaranth, pigweed, wild beet) were utilized by natives of early civilizations in the southwestern part of the United States. *Amaranthus tricolor* and *A. dubius* are extensively grown as vegetables in Asia, West Africa, and the Caribbean. *Amaranthus cruentus* is grown both for leaves and grain; its use for leaves is most prevalent in humid tropical Africa. *Amaranthus tricolor* is widely cultivated throughout Asia and the South Pacific as a vegetable. The leaves of *A. dubius* are considered to be a delicacy in many areas of the Caribbean. A dark-seeded strain of *A. cruentus* is commonly cultivated as a vegetable in West Africa.

Culinary Portrait

Grain Amaranth

Grain amaranth has a high-protein as well as a high-fat content, and can supply an excellent balance of carbohydrates, fats, and protein. It has been said that high-protein rice is the only other cereal-like plant known to be as useful in satisfying protein and energy needs. Because of its nutritional qualities, grain amaranth offers excellent possibilities for improving human nutrition, especially in Third World countries. Grain amaranth is easy to digest, and is customarily given to those who are recovering from an illness or a fasting period. Grain amaranth is also being used as a food for people with allergies to cereals. Amaranth seeds may be prepared as a porridge simply by cooking them for about 30 minutes in 2–3 parts of water. In Nepal, amaranth seeds are often eaten as a gruel. The grain can also be sprouted, and consumed like alfalfa sprouts. Still other methods of preparation are to pop the seeds like popcorn, or flake them like oatmeal. In Mexico, grain amaranth is popped and mixed with a sugar solution to make a confection called *alegria* ("happiness"). A traditional Mexican drink called *atole* is made from milled and roasted amaranth seed. In India, *A. hypochondriacus* is known as "the King's grain" and is often popped to be used in confections. In various Third World countries, the seeds are commonly ground into flour and used for breadmaking. Amaranth flour does not contain gluten, and will not rise when baked unless mixed with wheat or rye flour. However, amaranth flour

Livid amaranth (*Amaranthus blitum*), from Hallier (1880–1888, vol. 9, plate 860).

can be used directly to make cookies, pancakes, waffles, and the like, and such pastries are unusually moist and sweet. More than 40 products containing amaranth are currently on the market in the United States. In North America, amaranth is being used in breakfast cereals, bars including granola and other snack foods, crackers, cookies, noodles, pancakes, multigrain breads, and other flour-based products. The bulk of amaranth products are sold in the health food sections of grocery stores, in specialty food stores, or through direct marketing. Amaranth flour stores longer than wheat flour, and should be kept in an opaque container in a cool, dry place. Amaranth seeds should be stored in the same way.

Vegetable Amaranth

The leaves of many cultivated species of *Amaranthus* are commonly eaten as potherbs or leafy vegetables, alone or in combination with other vegetables and (or) meat. Young leaves of the species used for grain are sometimes also used as a vegetable. Amaranth leaves are a good source of dietary fiber and contain high amounts of proteins, vitamins, and minerals. If fresh leaves are available, they may be stored like spinach, in a refrigerator for a few days.

Although vegetable amaranth is considered a delicacy or a food staple in many parts of the world, this is not the case in North America, and the product is most likely to be available merely as imported canned goods. In the United States, amaranth is seldom used as a vegetable, and production is carried out by just a few growers, especially in the Southwest. Consumption is limited largely to ethnic uses, and to the occasional home garden. There have been concerns about the presence of oxalates in vegetable amaranths, which can bind calcium, leading to a medical condition called hypocalcemia (literally, "insufficient calcium"). Oxalates (common in rhubarb and spinach) can make up 0.2–11.4% of the dry matter in vegetable amaranths. Levels of oxalates can be reduced by steaming or boiling leaves for 10 minutes.

Curiosities of Science and Technology

- In Elizabethan times (the period of Elizabeth I, 1533–1603), red amaranth flowers were soaked in water for an hour and given to drunkards as "red wine" in hopes of breaking their alcoholic addiction.
- Amaranth was used as a source of a red food dye. Later, a comparable dye named "E123 – Amaranth" (the terminology in the European Community), and also known as "FD&C Red 2" and "US red dye No. 2" was synthesized, and has been used in ice creams, gravy granules, jams, jelly, tinned fruit pie fillings and prawns, cake mixes, soups, trifles, and other prepared foods. It has been found to cause allergic and (or) intolerance reactions, similar to nettle rash, particularly among asthmatics and those with an aspirin intolerance. As well, it poses a danger of birth defects and possibly cancer (judged by test animals). It is now generally banned, but can be used for very limited food purposes (e.g., in caviar) in some countries. Natural amaranth dye is still used for coloring chicha (an alcoholic drink of South America) and ceremonial maize dishes of some Native Americans.
- Tumbleweeds are plant species that mature into a dried-up rounded tangle of stiff branches, which is rolled by the wind, scattering seeds. Tumbleweeds are particularly common on prairies, which provide the flat land required so that they can roll. Some *Amaranthus* species, including *A. albus* L. (tumble pigweed) and *A. graecizans* L. (spreading pigweed) introduced from tropical America, are tumbleweed pests of western US agricultural areas.

Pendant amaranth (*Amaranthus caudatus*), from Duthie (1893).

- Plants that furnish food for humans can also furnish food for livestock. However, plants that produce fodder for livestock are chosen for their ability to produce a large amount of plant material in a short period of time, and excellent taste is not required. In China, amaranth has been cultivated specifically for use as a forage for cattle. In most countries, the straw or remains after the edible parts of plants have been harvested for use as human food is often salvaged for use as livestock feed, and this is frequently the case for amaranth raised as a grain. However, some types of amaranth accumulate toxic levels of oxalate and nitrate when grown under conditions of stress, so care needs to be exercised that livestock are not poisoned.
- There are two, widespread classes of plants that differ in the way they carry out photosynthesis. Most familiar temperate zone plants are called C3 plants because the first stable compound formed when carbon dioxide is processed is a three-carbon compound, i.e., "C3." The C4 plants are so named because the first organic compound incorporating CO_2 is a four-carbon compound. Amaranth is one of many tropical plants having the C4 type of photosynthesis, which allows for increased growth if temperature and light intensity are sufficiently high.
- In Peru, ash from amaranth stalks is used for soaking maize and making "tokra balls," which are gnawed on when chewing narcotic coca (*Erythroxylon coca* L.) leaves. Ash releases alkaloids from the coca, producing the stimulant effects of cocaine.

Key Information Sources

Breene, W.M. 1991. Food uses of grain amaranth. Cereal Foods World, **36**: 426–430.

Coons, M.P. 1982. Relationships of *Amaranthus caudatus*. Econ. Bot. **36**: 129–146.

Feine, L.B., Harwood, R.R., Kauffman, C.S., and Senft, J.P. 1979. Amaranth, gentle giant of the past and future. *In* New agricultural crops. *Edited by* G.A. Ritchie. Westview Press, Boulder, CO. pp. 41–63.

Grubben, G.J.H. 1976. The cultivation of amaranth as a tropical leaf vegetable. Communication 67. Royal Tropical Institute, Amsterdam, Netherlands. 207 pp.

Grubben, G.J.H. 1993. *Amaranthus* L. *In* Plant resources of South-East Asia, 8, vegetables. *Edited by* J.E. Siemonsma and K. Piluek. Pudoc Scientific Publishers, Wageningen, Netherlands. pp. 82–96.

Grubben, G.J.H., and Sloten, D.H. van. 1981. Genetic resources of amaranths: a global plan of action. International Board for Plant Genetic Resources, Rome, Italy. 57 pp.

Huttner, R. (*Editor*) and National Research Council. 1984. Amaranth: modern prospects for an ancient crop. National Academy of Sciences, Washington, DC. 92 pp.

Itúrbide, G.A., and Gispert, N. 1994. Grain amaranths (*Amaranthus* spp.). *In* Neglected crops: 1492 from a different perspective. *Edited by* J.E. Hernández-Bermejo and J. León. Food and Agriculture Organization of the United Nations, Rome, Italy. pp. 93–101.

Jain, S.K., and Sutarno, H. 1996. *Amaranthus* L. (grain amaranth). *In* Plant resources of South-East Asia, 10, cereals. *Edited by* C.J.H. Grubben and S. Partohardjono. Backhuys Publishers, Leiden, Netherlands. pp. 75–79.

Karasch, M. 2000. Amaranth. *In* The Cambridge world history of food. *Edited by* K.F. Kiple and K.C. Ornelas. Cambridge University Press, Cambridge, UK. pp. 75–81.

Kauffman, C.S., and Weber, L.E. 1990. Grain amaranth. *In* Advances in new crops. *Edited by* J. Janick and J.E. Simon. Timber Press, Portland, OR. pp. 127–139.

Martin, F.W., and Telek, L. 1979. Vegetables for the hot, humid tropics. Part 6. Amaranth and Celosia, *Amaranthus* and *Celosia*. Agricultural Research, Science and Education Administration, U.S. Dept. of Agriculture, New Orleans, LA. 21 pp.

Myers, R.L. 1996. Amaranth: new crop opportunity. *In* Progress in new crops. *Edited by* J. Janick. ASHS Press, Alexandria, VA. pp. 207–220.

National Research Council. 1984. Amaranth, modern prospects for an ancient crop. National Academy Press, Washington, DC. 80 pp.

Oke, O.L. 1983. Amaranth. *In* Handbook of tropical foods. *Edited by* H.T. Chan, Jr. Marcel Dekker Inc., New York, NY. pp. 1–28.

Paredes-Lopez, O. (*Editor*). 1994. Amaranth: biology, chemistry, and technology. CRC Press, Boca Raton, FL. 223 pp.

Rodale Research Center and National Academy of Sciences. 1986. Grain amaranth: expanding consumption through improved cropping, marketing, and crop development. Proceedings of the third amaranth conference, Kutztown, PA, 11–13 Sept. 1984. Rodale Press, Emmaus, PA. 233 pp.

Sauer, J.D. 1967. The grain amaranths and their relatives: a revised taxonomic and geographic survey. Ann. Mo. Bot. Gard. **54**: 102–137.

Sauer, J.D. 1995. Grain amaranths. *In* Evolution of crop plants. 2nd edition. *Edited by* J. Smartt and N.W. Simmonds. Longman Scientific & Technical, Burnt Mill, Harlow, Essex, UK. pp. 8–10.

Saunders, R.M., and Becker, R. 1984. *Amaranthus*: a potential food and feed resource. Adv. Cereal Sci. Technol. (Am. Assoc. Cereal Chem., St Paul, MN), **6**: 357–396.

Sealy, R.L., McWilliams, E.L., Novak, J., Fong, F., and Kenerley, C.M. 1990. Vegetable amaranths: cultivar selection for summer production in the south. *In* Advances in new crops. *Edited by* J. Janick and J.E. Simon. Timber Press, Portland, OR. pp. 396–398.

Senft, J.P., Kauffman, C.S., and Bailey, N.N. 1981. The genus *Amaranthus*: a comprehensive bibliography. Rodale Press, Emmaus, PA. 217 pp.

Singh, B.P., and Whitehead, W.F. 1996. Management methods for producing vegetable amaranth. *In* Progress in new crops. *Edited by* J. Janick. ASHS Press, Arlington, VA. pp. 511–515.

Singh, H., and Thomas, T.A. 1978. Grain amaranths, buckwheat and chenopods. Revised edition. Indian Council of Agricultural Research, New Delhi, India. 70 pp.

Singhal, R.S., and Kulkami, P.R.1988. Review: amaranths—an underutilized resource. Int. J. Food Sci. Technol. **23**: 125–139.

Sreelathakumary, I., and Peter, K.V. 1993. Amaranth, *Amaranthus* spp. *In* Genetic improvement of vegetable crops. *Edited by* G. Kalloo and B.O. Bergh. Pergamon Press, New York, NY. pp. 315–323.

Stallknecht, G.F., and Schulz-Schaeffer, J.R. 1993. Amaranth rediscovered. *In* New crops. *Edited by* J. Janick and J.E. Simon. Wiley, New York, NY. pp. 211–218.

Tucker, J.B. 1986. Amaranth: the once and future crop. BioScience, **36**: 9–13.

U.S. National Research Council, Board on Science and Technology for International Development. 1984. Amaranth: modern prospects for an ancient crop. National Academy Press, Washington, DC. 81 pp.

Weber, L.E., and Applegate, W.W. 1990. Amaranth grain production guide. Rodale Research Center and American Amaranth Institute, Emmaus, PA. 36 pp.

Williams, J.T., and Brenner, D. 1995. Grain amaranth (*Amaranthus* species). *In* Cereals and pseudocereals. (Second in series Underutilized Crops.) *Edited by* J.T. Williams. Chapman & Hall, London, UK. pp. 129–186.

Specialty Cookbooks

Bumbarner, M.A. 1997. The new book of whole grains: more than 200 recipes featuring whole grains, including amaranth, quinoa, wheat, spelt, oats, rye, barley, and millet. St. Martin's Griffin, New York, NY. 299 pp.

Livingston, A.D. 2000. The whole grain cookbook: delicious recipes for what, barley, oats, rye, amaranth, spelt, corn, millet, quinoa, and more, with instructions for milling your own. Lyons Press, New York, NY. 302 pp.

ANISE AND STAR ANISE

These two species are combined in this chapter because they produce essentially the same flavor.

A. Anise

Family: Apiaceae (Umbelliferae; carrot family)

Names

Scientific name: *Pimpinella anisum* L. (*Anisum vulgare* Gaertn.)

- Anise acquired its name by confusion with dill, from the Greek word for dill, *anison* (*áneeson* or *áneeton*).
- Pronunciation: ANN-iss.
- Anise is also known as aniseed (applied both to the seeds and the whole plant), common anise, healbite, sweet Alice, and sweet cumin (not the same as true cumin).
- The vegetable Florence fennel (*Foeniculum vulgare* subsp. *azoricum* (Mill.) Thell.) is sometimes known as "anise" or "sweet anise," a practice that often leads to confusion with the herb discussed in this chapter.
- Anise should not be confused with anise tree, the genus *Illicium*, which contains species known as Chinese or Japanese anise, purple anise, and star anise (see Star Anise, below).
- Anise hyssop is *Agastache foeniculum* (Pursh) Kuntze, another unrelated food plant that can be confused with anise.
- The genus name *Pimpinella* is a corruption of the medieval Latin *pinella* or *bipinnela*, two-winged, an allusion to the fine divisions of the leaves. It has also been suggested that the word arose from the medieval Latin *pipinella*, and in turn from the old Latin *pepo*, a pumpkin, or from the Latin *pampinus*, meaning a tendril or young shoot of a vine, or vine leaf, but the justification for these latter explanations is unclear.
- *Anisum* in the scientific name *P. anisum* has the same derivation as the English word anise, as noted above.

Plant Portrait

Anise is an aromatic annual, with white or yellow–white flowers. It is typically 50–90 cm (20–36 inches) in height when cultivated, sometimes as small as 10 cm (4 inches) in the wild. The feathery, fern-like foliage is lobed and finely cut, and the small flowers and slender "seeds" (see *Schizocarp* in Glossary) are displayed in flat clusters. The wild range is unknown, since the plant is widely distributed as a result of escape from cultivation. The species may be native to Egypt, Syria, or Cyprus to Greece. Anise is now widely established in other parts of Asia, and in Europe and North America. It has been claimed that anise seed was found in Egypt as early as 1500 BC. Anise has been valued as a drug plant and condiment since antiquity, when it was used by the classical Egyptians, Greeks, Romans, and Hebrews. In many countries, it is valued as much medicinally as for culinary purposes. In addition to the culinary uses noted below, the essential oil is employed externally, in perfumes, soaps, and skin creams, and has such additional uses as a sensitizer for bleaching colors in photography. Anise is widely cultivated commercially, notably in southern Europe (Cyprus, Bulgaria, Germany, Italy, France, Spain), North Africa, Turkey, Syria, Russia, India, Pakistan, China, Japan, Malagasy Republic, Chile, Argentina, and Mexico. Mexico is the world's largest producer. There is limited cultivation in the United States.

Culinary Portrait

The seeds, which are the economically important part of the plant, have an agreeable odor and pleasant, spicy taste. The aroma and flavor of the seeds have been described as powerfully sweet, aromatic, licorice-like, warm, fruity, camphoraceous, and cooling, with little perceptible aftertaste. The taste is reminiscent of licorice, and indeed, anise seed is often used to impart a licorice taste. The sweet licorice flavor of anise seed is used to intensify sweetness in cakes, pastries, and cookies. The seeds are also an ingredient of such baked goods as bread and rolls, and of soups, stews, curries, marinades, cheese, applesauce, and baked apples. Anise is a common ingredient in Asian cuisine. Anise-seed sugar cookies called *biscochitos* are served traditionally during the Christmas season in Native American and Spanish-American communities of the Southwestern United States. Many meats are seasoned with anise seeds or extract, including bologna, frankfurters, knockwurst, dry sausage, mortadella, pepperoni, spiced luncheon loaf, and sausages. Alone or in combination with cinnamon and bay, anise seed compliments duck, pork, and fish. The seeds can be added to such cooked vegetables as beet, cabbage, and carrot, and are said to have a special affinity for garlic. Like most herbs and spices, it should be remembered that very little goes a long way. For example, 0.62 mL (1/8 teaspoon) of ground seed is sufficient to season 454 g (1 pound) of canned green beans. When purchasing anise seeds, it is prudent to buy small amounts, so that the spice will be used up promptly and remain flavorful. The seeds store best in an airtight container.

The flavor of anise leaves resembles that of the seeds, although the leaves are more delicate than the seeds and are not as widely used. Both raw and cooked leaves are considered delicious, and are employed as a seasoning, salad component, and garnish. Small amounts of chopped leaves can be added to cooked vegetables, soups, stews, sauces, meat dishes, salads, fruit salads, cream cheese, fish, and pies. Like the seeds, the leaves are occasionally used to brew a sweet, licorice-tasting tea. Leaves of anise are sometimes encountered in commercial herbal tea blends, and are in fact recommended by many herbal practitioners as a digestive for various ailments. Occasionally, the roots have been used to prepare wine.

Oil from the seeds provides the flavoring agent for many famous alcoholic beverages, such as anise liqueur or Anisette (French and North African), Arak or Arrak (eastern Mediterranean), Berger (French), Ojen (Spanish), Raki (Turkish), Ouzo (Greek), Aguardiente (South American), Allasch (Latvian), Sambuca (Italian), Mistra (Italian), Tres Castillos (Puerto Rican), and muscatel wine. The oil is also used in cooking, in industry, and in medicine. It is employed to flavor a remarkable range of products, from toothpaste to mouse bait. Cough syrups, cough drops, chewing gum, tooth powder, ice cream, pickles, and tobacco products often have been flavored with anise oil. Most of the "licorice" flavor in candy actually comes from anise.

Anise milk (aniseed milk) is another beverage that is popular, especially in the Netherlands, where it is traditionally associated with ice-skating. Anise milk is made by scalding 4 cups of milk with 1 tablespoon of crushed aniseed, adding 1/2 cup of sugar, simmering for 5 minutes, then adding 2 tablespoons of cornstarch dissolved in a small amount of water, and warming the brew for another 5 minutes.

Anise (*Pimpinella anisum*), from Köhler (1883–1914).

Anise (*Pimpinella anisum*), from Vilmorin-Andrieux (1885).

Culinary Vocabulary

- "Biscotto," based on Italian, from the Medieval Latin *bis coctus*, "twice cooked," is a crisp Italian cookie traditionally flavored with anise and often containing almonds or filberts.
- "Pastis" (pronounced pas-TEES) refers to anise-based, French, aperitif liqueurs, such as Pernod and Ricard.

Curiosities of Science and Technology

- Estrogens are the dominant female reproductive hormones. These are currently used to treat medical conditions related to deficiencies of the naturally produced sex hormones, for example, menstrual irregularities. Anise has been known to have estrogenic activity since antiquity, and has long been considered to be an aphrodisiac. Not only humans but also animals respond to estrogens. Cattle given estrogens mate more readily because of increased libido. Increased milk secretion is another result. The chemicals dianethole and photanethole in anise are similar to the female sex hormone estrogen. It is remarkable that the estrogenic effects of anise may have been appreciated thousands of years ago, possibly as a result of careful observations by herdsmen of the effects on their animals of eating anise.
- After rich meals, the classical Romans traditionally ate a spiced cake called *mustaceum*, containing cumin, new wine, fat, cheese, and grated bay bark, and also anise seeds, the last-mentioned possibly in the belief that it promoted digestion. This may have given rise to the modern tradition of a spicy cake at weddings. Somewhat reminiscent of the Roman practice, in modern times in some countries, such as Afghanistan and India, slightly roasted anise seeds are sometimes chewed after meals to sweeten the breath and aid digestion.
- Anise has been said to be as attractive to dogs as catnip is to cats, and so is a common ingredient in dog foods. It was a frequent practice in England to douse a captured fox in anise before it was released for hunts. In modern times, some people opposed to the blood sport of fox hunting have laid false trails with aniseed, blood, or urine to confuse the hounds and huntsmen. More legitimately, anise odor has been used in England and the United States as an artificial scent for "drag hunting" with foxhounds. A sack saturated with oil of anise is dragged to lure the hounds. Anise is also reputed to be good bait for fish.
- Components in the oil of anise (also of dill and parsley) can be chemically transformed in the living body, resulting in the production of dangerous, hallucinogenic amphetamines. Too little is ingested in the amounts consumed normally to represent a hazard.
- The longest word ever to occur in any language in a literary work was a fricassée with 17 sweet and sour ingredients, including brains, honey, vinegar, fish, pickles, and the anise-flavored alcoholic beverage ouzo (pronounced OO-zoh; this Greek brandy is clear, turning cloudy when water is added, as noted below). The word appeared in *The Ecclesiazusae*, a Greek comedy written by Aristophanes over 2000 years ago. The Greek word is 170 letters long, but transliterates into the following 182-letter word: *lopadotemachoselachogaleokranioleipsanodrimhypotrimmatosilphioparaomelitokatakechmenokichlepikossyphophattoperisteralektryonoptekephalliokigklopeleiolagoiosiraiobaphetraganopterygon.*
- When mixed with cold water, anise-based liqueurs turn white and cloudy. The reason this occurs is that anethole and other essential oils in these beverages are soluble in alcohol but not in water; so when water is added, they come out of solution and form tiny droplets that cloud the drink. This is reminiscent of milk, which is white and opaque because of a colloid suspension consisting of very small particles of casein proteins.

Key Information Sources

Albert-Puleo, M. 1980. Fennel and anise as estrogenic agents. J. Ethnopharmacol. **2**: 337–344.

Arslan, N., Gurbuz, B., Sarihan, O.E.B., and Gumuscu, A. 2004. Variation in essential oil content and composition in Turkish anise (*Pimpinella anisum* L.) populations. Turk. J. Agric. For. (Turkey), **28**: 173–177.

Capelletti, E.M. 1980. Botanical identification of anise and hemlock fruits in powdered drug samples. Planta Med. **39**: 88–94.

Cardenas, L.B., and Guzman, C.C. de. 1999. *Pimpinella anisum* L. *In* Plant resources of South-East Asia, 13, spices. *Edited by* C.C. de Guzman and J.S. Siemonsma. Backhuys, Leiden, Netherlands. pp. 180–183.

Embong, M.B., Hadziyev, D., and Molnar, S. 1977. Essential oils from spices grown in Alberta. Anise oil (*Pimpinella anisum*). Can. J. Plant Sci. **57**: 681–688.

Ernst, D. 1989. *Pimpinella anisum* L. (anise): cell culture, somatic embryogenesis, and the production of anise oil. Biotechnol. Agric. For. **7**: 381–397.

Hornok, L. 1992. Anise (*Pimpinella anisum* L.). *In* Cultivation and processing of medicinal plants. *Edited by* L. Hornok. John Wiley & Sons, Chichester, UK. pp. 143–147.

Icon Health Publications. 2003. Anise: a medical dictionary, bibliography, and annotated research guide to internet references. ICON Health Publications, San Diego, CA. 100 pp.

Kollmannsberger, H., Fricke, G., Paulus, H., and Nitz, S. 2000. The flavour composition of Umbelliferous fruits: 1. Anise (*Pimpinella anisum*). Adv. Food Sci. **22(1–2)**: 47–61.

Lawrence, B.M. 1983. Progress in essential oils—anise oil, bergamot oil. Perfum. & Flavor. **8(3)**: 65–74.

Lawrence, B.M. 1987. Progress in essential oils. [Anise oil, bergamot oil, cumin oil, cypress oil.] Perfum. & Flavor. **12(2)**: 67–72.

Miro, M. de. 2004. *Illicium*, *Pimpinella* and *Foeniculum*. CRC Press, Boca Raton, FL. 176 pp.

Özgüven, M. 2001. Aniseed. *In* Handbook of herbs and spices. *Edited by* K.V. Peter. CRC Press, Boca Raton, FL. pp. 39–51.

Szujko-Lacza, J. 1975. Nectary gland development and functional duration in *Pimpinella anisum* L. Acta. Biol. Budap. **26**: 51–65.

Szujko[ne]-Lacza, J. 1976. Anise, *Pimpinella anisum* L. Akademiai Kiado, Budapest, Hungary. 95 pp. [In Hungarian.]

Szujko-Lacza, J., and Szocs, Z. 1975. The architecture and the quantitative investigation of some characteristics of anise, *Pimpinella anisum* L. Acta Bot. **23**: 443–450.

Szujko-Lacza, J., Sen, S., and Horvath, I. 1979. Effect of different light intensities on the anatomical characteristics of the leaves of *Pimpinella anisum*. Acta Agron. Acad. Scient. Hung. **28**: 120–131.

B. (Chinese) Star Anise

Family: Illiciaceae (star anise family)

Names

Scientific name: *Illicium verum* Hook. f.

- The "star" in "star anise" is based on the star-like form of the fruits of the plant; the "anise" is for the resemblance of taste and chemistry with anise.
- Chinese star anise is also called aniseed stars, badian, badiana, Chinese anise, and (very undesirably, since it leads to confusion with the poisonous relative) Japanese star anise.
- The genus name *Illicium* is based on the Latin *illicere*, to allure, a reflection of the pleasant fragrance of several of the species, including Chinese star anise.
- *Verum* in the scientific name *I. verum* is Latin for true. In the past, it was not appreciated that the poisonous Japanese star anise (see below) was different from the edible Chinese star anise, and the "true" in the scientific name points out that this is the truly edible species.

Plant Portrait

Chinese star anise is a subtropical evergreen shrub or small tree growing as tall as 18 m (60 feet). The species appears to be native to southeastern China and northeastern Vietnam, although some have questioned whether genuinely wild plants still exist. Its fruits have been used as a spice in China for perhaps 3000 years, and the spice was introduced into Europe in the 17th century. The tree produces white to yellow flowers that mature to deep rose, pink, or purple and develop into anise-scented, star-shaped fruits. The fruits have 5–12 pointed, boat-shaped sections, about 8 on average, which become tough-skinned and rust colored at maturity, each section measuring up to 3 cm (1¼ inches) in length, and containing one shiny brown seed (or occasionally empty). The fruit is usually picked before it can ripen, and dried, and is available whole, in pieces, or ground to a red-brown powder. Chinese star anise like anise described above is used directly as a culinary spice, and is similarly the source of a distilled oil, anethole, which is employed in medicine, and in industry for flavoring. China is the main exporter, with some product also from Laos, the Philippines, and Jamaica. The annual world value of star anise essential oil is several million dollars annually, and mostly the oil, not the fruit, is imported by Western countries. Because Chinese star anise contains a chemical

Left, poisonous Japanese star anise (*Illicium anisatum*), with fruit, from Köhler (1883–1914); *right*, edible star anise (*I. verum*), from Curtis (1787–present, plate 7005).

used to make antiviral medicines, the species has recently become extremely important in this regard (see below).

Japanese anise (also called bastard star anise, Japanese star anise, sacred anise tree, and Chinese anise) is *I. anisatum* L. (*I. religiosum* Siebold & Zucc.). This native of China, Japan, and Korea is an evergreen shrub or small tree that grows as tall as 9 m (30 feet), and is sometimes cultivated in the warmest parts of the United States. Branches are commonly used to decorate Buddhist graves in temple grounds in Japan. Although the seeds are used in China to treat toothache and dermatitis, and the essential oil has been used to treat colic in children, the seeds are very toxic, and indeed have been used as fish poison. Using the seeds of Japanese anise for culinary purposes could be fatal. In the past, commercial star anise preparations were sometimes contaminated with the poisonous species. In 2001, France banned the use of "star anise" in drugs and herb teas following reports of convulsions due to the presence of such contamination. In 2003, the U.S. Food and Drug Administration issued an advisory against consuming "star anise tea," since it was found that numerous products being sold were contaminated with Japanese star anise, resulting in illness. Remarkably, the common names for the poisonous Japanese star anise and the edible Chinese star anise are confused in several standard horticultural references works, which could lead to accidental poisoning. Some other poisonous species of *Illicium* are native to, or cultivated in, warm climates, including the southern United States, so that attempting to harvest and consume the fruits of *Illicium* trees that one encounters is hazardous without expert identification.

Culinary Portrait

Star anise has a powerful, licorice-like, bouquet that is more pungent and stronger than that of anise. The flavor is like a bitter, harsher aniseed, providing an authentic touch in the preparation of certain Chinese dishes. Star anise is important in Asian cuisines, predominantly in China, which has a long tradition of using the spice. The Chinese use star anise much in the same way they use cinnamon, especially when preparing meat and poultry dishes. In Asia, the fruits are used as a condiment to flavor such foods as broths, marinades, sautéed vegetables, rice, simmered dishes made with soy, sauces, chicken, duck, pork, beef, fish, shellfish, pastry, and beverages (including coffee and tea).

In the West, star anise competes with anise for much of the same market, and it is used mainly as a cheap substitute for anise in mulled wine, desserts, and liqueurs. Many anise-tasting liqueurs use at least some star anise. Especially in France, the oil is used as a raw material in the production of alcoholic beverages, particularly the liquors Pernod and Anisette (Pernod Ricard is one of the major users). While star anise fruits keep well (indeed better than anise)—for over a year when stored in airtight containers in a cool location—powdered star anise does not store well, and should be purchased in very small quantities after checking the flavor. Whole fruits can be added directly to the cooking pot. Small amounts are used, as the spice is powerful. A few fruits are sufficient to flavor an entire dish. Generally, it is recommended that about a third of the amount of star anise is required in comparison to anise.

Culinary Vocabulary

- "Five-spice powder" (known in China as *wu xiang fen*, pronounced woo shang fen) is a well-known Chinese spice mixture, generally some combination of star anise, fennel, clove, cinnamon, Szechwan pepper, ginger, nutmeg, cassia (or cinnamon), and cloves. In China, it is often added to fried vegetables or meat, or used in meat coatings and marinades.

Curiosities of Science and Technology

- The letters of the genus *Illicium* are all Roman numbers (I = 1, L = 2000, C = 100, U(= V) = 5000, M = 1000).
- The toxic Japanese star anise (*I. anisatum*) has been used in Japan as a drug to produce insensitivity by those about to commit hara-kiri.
- In the past in Asia, it was common to chew Chinese star anise fruits after meals to promote digestion and sweeten the breath, a practice that has persisted to the present. A Chinese practice is to suck on a point of the star-shaped fruit to ensure fresh breath.
- In Asian medicine, Chinese star anise is used to treat indigestion, facial paralysis, arthritis, and intestinal cramps. There is some very minor use in Western medicine for treating intestinal cramps.
- Dogs and cats are familiar with the taste of star anise, since it is a popular additive in their pet foods. Anethole, the essential ingredient, has been said to be as attractive to dogs as catnip is to cats (see information for ANISE, above).
- In DNA-based evolutionary studies reported in October 1999 in the journal *Science*, it was concluded that Chinese star anise represents one of the three most ancient of the hundreds of families of flowering plants.
- Swiss drug-maker F. Hoffmann-La Roche Ltd. is the manufacturer of the antiviral drug Tamiflu (oseltamivir phosphate). Tamiflu, mainly applied in the past to prevent influenza, is considered to be the first line of defense for potential new pandemics, and accordingly has been the subject of numerous news reports. The drug is manufactured from shikimic acid by a complex multi-step chemical synthesis procedure. All plants synthesize shikimic acid because important substances essential for biological functions are synthesized from it. Industry has found many ways to synthesize shikimic acid, including from quinic acid from cinchona bark, from the bacterium *E. coli*, and from carbohydrates (e.g., D-arabinose and

Left, edible fruit of star anise (*Illicium verum*); *right*, poisonous fruit of Japanese star anise (*I. anisatum*), from Hare et al. (1908).

D-ribose). Nevertheless, about two-thirds of the shikimic acid used for Tamiflu is obtained from Chinese star anise. (Many news reports erroneously identify Japanese star anise as the source.)

Key Information Sources

(For additional references, see ANISE, above.)

Agarwal, S.K., Siddiqui, M.S., Jain, S.P., and Kumar, S. 1999. Chemotaxonomical study of Indian *Illicium griffithii* and *Illicium verum* fruits. J. Med. Aromat. Plant Sci. **21**: 945–946.

Bellenot, D., and Saltron, F. 2003. Caracterization de la badiane alimentaire et non alimentaire. Ann. Falsif. Expertise Chim. Toxicol. (France), **96**(**962**): 39–54. [Discusses identification of fruits of *Illicium* species, in order to detect fruits of poisonous species of star anise that are often present as contaminants in collections of fruits of Chinese star anise.]

Bernard, T., Perineau, F., Delmas, M., and Gaset, A. 1989. Extraction of essential oils by refining of plant materials II. Processing of products in the dry state: *Illicium verum* Hooker fruit and *Cinnamomum zeylanicum* Nees bark. Flavour Fragrance J. **4**: 85–90.

Chang, K.S., and Ahn, Y.J. 2002. Fumigant activity of (E)-anethole identified in *Illicium verum* fruit against *Blattella germanica*. Pest Manage. Sci. **58**: 161–166.

Cu, J.Q., Perineau, F., and Goepfert, G. 1990. GC/MS analysis of star anise oil. J. Essent. Oil Res. **2**: 91–92.

De, M., De, A.K., Sen, P., and Banerjee, A.B. 2002. Antimicrobial properties of star anise (*Illicium verum* Hook f). PTR Phytother. Res. **16**: 94–95.

Gerhard, U. 1979. Star anise, a nicely shaped spice. Fleischwirtschaft, **59**: 176–177. [In German.]

Herisset, A., Jolivet, J., and Rey, P. 1972. Differentiation of some essential oils with related structures especially through their UV, IR and Raman spectra, Part 5, oils of Chinese star anise *Illicium verum*, anise *Pimpinella anisum* and sweet fennel *Foeniculum dulce*. Plant. Med. Phytother. **6**: 137–148.

Ho, S.H., Ma, Y., and Huang, Y. 1997. Anethole, a potential insecticide from *Illicium verum* Hook f., against two stored product insects. Int. Pest Control, **39**(**2**): 50–51.

Ho, S.H., Ma, Y., Goh, P.M., and Sim, K.Y. 1995. Star anise, *Illicium verum* Hook f. as a potential grain protectant against *Tribolium castaneum* (Herbst) and *Sitophilus zeamais* Motsch. Postharvest Biol. Technol. **6**: 341–347.

Hoh, H.-C. 1940. The star anise tree in Kwangsi. Sunyatsenia, **4**: 272–289.

International Organization for Standardization. 1995. Star anise (*Illicium verum* Hook. f.). Specification. International Organization for Standardization, Geneva, Switzerland. 7 pp.

Kataoka, E., Tokue, C., and Tanimura, W. 1987. Findings on sterol, tocopherol and phospholipids contained in cardamon and star anise. J. Agric. Sci. Tokyo (Japan), **31**: 189–196. [In Japanese, English summary.]

Lawrence, B.M. 1992. Progress in essential oils. Star anise oil. Perfum. Flavor. **17**(**2**): 49–50.

Lee, S.W., Li, G., Lee, K.S., Jung, J.S., Xu, M.L., Seo, C.S., Chang, H.W., Kim, S.K., Song, D.K., and Son, J.K. 2003. Preventive agents against sepsis and new phenylpropanoid glucosides from the fruits of *Illicium verum*. Planta Med. **69**: 861–864.

Liu, H., and Yang, C.S. 1989. Pollen morphology of Illiceaceae and its significance in systematics. Chinese J. Bot. **1**(**2**): 104–115.

Liu, Y., 1999. Study on mechanical drying methods of the star anise. J. Guangxi Agric. Biol. Sci. (China), **18**: 132–135. [In Chinese, English summary.]

Lô, V.N. 1999. *Illicium verum* Hook. f. *In* Plant resources of South-East Asia, 13, spices. *Edited by* C.C. de Guzman and J.S. Siemonsma. Backhuys, Leiden, Netherlands. pp. 130–134.

Minodier, P., Pommier, P., Moulene, E., Retornaz, K., Prost, N., and Deharo, L. 2003. Star anise poisoning in infants. Arch. Pediatr. **10**: 619–621. [In French.]

Nakamura, T., Okuyama, E., and Yamazaki, M. 1996. Neurotropic components from star anise (*Illicium verum* Hook. fil.). Chem. Pharm. Bull. (Tokyo), **44**: 1908–1914.

Okuyama, E., Nakamura, T., and Yamazaki, M. 1993. Convulsants from star anise (*Illicium verum* Hook. f.). Chem. Pharm. Bull. (Tokyo), **41**: 1670–1671.

Saltron, F., Langella, C., and Guerere, M. 2001. Evidence for the contamination of Chinese star anise by other *Illicium* species. Ann. Falsif. Expert. Chim. Toxicol. **95**: 397–402. [In French, English summary.]

Small, E. 1996. Confusion of common names for toxic and edible 'star anise' (*Illicium*) species. Econ. Bot. **50**: 337–339.

Small, E., and Catling, P.M. 2006. Blossoming treasures of biodiversity 19. Chinese star anise—defence against a potentially catastrophic global killer flu virus. Biodiversity, **7**(**1**): 56–60 + back cover.

Smith, A.C. 1947. The families Illiciaceae and Schisandraceae. Sargentia, **7**: 1–224.

APPLE

Family: Rosaceae (rose family)

Names

Scientific names: Apples are species and hybrids of *Malus*. Most modern apple varieties are placed in the hybrid *M. domestica* Borkh. ["*M. pumila*" and "*M. sylvestris*" of various authors; *M. sylvestris* Mill. var. *domestica* (Borkh.) Mansf.; *Pyrus malus* L.]. Several European species are the chief ancestors of modern apples, notably *M. pumila* Mill. (paradise apple), *M. sylvestris* Mill. (European crab apple), and *M. baccata* (L.) Borkh. (Siberian crab apple).

- The word "apple" is perhaps derived from the Old High German *apful*, or the Old Slavic *ablùko*.
- The "crab" in "crab apple" (crabapple) may be an allusion to the crab, from the French *crabbe* for this crustacean. The term "crab apple" is ambiguous; it is often applied to specific wild species, but also to any small sour apple, sometimes including wild varieties of the common domesticated apple.
- In old Europe, it was common practice to coin names for new edible plants that were introduced using familiar names, and this was particularly true for the well-known apple. Thus, depending on language, lemons were called "Persian apples," dates were "finger apples," potatoes were "apples of the earth," eggplants were "mad apples," pomegranates were "apples of Carthage," and tomatoes were "love apples" and "gold apples."
- Numerous plants unrelated to the apple have "apple" in their name, for example, Adam's apple [*Tabernaemontana divaricata* (L.) R. Br. ex Roem. & Schult.], bake apple (*Rubus chamaemorus* L.), custard apple (*Annona* species), and sugar apple (*Annona squamosa* L.).
- The genus name *Malus* is based on the Greek *melon*, a name that was applied to the apple as well as to other tree fruits with a fleshy exterior. The word also traces to an older language, Akkadian, where *malum* meant fullness or to be full.
- Words found in the scientific names of the principal apple species mentioned above: *pumila*, dwarf; *sylvestris*, growing in woods; *baccata*, berry-like (having fruits with a fleshy texture).

Plant Portrait

Apple trees vary in height to over 15 m (49 feet) and can live for more than a century, some surviving for more than 2 centuries. Depending on variety, apples range in size from a little larger than a cherry to as large as a grapefruit. The skin of an apple may be red, green, yellow, orange, or brown, and the flesh white, cream, pinkish, or greenish.

The apple has been grown by human beings since at least as far back as recorded history in both Europe and western Asia, and may have been the earliest of all fruits to be cultivated. Kazakhstan, dating back some 8500 years, is often said to be the birthplace of the modern apple. Lake Dwellers of central Europe grew apples in neolithic times, as reflected by the discovery of stored apples seeds. The Greeks and Romans were familiar with about two dozen varieties, while modern man is believed to have selected as many as 10 000, indeed some claim 15 000 (almost half of them American, as reflected in the expression "American as apple pie"). Although there are thousands of apple varieties, only about two dozen have commercial importance in the United States. Eight varieties account for 80% of American production: Red Delicious, Golden Delicious, Granny Smith, McIntosh, Rome Beauty, Jonathan, York, and Stayman. The Golden Delicious is very popular in Europe. In addition to the United States, large producers include Russia, China, Germany, and France. Apple cultivars are generally mutants that have been selected from great numbers of seedlings, and propagated and held true to type by budding or grafting. Trees raised from the seeds of apples rarely possess as desirable qualities as the maternal trees from which they were obtained. The apple is the most widespread of all fruits, accounting for about half of the world's output of deciduous tree fruit. Apples may be kept for 8 months in an atmosphere of reduced oxygen, regulated by carbon dioxide and low temperature; such controlled atmosphere storage prevents continued ripening and makes tree-fresh apples available through the winter months.

Varieties Popular in North America

- The McIntosh is red-skinned or greenish, blushed with red. This thin-skinned, all-purpose apple is tender and requires less cooking time than most other varieties. In 1796, John McIntosh left the Mohawk Valley of New York and settled in southern Ontario. About 15 years later, he discovered about 20 apple trees, which legend has it grew from seeds discarded by an Indian. One of the trees turned out to be the first McIntosh apple.

Picturesque apple tree (*Malus domestica*), from Foord (1906).

John's son Allen learned how to take shoots from the tree and graft them onto other trees, thus reproducing the McIntosh apple, possibly the most famous of all apple cultivars. The original tree was scorched during a fire in 1884, but survived until 1906. A monument marks the spot southeast of Ottawa where the tree lived for almost a century; and in 1996, the Canadian government issued a silver dollar featuring a bough full of McIntosh apples, commemorating the 100th year since John McIntosh arrived in Canada.

- Delicious-apple varieties account for half of the apples sold in the world. Red Delicious (or just Delicious), a red-skinned eating apple, is the leading variety, followed by Golden Delicious, a gold-skinned (i.e., pale to bright yellow) eating apple. Despite the similarity of names, the two varieties are unrelated. Delicious was discovered by a farmer named Hesse Hiatt in Peru, Iowa, in 1872, and Golden Delicious was found in 1914 on the Anderson Mullens farm in West Virginia. A commercial nursery, Stark Brothers, bought the rights to Delicious in 1895, and subsequently also purchased Golden Delicious, and renamed both varieties similarly. (The Red Delicious was originally named Hawkeye, and the Golden Delicious was originally called Mullen's Yellow Seedling.) The Golden Delicious has sweet, fine-grained, crisp flesh, which some find bland and mealy, but this apple retains its shape well when baked and also makes excellent applesauce. The Red Delicious is distinctive in appearance (it has a five-knobbed crown) and in flavor (described as "sugar and spice").
- The Red Astrachan is a red-skinned, tart, high-pectin apple suitable for pies, sauces, and jellies. It originated in Russia and was brought to the United States in 1870. This variety has become second only to Winesap as the most widely sold southern apple, and is one of the most widely distributed throughout the world. However, as is often the case, patriots often prefer the products of their own country, and so the foreign-sounding Red Astrachan was renamed the Abe Lincoln apple (it was also called Captain and Red Ashmore, especially by those who were not fond of Lincoln). The old name Red Astrachan, however, eventually triumphed in the United States.
- The Granny Smith, a green-skinned eating apple, originated in Australia about 1868 when gardener Maria Ann Smith (died 1870; her nickname was "Granny Smith") planted seeds she found in a gin barrel in New South Wales, and first grew the famous apple at Eastwood, Sydney. This popular apple is preferred by many cooks for its tart–sweet flavor and superb cooking qualities.

- The Jonathan apple is a highly aromatic, spicy fruit, the skin bright red with tiny white dots. In addition to being good for eating fresh, it makes excellent pies, applesauce, and apple butter (defined below), although its juiciness causes it to cook down considerably. It was discovered in 1800 in Woodstock, New York, and was originally called Rick after farmer Philip Rick, who found it. The variety was popularized by Jonathan Hasbrouk (an American judge who died in 1846; his first name became the name of the apple) and Judge J. Buel of Albany, New York.
- The Rome Beauty is a red-skinned baking apple that originated from a discarded tree given to Allanson Gillett (also spelled Gillet) by his brother Joel, in Proctorville, Ohio. The variety was named for the township of Rome where the Gillett farm was located.
- The York (especially the York Imperial) is a red-skinned baking apple, first discovered near York, Pennsylvania, about 1830 by a man named Johnson, who first called it Johnson's Fine Winter Apple. The name York was given to it by horticulturist Charles Downing.
- Stayman is a red-skinned eating apple, sometimes used for cooking. It was selected in Leavenworth, Kansas, in 1866, by Dr. J. Stayman, and is sometimes called Stayman's Winesap or Stayman Winesap.
- The Rhode Island Greening is a yellow–green-skinned baking or cooking apple. It is considered best for apple pie, and is commercially processed for applesauce and frozen pies. Accordingly, it is rarely seen in the market. It originated in Green's End, Rhode Island, about 1748, with the help of a tavern keeper named Green. Rhode Island adopted the Rhode Island Greening apple in 1991 as its official state fruit.

Some Other Varieties

- Various apple varieties have "pippin" in their names (e.g., Cox's Orange Pippin). In the past, the word "pippin" meant any apples grown from seed. The word originated in English in the 14th century, and was adopted from Old French *pepin*, "seed" or "seedling apple." The word "pippin" was first used in English to mean "seed." "Pip," meaning the internal seed of a fruit, is derived from "pippin."
- The Lady apple is a variety with small, red, strikingly beautiful apples suitable for decorative purposes or as a dessert. The taste is sweet and tart. This miniature apple is generally expensive, and of limited availability. This apple should not be confused with Pink Lady, a variety that arose as a cross between Golden Delicious and Lady Williams apples.
- The Gala apple, developed in New Zealand in the 1920s, has become one of the world's most widely grown varieties, and a mainstay of supermarket apple selections. Some forms of it are yellowish, others quite reddish. It is mostly used for eating as a dessert fruit, but is also recommended in salads, and for baking and applesauce.

Culinary Portrait

The flavor of apples can be sweet or tart, and varieties differ considerably in juiciness. Texture can be soft and smooth, crisp, mealy, crunchy, or hard. Most of an apple's fragrance cells are concentrated in the skin and, as they ripen, the epidermal cells develop more aroma and flavor. Apples usually contain over 80% water, and the chief food value is in the form of sugars that are produced from starches as the fruits ripen. Eaten raw, the fruits are a delectable snack or dessert. Apples are also baked, canned, stewed, frozen, dehydrated, and turned into pie filling, sauce, cider, vinegar, juice, paste (apple butter), pectin, jelly, and candy. Crab apples are especially rich in pectin, and are particularly suitable for making jellies. Indeed, crab apples are often added to other fruit jellies to improve their consistency. Crab apple jelly particularly complements cold meats and poultry. Apples are used to make an endless variety of desserts, including fritters, turnovers, charlottes, puddings, compôtes, and mousses. Apples are said to make the prince of pies. The fruit is also used in several specialty baked desserts, for example, apple grunt (also known as slump, a colonial American dessert made with fresh fruit topped with biscuit dough and steamed in a closed container), cobbler (a deep-dish fruit tart with a rich, sweet, biscuit-type dough covering the fruit), and pandowdy (an old-fashioned deep-dish New England fruit-pie-like dessert in which the crust is broken up partway through baking and pressed down into the fruit to absorb the juices—a technique called "dowdying"). There are countless recipes for combining apples with various cooked dishes, and numerous countries have their particular dishes, such as Austrian apple-and-raisin strudel, Normandy (France) apple tart, and British top-crusted pie (unlike American apple pie, which has pastry underneath and on top, British apple pie is normally baked in a deep pie dish with a crust on top only). Although most often used in sweet preparations (frequently flavored with cinnamon and vanilla), apples also are excellent in savory dishes, including meat, poultry, and fish. Apples also combine well with vegetables in salads.

Apple (*Malus domestica*), from Köhler (1883–1914).

Apple varieties are usually particularly suitable for some uses, and often unsuitable for others. Eating apples should be firm, juicy, tasty, and crisp. Pie apples should be drier, and slightly acidic. Oven-baking apples should be sweet, and not disintegrate easily. The Reverend W. Wilks is an interesting cooking apple that disintegrates to a yellow froth when cooked. Some varieties do not react well to baking, some actually becoming bitter. Good apples for jelly are often barely ripe, and are acidic, juicy, and high in pectin. Apples suitable for applesauce do not discolor easily. The McIntosh, Winesap, Jonathan, and Golden Delicious are familiar eating apples, which also happen to be good baking apples (and so are called "all-purpose apples"). However, Rome Beauty is much more suitable for baking than for eating raw, and Famed Delicious is excellent for eating raw but unsuitable for cooking. Gravenstein and Crispin are favorite pie apples, while the Yellow Newton is very good for making cider. Cortland is useful for salads because it stays snowy white when cut.

Apples are sometimes covered with wax or even shellac, and from time to time, concern is expressed about the possibility of pesticides leaving residues on the skin. Accordingly, washing under cold water or even accompanying this with scrubbing of the skin is desirable before consumption.

Cider is a wine produced from the fermented juice of the apple. It has an alcohol content of 3.5–7.5%. Sweet cider is made from apples with considerable sugar, while dry cider is produced from apples with plenty of malic acid. Bittersweet apples are high in tannins, which occur in the skins. This produces the mouth-puckering sensation of a zesty cider; the best cider apples are high in tannin. The juice of apples ferments by the action of yeast to produce alcohol. Vinegar may be manufactured by allowing vinegar bacteria to convert the alcohol of the hard cider into acetic acid. Apple cider vinegar is the most popular vinegar in the United States, and is considered by many to be an indispensable condiment.

Culinary Vocabulary

- A "cooking apple" in British English is equivalent to a "baking apple" in American English.

- In Britain, "minced meat" (or just "mince") is, as in North America, chopped meat or hamburger. However, "mincemeat" can refer to any combination of chopped meat, apples, raisins, spices, etc., which go into a mince pie.
- "Mock apple pie" was popular during the Great Depression, when apples were too expensive for many people to buy. The recipe traces to a pie made with hardtack (also known as sailor's biscuit and sea biscuit, a hard, coarse unleavened bread traditionally used as army or navy rations because of its long shelf life) during the American Civil War. The Nabisco company printed the recipe on its packages of Ritz Crackers during the Depression. The crackers were formed into a crust into which a cooked preparation of water, sugar, cream of tartar, lemon juice, and grated lemon rind was poured. Cinnamon, margarine, and a crust topping were added, and the pie was baked until golden brown. The crackers absorb the flavors and expand to produce an apple-like texture.
- "Apple sass" is a New England expression for "apple sauce."

Siberian crab apple (*M. baccata*), from Pallas (1784–1831).

- The word "cider" is derived from the Hebrew *shekar*, "strong drink of grain and honey," although in North America, cider is usually made from apples.
- In North America, unfermented apple juice is called "soft cider," while fermented juice is termed "hard cider" (with an alcohol content of 3–7%). "Sweet cider" is unfermented (alcohol-free) or partly fermented (producing a beverage with under 3% alcohol). The distilled liquor from cider is called "applejack" in the United States, "calvados" or apple brandy in Europe. Applejack can also be produced by adding extra sugar to the fermenting apples. "Ciderkin" is a diluted apple cider made from watered-down second and third pressings of the apples. "Cider royal" is cider mixed with brandy. "Stonewall," also known as "stonefence," a drink popular in the United States during the early 19th century, is made from apple cider or applejack and a liquor such as rum. "Essence of lockjaw" was a highly potent apple-based liquor that was made in colonial times. "Champagne cider" is a sort of sparkling cider that has been manufactured in England and Spain by inducing a secondary fermentation.
- "Bird's nest pudding" (also known as "crow's nest pudding") is a very old New England apple pudding.
- *Tart tatin* is a famous French upside-down apple tart prepared by covering the bottom of a shallow baking dish with butter and sugar, then apples and a pastry crust. The sugar and butter turn into a caramel that becomes the delicious topping when the tart is inverted onto a serving plate. Two French sisters living in the Loire Valley prepared and sold this dish for many years, and are commemorated in the longer French name *tarte des demoiselles Tatin*, "the tart of the young Tatin ladies."
- "Sea pie" is an old New England stew made with pork, veal, or fowl, mixed with sweet dried apples, molasses, and dumplings. The highly misleading name (sea pie in this sense is not a pie and does not contain sea foods) is said to have arisen because a sea captain provided the recipe. There is a traditional Quebec layered meat pie called *cipaille* (pronounced "sea pie"), which may have contributed to the name. In recent times, various "sea pie" recipes, in fact for pies made with seafood, have become available.
- "Schnitz" (pronounced schneetz) is dried, whole or quartered sweet apples used in Pennsylvania Dutch cookery (the "Dutch" in Pennsylvania Dutch is a linguistic transformation from *Deutsch*, i.e., German). Schnitz is generally available in areas with

large Amish populations. Schnitz and knepp (pronounced kae-mep), i.e., "apples and dumplings," is a classic Pennsylvania Dutch dish of dried apples simmered with ham and topped with spoonfuls of batter to produce dumplings. Other schnitz specialties of the Pennsylvania Dutch are dried-apple pie and dried-apple cake.

- Amish mothers traditionally bake special half-moon apple pies for children to eat on Sundays. These are known as "preaching pies" (also as "perfect pacifiers"), and are made with dried apples so they will not drip as they are eaten during Sunday services.
- "Apple brown betty" is a traditional American baked dessert made with sliced apples, bread crumbs, brown sugar, and often cinnamon and nutmeg. It is frequently served with heavy cream. (A "betty" is a baked pudding made with layers of sugared and spiced fruit and buttered breadcrumbs. Betties date back to colonial America.)
- "Apple butter" is an apple preserve made slowly by cooking down apples and spices into a thick paste that is used as a spread.
- A "Marlborough pie" is a single-crust apple pie flavored with various ingredients such as lemon juice, nutmeg, and sherry; but instead of sliced apples, it is made with applesauce, giving it a custard texture. It is thought that it originated in Massachusetts for use during Thanksgiving.

Curiosities of Science and Technology

- The first colonists in New England naively transplanted American wild crab apples to their gardens, hoping that cultivation would immediately turn them into large trees, producing succulent fruit as in Europe. Of course, the wild trees merely produced crab apples.
- Just as beer was commonly drunk during the Middle Ages in Europe because it was often more healthful than the contaminated water that was available, so alcoholic apple cider was often the only safe beverage available to North American colonists. Almost every village had a cider mill, and cider was the most popular drink in North America until the 19th century when beer became common.
- Moses Coates (1746–1816), a Pennsylvanian inventor, patented the apple parer in 1803.
- The wood of crab apple trees is renowned for its hardness, and in early times, it was a favorite material for cudgels.
- Henderson Luelling and William Meek from Iowa first brought commercial apple seedlings to the Pacific Northwest of the United States. Luelling's effort was particularly heroic: by himself, in 1847, he steered a covered wagon filled with earth in which apple seedlings were growing through hostile Indian territory.
- At the transition of the 19th and 20th centuries, "apple green" was a popular color because it was thought to be calming. Asylum walls were painted apple green to calm the patients, and husbands were advised to paint their kitchens apple green so their wives "would not insist on getting the vote."
- Twenty-five percent of the volume of an apple is air, which is why it floats.
- Why does the white flesh of apples and potatoes turn brown when exposed to air for a few hours? For the same reason that iron rusts—oxidation (a chemical combining with oxygen) occurs. When oxygen becomes available on account of removing the skin of the apple, the enzyme polyphenol oxidase (also known as tyrosinase) oxidizes compounds (phenolics) in the fruit, turning them brown. Browning of a peeled apple can be prevented simply by putting it in water, so that the enzyme does not have access to oxygen.
- In 1962, astronaut John Glenn ate the first meal in space—puréed applesauce squeezed from a tube.
- The apples from just one tree can weigh 363 kg (800 pounds).
- In northern Japan, apple farmers employ turkeys to protect their crops from marauding wild monkeys.
- A report issued by the British Dental Association in 2002 warned that because the sugar content of some modern apples in Britain, such as Pink Lady, Fuji, and Braeburn, was about 15% (higher than the 10–11% in popular apples a decade ago), to prevent dental decay it was advisable for children to eat such extra-sweet apples with meals, followed by brushing of teeth, rather than as between-meal snacks. A typical 140-g (5-ounce) apple was said to contain the equivalent of more than 4 teaspoons of sugar. Additionally, it was noted that those who cannot resist a late-night apple would be well advised to rinse their mouths with water and delay brushing for half an hour, to prevent brushing from damaging the tooth enamel, which may have been softened by acids from the apple. This alarmist report can be taken with a grain of salt. Apples have been called "nature's toothbrush," the

crisp texture cleaning teeth and stimulating gums. Moreover, much of the sugar in apples is in the form of fructose, the consumption of which has been shown to have several health benefits.

Key Information Sources

(For additional references, see PEAR.)

Brown, A.G. 1975. Apples. *In* Advances in fruit breeding. *Edited by* J. Janick and J.N. Moore. Purdue University Press, West Lafayette, IN. pp. 3–37.

Browning, F. 1998. Apples. North Point Press, New York, NY. 241 pp.

Dennis, F.G., Jr. 1986. Apple. *In* CRC handbook of fruit set and development. *Edited by* S.P. Monselise. CRC Press, Boca Raton, FL. pp. 1–44.

Ferree, D.C., and Warrington, I.J. (*Editors*). 2003. Apples: botany, production, and uses. CABI Publishing, New York, NY. 660 pp.

Finnigan, B.F., Colt, W.M., and Fallahi, E. 2000. Growing apples for local markets in cold climates. University of Idaho, Cooperative Extension System, Moscow, IA. 20 pp.

Glover, J. 2002. A cost of production analysis of conventional vs. integrated vs. organic apple production systems. Agricultural Research Center, College of Agriculture and Home Economics, Washington State University, Pullman, WA. 88 pp.

Jackson, J.E. 2003. Biology of apples and pears. Cambridge University Press, Cambridge, UK. 488 pp.

Juniper, B.E., Watkins, R., and Harris, S.A. 1998. The origin of the apple. Acta Hortic. **484**: 27–33.

Korban, S.S., and Chen, H. 1992. Apple. *In* Biotechnology of perennial fruit crops. *Edited by* F.A. Hammerschlag and R.E. Litz. CAB International, Wallingford, UK. pp. 203–227.

Kusumo, S., and Verheij, E.W.M. 1991. *Malus domestica* Borkh. *In* Plant resources of South-East Asia, 2, edible fruits and nuts. *Edited by* E.W.M. Verheij and R.E. Coronel. Pudoc, Leiden, Netherlands. pp. 200–203.

Lape, F. 1979. Apples & man. Van Nostrand Reinhold, New York, NY. 160 pp.

Marini, R.P. 2001. Training and pruning apple trees. Revised edition. Virginia Cooperative Extension, Blacksburg, VA. 8 pp.

Martin, A.A. 1976. All about apples. Houghton Mifflin, Boston, MA. 174 pp.

Morgan, J., Richards, A., and Dowle, E. 2002. The new book of apples. Revised edition. Ebury Press, London, UK. 316 pp.

Phillips, M. 1998. The apple grower: a guide for the organic orchardist. Chelsea Green Publishing Co., White River Junction, VT. 242 pp.

Pomology Research Station. 1976. Fundamentals of intensive apple production. An Foras Talúntais Dublin, Ireland. 219 pp.

Rosenstein, M. 1996. In praise of apples: a harvest of history, horticulture & recipes. Lark Books, New York, NY. 176 pp.

Twiss, S. 1999. Apples: a social history. National Trust, London, UK. 46 pp.

Watkins, R. 1985. Apple genetic resources. Acta Hortic. **159**: 21–30.

Way, R.D., Aldwinckle, H.S., Lamb, R.C., Rejman, A., Sansavini, S., Shen, T., Watkins, R., Westwood, M.N., and Yoshida, Y. 1990. Apples (*Malus*). *In* Genetic resources of temperate fruit and nut crops. *Edited by* J.N. Moore and J.R. Ballington, Jr. International Society for Horticultural Science, Wageningen, Netherlands. pp. 1–62.

Yepsen, R.B. 1994. Apples. W.W. Norton, New York, NY. 255 pp.

Specialty Cookbooks

(Note: There are hundreds of cookbooks dedicated to the apple. The following is just a sample.)

Alexander, S.K., and Fairbairn, K. 1984. All American apple cookbook. ABC Enterprises, Santa Ana, CA. 105 pp.

Blank, M., Sporrer, B., and Hrbkova, A. 2000. Over 100 delicious recipes: apples. DuMont Verlag, Cologne, Germany. 156 pp.

Haedrich, K. 2002. Apple pie perfect: 100 delicious and decidedly different recipes for America's favorite pie. Harvard Common Press, Boston, MA. 250 pp.

Hill, N. (Editor). 1995. The apple cookbook: more than sixty easy, imaginative recipes. Courage Books, Philadelphia, PA. 64 pp.

Jackson, L. 1996. Apples, apples, everywhere: favorite recipes from America's orchards. Images Unlimited Publishing, Maryville, MO. 252 pp.

Munson, S., and Nelson, J. 1989. Apple-lovers' cook book. Golden West, Phoenix, AZ. 119 pp.

Nibbler's Apple Orchard. 1981. From applesauce to stuffed opossum: 700 recipes for apple lovers. Cookbook Publishers, Aurora, OR. 343 pp.

Patent, Greg, and Patent, D.H. 1999. A is for apple: more than 200 recipes for eating, munching, and cooking with America's favorite fruit. Broadway Books, New York, NY. 276 pp.

Peterson, B.S. 2004. Mom 'n' Pop's apple pie: 1950s cookbook; over 300 great recipes from the Golden Age of American home cooking. 2nd edition. Last Gasp, San Francisco, CA. 112 pp.

S. Martinelli & Company. 1968. Apple cider recipes: 100th anniversary collection. S. Martinelli & Co., Watsonville, CA. 15 pp.

Virginia State Horticultural Society. 1988. The Virginia apple cook book. 200 delicious recipes made with health-giving Virginia apples. Virginia State Horticultural Society, Staunton, VA. 64 pp.

Woodier, O. 1984. Apple cookbook. Storey Books, North Adams, MA. 187 pp.

APRICOT

Family: Rosaceae (rose family)

Names

Scientific name: *Prunus armeniaca* L. (*Armeniaca vulgaris* Lam.)

- The word "apricot" has a complex history. The classical Romans called it *praecocum*, meaning "the precocious one," because it blossomed very early. Byzantine Greek acquired the word as *beríkokken*, and the Arabs turned the word into *birqūq*, with the Moors adding the definite article *al* (the), so that apricot was called *al-birqūq*. The Spanish converted the word into *albaricoque*, which the Italians transformed to *alberococco*, the base of the earliest English word for apricot, *abrecock*.
- In British English, the first syllable of "apricot" is pronounced like the word "ape"; in American English, it is pronounced like the first syllable of "apple".
- The Japanese apricot or Japanese flowering almond (*Prunus mume* Siebold & Zucc.) is a popular garden ornamental, and its fruit is much used as a pickle and source of sweet liquor, although the fresh fruit is not suitable for eating. (In Japan, the species is often called "Japanese plum.")
- Crosses between apricot and plums have yielded a number of hybrids, including some called "plumcots" (formed from PLUM + apriCOT), sometimes shortened to "pluots." In California, one common hybrid is sold as "aprium" (formed from APRIcot + plUM). [The names Pluot and Aprium are registered trademarks of Zaiger Genetics, Modesto, California.] Pluots are mainly plums, and have smooth skin like plums, while the aprium is mostly apricot, and has slightly fuzzy skin. Pluots and apriums are quite sweet, with a higher sugar content than normally found in plums or apricots.
- The Maypop (*Passiflora incarnata* L.), a wild edible plant of the eastern United States, is sometimes called wild apricot and apricot vine.
- *Armeniaca* in the scientific name *P. armeniaca* is Latin for belonging to Armenia, reflecting early belief that the apricot originated there.
- For the derivation of the genus name *Prunus*, see PLUM.

Plant Portrait

The apricot is a deciduous tree of medium size, usually held to not over 5.5 m (18 feet) in orchards by pruning, but capable of growing up to 10 m (33 feet). The flowers emerge before the leaves, and are white or pink. The tree grows wild in parts of central Asia, Korea, Siberia, and northern China. The apricot was developed as a crop in China, where it has been cultivated possibly for the last 4000 years, and over 2000 cultivars have been selected. It was then transported to Armenia, from where it came to Europe. Major producers are Turkey, Iran, Italy, Pakistan, France, Spain, Morocco, China, and South Africa. Most of the US production is in California, and most of the American crop is packed in tins or sold as dried fruit. Some apricots are grown commercially in Canada, mostly in British Columbia and Ontario. The fruit is generally round but somewhat compressed, and typically smaller than peaches. Sometimes apricots are slightly fuzzy, especially when young, but are nearly smooth when ripe. The color of the skin may be

Apricot (*Prunus armeniaca*), from Hallier (1880–1888, vol. 25, plate 2553).

Plumcot (apricot–plum hybrid), from *Revue de l'Horticulture Belge et Étrangère* (1911, vol. 37).

yellow, yellowish green, brownish orange, and blushed red (in Asia, white, black, gray, and pink apricots are known). The flesh is typically yellow or deep orange, but is sometimes almost white. The single, compressed stone is either clingstone (firmly attached to the flesh) or free, and is generally smooth like that of the plum, but in a few kinds is pitted much as in the peach. The stone contains a kernel, which is edible in some varieties, bitter in others. Like bitter almonds, bitter apricot kernels are poisonous, although roasting will denature the toxin. Apricots develop their flavor and sweetness on the tree, so they should be harvested just before the fruit is fully ripe. However, they are often picked unripe, the harder fruit better withstanding transportation. Unfortunately, of all the major summer deciduous fruits grown in North America, apricots are by far the most fragile, and so the best fresh apricots are usually obtained locally on the continent.

Culinary Portrait

Apricots are eaten fresh, canned, dried, and preserved, and are made into juice and baby food. They can be cooked, candied, or stewed, like peaches or nectarines, with which they are interchangeable in most recipes. The fruit is used in pies, cakes, sorbets, yogurt, crêpes, jams, and chutneys. In the Middle East, apricots are cooked with meat, going especially well with lamb. Apricots bruise easily, and therefore do not ship well, and they have a short shelf life. Fresh apricots are superior in flavor to dried, but canned apricots can be satisfactory. Good-quality apricots should be mature, plump, uniformly golden in color, juicy, and when ripe, yielding to gentle pressure on the skin. Poor-quality apricots may result from excessive maturity or immaturity. Overly matured fruit is dull looking, soft, shriveled, or mushy, while immature fruit is very firm, pale yellow, or greenish yellow. A slight amount of ripening is possible at room temperature, and when ready to eat (indicated by softness to the touch), the fruit should be promptly consumed or briefly refrigerated (no longer than a week). Lemon juice prevents the flesh from darkening after slicing. After the central stone is removed, the fruit can be frozen in slices or puréed. Dried apricots can easily be rehydrated, and are particularly popular with backpackers and hikers. Dried apricots are mildly laxative, and are best consumed in moderation. They should be soaked for several hours before cooking, as they can be more easily puréed. Apricots are also used in the making of liqueurs and brandy. Seed of Central Asian and Mediterranean apricots is generally "sweet," and can be used as a substitute for almonds (as typically done in the Middle East), or crushed for almond-like cooking oil. However, the seed of most apricots sold in Western countries contains toxic hydrocyanic acid, and should not be consumed.

Culinary Vocabulary

- "Apricot sheets," also known as apricot leather and apricot rolls, are a Middle Eastern specialty prepared by cooking down apricot purée with sugar and rolling it into sheets. In Western countries, they are often available in Middle Eastern specialty stores. Apricot sheets can be used to prepare beverages, sauces, and puddings.
- The Austrian dish known as *marillenknödel* has been described as one of the most delicious ways of eating apricots. It is an apricot stuffed with a lump of sugar, sealed in a thin coat of light dumpling mixture, poached, and finished in crispy, butter-fried bread crumbs and sugar.
- *Kamraddin*, a paste made from apricots, is popular in Arab countries.
- *Ramadaniya* (pronounced rah-mah-dahn-ee-yah) is a Middle Eastern compôte prepared with dried apricots, figs, plums, almonds, pistachios, pine nuts, orange-blossom water, and apricot paste.
- *Abricoter* (based on the French *abricot*, apricot) is a French culinary term used in making pastries, meaning to coat a cake or similar preparation with apricot jam that has been boiled down to thicken, flavored with a liqueur, and strained.
- "Apry" (pronounced AP-ree) is apricot brandy.
- A "Valencia" is a cocktail prepared with apricot brandy, orange juice, orange bitters, and sparkling wine.
- *Umeshu* is a Japanese liqueur prepared from unripe apricots.

Curiosities of Science and Technology

- Ancient Egyptians used apricot-pit charcoal as an ingredient in eyeliner and mascara.
- The chemical amygdalin, largely from apricot pits, came to be called the drug Laetrile or "Vitamin B17" (although not a vitamin), and elicited great interest as a cancer tumor destroyer during the 1970s. Amygdalin is present in the seeds of many plants of the rose family, including rose, almond, apple, peach, and wild cherry. The National Cancer Institute in the United States determined that Laetrile is an ineffective cancer treatment, and it was banned from interstate commerce. Laetrile is an example of a fraudulent, sordid, and ineffective cancer cure, and charlatans in other countries continue to exploit the desperation of people suffering from incurable cancer by offering Laetrile therapy. Toxic cyanide may be released from amygdalin through an enzymatic reaction when being digested; similarly, cyanide is liberated from the seeds of some plants of the rose family, such as apples. Indeed, cases of fatal cyanide poisoning after eating apple seeds have been reported. Self-treatment with apricot seeds can be deadly.
- Apricots are usually more colored (have a red cheek) on the side of the fruit that received the most sun.
- In 1891, a 34-year-old apricot tree in Calaveras County, California, was observed to have a trunk 2.3 m (7.5 feet) in circumference. It yielded 680 kg (1500 pounds) of fruit in one season.

Apricot (*Prunus armeniaca*), from *Revue de l'Horticulture Belge et Étrangère* (1909, vol. 35).

- The City of Patterson, California, is known as the Apricot Capital of the World and each June is the site of the largest apricot festival in North America. The city's design features circles and broad radial boulevards lined with apricot trees, patterned by the developer after the grand avenues of Paris and Washington, DC. The Patterson apricot variety accounts for 75% of all California apricots (and California grows 95% of the US supply of apricots).
- By 1920, California's apricot industry was centered in the Santa Clara Valley, an area 70 km (45 miles) southeast of San Francisco. After World War II, the valley, once known primarily for fruit farming, became the focus of a growing computer industry, and acquired the name "Silicon Valley." By the end of the century, California's apricot orchards were relocated to the San Joaquin Valley where they are predominately found today. Sprawling apricot orchards were uprooted to make way for electronics industry offices, laboratories, and factories. Silicon Valley is now recognized globally as the number-one address in high technology—the world's most significant concentration of high-tech industry and talent. Unfortunately, the change has been accompanied by environmental problems and the sacrifice of prime agricultural land. Another unfortunate outcome has been a lowering of the quality of American apricots, which are now grown on less-suitable land.
- Apricots were part of the astronauts' diet on the Apollo program (1967–1972, 17 flights, 11 manned), which witnessed 12 men walk on the moon.
- About 95% of the flesh of an apricot is water. When fresh apricots are converted to dried apricots, they lose about 85% of their weight.

Key Information Sources

Bailey, C.H., and Hough, L.F. 1975. Apricots. *In* Advances in fruit breeding. *Edited by* J. Janick and J.N. Moore. pp. 367–383.

Carlson, R.F., Hull, Jr., and Moulton, J.E. 1977. Growing apricots in Michigan. Michigan State University. Cooperative Extension Service, East Lansing, MI. 8 pp.

Gulcan, R., and Aksoy, U. (*Editors*). 1995. Proceedings of the tenth international symposium on apricot culture, Ege University, Faculty of Agriculture, Department of Horticulture, Izmir, Turkey, 20–24 September 1993. International Society for Horticultural Science, Wageningen, Netherlands. 661 pp.

Hesse, C.O. 1952. Apricot culture in California. College of Agriculture, University of California, Berkeley, CA. 58 pp.

ICON Health Publications. 2003. Apricots: a medical dictionary, bibliography, and annotated research guide to internet references. ICON Health Publications, San Diego, CA. 88 pp.

Karayiannis, I. (*Editor*). 1999. Proceedings [11th] international symposium on apricot culture, Veria-Makedonia, Greece, 25–30 May, 1997. International Society of Horticultural Sciences, Leuven, Belgium. 2 vols.

Logsdon, G. 1978. If you can grow peaches, you can grow apricots. Org. Gard. **25**(**12**): 94–102.

Mehlenbacher, S.A., Cociu, V., and Hough, L.F. 1990. Apricots (*Prunus*). *In* Genetic resources of temperate fruit and nut crops. *Edited by* J.N. Moore and J.R. Ballington, Jr. International Society of Horticultural Sciences, Wageningen, Netherlands. pp. 65–107.

Morozov, A.V., and Paunovic, S.A. (*Editors*). 1978. Proceedings of the 6th international symposium on apricot culture and decline, Yerevan, Armenian SSR, USSR, 4–8 July 1977. Armenian Scientific Institute of Viticulture, Oenology and Fruit Growing, Armenia, USSR. 354 pp.

Nyujtó, F., and Klement, Z. (*Editors*). 1985. Proceedings of the eighth international symposium on apricot culture and decline, Kecskemét, Hungary, July 15–21, 1985. International Society for Horticultural Science, Wageningen, Netherlands. 533 pp.

Nyujtó, F., and Klement, Z. (*Editors*). 1991. Proceedings of the ninth international symposium on apricot culture, Caserta, Italy, July 9–15, 1989. International Society for Horticultural Science, Wageningen, Netherlands. 2 vols.

Organisation for Economic Co-operation and Development. 1994. Apricots. Organisation for Economic Co-operation and Development, Paris, France. 77 pp. [In English and French.]

Ramming, D.W. 1977. Growing apricots for home use. U.S. Dept. of Agriculture Home and Garden Bulletin. No. 204. Revised edition. U.S. Dept. of Agriculture, Washington, DC. 8 pp.

Simon, P., and Weiland, G. 1980. Bibliography of the international literature on apricots. Universitatsbibliothek der Technischen Universitat Berlin, Germany. 30 pp. [In German.]

Specialty Cookbooks

Adams, G. 1985. The complete apricot cookbook. Adams Place, Kennewick, WA. 104 pp.

California Apricot Advisory Board. 1990. California's apricot growers' favorite recipes. California Apricot Advisory Board, Walnut Creek, CA. 35 pp.

Gennis, R. 1989. Lots of ®cots: (cooking with apricots). 2nd edition. Ben Ali Books. 62 pp.

Los Altos Quote Club. 1950s. A harvest of apricot recipes. Foothill Print & Publishing, Los Altos, CA. 43 pp.

Maurer, E.L. 1968. Dried apricot recipes from the Maurer orchard. C.A. and E.L. Maurer, Santa Clara, CA. 75 pp.

Rubin, C., and Rubin, J. 1974. Apricot cookbook. Emporium Publications, Newton, MA. 91 pp.

ARTICHOKE (GLOBE ARTICHOKE)

Family: Asteraceae (Compositae; sunflower family)

Names

Scientific name: *Cynara scolymus* L. [*C. cardunculus* L. subsp. *cardunculus*, *C. cardunculus* subsp. *scolymus* (L.) Beger, *C. cardunculus* var. *scolymus* (L.) Fiori].

- "Artichoke" is usually interpreted as a corruption of its Arabic name, *al'qarshuf*, taken up in English from the Italian dialect word *articiocco*. An alternative, perhaps more colorful explanation holds that the name is derived from the old North Italian (Ligurian) term *cocal*, meaning a pine cone, with which the artichoke seemed comparable.
- The "globe" in globe artichoke came into being in the mid 19th century to distinguish the vegetable from the Jerusalem artichoke (*Helianthus tuberosus* L.) and Chinese artichoke (*Stachys affinis* Bunge). Before the flowers open, the floral head is spherical (globe-shaped).
- The globe artichoke is also known as French artichoke and green artichoke.
- *Cynara*, the genus name, has been traced to the classical Greek *kynara* (or *kinara*, related to *kynaros akantha*, the Greek for a spiny plant). One interpretation holds that the word is related to the Greek *kuon*, a dog—a pejorative reference to the objectionable thistle-like spines (the spines have been likened to a dog's teeth). Alternatively, the genus name has been interpreted as based on the Latin *cinara*, for a kind of plant native to the island of Cinara (now Zinara) in the Aegean Sea.
- *Scolymus* in the scientific name is from the Greek *skolos*, a thorn, another reference to the spiny nature of the plant.

Plant Portrait

Globe artichoke or, more simply, artichoke, is a thistle-like, perennial herbaceous plant 1–1.5 m (3–5 feet) tall, with bristly leaves. In warm climates, the plants are grown as perennials, but in cooler regions, they may be grown as annuals. The artichoke is not known in the wild. It originated in the central and western Mediterranean region and was carried to Egypt and farther eastward 2000–2500

Globe artichoke *(Cynara scolymus)*, from Nicholson (1885–1889).

years ago. The species is closely related and probably derived from cardoon (*Cynara cardunculus* L.), and the latter may have been eaten first by the ancient Romans and Greeks. Artichoke essentially disappeared from use in Europe after the fall of the Roman Empire and resurfaced around 1400 in Naples and Sicily. It spread through the rest of Europe and to America, but became popular mostly in southern Europe. Although globe artichoke is now grown in many parts of the world, nearly 95% of world production is from areas bordering the Mediterranean, with Italy, Spain, and France accounting for 80% of the crop, and significant cultivation also in Egypt and Israel. In Europe, there are numerous varieties, the edible portions ranging considerably in size. The major growing area in North America is California. The main variety in California is the cultivar Green Globe, which is grown not from seed (which would not produce true-to-type plants) but from portions of the underground part of the plant.

The edible region of the artichoke is part of a head of small flowers enclosed by bracts (small, scaly leaves). The bases of the bracts cradle a small portion of edible material, which is eaten, but most of the edible material occurs at the fleshy end of the stem on which the head of flowers develops. If the immature heads of flower buds are not removed, they mature into flowers but usually produce few viable seeds. Mature plants develop 5–8 marketable artichokes. The largest artichokes grow at the top end of the stalk, the smallest at the bottom. Harvested artichokes are globular to cone-shaped, green, and often tinged with purple. "Winter-kissed" artichokes in North America are available in the fall and winter. They can be darker, with bronze-tipped leaves or a whitish tint from exposure to low temperatures, and are considered more tender and flavorful by some connoisseurs.

Culinary Portrait

Nestled in the center of a mature artichoke is an inedible, thistle-like, hairy "choke," which is easily removed after cooking. Mature artichokes are not eaten raw, except for the small purple Provençal variety, which has an undeveloped choke. Small or baby artichokes, no larger than 5 cm (2 inches) in diameter, often have no choke at the center, and are also sometimes consumed raw. When very young and lacking a choke, artichokes are said to have a completely edible "heart," made up of completely edible bracts and the fleshy end of the stem. The hearts can be marinated or battered and fried. They are also canned or frozen and used as hors d'oeuvres. Baby artichokes tend to be available in North America in the spring. Mature artichokes are normally cooked and served hot or cold with melted butter, salad dressing, or a sauce. They may be steamed, boiled, or stuffed. The nutty flavor of the material at the base of the bracts is prized.

Unwashed artichokes can be refrigerated for up to a week, if dampened with a few drops of water and sealed in an airtight bag. To process a large number of artichokes, some cooks wear waterproof gloves to prevent staining of the hands, and to prevent the bitter juice from being transferred to other food. To prepare, wash well, trim away all but 2.5 cm (1 inch) from the stem, and remove the bottom bracts. For very tender artichokes, cut away the top third of the artichoke and the tops of the remaining bracts using a stainless steel knife, and rub lemon on the cut surfaces. Always use a steel knife and a stainless steel or glass pot (iron and aluminum turn artichokes an unappetizing blue or black). The artichokes can be (1) steamed for an hour, upside down; (2) cooked in a pressure cooker for 8–10 minutes; or (3) placed in a large pot (not cast iron or aluminum) of water, with added lemon juice or a little vinegar, and seasonings, the water brought to a boil and simmered for 25–45 minutes (when ready, a central bract will pull out easily), removed and drained for a few minutes before serving. A large (225 g or 8 ounces) artichoke can also be placed in a small microwave-safe bowl, adding 2.5 cm (1 inch) of water, covering with wet paper towels, and cooking on high for 4 minutes.

Artichoke etiquette is challenging, and the following are guidelines. Pluck the bracts with your fingers. Pull off a bract, holding it by the pointed end. Put the other end in your mouth and pull it between your teeth, scraping its length (the amount of edible material increases toward the center of the artichoke). Put a small part of the edible portion of the bract in a dip such as a vinaigrette or mayonnaise, if provided, and scrape with your teeth as indicated above. Too much dip may prevent tasting the artichoke. Near the center, the bracts become almost white with purple tips. These require caution, as their purple ends are prickly (and are normally thrown or cut away). After the bracts are pulled off, the base remains, which is the best part. Scoop away the fuzzy choke with your knife or spoon (a properly prepared artichoke will already have had this choke removed). Using a knife and fork, cut off bites from the base, like pieces of prime fillet.

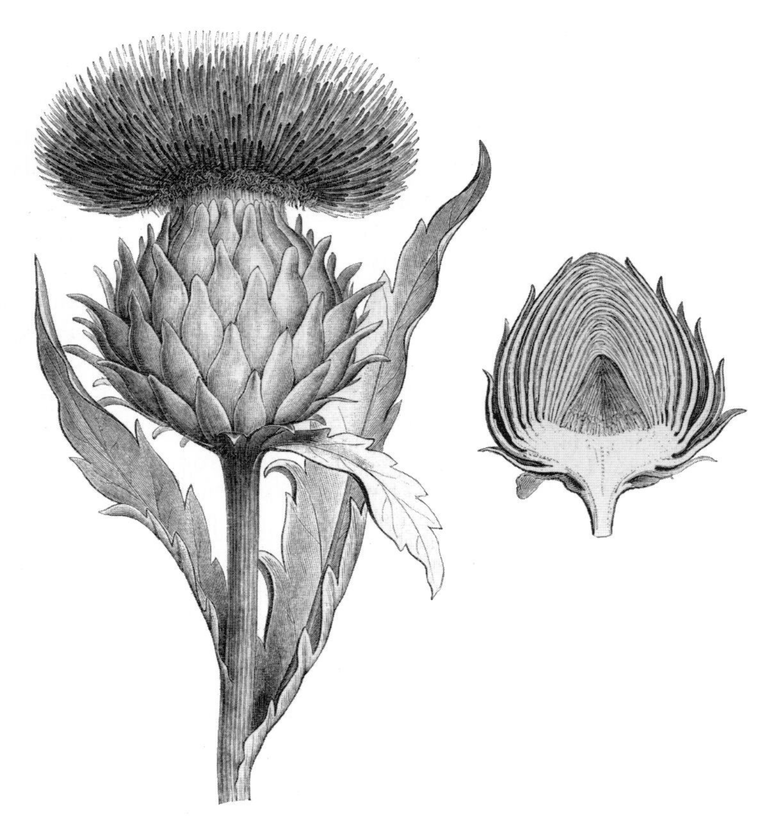

Globe artichoke *(Cynara scolymus)*. Flowering head at left from Engler and Prantl (1889–1915); long section at right from Baillon (1876–1892).

Culinary Vocabulary

- Cynar is an artichoke-flavored apéritif and digestive, made and quite popular in Italy.
- "Eggs Sardou" is a brunch dish made with an artichoke heart topped with anchovy, a poached egg, hollandaise sauce, chopped ham, and a garnish of sliced truffle.
- *Sauce vierge* is a French sauce prepared from béchamel sauce (a creamy mix of thickened milk, fat, and flour), artichoke purée, and whipped, unsweetened cream.

Curiosities of Science and Technology

- Artichoke has a long history of use as a medicine. Antonius Musa, physician to Roman emperor Augustus (63 BC–14 AD), claimed that artichoke was effective against 48 diseases, and decreased the intoxicating effects of wine. In Europe, artichoke has sometimes been used in traditional medicine to treat jaundice and coughs, and a variety of other conditions. Over-the-counter artichoke extracts are available in North America, and are intended to treat appetite loss and liver and gallbladder problems. Modern research has led to claims that extracts of artichoke have medicinal value for affecting gastrointestinal activity, blood-clotting time, capillary resistance, heart activity, and neutralization of some toxins.
- Artichokes were once prescribed as a beauty treatment.
- Castroville (south of San Francisco) is the self-proclaimed "Artichoke Center of the World." There is an "Artichoke Inn" and a "Giant Artichoke Restaurant." A large roadside exhibit in the city is said to be "the world's largest man-made artichoke." In fact, Italy, Spain, France, and Greece produce much more artichoke than the United States. It is true that Castroville is the artichoke center of the New World, and the cooling fogs from the Pacific Ocean that creep over the city have been credited with producing artichokes that are unsurpassed in quality.
- Artichoke flowers curdle milk.

- The "artichoke effect" alters taste perception in some people. Chlorogenic acid and cynarin in artichokes inhibit the taste buds. Drinking water, milk, or wine results in a return of normal taste sensation. People with the genes that predispose them to the artichoke effect may have the illusion that drinks taste exceptionally sweet just after they have eaten an artichoke (actually, the drink has simply returned their taste buds to normal function). To avoid the possibility that the artichoke effect could interfere with appreciation of the taste of wine, some wine connoisseurs will not eat artichokes when wine is served with the meal.

Key Information Sources

Ancora, G. 1986. Globe artichoke. Biotechnol. Agric. For. **2**: 471–484.

Foury, C. 1984. Floral biology of the artichoke (*Cynara scolymus* L.) and its application in breeding work, I. Data on floral biology. Amerind Publishing Co. Pvt. Ltd., New Delhi, India. 31 leaves.

Gerakis, P.A., and Honma, S. 1969. Response of globe artichoke (*Cynara scolymus* L.) to various nutritional environments in solution culture and to N, P, and K fertilizer in organic soil. Soil Sci. **108**: 290–295.

Gerakis, P.A., Markarian, D., and Honma, S. 1969. Vernalization of globe artichoke, *Cynara scolymus* L. J. Am. Soc. Hortic. Sci. **94**: 254–258.

Lanteri, S., Saba, E., Cadinu, M., Mallica, G.M., Baghino, L., and Portis, E. 2004. Amplified fragment length polymorphism for genetic diversity assessment in globe artichoke. Theor. Appl. Genet. **108**: 1534–1544.

Larkcom, J. 1979. Globe artichokes. Cultivation. Gard. (London), **104**: 107–111.

Martindale, W.L. 1971. Globe artichokes. J. Agric. (Melbourne), **69**: 304–306.

Pecaut, P. 1993. Globe artichoke, *Cynara scolymus* L. *In* Genetic improvement of vegetable crops. *Edited by* G. Kalloo and B.O. Bergh. Pergamon Press, New York, NY. pp. 737–746.

Quirce, S., Isabel-Tabar, A., Maria-Olaguibel, J., and Cuevas, M. 1996. Occupational contact urticaria syndrome caused by globe artichoke (*Cynara scolymus*). J. Allergy Clin. Immunol. **97**: 710–711.

Romano, C., Ferrara, A., and Falagiani, P. 2000. A case of allergy to globe artichoke and other clinical cases of rare food allergy. J. Invest. Allergol. Clin. Immunol. **10**: 102–104.

Rottenberg, A., and Zohary, D. 1996. The wild ancestry of the cultivated artichoke. Genet. Resour. Crop Evol. **43**: 53–58.

Rottenberg, A., Zohary, D., and Nevo, E. 1996. Isozyme relationships between cultivated artichoke and the wild relatives. Genet. Resour. Crop Evol. **43**: 59–62.

Ryder, E.J, De Vos, N.E., and Bari, M.A. 1983. The globe artichoke (*Cynara scolymus* L.). Hortscience, **18**: 646–653.

Sonnante, G., De Paolis, A., Lattanzio, V., and Perrino, P. 2002. Genetic variation in wild and cultivated artichoke revealed by RAPD markers. Genet. Resour. Crop Evol. **49**: 247–252.

Suslow, T., and Cantwell, M. 1997. Recommendations for maintaining postharvest quality: globe artichoke. Perishables Handling Quart. **92**(Nov.): 27–28.

Thomsen, C.D., and Barbe, G.D. 1986. Artichoke thistle in California. Calif. Weed Conf. **39**: 228–231.

Udell, R. 1982. How to eat an artichoke and other trying, troublesome, hard-to-get-at foods. Putnam, New York, NY. 127 pp.

Walker, P. 1972. Further notes on globe artichokes. R. Hortic. Soc. J. **97**: 490–492.

Welbaum, G.E. 1994. Annual culture of globe artichoke from seed in Virginia. HortTechnology, **4**: 147–150.

Welbaum, G.E., and Warfield, S.C. 1992. Growing globe artichokes from seed. Acta Hortic. **318**: 111–115.

Zohary, D., and Basnizky, J. 1975. The cultivated artichoke–*Cynara scolymus.*–Its probable wild ancestors. Econ. Bot. **29**: 233–235.

Specialty Cookbooks

Castelli, A.C., and Castelli, C.A. 1995. Artie and Chuck's Cynara erotica: an introduction to the sensuous artichoke. A.C. Castelli Associates, New York, NY. 118 pp.

Castelli, A.C., and Castelli, C.A. 1998. The sensuous artichoke: magic of the artichoke. A.C. Castelli Associates, Riverdale, Bronx, NY. 294 pp.

Comfort, M., Griffee, N., Walker, C., and California Artichoke Advisory Board. (*Editors*). 1998. California artichoke cookbook. Celestial Arts, Berkeley, CA. 96 pp.

Holtman, K, and Borovoy, J. 1995. The amazing California artichoke. California Artichoke Advisory Board, Castroville, CA. 92 pp.

Rain, P. 1985. The artichoke cookbook. Celestial Arts, Berkeley, CA. 174 pp.

AVOCADO

Family: Lauraceae (laurel family)

Names

Scientific name: *Persea americana* Mill.

- West Indian avocado—*P. americana* Mill. var. *americana* (*P. gratissima* C.F. Gaertn., *P. leiogyna* S.F. Blake)
- Mexican avocado—*P. americana* var. *drymifolia* (Schltdl. & Cham.) S.F. Blake
- Guatemalan avocado—*P. americana* var. *nubigena* (L.O. Williams) L.E. Kopp (*P. nubigena* L.O. Williams var. *guatamalensis* L.O. Williams)
- The word "avocado" appears to have originated by attempts to speak phonetically the Aztec (Nahuatl) name *ahuacatl* (testicle, for the shape of the fruit) or the Spanish derivative *aguacate*. This was so unpronounceable to early Americans that they preferred the name "alligator pear," a name generally interpreted as coined because of the somewhat similar color and roughness of the fruit surface with the alligator's skin. Alternatively, it has been claimed that the word "avocado" originated as a mispronunciation of the original Aztec name. The word "avocado" appeared for the first time in the plant catalog of Sir Hans Sloane, secretary and later president of the British Royal Society in 1696. The Aztecs added their word for sauce, *molli*, to *ahuacatl*, for "avocado sauce," and the Spaniards turned this into *ahuacamolli*, which was altered to "guacamole."
- The name "alligator pear" is sometimes also applied to the chayote (*Sechium edule* (Jacq.) Sw.).
- In addition to the name alligator pear, the avocado is known as avocado pear, and Trapp avocado.
- During British colonial days, avocados were a rarity in Britain, and were occasionally sent home by colonial servants overseas by ship, which resulted in nicknames such as butter pear, midshipman's butter [midshipman: one in training for a naval commission, especially a student], and subaltern's butter. On ships, avocado was actually used as a spread. In northern Mexico, avocado is mashed and used as a spread called *Mantequilla de Pobre* (poor man's butter), and indeed, this name is used in various areas of the tropics.
- In Hawaii, the West Indian avocado is called the summer avocado, while Guatemalan and hybrids of Guatemalan and Mexican avocados are called winter avocados.
- Hass (Haas), the most popular variety of avocado (it has black or brown skin, while most other varieties commonly sold in North American supermarkets have glossy green skin), was discovered by a California postman, Rudolph Hass, in 1926 in his La Habra backyard, where the original tree is still standing. The Hass avocado is also called the California avocado.
- The Bacon avocado does not taste like bacon. It is named for its originator, James E. Bacon, who found it on his ranch in Buena Park, California, and subsequently introduced it to the trade in the late 1920s. It is exceptionally resistant to cold, but is not as tasty as the more popular Hass and Fuerte avocados.
- The genus name *Persea* is based on the Greek *persea*, the name of an Egyptian tree (*Cordia myxa* L.).

Plant Portrait

The avocado tree is an evergreen, 12–24 m (40–80 feet) tall. The fruits take 9–15 months to mature. They range in length from 2.5 cm (1 inch) to 30 cm (1 foot), and in weight from 28 g (1 ounce) to 1.8 kg (4 pounds). A large pit at the center of the fruit usually detaches easily from the flesh. A milky substance covering the pit turns reddish when exposed to air and can stain fabrics. The flesh is rich, butter-textured, yellowish green, with a slightly nutty taste. The avocado is a native of Mexico and Central America. It probably originated in southern Mexico, but was cultivated from the Rio Grande to central Peru before the arrival of Europeans. The major civilizations of the New World—Maya, Aztecs, and pre-Incan people of Peru—all cultivated avocado. Carbon dating has shown that avocados were eaten as early as 9000–10 000 years ago. In 1519, Spanish soldier of fortune Hernando Cortez entered Mexico City, and his followers were the first Europeans to discover the fruit. Avocados were brought to Great Britain in the 17th century, but were a rarity in Europe until Israel began large-scale cultivation after the Second World War. They are now raised commercially in Israel, Turkey, Spain, France, Chile, Brazil, South Africa, Mexico, Hawaii, Florida, Indonesia, Australia, California, and some Caribbean islands, particularly the Dominican Republic. Mexico is the world's largest grower, with about a quarter of global production. (The United States banned importation of avocados from Mexico from 1914 until 1997.) Ninety-five percent of all avocados grown in the United States come from California, the remainder mostly from Florida. Although the avocado is an expensive luxury in many places, it is a very cheap food in others. In the West Indies, where the very large-fruited West Indian avocado is grown, it is sometimes known as "poor man's breakfast." Americans today eat about 1 kg (2 pounds) of avocados per person each year. Chileans eat about five times as much on a per capita basis.

Three principal groups of avocado have been recognized.

- Mexican avocado: Fruits are small (170–280 g; 6–10 ounces), fast-ripening (6–8 months), with soft paper-thin skins that turn glossy green or less frequently black when ripe. Mexican types discolor quickly and should be consumed quickly.
- Guatemalan avocado: Produces medium (up to 340 g or 12 ounces), ovoid or pear-shaped, pebbled, brittle-skinned, green, slow-ripening (12–18 months) fruits that turn blackish green or purple–black when ripe. These are the highest in oil content.
- West Indian avocado: Produces enormous, smooth–leathery, round, glossy green fruits that are low in oil and weigh up to 1 kg (about 2 pounds).

There are hundreds of varieties, available in sundry shapes—round, pear-shaped, crooknecked (like a squash), colored green, purple, maroon, and jet black, and with skin texture smooth or pebbly. Two varieties are most commonly sold in stores: Haas (a Guatemalan type)—thick-skinned, blackish, with a small pit and buttery texture; and Fuerte (a Mexican type)—thin-skinned, green, and smooth. Most avocado varieties do not come true from seed and must be propagated vegetatively. Avocados raised from pits are unlikely to bear satisfactory fruits.

Culinary Portrait

Avocado flesh is mainly consumed fresh in salads, after discarding the inedible skin. The tannins in the fruit result in a bitter flavor when avocado is cooked over high heat, which must be avoided. Avocado should not be boiled, but may be added at the end of cooking. The high fat content (at 20%, more than 20 times the average for other fruits) combines well with acidic fruit and vegetables, such as citrus, tomatoes, and pineapple, as well as with acidic dressings. Avocado is used as a hot-dog topping, in soups, sandwiches, hors d'oeuvre, and soufflés. The fruit also is used to supply the fat content of frozen desserts such as ice cream and sherbets. The manner of consumption of avocados varies. Some countries prefer the fruit cooked, others raw. The hole left when the stone is removed is commonly filled with dressing, shrimps and other seafood, chicken, and other delicacies.

In North America, guacamole is the most familiar dish, and is a favorite dip with potato chips, tortilla chips, and similar products. Guacamole is traditionally a mixture of mashed avocado, minced chile, and chopped cilantro, onions, and tomato. Guacamole does not brown greatly if allowed to sit for 1–2 hours, but longer periods may result in more browning than acceptable to most people. Some varieties brown less rapidly. The most common California varieties, Hass and Fuerte, discolor rapidly. The Lerman avocado, developed in Israel, requires 12–18 hours to brown. Some lemon, lime, tomato juice, or olive oil can be added to retard the avocado from darkening when exposed to air. Saran Wrap pressed tightly over the guacamole works excellently (only areas where bubbles of air are trapped become brown).

Avocados are picked unripe, and require a period of ripening. They ripen best between 15 and 21 °C (60–70 °F). At higher temperatures, they tend to develop brown spots, poor flavor, and fungal decay.

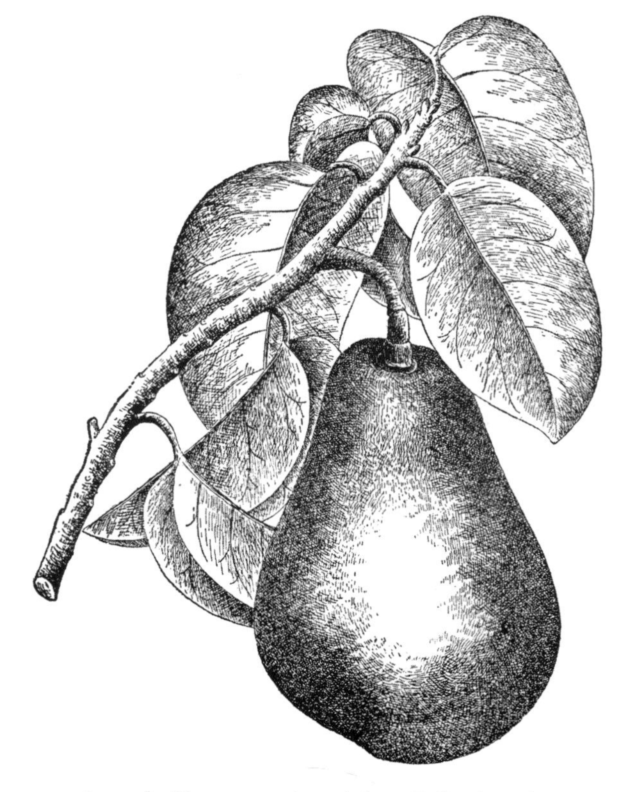

Avocado *(Persea americana)*, from Bailey (1916).

Accordingly, a sunny windowsill is inappropriate for ripening avocados. The fruit is very susceptible to chilling injury, and can become discolored, as well as developing an off-flavor, when stored in a refrigerator. However, avocados that have become soft naturally, and aren't bruised, can be stored in a refrigerator for weeks. Avocados are often cut in half lengthwise, preferably with a stainless steel knife. Should the flesh cling to the stone when the fruit is sectioned, the two halves may be gently twisted in opposite directions. When using half an avocado, retain the pit in the unused half, brush the cut area with lemon juice, cover tightly with plastic wrap, and keep for no more than 3 days in the refrigerator.

In Africa, avocado leaves are used to make a sparkling, slightly alcoholic drink, *babine*. In Mexico, avocado leaves are toasted and ground to produce a spice used to impart an anise–hazelnut flavor to many dishes. Sometimes even the fresh leaves are used as a spice, so that while home-grown avocados raised from seeds very rarely produce edible fruits, at least they can be used for a culinary purpose. However, it has been pointed out that the leaves of plants grown in a cold climate are milder than those grown in tropical regions.

Culinary Vocabulary

- The "cocktail avocado" (avocadito) has very small fruits, the size of plums (5–6.4 cm or 2–2½ inches long) and is completely seedless. Even smaller, finger-sized fruits that result when growing conditions are poor and pollination does not occur are often sold as cocktail avocados.
- The word "guacamole," referring to the most familiar preparation of avocado, is American-Spanish in derivation, based on a Native American word (Nahuatl, *ahuacamolle*), a compound of *ahuacatl*, avocado, + *molli*, sauce or stew.
- *Vitamina de abacate* (pronounced vee-tah-mee-nah day ah-bah-cah-tay) is a South American (notably Brazilian) beverage prepared with avocado, cold milk, sugar, and vanilla.

Curiosities of Science and Technology

- Archaeologists in Peru have found avocado seeds buried with mummies dating back to 750 BC.
- Some varieties of avocado (e.g., Fuerte) have flowers that open first as females, close in the afternoon, and open the next afternoon as males. Other varieties (e.g., Hass) have flowers that begin as female, but convert overnight and reopen as males the next morning. (For those who are interested, this is called "protogynous, diurnally synchronous dichogamy.")
- Many people have tried to grow an avocado from the pit, suspending it over a water-filled jar, with three toothpicks inserted on the sides. But should the pointy or broad end be down? The broad end should be down. [Well-ripened fruits are more likely than immature fruits to yield seeds that will sprout. Clean off adhering flesh with warm water. Maintain the water line at the middle point of the seed, or about an inch submerged. Keep the jar in a warm, sunny place away from drafts. If the seed fails to show any signs of growth after 6 or 7 weeks, it likely is a dud.]
- A mature avocado tree can produce 2 million flowers.
- A single avocado tree can produce up to 400 avocados every year. Avocado trees have been known to collapse under the weight of their fruit.
- Avocados on the tree are inhibited from ripening by a hormone supplied from the leaves, and so the fruit can be stored on the tree for up to 8 months. When the fruit is harvested, it begins to ripen. To speed up ripening of an avocado, place it in a paper bag at room temperature along with an apple or banana. Ethylene gas released by apples and bananas hastens ripening of various other fruits. To halt ripening of avocados, deprive them of oxygen (e.g., put them in a tightly sealed plastic bag).
- Although avocado trees are "evergreen," the leaves normally are retained only for 2–3 years.
- The largest number of culinary dishes made with avocado is found in Israel.
- Florida avocado varieties generally produce fruit with lower oil content compared with varieties grown in California.
- The avocado has a reputation as "fatty," and indeed is second only to olives among fruits in oil content. However, the oil is high in monounsaturated fatty acids, and therefore is relatively healthy, tending to reduce blood cholesterol.
- Ripe avocado has been recommended as a soothing material to be rubbed onto sunburned skin and winter-chapped hands (it will keep skin moist). It has even been suggested as a shaving cream!
- Avocados contain nearly twice the food energy of an equivalent weight of meat.
- Three-quarters of the average avocado is water.

Avocado *(Persea americana)*, from Engler and Prantl (1889–1915).

- Unlike most other fruits, the sugar content of avocados decreases rapidly during ripening. In Brazil, where the avocado is mashed as a dessert, sugar is added.
- The Calavo Growers' factory in Santa Paula, California, is the largest commercial supplier of guacamole in the United States, and Santa Paula advertises itself as the "Guacamole Capital of the World." The plant's annual production of guacamole and other avocado-based production is almost 10 million kg. About 40% of the world's production of guacamole comes from this factory.
- As noted above, browning or discoloration is a problem when an avocado is cut. It has been observed that the blossom end of a cut avocado browns more rapidly than the stem end.
- Many cookbooks claim that guacamole can be kept from browning by putting the avocado pit into the middle of the bowl. Harold McGee reported that experiments he conducted (*The Curious Cook*, 1990, Macmillan Publishing, New York) showed that when an avocado pit was placed in guacamole, the only place the guacamole remained green was where it actually contacted the pit, apparently because the contact prevented oxygen from reaching the guacamole. A light bulb immersed in the guacamole worked just as well.
- How do you tell a lawyer from an avocado? In French, the two words are the same (*avocat*), presenting a problem for automatic (computer) translation. Popular internet translation sites (such as Google and Bablefish.com) are unable to distinguish which word is appropriate. New software under development by various companies attempts to employ statistical methods instead of grammar rules and literal translations in order to determine the best word.

Key Information Sources

Ahmed, E.M., and Barmore, C.R. 1980. Avocado. *In* Tropical and subtropical fruits. *Edited by* S. Nagy and P.E. Shaw. AVI Publishing, Westport, CT. pp. 121–156.

Bergh, B., and Ellstrand, N. 1986. Taxonomy of the avocado. Calif. Avocado Soc. Yearb. **70**: 135–145.

Bergh, B.O. 1975. Avocados. *In* Advances in fruit breeding. *Edited by* J. Janick and J.N. Moore. Purdue University Press, West Lafayette, IN. pp. 541–567.

Bergh, B.O. 1986. *Persea americana*. *In* Handbook of flowering, vol. 5. *Edited by* A.H. Halevy. CRC Press, Inc., FL. pp. 253–268.

Bergh, B.O. 1995. Avocado. *In* Evolution of crop plants. 2nd edition. *Edited by* J. Smartt and N.W. Simmonds. Longman Scientific & Technical, Burnt Mill, Harlow, Essex, UK. pp. 240–245.

Bertoldi, J.R. 1989. Avocados: growing, harvesting, marketing. Observational Research Publications, Las Vegas, NV. 42 pp.

Bower, J.P., and Cutting, J.C. 1988. Avocado fruit development and ripening physiology. Hortic. Rev. **10**: 229–271.

Chalker, F.C., and Robinson, P.W. 1969. The avocado. 2nd edition. Bull. H63. New South Wales Dept. of Agriculture, Division of Horticulture, Sydney, Australia. 15 pp.

Furnier, G.P., Cummings, P.M., and Clegg, M.T. 1990. Evolution of the avocado as revealed by DNA restriction fragment variation. J. Hered. **81**: 183–188.

ICON Health Publications. 2003. Avocado: a medical dictionary, bibliography, and annotated research guide to internet references. ICON Health Publications, San Diego, CA. 108 pp.

Lovatt, C.J., Holthe, P.A., and Arpaia, M.L. (*Editors*). 1992. Proceedings of the second world avocado congress, April 21–26, 1991, Orange, California, U.S. University of California (Riverside), Riverside, CA. 2 vols.

Morton, J. 1987. Avocado. *In* Fruits of warm climates. *Authored by* J.F. Morton. Creative Resource Systems, Winterville, NC. pp. 91–102.

Muthukrishnan, C.R., and Khader, J.B.M.M.A. 1985. Avocado. *In* Fruits of India, tropical and subtropical. *Edited by* T.K. Bose. Naya Prokash, Calcutta, India. pp. 465–478.

Organization for Economic Co-operation and Development. 1995. Avocats [Avocados]. OECD, Washington, DC, and OECD Publications and Information Centre, Paris, France. 73 pp. [In English and French.]

Perper, H. 1965. The avocado pit grower's indoor how-to book. Walker, New York, NY. 62 pp.

Pliego-Alfaro, F., and Bergh, B.O. 1992. Avocado. *In* Biotechnology of perennial fruit crops. *Edited by* F.A. Hammerschlag and R.E. Litz. CAB International, Wallingford, UK. pp. 323–334.

Sanyal, D. 1999. Avocado. *In* Tropical horticulture, vol. 1. *Edited by* T.K. Bose, S.K. Mitra, A.A. Farooqui, and M.K. Sadhu. Naya Prokash, Calcutta, India. pp. 362–375.

Schroeder, C.A. 1968. Prehistoric avocados in California. Calif. Avocado Soc. Yearb. **52**: 29–34.

Scora, R.W., and Bergh, B.O. 1992. Origin of and taxonomic relationship within the genus *Persea*. *In* Proceedings of the 2nd world avocado congress, vol. 2. *Edited by* C.J. Lovatt and P. Holthe. University of California, Riverside, CA. pp. 505–514.

Smith, C.E. 1969. Additional notes on pre-Conquest avocados in Mexico. Econ. Bot. **23**: 135–140.

Storey, W.B., Bergh, B.O., and Zentmyer, G.A. 1986. The origin, indigenous range, and dissemination of the avocado. Calif. Avocado Soc. Yearb. **70**: 127–133.

Whiley, A.W. 1991. *Persea americana* Miller. *In* Plant resources of South-East Asia, 2, edible fruits and nuts. *Edited by* E.W.M. Verheij and R.E. Coronel. Pudoc, Leiden, Netherlands. pp. 249–254.

Whiley, A.W., Schaffer, B., and Wolstenholme, B.N. (*Editors*). 2002. The avocado: botany, production, and uses. CABI Publishing, New York, NY. 416 pp.

Specialty Cookbooks

Bauer, H., and Logerman, H.B. 1967. The avocado cookbook. Doubleday, Garden City, NY. 242 pp.

Carlisle, J. 1985. The avocado lovers' cookbook. Celestial Arts (in association with Blue Ribbon Publishers), Berkeley, CA. 140 pp.

Carr, A. 1988. A glut of avocados. Merehurst, London, UK. 92 pp.

Coker, B., and Schmidt, Z. 1990. Adventures in cooking avocados. Vantage Press, New York, NY. 79 pp.

Gordon, T. 1988. Avocado recipes, etc. 2nd edition. T. Gordon, Austin, TX. 128 pp.

Graham, J. 1983. The avocado cookbook. Wattle Books, Frenchs Forest, NSW, Australia. 96 pp.

Heaslip, C. 1984. The complete avocado cookbook. Bay Books, Sydney, Australia. 96 pp.

Hood, K.J.M. 2004. Avocado delights cookbook: a collection of avocado recipes. Whispering Pine Press, Spokane, WA. 320 pp.

Leneman, L. 1985. The amazing avocado (a Thorsons wholefood cookbook). HarperCollins, New York, NY. 96 pp.

Spain, H. 1979. The avocado cookbook. Creative Arts Book Co., Berkeley, CA. 146 pp.

Banana and Plantain

Family: Lauraceae (laurel family)

Names

Scientific name: *Musa* ×*paradisiaca* L. (a sterile hybrid of *M. acuminata* Colla and *M. balbisiana* Colla). A few cultivars belong to *M. acuminata*.

- The word "banana" originated from one of the languages of coastal West Africa, presumably in Guinea or Sierra Leone. It has been speculated that the English word "banana" comes from the Arabic *banana*, meaning finger or toe, and indeed, in the banana trade, a bunch of bananas is called a hand, while individual bananas are called fingers. Australians call bananas "ladyfingers" (a term that others apply to okra (*Abelmoschus esculentus* (L.) Moench).
- The word "plantain" (pronounced plahn-TAYNE) originated from the Spanish word *plantano* (plátano) for this vegetable. The banana type of plantain should not be confused with the common weedy plantains of the genus *Plantago* in which case the word plantain is derived from the Latin word *planta*, footprint, for the resemblance of the leaves to footprints (North American Indians called these weeds "white man's footprint," an indication that the *Plantago* weeds accompanied European settlements).
- The plantain is also called plantain banana and cooking banana.
- In Spanish and Portuguese, the fruit is *banana* and the tree is *banano*.
- The origin of the genus name *Musa* is problematical. Linnaeus (1707–1778), the father of biological nomenclature, created the genus name *Musa*, possibly basing it on the following: (*a*) one of several Arabic words for the fruit (*mouz*, *moz*, *mawz*, or *moaz*, derived from the Sanskrit *moka* or from the southern Arabian town Moka); (*b*) to Antonius Musa, a Roman medical man of the 1st century BC (physician to Octavius Augustus, first emperor of Rome, 63–14 BC); or (*c*) to honor the Muses (the nine sister goddesses in Greek mythology presiding over song, poetry, arts, and sciences).
- The Koran calls the banana the "Tree of Paradise" (i.e., the banana, not the apple, was the "forbidden fruit" present in the Garden of Eden), so Linnaeus coined the name *Musa paradisiaca*. Linnaeus also recognized a form of banana by the name *Musa sapientum*, (literally "wise banana") because Roman scholar Pliny the Elder (23–79) had noted how the sages (wise men) of India ate the fruit.
- According to Hindu legend, the banana was the forbidden fruit of the Garden of Eden, and Adam and Eve covered their nakedness with large banana leaves (rather than much smaller fig leaves). Accordingly, very early names for the banana were "apple of paradise" and "Adam's fig." The use of "fig" as a name for the banana also reflects Roman usage, and an early French name was *fig du paradis* ("fig of paradise").
- A "burro banana" or "box banana" is a banana variety grown in Mexico, with a flat, boxy shape and a tangy, lemon flavor.

Plant Portrait

Banana plants are tree-like plants that grow 2–9 m (6½–30 feet) in height, occasionally to 15 m (160 feet). The nonwoody "trunk" of a banana plant consists of tightly overlapping leaf sheaths. The leaves are huge—up to 4 m (13 feet) long and 1 m (3 feet) wide. Bananas originated in Malesia (Malaysia, Indonesia, the Philippines, Borneo, and Papua New Guinea), and have been consumed by humans for at least 4000 years. The fruit was well known to ancient Hindu, Greek, Roman, and Chinese cultures. Bananas and plantains are today grown in every humid tropical region and in selected areas in the subtropics, and are the fourth largest fruit crop of the world in value, and the largest in volume of world production. The fruit constitutes a crucial part of human diets in most tropical regions. The largest producers are Brazil and Uganda, with important crops also in India, the Philippines, Ecuador, Colombia, Honduras, Brazil, Tanzania, Rwanda, Indonesia, Thailand, Ivory Coast, and Vietnam. Small amounts are produced in the Hawaiian Islands, Florida, and along the coast of the Gulf of Mexico. The plant needs 10–15 months of frost-free conditions to produce a flower stalk. Each trunk produces one huge flower cluster, and the main stalk dies after bearing fruit. New stalks then grow from the rhizome (underground stem). Banana plants are extremely decorative, ranking next to palm trees for the tropical feeling they lend to the landscape. Dwarf varieties make good container specimens.

Over 500 varieties of bananas have been recognized. These have been classified into the following groups:

(*Musa ×paradisiaca*), from Lamarck and Poiret (1744–1829, plate 836).

- Fruit or dessert bananas are familiar yellow bananas, about 15–30 cm (6–12 inches) in length.
- Apple bananas are also yellow, but are smaller, 8–10 cm (3–4 inches), and ripen faster.
- Baby bananas, also yellow, measure 6–8 cm (2–3 inches) in length, and are exceptionally sweet.
- Red bananas have a green/red peel, pink flesh, and a typical banana taste despite their unusual appearance.
- Plantains are baking (cooking) bananas. These are starchy, 30–40 cm (12–16 inches) long, green, yellow, or reddish. They can not be eaten raw, and replace potatoes as a starchy food in tropical countries. Plantains are cultivated in India, Malaysia, Africa, and tropical America. Plantains are a staple in West and Central Africa, where people derive about a quarter of their energy requirements from this food.

Culinary Portrait

Dessert bananas are mostly eaten raw after they ripen (black spots on the skin indicate ripeness, not spoilage) at which point they have low starch, high sugar, and considerable flavor. By contrast, plantains convert less of their starch to sugar during ripening. They have high starch content and are eaten either when green or ripe, after boiling, frying, or roasting. Green bananas are sometimes used as a vegetable because they are less sweet and firmer than the ripe fruit. Bananas are especially desirable in diets where ease of digestibility, low fat, no cholesterol, high potassium, low sodium, and vitamin content are required. Bananas do not cause digestive disturbance, and readily neutralize free acid in the stomach. Banana diets are useful for babies, the elderly, and patients with gout, arthritis, and stomach problems (the fruit is mildly laxative when overripe). A small amount of banana is processed: canned as slices, dried as slices or flakes, frozen, or made into juice. Banana purée is canned or frozen for use in baby foods, baked goods, and drinks. Bananas are consumed to some extent in salads, desserts, and in a variety of baked confections, including pies, cakes, muffins, donuts, bread, and puddings. The fruit can be fried, made into fritters, or cooked with meats such as ham and bacon. Banana essence is a clear colorless liquid used in desserts, juices, and drinks. A flour can be made from dried, ripe fruit. In India and other tropical regions, virtually the entire banana plant may be eaten. After harvesting the fruit in some tropical regions, the banana stem, which has considerable starch, is boiled as a vegetable, or occasionally candied. In Southeast Asia, male flower buds or young flowers are also eaten as a vegetable, raw or cooked. In Central and East Africa, the juice from ripe bananas is fermented to make beer [consumption in Rwanda has reached 1.2 L (1¼ American quarts) per capita per day]. In some countries, banana sprouts are covered with a pot and grown in the dark to produce thick, long, white, edible spikes that resemble huge, white asparagus. Banana leaves are widely used as food wrappers for steaming.

Plantains have a mild flavor like squash and sweet potato (or like banana when very ripe), but must be cooked to be edible. Depending on recipe, either green or ripe plantains may be required. When ripe, they are brownish or black-skinned and yield to slight pressure. They can be roasted in their skins at 177 °C (350 °F) for 30–45 minutes, first making a lengthwise cut on one side to prevent bursting, and covering them with oil or butter. Alternatively, plantains can be peeled, removing all the fibrous strings—easily done when ripe, but the skin of green plantains is tenacious (one lengthwise incision, and three crosswise incisions, one in the center and the other two near the ends, will be helpful in removing the skin). It is sometimes suggested that when peeling plantains, the hands be moistened and rubbed with salt to prevent the juices from sticking to skin. Once peeled, the plantains can be halved lengthwise, and sautéed in oil until golden on both sides, or sliced and deep-fried like potato chips. Plantains are also boiled (about 25 minutes), steamed, and grilled until tender (about 45 minutes, 10 cm or 4 inches away from the heat). This vegetable is served as a side dish, incorporated into various culinary preparations (chunks are often added to soups and stews), stuffed or used as a stuffing, and sweetened and prepared as a dessert. Plantains should be stored at room temperature, but may be refrigerated briefly if very ripe.

Culinary Vocabulary

- The national snack of Nicaragua is the "nacatamal"—banana leaves surrounding layers of mashed potato mixed with lard, cornmeal, garlic, onions, and green peppers, and topped with pieces of pork, chile seasoning, tomato paste, and white rice.
- "Bananas Foster" is a dessert created in the 1950s at Brennan's Restaurant in New Orleans, in honor of a customer, Richard Foster. It consists of a bowl of lengthwise-sliced bananas sautéed quickly in a mixture of rum, brown sugar, and banana liqueur and served with vanilla ice cream.
- "Plantain flour" is prepared by drying and grinding plantains. It can be baked but is mostly used as a thickener for sauces.
- *Platanos a salteados* (pronounced plah-tah-noss ah sal-tae-ah-doss) is a South American dish of plantains, browned in chile butter and served with sour cream.
- The banana split was invented in 1904 at Strickler's Drug Store in Latrobe, Pennsylvania, by Dr. David Strickler, a pharmacist. The classic banana split is prepared by placing scoops of vanilla, chocolate, and strawberry ice cream in a row between the halves of a banana that has been split lengthwise. Pineapple topping is spooned over the vanilla ice cream, chocolate syrup over the chocolate, strawberry topping over the strawberry, and the confection is garnished with crushed nuts, whipped cream, and maraschino cherries. (Reference: Turback, M. 2004. The banana split book: everything there is to know about America's greatest dessert. Camino Books, Philadelphia. PA. 166 pp.)

Banana *(Musa ×paradisiaca)*, from Harter (1988).

Curiosities of Science and Technology

- Technical botanical terminology sometimes seems contradictory to how the same words are used in common speech, and trivia buffs often delight in taking advantage of this. For example, it is often pointed out that a banana "tree" is not a tree, it is a "herb." This is true in the sense that in technical language a herb is a nonwoody plant, but large banana plants have the appearance of trees even though they are nonwoody, and it is debatably an error to call a banana plant a tree. Moreover, one sense of the word "herb" is a plant that furnishes spices or medicinal substances, and this means that woody trees are often called herbs. Another trivia claim is that a banana is a "berry." In fact, the word "berry" is frequently misapplied to any pulpy or fleshy fruit. Technically, a berry is a baccate (apple-shaped), indehiscent (does not open by itself to distribute the seeds) fruit that arose from a single pistil (the female part of a flower), but since a banana is not apple-shaped, it is not technically a berry. A tomato, however, is a berry technically, but would certainly not be found in the berry section of a supermarket.
- The tongue and eyes of monkeys that eat too many unripe bananas turn green.
- Wild bananas rely on bats for pollination and seed dispersal.
- The country of Tonga once issued a stamp shaped like a banana.
- A biochemical oddity of the banana is that it synthesizes and stores in the peel several neurologically active compounds: serotonin, dopamine, and norepinephrine. This, along with a song entitled *Mellow Yellow*, popularized by the folk singer Donovan in 1966, seems to have encouraged people in the 1960s to start smoking banana peels in the expectation that they would become "high," and recipes for using banana peel as an intoxicant began to

circulate. It was claimed that a psychedelic substance called "bananadine" was present, but testing by the U.S. Food and Drug Administration revealed that the claim was a hoax.

- "Banana oil" (isoamyl acetate) is not obtained from bananas, although it smells like the fruit. It is a synthetic product manufactured from petroleum. It has several uses, for example, as a solvent in some cosmetic formulations, to treat the fabric from which model airplanes are made, and to test whether respirators are leaking (detected by the smell of the oil). Banana oil is poisonous (including the fumes) and flammable.
- The "banana seat," a long seat that was popular on children's bicycles of the 1970s, was created by a renowned American industrial designer, Viktor Schreckengost (1906–2008), and is considered to be the most celebrated and enduring bicycle seat style. The banana seat is so long that a back bar or sissy bar is required to support the back end. Riding double (with two riders) is possible on banana seats, but is not safe unless the rear rider has foot pegs.
- Sap from the banana plant is a tenacious stain that is extremely difficult to remove from hands and clothes.
- The "ice cream banana" is blue when immature, ripening to yellow like other bananas. It has a vanilla-custard taste and a marshmallow texture.
- Throughout the tropics, green banana leaves are used as umbrellas, plates, and wrapping material. Pads for carrying heavy loads on the head are made from dried leaves twisted into a ring.
- Bananas have been grown in Iceland, in soil heated by geysers.
- Mosquitoes are attracted to people who have recently eaten bananas.
- Bananas are one of the few fruits that ripen best off the plant. Fruits that mature on the plant tend to split open, and the pulp has a cottony texture and off-flavor.
- Banana plants developing in rich soils tend to produce large bunches of bananas, which can become so heavy that the plants fall over. To prevent this, some traditional farmers place prop poles under the base of growing bunches of bananas to keep the plants upright.
- The fruit on a bunch of bananas tend to ripen unevenly, those at the base of the bunch ripening first. This makes it difficult to harvest an entire bunch and use all of the bananas at once. In Polynesia and elsewhere, a curious procedure is used to force the bananas of a bunch to mature at the same time. A large pit is dug in the ground, lined with leaves, and filled with bunches of bananas with mature but unripe fruit. The pit is then covered with soil. Fires are built on the edges of the pit, and kept burning slowly, so that the bananas are kept very warm. When the pit is opened in 3–7 days, the bananas are uniformly and completely ripe. It has been speculated that the heat, exclusion of oxygen, and ethylene

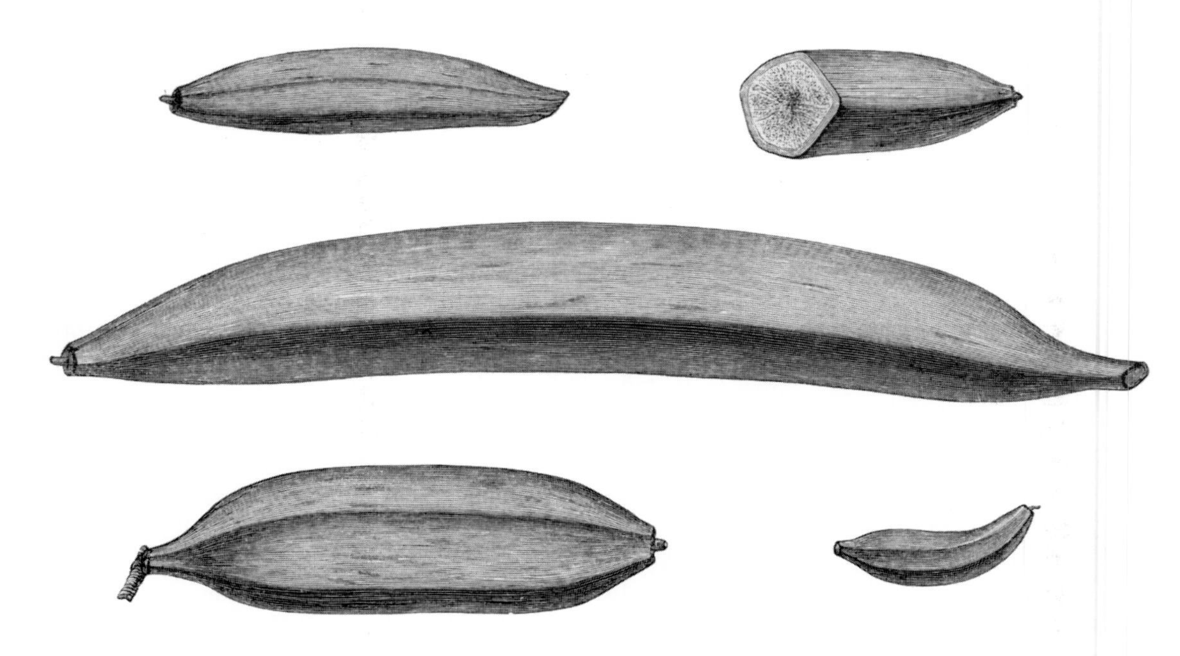

Varieties of banana (*Musa ×paradisiaca*), from Engler and Prantl (1889–1915).

(a natural ripeness promoter) produced by the bananas and retained in the closed pit combine to produce the uniform ripening.

- Bananas should not be refrigerated, although if faced with too many uneaten bananas, putting some in the refrigerator will prolong their life. In fact, little harm is caused to the taste, despite the skin turning an unsightly brown. Commercial (unripe) bananas are typically shipped in containers that maintain low temperature (12 to −25 °C, 54 to −13 °F) for up to 40 days in ethylene-free environments (ethylene is a natural promoter of ripening in most fruits). To produce uniform ripening, the temperature is raised and ethylene gas is sprayed on the bananas.
- The banana is an unusual crop from a genetic perspective. Most cultivated bananas are triploids (i.e., they have three sets of chromosomes), the result of past hybridization between plants with four sets of chromosomes and plants with two sets of chromosomes. As a rule, triploids are sterile, and unable to produce seeds. Accordingly, bananas are seedless. Most plants that are unable to produce seeds will not produce fruits either, but the banana plant is an exception. Since seeds for most banana cultivars are not available, banana plants are reproduced from the shoots that arise at the base of the plant. Because the plants are reproduced vegetatively, cultivars are very uniform, hence subject to diseases, as discussed next.
- In 2003, Belgian plant pathologist Emile Frison of the International Network for the Improvement of Banana and Plantain stated that the dominant commercial banana variety Cavendish might become extinct within 10 years. Limited genetic diversity of cultivated bananas (due to their asexual reproduction) makes them vulnerable to diseases. Indeed, the Cavendish's predecessor, the Gros Michel ("Big Mike"), was wiped out commercially by fungal diseases. The Cavendish banana, found mostly on Western supermarket shelves, has been under attack in some Asian countries by a new strain of Fusarium wilt known as "Panama disease." The magazine *New Scientist* stated "We may see the extinction of the banana, currently a lifesaver for hungry and impoverished Africans and the most popular product on the world's supermarket shelves." However, this considerably overstated the problem. While the Cavendish banana is grown by virtually all commercially important plantations, and is important in world trade, it accounts for only 10% of bananas produced and consumed globally.
- Bananas are notoriously susceptible to bruising, and to prevent this, a protective, ventilated case for single bananas called the "Banana Guard" was invented by Aberrant Designs of Vancouver, British Columbia (United States Patent 6612440, 2003).

Key Information Sources

Abdul, J.B.M.M., Chellappan, K., Pillai, O.A.A., and Chattopadhaya, P.K. 1985. Banana. *In* Fruits of India, tropical and subtropical. *Edited by* T.K. Bose. Naya Prokash, Calcutta, India. pp. 124–161.

Chattopadhyay, P.K. 1999. Banana. *In* Tropical horticulture, vol. 1. *Edited by* T.K. Bose, S.K. Mitra, A.A. Farooqui, and M.K. Sadhu. Naya Prokash, Calcutta, India. pp. 229–258.

Diekmann, M., and Putter, C.A.J. (*Editors*). 1996. *Musa* spp. 2nd edition. International Plant Genetic Resources Institute, Rome, Italy. 28 pp.

Espino, R.R.C., Jamaluddin, S.H., Silayoi, B., and Nasution, R.E. 1991. *Musa* L. (edible cultivars). *In* Plant resources of South-East Asia, 2, edible fruits and nuts. *Edited by* E.W.M. Verheij and R.E. Coronel. Pudoc, Leiden, Netherlands. pp. 225–233.

Gowen, S. 1994. Bananas and plantains. Chapman & Hall, New York, NY. 624 pp.

Haarer, A.E. 1964. Modern banana production. Leonard Hill, London, UK. 136 pp.

Hassan, A., and Pantasico, F.B. (*Editors*). Banana–fruit development, postharvest physiology, handling and marketing in ASEAN. ASEAN Food Handling Bureau, Kuala Lumpur, Malaysia. 147 pp.

Israeli, Y., and Lahav, E. 1986. Banana. *In* CRC handbook of fruit set and development. *Edited by* S.P. Monselise. CRC Press, Boca Raton, FL. pp. 45–73.

Jenkins, V.S. 2000. Bananas: an American history. Smithsonian Institution Press, Washington, DC. 210 pp.

Jones, D.R. 1999. Diseases of banana, abacá, and enset. CABI Publishing, Wallingford, Oxon, UK. 544 pp.

Lebot, V., Aradhya, K.M., Manshardt, R., and Meilleur, B. 1993. Genetic relationships among cultivated bananas and plantains from Asia and the Pacific. Euphtica, **67**: 163–175.

Lessard, W.O. 1992. The complete book of bananas. W.O. Lessard, Homestead, FL. 119 pp.

Loesecke, H.W. von. 1950. Bananas: chemistry, physiology, technology. 2nd edition. Interscience Publishers, New York, NY. 189 pp.

McClatchey, W.C. 2000. Bananas and plantains. *In* The Cambridge world history of food. *Edited by* K.F. Kiple and K.C. Ornelas. Cambridge University Press, Cambridge, UK. pp. 175–181.

Nasution, R.E. 1993. *Musa* L. *In* Plant resources of South-East Asia, 8, vegetables. *Edited by* J.E. Siemonsma and K. Piluek. Pudoc Scientific Publishers, Wageningen, Netherlands. pp. 215–217.

Novak, F.J. 1992. *Musa* (banana and plantains). *In* Biotechnology of perennial fruit crops. *Edited by* F.A. Hammerschlag and R.E. Litz. CAB International, Wallingford, UK. pp. 449–488.

Persley, G.J., and De Langhe, E.A. (*Editors*). 1987. Banana and plantain breeding strategies: proceedings of an international workshop held at Cairns, Australia, 13–17 October 1986. Australian Centre for International Agricultural Research, Canberra, Australia. 187 pp.

Persley, G.J., and George, P. (*Editors*). 1996. Banana improvement: research challenges and opportunities. World Bank, Washington, DC. 47 pp.

Philippine Council for Agriculture, Forestry and Natural Resources Research and Development. 1988. The Philippines recommend for banana. PCARRD Tech. Bull. Series 66/1988. Los Baños, Laguna, Philippines. 136 pp.

Reynolds, P.K. 1927. The banana: its history, cultivation, and place among staple foods. Houghton Mifflin, New York, NY. 181 pp.

Reynolds, P.K. 1951. Earliest evidence of banana culture. J. Am. Oriental Soc. 71(4; Suppl.). 28 pp.

Robinson, J.C. 1996. Bananas and plantains. CAB International, Wallingford, Oxon, UK. 238 pp.

Rowe, P., and Rosales, F.E. 1996. Bananas and plantains. *In* Fruit breeding, vol. 1. Tree and tropical fruits. *Edited by* J. Janick and J.N. Moore. John Wiley & Sons, New York, NY. pp. 167–211.

Rowe, P.R. 1984. Breeding bananas and plantains. Plant Breed. Rev. **2**: 135–155.

Seelig, R.A., and Bing, M.C. 1990*a*. Bananas. *In* Encyclopedia of produce. [Loose-leaf collection of individually paginated chapters.] *Edited by* R.A. Seelig and M.C. Bing. United Fresh Fruit and Vegetable Association, Alexandria, VA. 31 pp.

Seelig, R.A., and Bing, M.C. 1990*b*. Plantains. *In* Encyclopedia of produce. [Loose-leaf collection of individually paginated chapters.] *Edited by* R.A. Seelig and M.C. Bing. United Fresh Fruit and Vegetable Association, Alexandria, VA. 9 pp.

Shepherd, K. 1957. Banana cultivars in East Africa. Trop. Agric. Trin. **34**: 277–286.

Simmonds, N.W. 1962. The evolution of bananas. Tropical Science Series. Longmans, London, UK. 170 pp.

Simmonds, N.W. 1995. Bananas. *In* Evolution of crop plants. 2nd edition. *Edited by* J. Smartt and N.W. Simmonds. Longman Scientific & Technical, Burnt Mill, Harlow, Essex, UK. pp. 370–375.

Simmonds, N.W., and Shepherd, K. 1955. The taxonomy and origins of the cultivated banana. J. Linn. Soc. Bot. **55**: 302–312.

Simmonds, N.W., and Weatherup, S.T.C. 1990. Numerical taxonomy of the wild bananas. New Phytol. **115**: 567–571.

Soto, M.S. 1992. Bananos. Lithography Imprenta LIL, San José, Costa Rica. 649 pp. [In Spanish.]

Stover, R.H., and Simmonds, N.W. 1987. Bananas. 3rd edition. Longman Scientific & Technical, Harlow, Essex, UK. 468 pp.

Turner, D.W. 1994. Bananas and plantains. *In* Handbook of environmental physiology of fruit crops. Vol. 2. Subtropical and tropical crops. *Edited by* B. Schaffer and P.C. Anderen. CRC Press, Boca Raton, FL. pp. 37–64.

Turner, D.W. 1997. Bananas and plantains. *In* Postharvest physiology and storage of tropical and subtropical fruits. *Edited by* S.K. Mitra. CAB International, Wallingford, UK. pp. 47–83.

Valmayor, R.V., and Wagih, M.E. 1996. *Musa* L. (plantain and cooking banana). *In* Plant resources of South-East Asia, 9, plants yielding non-seed carbohydrates. *Edited by* M. Flach and F. Rumawas. Backhuys Publishers, Leiden, Netherlands. pp. 126–131.

Wainwright, H. 1992. Improving the utilization of cooking bananas and plantains. Outlook Agric. **21**: 177–181.

Wardlaw, C.W. 1972. Banana diseases; including plantains and abaca. 2nd edition. Longman, Londong, UK. 878 pp.

Specialty Cookbooks

Banana Export Industry Foundation. 1976. 100 + 1 banana recipes. New Day Publishers, Quezon City, Philippines. 110 pp.

Berry, A. 1995. Go bananas! cookbook. Runaway Press, Nashville, TN. 111 pp.

CQ Products. 2003. Banana cookbook: 101 recipes with bananas. CQ Products, UK. 110 pp.

Feldman, E. 1978. Going bananas: the complete banana cookbook. Universe Books, New York, NY. 112 pp.

Hood, K.J.M., et al. 2003. Banana delight cookbooks: a collection of banana recipes. Whispering Pine Press, Spokane, WA. 320 pp.

Jamaica Information Service. 1960. Banana recipes. Jamaica Information Service, Kingston, Jamaica. 32 pp.

Lindquist, C. 1993. The banana lover's cookbook. St. Martin's Press, New York, NY. 104 pp.

Lorenz Books. 1997. Bananas: a book of recipes. Lorenz Books, New York, NY. 64 pp.

Quick, S. 2000. Go bananas!: 150 recipes for America's most versatile fruit. Broadway Books, New York, NY. 193 pp.

Shepard, R.E. 1986. Banana cookbook. Macmillan, London, UK. 96 pp.

Thurlow, C. 1995. The amazingly simple banana diet. Alma House, London, UK. 96 pp.

United Fruit Company. 1949. Favorite banana recipes from famous New England inns: presented by Chiquita Banana. United Fruit Co., New York, NY. 15 pp.

United Fruit Company. 1953. Chiquita Banana presents 18 recipes from her minute movies. United Fruit Company, New York, NY. 18 pp.

United Fruit Company. 1954. Bananas: recipes for institutional service and menus. United Fruit Company, New York, NY. 52 pp.

VeuCasovic, E.M. 1982. Bananas by the bunch: a book of favorite recipes. E.M. VeuCasovic, Bexar, AR. 116 pp.

BARLEY

Family: Poaceae (Gramineae; grass family)

Names

Scientific name: *Hordeum vulgare* L.

- The word "barley" traces back through Old English and French, and possibly other European languages for this ancient cereal.
- The introduction of beer into Europe began in 55 BC by Julius Caesar's Roman legions. As barley was used in beer, it was also the staple food of the Roman gladiators. Barley's Latin genus name *Hordeum* is related to the word for gladiators, *hordearii*.
- *Vulgare* in the scientific name *H. vulgare* is Latin for "common."

Plant Portrait

Barley is an annual cereal grass, 30–120 cm (1–4 feet) tall. Barleys are divided into groups by the number of rows of kernels on the fruiting stem: two, four, or six, and are further classified into those with the hulls tightly attached to the grains, loosely attached, or lacking hulls. The grains can be red, white, black, purple, or blue. Some kinds are associated with special uses. For example, two-rowed varieties are usually used in the preparation of malt for brewing beer. One of the most ancient of cultivated grains, barley is probably native to the Middle East, from Afghanistan to northern India. It was known to the ancient Greeks, Romans, Chinese, and Egyptians, and was the principal bread material in Europe as late as the 16th century. Barley adapts well to a variety of climates, and is now widely cultivated in all temperate regions. The chief producing nations are the United States, Canada, the CIS (Russia and most other countries of the former USSR), China, India, and countries bordering the Mediterranean. Barley is the world's fourth most important cereal crop (after wheat, rice, and corn). In the Western Hemisphere, barley is used mainly as livestock feed, but also in making bread, beer, and whiskey. China overtook the United States as the world's larger beer consumer in 2003, and accordingly has become the larger consumer of malting barley.

Culinary Portrait

Barley is used for specialty food products, particularly by the brewing and malting industries as one of the primary components of beer and whiskey. High-protein barleys are generally valued for human food as well as for feeding livestock, and for producing malt for whiskey, while starchy barley is preferred for producing malt for beer. "Hulled barley" has had the outer husk removed but retains most of its bran, and most of its nutrients, which are found near the outer layer of the grain. Less nutritious but more edible is pearl barley, which is made by using abrasive disks to grind the hulls and bran off the kernels. Pearl barley is often used in soups, stews, puddings, and desserts because the grains tend to swell, acting as a thickener, and because the flavor is nutty and pleasant. Barley has less protein than wheat, and much less gluten, so it is not suitable for preparing porous, light bread. Barley flour made from whole grain is darker than whole-grain wheat flour. The dark breads of northern and central Europe and Russia are made from barley, with some wheat added to lighten the product. Flat, unleavened breads of North Africa and the Middle East, such as pita, are also made from barley. Wheat is usually also added to barley when it is used to make cookies, crêpes, and cakes. In Europe, North Africa, and the Middle East, barley is also commonly used in porridge. "Barley flakes" are prepared and used like rolled oats (see OAT).

To cook hulled barley, add 3–4 cups of cooking liquid for each cup of grain and maintain over low heat for about 1 hour. Both hulled and pot barley (defined below) should be soaked for several hours before cooking. Pearled barley does not require presoaking, and needs only about 30 minutes of cooking. Pearled barley can be mixed with rice and cooked.

"Malt," in general, is germinated grain, but is usually barley. The seeds are germinated in large vats for 2–3 days, and then allowed to sprout on special growing surfaces after which the seedlings are kiln-dried and screened. During this process, most of the starch in the seeds is converted to maltose (malt sugar), and considerable diastase (an enzyme promoting conversion of starches into sugar) is produced. Diastase is important in brewing alcoholic beverages because yeasts used in brewing need sugars to convert into alcohol, and some starting materials such as potatoes contain mainly starch, not sugar. The enzyme changes the starch into sugar, which the yeast converts into alcohol. Raw malt is not widely consumed, but the extract of malt is very nutritious, and the diastase present promotes digestion of starches. Accordingly, extract of malt has long enjoyed a reputation as a health food. "Malted milk," made with dried extracts of malted wheat and barley mixed with powdered milk, is a popular base for ice cream, pastries, and other sweet confections, and is also used to flavor breakfast cereals. Malt produced from roasted barley is employed as a substitute for coffee. "Malt liquor" refers to fermented beverages made from malted barley. In the United States, malt liquor refers to a high-alcohol beer. "Malt vinegar" is made from unhopped beer; "brown malt vinegar" has caramel added.

Culinary Vocabulary

- "Liquid bread" is a quaint old phrase for beer, reflecting the common use of barley in preparing both beer and bread.
- "John Barleycorn" is a nickname for alcoholic beverages, particularly for spirits made from malt (i.e., barley). The term was first recorded in 1620.
- "Barley-broo" is a Scottish term for whiskey, beer, and ale.
- "Scotch whisky," which is distilled from barley, is spelled without the "e" (i.e., not "whiskey"). ["Whisky" is the usual spelling in Scotland and is common in Canada, while "whiskey" is employed in the US and Ireland.]
- "Pot barley" (Scotch barley) is barley with the hull or outer husk removed, but not as polished as pearl barley. The abrasive process results in the grain losing a proportion of its vitamins and minerals, and almost all of its bran. Pot barley is used much like pearl barley. Its nutty flavor (more so than that of pearly barley) complements stews and soups, and it is recommended for Scotch broth.
- "Barley sugar" has three different meanings: (1) a candy, once made by combining boiled sugar, barley water, and lemon juice to produce an amber-colored twisted stick (it is still manufactured today, but made with synthetic flavoring); (2) a confection made by heating white sugar to the melting point, when it forms small grains resembling barley; (3) in British slang, twisting someone's arm painfully behind his back. The long twisted sticks of "barley sugar" (meaning 1) resulted in the term "barley sugar" being applied in architecture and design to twisted columns and pillars.
- "Barley water" has several meanings: (1) (especially in Great Britain) a boiled preparation of pearl barley flavored with lemon or orange juice, often mixed with milk, and served as a nourishing dietary supplement for invalids and infants; (2) (especially in Korea) a drink made by boiling barley in water; (3) a slightly fermented beverage made by steeping barley in water.
- "Pale malt" is air-dried, malted barley, a process that minimizes color development, as opposed to kiln-dried barley.
- "Orgeat" (pronounced OHR-zhat), a word derived from the Latin *hordeum* (see genus name *Hordeum*, above) has two meanings: (1) a barley–almond mixture, similar to English barley water (described above), and (2) a sweet syrup made from almonds, sugar, and rosewater or orange-flower water, and used to flavor cocktails and baked goods.
- "Barley wine" is an archaic English term for a dark, strong ale (or, more rarely, beer). British brews manufactured with barley are usually made without hops, and so technically are ales, not beer. Barley wine is also a sweet wine that was manufactured by the ancient Babylonians, Sumerians, and Egyptians, containing as much as 12% alcohol. Today, barley wine is likely to be called malt liquor.
- "Barley broth" (often called Scotch broth) is a Scottish soup made with lamb or mutton, barley, and various vegetables.
- "Barley tea," a beverage made from roasted barley, is known as *mugi cha* in Japan and *poricha* in Korea, the two countries where it is primarily consumed.

Barley *(Hordeum vulgare)*, from Baillon (1876–1892).

Curiosities of Science and Technology

- In 2004, the earliest evidence of milling of flour and baking of bread was published by Piperno et al. (2004). Ohalo II is a well-known, ancient human settlement, dated at 23 000 years before the present era, on the shore of the Sea of Galilee in Israel. On the floor of a hut, archeologists discovered a large, flat stone used to grind wild grass grains, revealed by the presence of starch grains from close wild relatives of domesticated wheat and barley. Additional evidence from an oven-like hearth suggests that dough may have been baked much as modern nomadic tribes in the region still prepare seed cakes. Such grinding and baking of cereal grains is considered to represent an important nutritional advancement in human history.
- Many ancient cultures brewed beer with barley. A Mesopotamian table dated at 7000 BC included a recipe for beer. Clay tables dated at 6000 BC in Babylon have a scene depicting the preparation of beer.
- The oldest known recipe for barley wine was found inscribed on a brick of a Babylonian library building, dated at 2800 BC.
- Egyptian hieroglyphics engraved inside the pyramids depict brewers at work and various recipes of beer so the Pharaoh would know how to make beer in the afterlife.
- Some Egyptian mummies were found to have necklaces made from barley seed.
- In ancient Greece, athletes in training consumed barley mush because it was considered to be the mildest of all cereals.
- At the beginning of the 3rd century, tribute to Rome from Carthage included 500 000 bushels of wheat and 300 000 bushels of barley.
- In 1324, King Edward II of England decreed that an inch was equal to the length of three barley grains ("barleycorns") from the center of the barley ear, placed end to end. The same decree specified that the longest normal human foot measured 39 barleycorns in length. In 1888, the US shoe industry accepted this old standard, establishing its size 13 (i.e., 39 barleycorns or 13 inches) as the largest, regularly manufactured shoe.
- Osteoporosis, or porous bone, is a disease characterized by low bone mass and structural deterioration of bone tissue, leading to bone fragility and an increased susceptibility to fractures of the hip, spine, and wrist. Of the 10 million Americans estimated to have osteoporosis, 8 million are women and 2 million are men. One in two women and one in four men age 50 and over will have an osteoporosis-related fracture in their lifetime. In 2002, researchers at King's College and St. Thomas's Hospital in London announced that they had found that intake of the element silicon was directly related to bone strength, and that consumption of it in early years was important in preventing the development of osteoporosis in later life. Silicon is absorbed from the soil by plants, especially by cereals such as wheat and barley. It is particularly deposited in the bracts (husks) of barley, and so beer made from barley is one of the richest sources of silicon in the modern diet. A pint of beer provides 12 mg of silicon, 40% of the average daily intake. Moreover, the absorption by the body of silicon has been found to be superior from beer than from other foods.

Barley *(Hordeum vulgare)*, from Hallier (1880–1888, vol. 8, plate 787).

Key Information Sources

Bamforth, C.W. 2003. Barley and malt starch in brewing: a general review. Tech. Q. Master Brew. Assoc. Am. **40**: 89–97.

Baum, B.R., Nevo, E., Johnson, D.A., and Beiles, A. 1997. Genetic diversity in wild barley (*Hordeum spontaneum* C. Koch) in the Near East: a molecular analysis using Random Amplified Polymorphic DNA. Genet. Resour. Crop Evol. **44**: 147–157.

Bhatty, R.S. 1986. The potential of hull-less barley—a review. Cereal Chem. **63**: 97–103.

Bothmer, R. von, Hintum, T. van, Knüpffer, H., and Sato, K. (*Editors*). 2003*a*. Diversity in barley (*Hordeum vulgare*). Elsevier Science Ltd., Amsterdam, Netherlands. 280 pp.

Bothmer, R. von, Jacobsen, N., Baden, C., Jorgensen, R.B., Linde-Laursen, I., Baden, C., and Jacobsen, N. 2003*b*. *Hordeum*. Springer-Verlag, Heidelberg, Germany. CD-ROM.

Ceccarelli, S., and Grando, S. 1996. *Hordeum vulgare* L. *In* Plant resources of South-East Asia, 10, cereals. *Edited by* C.J.H. Grubben and S. Partohardjono. Backhuys Publishers, Leiden, Netherlands. pp. 99–102.

Briggs, D.E. 1978. Barley. Wiley, New York, NY. 612 pp.

Brookes, P.A., Lovett, D.A., and MacWilliam, I.C. 1976. The steeping of barley. A review of the metabolic consequences of water uptake, and their practical implications for malting. J. Inst. Brew. (London), **82**: 14–26.

Cook, A.H. (*Editor*). 1962. Barley and malt: biology, biochemistry, technology. Academic Press, New York, NY. 740 pp.

Gales, K. 1983. Yield variation of wheat and barley in Britain in relation to crop growth and soil conditions—a review. J. Sci. Food Agric. **34**: 1085–1104.

Harlan, J.R. 1995. Barley. *In* Evolution of crop plants. 2nd edition. *Edited by* J. Smartt and N.W. Simmonds. Longman Scientific & Technical, Burnt Mill, Harlow, Essex, UK. pp. 140–147.

Henry, R.J. 1988. The carbohydrates of barley grains—a review. J. Inst. Brew. (London), **94**: 71–78.

Hockett, E.A. 2000. Barley. *In* Handbook of cereal science and technology. 2nd edition. *Edited by* K. Kulp and J.G. Ponte, Jr. Marcel Dekker, New York, NY. pp. 81–125.

ICON Health Publications. 2003. Barley: a medical dictionary, bibliography, and annotated research guide to internet references. ICON Health Publications, San Diego, CA. 284 pp.

Jakob, S.S., Meister, A., and Blattner, F.R. 2004. Considerable genome size variation of *Hordeum* species (Poaceae) is linked to phylogeny, life form, ecology, and speciation rates. Molec. Biol. Evol. **21**: 860–869.

Johnson, D.D. 1999. Grain quality in the Canadian barley sector: a review of regulations, industry practices, and policy issues. Dept. of Agricultural Economics, Agricultural Experiment Station, North Dakota State University, Fargo, ND. 31 pp.

MacGregor, A.W., and Bhatty, R.S. (*Editors*). 1993. Barley: chemistry and technology. American Association of Cereal Chemists, St. Paul, MI. 486 pp.

McCorriston, J. 2000. Barley. *In* The Cambridge world history of food. *Edited by* K.F. Kiple and K.C. Ornelas. Cambridge University Press, Cambridge, UK. pp. 81–89.

Nevo, E., Baum, B., Beiles, A., and Johnson, D.A. 1998. Ecological correlates of RAPD DNA diversity of wild barley, *Hordeum spontaneum*, in the Fertile Crescent. Genet. Resour. Crop Evol. **45**: 151–159.

Nishikawa, T., Salomon, B., Komatsuda, T., Bothmer, R. von, and Kadowaki, K. 2002. Molecular phylogeny of the genus *Hordeum* using three chloroplast DNA sequences. Genome, **45**: 1157–1166.

Piperno, D.R., Weiss, E., Holst, I., and Nabel, D. 2004. Processing of wild cereal grains in the Upper Paleolithic revealed by starch grain analysis. Nature, **430**: 670–673.

Pirie, R.B. 1992. Review of barley varieties—old and new. Brewer, **78**: 534–536.

Rasmusson, D.C. (*Editor*). 1985. Barley. American Society of Agronomy, Madison, WI. 522 pp.

Shewry, P.R. (*Editor*). 1992. Barley: genetics, biochemistry, molecular biology, and biotechnology. C.A.B. International, Wallingford, Oxon, UK. 610 pp.

Slafer, G.A., Molina-Cano, J.L., Savin, R., Araus, J.L., and Romagosa, I. (*Editors*). 2002. Barley science: recent advances from molecular biology to agronomy of yield and quality. Food Products Press, New York, NY. 565 pp.

Svitashev, S., Bryngelsson, T., Vershinin, A., Pedersen, C., Sall, T., and Bothmer, R. von. 1994. Phylogenetic analysis of the genus *Hordeum* using repetitive DNA sequences. Theor. Appl. Genet. **89**: 801–810.

United States. Science and Education Administration. 1979. Barley: origin, botany, culture, winter hardiness, genetics, utilization, pests. Revised edition. Science and Education Administration, U.S. Dept. of Agriculture, Washington, DC. 154 pp.

Wainwright, T., and Buckee, G.K. 1977. Barley and malt analysis—a review. J. Inst. Brew. (London), **83**: 325–347.

Wang, J. 2005. China: beer and malting barley. Bi-weekly Bull. (Agric. Agri-Food Can.), **18(10)**: 1–6.

Specialty Cookbooks

Allen, F., and Cantwell, D. 1998. Barley wine: history, brewing techniques, recipes. Brewers Publications (Classic Beer Style Series, 11), Boulder, CO. 208 pp.

Cottrell, E. 1974. The oats, peas, beans & barley cookbook. Woodbridge Press, Santa Barbara, CA. 271 pp.

Freeman, B. 1981. A book of Welsh bread: recipes for the old, traditional wholegrain wheat, barley and rye breads of Wales, adapted for baking today. Image Imprint, Cardiff, UK. 48 pp.

Hayes, J.L., and Leblang, B.T. 1995. Grains: seventy-six healthy recipes for barley, corn, rye, wheat, and other grains. Harmony Books, New York, NY. 137 pp.

Keane, M., and Chace, D. 1994. Grains for better health: over 100 delicious recipes using rice, wheat, barley, and other wholesome grains. Prima Publishing, Rocklin, CA. 224 pp.

Napier, R. 1995. Barley banquet: how to improve your life and health with special barley recipes. Green Gecko Publishing, Duncan, BC. 124 pp.

Sokolov, R.A. 1996.With the grain: 200 delectable recipes using wheat, corn, rice, oats, barley, and other grains. Knopf, New York, NY. 268 pp.

BEAN (COMMON BEAN)

Family: Fabaceae (Leguminosae; pea family)

Names

Scientific name: *Phaseolus vulgaris* L.

- The English word "bean" has been traced with certainty only within the confines of the English language, but probably came from German. The word was spelled the same and had the same meaning 1000 years ago in English. A variety of seed-like products of similar size and shape are also termed beans, such as coffee beans and jelly beans.
- The name "kidney bean" was first used by the English in 1551 to distinguish American common beans from Old World beans. It reflects the kidney-like shape, and a supposed strengthening effect on the kidneys.
- The name "string bean" was based on the stringy strands that were once present in the pod, but have now been removed by breeding to the point that they now develop only when the bean is overripe. The strands of inedible, tough fibers occur at the sutures of the two valves that make up the pod (additional information is provided below).
- The "army bean" was so named because of the widespread use of common beans by the military. Beans represent a way of feeding large numbers of people, at minimum expense, since this nutritious food is easily transported and keeps well.
- The "navy bean," a simply white common bean, derives its name from its use as a staple of U.S. Navy messes.
- The "snap bean" was so called because the tender pods can be easily snapped or broken into pieces before cooking.
- "Wax bean," applied most frequently to yellow edible-podded forms, is based on the waxy appearance of the pods.
- "Butter bean," applied to the seeds and also to yellow edible-podded varieties, is based on the buttery color. The term is also used for the lima bean, which see.
- The name "pinto bean" is derived from the Spanish word for "painted." Pinto beans have streaks of brownish pink on their skin, like the horse of the same name. The "crabeye bean" is an alternative name, presumably coined for a supposed resemblance to the beady, bean-shaped eye of the crab.
- "Frijol" is based on the American Spanish *frijol*, derived from an Old World name for the broad bean, which see.
- "Haricot" is a French word for bean, carried over into English, and now a general term referring to a wide variety of beans.
- The name "French bean" arose in England because the vegetable was first brought to northern Europe from Canada by French explorers. In American English, "French beans" are likely to be called string beans or snap beans.
- Other, self-explanatory names often encountered are green bean and salad bean.
- In 19th century America, dried beans were not considered to be *haute cuisine*, and were sometimes known by the euphemism "Alaska strawberries" (see STRAWBERRY for similar amusing terms).
- The genus name *Phaseolus* is derived from the Greek *phaseolos*, a bean. A more imaginative interpretation translates the Greek word as a small boat, supposedly the shape of the bean pod.
- *Vulgaris* in the scientific name *P. vulgaris* is Latin for common, of frequent occurrence, derived from the Latin *vulgus*, the common herd, or the common people.

Plant Portrait

The common bean is a herbaceous annual plant native to the Americas. It has been selected over the last 7000–8000 years from a wild-growing vine, and has been turned into a major food crop grown worldwide. Cultivation moved through the Americas, and beans had diversified into various selections by the time the Europeans arrived. By then, beans had become a staple food even in parts of Canada. Columbus is thought to have brought seeds back to Europe, and by the 17th century, their use was widespread in Eurasia. The common bean is either a climbing, twining vine 2–3 m (3–10 feet) long, or a dwarf bush, 20–60 cm (8–24 inches) high. The climbing forms are commonly called "pole beans." The flowers are shades of white, yellow, orange, red, and purple. Pods are typically 9–18 cm (3½–7 inches) long. Common beans are used either as fresh pods or as dried seeds that are cooked. Fresh beans are particularly important in North America and in developed countries in Europe.

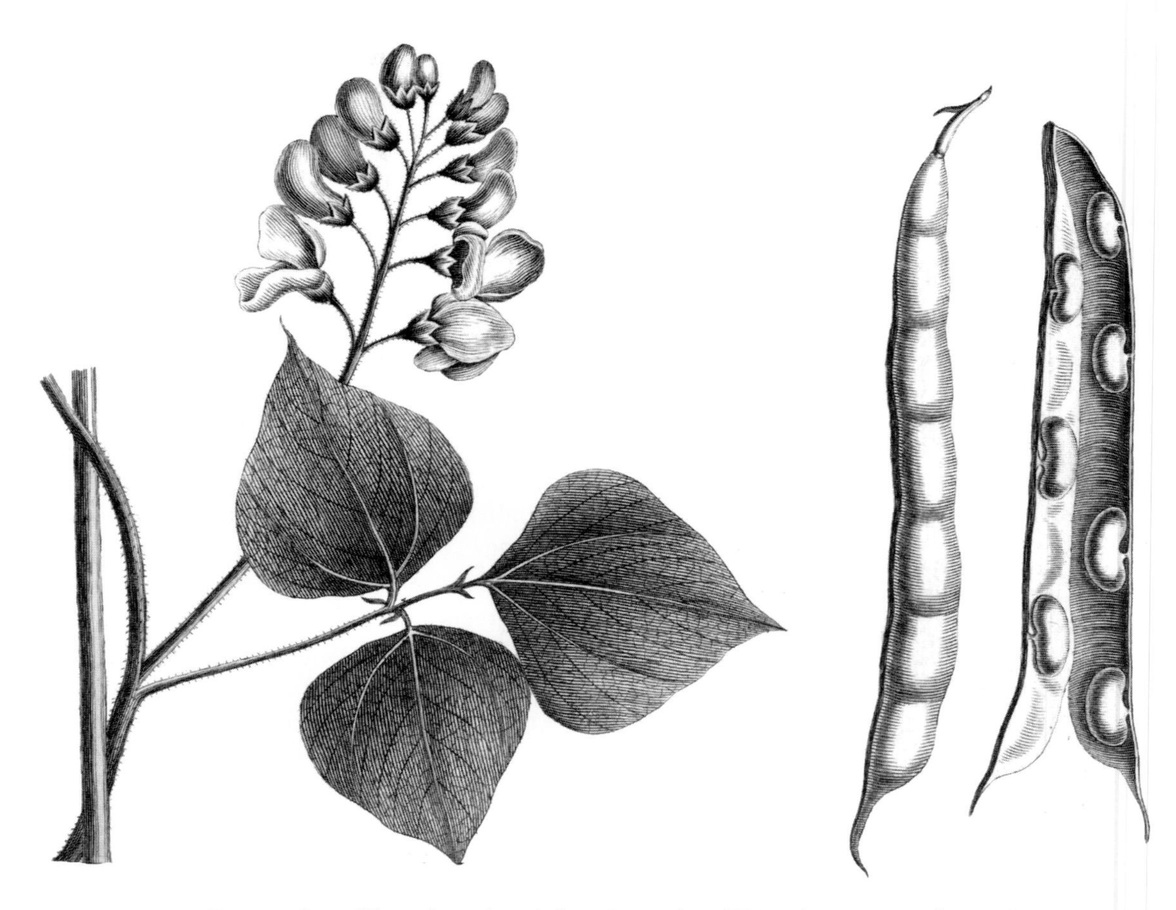

Common bean *(Phaseolus vulgaris)*, from Lamarck and Poiret (1744–1829, plate 610).

However, dry beans are far more important economically. Common bean is the most widely cultivated of all "beans" in temperate regions, but is also widely cultivated in semitropical countries. Major production areas are found from Mexico to South America, and in East Africa, where beans are a critical protein source, supplementing carbohydrate staple foods such as corn (maize). More than 14 000 cultivated varieties have been recorded. Selection of cultivated forms has resulted in hundreds of different kinds of market types, including the following categories.

A. Edible-podded beans, known as snap (stringless), French, or string beans, are raised for the edible pods, which are harvested while the pods are immature. These are grown in the garden both as bush and pole forms, with either green or yellow ("wax") pods. Some varieties in this class are also suitable for production of shelled beans (dry seeds).

B. Edible, shelled, unripe beans—grown for immature seeds shelled from the unripe pods. This kind of bean is infrequently encountered and is of limited importance.

C. Shell beans—grown for the dry, ripe seeds, used generally as a soup bean, a stewed bean, or as a baked bean. Such kinds are grown for "dry beans" (dried mature seeds) and when grown as large acreages are often called "field beans." A variety of types have been recognized, including the following, which are grown in North America:

1. Pea or navy: round, usually white seeds 8 mm (0.3 inch) long or less.
2. Great northern: an enlarged, lengthened kind of navy bean.
3. Pinto: like great northern in size, but plumper and colored pinkish buff blotched with dark brown.
4. Cranberry: a horticultural kind, plump, oblong, buff splashed with carmine.
5. Red kidney: long, broad-oval, kidney-shaped, varying from pink crimson to reddish brown.
6. White kidney: like red kidney, but white.

7. White marrow: short, plump, rounded, white seeds.
8. Black turtle soup (also black Spanish, tampico, and Venezuelan): small, elongated, flattened, round ends, jet-black in color.
9. Yellow eye: large, broad, plump, white, with distinct darkened area.
10. Jacob's cattle (also known as trout beans and Dalmatian beans, and also as torellen because they arrived in the US with German settlers who called themselves Torellen): plump, elongated, rounded ends, unusual coloration with brownish crimson round spots and blotches on a cream surface.
11. Soldier: intermediate in shape between navy and marrow; brownish red with blotchy markings.

Culinary Portrait

Fresh beans are steamed, boiled, baked, or fried, and used in vegetable dishes, salads, stews, and soups. Dried beans are soaked in water and then boiled or fried for use in a variety of dishes, such as casseroles and stir-fries. Both fresh and dried beans are canned and processed, and fresh beans are also frozen. Common products utilizing beans include pork and beans, refried beans, stews, soups, chile, bean flour, bean paste, fiber biscuits, and snack foods.

Beans are exceptionally advantageous to the human diet. They are low in fat and sodium, free of cholesterol, high in protein, and an excellent source of soluble and insoluble fiber, complex carbohydrates, and vitamins and minerals, especially B vitamins, potassium, and phosphorus. Flour made from dry beans is gluten free, and a nutritious option for people with celiac disease.

On the negative side, eating beans is well known to cause flatulence. Beans have complex sugars (oligosaccharides) that can not be metabolized by human digestive enzymes. Concentrations of these sugars in dried legumes cause the most problems, particularly in navy beans and lima beans. The sugars pass unaltered through the upper intestine, but in the lower intestine are decomposed by bacteria, which produce four main gases—carbon dioxide, hydrogen, methane, and nitrogen gas. Methane is odorless, and hydrogen, nitrogen, and carbon dioxide have little odor. The foul odor is mainly due to other substances, especially if they contain sulfur, such as eggs and cabbage. The problem is lessened by thorough cooking and by soaking the beans in several changes of water.

Culinary Vocabulary

- "Boston baked beans" are a mix of navy beans or pea beans (the latter preferred by New Englanders), salt pork, molasses, and brown sugar, baked in a casserole. The dish is so named because it was made and baked by Puritan Bostonian women on Saturday to be served for dinner that night. Because cooking was forbidden on the Sabbath, the leftover beans were served with brown bread for Sunday breakfast or lunch. This led to Massachusetts being called the "Baked Bean State."
- "Senate bean soup" is a bean soup served in the U.S. Senate dining room. It contains white beans, smoked ham hocks, mashed potatoes, onions, and garlic.
- The Dutch call their popular dish combining string beans and navy beans *blote billetjes in bet gras*, which translates as "bare buttocks in the grass."
- In addition to being the brand name of a commercial preparation that reduces intestinal gas, "Beano" is a slang abbreviation for "bean fest."
- "Refried beans" is a Mexican-American dish of mashed, cooked beans, typically served as a side dish or used as a filling for various tortillas. The English name is based on a mistranslation of the Spanish name *frijoles refritos*. The Spanish means well fried, not refried.
- "Ranch beans" is a pinto bean dish flavored with onions, garlic, and bacon.
- "Moors and Christians" is a dish combining black beans (representing the Moors, a dark-skinned Caucasoid people of northern Africa) and white rice (for lighter-skinned Christians).

Curiosities of Science and Technology

- In 1857, Charles Darwin, the "Father of Evolution," studied the relationship of bee pollinators and bean flowers, and concluded that "if every bee in Britain were destroyed, we should not again see a pod on our kidney beans." It isn't clear what variety Darwin studied because cultivars grown today have no or little need of bees to develop pods.
- Red beans and rice, a well-known Louisiana dish, are customarily served on Mondays in Louisiana, following a Sunday dinner of ham, and using the ham bone to flavor the bean–rice meal. This combination is believed to have originated with Louisiana black cooks (who called it "red and white"), but has

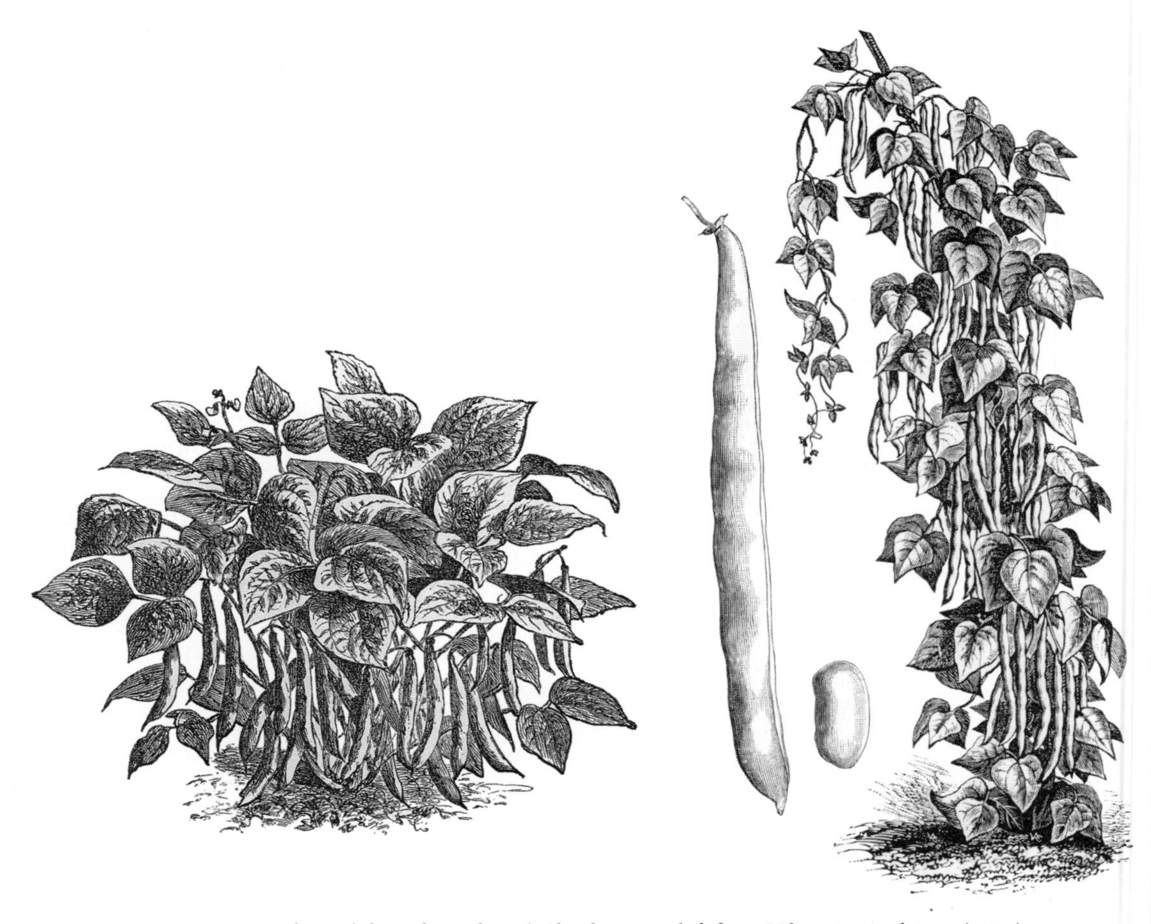

Common bean *(Phaseolus vulgaris)*. Shrub type at left from Vilmorin-Andrieux (1885), vine type at right from *Gartenflora* (1893, vol. 42).

become one of the mainstays of Cajun cooking. As is often the case with traditional dishes, there are culinary arguments about its preparation—what kind of red beans are best, and whether the beans and rice should be cooked separately or together. In any event, the basically vegetarian combination reflects a nutritional balance that occurs repeatedly in meat-restricted diets. Beans are a rich source of protein, particularly the essential amino acid lysine, but are usually deficient in the sulfur-containing amino acids methionine and cystine. By contrast, cereals like rice contain lower amounts of proteins, and these are deficient in lysine but adequate in sulfur-containing amino acids. Together, rice and beans provide a balance of essential protein components, which is necessary for survival, especially when adequate meat supplies are unavailable.

- The commercial product "Beano" ("Take Beano before and there'll be no gas") contains the enzyme alpha-galactosidase, extracted from the fungus *Aspergillus*. The enzyme neutralizes certain sugars that otherwise cause the infamous flatulence associated with eating beans. Three to eight drops of Beano are taken when first consuming gas-inducing food, but it must be placed in the mouth with food that is no more than lukewarm, or it will not work, as the enzyme becomes inactive at 54 °C (130 °F).
- Most dried beans will rehydrate to three times their dry size.
- Pods of wild plants of the pea family often dehisce suddenly because of the release of tensions built up in the parchment layer lining the inner pod. This layer is made up of oriented fibers, which twist in response to humidity, building up tension, and twisting the pod valves in opposite directions. At maturity, after low humidity has resulted in great tension developing, the sutures break, the valves of the pod twist quickly in opposite direction, and the seeds are ejected. In cultivated varieties, pods usually have a reduction of the parchment layer, and so reduced dehiscence. Some "stringless" types lack parchment completely, and are quite indehiscent.

- Starting in 1989, for several years in France there were numerous claims of consumers (all women) finding live toads in cans of beans (French beans, of course). Toads have a longstanding reputation of being able to survive for considerable periods when entombed, and seem to be the regular subject of such reports. The headline in one French newspaper was "Woman finds French beans in her can of toads." (A similar American newspaper article in 2003 asserted that a woman who found a frog in a can of peas was "hopping mad" and "nearly croaked.") Such claims of toads (dead or alive) in closed containers outnumber reports of mice, which should be (and no doubt are) found in food more frequently. This may simply be because the reputation for survival of toads predisposes stories about them to be taken more seriously. Psychological analysis of the reports in France concluded that women with phobias about amphibians had been sensitized by the first accounts of toads in beans, and responded by hysterically finding the subject of their fears.

Key Information Sources

Adam, M.W., Coyne, D.P. Davis, J.H.C., Graham, P.H., and Francis, C.A. 1985. Common bean (*Phaseolus vulgaris* L.). *In* Grain legume crops. *Edited by* R.J. Summerfield and E.H. Roberts. Collins, London, UK. pp. 433–476.

Davis, J.H.C. 1997. *Phaseolus* beans. *In* The physiology of vegetable crops. *Edited by* H.C. Wien. CABI International, Wallingford, Oxon, UK. pp. 409–428.

Debouck, D.G. 1994. Beans. (*Phaseolus* spp.). *In* Neglected crops. 1492 from a different perspective. *Edited by* J.E. Hernández-Bermejo and J. Léon. Food and Agriculture Organization of the United Nations, Rome, Italy. pp. 47–62.

Debouck, D.G., and Smartt, J. 1995. Beans. *Phaseolus* spp. (Leguminosae—Papilionoideae). *In* Evolution of crop plants. 2nd edition. *Edited by* J. Smartt and N.W. Simmonds. Longman Scientific & Technical, Burnt Mill, Harlow, Essex, UK. pp. 287–294.

Dev, J., and Gupta, V.P. 1997. Common bean: historic view and breeding strategy. Ann. Biol. (Hissar), **13**: 213–219.

Gepts, P. (*Editor*). 1988. Genetic resources of *Phaseolus* beans: their maintenance, domestication, evolution, and utilization. Kluwer Academic Publishers, Norwell, MA. 613 pp.

Gepts, P. 1998. Origin and evolution of common bean: past events and recent trends. Hortscience, **33**: 1124–1130.

Graham, P.H., and Ranalli, P. 1997. Common bean (*Phaseolus vulgaris* L.). Field Crops Res. **53(1–3)**: 131–146.

Kaplan, L. 1981. What is the origin of the common bean? Econ. Bot. **35**: 240–254.

Kaplan, L. 2000. Beans, peas, and lentils. *In* The Cambridge world history of food. *Edited by* K.F. Kiple and K.C. Ornelas. Cambridge University Press, Cambridge, UK. pp. 271–281.

Laing, D.R., Jones, P.G., and Davis, J.H.C. 1984. Common bean (*Phaseolus vulgaris* L.). *In* The physiology of tropical field crops. *Edited by* P.R. Goldsworthy and N.M. Fisher. John Wiley & Sons, Chichester, UK. pp. 305–351.

Maiti, R. 1997. *Phaseolus* spp. Bean Science. Science Publishers, Inc., Enfield, NH. 534 pp.

Maiti, R.K., Singh, V.P., Arreola, E.S., and Chirino, Y.S.U. 2002. Physiological, biochemical and molecular mechanisms of resistance of *Phaseolus* bean and other related crops to drought, high and low temperature and salinity: a review. Crop Res. (Hisar), **24**: 205–241.

Nwokolo, E. 1996. Common bean (*Phaseolus vulgaris* L.). *In* Food and feed from legumes and oilseeds. *Edited by* J. Smartt and E. Nwokolo. Chapman and Hall, London, UK. pp. 159–172.

Park, S.J. 1989. Growing field beans in Canada. Agric. Can. Publ. 1787/E. 30 pp.

Schoonhoven, A. van, and Voysest, O. (*Editors*). 1991. Common beans: research for crop improvement. C.A.B. International, Wallingford, UK. 980 pp.

Silbernagel, M.J. 1986. Snap bean breeding. *In* Breeding vegetable crops. *Edited by* M.J. Bassett. AVI Publishing Co., Westport, CT. pp. 243–282.

Singh, S.P. (*Editor*). 1999. Common bean improvement in the twenty-first century. Kluwer Academic Publishers, Dordrecht, Netherlands. 405 pp.

Singh, S.P. 2001. Broadening the genetic base of common bean cultivars: a review. Crop Sci. **41**: 1659–1675.

Singh, S.P., Gepts, P.L., and Debouck, D.G. 1991. Races of common bean (*Phaseolus vulgaris* L., Fabaceae). Econ. Bot. **45**: 379–396.

Skrypetz, S. 2006. Dry beans: situation and outlook. Bi-weekly Bull. (Agric. Agri-Food Can.), **19(19)**: 1–8.

Smartt, J. 1989. *Phaseolus vulgaris* L. *In* Plant resources of South-East Asia, 1, pulses. *Edited by* L.J.G. van der Maesen and S. Somaatmadja. Pudoc, Leiden, Netherlands. pp. 60–63.

Svetleva, D., Velcheva, M., and Bhowmik, G. 2003. Biotechnology as a useful tool in common bean (*Phaseolus vulgaris* L.) improvement: a review. Euphytica, **131**: 189–200.

Unk, J. 1984. A bab: *Phaseolus vulgaris*. Akadâemiai Kiadâo, Budapest, Hungary. 344 pp. [In Hungarian.]

Specialty Cookbooks

(Also see LENTIL and PEA.)

Bennett, V. 1967. The complete bean cookbook. Prentice Hall, Englewood Cliffs, NJ. 298 pp.

California Dry Bean Advisory Board. 1995. Beans, beans, beans: California's finest recipes. California Dry Bean Advisory Board, Dinuba, CA. 116 pp.

Choate, J. 1992. The bean cookbook: a celebration of the delicious legume from Hoppin' John to simple cassoulet. Simon & Schuster, New York, NY. 128 pp.

Cortez Chamber of Commerce. 1967. From the queen's kitchen: a collection of recipes from the pinto bean cooking contest, held each year during the Four Corners Harvest Festival in Cortez, Colorado. Revised edition. Cortez Chamber of Commerce, Cortez, CO. 116 pp.

Elliott, R. 1979. The bean book. Fontana, London. 294 pp.

Fischer, S. 1999. Bean lovers cookbook. Golden West Publishers, Phoenix, AZ. 109 pp.

Geil, P.B. 1996. Magic beans: 150 delicious recipes featuring nature's low-fat, nutrient-rich, disease-fighting powerhouse. Chronimed Publishing, Minneapolis, MN. 194 pp.

Hamlyn. 2001. The bean book: over 70 recipes using beans and other pulses. Hamlyn, London, UK. 144 pp.

Heebner, L. 1996. Calypso bean soup: and other savory recipes featuring heirloom beans from the West. Collins, San Francisco, CA. 100 pp.

Hériteau, J. 1978. The complete book of beans. Hawthorn Books, New York, NY. 194 pp.

Horsley, J. 1982. Bean cuisine. Prism Press, Stable Court, UK. 89 pp.

Idaho Bean Commission. 1982. A collection of Idaho's favorite bean recipes. Idaho Bean Commssion, Boise, ID. 101 pp.

Karn, M. 1991. Beans, beans: the musical fruit cookbook. 2nd edition. Ozark Postcard Publishers, Kimberling City, MO. 93 pp.

Mallos, T. 1980. The bean cookbook. Windward, Leicaster, UK. 128 pp.

Mayes, K., and Gottfried, S. 1992. Boutique bean pot: exciting bean varieties in superb new recipes. Woodbridge Press, Santa Barbara, CA. 208 pp.

Michigan Bean Commission. 1995*a*. Old favorites: time-honored and taste-tested recipes featuring Michigan dry beans. Michigan Bean Commission, St. Johns, MI. 17 pp.

Michigan Bean Commission. 1995*b*. Vegetarian favorites: a delicious collection of vegetarian recipes featuring Michigan dry beans. Michigan Bean Commission, St. Johns, MI. 29 pp.

Miller, A. 1997. The bean harvest cookbook. Taunton Press Newtown, CT. 169 pp.

Nebraska Dry Bean Growers Association. 1984. Nebraska bean cookbook. Nebraska Dry Bean Growers Association, Scottsbluff, NE. 56 pp.

Smith, W.J. 1991. The Little Mountain bean bible cookbook: over 1000 bean recipes. Little Mountains Publications, Bristol, VA. 404 pp.

White, S.C. 2004. The daily bean: 175 easy and creative bean recipes for breakfast, lunch, dinner, and...yes, dessert!: a cookbook. Lifeline Press, Washington, DC. 217 pp.

Upsom, N. 1982. The bean cookbook: dry legume cookery. Pacific Search Press, Seattle, WA. 153 pp.

BEET, SUGAR BEET, AND SWISS CHARD

Family: Chenopodiaceae (goosefoot family)

A. Beet and Sugar Beet

Names

Scientific name: *Beta vulgaris* L. subsp. *vulgaris*

- The English word "beet" is derived through the Old English *bete* from the Latin word for beet, *beta*. Some authorities, including Columella, a 1st century AD Roman who wrote about agriculture, have rather dubiously attributed the use of the Greek letter beta (ß) for the beet to a resemblance of the seed to this second letter of the Greek alphabet. (Note that beetroot was not available in ancient Greece and Rome, just foliage beets.)
- Mangel-wurzels or mangels are large varieties of beets used as fodder. "Mangel-wurzel" is from the German *mangold*, beet, and *wurzel*, root; mangel is short for mangel-wurzel.
- The genus name *Beta* is based on the Latin word for beet, *beta*.
- *Vulgaris* in the scientific name is Latin for common or ordinary.

Plant Portrait

The major types of beet include the garden beet, sugar beet, and fodder beet, all of which have enlarged underground storage organs, and are discussed in the following. Swiss chard, grown for its edible leaves and leaf stalks, is also a type of beet.

Beets are a relatively recent crop. Whereas Swiss chard, a close relative of the beet, was domesticated by classical Greek times, beetroot was not reliably reported until the 12th century. The wild sea beet, native to seashores of Eurasia and North Africa, is the ancestor of the modern beet. Through time, the root was selected from a long tapered form to a rounded shape. Beets can be grown as annuals, biennials, and perennials, but in North America, they are generally raised as annuals. The beet plant can flower in its first season, but this drains the edible parts of the plant of food reserves and makes the root woody and unpalatable. Cultivated varieties of beet have been selected to grow as biennials, so that premature flowering does not occur in the first year of growth. If permitted to grow for a second year, the plants flower and produce seeds. Beets produce an aggregated cluster of seeds called a seed ball, which makes planting difficult. However, so-called "monogerm" varieties have been bred, which produce seeds singly, making them much easier to plant. The beet has an enlarged root, consisting of a series of alternating bands of conducting and storage tissue, which are actually growth rings. Beetroots vary in color from red–purple, red, and pink to yellow and white. High temperatures result in poor coloration. The staining power of beet juice is substantial, and even gloss paint can be indelibly marked.

In the 18th century, procedures to produce sugar from beets were found. By 1801, the first sugar refinery for sugar beets was operating in Europe, and in 1897 in California. Major producers are Russia, Ukraine, Poland, France, Germany, China, and Turkey. Sugar beets are also raised in both the United States and Canada. By selection, the sugar content has been increased from 5% to more than 20%, which is about five times as much as found in most garden beets. Sugar beets are one of the most important industrial crops of the temperate and subtropical zones. Because sugar cane can not be grown in temperate climates, sugar beet is the major domestic source of sugar in many northern countries; indeed, beet sugar accounts for more than a third of the world's production of refined table sugar (sucrose). As with cane sugar, raw sugar from beet is brown and is made white by refining. Impurities not crystallized in the refining process produce molasses, which can be used in manufacturing fusel oil, alcohol, rum, and vinegar. The pulp remaining from the extraction process, and the tops of the plants, are useful as stock feed and as fertilizer.

Fodder beet (also called mangel-wurzel, mangel, and mangold) was developed by the 1700s for use as animal feed. The coarse-grained storage organs are large and frequently weigh more than 1 kg (over 2 pounds); they often rise some distance out of the ground. Colors vary as in the garden beet, but white is most commonly encountered. Fodder beets yield a greater crop per hectare than sugar beets but have less dry matter. The roots can be fed to cattle either whole or chipped. Also, they may be useful for ethanol

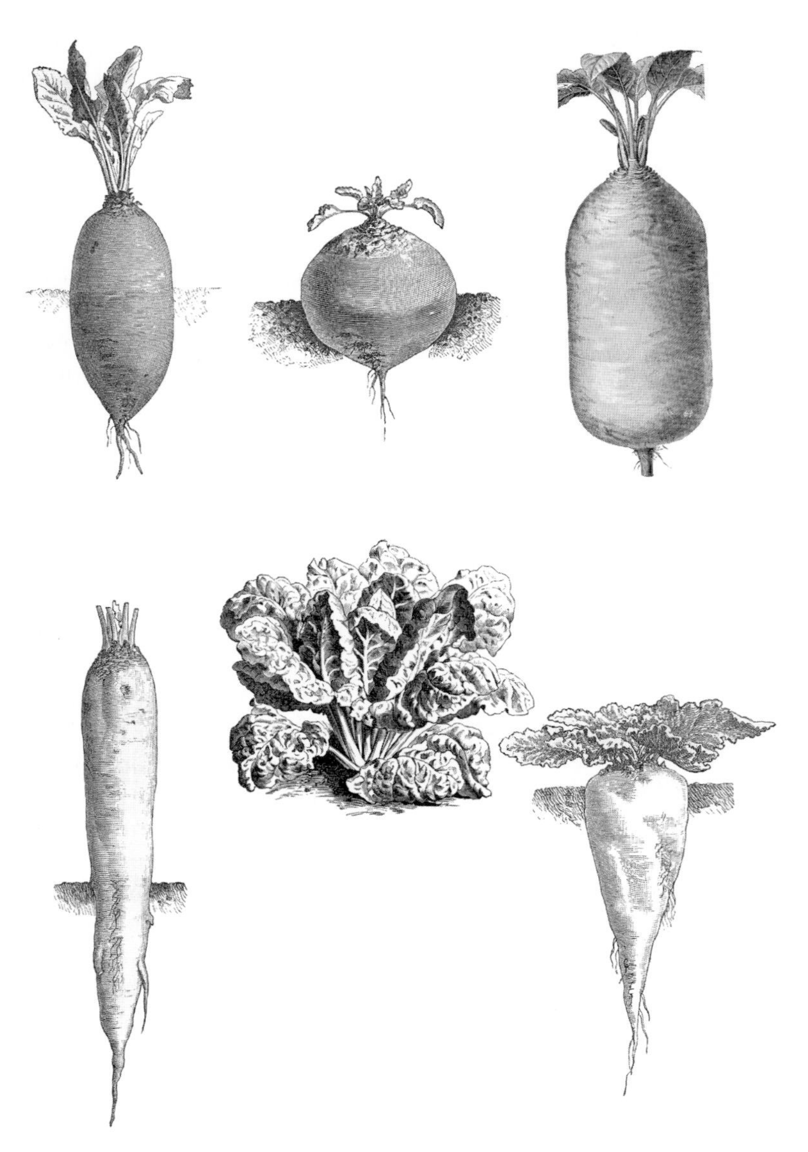

Varieties of beet (*Beta vulgaris* subsp. *vulgaris*). *Center*: variety with edible leaves, White-leaved-beet, from Vilmorin-Andrieux (1885). *Upper left*: fodder beet (for feeding livestock), Yellow Tankard Mangold, from Vilmorin-Andrieux (1885). *Upper center*: another fodder beet, Orange Globe Mangold, from Vilmorin-Andrieux (1885). *Upper right*: the variety Long Golden Yellow Guardian, from *Revue de l'Horticulture Belge et Étrangère* (1911, vol. 37). *Lower left*: Long Yellow beet, from Vilmorin-Andrieux (1885). *Lower right*: White Imperial sugar beet, from Vilmorin-Andrieux (1885).

production, yielding 2½ times more alcohol than corn and 50% more than sugar beets. A good domestic ale has been brewed from the mangold.

Culinary Portrait

Garden beets are a popular salad vegetable, and young beetroots are excellent for pickling. Both the root and tops can be steamed or boiled. The tops are cooked like spinach or Swiss chard. In the home kitchen, beets should be handled carefully before cooking to prevent bruises or nicks that result in bleeding into the water. For the same reason, the roots should also not be peeled before boiling or steaming. Most roots are processed commercially, either whole, sliced, or cubed, and canned. The juice is added as a colorant in many products. Beetroots are used in soups, stews, and casseroles and are added cold to salads. A pleasant wine has been made from beetroots. Beets have even been used as a substitute for coffee, after being dried, roasted, and ground to a powder.

Beets can be purchased bunched with or without tops. When buying beets, avoid very large or elongated roots, which may be fibrous. Choosing roots of uniform size is advisable when cooking whole beets, to promote uniform cooking. Fresh beets may be stored up to 4 weeks in a refrigerator, preferably with leaves or at least 5 cm (2 inches) of stem attached. Leaves are best stored unwashed in the refrigerator in a perforated plastic bag, and used within 5 days.

To retain color and prevent bleeding, beets are best cooked whole and unpeeled (but scrubbed delicately if necessary). The bright red color of beets becomes purple in alkaline ingredients such as baking soda, while salt makes the color paler (salt should be added only at the end of cooking). The color of faded beets can be brightened by adding an acidic ingredient such as lemon juice or vinegar. Gloves can be worn to avoid getting hands stained, or lemon juice may be used to remove stains from hands. Wax paper or newspaper can be spread over the work surface to avoid additional staining.

Canned beets are very popular, but unfortunately, the taste is markedly inferior to that of fresh beets, and experience with canned beets has resulted in many people having learned to dislike this vegetable. Additionally, overcooking of fresh beets takes away their natural juicy firmness and mild sweetness. Roasting in foil, rather than boiling, is one way of emphasizing the natural flavor of beets. After mild scrubbing with a brush, the ends are cut off and the beets are wrapped in foil and roasted at 200 °C (400 °F) until they are easily pierced, usually about 45 minutes. The skin may be removed with a vegetable peeler. The warm beets may then be sliced and flavored—for example, with balsamic vinegar, extra-virgin olive oil, and freshly ground pepper—and combined with greens such as arugula.

Culinary Vocabulary

- The British expression "fresh beetroot" is equivalent to the American "raw beets."
- The basis of the popular Russian and Polish soup *borscht* (borsch, borsht, borshch, borschok, typically pronounced BOHR-scht or BOHR-sh) is a beet and cabbage purée. Meat may be added, as well as a garnish of sour cream, and the soup may be served hot or cold. Borscht was well known to American Jews who emigrated from Russia and Poland in the 1930s, and as a result, the soup became popular in restaurants serving kosher food to a Jewish clientele. The "Borscht Circuit" refers to the summer resort hotels in the Catskill Mountains of New York state where American entertainers perform largely for Jewish audiences. The word "borscht" is a Russian word for "cow parsnip," an original constituent of the soup.
- "Flannel hash" is a New England specialty made by frying together chopped cooked beets, potatoes, onions, and crisp bacon until the ingredients are crusty and brown. Authentic red flannel hash calls for including beets as 85% of the volume.
- "Harvard beets" is a dish of beets cooked in a sweet and sour sauce of vinegar, sugar, and butter thickened with cornstarch. One explanation for the name is that it was created by a Harvard University student. A second explanation holds that the name is based on the deep crimson color of the cooked beets being similar to the color of the Harvard football team jerseys. A third explanation holds that the dish originated in the 17th century in a tavern in England named Harwood's. A Russian émigré, who was familiar with the tavern and the dish, subsequently opened a tavern in Boston with the same name, and his clients, familiar with the nearby Harvard University, mispronounced "Harwood beets" as "Harvard beets."
- *Rozl* or *rossel* (pronounced ross-ell) is a soured, deep-red beet juice employed as a flavorant in Jewish cuisine, especially for soups of the same name. *Rozlfleish* (pronounced row'zil-fly'sch) is a sweet-and-sour Jewish pot roast flavored with rozl and sugar with added onions and potatoes.
- "Red beet eggs" is a Pennsylvania Dutch dish prepared by steeping eggs in the liquid from beets pickled in vinegar, water, and brown sugar.
- "Black death vodka" is a Belgian-made vodka of Icelandic origin, prepared entirely from sugar beet.

Curiosities of Science and Technology

- In the past, various medicinal properties were attributed to beets. The sugar beet was said to be good for headache and all brain problems. The juice of red beets was claimed to be beneficial for jaundice and toothaches. Beet juice was recommended for anemia, yellow jaundice, and toothache. It was also put into the nostrils to clear ringing ears. Beet juice in vinegar was said to rid the scalp of dandruff, and was recommended to prevent falling hair.
- Russian peasants buried beets embedded with mosquitoes and flies in a ceremony meant to relieve them of insect bites. This odd procedure appears related to the use at the time of beet juice as an insect repellent.
- In Russia, beets were widely used as a rouge.

- Before pumpkins became common in Europe, jack-o'-lanterns in Ireland and Scotland were carved out of turnips, potatoes, and beets.
- Napoleon Bonaparte (1769–1821) instituted measures that promoted the beet sugar industry in France, following disruption of the supply of cane sugar from the West Indies, caused by the British naval blockade. This was the first major support provided by a government to promote the development of a new agricultural crop. The major beneficiary was Benjamin Delessert (1773–1847), a French industrialist who in 1812 developed the first commercially successful process to extract sugar from sugar beets. He is commemorated by the Boulevard Delessert in Paris.
- The Pennsylvania Dutch used beet juice to tint and flavor hard-boiled eggs.
- Fodder beets weighing over 50 kg (110 pounds) have been grown.
- The expression used by beet farmers "plant late to make red beets redder" is based on the fact that extra time in the field in later summer or early fall, when temperatures are cooler, tends to increase the intensity of color.
- Sugar beet is a temperate-climate crop, while sugar cane requires tropical to subtropical conditions. Sugar is a rather unique plant food in the respect that it originates both from temperate and tropical regions. Relatively few countries extend across a sufficient range of climatic zones to be able to produce both cane and beet sugar. Such countries include the United States and China.
- Utah declared the sugar beet to be its official "State Historical Vegetable" in 2002. Beginning in the 19th century and lasting for almost 100 years, the sugar beet industry was very important in Utah.
- Eating beets can result in scarlet urine and stools—harmless, although the bloody color may be alarming. Beet-colored urine is facetiously known in medical circles as "beeturia."
- The Chioggia (chiogga) or Candy Cane beet is a most unusual variety with concentric rings of red and white.

Key Information Sources

(Additional references on sugar beet are in SUGAR CANE.)

Bock, T.S. de. 1986. The genus *Beta*: domestication, taxonomy and interspecific hybridization for plant breeding. Acta Hortic. **182**: 335–343.

Boswell, V.R. 1954. Growing table beets. U.S. Government Printing Office, Washington, DC. 4 pp.

Coons, G.H. 1954. The wild species of *Beta*. Proc. Am. Soc. Sugar Beet Technol. **8**: 142–147.

Ecke, W., and Michaelis, G. 1990. Comparison of chloroplast and mitochondrial DNA from five morphologically distinct *Beta vulgaris* cultivars: sugar beet, fodder beet, beet root, foliage beet, and Swiss chard. Theor. Appl. Genet. **79**: 440–442.

Ford-Lloyd, B.V. 1995. Sugarbeet, and other cultivated beets. *In* Evolution of crop plants. 2nd edition. *Edited by* J. Smartt and N.W. Simmonds. Longman Scientific & Technical, Burnt Mill, Harlow, Essex, UK. pp. 35–40.

Ford-Lloyd, B.V., and Hawkes, J.G. 1986. Weed beets: their origin and classification. Acta Hortic. **182**: 399–401.

Ford-Lloyd, B.V., and Williams, J.T. 1975. A revision of *Beta* section *Vulgares* (Chenopodiaceae), with new light on the origin of cultivated beets. Bot. J. Linn. Soc. **71**: 89–102.

Geyt, J.P.C. van, Lange, W., Oleo, M., and Bock, T.S.M. de. 1990. Natural variation within the genus *Beta* and its possible use for breeding sugar beet: a review. Euphytica, **49**: 57–76.

Her Majesty's Stationery Office. 1970. Sugar beet cultivation. 3rd edition. H.M.S.O., London, UK. 114 pp.

ICON Health Publications. 2003. Beets: a medical dictionary, bibliography, and annotated research guide to internet references. ICON Health Publications, San Diego, CA. 140 pp.

Longden, P.C. 1993. Weed beet: a review. Aspects Appl. Biol. (U.K.), **35**: 185–194.

McGinnis, R.A. (*Editor*). 1982. Beet-sugar technology. 3rd edition. Beet Sugar Development Foundation, Fort Collins, CO. 855 pp.

Mukhopadhyay, A.N. 1987. CRC handbook of diseases of sugar beet. CRC Press, Boca Raton, FL. 177 pp.

Nathan, R.A. 1978. Fuels from sugar crops: systems study for sugarcane, sweet sorghum, and sugar beets. Technical Information Center, U.S. Dept. of Energy, Oak Ridge, TN. 137 pp.

Oyen, L.P.A. 1993. *Beta vulgaris* L. *In* Plant resources of South-East Asia, 8, vegetables. *Edited by* J.E. Siemonsma and K. Piluek. Pudoc Scientific Publishers, Wageningen, Netherlands. pp. 97–101.

Pink, D.A.C. 1993. Beetroot, *Beta vulgaris subsp.* vulgaris L. *In* Genetic improvement of vegetable crops. *Edited by* G. Kalloo and B.O. Bergh. Pergamon Press, New York, NY. pp. 473–477.

Schalit, M. 1970. Guide to the literature of the sugar industry. An annotated bibliographical guide to the literature on sugar and its manufacture from beet and cane. Elsevier, Amsterdam, Netherlands. 184 pp.

Scott, R.K., and Cooke, D.A. 1993. (*Editors*). The sugar beet crop: science into practice. Kluwer Academic Publishers, Dordrecht, Netherlands. 675 pp.

Van Geyt, J.P.C., Lange, W., Oleo, M., and De Bock, T.S.M. 1990. Natural variation within the genus *Beta* and its possible use for breeding sugar beet: a review. Euphytica, **49**: 57–76.

Vukov, K. 1977. Physics and chemistry of sugar-beet in sugar manufacture. Elsevier Scientific Publishing Co., New York, NY. 594 pp.

Wayland, B. 2003. The beet goes on. Harrowsmith Country Life, **27**(**173**): 24, 38–39.

Whitney, E.D., and Duffus, J.E. (*Editors*). 1986. Compendium of beet diseases and insects. American Phytopathological Society, St. Paul, MN. 76 pp.

Specialty Cookbooks

Bent County Democrat. 1924. Official beet sugar cook book, compiled by users of beet sugar. Wick & Clark Publishers, Las Animas, CO. 44 pp.

Blanksteen, J. 1974. Nothing beets borscht: Jane's Russian cookbook. Atheneum, New York, NY. 242 pp.

B. Swiss Chard

Names

Scientific name: *Beta vulgaris* L. subsp. *cicla* (L.) W.D.J. Koch

- "Chard," like the word "cardoon" (*Cynara cardunculus* L., another vegetable), is based on the French *carde*, from the Latin *carduus*, thistle. For both chard and cardoon, the enlarged midrib of the edible leaf, reminiscent of the thistle, is the basis of the word.
- The name "chard" is also sometimes given to blanched shoots of globe artichoke (*Cynara scolymus*, which see), to young flowering shoots of salsify (*Tragopogon porrifolius* L.), and to cardoon. The name "seakale beet" sometimes is used for Swiss chard, but this usage may produce confusion with the true sea kale (*Crambe maritima* L.).
- By the 19th century, seed catalogues were adding "Swiss" to the name "chard," presumably to distinguish it from other types of plant called chard. It is not clear why the term "Swiss" was chosen, although it has been suggested that it is because the Dutch associated the vegetable with Switzerland.
- "Silver beet" is another name for the Swiss chard (especially in Australia and New Zealand), and reflects the close relationship with the beet, as well as the silvery color of the leaves of Swiss chard.
- The names "rhubarb chard," "ruby chard," and "spinach rhubarb" are based on forms with scarlet leaf stalks.
- A related form with edible pulpy leaves but without swollen leafstalks is called leaf beets, beetgreen, or spinach beets. Swiss chard and spinach beets are often referred to as "foliage beets." Forms with unswollen leafstalks have been separated as *B. vulgaris* subsp. *cicla* var. *cicla* L., in contrast with Swiss chard, which is then called *B. vulgaris* subsp. *cicla* var. *flavescens* (Lam.) Lam. & DC.
- The best-known cultivar of Swiss chard, Lucullus, recalls the great Roman general Lucius Licinius Lucullus (about 117–57 BC), who very likely consumed the vegetable, since it was available in ancient Rome.
- See BEET AND SUGAR BEET, above, for a discussion of the genus name *Beta* and the word *vulgaris* in the scientific name.
- *Cicla* in the scientific name *B. vulgaris* L. subsp. *cicla* is derived from *sicula*, which refers to Sicily, one of the places where chard first grew.

Plant Portrait

Chard or Swiss chard is a biennial, cultivated as an annual. It is grown for its edible leaves (i.e., leaf blades) and leaf stalks, and lacks the enlarged edible root of its close relative the garden beet. The plant has swollen leaf stalks up to 60 cm (2 feet) in length, topped by leaf blades up to 15 cm (6 inches) wide. The midrib of the leaf blade is succulent, like the stalk. It is believed that wild beet leaves were harvested as a potherb several thousand years ago in the Mediterranean area, resulting in the selection of forms with especially edible leaves and leaf stalks. Some forms have flat leaves, others have curly foliage. Swiss chard is a very old vegetable. The ancient Greeks recognized red, white, and "black" (intense deep green) forms, and the Romans were also well acquainted with this vegetable. Swiss chard was introduced to North America by Europeans in 1806. In North America, it is grown mainly in home gardens, rather than commercially. Although less popular than its cousin, beetroot, Swiss chard is often consumed, its leaf stalks serving the role of asparagus, a much more seasonal vegetable. Swiss chard is a minor crop in North America. California, Florida, New Jersey, and Texas are the major regions of production. Canadian production, intended for local markets, is limited to small areas. Chard does not can well, and generally, the fresh product is found in commerce.

Stunningly beautiful forms of Swiss chard, available in a rainbow of colors, are now commonly cultivated as ornamentals rather than as vegetables.

Silvery Sea Kale chard

White Curled Swiss chard

Varieties of Swiss chard (*Beta vulgaris* subsp. *cicla*), from Vilmorin-Andrieux (1885).

Culinary Portrait

Although not well known, Swiss chard is considered by many to be one of the finest of vegetables. The stalks are cooked like celery and taste best after steaming. They can also be used in stir-fries. The leaf blades (often with the leaf stalks attached) are cooked like spinach (the taste is said to be similar), using only the water left on the leaves after washing. In fact, a combination of spinach and Swiss chard has a pleasant flavor. Leaves (leaf blades) and stalks are best cooked separately when the stalks are wider than 1.3 cm (1/2 inch). Stalks may be steamed for 8–15 minutes, while the leaf blades are steamed for 5–8 minutes. Young tender chard leaves can be eaten raw, adding a beet-like flavor to salads and sandwiches. Chard also is a good addition to soups and ragouts. Chard leaves (i.e., leaf blades) freeze well after blanching, but the stalks become soggy and rather unappealing. Numerous recipes for preparing chard are available on the Web. Note that the stalks and leaves of Swiss chard and leaf beets contain appreciable amounts of calcium oxalate, which can be toxic when consumed to excess. Unwashed chard can be stored in a perforated plastic bag in a refrigerator for up to 4 days.

Culinary Vocabulary

- As a result of the popularity of the edible swollen leaf stalk (i.e., petiole) of chard, the term "chard" is sometimes applied to any edible leaf stalk of leafy green vegetables. (The term "midrib" is sometimes also used in this sense, although this is incorrect, since the midrib refers to the main vein of the leaf blade.)

Curiosities of Science and Technology

- In ancient times, it was commonly believed that the shape and color of a plant were indicative of its medical virtues. Because of their blood-red color, early foliage beets were believed to be good for the blood and were eaten to improve "weak blood." The classical Greeks used foliage beets to "cool" blood.
- Swiss chard and the garden beet can be crossed to produce seeds, but since these biennial plants are grown as annuals and harvested in the first year, this is unlikely to occur in gardens. These plants produce wind-borne pollen in their second year, which can be carried over long distances, so seed producers need to ensure that their plants are well isolated from possible sources of contaminating

pollen. Hybrid seed would result in plants that are not useful either for their roots or their aboveground parts.

Key Information Sources

(For additional references, see Beet and Sugar Beet above.)

Acton, P. 1999. Chard: delicious, nutritious, and lovely. Org. Gard. **46**(**3**): 30–33.

Ali, I., Shah, F.H., Raoof, A., Ali, A., and Ejaz, M. 2003. Effect of N levels and sowing methods on the growth and yield of Swiss chard cv. Kalam selection. Balochistan J. Agric. Sci. (Pakistan), **4**: 11–15.

Andersen, C.R. 2002. Chard. Cooperative Extension Service, University of Arkansas, Fayetteville, AR. 2 pp.

Banadyga, A.A. 1977. Greens or "potherbs"—chard, collards, kale, mustard, spinach, New Zealand spinach. Agric. Inf. Bull. U.S. Dep. Agric. **409**: 163–170.

Bartsch, D., Brand, U., Morak, C., Pohl-Orf, M., Schuphan, I., and Ellstrand, N.C. 2001. Biosafety of hybrids between transgenic virus-resistant sugar beet and Swiss chard. Ecol. Appl. **11**: 142–147.

Gil, M.I., Ferreres, F., and Tomas-Barberan, F.A. 1998. Effect of modified atmosphere packaging on the flavonoids and vitamin C content of minimally processed Swiss Chard (*Beta vulgaris* subspecies *cycla*). J. Agric. Food Chem. **46**: 2007–2012.

Gonzalez-Mancebo, E., Alfaya, T., Pulido, Z., Leon, F., Cuevas, M., and de la Hoz, B. 2000. Swiss chard-induced asthma. Allergy (Copenhagen), **55**: 511–512.

Hoz, C.B. de la, Fernandez, R.M., Quirce, S., Cuevas, M., Fraj, J., Davila, I., Igea, J.M., and Losada, E. 1991. Swiss chard hypersensitivity clinical and immunologic study. Ann. Allergy, **67**: 487–492.

Jarmin, M.L., and Thorton, R.E. 1985. Table beet & Swiss chard seed production in the Pacific Northwest. Oregon State University Extension Service, Corvallis, OR. 4 pp.

Kugler, F., Stintzing, F.C., and Carle, R. 2004. Identification of betalains from petioles of differently colored Swiss chard (*Beta vulgaris* L. ssp. *cicla* [L.] Alef. cv. Bright Lights) by high-performance liquid chromatography–electrospray ionization mass spectrometry. J. Agric. Food Chem. **52**: 2975–2981.

Mayberry, K., Savio, Y., and Myers, C. 1991. Swiss chard. Crop Sheet SMC-033. *In* Specialty and minor crops handbook. *Edited by* C. Myers. The Small Farm Center, Division of Agriculture and Natural Resources, University of California, Oakland, CA. 2 pp.

Moreira, M. del R., Roura, S.I., and Valle, C.E. del. 2003. Quality of Swiss chard produced by conventional and organic methods. Lebensm.-Wiss. Technol. **36**: 135–141.

Pokluda, R., and Kuben, J. 2002. Comparison of selected Swiss chard (*Beta vulgaris* ssp. *cicla*) varieties. Hortic. Sci. UZPI (Czech Republic), **29**(**3**): 114–118.

Pyo, Y.H., Lee, T.C., Logendra, L., and Rosen, R.T. 2004. Antioxidant activity and phenolic compounds of Swiss chard (*Beta vulgaris* subspecies *cycla*) extracts. Food Chem. **85**: 19–26.

Roberts, C.R. 1972. Growing Swiss chard. Kentucky University Cooperative Extension Service, Kentucky University Dept. of Horticulture, Lexington, KY. 1 p.

Roura, S.I., Davidovich, L.A., and Valle, C.E. del. 2000*a*. Quality loss in minimally processed Swiss chard related to amount of damaged area. Lebensm.-Wiss. Technol. **33**: 53–59.

Roura, S.I., Davidovich, L.A., and Valle, C.E. del. 2000*b*. Postharvest changes in fresh Swiss chard (*Beta vulgaris*, type *cycla*) under different storage conditions. J. Food Qual. **23**(**2**): 37–147.

Rutzke, M., Gutenmann, W.H., Williams, S.D., and Lisk, D.J. 1993. Cadmium and selenium absorption by Swiss chard grown in potted composted materials. Bull. Envir. Contam. Toxicol. **51**: 416–420.

Zhang, P., Ganje, T.J., Page, A.L., and Chang, AC. 1988. Growth and uptake of selenium by Swiss chard in acid and neutral soils. J. Environ. Qual. **17**: 314–316.

BLACK-EYED PEA (COWPEA)

Family: Fabaceae (Leguminosae; pea family)

Names

Scientific name:

- *Vigna unguiculata* (L.) Walp. subsp. *unguiculata* [*V. sinensis* (L.) Savi ex Hassk.]
- The name "black-eyed pea," recorded in 1738, is based on varieties that have white seeds with a black spot. The dark spot occurs at the "hilum" of the seed (the point where the seed is attached to the pod), making the seed look like a dark eye. While the hilum is generally black, it may be brown, dark crimson, or red, depending on variety.
- Black-eyed peas are so commonly grown in the southern United States that they are often simply called "peas." Indeed, more often than not, in the Deep South, "peas" means black-eyed peas, not the common pea (*Pisum sativum*). Black-eyed peas are also often called "beans" (sometimes black-eyed beans) in the southern United States.
- It is frequently asked whether the black-eyed pea is a pea or a bean. Since both "pea" and "bean" are imprecise terms, which are both applied to the black-eyed pea, the question is semantic. A black-eyed pea is both a pea and a bean.
- The name "cowpea" originated in the United States, and is recorded as appearing first in print in 1798. This name is common in Britain, where the name "black-eyed pea" is rarely used.
- The name "crowder" (or crowder pea) was used for early varieties called "crowders," which had seeds crowded so closely in the pod that the ends of the seeds were flattened.
- "Southern pea" reflects the frequent cultivation of black-eyed peas in the southern United States.
- The black-eyed pea was once known by the French name *mogette*, a diminutive of *monge*, meaning a nun. It was so named because the black eye in the center seemed reminiscent of a nun's headdress.
- See MUNG BEAN for information on the genus name *Vigna*.
- *Unguiculata* in the scientific name is Latin for "with a small nail" or "with a small claw," a technical way of expressing that the petals of the flowers have small stalks.

Plant Portrait

The black-eyed pea or cowpea is an annual tropical or subtropical herb growing to 75 cm (2½ feet) or more in height, but sometimes bending over considerably. The pods may be up to 30 cm (1 foot) long, straight, curved, or coiled. The seeds are 2–12 mm (1/16–1/2 inch) long, and may be smooth or wrinkled, red, black, brown, green, white, solid in color or spotted. The seeds may be kidney-shaped, globular, or somewhat rectangular. The plant has been used since ancient times in Africa and Asia and is now widely cultivated for human food and animal feed and as a green manure, hay, or silage. It originated in tropical Africa where a wild form still occurs. Essentially wild forms are still grown in Ethiopia. The center of domestication was in west-central or south-central Africa, in the region from Nigeria to Zaire. The time of domestication is uncertain, but was likely thousands of years ago. It is thought that the history of the plant dates back 5000–6000 years ago in West Africa when it was closely associated with the cultivation of sorghum and pearl millet. This legume was known in Greece of 300 BC, and by the end of the Roman Era was a minor crop in southern Europe. Black-eyed peas are sensitive to cold and are killed by frost, and so are primarily grown in warm regions.

After the common bean (*Phaseolus vulgaris*), the cowpea is the most important pulse crop grown in Africa (pulses are edible-seeded species of the pea family). The largest acreage is in Africa, with Nigeria and Niger predominating, but Brazil, West Indies, India, the United States, Burma, Sri Lanka, Yugoslavia, and Australia all have significant production. Cowpea is widely known as a crop of the poor. It is grown by tens of millions of smallholder farms in the drier zones of Africa, in a great arc from Senegal eastward to Sudan and Somalia and southward to Zimbabwe, Botswana, and Mozambique. Two hundred million children, women, and men consume cowpea often, even daily when it is available.

The black-eyed pea was introduced to Jamaica by Spanish slave traders about 1765, who carried them as part of the ship stores to feed their tragic cargo. The slave industry also brought black-eyed peas to the

United States by way of the West Indies, particularly to the territory of Louisiana, where as a result many Creole dishes are made with them.

In North America, black-eyed peas are cultivated mostly in the Cotton Belt of the southeastern United States, mainly as a forage crop for livestock and as a green manure (i.e., a crop that is plowed under to enrich the soil), but also for green pods and green seeds for human consumption. Black-eyed peas are also raised in California for dry beans.

Culinary Portrait

In North America, the black-eyed pea is used primarily as a dry bean. Black-eyed peas are particularly popular in many southern US "soul foods," including the dish "Hoppin' John" (see Pigeon Pea). In the United States, black-eyed peas are often prepared in combination with pork. The seeds are very flavorful, and are used in soups, salads, casseroles, and fritters. In Asia, the dried seeds are commonly eaten whole, or cooked in soups and stews, or prepared as purées. Also in Asia, the young pods are consumed as a vegetable. Cooking is known to improve the nutritional value of the seeds, possibly because certain toxins (trypsin inhibitors) that are present are deactivated by heat. The seeds are cooked for about 1 hour, and if overcooked even slightly, they tend to become mushy. In a pressure cooker at 15 pounds of pressure, presoaked seeds may be ready in only 10 minutes, while unsoaked seeds require 10–20 minutes. In parts of Africa, the tuberous roots are also eaten. Sometimes, the seeds are germinated and consumed as sprouts.

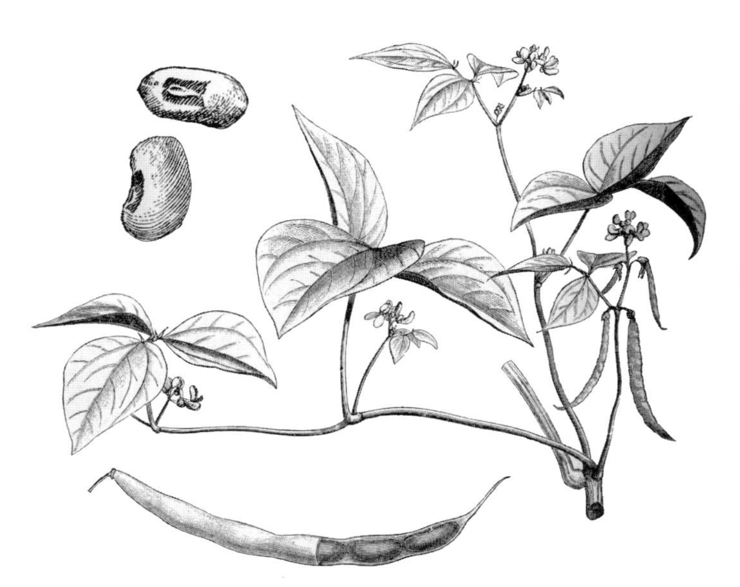

Black-eyed pea (*Vigna unguiculata* subsp. *unguiculata*), from Engler and Prantl (1889–1915). Seeds at upper left are from Bailey (1900–1902).

The tender leaves of cowpea plant are nutritious, although normally not consumed by humans in North America. In many areas of Africa and occasionally in Asia, however, fresh leaves are regularly harvested and consumed. The leaves contain 25% protein as a percentage of dry weight, and the protein quality is high. The US National Aeronautics and Space Administration was so impressed with the nutritional potential of cowpea leaves that they listed cowpeas as having potential in future space stations as food for astronauts.

Culinary Vocabulary

- "Texas caviar" is a dish (variously called a dip, relish, or salsa) of black-eyed peas and other vegetables, such as green and red peppers, and green onions. Numerous recipes are available on the Web.
- *Bollitos* (pronounced boh-yee-TOHS) is a Cuban dish made by deep-frying rolled balls of black-eyed peas and garlic.

Curiosities of Science and Technology

- Thomas Jefferson (1743–1826), the 3rd American president, is widely considered to have been the most significant individual who introduced new crops to the United States. He raised black-eyed peas at Monticello, his well-known estate in Virginia.
- In Africa, the Hausa and Edo tribes use cowpeas medicinally, one or two seeds ground and mixed with soil or oil to treat stubborn boils.
- Athens, Texas, claims to be the "Black-Eyed Pea Capital of the Planet." A Black-Eyed Pea Jamboree is held every July, featuring many foods using the legume—even martinis made with black-eyed peas soaked in crushed chiles.
- In 1988, Oklahoma declared that its "state meal" should consist of a choice of fried okra, squash, cornbread, barbecue pork, biscuits, sausage and gravy, grits, corn, strawberries, chicken-fried steak, pecan pie and black-eyed peas.
- A special part of African-American hoodoo (voodoo) practice is the mojo bag—a bag usually made of flannel, which contains one or more magical items. There are numerous combinations of objects that are placed in a mojo bag, depending on its purpose. For a wish to come true, a red flannel bag is anointed with Van Van (brand) oil and the following are added: seven wishing "beans" (normally black-eyed peas in the American South where hoodoo has been practised), a rabbit foot,

and a piece of parchment upon which the wish has been written in red ink. Some people write a different wish for each of the seven "beans."

Key Information Sources

Creighton Miller, J.C., Jr., Miller, J.P., and Fery, R.L. (*Editors*). 1990. Cowpea research: a U.S. perspective: proceedings of the second southernpea (cowpea) workshop, Feb. 6, 1989. Nashville, Tennessee. Texas Agricultural Experiment Station, Texas A & M University System, College Station, TX. 81 pp.

Fery, F.L. 2002. New opportunities in *Vigna*. *In* Trends in new crops and new uses. *Edited by* J. Janick and A. Whipkey. ASHS Press, Alexandria, VA. pp. 424–428.

Fery, R.L. 1980. Genetics of *Vigna*. Hortic. Rev. **2**: 311–394.

Fery, R.L. 1981. Cowpea production in the United States. HortScience, **16**: 473–474.

Fery, R.L. 1990. The cowpea: production, utilization, and research in the United States. Hortic. Rev. **12**: 197–222.

Hall, A.E., Cisse, N., Thiaw, S., et al. 2003. Development of cowpea cultivars and germplasm by the Bean/Cowpea CRSP. Field Crops Res. **82**: 103–134.

Ligon, L.L. 1958. Characteristics of cowpea varieties (*Vigna sinensis*). Oklahoma Agricultural Experiment Station, Stillwater, OK. 47 pp.

Lush, W.M., and Evans, L.T. 1981. Domestication and improvement of cowpea. Euphytica, **30**: 579–587.

Ng, N.Q. 1995. Cowpea. *Vigna unguiculata* (Leguminosae—Papilionoideae). *In* Evolution of crop plants. 2nd edition. *Edited by* J. Smartt and N.W. Simmonds. Longman Scientific & Technical, Burnt Mill, Harlow, Essex, UK. pp. 326–332.

Ng, N.Q., and Monti, L.M. (*Editors*). 1990. Cowpea genetic resources: contributions in cowpea exploration, evaluation and research from Italy and the International Institute of Tropical Agriculture. International Institute of Tropical Agriculture, Ibadan, Nigeria. 200 pp.

Nwokolo, E., and Ilechukwu, S.N. 1996. Cowpea (*Vigna unguiculata* (L.) Walp.). *In* Food and feed from legumes and oilseeds. *Edited by* J. Smartt and E. Nwokolo. Chapman and Hall, London, UK. pp. 229–242.

Oghiakhe, S. 1997. Trichomes and resistance to major insect pests in cowpea, *Vigna unguiculata* (L.) Walp.: a review. Discovery Innovation, **9**: 173–178.

Rawal, K.M. 1975. Natural hybridization among wild, weedy and cultivated *Vigna unguiculata* (L.) Walp. Euphytica, **24**: 699–707.

Singh, B.B. (*Editor*). 1997. Advances in cowpea research: proceedings of the second world cowpea research conference, held at Accra, Ghana, 3–7 Sept. 1995. International Institute of Tropical Agriculture, Ibadan, Nigeria. 375 pp.

Singh, S.R., and Allen, D.J. 1979. Cowpea pests and diseases. International Institute of Tropical Agriculture, Ibadan, Nigeria. 113 pp.

Singh, S.R., and Rachie, K.O. (*Editors*). 1985. Cowpea research, production and utilization. John Wiley & Sons, Chichester, Essex, UK. 460 pp.

Som, M.G., and Hazra, P. 1993. Cowpea, *Vigna unguiculata* (L.) Walp. *In* Genetic improvement of vegetable crops. *Edited by* G. Kalloo and B.O. Bergh. Pergamon Press, New York, NY. pp. 339–354.

Steele, W.M., Allen, D.J., and Sumerfield, R.J. 1985. Cowpea (*Vigna unguiculata* (L.) Walp). *In* Grain legume crops. *Edited by* R.J. Summerfield and E.H. Roberts. Collins, London, UK. pp. 520–583.

Summerfield, R.J., Huxley, P.A., and Steele, W.M. 1974. Cowpea (*Vigna unguiculata* (L.) Walp). Field Crop Abstr. **27**: 301–312.

Whittle, E.C. 1984. The heat-loving peas. Org. Gard. **31**: 74–77.

Wien, H.C., and Summerfield, R.J. 1984. Cowpea (*Vigna unguiculata* (L.) Walp). *In* The physiology of tropical field crops. *Edited by* P.R. Goldsworthy and N.M. Fisher. John Wiley & Sons, Chichester, UK. pp. 353–383.

Specialty Cookbooks

Athens (Texas) Chamber of Commerce. 1971/1972. Reci peas from the Black-eyed Pea Jamboree, Athens, Texas. Charles, the Printer, Athens, TX. 76 pp. [1978 version, 158 pp., sponsored by Jaycee-ettes.]

Carver, G.W. 1925. How to grow the cow pea and forty ways of preparing it as a table delicacy. Tuskegee Normal and Industrial Institute, Tuskegee, AL. 22 pp. [See PEANUT for information about George Washington Carver.]

Dovlo, F.E., Williams, C.E., and Zoaka, L. 1976. Cowpeas: home preparation and use in West Africa. International Development Research Centre, Ottawa, ON. 96 pp.

Jones, B.C. 1991. Black-eyed peas to pralines: Texas style. Barbara C. Jones, Bonham, TX. 158 pp.

Khon Kaen University. 1981. Cowpea dishes: the new development for home use. Faculty of Agriculture, Khon Kaen University, Khon Kaen, Thailand. 27 pp.

Langworthy, C.F., and Hunt, C.L. 1917. Use of corn, kafir, and cowpeas in the home. U.S. Dept. of Agriculture, Washington, DC. 15 pp.

Bottle Gourd

Family: Cucurbitaceae (gourd family)

Names

Scientific name: *Lagenaria siceraria* (Molina) Standl. [*Cucurbita siceraria* Molina, *L. leucantha* (Duchesne ex Lam.) Rusby, *L. vulgaris* Ser.]

- "Bottle gourd" is so named because it is a gourd that has been extensively used to make containers, including bottles.
- The bottle gourd is also known as birdhouse gourd, bottle squash, calabash, calabash gourd, Chinese preserving melon, cucuzzi (often denotes a long club-shaped type), Guinea bean, Italian edible gourd, New Guinea bean, spaghetti squash, suzza melon, Tasmania bean, trumpet gourd, white-flowered gourd, and zucca (melon).
- "Gourds" have been defined as members of the Cucurbitaceae, with durable, hard-shelled fruit grown for ornament, utensils, or general interest (a few kinds occasionally called gourds have soft flesh and nondurable rinds). The two species most commonly encountered in North America as gourds are *L. siceraria* and *Cucurbita pepo* L. subsp. *ovifera* (L.) D.S. Decker, the white-flowered and yellow-flowered gourds, respectively. The fruits of the yellow-flowered gourd are never as palatable as those of the white-flowered gourd. The flowers of the white-flowered gourd open at night, whereas those of the yellow-flowered open about daybreak and usually wilt by noon.
- The word "calabash" is from the French *calebasse*, gourd, from Spanish *calabaza*, from Catalan *carabaça*, and perhaps from Arabic *qar'a ybisa*, dried gourd. "Calabash" is also used to denote the fruit of a tropical tree, *Crescentia cujete* L., the calabash tree, whose fruits resemble gourds. The sweetish fruits and seeds have been eaten, but are used primarily, and rather widely, for utensils.
- The Jewish prophet Jonah became furious when a worm gnawed its way into his gourd and it withered (Jonah 4: 6–10). Jonah's fury at the loss of his gourd is commonly interpreted as a parable of his anger with God. "Jonah's gourd" is generally believed to be the castor oil plant (*Ricinus communis* L.), but it has been suggested that it is the bottle gourd. This makes more sense, as the castor oil bean pods are much smaller and indeed poisonous.
- The genus name *Lagenaria* is based on the Latin *lagena*, a bottle.
- *Siceraria* in the scientific name *L. siceraria* is based on the Latin *sicera* (from the Greek *sikera*), a spirituous drink, + *aria*, like, i.e., "like an alcoholic beverage."

Plant Portrait

The bottle gourd is a vigorous, annual, running or climbing vine reaching up to 10 m (33 feet) in tropical areas. The fruit is green initially, turning whitish or yellowish at maturity. Fruits are very variable, with lengths of 10 cm (4 inches) to more than 2 m (6½ feet), weights ranging up to 50 kg (110 pounds), and such shapes as straight, crooknecked, club-form, dumbbell-form, twisted, coiled, disk-like, and globular. The flesh is white, pulpy, and many-seeded. The fruit may have a sterile (seedless) neck from 8 to 40 cm (3–16 inches) in length and 2.5–5 cm (1–2 inches) in width. Wider necks usually contain seeds, and the neck may have a seed-containing bulge. The long, narrow forms are best for vegetables, while the round types make good containers.

The species probably originated in tropical Africa and was distributed in early times to South America, possibly by floating. Tests have demonstrated that the mature gourds will float for the better part of a year in seawater while the seeds remain viable. Archaeological evidence has shown that the bottle gourd was in Peru around 12 000 BC, Thailand about 8000 BC, Mexico about 7000 BC, and Egypt about 3500 BC. Bottle gourd is widely grown in tropical areas as a vegetable and for useful gourds. This species is adapted to a hot, dry climate and is much more important in tropical and semitropical regions than in North America, where it is rarely grown commercially and is mostly encountered in home gardens as a curiosity.

The hollowed-out gourds have been employed in numerous ways since antiquity. They have been used extensively as containers for holding and carrying water, foods, and other substances. Some gourds make excellent spoons or "dippers" for food and water. Gourds have been turned into musical instruments, pipes, floats for fishnets, birdhouses, and decorative masks and charms. For those who would like to preserve bottle gourds grown in their garden, the following steps should be followed. After

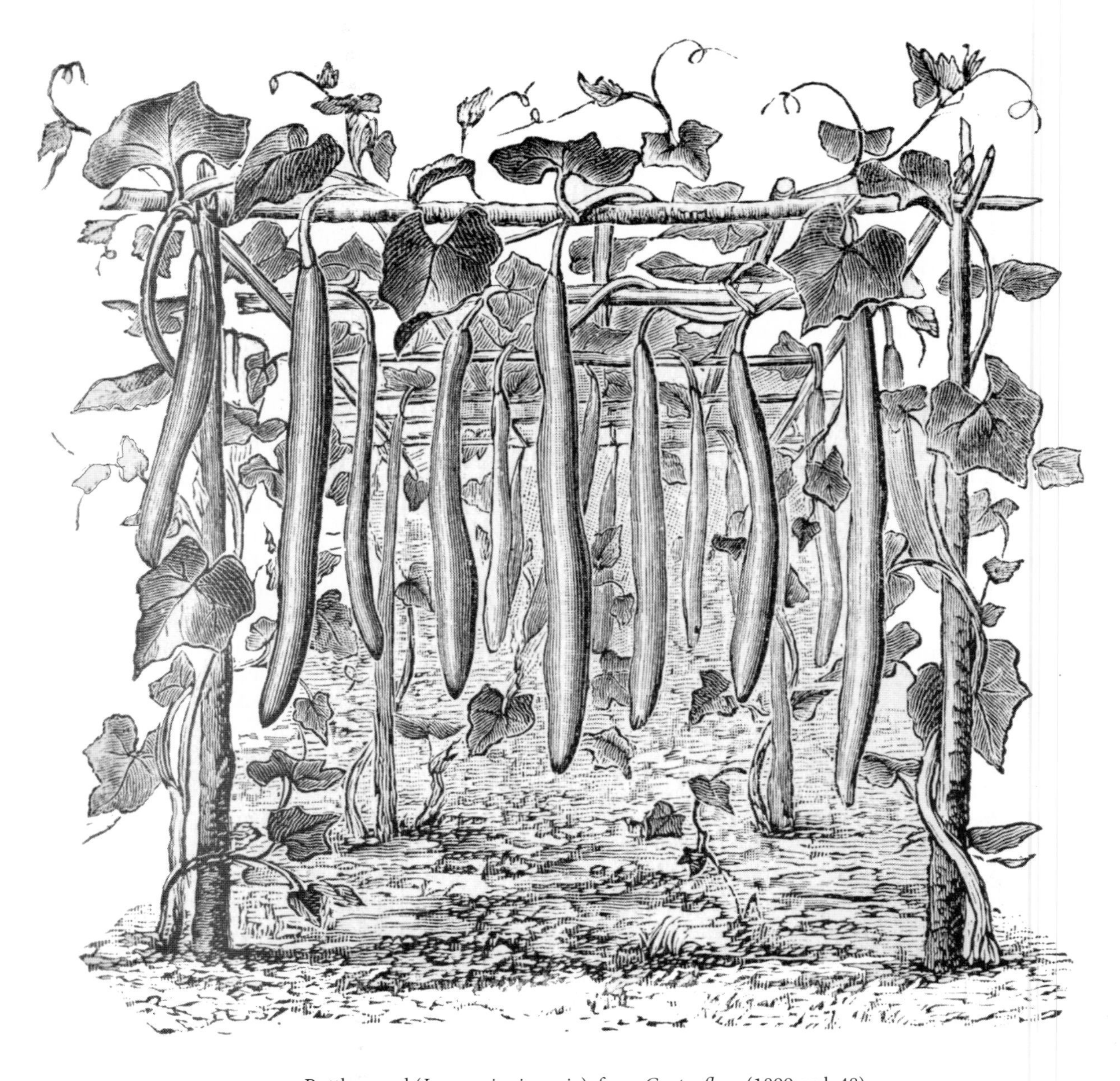

Bottle gourd (*Lagenaria siceraria*), from *Gartenflora* (1899, vol. 48).

harvesting, the gourds should be washed in warm soapy water and rinsed in a strong solution of a nonbleaching disinfectant such as borax. They are best hung from strings in a well-ventilated room for 4–6 weeks, or up to 6 months in some cases. Direct sunlight should be avoided during curing. Ripening may be hastened by piercing the ends of the gourds with a darning needle. Any moldy areas may be washed off, or the gourd, if badly molded, should be discarded. Properly cured bottle gourds should last for years. Once cured, the gourds may be soaked in hot water for several hours to loosen the outer skin and scrubbed with steel wool or fine sandpaper to prepare the surface for coating with a wax paste, shellac, varnish, or paint. The gourds make attractive home decorations. Bottle gourds may be carved into various shapes such as dippers, scoops, storage vessels, and birdhouses. Designs can be cut into the surface or burned on with an electric wood burner.

Culinary Portrait

Young bottle gourd fruits of certain varieties are slightly sweet, tender, free of bitterness, and as nutritious as the popular summer squashes. Other varieties are bitter and may contain poisonous substances. The immature fruits of suitable varieties can be cooked and used in vegetable dishes. The flesh is white, firm, and has an excellent texture and a mild taste. Bottle gourd fruit is best used like summer squash such as zucchini (some varieties are said to be indistinguishable in taste from zucchini). It can be sliced, coated with flour, and fried like eggplant, or boiled, diced, and creamed. The young leaves or immature shoots may also be hand-picked and cooked. In western Africa, the seeds are added to soups or used as a snack food. The immature flesh has been used to make icing for cakes.

Curiosities of Science and Technology

- The bottle gourd is the only cultivated crop known to have existed in both the Old and New Worlds in pre-Columbian times.
- In pre-Columbian Peru, the surgical operation of trepanation (removing a piece of the skull) was performed after injuries. A piece of gourd was sometimes used to replace the portion of skull that was removed.
- In China, doctors and pharmacists often hang a bottle gourd outside their place of business as a sign of their profession. Belief in the curative powers of the bottle gourd is quite old in China. At times in the past, people carried charms in the shape of bottle gourds to ward off diseases.
- The shape of the typical Chinese porcelain soupspoon is based on the curvature and shape of the bottle gourd.
- The Chinese used to place crickets in small cages to enjoy their singing. About the year 1000, they began to hold cricket fights, and gourds were used to house the crickets over winter.
- Another old use for gourds was as flasks for dispensing gunpowder.
- Calabash pipes for smoking tobacco are made in part from the bottle gourd. The pipes were once supplied

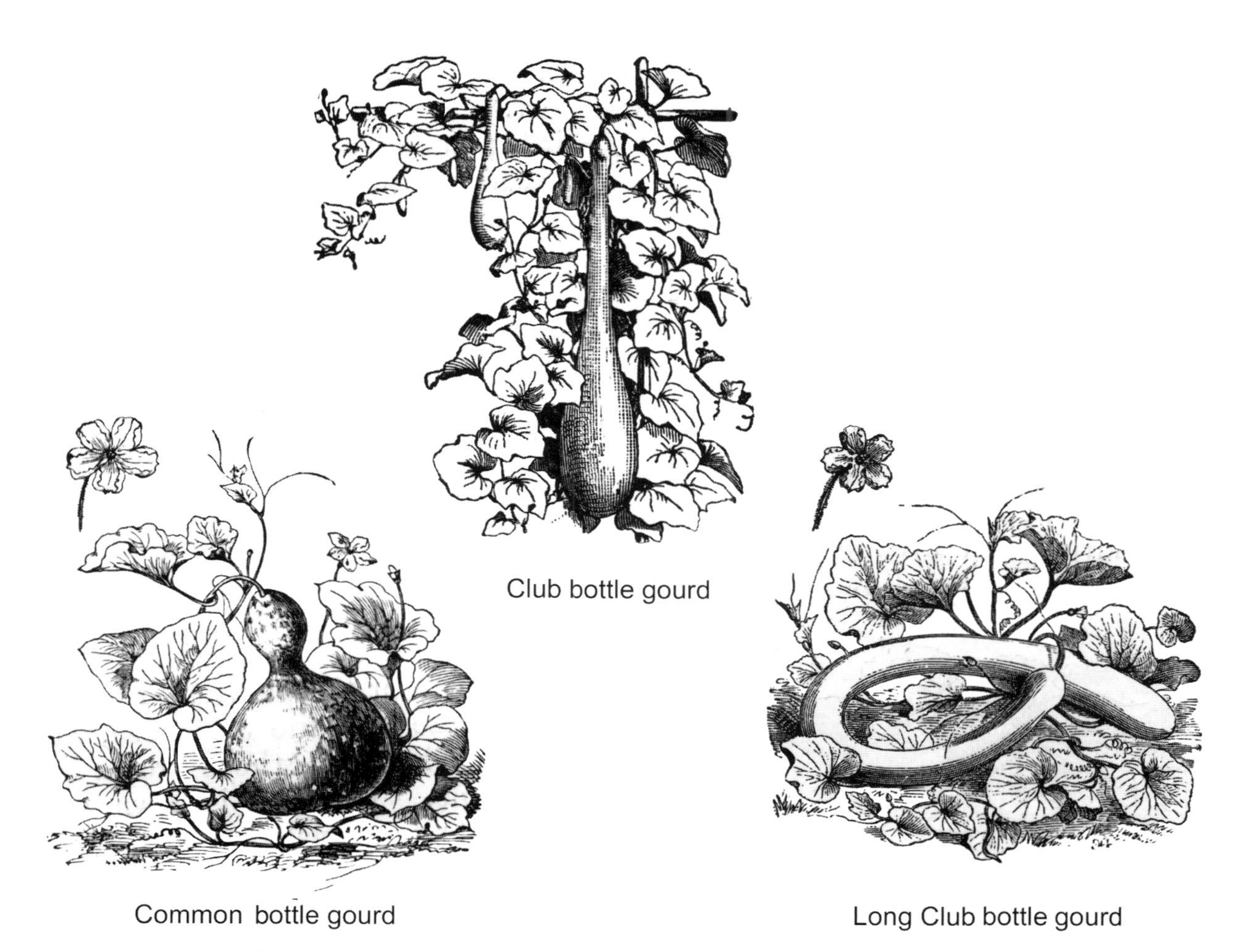

Varieties of bottle gourd (*Lagenaria siceraria*), from Vilmorin-Andrieux (1885).

entirely from South Africa, where cultivars with straight necks were artificially forced to grow into the desired shape. The pipe is formed by the neck end of the bottle gourd. A bowl of clay, porcelain, or meerschaum is put in the large end of the neck, and a stem is added at the smaller (lip) end.

- While the blossoms of bottle gourd are pollinated primarily by bees, it appears that nectar-collecting bats may also contribute to its pollination in the native tropical habitat. The flowers have a musk-like odor, typical of many nocturnal flowers visited by bats.
- Designs lightly scratched into the skin of growing bottle gourd fruit will develop into scars that remain intact in the mature fruits. Some artisans take advantage of this to produce ornately decorated fruit. Another technique is to constrict the young fruit with bands to produce particular shapes as the fruits mature.
- Names for star constellations differ considerably, depending on the country. The "Big Dipper" assemblage of stars in the night sky is also known as the "Drinking Gourd."

Key Information Sources

Calabrese, N., Di-Venere, D., and Linsalata, V. 1999. Technological and qualitative aspects of calabash gourd, *Lagenaria siceraria* (Molina) Standley, for processing. Acta Hortic. **492**: 179–186.

Damato, G., Manolio, G., and Bianco, V.V. 1998. Sowing dates, nitrogen rates, pruning and yield of *Lagenaria siceraria* (Molina) Standl. in Southern Italy. Acta Hortic. **467**: 295–303.

Decker-Walters, D., Staub, J., Lopez-Sesse, A., and Nakata, E. 2001. Diversity in landraces and cultivars of bottle gourd (*Lagenaria siceraria*; Cucurbitaceae) as assessed by random amplified polymorphic DNA. Genet. Resourc. Crop Evol. **48**: 369–380.

Decker-Walters, D.S., Wilkins-Ellert, M., Chung, S.-M., and Staub, J.E. 2004. Discovery and genetic assessment of wild bottle gourd [*Lagenaria siceraria* (Mol.) Standley; Cucurbitaceae] from Zimbabwe. Econ. Bot. **58**: 501–508.

Dodge, E.S. 1943. Gourd growers of the South Seas; an introduction to the study of the *Lagenaria* gourd in the culture of the Polynesians. The Gourd Society of America, Boston, MA. 119 pp.

Heiser, C.B., Jr. 1973. Variation in the bottle gourd. *In* Tropical forest ecosystems in Africa and South America: a comparative review. *Edited by* B.J. Meggers, E.S,. Ayensu, and W.D. Duckworth. Smithsonian Institution Press, Washington, DC. pp. 121–128.

Heiser, C.B, Jr. 1979. The gourd book. University of Oklahoma Press, Norman, OK. 248 pp.

Heiser, C.B., Jr. 1989. Domestication of Cucurbitaceae: *Cucurbita* and *Lagenaria*. *In* Foraging and farming: the evolution of plant exploitation. *Edited by* D.R. Harris and G.C. Hillman. Unwin Hyman, London, UK. pp. 471–480.

Kalloo, G. 1993. Bottle gourd, *Lagenaria siceraria* (Mol.) Standl. *In* Genetic improvement of vegetable crops. *Edited by* G. Kalloo and B.O. Bergh. Pergamon Press, New York, NY. pp. 247–249.

Martin, F.W. 1979. Sponge and bottle gourds, *Luffa* and *Lagenaria*. Agricultural Research (Southern Region), Science and Education Administration, U.S. Dept. of Agriculture, Mayagüez Institute of Tropical Agriculture (Puerto Rico), New Orleans, LA. 19 pp.

Myers, C., and Perry, E. 1991. Bottle gourd, calabash gourd, cucuzzi. Crop Sheet SMC-006. *In* Specialty and minor crops handbook. *Edited by* C. Myers. The Small Farm Center, Division of Agriculture and Natural Resources, University of California, Oakland, CA. 2 pp.

Ndukwu, B.C., Obute, G.C., and Obute, G.C. 2002. Morphological and ethnobotanical consideration of the genus *Lagenaria* Ser. (Cucurbitaceae) in the Niger Delta area. J. Econ. Taxon. Bot. **26**: 751–757.

Nwokolo, E. 1996. Bottle gourd, buffalo gourd and other gourds. *In* Food and feed from legumes and oilseeds. *Edited by* E. Nwokolo and J. Smartt. Chapman and Hall, London, UK. pp. 290–297.

Pitchaimuthu, M., and Sirohi, P.S. 1997. Genetic analysis of fruit characters in bottle gourd (*Lagenaria siceraria* (Mol.) Standl). J. Genet. Breed. **51**: 33–37.

Richardson, J.B. 1972. The pre-Columbian distribution of the bottle gourd (*Lagenaria siceraria*): a re-evaluation. Econ. Bot. **26**: 265–273.

Singh, C.B., and Singh, S.N. 1970. Studies in the floral biology of white bottle gourd. Allahbad Farmer, **44(3)**: 133–138.

Singh, R.D., and Singh, J.P. 1998. Improvement and cultivation: *Lagenaria* and *Luffa*. *In* Cucurbits. *Edited by* N.M. Nayar and T.A. More. Science Publishers, Enfield, NH. pp. 199–203.

Speck, F.G. 1941. Gourds of the southeastern Indians: a prolegomenon on the *Lagenaria* gourd in the culture of the southeastern Indians. The New England Gourd Society, Boston, MA. 113 pp.

Towle, M.A., and Schultes, R.E. 1952. The pre-Columbian occurrence of *Lagenaria* seeds in coastal Peru. Bot. Mus. Leafl. Harv. Univ. **15(6)**: 171–190.

Whitaker, T.W. 1948. *Lagenaria*: a pre-Columbian cultivated plant in the Americas. Southwestern J. Anthropol. **4:(1)**: 48–69.

Widjaja, E.A., and Reyes, M.E.C. 1993. *Lagenaria siceraria* (Molina) Standley. *In* Plant resources of South-East Asia, 8, vegetables. *Edited by* J.E. Siemonsma and K. Piluek. Pudoc Scientific Publishers, Wageningen, Netherlands. pp. 190–192.

BRAZIL NUT

Family: Lecythidaceae (monkey-pot family)

Names

Scientific name: *Bertholletia excelsa* Bonpl.

- "Brazil nut" is obviously named for the country of Brazil that furnishes most of the world's supply.
- The Brazil nut is also known as cream nut (for the color of the nuts) and Para nut (for the Brazilian state of Pará, a major area of production).
- In Brazil, the nut is called *castanha*, Portuguese for "chestnut." In France, Brazil nuts have been called "American chestnuts."
- The genus name *Bertholletia* was coined by German botanist–explorer Alexander von Humboldt (1769–1859) and his French colleague Aimé Bonpland (1773–1858; note the latter's abbreviation, Bonpl., in the full scientific name, *Bertholletia excelsa* Bonpl.). They were the first Europeans to collect and scientifically study the Brazil nut. (Humboldt was also the first European to observe how the poison curare was made.) The genus was named for Humboldt's friend, the chemist and salon host Claude Louis Berthollet (1748–1822).
- *Excelsa* in the scientific name *B. excelsa* is Latin for tall, pointing out that this is one of the tallest of trees.
- The Lecythidaceae is called the monkey-pot family because the genus for which the family is named, *Lecythis*, is best known for species called monkey-pot. Sugar was placed in chained, empty fruits (the "monkey-pots") to trap monkeys, which would not open their paws once they had seized some sugar, and so could not withdraw their arms and escape.

Brazil nut (*Bertholletia excelsa*), from Martius (1831, vol. 2, plate 1.1). A strangler fig has attached itself to the trunk, and several vines hang down from the branches.

Plant Portrait

The Brazil nut tree is a giant evergreen tree, 23–30 m (75–100 feet) tall, sometimes reaching 50 m (164 feet) in height, and usually rising well above the surrounding jungle. Trunk diameters reach 1.2–2.4 m (4–8 feet), rarely 5 m (16 feet). The trees often achieve an age of 1000 years, and one tree was found to be 1600 years old. The fruits are large, spherical, dark brown, 10–15 cm (4–6 inches) in diameter, and weigh 0.5–2.5 kg (1–5½ pounds). They resemble large, husked coconuts, with a hard, woody exterior. Inside are 10–25 Brazil nuts, 3.8–6.4 cm (1½–2½ inches) long, three-sided, arranged like the sections of an orange. Once the nuts are broken out of the fruits, each nut is still encased in its own shell, and Brazil nuts are marketed shelled or unshelled. The shell is about twice as heavy as the white-fleshed kernel it contains. The kernel is enclosed in a thin layer of brownish skin. Brazil nuts are almost all collected from wild, not planted trees. The trees are native to a huge area of the Amazon Basin that includes Brazil and the adjoining countries of Bolivia, Peru, Colombia, the Guianas, and Venezuela. For centuries, the indigenous tribes of the rainforest have relied on Brazil nuts as an important and significant staple of their diet. Peoples like the Yanomami eat the nuts raw or grated and mixed into a manioc porridge. Spanish colonists learned of the Brazil nut in the 16th century, and Dutch traders exported small amounts to Europe in the 17th century. More substantial exports to Europe occurred in the 18th century, and in 1810, the first shipment was

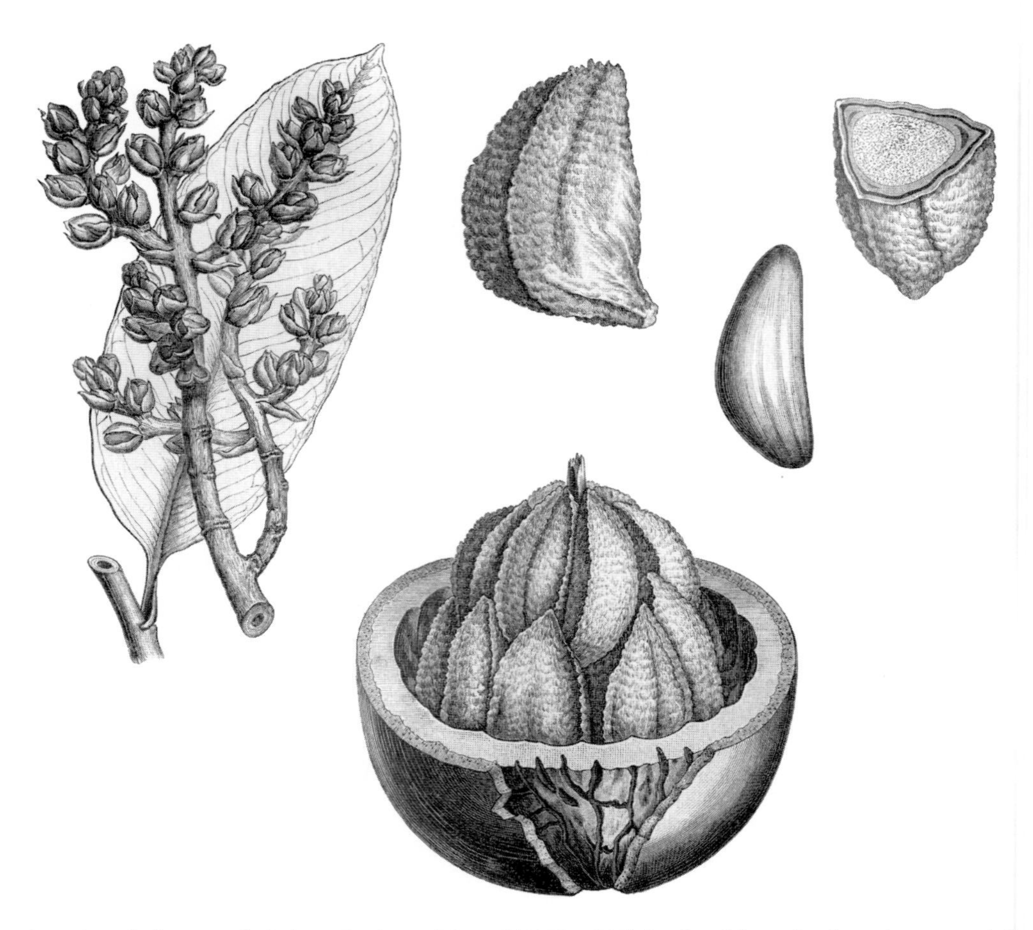

Brazil nut (*Bertholletia excelsa*), from Engler and Prantl (1889–1915). Leaf and flowering branch at upper left, opened fruit at bottom, views of kernel and its shell at upper right.

made to the United States. Today, most of the Brazil nut crop is exported to the United States and Europe, where the nuts are a popular delicacy. In 2003, the European Community banned the importation of Brazil nuts, based on concern about contamination with aflatoxin, a carcinogenic substance produced by molds. Attempts to establish plantations in countries outside of South America have not proven successful. Brazil nut is the only commercial nut produced exclusively in Amazon forests, and the only internationally traded nut collected almost entirely from the wild. An article in *Science Magazine* (19 Dec. 2003) suggested that the annual harvest of at least 45 000 tonnes (50 000 tons) from the Amazon region is not sustainable, and conservation measures need to be enacted.

Culinary Portrait

Brazil nuts have crunchy, yellowish kernels that taste like coconut. They are consumed raw, roasted, salted, and in baked goods, confectionery products (especially as chocolate-covered candies), and ice cream—all foods that are well complemented by the flavor. Because of the similar taste to coconut and macadamia nut, Brazil nut can replace them in most recipes. The nuts are popular in shelled nut mixtures. Brazil nuts are about two-thirds fat or oil (hence, they have a high caloric content), and are best kept cool and protected from air in sealed, light-proof, vacuum-packed containers to prevent rancidity. Once blanched, crushed, or sliced, contact with air tends to stimulate rancidity. Staleness is a problem, and the nuts should be purchased from a reliable establishment where there is good turnover, and eaten promptly. In Brazil, where the supply is considerable, the nuts are used in a variety of culinary dishes, for example, as a ground stuffing for avocados, as an accompaniment to meat and poultry dishes, and as a garnish for vegetables. Brazil nut oil is used in Brazil as a substitute for olive oil, and there is some export. Because the shells are so hard, shelled nuts are preferably purchased. To facilitate cracking nuts in the shell, they may first be frozen, allowed to defrost slightly, and then broken open with a hammer. Alternatively, they may first be steamed for a few minutes, and then broken open with a hammer.

Curiosities of Science and Technology

- Collecting Brazil nuts is hazardous. The nuts can weigh as much as 2.7 kg (6 pounds). At maturity, the heavy fruits fall to the ground from heights of 30 m (100 feet) or more, with a velocity of about 275 m (900 feet) per minute, sometimes killing people. In the Amazon Basin, one often sees injured people and animals near groves of Brazil nuts. At times, the falling nuts hit with such impact that they literally plant themselves in the ground. Laborers make it a point never to be under the trees on windy or rainy days, when falling nuts are common. Some even carry shields over their heads. But falling nuts aren't the only problem. Workers are exposed to vipers, jaguars, tyrannical bosses and traders, death from drowning and armed skirmishes over possession of trees, and diseases like malaria and leishmaniasis (caused by protozoan parasites of the genus *Leishmania*, the disease is transmitted by sand flies, occurs mostly in the tropics and subtropics, and is generally recognized by nonfatal, disfiguring lesions). In 1985, in Macapa, Brazil, six collectors were killed and 12 wounded in a fight over Brazil nuts.
- Agoutis are cat-size brown rodents. They love Brazil nuts, gather up all they can find, and bury them individually, just as squirrels bury acorns for future food. The agouti is virtually the only animal in the Amazon Basin that has teeth strong enough to open the thick husk and liberate the seeds so they can sprout. Brazil nut gatherers often find that the agoutis have taken most of the nuts that they had hoped to collect.
- The Brazil nut poison (arrow) frog (*Dendrobates castaneoticus*) is a little black frog with a white stripe pattern and orange spots. After agoutis have opened Brazil nut fruits, the empty shells often collect rainwater. The tadpoles of the frog frequently live in the rainwater accumulated in Brazil nut fruit shells. Poison arrow frogs (also called poison dart frogs) are small, brightly colored frog species of tropical rainforests of South and Central America. Poison from their skin is applied to the tips of hunting arrows and blow-gun darts by some South American Indians. Curiously, the skin of captive poison frogs (in zoos and in the pet trade) often loses its toxicity, the poison apparently dependent on the frogs continuing to consume their natural diet.
- Brazil nut trees normally yield about 30–50 kg (66–110 pounds) of fruits annually, but some trees have been known to produce over 2000 kg (4500 pounds) in a year.
- The Amazon Basin, the home of the Brazil nut, is eight times larger than California. It is nearly 100 times larger than the Netherlands.
- Selenium, although extremely toxic, is essential in minute amounts for human nutrition. Selenium may deter ovarian cancer by enhancing the activity of the antioxidant vitamin E (antioxidants counter the effects of "free radicals," harmful products of metabolism that are thought to predispose people to disease and aging). In areas where the soil content of selenium is high, so that crops absorb much of the element, it has been observed that there are lower incidences of some types of cancer. Brazil nuts are extraordinarily rich in selenium. The nuts can contain 250 times as much selenium as in most foods, depending on the soil where they are grown. One single Brazil nut can exceed the US *Recommended Daily Allowance* of selenium. But while there does seem to be a link of selenium consumption and lower risk of cancer, too much can be toxic and cause balding. Some nuts of relatives of the Brazil nut in the Lecythidaceae family contain so much selenium that overeating them can lead to hair loss.

Brazil nut (*Bertholletia excelsa*), from Humboldt and Bonpland (1807). These fruits are 10–15 cm (4–6 inches) in diameter.

Brazil nuts *(Bertholletia excelsa)*. Photograph by Eric Johnson.

- Almost 70% of a Brazil nut is oil or fat, and with such a high oil content, the nuts will burn like miniature candles when lit. In fact, the oil is extracted from the nuts and used by indigenous and rural peoples of the Amazon Basin for oil for lamps.
- The empty fruit pods of Brazil nuts, often called "monkey pots," are sometimes used to carry around small smoky fires to discourage attacks of black flies.
- In the Amazon region, Brazil nuts have been used as money.
- Genetic engineering can improve the nutritional quality of crops, but there are some potential dangers. The Brazil nut is rich in the amino acid methionine, an essential protein constituent that is present in low amounts in some crops, such as soybean. However, Brazil nut protein is the cause of allergic reactions in people sensitive to these nuts. In the early 1990s, experimental soybeans, in which a gene from the Brazil nut had been implanted to improve the protein quality, were found to be capable of provoking serious allergic reactions in individuals sensitive to Brazil nuts but not to regular varieties of soybeans. The modified soybean was never marketed, and no one actually became sick. Nevertheless, this experiment has become a classic example of the possible dangers of genetically engineering food plants.

Key Information Sources

Asero, R., Mistrello, G., Roncarolo, D., and Amato, S. 2002. Allergy to minor allergens of Brazil nut. Allergy (Copenhagen), **57**: 1080–1081.

Borja, J.M., Bartolome, B., Gomez, E., Galindo, P.A., and Feo, F. 1999. Anaphylaxis from Brazil nut. Allergy (Copenhagen), **54**: 1007–1008.

Camargo, P.B. de, Salomao, R. de P., Trumbore, S., and Martinelli, L.A. 1994. How old are large Brazil-nut trees (*Bertholletia excelsa*) in the Amazon? Sci. Agric. (Brazil), **51**: 389–391.

Clement, C.R. 1993. Brazil nut (*Bertholletia excelsa*, Lecythidaceae). *In* Selected species and strategies to enhance income generation from Amazonian forests. *Edited by* J.W. Clay. Forestry Dept., Food and Agriculture Organization, Rome, Italy. pp. 115–127.

Cunha, R., Prado, M.A do, De Carvalho, J.E.U., and De Goes, M. 1996. Morphological studies on the development of the recalcitrant seed of *Bertholletia excelsa* H.B.K. (Brazil nut). Seed Sci. Technol. **24**: 581–584.

Ip, C., and Lisk, D.J. 1994. Bioactivity of selenium from Brazil nut for cancer prevention and selenoenzyme maintenance. Nutr. Cancer, **21**(**3**): 203–212.

Kainer, K.A., De Matos-Malavasi, M , Duryea, M.L., and Rodrigues Da Silva, E. 1999. Brazil nut (*Bertholletia excelsa*) seed characteristics, preimbibition and germination. Seed Sci. Technol. **27**: 731–745.

LaFleur, J.R. 1992. Marketing of Brazil nuts: a case study from Brazil. Food and Agriculture Organization of the United Nations, Rome, Italy. 61 pp.

Melo, V.M.M., Xavier, F.J., Lima, M.S., and Prouvost, D.A. 1994. Allergenicity and tolerance to proteins from Brazil nut (*Bertholletia excelsa* H.B.K.). Food Agric. Immunol. **6**: 185–195.

Miller, C.J. 1990. Natural history, economic botany, and germplasm conservation of the Brazil nut tree (*Bertholletia excelsa*, Humb. & Bonpl.). M.Sc. Thesis, University of Florida. 151 pp.

Mori, S.A., and Prance, G. 1990. Taxonomy, ecology, and economic botany of the Brazil nut (*Bertholletia excelsa* Humb. & Bonpl.: Lecythidaceae). Adv. Econ. Bot. **8**: 130–150.

Myers, G.P., Newton, A.C., and Melgarejo, O. 2000. The influence of canopy gap size on natural regeneration of Brazil nut (*Bertholletia excelsa*) in Bolivia. For. Ecol. Manag. **127(1–3)**: 119–128.

Nelson, B,W., Absy, M.L., Barbosa, E.M., and Prance, G.T. 1985. Observations on flower visitors to *Bertholletia excelsa* H.B.K. and *Couratari tenuicarpa* A.C. Sm. (Lecythidaceae). Acta Amazonica, **15**(1–2 Suppl.): 225–234.

O'Malley, D.M., Buckley, D.P., Prance, G.T., and Bawa, K.S. 1988. Genetics of Brazil nut (*Berholletia excelsa* Humb. & Bonpl.: Lecythidaceae). 2. Mating system. Theor. Appl. Genet. **76**: 929–932.

Peres, C.A., and Baider, C. 1997. Seed dispersal, spatial distribution and population structure of Brazilnut trees (*Bertholletia excelsa*) in southeastern Amazonia. J. Trop. Ecol. **13**: 595–616.

Peres, C.A., Baider, C., Zuidema, P.A., Wadt, L.H.O., Kainer, K.A., et al. 2003. Demographic threats to the sustainability of Brazil nut exploitation. Science, **302**: 2112–2114.

Peres, C.A., Schiesari, L.C., and Dias-Leme, C.L. 1997. Vertebrate predation of Brazil-nuts (*Bertholletia excelsa*, Lecythidaceae), an agouti-dispersed Amazonian seed crop: a test of the escape hypothesis. J. Trop. Ecol. **13**: 69–79.

Reilly, C. 1999. Brazil nuts—the selenium supplement of choice? BNF Nutr. Bull. **24**: 177–184.

Schreiber, W.R. 1942. The Amazon Basin Brazil nut industry. U.S. Dept. of Agriculture, Office Office of Foreign Agricultural Relations, Washington, DC. 36 pp.

Small, E., and Catling, P.M. 2005. Blossoming treasures of biodiversity 17. Brazil nut —a key resource in a globally important ecosystem. Biodiversity, **6(3)**: 49–54.

Zuidema, P.A., and Boot, R.G.A. 2002. Demography of the Brazil nut tree (*Bertholletia excelsa*) in the Bolivian Amazon: impact of seed extraction on recruitment and population dynamics. J. Trop. Ecol. **18**: 1–31.

Specialty Cookbooks

Brazil Nut Advertising Fund. 1950s. Kernel Nut of Brazil says: let my company help your company make more money with these new Brazil nut candy formulas. Brazil Nut Advertising Fund, New York, NY. 56 pp.

Brazil Nut Association. 1940–1949. Kernel Nut of Brazil presents a parade of commercial formulas for Brazil nut candies: prepared for wholesale and retail candy manufacturers. Brazil Nut Association, New York, NY. 66 pp.

Brazil Nut Association. 1950s. A parade of Brazil nut recipes. Brazil Nut Association, New York, NY. 31 pp.

BROAD BEAN

Family: Fabaceae (Leguminosae; pea family)

Names

Scientific name: *Vicia faba* L.

- The word "bean" is based on a root common to Old English, Old High German, and Old Norse, which apparently once referred to the broad bean. Today, a wide variety of seeds (and sometimes fruits), especially of the legume family, are called beans.
- The English name "broad bean" is based on the substantial breadth of the seeds of the largest varieties.
- "Fava" is an English corruption of the Latin *faba*, bean. The use in English of "faba bean" is said to have originated in Canada. The expression "faba bean" is tautological (i.e., it repeats a word or thought, literally translating as "bean bean").
- Alternative names for the broad bean include faba bean, fava bean, English bean, European bean, horse bean, pigeon pea, tick bean, tick pea, and Windsor bean. Small-seeded forms, with seeds less than 15 mm (0.6 inch) long, are often called pigeon pea (they are fed to pigeons) and tickpea. However, see PIGEON PEA, which has priority for this name. Medium-seeded forms, with seeds 15–19 mm (0.6–0.75 inch long), are often called horsebean (a reflection of being fed to horses).
- Outside North America, the name "broad bean" almost exclusively denotes *V. faba* used for human food.
- In Europe, the smaller beans are often known as the field bean, or sometimes common bean. However, in North America, the terms "field bean" and "common bean" are predominantly used for *Phaseolus vulgaris*, including the green, yellow, and pole beans familiar to home gardeners.
- In Quebec, a medium-sized type of faba bean called *la gourgane* is grown for human consumption, particularly in the St. Jean area of the province, especially for soup.
- The genus name *Vicia* is the ancient Roman (Latin) name for a plant of uncertain identity, possibly the vetch.
- The 300 or more families of flowering plants are named after a genus that they contain. For example, the hemp family (Cannabaceae) is named after the genus *Cannabis* (*Cannabis* + suffix aceae). For most families, this naming convention is mandatory, but for eight important families, some older names can be used alternatively. For example, the pea family is known by its traditional name, Leguminosae, and also as the Fabaceae. The name Fabaceae is based on the genus name *Faba*. *Vicia faba* is often placed in the genus *Faba* rather than in the genus *Vicia*.
- *Faba* in the scientific name *V. faba* is an old name of Etruscan (Italian) origin for a plant of uncertain identity. As noted above, *faba* is also an old Roman word for "bean."

Plant Portrait

The broad bean (also commonly known as the fava bean and faba bean) is an annual (occasionally biennial) herbaceous plant, 30–180 cm (1–6 feet) tall, with fragrant flowers 2–3 cm (about an inch) long, marked with white, black, dark brown, or violet. The pods vary considerably in length, depending on variety, and may be as short as 5 cm (2 inches) or as long as 30 cm (1 foot). The flattish seeds may be dark brown, brown, reddish, pinkish yellow, green, or a variegated combination of colors. The plant is extensively cultivated as food for humans in the north temperate zone, and at higher altitudes during the cool season, in some subtropical regions. It appears that broad bean originated in the Near East, with selections radiating outwards geographically. Evidence of usage has been found at Iron Age sites in Europe, and there are also findings of seeds from around 1800 BC in ancient Egypt. Broad bean was first used in China about 1200. It was carried to Mexico and South America by Spanish explorers. Large-seeded forms (seeds longer than 20 mm or 3/4 inch), usually known as fava bean, faba bean, and Windsor bean, are most commonly used for human food, while the smaller-seeded forms are used mostly for animal feed. The seeds of the type used for human food are usually the size of lima beans, and have been employed like limas in those parts of western Europe and North America that have a climate too chilly to germinate limas and a growing season too short or too sunless to mature either pole beans or limas. China is the main grower, accounting for about 40% of world production. Other major producers include Egypt, Ethiopia, Morocco, Germany, the United Kingdom, and Australia.

Broad bean (*Vicia faba*), from Thomé (1903–1905, vol. 3).

Culinary Portrait

The seeds of broad bean have a strong, gamy flavor, and are not as sweet as limas. The nutritional value of broad bean is considerable, and it has been called "poor mans meat." Both whole pods and shelled seeds of broad bean can be used for human food. Broad beans can be harvested at an immature stage, and the whole pods cooked. When young and tender, they can be eaten raw in salads. As the pod ages, the seeds should be shelled, and because some people find the thick, mature skins difficult to chew and bitter, may even need to be skinned before cooking. The skins come off easily when the seeds are soaked for 12–24 hours (the water should be changed frequently); or when plunged into boiling water for several minutes, drained, and rinsed under cold water. Dried beans may be cooked for about 2½ hours; skinned, dried beans that have been soaked for 8–12 hours cook in about 1½ hours; and in a pressure cooker, unsoaked beans require 25 minutes, or 20 minutes if they have been soaked. The immature pods are canned and frozen in northwest Europe, where problems of mechanized harvesting limit the crop. The ripe seeds are used in stews and soups and may be deep-fried, prepared as fritters, ground for meal, boiled, creamed, or puréed. The beans can be dried for storage. Mature beans can be sprouted by soaking them in water, and allowing the seeds to germinate. The sprouts may be added to soups. In Latin America, seeds are roasted and used like peanuts, even though they are much harder to chew than the latter. Recently, there has been interest in developing protein concentrates from broad bean, for use in the manufacture of textured protein to replace meat.

Broad beans contain anti-nutritional constituents that can cause significant problems in some people. The toxic constituents are called hemagglutinins and trypsin inhibitors. Long-term ingestion of hemagglutinin can seriously impair intestinal absorption, while ingestion of trypsin inhibitors may increase the requirement for vitamin B12. The presence of indigestible sugars called oligosaccharides (raffinose and stachiose) leads to flatulence.

The most serious effects result from the presence of chemicals called beta-glycosides (vicine and covicine), which lead to a hemolytic (red blood cell destruction) syndrome in some individuals. The condition, known as "favism," is characterized by weakness, pallor, jaundice, and hemoglobinuria (blood in urine). Death can result from renal failure. This syndrome occurs in individuals who are deficient in a certain enzyme (glucose-6-phosphate dehydrogenase or more simply G6PD), but it may not be the only deficiency associated with favism. In those susceptible to favism, the G6PD activity is only 0–6% of normal. Certain racial or ethnic groups, including Oriental Jews, Mediterranean Europeans, Arabs, Asians, and Blacks, have higher incidence of favism than other groups. Favism occurs in about 1% of Whites and 15% of Blacks. In cases of extreme sensitivity, even the pollen induces an attack. Favism does not occur in people without the genetic predisposition. Deficiency of the enzyme is thought to be linked to protection against malaria.

Curiosities of Science and Technology

- Many ancient cultures seem to have developed a love–hate relationship with the broad bean. On the negative side, several students of the Greek philosopher and mathematician Pythagoras (who died about 497 BC) are said to have allowed themselves to be slaughtered by soldiers rather than escape through a field of beans. Pythagoras himself is thought to have refused to enter a bean field while attempting to escape from pursuing Crotons, which led to his capture and execution. Upper-class Greeks and Romans regarded beans as harmful and thought that eating them would cloud their vision. The Egyptians considered the broad bean to be unclean. It is possible that this aversion to broad beans traces to the harmful effects of favism because about 1% of people of Greek and Italian descent are susceptible to this condition, as are people of Semitic descent. By contrast, the broad bean was respected by others. The Fabii, one of the noble families of Rome, derived their name from the faba bean. During the Roman feast of Fabaria, beans were offered symbolically.
- The ancient Romans used beans for balloting, both in their elections and in courts. Black beans stood for opposition or guilt, and white for agreement or innocence.
- Many species of *Vicia*, including the broad bean, possess nectaries (nectar-producing glands) on the stipules (bracts at the base of the leaves). In most plants, nectaries are only found in flowers, where they attract pollinators. Ants have often been observed collecting nectar from the leaf nectaries of the faba bean. It is believed that these ants have established a symbiotic (mutually beneficial) association with the faba bean, the ants protecting the plants against caterpillars while obtaining nourishment from the nectar.

Key Information Sources

Abdallah, M.M.F. 1979. The origin and evolution in *Vicia faba* L. *In* Proceedings of the first Mediterranean conference on genetics, Cairo, Egypt, March, 1979. *Edited by* Jam`iyah al-Misriyah li-`Ulum al-Wirathah. Egyptian Society of Genetics, Cairo, Egypt. pp. 713–746.

Adsule, R.N., and Akpapunam, M. 1996. Faba bean (*Vicia faba* L.). *In* Food and feed from legumes and oilseeds. *Edited by* J. Smartt and E. Nwokolo. Chapman and Hall, London, UK. pp. 197–202.

Bond, D.A. 1995. Faba bean. *In* Evolution of crop plants. 2nd edition. *Edited by* J. Smartt and N.W. Simmonds. Longman Scientific & Technical, Burnt Mill, Harlow, Essex, UK. pp. 312–316.

Bond, D.A., Jellis, G.J., Rowland, G.G., Le Guen, J., Robertson, L.D., Khalil, S.A., and Li-Juan, L. 1993. Present status and future strategy in breeding faba beans (*Vicia faba* L.) for resistance to biotic and abiotic stresses. Euphytica, **73**: 151–166.

Bond, D.A., Lawes, D.A.. Hawtin, G.C., Saxena, M.C., and Stephens, J.H. 1985. Faba bean (*Vicia faba* L.). *In* Grain legume crops. *Edited by* R.J. Summerfield and E.H. Roberts. Collins Professional and Technical Books, London, UK. pp. 199–265.

Crofton, G.R.A. 1997. The principal seed characters of field beans (*Vicia faba* L. (Partim)) in relation to variety classification. Plant Var. Seeds, **10**(**2**): 81–94.

Cubero, J.I. 1973. Evolution trends in *Vicia faba*. Theor. Appl. Genet. **43**: 59–65.

Cubero, J.I. 1974. On the evolution of *Vicia faba* L. Theor. Appl. Genet. **45**: 47–51.

Duc, G. 1997. Faba bean (*Vicia faba* L.). Field Crops Res. **53(1–3)**: 99–109.

Hebblethwaite, P.D. (*Editor*).1983. The faba bean (*Vicia faba* L.): a basis for improvement. Butterworths, London, UK. 573 pp.

Hebblethwaite, P.D. (*Editor*). 1984. *Vicia faba*: agronomy, physiology, and breeding. [Proceedings of a seminar in the CEC Programme of Coordination of Research on Plant Protein Improvement, held at the University of Nottingham, UK, 14–16 September 1983]. M. Nijhoff for the Commission of the European Communities, The Hague, Netherlands. 333 pp.

Drew, N., and Bretag, T. 1994. Irrigated faba bean growers handbook. Victorian Dept. of Agriculture, Kerang, Australia. 24 pp.

Filippetti, A., and Ricciardi, L. 1993. Faba bean, *Vicia faba* L. *In* Genetic improvement of vegetable crops. *Edited by* G. Kalloo and B.O. Bergh. Pergamon Press, New York, NY. pp. 355–385.

Grashoff, C. 1992. Variability in yield of faba beans (*Vicia faba* L.). Landbouwuniversiteit te Wageningen, Wageningen, Netherlands. 128 pp.

Hanna, A.S., and Lawes, D.A. 1967. Studies on pollination and fertilization in the field bean (*Vicia faba* L.). Annu. Appl. Biol. **59**: 289–295.

Jansen, P.C.M. 1989. *Vicia faba* L. *In* Plant resources of South-East Asia, 1, pulses. *Edited by* L.J.G. van der Maesen and S. Somaatmadja. Pudoc, Leiden, Netherlands. pp. 64–66.

Ladizinsky, G. 1975. On the origin of the broad bean, *Vicia faba* L. Isr. J. Bot. **24**: 80–88.

Lang, L. 1993. Faba bean in China: state-of-the-art review. International Center for Agricultural Research in the Dry Areas, Aleppo, Syria. 144 pp.

Piggin, C. (*Editor*). 1994. Research and development prospects for faba bean: report of a workshop held in Melbourne, Australia, 28–29 March 1994. Australian Centre for International Agricultural Research, Canberra, Australia. 44 pp.

Platford, G., Rogalsky, J.R., Small, D., Furgal, J.F., Campbell, L.D., Devlin, T.J., Ingalis, J.R., and Stothers, S.C. 1983. Fababean production and use in Manitoba. Manit. Agric. Publ. 8358. 20 pp.

Sokolov, R. 1984. Broad bean universe. Nat. Hist. **12**: 84–87.

University of Gottingen. 1986. *Vicia faba*: cultivation, breeding and nitrogen fixation: proceedings of a workshop in the CEC programme of coordination of agricultural research, July 16–19, 1986, University of Gottingen, Germany. [Vortrage fur Pflanzenzuchtung Heft 11.] Saatgut-Treuhandverwaltungs, Bonn, Germany. 234 pp.

Witcombe, J.R., and Erskine, W. 1984. Genetic resources and their exploitation: chickpeas, faba beans, and lentils. Martinus Nijhoff/Junk Publishers, The Hague, Netherlands. 256 pp.

Zohary, D. 1977. Comments on the origin of cultivated broad bean. Isr. J. Bot. **26**: 39–40.

Specialty Cookbooks

(See Bean, Common; especially cookbooks from the UK.)

BROCCOLI AND BROCCOLI RAAB

Family: Brassicaceae (Cruciferae; mustard family)

Names

Scientific names:

Broccoli—*Brassica oleracea* L. var. *italica* Plenck

Broccoli raab—*B. ruvo* L.H. Bailey (*B. rapa* L. "Broccoletto Group"; *B. rapa* "Ruvo Group")

- The name "broccoli" is from the Italian *broccolo*, which refers to the flowering top of a cabbage (broccoli literally means "little shoots"). The word came originally from the Latin *brachium*, meaning arm or branch. Broccoli in Italian is plural, and in English, the plural of broccoli is broccoli.
- Other names for different types of broccoli: bouquet broccoli, heading broccoli, true-heading broccoli, sprouting broccoli, green sprouting broccoli, purple sprouting broccoli, white sprouting broccoli (early and late forms are sometimes distinguished).
- In Britain, the word "broccoli" is sometimes used in a seasonal context to distinguish different groups of cauliflower.
- "Calabrese" (which means "Calabrian" in Italian, named for Calabria in southwest Italy) is a (typically annual) form of broccoli with tightly packed flowers heads, which are either green or purple. The Italian pronunciation of the name is cah-la-brae-sae. Other names for this vegetable include asparagus broccoli, sprouting broccoli, and true-heading broccoli. Considerable amounts are produced in California. Depending on where one is in the world, calabrese may simply be called broccoli.
- "Chinese broccoli" (or Chinese kale) is *B. alboglabra* L., which is treated in KALE AND COLLARD.
- The hybrid of broccoli and cauliflower is known as broccoflower.
- Broccoli raab (also spelled rabe) is also known as bitter broccoli, Italian turnip broccoli, ruvo kale, and spring raab.
- Several other vegetables are highly reminiscent of broccoli. Particularly interesting is "Rapini" (or "rappini"), also known by the cultivar name Rapine. Rapini has been placed in *Brassica rapa* subsp. *rapa* (which includes the turnip).
- See CABBAGE for the derivation of the scientific name *Brassica oleracea*.
- *Italica* in the scientific name *B. oleracea var. italica* means Italian.

Plant Portrait

Broccoli is quite similar to cauliflower (which see). Both annual and perennial forms are grown. The flower head of cauliflower is blue–green, white, or purple, and this marketable part is called the "curd." Broccoli is the immature flower head of the plant, which includes the stalks and young flower buds. It differs from early cauliflower cultivars, which develop undifferentiated flowers that abort prior to flowering and production of the curd. Broccoli likely originated from wild cabbage (*Brassica oleracea*), although some authorities believe that both broccoli and cauliflower may be descended from the wild species *B. cretica* Lam. Since Roman times, the two crops have frequently not been distinguished. Even in modern times, the two are unclearly identified, depending where the crop is grown. Accurate descriptions of broccoli were not prepared until the

Broccoli (*Brassica oleracea* var. *italica*) plant in early flower, by B. Flahey.

Broccoli (*Brassica oleracea* var. *italica*), by B. Flahey.

Middle Ages, when the English referred to sprouting cauliflower or Italian asparagus. American writers in 1806 called it green broccoli. Americans broadened the popularity of this vegetable, which otherwise was limited largely to Italy. Today, almost all of the commercial broccoli grown in the United States is produced in California.

Broccoli raab includes kinds of fast-growing annuals that form loose-flowering heads in the axils of the lower leaves. In this type of vegetable, the curd (floral head) is much less developed than in broccoli, and the leaves and attached succulent stem form most of the vegetable. Broccoli raab has been consumed for centuries in the Mediterranean area, especially in Italy, and although introduced by Italian immigrants to the United States around the beginning of the 20th century, it is relatively unknown in North America.

Culinary Portrait

Broccoli is a well-known vegetable, eaten raw or cooked in salads and a variety of prepared meals. When purchasing broccoli, the buds should be tightly closed (open flowers indicate the vegetable is old), and the stalks and branches should be tender and firm (easily pierced with a fingernail). Reject broccoli that has yellowed or wilted leaves. Broccoli is usually boiled, and served with butter or a sauce such as hollandaise. Cooking can be speeded up by splitting the stalk in two. Even better, since broccoli buds cook more quickly than the stems, divide the tops and peel and section the stalks into short lengths; and first cook the stem portions in boiling salted water for 5 minutes, then add the top sections and cook for another 5 minutes. The

Broccoli raab (*Brassica ruvo*), from Vilmorin-Andrieux (1885).

strong cooking odor can be reduced by adding a piece of bread to the cooking water. It has been recommended that when steaming broccoli, the lid should be lifted two or three times to allow the gases to escape—a procedure said to keep the color bright green.

Broccoli raab has a more pungent and bitter taste than broccoli, and is not to everyone's taste. It is a favorite in Italian cooking, where it adds a zesty flavor to bland foods such as mild potato and pasta dishes and holds its own with highly seasoned foods such as spiced sausage. Broccoli raab cooks faster than broccoli (2–6 minutes) and turns soft suddenly. To reduce the pungent flavor, it can be blanched for 1 minute in boiling salted water, drained, and then cooked. The stalks (which taste better than the leaves) require slightly more cooking than the leaves and flower clusters, and may be cooked first for 1 minute, then the other parts added. Steaming is not recommended, as the cooked vegetable will tend to retain more bitterness.

Curiosities of Science and Technology

- In the past, titles of scientific publications were often very much more detailed than at present. Consider Stephen Switzer's book, published in London, England, in 1727: "The practical kitchen gardiner, or, A new and entire system of directions for his employment in the melonry, kitchen-garden, and potagery, in the several seasons of the year: being chiefly the observations of a person train'd up in the neat-houses or kitchen-gardens about London: illustrated with plans and descriptions proper for the situation and disposition of those gardens: to which is added, by way of supplement, the method of raising cucumbers and melons, mushrooms, borecole, broccoli, potatoes, and other curious and useful plants, as practised in France, Italy, Holland and Ireland: and also, an account of the labours and profits of a kitchen-garden, and what every gentleman may reasonably expect therefrom every month of the year: in a method never yet attempted."
- A record-setting 15.9-kg (35-pound) broccoli was grown in 1993 by John V.R. Evans of Alaska.

Key Information Sources

(See additional references in CAULIFLOWER and CABBAGE.)

Bouquet, A.G.B. 1929. Cauliflower and broccoli culture. Orange Judd, New York, NY. 125 pp.

Burt, J.R., and Hegney, M. 1989. Proceedings of the Cauliflower and Broccoli Growers seminar Manjimup, Western Australia, Aug. 9, 1989. Western Australian Dept. of Agriculture, Perth, WA, Australia. 53 pp.

Downard, G.C. 1992. Big broccoli book. Random House, New York, NY. 97 pp.

Gray, A.R. 1982. Taxonomy and evolution of broccoli (*Brassica oleracea* var. *italica*). Econ. Bot. **36**: 397–410.

Gray, A.R. 1989. Taxonomy and evolution of broccolis and cauliflowers. Baileya, **23**: 28–46.

Gray, A.R. 1993. Broccoli, *Brassica oleracea* L. (Italica group). *In* Genetic improvement of vegetable crops. *Edited by* G. Kalloo and B.O. Bergh. Pergamon Press, New York, NY. pp. 61–86.

Great Britain Ministry of Agriculture, Fisheries and Food. 1952. Winter (broccoli) and summer cauliflowers. Bulletin 131, 3rd edition. Her Majesty's Stationary Office, London, UK. 20 pp.

Maier, A. 1976. How to select broccoli. Cooperative Extension Service, North Dakota State University, Fargo, ND. 4 pp.

Mitchell, D.R. 1998. The broccoli sprouts breakthrough: the new miracle food for cancer prevention. St. Martin's Paperbacks, New York, NY. 165 pp.

Organisation for Economic Co-operation and Development. 2000. Broccoli. OECD, Paris, France. 70 pp. [In English and French.]

Steffens, G. McM. 1985. The National Gardening book of cauliflower, broccoli & cabbage. Revised edition. National Gardening Association, Burlington, VT. 27 pp.

United States Department of Agriculture. 1984. Growing cauliflower and broccoli. Farmer's Bulletin 2239 (revised). U.S. Dept. of Agriculture Extension Service, Washington, DC. 14 pp.

Specialty Cookbooks

Brody, L. 1993. Broccoli by Brody: recipes for America's healthiest vegetable. W. Morrow, New York, NY. 116 pp.

Haspel, B., and Haspel, T. 1999. The dreaded broccoli cookbook: a good-natured guide to healthful eating, with 100 recipes. Scribner, New York, NY. 317 pp.

Holt, T. 1993. Broccoli power. Dell Publishing Co., New York, NY. 161 pp.

Katzen, M. 2000. The new enchanted broccoli forest. Ten Speed Press, Berkeley, CA. 303 pp.

Oklahoma State Department of Agriculture. 1983. Oklahoma broccoli: recipes. Dept. of Agriculture, Marketing Industry Division, Oklahoma City, OK. 6 pp.

Pazienza, A. 1973. Cooking broccoli and cauliflower. Abson Books, Bristol, UK. 33 pp.

Price, B.J. 1994. 101 ways to fix broccoli. Carlton Press, New York, NY. 112 pp.

Wasserman, D., and Stahler, C. 1985. I love animals and broccoli. Baltimore Vegetarians, Baltimore, MD. 40 pp.

BRUSSELS SPROUTS

Family: Brassicaceae (Cruciferae; mustard family)

Names

Scientific name: *Brassica oleracea* var. *gemmifera* Zenker (*B. oleracea* Brussels Sprouts Group)

- "Brussels sprouts" is named after Brussels in Belgium, where it probably first attained importance.
- "Brussels sprouts," and much less commonly "Brussels sprout," are used interchangeably.
- "Brussel's sprouts" is incorrect.
- In German, Brussels sprouts is known as *Rosenkohl*—"rose cabbage."
- See CABBAGE for the derivation of the scientific name *Brassica oleracea*.
- The Latin variety name *gemmifera* is from the Latin *gemma*, bud, + *fero*, to bear, in allusion to the bud-like "sprouts."

Plant Portrait

The vegetable Brussels sprouts resembles miniature cabbage heads (called "sprouts") 2.5–4 cm (1–1½ inches) in diameter. These are produced in the axils of the leaves on the main stem. Sprouts first appear in the lower leaf axils and develop in higher ones in sequence towards the stem tip. The smaller sprouts are more tender. It is claimed that primitive types of Brussels sprouts were grown in Italy in Roman times, and possibly as early as the 1200s in Belgium. However, the modern, familiar Brussels sprouts was first cultivated in large quantities in Belgium and France. The earliest record of modern types dates back about 500 years. This vegetable likely arose from Savoy cabbage (see CABBAGE). When the head of Savoy cabbage is removed, the plant develops little heads in the leaf axils. Writings from 1587 described the "new cabbage" and associated it with the city of Brussels. Its use spread across temperate Europe where it became popular because it continuously produced miniature cabbages throughout the growing season. Brussels sprouts was widely consumed in the 1800s in Europe, but not until 1925 was it grown commercially in North America. The most important commercial areas in the world for this minor vegetable are Europe, Japan, and North America. Most commercial cultivation in the United States is concentrated in California, with limited production in New York. Canadians are fonder of Brussels sprouts than Americans, and while Canada produces about half of its consumption of this crop domestically, much is imported from California.

Culinary Portrait

Brussels sprouts are not eaten raw. This well-known vegetable is cooked and most commonly served as a separate dish, typically with butter or a béchamel sauce. The sprouts are also added to soups, stews, stir-fries, and casseroles, and may be puréed with potatoes or served cold in a salad.

Brussels sprouts contain high levels of antioxidants, fiber, potassium, iron and, like other members of the cabbage clan, cancer-inhibiting chemicals. Unfortunately, they also have high amounts of sulforaphane, which stinks when Brussels sprouts (and other members of the cabbage family) are aged or overcooked. To avoid this problem, buy fresh sprouts and prepare them properly. Purchase fresh, small sprouts that are evenly sized, compact, firm to the touch, with dark green leaves. The smaller the sprouts, the more delicate the flavor (the larger heads have a more cabbage-like taste). Avoid those with yellowed, withered, or loose leaves, and those in which the stem end, where they were cut from the stalk, has dried out. Prepare and eat as soon as possible, within 3 days of purchase. Store the sprouts unwashed, in an open paper or plastic bag in the refrigerator. To clean, soak in cold water for several minutes and drain. Trim the bottom of the stems, and remove and discard the tough outermost leaves. Sprouts can be cut in half or quartered to promote quick and even cooking. Some chefs cut a 6 mm (1/4 inch) "X" in the stem in the belief that it assists in even cooking (many chefs consider this to be an old wives' tale). The sprouts can be steamed (15 minutes) or braised (15 minutes), boiled (8–12 minutes), blanched, roasted, or sautéed. Steam or boil uncovered, in a small amount of water (e.g., 2 cm or 3/4 inch) for a short time until just tender. When sprouts are overcooked, the amount of foul-smelling hydrogen sulfide can increase dramatically, and the vegetable becomes pasty.

Culinary Vocabulary

- The French expression *à la bruxelloise* refers to a French garnish for small joints of meat, made with stewed Brussels sprouts, chicory, and potatoes.

Curiosities of Science and Technology

- Some gardeners in colder climates leave their Brussels sprouts on the stalks until the first frost arrives, since frost or cool weather sweetens this vegetable.
- Another gardening trick with Brussels sprouts is to break off all the lower branches to a height of 15–20 cm (6–8 inches) as soon as the first sprout begins to form. This encourages the development of taller plants with more and earlier sprouts.
- The problem with Brussels sprouts and other members of the cabbage clan is bitterness. There are two dozen kinds of bitter receptors in the taste buds of humans, each sensitive to slightly different substances. This complicated system is thought to have evolved to guard against poisons, since plant poisons almost always taste bitter. Given current concerns over the presence of agricultural toxins in food plants, it is startling to learn that at least 10 000 times the weight of natural poison is consumed in food compared to the amount of man-made pesticide residues. Indeed, about 5–10%

Brussels sprouts (*Brassica oleracea* var. *gemmifera*), from Nicholson (1885–1889).

of the dry weight of plants that people consume is made up of natural toxic chemicals that plants manufacture to protect themselves against microorganisms, insects, and animal predators. While many of the thousands of chemicals in plants can be poisonous, their concentrations are for the most part too low to be of concern. The German alchemist and physician Paracelsus (1493–1541) taught that the only difference between a medicine and a poison is the dose, and in small doses, bitter substances can be very beneficial. The ancient Greek physician Hippocrates (460?–377 BC), considered to be the Father of Medicine, said, "Let food be thy medicine," and this certainly applies to some bitter foods. Researchers are making progress developing "bitter blockers," additives that could fool the taste buds into ignoring bitter elements in food, beverages, and medicine. In the meantime, for many the only way to consume foods that taste bad is to combine them with foods that taste good.

Key Information Sources

(See additional references in CABBAGE.)

Abuzeid, A.E., and Wilcockson, S.J. 1989. Effects of sowing date, plant density and year on growth and yield of Brussels sprouts (*Brassica oleracea* var. *bullata* subvar. *gemmifera*). J. Agric. Sci. **112**: 359–375.

Babik, I., Rumpel, J., and Elkner, K. 1996. The influence of nitrogen fertilization on yield, quality and senescence of Brussels sprouts. Acta Hortic. **407**: 353–359.

Bedford, L.V. 1989. Sensory appraisal of Brussels sprouts for the freezing industry. Acta Hortic. **244**: 57–64.

Beecher, C.W. 1994. Cancer preventive properties of varieties of *Brassica oleracea*: a review. Am. J. Clin. Nutr. **59(5S)**: 1166S–1170S.

Burge, W. 1996. Sweeten up your fall with Brussels sprouts. Org. Gard. **43(6)**: 51–53.

Doorn, H.E. van, Kruk, G.C. van der, Holst, G.J. van, Raaijmakers-Ruijs, N.C.M.E., Postma, E., Groeneweg, B., and Jongen, W.H.F. 1998. The glucosinolates sinigrin and progoitrin are important determinants for taste preference and bitterness of Brussels sprouts. J. Sci. Food Agric. **78**: 30–38.

Everaarts, A.P., and Sukkel, W. 1999. Bud initiation and optimum harvest date in Brussels sprouts. Sci. Hortic. (Amsterdam), **81**: 361–367.

Fenwick, G.R., Griffiths, N.M., and Heaney, R.K. 1983. Bitterness in Brussels sprouts (*Brassica oleracea* L. var. *gemmifera*): the role of glucosinolates and their breakdown products. J. Sci. Food Agric. **34**: 73–80.

Great Britain. Agricultural Development and Advisory Service. 1979. Brussels sprouts: production, harvesting, and marketing of Brussels sprouts for the fresh market and for processing. Great Britain Agricultural Development and Advisory Service, Her Majesty's Stationary Office, London, UK. 68 pp.

Heaney, R.K., and Fenwick, G.R. 1980. The glucosinolate content of *Brassica* [Brussels sprouts] vegetables. A chemotaxonomic approach to cultivar identification. J. Sci. Food Agric. **31**: 794–801.

Kronenberg, H.G. 1975. A crop geography of late Brussels sprouts. Neth. J. Agric. Sci. **23**: 291–298.

Lipton, W.J., and Mackey, B.E. 1987. Physiological and quality responses of Brussels sprouts to storage in controlled atmospheres. J. Am. Soc. Hortic. Sci. **112**: 491–496.

Moore, J.F. 1953. Growing Brussels sprouts in Western Washington: adapted varieties, cultural practices, and economic aspects. State College of Washington, Institute of Agricultural Science, Western Washington Experiment Station, Puyallup, WA. 10 pp.

Ockendon, D.J., and Smith, B.M. 1993. Brussels sprouts, *Brassica oleracea* var. *gemmifera* DC. *In* Genetic improvement of vegetable crops. *Edited by* G. Kalloo and B.O. Bergh. Pergamon Press, New York, NY. pp. 87–112.

Paolini, M. 1998. Brussels sprouts: an exceptionally rich source of ambiguity for anticancer strategies. Toxicol. Appl. Pharmacol. **152(2)**: 293–294.

Sciaroni, R.H. 1953. Brussels sprouts production in California. Calif. Agric. Exp. St. Circ. 427. Division of Agricultural Sciences, University of California, Berkeley, CA. 15 pp.

Scott, B.S. 1984. Brussels sprouts: a winter treat [North Carolina]. Organic Gard. **31(4)**: 40–42.

Welbaum, G.E. 1993. Brussels sprouts as an alternative crop for southwest Virginia. *In* New crops. *Edited by* J. Janick and J.E. Simon. Wiley, New York, NY. pp. 573–576.

Whitwell, J.D., Senior, D., and Morris, G.E.L. 1981. Effects of variety, plant density, stopping time and harvest date on drilled Brussels sprouts for processing. Acta Hortic. **122**: 151–165.

Wien, H.C., and Wurr, D.C.E. 1997. Cauliflower, broccoli, cabbage and Brussels sprouts. *In* The physiology of vegetable crops. *Edited by* H.C. Wien. CABI, Oxford, UK. pp. 511–552.

Specialty Cookbooks

(See additional references in BROCCOLI and CABBAGE.)

Brussels Sprout Marketing Program. n.d. Classic recipes for California Brussels sprouts. 9 pp. [Available from library, University of California, Berkeley.]

CABBAGE

Family: Brassicaceae (Cruciferae; mustard family)

Names

Scientific names: Cabbages are species of *Brassica*. These are categorized in the following two groups:

Western Cabbages

- Cabbage—*B. oleracea* L. var. *capitata* L.
- Savoy cabbage—*B. oleracea* var. *sabauda* L.

Oriental Cabbages

- Pak-choi (bok-choi, bok-choy, celery mustard, Chinese cabbage, Chinese mustard, mustard cabbage, pak-choi, pak-choy, white celery mustard)—*B. rapa* L. subsp. *chinensis* (L.) Hanelt
- Pe-tsai (celery cabbage, Chinese cabbage, pai-tsai)—*B. rapa* subsp. *pekinensis* (Lour.) Hanelt
- Chinese savoy (broad-beak mustard, ta ko tsai, taatsi)—*B. rapa* subsp. *narinosa* (L.H. Bailey) Hanelt
- Mizuna—*B. rapa* subsp. *nipposinica* (L.H. Bailey) Hanelt

- "Cabbage" is from the Middle English *caboche*, which is from the Old North French *caboche*, meaning "head." The English expression "cabbage head" is tautological (i.e., it repeats a word or thought, literally translating as "head head").
- "Red-in-snow cabbage" (snow cabbage) is not cabbage; it is a form of Chinese amaranth (*A. tricolor* L.; see AMARANTH).
- "Cabbage trees" refer to species of *Cordyline*.
- Savoy is a former Italian duchy in the Western Alps, ceded to France in 1860. The name "Savoy cabbage" is a translation of the French *chou de Savoie*, cabbage of Savoy. (Note that Savoy is capitalized in "Savoy cabbage" but not in "Chinese savoy.")
- "Pak-choi," "bok-choi," and "pe-tsai" are all based on Chinese dialects for "white vegetable." This has often led to confusion, with the names sometimes applied interchangeably to the same or different vegetables.
- The genus *Brassica* is based on the Latin name of the cabbage. It has been suggested that the word was used by the playwright Plautus in the 3rd century BC, and had the original Latin meaning "cut off the head," presumably in relation to the cabbage head.
- *Oleracea* in the scientific name *B. oleracea* is based on the Latin *oleracea*, meaning having the nature of a potherb or kitchen vegetable.

Drumhead cabbage

Early Étampes cabbage

Varieties of Western cabbage (*Brassica oleracea* var. *capitata*), from Vilmorin-Andrieux (1885).

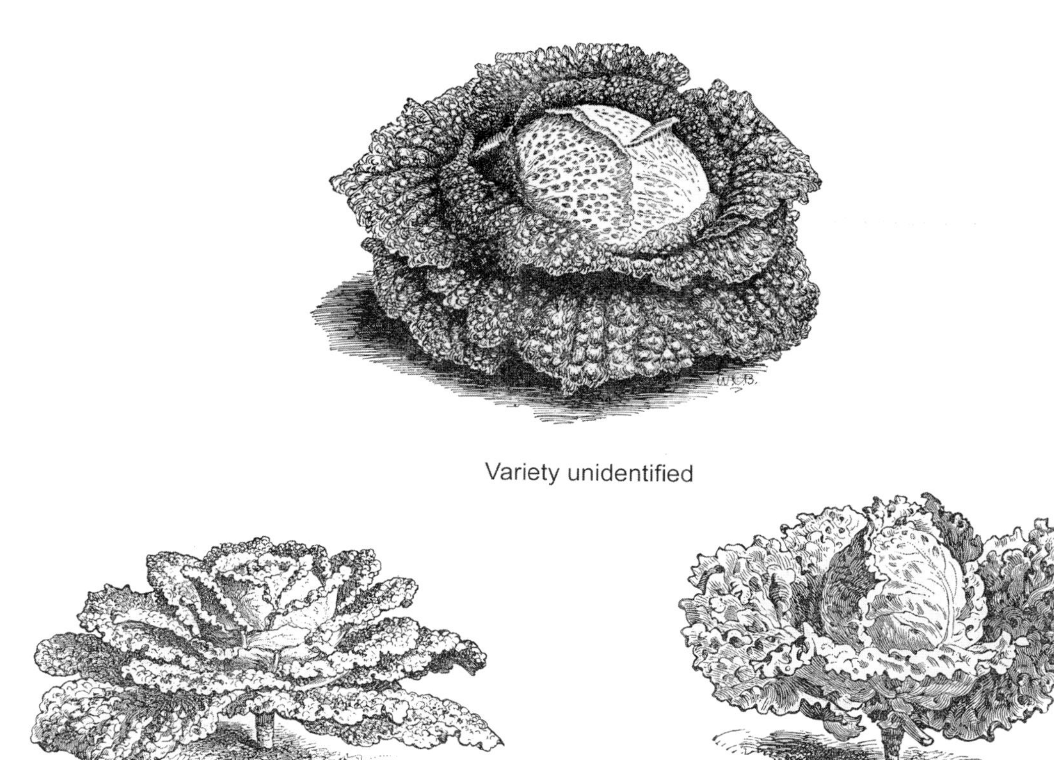

Varieties of Savoy cabbage (*Brassica oleracea* var. *sabauda*). Unidentified variety from Bailey (1900–1902), the remaining two from Vilmorin-Andrieux (1885).

- The Latin variety name *capitata* is based on the Latin word for having a head, an allusion to the head of the cabbage.
- *Rapa* in the scientific name *B. rapa*, and in the name "rape" for oilseed forms of *B. rapa*, is based on the Latin *rapa*, turnip.

Plant Portrait

Cabbage crops are grown worldwide. Approximately 80% of the total world's harvest comes from Asia (especially India and China) and Europe (especially Italy and France).

Western Cabbages

Western cabbages originated from the wild form of *B. oleracea*, which is found along the coasts of Europe. The wild ancestor is a perennial herb with a strong taproot and a stem that becomes woody at the base. It occurs on limestone and chalk cliffs, and on grassy slopes and among shrubs. The most likely area of origin of the cultivated forms is the Mediterranean. The Greek writer Theophrastus (372–286 BC) described a simple type of cultivated cabbage. Both regular-leaved cabbage and Savoy or Milan (crinkle-leaved) cabbage may have been familiar to ancient Romans. Cabbage became popular when the ancient Romans, or more likely the Celts, spread it throughout Europe and Britain. By the Middle Ages, white, red, and Savoy cabbage were known in the Mediterranean and much of Europe. Jacques Cartier brought cabbage to North America in 1541. The first written record of cabbage in the United States is 1669. By the 1700s, both indigenous people and the colonials were growing this vegetable. Western cabbages are biennial plants, but are harvested in the first season. The first year of growth produces the tightly closed ball of leaves called the "head," usually marketed with a weight of 1–3 kg (2–7 pounds) and a diameter of 10–20 cm (4–8 inches). The inner leaves, protected from sunlight, are paler than the outer leaves. Varieties have been selected for fresh market, late or storage market, and for processing. Varieties have also been selected that have differently shaped heads, including round or ball-shaped, pointed, conical, and drumhead. Leaf color may be green, purple, red, or white. Leaves may be smooth, or crinkled as in the Savoy cabbages. The "blistering" or "puckering" of

Savoy cabbage leaves is due to the veins growing less rapidly than other leaf tissues. Cabbage is often considered to be Russia's national food. Russians often consume cabbage at several meals of the day, and eat seven times as much cabbage as the average North American.

Oriental Cabbages

The Oriental cabbages, particularly pe-tsai, are cultivated worldwide and are extremely important in several Asian countries. They are indispensable leaf vegetables in China, Japan, Indonesia, and Malaysia. In China, they make up a quarter of all expenditures for vegetables. These cabbages are known only in cultivation, and probably arose in China from domesticated oilseed forms (known as "rape") of the species *Brassica rapa*, by selection for leaf characteristics. A cabbage-like plant was mentioned in writings from about 300 AD in China. Chinese cabbages were introduced to North America during the 1800s. Unlike Western cabbages, the Oriental cabbages are annuals, although pak-choi may also be a biennial. There are many forms available, and distinguishing them can be difficult (see the illustrations). In China, pe-tsai is referred to as "large white cabbage," whereas pak-choi, Chinese savoy, and occasionally mizuna are termed "small white cabbages." The small white cabbages do not have a strict cabbage-like head. Pak-choi includes selections with loosely packed leaves. Chinese savoy has a flattened rosette of leaves, at least early in its development. Mizuna more closely resembles mustard greens (*B. juncea*; see Mustard) than it does cabbages. There are two types of subsp. *nipposinica*. The "Mizuna Group" consists of deeply dissected, bipinnate leaves; the "Mibuna Group" has slender, entire leaves. Pe-tsai is a heading type that has a more delicate flavor than cabbage. It is used raw in salads or boiled, stir-fried, pickled, made into sauerkraut, and used in soups. Pe-tsai resembles ordinary cabbage in its production of a tightly curled head, but is distinguished by its elongated shape, reminiscent of romaine lettuce. It has a crisp texture, owing to a high water content, less fiber than most cabbages, and a somewhat mustard-like flavor, which is more pleasant than that of most Western cabbages. Pak-choi, which resembles celery and Swiss chard, has both edible leaves and leaf stalks. The leaves can be boiled and the stalks boiled or used in stir-fries. Pak-choi is probably the most familiar of Chinese greens, the leaves and white chunks of stem being staples in Oriental restaurants. Chinese savoy and mizuna are both grown for their leaves that are used raw in salads or boiled.

Culinary Portrait

Cabbages are versatile, and used as food in many ways: in salads and pickles, as raw and cooked vegetables, and as a base for soups, stews, stir-fries, and stuffings. Cabbage blends well with carrots, onion, potatoes, and meats such as bacon and sausages. Coleslaw is a salad of shredded or chopped cabbage, and its taste is improved when chilled in the refrigerator for at least 30 minutes before serving. Cabbage is commonly boiled, braised, steamed, and sautéed, and sometimes microwaved. Cooked cabbage leaves are stuffed with various combinations of meat, sauce, and other vegetables to make tasty cabbage rolls. Cabbage is the main constituent of the popular German dish sauerkraut. Sauerkraut is finely cut cabbage that has been fermented in brine. "Late cabbage," used for sauerkraut, had its origin in Germany. Yeast is universally present in fermenting cabbage. The sour taste of sauerkraut is caused by lactic acid formed by the action of lactic-acid bacteria on the sugar in the cabbage juice. Fermented cabbage is easier to digest than raw cabbage, retains much of the vitamin content of the raw vegetable, and many prefer its texture and flavor.

Because cabbage is cheap, it became and remains a popular vegetable in areas that were once, or are still, economically depressed. Partly for this reason, cabbages are often scorned by gourmets. The vegetable also stinks when boiled, because it gives off sulfurous vapors reminiscent of rotten eggs and ammonia, which has also long discouraged many from eating this very nutritious food. Much of the strong odor is given off in the first 5 minutes of cooking. The smell can be reduced by slicing cabbage into thin strips and stir-frying these so that the coating of hot oil seals in the odors (and indeed the flavor). Another technique of minimizing odor is to place a piece of bread in the cooking water. Overcooked cabbage tends to lose flavor, nutrients, and color, and becomes pasty. It has been recommended that only a little water be present in the cooking pot (2 cm or 3/4 inch), and cooking be brief (once the water is boiling, 4–8 minutes for shredded cabbage, 10–15 minutes for cabbage cut into quarters).

Red cabbage loses its red color in alkaline conditions, becoming mauve, blue, or dirty green. Accordingly, traditional recipes for red cabbage include acidic fruit, vinegar, or wine to reduce the alkalinity. This not only preserves the color but also the flavor. To prevent discoloration, a stainless steel knife should be used to cut red cabbage. Red cabbage is often pickled.

Cabbage tends to wilt at room temperate. It is best stored in a refrigerator in a closed container or plastic bag, preferably just above freezing. Unless well wrapped, the odor of cabbage, especially when cut, tends to transfer to other foods in a refrigerator. If a plastic bag in which cabbage is stored in a refrigerator is too tightly sealed, the vegetable may start to rot, but if it is not adequately sealed, the flavor of other foods may be affected.

Just prior to use, the stem end of the head of cabbage is trimmed away; the outer, coarser leaves may be discarded; the cabbage is rinsed under cold water, and patted dry. To prevent excessive loss of vitamin C, cutting cabbage with a knife has been recommended in preference to using a chopper, followed by use within several hours.

Cabbage and related vegetables (i.e., belonging to the genus *Brassica*, such as cauliflower, broccoli, Brussels sprouts, kohlrabi, rutabaga, and turnip) contain very high levels of antioxidant chemicals (such as dithioltiones and glucosinolates) and anticancer compounds (including isothiocyanates, coumarins, and phenols). The anticancer properties of cabbage are so well established that it has been recommended that there should be increased consumption of cabbages and other crucifer crops (i.e., those belonging to the Cruciferae or cabbage family). Cabbage is so rich nutritionally that it has been called "man's best friend in the vegetable kingdom."

Culinary Vocabulary

- The tender center of some selections of pak-choi (bok-choi) is called *sum* ("heart of cabbage"), and is sometimes sold separated under the names pak-choi sum or bok-choi sum (or simply choy sum).
- "Cole slaw" (or "cold slaw") was so named at the end of the 18th century from the Dutch *kool sla*, cabbage + salad. Probably, the name arose in the United States from Dutch settlers. The original Dutch preparation was likely served hot. It has been said that the tradition of always serving cold cole slaw in North America arose because Americans assumed that "cole" in "cole slaw" meant "cold." Occasionally, the same dish served hot was called "hot slaw."
- *Bigos* (pronounced BEE-gohs), the Polish national dish, is a stew made with alternating layers of sauerkraut and meat. It is also known as hunter's stew. According to legend, hunting parties once brought along servants with large kettles, who kept these continuously heated while adding wild meats, as they were caught, between layers of sauerkraut. The dish was kept simmering for days, eaten daily but considered best on the last day of the hunt.
- *Choucroute* (pronounced shoo-CROOT) is a French sauerkraut dish characteristic of the Alsace region, made from cabbage cooked with goose fat, onions, juniper berries, and white wine. *Choucroute à l'ancienne,* also from France's Alsace district, adds carrots, peppercorns, caraway seeds, garlic cloves, bay leaves, and thyme, and complements the dish with sausages and smoked pork loin. *Choucroute garni* supplements choucroute with potatoes, smoked pork, pork sausages, ham, or goose.
- The "Reuben sandwich" is a combination of sauerkraut, Emmental (i.e., Swiss) cheese, and corned beef between two slices of rye bread, fried in butter. Its invention is most reliably credited to Reuben Kulakofsky (1873–1960) in 1925 while working as a chef at the Blackstone Hotel in Omaha, Nebraska.
- "Bubble and Squeak" is a potato and cabbage dish, named for the bubbling and squeaking that occurs while these vegetables are being fried.

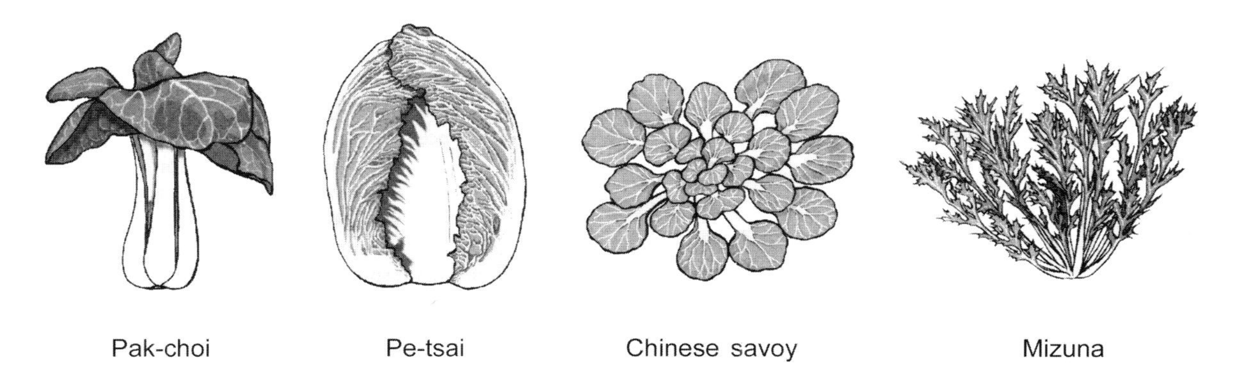

Main types of Oriental cabbages, by M. Jomphe.

Pak-choi (*Brassica rapa* subsp. *chinensis*). *Left*, from Vilmorin-Andrieux (1885); *right*, from Bailey (1900–1902).

- "Colcannon" (pronounced kuhl-CAN-uhn, based on the Gaelic *cal ceannan*, literally "white-headed cabbage") is an Irish dish similar to bubble and squeak (the word "cale-cannon," referring to similar dishes, is derived from colcannon). Colcannon is made of mashed potatoes and cooked cabbage mixed with butter and milk and flavored with chives and parsley. This stew-like preparation is a "national food" in Ireland, where it is typically part of Saint Patrick's Day food festivals. By contrast, corned beef and cabbage is much more commonly associated with the US version of Saint Patrick's Day. The word "colcannon" also refers to a Scottish Highland dish of mashed cabbage, potatoes, turnips, and carrots cooked with butter.
- A Scottish dish of potatoes and cabbage mashed together and browned in the oven has the amusing name of "rumbledethumps."
- "Shchi" (often "schi" in English) is an extremely popular soup in Russia, normally featuring cabbage, with the addition of whatever other vegetables are available, to which some pork or beef broth may be added. *Shchi soldataki*, soldier's shchi, uses sauerkraut instead of cabbage.
- *Kim chee*, *kimchee*, or *kimchi* (pronounced gimchee), also known as Korean cabbage pickle, is a very spicy Korean pickled cabbage (usually napa cabbage), usually made with very hot chile pepper, garlic, green onion, ginger, and other spices. There are many different recipes. By decree of the South Korean government, kim chee is a National Treasure. It is available in cans, sometimes also in plastic bags, in many North American markets.
- The phrase *à la limouisine* refers to a French method of cooking red cabbage with bacon fat, and garnishing with braised chestnuts. Meat and poultry so prepared are also designated by this expression.

Curiosities of Science and Technology

- Both the ancient Egyptians and Romans thought that cabbage could prevent inebriation, and ate it before banquets to keep from becoming drunk. The Greek philosopher Aristotle (384–322 BC), considered one of the wisest people who ever lived, also followed this practice. The Romans also ate cabbages to cure hangovers. A Russian cure for hangover is similar: drinking pickled cabbage water the morning after. An old Ukrainian cure for a hangover is to consume chilled sauerkraut. Backing up these practices is the modern finding that certain chemicals (phenols) in cabbages enhance detoxification enzymes.
- In 1772, English mariner Captain James Cook (1728–1779) learned that sauerkraut had healthful properties, and he ordered 12 tons for his second great voyage exploring the Pacific. In search of the great southern continent then believed to exist, the voyage skirted Antarctic ice fields, visited Tahiti and New Hebrides, and discovered New Caledonia. The vitamin C value of the sauerkraut

contributed to the fact that during the 3-year journey, only one of the 118-man crew perished because of scurvy.

- The common cabbage is 90% or more water.
- Despite its northern location, the long hours of sunlight in the summer can produce giant vegetables in Alaska. One farmer in the Matanuska Valley grew a 45-kg (100-pound) cabbage.
- Pewter pieces can be cleaned by rubbing them with cabbage leaves.
- "Skunk cabbages" are not related to true cabbages. "Eastern skunk cabbage," also known as swamp cabbage, is *Symplocarpus foetidus* (L.) Salisb. ex W.P.C. Barton. The species occurs in eastern North America and Asia. The "skunk" in the name comes from the skunk-like odor when the plant is bruised, and the "cabbage" from the clump of leaves that are formed after flowering, resembling a cabbage. The putrid smell serves to attract insects to pollinate the plant. Flour from the roots of this wild plant was used by eastern American Indians to prepare bread, and the leaves were used as a vegetable. The "western skunk cabbage" is *Lysichitum americanum* Hultén & St. John. It is found from Alaska south to Northern California, and east to Montana and Idaho, and was consumed by Indians in western North America much like the eastern skunk cabbage was eaten. Although some wild food collectors still dine on skunk cabbages, the plants can't be recommended for their culinary virtues. They contain microscopic bundles of needle-like crystals of calcium oxalate in their stems, leaves, and underground parts. While these are apparently denatured somewhat with cooking and (or) drying, there is a possibility of irritation of the mouth, tongue, and throat.
- Germplasm banks are collections of living material, in most cases seeds, particularly of economically important plants, which are preserved on a long-term basis so that new plants can be grown when needed for research and breeding. The largest and most important germplasm bank in the world is the US National Plant Germplasm System, under the control of the U.S. Department of Agriculture. This has more than 400 000 collections, numbered consecutively starting with P.I.1 ("Plant Introduction 1"), which was catalogued in 1898. The collection P.I.1 is a cabbage variety introduced from Russia.

Key Information Sources

Chen, C. 1984. Morphology of Chinese cabbage. Science Publishing Company, Beijing, China. 116 pp.

Chiang, M.S., Chong, C., Landry, B.S., and Crête, R. 1993. Cabbage, *Brassica oleracea* subsp. *capitata* L. *In* Genetic improvement of vegetable crops. *Edited by* G. Kalloo and B.O. Bergh. Pergamon Press, New York, NY. pp. 113–155.

Dahlen, M., and Phillips, K. 1983. A popular guide to Chinese vegetables. Crown Publishers, Inc., New York. 113 pp.

Dickson, M.H., and Wallace, D.H. 1986. Cabbage breeding. *In* Breeding vegetable crops. *Edited by* M.J. Bassett. AVI Publishing Co., Westport, CT. pp. 395–432.

Gómez-Campo, C. (*Editor*). 1999. Biology of *Brassica* coenospecies. Elsevier, New York, NY. 489 pp.

Hodgkin, T. 1995. Cabbages, kales, etc. *In* Evolution of crop plants. 2nd edition. *Edited by* J. Smartt and N.W. Simmonds. Longman Scientific & Technical, Burnt Mill, Harlow, Essex, UK. pp. 76–82.

Jansen, P.C.M., Siemonsma, J.S., and Narciso, J.O. 1993. *Brassica oleracea* L. *In* Plant resources of South-East Asia, 8, vegetables. *Edited by* J.E. Siemonsma and K. Piluek. Pudoc Scientific Publishers, Wageningen, Netherlands. pp. 108–111.

Jiang, M.C. 1981. Cultivation of Chinese cabbage. Agriculture Publishing Company, Beijing, China. 260 pp.

Kalloo, G., and Rana, M.K. 1993. Chinese cabbage, *Brassica pekinensis*, *B. chinensis*. *In* Genetic improvement of vegetable crops. *Edited by* G. Kalloo and B.O. Bergh. Pergamon Press, New York, NY. pp. 179–186.

Kraus, J.E. 1940. Chinese cabbage varieties: their classification, description, and culture in the Central Great Plains. Circular 571. U.S. Dept. of Agriculture, Washington, DC. 20 pp.

Kuo, C.G., and Toxopeus, H. 1993. *Brassica rapa* L. group Chinese Cabbage. *In* Plant resources of South-East Asia, 8, vegetables. *Edited by* J.E. Siemonsma and K. Piluek. Pudoc Scientific Publishers, Wageningen, Netherlands. pp. 127–130.

Li, C. 184. Chinese cabbages of China. Agriculture Publishing company, Beijing, China. 234 pp.

Lin, W. 1980. A study on the classification of Chinese cabbages. Acta Hortic. Sinica, 7: 21–28.

Myers, C. 1991. Nappa cabbage, Chinese cabbage, celery cabbage, pe-tsai. Crop Sheet SMC-024. *In* Specialty and minor crops handbook. *Edited by* C. Myers. The Small Farm Center, Division of Agriculture and Natural Resources, University of California, Oakland, CA. 4 pp.

Nieuwhof, M. 1969. Cole crops: botany, cultivation, and utilization. L. Hill, London, UK. 353 pp.

Opeña, R.T., Kuo, C.G., and Yoon, J.Y. 1988. Breeding and seed production of Chinese cabbage in the tropics and subtropics. Tech. Bull. 17. Asian Vegetable Research and Development Center, Shanhua, Tainan, Taiwan. 92 pp.

Prakash, S., and Hinata, K. 1980. Taxonomy, cytogenetics and origin of crop brassicas, a review. Opera Bot. **55**: 1–57.

Pritchard, M.K., and Becker, R.F. 1989. Cabbage. *In* Quality and preservation of vegetables. *Edited by* N.A. Eskin. CRC Press, Inc., Boca Raton, FL. pp. 265–284.

Seelig, R.A., and Bing, M.C. 1990. Cabbage. *In* Encyclopedia of produce. [Loose-leaf collection of individually paginated chapters.] *Edited by* R.A. Seelig and M.C. Bing. United Fresh Fruit and Vegetable Association, Alexandria, VA. 13 pp.

Talekar, N.S., and Griggs, T.D. (*Editors*). 1981. Chinese cabbage. Proceedings of the first international symposium of Chinese cabbage. Asian Vegetable Research and Development Center, Shanhua, Tainan, Taiwan. 489 pp.

Tay, D.C.S., and Toxpeus, H. 1993. *Brassica rapa* L. cv. Group Pak Choi. *In* Plant resources of South-East Asia, 8, vegetables. *Edited by* J.E. Siemonsma and K. Piluek. Pudoc Scientific Publishers, Wageningen, Netherlands. pp. 130–134.

Thompson, K.F. 1976. Cabbages, kales, etc. *In* Evolution of crop plants. *Edited by* N.W. Simmonds. Longman, London, UK. pp. 49–52.

Tsunoda, S., Hinata, K., and Gómez-Campo, C. (*Editors*). 1980. Brassica crops and wild allies: biology and breeding. Japan Scientific Societies Press, Tokyo, Japan. 354 pp.

Vossen, H.A.M. van der. 1993. *Brassica oleracea* L. cv. Groups White Headed Cabbage, Red Headed Cabbage, Savoy Headed Cabbage. *In* Plant resources of South-East Asia, 8, vegetables. *Edited by* J.E. Siemonsma and K. Piluek. Pudoc Scientific Publishers, Wageningen, Netherlands. pp. 117–121.

Wien, H.C. 1997. Cauliflower, broccoli, cabbage and Brussels sprouts. *In* The physiology of vegetable crops. *Edited by* H.C. Wien. CABI International, Wallingford, Oxon, UK. pp. 511–552.

Specialty Cookbooks

Cooper, M. 2003. The ultimate cabbage soup diet: lose up to 10 pounds in seven days—and never feel hungry. John Blake, London, UK. 192 pp.

Daniels, M. 2001. The cabbage soup diet recipe book. John Blake, London, UK. 197 pp.

Haslinger, A.L. 1974. The cabbage cookbook. Arco Publishing Co., New York, NY. 156 pp.

Hendrickson, A., and Hendrickson, J. 1989. Broccoli & company: over 100 healthy recipes for broccoli, Brussels sprouts, cabbage, cauliflower, collards, kale, kohlrabi, mustard greens rutabaga, and turnip. Storey Communications, Pownal, VT. 140 pp.

Hiatt, J.M., 1989. Cabbage: cures to cuisine. Naturegraph Publishers, Happy Camp, CA. 127 pp.

Kaufmann, K.N., and Schoenech, A. 1998. The cultured cabbage: rediscovering the art of making sauerkraut. Revised edition. Alive Books, Burnaby, BC. 79 pp.

Paananen, E. 1984. Of cabbages and the king: the world cabbage cookbook. Betterway Publications, White Hall, VA. 189 pp.

Telian, J., and Zantman, H. 2002. The savoy cabbage cookbook. Spearhead, Claremont, South Africa. 111 pp.

Turgeon, S. 1977. Of cabbages and kings cookbook: an uncommon collection of recipes featuring that family of vegetables which includes broccoli, Brussels sprouts, cauliflower, collards, turnips, kale, and kohlrabi. Curtis Publishing Co. Indianapolis, IN. 135 pp.

Cardamom (Cardamon)

Family: Zingiberaceae (ginger family)

Names

Scientific name: *Elettaria cardamomum* (L.) Maton (*Amomum cardamomum* L.)

- The English name "cardamom" (or "cardamon") is based on a blend of the Greek *kardamōmon*, the name for some peppery herb, and *amōmon*, an Indian spice plant.
- In North America, "cardamom" is the preferred term, although sometimes "cardamon" is used. In England, the Middle East, and the Far East, the plural (cardamoms) is generally employed.
- Cardamom is also known as Malabar cardamom and Ceylon cardamom. "Malabar cardamom" has flowering and fruiting stems that trail on the ground, while "Mysore cardamom" has flowering stems that hold the flowers and fruits (which are larger) off the ground. The latter type is often preferred because the fruits are less likely to be damaged or spoiled because they lie on the ground.
- "Black cardamom" and "brown cardamom" are collective names for several relatives of cardamom, often of inferior quality. One of these is black cardamom (Indian cardamom, Nepal cardamom), *Amomum subulatum* Roxb., a species of the Eastern Himalayas, produced mainly in Nepal and Sikkim. "Wild cardamoms" are also species used as substitutes or adulterants. Siam cardamom is *Amomum krervanh* Pierre ex Gagnep. (*A. testaceum* Ridley), a native of Southeast Asia, used in Thailand and Cambodia as a cardamom substitute. Round or Jawa cardamom is *A. compactum* Sol. ex Maton (*A. kepulaga* Sprague & Burkill), from Indonesia. Grains of Paradise, also known as Melegueta pepper, is *Aframomum melegueta* K. Schum.), from Africa's west coast, which has seeds approximately the size and shape of cardamom seeds, but reddish brown in color (or pale gray when powdered); the seeds are hard to obtain in Western countries, but are sometimes used as a reasonable substitute for cardamom in the Old World.
- "Small cardamom" is *Elettaria cardamomum*, contrasted with "large cardamom," a phrase used for *Aframomum* and *Amomum* species.
- The genus name *Elettaria* is based on the local name of the spice in a Southeast Asian language (perhaps like the Hindi *elaichi*).
- For the derivation of *cardamomum* in the scientific name *E. cardamomum*, see the information above for the word "cardamom."

Plant Portrait

Cardamom is a perennial, clump-forming, tropical herb, native to monsoon forests of southern India and Sri Lanka (Ceylon). It develops leafy shoots 1.8–5.5 m (6–18 feet) high, and flowering stems 60–120 cm (2–4 feet) tall that arise from its base. The orchid-like flowers are 2.5–3.8 cm (1–1½ inches) long, white or pale green with a violet central lip. The fruit is three-sided, oval, 1.3–1.9 cm (1/2–3/4 inch) long, with 15–20 hard, reddish brown to brownish black, angular seeds with a strong aromatic odor and flavor. The spice consists of the whole fruits or just the hulled seeds. Whole seedpods are usually white (bleaching to a creamy white color may result from the fumes of burning sulfur), but are also sold in black form (the result of sun-drying, producing strong flavor) and green (the result of drying in indoor kilns, producing a milder flavor, often considered superior).

Cardamom was mentioned in an Egyptian papyrus dated at 1550 BC as having numerous medical properties, and it has been used in India as medicine at least since the 4th century BC. The spice is believed to have been grown in the garden of the king of Babylon in 721 BC. The classical Greeks and Romans imported it from India. Today, about 80% of the world's supply is produced in India, with lesser amounts supplied by Guatemala, Tanzania, Malaysia, Cambodia, Nepal, Sri Lanka, Mexico, and Thailand. Cardamom is so important in India that it has given its name to the Cardamom Hills of the Kerala region (a state in the southern tip of India); to Cardamom County, a resort area in the Cardamom Hills; and to the town of Bodinayakanur, which is called "Cardamom City" in recognition of the large quantity of the spice that is traded there. Cardamom is much more popular in Asia than in North America and most of Europe, although it does have considerable use in Scandinavian countries. Extracted essential oil from cardamom is occasionally used medicinally to reduce gas and to mask the bad taste of drugs. The essential oil is also used to flavor tobacco, and as an ingredient in perfume. However, the oil is expensive and so has limited use.

Cardamom (*Elettaria cardamomum*), from Köhler (1883–1914).

Culinary Portrait

Cardamom is sweetly, very pleasantly aromatic, with a warm, slightly pungent flavor somewhat reminiscent of camphor. It is used to flavor cakes, cookies, muffins, buns, and other baked goods, curries, rice, vegetables, meats, and marinades. In most recipes, cardamom can replace ginger and cinnamon. To release the flavor, cardamom pods should be crushed before use. Powdered cardamom is desirable for baking. Essential oil from cardamom is used to a small extent to flavor liqueurs and commercial meats.

When kept in the seedpods, the seeds retain their taste better, and so it is preferable to buy entire, not ground cardamom. Aside from the fact that the flavor is retained better when the seeds are kept in the whole pods, there is another reason for keeping the seeds within their capsules. Cardamom is very expensive, and when the seeds are removed from the pod, this spice is easily adulterated with pepper, mustard seed, or cheaper grades of cardamom. Even when left in the pods, in a year the seeds lose about 40% of their essential oil, which is responsible for much of the flavor. Storing in a sealed container is advisable. The shelf life of powdered cardamom is only a few months.

Cardamom is employed principally in Indian and Middle Eastern cuisines. In Arab countries, particularly Saudi Arabia, it is the most popular of spices. Indian curry contains cardamom as a main ingredient. Cardamom is particularly used in preparing cardamom coffee (*gahwa*), often served ritually to guests from a brass pot with an elongated curved spout. Bedouins (Arabic nomads) sometimes have coffee pots designed to hold several cardamom capsules in their spouts in order to flavor the coffee as it is poured. As much as 2 teaspoons of cardamom may be used for a small cup of coffee, producing a beverage that is very heavily spiced.

Culinary Vocabulary

- "Glogg" (glögg, gloog) is a traditional Christmas drink in Sweden as well as for people of Swedish ancestry elsewhere. It is made with red wine, sugar, raisins, cinnamon sticks, cloves, and cardamom seeds—mixed and heated for about an hour, allowed to sit for a day, sieved and reheated before serving, with liquor and a garnish of almonds, raisins, or orange peel sometimes added.

Curiosities of Science and Technology

- Cardamom is the third most costly spice, following saffron and vanilla. In Asia, it has been said that while black pepper is the king of spices, cardamom is the queen.
- An old Asian tradition is chewing fennel and cardamom seeds after meals to sweeten breath and act as a toothbrush.
- In the story *Beyond the Pale*, by English writer Rudyard Kipling (1865–1936), a man in India receives an "object letter," i.e., a collection of several objects, each with symbolic value, which needs to be translated according to local knowledge in order to be understood. The objects included 11 cardamom seeds (one cardamom seed signified jealousy). The message was "A widow desires you to come at eleven o'clock to the gully with the heap of cattle food."
- In Scandinavia, cardamom seeds are sometimes chewed to mask alcoholic breath.
- About 110 000 cardamom seeds weigh 1 kg (there are about 50 000 seeds in a pound).
- It has been suggested that the popularity of cardamom in Arab countries is due to the cooling effect

Cardamom (*Elettaria cardamomum*), from Jackson (1890).

of the spice, which is welcome in the hot climate, and the beliefs that it is good for digestion and is a powerful aphrodisiac. According to a saying, "a poor man in Saudi Arabia would rather give up his rice than his cardamom."

Key Information Sources

Burtt, B.L., and Smith, R.M. 1983. Zinbiberaceae. *Elettaria*. *In* A revised handbook to the flora of Ceylon. Vol. 4. *Edited by* M.D. Dassanayake and F.R. Fosberg. Amerind Publishing Co., New Delhi, India. pp. 528–530.

Grant, A.W.B. 1982. Cardamom. 2nd edition. Dept. of Primary Industry, Port Moresby, Papua, New Guinea. 35 pp.

Institut de Recherches Agronomiques Tropicales de des Cultures Vivrières. [IRAT], and Centre de Coopération Internationale en Recherche Agronomique pour le Développement [CIRAD], Nogent-sur-Marne, France. 1988. Cardamom bibliography. CIRAD and IRAT, Montpellier, France. 365 pp. [In English and French.]

Jos, C.A. 1982. Story of queen of spices. Cardamom (India), **14(6)**: 3–13.

Korikanthimath, V.S. 1995. Economics of sustained production of cardamom. J. Spices Aromat. Crops, **4(2)**: 119–128.

Korikanthimath, V.S. 2001. Cardamom (small). *In* Handbook of herbs and spices. *Edited by* K.V. Peter. CRC Press, Boca Raton, FL. pp. 123–133.

Nair, K.N., Narayana, D., and Sivanandan, P. 1989. Ecology or economics in cardamom development. Oxford & IBH Publishing Co., New Delhi, India. 119 pp.

Narasimhan, S., Raghuveer, K.G., and Lewis, Y.S. 1982. Cardamom—production, technology, chemistry, and quality. CRC Crit. Rev. Food Sci. Nutr. **16**: 229–236.

Nayar, N.M., and Ravindran, P.N. 1995. Herb spices. *In* Evolution of crop plants. 2nd edition. *Edited by* J. Smartt and N.W. Simmonds. Longman Scientific & Technical, Burnt Mill, Harlow, Essex, UK. pp. 491–494.

Parameswar, N.S. 1973. Floral biology of cardamom (*Elettaria cardamomum* Maton). Mysore J. Agric. Sci. **7**: 205–213.

Peter, K.V., and Kandiannan, K. 1999. Cardamom. *In* Tropical horticulture, vol. 1. *Edited by* T.K. Bose, S.K. Mitra, A.A. Farooqui, and M.K. Sadhu. Naya Prokash, Calcutta, India. pp. 657–670.

Ramanatha Menon, M. 1988. Diseases of cardamom: a review. Kerala Agricultural University, Trichur, India. 38 pp.

Ravindran, P.N., and Madhusoodanan, K.L. 2002. Cardamon: the genus *Elettaria*. Taylor and Francis, New York, NY. 374 pp.

Sahadevan, P.C. 1965. Cardamom. [Trivandrum] Agricultural Information Service, Dept. of Agriculture, Kerala, India. 90 pp.

Samarawira, I. 1972. Cardamon. World Crops, **24**: 76–78.

Sarna, Y.R., Devasahayam, S., and Anandaraj, M. 1992. Problems and prospects: proceedings of national seminar on black pepper and cardamom, 17–18 May 1992, Calicut, Kerala, India. Indian Society for Spices, Calicut, India. 90 pp.

Shankaracharya, N.B., and Natarajan, C.P. 1971. Cardamom: chemistry, technology and uses. Indian Food Packer, **25**: 28–36.

Subbarao, G. 1980. Bibliography on cardamom (*Elettaria cardamomum* (L.) Maton). Central Plantation Crops Research Institute, Kasaragod, Kerala, India. 97 pp.

Wardini, T.H., and Thomas, A. 1999. *Elettaria cardamomum* (L.) Maton. *In* Plant resources of South-East Asia, 13, spices. *Edited by* C.C. de Guzman and J.S. Siemonsma. Backhuys, Leiden, Netherlands. pp. 116–120.

Wijesekera, R.O.B., Nethsingha, C., Paskaranathan, U., and Rajendran, P. 1975. Cardamom. National Science Council of Sri Lanka, Colombo, Sri Lanka. 54 pp.

Specialty Cookbooks

Morris, S. 1998. Cardamom & coriander: a celebration of Indian cooking. Metro, London, UK. 220 pp.

Rich, J.M. 1993. Cooking with cardamom: the exotic mystery spice. J.M. Rich, Clearwater, FL. 81 pp.

CARROT

Family: Apiaceae (Umbelliferae; carrot family)

Names

Scientific name: *Daucus carota* L.

- "Carrot" and *carota* in the scientific name of the carrot are based on the classical Greek word for carrot, *karoton*.
- "Carrot," or similar words, often have nothing to do with the vegetable called carrot. The points of the hoops stuck in the ground used in croquet are called carrots. *Carat* is the unit of weight for precious stones, equal to 200 mg. *Karat* is 1/24 part of otherwise pure gold. *Caret* is a proofreader's insertion mark.
- Wild carrot is called Queen Anne's lace, generally interpreted as commemorating Queen Anne (1655–1714), who is believed to have pricked her finger while making lace, drawing a drop of blood. The foliage is indeed lacy, and at the time of Queen Anne, lace collars were popular. A dark reddish flower characteristically occurs in the center of the umbel of wild carrot flowers, and this is said to be the blood of the Queen. An ancient folk remedy to treat epilepsy was to eat the reddish flower. The reddish flower may serve to draw insects to the flowers, needed for pollination. An alternative interpretation of the name "Queen Anne" is that is refers to St. Anne, the mother of the Virgin Mary and the patron saint of lace makers.
- The wild carrot (Queen Anne's lace) is sometimes called bird's nest because of the shape of the fruiting flower head.
- According to an old superstition, your mother would die if wild carrot was brought into the house. Wild carrot is usually called Queen Anne's lace, but because of this old belief it has also been called "mother die."
- The "Peruvian carrot," *Arracacia xanthorryza* Bancr., more commonly known as arracha and Peruvian parsnip, and sometimes called white carrot, is a root vegetable of the carrot family, cultivated in northern South America and the West Indies.
- The natural sweetness of the carrot is reflected in an early Celtic name for the vegetable: "honey underground," a name that was used for some time in Ireland.
- The genus name *Daucus* has been said to be derived from the Greek *daio*, to make hot, an allusion to its supposed effect in past medicinal practice.

Plant Portrait

The cultivated carrot is one of the most important root vegetables grown in temperate regions of the world. It was derived from the wild carrot, which has whitish roots. Early writings in classical Greek and Roman times refer to edible white roots, but these may have been parsnips. There are white-rooted carrots in existence today, used as fodder for livestock. The earliest vegetable definitely known to be a carrot dates to 10th century Persia and Asia Minor, and was quite unlike the familiar orange-rooted carrot of today. This Asian type of carrot has purplish or yellowish roots. The purple is due to purple pigments called anthocyanins, which are water soluble like the purple coloring of beet, and stain skin and clothing. These old-fashioned types of purple and yellow carrots are no longer grown. The familiar, orange-rooted type of carrot is orange because of orange pigments, known as carotenes, and these are converted to vitamin A by the human body. The orange carrot seems to have first become well known about 1600 in the Netherlands. Before then, orange carrots were completely unknown (this is quite clear when one looks at carrot colors in old paintings before and after 1600). The cultivated carrot is remarkably diverse, and comes in an astonishing range of shapes. Size varies from large roots of the order of 200 g (7 ounces) to "baby" carrots of 5–10 g (0.18–0.35 ounce). Carrots are sweet, most of the sugar located in the outer ring of flesh circling the core. With maturity, simple sugar (glucose) is converted to table-sugar (sucrose), and the carrot becomes sweeter.

Culinary Portrait

Carrots are not everyone's favorite food. The pejorative slang phrase "rabbit food" is applied to carrots more than any other vegetable. They are valued in dieting, but the expression "carrot-stick dieting" is sometimes used to refer to a rather monotonous weight-reducing regime. One dieting cookbook is entitled *I'm Sick of Carrot Sticks*. Despite such viewpoints, the carrot is a star of the vegetable world.

The carrot is widely consumed, its orange–yellow color adding attractiveness to foods. It is available fresh, canned, frozen, and dehydrated, and is consumed raw, as juice, or cooked in various ways. Carrots are served raw as appetizers and in salads, and cooked carrots are served separately as side dishes (typically whole and glazed, or with a cream sauce or butter, or puréed), or combined with innumerable other foods. Carrots are found in stews, soups, quiches, soufflés, and omelets. They are also used in the baking of some cakes and cookies. Valuable nutrients are near the thin outer skin of carrots, which are therefore best scraped with a hard brush rather than peeled, especially when young (old carrots may need to be peeled). If the top end of a carrot is exposed to sunlight, it becomes green and bitter, and this should be cut away. Young carrots are best served raw or cooked for a short time. Older carrots withstand long, slow cooking better, and also retain color and flavor. Carrots are used as a flavoring ingredient of soups and sauces and are a chief component of mixed vegetable dishes. Carrot syrup is sometimes employed as a sweetening agent. Alcoholic tincture of carrot seeds is incorporated into French liqueurs, and the roots are occasionally used to make wine. Roasted roots have sometimes been used as a substitute for coffee. Carrot tops may be added to salads, soups, and stews.

Carrots store relatively well. Young carrots may be kept for up to 2 weeks in a refrigerator, older ones for up to 3 weeks. The roots are best wrapped in paper toweling (to prevent condensation wetting the carrots, which will spoil them) and placed in a perforated plastic bag before refrigeration. Carrots can also be stored in a dark, cold (just above freezing) place with humidity over 90% (but with some ventilation). Old traditional ways of storing carrots for months were to simply leave them in the soil where they grew (which works as long as the plants are mulched and the temperature does not drop too low) or to cover them, unwashed, with sand in a cold room.

A culinary trick used in slicing carrots and other long, round root vegetables is to first cut a thin strip off one side, and lay it on that flat side so that it won't slip while being sliced crosswise. Carrot curls are made by first peeling carrots, then using a vegetable peeler to shave strips lengthwise from long, straight carrots, working spirally around the carrot. The curls can be wound around a finger until completed. Their flavor is enhanced by placing them on ice and keeping them refrigerated for an hour before serving.

Carrot tops (i.e., leaves) are best chopped off before being shipped to market because they sap moisture from the roots and lose it to the air, causing the roots to shrivel. The tops may be left on to create an attractive appearance in the store, but should be chopped off before storing so that they will remain fresh longer.

Culinary Vocabulary

- The French word *Crécy* (pronounced kray-cee) indicates that carrots are being used in a dish. *À la Crécy* refers to a French garnish made of julienned carrots (i.e., cut into matchstick shape) or, more generally, a French method of preparation in which carrots are used. *Purée Crécy* is puréed carrots. *Consommé Crécy* is a rich beef broth garnished with julienned carrots. *Omellete Crécy* is an omelet made with carrots. Crécy is a village in France, which once produced carrots famous for their fine quality. In 1346, British troops are believed to have been fed carrot soup during a battle with the French; it is thought that ever since, the word Crécy has been associated with carrots.
- *Carrottes à la Vichy* is a French dish of carrots cooked in Vichy water (a French mineral water from the town of Vichy).
- "Tzimmes" is a Jewish dish of root vegetables and other components, which typically features sliced carrots sweetened with honey. It is often served at Rosh Hashana, the Jewish New Year, with the symbolism that the round sections of carrots resemble coins, indicative of prosperity.

Curiosities of Science and Technology

- In the Middle Ages, women decorated their hair with feathery carrot leaves. At the court of English King James I (1566–1625), carrot foliage was worn as hat plumes.
- Originally, a "carrot and stick" was a device fastened to the necks of horses and mules just beyond their ability to grasp it in their mouths, and designed to motivate them to move forward. The expression "carrot and stick" also implies motivating by offering carrots as a reward, but coupled with the threat of being beaten by a stick. This use of the word "carrot" to mean "inducement" was first recorded in the 1890s. The idea that the stick represented the opposite of a reward, i.e., a punishment, is something that people added over time.
- The birds-nest-shaped fruit cluster of carrot has a remarkable mechanism for seed dispersal. The stalks are hygroscopic (readily taking up and retaining moisture), so that when conditions are dry and suitable for seed dispersal, they bend outward, exposing the fruits to wind and animals; when conditions are wet, they bend inwards, forming the familiar bird's nest structure, which protects the seeds.

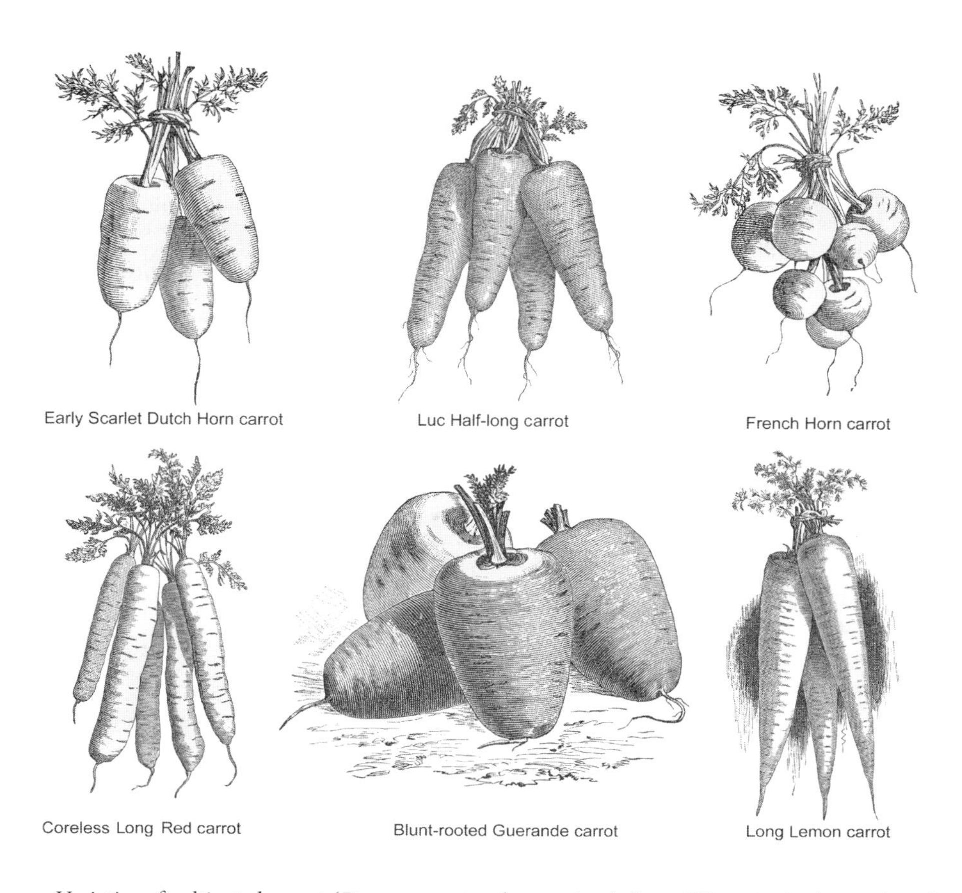

Varieties of cultivated carrot (*Daucus carota* subsp. *sativus*), from Vilmorin-Andrieux (1885).

- Very high consumption of carrots can give the skin an orange tone, particularly on the palms of hands and feet. Although indicative of an unbalanced diet, this is not harmful, and it is believed to be very difficult to consume too much beta-carotene, the precursor of vitamin A, by eating carrots because the body doesn't convert the excess beta-carotene to vitamin A. However, it is certainly possible to consume too much vitamin A, which may be toxic at only 15 times the recommended dietary allowance.
- Carrots provide 30% of the vitamin A content of the US diet.
- Vision is extremely important to humans, and the bright orange, yellow, and red colors of fruits and vegetables have been selected because they are attractive, not because the pigments causing the colors are often beneficial nutritionally. Carotenoid pigments as found in carrots are not only attractive to humans, but also to birds (where they produce yellow and red feathers), fish (e.g., goldfish and salmon), and lower animals, where the carotenoids can be combined with proteins to produce blue, green, and purple colors.
- The natural colors of vegetables and fruits are known to be based on pigmented chemicals that help prevent many serious illnesses such as cancer and heart disease, as well as defending the body against ageing. A "traffic-light-diet," with plenty of red, orange, and green fruits and vegetables, is especially healthy. For examples,
  - Red: sweet pepper, raspberry, red currant, strawberry, grape, tomato, and carrot.
  - Yellow/orange: orange, mango, plum, melon, pumpkin, sweet potato, and of course, carrot.
  - Green: cabbage, Brussels sprouts, broccoli, lettuce, pea, and spinach.
- A carrot is approximately 88% water.
- A neurotoxin, carotatoxin, occurs naturally in carrots, but in such small concentration that it has little significance for normal consumers of carrots.
- Carrots are used in some commercial breast creams because beta-carotene is a phytoestrogen, a substance similar to female hormones, and so hypothesized to stimulate breast development.

Wild carrot or Queen Anne's lace (*Daucus carota* subsp. *carota*), from Hallier (1880–1888, vol. 27, plate 2833).

- Growing clones (copies) of plants and animals, starting with samples of cells, is now a widely accepted fact of science, but it was not always so. Until the late 1950s, biologists thought that mature cells would not divide and grow to form complete organisms. In 1958, F.E. Steward, a biologist working in the Lab of Cell Physiology, Growth, and Development at Cornell University in Ithaca, New York, succeeded in growing complete carrot plants, starting by culturing a carrot root cell in a tube containing cell nutrients. Carrot tissue is now extremely popular "test tube" culture material. Human and carrot cells have been fused.
- It has been calculated that one carrot provides enough energy to walk 1.6 km (1 mile).
- Carrots are said to be a good accompaniment to beef because the mildly laxative effect of carrots can counteract the sometimes constipating effect of beef.
- In some of its legislation, the European Community has classified the carrot as a fruit, not a vegetable.

Key Information Sources

Babb, M.F., Kraus, J.E., and Magruder, R. 1950. Synonymy of orange-fleshed varieties of carrots. Circ. 833. U.S. Dept. of Agriculture, Washington, DC. 100 pp.

Banga, O. 1963. Main types of the Western carotene carrot and their origin. W.E.J. Tjeenk Willink, Zwolle, Netherlands. 153 pp.

Brown, J.G. 1950. Diseases of carrots. Agricultural Experiment Station, University of Arizona, Tucson, AZ. 19 pp.

Crête, R. 1980. Diseases of carrots in Canada. Revised edition. Agriculture Canada, Ottawa, ON. 26 pp.

Crosby, D.G., and Aharonson, N. 1967. The structure of carotatoxin, a natural toxicant from carrot. Tetrahedron, **23**: 465–472.

Dale, H.M. 1974. The biology of Canadian weeds. 5. *Daucus carota*. Can. J. Plant Sci. **54**: 673–685.

Dowker, B.D. 1971. Variation studies in carrots as an aid to breeding. I. Concepts. J. Hortic. Sci. **46**: 485–497.

Dowker, B.D., Fennell, J.F.M., and Jackson, J.C. 1976. Variation studies in carrots as an aid to breeding. III. Size and shape characters. J. Hortic. Sci. **51**: 235–244.

Esau, K., Hewitt, W.B., and Brooks, R.M. 1949. Developmental anatomy of the fleshy storage organ of *Daucus carota*. Hilgardia, **13**: 175–306.

Heywood, V.H. 1983. Relationships and evolution in the *Daucus carota* complex. Isr. J. Bot. **32**: 51–65.

Mackevic, V.I. 1929. The carrot of Afghanistan. Bull. Appl. Bot. Genet. Plant Breed. **20**: 517–557.

Magruder, R. 1940. Descriptions of types of principal American varieties of orange-fleshed carrots. U.S. Dept. of Agriculture, Washington, DC. 48 pp. + plates.

Organisation for Economic Co-operation and Development. 2000. Carrots. O.E.C.D., Paris, France. 69 pp. [In English and French.]

Peterson, C.E., and Simon, P.W. 1986. Carrot breeding. *In* Breeding vegetable crops. *Edited by* M.J. Bassett. AVI Publishing Co., Westport, CT. pp. 321–356.

Quagliotti, L. 1967. Effects of different temperatures on stalk development, flowering habit and sex expression in the carrot. Euphytica, **16**: 83–103.

Riggs, T.J. 1995. Carrot. *In* Evolution of crop plants. 2nd edition. *Edited by* J. Smartt and N.W. Simmonds. Longman Scientific & Technical, Burnt Mill, Harlow, Essex, UK. pp. 477–480.

Roxas, V.P. 1993. *Daucus carota* L. *In* Plant resources of South-East Asia, 8, vegetables. *Edited by* J.E. Siemonsma and K. Piluek. Pudoc Scientific Publishers, Wageningen, Netherlands. pp. 167–171.

Rubatzky, V.E., Quiros, C.F., and Simon, P.W. 1999. Carrots and related vegetable Umbelliferae. CABI, Wallingford, Oxon, UK. 294 pp.

Shimmin, T. (*Editor*). 2000. Proceedings of carrot conference Australia: Perth, Western Australia, Oct. 2000. Carrot Association for Research and Development (WA Inc.), Perth, WA, Australia. 80 pp.

Simon, P.W. 1993. Carrot, *Daucus carota* L. *In* Genetic improvement of vegetable crops. *Edited by* G. Kalloo and B.O. Bergh. Pergamon Press, New York, NY. pp. 479–484.

Small, E. 1978. A numerical taxonomic analysis of the *Daucus carota* complex. Can. J. Bot. **56**: 248–276.

Stein, M., and Nothnagel, T. 1995. Some remarks on carrot breeding (*Daucus carota sativus* Hoffm.). Plant Breed. **114**: 1–11.

Walde, S.G., Math, R.G., Chakkaravarthi, A., and Rao D.G. 1992. Preservation of carrots (*Daucus carota* L.) by dehydration techniques—a review. Indian Food Packer, **46**(**6**): 37–42.

Whitaker, T.W. 1970. Carrot production in the United States. U.S. Agricultural Research Service, Washington, DC. 37 pp.

Specialty Cookbooks

Barry, L., and Barry, B. 1981. Carrot cookbook. Barry Publishing Co., Prescott, AR. 92 pp.

Central Florida Wildlife Adoption Society. 1984. The gourmet carrot cookbook. Central Florida Adoption Society, Howey-in-the-Hills, FL. 98 pp.

Downard, G. 1995. The big carrot book. Random House, New York, NY. 94 pp.

Hendrickson, A., and Hendrickson, J. 1986. The carrot cookbook. Storey Communications, Pownal, VT. 171 pp.

Hill, N. 1995. The carrot cookbook: more than sixty easy, imaginative recipes. Courage Books, Philadephia, PA. 64 pp.

Jean, N. 1982. The classic carrot cookbook for 24-carat cooks: a collection from the kitchens of garden club members and their friends. Arizona Federation of Garden Clubs, Tempe, AZ. 136 pp.

Lukin, A. 1989. The big carrot vegetarian cookbook: recipes from the kitchen of The Big Carrot. Second Story Press, Toronto, ON. 141 pp.

Saling, A. 1975. The carrot cookbook. Pacific Search, Seattle, WA. 158 pp.

Siegel, H. and Gillingham, K. 2000. The totally carrot cookbook. Celestial Arts, Berkeley, CA. 95 pp.Cassava

CASSAVA

Family: Euphorbiaceae (spurge family)

Names

Scientific name: *Manihot esculenta* Crantz [*Jatropha manihot* L., *M. aipi* Pohl, *M. dulcis* (J.F. Gmel.) Pax, *M. melanobasis* Müll. Arg., *M. utilissima* Pohl]

- "Cassava" is derived from the Spanish *cazabe*, cassava bread, which in turn comes from the Taino of Haiti (language of the Taino Amerindians) word *cacabi* or *cāsavi*.
- Cassava is also called Brazilian arrowroot, manioc, tapioca (plant), and yuca. Cassava is known as manioc in French-speaking areas, and tapioca and yuca in Latin America.
- The name "manioc" as well as the genus name *Manihot* are based on *mandioca*, the Tupi language of the Brazilian Amazon. The word is related to *Mani-óca*, a legendary Indian woman whose body grew the manioc plant according to Brazilian Indian legends.
- "Tapioca" is derived through Spanish from the Tupi (language used in Amazon Basin of South America) word *typóca*, meaning "juice to be removed from the heart (of the root)." This reflects the mode of preparation: starch was produced by crushing the roots, steeping in water, and squeezing the liquid out.
- "Yuca" (with one *c*) is the common word for cassava in Latin America, and because most cassava in North America is imported from South America, it is commonly called yuca in North America. In most of the English-speaking word, cassava is the accepted name. The word "yuca" originated from the same word as did cassava (see above).
- Yuca should not be confused with yucca (with two *c*'s). Yucca species are in the genus *Yucca* of the family Agavaceae. The most well-known *Yucca* is *Y. brevifolia* Engelm., the Joshua tree.
- *Esculenta* in the scientific name *M. esculenta* means edible, good to eat.

Plant Portrait

Cassava is a large semi-woody shrub or small tree, 1.3–5 m (4–16 feet) tall. It is considered to be native to Brazil and Paraguay (although wild forms are not known), and is now extensively cultivated in tropical and sometimes also in subtropical regions as an important food and for the production of starch. Cassava is a major supplier of low-cost carbohydrates for populations in the humid tropics, the most important tropical root crop, and the world's fourth most important source of calories (after rice, sugar cane, and maize). However, the nutritional value is limited, especially as processing methods noted below can leach away or destroy vitamins, and cassava must be supplemented with other foods to prevent malnutrition. The largest producers are Brazil, Thailand, Nigeria, Zaire, and Indonesia. The crop is grown to a small extent in the southern United States, mainly for stock feed. There is substantial production in Puerto Rico and Hawaii. Cassava is raised mostly for its edible tubers (underground storage stems), which are fleshy in cultivated varieties, up to 2.5 m (8 feet) long and 10–15 cm (4–6 inches) in diameter, mostly weighing 4–7 kg (9–15 pounds), sometimes up to 40 kg (88 pounds). The tuberous roots are clustered at the stem base. In specialty markets that carry cassava, the roots may be shaped like clubs or slim sweet potatoes. Under the bark-like covering, the interior of the roots is hard and white. Plants are propagated by stem pieces, somewhat like sugar cane, and are frequently grown as annuals, although some cultivars require 18 or more months of growth before they can be harvested. In Africa, the tubers are sometimes left in the field for years, as insurance against famine, and although they become rather fibrous, they grow very large. There are about 200 recognized cultivated varieties, commonly placed in two basic types, sweet and bitter. Sweet cassavas are more widely distributed in South America than bitter varieties, and were cultivated by older civilizations, as long as 4000 years ago in Peru and 2000 years ago in Mexico.

Culinary Portrait

In countries where cassava is a staple, there are as many methods of preparing it as there are for potatoes. This vegetable can be boiled, fried, puréed, used in soups, stews, made into bread, etc. Recipes are available on the Web.

Cassava starch is known as tapioca, and is the major form of cassava encountered outside of the tropics. "Tapioca pudding" can be made with starch from either cassava or sago palm (*Metroxylon sagu* Rottb.), and sometimes cassava starch is called sago starch. Tapioca starch is also confused with the rather similar arrowroot starch (from *Maranta arundinacea*

L.; cassava starch is sometimes called "Brazilian arrowroot"), and much of the commercial "arrowroot starch" available is actually cassava starch. Tapioca is sold in flake or flour form and as pellets (pearl tapioca). When cooked, it becomes transparent and increases in size, and is used to thicken puddings and soups. To prevent lumps from forming, tapioca should be stirred constantly while being prepared. Tapioca is bland, tending to absorb flavors, and is particularly useful to thicken soups, stews, pies, fruits, puddings, and sauces.

In addition to tapioca, in countries where cassava is grown in large quantities, it is often processed into flour, syrup, vinegar, or alcoholic beverages. Cassava leaves are employed in West African cooking to flavor sauces that are consumed over rice. Dried or (preferably) frozen cassava leaves are often available in Caribbean, African, and Brazilian groceries.

Cassava (*Manihot esculenta*), from Engler and Prantl (1889–1915).

There is poisonous hydrogen cyanide (HCN, also known as hydrocyanic acid and prussic acid) in the skin of sweet varieties, which must be peeled before eating. In bitter varieties, the HCN is distributed throughout the root, which must be cooked before using. So-called sweet varieties are not necessarily less toxic than bitter varieties, and it is recommended that no variety of cassava be consumed fresh. Cassava marketed in North America is generally low in HCN, and more than 1 kg (about 2 pounds) of unprocessed tubers would have to be consumed before a lethal dose would be reached.

Commercial sweet cassava tubers are very perishable, with a shelf life of only a few days, although special techniques can be used to preserve fresh cassava for up to several weeks. By contrast, mature bitter varieties will store for several months. Quality of fresh cassava is generally poor unless bought immediately from a local market, and most purchases are by ethnic groups familiar with it. Frozen, peeled tubers are marketed in North America, packaged like frozen French fries. In most Western countries, frozen cassava, which is generally partly cooked before being frozen, is the best choice. Fresh tubers in North American markets will probably be covered by wax to prevent drying out and oxidation. Cutting the tuber to ensure it is white is recommended to be sure that it is fresh; if the store will not permit this, or does not offer this demonstration, it is risky to buy cassava. Once cut, cassava can be wrapped in plastic and stored for up to 4 days in a refrigerator, although the flesh discolors on standing.

Culinary Vocabulary

- "Instant tapioca" has been processed into flakes, flour, or granular form, and has been precooked before being marketed, so that only 10 minutes are required for preparation.
- "Cassareep" (pronounced KAS-sah-reep), a staple condiment of West Indian cuisine, is a syrupy, dark brown liquid prepared by boiling down and seasoning the bitter juice of the cassava root with brown sugar and spices. The word cassareep is based on a local Cariban language. Cassareep is the basis of many West Indian sauces, and is considered indispensable in West Indian pepper pot (a meat stew of oxtails, beef or pork and chicken, served with rice). The grated cassava that is left after the juice

is squeezed out to make cassareep is often baked to make flat cakes called "cassava bread" or, alternatively, is sun-dried to produce "cassava meal," most of which is used to produce tapioca.

- "Farofa" is a Brazilian condiment made with cassava flour, and other ingredients such as palm oil, onions, and chiles. It is sprinkled over beef, beans, and other dishes.
- "Foo-foo" (foofoo, foufou, foutou) is an extremely thick porridge made from ground cassava, corn, rice, and (or) yams. It is a staple of Central and West African cuisines. Cuban dictator Fidel Castro (1926–) described it as a "great delicacy" that was commonly consumed by his guerrillas while they were hiding in the hills before the revolution. The Caribbean counterpart of foo-foo is coo-coo [coo coo, coocoo; called "funchi" (pronounced "foon-jee") and "fungi" in the Netherlands, Antilles, and the Virgin Islands], a pasty dish of steamed cornmeal and okra garnished with vegetables and herbs. Different versions of coo coo use as a major constituent any of breadfruit, cassava, coconut, okra, or plantain. The North African counterpart is couscous, a meat or vegetable stew mixed with bits of semolina flour.
- Deep-fried root pieces of cassava are offered in Miami Latin restaurants under the name of "Miami fries."
- "Cavim" is a popular Brazilian brandy made partly from cassava.
- "Tiquira" (pronounced te-kee-rah) is a Brazilian beverage, with a very high alcohol content, made from fermented tapioca roots.
- "Gari" is a starch made from cassava. This is in the form of a coarse flour, and is popular in West African and Latin American markets.

Curiosities of Science and Technology

- In Central African Republic, Congo, Mozambique, and Zaire, the average person consumes more than 200 kg (220 pounds) of cassava annually.
- Konzo disease occurs among populations that consume large amounts of cassava, and fail to sufficiently detoxify it. The disease is marked either by acute poisoning with severe sudden illness and death or by a later onset of a suddenly appearing paralysis. Outbreaks of konzo have been identified in certain rural areas of Africa: Central Africa Republic, Mozambique, Tanzania, and Zaire. Konzo mainly affects children and women of childbearing age. The name Konzo was assigned to the disease in the first report from the southern Bandundu region in Zaire.
- Carribean Arawak Indians reportedly committed suicide by eating raw cassava rather than face slavery under post-Columbian Spanish invaders.
- The cassava plant grows naturally in the tropics, so it does not need its large storage roots in order to survive a cold season. The storage roots seem designed by nature to survive two natural climatic stresses: occasional droughts, and storms with high winds, which can destroy the aboveground parts of the plant.
- Headline: "1,500-Ton Tapioca Pudding Nearly Sinks Ocean Liner" (*The Capital Times*, Madison WI, 14 September 1972). The 12 165-ton Swiss freighter Cassarate arrived at Cardiff Wales with a fire that had smoldered in timber stacked in the upper holds for the last 25 days. The crew had kept the fire from becoming serious by continually hosing down the timber, but the water seeped down to the lower holds where 1500 tons of tapioca were stored. The water swelled the tapioca. Then the heat from the flames started to cook the sticky mess. The swelling tapioca—enough to serve a million plates—could have buckled the ship's steel plates, but fortunately, the huge tapioca pudding, estimated at 500 truck loads, was removed before this happened.
- Licking postage stamps can be nutritious, as tapioca is often used in the glue on the back. The U.S. Postal Service has used a stamp adhesive made of sweet potato, corn, and cassava. The base takes into account religious and vegetarian dietary strictures.

Key Information Sources

Anazonwu-Bello, J. 1977. Cassava products and recipes. Ministry of Agriculture and Natural Resources, Enugu, Nigeria. 24 pp.

Byrne. D. 1984. Breeding cassava. Plant Breed. Rev. **2**: 73–134.

Cock, J.H. 1982. Cassava: a basic energy source in the tropics. Science, **218**: 755–762.

Cock, J.H. 1984. Cassava. *In* The physiology of tropical field crops. *Edited by* P.R. Goldsworthy and N.M. Fisher. John Wiley & Sons, Chichester, UK. pp. 529–549.

Cock, J. 1985. Cassava, new potential for a neglected crop. Westview Press, London, UK. 191 pp.

Cock, J., and Reyes, J.A. 1985. Cassava: research, production and utilization. United Nations Development Programme, New York, NY. 745 pp.

Eterjere, E.O., and Bhat, R.B. 1985. Traditional preparation and uses of cassava in Nigeria. Econ. Bot. **39**: 157–164.

Grace, M.R. 1977. Cassava processing. Revised edition. Food and Agriculture Organization of the United Nations, Rome, Italy. 155 pp.

Gulick, P., Hershey, C., and Esquinas-Alcazar, J.T. 1983. Genetic resources of cassava and wild relatives. International Board for Plant Genetic Resources, Rome, Italy. 56 pp.

Harris, N.V. 1985. Economics of growing cassava for ethanol. Department of Resources and Energy, Canberra, Australia. 96 pp.

Hendershott, C.H. 1972. A literature review and research recommendations on cassava (*Manihot esculenta* Crantz). University of Georgia, Athens, GA. 326 pp.

Henry, G, and Gottret, V. 1996. Global cassava trends: reassessing the crop's future. Centro Internacional de Agricultura Tropical, Cali, Columbia. 45 pp.

Hershey, C.H. 1993. Cassava, *Manihot esculenta* Crantz. *In* Genetic improvement of vegetable crops. *Edited by* G. Kalloo and B.O. Bergh. Pergamon Press, New York, NY. pp. 669–691.

Hillocks, R.J., Thresh, J.M., and Bellotti, A. 2002. Cassava: biology, production and utilization. CABI Publishing, Wallingford, UK. 332 pp.

International Institute of Tropical Agriculture. 1990. Cassava in tropical Africa: a reference manual. International Institute of Tropical Agriculture, Ibadan, Nigeria. 176 pp.

Jennings, D.L. 1970. Cassava in Africa. Field Crop Abstr. **23**: 271–278.

Jennings, D.L. 1995. Cassava, *Manihot esculenta* (Euphorbiaceae). *In* Evolution of crop plants. 2nd edition. *Edited by* J. Smartt and N.W. Simmonds. Longman Scientific & Technical, Burnt Mill, Harlow, Essex, UK. pp. 128–132.

Jennings, D.L., and Hershey, C.H. 1985. Cassava breeding: a decade of progress from international programmes. *In* Progress in plant breeding, vol. 1. *Edited by* G.E. Russell. Butterworth-Heinemann, London, UK. pp. 89–116.

Karasch, M. 2000. Manioc. *In* The Cambridge world history of food. *Edited by* K.F. Kiple and K.C. Ornelas. Cambridge University Press, Cambridge, UK. pp. 181–187.

Lancaster, P.A., and Brooks, J.E. 1983. Cassava leaves as human food. Econ. Bot. **37**: 331–348.

Leatherdale, D. 1977. Cassava thesaurus. Cassava Information Center, Centro Internacional de Agricultura Tropical, Cali, Colombia. 146 pp.

Manrique, L.A. 1997. Cassava: production principles and practices. Manrique International Agrotech, Honolulu, HI. 248 pp.

Odigboh, E.U. 1983. Cassava: production, processing, and utilization. *In* Handbook of tropical foods. *Edited by* H.T. Chan, Jr. Marcel Dekker Inc., New York, NY. pp. 145–200.

Phillips, T.P. 1974. Cassava utilization and potential markets. International Development Research Centre, Ottawa, ON. 182 pp.

Renvioze, B.S. 1973. The area of origin of *Manihot esculenta* as a crop plant: a review of the evidence. Econ. Bot. **6**: 352–360.

Rogers, D.G., and Appan, S.G. 1973. *Manihot, Manihotoides* (Euphorbiaceae). Published for Organization for Flora Neotropica *by* Hafner Press, New York, NY. 272 pp.

Rogers, D.J. 1963. Studies of *Manihot esculenta* and related species. Bull. Torrey Bot. Club, **90**: 43–54.

Rogers, D.J. 1965. Some botanical and ethnological considerations of *Manihot esculenta*. Econ. Bot. **19**: 369–377.

Rogers, D.J., and Fleming, H.S. 1973. A monograph of *Manihot esculenta* with an explanation of the taximetrics methods used. New York Botanical Garden, Bronx, NY. 113 pp.

Rosling, H. 1988. Cassava toxicity and food security: a review of health effects of cyanide exposure from cassava and of ways to prevent these effects. 2nd edition. International Child Health Unit, University Hospital, Uppsala, Sweden. 40 pp.

Sarma, J.S., and Kunchai, D. 1991. Trends and prospects for cassava in the developing world. International Food Policy Research Institute, Washington, DC. 64 pp.

Silvestre, P. 1989. Cassava. Macmillan, London, UK. 82 pp.

Thampan, P.K. 1979. Cassava. Kerala Agricultural University, Trichur, Kerala, India. 242 pp.

Veltkamp, H.J., and Bruijn, G.H. de. 1996. *Manihot esculenta* Crantz. *In* Plant resources of South-East Asia, 9, plants yielding non-seed carbohydrates. *Edited by* M. Flach and F. Rumawas. Backhuys Publishers, Leiden, Netherlands. pp. 107–113.

Specialty Cookbooks

Anazonwu-Bello, J. 1977. Cassava products and recipes. Ministry of Agriculture and Natural Resources, Enugu, Nigeria. 24 pp.

Onabolu, A. 1988. Sweet cassava: a food for all seasons. United Nations Children's fund, Household Food Security and Nutrition Program, Ibadan, Nigeria. 60 pp.

Cauliflower

Family: Brassicaceae (Cruciferae; mustard family)

Names

Scientific name: *Brassica oleracea* L. var. *botrytis* L. (*B. oleracea* Cauliflower Group)

- The word "cauliflower" is a combination of the Latin *caulis* (stem) and *flore* (flower), which reflects the edible head of flowers.
- See CABBAGE for the derivation of the scientific name *Brassica oleracea*.
- *Botrytis* in the Latin name of cauliflower, *B. oleracea* var. *botrytis*, is the Greek word for a cluster or bunch of grapes, an imaginative allusion to the appearance of the flower head.

Plant Portrait

Cauliflower and broccoli (which see) are quite similar. Cauliflower is the flower head of a type of cabbage in which the flowers haven't matured but have become quite fleshy; broccoli is the same thing but includes the fleshy stalks and flower buds that are younger than the flowers of cauliflower. As noted in the chapter on broccoli, there is considerable confusion between many cultivars of cauliflower and those of broccoli. For example, the cultivar Romanesco, developed in central-western Italy near Rome, is revered in Europe for its nut-like texture and sweet flavor. Although often sold as a "broccoli" in garden catalogs, it is a cauliflower. Names such as "white-sprouting broccoli" and "heading broccoli" are still applied to forms of cauliflower in Europe.

During the first year of growth, when they are harvested, cauliflower heads remain functionless, except of course for being eaten. The flowers can be made to become fertile and produce seeds by exposure to low temperatures in storage. Most cultivars of cauliflower require blanching to become a brilliant white rather than a dull yellow. This is achieved by covering the flowering heads to prevent the formation of green chlorophyll that sunlight would cause. Blanching is done by tying the upper leaves over the heads while they are the size of a teacup. Purple- and green-headed cultivars are not blanched.

Cauliflower is known only in cultivation. Its origin as a crop is obscure and is intertwined with that of broccoli. Likely it originated directly from the wild cabbage, *B. oleracea* var. *oleracea*, rather than from other cultivated forms. Records of cauliflower go back to the 6th century BC. By the 12th century in Spain, the Moors had selected three different types. By the Middle Ages, both purple and white types were known, including the Erfurt cauliflower from which the modern North American Snowball varieties were developed. Major cauliflower-producing nations include China, India, France, Italy, and the United States.

Culinary Portrait

Cauliflower has been described as the aristocrat of the cabbage clan. It is occasionally employed as a raw vegetable in salads or by itself, but more frequently is served *au gratin* (prepared with a topping of buttered crumbs or grated cheese) or cooked as a vegetable and added to stews, casseroles, soups, pasta, omelets, and meat and seafood dishes. It is also a common component of mixed pickles. In most recipes, cauliflower can be interchanged with broccoli. Cauliflower tends to darken as it cooks, which can be prevented by adding half a cup of milk to the cooking water. Adding a piece of bread to the cooking liquid absorbs some of the odor, a practice common to most of the other close relatives of the cabbage. To preserve its flavor and nutrients, cooking should stop as soon as the vegetable is tender. Cauliflower cooks quickly, and when overcooked tends to fall apart and become pasty. Cauliflower has been described as "vegetable liver," a reflection of its high content of iron. It also has considerable calcium, and like other members of the cabbage clan, is not only nutritionally rich but contains health-promoting chemicals (see CABBAGE).

Heads of cauliflower deteriorate rapidly, and older stock in markets may have had wilted leaves trimmed away to mask the appearance of being overly mature. When purchasing cauliflower, select heads that look very fresh and firm, with no brownish areas resulting from age and disease.

Culinary Vocabulary

- The French expression *à la Dubarry* (*à la du Barry*), pronounced doo-bah-ree, refers to any French dish prepared or garnished with cauliflower, particularly cooked cauliflower served with cheese sauce. The Countess Du Barry (Jeanne du Barry, 1743–1793, French courtier and mistress of King Louis XV, guillotined during the French Revolution), who is commemorated by the term, was presumably fond of cauliflower. Crème Dubarry is a creamy cauliflower soup.

Varieties of cauliflower (*Brassica oleracea* var. *botrytis*), from Vilmorin-Andrieux (1885).

- The French culinary expression *à la polonaise* refers to a method of food preparation associated with Polish cuisine. Dishes, especially cauliflower and asparagus, are boiled and sprinkled with chopped hard-boiled egg yolk, breadcrumbs, parsley, and melted butter.
- An Italian Christmas Eve is known as "Fish Night" because the dinner features seven kinds of fish, symbolizing the seven phases of the moon, and the lucky number seven. As well, three kinds of vegetables are served—artichoke hearts, broccoli, and cauliflower—representing the Trinity in Christian belief.

Curiosities of Science and Technology

- According to the Guinness World Record Holder 2000 Calendar, Karen Roman grew the world's largest cauliflower, weighing 10 kg (22 pounds).
- In 1970, an orange cauliflower mutation was discovered by a farmer in the Holland Marsh, north of Toronto, Canada. Three decades of research and development followed, particularly by Michael Dickson, a vegetable breeder at the New York State Agricultural Experiment Station affiliated with Cornell University. The resulting orange cauliflower is called Citrus. Its orange color is due to beta-carotene, the same chemical that produces orange color in oranges and carrots. The flowering head does not require shading from light, unlike most white cauliflowers, that do need to be protected from light to develop their color.
- "Cauliflower ear" (auricular hematoma) is an ear swollen and deformed by trauma or friction and is common in boxing, wrestling, rugby, judo, and other contact sports. The expression was first recorded in 1909. "Cauliflower ear corn" is a weird type of corn that has been produced several times in genetic experiments in which the corncob turns into a proliferating cauliflower-like mass of tissue.

Key Information Sources

(See additional references in Broccoli.)

Andersen, C.R. 2000. Cauliflower. Cooperative Extension Service, University of Arkansas, Fayetteville, AR. 4 pp.

Anonymous. 1964. Cauliflowers and related crops. Bull. 131. 4th ed. Her Majesty's Stationary Office, London, UK. 39 pp.

Anonymous. 1982. Cauliflowers. Grower, London, UK. 87 pp.

Carby-Samuels, H.R. 1985. A study of cauliflower production & marketing in Ontario. Regional Development Branch, Agriculture Canada, Toronto, ON. 39 pp.

Crisp, P. 1982. The use of an evolutionary scheme for cauliflower in the screening of genetic resources. Euphytica, **31**: 725–734.

Crisp, P., and Tapsell, C.R. 1993. Cauliflower, *Brassica oleracea* L. *In* Genetic improvement of vegetable crops. *Edited by* G. Kalloo and B.O. Bergh. Pergamon Press, New York, NY. pp. 157–178.

Gardner, R. 1976. Cauliflowers—the aristocrats of the cabbage clan. J. Agric. Victoria (Australia), **74**: 75–76.

Gray, A. 1989. Green curded cauliflowers. Garden (U.K.), **114**: 31–33.

Gray, A.R., and Crisp, P. 1977. Breeding systems, taxonomy and breeding strategy in cauliflower (*Brassica oleracea* var. *botrytis*). Euphytica, **26**: 369–375.

Huyser-Honig, J. 1995. 12 easy-grow cauliflowers. Org. Gard. **42(9)**: 34–38.

Koike, S.T., and Chaney, W.E. 1997. Cauliflower production in California. Division of Agriculture and Natural Resources, University of California, Oakland, CA. 3 pp.

Maier, A. 1976. How to select cauliflower. Cooperative Extension Service, North Dakota State University, Fargo, ND. 4 pp.

Organisation for Economic Co-operation and Development. 1969. Cauliflowers. Organisation for Economic Co-operation and Development, Paris, France. 57 pp.

Smyth, D.R. 1995. Origin of the cauliflower. A mutant gene that is responsible for generating cauliflower-like heads in the model laboratory plant *Arabidopsis* has been cloned, and the same gene has been shown to be mutant in edible cauliflowers. Curr. Biol. (U.K.), **5**: 361–363.

Swarup, V., and Chatterjee, S.S. 1972. Origin and genetic improvement of Indian cauliflower. Econ. Bot. **26**: 381–392.

University of Tennessee. 1947. Cauliflower: an annotated bibliography from selected references. Agricultural Experiment Station, University of Tennessee, Knoxville, TN. 18 pp.

Vossen, H.A.M. van. 1993. *Brassica oleracea* L. cv. Groups Cauliflower & Broccoli. *In* Plant resources of South-East Asia, 8, vegetables. *Edited by* J.E. Siemonsma and K. Piluek. Pudoc Scientific Publishers, Wageningen, Netherlands. pp. 111–115.

Waters, L. 1983. Growing broccoli and cauliflower in Minnesota: a production guide. University of Minnesota, Agricultural Extension Service, St. Paul, MN. 7 pp.

Watts, L.E. 1963. Investigations into the breeding system of cauliflowers. *Brassica oleracea* var. *botrytis*. I. studies of self-incompatibility. Euphytica, **12**: 323–340.

Whitwell, J.D., Jones, G.L., and Williams, J.B. 1982. Cauliflowers. ADAS/MAFF Reference Book 131. Growers Books, London, UK. 87 pp.

Wurr, D.C.E. 1990. Prediction of the time of maturity in cauliflowers. Acta Hortic. **267**: 387–394.

Specialty Cookbooks

(See additional references in the chapters on BROCCOLI and CABBAGE.)

Oklahoma State Dept. of Agriculture. 1983. Oklahoma cauliflower: recipes. Oklahoma State Dept. of Agriculture, Marketing Industry Division, OK. 6 pp.

Walsh, B.H. 1976. Using Florida vegetables: broccoli and cauliflower. Florida Cooperative Extension Service, Institute of Food and Agricultural Sciences, University of Florida, FL. 4 pp.

CHERRY

Family: Rosaceae (rose family)

Names

Scientific names: *Prunus* species

- Sweet cherry—*P. avium* (L.) L. [*Cerasus avium* (L.) Moench]
- Sour cherry—*P. cerasus* L. (*Cerasus vulgaris* Mill.)

- "Cherry" is from the Latin *cerasus*, cherry tree, which is from the Greek word for the cherry tree, *kerasos*, derived ultimately from *karsu*, from the Akkadian language of Mesopotamia. (Also see *Cerasus*, below.)
- The sweet cherry is also called gean and mazzard cherry.
- "Sweet cherry" and "sour cherry" are so named for the taste of the fruit. The sour cherry is also called the "tart cherry" and "pie cherry."
- See PLUM for the derivation of the genus name *Prunus*.
- The Bing cherry, the most famous sweet cherry variety, was developed in 1875 by Seth Luelling, a grower in Milwaukie, Oregon. The Bing was named for Luelling's Manchurian foreman.
- Montmorency (for which the Montmorency sour cherry is named) was a prerevolutionary French dukedom in the southwest of France, famous for its cherries, and according to legend, for a duke who insisted on having them with everything. "Duck Montmorency" is a culinary preparation of duck made, of course, with cherries.
- "Cherry" is part of the name of various plant species that are unrelated to cherry species. For example, "cherry currant" refers to a variety of the red currant (see CURRANT), "cherry pepper" refers to various peppers (see CHILE PEPPER), "cherry tomato" refers to small tomatoes (see TOMATO), and "cherry orange" is the kumquat (*Fortunella* species of the citrus family).
- *Avium* in the scientific name *P. avium* is Latin for "of the birds," referring to the tree's attractiveness to birds. Wild cherries are often called "bird cherry."
- *Cerasus* in the scientific name *P. cerasus* is the Latin name for a cherry tree brought into Europe from Cerasus (also spelled Kerasoon, Giresun, or Kerasun), a city and port in Northeastern Turkey.

Plant Portrait

The history of cherry cultivation is lost in antiquity. Cherry stones have been found in the deposits of the ancient Swiss Lake Dwellers, in the mounds and cliff-caves of prehistoric inhabitants of America, and in the ancient rubbish heaps of Scandinavia, indicating that the plants were cropped, although not necessarily cultivated. There is no evidence that cherries were grown by the ancient Egyptians, Syrians, or Babylonians. However, the ancient Greeks and Romans grew domesticated cherries, and it is probable that domestication took place at some time between 2000 and 3000 years ago in the region of Asia Minor or Greece. Interest in cherries declined following the fall of the Roman Empire, but in the Middle Ages, cherry cultivation became popular once again in Europe, and the practice was exported to the New World. The cherry is one of the most important fruit crops grown in the cooler temperate regions of the world. Although several other cherries are cultivated for their fruit, the sweet and sour cherries are the most important commercially.

The sweet cherry tree is native to Europe, western Asia, and North Africa. More than 1000 sweet cherry fruit cultivars have been named. In addition, there are also numerous ornamental forms with double flowers, dissected leaves, or pendent branchlets. North American sweet cherries are grown primarily in the Pacific Coast states, but Michigan produces about 20% of the annual crop. Most sweet cherries are self-unfruitful, i.e., they must have another sweet cherry tree for pollination in order to set fruit. Sweet cherries are not as hardy as sour cherries, and generally are just a little hardier than the peach.

The sour cherry is thought to occur wild in southwestern Asia and southeastern Europe, although some authorities think the species exists only in cultivation. It is hardier than the sweet cherry, and some cultivars are even hardier than the apple. There are hundreds of varieties, including ornamental forms. Montmorency, a 400-year-old French variety, makes up most of the United States sour cherry industry. Michigan produces almost three-quarters of the sour cherries grown in the United States. The sour cherry is one of the few fruit trees suited for shady conditions. Unlike the sweet cherry tree, sour cherries are usually self-fertile, and can set fruit without the need for pollen from another tree.

Sweet cherry (*Prunus avium*), from Köhler (1883–1914).

Culinary Portrait

Most cherries used for purposes other than eating out of hand are sour cherries, while sweet cherries are especially popular for eating raw. Tart cherries are seldom sold fresh; they are generally canned or frozen shortly after harvesting for use in products throughout the year. Sweet cherries should be picked when mature, as they do not ripen after they are harvested. Unfortunately, when harvested when fully mature and at their best, the fruit deteriorates before reaching distant markets, so that store-bought sweet cherries rarely taste as good as home-grown fruit. The fruit spoils rapidly at room temperature, but may be kept in the refrigerator for several days. Cherries are best stored in a perforated plastic bag, but away from strong-smelling foods, as they tend to absorb odors. Culinary dishes made with sweet cherries are often not as good as those prepared with sour cherries, since many sweet cherries lack acidity. If sweet cherries are used in making jams and jellies, they are best combined with currant juice or other fruit juice that provides acidity. The pits of cherries can be removed by using a cherry pitter tool, or more laboriously by cutting the fruit in half and plucking out the pit.

Aside from eating fresh, cherries are prepared in a variety of ways. They may be cooked, candied, dried, preserved, in syrup, or made into alcoholic beverages, including liqueurs. Cherries are widely used in the making of pies, jams, preserves, and pastries. Candied cherries are an essential ingredient of fruitcake. Cherries can be used to accompany poultry and wild game.

Cherries in baked goods sometimes produce blue discoloration around the fruit, owing to a chemical reaction between the cherries and baking powder or baking soda. This can be prevented by substituting buttermilk or sour cream for milk in the recipe or adding an acidic liquid such as lemon juice.

One type of sour cherry is called the marasca cherry. Its fruits are much smaller, darker, and more acidic than most other sour cherries, and are employed to produce maraschino, a liquor used for the manufacture of maraschino cherries. Maraschino cherries were once made by soaking cherries in 70-proof Maraschino cordial, produced from the juice of the Italian Dalmatian marasca wild cherry. Today, the cherries are soaked in sodium metabisulfate, calcium chloride, citric acid, corn syrup, and fructose

Old sweet cherry tree (*Prunus avium*), from Bailey (1900–1902).

solutions, and artificially flavored and colored red or green, which doesn't seem nearly as appealing as the original method.

Cherry soup is a traditional summer soup made since the late 1700s in rural Pennsylvania and in the American Midwest, where many German immigrants settled. It is a combination of pitted sweet cherries, dry wine, sugar, and spices such as cinnamon and cloves. The soup is served during hot summer days—either hot or cold—often with dumplings.

Culinary Vocabulary

- "Kirsch" (based on the German word for cherry) or kirschwasser (German for "cherry water") is a German clear brandy made generally from double-distilled cherry juice from morello cherries (a sour type, with dark skin and juice). Kirsch has been interpreted in a more general sense to refer to any clear, white brandy distilled from European cherries and their pits, produced not only in Germany but also in Switzerland and Alsace. The pits are left in the mash used to distill the brandy, producing a bitter almond flavor. Kirsch is dry, and should not be confused with the sweet, dark, sticky "cherry brandy."
- A "Forester" is a cocktail prepared with bourbon, cherry liqueur, and lemon juice, and garnished with a maraschino cherry.
- "Cherries Jubilee" is a famous dessert [said to have been created by the famous French chef Auguste Escoffier (1846–1935) on the occasion of his 50th birthday], made with black cherries flambéed with kirsch or brandy, to which vanilla ice cream is added.
- "Black Forest Cake" is a layered chocolate cake with kirsch, whipped cream, sour cherries, and chocolate curls. It originated in Swabia in the Black Forest region of southern Germany, known for both the quality of its pastries, and its kirsch. The cake is thought to date back to the late 16th century in the Black Forest Region—*Der Schwarzwald* in German, hence the German name *Schwarzwälderkirschtorte*, literally "Black Forest cherry cake."
- "Cherry bounce" in Britain refers to cherry brandy, and in America, to a homemade cherry-flavored liqueur made from rum or whisky and sometimes cider.
- "Peter Herring" (once called "Cherry Herring") is a Danish, cherry-flavored liqueur, prepared from cherry juice and pits.
- A "bachelor's button" is a cookie topped with a maraschino cherry, the confection made to resemble a button. (Several species of plants are also known as bachelor's button.)

- A "hot cherry" is a round, pungent chile pepper mostly used for pickling.

Curiosities of Science and Technology

- In common language, colors are often difficult to describe and distinguish. "Cherry" is part of the names of various, rather subjectively defined colors, including cherry (a moderate red), cherry blossom (also a moderate red, but paler), cherry pink (pink like cherry blossoms), cherry rose (a deep pink), and cherry wine (a deep red).
- Southwest injury or sunscald is particularly damaging to cherry trees. Bright sunshine during cold periods in midwinter warms the bark on the trunks on the southwest side during the day, and this may be followed by rapid chilling when the sun drops towards the horizon. Bark temperatures may fall from 20 °C to −18 °C (68 °F to 0 °F). This injures the bark and adjacent areas, which frequently die, leaving bare wood areas. Painting with white latex base paint and shading the trunks can prevent this kind of injury.

Sour cherry (*Prunus cerasus*), from Hallier (1880–1888, vol. 25, plate 2558).

- A major hazard of growing cherries is fruit splitting as a result of prolonged rain during the harvest. Sweet cherries absorb water by osmosis when in contact with rain water, and swell until the skin ruptures, unless natural drying is fairly rapid. Entire commercial harvests have been lost because of this. The problem rarely affects sour cherries.
- Cherries should be picked with the stems attached. Cherries with their stalks attached have a much longer shelf life.
- It has been claimed that along with the ornamental cherry trees that were introduced from Japan to beautify the Washington, DC, area in 1912 came the Oriental fruit moth (*Grapholita molesta*), which loves peaches, and has been a major pest of peaches in North America ever since. British Columbia remains the only commercial fruit-growing area of North America that is free of this pest.
- Candy-maker James Smith, along with his two sons William and Andrew, originally from Scotland, moved from St. Armand, Quebec, to Poughkeepsie, New York, in 1847 and started the Smith Brothers firm. Drawings of the two bearded brothers came to grace the packages of cough drops, and by chance, the word "Trade" appeared under the picture of William and the word "Mark" under that of Andrew, and the Smith Brothers became known to generations as Trade and Mark. Andrew died in 1895, and William continued as President of Smith Brothers almost up to his death in 1913. He was succeeded by his son, Arthur, who added the famous Smith Brothers wild cherry flavor in 1948. The ingredients that produced the "cherry" flavor are unclear, but it appears that bark extracts from wild cherry trees were added. In fact, boiled preparations of bark from various native and introduced species of cherry have long been used by several North America Indian groups as a remedy for coughs and colds.
- Reportedly, overripe cherries left on a tree may ferment under hot conditions, producing intoxication in birds that consume them in large quantities. Also reportedly, maraschino cherry chocolate confections made with liquor have intoxicated children who similarly consumed them to excess.

- There are about 7000 cherries on an average sour cherry tree. It takes about 250 cherries to make a cherry pie, so each tree potentially could produce enough cherries for 28 pies.
- The average American consumes about 0.45 kg (1 pound) of sour cherries annually.

Key Information Sources

Chadbund, G. 1972. Flowering cherries. Collins, London, UK. 160 pp.

Christensen, J.V. 1990. A review of an evaluation of 95 cultivars of sour cherry. Tidsskrift Planteavl. **94(1)**: 51-64.

Downing, D.L., Huehn, W.G., and LaBelle, R.L. 1971. Handling of red tart cherries for processing—a review. N. Y. Food Life Sci. Bull. **11**(May): 1–6.

Fischer, M. 1996. Resistance breeding in sweet cherries. Acta Hortic. **410**: 87–96.

Fogle, H.W. 1975. Cherries. *In* Advances in fruit breeding. *Edited by* J. Janick and J.N. Moore. Purdue University Press, West Lafayette, IN. pp. 348–366.

Fogle, H.W., Cochran, L.C., and Keil, H.L. 1974. Growing sour cherries. U.S. Dep. Agric. Handbook No. 451. U.S. Dept. of Agriculture, Washington, DC. 36 pp.

Fogle, H.W., Snyder, J.C., Baker, H., Cameron, H.R., Cochran, L.C., Schomer, H.A., and Yang, H.Y. 1973. Sweet cherries: production, marketing, and processing. U.S. Dep. Agric. Handb. No. 442. U.S. Dept. of Agriculture, Washington, DC. 94 pp.

Grubb, N.H. 1949. Cherries. C. Lockwood, London, UK. 186 pp.

Gruppe, W. (*Editor*). 1985. International workshop on improvement of sweet and sour cherry varieties and rootstocks: new developments and methods (Grünberg/Giessen and Weihenstephan/Munich, Germany, 2–6 July 1984). International Society for Horticultural Science, Wageningen, Netherlands. 380 pp.

Hedrick, U.P., Howe, G.H., Taylor, O.M., Tubergen, C.B., and Wellington, R. 1915. The cherries of New York. J.B. Lyon Co., Albany, NY. 371 pp.

Iezzoni, A., Schmidt, H., and Albertini, A. 1990. Cherries (*Prunus*). *In* Genetic resources of temperate fruit and nut crops. *Edited by* J.N. Moore and J.R. Ballington, Jr. International Society for Horticultural Science, Wageningen, Netherlands. pp. 111–173.

Jefferson, R.E., and Fusonie, A.E. 1977. The Japanese flowering cherry trees of Washington, D.C.; a living symbol of friendship. Natl. Arbor. Contrib. No. 4. U.S. Dept. of Agriculture, Washington, DC. 66 pp.

Kappel, F., and Lay, W. 1997. Sweet cherry breeding in Canada from the early 1900s to 1994. Fruit Var. J. **51**: 233–238.

Kuitert, W., and Peterse, A.H. 1999. Japanese flowering cherries. Timber Press, Portland, OR. 395 pp.

Leather, S.R., and Bland, K.P. 1999. Insects on cherry trees. Richmond Publishing Co., Slough, UK. 82 pp.

Looney, N.E., and Webster, A.D. (*Editors*). 1996. Cherries: crop physiology, production and uses. CABI, Wallingford, UK. 513 pp.

Marshall, R.E. 1954. Cherries and cherry products. Interscience Publishers, New York, NY. 283 pp.

Ministry of Agriculture, Fisheries and Food. 1981. Cherries. Ministry of Agriculture, Fisheries and Food, Alnick, Northumberland, UK. 29 pp.

Sekse, L. 1995. Fruit cracking in sweet cherries (*Prunus avium* L.). Some physiological aspects—a mini review. Sci. Hortic. **63**: 135–141.

Timon, B., Gergely, I., and Hampson, C.R. (*Editors*). 1996. Proceedings international cherry symposium, Budapest, Hungary, 14–18 June 1993. International Society for Horticultural Science, Leuven, Belgium. 565 pp.

Ystaas, J., and Callesen, O. (*Editors*). 1998. Proceedings third international cherry symposium, Ullensvang, Norway and Aarsley, Denmark, 23–29 July, 1997. International Society for Horticultural Science, Leuven, Belgium. 2 vols.

U.S. Dept. of Agriculture. 1977. Growing cherries east of the Rocky Mountains. Farmer's Bull. 2185. Revised edition. U.S.D.A., Agricultural Research Service, Washington, DC. 31 pp.

Specialty Cookbooks

Bader, M. 1995. Cherry creations: the ultimate cherry cookbook. Northstar Publishing Co., Las Vegas, NV. 224 pp.

Cherry Cookbook Committee. 1976. 150 cherry recipes. Cherry Cookbook Committee, National Cherry Festival, Traverse City, MI. 106 pp.

Cherry Marketing Institute. 1994. A celebration of cherries: delightful recipes for dining adventure. Cherry Marketing Institute, Okemos, MI. 37 pp.

Conyers Cherry Blossom Festival Foundation. 1991. A taste of the Cherry Blossom Festival. Illustrated Art & Design, Conyers, GA. 173 pp.

Eberly, C. 1984. 101 cherry recipes. Eberly Press, Lansing, MI. 47 pp.

Eschmann, C. 1984. Cherries galore: a cookbook. Macon Telegraph and News, Macon, GA. 206 pp.

Gohlke, A. 1982. Cherry delights. Reiman Publications, Milwaukee, WI. 64 pp.

Hill, J., and Littler, J. 1993. Cherries galore: a Door County cherry recipe collection. Hill Orchards of Scandia Road, Siste Bay, WI. 252 pp.

Morgenroth, B. 1998. The best 50 cherry recipes. Bristol Publishing Enterprises, San Leandro, CA. 78 pp.

Rubin, C., and Rubin, J. 1974. Cherry cookbook. Emporium Publications, Newton, MA. 91 pp.

Whitaker, T. 1976. Light-sweet cherries: over 100 recipes and serving ideas. Whitaker, Silveton, OR. 88 pp.

CHICKPEA (GARBANZO BEAN)

Family: Fabaceae (Leguminosae; pea family)

Names

Scientific name: *Cicer arietinum* L.

- Chickpeas (chick peas) have no connection with chickens. The "chick" in chickpea arose from the Latin *cicer* for chickpea, changing first to the Middle English *chiche*.
- The chickpea is better known in North America as garbanzo bean. "Garbanzo" is the Spanish word for the chickpea, and came from the Latin *ervum*, bitter vetch.
- Other names for the chickpea or garbanzo: Bengal gram, channa (for a very widely planted variety in India), chich, chich-pea, dhal, Egyptian pea, and gram [perhaps better reserved for the mung bean, *Vigna radiata* (L.) R. Wilcz., which see for the derivation of "gram"].
- The genus name *Cicer* is from the ancient Latin *cicer*, probably from the Greek *krios*, likely designating the chickpea.
- *Arietinum* in the scientific name *C. arietinum* is from the Latin *aries*, a ram, from a fancied resemblance of the seeds to a ram's head (it is claimed that the larger varieties of seeds when viewed from the side suggest a ram's head with large curly horns). "Ram's head pea" is an occasional English name for the chickpea.

Plant Portrait

The chickpea is an upright annual herb, 20–100 cm (8–40 inches) tall. The flowers are white, pink, purplish, or blue, 8–12 mm (0.3–0.5 inch) long. These produce pods 14–35 mm (0.6–1.4 inches) long, with one or two (sometimes up to four) seeds. The seeds are round or elongated, smooth, wrinkled, or warty, cream, white, yellow, brown, red, black, or green, and vary from the size of small peas to that of large lima beans. "Kabuli" chickpeas of Mediterranean and Middle Eastern origin have very large seeds (and are the type known as garbanzo beans); "Desi" chickpeas of Indian distribution have smaller seeds but are high-yielding in India and many other locations. Kabuli seeds in North America are typically cream-colored and have a thin seed coat, while Desi type seeds have a thick seed coat and a darker color. The chickpea seems to have arisen from a wild species of southeastern Turkey. Chickpeas were used in Turkey at least by 5500 BC, and were cultivated in the Mediterranean area since 3000 or 4000 BC. This bean has been grown in India since 2000 BC, and in Ethiopia at least since 1000 BC. Chickpeas were known to the ancient Egyptians, Hebrews, and Greeks. In the 16th century, the chickpea was brought to South and Central America. Expatriate Indians introduced it into eastern and southern Africa and the West Indies.

Today, the chickpea is widely grown in the Mediterranean area, in Asia (including India, Pakistan, Syria, Turkey, and Iran), and to a lesser extent in Australia, Myanmar, Ethiopia, Mexico, Argentina, Chile, and Peru. India produces about two-thirds of the world's supply. More than 90% of the world crop is consumed in the countries where it is produced. The chickpea is adapted to a tropical climate of moderate temperature and rainfall, and can be grown well in regions of North America. In recent years, cultivation has been increasing in North America. The top three exporting countries are Australia, Mexico, and Canada. In the United States, chickpeas are grown commercially in Idaho, California, North Dakota, Washington, South Dakota, and Montana. About three-quarters of world production is of the Desi type. The Kabuli type predominates in American production, while both types are grown in Canada. In addition to food use, chickpea starch is employed for textile sizing and preparation of adhesives.

Culinary Portrait

Chickpeas are used almost exclusively as a human food, consumed fresh, or dried and cooked. Dried chickpeas must be soaked before being boiled, and this is typically carried out overnight. Old-fashioned varieties required boiling for up to 4 hours in an open container, or cooking for an hour in a pressure cooker, although modern varieties can be tender in only 20 minutes in a pressure cooker (it has been said that it is almost impossible to overcook chickpeas). Desi type seeds must be hulled, and are used whole, split, or milled. In India and surrounding countries, hulled, split Desi chickpeas are used to produce "dhal," commonly eaten as a thick soup. The seeds can also be ground into a fine flour called *besan*, used for preparing cooked foods. Besan is used in many ways, including mixed with wheat flour to make roti or chapatti, and also for sweets and snacks. Chickpeas are extensively used for making dishes such as couscous, hummus, and felafel in the Mediterranean region, the Middle East, and parts of Asia. Recently, these dishes have been popularized in North America. Couscous refers to small, spherical bits of semolina dough, which are steamed and served like rice, mixed with a stew or sauce. Felafel is a mixture of ground chickpeas, garlic, and a herb mixture, formed into balls or patties and fried. (In Egypt, where felafel supposedly originated, it is made with broad beans, but in Syria, Lebanon, and Israel, it is made with chickpeas.) Hummus (hommus) is made by mashing chickpeas and mixing with olive or sesame oil and spices, and serving usually as a dip. In North America, Kabuli type seeds are valued for use as an ingredient in salad bars at the green–immature stage for use as "garbanzo beans." Kabuli seeds are also used in preparing a wide variety of snack foods, soups, sweets, and condiments. Roasted seeds and roots are occasionally employed as a coffee substitute. In Asia, sprouted seeds are consumed as a vegetable or salad, and young plants and pods as greens.

Celiac disease is a condition where glutens and related proteins consumed from the closely related cereals wheat, rye, barley, and triticale damage the small intestines (see WHEAT for additional information). Because chickpea is gluten free, it can be used in food products developed for people who are allergic to gluten.

Culinary Vocabulary

- *Pastéis de grao* (pronounced pass-ta-ess da gra-oh) is a small Portuguese cake prepared with chickpea flour.
- *Socca* (pronounced soh-kah) are thick, coarse chickpea flour pancakes originating from Nice, France.

Curiosities of Science and Technology

- In ancient times, chickpeas were believed to be a powerful aphrodisiac, especially for men. The Romans fed them to their stallions in the belief that it would make them more vigorous. However, Arabian herbalists recommended a decoction (boiled mixture) of chickpeas and honey as an antiaphrodisiac.

Chickpea (*Cicer arietinum*), from Hallier (1880–1888, vol. 24, plate 2472).

- In 1282, a bloody uprising against French rule (by Charles I of Anjou) occurred in Sicily. The Sicilian rebels were ordered to massacre every French person on sight. To decide if someone was French, everyone was commanded to say *cecceri*, an Italian dialect expression for chickpeas. Those who did not pronounce the word correctly were executed. Such a linguistic clue is called a shibboleth, after the biblical story of how the Gileadites identified their enemies, the Ephraimites, by having them pronounce the Hebrew word *shibboleth* (meaning either "rushing water" or "an ear of wheat"). The Ephraimites could only say "sibboleth." Today, in English, shibboleth means any pronunciation, custom, or practice that identifies someone as an outsider.

- Chickpea plants are covered by miniature glands that secrete malic acid (about 94%) and oxalic acid (about 6%). The acids have been harvested in India by spreading thin muslin over the crop at night and wringing out the soaked cloths afterwards. The resulting preparation was recommended in the past as an aphrodisiac, and to treat bronchitis, cholera, constipation, flatulence, snakebite, and warts. The exuded material is now used as a sort of vinegar. Walking through a chickpea field may result in soiling of trousers and shoes from the malic and oxalic acids that are exuded by the plants. Hand-harvesting of chickpea must be done with gloves for protection from the acid.

Key Information Sources

Berrada, A., Stack, M.W., Riddell, B., Brick, M.A., and Johnson, D.L. 1999. Chickpea: a potential crop for southwestern Colorado. *In* Perspectives on new crops and new uses. *Edited by* J. Janick. ASHS Press, Alexandria, VA. pp. 206–213.

Bestow, S. 1996. Chickpea management guide. Agriculture Western Australia, Perth, WA, Australia. 18 pp.

International Crops Research Institute for the Semi-arid Tropics. 1980. Proceedings of the international workshop on chickpea improvement, Hyderabad, India, 28 Feb.–2 Mar. 1979. International Crops Research Institute for the Semi-arid Tropics (ICRISAT), Patancheru, Andhra Pradesh, India. 298 pp.

Joshi, P.K. 2001. The world chickpea and pigeonpea economies: facts, trends, and outlook. ICRISAT, Andhra Pradesh, India. 62 pp.

Kupicha, F.K. 1977. The delimitation of the tribe Vicieae and the relationships of *Cicer* L. Bot. J. Linn. Soc. **74**: 131–162.

Ladizinsky, G. 1995. Chickpea. *In* Evolution of crop plants. 2nd edition. *Edited by* J. Smartt and N.W. Simmonds. Longman Scientific & Technical, Burnt Mill, Harlow, Essex, UK. pp. 258–261.

Ladizinsky, G., and Adler, A. 1976*a*. The origin of chickpea *Cicer arietinum*. Euphytica, **25**: 211–217.

Ladizinsky, G., and Adler, A. 1976*b*. Genetic relationships among the annual species of *Cicer* L. Theor. Appl. Genet. **48**: 197–204.

Loss, S.P., Brandon, N., and Siddique, K.H.M. 1998. The chickpea book: a technical guide to chickpea production. Agriculture Western Australia, South Perth, WA, Australia. 75 pp.

Maiti, R.K., and Wesche-Ebeling, P. 2001. Advances in chickpea science. Science Publishers, Enfield, NH. 360 pp.

McKay, K. 2002. Growing chickpea in the northern Great Plains. NDSU Ext. Serv. Circ. A-1236. North Dakota State University, Fargo, ND. 8 pp.

Pundir, R.P.S., and van der Maesen, L.J.G. 1983. Interspecific hybridization in chickpea. Int. Chickpea Newsl. **8**: 4–5.

Saxena, M.C., and Singh, K.B. (*Editors*). 1987. The chickpea. CABI, Wallingford, UK. 409 pp.

Saxena, N.P 1984. Chickpea. *In* The physiology of tropical field crops. *Edited by* P.R. Goldsworthy and N.M. Fisher. John Wiley & Sons, Chichester, UK. pp. 419–452.

Singh, K.B., and Saxena, M.C. 1999. Chickpeas. Macmillan Education, London, UK. 134 pp.

Singh, K.B., and van der Maesen, L.J.G. 1977. Chickpea bibliography, 1930 to 1974. ICRISAT, Hyderabad, India. 223 pp.

Singh, K.B., Malhotra, R.S., and Muehlbauer, F.J. 1984. An annotated bibliography of chickpea genetics and breeding, 1915–1983. ICARDA and ICRISAT, Aleppo, Syria. 195 pp.

Singh, K.B., Malhotra, R.S., Halila, M.H., Knights, E.J., and Verma, M.M. 1994. Current status and future strategy in breeding chickpea for resistance to biotic and abiotic stresses. *In* Expanding the production and use of cool season food legumes. *Edited by* F.J. Muehlbauer and W.J. Kaiser. Kluwer Academic Publishers, Dordrecht, Netherlands. pp. 572–591.

Smithson, J.B., Thompson, J.A., and Summerfield, R.J. 1985. Chickpea (*Cicer arietinum* L.). *In* Grain legume crops. *Edited by* R.J. Summerfield and E.H. Roberts. Collins, London, UK. pp. 312–390.

Sotelo, A., and Adsule, R.N. 1996. Chickpea (*Cicer arietinum* L.). *In* Food and feed from legumes and oilseeds. *Edited by* J. Smartt and E. Nwokolo. Chapman and Hall, London, UK. pp. 82–89.

Van der Maesen, L.J.G. 1972. *Cicer* L., a monograph of the genus, with special reference to the chickpea (*Cicer arietinum* L.) its ecology and cultivation. H. Veenman & Zonen, Wageningen, Netherlands. 342 pp.

Van der Maesen, L.J.G. 1989. *Cicer arientinum* L. *In* Plant resources of South-East Asia, 1, pulses. *Edited by* L.J.G. van der Maesen and S. Somaatmadja. Pudoc, Leiden, Netherlands. pp. 42–43.

Van Rheenen, H.A., and Saxena, M.C. (*Editors*). 1990. Chickpea in the nineties. International Crops Research Institute for the Semi-arid Tropics, Patancheru, India. 403 pp.

Specialty Cookbooks

Saskatchewan Pulse Crop Development Board. 1994. Discover the pulse potential: peas, beans, lentils, chickpeas: includes nutritional analysis & diabetic food choices. Saskatchewan Pulse Crop Development Board, Regina, SK. 172 pp.

Schultz, J.S.F.T.W., and Regardz, B. 1988. Chickpea cookbook. India Joze Restaurant Publishing, Santa Cruz, CA. 106 pp.

CHILE PEPPER AND SWEET PEPPER

Family: Solanaceae (potato family)

Names

Scientific names: *Capsicum* species

- Chile and sweet pepper—*C. annuum* L.
- Tabasco pepper—*C. frutescens* L.
- Habañero pepper (bonnet pepper, Scotch bonnet, squash pepper)—*C. chinense* Jacq.
- Peruvian pepper (aji, cherry capsicum, pimentchien, uchu)—*C. baccatum* L. var. *pendulum* (Willd.) Eshbaugh
- Apple pepper (apple chile, rocoto, locoto)—*C. pubescens* R.& P.

- Chile or *Capsicum* peppers should not be confused with the commonly used spice also known as "pepper," obtained from *Piper nigrum* (see PEPPER).
- The country Chile is not responsible for the name chile pepper. The name for the country of Chile is derived from the South American Araucanian Indian tribe's word for "the end of the earth."
- The word "chile" comes from *chilli,* the Mexican Indian (Nahuatl) tribal name for peppers. Nahuatl, the native language of the ruling Aztecs, was the *lingua franca* of the central highlands of Mexico at the time of the arrival of the Spaniards, and is still spoken in a modified form today by 1.5 million people in Mexico. The Anglicized form "chile" in "chile pepper" is also commonly spelled "chili" and "chilli," and occasionally "chillie," "chili," and even "chilly" (although a "chilly hot pepper" is an oxymoron, i.e., a contradiction of terms). "Chile" and "chili" are general spellings in the United States. "Chilli" (plural, chillies) is commonly used in Britain, India, Africa, and the countries of the Far East, but not in the United States. The Portuguese took "chilli" peppers to India in the 16th century, where the spelling "chilli" has persisted. The same is true for the Philippines and other areas of the Pacific and Far East. In Australia and India, "chillie" and "chilly" are also generally used for hot peppers. Most of the English-speaking world uses the spelling "chilli." However, in North America, the Spanish *chile* and the anglicized *chili* are by far the most popular terms; for consistency, *chile* is used in this book. The term *chili* is also commonly applied to a variously concocted stew of meat or beans in a red pepper sauce; some pepper specialists prefer to reserve the term *chili* for this dish. "Chili" is short for "chili con carne" (or "chile con carne"), a Spanish-American term meaning "chile with meat."
- New Mexicans insist on the spelling *chile* for both the whole fruit and the red stew, contending that both the stew and the same cultivars grown in New Mexico are vastly superior to those of Texas and California. Chili has been the state dish of Texas (where it is affectionately known as "a bowl of red").
- Several chile cultivars have interesting names. The cultivar Cascabel in Spanish means "rattle," referring to the rattle the fruits make when dried. The cultivar Guajillo is very popular in parts of Mexico where it is known as *chile travieso*, meaning "naughty chile" owing to hot flavor. One American seed supplier sells a pepper with hot, small, orange fruits whose characteristic size and shape have resulted in it being called Rat Turd pepper. The Peter Pepper, also known as the Penis Pepper, is an ornamental form, with fruits decidedly phallic in appearance. A fiery Brazilian pepper is known as *mata-frade*, literally, "it kills the friar."
- "Cayenne" in "cayenne pepper" is based on the Tupi language (of Brazil) term for "hot pepper," *quiínia*. Cayenne, a municipality and the capital in French Guiana, on the northern coast of South America, acquired its name when Europeans mistakenly associated the hot pepper with this geographical area. That is, the area was named after the pepper, not vice-versa.
- Jalapeño peppers are elongated, dark green, and highly pungent. The name of the jalapeño chile traces to the town of Jalapa in Veracruz, east-central Mexico.
- Pimiento (often spelled pimento) peppers are non-pungent fruits that are often heart-shaped. The thick, red, fleshy fruit wall is used in cooking, and is also called "pimiento." This is used fresh or processed in canning, stuffing olives, cheeses, and in salads for color and flavor. The use of the term "pimiento" should not be confused with allspice

(*Pimenta dioica*, (L.) Merr.), a different species that is often referred to as pimiento in the world spice trade. The word "pimiento" is derived from the Castilian Spanish word for pepper, *pimienta*, which traces to the Latin *pigmentum*, meaning "coloring matter."

- "Paprika" is derived from essentially the same word in Serbian or Hungarian for pepper. Both cultures make considerable use of medium–hot chile powder.
- In the Midwest of the United States, sweet bell peppers are sometimes called mangos. Some produce wholesalers and retailers label boxes "mango fruit" or "mango peppers" to avoid confusion. However, to confuse matters further, California specialty produce retailer Frieda's Inc. introduced "Dried Chile Mangoes," which are dried mango slices dusted with sugar and spicy chiles.
- "Christmas peppers" are ornamental chiles. Ornamental chiles grown today may have all the colors of the rainbow, and four or five differently colored fruits on the same plant. Since they are often marketed during the holiday season, covered with bright red fruits, they are sometimes called "Christmas peppers."
- The "chilipepper" is a North American groundfish (*Sebastes goodei*).
- The "tomato pepper" or "topepo" is allegedly (but doubtfully) a hybrid of *C. annuum* and the tomato (*Lycopersicon esculentum* Mill.).
- The genus name *Capsicum* comes from the Greek *kapto*, to bite, in reference to the hot taste of the fruit. Another explanation: the name comes from the Latin *capsa*, meaning "container-like," an allusion to the shape of the fruit.
- One would expect that the name *C. annuum* would designate a plant that is naturally (not just grown as) an annual. *Capsicum annuum* is in fact a perennial, and in the tropics will grow into small to medium-sized trees as tall as 4 m (13 feet). The *International Code of Botanical Nomenclature* does not insist that the names of plants accurately describe them, and there are many cases of names that seem inappropriate.
- The French botanist who named *C. chinense* in 1776 obtained his seeds from China, hence the "*chinense*" in the scientific name, despite the fact that it originated in the New World like all *Capsicum* species.
- *Capsicum chinense* is most commonly called the "Habañero pepper." Habañero is a Mexican name, and means "of Havana" or "from Havana." Several similar sounding words could be confused with habañero. The word "habanero" means a native or inhabitant of Havana, Cuba. The habanera is a slow, voluptuous Cuban dance, or a slow Cuban dance tune. Habenaria refers to any plant or flower of the orchid genus *Habenaria*.
- *Capsicum frutescens* acquired its common name, Tabasco pepper, from the state of Tabasco and the Tabasco River in Mexico (tabasco means "damp earth" in a native American language). An American soldier returning from the Mexican War of 1846–1848 brought seeds of the species from Tabasco to Louisiana. *Frutescens* in the scientific name is Latin for shrubby, which is descriptive of the species.
- *Baccatum* in the name *C. baccatum* is Latin for berry-like or pulpy, referring to the fruit.
- *Pubescens* in the name *C. pubescens* is Latin for hairy, a reference to the pubescence of the plant.

Plant Portrait

There are about 25 species of *Capsicum*, which are perennial shrubs native to Mexico, Central and South America. The following five species have been domesticated and are cultivated for their fruits.

Common chile pepper (*C. annuum*) is a herbaceous annual in temperate regions but may become a woody biennial in warmer climates. It is the most widely used pepper in the world and includes sweet peppers, cayenne, pimento, and pungent chile peppers. Various fruit types and degrees of pungency have been selected. There are far more cultivars of *C. annuum* than of all the remaining species of *Capsicum* combined.

Tabasco pepper (*C. frutescens*) is a subshrub, growing 1–1.5 m (3–5 feet) high, with small, oblong, bright red fruit 2–3 cm (about an inch) long, pointed at the tip. The species is widely cultivated in tropical, subtropical, and warm temperate countries, notably in India, tropical America, and the southern United States. Major producers of tabasco pepper, in decreasing order of importance, are India, Sierra Leone, Zanzibar, Japan, Mexico, and the United States. In the United States, production is in Louisiana. Cultivars include the "tabasco" peppers, which are used in commercial production of hot sauces, notably in the Gulf American states. Tabasco sauce has long been an accompaniment to Louisiana oysters. *Tabasco* is a registered trademark for the hot sauce manufactured by the McIlhenny Company, which has been involved in a number of legal challenges regarding whether companies using the word "Tabasco" have committed infringements on its trademark.

Chile pepper (*Capsicum annuum*), from Köhler (1883–1914).

Habañero pepper (*C. chinense*) is a shrubby species, widely grown in South America, and popular in many Caribbean countries, particularly Cuba and Jamaica. *Capsicum chinense* is very closely related to *C. frutescens*. The fruits are usually very pungent, although some nonpungent forms have been found in Argentina. *Capsicum chinense* furnishes exceptionally, indeed dangerously hot chile peppers. The volcanic flavor of the best-known cultivar, Habañero, is blisteringly hot. *Capsicum chinense* is cultivated in the West Indies, the Amazon basin, northern South America, Peru, Bolivia, and Africa.

Peruvian pepper (*C. baccatum*) is cultivated in South America, and is also grown to a limited extent in the United States. Both wild and domesticated forms are known, with pungent (or very rarely sweet), and large and small fruits, often marketed as chiles.

Apple pepper (*C. pubescens*) is cultivated at higher elevations from Bolivia to Mexico, and in southern California, but has no commercial potential in temperate climates.

Capsicum peppers include both sweet (vegetable) forms and hot (spice or "pungent") types, and both classes are known within several of the cultivated species. Although the archetypal hot pepper is long, thin, and bright red, peppers of other colors and shapes can also be devastatingly pungent. A general but fallible rule is that the smaller the fruit, the hotter it is.

The fruits of peppers are highly perishable, and not surprisingly, ancient remains are few. However, there is evidence that chiles were eaten by native peoples in Mexico as early as 7000 BC and started to be cultivated between 5200 and 3400 BC. Archaeologists have found Peruvian embroidery depicting *Capsicum* from 400 AD, and pottery and carvings from 800 AD in the Americas. It has been said that during Columbus' first expedition (1492–1493) he observed people using colorful red fruits with much of their foods. This seasoning was deemed stronger than the black pepper of Asia (*Piper nigrum*) that he had sought. Columbus took samples back to Spain and named it "red pepper." His choice of pepper as a name was obviously guided by his expectation of finding black pepper, and indeed the need to justify his expedition by such a discovery. During Columbus' second expedition, the physician Dr. Diego Alvarez Chanca observed that the natives used chile peppers as a condiment and medicine. In fact, the Western Hemisphere indigenous peoples had been cultivating this pungent food additive for many centuries prior to Spanish occupation. The Peruvian Incas extensively used colorful, fiery fruits. Similar fruits were employed by the Mayans for cramps and diarrhea. The use of *Capsicum* quickly spread once it was introduced in Europe. Later, some of the milder forms were adopted. Chile pepper was eventually reintroduced to the northern part of the Western Hemisphere with European colonization.

Chile pepper is said to be the most used spice in the world. Various explanations of why this is so have been advanced, including the following: value as an appetite stimulant, ability to mask spoilage, high vitamin content, and causing salivation, which may facilitate digestion of high starch diets. Perhaps the most interesting explanation is that chiles cause sweating, which in tropical areas of the world would facilitate loss of excessive heat. About one-quarter of the world's adult population regularly eats chile pepper as a flavoring in food, and for many people, hot chile pepper may constitute about 1% of the total diet. The largest cultivation areas of chile peppers are in tropical and subtropical Asia, from the Indian subcontinent through Southeast Asia to Japan, and in East Africa. Central and South America and the Caribbean produce important crops as well. Major producers of sweet pepper varieties are China, Turkey, Nigeria, Spain, Mexico, and Romania. A large proportion of chile pepper production is consumed domestically. In the United States, major cultivation areas of chile peppers include California, New Mexico, Texas, North and South Carolina, and Louisiana.

Capsaicin (pronounced cap-SAY-a-sin) is the most significant chemical compound responsible for the capacity of chile to cause a burning, painful sensation, both externally and internally. In 1912, a method was devised by chemist Wilbur Scoville to compare the pungency of peppers in "Scoville heat units." Repeated exposure to capsaicin desensitizes people, allowing them to tolerate ever-increasing doses. For this reason, measuring the Scoville Heat Unit rating of a given chile pepper used to be performed with subjects who were not regular pepper eaters. What is considered mild to a "hot chile junky" could be intolerably fiery to those not habituated to the spice. Today, machines rather than people measure a pepper's heat by finding the capsaicin level and converting this to Scoville Units (pure capsaicin equals 16 000 000 Scoville Units). Bell (sweet) peppers are rated at 0, jalapeño and cayenne at 2000–35 000, and tabasco (*C. frutescens*) at 30 000–80 000. The highest ratings of *C. annuum* is around 100 000. The *C. chinense* cultivar Habañero has an impressive rating of 200 000–300 000. High-pungency flavoring extracts (oleoresins) measuring up to 1 000 000 have been made from African-type chiles. There has been a lively debate as to whether or not the Habañero cultivar, which has been called "the king of sting," is or is not the hottest of the chiles. Some people contend that it is. Others claim that many of the hot South American chile peppers will cause the Mexican Habañero to pale by comparison. The varieties of *C. chinense* grown in Africa are also reportedly the most pungent of all. The *Guinness World Records 2001* book states that the world's hottest spice is Red Savina Habañero, developed by GNS Spices of Walnut, California, with a rating of 350 000–570 000. In 2006, it was widely reported in the media that Michael and Joy Michaud of "Peppers by Post" in West Dorset, England, had developed a cultivar of *C. chinense* called Dorset Naga, which produced a rating of up to 970 000.

The most pungent types of chile are used to prepare pharmaceuticals. These include stimulant and counter-irritant balms applied externally, and formulations taken internally to treat flatulence and arouse appetite. Capsaicin ointment has proved useful for reducing pain due to shingles and stump pain following amputation. Chile extracts are found in North America in some commercial laxatives, external analgesic preparations, and liniments. Capsaicin is the active ingredient in two popular rubdown liniments used for sore muscles, Heet and Sloan's Liniment.

Culinary Portrait

Sweet peppers include the bell peppers, so named for the bell-like shape of many of the cultivars. Sweet peppers are typically forms with large, smooth, thick flesh of various colors, most commonly green and red. Red and orange sweet peppers purchased in supermarkets tend to be sweeter than green sweet peppers. Sweet peppers are eaten raw or cooked, and are used not only as a vegetable but also as condiments to flavor other foods. Some people find that sweet peppers give them indigestion unless they are peeled. Raw sweet pepper is often served with a dip or as an appetizer, and is used in salads. Sweet peppers are added to soups, stews, omelets, tofu, rice, pasta, pizza, and many other dishes, as well as processed meats. Large, thick-fleshed sweet varieties are often stuffed with meat and cooked. Sweet peppers become sweeter when cooked, but if overcooked they lose flavor and nutrients. Cooking brown, black, and purple peppers may cause them to turn green. Sweet peppers can be stored, unwashed, in a perforated plastic bag in a refrigerator for about a week.

Hot peppers are used particularly as flavoring and seasoning for various foods, and in condiments. They are also very commonly employed as a source of yellow, orange, and red color in food products. Peppers differ in aroma, and vary in taste from mild to fiery. Cayenne pepper and "red pepper" products are ground from high- and medium-pungency chiles, and when made into sauces, they add color and flavor to spaghetti, pizza, gravies, and stews. Cayenne pepper should be used with discretion, a mere pinch usually sufficing to season an entire dish. Similarly, very little tabasco sauce (1–3 drops) usually is enough to season a dish. Ground chile is made from macerated whole fruits, including the fruit wall, seeds, and placenta (the part of the ovary to which the seeds are attached). Chile powder is a commercial blend of several finely ground spices including chile, cumin (*Cuminum cyminum* L.), oregano (*Origanum* species), and garlic (*Allium sativum*). Both mixtures are used extensively in many different ethnic dishes. These include Mexican, Indian (and other Asian), and certain Italian dishes, for example, as flavorings for meat and vegetable soups, pasta, pizza, curries, chili con carne, tacos, burritos, and processed meats. Paprika, a nonpungent (for the most part) bright red powder, is made from the fruit wall of certain highly colored, sweet to mildly pungent cultivars and is used mainly as a coloring agent in food preparations such as cheese, egg, and fish dishes, salads, sauces, and rice. "Hungarian paprika" can be notably hot. Paprika should not be cooked too long, as this results in deterioration of the flavor and color. The fresh, fleshy fruit wall of pimiento peppers is used to stuff olives, and to flavor cheese and processed meats. Several cultivars are used specifically in given ethnic dishes, for examples, Ancho and New Mexican in chile rellenos, Pasilla for mole sauce, and Serrano for pico de gallo. In West Indian markets of some large North American cities, a seasoning mixture called *sofrito* (a blend of fetid herbs and *C. chinense* chiles) is sold. The oleoresins obtained from the various hot peppers are also employed in the food industries. These are added to meats, sausage, catsup, sauces, ginger ale, rum, and even chewing and smoking tobacco.

Hot peppers are widely used by poor peoples around the world because they add flavor very cheaply to meals, and as a consequence, chile pepper seasoning is characteristic of the cuisines of many nations of Asia and South America. Ethnic food of these regions, for example, that of India and Mexico, is often heavily spiced with chile pepper, and many Westerners attempting to reproduce such dishes think that the food will only be "authentic" by adding the same, large amounts of hot peppers. What is generally not realized is that, in cultures that have become habituated to chile pepper, people have become used to large doses, and the food consequently does not taste nearly as hot to them as it does to those unused to the high concentrations. To reproduce an "authentic" taste sensation for those who are not used to very high levels of chile pepper, it is quite wrong to add it until the meal is blindingly hot.

Fresh hot peppers can be stored, wrapped in plastic and unwashed, in a refrigerator, for up to a week. Dried chile peppers can be stored for up to 8 months in a plastic bag in a refrigerator. Paprika should be stored in an airtight container in the refrigerator (it is one of the few spices that keep better refrigerated). Tabasco sauce keeps indefinitely at room temperature.

Patients with duodenal ulcers are often advised to consume a bland diet, with little spice, especially chile. At least one study has contradicted this conventional advice, concluding that bland food is unlikely to be useful, and that consumption of chile does not retard the healing of duodenal ulcers. Indeed, there is evidence that capsaicin in chile may have a protective effect against peptic ulcer disease. Nevertheless, high doses of red pepper are an extreme irritant to the esophagus and stomach and, perhaps, kidneys. Individuals with intestinal or kidney ailments are advised not to use hot red pepper.

Warning: Hot peppers can irritate skin and eyes; in large doses, chiles may cause stomach and intestinal irritation.

Precautions: When handling hot peppers, avoid touching eyes, sensitive areas, or cuts. Wear rubber gloves, especially when preparing large quantities. Wash hands well, also knife and chopping board, in chlorine and water (although soap and water will not get rid of all of the capsaicin). If gloves have not been worn, a nail brush may be necessary to remove the oil from under the fingernails.

What to do about a skin reaction: Because it is essentially insoluble in cold water and only slightly soluble in hot water, traces of capsaicin on the hands can be transferred accidentally to sensitive mucous membranes well after the initial contact. Romantic activity after handling chile could be hotter than anticipated! Following work with hot peppers, it is advisable to wash one's hands before as well as after using the rest room! Cold water provides only temporary relief. As normal washing with soap will not remove capsaicin well, it has been recommended that when necessary, this chemical be removed from the hands by washing with vinegar (capsaicin is soluble in vinegar) or a mild chlorine household bleach solution (one part common bleach to five parts of water; a stronger solution does not seem to improve removal of the capsaicin). Alcohol on the skin may also be effective.

What to do about an oral reaction: A sore mouth can be soothed with beans, plain cooked rice, bread or other starchy foods. Drinking water, especially ice water, can provide some psychological relief, but does not help much because of the insolubility of capsaicin in water. Drinking alcohol (especially beer) or consuming milk fat (such as whole milk) is generally recommended to ease the pain in the mouth. Casein-containing foods, such as milk, cheese, butter, yogurt, sour cream, and ice cream reduce the effects of capsaicin through a detergent effect. Sugar and chocolate are also reported to be helpful.

Culinary Vocabulary

- "Hot pepper oil" is vegetable oil flavored with chile peppers and used in Chinese and Italian cuisine.
- "Piripiri" (= piri-piri, pronounced pee-ree-pee-ree), a Portuguese word much used in Africa, refers to small, very hot, red chiles, as well as the national dish of Mozambique (made up of seafood, fowl, or meat that has been marinated in lemon juice, garlic, and chiles, then grilled and served with rice). The name has also been applied to any chile-based sauce served as a condiment. As well, in Swahili, piripiri is a slang term for "penis."

- "Abruzzese" (pronounced ah-BROOZ-dzee) refers to Italian dishes prepared with the generous use of hot chiles, a tradition associated with Italy's Abruzzi region.
- The expression *à la hongroise* is French for Hungarian style, and indicates a dish almost always with the national spice, paprika (also onions, sour cream, and perhaps bell pepper, cabbage, and (or) leek).
- "Chile (chili) powder" is a confusing term, often referring to a blend of dried, pulverized chiles and such spices as garlic, cumin, and oregano. In Asia, chile powder tends to be much hotter than Latin American chile powder.
- The traditional Hungarian dish "goulash," a paprika-seasoned mixture of meat and vegetables, is based on the Hungarian word gulyás, herdsman's stew.
- The word "salsa" (from the Spanish for "sauce," and pronounced SAHL-sah) generally refers to a Mexican cold sauce made from such ingredients as tomatoes, cilantro, chiles, and onions. Similar vegetable and spice purées are common to many cultures (and known by such names as chutneys and chow-chows). Tomatillo (*Physalis philadelphica* Lam.) is often added because the fruits contain a pectin-like substance that thickens the salsa when refrigerated. A very wide variety of types of salsa with different ingredients are prepared today, but the most common ingredient remains chile pepper. In Mexico and the southwestern United States, salsa is used in popular dishes such as tacos, enchiladas, tostadas, gazpacho, moles, and guacamole. (In Italian usage, salsa is a general term for pasta sauces; in Portuguese, salsa refers to parsley.) In dollar value, salsa sales are higher than sales for ketchup in the United States.

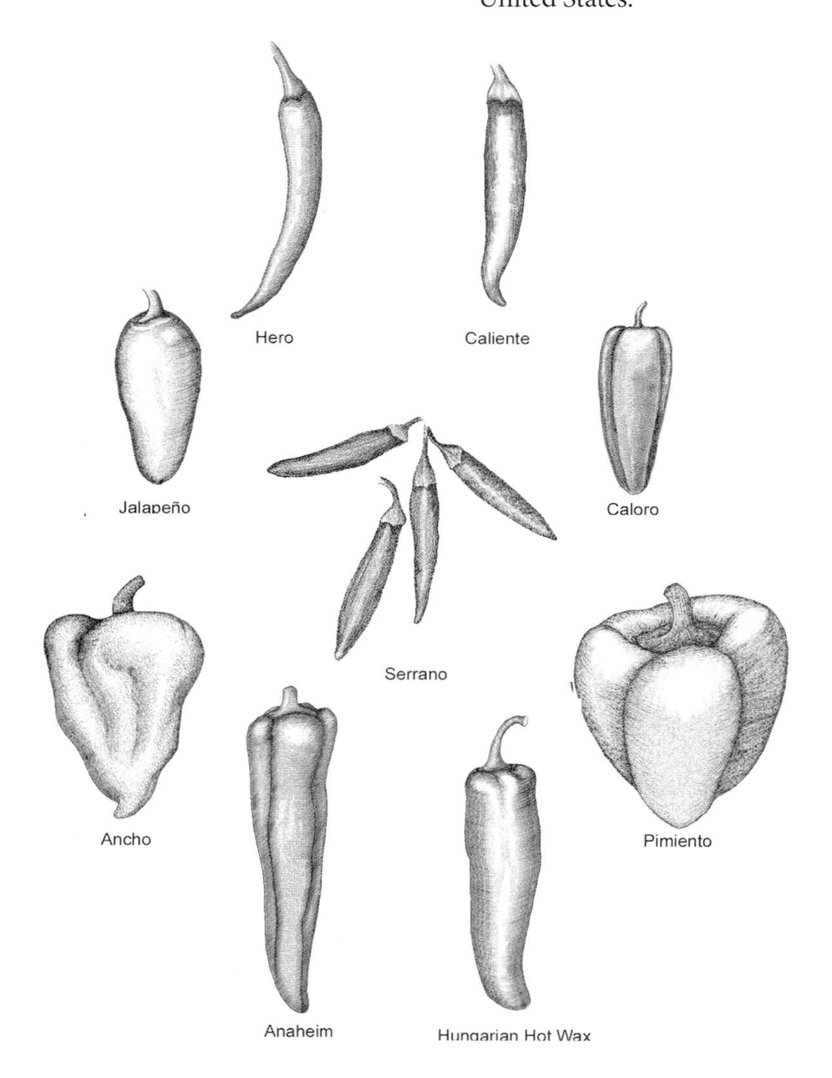

Varieties of chile pepper (*Capsicum annuum*), by M. Jomphe.

- "Bullshot" is a beverage made of two parts beef bouillon, one part vodka, and dashes of Worcestershire sauce, bitters, and tabasco sauce.
- A "Cajun martini" is a cocktail prepared with vodka (usually pepper-flavored) and dry vermouth, garnished with a jalapeño pepper.
- The food phrase "à la king" refers to a dish, usually chicken or turkey, in a rich cream sauce containing mushrooms, pimientos, green peppers, and sometimes sherry. Creation of the dish is variously credited to the following: chef George Greewald of New York's Brighton Beach Hotel, supposedly made for the proprietors, Mr. And Mrs. E. Clark King III; Foxhall P. Keene, the son of Wall Street broker and sportsman James R. Keene, at Delmonico's restaurant in New York City; and James R. Keene at Claridge's restaurant in London, after Keene's horse won the 1881 Grand Prix.
- *Chiles rellenos* (Mexican Spanish for "stuffed peppers") is a Mexican dish of fried, stuffed chile peppers, which is popular in restaurants that serve Mexican food.
- "Hunan hand" was so named after a man received serious capsaicin burns on his fingertips by chopping up red peppers for a Hunan Chinese recipe.
- "Fisherman's joke" is an Italian dish (*scherzo del pescatore*), referring to a platter of cooked small squids, which are customarily eaten by popping them whole into the mouth. The "joke" is that hidden in the platter is one squid that has been filled with hot pepper.

Curiosities of Science and Technology

- Conquistador Bernal Díaz (1496–1584), a participant in the conquest of Mexico in 1819, recorded that the Aztecs consumed the arms and legs of their sacrificial victims with a sauce made of chile peppers, tomatoes, wild onions, and salt—ingredients that are nearly the same as used in the salsa in most Mexican homes today.
- The Mayans are said to have hurled chile by catapult to blind and stun their enemies.
- Indigenous New World people are reported to have fought the Spanish invaders by using irritating, burning chile smoke.
- The Carib Indians prepared a "pepper-pot," a kind of camper's stew in which various ingredients are constantly added, so that the pot is never empty.
- The first commercial chile powder was produced in 1894 by café owner William Gebhardt, a German immigrant to New Braunfels, Texas. He is also credited with producing the first canned chili con carne at his factory in 1911. (As noted above, chile con carne is the same as chili.) Gebhardt's brand of chile powder is still one of the most popular, and is specified in many recipes.
- Peppers are high in vitamin C (ascorbic acid). The Hungarian Albert Szent-Györgyi was awarded the Nobel Prize for physiology and medicine in 1937 for isolating vitamin C in 1928 from paprika fruits and showing that they were one of the richest sources available for this vitamin.
- In 1983, to prevent rambunctious teenagers from sucking tokens out of subway turnstiles, New York transit officials dusted ground hot pepper on token slots.
- Shri Kishan Joshi of Almora, India, in 1985–1986, grew a chile plant 6.6 m (21 feet 7 inches) tall. Dr. P.W. Bosland (personal communication) reports having seen *C. annuum* "trees" 10 m (33 feet) tall in the tropics. Dr. W.H. Eshbaugh (personal communication) has seen plants of *C. pubescens* straggling to lengths up to 18 m (58 feet) in Bolivia.
- Southern Columbian natives mixed powdered chile pepper fruits with coca leaf (from *Erythroxylum* species, particularly *E. coca* Lam.) and snuffed the mixture, knowing that the mucous membrane irritation from the pepper accelerates the uptake of the narcotic.
- Peruvian Amerindian nursing mothers applied pepper juice to their nipples to wean their infants; similarly, some parents today use "Nothumb," a chile preparation applied to children's thumbs to keep them from sucking them.
- Koreans have been observed to sprinkle red pepper in flannel bellybands wrapped around their middles in cold weather as a way of keeping warm.
- Red chile mash added to chicken feed darkens the color of the egg yolks, making the eggs more saleable.
- Pink flamingos in captivity lose the brilliance of their plumage, but feeding chile pepper mash puts the birds back in the pink.
- The bite in ginger ale and the peppery flavor of old time "red hots" is often the result of added capsaicin.
- Mixtures containing *Capsicum* extracts are marketed for placing in stockings to keep feet warm on cold days. One such commercial foot warmer powder is known as Fire Feet.

Tabasco pepper (*Capsicum frutescens*), from Duthie (1893).

- The wild forms of *C. annuum* are called bird peppers, a reference to the attractiveness of the red fruits of *Capsicum* species to birds. Very hot chiles used in Asia, and occasionally elsewhere, are also sometimes called "bird peppers" (and "bird's eye chiles"). In the wild, pepper seeds are commonly distributed by birds, many of them passing unharmed through the birds' guts. Seeds of most wild species are slow to germinate because of germination inhibitors, which are effectively removed by passing through a bird's digestive system.
- Birds apparently don't have taste buds that distinguish the hotness of capsaicin, as reflected by their consumption of chile peppers. "Squirrel Away" is a commercial birdseed supplement made with chile powder from *C. chinense*. Used at wild bird feeders, it discourages squirrels from foraging on the birdseed at the feeders.
- Before the Spanish occupation, chiles were used as currency by some native New World peoples.
- Mexican immigrants to West Texas have been observed using pepper to assist women in giving birth. A mother-to-be is stretched out on the floor with a rope around her middle, with a corncob used as a turnbuckle. As labor progresses, the midwife tightens the rope by turning the corncob. When birth seems imminent, a final turn of the corncob is made; at the same time, a handful of tobacco snuff mixed with chile powder is thrown into the face of the expectant woman, who sneezes violently, and (it is expected) expels the baby.
- A sailing enthusiast had the idea of trying the extra hot Habañero cultivar to see what affect it would have on barnacles and tubeworms from growing on the underwater portion of boats. This insight led to the development of an epoxy-based

experimental marine paint called Barnacle Ban. This includes derivatives from the cultivar Habañero, and is so hot it will blister fingers. It is claimed to prevent mussels (especially zebra mussels) and barnacles from attaching themselves to boats, buoys, and intake pipes.

- Chile is used in China to treat toothache. Moreover, a folk remedy in North America for relieving toothache involves soaking a piece of cotton in tincture of cayenne and placing it inside the tooth cavity. Oil of cloves is famous for its ability to reduce the pain of a toothache, because of the active ingredient eugenol. Capsaicin is rather similar chemically to eugenol, which may explain why the effect of chewing a chile pepper over an aching tooth has been said to be similar to the effect of oil of cloves.
- A research study found that 75% of headache patients felt relief when they rubbed capsaicin on their nose.
- The scientific journal *Toxicon* reported that drinking 1.7 L (1.5 quarts) Louisiana-style hot sauce in one sitting will cause death by respiratory failure if body weight is 63.5 kg (140 pounds) or less.

Key Information Sources

Andrews, J. 1986. Peppers. The world's favorite all-American spice. Pac. Discovery, **39**(**1**): 20–33.

Andrews, J. 1988. Around the world with the chili pepper: post-Columbian distribution of domesticated capsicums. J. Gastronomy, **4**(**3**): 21–35.

Andrews, J. 1995. Peppers. The domesticated capsicums. 4th edition. University of Texas Press, Austin, TX. 170 pp.

Andrews, J. 2000. Chilli peppers. *In* The Cambridge world history of food. *Edited by* K.F. Kiple and K.C. Ornelas. Cambridge University Press, Cambridge, UK. pp. 281–288.

Berke, T.G., and Shieh, S.C. 2001. Capsicum, chillies, paprika, bird's eye chilli. *In* Handbook of herbs and spices. *Edited by* K.V. Peter. CRC Press, Boca Raton, FL. pp. 111–122.

Bosland, P.W. 1992. Chiles: a diverse crop. HortTechnology, **2**(**1**): 7–10.

Bosland, P.W. 1993. *Capsicum*: a comprehensive bibliography. 2nd edition. The Chile Institute, Las Cruces, NM. 255 pp.

Bosland, P.W. 1994. Chiles: history, cultivation, and uses. *In* Spices, herbs and edible fungi. *Edited by* G. Charalambous. Elsevier, London, UK. pp. 347–366.

Bosland, P.W., Bailey, A.L., and Inglesias-Olivas, J. 1990. *Capsicum* pepper varieties and classification. Coop. Ext. Serv. Circ. 530. College of Agriculture and Home Economics, New Mexico State University, Albuquerque, NM. 13 pp.

DeWitt, D. 1991. Chile peppers: a selected bibliography of the capsicums. The Chile Institute, Las Cruces, NM. 74 pp.

DeWitt, D., and Bosland, P.W. 1993. The pepper garden. Ten Speed Press, Berkeley, CA. 240 pp.

Eshbaugh, H.W. 1980. The taxonomy of the genus *Capsicum* (Solanaceae). Phytologia, **47**: 153–166.

Eshbaugh, W.H. 1993. Peppers: history and exploitation of a serendipitous new crop discovery. *In* New crops. *Edited by* J. Janick and J.E. Simon. John Wiley and Sons, New York, NY. pp. 132–139.

Eshbaugh, W.H., Guttman, S.I., and McLeod, M.J. 1983. The origin and evolution of domesticated *Capsicum* species. J. Ethnobiol. **3**: 49–54.

Govindarajan, V.S. 1985. Capsicum—production, technology, chemistry and quality. Part I. History, botany, cultivation and primary processing. CRC Crit. Rev. Food Sci. Nutr. **22**: 109–176.

Govindarajan, V.S. 1986*a*. *Capsicum*—production, technology, chemistry, and quality—Part II. Processed products, standards, world production and trade. CRC Crit. Rev. Food Sci. Nutr. **23**: 207–288.

Govindarajan, V.S. 1986*b*. *Capsicum*—production, technology, chemistry, and quality. Part III. Chemistry of the color, aroma, and pungency stimuli. CRC Crit. Rev. Food Sci. Nutr. **24**: 244–355.

Govindarajan, V.S., and Sathyanarayana, M.N. 1991. Capsicum—production, technology, chemistry, and quality. Part V. Impact on physiology, pharmacology, nutrition, and metabolism; structure, pungency, pain, and desensitization sequences. CRC Crit. Rev. Food Sci. Nutr. **29**: 435–474.

Greenleaf W.H. 1986. Pepper breeding. *In* Breeding vegetable crops. *Edited by* M.J. Bassett. AVI Publishing Co., Westport, CT. pp. 67–134.

Heiser, C.B., Jr. 1995. Peppers. *Capsicum* (Solanaceae). *In* Evolution of crop plants. 2nd edition. *Edited by* J. Smartt and N.W. Simmonds. Longman Scientific & Technical, Burnt Mill, Harlow, Essex, UK. pp. 449–451.

Heiser, C.B. [Jr.], and Pickersgill, B. 1969. Names for the cultivated *Capsicum* species (Solanaceae). Taxon, **18**: 277–283.

Heiser, C.B., Jr., and Smith, P.G. 1953. The cultivated capsicum peppers. Econ. Bot. **7**: 214–227.

International Board of Plant Genetic Resources. 1983. Genetic resources of *Capsicum*. IBPGR, Rome, Italy. 49 pp.

Jesiolowski, J. 1993. Grow red hot chile peppers. Org. Gard. **40**(**2**): 27–32.

Kang, J.Y., Yeoh, K.G., Chia, H.P., Lee, H.P., Chia, Y.W., Guan, R., and Yap, I. 1995. Chili—protective factor against peptic ulcer? Dig. Dis. Sci. **40**: 576–579.

Lippert, L.F., and Scharffenberg, R.S. 1964. Garden pepper (*Capsicum* sp.). Vegetable crops bibliographies, vol. 1. Bibliographic Associates, West Covina, CA. 258 pp.

Locock, R.A. 1985. *Capsicum*. Can. Pharm. J. **118**: 517–519.

Magra, J.A. 1975. *Capsicum*. Crit. Rev. Food Sci. Nutri. **6**: 177–199.

Martin, F.W., Santiago, J., and Cook, A.A. 1979. Vegetables for the hot, humid tropics. Part 7. The peppers, *Capsicum* series. Agricultural Research, Science and Education Administration, U.S. Dept. of Agriculture, New Orleans, LA. 18 pp.

McLeod, M.J.S., Guttman, S.I., and Eshbaugh, W.H. 1982. Early evolution of chili peppers (*Capsicum*). Econ. Bot. **36**: 361–368.

Munting, A.J. 1974. Development of flower and fruit of *Capsicum annuum* L. Acta Bot. Neerl. **23**: 415–432.

Naj, A. 1993. Peppers. Vintage Books, New York, NY. 245 pp.

Palevitch, D., and Craker, L.E. 1993. Nutritional and medical importance of red peppers. Herb Spice Med. Plant Dig. **11**(**3**): 1–4.

Pickersgill, B. 1971. Relationships between weedy and cultivated forms in some species of chili peppers (Genus *Capsicum*). Evolution, **25**: 683–691.

Pickersgill, B. 1984. Migrations of chili peppers, *Capsicum* spp., in the Americas. *In* Pre-Columbian plant migration, *Edited by* D. Stone and R.M. Bird. Peabody Museum of Archaeology and Ethnology, Harvard University, Cambridge, MA. pp. 106–123.

Pohronezny, K. 2003. Compendium of pepper diseases. American Phytopathological Society Press, St. Paul, MN. 63 pp.

Poulos, J.M. 1993. *Capsicum* L. *In* Plant resources of South-East Asia, 8, vegetables. *Edited by* J.E. Siemonsma and K. Piluek. Pudoc Scientific Publishers, Wageningen, Netherlands. pp. 136–140.

Rylski, I. 1985. *Capsicum*. *In* Handbook of flowering. Vol. II. *Edited by* A.H. Halevy. CRC Press, Boca Raton, FL. pp. 140–146.

Rylski, I. 1986. Pepper (*Capsicum*). *In* CRC handbook of fruit set and development. *Edited by* S.P. Monselise. CRC Press, Boca Raton, FL. pp. 341–354.

Smith, P.G., and Heiser, C.B., Jr. 1951. Taxonomic and genetic studies on the cultivated peppers *C. annuum* L. and *C. frutescens* L. Am. J. Bot. **38**: 367–368.

Smith, P.G., Villalon, B., and Villa, P.L. 1987. Horticultural classification of peppers grown in the United States. Hortscience, **22**: 11–13.

Stock, M.T., and Hunter, K. 1995*a*. The healing powers of peppers. Part 1: cayenne as a curative. Chile Pepper [Mag.], **9**(**3**): 18–21, 37, 39–41.

Stock, M.T., and Hunter, K. 1995*b*. The healing powers of peppers. Part 2. Chile Pepper [Mag.], **9**(**4**): 20–23, 52.

Wien, H. C. 1997. Peppers. *In* The physiology of vegetable crops. *Edited by* H.C. Wien. CABI International, Wallingford, Oxon, UK. pp. 259–293.

Specialty Cookbooks

(Note: There are dozens of cookbooks dedicated to chile peppers. The following is a selection.)

Andrews, J. 1993. Red hot peppers. Macmillan Publishing Co., New York, NY. 242 pp.

Andrews, J. 1999. The pepper trail: history & recipes from around the world. University of North Texas Press, Denton, TX. 261 pp.

Automobile racing drivers—United States. 1998. NASCAR cooks with Tabasco brand pepper sauce: celebrating NASCAR's 50th anniversary, featuring the hottest collection of favorite recipes from 50 NASCAR celebrities. HarperCollins Publishers, New York, NY. 96 pp.

Belsinger, S., and Dille, C. 1994. The chile pepper book: a fiesta of fiery, flavorful recipes. Interweave Press, Loveland, CO. 96 pp.

Ciletti, B. 1998. The pepper harvest cookbook. Taunton Press, Newtown, CT. 176 pp.

Creasy, R. 2000. The edible pepper garden. Periplus, Boston, MA. 105 pp.

Dewitt, D. 1999. The chile pepper encyclopedia. Everything you'll ever need to know about hot peppers, with more than 100 recipes. William Morrow & Co., New York, NY. 338 pp.

DeWitt, D., and Gerlach, N. 1984. The fiery cuisines. St. Martin's Press, New York, NY. 229 pp.

Dewitt, D., and Gerlach, N. 1990. The whole chile pepper book with over 180 hot and spicy recipes. Little Brown & Co., Boston, MA. 373 pp.

Dewitt, D., and Gerlach, N. 1995. Heat wave! the best of Chile Pepper Magazine: 200 great recipes from hot & spicy world cuisines. Crossing Press, Freedom, CA. 179 pp.

Fleetwood, J. 2001. Red hot chili pepper: bring a touch of fire to your cooking with this sizzling collection of more than 140 chilli-hot recipes. Lorenz, London, UK. 256 pp.

Hazen-Hammond, S. 1993. Chile pepper fever: mine's hotter than yours. Voyageur Press, Stillwater, MN. 127 pp.

McIlhenny Co. 1990. What's hot: hot tips (mild ones, too) on how to deliciously use the 720 drops in a 2 oz. bottle of Tabasco® pepper sauce. McIlhenny Co., Avery Island, LA. 14 pp.

McIlhenny, P., and Hunter, B. 1993. The tabasco cookbook: 125 years of America's favorite pepper sauce. Clarkson Potter, New York, NY. 144 pp.

McLaughlin, M. 1986. The Manhattan Chili Co. Southwest American cookbook. Crown Publishers, New York, NY. 120 pp.

Siegal, H., and Gillingham, K. 1994. The totally chile pepper cookbook. Celestial Arts, Berkeley, CA. 95 pp.

Steer, G. 1995. Great chili pepper cookbook. Book Sales Inc., Edison, NJ. 128 pp.

CHOCOLATE (CACAO)

Family: Sterculiaceae (cacao or sterculia family)

Names

Scientific name: *Theobroma cacao* L.

- The word "chocolate" has been claimed to be based on the South American Indian (Náhuatl) word, *xocoatl*, for the cacao bean. *Chocolatl*, the beverage form of chocolate that was consumed by the Indians, has been interpreted as derived from *xocoatl*, meaning "bitter drink." Some linguists have proposed other origins. One theory is that the Spaniards coined chocolate as a new word, combining the Mayan word for hot, *chocol*, with the Aztec word for water, *atl*.
- The Nahuatl word *xocoatl* (mentioned above) is translated as both "cocoa" and "cacao" in English. "Cacao" is the version found in the species name *T. cacao*, and is used in such expressions as "cacao tree" (in which case cacao is pronounced kah-KAY-oh or kah-KAH-oh). It has been claimed that the word *cacao* is derived from *cacahuaquchtl*, the Mayan Indian word for cacao tree, and indeed the tree sacred to the Mayan gods.
- Because of a mistake in spelling, probably made by English importers many years ago, cacao beans became known as cocoa beans in English-speaking countries. As a result, many people think that the beans come from the coconut palm tree instead of the cacao tree. "Coco butter" is a misspelling of "cocoa butter," which is obtained from cacao beans, not from coconuts. "Coco bean" is a variety of the common bean (which see), *Phaseolus vulgaris*.
- According to Wessel and Toxopeus (2000), "In the current English literature the terms 'cacao' for the plant and 'cocoa' for its products have generally been replaced by 'cocoa' to encompass both meanings."
- The genus name *Theobroma* is from the Greek *theos*, god, + *broma*, food, i.e., "food of the gods." Of course, chocolate also has the power to gladden mortal hearts.

Plant Portrait

The cacao or chocolate tree is a small tropical evergreen tree, reaching a height of 3.7–7.5 m (12–25 feet), occasionally growing to 15 m (49 feet). This native of South America and the West Indies is the source of the world's chocolate and cocoa. Cacao trees are adapted to the shade of taller trees, and are grown especially in plantations of rubber trees, coconuts, and bananas. The tree is unusual in bearing its flowers and pods (which are shaped like American footballs) on the main trunk, as well as on the larger branches. Trees generally develop 20–40 pods, which yield 1–3 kg (2–6.7 pounds) of dried beans each year. The pods are 20–30 cm (8–12 inches) long, yellow or red, and contain white, mucilaginous, mildly sweet, buttery pulp and 20–40 flat seeds (generally called beans). The seeds are about 2.5 cm (1 inch) long, reddish brown externally, dark brown internally. The cacao tree was extensively cultivated from Mexico to South America long before the discovery of the New World. At present, cacao trees are grown mainly in West Africa (the source of two thirds of world production), Asia (about 18% of world production) and in the Americas (about 15% of world production). Côte d'Ivoire (Ivory Coast) produces more than 40% of the world's cocoa beans, and it has been estimated that 10 million people work in the cocoa industry in that country alone. Ghana and Indonesia rank next among major world producers, followed by Brazil, Nigeria, and Malaysia. Cross-pollination in cacao is necessary to produce pods. In its native area, cacao is insect pollinated, various authorities crediting pollination particularly to midges, but also to some extent to aphids, thrips, ants, and bees. Outside of the native South American range, the lack of adequate pollination and pod production is sometimes a serious problem.

The cocoa or chocolate nuts of commerce are the seeds taken from the fruit and deprived of their slimy covering. There are many varieties, but the international cocoa community generally classifies cocoa beans into two broad types. *Forastero cocoa* or "bulk cocoa" has highly pigmented beans used in the manufacture of cocoa butter and high-volume chocolate lines, and provides more than three-quarters of world production. The Forastero is found in West Africa, Brazil, and Ecuador. The second type is *criollo cocoa*, mainly grown in Central and northern South America, and the white or pale violet beans are used to manufacture chocolate of the highest quality. This type is quite fragile and therefore not much cultivated.

The "Trinitario" is a hybrid of the two previous groups. It provides about 20% of the world production. The flavor of cocoa liquor, used in the manufacture of chocolate, is determined not only by the variety but also by where the trees are grown and how the beans are processed. The seeds are prepared for commerce either by simple drying, in which case they retain their bitterness, or by various other processes that remove the bitterness. Chocolate and cocoa are made from cocoa beans by fermentation and roasting. Cocoa butter, a liquid also called cocoa liquor, is extracted from the beans and is used in a number of products, ranging from cosmetics to pharmaceuticals, but mainly in the manufacture of chocolate confectionary.

Chocolate making involves the following steps. After cocoa pods are harvested, the beans inside are scraped out and fermented. The fermented beans are sun-dried, and shipped to manufacturers, who blend and roast them. Next, the beans are crushed, their shells or husks are removed, and the cleaned cocoa kernels, called nibs, are milled into a thick paste called "chocolate liquor." The chocolate liquor may be refined additionally by pressing out the cocoa butter, leaving dry cocoa powder. Various types of chocolate are prepared by adding sugar, milk solids, vanilla, and other flavorings. The mixture is finally blended and milled until smooth, and the finished chocolate is poured into molds, hardened, and wrapped and shipped to stores.

Sweetness of chocolate depends on the content of chocolate liquor (which is naturally bitter) and the additives (particularly sugar) that manufacturers place in their chocolates. "Unsweetened chocolate" (also called bitter chocolate and baking chocolate, and generally considered ideal for baking purposes) is hardened chocolate liquor containing no sugar. It may contain artificial or natural vanilla. "Bittersweet chocolate," depending on national regulations, contains 35–50% chocolate liquor, but the degree of sweetness depends on the brand; one would expect minimal amounts of sugar, but there may be 35–50% sugar present, overlapping with semisweet and sweet chocolates. "Semisweet" chocolate contains 15–35% chocolate liquor and has "moderate" amounts of sugar. "Sweet" chocolate has a minimum of 15% chocolate liquor, and generally has the highest amounts of sugar. Because of differences in the amounts of sugar different manufacturers add, the overall sweetness of chocolate is sometimes not well indicated by the terms bittersweet, semisweet, and sweet.

The use of cocoa for eating and drinking probably dates back several thousand years. The ancient Olmecs (1500–300 BC) were the first to cultivate cocoa, in the lush jungles of Central and South America. Cocoa was used to prepare a ritual drink reserved for the kings of the Mayan empire (200 BC–1550 AD). For most of history, chocolate was consumed only as a liquid. The New World tradition of drinking spiced liquid cocoa was brought back to Spain in the 16th century by the Conquistadores. For nearly a century, liquid chocolate made from cocoa, sugar, cinnamon, and vanilla was the exclusive drink of the Spanish Royal Court, but by the early 18th century, cocoa houses were popular in major European cities. Solid or "Dutch chocolate" was invented by a Dutch chemist, C.J. van Houten, in 1828. He discovered that processing chocolate liquor with alkali neutralizes various acids and so tends to increase the chocolate flavor, reduce bitterness, and darken the chocolate. Today, most chocolates are prepared with alkali, a procedure still called "Dutching."

Culinary Portrait

There is probably no other food than chocolate that holds so universal an appeal to people of all ages. The cocoa bean is the source of a wide range of foods and delicacies, the basis of a veritable cornucopia of delights that has not only enriched our culinary lives, but also has captured the imaginations and souls of people for generations. Valentine's Day (February 14) is marked by gifts of heart-shaped boxes of chocolates, a tradition that was well established by the 1890s. Easter is characterized by chocolate bunnies and eggs. At least to the sweeter sex, a gift of chocolate is almost always welcome ("sweets to the sweet," as the saying goes). Many cookbooks and cooking courses deal exclusively with chocolate, and more than a million recipes using chocolate are available on the Web, ranging from the preparation of simple classic desserts to exquisite complex creations that are works of art. The topic is so vast that only a few elementary culinary observations can be made here.

Chocolate or cocoa is a principal flavoring for desserts, puddings, baked goods, icing, and ice cream, and combines well with fruits, nuts, coffee, spirits, and mint. Chocolate is used in cakes, cookies, and a wide variety of confections—indeed every imaginable kind of dessert. In addition to sweet preparations, chocolate is also used, especially in northern Europe, to flavor savory dishes, particularly sauces for game.

When melting chocolate, cut it into small chunks; never put a big block in a pan. Chocolate melts most quickly and easily when it is grated or chopped. Use a large, sharp knife to shave off pieces, and a cutting

Chocolate (*Theobroma cacao*), from Harter (1988).

board. An uncovered double boiler is frequently used to melt chocolate. In melting chocolate, it takes only seconds to go from just right to burned. The temperature of the chocolate should never be greater than 49 °C (120 °F), to avoid altering its flavor. Unsweetened, sweet, semisweet, and bittersweet chocolate require stirring every minute or so, while milk and white chocolate require constant stirring. Chocolate doesn't look melted until stirred. No water should ever contact chocolate while it is melting, as this causes lumps to form.

"Chocolate syrup," used to garnish and flavor pastries and dessert treats, is not melted chocolate. It is a mixture of chocolate and cream or butter, often sweetened and flavored. "Chocolate milk" is simply whole, pasteurized, homogenized milk with 1.5–2% liquid chocolate, and sometimes added sweetener. "Chocolate plastic" (chocolate modeling paste, modeling chocolate) is a pliable decorating paste made from a mixture of chocolate and corn syrup. Couverture chocolate (also known as coating chocolate and enrobing chocolate) contains 32–39% cocoa butter. The extra cocoa butter gives it the right texture and consistency for hand-dipping chocolates and for enrobing them (pouring over the candies to coat them).

Solid chocolate is the most suitable form for storing. Chocolate maintains its freshness best if it is well wrapped in foil and stored in a fairly cool, dry place, between 13 and 15 °C (55–60 °F). At very high temperatures, the cocoa butter melts into the wrapper and the chocolate loses flavor. Never refrigerate or freeze chocolate, since the moisture that forms on the surface makes it likely to lump when melted. Excessive moisture may produce mustiness. Sugar bloom is a condition associated with high humidity or heat during storage, in which sugar dissolves out and recrystallizes on the surface, and pale spots ("bloom") cover the surface (this does not necessarily indicate spoilage). Recommended shelf life of unsweetened and dark chocolate is generally a year, and 7–8 months for milk and white chocolate. However, if the chocolate grays in color, cocoa butter has merely risen to the top and neither flavor nor quality is impaired.

Cocoa (or "unsweetened cocoa" or cocoa powder) is chocolate that has had most of the fat removed and has been pulverized into a powder ("cocoa" also refers to a hot beverage made with cocoa, a sweetener, and milk; and to mixes (also known as "instant cocoa"), almost always with sugar added, that are used to make beverage cocoa). Because cocoa contains far less cocoa butter and fat (ranging from 14 to 25%) than chocolate, health conscious cooks often prefer to use cocoa. In recipes, unsweetened regular (nonalkalized) cocoa always requires adding baking soda to help neutralize the mix and darken the color (most American brands are nonalkalized). Unsweetened Dutch-processed cocoa (which has a richer, more intense flavor) contains an alkali (usually potassium carbonate) that neutralizes acids, and is used in recipes that do not call for baking soda (most European and European-style cocoas are Dutch processed). Cocoa should be sifted before adding to a recipe to prevent lumps from forming. Refrigerating or freezing cocoa makes it difficult to sift, and it is best stored at room temperature in an airtight container, away from garlic, onions, or other strong-flavored foods, as their aroma could be absorbed.

"Drinking chocolate" is simply chocolate dissolved in liquid for the purpose of drinking. Improved drinking chocolate dates to the 19th century, when the technology was invented to remove most of the fat from chocolate (just as is done with cocoa). Cocoa is in fact drinking chocolate, but nonalkalized cocoa is considered superior in taste, and the phrase "drinking chocolate" is sometimes reserved for the latter form of cocoa. Cocoa produced by the alkali process blends more easily with milk, but is not as good tasting.

White chocolate is a combination of cocoa butter, sugar, butterfat, milk solids, lecithin, and flavorings. White chocolate has far less chocolate then conventional chocolate, and gets its mild chocolate flavor from cocoa butter, not from other components of the chocolate bean. The ivory color is that of the cocoa butter. White chocolate contains no cocoa solids, and is not considered "real chocolate" by many people, but is nevertheless liked for its smooth, creamy taste. Bright white "white chocolate" probably contains no cocoa butter and in most places in North America cannot legally be called chocolate.

Culinary Vocabulary

- "Alkalized cocoa" is another expression for Dutch-processed cocoa, which as noted above is cocoa that has been treated with an alkali solution. This raises the pH level (making for less acidity), and results in a milder, darker powder than untreated cocoa.
- "German chocolate cake" isn't German. The original recipe was published in 1957 by a Texas newspaper, *The Dallas Morning News*, and referred to Baker's "German's sweet chocolate," a dark, sweet, baking chocolate created by Englishman Samuel German, working as a confectioner for the Baker chocolate company in 1852. In most recipes and products today, the "'s" has been dropped from "German's," giving the false idea that the recipe was German in origin.
- "Dietetic chocolate" is not low-calorie chocolate, as the name suggests. It is chocolate sweetened with sorbitol or mannitol instead of sugar, for the use of diabetics and others on sugar-restricted diets. Extra fat is often added, and the flavor is not to the taste of many people.
- "Milk chocolate" was widely introduced in 1875, the result of a collaboration between Henry Nestlé, a producer of evaporated milk, and Daniel Peter, a Swiss chocolate maker. The idea of mixing chocolate with milk was first proposed in the 18th century by Sir Hans Sloane (1660–1753), personal physician to Queen Anne (1665–1714). The Swiss dominated the market for milk chocolate until the 20th century. Milk chocolate is now preferred by 80% of the world's population.
- "Mud pie" is a dessert, typically consisting of a chocolate cookie crust filled with chocolate and other flavors of ice cream and drizzled with chocolate sauce. A "Mississippi mud pie" is a chocolate pie (or cake), supposedly named because it looks like the thick mud along the banks of the Mississippi River. It is generally made with cocoa powder and chopped pecans, baked with a marshmallow topping, and covered with a cocoa and pecan icing.
- A "black bottom" refers either to a sundae made with chocolate ice cream and chocolate syrup, or a pie made with a chocolate custard topped with rum custard and whipped cream. The term "black bottom" originally meant a low-lying area inhabited by black people, and either this meaning was changed to indicate culinary usages, or the culinary terms arose from the black bottom dance of 1926 and following years.
- The British expression "plain chocolate" is equivalent to the American "semi-sweet chocolate."
- "Chocolate velvet cake" is a very rich, dense fudge-like chocolate cake. It was created in 1949 by pastry chef Albert Kumin of the Four Seasons restaurant chain.

- "Sachertorte" is said to be Vienna's most famous cake. This "arch-chocolate cake" was invented by Franz Sacher, chief pastry cook of Austrian statesman Prince von Metternich (1733–1859), on the occasion of the Congress of Vienna (1814–1815; a pivotal set of meetings of the great powers of Europe, spearheaded by Metternich, which led to unforeseen results, including it is said the World Wars). Sachertorte is a chocolate sponge topped with rich chocolate icing, with two layers separated by jam.
- An "egg cream" contains neither eggs nor cream. This frothy New York City favorite since the 1930s is made from chocolate syrup, ice-cold milk, and seltzer water. When properly mixed, a foamy head that looks like beaten egg whites tops the beverage (hence the name "egg cream"). Because fountain seltzer, not bottled club soda, is important, egg creams are difficult to make at home.
- A "Brandy Alexander" is a cocktail made from brandy, cream, and a chocolate liqueur, and typically served after dinner. If gin is substituted for brandy, the beverage is known as a "Panama." This drink, known since the Prohibition era, was long considered suitable for ladies and those not used to drinking.
- "Devil's food" refers to a cake, muffin, or cookie made with dark chocolate. The term is based on the humorous idea that such preparations are so rich and delicious that they must be sinful, and so associated with the Devil. Dark devil's food cake contrasts with snowy white angel food cake, which appeared earlier (the first devil's food recipe appeared in 1905), and the two kinds of baked goods reflect the widespread conception that white is good and black is bad. Devil's food cake (also known as red velvet cake, red devil's cake, Waldorf Astoria cake, and $100 dollar cake) is traditionally reddish. Originally, the cake acquired its red color from a chemical reaction between early varieties of cocoa and baking soda (which also gave the cake a soapy taste). Today, red food dye is used to give the cake an authentic appearance. In the 1970s, the cakes became less popular when red dye was linked to cancer. According to an old (and unreliable) story, the name "$100 dollar cake" traces to the request by a customer of the recipe for the cake, following which he was presented with a bill for $100. The name "Waldorf Astoria cake" traces to an unverified claim that in the 1950s the cake originated at the Waldorf-Astoria Hotel.
- "Boston cream pie" dates back to early 19th century Boston, when it was known as "Boston pie," and was simply a plain two-layer sponge cake filled with a vanilla custard. About the mid 1850s, someone (just who is uncertain) at Boston's Parker House Hotel added a chocolate glaze topping, and the dessert became known as Parker House chocolate pie and Boston cream pie. By modern standards, this "pie" is a "cake," and the word "pie" in "Boston cream pie" may be due to the past practice of baking cakes in pie tins, or perhaps because the original version was low, like a pie.

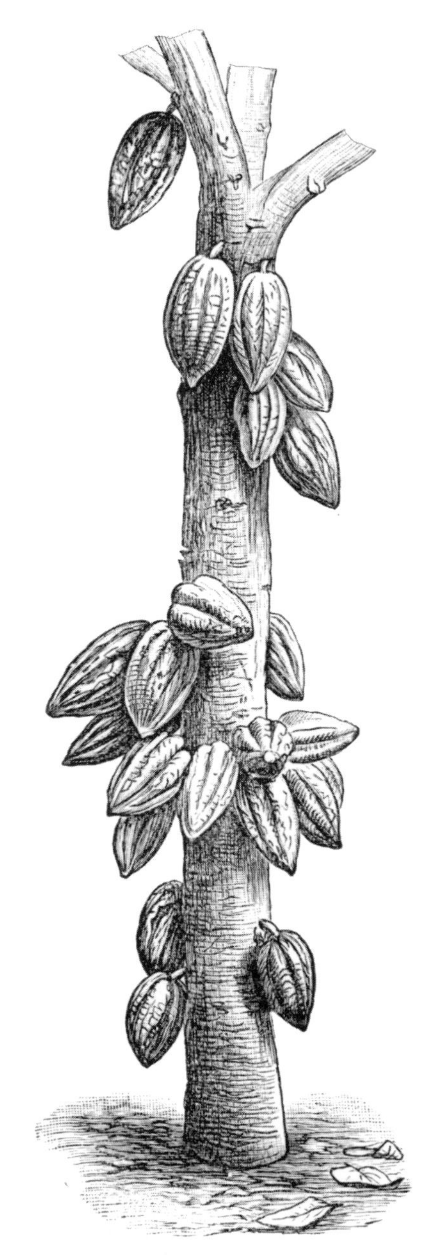

Trunk of chocolate tree (*Theobroma cacao*), from Strasburger (1930).

- "Whoopie pie" is a Pennsylvania German cupcake prepared with leftover chocolate cake batter and white icing.
- A "Bourbon biscuit" is an oblong chocolate cookie with a chocolate cream filling. It is said to commemorate the Bourbon family, the rulers of France from 1272 to 1792.
- A "chocolate truffle" (often shortened to truffle) is a rich, creamy candy made with butter, cream, flavorings, and sometimes nuts, formed into small balls, and coated with cocoa powder or melted chocolate. The chocolate truffle was named after the fungus of the same name because early versions of the confection seemed reminiscent in appearance.
- A "Happy Marriage" is a beverage prepared with equal parts of hot chocolate and coffee.

Curiosities of Science and Technology

- Cacao beans were both a valuable commodity, and a major form of currency and tribute payment in the Aztec empire (1376–1520), which included Mexico, Honduras, Guatemala, and El Salvador when the Spaniards arrived. The small seeds were used as coins, a practice that persisted in some areas until the 20th century. (This is an example of money growing on trees!) In Aztec times, about 100 beans would buy a slave. A rabbit cost 10 beans, a pumpkin 4. While on campaigns, Aztec soldiers carried tablets of ground chocolate, to be stirred into water as "instant chocolate."
- Napoleon Bonaparte (1769–1821) is said to have carried chocolate with him on his military campaigns, and eaten it whenever he needed a boost of energy. Similarly in more recent times, Edmund Hillary (1919–2008) and his teammates devoured chocolate while struggling up Mount Everest. American and Russian space flights have carried chocolate aboard, and armies have often used it for quick energy.
- In 1908, Thodor Tobler, a designer who operated a successful sweets company in Switzerland, adopted a triangle shape to distinguish his chocolate bar from that of his competitors. Information provided by the Tobler company contends that the well-known triangular Toblerone chocolate, honey, and almond bar takes its shape from the Matterhorn mountain. According to another widely circulated story, however, Tobler was inspired to adopt his unconventional shape for his chocolate bar after viewing the Folies Bergère dance troupe in Paris forming a triangular human pyramid.
- In 1920, Harry Burt (? –1926), a candy maker and owner of an ice cream parlor in Youngstown, Ohio, became the first person to put a chocolate-covered ice cream bar on a stick. This was called the "Good Humor Ice Cream Sucker," and was sold by vendors driving white trucks. The Good Humor ice cream company subsequently became an American tradition that has lasted to the present.
- In the 1940s and 1950s, the Hershey Company extracted theobromine from its cocoa beans and sold it to Coca Cola, which used it to pep up its drinks.
- The first suntan cream was invented by Miami Beach pharmacist Benjamin Green, when he cooked cocoa butter in a coffee pot. His invention was introduced as Coppertone Suntan Cream in 1944.
- In Alfred Hitchcock's 1960 movie *Psycho*, chocolate syrup was used to simulate blood in the famous shower scene. Psycho, the mother of all slasher films, topped the American Film Institute's 2001 list of "the most thrilling 100 movies" ever made.
- Eskimo Pie, the first patented commercial chocolate-covered ice cream bar, was invented by Christian Kent Nelson (1893–1992) in his home laboratory in Onawa, Iowa, in 1920. He first called them "I-Scream Bars."
- Recalling the days when the Aztecs served chocolate in gold vessels, some specialty chocolate houses in Europe apply a gold leaf to their confections as an expensive and rather decadent novelty. Despite the high cost, the gold leaf is extremely thin and the minute amounts ingested are believed to pass harmlessly through the digestive system within a day or two. (Danziger Goldwasser is a liqueur, mostly made in Germany, which contains flecks of gold leaf in suspension, which swirl about colorfully when the liquid is shaken or poured. It dates from the days when gold was considered a cure for all diseases.)
- In hot cocoa, the cocoa particles are not dissolved in the water, but rather are suspended, and with time the particles drop to the bottom of the cup.
- The melting point of cocoa butter is just below the human body temperature, explaining why it melts in your mouth.
- A century-old bar of chocolate was recovered from the expedition ship Discovery, which carried British explorer Robert F. Scott (1868–1912) on his 1901–1905 Antarctic expedition attempting to reach the South Pole. Cadbury, the maker, stated

that it could still be eaten. The 10-cm (4-inch) bar was sold by Christie's auction house in London in 2001 for about $700.

- Unlike most cats, dogs love the taste and smell of chocolate. Chocolate can be lethal to dogs, owing to theobromine, which stimulates the cardiac muscle and the central nervous system, irritates the gastrointestinal tract, and in some dogs, causes internal bleeding. As little as 57 g (2 ounces) of milk chocolate can be poisonous for a 4.5-kg (10-pound) puppy. Within 2 hours of ingestion, try inducing vomiting unless your dog is markedly stimulated, comatose, or has lost the gag reflex. If your dog has eaten a considerable amount of chocolate, take it to the vet without delay. Symptoms can include vomiting, hyperactivity, hyperventilation, dehydration, seizures, and coma. Symptoms may not show up for several hours after ingestion but usually are apparent within 12 hours. In the absence of major symptoms, administer activated charcoal. In a pinch, very thoroughly burnt toast will do. The darker the chocolate and the smaller the animal, the more dangerous the situation. Baking chocolate is the strongest, followed by dark chocolate and then milk chocolate. Baking chocolate has 10 times as much theobromine as milk chocolate, while white chocolate has negligible amounts. (Carob, described next, can be substituted for chocolate to make chocolate-like treats suitable for canines.)
- The carob tree (*Ceratonia siliqua* L.) is native to the Middle East, and has been cultivated for thousands of years in the Mediterranean region. Its pods yield carob, which is considered to have healthful attributes as a food, and to be a good substitute for chocolate. These properties have generated a commercial market for carob, especially since the 1980s. Carob can replace chocolate in recipes (1 part chocolate is equivalent to 1¾ parts carob), but is generally not considered as flavorful. Carob dissolves less readily than chocolate (hot water is therefore recommended), melts at a lower temperature, and liquefies more readily; these properties can present problems when preparing some dishes. Carob chocolates are widely sold on the basis that they have fewer calories and are virtually fat free, but in fact, on a weight basis, fat and calorie content of carob and chocolate are comparable. Moreover, carob's fat is mostly saturated, and many carob chocolates have had considerable fat and sugar added to enhance taste.
- The candies most likely to cause tooth decay are dark chocolate and fudge. Nevertheless, some research groups have found that cocoa powder contains a substance that may actually inhibit tooth decay.
- It is a widespread myth that chocolate aggravates acne.
- A "chocoholic" is someone addicted to chocolate. This slang term is normally used facetiously, and genuine chocoholism or a pathological addiction to chocolates is extremely rare. Various explanations of the attractiveness of chocolate, none completely accepted, have been advanced, including the following. (1) Theobromine in chocolate is chemically related to caffeine, which is also rather addictive. (2) The sugar in chocolate releases serotonin, a brain chemical that produces a sense of well-being. (3) Chocolate contains the same chemical, phenylethylamine, which your brain produces when you fall in love. (4) The smooth, rich taste provides sensory pleasure to the taste buds. (5) Fatty acids are present that are chemically and pharmacologically related to some marijuana compounds (one study suggested that to achieve the same effect as a marijuana joint, 9 kg or 20 pounds of chocolate would have to be eaten). (6) Many people use chocolate as a reward and learn to associate the product with positive self-esteem. Studies have shown that chocolate often helps counteract depression.
- In experiments conducted by Adam Drewnowski of the University of Washington, when chocolate lovers were given a drug, naloxone, which blocks addictive drugs from stimulating receptors in the brain to release natural opioids and dopamine (neurochemicals that trigger feelings of pleasure or well-being), they no longer craved chocolate.
- The craving for chocolate has been explained as a result of the survival value of high-fat, high-calorie food in the prehistoric past. Unfortunately, modern people, surrounded by such "high-density" foods, remain hard-wired to consume them, wreaking havoc on waistlines. The following advice has been offered to curb overeating chocolate and similar fattening treats:
    - When a craving arises, don't indulge immediately; wait 20 minutes; cravings often subside.
    - Keep chocolate and other craved foods out of the house.

- Complete self-denial can lead to obsession, so it is preferable to indulge occasionally.
- Eat treats slowly, savoring every morsel.
- Buying small, expensive portions of the highest-quality treats can be a way of limiting the volume of consumption.
- When cravings occur regularly, find low-calorie ways of satisfying them. Sweet tooths can be sated with hot cocoa or fat-free foods.

- In 1992, the Nabisco Company introduced the Snackwell's line of low-fat cookies and crackers, including several types of chocolate cookies that quickly became extremely popular. The "Snackwell Effect" refers to an overconsumption of low-fat food associated with the belief that the food isn't fattening. To Nabisco's credit, some Snackwell's products were labeled "NOT FOR WEIGHT CONTROL." The similar false sense of security associated with "light" foods has also led to harmful binging.
- In 2001, to commemorate the 100th anniversary of the Swiss Chocolate Federation, Switzerland issued a postage stamp that smelled and tasted like chocolate.
- An engineering graduate student at Michigan State University discovered that a melted Hershey chocolate bar is an "electrorheological fluid," commonly called a "smart fluid." When a high-voltage electrical field is applied to the melted chocolate, it instantaneously becomes a stiff gel, and the reverse occurs when the power is turned off. This property of smart fluids is known as the "Winslow effect," and it occurs in several liquids that are, like chocolate, suspensions of droplets in an oily fluid. Such smart fluids may be used one day in computer-controlled shock absorbers. Newspapers reported this story in 1996 under the headline "Chocolate shock absorbers?"

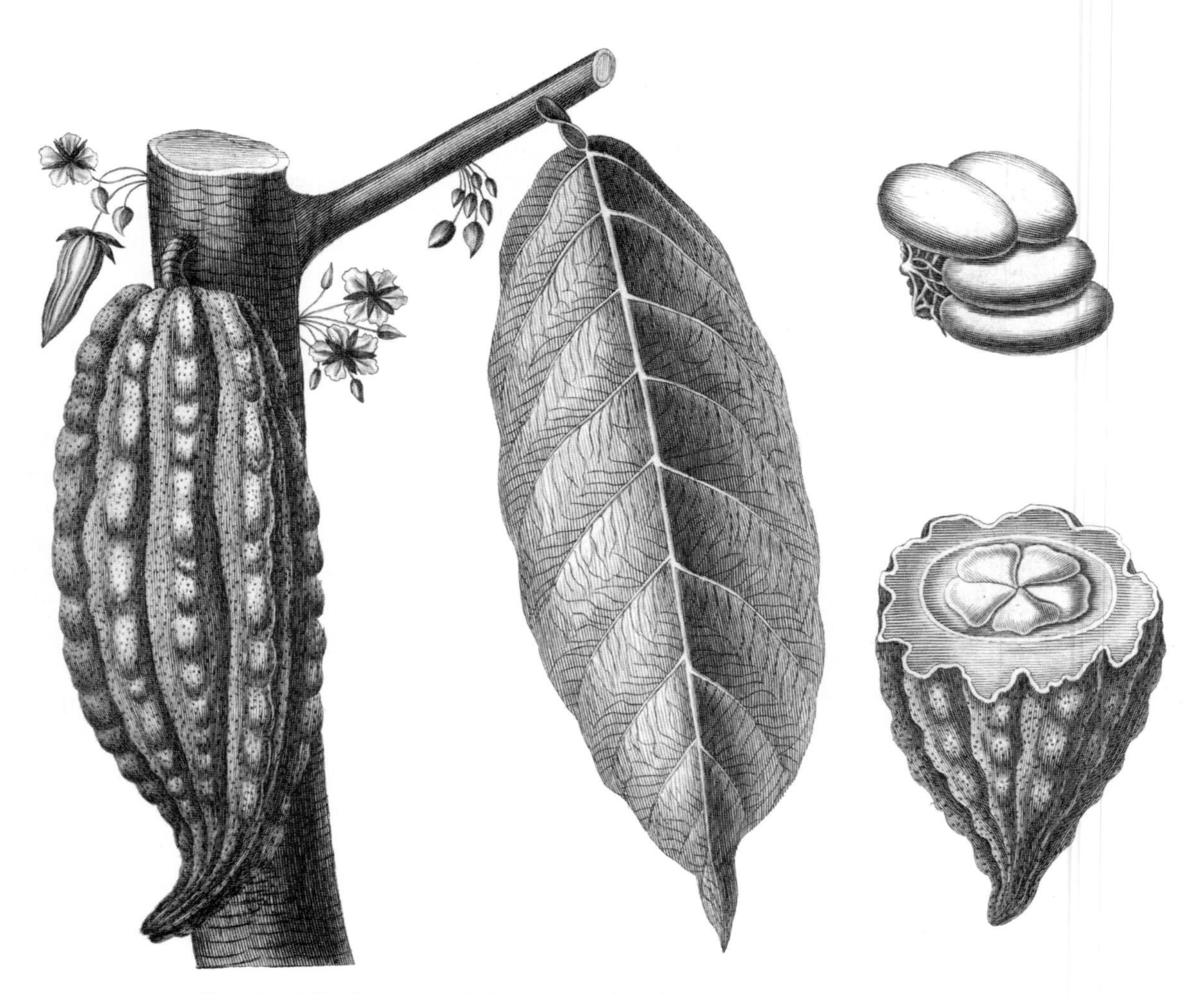

Chocolate (*Theobroma cacao*), from Lamarck and Poiret (1744–1829, plate 635).

- Cocoa beans are not only used to manufacture chocolate, but also as a source of butterfat, a natural ingredient for the cosmetics and pharmaceutical industries. If you are worried about dry skin and wrinkles, cocoa makes a fine facial pack to rejuvenate your skin. Mix a little dairy cream and olive oil with about 50 mL (1/4 cup) of cocoa powder to make a creamy paste (this should be enough for three or four facials; store unused part in refrigerator). Apply to the face (à la mud pack) and leave on for about 10 minutes before washing off with warm water. Regular treatments are claimed to keep your face youthful and feeling softer.
- Chocolate bars get their crisp structure and glossy shine from having the cocoa fat crystals aligned in a particular formation. When chocolate is melted (for example, as a result of being left in a hot car), the formation is lost and the texture and appearance are much less attractive. The same is true for other fats, such as butter. Melted butter that solidifies never looks as smooth and waxy as it did when it came out of the package. Chocolate makers achieve the desired crystalline arrangement by a process called tempering, involving melting, pouring onto a cold table, and stirring at precise temperatures.
- A report in 2004 in *New Scientist* noted that Katri Raikkonen at the University of Helsinki, Finland, and her colleagues found that the amount of chocolate that women consumed during pregnancy was related to increased smiling and laughing in the babies that were born. It is thought that eating chocolate reduced stress in the expectant mothers, and that this in turn affected the babies.
- Recent research has shown that drinking the traditional cup of cocoa at bedtime can substantially lower blood pressure, so much so that the result is a reduction of risk of stroke by about 20%, coronary heart disease by 10%, and overall mortality by 8%. Dark chocolate also seems to have the same effect, apparently because of the presence of polyphenols or flavonoids. (Reference: Taubert, D., Roesen, R., and Schömig, E. 2007. Effect of cocoa and tea intake on blood pressure: a meta-analysis. Arch. Intern. Med. **167**: 626–634.)

Key Information Sources

Bartley, B.G.D. (*Editor*). 2005. The genetic diversity of cacao and its utilization. CABI Publishing, Wallingford, Oxon, UK. 368 pp.

Beckett, S.T. 1999. Industrial chocolate manufacture and use. 3rd edition. Blackwell Science, Malden, MA. 488 pp.

Beckett, S.T. 2000. The science of chocolate. Royal Society of Chemistry, Cambridge, UK. 175 pp.

Bloom, C. 1998. All about chocolate: the ultimate resource to the world's favorite food. Macmillan, New York, NY. 370 pp.

Coady, C. 1995. The chocolate companion: a connoisseur's guide to the world's finest chocolates. Simon & Schuster, New York, NY. 192 pp.

Coe, S.D., and Coe, M.D. 1996. The true history of chocolate. Thames and Hudson, New York, NY. 280 pp.

Cook, L.R. 1963. Chocolate production and use. Magazines for Industry (Catalog and Book Division), New York, NY. 463 pp.

Cuatrecasas, J. 1964. Cacao and its allies: a taxonomic revision of the genus *Theobroma*. Contrib. U.S. Nat. Herbarium, **35**: 379–614.

Farooqi, A.A., and Sreeramu, B.S. 1999. Cocoa. *In* Tropical horticulture, vol. 1. *Edited by* T.K. Bose, S.K. Mitra, A.A. Farooqui, and M.K. Sadhu. Naya Prokash, Calcutta, India. pp. 503–520.

Fuller, L.K. 1994. Chocolate fads, folklore & fantasies: 1,000 + chunks of chocolate information. Haworth Press, New York, NY. 276 pp.

Heijbroek, A.M.A., and Konijn, R.J. 1995. The cocoa and chocolate market. Rabobank, Utrecht, Netherlands. 64 pp.

Kennedy, A.J. 1995. Cacao. *Theobroma cacao* (Sterculiaceae). *In* Evolution of crop plants. 2nd edition. *Edited by* J. Smartt and N.W. Simmonds. Longman Scientific & Technical, Burnt Mill, Harlow, Essex, UK. pp. 472–475.

Kennedy, A.J., Lockwood, G., Mossu, G., Simmonds, N.W., and Tan, G.Y. 1987. Cocoa breeding: past, present and future. Cocoa Growers' Bull. **38**: 5–22.

Knight, I. 1999. Chocolate and cocoa: health and nutrition. International Cocoa Research and Education Foundation, London, UK. 342 pp.

MacLeod, M.J. 2000. Cacao. *In* The Cambridge world history of food. *Edited by* K.F. Kiple and K.C. Ornelas. Cambridge University Press, Cambridge, UK. pp. 635–641.

Marcus, A. 1979. The chocolate bible. Putnam, New York, NY. 279 pp.

Minifie, B.W. 1989. Chocolate, cocoa, and confectionery: science and technology 3rd edition. Van Nostrand Reinhold, New York, NY. 904 pp.
Presilla, M.E. 2001. The new taste of chocolate: a cultural and natural history of cacao with recipes. Ten Speed Press, Berkeley, CA. 198 pp.
Soria, J. de V. 1970. Principal varieties of cocoa cultivated in Tropical America. Cocoa Growers' Bull. **15**: 12–21.
Toxopeus, H. 1969. Cacao. *In* Outlines of perennial crop breeding in the tropics. *Edited by* F.P. Ferwerda and F. Wit. Misc. Paper 4. Wageningen Agricultural University, Wageningen, Netherlands. pp. 79–109.
Wessel, M., and Toxopeus, H. 2000. *Theobroma cacao* L. *In* Plant resources of South-East Asia, 16, stimulants. *Edited by* H.A.M. van der Vossen and M. Wessel. Backhuys, Leiden, Netherlands. pp. 113–121.
Wieland, H. 1972. Cocoa and chocolate processing. Noyes Data Corp., Park Ridge, NH. 305 pp.
Wood, G.A.R., and Lass, R.A. (*Editors*). 1985. Cocoa. 4th edition. Longman Scientific & Technical, Harlow, UK. 620 pp.
Young, A.M. 1994. The chocolate tree: a natural history of cacao. Smithsonian Institution Press, Washington, DC. 200 pp.

Specialty Cookbooks

(Note: There are hundreds of cookbooks dedicated to chocolate. The following are selected examples.)

Beach, N. 1995. The Ghirardelli chocolate cookbook. Ten Speed Press, Berkeley, CA. 136 pp.
Berger, L. 1984. The gourmet's guide to chocolate. Quill, New York, NY. 128 pp.
Desaulniers, M. 1992. Death by chocolate: the last word on a consuming passion. Rizzoli, New York, NY. 143 pp.
Edelston, K. (*Editor*). 2003. Chocolate: over 250 recipes for cakes, cookies, desserts, party food, and drinks. Thunder Bay Press, San Diego, CA. 399 pp.
Finsand, M.J. 1984. The diabetic chocolate cookbook. Sterling, New York, NY. 160 pp.
Gillespie, G.R. 1996. 1001 chocolate treats. Black Dog & Leventhal, New York, NY. 423 pp.
González, E. 1998. The art of chocolate: techniques & recipes for simply spectacular desserts & confections. Chronicle Books, San Francisco, CA. 165 pp.
Hasson, J. 2003. 125 best chocolate chip recipes. R. Rose, Toronto, ON. 192 pp.
Malgieri, N. 1998. Chocolate: from simple cookies to extravagant showstoppers. HarperCollins, New York, NY. 464 pp.
McFadden, C., and France, C. 1997. The ultimate encyclopedia of chocolate: with over 200 recipes. Smithmark Publishers, New York, NY. 256 pp.
Szogyi, A. (*Editor*). 1997. Chocolate: food of the gods. Greenwood Press, Westport, CN. 228 pp.

CINNAMON AND CASSIA

Family: Lauraceae (laurel family)

Names

Scientific names: *Cinnamomum* species

- Cinnamon—*C. verum* J. Presl (often called *C. zeylanicum* Blume)
- Cassia—*C. aromaticum* Nees ["*C. cassia*" of authors; other species of *Cinnamomum* are sometimes sold as cassia]

- The genus name *Cinnamomum* and the English word "cinnamon" both derive from the Latin *cinnamomum*, which comes from the Greek *kinnámoomon*, which in turn is probably based on Semitic words, such as the Hebrew *quinamom* and the Arabic *amomon*, meaning fragrant spice plant.
- Cinnamon is widely known as Ceylon cinnamon.
- Cassia is also known as cassia bark, bastard cinnamon, Chinese cassia, Chinese cinnamon, and false cinnamon.
- *Verum* in the scientific name *C. verum* is Latin for true, hence the alternative common name "true cinnamon." *Aromaticum* in the scientific name *C. aromaticum* means aromatic.
- The Dutch, French, Spanish, and Italian names are derived from the Latin *canella,* meaning "small tube" or "pipe," an apt description of a cinnamon stick.

Plant Portrait

Cinnamon has a long history. It is mentioned in the Bible and in Sanskrit writings, and was familiar to the ancient Greeks and Romans. In early European times, most spices were supplied by Arab traders, who guarded their monopoly with fantastic tales about how dangerous or difficult it was to obtain the materials. Cinnamon sticks were alleged to be parts of the nests of huge birds, found on sheer cliffs, which could not be reached. To get the cinnamon, the Arabs alleged that they provided the birds with parts of dead oxen and asses, until their nests became so heavy that they fell to the ground, and the cinnamon could be collected. A different version of this tale required shooting lead arrows into the nest, until the accumulated weight brought it down. Another story told how cinnamon could only be collected in valleys guarded by poisonous snakes. Wars were fought over cinnamon. In the 17th century, the Dutch seized the Portuguese-controlled island of Ceylon, the world's largest cinnamon supplier. The Dutch also bribed the local king of an area along the coast of India where cinnamon was grown, to destroy that supply, so that Holland would have a monopoly on the prized spice. Then the French seized Ceylon, and finally, in 1795, England captured the island. But by 1833, the cinnamon monopoly began to fail as other countries took up the cultivation of the spice.

The true cinnamon is an evergreen tropical tree. Cinnamon trees are native to Ceylon (Sri Lanka) and southwest India, and are grown in other tropical areas, notably in the Seychelles and the Malagasy Republic. Wild cinnamon trees are known to grow to a height of 17 m (56 feet), but when cultivated, they are grown as short bushes to make harvesting easier. The trees have a productive life span of about 40–50 years.

Unlike most edible plant foods and spices, cinnamon is tree bark. The bark is stripped off, the outer bark removed, and the inner bark dried in long strips to form "quills," which are pale brown and curl into semi-tubes when dry, and can be enclosed inside each other for packing. Cinnamon is also ground to powder. Essential oil is produced from the leaves and bark, and is used in the food, pharmacy, cosmetics, and perfumery industries.

Most commercial ground cinnamon is actually cassia or a combination of cinnamon and cassia. Cassia is a tree, native to Burma, and is cultivated and harvested in the same way as true cinnamon. "Cinnamon" from cassia is mentioned in Chinese writings dating to 2800 BC.

Culinary Portrait

Most cinnamon sold in North America is cassia, and most people prefer it, having become familiar with its taste. By contrast, much of the cinnamon sold in Europe is true cinnamon. How to distinguish true cinnamon from cassia: Cinnamon quills or sticks are curled in telescopic form, while those of cassia curl inward from both sides, like a scroll. Ground spice is more difficult to distinguish. True ground cinnamon is tan in color with a warm, sweet flavor, whereas ground cassia is reddish brown, usually coarser in texture, with a more bitter, stronger flavor and a more aromatic bouquet. Cinnamon is most commonly sold as sticks and powder. Ground cinnamon is more flavorful than cinnamon sticks, but does not store as long. Cinnamon should be stored in a sealed container in a cool, dry, dark place.

True cinnamon (*Cinnamomum verum*), from Köhler (1883–1914).

Cinnamon is used in bakery products, sauces, puddings, pancakes, pickles, curry powder, yogurt, candies, and various other confectionery. The spice is an essential ingredient in chicken and beef dishes of Morocco and Greece, and is commonly employed with meats, especially lamb, in the Middle East. In Europe and North America, cinnamon tends to be used mostly in sweets. Cinnamon sticks make tasty stir sticks for coffee, tea, and mulled cider.

Cassia (or "cinnamon" from the cassia tree) has a stronger flavor than cinnamon, so less is required in recipes. Cassia is usually more suitable for savory dishes than for sweets. Dried cassia buds that look like cloves are used in the East for pickles, curries, candies, and spicy meat dishes. Cassia (as "cinnamon") is often added to baked zucchini, mulled wine, and a variety of dishes such as soups, meats, tomato sauces, vegetables, stews, couscous, pasta, and marinades. Cinnamon-flavored cassia flowers in brine are used to perfume sweets, fruits, teas, and wines. Cassia leaves are sometimes employed like bay leaves, notably in India to flavor curries.

Culinary Vocabulary

- "Snickerdoodles" is a southern United States word for cinnamon sugar cookies. More generally, snickerdoodles are crisp or soft cookies with a characteristically crackly surface, usually flavored with cinnamon and nutmeg, and coated in sugar before baking. The cookies are thought to have originated in 19th century New England, and the name "snickerdoodle" (pronounced SNIHK-uhr-doo-dl) seems to be whimsical.
- A "railroad cookie" is one swirled on the inside with cinnamon and brown sugar. The treat has been known at least since the late 19th century, and the name may have arisen by the resemblance of the dark tracks of filling to railroad tracks.
- *Zimtsterne* (pronounced zeemt-star-nay) are German Christmas cookies, also known as "cinnamon stars." They are prepared with almonds, cinnamon-flavored meringue, and flavorings.

- "Cassia blossom wine" (Guilin *Gui Hua Jiu*, *Gui Hua Jiu*), from the southwestern province of Guangxi, is flavored with *Osmanthus fragrans* Lour. (sweet osmanthus, fragrant olive), an Asian plant used for fragrance and taste, to make an elegantly perfumed light and semisweet rice wine. *Osmanthus fragrans* is also used to flavor "cassia blossom jam" (Gui Hua jam), a sugary paste sometimes available in Oriental grocery stores. The traditional English translation of "Gui Hua" as "cassia" is misleading, as true cassia is not used. (Information provided by Dr. T. Li.)

Curiosities of Science and Technology

Cinnamon

- Ancient Egyptians used cinnamon in embalming mummies.
- A beverage prescribed for colds by ancient Roman physicians—hot liquor and stick cinnamon—is still employed today.
- Pliny the Elder (23–79) of Roman times recorded that cinnamon was 15 times as valuable as silver on a weight basis.
- Cinnamon has been used for both sacred and profane purposes. In Palestine, cinnamon was used for rites of the Tabernacle, and in medieval Europe, it was always added to the incense burned in churches.
- Cinnamon trees grow so slowly that in Sri Lanka and other countries they do not rate valuable field space that can be used by fast-growing crops. As a result, the cinnamon trees are often planted in straggling rows on street shoulders.
- Despite their name, black bears are not always black; they may be cinnamon, blonde, pale blue (near Yakutat, Alaska), or even white (British Columbia).
- The top 10 foods found in American kitchens are chili powder, cinnamon, ketchup, margarine, mustard, potatoes, seasoned salt, soy sauce, spaghetti, and vegetable oil. (Surprisingly, coffee, milk, and tea failed to make the list.)

Cassia (*Cinnamomum aromaticum*), from Köhler (1883–1914).

- In the language of flowers, popular in Victorian England as a secret code of communicating information within written letters, "cinnamon" meant "my fortune is yours."

Cassia

- Reflecting old class distinctions, 15th century English author John Russell, in his *Bok* [Book] *of Nurture*, wrote, "Synamone [cinnamon] is for lordes, canelle [cassia] for common people."

Key Information Sources

Ceylon Industrial Development Board. 1973. Cinnamon: cultivation and processing. Industrial Development Board, Ceylon. 8 pp.

Dao, N.K., Hop, T., and Siemonsma, J.S. 1999. *Cinnamomum* Schaeffer. *In* Plant resources of South-East Asia, 13, spices. *Edited by* C.C. de Guzman and J.S. Siemonsma. Backhuys, Leiden, Netherlands. pp. 94–99.

Ferguson, A.M., and Ferguson, J. 1900s. All about cinnamon; including practical instructions for planting, cultivation and preparation for market, with information from a variety of sources. A.M. & J. Ferguson, Colombo, Ceylon. 43 pp.

Flach, M., and Siemonsma, J.S. 1999. *Cinnamomum verum* J.S. Presl. *In* Plant resources of South-East Asia, 13, spices. *Edited by* C.C. de Guzman and J.S. Siemonsma. Backhuys, Leiden, Netherlands. pp. 99–104.

International Organization for Standardization. 1997*a*. Cinnamon, Sri Lankan type, Seychelles type and Madagascan type (*Cinnamomum zeylanicum* Blume). International Organization for Standardization, Geneva, Switzerland. 8 pp.

International Organization for Standardization. 1997*b*. Cassia, Chinese type, Indonesian type and Vietnamese type (*Cinnamomum aromaticum* (Nees) syn. *Cinnamomum cassia* (Nees) ex Blume, *Cinnamomum burmanii* (C.G. Nees) Blume and *Cinnamomum loureirii* Nees). International Organization for Standardization, Geneva, Switzerland. 7 pp.

Jayaprakasha, G.K., Rao, L.J., and Sakariah, K.K. 1997. Chemical composition of the volatile oil from the fruits of *Cinnamomum zeylanicum* Blume. Flavour Fragrance J. **12**: 331–333.

Joy, P.P., Thomas, J., Mathew, S., and Ibrahim, K.K. 1998. Growth, leaf oil yield and quality investigations in cinnamon (*Cinnamomum verum*). J. Med. Aromatic Plant Sci. **20**: 401–406.

Kaul, P.N., Bhattacharya, A.K., Rajeswara-Rao, B.R., Syamasundar, K.V., and Ramesh, S. 2003. Volatile constituents of essential oils isolated from different parts of cinnamon (*Cinnamomum zeylanicum* Blume). J. Sci. Food Agric. **83**: 53–55.

Kostermans, A.J.G.H. 1998. *Cinnamomum* (Lauraceae). Reinwardtia (Indonesia), **11**(**3**): 195–213.

Manning, C.E.F. 1970. The market for cinnamon and cassia and their essential oils. Tropical Products Institute, London, UK. 74 pp.

Nayar, N.M., and Ravindran, P.N. 1995. Tree spices. *In* Evolution of crop plants. 2nd edition. *Edited by* J. Smartt and N.W. Simmonds. Longman Scientific & Technical, Burnt Mill, Harlow, Essex, UK. pp. 495–497.

Peter, K.V., and Kandiannan, K. 1999. Cinnamon. *In* Tropical horticulture, vol. 1. *Edited by* T.K. Bose, S.K. Mitra, A.A. Farooqui, and M.K. Sadhu. Naya Prokash, Calcutta, India. pp. 717–723.

Pillai, K.S. 1965. The cinnamon. Farm Bulletin No. 3. Agricultural Information Service, Dept. of Agriculture, Kerala, India. 40 pp.

Ravindran, P.N., Babu, K.N., and Shylaja, M. 2003. Cinnamon and cassia: the genus *Cinnamomum*. Taylor & Francis, London, UK. 440 pp. [Medicinal aspects.]

Senanayake, U.M., Edwards, R.A., and Lee, T.H. 1976. Cinnamon. Food Technol. Aust. **28**: 333–338.

Ter Heide, R. 1972. Qualitative analysis of the essential oil of cassia (*C. cassia* Blume). J. Agric. Food Chem. **20**: 747–751.

Thomas, J., and Duethi, P.P. 2001. Cinnamon. *In* Handbook of herbs and spices. *Edited by* K.V. Peter. CRC Press, Boca Raton, FL. pp. 143–153.

Vernin, G., Vernin, C., Metzger, J., Pujol, L., and Parkanyi, C. 1994. GC/MS analysis of cinnamon and cassia essential oils: a comparative study. Dev. Food Sci. **34**: 411–424.

Wijesekera, R.O.B., Jayewardene, A.L., and Rajapakse, L.S. 1974. Volatile constituents of leaf, stem and root oils of cinnamon (*Cinnamomum zeylanicum*). J. Sci. Food Agric. **25**: 1211–1220.

Yazaki, K., and Okuda, T. 1993. *Cinnamomum cassia* Blume (cinnamon): in vitro culture and the production of condensed tannins. Biotechnol. Agric. For. (Berlin), **24**: 122–131.

Specialty Cookbooks

Ostergren, C. 1989. Just cinnamon rolls: recipes from country women across North America. Mashue Printing, Midland, MI. 54 pp.

Pappas, L.S. 1994. Cinnamon. Chronicle Books, San Francisco, CA. 72 pp.

Welsh, F.M. 1982. The cinnamon cook book. Welsh Enterprises, US. 147 pp.

COCONUT

Family: Arecaceae (Palmae; palm family)

Names

Scientific name: *Cocos nucifera* L.

- "Coconut" is derived from the Portuguese (and Spanish) word *coco* (as explained in the following) + nut.
- The genus name *Cocos* probably derives from the Portuguese word *coco*, grimace or grin, because a coconut has three germinating pores, resembling a spooky face (see the illustration on page 183).
- Coconut has nothing to do with the coca plant (*Erythroxylum coca* Lam.), the leaf of which yields cocaine.
- *Nucifera* in the scientific name *C. nucifera* is Latin for nut-bearing.

Plant Portrait

Coconuts are the fruits of the coconut palm, a picturesque slender tree that grows to 24 m (80 feet) tall, sometimes to 30 m (100 feet; a coconut palm in Ceylon reached 36 m or 117 feet in height). The trunk may be as wide as 60 cm (2 feet). The top of the tree is crowned with a thatch of feather-like leaves 2.7–5.6 m (9–18 feet) long. Dwarf varieties grow to a height of 7.6–9 m (25–30 feet), and begin to flower after 3 years at a height of only 1 m (about 1 yard). The life span of these small trees is only about 30 years, but they are valued because they bear early and are resistant to lethal yellowing disease, which is a very serious problem of coconuts. Coconut palms grow naturally on tropical and subtropical seacoasts throughout Asia, Africa, Latin America, and the Pacific region. They do not like dry climates, and in the United States, the plants survive well only in southern Florida and the southernmost tip of Texas. Because the coconut floats and is distributed widely by sea, its center of origin is unclear—perhaps the Indo-Malaysian region or the Pacific islands. This is one of the most valuable of tropical plants, and indeed one of the 10 most useful trees in the world. Although coconuts generally grow on tropical seacoasts, inland plantations have been established. The fruits are a major crop in the humid areas of many tropical countries, particularly in the Philippines, Indonesia, India, Ceylon, Mexico, Brazil, Mozambique, Tanzania, and Ghana. Coconuts are also very important to the economies of small islands in the Pacific, Indian Ocean, and the Caribbean. The trees can start bearing at 6 years of age, and can live for 100 years. Coconut palms can produce clusters of 6–12 nuts as often as 12 times yearly. Good trees in their prime produce about 75 coconuts annually, but generally, trees yield about 25 each year. A coconut requires 6–12 months to ripen. Coconuts may be left to mature and fall off naturally, but are usually harvested by climbers, often by trained monkeys, or cut off with knives attached to the ends of long bamboo poles.

Coconuts sold in stores have been de-husked, so that their enormous size is generally not appreciated. The nuts are generally as large as a man's head, and consist of several layers. The outer layer is a fibrous husk, which is green when immature and brown at maturity. Inside this is a hard shell, 15–20 cm (6–8 inches across), and inside the shell is a relatively thin, white, fleshy layer consisting of white, oily "meat." The interior of the nut is hollow and partly filled with watery, sweet "coconut milk," especially in unripe nuts. As noted below, commercial "coconut milk" is something else.

Extracted fiber from the fruit husk is known as coir, and is used to make ropes, mats, fishing lines, carpets, baskets, brooms, and brushes. The fiber is also used for thatching roofs and for making charcoal. Coconut leaves are often woven into baskets, platter, mats, and hats. Coconut shells can be used as bowls and cooking utensils. In addition to culinary uses noted below, coconut oil is employed in making soap, detergents, shampoos, face creams, shaving cream, perfumes, candles, synthetic rubber, hydraulic brake fluid, and many other manufactured products.

Culinary Portrait

Coconut is widely employed in cooking. It is added to numerous dishes, including soups, appetizers, main courses, and desserts. Coconut is used in curries, sauces, rice dishes, marinades, stews, puddings, and in combinations with meat, poultry, and seafood. Dried coconut is sometimes used primarily for its decorative effect.

Natural "coconut milk" (coconut juice is a more appropriate term) in regions where coconuts are grown is a refreshing, sugary drink obtained directly from fresh coconuts. "Coconut syrup" is made by evaporating coconut juice and adding sugar. "Coconut honey" is a concentrated form of coconut syrup. In temperate regions, "coconut milk" is usually obtained by squeezing liquid out of grated coconut meat or by pouring boiling water over grated coconut meat and squeezing the mixture. Coconut milk resembles cow's milk somewhat, and occasionally has been used as a substitute. "Coconut cream" is simply a more concentrated form of coconut milk. Very young coconuts, which are too perishable to ship, are an exceptional treat, containing "spoon coconut"—a soft, gelatinous meat with the consistency of melon, which is eaten with a spoon from the shell, and possesses a better flavor than dried coconut. As the nut ripens, the meat becomes solid, coating the inner wall, and there is less water. The meat of the coconut, known as copra, is the world's chief source of vegetable fat. Copra contains about 60% oil. This coconut oil (which is also called coconut butter, since in the unrefined state it is a white, fatty solid at room temperature) is used for margarine and especially as a cooking oil. The oil is not suitable for use as a salad oil. Desiccated coconut is dried coconut meat, like copra, but is ground before drying.

Sap from the tree (often milked from the flower stalks before the flowers open) is evaporated down to make sugar, or fermented to produce alcohol, the mildly alcohol (up to 8%) beverage called toddy, and a potent spirit called arrack (also spelled "arak"). Arrack is made in the Far East and Middle East, from fermented rice and molasses to which the fermented

Coconut (*Cocos nucifera*), from Köhler (1883–1914).

sap of the coconut palm (or other palms) is frequently added. Indonesia is said to make the best arrack. Toddy can be converted to vinegar.

A good-quality coconut should feel heavy, and when shaken should produce the impression that it is full of liquid, as the milk is sloshed around. Avoid moldy coconuts, or those with wet "eyes." To extract the liquid, use an ice pick to poke two holes in two of the three eyes, and drain. To remove the thick outer covering, place the coconut in an oven at 200 °C (400 °F) for 20 minutes, allow to cool slightly, hold over a large bowl, and strike it all the way around with the back of a knife or cleaver until it splits open. Alternatively, some people strike the coconut with a mallet or hammer, or simply drop it onto a hard floor that will withstand damage. Instead of heating the coconut, an alternative is to place it in a freezer (at 0 °C or 32 °F) for an hour. The flesh should taste sweet, but if not, the material should be discarded or returned. A clean screwdriver can be employed to chip off most of the meat from the shell, and a vegetable peeler can be employed to scrape away the brown, thin skin that remains attached to the pieces of flesh. A food processor, blender, or fine grater can be used to shred the coconut.

Alternatively, dried, shredded coconut can simply be purchased for most cooking needs, and when a superior taste is desired, some supermarkets stock freshly frozen shredded coconut. Dried, grated coconut is often coated with sugar, and for those wishing to avoid the sugar, it may be necessary to purchase unsweetened coconut in natural-food stores. Unopened coconuts, depending on how fresh they are when purchased, can be stored for as long as 4 months at room temperature, and longer in a refrigerator. Whether or not the pulp inside a coconut has turned rancid can only be determined by opening the shell. Extracted coconut meat should be consumed quickly, but may be refrigerated for several days. Dried coconut stores indefinitely in a cool, dry place, protected from air and insects in a tightly sealed container.

The terminal growing bud of palm trees, nestled in the crown of the tree, can be used to produce an unusual, expensive, salad vegetable with a crisp, crunchy texture and taste reminiscent of white asparagus and bamboo shoots. This looks like a bundle of tightly packed, yellow–white, cabbage-like leaves. Because harvesting this "palm heart" kills the tree, it is generally collected from trees felled for lumber. Because the tree has to be sacrificed, this tasty part of the plant has been called "millionaire's salad."

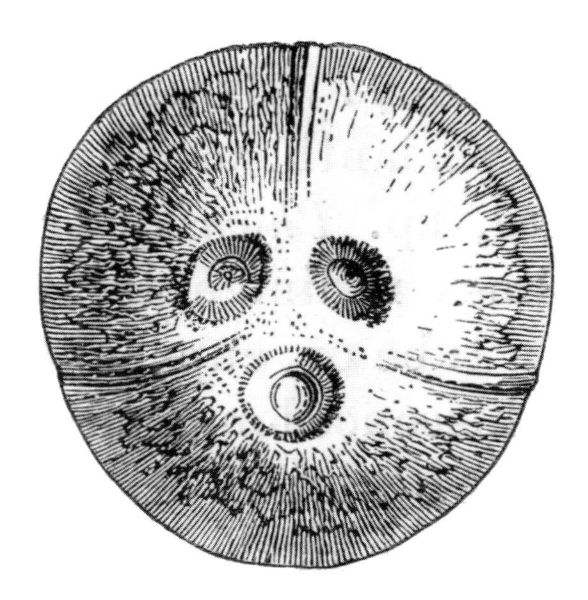

Coconut (*Cocos nucifera*), from Maout & Decaisne (1876), illustrating the three "eyes" at the base of a coconut.

Culinary Vocabulary

- See CHOCOLATE (cacao) for confusion between "cocoa" and "coco," which has led some people to think that chocolate comes from coconuts. "Coco butter" is a misspelling of "cocoa butter," which comes from chocolate seeds, not from coconuts.
- A "snow ball" (sno' ball) is a dessert made by rolling a scoop of ice cream in shredded coconut and topping it with chocolate syrup.
- *Cocada* (pronounced koa-KAH-dah) is a Spanish sweet coconut custard.
- *Cocalán* (pronounced coe-cah-lahn) is a honey-like syrup produced in Chile from palm sap.
- Cocoribe is an American liqueur prepared from coconuts and light Virgin Islands rum.
- "Latick" is a brown, curd-like material that remains after coconut milk is heated to the point of oil separation. It is employed as a topping for Filipino snack foods and desserts.
- "Sad cake" is an American Southern dessert, prepared with coconut, pecan, and raisins. Its name comes from the sad appearance when the cake sinks in the middle after baking.

Curiosities of Science and Technology

- Cadang-cadang disease, which occurs in coconut trees, is named after the noise the coconuts make when they fall to the ground.

- The outer part of the trunk of the coconut palm furnishes a construction lumber known as porcupine wood for houses and furniture; the name "porcupine wood" is based on the surface pattern, suggestive of porcupine quills.
- Coconut milk is so pure and sterile that during World War II it was used in emergencies instead of sterile glucose solution, and put directly into a patient's veins.
- Because coconut water is naturally sterile, it is used as a diluent, in combination with egg yolk, in artificial insemination.
- A single coconut has as much protein as a quarter pound of beefsteak.
- A freshly harvested coconut from a bunch that has been in the sun has a natural effervescence, and when opened will hiss while the gas is released.
- The husks of coconuts are generally inedible, but some rare plants produce fruit with a spongy, easily cut husk that is sweet to chew like sugar cane.
- The milky tissues within a coconut are intended to feed the growing embryo. However, the embryo's leaves and root develop mostly outside of the fruit. A connection (a sort of very short umbilical cord called a haustorium) allows the young plantlet to absorb the nourishment of the coconut. As the feeding process continues for up to 15 months, the connection can become quite large. The connection is edible, and is called a "coconut apple." Because it is still within the coconut, it is commonly obtained by grasping the leaves of a sprouted coconut, uprooting it, and swinging it against the trunk of a nearby tree to split open the fruit. The "apple" is then picked out and consumed.
- Coconut palms are often grown as ornamental trees in tropical regions. Unfortunately, the nuts are so heavy that they sometimes cause injuries to humans, cars, and rooftops when they fall. Coconuts are as big as a man's head and 1–2 kg (about 2–4½ pounds) in weight, and some people have been killed by falling coconuts. Coconut palm trees have been removed from some resorts in Florida and the Caribbean in order to avoid lawsuits from injured guests.
- The Macapuno (makapuno) coconut of the Philippines is famous for having no milk cavity. Its jelly-like flesh fills the middle of the coconut, and can be eaten with a spoon. Such coconuts also occur in other coconut-growing countries. The embryo is normal, but can only be germinated under the artificial conditions of a tissue culture laboratory.

Coconut (*Cocos nucifera*), from Hooker (1850), showing the collecting of toddy (palm sap) from the male inflorescences. The liquid is used to produce sugar and alcoholic beverages.

- In Hawaii, the swollen base of the trunk of coconut palms is hollowed out and turned into hula drums.
- Scientists are not immune to hoaxes, and (mostly in the past) some were taken in by the following story. The "coconut pearl" has been claimed to be the rarest and most valuable "botanical jewel" in the world. Like the pearls of oysters, this is said to be a shiny calcareous sphere that occasionally forms inside a coconut. The allegation that there really are coconut pearls is extremely doubtful, although there are instances of pearls that are believed to have originated from coconuts, including the "Maharaja coconut pearl," once displayed in the shell of a coconut at the Fairchild Tropical Garden in Coral Gables, Florida (a research center with an outstanding reputation).

- The coconut crab (*Birgus latro*), a relative of the hermit crab, is the world's largest crab. An adult may stretch over a meter (more than a yard) from claw to claw, weigh up to 4 kg (9 pounds), and live for over 30 years. The crab occurs in the tropical Indian and Pacific Oceans, but because of excessive harvesting, it has disappeared over much of its range. Coconut crabs eat fruit, rotting leaves, animals, and most notably, coconuts, which they open with their massive claws. Although they can climb trees to reach fruit, and are reputed to cut off coconuts, in fact they only eat fallen coconuts. When it is in its larval stages, it is believed that the crab can be transported long distances on floating coconuts. When young and their abdomens are still soft, the crabs use abandoned shells as armor, in the manner of hermit crabs. As they get bigger and run out of suitable shells, some may use discarded coconut shells. The alternative name of the coconut crab, "robber crab," is due to its attraction to shiny pots and trinkets, which it steals from tents and houses.
- In England, coconuts are split open and hung outdoors for birds to feed on, especially in winter.
- So-called "coconut-banana honey," said to be made by bees frequenting both banana and coconut groves, has sold as a luxury food in Paris at $90 for a 400-g (14-ounce) jar.
- Until the middle of the 20th century, all floating soaps were made from coconut oil. Then it was discovered that most soaps could be made to float simply by pumping air through the mixture during manufacture.
- Most soaps do not lather well in hard water, but because of its high lauric acid content, soaps made from coconut oil are quick-lathering even in hard water.

Key Information Sources

Aroksaar, R., and Alalamua, H. (*Editors*). 1978. Coconut intercropping and coconut by-products: a bibliography of published literature, 1966–1978. South Pacific Commission, Noumea, New Caledonia. 51 pp.

Banzon, J.A., and Velasco, J.R. 1982. Coconut production and utilization. Philippine Coconut Research and Development Foundation, Pasig, Metro Manila, Philippines. 351 pp.

Banzon, J.A., et al. 1990. Coconut as food. Philippine Coconut Research and Development Foundation, Diliman, Quezon City, Philippines. 239 pp.

Benbadis, A.K. 1992. Coconut and date palm. *In* Biotechnology of perennial fruit crops. *Edited by* F.A. Hammerschlag and R.E. Litz. CAB International, Wallingford, UK. pp. 383–400.

Child, R. 1974. Coconuts. 2nd edition. Longmans, London, UK. 335 pp.

Farooqi, A.A., and Sreeramu, B.S. 1999. Coconut. *In* Tropical horticulture, vol. 1. *Edited by* T.K. Bose, S.K. Mitra, A.A. Farooqui, and M.K. Sadhu. Naya Prokash, Calcutta, India. pp. 538–562.

Frémond, Y., Ziller, R., and de Nucé de Lamothe, M. 1968. The coconut palm. International Potash Institute, Berne, Switzerland. 227 pp.

Green, A.H. (*Editor*). 1991. Coconut production: present status and priorities for research. World Bank, Washington, DC. 150 pp.

Grimwood, B.E. [and contributions by Ashwood, F., et al.]. 1975. Coconut palm products: their processing in developing countries. Food and Agriculture Organization of the United Nations, Rome, Italy. 261 pp.

Harries, H.C. 1977*a*. The Cape Verde region (1499 to 1549); the key to coconut culture in the Western Hemisphere? Turrialba, **27**: 227–231.

Harries, H.C. 1977*b*. Coconut (*Cocos nucifera* L.) [Pests, diseases, quarantine regulations]. *In* Plant health and quarantine in international transfer of genetic resources. *Edited by* W.B. Hewitt and L. Chiarappa. CRC Press, Cleveland, OH. pp. 125–136.

Harries, H.C. 1978. The evolution, dissemination and classification of *Cocos nucifera*. Bot. Rev. **44**: 265–320.

Harries, H.C. 1981*a*. Practical identification of coconut varieties. Oleagineux, **36**: 63–72.

Harries, H.C. 1981*b*. Germination and taxonomy of the coconut *Cocos nucifera*. Ann. Bot. **48**: 873–883.

Harries, H.C. 1982. Coconut varieties. Indian Coconut J. **12(11)**: 3–10.

Harries, H.C. 1992. Biogeography of the coconut *Cocos nucifera* L. Principes, **36**: 155–162.

Harries, H.C. 1995. Coconut (*Coccos nucifera*). *In* Evolution of crop plants. 2nd edition. *Edited by* J. Smartt and N.W. Simmonds. Longman Scientific & Technical, Burnt Mill, Harlow, Essex, UK. pp. 389–394.

Harries, H.C. 2000. Coconut. *In* The Cambridge world history of food. *Edited by* K.F. Kiple and K.C. Ornelas. Cambridge University Press, Cambridge, UK. pp. 388–397.

Harries, H.C. 2002. The "Niu" Indies: Long lost "home" of the coconut palm. Palms, **46(2)**: 97–100.

Harries, H.C. 2004. Fun made the fair coconut shy. Palms, **48(2)**: 77–82.

Holzapfel, C., and Holzapfel, L. 2003. Coconut oil: for health and beauty. Book Publishing Co., Summertown, TN. 127 pp.

Mahabale, T.S. 1976. The origin of the coconut. The Palaeobotanist, **25**: 238–248.

Mahindapala, R., and Pinto, J.L.J.G. 1991. Coconut cultivation. Coconut Research Institute, Lunuwila, Sri Lanka. 162 pp.

Mandal, R.C. 2000. Coconut: production and protection technology. Agrobios, Jodhpur, India. 168 pp.

Maynard, R. 1986. The coconut book. Grove Press, New York, NY. 174 pp.

Mayuranathan, P.V. 1938. The original home of the coconut. J. Bombay Nat. Hist. Soc. **40**: 174–182, 776.

Melendez-Ramirez, V., Parra-Tabla, V., Kevan, P.G., Ramirez-Morillo, I., Harries, H., Fernandez-Barrera, M., and Zizumbo-Villareal, D. 2004. Mixed mating strategies and pollination by insects and wind in coconut palm (*Cocos nucifera* L. (Arecaceae)): importance in production and selection. Agric. For. Entomol. **6**: 155–163.

Menon, K.P.V., and Pandalai, K.M. 1958. The coconut palm, a monograph. Indian Central Coconut Committee, Ernakulam, India. 384 pp.

Meyer, C. 1976. Coconut, the tree of life. Morrow, New York, NY. 96 pp.

Ohler, J.G. 1984. Coconut, tree of life. Food and Agriculture Organization of the United Nations, Rome, Italy. 446 pp.

Ohler, J.G. (*Editor*). 2000. Modern coconut management: palm cultivation and products. Intermediate Technology, London, UK. 458 pp.

Ohler, J.G., and Magat, S.S. 2001. *Coccos nucifera* L. *In* Plant resources of South-East Asia, 14, vegetable oils and fats. *Edited by* H.A.M. van der Vossen and B.E. Umali. Backhuys Publishers, Leiden, Netherlands. pp. 76–84.

Onwudike, O.C. 1996. Coconut (*Cocos nucifera* L.) kernel, oil and meal. *In* Food and feed from legumes and oilseeds. *Edited by* J. Smartt and E. Nwokolo. Chapman and Hall, London, UK. pp. 306–317.

Piggott, C.J. 1964. Coconut growing. Oxford University Press, London, UK. 109 pp.

Silas, E.G., Aravindakshan, M., and Jose, A.I. 1991. Coconut breeding and management. Kerala Agricultural University, Vellanikkara, Trichur, India. 380 pp.

Thampan, P.K. 1975. The coconut palm and its products. Green Villa Publishing House, Cochin, India. 302 pp.

Thangaraj, T., and Muthuswami, S. 1985. Coconut. *In* Fruits of India, tropical and subtropical. *Edited by* T.K. Bose. Naya Prokash, Calcutta, India. pp. 345–387.

Woodroof, J.G. 1979. Coconuts: production, processing, products. 2nd edition. AVI Publishing Co., Westport, CN. 307 pp.

Zizumhbo-Villarreal, D., Hernandez-Roque, F., and Harries, H.C. 1993. Coconut varieties in Mexico. Econ. Bot. **47**: 65–78.

Specialty Cookbooks

Escudero, C.A. 1980. The coconut cookbook. Mr. & Mrs. Publishing Co., Manila, Philippines. 100 pp.

Fife, B. 2004. Coconut lovers cookbook. Piccadilly Books, Colorado Springs, CO. 224 pp.

General, H.C. 1994. The coconut cookery of Bicol. Bookmark, Metro Manila, Philippines. 124 pp.

Hood, K.J.M., et al. 2004. Coconut delights cookbook: a collection of coconut recipes. Whispering Pine Press, Spokane Valley, WA. 320 pp.

MacBean, V. 2001. Coconut cookery: a practical cookbook encompassing innovative uses of the tropical drupe *Cocus nucifera*, accompanied by assorted information and anecdotes ranging from hard data to the frankly frivolous. Frog Ltd., Berkeley, CA. 198 pp.

Persley, G.J., Foale, M.A., and Wright, B. 1990. Coconut cuisine: a taste of the tropics. Inkata Press, Melbourne, Australia. 57 pp.

Rohini Iyer, R., and Thampan, P.K. 1991. Coconut recipes around the world. Central Plantation Crops Research Institute, Kasaragod, Kerala, India. 173 pp.

COFFEE

Family: Rubiaceae (madder family; more familiarly, this is occasionally called the "coffee family")

Names

Scientific names: *Coffea* species

- Arabian coffee—*C. arabica* L.
- Robusta coffee—*C. canephora* Pierre ex A. Froehner (*C. robusta* L. Linden)
- Liberian coffee—*C. liberica* W. Bull ex Hiern (*C. excelsa* A. Chev., *C. dewevrei* De Wild. & T. Durand)

- The word "coffee" may be based on the town of Kaffa in Ethiopia, where it is believed coffee originated. The genus name *Coffea* has the same derivation. An alternative theory is that the word was derived from the Turkish *kahveh* or *qahwah*, meaning "a wine tonic that restores the appetite."
- Arabian coffee is also called arabica coffee and, simply, coffee.
- *Arabica* in the scientific name *C. arabica* means Arabian, reflective of the long association of Arabia with coffee production.
- *Canephora* in the scientific name *C. canephora* is based on the Greek *kaneon*, a wicker basket, or *kanes*, a basket or mat of reeds, + *phora*, bearing, i.e., "basket-bearing" or "reed-bearing," presumably because the branches suggested this appearance to the author who coined the name.
- *Coffea canephora* was discovered in the Congo in 1898, and is sometimes called Congo coffee tree. Unfortunately, this is easily confused with the name "Congo coffee," used for *C. congensis* A. Froehner.
- Robusta coffee is also called Rio Nunez coffee, after the region of Coastal Guinea in West Africa.
- Robusta coffee is so named because of the old scientific name, *C. robusta* (the species is now known as *C. canephora*), the Latin *robusta* meaning robust, for the large size of the tree.
- Liberian coffee is also known as liberica coffee.
- Exelsa coffee, once recognized as a separate species (*C. dewevrei* De Wild. & T. Durand) is now classified as a kind of Liberian coffee [*C. liberica* var. *dewevrei* (De Wild. & T. Durand) Lebrun]. It is found in western and west-central tropical Africa, and is used to produce only a very small amount of the world's coffee.
- *Liberica* in the scientific name *C. liberica* is Latin for Liberian, referring to the country in West Africa.
- *Coffea liberica* is also called liberica coffee.

Plant Portrait

Arabian coffee (*C. arabica*) is an evergreen shrub or small tree growing up to 4.6 m (15 feet) in height. The leaves are dark and glossy; the flowers are white, and produced in clusters; and the fruit is a yellow to crimson berry that becomes black on drying. Three-quarters of world coffee production is arabica coffee. Despite its name, the plant originated in Ethiopia, not Arabia. It is thought to have been introduced into Arabia prior to the 15th century. The species was planted in Java in 1690, and cultivation soon spread throughout the West Indies, Central America, and certain regions of South America. Later, it reached India and Sri Lanka. Subvarieties include Moka, Maragogipe, San Ramon, Columnaris, and Bourbon. Arabica coffees from Brazil are called Brazilian Coffees; those from Colombia, Venezuela, Peru, Guatemala, Salvador, Haiti, and Santo Domingo are called milds. Arabica coffee is full-bodied, sharp in taste, with a relatively low caffeine content. Arabica beans are slightly elongated, with greenish blue shades.

Robusta coffee (*C. canephora*) can be a tree over 12 m (about 40 feet) high, but also grows as a shrub. The seeds are smaller and rounder, with a straight crease. The species is native to Western Africa. It is cultivated in Africa, Asia, and Indonesia, where the climate is unsuitable for *C. arabica*. The plants produce about one-quarter of total world coffee production. Because of their high content of caffeine (about twice as much as arabica) and strong character, robustas are used mostly in specialty blends. The two main varieties are Kouillou and Robusta. The taste is generally considered less refined and harsher than arabica coffee, and the beans fetch a lower price.

Liberian coffee (*C. liberica*) is a tree native to tropical Africa, Liberia, and the Ivory Coast, and is also grown in Guinea, Ghana, Nigeria, Cameroon, Zaire, and the north of Angola. The tree reaches a height of 20 m (66 feet), and has very large leaves, and berries twice the size of the arabicas. It grows well in high altitudes and is resistant to droughts and diseases. Its flavor is not greatly valued, and it is being slowly replaced by robusta coffee.

Plantation coffee shrubs are 2–5 m tall (6½–16½ feet). Red pulpy berries are produced, about the size of big cherries, each containing two seeds ("beans") coupled together (but see information below regarding "peaberries"). Coffee fruit are hand-harvested. Over 50 countries grow coffee, all of them near the equator, between the tropics of Cancer and Capricorn. Brazil is the largest producer (responsible for about a third of world production), followed by Columbia. Other top 10 coffee-producing counties include Indonesia, Mexico, Ethiopia, Guatemala, India, Côte d'Ivoire (Ivory Coast), Uganda, and Vietnam.

Coffee has been known for almost 3000 years, although modern habits of consumption are relatively recent. Coffee roasting started in the 14th century, and became prevalent in Turkey in the mid 16th century. The brew was outlawed by the Ottoman Sultan in Istanbul in 1543. Nevertheless, in 1554 (or as early as 1474 according to some), the first coffeehouse was set up in Istanbul, and coffee spread throughout the Ottoman Empire, and to European countries, Western Asia, and India. Today, coffee is a giant global industry, estimated to employ about 23 million people. Over 400 billion cups are consumed every year, making coffee the world's most popular beverage. Scandinavians are the world's greatest coffee drinkers (Finns consume about 4.7 cups per persons per day, Norwegians about 4.3, Danes about 4.1). Americans drink about 1.8 cups per person per day, and the United States purchases more of the world's coffee than any other nation (and consumes about 450 million cups a day).

The history of coffee growing is marked by theft. Until late in the 17th century, the Arabs maintained a monopoly over the supply of coffee by forbidding seeds to be taken out of Arabia. Then Dutch spies stole some plants, and cultivated them in Java. In 1714, a coffee seed from Java was used to establish a plant in the botanical garden in Paris—the first coffee plant in France. That plant was destined to become the mother of all coffee plants in Latin America, but not before additional thefts took place. In 1723, a young French naval officer, Gabriel Mathieu de Clieu, led a gang of masked men over the garden walls and stole the 10-year-old coffee plant. De Clieu sailed for the West Indies with the potted plant in a glass box, intending to establish a coffee plantation in Martinique, where he lived. During the trip, a Dutch agent tried to destroy the plant, and was prevented from doing so by hand-to-hand combat with De Clieu. The ship became becalmed, and as drinking water grew scarce, the captain strictly rationed the water supply, refusing to issue any for the plant, forcing De Clieu to share his meager water ration with the plant! Both the coffee plant and

Arabian coffee (*Coffea arabica*), from Köhler (1883–1914).

De Clieu survived, and the young tree was planted in Martinique with an armed guard around it until it was large enough to produce additional plants. From these, coffee spread throughout the West Indies. The coffee plants were well guarded to protect the early plantations from competitors, but cuttings were stolen by Lieutenant Colonel Francisco de Melo Palheta of the Brazilian army, who was sent to French Guinea for the purpose, and the cuttings were used to establish one of the greatest coffee empires in the world in Brazil.

Coffee beans sold on the world market are "green," i.e., unroasted, and in this form, they can be stored for several years without loss of flavor (some coffees become more flavorful with age). Coffees from different sources are blended together to produce a consistent, desired flavor. Roasting is the key process that brings out coffee's flavor and aroma. The beans are heated at high temperatures in ovens, then quickly cooled and coated with resin, gum arabic, or sugar to make them shiny and preserve their flavor. During roasting, the greenish gray color changes to brown because of caramelization of sugars. The longer the roasting and the higher the temperature, the darker the color becomes. Longer roasting also makes for less acidic beans, but they become more bitter. The aroma that develops during roasting is partly due to the caramel that is produced, but mainly to the essential oils and fatty acids that rise to the surface of the beans.

Culinary Portrait

Although coffee is widely used as a flavoring, for example, in ice cream, pastries, candies, and liqueurs, it is primarily roasted, ground, and brewed. There is an ongoing debate regarding whether coffee or tea (which see) is the most important beverage in the Western World.

Coffee can be brewed by various methods, each best suited to a particular grind. Stronger and more flavorful brews are produced by using finer grinds, and as fewer beans are required, finer grinds are more economical.

The simplest but least satisfactory way of preparing coffee is "open pot brewing" or "steeping." Water is brought to a boil in a container, regular grind coffee is added, the heat is turned off, the container is covered and allowed to stand for 5 minutes after which several spoonfuls of cold water are added to settle the grounds and the coffee is drunk.

"Drip coffee" brings coffee into contact with water for as short a time as possible, extracting a maximum of flavor and aroma but a minimum of bitter taste. Drip pots can use a perforated top or a mesh filter made of paper or cloth and placed in a special conical basket. Coffee is placed in the top part of the pot, hot water is poured over it (once only), and trickles down to the lower part of the pot. If the grind is too fine, the filter pores become clogged; if the grind is too coarse, the brew will be bland.

"Percolation," the most automated method, is preferred in North America, but some experts dislike it in the belief that it promotes bitterness. A perforated basket containing coffee grounds is placed on a long stem. When the water boils, it rises through the stem by means of pressure or gravity, and falls on the grounds, seeps through them and back into the pot, the cycle continuing until the coffee reaches the desired strength.

The average person can prepare a perfect cup of coffee by following these steps. Use clean equipment. (Purists recommend using porcelain or earthenware coffee pots rather than those made of metal.) Grind recently roasted, high-quality, whole arabica beans. Take pure, fresh, cold water and bring it to a near boil. Alternatively, although experts claim boiled water will lose its oxygen and taste flat, bring water to a boil and wait a minute or so; the ideal brewing temperature is a few degrees below boiling. Purists recommend using nonchlorinated water, and avoiding water that is high in calcium, sulfur, or iron. For each cup, combine 170 g (170 mL, or 6 ounces) of the almost boiling water with 2 tablespoons of coffee for 4–5 minutes. Filter the coffee, and pour it into a cup. If desired, add sugar, cream, whipped cream, cinnamon, cardamom, nutmeg, citrus, ground nuts, or chocolate. Flavoring syrups should be added immediately after the coffee is brewed, not during brewing (or preflavored beans can be purchased). Drink immediately.

Coffee should not be reheated. Metallic coffeemakers and metal cups tend to deteriorate the taste. Washing coffeemakers is a good idea, as the oily residue can also result in unpleasant-tasting coffee. To clean an electric coffee pot, boil some water in it to with 1 teaspoon of liquid detergent added. All traces of dishwashing detergent should be rinsed away.

To make iced coffee, prepare a much more concentrated brew, since it will be diluted by the ice. Alternatively, pour brewed coffee into ice cube trays, freeze, and add to regular-strength coffee. Coffee left over from a meal can be saved as ice cubes for iced coffee.

How recently coffee beans were roasted is critical to coffee quality. The finest coffee is rarely heavily roasted, as this would spoil the subtle flavor. The flavor of the beans is developed only after they are roasted, but deteriorates quickly. While most people are content to buy supermarket coffee beans, purchasing coffee from a local, professional, specialty roaster is the best way to ensure getting really fresh beans. If possible, buy only the quantity that will be used up in the coming week. If unavoidable, coffee beans can be stored in an airtight container, with as little air inside as possible, in a freezer, and will remain in good condition for months. The beans do not have to be defrosted before grinding.

Expensive (burr-type) grinders work faster but the cheaper (blade-type) kind achieve the same quality. A medium grind is best for drip methods that provide for dissolving the flavorful coffee components in 5 minutes; longer and the brew becomes bitter. Since the contact is brief, a fine, powdery grind is best for espresso or vacuum brewing.

"Espresso coffee" is made with very finely ground, dark roasted coffee beans that are held in hot water in the tank of an espresso machine. Unlike most other methods of preparation of coffee that rely on brewing by percolating hot water over, or dripping through, loose grounds, in an espresso machine the water is forced under pressure through tightly packed grounds. Very expensive machines are available for preparing espresso coffee. For most people, it is much simpler, much less expensive, and much more likely that really great espresso will be consumed if it is simply bought at a coffeehouse. In fact, in Italy, espresso is mostly drunk in cafés, not at home. At most espresso bars and upscale coffee shops, espresso is produced one cup at a time, and usually served in a preheated demitasse (i.e., small) cup, along with a sugar cube or twist of lemon to reduce the bitterness.

"Turkish" or "Middle Eastern" coffee employs pulverized coffee ground and a long-handled eastern coffee pot called a *kanika*. Ibrik coffeemakers are widely used to make Turkish coffee. Coffee, water, and sugar are combined in the pot, boiled three times, and poured into tiny cups. The grounds are allowed to settle in the cup, and the tiny amount of strong, sweet coffee is carefully sipped. Coffee prepared this way is rich and full-bodied.

Coffee connoisseurs judge the brew with respect to four basic components: aroma, body, acidity, and flavor. Obviously, aroma refers to fragrance and flavor to the taste. Body means the feel or "weight" of the coffee in the mouth, how it rolls around the tongue and fills the throat on the way down. Acidity is not literally the pH level, but the sparkle, brightness, or tang that adds zest to the cup.

Once a container of coffee is opened and exposed to air, it deteriorates rather quickly. It is best to buy only as much as can be used up in a week. Coffee can be stored in a clean, airtight container in the refrigerator.

The stimulating effects of coffee are due to some extent to theobromine, but mainly to caffeine. Caffeine is a drug. It is the world's most popular stimulant, and the health aspects of it have long been a concern. The addictive character of caffeine is evidenced by the slight withdrawal symptoms and edginess that may be experienced before the first coffee of the day, and many people have come to depend on a "fix" of caffeine as a kick start. Caffeine stimulates the cerebral cortex, allaying fatigue and drowsiness. Other illustrations that caffeine is a drug are the facts that many people become jittery, or get a "buzz," or make frequent visits to the bathroom if they drink too much coffee, which obviously affects the brain and the kidneys. Coffee also produces diarrhea in some persons. It has been said that for most people, up to 400 mg of caffeine a day—the equivalent of 4 or 5 cups of coffee—should not raise health concerns. However, individuals vary in their sensitivity. Children are more susceptible. Many over the age of 30 find that caffeine keeps them awake at night (indeed, people driving long distances commonly use coffee or other caffeine-containing beverages to keep awake). Consuming more than 7 or 8 cups of coffee of average caffeine content in a day could lead to extreme effects, known as caffeinism. Symptoms could include missed heartbeats, palpitations, mood swings, sweating, and even panic attacks. Drinking more than 11 cups a day could be toxic. Because of suspicion that excess caffeine could trigger a miscarriage, some specialists recommend keeping caffeinated coffee consumption to no more than 3 cups a day when pregnant; some recommend avoiding caffeine completely. Excessive caffeine is also thought to increase the risk of osteoporosis (brittle, crumbly bones) and fractures.

Culinary Vocabulary

- Coffee was introduced to Indonesia by the Dutch in 1696. Java became a major coffee producing area, so much so that the name of this port became a common name for coffee.
- "Mocha" today has several meanings: (*a*) a coffee from Ethiopia (the word originally referred to bitter Ethiopian coffee); (*b*) a Yemeni coffee tasting like chocolate; (*c*) a beverage made with coffee and cocoa mixed together with water or milk; and (*d*) a flavoring combining coffee and chocolate (especially delicious because the coffee accents the flavor of the chocolate). Mocha was named for the southern port in Yemen on the Red Sea from which coffee was exported in large quantities. A more

colorful explanation is that several people were banished to the desert to die of starvation. In desperation, they boiled and ate the fruit of an unknown plant, which helped them survive. This was interpreted as a religious sign by nearby residents, who honored the plant and its beverage by giving it the town's name, Mocha.

- "Jamoke" (jamoca), based on "java" + "mocha," is an early 20th century term for coffee in general.
- A "café" (often rendered "cafe" in the United States and elsewhere) is a restaurant, barroom, cabaret, or nightclub. The word originated from the French, *café*, coffee, because the first cafés were coffeehouses, established during the 15th century in the Arabian cities of Mecca and Cairo. The café has always been a place where people can not only drink coffee, but also enjoy food, read, socialize, or just watch the world go by.
- "Cafeteria" was originally American Spanish for "coffee-house," based on the Spanish *caferero*, "coffee seller."
- After the American Revolution, coffee was popular but very expensive in the United States. Among the mock coffees that were served were "Boston coffee" (made from rye), "Canadian coffee" (made from peas), and "crumb coffee" (also known as "crust coffee"; made from burnt bread).
- Admiral Josephus "Joe" Daniels, Chief of U.S. Naval Operations, forbade alcohol onboard ships, except for special occasions. When coffee replaced the previously more available alcohol, it was called "cup of Joe."
- It has been claimed that the "bean" in "coffee bean" is a corruption of the Arabic for the coffee plant, *bunn*. This seems far-fetched.
- The "elephant bean" is a type of hybrid arabica coffee bean that is about a third larger than other coffee beans. It arose in Bahia State, Brazil, and was first identified in 1870. Before World War I, it was popular in many countries, including Germany where it was the only coffee served in the Kaiser's household. The best elephant beans are grown in Mexico and Guatemala. The coffee has a smooth taste with no bitterness and light acidity, but has been falling out of favor.
- George Washington coffee, the first major brand of instant coffee, was so popular during the First World War that soldiers asked for "a cup of George."
- "Calliope coffee" is a gelatin dessert prepared with strong coffee. Garnished with whipped cream, it was popular on Mississippi River showboats during the early 20th century.
- Types of coffee preparations:
  - Café Amaretto: hot, black coffee made with amaretto cloves, sugar, orange zest, lemon zest, and cinnamon sticks.
  - Café au lait (pronounced ka-FAY oh LAY): a coffee of French origin, made with equal parts hot, strong coffee and hot milk (compare caffè latte, and coffee milk, below).
  - Café brûlot (pronounced ka-FAY broo-LOW, French for "burnt brandy"): a traditional New Orleans combination of dark, strong coffee and brandy, flavored with citrus rind, spices, and sugar, and usually served flaming after dinner.
  - Café Carioca (pronounced kah-FEH ca-ree-O-ca): a coffee of Portuguese origin, made by diluting coffee with hot water.
  - Café continental: hot coffee mixed with coriander, sugar, and warmed sweet red wine, and served in a mug garnished with a quartered slice of orange.

Arabian coffee tree (*Coffea arabica*), from Baillon (1876–1892).

- Café de olla: a Mexican coffee flavored with cloves, cinnamon, and brown sugar.
- Café Diable (pronounced ka-fay dee-ah-blae): hot black coffee with Cognac, Cointreau, curaçao, cloves, coffee beans, and cinnamon sticks.
- Café double (pronounced ka-FAY dewbl): coffee beverage of French origin, made double strength.
- Café filtre (pronounced ka-FAY filt'r): coffee made by pouring hot water through ground coffee beans held in a filtering device fitted over a cup or pot.
- Café Normande (pronounced ka-FAY nohr-mahnd): coffee preparation of French origin, made with coffee and Calvados, served hot.
- Café Royale (pronounced ka-FAY roi-yahl): hot, black coffee made with bourbon and sugar.
- Café liégeois is a beverage made with cold coffee and ice cream.
- Cafè Americano (pronounced kahf-AY a-mer-i-CAH-no): Italian coffee made with about one-fourth espresso and three-fourths hot water.
- Caffè corretto (pronounced kahf-AY cohr-ett-toh): Italian espresso with a small amount of brandy or liqueur, usually served in a small cup.
- Caffè latte (pronounced kahf-AY LAH-tay): an Italian coffee preparation, with one-third or less espresso and two-thirds or more steamed milk, sometimes served with foam on top, and usually served in a tall glass.
- Caffè mocha (pronounced kahf-AY MO-kah): coffee made with one-third espresso, two-thirds steamed milk, and chocolate syrup, and usually topped with whipped cream sprinkled with cocoa powder, and served in a tall glass.
- Caffè ristretto (prounced kahf-AY ree-strai-toh): Italian beverage made with one part coffee and one part water—one of the most concentrated espresso drinks—usually served in a small cup.
- Cappuccino: a form of espresso (discussed below) popular in Europe. This consists of about one-third coffee and two-thirds steamed milk, served in a heated cup and topped with frothed milk. In Italy, cappuccino is made by forcing steam through milk or cream to produce a creamy topping, but in North America, it is often served simply with whipped cream on top. In coffee bars of the 1950s, cappuccino was known as "frothy coffee." In North America, cappuccino is frequent garnished with cinnamon, nutmeg, or chocolate. The Italian term cappuccino means "capuchin," and was presumably applied to coffee because of the similarity of the beverage and the pale brown robes worn by monks of the *Order of Friars Minor Capuchins* (an independent branch of the Franciscans). (Another interpretation holds that the drink is named after a Capuchin monk in whose garden coffee was grown in 1774 in Brazil.) The word "capuchin" ultimately comes from the Latin *cappa*, "hood," which is based on *caput*, "head."
- Espresso: an Italian coffee-brewing method in which hot water is forced through finely ground and packed coffee, usually dark-roasted beans, under high pressure, and the resulting beverage is usually served in a small cup (see below for additional information). In Italy, espresso is served without milk or cream, and in English, the term espresso is often interpreted as "black coffee" (as opposed to cappuccino, which contains milk). Espresso is often incorrectly spelled "expresso," leading to the incorrect conclusion that the beverage is made rapidly (the Italian word *espresso* means both "very quick" and "pressed out," but the coffee is named for the latter meaning).
- Espresso con panna (pronounce ess-PRESS-o cone PA-na): espresso with a dollop of whipped cream.
- Espresso lungo (pronounced ess-PRESS-o loon-go): espresso made with almost twice as much water as regular espresso, popular in the US.
- Espresso machiatto (pronounced es-PRESS-o mock-e-AH-toe): espresso with a dollop of steamed milk, usually served in a small cup.
- Espresso powder: a powder made from dried roasted espresso beans, used to give a rich coffee flavor and aroma to pastries, desserts, and confections.
- Espresso ristretto: *ristretto* means "restricted," referring to stopping the flow of water at about 29 mL (1 ounce) to produce a more concentrated brew than normal espresso.
- Belgian coffee: coffee in which a chunk of chocolate is placed to melt.
- Irish coffee: an alcoholic drink, typically made with strong, black coffee, sugar, and Irish whiskey, topped with whipped cream.

- A "Black Russian" is a cocktail made with 2 parts vodka and 1 part coffee-flavored liqueur (such as Kahlúa) served over ice. A "white Russian" is essentially the same thing, with cream. A "Coffee Cooler," almost identical, is prepared with iced coffee, vodka, Kahlúa, and heavy cream, and served with a topping of coffee ice cream.
- "Jamaican Coffee" is a cocktail prepared with hot black coffee, dark Jamaican rum, cognac, Tia Maria (a dark brown coffee-flavored Jamaican liqueur), and additions such as whipped cream, cinnamon, and ground ginger.
- Fineness of coffee grinds:
  - Coarse (loosely, "regular"): large particles, suitable for percolators, plunger pots, urns, and cold-water extraction.
  - Medium: finer than coarse, but still suitable for the same uses.
  - Drip: finely ground for cone-shaped filters and drip pots. (Sometimes "drip" is synonymous with "medium.")
  - Espresso: very finely ground, suited for espresso machines. ("Fine," with a texture close to powder, is often equated with espresso grind.)
  - Turkish: pulverized like flour, for use in traditional Turkish brewing.
- Kinds of roast:
  - American: (= regular roast): somewhat heavier than a "light roast."
  - Italian: glossy and dark brown; preferred for espresso.
  - Heavy (= "Dark French"): extremely dark.
  - Viennese: a blend of one-third heavy and two-thirds regular. (This should not be confused with "Viennese coffee," which is strong, hot coffee topped with whipped cream, served in a tall glass.)
  - European: a blend of one-third regular and two-thirds heavy.
- A "coffee-cup-full" is a traditional measure of volume, approximate 180 mL (6 American fluid ounces, compared to the standard American cup of 8 fluid ounces). An "espresso cup" is a small cup with a handle and a capacity of about 43 mL (1.5 fluid ounces). An "approved coffee measure" (also known as a "coffee scoop") is a measuring spoon developed by the Coffee Brewing Institute to measure 2 level tablespoons.
- "Demi-tasse" (demitasse) refers either to a small cup used for coffee, or strong, black coffee served after dinner, especially in France.
- "Boston coffee," a phrase that is rarely encountered today, refers to coffee made by first adding cream and sugar to a cup and then pouring in fresh percolated coffee (rather than adding coffee first and mixing in cream and sugar last). There are those who claim that this sequence of preparation makes for a superior coffee. The connection with Boston is obscure.
- "Thai coffee" is coffee with sweetened condensed milk.
- "Red-eye gravy" (redeye gravy, red ham gravy) is made of ham drippings often flavored with coffee. Several explanations have been advanced to explain the name. One explanation holds that the name is due to the reddish brown small circular droplets of ham fat in the gravy resembling eyes. Another theory holds that the heavier ingredients settle in the bottom of the bowl in the shape of a red eye, giving the gravy its name. A much more colorful explanation holds that the name is due to American general and president Andrew Jackson (1767–1845), who once asked his cook to prepare lunch. The cook had been drinking corn whiskey and his eyes were red. General Jackson told him to prepare country ham with gravy as red as his eyes. Ham gravy became known as "red eye gravy" from then on.
- "Babka" (based on diminutive of the Polish *baba*, old woman) is a coffee cake flavored with orange rind, rum, almonds, and raisins.
- *Pousse-café*, colloquial French for "push-coffee," is a phrase that came into refined (or pretentious) English. In France and elsewhere, the term refers to a small drink of brandy or liqueur consumed at the end of a meal as a "coffee chaser." In the United States, an elaborate cocktail is often served under the heading of pousse-café, prepared by pouring successive layers of cordials on top of each other in a slender cordial glass to create a rainbow of colors (the heavier ones, i.e., those with a higher specific gravity, placed at the bottom).
- "Coffee milk" is a blend of a little coffee or coffee syrup in a glass of sweetened milk. It is the coffee equivalent of "cambric tea" (see TEA). In 1993, it was made "The Official State Drink" of Rhode Island, where it is usually found. Coffee milk has been known in Rhode Island since the early 1920s. When ice cream is added, coffee milk is called a "coffee cabinet" or "coffee cab." (The "cabinet" in the name is apparently due to the unknown originator keeping

Arabian coffee (*Coffea arabica*), from Baillon (1876–1892).

his blender in a kitchen cabinet.) An independent meaning of "coffee milk" sometimes encountered is that it is simply coffee served with hot milk.

- "Moose milk" is an alcoholic beverage, popular in the 1930s among those who invented the sport of frostbite dinghy racing. It is made of one part coffee, two parts milk, and five or more parts rum, and can be served hot or cold.
- American cowboys had several slang names to describe their coffee, including the following: Arbuckle's (for the brand that was most used); belly-wash (weak coffee; the same term has been used by loggers to mean a soft drink); black jack; black water; cowboy coffee (black coffee without sugar); Indian coffee (a weakened, reboiled beverage made from used grounds, not considered fit for cowboys but allegedly good enough for Indian visitors); and six-shooter coffee (so strong it could float a six-shooter pistol).
- "Draw one" and the number 44 are old lunch-counter slang for an order for coffee. "Draw one in the dark" meant a cup of black coffee. A "pair of drawers" was 2 cups of coffee. A "belly warmer" was a cup of coffee. "Sinkers and suds" meant doughnuts and coffee. "Blonde" meant coffee with cream. "Blonde and sweet" was coffee with sugar and cream. "High-test" meant regular coffee, in contrast to decaffeinated.
- World War II American GIs call mess hall coffee "battery acid."
- Coffee berries may contain one or two seeds (i.e., beans). Fruit that contain just one seed are usually found on the ends of branches. "Peaberry" coffee is made with berries containing just one seed. Peaberry coffee is favored by some connoisseurs, although this may simply be due to the old way of roasting coffee in a pan, because the round beans tended to roast more evenly when shaken in the pan.
- "Dyspepsia coffee" is prepared with equal parts of ground coffee and cornmeal, moistened with molasses, browned in an oven, and used like coffee, as an alleged cure for indigestion.
- "Freshly brewed coffee" is generally defined in the food and beverage industry as coffee that was prepared no more than 1 hour ago.

Curiosities of Science and Technology

- The world's first coffeehouse was apparently the Kiva Han, constructed in Constantinople about 1475.
- In the 17th century, coffeehouses in the Middle East were centers of political intrigue. Mahomet

Kolpili, the grand vizier of Turkey became so concerned about the 600 coffeehouses in Istanbul that he decreed a penalty of cudgeling for operating a coffeehouse. For a second offence, the perpetrator was sewn into a leather bag and thrown into the Bosporus Strait (which is beside the city, and separates the European and Asian parts of Turkey).

- Turkish law once made it legal for a woman to divorce her husband if he failed to provide her with her daily quota of coffee.
- In 1652, the first coffeehouse opened in England. Coffeehouses became popular forums for learned discussion, and were dubbed "penny universities," a penny being the price of a cup of coffee. Coffee had the virtue of stimulating the body while clearing the mind, and was therefore conducive to intellectual discussions, unlike alcoholic beverages in pubs, which led to fights and orgies.
- In 1668, Edward Lloyd's coffeehouse opened on Lombard Street, London, England, and became frequented by merchants and maritime insurance agents. It eventually became Lloyd's of London, the world's best-known insurance company.
- In 1683, the Turkish Army surrounding Vienna was expelled from Austria. Franz Georg Kolschitzky, a Viennese who had lived in Turkey, recognized the coffee that was left behind, claimed it, and opened central Europe's first coffeehouse. He also established the habit of refining the brew by filtering out the grounds, sweetening it, and adding a dash of milk.
- In 1702, the first coffeehouse in the United States was established in Philadelphia.
- Swedish King Gustav III (1746–1792) is reputed to have commuted the death sentence of twin brothers convicted of murder provided that they agree to be guinea pigs in a scientific experiment designed to test which of coffee drinking or tea drinking was superior. One brother could drink tea but not coffee, while the other could only drink coffee. The tea drinker died first, at the age of 83. Sweden developed into one of the top coffee-drinking nations of the world.
- In 1822, the first crude espresso machine was invented in France. Early espresso machines sometimes exploded. The Italians perfected the device, and there are now over 200 000 espresso bars in Italy. More than 4 500 000 metric tons (one billion pounds) of espresso coffee are consumed each year in Italy.
- In 1865, James Mason patented the coffee percolator.
- In early London, England, bags of coffee were first sold at "candle auctions." While a candle stub flickered, bids on coffee were accepted by an auctioneer. As soon as the candle sputtered out, the last bidder received the lot.
- During the 19th century in the United States, primitive transportation and packaging resulted in rather low-quality coffee in much of the country, and there was often little appreciation of how coffee should look and taste. To improve the taste, some brewed coffee with an egg to give the drink a rich yellow color (merchants added rust or beef blood to make coffee look better). Another popular method to improve the taste was to add the uncooked skin of a mild codfish.

Liberian coffee (*Coffea liberica*), from *Gartenflora* (vol. 29, 1880).

- Coffee rust (coffee leaf rust, caused by the rust fungus *Hemileia vastatrix*), is an extremely serious disease of coffee plants, often wiping out entire plantations. In the late 19th century, coffee rust seriously damaged the coffee plantations of Sri Lanka, Java, and Malaya, leading to the cultivation of rubber as an alternative crop in these countries. In Sri Lanka, tea also became a major new replacement crop.
- In 1908, Melitta Bentz, in Germany, invented the world's first drip coffee maker. She pierced holes in a tin container, put a circular piece of her son's notebook paper in the bottom of it, and put this creation over a pot. Earlier coffee filters were made of horsehair.
- In 1886, former wholesale grocer Joel Cheek named his popular coffee blend "Maxwell House," after the hotel in Nashville, Tennessee, where it was served. By the time president Theodore Roosevelt visited Nashville in 1907, the Maxwell House brand had established strong presence in the Southeast. On finishing a cup of the coffee, Roosevelt was reported to say, "Good. Good to the last drop," which became an advertising slogan. Coca-Cola actually used the statement first, and in any event, the story was probably made up for advertising purposes.
- American inventor and salesman Charles W. Post (1854–1914) in 1895 marketed Postum, a coffee substitute made of wheat, bran, and molasses. By 1900, a host of competitors were undermining his business by selling their own brands of coffee substitutes. Alarmed, Post simply packaged some of his Postum as a "new" product called "Monk's Brew," and sold it so cheaply (5 cents a box compared to 25 cents for the same amount of Postum) that his competitors were driven out of business. Post then retrieved carloads of unsold boxes of Monk's Brew, and repackaged them as Postum.
- The first soluble instant coffee was invented by Japanese-American chemist Satori Kato of Chicago in 1901. ("Coffee powder," quite inferior to this, was available in the 18th century.)
- It has been claimed that the coffee break, as known in the United States, was invented in New Orleans about the 1920s.
- German researchers perfected the process of removing caffeine from the beans without destroying the flavor in 1903. Decaffeinated coffee was marketed under the brand name Sanka (from the French *sans caffeine*, "without caffeine"), which was introduced to the United States in 1923.
- An approximate equivalent of Irish coffee, mixing caffeine equal to two strong cups of coffee with the equivalent of one shot of liquor, was found to be effective in limiting damage due to ischemic stroke (the kind due to blood clots in the brain) in rats if administered intravenously within 3 hours. The experimental drug, called caffeinol, appears to be safe in humans (the study was published in April 2003, in *Stroke*: Journal of the American Heart Association). However, there is no evidence that drinking Irish coffee will prevent a stroke in the first place.
- How is decaffeinated coffee made? Several methods are in use. Water extraction is widely practiced, consisting of soaking the beans to dissolve the caffeine, extracting the caffeine from the water with activated carbon, and then resoaking the beans in the decaffeinated water to reabsorb the flavor compounds that were lost in the initial extraction. Some solvents previously used to extract caffeine have been mostly discontinued because of suspicion that they could cause cancer. In the widely used carbon dioxide process, green beans are soaked in highly compressed carbon dioxide, which extracts the caffeine. Whatever the method, not all of the caffeine is removed (generally more than 97% is removed), there is some change in flavor, and the extracted caffeine is generally used in other commercial preparations.
- In 1938, the Nestlé company invented freeze-dried coffee, which was marketed as Nescafé. Today, Nescafé Instant is the best-selling coffee in the world. The process of preparing freeze-dried coffee begins with brewing extra-strong coffee. This is cooled to a slush of coffee and ice crystals, and placed in a vacuum chamber where the ice is removed. The concentrated liquid coffee is then dried at below-freezing temperatures, producing the distinctive chips of freeze-dried coffee.
- About 10 000 arabica beans are required to make a kilogram of coffee (4000 beans for a pound). This is equivalent to the yearly production of two trees (one tree producing about a pound of coffee annually).
- The 31 August 1990 issue of the journal *Science* noted that of the over 800 chemicals present in roasted coffee, the safety of only 21 had been tested on animals, and 16 of these were capable of causing cancer in rats.
- In 1992, 79-year-old Stella Liebeck of Albuquerque, New Mexico, was in the passenger seat of her grandson's car when a Styrofoam cup of McDonald's coffee that she was holding between

her knees spilled into her lap. A doctor determined that she suffered third-degree burns, and she was hospitalized for 8 days. Hundreds of claims had previously been made against McDonald's, but the resulting legal case caught the public's attention, many perceiving that this was simply a comical example of excessive litigiousness. Testimony indicated that the restaurant had held its coffee at between 82 °C and 88 °C (180–190 °F) on the theory that this resulted in the best flavor, although far too hot to drink until cooled. Other establishments sell coffee at substantially lower temperatures, and coffee served at home is generally 57–60 °C (135–140 °F). The jury awarded Liebeck $200,000 in compensatory damages, reduced to $160,000 because it was found that she was 20% at fault in the spill. The jury also awarded Liebeck $2.7 million in punitive damages, an amount subsequently reduced in a secret settlement. Coffee temperature at many McDonald's restaurants also was subsequently reduced. The case is notable in that it stimulated numerous other claims by consumers alleging damages caused by large corporations.

- The Italian anise-flavored liqueur Sambuca is often served flaming, with three coffee beans floating on top (colorfully described in Italian as *tre mosche*—"three flies").
- A range of exotic flavors results from fermenting beans in the stomach of civets, parrots, monkeys, and other animals. Kopi Luwak, the world's rarest coffee (only about 227 kg or 500 pounds are produced annually), costing about $660 per kg ($300 per pound), is literally the end product of a cat-like marsupial, the palm civet (*Paradoxurus hermaphroditus*) or luwak in Indonesian, which loves eating coffee berries. The enzymes in the animal's stomach add a unique flavor during the fermentation process. Some have dubbed Kopi Luwak coffee as "crappacino."
- Today, coffee is second only to oil as a world trade item. On occasion, the World Bank has accepted coffee in lieu of currency.
- More pesticides are used on coffee than on any other product consumers eat or drink. However, tobacco and cotton require higher levels of pesticides.
- The lids for take-out polystyrene coffee cups have been the subject of extraordinary technological development. The Solo Company "Traveler" design, which was adopted by the Starbuck's chain, represented a great step forward in the 1980s. This was invented by Jack Clements of Ada, Oklahoma, and provides an area for the drinker's lip and nose by raising the level of the lid above the cup. The lid also offered a kind of overflow reservoir, useful for foam-topped gourmet coffees. Dozens of patented polystyrene coffee-cup lids are now available, and major food chains generally have their own design. Quite recently, engineer Al Bibeau and other designers at Sweetheart's Owings Mill, Maryland, Factory and Engineering Center, trademarked a lid called "Gourmet," which resembles a miniature modern civic arena, with gently angled, curving lid, and has been called the most visually elegant of lids.
- The "Java Log" is a 2-kg (5-pound) synthetic log for fireplaces made partly out of coffee grounds. It was invented by Rod Sprules and his wife Joanne Johnson of Ottawa, Canada. It is said that "some people smell the coffee, others don't."
- In 2001, Brazil issued a coffee-scented stamp to promote its coffee.
- Silver polish can be used to remove stubborn coffee stains from plastic cups. Rubbing with a damp cloth dipped in baking soda may also work. Egg yolk mixed with warm water has been employed to remove coffee and tea stains from fabrics.
- The risk of gout has been found to be about 59% lower for men who drink 6 cups of coffee or more per day compared to men who drink no coffee. Gout is a painful, common form of inflammatory arthritis owing to accumulation of uric acid crystals that often settle in the large joint of the big toe. The disease affects about 10% of men over 50 and 10% of women over 60. Drinking tea, and indeed consumption of caffeine, appears not to have any effect on gout. (Reference: Choi, H.K., Willett, W., Curhan, G. 2007. Coffee consumption and risk of incident gout in men: a prospective study. Arthritis & Rheumatism, **56**: 2049–2055.)

Key Information Sources

Ahmad, J., and Vishveshwara, S. 1980. *Coffea liberica* Bull ex Hiern: a review. Indian Coffee, **44**: 29–36.

Cheney, R.H. 1925. Coffee: a monograph of the economic species of the genus *Coffea* L. New York University Press, New York, NY. 244 pp.

Clarke, R.J., and Macrae, R. (*Editors*). 1985–1988. Coffee. Elsevier Applied Science Publishers, New York, NY. 6 vols. [1, Chemistry; 2, Technology; 3, Physiology; 4, Agronomy; 5, Related beverages; 6, Commercial and technico-legal aspects].

Clarke, R.J., and Vitzthum, O.G. 2001. Coffee: recent developments. Blackwell Science, Malden, MA. 246 pp.
Clifford, M.N., and Willson, K.C. 1985. Coffee: botany, biochemistry, and production of beans and beverage. AVI Publishing, London, UK. 457 pp.
Coste, R., and Cambrony, H. 1992. Coffee: the plant and the product. Macmillan, London, UK. 328 pp.
David, J., Schapira, D., and Schapira, K. 1996. The book of coffee and tea: a guide to the appreciation of fine coffees, teas, and herbal beverages. 2nd rev. edition. St. Martin's Griffin, New York, NY. 313 pp.
Davis, A.P., Chester, M., Maurin, O., and Fay, M.F. 2007. Searching for the relatives of *Coffea* (Rubiaceae, Ixoroideae): the circumscription and phylogeny of Coffeeae based on plastid sequence data and morphology. Am. J. Bot. **94**: 313–329.
Davis, A.P., Govaerts, R., Bridson, D.M., and Stoffelen, P. 2006. An annotated taxonomic conspectus of the genus *Coffea* (Rubiaceae). Bot. J. Linn. Soc. **152**: 465–512.
Dicum, G., and Luttinger, N. 1999. The coffee book: anatomy of an industry from the crop to the last drop. New Press, New York, NY. 196 pp.
Farooqi, A.A., and Sreeramu, B.S. 1999. Coffee. *In* Tropical horticulture, vol. 1. *Edited by* T.K. Bose, S.K. Mitra, A.A. Farooqui, and M.K. Sadhu. Naya Prokash, Calcutta, India. pp. 474–502.
Flament, I., and Bessière-Thomas, Y. 2002. Coffee flavor chemistry. Wiley, New York, NY. 410 pp.
Pendergrast, M. 1999. Uncommon grounds: the history of coffee and how it transformed our world. Basic Books, New York, NY. 458 pp.
Robinson, J.B.D. 1986*a*. Tabular descriptions of crops grown in the tropics: 11. Arabica coffee (*Coffea arabica* L.). CSIRO, Institute of Biological Resources, Division of Water and Land Resources, Canberra, Australia. 53 pp.
Robinson, J.B.D. 1986*b*. Tabular descriptions of crops grown in the tropics: 12. Robusta coffee (*Coffea canephora* Pierre ex Froehner). CSIRO, Institute of Biological Resources, Division of Water and Land Resources, Canberra, Australia. 52 pp.
Söndahl, M.R., and Lauritis, J.A. 1992. Coffee. *In* Biotechnology of perennial fruit crops. *Edited by* F.A. Hammerschlag and R.E. Litz. CAB International, Wallingford, UK. pp. 401–420.
Sosef, J.S.M., and Boer, E. 2000. *Coffea liberica* Bull ex Hiern. *In* Plant resources of South-East Asia, 16, stimulants. *Edited by* H.A.M. van der Vossen and M. Wessel. Backhuys, Leiden, Netherlands. pp. 74–78.
Spiller, G.A. 1998. Caffeine. CRC Press, Boca Raton, FL. 374 pp.
Topik, S.C. 2000. Coffee. *In* The Cambridge world history of food. *Edited by* K.F. Kiple and K.C. Ornelas. Cambridge University Press, Cambridge, UK. pp. 641–653.
Ukers, W.H. 1948. The romance of coffee; an outline history of coffee and coffee-drinking through a thousand years. Tea and Coffee Trade Journal Co., New York, NY. 280 pp.
Uribe Compuzano, A. 1954. Brown gold; the amazing story of coffee. Random House, New York, NY. 237 pp.
Van der Vopssen, H.A.M., Mawardi, Soenaryo [no initial], and Mawardi, S. 2000. *Coffea* L. *In* Plant resources of South-East Asia, 16, stimulants. *Edited by* H.A.M. van der Vossen and M. Wessel. Backhuys, Leiden, Netherlands. pp. 66–74.
Weinberg, B.A., and Bealer, B.K. 2001. The world of caffeine: the science and culture of the world's most popular drug. Routledge, New York, NY. 394 pp.
Wellman, F.L. 1961. Coffee; botany, cultivation and utilization. Leonard Hill, London, UK. 488 pp.
Wrigley, G. 1988. Coffee. Longman Scientific & Technical, Harlow, Essex, UK. 639 pp.
Wrigley, G. 1995. Coffee. *In* Evolution of crop plants. 2nd edition. *Edited by* J. Smartt and N.W. Simmonds. Longman Scientific & Technical, Harlow, Essex, UK. pp. 438–449.

Specialty Cookbooks

(Note: There are hundreds of books dedicated to preparing coffee, and several to cooking with coffee. The following are examples.)

Albright, B., and Uher, J. 1999. Maxwell House coffee drinks & desserts cookbook: from lattes and muffins to decadent cakes and midnight treats: original recipes. Clarkson Potter, New York, NY. 208 pp.
Calvert, J.S. 1994. Coffee: the essential guide to the essential bean. Hearst Books, New York, NY. 143 pp.
Castle, T.J. 1991. The perfect cup: a coffee lover's guide to buying, brewing, and tasting. Aris Books, Reading, MA. 244 pp.
Darroch, N. 1978. Cooking with coffee. Hart Publishing Co., New York, NY. 112 pp.
Davids, K. 2001. Espresso: ultimate coffee. 2nd edition. St. Martin's Griffin, New York, NY. 180 pp.
Foster, C. 1992. Cooking with coffee. Simon & Schuster, New York, NY. 160 pp.
Kummer, C. 1995. The joy of coffee: the essential guide to buying, brewing, and enjoying. Chapters Publications, Shelburne, VT. 288 pp.
Olsen, D., Carroll, J.P., and Brody, L.1994. Starbucks passion for coffee: a Starbucks coffee cookbook. Sunset Books, Menlo Park, CA. 96 pp.
Perry, S., and Gowans, E. 1991. The complete coffee book: a gourmet guide to buying, brewing, and cooking. Chronicle Books, San Francisco, CA. 96 pp.
Tekulsky, M. 1993. Making your own gourmet coffee drinks; espressos, cappuccinos, lattes, mochas, and more! Crown Publishers, New York, NY. 96 pp.
Ward, M. 1996. The top 100 international coffee recipes: how to prepare, serve, and experience great cups of tasty, healthy coffee for all occasions. Lifetime Books, Hollywood, FL. 184 pp.

COLA

Family: Sterculiaceae (cacao or sterculia family)

Names

Scientific names: *Cola* species

- Abata cola—*C. acuminata* (P. Beauv.) Schott & Endl.
- Bitter cola—*C. nitida* (Vent.) Schott & Endl.

- The name "cola" is of West African origin, similar to the Temne word *kla*, kola nut.
- Cola is also spelled "kola," and the latter name is used for both of the above species.
- The cola nut has also been known as bissy nut (or bissy-bissy), guru (goora) nut, ombene, and temperance nut.
- "Abata" in "abata cola" is an African name for *C. acuminata*.
- *Garcinia kola* Heckel of the Clusiaceae (Guttiferae) family is also known as bitter kola (and false kola), and is chewed as a social masticatory in Africa. These similarities in name and usage can lead to considerable confusion.

Abata cola (*Cola acuminata*) in flower, from Köhler (1883–1914).

- *Acuminata* in the scientific name *C. acuminata* is Latin for tapering to a narrow point, which is descriptive of the shape of the apex of the leaves.
- *Nitida* in the scientific name *C. nitida* is Latin for shiny, descriptive of the leaves.

Plant Portrait

Cola acuminata and *C. nitida* are the two most commonly cultivated cola species. These are tropical, African, evergreen trees, quite similar to each other. *Cola acuminata* is 7–10 m (23–33 feet), occasionally up to 13 m (43 feet) in height, while *C. nitida* is 8–12 m (26–39 feet), occasionally as tall as 25 m (82 feet). The fruit is a leathery or woody oblong pod, 8–20 cm (3–8 inches) long. The seeds are nearly globose, about 2.5 cm (1 inch) in diameter. The cola "nuts" of commerce are these reddish, fragrant, nut-like seeds, which are rich in the chemicals caffeine and theobromine. The seeds are chewed as a stimulant by local populations, a practice carried out in West Africa for thousands of years. The seeds have been traded to other countries since at least the 14th century, and are exported for commercial use in soft drinks and pharmaceutical medicines. The species are cultivated in West Africa, and also in the West Indies and other tropical countries. Cola nuts are the most important "nut" crop in Africa next to the oil palm.

In Africa, the use of cola has been taken up as a "social lubricant," particularly by Islamic people, who cannot drink alcohol according to their religion. Fresh cola nuts are chewed as a ceremonial greeting, and used to sweeten invitations and indicate congratulations. A host who fails to give cola nuts to departing guests in some cultures commits a serious social blunder. Cola nuts are commonly given as tokens of friendship.

Culinary Portrait

In Western countries, the chief use of cola seeds is as a source of extracts to flavor cola drinks, especially carbonated and flavored soft drinks, which by extension are called colas. Outside of Africa, the French are the chief consumers of cola nuts in cuisine, particularly in desserts, offsetting the bitter taste with sweet substances. Cola particularly seems to complement chocolate in creamy desserts.

Abata cola (*Cola acuminata*) in fruit, from Köhler (1883–1914).

Cola seeds and extracts are widely used in tropical countries as a stimulant and appetizer. Fresh nuts are difficult to find outside of the tropical areas where they are grown, but are sometimes sold at African markets in international cities, like Washington, DC. Cola nuts are in demand for use as a masticatory, i.e., to chew like gum. They can also be eaten and used as condiments. Such uses are largely confined to those who have become habituated during childhood. The fresh nut tastes bitter, but produces a curious physiological change: after eating, a sense of well-being spreads through the body and a sweet taste is present in the mouth, which makes food and beverages taste sweet. Older nuts lose their stimulant value and habitual users seek fresh nuts.

Because of the caffeine content, cola nut should not be consumed by women who are pregnant or nursing, those suffering from insomnia or anxiety problems, and patients with a history of high blood pressure, heart trouble, palpitations, seizures, insomnia, heart disease, high cholesterol, stroke, or stomach or duodenal ulcers. Many doctors advise that children should consume no more than two cans of cola a day. Tumors of the mouth and gut occur relatively often where cola nut chewing is common, especially in northern Nigeria, and it has been suggested that cola nuts are carcinogenic. For additional information on the health effect of caffeine, see COFFEE.

Culinary Vocabulary

- "Sudan coffee" is a caffeine-rich drink made in Africa from roasted, ground cola nuts.
- "Cuba Libre" is a cocktail prepared with cola, rum, and lime juice, and garnished with a wedge of lemon or lime.
- "Dope" is a term often used to refer to illicit drugs, especially marijuana. In the American South, the word has been used to refer to cola drinks because of the stimulating effects of the caffeine content.

Curiosities of Science and Technology

- Cola trees are native to Africa, and Africans have developed diverse uses for the plants: the pods are employed to ease labor pains; the seeds to relieve diarrhea, nausea, and hangover; the bark to heal wounds; the wood to build houses and boats, and carve figurines; and the roots and twigs to clean teeth and sweeten breath.
- Soldiers in Africa have used cola nuts to suppress the need to eat and drink, and provide energy, during long marches. In the 1890s, the Surgeon General of the U.S. Army recommended that infantrymen on forced marches chew cola beans to improve their performance. An early Coca-Cola advertisement claimed that English newspaperman Henry Morton Stanley (1841–1904) might not have found the great Scottish missionary–explorer David Livingstone (1813–1873) and uttered the phrase, "Dr. Livingstone, I presume," were it not for the cola nut, which energized his native bearers as they trekked through Darkest Africa.
- Dried cola nuts have been used as money in Africa.
- Carbonated cola drinks owe their existence to the chemist Joseph Priestley (1733–1804, co-discoverer of oxygen). Using carbon dioxide generated in a brewery, in 1772 he devised a way of impregnating water with the gas, and he subsequently marketed his discovery as "Pyrmont water," alleging it had curative properties. His invention has kept the world burping for over 2 centuries.
- In 1886, John S. Pemberton, an Atlanta druggist seeking a headache and hangover remedy, added cola nut extract to coca extract and produced Coca-Cola, destined to become the most popular nonalcoholic beverage in the Western World, after water. He advertised it as follows: "This 'intellectual beverage' and temperance drink contains the valuable tonic and nerve stimulant properties of

Bitter cola (*Cola nitida*), from Chevalier and Perrot (1911). A seed and a germinating seed are shown at the bottom.

the coca plant and cola (or kola) nuts, and makes not only a delicious, exhilarating, refreshing, and invigorating beverage...but a valuable brain tonic, and a cure for all nervous affections." The original beverage included cocaine leaves and cola beans, both containing stimulants. Coca-Cola did not become completely cocaine free until 1929.

- In 1906, the name "Coca-Cola" was the subject of a lawsuit under the US Pure Food and Drug Act. The company was accused of having a misleading brand because the drink contained neither coca nor cola. In 1909, the issue reached the Supreme Court. Coca-Cola argued that since Grape-Nuts contained neither grapes nor nuts, and since butternuts contained no butter, it did not matter if Coca-Cola did not contain either coca or cola. The company lost its case, and in 1916 was required to put in at least trace amounts of both coca and cola in order to continue to use the name. To avoid losing its trademark, trace amounts were added to the formula.
- In 1898, pharmacist Caleb D. Bradham invented Pepsi-Cola in New Bern, North Carolina. The beverage was first known as "Brad's drink," and was originally used as a cure for peptic ulcers.
- "Coca-Cola" is the best-known trademark, recognized by 94% of the world's population. In the 1930s, Coca-Cola took the position that the word "Cola" was part of its trademark, and sued Pepsi-Cola for infringing on its trademark. The court ruled that the "C" for copyright appeared only in the bottom of the "Coca" part of the word "Coca-Cola" on its products, and so anyone could use "Cola."
- Early cola beverages were sold in glass bottles stoppered with corks. Some of the first bottles used for the purpose had hemispherical bases, so that they could not be stored upright, and as they were stored on their side, this kept the cork moist and expanded for better sealing (side-storage is followed to this day for cork-stoppered wines). Because they could not stand up, the round-based containers were called "drunken bottles." They were replaced with flat-bottomed bottles, but these too were expected to be stored on their sides. Screw stoppers were introduced in the 1870s, making it unnecessary to continue side storage.
- Caffeine is part of many cola drinks, including Coca-Cola. During World War II, caffeine was scarce, and chemists suggested employing uric acid, a close chemical relative of caffeine, which can easily be converted into caffeine, to solve Coca-Cola's shortage of caffeine. They proposed using bat guano (feces), which is extremely rich in uric acid, as their source. The plan was abandoned when it was realized that Coca-Cola would be accused of adding bat excrement to their beverage.

Abata cola (*Cola acuminata*) in a forest in Guinea, Africa, from Jumelle (1901).

- There are more than 1000 Coca-Cola bottling plants in the world. Over 7000 soft drinks from Coca-Cola are drunk each second of the day, accounting for nearly half of all the soft drinks consumed in the world.
- Pouring cola drinks into a toilet has been recommended as a means of giving the porcelain a brilliant shine.
- Jell-O (of Kraft General Foods) produced a cola-flavored gelatin in 1942, but the next year it was discontinued.

Key Information Sources

Abaka, E. [same author as Abaka, E.K., below]. 2000. Kola nut. *In* The Cambridge world history of food. *Edited by* K.F. Kiple and K.C. Ornelas. Cambridge University Press, Cambridge, UK. pp. 684–692.

Abaka, E. 2004. Kola is God's gift: agricultural production, export initiatives & the kola industry in Asante & the Gold Coast c. 1820–1950. James Currey, Oxford, UK. 256 pp.

Abaka, E.K. 1998. Eating kola: the pharmacological and therapeutic significance of kola nuts. Ghana studies. Vol. 1. African Studies Program, University of Wisconsin, Madison, WI. pp. 1–11.

Adeyeye, E.I., and Ayejuyo, O.O. 1994. Chemical composition of *Cola acuminata* and *Garcinia kola* seeds grown in Nigeria. Int. J. Food Sci. Nutr. **45**: 223–230.

Alaribe, A.A.A., Ejezie, G.C., and Ezedinachi, E.N.U. 2003. The role of kola nut (*Cola nitida*) in the etiology of malaria morbidity. Pharm. Biol. **41**: 458–462.

Beattie, G.B. 1970. Soft drink flavours, their history and characteristics. I. Cola or 'kola' flavours. Flavour Ind. **1**(**6**): 390–394.

Brown, D.A.L. 1970. A review of germination of kola seed (*Cola nitida*) (Vent.) Schott & Endl. Ghana J. Agric. Sci. **5**: 179–186.

Cocoa Research Institute of Nigeria. 1971. Progress in tree crop research in Nigeria: cocoa, kola, and coffee. Cocoa Research Institute of Nigeria, Ibadan, Nigeria. 183 pp.

Eijnatten, C.L.M. van. 1969*a*. Kola, its botany and cultivation. Koninklijk Instituut voor de Tropen, Amsterdam, Netherlands. 100 pp.

Eijnatten, C.L.M. van. 1969*b*. Kolanut (*Cola nitida* and *C. acuminata*). *In* Outlines of perennial crop breeding in the tropics. *Edited by* F.P. Ferwerda and F. Wit. Misc. Paper 4. Landbouwhogeschool, Wageningen, Netherlands. pp. 289–307.

Eijnatten, C.L.M. van. 1976. Kola: *Cola* spp. (Sterculiaceae). *In* Evolution of Crop Plants. *Edited by* N.W. Simmonds. Longman Scientific & Technical, Burnt Mill, Harlow, Essex, UK. pp. 284–285.

Einjnatten, C.L.M. van, and Roemantyo [no initial]. 2000. *Cola* Schott & Engl. *In* Plant resources of South-East Asia, 16, stimulants. *Edited by* H.A.M. van der Vossen and M. Wessel. Backhuys, Leiden, Netherlands. pp. 78–83.

Emmins, C. 2000. Soft drinks. *In* The Cambridge world history of food. *Edited by* K.F. Kiple and K.C. Ornelas. Cambridge University Press, Cambridge, UK. pp. 702–712.

Frederick Stearns & Co. 1894. Kola. Stewart, Detroit, MI. 78 pp.

Heckel, É.M., Neish, J., and Chittendon, R.H. 1895. Kola: a study of its history, physiological and therapeutic action, pharmacy and uses in medicine. Johnson & Johnson Publishing, New Brunswick, NJ. 69 pp.

Ibikunle, B.O., and Mackenzie, J.A. 1974. Germination of kola (*Cola nitida* (Vent) Schott Endl). Turrialba, **24**: 187–192.

Idigo, A.C. 2002. Oji: *Cola acuminata*—Oji Igbo: the cornerstone of Igbo traditional ceremonies. Snaap Press, Enugu, Nigeria. 62 pp.

Mason, J. 1982. Dida obi (kolanut divination). Yoruba Theological Archministry, New York, NY. 20 pp.

Nzekwu, O. 1961. Kola nut. Nigeria Magazine (Lagos), **71**: 298–305.

Odegbaro, O.A. 1973. Regeneration of old kola trees, *Cola nitida* (Vent) Schott and Endlicher, by coppicing. Turrialba, **23**: 334–340.

Ogutuga, D.B.A. 1975. Chemical composition and potential commercial uses of kola nut, *Cola nitida*, Vent. (Schott and Endlicher). Ghana J. Agric. Sci. **8**: 121–125.

Oladokun, M.A.O. 1989. Nut weight and nutrient contents of *Cola acuminata* and *C. nitida* (Sterculiaceae). Econ. Bot. **43**: 17–22.

Oladokun, M.A.O., and Adedipe, N.O. 1988. Some comparative morpho-physical features of kola, *Cola* spp. Philippine J. Crop Sci. **13**(**2**): 69–76.

Oludemokun, A.A. 1982. Processing, storage and utilization of kola nuts, *Cola nitida* and *C. acuminata*. Trop. Sci. (U.K.), **24**(**2**): 111–117.

Parke, Davis & Company. 1895. The pharmacology of kola. 3rd edition. Parke, Detroit, MI. 43 pp.

Quarcoo, T. 1973. A handbook on kola. Cocoa Research Institute of Nigeria, Ibandan, Nigeria. 90 pp.

Riedel, H.R. 1988. The use of cola nuts in confectionery. Confectionery Prod. **54**(**10**): 629–631, 633.

Russell, T.A. 1955. The kola of Nigeria and the Cameroons. Trop. Agric. (Trinidad), **32**: 210–240.

Sundström, L.1966. The cola nut: functions in West African social life. Stud. Ethnographica Upsaliensia (Uppsala), **26**(**2**): 135–142.

Specialty Cookbooks

Graham, E.C., and Ralph Roberts, R. 1998. Classic cooking with Coca-Cola. Hambleton-Hill Publishing, Nashville, TN. 195 pp.

Publications International, Ltd. 2000. Coca-Cola cool recipes. Publications International, Ltd., Lincolnwood, IL. 94 pp.

Corn (Maize)

Family: Poaceae (Gramineae; grass family)

Names

Scientific name: *Zea mays* L.

- The English word "corn" has been traced to an Indo-European word of uncertain spelling ("grn"?), meaning "small nugget." *Granum*, meaning grain in Latin, has the same root. By extension, this came to be applied to the grain of cereals (a cereal is any edible grass seed).
- The concept of "corn" as a "small particle" explains why "corn" is in some words that have nothing to do with corn (maize). A "peppercorn" is a dried pepper berry. "Corned beef" is cooked beef preserved with "corns" or grains of salt.
- In the Old World, the name "corn" has been used for the leading cereal crop of a major region. In England, corn means wheat; in Scotland and Ireland, oats. In North America, corn means corn, although wheat is the leading cereal.
- The word "maize" traces to Columbus' 1492 expedition to the New World. Columbus reached some island (not determined with certainty) in the northern Antilles, near San Salvador, populated by Tahino people, who used the word *mahis*, meaning "source of life," for their staple cereal, corn. The Spanish distributed corn along with the original name. *Mahis* was transmuted phonetically into *maíz* in Spanish and maize in English. *Mays* in the scientific name is also based on the Indian name.
- Corn was a novelty in Europe in the first half of the 16th century, and it was often not appreciated that it originated in the New World. Such names as Welsh corn, Asiatic corn, and Turkish corn were used. "Turkish" was commonly applied to new foods because this was the time that the Turks had invaded Europe, bringing new foods with them. This is the reason that the "turkey" to this day bears the name of the country Turkey, although the bird comes from North America.
- Variety Massangeana of the popular house plant *Dracaena fragrans* (L.) Ker-Gawl is widely known as "corn plant" or "cornstalk plant" because of a superficial resemblance to *Zea mays*. It is not edible and has been the cause of poisoning of cats and dogs.
- The genus name *Zea* is based on the Greek word *zea*, referring to some kind of cereal (not *Zea mays*).

Plant Portrait

Corn, also known as sweet corn, maize, and Indian corn, is an annual herbaceous grass, varying in height from 0.5 to 7 m (2–22 feet). Corn plants have separate male and female flowers on the same plant. The male flowers, corresponding to the "tassels," are usually located at the top of the main shoot. The female flowers, making up the "ear," are located in the axils of lower leaves. Ears of corn usually have an even number of rows, normally between 8 and 24. Cultivated corn includes types used for human food, as well as for animal feed and silage. Wild forms are known from Mexico and Guatemala and, like three exclusively wild species of *Zea*, are termed "teosinte." Domesticated corn probably originated in southern or central Mexico. Well-known evidence for early use of the genus *Zea* has come from 7000-year-old dry caves in Mexico's Tehuacán Valley, and still earlier finds have been uncovered. Corn, long used as a source of flour by native Americans in North and South America, remains a staple in the diet of many native groups.

Sweet-tasting corn, which arose from time to time, was preserved by native Americans, many of whom used corn daily for food. Sweet corn is distinguished from other corns by genes for high sugar content. In 1779, a sweet corn was brought back from an expedition against the Six Nations Indians, from the Susquehanna region of Pennsylvania. It was called "papoon" by the Indians. The first commercial cultivar of sweet corn, developed from this source, was released in the northeastern United States in 1832. Since then more than 2000 cultivars have been selected, first as predominantly white-kerneled varieties and, after the release of Golden Bantam in 1902, increasingly yellow-kerneled cultivars.

Most corn grown in developed nations today is hybrid corn. Corn hybrids often produce higher yields and are relatively uniform, making harvesting and processing easier. Modern corn hybrids also may respond better to fertilization. Extensive research has resulted in the hybrid varieties used today, which are generated from well-defined inbred lines. Farmers are dependent on seed supply companies for hybrid corn seed because the kernels produced on hybrid corn plants are genetically heterogeneous, and will not produce the uniform, desired characteristics of these plants.

Several basic categories of corn are recognized. "Floury" varieties have soft, mealy starch. "Sweet" corn has a large amount of sugar. "Waxy" corn has starch composed of amylopectin. "Flint" corn has hard starch. "Pop" corn is a very hard type of flint corn. "Dent" corn has a softer kernel. Dent corn is the most important commercial type of corn grown in the United States, and indeed accounts for 95% of all of the world's corn. It is mostly yellow or white, the kernel forming a dent on the crown at maturity (the small indentation is due to shrinkage of the endosperm, i.e., the starchy core). However, dent corn is usually not eaten fresh, but is mostly used as livestock feed, in food processing industries, and to prepare starch and alcohol. Hybrid sweet corn, the predominant type eaten fresh, was developed in the 1930s, which has led to high-quality, uniform, and productive corn cultivars. In the 1970s, the "supersweet" corns were introduced. Today, three classes of sweet corn are recognized, normal, sugar-enhanced, and supersweet. Each contains three color groups: yellow, yellow-and-white (peaches and cream), and white. The sugar content in the normal or less-sweet group is between 9 and 16%, compared to 14–44% in sweeter varieties.

Of all the crops in the world, the value of corn is exceeded only by wheat, and possibly also rice. Corn is extensively grown in temperate and semitropical areas. It is the largest crop in the United States, both in terms of area planted and the value of the crop produced. The United States produces about 40% of the world's total corn crop, followed by China, with 20%. The US "Corn Belt" is a region of the Midwest, including the north-central plains. It is centered in Iowa and Illinois, and extends into southern Minnesota, southeastern South Dakota, eastern Nebraska, northeastern Kansas, northern Minnesota, Indiana, and western Ohio. Corn acreage once was larger than for any other crop, and the region still produces much of the US corn crop, but other crops, including soybean and alfalfa, are now also important. The region is now commonly called the Feed Grains and Livestock Belt. The top three corn-producing states, Iowa, Illinois, and Nebraska, produce almost 50% of the corn in the country. Only about 10% of corn grown in the United States is eaten as a vegetable. About 50% is used as livestock feed, some cultivars having been especially selected for the purpose. The remaining 40% is used to make hundreds of products, ranging from alcohol to industrial compounds. On a world basis, about one-fifth of all corn grown is eaten directly by people, two-thirds is consumed by livestock, and about one-tenth is used as a raw material in manufactured goods. Corn is one of the most widely used crop plants. In addition to foods, corn constituents are used to manufacture furfural, acetone, maltose, butyl alcohol, paper, adhesives, textiles, charcoal, medicines, fuels, and other products.

Culinary Portrait

Corn has been used as human food for thousands of years, usually ground as flour for breads, cakes, and other staple foods. When immature, it was commonly used as a vegetable by eastern American woodland Indians. Such consumption of whole immature cobs is seen currently in some parts of the world, such as in Southeast Asia, and the small whole cobs are often served as a cocktail or gourmet dish. Sweet corn is an important fresh and processed vegetable in North America. Native Americans were also familiar with popping corn, which is now a familiar snack food. Tortillas—thin, unleavened bread cakes—are traditionally made in Mexico simply by boiling kernels and mashing them into a paste, which is used as a starting dough. Today, corn is processed to provide such items as corn chips, tortillas, breakfast cereals, soft drinks, ice cream, peanut butter, salad dressing, and gelatin. Some principal edible products include corn oil (extracted from the germ, i.e., the embryo of the grain), cornmeal, cornstarch, corn sugar, and corn syrup. Corn oil is polyunsaturated, therefore desirable for low-cholesterol diets. It is excellent for frying and for making pastries and margarine, and is occasionally used as a salad oil. Cornmeal is ground grains of corn, and is used to make bread, cakes, and pastries. Several European mush dishes are made from cornmeal, including Romanian *mamaliga*, Italian *polenta*, and French *armotte*. The African "mealie porridge" is also a cornmeal mush. Cornstarch is a fine white powder extracted from corn. It has little taste and is chiefly used to thicken liquids, for example, in preparing soup, pudding, sauces, and pie fillings, and is employed in many Chinese dishes. Unlike ordinary flour, cornstarch does not contain gluten, and so has less of a tendency to form lumps, and makes a better thickener than flour. (To prevent lumps from forming when adding cornstarch to a hot mixtures, it may first be dissolved in a cold liquid.) Cornstarch lightens the texture of cakes and pastries when mixed at a ratio of one part to four parts flour. Cornstarch also is used in glazes, as a coating for foods that are deep-fried, and as a binder with minced meats. Some cooks advise that cornstarch should first be cooked for at least 1 minute to rid it of bitterness. "Modified cornstarch" has been treated to alter its culinary properties. Corn sugar is produced from cornstarch, and is used principally in industrial processes, such as manufacturing beer, vinegar, and caramel. Corn syrup is obtained by heating cornstarch with certain acids. It is used to sweeten numerous other products. Corn syrup does not crystallize or disintegrate or become grainy when cooked, and stays soft and elastic, and for all of these reasons, it is especially useful in the candy industry.

People have become so used to convenience that they now shy away from foods that need preparation before cooking. For some, even husking a cob of corn is too much hassle. But for those willing to prepare corn properly, the following tips are offered. Completely ripe corn has bright green, moist husks, and one should be able to feel the individual kernels inside by pressing gently against the husk. Corn is best cooked and eaten immediately after picking, but can be kept moist in a plastic bag and refrigerated for up to 3 days (corn can lose half its flavor in 1 day at room temperature during a hot summer day). *To boil*, remove husk and bits of silk, trim away the stem end and remove any disfigured kernels. Drop the cobs in a large pot of water brought to a boil on high heat. (Some cooks add a tablespoon of sugar to the pot to improve the taste. Adding salt, however, has been discouraged in the belief that it makes the corn tougher.) After the water returns to a boil, cook 3 minutes for young, pale cobs or 4–5 minutes for mature cobs. Remove the cobs immediately and keep them warm (for example, place in linen napkins on a platter). *To pressure cook (steam)*, use 1 cup of water, cook for 3–5 minutes. *To barbecue*, leave the husk on the cob, place the whole cobs on the barbecue over medium heat. Cover with a barbecue lid and cook 5 minutes. To check doneness, remove a strip of husk. When complete, remove the cobs, cool them for several minutes, and pull off the husks with the silk. Alternatively, the corn can be husked and wrapped in aluminum foil before being barbecued. *To bake*, cook the corn in their husks for 35 minutes at 218 °C (425 °F). Alternatively, remove the husk and wrap in aluminum foil. *To microwave*, leave the husk on the cob, place the cobs in the microwave and cook one or two cobs on high for 2 minutes per cob, and let stand 2 minutes per cob. Corn is traditionally flavored with butter and salt.

Corn kernels can be removed from the cob (before or after cooking), and added to mixed vegetables, soups, stews, and salads.

Corn whisky is prepared from a mash of 80% or more corn. Corn is also the principal grain in the mash for Bourbon whisky, but by law in the United States, the mash need only be 51% corn. Bourbon is considered to be the all-American whiskey. Several Kentuckians have been credited with its invention, perhaps most often a Baptist minister, the Reverend Elijah Craig, in 1789.

Caramel is produced by cooking sugar until it becomes a thick, dark liquid, the color ranging from golden to dark brown. Caramel corn is corn covered with caramel. This traditional snack food is still consumed at public entertainment events, but has lost some of its former popularity, compared to buttered or butter-flavored popcorn. Cracker Jack, a mixture of caramelized corn and peanuts, was created in Chicago in 1872 by brothers Fred and Louis Ruekheim, German immigrants.

Corn husks are the best-known type of food wrapper. Mexican cuisine uses a great range of leaves and husks as wrappers in which to cook foods, but the corn husk is the most important. Corn husks are usually used to make masa-filled tamales (masa is corn dough), but can be used to wrap different vegetables and fish.

One of the most unusual of foods is the Mexican preparation *cuitlacoche* (also *huitlacoche*; derived from the Aztec word *cuitlatl*, waste, + *cochi*, black). This is based on galls or tumors from corn ears infected with corn smut, a plant disease caused by the fungus *Ustilago maydis*. Considered a great delicacy in Mexico, the prepared tumors are used to flavor dishes. In Mexico, the tumors are picked before they can ripen and distribute their spores, and the resulting black fluid that is produced on cooking is said to have a delicious mushroom flavor. The tumors are thought to be poisonous when ripe.

The disease pellagra was formerly associated with poor people who principally ate corn. Pellagra is due to a deficiency of niacin, also known as vitamin B3, which is substantially lacking in corn, and needs to be provided by other foods. Niacin-rich foods include liver, whole grains, eggs, avocados, peanuts, fish, and meats. Niacin maintains normal function of the skin, nerves, and digestive system, as well as providing other benefits. Normally, 50–80% of the niacin in corn can not be taken up by people eating it. However, Native Americans learned to add lime, caustic soda, or ashes to corn while preparing it, which considerably increases the amount of niacin assimilated. Today, pellagra is very rare in the Western World, found principally in alcoholics and people suffering from severe gastrointestinal problems. Efforts are underway to breed corn with higher levels of niacin, as there are still areas of the world where excessive dependence on corn leads to the development of pellagra.

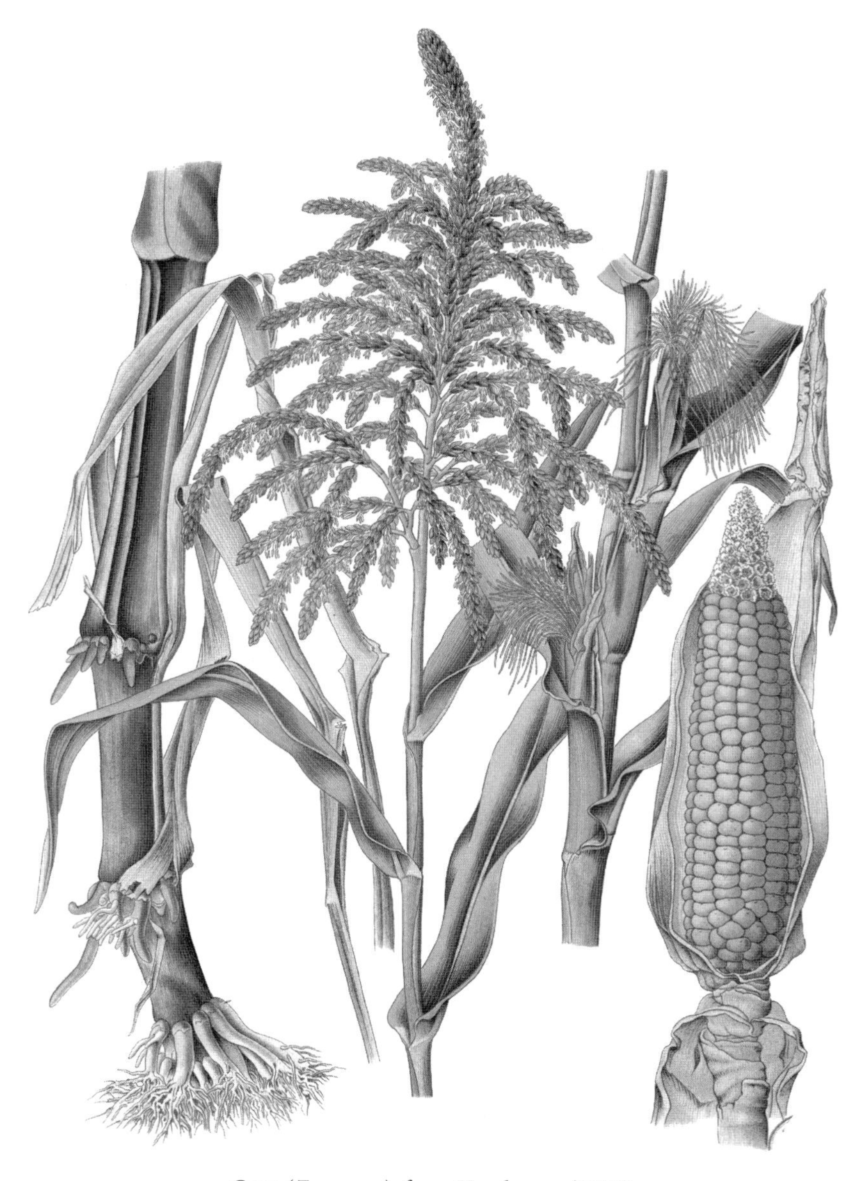

Corn (*Zea mays*), from Henderson (1890).

Culinary Vocabulary

- "Grit," meaning "very small, tough grains," is derived from "grn," as discussed above for the origin of the word corn. The word is related to grist, as in grist mill, and traces through *grytte*, a Middle English word meaning coarse or coarsely ground. "Grits" as a general word means coarsely ground, hulled grain. In the Deep South of the United States, where grits are a staple, the word specifically means coarsely ground hulled corn (also known as hominy and hominy grits), generally cooked as a cereal or a side dish to the Southern breakfast of eggs, bacon or country ham, red-eye gravy, and biscuits.
- "Hominy," another name for grits, is a word of native American origin. The Algonquian *appuminnéonash*, "parched corn," has been suggested as the source; so has the Algonquin word *tackhummin*, "hulled corn."
- Grits is plural (according to a Southern saying, "no one can eat just one"), but in practise is often used as a singular word ("grits is good").
- "Corn" is an obsolete American word for moonshine or white lightning, i.e., illicitly distilled liquor made from corn, sugar, and water.
- "Cornflour" is generally interpreted as a very fine flour ground from corn, although sometimes finely ground flour of other cereals, such as rice, is also called cornflour. In Britain, the term cornstarch may mean cornflour.

- "Corn germ" is the embryo of the corn kernel. It is quite fatty, containing almost half of the calories of corn, and because the fat tends to become rancid on storage, products like cornmeal and cornflour are usually made from degermed corn. Corn germ is sometimes available in stores, generally in hermetically sealed packages that need to be refrigerated once opened. Corn germ is crunchy, has a flavor reminiscent of hazelnut, and may be used to enrich a variety of dishes.
- "Cajun popcorn" is not popcorn. It is an appetizer prepared with spicy, shelled, battered, deep-fried crawfish or shrimp.
- "Johnnycake," in America where it originated, is a griddle cake made of cornmeal, salt, and boiling water or cold milk (an independent meaning of johnnycake is a Caribbean breakfast food made from flour, water, salt, and baking powder, shaped into balls and fried). The name may be derived from an Indian word for flat cornmeal cakes, *joniken*; or from "Shawnee cake," after the Indian tribe of the Tennessee Valley; or from the Dutch *pannekoeken*. It has also been claimed that this flat white cornmeal bread was named after the "Johnny Rebs" (i.e., Confederate soldiers) of the American Civil War because they ate so much corn pone. However, a hard, cornmeal-based bread called "jonny cake" was also eaten during the American Revolutionary War (1775–1783). Johnnycake is also called "journey cake"—some say because Johnny is pronounced like journey in New England, others say because it could be easily prepared during a journey. "Hoe cake" or "Johnny hoe cake" is a variation that is said to have originally been cooked on a hoe.
- "Polenta" (pronounced poh-LEHN-tah; the Italian word for cornmeal) is a traditional North-Italian dish made by cooking cornmeal with a liquid until it forms a soft mass, grilled or fried, cut into squares, and eaten hot or cooled. This cornmeal mush or porridge is much like the type of corn preparation made by Indigenous North Americans for thousands of years. Today, this extremely versatile food is served in countless ways, for example, topped with melted mozzarella cheese, or seasoned with a tomato sauce, or covered with ground meat. In the American South, a dish called polenta is made with grits instead of cornmeal.
- Extremely popular in the southern United States, "corn pone" is an eggless cornbread that is shaped into small ovals and fried or baked. The pone comes from a Native American word for "bread," related to the Delaware Indian word *äpân*, "baked." In 1960, Lyndon B. Johnson (1908–1973; president 1963–1969) named his campaign train, which helped hold the South for the Kennedy–Johnson ticket, "The Corn Pone Special."
- Bourbon whisky, made principally from corn, commemorates Bourbon County, originally in Virginia and later part of Kentucky. The Bourbons were European monarchs, and some of their subjects settled in the Americas.
- "Tiswin" is a fermented beverage made by Apaches from corn, brown sugar, and spices. New Mexican Indians call the drink "tiswino" and "tesquino."
- "Barefoot bread" is a southern United States expression for corn pone.
- "Anadama bread" is a colonial American yeast bread flavored with cornmeal and molasses. According to a legend explaining the origin of the name, a Gloucester, Massachusetts fisherman's wife named Anna fed her husband only cornmeal and molasses, until one night he became angry, added yeast and flour, and prepared a bread in the oven while muttering "Anna, damn her!" Another, more kindly legend holds that the name is related to a sea captain's wife who was famous for her corn-meal-and-molasses bread, and whose gravestone was inscribed with the commemoration "Anna was a lovely bride, but Anna, damn'er, up and died."
- The southern US specialty "hush puppies" are deep-fried small cornmeal dumplings flavored with chopped scallions. Served with fried catfish, cornmeal hush puppies are considered one of the South's culinary delicacies. The unusual name of hush puppies is said to have arisen following the Civil War when food was in short supply and cooks used to toss scraps of the fried batter to whimpering dogs while saying "Hush, puppy!" An alternative explanation is based on consumption of salamanders in the southern United States, which were called "water puppies" and "midpuppies" because of their short, stout legs. These, it is claimed, were deep-fried with cornmeal dough. They were given the euphemistic name hush puppies to mask their lowly nature, and subsequently, the name persisted after the salamanders were no longer consumed.
- The highly popular "corn dog" is prepared by dipping hot dogs in a cornmeal batter, quickly deep-frying, and serving on a stick with mustard or other condiment. The corn dog was first marketed at the 1942 Texas State Fair, by Neil and Carl Fletcher, and was called the "Fletchers' Original State Fair Corny Dog."

- A "corn dodger" (= dodger) is a small oval cake of baked or deep-fried corn bread, traditionally made in the southern United States. It is made from cornmeal or cornflour (i.e., finely ground cornmeal), water or milk, oil or lard, and usually sugar. These were sometimes so hard when first baked that if thrown, the intended victim "dodged" to avoid injury, resulting in the name.
- "Ceviche," an Ecuadorian seafood dish marinated in lemon and onions, is traditionally served with popcorn.
- "Choclo" (pronounced choh-cloh) refers to morsels of corn on the cob that are added to soups and stews in South America.
- The expression "to shell corn" is odd when one considers that corn does not have shells. It has been speculated that this usage traces to the times when North American Indians stripped off corn kernels with mussel shells.
- "Cornhusker" is an especially moist, soft cheese made in America from cow's milk. Similar to Colby, it was developed by the Nebraska Agricultural Experiment Station in 1940.
- "Seven-minute frosting" is a fluffy meringue frosting prepared with beaten egg white, sugar, and corn syrup.
- "Potlikker" (pot liquor) is the liquid remaining when green, peas, pork, and other items are boiled, and this vitamin-rich broth has long been popular in the southern United States, especially in combination with corn pone, which serves as a sponge to mop it up (reminiscent of coffee and doughnuts). Over the years, an argument has persisted whether corn pone should be dunked into the broth to produce the best results, or whether corn pone should be crumbled into the potlikker. At the 1932 Democratic National convention, Franklin D. Roosevelt (1882–1945, 32nd president of the United States) joked that the issue should be referred to the platform committee to decide on whether a vote was in order.

Curiosities of Science and Technology

- Chicha is an alcoholic drink of South America, particularly consumed in the Andes, where brewing of the beverage started with the Incas. The Incas used chicha in paying homage to the gods. Virgin women, in the tradition of modern Catholic nuns and the vestal virgins of ancient Rome, lived next to the temples where they brewed chicha and served it to the sun in a gold cup every morning, as well as to the priests. The women made the chicha by chewing the corn and spitting it into containers. It has been suggested that the enzyme from their saliva helped the grain to ferment, and it was believed that older women's saliva made for a tastier and more potent brew. In recent times, people have hired themselves out as specialists in chewing corn for the production of chicha.
- Corn was used as currency by Indians of Peru.
- The sugar-filled leaves of the corn plant were used by Indians as chewing gum.
- Corn, squash and beans were known as the "Three Sisters" by the Native Americans—sisters who should be planted together. It is now clear that collectively these three vegetables provide a superb balance of the nutrients required by humans.
- Blue is unusual as a food color, but the ancient southwestern US Pueblo tribes were fascinated with blue corn. Anthropologists have speculated that this was because the Indians associated the corn with spiritual aspects of the blue sky.
- Piki bread, a thin cornmeal bread often made from blue corn batter on a hot stone, was the bread of several southwestern US Indian tribes. Before being considered suitable for marriage, young women were once required to demonstrate that they had mastered the ability to prepare this bread.
- Today, there is often little appreciation of how extensive First Nation villages were in North America. French explorer Samuel de Champlain (1567?–1635), on wandering about Huronia (the area occupied by the Hurons, north of Toronto, Canada), recorded that he became lost far more frequently in cornfields than in the forests of the region.
- Early colonial farmers in North America fertilized corn by adding a fish to each hill, a practice originated by Native Americans.
- Sybilla Masters (?–1720), who divided her time between England and the American colonies, is known as the "first American woman inventor." She and her husband Thomas applied for a British patent for her invention for a new corn mill (the United States did not have its own patent office until 1793). They received British patent number 401 of 1715 (actually secured in 1716), filed in Thomas's name because women could not be patent holders. She used hammers instead of grinding wheels to make hominy meal from corn, and called the product "Tuscarora rice." (Tuscarora is a town in Pennsylvania, commemorating the Tuscarora Indians of the Iroquois Nation; the Masters were

Corn (*Zea mays*), from Henderson (1890).

merchants in Pennsylvania). The patent is believed to be the first British patent granted to a resident of the American colonies, and may have been the first patent ever granted anywhere for an invention that was made by a woman. The mill worked well, but the enterprise failed because the English would not develop a taste for southern grits. In the American colonies, the Masters sold the cornmeal as a cure for consumption (tuberculosis).

- The first certain account of plant hybridization was provided in a letter written in 1716 by Cotton Mather in which he discussed odd character combination in corn. Mather, a Puritan clergyman in New England, noticed that the windward side of a neighbor's corn (the side most exposed to pollen from other fields) developed multicolored cobs, while the rest of the planting developed yellow cobs.
- In the 1800s, American farmers often used a combination of cornmeal with soap to wash dirt from their hands. The coarse cornmeal helped loosen the dirt and it was believed to reduce chapping.
- It has been claimed that corn facilitated rapid, large-scale European settlement of North America. In contrast to wheat, barley, and rye, a single kernel of corn produces a much greater return than a single seed of the European grains. This large yield per plant made it worthwhile to tend each plant carefully in hills, a practice that is much easier than fully plowing a field, especially for pioneers starting with uncleared land.

- The corncob pipe was patented in 1878 by Henry Tibbs.
- Corn "pops" when heated because the outer part of the storage carbohydrates making up the kernel is hard while the center is soft. When heated, the small amount of water inside the soft, central, starchy part turns to steam, becomes inflated, and bursts, causing the seed to explode with a pop and turn inside out. The volume of modern kinds of popcorn can increase more than 40 times when popped.
- Popcorn was used by the Aztecs as decoration for ceremonial headdresses and necklaces for people, as well as for statues of their gods. Kernels of popcorn have been discovered in the graves of pre-Columbian Indians.
- Early Native Americans spread oil on an ear of corn, which they laid near a fire, causing the kernels to pop attached to the ear, producing "popcorn-on-the-cob."
- Although popcorn is now viewed as a snack food, during colonial times in North America popcorn was commonly used as a breakfast food, served with sugar and milk or cream.
- Many insects have evolved in nature to be specialist feeders on certain plants. However, the names assigned to some insect specialists of crops can obscure the natural feeding habits of these insects. The European corn borer is the most devastating insect pest of corn. However, corn is native to the Americas, while the corn borer is native to the Old Word, where it is a specialist feeder on hops (*Humulus lupulus* L.) and hemp (*Canabis sativa* L.). The corn borer did not meet and take an immediate liking to corn until Europeans brought the latter back to the Old World in the 16th century. The corn borer was brought to the Americas in the 20th century, so that corn grown in the Americas was not attacked by the insect until about a century ago.
- In 1945, Percy Spencer discovered that when popcorn was placed under microwave energy, it popped. This ultimately led to the invention of the microwave oven.
- The average ear of corn has 800 kernels in 16 rows. A kilogram of corn has about 3000 kernels (a pound has approximately 1300).
- Corn has been used in the production of alcohol for many years. Native Americans appear to have used corn to brew beer before Europeans arrived in the Americas. The 1792 Whiskey Rebellion in the United States resulted when efforts were made to tax corn whiskey. At the time, farmers commonly converted corn into whiskey, in part because it was much easier than grain to transport to customers.
- Barbara McClintock received the Nobel Prize in 1983 for her work with the complex color patterns of corn, studies that revealed moveable genetic elements termed "jumping genes." It has been said that scientists understand more about corn genetics than the genetics of any other flowering plant.
- Corn was grown in outer space, on the Atlantis shuttle that circled the earth for 5 days in 1989. In the absence of light, corn seedlings grew sideways in space, reflecting the importance of gravity on earth for normal plant growth. The roots did not grow downwards, and the shoots tied themselves into knots. Had light been provided, the shoots probably would have grown towards the light.
- At the 1994 winter Olympics in Lillehammer, Norway, food was served on plates made from potato starch and eaten with cutlery made out of corn.
- Corn pollen grains are among the largest in the grass family, and are produced in prodigious amounts. Two to five million pollen grains are released per plant. It has been estimated that an average of 6600 pollen grains are developed from each square centimeter (42 500 per square inch) in a cornfield.
- Simply because of variation in the size of the cob, a slender ear of corn may have as many kernels as a thicker ear.
- The famous cereal developer Dr. John Harvey Kellogg (1852–1943) developed Kellogg's Toasted Corn Flakes, which revolutionized North American breakfast habits, and remains the best-selling cereal today. Dr. Kellogg's thought that people needed dry, brittle food to chew in order to keep their teeth healthy, but when one of his patients broke her false teeth on a piece of zwieback (a special egg bread made into rusks, which takes its name from the German meaning twice-baked), he invented the more easily chewed corn flakes. Corn flakes were so popular in the early 1900s that over 40 companies produced the breakfast cereal in the Battle Creek region of Michigan.
- A patient of Dr. Kellogg, Charles W. Post (1854–1914), was inspired by his example, and invented Grapenut Flakes (i.e., Grape-Nuts, actually made from wheat and barley), and later a corn-flakes cereal called Post Toasties. Although first marketed by General Foods in 1908 as Post Toasties, it was first released in 1904 as "Elijah's Manna." The biblical name aroused religious opposition, and so it was changed.

- About 8% of the weight in a box of corn flakes is corn.
- Before 1970, hybrid corn in North America was dangerously uniform. To generate hybrid seeds, plant breeders used a Texas corn line that could not produce fertile pollen to eliminate the labor costs of detasseling (removing the male flowers) to prevent self-fertilization. This work resulted in complete and inexpensive control of hybrid crosses in corn breeding programs. By the late 1960s, virtually all commercial corn sold in the United States was based on cultivars with the Texas corn line. In the summer of 1970, a mutant form of a blight fungus spread north across the United States at the rate of 80 km or more a day, attacking all corn possessing the genes from the Texas corn line. This near-disaster has led to the use of more diverse varieties of corn.
- Most major soft drinks today are sweetened with corn syrup, not sugar.
- There are two, widespread classes of plants that differ in the way they carry out photosynthesis. Most familiar temperate zone plants are called C3 plants because the first stable compound formed when carbon dioxide is processed is a three-carbon compound, i.e., C3. C4 plants are so named because the first organic compound incorporating CO_2 is a four-carbon compound. Many tropical plants have the C4 type of photosynthesis, which allows for increased growth if temperature and light intensity are sufficiently high. Corn is a C4 plant. However, sweet corn, which is mainly grown at high latitudes, does not have as well developed a C4 system as corn varieties grown in more southern areas.
- The fact that corn has a different photosynthetic physiology than most other North American food plants has provided a technique for dating human bones that is different from the usual radiometric dating technique. Because of its distinctive set of photosynthetic enzymes, corn takes up a different proportion of the isotopes of carbon available (C12 or C13) in atmospheric carbon dioxide than do most other food plants eaten in temperate regions. When people or animals eat corn as a major part of their diet, their bones have a carbon isotope ratio that is different from that of the bones of non-corn eaters. Because these differences can be determined from skeletal remains, archeologists can determine historical patterns of corn consumption.
- A shoe horn can be used to remove kernels of corn from the cob.
- In 1997, Wang Litian, an amateur Chinese plant breeder, produced two new black varieties of corn, named Black Pearl and Black Guanyin.
- In 1995, Britain's Institute of Food Research spent about $300,000 for research to examine why corn flakes and other cereals become soggy in milk. For this, it was awarded "the IgNobel physics prize for research which cannot or should not be reproduced." However, the project does not deserve to be ridiculed. In 2002, Monsanto, the biotechnology giant, embarked on a multimillion-dollar research and breeding program to create a strain of corn that resists moisture. The goal: discover plant genes that produce high levels of wax, and transfer these into corn grown for breakfast cereals to produce corn flakes that remain crunchy even when drenched in milk. In fact, manufacturers presently add plant waxes in an attempt to keep the cereal from soaking up milk and becoming soggy.
- In 2006, the French National Institute for Agronomical Research in Nantes reported that its scientists had identified the factors that affect the crunchiness of corn flakes. It has been known for many years that Argentinian Plata corn produces corn flakes that resist sogginess and make pleasant sounds when consumed. Researchers found that the internal structure of the corn, particularly the interface between proteins and starch, is the key consideration to producing corn that does not become soggy in milk and produces a satisfying crunch in the mouth. It was also discovered that miniature bubbles form in each Argentinian Plata corn flake during cooking and that when crushed they produce a pleasing crunchiness.

Key Information Sources

Aldrich, S.R., Scott, W.O., and Leng, E.R. 1975. Modern corn production. 2nd edition. A & L Publications, Champaign, IL. 378 pp.

Alexander, D.E. 1989. Maize. *In* Oil crops of the world. *Edited by* G. Robbelen, R.K. Downey, and A. Ashri. McGraw-Hill, Toronto, ON. pp. 431–437.

Berger, J. 1962. Maize production and the manuring of maize. Centre d'Étude de l'Azote, Geneva, Conzett and Huber, Zurich, Switzerland. 315 pp.

Fischer, K.S., and Palmer, A.F.E. 1984. Tropical maize. *In* The physiology of tropical field crops. *Edited by* P.R. Goldsworthy and N.M. Fisher. John Wiley & Sons, Chichester, UK. pp. 213–248.

Food and Agriculture Organization of the United Nations. 1992. Maize in human nutrition. FAO, Rome, Italy. 160 pp.
Francis, A. 1990. The Tripsacinae: an interdisciplinary review of maize (*Zea mays*) and its relatives. Acta Bot. Fenn. **140**: 1–51.
Freeling, M., and Walbot, V. 1994. The maize handbook. Springer-Verlag, New York, NY. 759 pp.
Fussell, B.H. 1992. The story of corn. Knopf, New York, NY. 356 pp.
Goodman, M.M. 1988. The history and evolution of maize. CRC Crit. Rev. Plant Sci. **7**: 197–220.
Goodman, M.M. 1995. Maize. *In* Evolution of crop plants. 2nd edition. *Edited by* J. Smartt and N.W. Simmonds. Longman Scientific & Technical, Burnt Mill, Harlow, Essex, UK. pp. 192–202.
Häfliger, E. (*Editor*). 1979. Maize. Ciba-Geigy, Basle, Switzerland. 105 pp.
Huelsen, W.A. 1954. Sweet corn. InterScience Publishers, New York, NY. 409 pp.
Inglett, G.E. 1970. Corn: culture, processing, products. Avi Publishing Co., Westport, CT. 369 pp.
Johnson, D.L., and Jha, M.N. 1993. Blue corn. *In* New crops. *Edited by* J. Janick and J.E. Simon. Wiley, New York, NY. pp. 228–230.
Johnson, L.A. 2000. Corn: the major cereal of the Americas. *In* Handbook of cereal science and technology. 2nd edition. *Edited by* K. Kulp and J.G. Ponte, Jr. Marcel Dekker, New York, NY. pp. 31–80.
Jugenheimer, R.W. 1976. Corn: improvement, seed production, and uses. Wiley, New York, NY. 670 pp.
Kaukis, K., and Davis, D.W. 1986. Sweet corn breeding. *In* Breeding vegetable crops. *Edited by* M.J. Bassett. AVI Publishing Co., Westport, CT. pp. 475–519.
Koopmans, A., Have, H. ten, and Subandi [no initial]. 1996. *Zea mays* L. *In* Plant resources of South-East Asia, 10, cereals. *Edited by* C.J.H. Grubben and S. Partohardjono. Backhuys Publishers, Leiden, Netherlands. pp. 143–149.
Mangelsdorf, P.C. 1974. Corn: its origin, evolution, and improvement. Belknap Press of Harvard University Press, Cambridge, MA. 262 pp.
Martin, F.W., Daloz, C., and Carmen Vélez, M. del. 1981. Vegetables for the hot humid tropics, part 8, Vegetable corns, *Zea mays*. Mayaguez Institute of tropical Agriculture, Mayaguez, Puerto Rico. 14 pp.
Messer, E. 2000. Maize. *In* The Cambridge world history of food. *Edited by* K.F. Kiple and K.C. Ornelas. Cambridge University Press, Cambridge, UK. pp. 97–112.
Moris, C. 1979. Maize beer in the economics, politics, and religion of the Inca empire. *In* Fermented food beverages in nutrition. *Edited by* C.F. Gastineau, W.J. Darby, and T.B. Turner. Academic Press, New York, NY. pp. 21–34.
Small, E., and Cayouette, J. 1992. Biodiversity diamonds—the example of wild corn. Canadian Biodiversity, **2**(**3**): 24–28.
Smith, C.W., Betrán, J., and Runge, E.C.A. 2004. Corn: origin, history, technology, and production. Wiley, Hoboken, NJ. 949 pp.
Sprague, G.F., and Dudley, J.W. 1988. Corn and corn improvement. 3rd edition. American Society of Agronomy, Madison, WI. 986 pp.
Staller, J.E., Tykot, R.H., and Benz, B.F. (*Editors*). 2006. Histories of maize: multidisciplinary approaches to the pre-history, linguistics, biogeography, domestication, and evolution of maize. Elsevier, Amsterdam, Netherlands. 678 pp.
Tracy, W.F. 1993. Sweet corn, *Zea mays* L. *In* Genetic improvement of vegetable crops. *Edited by* G. Kalloo and B.O. Bergh. Pergamon Press, New York, NY. pp. 777–807.
Walden, D.B. 1978. Maize breeding and genetics. Wiley, New York, NY. 794 pp.
White, P.J., and Johnson, A. 2003. Corn: chemistry and technology. 2nd edition. American Association of Cereal Chemists, St. Paul, MN. 892 pp.
Wiley, R.C., Schales, F.D., and Corey, K.A. 1989. Sweet corn. *In* Quality and preservation of vegetables. *Edited by* N.A. Eskin. CRC Press, Inc., Boca Raton, FL. pp. 121–157.
Wolfe, D.W., Azanza, F., and Juvik, J.A. 1997. Sweet corn. *In* The physiology of vegetable crops. *Edited by* H.C. Wien. CABI International, Wallingford, Oxon, UK. pp. 461–478.

Specialty Cookbooks

(Note: There are numerous books on cooking corn of which the following are examples.)

Barrett, M.M. 1999. Corn lovers cook book. Golden West, Phoenix, AZ. 86 pp.
Buff, S. 1993. Corn cookery. Lyons & Burford, New York, NY. 222 pp.
Fussell, B.H. 1995. Crazy for corn. HarperPerennial, New York, NY. 237 pp.
Giedt, F.T. 1995. Popcorn! Simon and Schuster, New York, NY. 110 pp.
Mack, P. 1999. Corn. Record Books, Hackensack, NJ. 144 pp.
McNair, J.K. 1990. James McNair's corn cookbook. Chronicle Books, San Francisco, CA. 96 pp.
Rhijn, P. van, and Portilla, M.L. 1993. Maize cookery: corn-based recipes from Mexico. Publisher unknown, Mexico. 191 pp.
Siegel, H., and Gillingham, K. 1994. The totally corn cookbook. Celestial Arts, Berkeley, CA. 95 pp.
Tannis, D. 1995. Corn. Collins, San Francisco, CA. 95 pp.
Woodier, O. 2002. Corn: roasted, creamed, simmered + more. Storey Books, North Adams, MA. 187 pp.

Cotton

Family: Malvaceae (mallow family)

Names

Scientific names: *Gossypium* species

- American upland cotton—*G. hirsutum* L. (*G. mexicanum* Tod.)
- Sea island cotton—*G. barbadense* L. (*G. peruvianum* Cav., *G. vitifolium* Lam.)
- Tree cotton—*G. arboreum* L. (*G. nanking* Meyen, *G. obtusifolium* Roxb. ex G. Don)
- Levant cotton—*G. herbaceum* L.

- The word "cotton" comes from the Arabic word *qutun* or *kutun*, used to describe any fine textile.
- American upland cotton, *G. hirsutum*, is also known as American cotton, Bourbon cotton, cotton belt cotton, upland cotton, and West Indian cotton.
- *Gossypium barbadense* had spread to the West Indies by the time the Spanish arrived in the New World, and then to the southern United States, including the barrier islands off Georgia. People noticed its association with all of these Sea Islands, and so it came to be called "sea island cotton."
- *Gossypium barbadense* was introduced into Egypt about 1817 by a Frenchman. Its popularity there led to its being called "Egyptian cotton." "Pima" is another name for the species. Pima cotton traces to growth of the species in Pima, Arizona, from an Egyptian selection.
- Additional names for sea island cotton are American Pima cotton, American-Egyptian cotton, Brazilian cotton, Egyptian cotton, Gallini cotton, kidney cotton, long-staple cotton, and Peruvian cottton.
- Levant cotton, *G. herbaceum*, is also known as African cotton and West Asian cotton.
- Tree cotton, *G. arboreum*, is also known as Indian cotton and Pakistani cotton.
- The genus name *Gossypium* is a late (Medieval period) Latin name for the cotton plant.
- *Hirsutum* in the scientific name *G. hirsutum*, the most important species of cotton, is Latin for hairy.
- *Barbadense* in the scientific name *G. barbadense* means "of Barbados." Barbados had the first commercial slave plantation cultivating and exporting cotton, the trade based on *G. barbadense*.
- *Arboreum* in the scientific name *G. arboreum* means tree-like.
- *Herbaceum* in the scientific name *G. herbaceum* means herbaceous.

Plant Portrait

There are more than three dozen species of *Gossypium* in the tropics and subtropics of the Old and New Worlds, and several of these have been used since prehistory as sources of textile fiber. Cotton has been spun, woven, and dyed for thousands of years in both the Old and New Worlds. It was used for clothing by people of ancient India, Egypt, and China. In the 1st century, Arab traders brought fine muslin and calico to Italy and Spain. The Moors introduced the cultivation of cotton into Spain in the 9th century. Native Americans also spun and wove cotton into fine garments and dyed tapestries. Cotton fabrics found in Peruvian tombs are said to belong to a pre-Inca culture. Cotton is the world's most important vegetable fiber and makes up almost half of all the fiber sold in the world. The seed fiber is the raw material for a large volume of textile products. However, as noted below, cotton is also the source of an extremely important edible vegetable oil. Today, cotton is grown in over 90 countries, predominantly in China, the United States, Uzbekistan, India, Pakistan, Turkey, and Egypt.

American upland cotton, the most important of the four cultivated cotton species, is thought to have originated in Central America, where perennial forms are known. It was domesticated at least by 3400 BC in Mexico. Upland cotton is raised as an annual subshrub, which grows up to 1.5 m (4 feet) in height. Hundreds of cultivars are known. The capsules split open as they ripen, at which time they are known as bolls. Each capsule contains up to 50 seeds, which are attached to fibers, known as lint before they are processed into commercial textiles. The fibers and seeds are harvested from the bolls. Upland cotton was taken from Mexico to the United States about 1700. During the American Civil War, it was introduced into most tropical and subtropical countries of the world. It is now the source of all commercial cotton crops of Africa outside the Nile Valley, all those of South America except in Peru and northern Brazil, of the modern Russian crop, and much of that of northern India and Pakistan, the Philippine

Islands, and the Cotton Belt of the United States. American upland cotton is the type most commonly cultivated in the United States. Today, the leading cotton states are Texas, California, Mississippi, Arizona, and Louisiana.

Sea island cotton, a New World species that, it has been claimed, was being used in Peru of 8000 BC, produces exceptionally long, high-quality fibers, but is not as resistant to the boll weevil. Both American species produce much longer fibers than the Old World species of *Gossypium*, and this explains why they are the most important cotton species today. Sea island cotton is grown mostly in Asia, with limited cultivation in the United States.

Levant cotton, *G. herbaceum*, is native to sub-Saharan Africa and Arabia. It was probably domesticated in Ethiopia or southern Arabia. Today, this species is cultivated in Asia.

Gossypium hirsutum
(American upland cotton)

Gossypium barbadense
(sea island cotton)

Gossypium herbaceum
(Levant cotton)

Gossypium arboreum
(tree cotton)

The four major species of *Gossypium* used as sources of cotton, from Engler and Prantl (1889–1915).

Tree cotton, *G. arboreum*, is native to Northwest India and Pakistan and was being used to produce cotton textiles at least by 2000 BC. As with Levant cotton, tree cotton is cultivated in Asia.

Cotton is often called "King Cotton" because of its predominance in societies that have grown it, or manufactured textiles from it. It has been a cruel monarch, associated with terrible exploitation, wars, and social disruption. Britain's need for vast African and Indian markets for its cotton-manufactured products influenced its role as an imperial sea power. Beginning in North America in the Jamestown colony of 1607, cotton cultivation became the basis of the slave-labor economy of the Deep South and a principal economic cause of the Civil War. Cotton really became king in the early 1800s because of the invention of the cotton gin, which accelerated the Industrial Revolution, led to a huge resurgence of slavery, the isolation of the American South, the American Civil War, and migration of blacks northward in the 20th century. In England, cotton played a major role in the Industrial Revolution, and although the slave labor in the New World associated with cotton was extremely cruel, cruelty was also evident in the Old World. Women and children worked 16-hour days in English cotton mills, children entering at the age of 7, and people were commonly chained to the mills so their bodies wouldn't fall into the machinery.

Cottonseed oil, a by-product of the cotton industry, has been important since the late 19th century, when David Wesson (1861–1934, founder of Wesson Oil) of the Southern Cotton Oil Company introduced Wesson Oil in 1899 after perfecting a technique for deodorizing cotton oil. This revolutionized the cooking-oil industry, making cottonseed oil the first vegetable oil used in the United States, and the world's leading vegetable oil until the 1940s. In the United States, the cotton oil crop ranks third in volume behind soybean and corn oil, and represents about 5% of the total domestic fat and oil supply. American upland cotton is the predominant species used as an oil crop. The oil content of cotton seeds is about 20% (the kernels of the seeds, which make up half the weight of the seeds, contains 28–40% oil). The oil is obtained from the seeds by pressing or by extractive solvents. There are several uses for cotton oil in addition to its culinary use, described below. Paint makers use it to some extent as a semidrying oil. Less-refined grades are used in the manufacture of cosmetics, soap, candles, detergents, artificial leather, oilcloth, and many other commodities. After the oil is expressed from the seeds, the residue, called cottonseed cake or meal, is an important protein concentrate for livestock. A toxic substance, gossypol, must be removed before cottonseed can be used for food or feed. Gossypol is used in China as a male contraceptive.

Culinary Portrait

Cottonseed oil is used in North America as a salad or cooking oil. As a salad oil, it is employed in mayonnaise, salad dressings, sauces, and marinades. As a cooking oil, it is used for frying in both commercial and home cooking. In Japan, cottonseed oil is esteemed for making tempura, i.e., small portions of shrimp, vegetables, and other foods dipped in a thick batter and deep-fried. As a shortening or margarine, the oil is ideal for baked goods and cake icings. Cottonseed oil is also employed as a food packing, for example, as a medium in canned sardines. It has a high saturated fat level, much higher than the oils of corn, canola, soybean, sunflower, peanut, and safflower, and therefore is considered less healthy.

Culinary Vocabulary

- "Crisco," Proctor and Gamble's brand of solid shortening, introduced in 1911, is a near-acronym for "crystallized cottonseed oil." Crisco was originally made with hydrogenated cottonseed oil, but is now produced primarily from soybean. (One earlier name, Krispo, was abandoned for trademark reasons, and another name, Crist, was also abandoned because of its similarity to Christ. It has been said that these early names were chosen by Proctor & Gamble employees to suggest the hissing sound of the shortening in the pan, and that "Crisco" in fact is a combination of the names.)

Curiosities of Science and Technology

- Just before Columbus sailed the ocean blue, most Europeans were wearing a heavy woolen–linen mix made from flax and wool, a fabric called "linsey-woolsey." Upon landing in the Bahamas on 12 October 1492, Columbus was given thread made by the local Indians from American upland cotton, which was to become the predominant cotton species of the world, the predominant fabric for making clothing, and indeed the most important nonfood plant commodity with the exception of wood.
- The introduction of American upland cotton to Europe was a threat to the established wool industry. In 1666, there were protectionist "anti-calico" laws forbidding the wearing of cotton.

- Denim, primarily a cotton fabric, was first made in the 16th century at Nimes, France. The word "denim" comes from the name Nimes.
- Jeans have referred to durable work clothing made of cotton for many years. In the 16th century, sailors at the port of Genoa, Italy, wore denim trousers. The word "jeans" comes from Genes, the French word for Genoa.
- Poor people who could not afford silk tried to make cotton cloth look silky by beating on the fabric with sticks to soften the fibers, and rubbing it against a large stone to make it shiny. The shiny cotton was called "chintz." Because chintz was a cheaper copy of silk, calling something "chintzy" has come to mean it is cheap and of limited quality.
- American inventor Eli Whitney (1765–1825) perfected the cotton gin in 1792, a simple device that quickly removed the tiny seeds from cotton. Prior to his invention, a slave produced about 1 kg (2.2 pounds) of lint in 2 days. The cotton gin increased the yield of a laborer 100 times, which caused the cotton-producing US states to increase their productivity 10 times. Whitney made no money from the cotton gin because he did not have a valid patent.
- In the Old American South, cotton was chewed as a headache remedy.
- In the 1923, Leo Gerstenzang, a Polish-born American, noticed that his wife would wrap a wad of cotton around a toothpick to clean their baby's ears. He then manufactured a ready-to-use cotton swab, first called "Baby Gays," and later Q-Tips (the "Q" stood for quality). Doctors today advise that Q-tips should not be used to clean inside the ears.
- How does one judge the quality of bed sheets? Most shoppers think that thread count per inch is the definitive criterion; it is not. Linen (from flax), for example, is loosely woven but incredibly durable, and flannel has low thread counts of very fat threads. Several factors have a far greater effect on comfort and durability, particularly the quality of the fabric. In the case of cotton, yarn spun from long fibers has fewer connecting points, and sheets manufactured from it are smoother and stronger. The words "Egyptian cotton" and "Supima" are indications of top-quality fibers at least 2.5 cm (1 inch) long. Pima is the generic term for such long-fiber cottons grown in the United States, Australia, Israel, and Peru, while Supima is a licensed trademark for products made with 100% American Pima.
- A "hank" is a length of thread or yarn, specific to the type of fiber. For linen, a hank is 274 m (300 yards), while for cotton it is 768 m (840 yards).
- "Colored cotton" is natural cotton that has been bred to produce light green, light brown, and rusty red bolls. The colored fluff is processed into cloth that is gaining popularity with environmentally conscious consumers and people who are allergic to the dyes in regular cotton. Colored cotton is twice as expensive as white cotton.
- "Gun cotton" is a highly explosive preparation made by saturating cotton or other cellulose material with nitric and sulphuric acids.
- Cotton is treated to change its properties. "Mercerized cotton" is soaked in caustic soda for about 2 hours, making it shiny. "Sanfordizing," invented in 1970, involves ammonia processing, which prevents shrinkage after washing. "Permanent press" is achieved with chemicals applied to cross-link the cellulose fibers, so the fabric returns to its shape.
- More than 75% of the people in the world wear cotton clothing. The average American uses up 6.4 kg (14 pounds) of cotton annually.
- Wool clothing is notorious for being attacked by moth larvae and silverfish. Insects love wool because animal fibers are largely protein. By contrast, cotton fibers are mostly cellulose, which has virtually no food value for many insects. (Termites, like cows, have their own internal microorganisms, which can digest cellulose for their host.)
- Major league baseballs have a small cork core wound with 289 m (316 yards) of wool yarn, covered with two pieces of cowhide held together with 216 raised red cotton stitches.
- Most paper currency is made out of cotton, not paper based on wood pulp fiber, which is simply too fragile.
- In Ethiopia, cotton cloth was once commonly used as money.

Key Information Sources

Bailey, A.E. 1948. Cottonseed and cottonseed products; their chemistry and chemical technology. Interscience, New York, NY. 936 pp.

Baumann, L.A., and Whitten, M.E. 1970. Effects of storage temperatures on quality of cottonseed salad oil. U.S. Agricultural Research Service, Washington, DC. 22 pp.

Dani, R.G. 1990. Genetic research of cottonseed oil: a review. Coton Fibres Trop. (France), **45**: 71–75.

Dodge, B.S. 1984. Cotton, the plant that would be king. University of Texas Press, Austin, TX. 175 pp.
Endrizzi, J.E., Turcotte, E.L., and Kohel, R.J. 1985. Genetics, cytology, and evolution of *Gossypium*. Adv. Genet. **23**: 271–375.
Food and Agriculture Organization, and World Health Organization. 1970. Recommended international standard for edible cottonseed oil. FAO/WHO Codex Alimentarius Commission, Joint FAO/WHO Food Standards Programme, FAO, Rome, Italy. 20 pp.
Fryxell, P.A. 1976. A nomenclator of *Gossypium*: the botanical names of cotton. Agricultural Research Service, U.S. Dept. of Agriculture, Washington, DC. 114 pp.
Fryxell, P.A. 1979. The natural history of the cotton tribe (Malvaceae, tribe Gossypieae). Texas A&M University Press, College Station, TX. 245 pp.
Gelmond, H. 1979. A review of factors affecting seed quality distinctive to cotton seed production. Seed Sci. Technol. **7**: 39–46.
Harrison, J.R. 1977. Review of extraction process: emphasis cottonseed. Oil Mill Gazetteer, **82**(**4**): 16–17, 20–24.
Hearn, A.B., and Constable, G.A 1984. Cotton. *In* The physiology of tropical field crops. *Edited by* P.R. Goldsworthy and N.M. Fisher. John Wiley & Sons, Chichester, UK. pp. 495–527.
Hutchinson, J.B. 1947. The evolution of *Gossypium* and the differentiation of the cultivated cottons. Oxford University Press, London, UK. 160 pp.
James, C. 2002. Global review of commercialized transgenic crops: 2001 feature: Bt cotton. International Service for the Acquisition of Agri-Biotech Applications, Ithaca, NY. 184 pp.
Jones, L.A., and King, C.C. 1990. Cottonseed oil. National Cottonseed Products Association/Cotton Foundation, Memphis, TN. 60 pp.
Kohel, R.J., and Lewis, C.F. (*Editors*). 1984. Cotton. American Society of Agronomy, Madison, WI. 605 pp.
Lane, R.P. 1934. Cotton and cottonseed: a list of the publications of the United States Department of Agriculture on these subjects, including early reports of the United States Patent Office. U.S. Dept. of Agriculture, Washington, DC. 149 pp.
Murti, K.S., and Achaya, K.T. 1975. Cottonseed chemistry and technology in its setting in India. Council of Scientific & Industrial Research, New Delhi, India. 348 pp.
Pandey, S.N. 1998. Cottonseed and its utilization. Indian Council of Agricultural Research, Directorate of Information and Publication on Agriculture, New Delhi, India. 212 pp.
Saunders, J.H. 1961. The wild species of *Gossypium* and their evolutionary history. Oxford University Press, London, UK. 62 pp.
Taylor, B.F. 1936. Early history of the cotton oil industry in America. Taylor, Columbia, SC. 20 pp.
Verdery, M.C. 1971. Technology for the production of protein foods from cottonseed flour. Food and Agriculture Organization of the United Nations, Rome, Italy. 42 pp.
Wendel, J.F. 1995. Cotton. *In* Evolution of crop plants. 2nd edition. *Edited by* J. Smartt and N.W. Simmonds. Longman Scientific & Technical, Burnt Mill, Harlow, Essex, UK. pp. 358–366.
Wrenn, L.B. 1995. Cinderella of the new South: a history of the cottonseed industry, 1855– 1955. University of Tennessee Press, Knoxville, TN. 280 pp.

Specialty Cookbooks

Simmons, R. (*Editor*). 1984. Cottonseed cookery. Food Protein Research and Development Center, Texas Engineering Experiment Station, Texas A&M University System, College Station, TX. 48 pp.
Texas Woman's University. 1984. Cooking with cottonseed. College of Nutrition, Textiles and Human Development, Texas Woman's University, Denton, TX. 27 pp.

CUCUMBER

Family: Cucurbitaceae (gourd family)

Names

Scientific names: *Cucumis* species

- Cucumber—*C. sativus* L.
- West Indian gherkin—*C. anguria* L.
- Chinese cucumber—*C. melo* L. var. *conomon* (Thunb.) Mak.

- The word "cucumber" is based on the ancient Roman name for the cucumber, *cucumis*, which is also the basis of the genus name *Cucumis*.
- The Chinese cucumber is also known as Oriental pickling cucumber, Oriental pickling melon, pickling melon, and sweet melon.
- The West Indian gherkin is also known as bur(r) cucumber, bur(r) gherkin, Jamaican cucumber, Jamaican gherkin, gooseberry gourd, pepino, pepinillo, and West Indian gourd.
- The name "pepino," an alterative name for the West Indian gherkin, is also used for the Peruvian pepino (*Solanum muricatum* Aiton).
- The term "cowcumber," sometimes used today disparagingly, was the usual way 17th century English writers referred to the cucumber.
- For information on "cucumber pickle," "pickled cucumber," and "preserved cucumber," all of which are alternative names for "tea melon," see MELON, CULINARY VOCABULARY.
- *Anguria* in the scientific name *C. anguria* is based on the Greek *angyria*, a cucumber.
- See MELON (which is *C. melo*) for information on the derivation of *melo* in the scientific name *C. melo*.
- *Sativus* in the scientific name *C. sativus* means sown or cultivated.

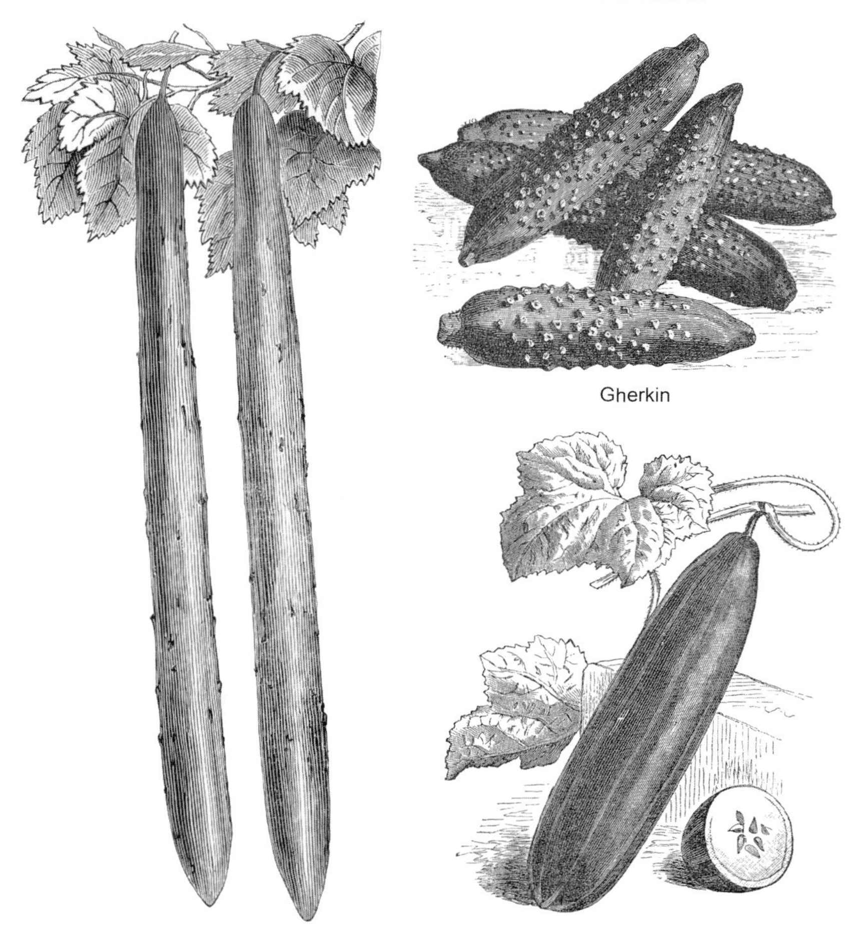

Cucumber (*Cucumis sativus*). English cucumber from *Revue de l'Horticulture Belge et Étrangère* (1911, vol. 37). Remaining illustrations from Vilmorin-Andrieux (1885).

Chinese (snake) cucumber (*Cucumis melo* var. *conomon*), from Vilmorin-Andrieux (1885).

Plant Portrait

Cucumber plants are vines with large, yellow flowers. The plants usually trail on the ground in the open, but sometimes climb on trellises in greenhouses, clinging by means of tendrils. Some so-called "bush-type" cultivars do not trail far over the ground. The English forcing cucumber sets fruits without the requirement for sexual fertilization, and the fruits are seedless. Sexual expression in cucumbers is complex. Some cucumbers have male and female flowers on the same plant. Others, including most field-grown varieties, mostly have female flowers, but with a few male flowers. Field cucumbers are insect-pollinated, and in commercial production, beehives are often provided. The fruits may differ in color (usually yellow or green), shape (usually cylindrical or oblong), and flesh characteristics. The skin may be smooth, ridged, rough, or spiny, and most varieties have edible seeds.

The cultivated cucumber is an ancient, Old World vegetable that probably originated in India and spread westward to ancient Egypt and Greece. Wild cucumbers that likely gave rise to the modern cultivated forms have been found in the Himalayan region of Asia. The cucumber was introduced into China in pre-Christian times, probably during the Han Dynasty (206 BC –220 AD). Columbus brought cucumber seeds to Haiti on his second voyage (in 1494), and their use in North America spread from there. Cucumbers are an important crop worldwide, common in home gardens, on truck farms, and as a greenhouse crop. "Burpless" cucumbers have been selected to prevent burping of gas, which affects some individuals after eating cucumbers. Pickling varieties are mostly picked when small and immature.

The Chinese cucumber is an annual, herbaceous vine. It is a native of Asia and was described in Chinese literature as long ago as 560 AD. The Chinese cucumber is grown in many parts of Asia and, to a lesser extent, in tropical areas around the world. However, this exotic is only occasionally grown in North America. It should not be confused with the familiar field cucumbers of commerce that belong to *C. sativus*, although it is similar in shape and color. Chinese cucumbers can grow much longer than field cucumbers (up to 50 cm or 20 inches in length), and in Japan, they are marketed as "yard-long cucumbers." The Chinese cucumber should also not be confused with two other somewhat similar vegetables, the "snake melon" (see Melon), and the "snake cucumber" (see below). The fruit has a thin, light-green skin, the flesh is crisp with a delicate flavor, and there are few seeds. The fruit is also burpless.

The West Indian gherkin is also sometimes cultivated in North America as a novelty. The fruit is generally oval in shape, 2.5–7.5 cm (1–3 inches) long, much more warty than the common cucumber, and has a large cavity with many seeds.

There are three basic processes for preparing pickles. The first simply requires storing the cucumbers in salt brine, sometimes for as long as 3 years. To remove the excess salt, the pickles are then rinsed, placed in fresh water, and heated. They may be sliced or left whole, and are packed in a finishing "liquor." A second process similarly controls natural fermentation with brine, but is done under refrigeration, which results in very flavorful cucumber-like pickles, but with a relatively short shelf life (3 months or so). In the *fresh pack* process, the cucumbers are simply placed in flavored brine or syrup and quickly pasteurized. This produces less-salty pickles.

Culinary Portrait

Cucumbers are generally eaten raw, peeled or unpeeled, and added to salads. It is not necessary to peel cucumbers, but old or large specimens and those with a waxy coating are often best peeled. Cucumbers are also cooked and used in soups, stews, and vegetable dishes. Cucumbers can be sautéed, steamed, and indeed prepared by most of the cooking methods suitable for squash. Pickling varieties are used in relishes, chutneys, and pickles. Medium-sized cucumbers are preferable to very large ones, which are often bitter or bland in taste, with numerous hard seeds.

Chinese cucumbers provide long, tasty fruit suitable for salads, vegetable dishes, and pickles. In Japan, the fruits are used for making expensive pickles, called *tsukemono*, after removing the skin and seeds and preserving in sake.

West Indian gherkin fruits are used mostly in pickles but are also eaten as a cooked vegetable and incorporated into curries. In North America, the "gherkins" of commerce usually are small or immature common cucumbers.

Culinary Vocabulary

- A "gherkin" is a small pickling or pickled fruit of the cucumber. The word is from the plural of the Dutch *gurk*, cucumber, which is ultimately derived from the late Greek *aggourion*, a diminutive of *agourous*, not ripe, green, i.e., something which is small and immature.
- The British expression "pickled cucumber" is equivalent to the American "dill pickle."
- "Bread and butter pickles" are sweet pickles made from sliced cucumbers, cured with spices such as mustard and celery seeds and onions. The name may have originated during the Depression Era, when pickles were an important dietary staple. The phrase "bread and butter" has long been applied to simple, basic, essential, common things of life (although the finest center-cut chuck steaks are also sometimes known as "bread and butter").

Curiosities of Science and Technology

- Cucumber has long been employed for its purported cosmetic properties. Egyptian Queen Cleopatra (69–30 BC) ate pickles in the belief that they enhanced her beauty. In 1887, the legendary actress Sarah Bernhardt (1844–1923) packed 30 jars of cucumber cold cream to keep her complexion perfect on a European tour. Cucumber flesh is sometimes rubbed directly on skin to keep it soft and white. It is supposedly cooling and soothing to irritated skin, such as that caused by sunburn. Cosmetic soaps and bubble baths are sometimes made with cucumber.

West Indian gherkin (*Cucumis anguria*). *Left*, from Bailey (1910); *right*, from Vilmorin-Andrieux (1885).

- The Roman Emperor Tiberius (42 BC–37 AD) had cucumbers on his table daily, summer and winter, making it necessary to grow them out of season using artificial methods.
- The Italian navigator Amerigo Vespucci (1454–1512), who contributed his name to "America," supplied his ships with pickles, believing that they helped sailors survive scurvy. In his early years, he was a pickle vendor, supplying seafarers with the product. Pickles contain vitamin C, which combats scurvy, although probably in insufficient amounts to make much difference.
- In China, it was once fashionable to attach stones to the bottoms of cucumbers growing on trellises so that they would grow straight.
- American businessman Henry John Heinz (1844–1919) was a marketing genius. The Heinz display at the Chicago World's Fair of 1893 was a major attraction to the 1 million visitors, many of whom picked up a tiny green metallic reproduction of a Heinz pickle to attach to their key chains or charm bracelets. Heinz erected a six-story illuminated green pickle at the intersection of Fifth Avenue and 23rd Street, which displayed the Heinz name for several years until 1906.
- The expression "cool as a cucumber" apparently comes from English physicians advising patients to lie on cucumbers, in the mistaken belief that this would lower their fever.
- According to *The Food of the Western World*, a dictionary of gastronomy on food from North America and Europe by Theodora Fitzgibbon, when cutting cucumber, one should always start from the pointed flower end, as this prevents any bitter taste. (Perhaps the idea here is that the relatively tender flesh at the flower end should be consumed first, as it will tend to store less well than the rest of the cucumber.)
- The craving for pickles (as well as pretzels and anchovies) by pregnant women has been explained as a consequence of a greater need for salt, which is required to meet the 40% increase in blood needed by the placenta (blood is salty), as well as the salty fluid bathing the fetus.
- Homemade "refrigerator dill pickles" are traditionally prepared by placing cucumbers in a brine mixture along with vinegar, dill, and other spices in jars left at room temperature for about a week. The jars are then kept in a refrigerator for several more weeks of slow fermentation. Mark Harrison, a food scientist at the University of Georgia, reported in 2004 that the procedure is so vulnerable to potentially deadly bacterial contamination that people should simply not try to make this popular pickle. *Listeria* is of particular concern, since home recipes often do not include enough vinegar to control this bacterium. In commercially manufactured pickles, the presence of lactic-acid bacteria generally serves to destroy the *Listeria*.
- Very reminiscent of the cucumber is the snake cucumber, *Trichosanthes cucumerina* L. var. *anguina* (L.) Haines (*T. anguina* L.), also known as club gourd, serpent cucumber, snake gourd, vegetal snake, and viper gourd. This is a tropical annual, trailing or climbing vine, native to India and Pakistan, and grown in Southeast Asia, China, Japan, West Africa, Latin America, the Caribbean, and tropical Australia. The fruits produced are very slender, as long as 2 m (6½ feet), and as thick as 10 cm (4 inches). However, they are usually harvested when immature, 30–40 cm (12–16 inches) long, while they are tender, and are usually cooked, especially in soup. The mature fruits may remain straight, or curl and resemble a snake.

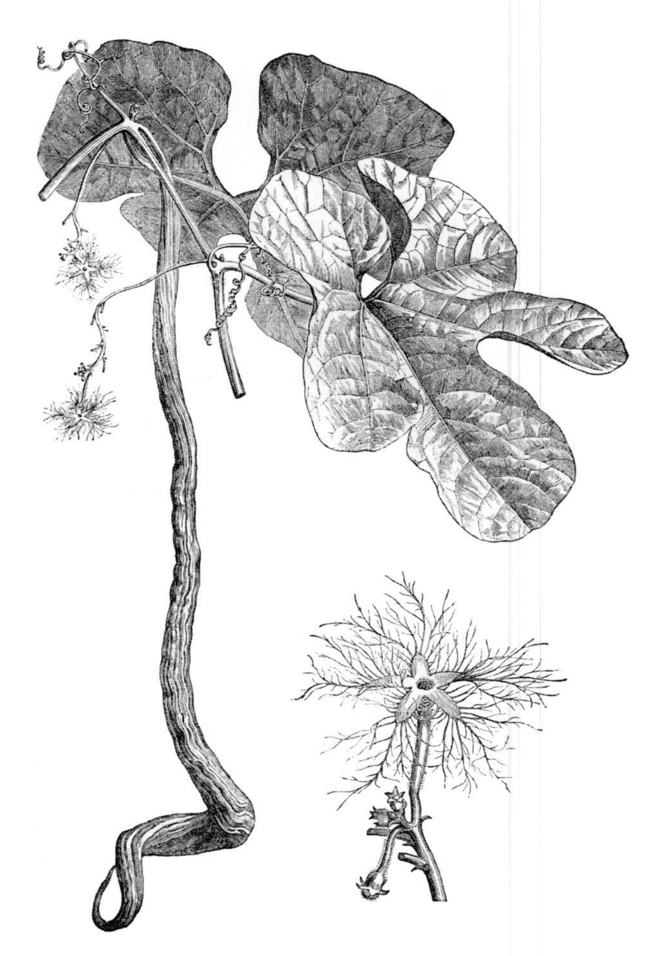

Snake gourd (*Trichosanthes anguina*), from Paillieux and Bois (1892).

Key Information Sources

Bates, D.M. 1995. Cucumbers, melons and water-melons. *In* Evolution of crop plants. 2nd edition. *Edited by* J. Smartt and N.W. Simmonds. Longman Scientific & Technical, Burnt Mill, Harlow, Essex, UK. pp. 89–96.

Daeschel, M.A., and Fleming, H.P. 1987. Achieving pure culture cucumber fermentations: a review. Developments Ind. Microbiol. **28**: 141–148.

Drimtzias, E.N. 1990. A critical review on cultivation and nutrition of cucumber and tomato with emphasis on soilless culture. KVL, Copenhagen, Denmark. 67 pp.

Ennis, D.M., and O'Sullivan, J. 1979. Cucumber quality—a review. J. Food Sci. **44**: 186–189, 197.

Garrett, J.T. 1968. Growing pickling cucumbers. Mississippi Cooperative Extension Service, Mississippi State University, MI. 8 pp.

Gildemacher, B.H., and Jansen, G.J. 1993. *Cucumis sativus* L. *In* Plant resources of South-East Asia, 8, vegetables. *Edited by* J.E. Siemonsma and K. Piluek. Pudoc Scientific Publishers, Wageningen, Netherlands. pp. 157–160.

Gildemacher, B.H., Jansen, G.J., and Chayamarit, K 1993. *Trichosanthes* L. *In* Plant resources of South-East Asia, 8, vegetables. *Edited by* J.E. Siemonsma and K. Piluek. Pudoc Scientific Publishers, Wageningen, Netherlands. pp. 271–274.

Grower Books. 1980. Cucumbers. Grower Books, London, UK. 72 pp.

Hughes, G.R., Averre, C.W., and Sorensen, K.A. 1988. Growing pickling cucumbers in North Carolina. N.C. Agricultural Extension Service, North Carolina State University, Raleigh, NC. 15 pp.

Kirkbride, J.H., Jr. 1988. Botanical origin of the West Indian gherkin (*Cucumis anguria* L. var. *anguria*). Am. J. Bot. **75**(6, part 2): 186.

Kirkbride, J.H. 1993. Biosystematic monograph of the genus *Cucumis*: botanical identification of cucumbers and melons. Parkway Publishers, Boone, NC. 159 pp.

Lower, R.L., and Edwards, M.D. 1986. Cucumber breeding. *In* Breeding vegetable crops. *Edited by* M.J. Bassett. AVI Publishing Co., Westport, CT. pp. 173–207.

Maynard, D., and Maynard, D.N. 2000. Cucumbers, melons, and watermelons. *In* The Cambridge world history of food. *Edited by* K.F. Kiple and K.C. Ornelas. Cambridge University Press, Cambridge, UK. pp. 298–313.

Naudin, C. 1962. A monograph of species and varieties of genus *Cucumis*. Ann. Sci. Nat. (4th Ser.), **18**: 159–209.

Niemela, S.I., and Laine, J.J. 1975. Effects of cucumber size and brine composition on the quality of dill pickles. J. Food Sci. **40**: 684–688.

Papadopoulos, A.P. 1994. Growing greenhouse seedless cucumbers in soil and in soilless media. Agriculture and Agri-Food Canada, Ottawa, ON. 126 pp.

Pierce, L.K., and Wehner, T.C. 1990. Review of genes and linkage groups in cucumber. HortScience, **25**: 605–615.

Raj, N.M., Prasanna, K.P., and Peter, K.V. 1993. Snake gourd, *Trichosanthes anguina* L. *In* Genetic improvement of vegetable crops. *Edited by* G. Kalloo and B.O. Bergh. Pergamon Press, New York, NY. pp. 259–264.

Schultheis, J.R. 1998. Commercial production of pickling & slicing cucumbers in North Carolina. N.C. Cooperative Extension Service, North Carolina State University, Raleigh, NC. 43 pp.

Tatlioglu, T. 1993. Cucumber, *Cucumis sativus* L. *In* Genetic improvement of vegetable crops. *Edited by* G. Kalloo and B.O. Bergh. Pergamon Press, New York, NY. pp. 197–234.

Wehmer, T.C., and Robinson, R.W. 1991. A brief history of the development of cucumber cultivars in the U.S. Cucurbit. Genet. Coop. Rep. **14**: 1–4.

Whitaker, T.W., and Davis, G.N. 1962. Cucurbits. Interscience Publishers Inc., New York, NY. 249 pp.

Wittwer, S.H., and Honma, S. 1979. Greenhouse tomatoes, lettuce and cucumbers. Michigan State University Press, East Lansing, MI. 225 pp.

Specialty Cookbooks

(For additional references, see MELON.)

National Gardening Association. 1987. Book of cucumbers, melons, & squash. Revised edition. Villard Books, New York, NY. 86 pp.

New York City. 1946. Forty ways to prepare cucumbers. Revised edition. City of New York, Department of Markets, Division of Consumers' Service and Research, New York, NY. 8 pp.

Newcomers Covina. 1960s. Newcomers cucumbers: a collection of favorite recipes. Newcomers Covina, West Covina, CA. 68 pp.

Oklahoma State Dept. Agriculture. 1983. Oklahoma cucumber recipes. State Dept. of Agriculture, Marketing Industry Division, Oklahoma City, OK. 6 pp.

South Carolina Department of Agriculture. 1977. Cool crisp cucumber ideas. S.C. Dept. of Agriculture, Columbia, SC. 5 pp.

CURRANT

Family: Saxifragaceae (saxifrage family)

Names

Scientific names: *Ribes* species

- Red currant (common currant, garden currant)—*R. rubrum* L. [*R. silvestre* (Lam.) Mert. & Koch., *R. sativum* (Rchb.) Syme]; the predominant red currant grown in North America, also grown in Eurasia.
  - Northern red currant (downy currant, Nordic currant, red currant)—*R. spicatum* E. Robson (commonly misidentified as *R. rubrum*); much more commonly grown in Europe than in North America.
  - White currants—selected (albino) mutations, mostly hybrids of the above two species.
  - Rock currant (rock red currant)—*R. petraeum* Wulfen; rarely grown, but a parent of some cultivated hybrids.
  - Black currant (European black currant, cassis)—*R. nigrum* L. (not to be confused with an American black currant species that is occasionally grown).
  - Golden currant (buffalo currant, Missouri currant, clove currant)—*R. odoratum* H.L. Wendl. The only one of this list of species that grows wild in the United States, the fruits are more often black than golden, not grown often.
- Red currant (*R. rubrum*) in northern European countries was often known as "wine berry" for its use in preparing wine.
- "Jostaberries" are hybrids of black currant and the American gooseberry, *R. hirtellum*, produced in Germany during the 1930s–1950s. They have been commercially promoted in recent years in the United States, but have been disappointing in flavor to date.
- The genus name *Ribes* is said to derive from the Arabic *ribas*, rhubarb, presumably applied to the

Red currant (*Ribes rubrum*), from *Gartenflora* (1867, vol. 16), showing seven different 19th century cultivated varieties.

European black currant (*Ribes nigrum*), from Morren (1851–1885, vol. 1), showing three different 19th century cultivated varieties.

gooseberry because of a supposed similarity in taste that was noted by Arabs while they expanded into Spain centuries ago. Alternatively, the name has been interpreted as originating from the Old Danish colloquial *ribs*, red currant, and as the Latinized form of *riebs*, an old German word for currant.

- Explanations for the Latin words used in the scientific names of the species listed above: *rubrum* (red, for the color of the berries), *spicatum* (spike-like, describing the cluster of flowers), *petraeum* (growing among rocks), *nigrum* (black, for the color of the berries), and *odoratum* (fragrant, for the odor of the plant).

Plant Portrait

About 100 of the species in the genus *Ribes* are called "currants," the remaining 50 or so are "gooseberries." Varieties (many of hybrid origin) of those listed above, especially *R. rubrum*, are the most important commercial sources of currant fruit. Currant species are shrubs of northern areas. They are similar to gooseberry shrubs, but usually lack the spines and bristles of the latter (although spineless gooseberry varieties have been selected). The two important commercial species, red and black currant, are from Europe. The berries are small, red, black, white, or golden. Red currants arose in the Old World. They seem to have been unknown to the Greeks and Romans, and were recorded as cultivated in Europe

White currant (*Ribes rubrum* × *R. nigrum*), from Henderson (1890).

apparently no earlier than the 16th century, although in fact it is probable that they were grown earlier. Red currants were introduced to North America in the 17th century. As with other fruit crops, selection has been primarily for size and flavor of fruit, and productivity of the plants under cultivation. The European black currant is a recently domesticated plant, having been taken into cultivation in northern Europe within the last 400 years. The fruit was used medicinally and for making wine several hundred years ago, the flavor of the fresh fruit not proving attractive. The black currant was brought to North America in the early 1600s.

About 1892, white-pine blister rust was imported to North America. This fungus, which occupies the black currant as an alternate host, led to severe infection of five-needle (white) pine trees, leading to legislation preventing the cultivation of currants and gooseberries over wide areas of North America, and indeed to programs of eradication. The result was to limit the commercial significance of both currants and gooseberries in North America, most particularly in the United States. Today, red currants are grown in small quantities in the United States and Canada, while there is little cultivation of the black currant, which is especially sensitive to white-pine blister rust. Red currants are grown mostly in Germany and Poland, with lesser amounts in other European countries. There is substantial production of black currants in Great Britain, Poland, Germany, and other European countries. The fruit of the black currant is exceptionally high in vitamin C, and has been considered a valuable antiscorbutic (a substance that prevents scurvy) for many years.

A cultivar of the rock currant, Jahns Prairie (named for Otto L. Jahn, the first curator of the National Clonal Germplasm Repository in Covallis, Oregon) was released jointly by the U.S. Department of Agriculture and Agriculture and Agri-Food Canada in the 1990s. This variety, selected by Dr. Jahn from a native population of the species growing in Alberta, is disease resistant and produces high-quality, dark red, dessert gooseberries.

Culinary Portrait

Most fresh currants are used in jams, jellies, and preserves, and also for their juice. Red currants can be eaten raw, either directly or in salads, but because of their tartness they are usually cooked. Red-currant jelly is considered a delicacy, and indeed the standard par excellence in jellies. Red and white currants are used to make the prized French Bar-le Duc jelly (see below for additional information). Red currants are also the key ingredient of Cumberland sauce (also see below for more information). In Europe, red-currant jelly is used in cooking of pastries, and for making glazes and sauces for meats and poultry. Red-currant jelly is a traditional accompaniment of roast venison and mutton. Red currants can also be added to cakes, pies, puddings, and mixed-fruit desserts. Many find the taste of black currants objectionable, limiting the popularity of this fruit in North America. Black currants are rarely used as a dessert fruit. In the Old World, the berries were employed for canning, juice, jams, jellies, and pie fillings, for black-currant crème, and liqueurs such as Cassis (a name not only used for liqueurs flavored with black-currant juice, but also for certain French wines from the Côtes de Provence region), and for converting white wine to "rosé." "Crème de Cassis" is applied to a reddish purple liqueur made from black currants, but sometimes also to a sweet syrup made from currants, which may or may not contain alcohol. Fresh currants should be consumed as soon as possible, but may be stored, unwashed, in a plastic bag in a refrigerator for a few days. Currant berries are best cooked slowly for 3–5 minutes, using a small quantity of water or juice, and adding sugar as desired after cooking.

Culinary Vocabulary

- "Dried currants" are not dried currant berries, but are small seedless raisins, the dried fruit of several grape varieties grown in the Middle East. The name "currant" in this sense is an allusion to the Greek port of Corinth (*courante* in Middle English), the center of export. The name was applied to certain species of *Ribes* because of the similarity between their fruit clusters and those of the grapes called currants. The word "currant" has been used for fruit from *Ribes* only since about 1550, while it was used for dried raisins long before this date.
- "Zante currants" are Black Corinth grapes (see Grape for additional information).
- "Brownies" in North America are small, rich, chewy, squares of chocolate cake, usually with nuts. In Australia and New Zealand, the term may refer to a kind of bread made with currants.
- "Cumberland sauce" is a piquant English sweet-and-sour sauce featuring red-currant jelly, and usually also with port, lemon, orange juice and zest, and mustard. It is usually served with duck, venison, or other game, typically cold. It has been speculated that the name traces to Ernest, Duke of Cumberland, brother of George IV, who in 1837 became the last independent king of Hanover, Germany; however, most authorities discount this explanation of the name, which remains of obscure origin.
- "Eccles cake" is a small round cake (a sort of cookie) prepared with buttery, flaky pastry, and filled with currants. The name comes from the town of Eccles in Greater Manchester, England, and is said to have been invented by the culinary writer Elizabeth Raffald (1733–1781).
- "Election cake" is a classic English fruitcake or plum cake, originally made with molasses, spice, raisins, and currants, and later, brandy. It became a popular tradition to serve these huge cakes, each weighing about 5.4 kg (12 pounds), while waiting for election results in New England.
- Bar-le-Duc jelly (Lorraine jelly, confiture de groseille) is a famous currant preserve (mislabeled a "jelly," since as noted in the following, whole berries are present), originally from the town of Bar-le-Duc in northwestern France. It was once made from white currants, with the tiny seeds (on average, seven per berry) removed by hand, often using a sharpened goose feather, and the fruit immediately immersed in burning-hot, clear, white syrup, a method that preserved the flavor and bright color of the whole berries. The process is thought to have been created by the 14th century provisions chief of Duke Robert of the old province of Lorraine. The choice preserve is now produced from red and white currants as well as other berry fruits, and the seeds are not generally removed manually.
- A "garibaldi" is a British biscuit with a layer of currants inside, which used to be popular in the first half of the 20th century. It commemorated Giuseppe Garibaldi (1807–1882), an Italian military and national leader. Its colloquial nickname was "squashed-fly biscuit," named for the appearance of the currants.
- "Plum duff" is an old-fashioned English boiled suet pudding flavored with currants or raisins.

Curiosities of Science and Technology

- The black currant has had its detractors for a long time. In an old herbal (plant medicinal book), the fruits were described as "of a stinking and somewhat loathing savour," and indeed, few attempt to eat the fresh fruit. Perhaps in accord with the adage "medicine has to taste bad to be good," black currant has often been used medicinally. Black currants have long been a popular home remedy against colds, and the exceptionally high vitamin C content (as much as five times the concentration in citrus fruits) is provocative in view of the controversial claims that this vitamin is medicinally effective against colds.
- The "Victorian language of flowers" was a secret coded language in Victorian times, with flowers and plants symbolic of certain messages, so when the flower or plant was mentioned in a letter, those who knew the code could understand the hidden information. "Currant" meant "thy frown will kill me," while "branch of currants" stood for "you please all."
- As noted above, currant cultivation in the United States has been discouraged because the plants serve as the alternate host for a fungus that infects pines. At one time, white-pine rust from currants destroyed 650 million board feet of white pine yearly. (One board foot = 2.4 L = 144 cubic inches of material, i.e., 1 foot × 1 foot × 1 inch.)

Rock currant (*Ribes petraeum*), a rarely grown species that is a parent of some cultivated hybrids, from Lamarck and Poiret (1744–1829, plate 146).

- During the Second World War, it was part of the war effort in England to plant black currants, which are extremely high in vitamin C, in an effort to replace oranges as the major source of the vitamin.
- Currants (and also their relatives the gooseberries) bloom early in the spring, and severe frosts can injure blossoms and young, developing berries. Commercial planters often spray water over the plants when temperatures reach −1 °C (30 °F). The conversion of water to ice on the plants releases heat, which protects the blossoms and berries.

Key Information Sources

Audette, N., and Lareau, M.J. 1997. Currants and gooseberries culture guide. Conseil des productions végétales du Québec and Horticultural R & D Centre, QC. 43 pp.

Barney, D.L., and Hummer, K.E. 2005. Currants, gooseberries, and jostaberries: a guide for growers, marketers, and researchers in North America. Food Products Press, Binghamton, NY. 299 pp.

Barre, D.E. 2001. Potential of evening primrose, borage, black currant, and fungal oils in human health. Ann. Nutr. Metab. **45**(**2**): 47–57.

Beecham Foods. 1973. Black currants: a grower's guide. Revised edition. Beecham Foods, Brentford, UK. 98 pp.

Berger, A. 1924. A taxonomic review of currants and gooseberries. New York State Agricultural Experiment Station, Geneva, NY. 118 pp.

Bird, R., and Whiteman, K. 2002. Growing berries and currants: a directory of varieties and how to cultivate them successfully. Lorenz, London, UK. 64 pp.

Brennan, R.M. 1990. Currants and gooseberries (*Ribes*). *In* Genetic resources of temperate fruit and nut crops. *Edited by* J.N. Moore and J.R. Ballington, Jr. International Society for Horticultural Science, Wageningen, Netherlands. pp. 459–488.

Bunyard, E.A. 1921. A revision of the red currants. J. Pomol. **2**: 38–55.

Crawford, M. 1997. Currants and gooseberries: production and culture. Agroforestry Research Trust, Totnes, UK. 48 pp.

Crowther, D.S. 1956. Currants and gooseberries. W.H. & L. Collingridge, Ltd., London, UK. 92 pp.

Darrow, G.M. 1919. Currants and gooseberries. U.S. Dept. of Agriculture, Washington, DC. 40 pp.

Darrow, G.M., and Detwiler, S.B. 1946. Currants and gooseberries: their culture and relation to white-pine blister rust. Revised edition. U.S. Dept. of Agriculture, Washington, DC. 36 pp.

Harmat, L., Porpaczy, A., Himelrick, D.G., and Galletta, G.J. 1990. Currant and gooseberry management. *In* Small fruit crop management. *Edited by* G.J. Galletta and D.G. Himelrick. Prentice Hall, Englewood Cliffs, NJ. pp. 245–272.

Hatton, R.G. 1920. Black currant varieties. A method of classification. J. Pomol. **1**: 68–80, 145–154.

Janczewski, E. de, and Glinka, R. von. 1907. Monograph of the currants, *Ribes*. Mem. Soc. Phys. Hist. Nat Genève, **35**: 199–517. [In German; W. Kündig & fils, Geneva, Switzerland.]

Kampuss, K., and Pedersen, H.L. 2003*a*. A review of red and white currant (*Ribes rubrum* L.): research and literature. Small Fruits Rev. **2**(**3**): 23–46.

Kampuss, K., and Pedersen, H.L. 2003*b*. A review of red and white currant cultivars. Small Fruits Rev. **2**(**3**): 47–101.

Keep, E. 1975. Currants and gooseberries. *In* Advances in fruit breeding. *Edited by* J. Janick and J.N. Moore. Purdue University Press, West Lafayette, IN. pp. 197–268.

Keep, E. 1995. Currants. *In* Evolution of crop plants. 2nd edition. *Edited by* J. Smartt and N.W. Simmonds. Longman Scientific & Technical, Burnt Mill, Harlow, Essex, UK. pp. 235–239.

Knight, R.L., and Keep, E. 1955. Abstract bibliography of fruit breeding and genetics to 1955; *Rubus* and *Ribes*, a survey. Commonwealth Agricultural Bureaux, East Malling, Kent, UK. 254 pp.

Knight, R.L., Parker, J.H., and Keep, E. 1972. Abstract bibliography of fruit breeding and genetics, 1956–1969. *Rubus* and *Ribes*. Tech. Commun. No. 32. Commonwealth Bureau of Horticulture and Plantation Crops, Slough, UK. 449 pp.

Miles, N.W. 1983. Currants and gooseberries. Ontario Ministry of Agriculture and Food, Toronto, ON. 4 pp.

Olander, S. 1998. A bibliography of black currants. Swedish University of Agricultural Sciences, Uppsala, Sweden. 42 pp.

Oldham, C.H. 1956. Black and red currants. Publisher not stated, Littlebury, UK. 136 pp.

Todd, J.C. 1962. Black currant varieties: their classification and identification. Ministry of Agriculture, Fisheries and Food, Tech. Bull. 11. Her Majesty's Stationary Office, London, UK. 94 pp.

United Kingdom Ministry of Agriculture, Fisheries and Food. 1975. Red currants. Advisory leaflet 521. Revised edition. Great Britain. Ministry of Agriculture, Fisheries and Food, Pinner, UK. 8 pp.

United Kingdom Ministry of Agriculture, Fisheries and Food. 1976. Black currants. Advisory leaflet 543. Revised edition. Great Britain. Ministry of Agriculture, Fisheries and Food, Pinner, UK. 9 pp.

Williamson, B., Brennan, R.M., and Gordon, S.L. (*Editors*). 2002. Proceedings of the eighth international *Rubus* and *Ribes* symposium, Dundee, Scotland, July 2001. International Society for Horticultural Science, Leuven, Belgium. 2 vols.

Specialty Cookbooks

Hendrickson, R. 1981. The berry book. The illustrated home gardener's guide to growing and using over 50 kinds and 500 varieties of berries. Doubleday & Company, Garden City, NY. [Chapter 12: The electrifying currant. pp. 184–193.]

Pellowe, S. 1989. Saffron & currants: a Cornish heritage cookbook. Renard Productions, Aurora, IL. 51 pp.

Senn, C.H. 1906. Currants: a few tasty recipes. Publisher not stated, London?, UK. 23 pp.

Senn, C.H. 1915. Home recipes for cooking currants. Publisher not stated, London?, UK. 32 pp.

Senn, C.H. 1919. Currant cookery. Publisher not stated, London?, UK. 24 pp.

Wayne, S. 1930–1939? The currant recipe book. Advertising Dept., Central Currant Office, London, UK. 32 pp.

DATE

Family: Arecaceae (Palmae; palm family)

Names

Scientific name: *Phoenix dactylifera* L.

- The English word "date" is derived from the Greek word *dactylus*, meaning finger, the same word in the scientific name *P. dactylifera*. Before it was adopted in Greek, the word was used in ancient Hebrew and Syria as the name of the date palm. The use of the word "finger" for the date palm has been variously interpreted—as a reference to the tall, slender form of the tree, or to the finger-like shape of the fruit clusters, or the appearance of the leaves.
- The word "date," meaning time on a calendar, is unrelated, deriving from the Latin *datum*, meaning given. In Roman times, the date was indicated by the expression *Data Romae*, literally "given (or issued) at Rome"; hence, "date" came to mean day of the calendar year.
- The "Indian date" is the tamarind (*Tamarindus indica* L.).
- The classical Greek philosopher and botanist Theophrastus (about 372–287 BC) used the name phoenix for the date palm. The genus name *Phoenix* relates directly or indirectly to the Phoenix, a legendary bird of ancient Greece that built its own funerary pile, fanned the fire with its wings, and eventually arose again from its own ashes. The Phoenix was also a legendary bird of ancient Egypt (also known as the Roc and the sun bird), found on mural paintings that decorated the tombs of the kings and the nobles, and both the bird and the date palm were called by the same name (Bennu). The Phoenix was born and reborn from a date palm, and the name Phoenicia (modern Lebanon and Syria) was derived from it.

Plant Portrait

Dates are the fruit of a tall palm tree, ranging up to 30 m (100 feet). In nature, there are equal numbers of male and female date palms, with pollination occurring by wind. In early times, Arabs learned how to pollinate the trees by hand, so that it was only necessary to have one male plant for numerous females. The date fruit looks like an elongated olive, varying in color when mature, from tan to a dark, brownish yellow. Each fruit contains a single hard seed, which is deeply grooved on one side. This slender pit is often removed from dates sold in stores. Dates are one of the world's oldest crops, probably originating in western Asia and North Africa. Date palms were cultivated as far back as 4000 BC in southern Iraq. Today, date trees are raised commercially in the Old World from Morocco to India, and the leading producers are Iraq, Saudi Arabia, Algeria, Iran, and Egypt. In the United States, date palms are grown in the southern deserts of California and Arizona. Although some date trees are established by seeds, date cultivars are propagated vegetatively, using branches that form at the base of the trunk. There are some 1500 cultivars of dates. An average mature tree produces 45 kg (100 pounds) of fruit annually, a good tree 68 kg (150 pounds), and some improved cultivars have yielded over 100 kg (220 pounds). Dates are generally harvested before they are mature, then ripened in storage, like bananas. There are three basic types of dates: soft, semidry, and dry. Soft dates, grown in large quantities in Iraq, are popular in the Middle East and are frequently sold in pressed masses and eaten raw. Semidry dates are grown in North African countries and the United States. The semidry Algerian variety Deglet Noor is the world's most widely consumed date. Semidry dates store well, but need to be softened in water before eating, and are generally sold in the Middle East.

Date trees have commonly been planted in oases, where they literally have saved travelers from starvation, and so have been called "The Tree of Life." Date palms are also known as the "Bride of the Desert," because they add beauty to barren lands.

Culinary Portrait

About 75% of a dry date is sugar (mostly glucose and fructose), so that the fruit is often consumed out of hand, sometimes stuffed with nuts or marzipan or rolled in sugar or syrup. In North America, dates are usually used in sweet foods, such as cakes, cookies, muffins, squares, and cereals. In Middle Eastern cooking, dates are added to a wider variety of dishes, including salads and couscous, and are stuffed, candied, and employed to manufacture spirits. In India, dates are used in curries and chutneys. Dates can be hydrated by soaking them in water for several hours. Date trees are also tapped like maple sugar trees to produce a sugary sap that can be boiled down to produce sugar. As well, the sap is fermented to make palm

Date (*Phoenix dactylifera*), from Figuier (1867).

wine. Soft dates should be stored in the refrigerator. Semisoft types will keep longer if refrigerated, but will store well if kept tightly wrapped at room temperature. Airtight containers of dates will stay moist in a refrigerator for up to 8 months. If not well sealed, dates have a tendency to absorb flavors and odors of other foods.

Culinary Vocabulary

- "Date syrup" is made by boiling dates in water for several hours. The thick syrup is used in Middle Eastern cooking much as corn syrup or molasses is employed in Western cooking.
- "Sea dates," marine mollusks of the species *Lithophaga lithophaga* L., are so named because they are the size, shape, and color of dates. They have a delicate, sweet taste, and are usually eaten raw in Mediterranean countries. Sea dates are also called European date mussels.

Curiosities of Science and Technology

- The art of the world's two most ancient cultures, Egypt and Sumer, is full of palm-tree images. For the Egyptians, the palm was associated with the god Heh, the embodiment of eternity. For the Sumerians, it was the symbol of the garden of paradise.
- The ancient Egyptians washed the bodies of the dead with date wine during mummification.
- There is an astonishing sequence of adopting the palm leaf as a symbol. The Egyptians held the date palm to be the symbolic tree of the year, since it produced a new branch every month. After the exodus of the Jews from Egypt, the date palm became the sacred emblem of Judea. Next, in 53 BC, the Roman legions triumphed over Judea, and used the palm leaf as an emblem of their triumph and a symbol of their plunder and destruction of Egypt. In 29 AD, Christians accepted the palm leaf as symbolic of the triumphant entry of Christ into Jerusalem. When Christians were forced to find shelter in the Catacombs, the palm leaf became the emblem of the martyrs. There are 60 references to date palms in the Old Testament. Date palm fronds are used on Palm Sunday, commemorating the entry of Jesus in Jerusalem (Lion's Gate or east entrance to Jerusalem through which Jesus is supposed to have entered the city). In Italy, some groves of date palms are maintained solely to supply young leaves for religious use on Palm Sunday. In areas of Spain, only the leaves of male palms are used for this purpose.
- According to an Arab saying, there are as many uses for dates as there are days in the year. Every part of the tree has its uses. The wood and leaves provide timber and fabric for houses and fences; the leaves are used for making ropes, cord, baskets, crates, mats, and furniture; the palm heart is eaten in salads; flour is made from the pithy core of the trunk; oil from the seeds is used to make soap; and the unused portions of the plants are used as fuel.
- There are over 100 million date palm trees in the world.
- Male and female flowers of date trees are on separate plants (rarely, some plants have flowers of both sexes). Although the flowers are wind-pollinated, in commercial practice most dates are produced by hand pollination. This is traditionally accomplished by cutting off a bunch of male flowers and tying them to a bunch of female flowers, and letting nature take its course. In date-tree plantations, only one male is required for every 50 females.

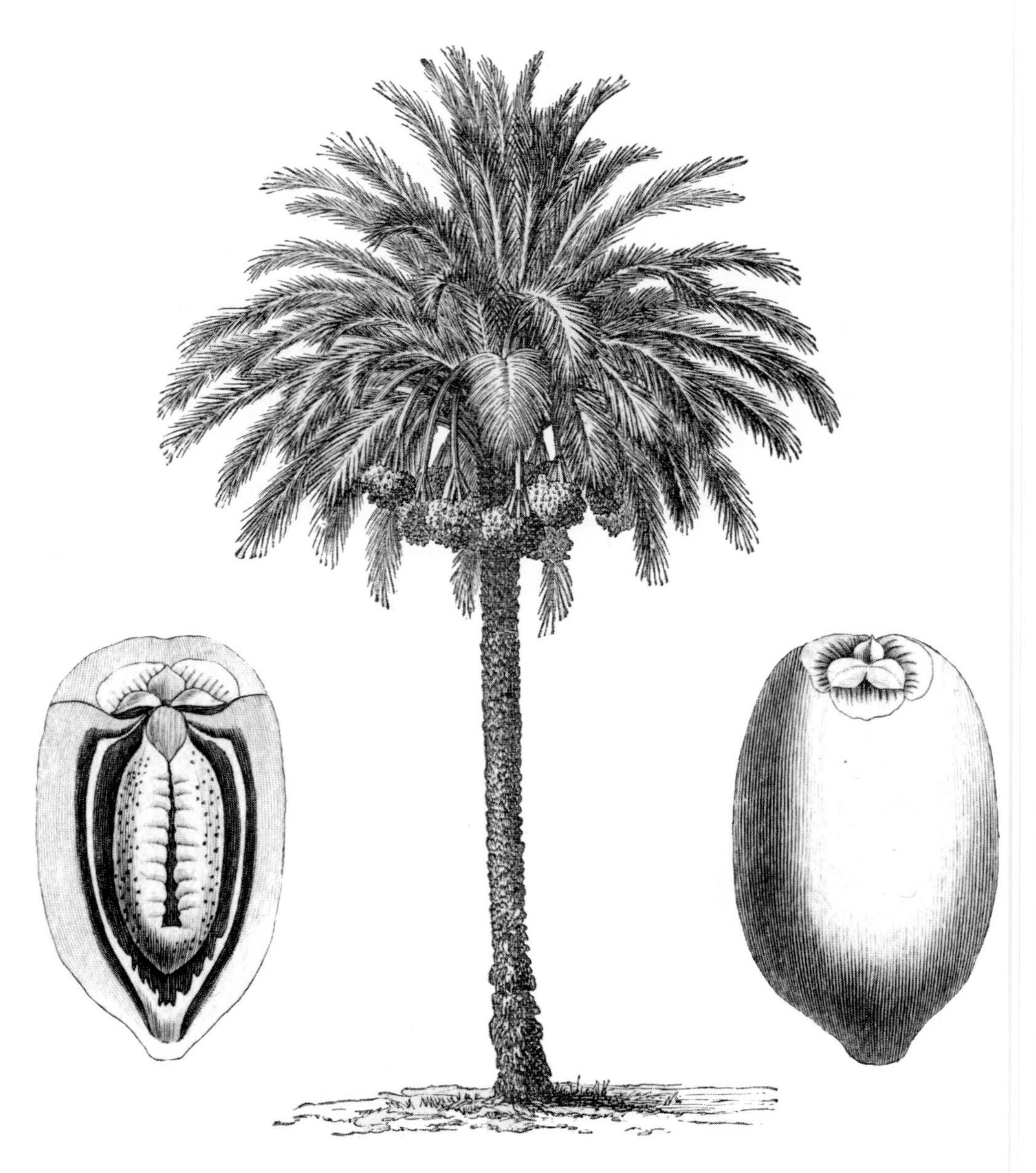

Date (*Phoenix dactylifera*). Tree in center from Maout and Decaisne (1876). Long section of fruit (*left*) and intact fruit (*right*) from Lamarck and Poiret (1744–1829, plate 893).

- Indio, California, and the surrounding Coachella Valley produce most of the dates consumed in the United States. The climate is similar to that of the Sahara Desert, so the region is ideal for growing dates. Arabic names became popular for towns of the region, such as Mecca and Oasis, with streets such as Arabia and Deglet Noor. An annual "National Date Festival" has been held since 1921 at Indio, with people dressed in Arabian costumes, camel races, and numerous date products including pies, cakes, cookies, date butter spreads, milkshakes, and chocolate-covered dates.

Key Information Sources

Abou-Aziz, A.B., Abdel-Wahab, F.K., El-Nabawy, S.M., and Abdel-Kader, A.S. 1975. Keeping quality of fresh date fruits as affected by cultivar and storage temperatures. Egypt. J. Hortic. **2**: 67–74.

Al-Khalifah, N.S., and Askari, E. 2003. Molecular phylogeny of date palm (*Phoenix dactylifera* L.) cultivars from Saudi Arabia by DNA fingerprinting. Theor. Appl. Genet. **107**: 1266–1270.

Asif, M.I.A., and Al-Ghamdi, A.S. 1986. Bibliography of date palm (*Phoenix dactylifera* L.). Date Palm Research Center, King Faisal University, Al Hassa, Saudi Arabia. 217 pp.

Barrett, H.C. 1973. Date breeding and improvement in North America. Fruit Var. Hortic. Dig. **27(3)**: 50–55.

Barreveld, W.H. 1993. Date palm products. Food and Agriculture Organization of the United Nations, Rome, Italy. Various paginations.

Carpenter, J.B., and Ream, C.L. 1976. Date palm breeding, a review. Date Growers' Inst. Rep. **53**: 25–33.

Corniquel, B., and Mercier, L. 1994. Date palm (*Phoenix dactylifera* L.) cultivar identification by RFLP and RAPD. Plant Sci. (Ireland), **101**: 163–172.

De Mason, D.A., and Stolte, K.W. 1982. Floral development in *Phoenix dactylifera*. Can. J. Bot. **60**: 1439–1446.

Devanand, P.S., and Chao, C.T. 2003. Genetic variation within 'Medjool' and Deglet Noor' date (*Phoenix dactylifera* L.) cultivars in California detected by fluorescent-AFLP markers. J. Hortic. Sci. Biotechnol. **78**: 405–409.

Djerbi, M. 1983. Diseases of the date palm (*Phoenix dactylifera* L.). Food and Agriculture Organization, Regional Project for Palm and Dates Research Centre in the Near East and North Africa, Baghdad, Iraq. 124 pp.

Dowson, V.H.W. 1976. Bibliography of the date palm. Field Research Projects, Coconut Grove, FL. 139 pp.

Elhoumaizi, M.A., Saaidi, M., Oihabi, A., and Cilas, C. 2002. Phenotypic diversity of date-palm cultivars (*Phoenix dactylifera* L.) from Morocco. Genet. Resour. Crop Evol. **49**: 483–490.

International Date Palm Forum [Markaz al-Imarat lil-Dirasat wa-al-Buhuth al-Istiratijiyah]. 2003. The date palm: from traditional resource to green wealth. Emirates Center for Strategic Studies and Research, Abu Dhabi. 433 pp.

Long, E.M. 1943. Development anatomy of the fruit of the Deglet Noor date. Bot. Gaz. **104**: 424–426.

Morton, J. 1987. Date. *In* Fruits of warm climates. *Authored by* J.F. Morton. Creative Resource Systems, Winterville, NC. pp. 5–11.

Nixon, R.W., and Carpenter, J.B. 1978. Growing dates in the United States. Revised edition. U.S. Dept. of Agriculture, Science and Education Administration, Washington, DC. 63 pp.

Oudejans, J.H.M. 1969. Date palm. *In* Outlines of perennial crop breeding in the tropics. *Edited by* F.P. Ferwerda and F. Wit. Misc. Paper 4. Landbouwhogeschool, Wageningen, Netherlands. pp. 243–257.

Pareek, O.P. 1985. Date palm. *In* Fruits of India, tropical and subtropical. *Edited by* T.K. Bose. Naya Prokash, Calcutta, India. pp. 566–580.

Poponoe, P.B. 1973. The date palm. Field Research Projects, Coconut Grove, FL. 247 pp.

Reuveni, O. 1986. Date. *In* CRC handbook of fruit set and development. *Edited by* S.P. Monselise. CRC Press, Boca Raton, FL. pp. 119–144.

Samarawira, I. 1983. Date palm, potential source for refined sugar. Econ. Bot. **37**: 181–186.

Vandercook, C.E., Hasegawa, S., and Maier, V.P. 1980. Dates. *In* Tropical and subtropical fruits: composition, properties and uses. *Edited by* S. Nagy and R.E. Shaw. AVI Publishing Co., Westport, CT. pp. 506–541.

Wrigley, G. 1995. Date palm. *In* Evolution of crop plants. 2nd edition. *Edited by* J. Smartt and N.W. Simmonds. Longman Scientific & Technical, Burnt Mill, Harlow, Essex, UK. pp. 399–403.

Zaid, A., and Arias-Jiménez, E.J. 2002. Date palm cultivation. Food and Agriculture Organization of the United Nations, Rome, Italy. 292 pp.

Specialty Cookbooks

California Date Exchange. 1933. How to serve California dates. California Date Exchange, Indio, CA. 16 pp. [California Date Growers Association. 1932. How to serve fresh dates. California Date Growers Association, Indio, CA. 16 pp.]

Heetland, R.I. 1993. Date recipes. Revised edition. Golden West Publishing, Phoenix, AZ. 126 pp.

Henderson, S., Finger, J., and Jim, D. 1968. Fresh date cookbook. Henderson Publications, Los Angeles, CA. 32 pp.

Hills Brothers Company. 1928. Dates in the healthful diet. Hills Brothers Company, New York, NY. 42 pp.

Iraqi Date Association. 1953. Iraqi dates; culture, industry, trade, and delicious recipes. Iraqi Date Association, Baghdad, Iraq. 95 pp.

Khazzam, J.S., Inc. 1984. Cooking with dates. J.S. Khazzam, Toronto, ON. 40 pp.

Metzler, M.S. 1929. Date recipes: teaches the use of Coachella Valley dates, both fresh and cooked. Press of Coachella Valley Submarine, Coachella, CA. 61 pp. [Metzler, M.S. 1944. Date recipes: from the American home of the date, grown in the date gardens of Coachella Valley. Valerie Jean Date Shop, Thermal, CA. 64 pp.]

Parfitt, J., and Valentine, S. 1995. Dates. MCBS in Association with Summertime Publishing, Mina Al-Fahal, Sultanat of Oman. 93 pp.

Petinak, M.J. 1941. Dates as food, how to use them. 2nd edition. M.J. Petinak, Los Angeles, CA. 64 pp.

EGGPLANT

Family: Solanaceae (potato family)

Names

Scientific name: *Solanum melongena* L.

- The name "eggplant" (or egg-plant) was coined because of the resemblance to eggs of the small-fruited forms of China, which were the first varieties known to English-speaking people. Some white-fruited ornamental eggplants available today have strikingly egg-like fruit.
- In tropical countries, some species of *Solanum* other than *S. melongena* are called eggplant.
- In Europe, eggplant is usually known as aubergine, the name in both English and French. "Aubergine" is derived from the Arabic *bathinjan*, which in turn comes from the Persian *badnjan*, and ultimately from the Sanskrit *vatingana*, apparently meaning "antifart vegetable," for its presumed antiflatulent effect.
- In Indian and African English, "brinjal" is the name of the eggplant. The word originates from the Portuguese *beringela*, which traces to the same roots as the word aubergine.
- British traders bringing eggplants to the London market from West Africa in the 17th century coined the name "Guinea squash."
- Like the tomato, the eggplant was once called the "love apple" in Europe and was thought to be an aphrodisiac. Eggplants were brought to America by Spaniards as *berengenas*, meaning "apples of love."
- A "hairy eggplant" is an eggplant variety with orange skin covered with fine dark hairs. It is known as *mauk* in Thailand, where it is used raw in salads and minced in sauces, producing a bitter flavor.
- See Potato for information on the genus name *Solanum*.
- *Melongena* in the scientific name *S. melongena* is from the Italian *melanzane*, mad apple, reflecting the 16th century European idea that the fruit causes insanity.

Plant Portrait

The eggplant is a perennial that is grown as an annual, and although considered to be a herb, becomes somewhat woody and shrubby at maturity. The stems may become longer than 2 meters (6½ feet), but are commonly no longer than a meter (about a yard). Eggplant originated in Southeast Asia more than 4000 years ago. Large-fruited varieties came from India, and smaller-fruited forms from China (the fruit of the "pea aubergine" is the size of a large cherry). Eggplant has been known in China since the 5th century BC but was not present in ancient Greece or Rome. Seeds of eggplant were brought westward from China about 1500 years ago. The Moors are thought to have introduced eggplant to Spain around 1200 AD. By the Middle Ages, all the colors of fruits that occur today were available. Several varieties were known in Europe and North America by the mid 1800s. The plant is subtropical, requiring warmth and protection from cold, and so eggplant is a very important crop in subtropical Africa, Asia, and America. Eggplants are particularly popular in Asia, but they are also a staple in North Africa, from Egypt to Morocco. The main European producers are Italy, Spain, and France, while China, Turkey, and Japan are principal Asian growers. Eggplant is a minor crop in North America. Fruit types are varied, including a small "green grape" type from Thailand; "Easter-egg" types, with variously colored fruits on a single plant; Italian "bella" types, with delicate pink and white shading on round fruits; the standard commercial class, with black 1.4-kg (3-pound) fruits; and the increasingly popular, slim, purple–black Japanese or Oriental types. Shapes vary from round to snake-like, and color variations include white, red, green, purple, pink, and glossy solid black. Even striped forms are known. Although eggplant is mainly used for food, ornamental eggplants have also been bred.

Culinary Portrait

The eggplant is a well-known cooked vegetable, with a distinctive taste that is frequently either loved or hated. Eggplant is rarely eaten raw, although some Asian varieties are consumed uncooked in salads. This vegetable can be baked in a conventional or microwave oven, grilled, sautéed, boiled, marinated, or fried. Long, slender eggplants tend to be less moist and are good for frying. Fat, rounded eggplants tend to be juicier and are better suited for baking and casseroles. Eggplant is used in cuisines around the world. Ratatouille, a classic French dish, is a vegetable stew, usually made with eggplant, tomatoes, zucchini, peppers, and onions, seasoned with herbs and garlic. The Middle Eastern baba gannouj is a purée of roasted

Eggplant (*Solanum melongena*), from Henderson (1890).

eggplant and tahini, flavored with garlic and lemon juice. The popular Greek dish moussaka consists of layers of seasoned ground lamb and eggplant, and is usually topped with a custard or cheese sauce and baked. Eggplant is famous in Italian cuisine, especially as eggplant parmagiana, an Italian-American dish of breaded, sautéed eggplant baked with tomato sauce. Involtini (Italian for "rolled up") refers to wraps, typically made in southern Italy, with eggplant slices wrapped around a mixture of basil, pine nuts, breadcrumbs, garlic, provolone and Parmesan cheese, and egg, and baked in tomato sauce sprinkled with mozzarella. Caponata is a dish of eggplant and other vegetables, pine nuts, and anchovies, cooked in olive oil and served at room temperature, often as an appetizer. Stuffed eggplants are widely popular. In India, eggplant is found in curries and pickles, and pickled tiny eggplants are sometimes available in Western markets.

Tips: Buy only dark, almost black, eggplants that are firm and shiny; a spongy eggplant is either old or has too many seeds. Eggplants will become soft, wrinkled, and bitter if left at room temperature for several days, and such shriveled fruits should not be purchased. Ripeness can be checked by pressing the skin lightly with a finger: if the imprint remains visible, the eggplant is ripe, but if the flesh springs back, it is not ready to be consumed. Young, small eggplants tend to have fewer seeds, and their skin is less bitter and more tender. Store eggplants at 8–12 °C (46–54 °F), with a relative humidity of 90–95%, but for only a few days before eating (a perforated plastic bag is recommended for storage in a refrigerator; if eggplants have been wrapped in plastic film when purchased, remove and discard the film). Handle with care, as eggplants bruise easily. Peeling an eggplant before cooking is unnecessary, and usually undesirable, since the skin is tasty; however, very large, older eggplants develop a tough exterior, which may be peeled away. Cut eggplants just before using, as their flesh darkens rapidly when exposed to air. Brushing the flesh with a little lemon juice can prevent discoloration if the dish needs to stand for some time before serving. Eggplants were once salted after cutting and allowed to sit for about a half hour, then rinsed and patted dry, a process designed to eliminate some of the acrid taste. Since much of the bitter taste is localized near the skin, peeling was used to achieve the same goal. Such procedures are only necessary with older, larger, overripe eggplants. Because of the very spongy texture, eggplants seem to drink up several times their weight in oil during frying, although eventually the tissues collapse, releasing much of the absorbed oil. Coat slices of eggplant well with a batter or crumb crust to prevent the oil from penetrating the flesh when deep-frying or pan-frying. When baking eggplant in an oven, pierce the whole, unpeeled fruit in several places to let steam escape. Incisions can be made in the flesh to promote uniform cooking.

Culinary Vocabulary

- "Baba gannouj" (baba ghannouj, baba ghanoush, baba ghanouj), the Middle Eastern eggplant dish described above, is a loose translation of the Lebanese "spoiled old daddy."
- "Poor-man's-caviar" is another colorful name for the eggplant. Baba gannouj has been listed in restaurants in Western countries as "eggplant caviar," and sometimes a thick dip or spread called eggplant caviar is made with puréed, roasted eggplant, onion, olive oil, and seasonings. "Caviar" is in these names because the small, granular seeds are vaguely reminiscent of genuine caviar, i.e., the eggs of the sturgeon.
- "Vegetable liver" is a Jewish-American dish made with eggplant and seasonings. It substitutes for chopped chicken liver, which Kosher dietary laws forbid at certain times. The expression "vegetable liver" has also been used to indicate other vegetable dishes with the texture and appearance of liver.
- A succulent Turkish eggplant dish is known as *immam bayildi*, literally "the priest fainted," because, it is said, an imam consuming it fainted from pleasure. Alternatively, he is claimed to have fainted because his daughter used up an expensive jar of olive oil making the dish; or when he learned of the cost of the ingredients.
- *Hünkâr beğendi*, literally "the king liked it," is a Turkish and Greek specialty consisting of a lamb stew served over an eggplant purée.
- Greeks make an eggplant jam called *gliki melitzanis*.

Curiosities of Science and Technology

- In the 5th century, Chinese high-society women made a black dye from dark eggplant skins to stain and polish their teeth to a black luster.
- Eggplants, like tomatoes and potatoes, are in the Solanaceae (potato family), which contains some species (called nightshades) that have poisonous fruits. Familiarity with these poisonous species led many Europeans to conclude that the rather similar eggplant must also be poisonous. The 16th century Italian physician Castore Durante wrote that eating too much eggplant caused melancholy,

Varieties of eggplant (*Solanum melongena*), from Vilmorin-Andrieux (1885).

cancer, leprosy, headaches, hardening of the liver and spleen, and long fevers, and was even bad for the complexion.

- The cuisine of the Ottoman Empire was obsessed with making stuffed vegetables called "dolmas." To this day in Turkey and nearby countries, eggplants are hollowed out and the flesh consumed immediately, while the eggplant shells are hung from clotheslines to dry.
- Thomas Jefferson (1743–1826; 3rd president of the United States) has been credited with the introduction of eggplant to North America. Indeed, he is widely viewed as the foremost advocate of introducing new crops to North American agriculture.
- A proposed explanation for the lack of popularity of the eggplant (at least in North America) is the color: instinct warns against purple food. Also, the spongy, corpse-like look of eggplant flesh has been claimed to repel some people.
- Eggplant fried good and brown is known for its concentrated flavor, so much so that in Egypt, street merchants announce eggplants for sale by crying, *Ya <arus el-qaleyya* ("the bride of the frying pan").
- Eggplants commonly sold in North America contain about 93% water.
- It is widely alleged that "male eggplants" can be identified by their "outie bellybuttons," while "female eggplants have "innie bellybuttons." Male eggplants are said to have a shallow round dimple in the end opposite the stem (i.e., the blossom end), and the blossom end is rounder and smoother, while the blossom end of a female eggplant is described as having an indentation that is deep and elliptical (shaped like a dash). Female eggplants are supposed to have more seeds than the males, and the males are said to be sweeter because they do not have as many bitter seeds. In addition, some have claimed that females have a brighter sheen to their skin. Botanically, there is no such thing as a male eggplant or a female eggplant (eggplants produce hermaphroditic flowers, which can self-fertilize). However, there is a biological explanation regarding why eggplants with an indentation at the blossom end tend to have fewer seeds. This is the result of poor pollination, producing a condition known as "catfacing" (which also occurs in tomatoes). While fewer seeds may be present, the quality may be less desirable.

Key Information Sources

Adams, W.D. 1975. The enticing eggplant. Am. Hortic. Mag. **54(2)**: 22–24.

Anonymous 1987. Aubergines. OECD, Paris, France. 53 pp.

Carmen, D.P. del, Nuevo, P.A., Amatorio, E.Q., Azucena, M.L.A., Esguerra, E.B., Maunahan, M.V., and Bautista, O.K. (*Editors*). 2001. Postharvest handling of eggplant in the Philippines. Proceedings of the 1st International Postharvest Horticulture Conference 2001. Foundation for Resource Linkage and Development, Inc., Philippines, and University of Los Banos, College, Laguna, Philippines. 280 pp.

Choudhury, B. 1995. Eggplant. *In* Evolution of crop plants. 2nd edition. *Edited by* J. Smartt and N.W. Simmonds. Longman Scientific & Technical, Burnt Mill, Harlow, Essex, UK. pp. 464–465.

Isshiki, S., Suzuki, S., and Yamashita, K. 2003. RFLP analysis of mitochondrial DNA in eggplant and related *Solanum* species. Genet. Resourc. Crop Evol. **50**: 133–137.

Kalloo, G. 1993. Eggplant, *Solanum melongena* L. *In* Genetic improvement of vegetable crops. *Edited by* G. Kalloo and B.O. Bergh. Pergamon Press, New York, NY. pp. 587–604.

Khan, R. 1979. *Solanum melongena* and its ancestral forms. *In* The biology and taxonomy of the Solanaceae. *Edited by* J.G. Hawkes, R.N. Lester, and A.D. Skelding. Academic Press, New York, N.Y. pp. 629–636.

Lester, R.N., Jaeger, P.M.L., Bleijendaal-Spierings, B.H.M., Bleijendaal, H.P.O., and Holloway, H.L.O. 1990. African eggplants—a review of collecting in West Africa. Plant Genet. Resour. Newsl. **81/82**: 17–26.

Martin, F.W., and Pollack, B.L. 1979. Vegetables for the hot humid tropics, part 5, eggplant, *Solanum melongena*. Mayaguez Institute of tropical Agriculture, Mayaguez, Puerto Rico. 18 pp.

Martin, F.W., and Rhodes, A.M. 1979. Subspecific grouping of eggplant cultivars. Euphytica, **28**: 367–383.

Narikawa, T. 1988. Potential of wild relatives of solanaceous vegetables with reference to breeding resistant eggplant. *In* Crop genetic resources of East Asia: proceedings of the international workshop on crop genetic resources of East Asia, 10–13 November 1987, Tsukuba, Japan. *Edited by* S. Suzuki. International Board for Plant Genetic Resources, Rome, Italy. pp. 161–167.

Nothmann, J. 1986. Eggplant. *In* CRC handbook of fruit set and development. *Edited by* S.P. Monselise. CRC Press, Boca Raton, FL. pp. 145–152.

Nothmann, J., Rylski, I., and Spigelman, M. 1983*a*. Color and variations in color intensity of fruit of eggplant cultivars. Sci. Hortic. **4**: 191–197.

Nothmann, J., Rylski, I., and Spigelman, M. 1983*b*. Floral morphology and position, cluster size and seasonal fruit set in different eggplant cultivars. J. Hortic. Sci. **58**: 403–409.

Omidiji, M.O. 1976. Evidence concerning the hybrid origin of the local garden egg (*Solanum melongena* L.). Niger. J. Sci. **10**: 123–132.

Sakata, Y., and Lester, R.N. 1994. Chloroplast DNA diversity in eggplant (*Solanum melongena*) and its related species *S. incanum* and *S. marginatum*. Euphytica, **80**: 1–4.

Sakata, Y., Nishio, T., and Matthews, P.J. 1991. Chloroplast DNA analysis of eggplant (*Solanum melongena*) and related species for their taxonomic affinity. Euphytica, **55**: 21–26.

Sambandam, C.N. 1964. Natural cross pollination in eggplant (*Solanum melongena*). Econ. Bot. **18**: 128–131.

Seelig, R.A., and Bing, M.C. 1990. Eggplant. *In* Encyclopedia of produce. [Loose-leaf collection of individually paginated chapters.] *Edited by* R.A. Seelig and M.C. Bing. United Fresh Fruit and Vegetable Association, Alexandria, VA. 14 pp.

Smith, D. 1979. Peppers and aubergines. Grower Books, London, UK. 91 pp.

Sutarno, H., Danimihardja, S., and Grubben, G.J.H. 1993. *Solanum melongena* L. *In* Plant resources of South-East Asia, 8, vegetables. *Edited by* J.E. Siemonsma and K. Piluek. Pudoc Scientific Publishers, Wageningen, Netherlands. pp. 255–258.

Specialty Cookbooks

Kéhayan, N. 1995. Essentially aubergines: a culinary tale. Grub Street, London, UK. 224 pp. [Reprinted 2002; also published in 1996 as *Essentially Eggplants* by Perseus Publishing.]

Moon, R. 1998. The aubergine cookbook: classic and contemporary recipes for today's healthy diet. Apple, London, UK. 128 pp.

Pazienza, A. 1975. Cooking aubergines, egg plants. Abson books, Abson, Wick, Bristol, UK. 35 pp.

Protheroe, K., and Van der Linde, M. 1999. The lean aubergine. Struik, Cape Town, South Africa. 207 pp.

Ramsay, G., and Denny, R. 1996. Gordon Ramsay's passion for flavour: recipes from the aubergine. Conran Octopus, London, UK. 192 pp.

Smalheiser, S. 2000. Simply eggplant: kosher recipes from around the world. Gefen Publishing House, Hewlett, NY. 179 pp.

Upsom, N. 1979. The eggplant cookbook. Pacific Search Press, Seattle, WA. 154 pp.

Wilton-Smith, J., and Donkin, M.E. 1986. Aubergine: a miscellany of gastronomic delights and horticultural information. Stylus, Salisbury, UK. 24 pp.

FIG

Family: Moraceae (mulberry family)

Names

Scientific name: *Ficus carica* L.

- The English word "fig" and the genus name *Ficus* are assumed to trace to the Latin word for the fig, *ficus*.
- "Capri" in "caprifig" means "goat," an allusion to the fact that this type of fig, a native of Turkey, was fed to goats.
- "Banana figs" are dried bananas ("banana fig" also refers to a species of fig that grows in Australia, *Ficus pleurocarpa* F. Muell., which has yellow, banana-shaped, edible fruits).
- The Mission fig, a popular variety in the United States, acquired its name after being bought to America by Spanish missionaries, dating back to the times of the conquistadors.
- *Carica* in the scientific name *F. carica* refers to Caria, an ancient region of southwestern Asia Minor where figs seem to have been extensively cultivated.

Plant Portrait

The fig is a small, picturesque deciduous tree, generally 3–9 m (10–30 feet) high, with numerous spreading branches and a trunk rarely more than 17.5 cm (7 inches) in diameter. Fig trees can live as long as 100 years and (very rarely) grow as tall as 30 m (100 feet). The fruit of the fig is generally pear-shaped, 2.5–6 cm (1–2½ inches) wide, 2.5–10 cm (1–4 inches) long, and varies in color from yellowish green to coppery, bronze, or dark purple, and may also be white, brown, or black. The skin of the fig is thin and tender, and the fleshy wall is whitish, pale yellow, or amber, or more or less pink, rose, red, or purple. The flesh is juicy and sweet when ripe, gummy with latex when unripe. Seeds may be large, medium, small, or minute and range in number from 30 to 1600 per fruit.

The fig is native to Western Asia and was distributed by man throughout the Mediterranean area. Figs were collected from wild trees by the ancient Egyptians as long as 6000 years ago, and the trees have been cultivated for thousands of years. Figs were introduced into England in the 16th century, and to the Americas after the discovery of the New World. They grow best in semiarid tropical and subtropical regions of the world. There are hundreds of varieties. Today, the United States, Turkey, Greece, and Spain are the leading producers. Four principal varieties of figs are grown in California: the amber-colored Calimyrna; the dark, purplish Mission; the Adriatic; and the Kadota. In Europe, western Asia, northern Africa, and California, commercial canning and drying of figs are important industries.

The fig fruit is not produced conventionally from the ovary of a single flower like most fruits. It is actually a fleshy, flask-shaped, modified stem lined on the inside with hundreds of tiny one-seeded fruits. Technically a fig is a type of "multiple fruit" called a "syconium." Before the female flowers are fertilized to produce the tiny fruits, tiny flowers are massed on the inside wall of the early fig structure. In the common fig, the flowers are all female and need no pollination. There are three other types: the caprifig, which has male and female flowers and requires visits by the females of a tiny wasp, which crawls through a small hole at the end of the fig to reach the tiny flowers; the Smyrna fig, needing cross-pollination by caprifigs in order to develop normally; and the San Pedro fig, which is intermediate, its first crop independent like the common fig, its second crop dependent on pollination. Fruit that has been pollinated produces tiny seeds, and in the case of Smyrna figs, this is said to give them a more nutty flavor.

Culinary Portrait

In warm, humid climates, figs are generally eaten fresh and raw without peeling, and they are often served with cream and sugar. The skin may be peeled back from the stem end to expose the flesh for eating out of hand. Fresh figs are very perishable, with a shelf life of only 1–2 days when removed from cold, moist storage. If placed in a refrigerator (only for a day or two), they should be wrapped well to prevent absorption of odors. They will store for a month at just above freezing, and for several months when frozen. Fresh figs in the market have usually been picked rather unripe to prolong their shelf life, but this is at the cost of taste. Dried figs, like the fresh fruit, are often eaten out of hand. Dried figs can be stuffed with almonds, other nuts, or sections of orange. The dried fruit can be used in place of prunes in most recipes. Fresh or dried, figs are an excellent appetizer and accompaniment of cheese, ham, and other piquant foods. Figs may be stewed or cooked in various ways, as in pies, puddings, cakes, bread or other bakery products, or prepared as jam, marmalade, or paste. The fruit blends well with poultry and wild game. Figs are occasionally used in the Old World as a flavoring for coffee, after roasting and drying. Another occasional Old World use of figs is as a base for preparing alcoholic beverages (*boukha* is a popular brandy made of figs in North Africa). Another alcoholic way of consuming figs is to soak them in port, sherry, or whisky. Fig syrup is well known as a laxative.

Culinary Vocabulary

- "Fig Sue" is a dish that was once commonly consumed in England. It is composed of figs boiled in ale with pieces of bread. Its origin is religious, tracing to the New Testament account of a fig tree without ripe fruit that Jesus came upon on the outskirts of Jerusalem. Jesus said, "let no fruit grow on thee henceforward for ever," and the tree withered away (Matthew 21: 18–19). Jesus' cursing the fig tree for its fruitlessness was an allusion to the fruitlessness of Israel of the times. To commemorate this event, Palm Sunday was once called "Fig Sunday" in England, and Fig Sue was ritually eaten.
- "Foie gras" (French for "fat liver") is the liver of a goose that has been artificially fattened, usually by force feeding (a cruel procedure according to many). A single liver can weigh a kilogram (over 2 pounds), or a tenth of the weight of the whole goose. Usually served as a pâté (about 75% foie gras, with pork fat, onion, and mushrooms), it is considered a great delicacy and very expensive. Roman consul Quintus Caecilius Metullus Pius Scipio, who lived in the 2nd and 1st centuries BC, is credited with inventing foie gras, using figs as the fattening food (corn is usually used today). The French word for liver, *foie*, in fact traces to the Latin word for fig, *ficus*, because of this use.
- "Figpeckers" are Old World warblers (especially species of *Sylvia*) that commonly feed on figs or grapes. In some parts of Europe, these songbirds are used as food, typically enveloped in vine leaves and grilled.
- "Figgy pudding," as mentioned in the traditional carol "We Wish You a Merry Christmas," is made with raisins, not figs. "Figs" was an English dialect word for raisins.
- The fruit most often mentioned in the Bible is the fig. Therefore, it is not surprising that figs are a regular ingredient in "scripture cake," which is made with ingredients listed in the Bible (for example, almonds, butter, citron, eggs, honey, raisins, salt, sugar ("sweet cane"), water, "flour," and "spices"). This cake was popular in the American colonies, and was traditionally served among Baptists. One "recipe" that requires knowledge of the Bible calls for 4½ cups I Kings 4:22, 1½ cups Judges 5:25, 2 cups Jeremiah 6:20, 2 cups I Samuel 30:12, 2 cups Nahum 3:12, 1 cup Numbers 17:8, 2 tablespoons I Samuel 14:25, season to taste with II Chronicles 9:9, 6 (eggs!) of Jeremiah 17:11, a pinch of Leviticus 2:13, ½ cup Judges 4:19, and 2 teaspoons Amos 4:5.
- Fig paste with added wheat and corn flour, whey, syrup, oils, and other ingredients forms the filling for the well-known bakery product "Fig Newton." Fig Newtons were first made by the Kennedy Biscuit Works of Cambridgeport, Massachusetts (now Nabisco). The company already had several lines of cookies named after regional areas, and apparently named Fig Newtons after Newton, Massachusetts, a town (now a suburb of Boston) near the manufacturing plant. The Fig Newton owes its existence to a machine developed by James Mitchell, which combines a hollow cookie crust with a jam filling. The Mitchell invention was first used to make Fig Newtons in 1892.

Curiosities of Science and Technology

- An old Egyptian painting produced 4000 years ago shows trained baboons harvesting figs.
- Grape vines and fig trees are ancient staple crops that have been associated with each other. Both species are biblical symbols of peace and plenty. In ancient Syria and Judea, grape vines were trained particularly on the fig tree. The association of the two is reflected in the biblical proverb: "They shall sit every man under his vine and his fig tree, and none shall make them afraid" (Old Testament, Micah IV/4).
- In classical Greece, athletes were sometimes fed almost entirely on figs, in the belief that the fruit increased their strength and swiftness. In the original Olympic games, winning athletes were crowned with fig wreaths and given figs to eat.
- The ancient city of Attica was famous for its luscious figs. Solon (ca. 630–ca. 560 BC), the ruler of Attica, made it illegal to export figs out of Greece.
- In past times in the Middle East, dried figs were often threaded onto strings for convenient carrying on long arduous journeys across the desert. The figs provided a nutritious, high-protein, high-carbohydrate diet in a region where food was scarce. Today, figs are similarly useful for backpackers hiking in the wilderness.

Fig (*Ficus carica*), from Thomé (1903–1905, vol. 2).

- In the Old Testament, the fig tree of Paradise rescues Adam and Eve from their nakedness. According to legend, the fig was used because it had the largest leaves of the plants in Paradise. Fig leaves covering the genitalia of the first humans led to the metaphor of "fig leaf" as something that conceals an embarrassing truth or excuses some unwise action. A latex is exuded from different parts of the tree, including the unripe fruits, and this may be severely irritating to the skin if not removed promptly. For this reason, it is not a good idea to actually wear fig leaves over naked tender skin. The latex is powdered for use in coagulating milk to make cheese. Indeed, in ancient Greece, shepherds employed fig branches to stir cheese, with the result that the latex present accelerated the coagulation of the cheese. A protein-digesting enzyme in the latex, ficin, is used to tenderize meat. In tropical America, the latex is often used for washing dishes, pots, and pans. It was an ingredient in some of the early commercial detergents for household use but was abandoned after many reports of irritated or inflamed hands in housewives. The latex is used in folk medicine, applied on warts, skin ulcers, and sores, and taken as a purgative (to induce vomiting) and vermifuge (to get rid of intestinal worms), but its use is very risky.

- In Europe, poisoned figs became a customary way of poisoning one's enemies, so much so that "Italian fig" and "Spanish fig" were phrases that implied poisoned fruit intended for the purpose, whether they were figs or some other fruit. More recently, these phrases refer to an obscene gesture.
- In the Mediterranean area, travelers traditionally are allowed to help themselves to a fruit as they pass a fig tree. This hospitality does not extend to picking figs into a basket, which is interpreted as theft.
- There are about 1000 species in the fig genus *Ficus*. Perhaps the most significant besides the common fig is the bo tree, *Ficus religiosa* L., of Burma, Ceylon, and India, also called the bodhi tree, *bodhi* being the Sanskrit word for enlightenment or wisdom. This tree is sacred to both Hindus and Buddhists. Another remarkable species is the banyan tree or East Indian fig tree, *Ficus benghalensis* L., considered sacred in India because the deity Vishnu was born beneath it. Hindus also regard the banyan as sacred, for it is said that Buddha once meditated beneath one. "Banyan" comes from the "banians," Hindu merchants who set up markets in the shade of these enormous trees. One of the largest trees on record grew at the Calcutta Botanic Garden, and was listed in the 1985 and 1997 *The Guinness Book of World Records* as the world's largest (spreading) tree crown, with 1000 prop roots and said to cover an area of up to 1.6 ha (4 acres). However, a banyan in Andhra Pradesh (called the Thimmamma Marimanu) is said to cover 2.1 ha (5.2 acres). The canopy of some banyans provides shade for entire villages. A remarkable story comes from the Transvaal region of South Africa, where a traveler reported a village with 17 conical huts built above the reach of lions on the limbs of one gigantic banyan-type fig tree species.
- Strangler figs are the "boa constrictors of the plant world." In tropical rain forests, their seeds germinate high up in the canopy, where they can receive light. The young strangler fig plants send numerous aerial roots down to the ground, far below. These roots gradually wrap around the trunk of the supporting tree, and gradually shade it to death. The ancient Aztecs and Mayans made paper from the bark of strangler figs, much as the ancient Egyptians made papyrus paper.
- Figs are claimed have the highest dietary fiber content of any common fruit, nut, or vegetable.
- Often said to be the sweetest of fruits, dried figs may be 50% sugar. In fact, before refined sugar, figs were the most important sweetening agent except for honey.

Key Information Sources

Aksoy, U., Ferguson, L, and Hepaksoy, S. (*Editors*). 1998. Proceedings first international symposium on fig. Izmir, Turkey, 24–28 June 1997. International Society for Horticultural Science, Leuven, Belgium. 325 pp.

Beck, N.G., and Lord, E.M. 1988. Breeding system in *Ficus carica*, the common fig. I. Floral diversity. Am. J. Bot. **75**: 1904–1912.

Bolin, H.R., and King, A.D., Jr. 1980. Figs. *In* Tropical and subtropical fruits: composition, properties and uses. *Edited by* S. Nagy and R.E. Shaw. AVI Publishing Co., Westport, CT. pp. 492–505.

Condit, I.J. 1933. Fig culture in California. University of California Press, Berkeley, CA. 69 pp.

Condit, I.J. 1947. The fig. Chronica Botanica, Waltham, MA. 222 pp.

Condit, I.J. 1955. Fig varieties: a monograph. Hilgardia, **23**: 323–538.

Condit, I.J., and Enerud, J. 1956. A bibliography of the fig. Berkeley, CA. 663 pp.

Corrales, M.L., and Garcia, M.J.B. (*Editors*). 2003. Proceedings of the 2nd international symposium on fig, Caceres, Spain, 7–11 May, 2001. International Society for Horticultural Science, Leuven, Belgium. 312 pp.

Crane, J.C. 1986. Fig. *In* CRC handbook of fruit set and development. *Edited by* S.P. Monselise. CRC Press, Boca Raton, FL. pp. 153–165.

Ferguson, L., Michailides, T., and Shorey, H.H. 1990. The California fig industry. Hortic. Rev. **12**: 409–490.

Galil, J., and Neeman, G. 1977. Pollen transfer and pollination in the common fig (*Ficus carica* L.). New Phytol. **79**: 163–171.

Hill, D.S. 1967. Figs (*Ficus* spp.) and fig-wasps (Chalcidoidea). J. Nat. Hist. **1**: 413–434.

Krezdorn, A.H., and Adriance, G.W. 1968. Fig growing in the South. Agric. Handb. 196. U.S. Dept. of Agriculture, Washington, DC. 26 pp.

Lyons, C.G., and McEachern, G.R. 1987. Home fruit production—figs. Texas Agricultural Extension Service, Texas A & M University System, College Station, TX. 7 pp.

Morton, J. 1987. Fig. *In* Fruits of warm climates. *Authored by* J.F. Morton. Creative Resource Systems, Winterville, NC. pp. 47–50.

Obenauf, G. 1978. Commercial dried fig production in California. Division of Agricultural Science, University of California, Berkeley, CA. 30 pp.

Sadhu, M.K. 1985. Fig. *In* Fruits of India, tropical and subtropical. *Edited by* T.K. Bose. Naya Prokash, Calcutta, India. pp. 505–519.

Serafimova, R., and Penev, S. 1980. The fig. K.G. Danov, Plovdiv, Bulgaria. 144 pp. [In Bulgarian, English summary.]

Storey, W.B. 1975. Figs. *In* Advances in fruit breeding. *Edited by* J. Janick and J.N. Moore. Purdue University Press, West Lafayette, IN. pp. 568–589.

Storey, W.B. 1977. The fig (*Ficus carica* Linnaeus): its biology, history, culture, and utilization. Jurupa Mountains Cultural Center, Riverside, CA. 94 pp.

United States Dept. of Agriculture. 1979. Growing figs in the South for home use. U.S. Dept. of Agriculture, Science and Education Administration. Washington, DC. 12 pp.

Valdeyron, G., and Lloyd, D.G. 1979. Sex differences and flowering phenology in the common fig *Ficus carica*. Evolution, **33**: 673–685.

Zohary, D. 1995. Fig. *In* Evolution of crop plants. 2nd edition. *Edited by* J. Smartt and N.W. Simmonds. Longman Scientific & Technical, Burnt Mill, Harlow, Essex, UK. pp. 366–370.

Specialty Cookbooks

Allardice, P. 1993. A fresh look at figs: traditions, myths and mouth-watering recipes. Hill of Content, Melbourne, Australia. 99 pp.

California Dried Fig Advisory Board. 1953–1973. 48 family favorites with California figs. California Dried Fig Advisory Board, Fresno, CA. 19 pp.

Simmons, M. 2004. Fig heaven: 70 recipes for the world's most luscious fruit. William Morrow, New York, NY. 166 pp.

FILBERT AND HAZELNUT

Family: Corylaceae (hazel family; this family is often placed within the birch family, the Betulaceae)

Names

Scientific names: *Corylus* species

European

- European filbert (European hazel, European hazelnut, Barcelona nut, cobnut)—*C. avellana* L.
- Turkish filbert (Turkish hazel, Turkish tree hazel, Byzantine hazel, Constantinople hazel)—*C. colurna* L.
- Giant filbert (Lambert's filbert, Spanish nut)—*C. maxima* Mill.

American

- American hazelnut (American filbert)—*C. americana* Marshall
- Beaked hazelnut (wild filbert, beaked filbert)—*C. cornuta* Marshall

- European nuts of the genus *Corylus* are often called filberts, while the nuts of native species of *Corylus* of North America tend to be called hazelnuts, and this practice is followed here. Some authors have employed the name filbert for any superior nuts from cultivated *Corylus*, but this usage has not been consistent. Another practice has been to call cultivated varieties filberts, and wild types, hazels. Hazelnut is currently preferred by the North American marketing industry. For centuries, it was common practice in England to term nuts with long husks as filberts, and those with short husks as hazels. Sometimes this practice is still followed in Great Britain, but because of hybridization between the two kinds, such terminology is obsolete.
- The name "filbert" is said to have originated from the German *Vollbart*, meaning full beard, referring to the long husk of the nut. It has also been associated with St. Philbert (Phillibert, about 684), a Frankish abbot of Jumièges in Normandy, whose feast date is August 22, when the nuts of European hazelnut have matured. Still another suggestion is that "filbert" traces to the medieval King Philibert of France, who introduced new cultivated hazels.
- The word "hazel" is derived from *haesel*, the Anglo-Saxon word for hood or bonnet, descriptive of the husk enclosing the fruit. Alternatively, the root of *haesel* has been traced to the older Anglo-Saxon *haes*, or German *heissen*, "to give orders," the hazel rod representing a sign of authority for shepherd chieftains.
- There are several interpretations of "cobnut," used for the European filbert in Britain. The word "cobnut" dates at least to the 16th century, when English children played the game "cobnut" with nuts of *Corylus*. The winning nut in this game was the "cob." A "cob" is also the name of a short, stout, English workhorse, and is said to be applied to the nuts because their brown color and stocky shape were considered reminiscent of the horses (a rather fanciful idea). Still another explanation is that the name "cobnut" may derive from the 15th century "cobill nut," the later name perhaps pointing out the similarity of the nut with a cobble or rounded stone. In Britain, "cob nuts" have been distinguished from "filberts" by the husk of the former only partially

European filbert (*Corylus avellana*), from Morren (1851–1885, vol. 3).

European filbert (*Corylus avellana*), from Thomé (1903–1905, vol. 2, plate 173), showing male branch above, fruiting branch below.

covering the nut, while in the latter the husk completely surrounds the nut.

- "Trazels" are hybrids of Turkish filbert (*C. colurna*) with the European hazelnut (*C. avellana*).
- The name "filazel" is applied to hybrids of beaked hazelnut (*C. cornuta*) and European hazelnut (*C. avellana*).
- Hybrids of American hazelnut (*C. americana*) and European hazelnut (*C. avellana*) have been referred to as "hazelberts."
- Contorta is an ornamental variety of *C. avellana* that has corkscrew twisted and spiraled branches. It is also known as "Harry Lauder's Walking Stick," named after the Scottish Vaudeville comedian (1870–1950) who used a wild-looking, bent, gnarly cane.
- The genus name *Corylus* originates from the Greek word *korys*, referring to a helmet or hood, descriptive of the husk that encloses the fruit.
- *Avellana* in the scientific name *C. avellana* is said to be derived either from Abellina in Asia (allegedly the present valley of Damascus in Asia) or the town of Avellino near Pompeii and Mt. Vesuvius in Italy where hazel plantations were established. (Archeological studies have shown that at the time of the eruption of Mount Vesuvius in 79 AD, the inhabitants of Pompeii were eating hazelnuts obtained from extensive plantings near the mountain.) In the first century, the Roman food-specialist Apicius concocted a filbert recipe that he called *nux abellana* ("nut of Abellana").
- *Maxima* in the scientific name *C. maxima* is Latin for very large, a reference to the large size of the plant.
- *Colurna* in the scientific name *C. colurna* is the ancient name for this species.
- *Americana* in the scientific name *C. americana* is the Latinized form of "American."
- *Cornuta* in the scientific name *C. cornuta* is Latin for horned or beaked, a reference to the long, horn-like husk around the fruit.

Plant Portrait

Hazelnut or filbert is second only to almond in world nut production. The main source of the nuts of commerce is the European filbert, a thicket-forming shrub or small tree up to 6 m (20 feet) tall, native to most of Europe. European filberts are thought to be derived, at least in part, by hybridization with the giant filbert, a deciduous shrub or small tree up to 10 m (33 feet) in height that occurs in southeastern Europe and western Asia. The wild giant filbert is often used as a source of nuts, and these are sometimes sold in commerce as the European filbert. Some of the crop from Turkey (where most of the world's filberts are raised) is from hybrids of the European and Turkish filberts. The Turkish filbert is a deciduous shrub or small tree rarely up to 25 m (82 feet) tall, native to southern Europe eastward to Turkey. The Spanish cultivar Barcelona, a European filbert, is predominant in North America. It is cultivated in the Pacific Northwest, including the coastal valleys of the Cascade Range of Oregon and Washington, and the Fraser Valley in southern British Columbia. The American crop provides about 3% of the world supply, and the United States imports about 50% of the hazelnuts it consumes annually. About 70% of the world's filberts come from small Turkish farms on the southern coast of the Black Sea, about 20% comes from coastal regions of Italy, and about 7% from Spain's Mediterranean coastal areas. Filberts are also produced to a minor extent in other areas of Europe, northeastern Asia, Australia, New Zealand, South Africa, and eastern North America.

The two North American species of hazelnut are deciduous shrubs, growing 1–3 m (3–10 feet) tall. American hazelnut grows wild from eastern Canada to Florida. The beaked hazelnut grows wild across Canada, in the northeastern United States, and on the west coast from California northward. These native North American species have been hybridized with the European species to produce hardy varieties of the latter. The nuts of the American species are too small to compete with the European filberts, but there are a few varieties of American hazelnut that are much improved compared to the wild species. The uninitiated wild-nut collector should be warned that the husk of the beaked hazel is covered with tiny, skin-piercing hairs, which can be quite irritating to the hands.

Hazelnuts were eaten by early European man during the Mesolithic period of the Stone Age 10 000 years ago, as indicated by presence of the nuts in many archeological sites. Old Chinese manuscripts indicate that filberts were harvested nearly 5000 years ago, and they were cultivated by the Greeks and Romans 2000–2500 years ago. North American wild hazelnuts were

Turkish filbert (*Corylus colurna*), from Loudon (1844, vol. 8).

Fruit ("nuts") of American hazelnut (*Corylus americana*), from United States Department of Agriculture, Division of Pomology (1896, plate 13). A form with an open involucre is shown at left, and a form with a closed involucre at right.

widely collected by native people throughout their areas of occurrence. Early North American explorers and settlers also ate the wild nuts.

Culinary Portrait

Hazelnuts or filberts are consumed whole, ground, or chopped. The texture and flavor are highly appealing. The nuts are eaten extensively during the Christmas season, but are also used as a substitute for flour, as a spread like butter, in toppings and coatings, and in cereals, garnishes, soups, cookies, ice cream, and chocolate. The nuts are popular in stuffed vegetable, cheese, and other savory vegetarian dishes, and are also made into milks for desserts, and sauces and stuffings for meat, game, and fish. A superior, well-flavored salad oil (not recommended for frying) is extracted from the nuts, sometimes with over 90% unsaturated fatty acids. Like the nuts, it must be stored carefully to prevent rancidity. In western Europe, filberts are particularly important in the confectionery business. Both praline mixtures made from ground hazels and sugar, and a paste made from ground blanched filberts and chocolate are widely used in areas where rival almonds and walnuts are available later in the season; and a whole hazelnut is as likely as the latter to be found in the famous chocolate pralines of Belgium or Switzerland. The filbert is also one of the nuts commonly used in European sugar paste or nougat confections, particularly in Spain. Filberts are often substituted for almonds in the Mediterranean region.

The nuts are marketed shelled or unshelled, and may be available whole or ground, and often roasted and salted. Roasting can be carried out by placing the nuts on a cookie sheet and baking in an oven at 93–141 °C (200–285 °F) until golden brown, stirring occasionally. After roasting but while the nuts are still hot, the thin brown skin can be peeled by rubbing with a thick cloth. The nuts are less perishable than such fatty nuts as Brazil nuts, macadamias, and pecans, but nevertheless become bitter and dry out relatively quickly if not stored properly. Fresh nuts can be kept in a cool dry place for a month. Shelled nuts can be refrigerated for up to 4 months, or frozen for up to a year.

Culinary Vocabulary

- Nutella™ is a hazelnut spread made with skim milk and cocoa virtually unknown in the United States, but familiar in Europe and Canada.
- The Italian expression *brutti ma buoni* (pronounced BROO-tee mah B'WHO-nee), meaning "good but ugly," refers to a meringue cookie prepared with hazelnuts or almonds.

Curiosities of Science and Technology

- After melting of the continental glaciers 10 000–7000 years ago, hazelnut was so abundant in Europe that scientists specializing in studying the past call the period the "Hazel Maximum."

Beaked hazelnut (*Corylus cornuta*), from *Garden and Forest* (1895, vol. 8).

- The famous "Ice Man" (nicknamed Ötzi, after the Ötztaler Alps), found in 1991 in the Tyrolean Alps of Italy and dated to the late Stone Age over 5000 years ago, used hazelwood to reinforce his skin quiver and backpack.
- In Greek mythology, Jupiter had two sons, Apollo (the god of harmony) and Mercury (the god of eloquence), who one day exchanged gifts. Apollo received a tortoise-shell lyre, whose music was intended to enhance the artistic spirit of mankind. Mercury got a winged wand made of hazel, supposed to assist people in expressing their thoughts through words. The symbolic winged hazel rod, entwined with two twisting serpents, became the symbol of communication, and is still used as such. It was said that whomever Mercury touched with his hazel rod gained the gift of eloquence.
- European hazelnut has had numerous medicinal uses dating from the Greek and Roman empires, including treatment of baldness, the common cold, and stomach and sexual problems. Eating 12 filberts before going to bed was employed as a cure for bed-wetting in Spanish traditional medicine. American hazelnut was used to treat ulcers and tumors by the Hurons in southern Ontario and has also been used by native people as an analgesic. Decoctions containing hazelnut were employed by the Chippewa to treat convulsions and hemorrhage and to heal wounds. There is little use of hazelnuts in modern medicine.
- Rhabdomancy, the art of finding underground riches and subterranean springs with a forked branch, was largely practiced with hazelnut twigs. Use of the Y-shaped branch, called dowsing or divining, revealed the location of treasure, minerals, ore, and water. Divining rods were used by the ancient Greeks, and were very popular during the Middle Ages. The Father of biological nomenclature, Linnaeus (1707–1778), admitted to being half a convert to their use. The natural affinity of the hazel for damp places may be at the root of the idea that hazel indicates a water source.
- In the 17th and 18th centuries, hazels became popular among the wealthy for ornamental purposes. They were planted to provide privacy and color in gardens, as ornamental hedges and, in France, were trained into arbors known as "nut walks."

- The fine oil of filberts has a reputation for improving the tone of violins when used in their varnish.
- Foreign names for filberts or hazels are often related to place-names in Europe. For example, filberts are called *avelines* in France because they were cultivated for centuries around the Italian city of Avellino. The Gaelic word for hazel is Coll. It appears frequently in place-names in the west of Scotland, such as the Isle of Coll and Bar Calltuin. It also appears in the name of Clan Colquhoun whose clan badge is the hazel. The longest place-name in Britain (located in Wales) is Llanfairpwllgwyngyllgogerychwyrndrobwll llantyssiliogogogoch, which means (translated from the Welsh) "St. Mary's Church in the hollow of the white hazel near to the rapid whirlpool of Llantysilio of the Red Cave."
- A survey showed that hazelnuts are as popular as almonds in chocolate, but less popular than pecans in ice cream and walnuts in brownies.
- According to *Ripley's Believe It or Not!*, Constantin Boym and Laurene Leon of New York City invented edible almond- and hazelnut-flavored pencils.

Key Information Sources

Axer, J. 1980. Filberts and the world of nuts. Proc. Nut Grow. Soc. Oreg. Wash. B. C. **1980**: 83–87, 90–96.

Baron, L. 1985. Growing hazelnuts in Oregon. Revised edition. Extension Service, Oregon State University, Corvallis, OR. 19 pp.

Bergougnoux, F., Germain, E., and Sarraquigne, G.E. 1978. Le noisetier, production et culture. Invuflec, Paris, France. 164 pp. [The hazelnut tree: production and cultivation. In French. New edition: Germain, É., Sarraquigne, J.-P., Breisch, H., Hutin, C., and Léglise, de Taffin, H. 2004. Le noisetier. Invuflec, Paris, France. 296 pp.]

Brink, V. 1964. Success with filberts. Org. Gard. Farming, **11**(**9**): 28–31.

Catling, P.M. and Small, E. 2000. Poorly known economic plants of Canada—25. American hazelnut, *Corylus americana*, and beaked hazelnut, *C. cornuta*. Bull. Can. Bot. Assoc. **33**: 16–20.

Cho, H.-M. 1988. Potential use of Korean filbert (*Corylus* spp.) landraces. *In* Crop genetic resources of East Asia. *Edited by* S. Suzuki. International Board for Plant Genetic Resources, Rome, Italy. pp. 169–181.

Commonwealth Agricultural Bureaux. 1980. Growing *Corylus* spp. (hazels, cobnuts, filberts), 1972–1979. Annotated Bibliographies. Plant series no. 8/80. Commonwealth Agricultural Bureaux, Farnham Royal, UK. 21 pp.

Debor, H.W. 1978. Bibliography of the international literature on filberts. Technische Universität, Berlin, Germany. 129 pp. [In English and German.]

Farris, C.W. 1972. The tree hazels. North. Nut Grow. Assoc. Annu. Rep. **63**: 24–26.

Farris, C.W. 1978. The trazels. North. Nut Grow. Assoc. Annu. Rep. **69**: 32–34.

Farris, C.W. 1988. Potential for genetic improvement in hazelnuts. North. Nut Grow. Assoc. Annu. Rep. **79**: 46–47.

Gellatly, J.U. 1966. Tree hazels and their improved hybrids. North. Nut Grow. Assoc. Annu. Rep. **57**: 98–101.

Great Britain. Ministry of Agriculture, Fisheries and Food. 1966. Cob-nuts and filberts. Advis. Leafl. 400. Ministry of Agriculture, Fisheries and Food, London, UK. 6 pp.

Jona, R. 1986. Hazelnut. *In* CRC handbook of fruit set and development. *Edited by* S.P. Monselise. CRC Press, Boca Raton, FL. pp. 193–216.

Lagerstedt, H.B. 1975. Filberts. *In* Advances in fruit breeding. *Edited by* J. Janick and J.N. Moore. Purdue University Press, West Lafayette, IN. pp. 456–489.

Lagerstedt, H.B. 1979. Filberts. *In* Nut tree culture in North America. *Edited by* R.A. Jaynes. Northern Nut Growers Association, Hamden, CT. pp. 128–147.

Lagerstedt, H.B. 1983. The American nut industry—filberts (*Corylus avellana*), cultivars, culture production. North. Nut Grow. Assoc. Annu. Rep. **74**: 179–185.

Mehlenbacher, S.A. 1990. Hazelnuts (*Corylus*). *In* Genetic resources of temperate fruit and nut crops. *Edited by* J.N. Moore and J.R. Ballington, Jr. International Society for Horticultural Science, Wageningen, Netherlands. pp. 791–836.

Miller, R.C., and Devlin, K.A. 1948. Processing filbert nuts. Oregon State College, Agricultural Experiment Station, Corvallis, OR. 16 pp.

Organisation for Economic Co-operation and Development.1981. Unshelled sweet almonds, unshelled hazel nuts. Organization for Economic Co-operation and Development, Paris, France, and Washington, DC. 77 pp.

Slate, G.L., and Lewis, G. 1930. Bull. N. Y. State Agric. Exp. Stn. No. 588. New York State Agricultural Experiment Station, Geneva, NY. 32 pp.

Tesche, W.C. 1956. The walnut and filbert industries of the Mediterranean Basin. United States Dept. of Agriculture, Foreign Agricultural Service, Washington, DC. 48 pp.

Thompson, M.M. 1977. Inheritance of nut traits in filbert (*Corylus avellana* L.). Euphytica, **26**: 465–474.

Specialty Cookbooks

Oregon Filbert Commission, and Nut Growers Society of Oregon and Washington. 1973. A treasury of prize winning filbert recipes. 3rd edition. Oregon Filbert Commission, Tigard, OR. 128 pp.

Fonio

Family: Poaceae (Gramineae; grass family)

Names

Scientific names: *Digitaria* species

- Fonio (white fonio)—*D. exilis* (Kippist) Stapf [*Paspalum exile* Kippist, *Panicum exile* (Kippist) A. Chev., *Syntherisma exilis* (Kippist) Newbold]
- Black fonio—*D. iburua* Stapf [*Syntherisma iburua* (Kippist) Newbold]

Vocabulary

- The name "fonio" originated as a French word, based on a native North African name for the plant and the grain produced from it.
- Both white fonio and black fonio produce white and dark seeds; black fonio is "blacker" in the sense that its flowering and fruiting heads are reddish or dark brown.
- In English, fonio has usually been called "hungry rice," a misleading, derogatory name invented by Europeans ignorant of the nature of the crop. Fonio has no relation to rice, although the two are both in the grass family. The use of "hungry" is disparaging, implying that one has to be hungry to eat this food, which in fact is good tasting.
- Other English names for fonio include acha, hungry millet, hungry koos, findi, fonio millet, and fundi millet.
- White fonio has been called "true fonio."
- Black fonio is also called black acha.
- In Nigeria, white fonio is often called acha, while black fonio is called ibura.
- Fonio is a special "millet," i.e., a minor true cereal (see MILLET for additional information on this category of food plants).
- The genus name *Digitaria* is based on the Latin *digitus*, a finger, + *aria*, like, for the finger-like flowering branches.
- *Exilis* in the scientific name *D. exilis* is Latin for slender, a description of the plant.
- *Iburua* in the scientific name *D. iburua* is based on the Nigerian name for black fonio, *ibura*.

Plant Portrait

The two species of fonio are indigenous, annual, West African grasses with small but abundant seed. The seeds of both range from very white to fawn yellow or purplish. White fonio is the most widely used of the two species, and is grown from Senegal to Chad, particularly on the upland plateau of central Nigeria. White fonio is generally up to 75 cm (2½ feet) tall, while black fonio is taller, sometimes reaching 1.4 m (4½ feet). Black fonio was first recognized in the literature fairly recently, in 1911, and its cultivation is restricted to the Jos-Bauchi Plateau of Nigeria and to northern regions of Togo and Benin. Although much less grown, black fonio is not considered inferior and may have considerable potential in countries outside of Africa. Fonio is grown mostly on small farms for home consumption in 15 African countries, mainly across the dry grasslands of West Africa. It is said to be Africa's oldest cereal, and has been cultivated for thousands of years. It is also said to be the world's faster maturing cereal, some varieties producing grain

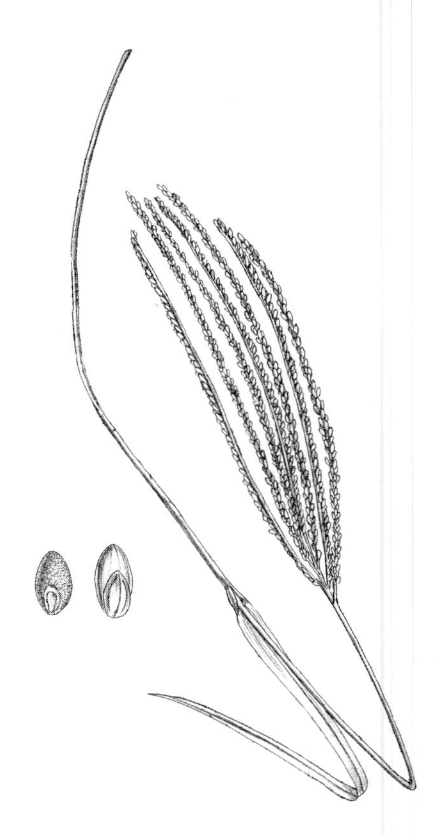

White fonio (*Digitaria exilis*), redrawn from Prain (1916) by B. Brookes.

Black fonio (*Digitaria iburua*), redrawn from Prain (1916) by B. Brookes.

just 6 weeks after planting, although other varieties take as long as 6 months. It was once the major food of West Africa, and today is still either a staple or a major part of the diet of certain regions of western Africa (such as Mali, Burkina Faso, Guinea, and Nigeria). The crop currently is consumed by 3–4 million people. Fonio is rarely cultivated outside of Africa, although it is occasionally grown commercially on a small scale and marketed in North America.

Culinary Portrait

In Africa, fonio is used in various ways: as a porridge, in soups, ground and mixed with other flours to make breads, popped, brewed for beer, and as couscous with fish or meat. It has been described as a good substitute for semolina (the wheat product used to make spaghetti and other pastas). Although fonio is a poor people's food in Africa, it is one of the world's best-tasting and most nutritious cereals. The protein in the grain is of particular value, high in two sulfur-containing amino acids (methionine and cystine), which are vital to human health, but deficient in today's major cereals (wheat, rice, corn, sorghum, barley, and rye). Although almost unknown in Western nations, fonio is occasionally consumed as a gourmet grain.

Curiosities of Science and Technology

- As with many food plants used predominantly in poor countries, fonio has suffered from the reputation of being fit for consumption only by peasants and the poor. Some crops now commonly consumed in Western nations once similarly were subjected to such discrimination. In Great Britain, the English refused to eat potatoes, the food of the Irish poor. In the United States, peanuts were considered to be a "slave food" until the late 19th century.
- The genus *Digitaria* is infamous for crabgrass—species that are serious, annual weeds in lawns and agricultural fields (some "crabgrass" species are in other genera of grasses). Crabgrass is hard to remove from lawns because its low growth escapes the mower, very few herbicides can control it without damaging the lawn, and it grows aggressively. It is often said that, practically speaking, crabgrass can never be completely eradicated.

Crab grass (*Digitaria sanguinalis*), from Hitchcock and Chase (1951).

Key Information Sources

Ayo, J.A. 2004. Effect of acha (*Digitaria exilis* Staph) and millet (*Pennisetum typhodium*) grain on kunun zaki (African traditional beverage). Brit. Food J. **106**: 512–519.

Ayo, J.A., and Nkama, I. 2003. Effect of acha (*Digitaria exilis* Staph.) grain flours on the physical and sensory quality of biscuit. Nutr. Food Sci. **33**: 125–130.

Carcea, M., and Acquistucci, R. 1997. Isolation and physicochemical characteristics of fonio (*Digitaria exilis* Stapf) starch. Starke, **49**: 131–135.

De Lumen, B.O., Thompson, S., and Odegard, W.J. 1993. Sulfur amino acid-rich proteins in acha (*Digitaria exilis*), a promising underutilized African cereal. J. Agric. Food Chem. **41**: 1045–1047.

Gueye, M.T., and Delobel, A. 1999. Relative susceptibility of stored pearl millet products and fonio to insect infestation. J. Stored Prod. Res. **35**: 277–283.

Haq, N., and Dania-Ogbe, F. 1995. Fonio (*Digitaria exilis* and *D. iburua*). *In* Cereals and pseudocereals. *Edited by* J.T. Williams. Chapman & Hall, London, UK. pp. 225–245.

Hilu, K.W., M'Ribu, K., Liang, H., and Mandelbaum, C. 1997. Fonio millets: ethnobotany, genetic diversity and evolution. S. Afr. J. Bot. **63**: 185–190.

Irving, D.W., and Jideani, I.A. 1997. Microstructure and composition of *Digitaria exilis* Stapf (acha): a potential crop. Cereal Chem. **74**: 224–228.

Jideani, A.I., and Akingbala, J.O. 1993. Some physicochemical properties of acha (*Digitaria exilis* Stapf) and iburu (*Digitaria iburua* Stapf) grains. J. Sci. Food Agric. **63**: 369–374.

Jideani, A.I., Apenten, R.K.O., and Muller, H.G. 1994. The effect of cooking on proteins from acha (*Digitaria exilis*) and durum wheat. J. Sci. Food Agric. **65**: 465–476.

Jideani, I., Takeda, Y., and Hizukuri, S. 1996. Structures and physiochemical properties of starches from acha (*Digitaria exilis*), Iburu (*D. iburua*), and tamba (*Eleusine coracana*). Cereal Chem. **73**: 677–685.

Jideani, I.A. 1990. Acha, *Digitaria exilis*, the neglected cereal. Agric. Int. **42**: 132–134, 143.

Jideani, I.A. 1999. Traditional and possible technological uses of *Digitaria exilis* (acha) and *Digitaria iburua* (iburu): a review. Plant Foods Hum. Nutr. **4**: 363–374.

Jideani, I.A., Owusu, R.K., and Muller, H.G. 1994. Proteins of acha (*Digitaria exilis* Stapf): solubility fractionation, gel filtration and electrophoresis of protein fractions. Food Chem. **51**: 51–59.

Kuta, D.D., Kwon-Ndung, E., Dachi, S., Ukwungwu, M., and Imolehin, E.D. 2003. Potential role of biotechnology tools for genetic improvement of "lost crops of Africa:" The case of fonio (*Digitaria exilis* and *Digitaria iburua*). Afr. J. Biotech. **2**: 580–585.

Kwon-Ndung, E.H., Misari, S.M., and Dachi, S.N. 1998. Collecting germplasm of acha, *Digitaria exilis* (Kipp. Stapf.) accessions in Nigeria. Plant Genet. Resour. Newsl. **116**: 30–31.

Lasekan, O.O., and Feijao-Teixeira, J.P. 2004. Aroma compounds of malted acha (*Digitaria exilis*, Stapf). J. Food Qual. **27**: 153–161.

Lasekan, O.O., Teixeira, J.P.F., and Salva, T.J.G. 2001. Volatile flavour compounds of cooked acha (*Digitaria exilis* Stapf). Food Chem. **75**: 333–337.

Nzelibe, H.C., and Nwasike, C.C. 1995. The brewing potential of 'acha' (*Digitaria exilis*) malt compared with pearl millet (*Pennisetum typhoides*) malts and sorghum (*Sorghum bicolor*) malts. J. Inst. Brewing, **101**: 345–350.

Nzelibe, H.C., Obaleye, S., and Onyenekwe, P.C. 2000. Malting characteristics of different varieties of fonio millet (*Digitaria exilis*). Eur. Food Res. Technol. **211**: 126–129.

Obizoba, I.C., and Anyika, J.U. 1994. Nutritive value of baobab milk (gubdi) and mixtures of baobab (*Adansonia digitata* L.) and hungry rice, acha (*Digitaria exilis*) flours. Plant Foods Hum. Nutr. **46**: 157–165.

Temple, V.J., and Bassa, J.D. 1991. Proximate chemical composition of acha (*Digitaria exilis*) grain. J. Sci. Food Agric. **56**: 561–563.

Wet, J.M.J. de. 1995. Minor cereals. *In* Evolution of crop plants. 2nd edition. *Edited by* J. Smartt and N.W. Simmonds. Longman Scientific & Technical, Burnt Mill, Harlow, Essex, UK. pp. 202–208.

GARLIC

Family: Liliaceae (lily family); sometimes placed in Alliaceae (onion family)

Names

Scientific names: garlic species belong to the genus *Allium*. These include the following:

- Garlic (common garlic)—*Allium sativum* L. var. *sativum*
- Rocambole—*A. sativum* var. *ophioscorodon* (Link) Döll (*A. sativum* Ophioscorodon Group)
- Canada garlic—*A. canadense* L.

- The word "garlic" is derived from the Anglo-Saxon *gar*, lance or spear (after the shape of the stem), and *leac*, leak or potherb. The name "garfish," referring to a fish with a long-pointed snout, is also based on the similarity to a spear.
- Obsolete names for garlic include Devil's posy, heal-all, poor man's treacle, rustic's treacle, and stinking rose.
- The English word *treacle* doesn't just mean molasses. It is derived from the Greek *theriake*, an antidote against a poisonous bite, and indeed refers to a medicinal compound formerly widely used against poison. The extensive use of garlic as medicine by the peasantry of the British Isles led to the English names rustic's treacle and poor man's treacle.
- For thousands of years in India, garlic was considered a good treatment for leprosy. After Britain colonized India, leprosy became known as *peelgarlic* because lepers spent much time peeling and eating garlic.
- Rocambole is also known as ophio garlic, serpent garlic, Spanish garlic, and top-setting garlic.
- Canada garlic (*A. canadense*) is also known as American wild onion, Canadian garlic, meadow garlic, meadow leek, rose garlic, wild garlic, and wild onion. The name "wild garlic" is ambiguous, since it is applied to other species, such as *A. oleraceum* L. and *A. vineale* L., Eurasian species naturalized in North America.
- "Tahiti garlic" or "elephant garlic" is not really a garlic, it is an onion (which see).
- Some African species of *Tulbaghia* are known and sometimes eaten as "garlic." In northern North America, some of these are grown in pots as ornamentals, while farther south, they can be grown outdoors. *Tulbaghia violacea* Harv., society garlic, produces garlic-flavored leaves and flowers that can be used in soups, salads, and other foods.
- The genus name *Allium* is the classical Latin name for garlic.
- *Sativum* in the scientific name of common garlic is Latin for planted or cultivated.

Plant Portrait

Garlic is a herbaceous perennial known, for the most part, only in cultivation. Most clones are sterile, and must be propagated from bulbs. The garlic "bulb" is composed of swollen modified leaves like that of the onion. Unlike the onion, the leaves of the garlic bulb are organized into segments known as cloves or kernels. A bulb can contain from six to more than two dozen cloves. The cloves are joined to the central axis of the plant, and may be encased in a membranous (parchment-like) sheath, which can be white, violet, or reddish, depending on variety. In commerce of the world, two general types of garlic have been recognized. "Creole garlic" has large cloves that are dark in color by comparison with "Italian garlic," which has small pinkish cloves with a stronger flavor. It is thought that garlic originated from a wild garlic species, perhaps in the desert of Kirghiz in western Asia. A garlic recipe written in cuneiform characters is known from about 3000 BC. Garlic was consumed over 4000 years ago by the early Chinese, Sumerians, and Indians, as well as the ancient Egyptians. Garlic was also used later by the Romans. It has been suggested that garlic acquired an early reputation for improving physical performance because it is beneficial in amoebic dysentery, a condition common all over the East. Because of the antifungal and antibacterial activity of garlic, its addition to meat products and meat stews in the Middle Ages probably assisted in the control of microbial spoilage. Hernando Cortez (1485–1547), the Spanish conqueror of Mexico, brought garlic to North America. Garlic consumption is relatively low in northern Europe, the United States, and Canada, and high in southern Europe, North Africa, parts of Asia, and Latin America. Leading producers are China, Korea (Rep.), India, Spain, Egypt, the United States, Indonesia, and Tajikistan. California supplies most of the US need for garlic.

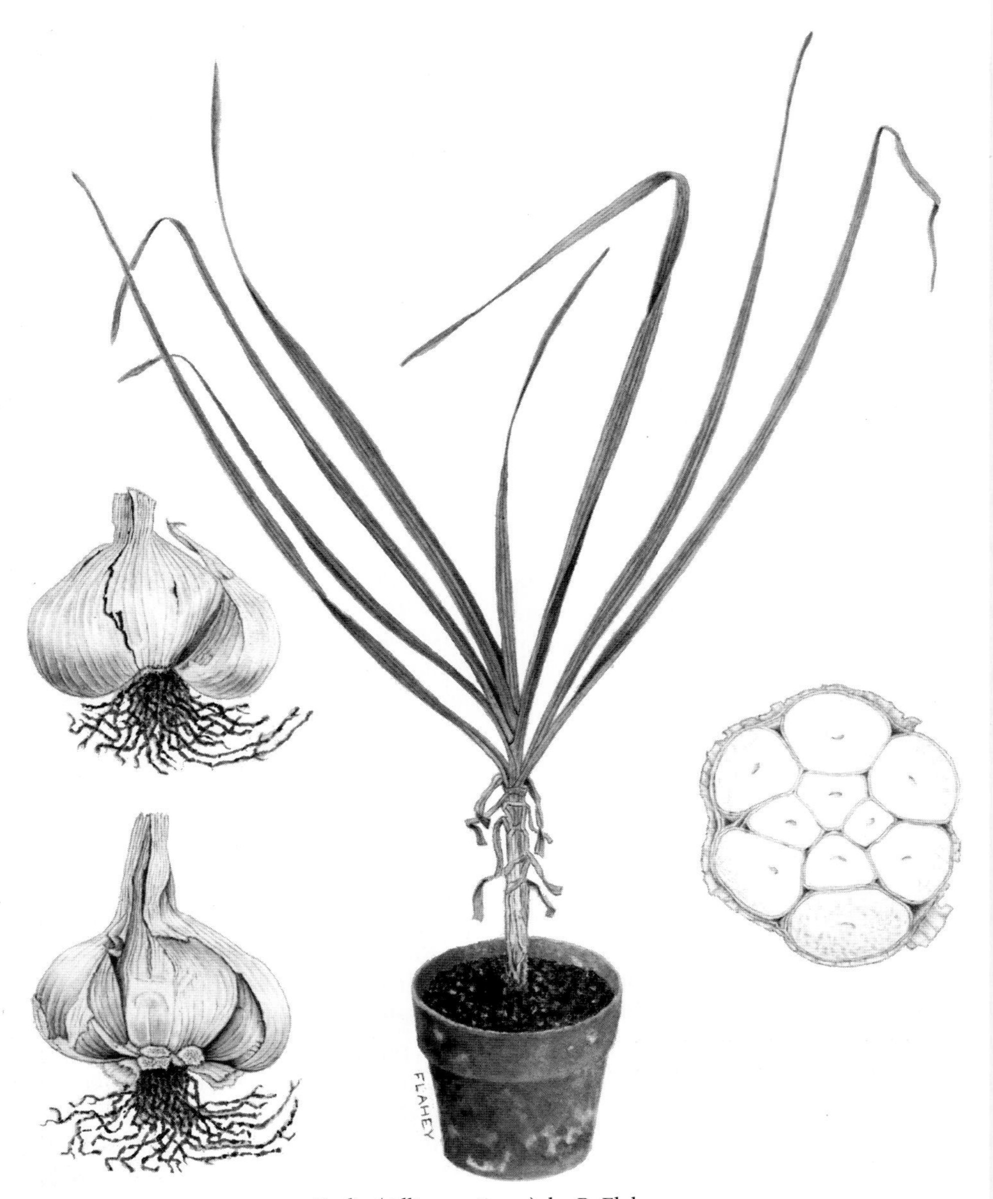

Garlic (*Allium sativum*), by B. Flahey.

One of the most distinctive varieties of garlic is rocambole. This has distinctive flower stalks that twist and bend, first towards the soil surface and then skyward, with pea-sized bulbils produced at the end of the stalk. The bulbs and bulbils are mildly garlic-flavored and are used to season food. The young leaves are delicately chive-like in taste. Several varieties are becoming popular as gourmet items.

Canada garlic is usually collected as a wild food, but it is also sometimes cultivated. It occurs in thickets, woods, and meadows from Ontario to New Brunswick in the north, south to Texas and Florida. The bulbs of Canada garlic are sweet and palatable after boiling.

Culinary Portrait

Although garlic can be consumed as a vegetable, it is primarily used as a condiment. If used sparingly, garlic enhances the flavor of many foods. Some people experience digestive difficulties with garlic, and others may be allergic to it, developing a skin rash or irritation, but for most people, enjoying garlic is simply a matter of experience. Garlic is eaten fresh, dried, or powdered. Fresh garlic is widely employed in cooking and as a flavoring, either whole or grated. Garlic is added to appetizers, stir-fries, pasta, meat dishes, poultry, seafood dishes, stews, casseroles, marinades, soups, salads, salad dressings, and sauces. Meat dishes, such as leg of lamb, can be flavored by inserting slices of garlic into incisions. For a mild flavor, the inside of salad bowls can simply be rubbed with the peeled half of a raw clove. Garlic sauces and preparations are a specialty of the Mediterranean region, and notably include *aioli* (*aïoli*), pesto, persillade, skorthalia, and gremolada. Aioli, a garlic mayonnaise, originates from provincial France. This is a thick, golden, strong-tasting mayonnaise made with eggs, olive oil, and crushed garlic. Skorthalia (skordalia) is a Greek sauce made from garlic, lemon, and nuts, and gremolada is an aromatic garnish of chopped parsley, garlic, and lemon zest. Garlic can be used with any vegetable compatible with onion. The green leaves of fresh garlic can be used like chives. The cloves can be steamed or baked whole. Some authorities maintain that garlic should not be cut, but crushed with a strong blow from the flat of a knife blade, or in a garlic press. However, aïoli needs to be pounded to a cream in a mortar, and other garlic preparations are commonly cut. To release as much flavor as possible, most authorities recommend mashing or chopping fresh garlic; alternatively, for a mild, sweet flavor, garlic can be cooked. If oil or fat used to fry garlic is too hot, an acrid flavor is developed. Garlic becomes milder with cooking, but turns bitter if burned. When garlic is sautéed until it turns brown, its flavor is destroyed and it adds bitterness to foods with which it is combined.

Fresh garlic can be peeled without the skin sticking to one's fingers in several ways (first cut off the tough piece at the bottom):

- Stove Top Method (for large quantities): separate head of garlic into individual, unpeeled cloves. Drop cloves into boiling water, cover, cook 2 minutes, drain in colander, rinse with cold water, let stand until cool enough to handle, peel.

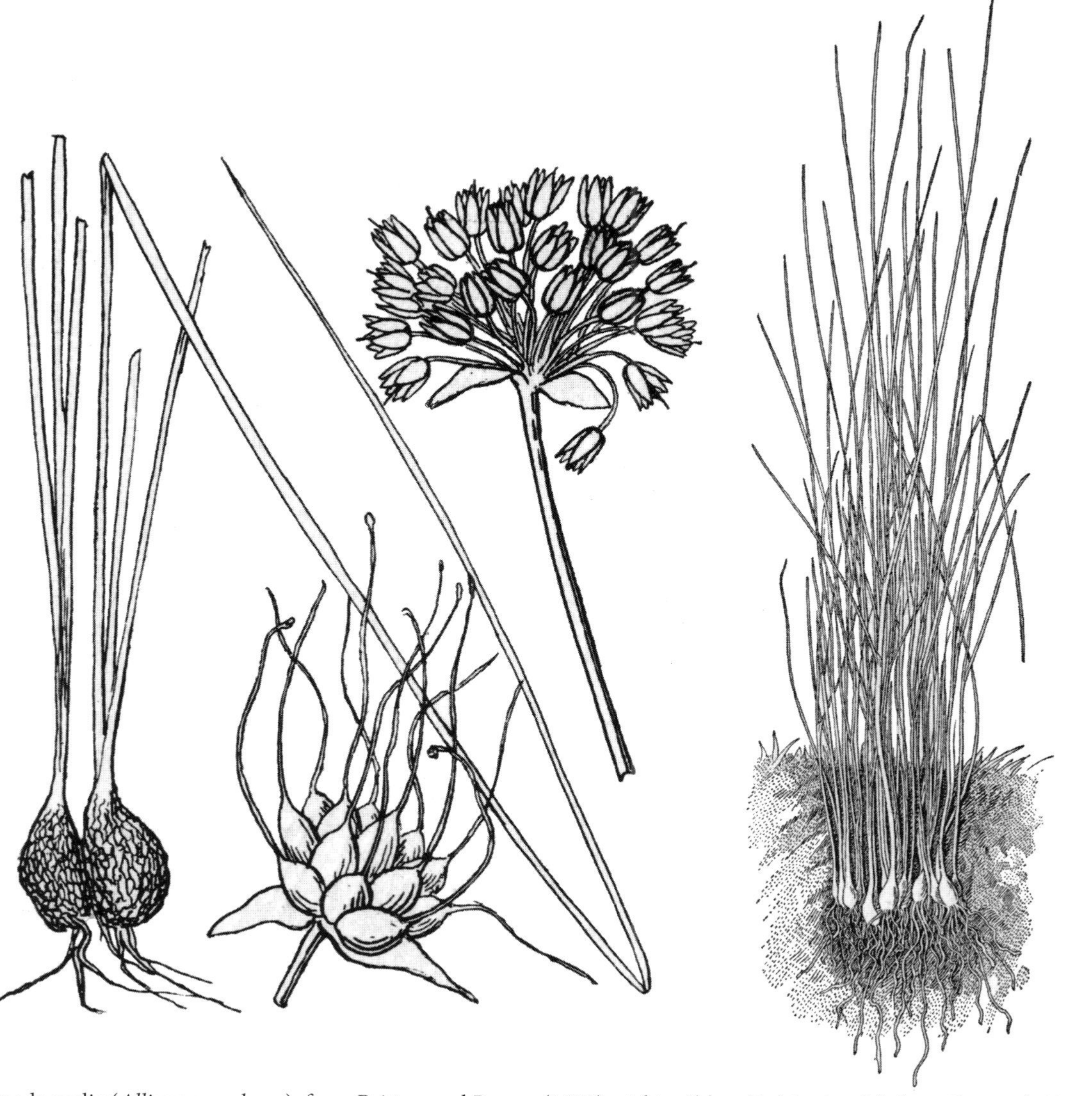

Left, Canada garlic (*Allium canadense*), from Britton and Brown (1898); *right*, wild garlic (*A. vineale*), from Dewey (1897).

- Microwave Method: place whole head of garlic on paper plate, microwave on high for 1 minute (rotating plate, if stationary, for 30 seconds), leave in microwave oven 1 minute, remove and let stand until cool enough to handle, then peel.
- The Knife Method: place clove of garlic on cutting board, place the flat side of a chef's knife on top of the clove, rap the blade sharply with fist to break the skin for easy removal. Too much pressure will smash the clove.

Garlic bread is a traditional item served with Italian meals. Authentic garlic bread is prepared by rubbing toast with a garlic clove, although in recent times, garlic powder and chopped garlic are used for the purpose. Commercially prepared garlic bread is now widely available in supermarkets.

Garlic salt is a free-flowing, blended, dry mixture of noniodized salt (not more than 81%), garlic powder (18–19%), and an anticaking agent such as calcium stearate or tricalcium phosphate. Powdered garlic is extremely hygroscopic, taking up moisture unless stored in tightly sealed containers. Unless kept in a closed container, garlic salt also tends to become lumpy and hard. Starch is sometimes added to help prevent caking. Garlic salt is widely used by the meat packing industry for flavoring, and in the home kitchen for various culinary purposes. The high salt content means that it should not be used by persons who must avoid high sodium intake. A salt substitute recommended by the American Heart Association is made up of one part each of basil, black pepper, garlic powder, mace, marjoram, onion powder, parsley, sage, savory, and thyme, and one-half part of cayenne pepper.

Because garlic can retard clotting time, individuals regularly taking aspirin or other anticoagulant drugs, or susceptible to low blood clotting conditions, should avoid consuming large amounts.

Garlic keeps well, but the bulbs should be thoroughly dry. Individual bulbs tend to lose moisture rapidly, and it is preferable to store a group of bulbs together. The bulbs can be stored in a cold (just above freezing) or cool (18 °C or 64 °F), well-ventilated dark area. Garlic tends to sprout if stored in the refrigerator, and spreads odor to other foods. White garlic is said to keep for about 6 months, pink for nearly a year. When bulbs become soft or spots appear on them, they should be discarded.

Culinary Vocabulary

- A "clove" of garlic is based on the same Germanic source as the English verb cleave, and has nothing to do with the spice called clove (*Syzgium aromaticum* (L.) Merr. & L.M. Perry). Still another independent meaning of clove is the following: a British unit of weight for goods such as cheese, equal to 3.5 kg (8 pounds).
- "Alliaceous" (pronounces ALL-ley-AY-shus) means smelling like garlic or onions. The word can apply to anything, but is particularly used for food and breath.

Curiosities of Science and Technology

- Garlic is one of the oldest medicines. About 1874, the German Egyptologist Georg Ebers discovered the world's most ancient surviving medical text, a papyrus about 20 m (66 feet) long dating to about 1550 BC. This has come to be known as the Codex Ebers. It contains 876 therapeutic formulae of which 22 are garlic prescriptions, including uses for heart problems, headache, bites, worms, and tumors. The Roman scholar Pliny (23–79) declared garlic to be a remedy for 61 ailments (including vampires, hair loss, and warts). There is some scientific evidence supporting past wisdom of using garlic to treat the following: general heart problems, hypertension, blood clots, general wounds, gangrene, plague, cholera, typhus, meningitis, amoebic dysentery, intestinal worms, tumors, hypoglycemia, and stomach and intestinal ailments. Today, garlic is believed to have useful medicinal potential as an antimicrobial agent and for treating atherosclerotic disease and tumors.
- While Major Stephen Long was exploring Nebraska and Colorado in the 1820s, about 100 men in his party died of scurvy. Local Indians provided a wild garlic to the survivors, who did not perish owing to the high vitamin C content of the herb. A decade later, Prince Maximilian of Wied was conducting a 2-year exploration of the Missouri River Valley, and also fell ill with scurvy. A cook who had been with Major Long remembered the garlic cure, acquired wild garlic from Indians, and healed the Prince.
- Garlic juice was considered a valuable antiseptic during World War I, used to combat typhus and amoebic dysentery, and especially to disinfect wounds. The British government ordered tons of the bulbs for such use. The raw juice was diluted with water and put on swabs of sterilized sphagnum moss. This treatment is said to have saved the lives of thousands of men by preventing infection of wounds. Modern antibiotics, particularly penicillin, replaced garlic for these uses by World War II, but Russia had an insufficient supply, and Red Army physicians had to rely heavily once again on

garlic to treat battle wounds to prevent septic poisoning or gangrene. Hence, garlic came to be called "Russian penicillin."

- Research in the Veterans Administration Hospital in Minneapolis provided evidence that garlic breath often lasts for several days.
- Studies in the early 1990s by J.A. Mennella and G.K. Beauchamp showed that lactating mothers who ate garlic produced milk with a garlic odor. However, it was found that infants remained attached to the breast longer and ingested more milk when it smelled of garlic.
- Many suggestions have been made as to how to remove garlic odor from the mouth. Rinsing the mouth and brushing the teeth with soap and water and even 30% alcohol have proven unsuccessful. Home remedies have included strong coffee, honey, yogurt, or milk, and the chewing of coffee beans, cloves, cardamom seeds, parsley, chervil, mint-flavored chewing gum, and even slowly eating a grated potato with a few spoonfuls of honey. The most widely cited method is to chew parsley. Complete deodorization has been achieved by rinsing the mouth and cleaning the teeth and tongue with chloramine, apparently owing to the liberated chlorine. A much more palatable cure is used by the French, who believe that red wine is especially effective; whether or not this works, the problem may seem less important.
- Allicin, the substance responsible for the characteristic garlic odor, enters the blood stream when ingested, reaching all parts of the body. Elimination is not only by air exhaled from the lungs, but also via the skin, so that sweat acquires the odor of garlic.
- An interspecific hybrid between garlic and common onion was reported in 1992. This was an unusual achievement because the two are quite distantly related, and fertile clones of garlic that could be used to make the hybrid are quite rare. The hybrid, which was sterile and incapable of setting seed, produced characteristically onion chemicals as well as characteristically garlic chemicals, and was intermediate in odor. This new vegetable resembled a bulbing onion.
- Garlic juice has been added to oil and sprayed on ponds to destroy mosquito infestations.
- Extracts from garlic have been found to be very toxic to snails that serve as intermediate hosts for parasites attacking cattle and other livestock in parts of northern India.

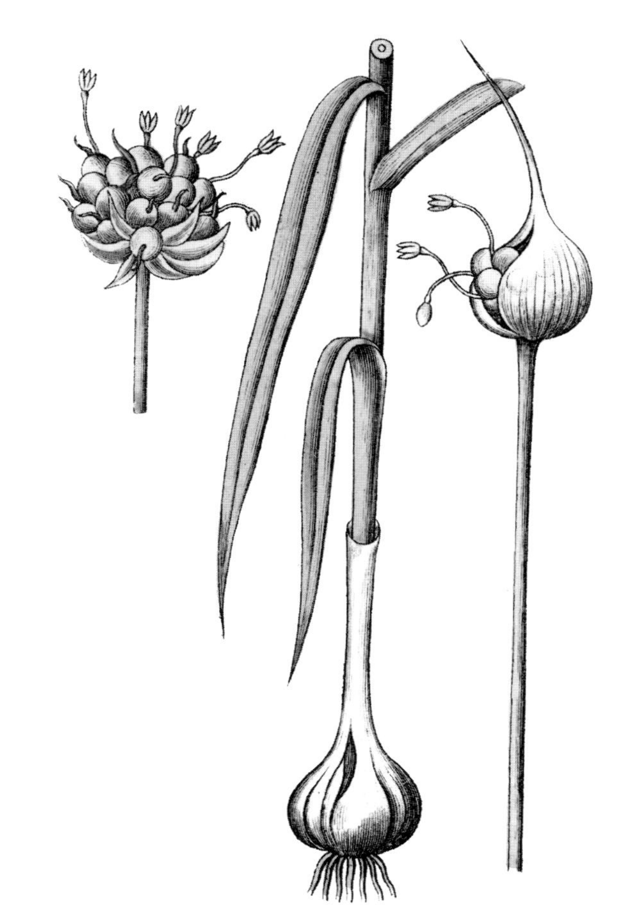

Garlic (*Allium sativum*), from Hallier (1880–1888, vol. 3, plate 254).

- Rather surprisingly, garlic derivatives have been used in perfumery.

Key Information Sources

(See Onion for additional references.)

Allison, L. 1993. The magic of garlic. Cool Hand Communications, Boca Raton, FL. 127 pp.

Batchvarov, S. 1993. Garlic, *Allium sativum* L. *In* Genetic improvement of vegetable crops. *Edited by* G. Kalloo and B.O. Bergh. Pergamon Press, New York, NY. pp. 15–27.

Bergner, P. 1996. The healing power of garlic. Prima Publishing, Rocklin, CA. 290 pp.

Binding, G.J. 1970. About garlic. Thorsons, London, UK. 64 pp.

Brewster, J.L. 1997. Onions and garlic. *In* The physiology of vegetable crops. *Edited by* H.C. Wien. CAB International, Wallingford, Oxon, UK. pp. 581–619.

Dorant, E. 1995. Do onion, leek or garlic pills help against cancer? Voeding, **56(5)**: 14–15.

Engeland, R.L. 1991. Growing great garlic: the definitive guide for organic gardeners and small farmers. Filaree Productions, Okanogan, WA. 213 pp.

Etoh, T. 1985. Studies on the sterility of garlic, *Allium sativum* L. Mem. Fac. Agric. Kagoshima Univ. **21**: 77–132.

Fenwick, G.R., and Hanley, A.B. 1985*a*. The genus *Allium*—Part 1. CRC Crit. Rev. Food Sci. Nutr. **22**: 199–271.

Fenwick, G.R., and Hanley, A.B. 1985*b*. The genus *Allium*—Part 2. CRC Crit. Rev. Food Sci. Nutr. **22**: 273–377.

Fenwick, G.R., and Hanley, A.B. 1985*c*. The genus *Allium*—Part 3. CRC Crit. Rev. Food Sci. Nutr. **23**: 1–73.

Hanley, A.B., and Fenwick, G.R. 1985. Cultivated alliums. J. Plant Foods, **6**: 211–238.

Harris, L.J. 1980. The book of garlic. 3rd edition. Panjandrum/Aris Books, Los Angeles, CA. 286 pp.

Heinerman, J. 1994. The healing benefits of garlic: from pharoahs to pharmacists. Keats Publishing, New Canaan, CN. 196 pp.

Koch, H.P., and Lawson, L.D. 1996. (*Editors*). Garlic: the science and therapeutic application of *Allium sativum* L. and related species. 2nd edition. Williams & Wilkins, Baltimore, MD. 329 pp.

Lawson, L.D. 1998. Garlic: a review of its medicinal effects and indicated active compounds. *In* Phytomedicines of Europe: their chemistry and biological activity. *Edited by* L.D. Lawson and R. Bauer. American Chemical Society, Washington, DC. pp. 176–209.

Mann, L.K. 1952. Anatomy of the garlic bulb and factors affecting bulb development. Hilgardia, **21**: 195–251.

Meer, Q.P. van der, and Permadi, A.H. 1993. *Allium sativum* L. *In* Plant resources of South-East Asia, 8, vegetables. *Edited by* J.E. Siemonsma and K. Piluek. Pudoc Scientific Publishers, Wageningen, Netherlands. pp. 77–80.

Mitchell, J.C. 1980. Contact sensitivity to garlic (*Allium*). Contact Dermatitis, **6**: 356.

Nagpurkar, A., Peschell, J., and Holub, B.J. 2000. Garlic constituents and disease prevention. *In* Herbs, botanicals & teas. Technomic, Lancaster, PA. pp. 1–21.

Organisation for Economic Co-operation and Development. 1979. Garlic. Revised edition. OECD, Paris, France. 45 pp. [In English and French.]

Pandey, U.B. 2001. Garlic. *In* Handbook of herbs and spices. *Edited by* K.V. Peter. Woodhead Publishing, Cambridge, UK. pp. 180–194.

Rabinowitch, H.D., and Currah, L. (*Editors*). 2002. *Allium* crop science: recent advances. CABI Publishing, New York, NY. 515 pp.

Ramanujam, K. 1962. Garlic in the treatment of acute leprosy neuritis. Lepr. India, **34**: 174.

Reuter, H.D., and Sendl, A. 1994. *Allium sativum* and *Allium ursinum*: chemistry, pharmacology and medicinal applications. Econ. Med. Plant Res. **6**: 56–113.

Rosin, S., Tuorila, H., and Uutela, A. 1992. Garlic: a sensory pleasure or a social nuisance? Appetite, **19**: 133–143.

Silagy, C., and Neil, A. 1994. Garlic as a lipid lowering agent—a meta-analysis. J. R. Coll. Physicians Lond. **28**: 39–45.

Warshafsky, S., Kamer, R.S., and Sivak, S.L. 1993. Effect of garlic on total serum cholesterol. A meta-analysis. Ann. Intern. Med. **119**: 599–605.

Whitaker, J.R. 1976. Development of flavor, odor, and pungency in onion and garlic. Adv. Food Res. **22**: 73–133.

Specialty Cookbooks

(Note: There are at least a hundred cookbooks dedicated to garlic. The following are examples.)

Aaron, C. 1997. The great garlic book: a guide with recipes. Ten Speed Press, Berkeley, CA. 146 pp.

Baker, S., and Sbraga, M. 1997. The unabashed garlic & onion lover's international cookbook. Avery Publishing, Garden City Park, NY. 197 pp.

Belsinger, S., and Dille, C. 1993. The garlic book: a garland of simple, savory, robust recipes. Interweave Press, Loveland, CO. 71 pp.

DiResta, D. 2000. The garlic cookbook. Revised edition. Bristol Publishing Enterprises, San Leandro, CA. 155 pp.

Ferrari, L. 1994. The good-for-you garlic cookbook: over 125 deliciously healthful garlic recipes. Prima Publishing, Rocklin, CA. 191 pp.

Friedlander, B., and Cato, B. 1976. The great garlic cookbook. Macmillan, New York, NY. 160 pp.

Gilroy California. 1987. The complete garlic lovers' cookbook: from Gilroy, garlic capital of the world. Celestial Arts, Berkeley, CA. 349 pp.

Gilroy Garlic Festival. 1993. Garlic lover's greatest hits, 1979–1993: 15 years of prize-winning recipes. Celestial Arts, Berkeley, CA. 129 pp.

Griffith, L., and Griffith, F. 1998. Garlic, garlic, garlic: exceptional recipes from the world's most indispensable ingredient Houghton Mifflin, Boston, MA. 432 pp.

Holder, K., and Duff, G. 1996. A clove of garlic: garlic for health and cookery: recipes and traditions. Chartwell Books, Edison, NJ. 128 pp.

Kreitzman, S. 1984. Garlic: 101 savory and seductive recipes, along with fascinating facts and folklore. Harmony Books, New York, NY. 128 pp.

Periplus Editions. 2000. Garlic: garlic recipes by leading chefs from around the world. Periplus, Boston, MA. 159 pp.

Perry, S. 2004. Everything tastes better with garlic: positively irresistible recipes. Chronicle Books, San Francisco, CA. 131 pp.

Renoux, V. 2004. For the love of garlic: the complete guide to garlic cuisine. Square One Publishers, Garden City Park, NY. 208 pp.

GRAPE

Family: Vitaceae (grape family)

Names

Scientific names: *Vitis* species

- Wine grape—*Vitis vinifera* L
- Fox grape—*Vitis labrusca* L.
- Muscadine grape—*V. rotundifolia* Michx.

- The English word "grape" is from the Old French words *grape*, *grappe*, and *crape*, meaning "bunch of grapes." The word was derived in the 13th century from the French term *grape*, a small bill-hook used to cut off grapes from the vine; this usage tracing to the Old High German *krapfo*, "a hook."
- The scientific name *Vitis* is the classical Latin word for vine.
- *Vinifera* in the scientific name of the European grape is Latin for wine-bearing. *Vitis vinifera* is known simply as the grape, but also as common grape, Old World grape, wine grape, and raisin grape.
- The fox or skunk grape is said to be named for the "foxy," aromatic, or musky odor. By 1864, Webster's Dictionary defined "foxy" as having the coarse flavor of the fox grape. Whether the grapes actually smell like a fox has been questioned, and in any event, very few people today have been close enough to a fox to smell one. It has also been said that "fox" in the name was simply a way of indicating that the species is wild. The fox grape is also known as the northern fox grape and skunk grape.
- *Labrusca* in the scientific name of the fox grape is an early Latin name for European wild grapes.
- "Muscadine" is probably a variant of "muscatel," from Old French *muscadel*, resembling musk (tracing to the Latin *muscus*, musk), and applied to the Muscat grape used for wine and raisins. The muscadine grape is also known as the bullace, scuppernong, and southern fox grape.
- "Catawba" was the name of an Indian tribe of the Carolinas who were part of the Sioux nation. The word was originally *katapa*, meaning "to be separate," reflecting the separation of the Catawba from the other Sioux, who resided mostly in the west on the Great Plains. The name was passed on to the Catawba River, part of the original homeland of the tribe, and eventually to the Catawba grape.
- *Rotundifolia* in the scientific name of the muscadine grape is Latin for "rounded-leaved."
- In 1876, William Thompson, a Scottish immigrant to the United States, grew the Sultana grape, a seedless Turkish grape variety that was sweet, thin-skinned, and seedless. The Sultana was renamed the Thompson Seedless, and now makes up 95% of California's raisins. The original name, Sultana, is still popular outside of North America.
- Scuppernong is the best-known variety of the muscadine grape, *V. rotundifolia*. The name "Scuppermong" comes from the Scuppermong River, and is derived from the Algonquin Indian name *Askuponong*, interpreted as "place of the sweet bay tree" or "place of the magnolia." The first English settlement in the New World was in 1586 at Roanoke Island, off the northeast coast of North Carolina, and the colonists are credited with discovering the original vine growing there around 1584–1585. During the American Civil War, troops garrisoning Roanoke Island noted the vine was still there and growing strong. Today, the vine is more than 400 years old, has a trunk 0.6 m (2 feet) thick and covers 0.2 ha (half an acre), and is a tourist attraction. Over the years, many cuttings were taken to grow the grape commercially.

Plant Portrait

"Grapes are the highest value commodity that we know of in agriculture" (Christian Butzke, Purdue University; reported in 2006 by Bloomberg News Service), so it is not surprising that they are the subject of very specialized areas of science and technology. Viticulture is the cultivation and production of grapes; enology is the science of wine and wine making; ampelography (pronounced am-peh-LAW-gra-fee; from the Greek *ampelos*, vine, + *graphe*, writing) is the study and classification of grapevines and grapes (which today includes DNA fingerprinting).

There are about 60 species of the genus *Vitis*, all of which produce edible, although not necessarily palatable, fruits. The woody vines of grape species climb by means of tendrils. The principal commercial grape species are discussed below. Many grape cultivars are of hybrid origin, and not simply referable to any one species. Also, many wild species, and hybrids, are used as rootstocks on which the fruit-bearing portions of the principal species are grafted. Modern cultivars of *Vitis* are clones, essentially genetically uniform, and are propagated vegetatively, like apple and potato varieties.

Wine grape (*Vitis vinifera*), from Köhler (1883–1914).

Seedless clones are of course easily reproduced without seeds. Grapes are an unusual fruit in that they must be allowed to ripen fully before harvesting, as further ripening does not occur after the fruit is picked. Accordingly, many attempts to grow grapes in regions to which they are not climatically suited end in disaster. Nevertheless, there are 2000 or more wineries in North America, and almost every state of the United States and province of Canada produce some type of wine. Mexico makes some commercial wines, but its grapes are used mostly for producing brandy.

Wine grape

This species is by far the most important, accounting for 95% of grape production. *Vitis vinifera* is somewhat smaller and less high climbing than most American species of *Vitis*, but nevertheless can easily spread or climb 20 m (66 feet) from the root system. The plant is a long-lived woody vine, and in commercial production, it is drastically restricted in size by pruning, in order to facilitate harvesting. Pruning of grapes grown for wine production also is intended to limit the number of bunches of grapes, and thus concentrate the flavor. A single bunch of grapes may contain only several grapes, or as many as 300. The plant is generally supported on a trellis, but some wine types are grown with a short, self-supporting trunk. Individual fruits are 1–3 cm (1/3–1¼ inches) in diameter, and colors include light green, yellow–green, green (known as "white" in Europe), blue–black (purple), red, and black. The skin has a thin, powdery coating called the "bloom." The common grape is one of the most ancient of harvested plants. It probably originated in the Mediterranean area and was cultivated in the Near East perhaps as long ago as 6000 years. Making wine from grapes is a Western practice. Evidence of wine making dates back to the Bronze Age in Europe, approximately 3000 BC. Currently, European grapes are commercially cultivated around the world, mostly in Western Europe, the Balkans, California, Australia, South Africa, and parts of South America. France, Italy, Spain, and the US produce and consume more than half of the wine produced in the world. Some wild fruit is capable of producing wine as good as made from many modern cultivars, so it is not surprising that *V. vinifera* grapes are ideally suited to wine making, and on the whole, other species produce inferior wines. Thousands of wine grape cultivars are known, although less than 100 are grown in substantial quantities. Cabarnet Sauvignon is the most important red wine grape in the world (Merlot, Pinot Noir, and Syrah (Shraz) are other leading red wine grapes); Chardonnay is the world's most important white wine grape (Riesling and Sauvignon Blanc are other leading white wine grapes). The European grape is not well adapted to cultivation in North America except for part of California, but hybridization with American species has increased its range of cultivation. Many of the hybrids are called "French hybrids" because French breeders did much of the work.

Fox grape

The fox grape is native to the eastern United States. The grape varieties developed from this species are hardier and more disease resistant than the Old World grape. The berries may be green, red, purple–black, or reddish brown, and are 1.3–2.5 cm (1/2–1 inch) in diameter. They are sweet or astringent, and "foxy" in flavor. Generally, fox grape cultivars are hybrids with other species, especially the common grape, or at least have some other species in their parentage. Concord is the best-known variety, and Catawba, Delaware, and Niagara are other well-known varieties. Concord was originated by Ephraim W. Bull (1805–1985) in Concord, Massachusetts. It grew from a seed planted in 1843, which bore fruit in 1849. Prior to the appearance of the Concord, grape growing in Eastern America had limited success. Except for cultivars of the common grape, the Concord is the most important grape cultivar in the New World. Grape juice and jelly are typically made from Concord grapes. Catawba was first discovered in woods near the Catawba River in North Carolina in 1901. It is a hybrid of the fox grape and the wine grape, and produces a dark, rich, red wine with a taste reminiscent of muscat. The cultivation of Catawba helped to establish grape growing in the Atlantic states.

Muscadine grape

The muscadine grape grows as a wild plant in the southeastern United States and in Mexico. The berries are 2–4 cm (3/4–1½ inches) in diameter, and have a thick, tough skin. The grapes tend to be dull purple, but a range of other colors has been recorded. As with all American species, the wild fruit was harvested by Native Americans. The species was domesticated by settlers in the New World, and a number of cultivars are now grown in the Cotton Belt of the United States. Commercial production of muscadine grapes is essentially limited to the US Southeast. Scuppernong is the oldest and best-known variety, and indeed its name has become a general name for all muscadine grapes. Several dozen cultivars are now available. Like the fox grape, muscadine grapes may be strongly musk-flavored. However, some modern cultivars have a unique fruity flavor with very little muskiness. The fruit of the muscadine grape is eaten fresh and is commonly made into juice, wines, pies, jellies, and other processed products.

Wine has been enormously important to society, perhaps most interestingly in its relationship to religion. During the Roman Era, grape growing became associated with Christianity, wine being an essential ingredient in the consecration of the Mass. Not surprisingly, Catholic monasteries were very important centers of wine grape growing in the Middle Ages. However, there is a schizophrenic relationship between wine (an alcoholic beverage) and Christianity. Wine is mentioned 185 times in the Old Testament and 26 times in the New Testament. The Bible warns against the evil of drunkenness (e.g., Proverbs 20:1, Habakkuk 2:15). Nevertheless, Jesus served wine at the Last Supper, and performed his first miracle by turning water into wine during the wedding reception at Cana (John 2:1–11). According to section 924 of the Roman Catholic church law code, wine served at mass must be natural, i.e., alcoholic. The fact that Catholic parishioners frequently receive only bread, and not wine at the altar, became an important issue of dispute with Protestants in the 16th century Reformation. Most American Protestant churches use grape juice in communion services, as does the United Church in Canada. The zealously anti-alcohol Mormon religion requires that water be consumed during communion to avoid even the slightest risk of fermentation. Some Protestant theologians do not accept the view that Christ associated himself with wine, a sinful alcoholic beverage, and interpret favorable Bible mention of "wine" as actually referring to grape juice. Muslims do not drink wine because the Koran considers it a "satanic device." Nevertheless, there are said to be "rivers of wine" in paradise, and the devout are rewarded with it in the afterlife.

Jewish dietary laws are particularly strict with regard to prepared foods made with grapes, and accordingly when they are kosher (i.e., meet prescribed standards for consumption by religious Jews). These restrictions derive from Jewish laws against using products of idolatry. Wine was commonly used in the rituals of ancient pagan religions, and accordingly, use of wines and other grape products made by non-Jews was, and remains, prohibited. Even fruit drinks or fruit-flavored drinks sweetened with grape juice are forbidden unless prepared by Jews under supervision of a rabbi, and certified to be kosher. Kosher baking powder is extremely uncommon because baking powder is made with cream of tartar, a by-product of wine making. However, the dietary restrictions do not apply to whole grapes, and a very restricted range of simple products made from whole grapes, such as some fruit cocktails and grape jams made from whole grapes. However, grape jams produced from grape pulp or grape juice (which fall into the category of grape products) are only kosher when they meet dietary law requirements and receive official approval.

Culinary Portrait

Table Grapes

Grapes used fresh should have attractive appearance (large, bright, and sometimes unusually shaped berries) and, for storage or shipping, have firm, tough, fairly thick skin. Seedless table grapes have become very popular in recent decades. Grapes can be eaten out of hand, added to salads, used as an accompaniment to poultry, meat, and seafood, made into sauces, jam, or jelly, or used as a component of a very wide range of dishes, including pastries, stuffings, curries, and stews. Grapes tend to shrivel and ferment at room temperature, and are best stored in a perforated plastic bag in a refrigerator. If only part of a bunch of grapes is being consumed, it is preferable to cut off small clusters from the main stem with scissors, instead of pulling off individual grapes, as this tends to cause drying out of the remaining bunch when it is stored.

Juice Grapes

Juice grapes are used to produce unfermented juice. Grape juice is pasteurized in North America, which causes most vinifera cultivars to lose their fresh flavor. The strongly flavored American variety Concord retains its flavor very well and so is widely used for juice.

Raisins

Grapes used for drying are table grapes, rather than wine grapes. Raisin grapes are produced primarily from the varieties Thompson Seedless, Malaga, Sultanina, Black Corinth, and Muscat. Desirable characteristics of raisin grapes include the following: seedlessness; pleasing, rich flavor; soft texture; high sugar content; tender skin; and little tendency to become sticky. The large-fruited Muscat has seeds, but these are removed after drying. Corinth grapes, also known as Zante currants, are tiny, black, seedless grapes, the raisins of which are very popular in making pastries. "Golden raisins" are produced by treating Thompson grapes with sulfites before they are dried, resulting in conservation of their light color, which varies from amber to golden yellow. Raisins are eaten directly, added to nut and dried-fruit mixtures, mixed with cereals, salads, sauces, poultry stuffing, meat loaf, puddings, and other dishes, and cooked in pastries, breads, pudding, and other confections.

Edible Leaves

Grape leaves have been a minor food throughout recorded history, and young, tender leaves are still consumed, especially in Greece, Romania, Turkey, and North Africa. Grape leaves are consumed raw in salads, or stuffed with rice, meat, and (or) other foods and baked. Middle Eastern countries have their own variations of stuffed grape leaves, of which perhaps the best known is the Greek *dolmades* (often called "dolmas," although this is a more generic term, referring to any fruits, vegetables, or leaves that are prepared as stuffed dishes). Canned grape leaves are usually available in markets specializing in Middle Eastern foods.

Edible Seeds

Grape seeds are edible, and in some cases, the oil is pressed out and used for cooking. Recently, grape-seed extracts have become popular in dietary supplement products. These extracts have been used to treat a variety of conditions, and seem to be most effective for preventing degeneration of the vascular system.

Miscellaneous Old Uses

In early times, grapes were used to make some food products that are no longer significant. One of these was grape sugar, a sweetener that was second only to honey. Grape sugar remained important until after the Crusades (11th–13th centuries), when cane sugar began to be imported from the Near East. During the Middle Ages, "verjuice," the juice of unripe

Labrusca grape (fox grape, *Vitis labrusca*), from Bailey (1900–1902).

fruit (particularly grapes), was a staple cooking additive, employed to add tartness.

Wine

Although wines can be made from various fruits and even other materials (mead, for example, is a wine made from honey), wine is mostly fermented grape juice. Today, most grapes are consumed as wine and spirits, and in the light of this, it is perhaps not surprising that world production of grapes exceeds that of any other fruit. Table (dessert) grapes, raisin grapes, wine grapes, and juice grapes are often different, specialized cultivars. Cultivars suitable for table grapes are usually not suitable for making wine, and vice versa. All mature grapes may be fermented into wine, but only certain cultivars produce wine of high quality, characterized by distinctive flavor, bouquet, and general excellence. Table or dry wines are produced from grapes of high acidity and moderate sugar content, whereas dessert or sweet wines are usually made of grapes high in sugar and relatively low in acid. High yield and high quality of the resulting wine are the standards by which wine grapes are judged.

Wine is a product of immense cultural appeal, and some comments about it are in order. Wines are produced by fermentation of the sugar in grapes, largely by yeasts that are present on the skins. As well, artificially cultured wine yeasts are sometimes added to the crushed grapes in a fermentation vat. In fermentation, sugar is converted to alcohol and carbon dioxide. Wine quality is determined by cultivar, climate, soil, and the skill of the grape grower and wine producer. "Dryness" or sweetness of wines is the result of the amount of sugar present. Red wines owe their color to the pigment of black (blue) grapes, which is not removed from the juice during fermentation. In rosé wines, which are pink, the red skin of the grapes is left in the fermenting juice for only a limited period. White wines are derived from "white" (i.e., green) grapes, or from black grapes deprived of their pigmented skin. Natural or "table" wine (such as Bordeaux and Chablis) is produced by natural fermentation with just a slight addition of yeast and sugar. The alcoholic content of true wines varies from 7% to about 16%. Some wines (such as sherry and port) are fortified by the addition of alcohol (usually in the form of grape brandy), and alcohol content is 15–22% by volume. Sparkling wines are manufactured by corking bottles before fermentation has ceased (so-called double-fermentation), allowing carbon dioxide to accumulate (regular or nonsparkling wines are called "still wines"). Champagne is the best example of a naturally sparkling wine. Artificial sparkling wines simply have carbon dioxide added. Brandies, such as Cognac, are distillates from fermented grape juice, or in some cases are produced from dried grapes. Brandy may be used in preparing liqueurs and fortified wines.

Any wine can be served with any dish, but it is also true that chocolate mousse can be served with ketchup. (Champagne is said to be the only wine that can be served with any food at any time.) Matching wines and food is the subject of considerable snobbery, but most tastes will be satisfied by following some simple rules. Basic tips for serving wine with food follow.

- Sparkling wines are best as an apéritif, i.e., served before the main course. Sweet sparkling wines are also excellent at the end of a meal.
- Chocolate can dull the taste of wine, even making it bitter, unless the wine selected is sweeter than chocolate. Ice wines and port stand up well to the richness of chocolate.
- When strong sauces and strongly flavored dishes are served, full-flavored, full-bodied wines are most appropriate.
- Curry powder and very spicy food are too overpowering to be served with wine. Beer is better.
- Delicate foods and light dishes can be overwhelmed by strong wines. Lighter, more fruity wines are preferable.
- Meat-textured fish such as tuna and shark are best paired with a light red wine, as the texture of the fish overwhelms most white wines.
- Pickles, artichokes, bananas, and citrus fruits clash with most wines.
- Salads and dishes with vinegar reduce the taste of most wines. Mineral water is preferable for salads.
- Smoked fish and meat, asparagus, and eggs are difficult to match with wine.
- When serving more than one wine at a meal, several recommendations apply: choose whites before reds, dry before sweet, light before heavy, simple before complex, and young before old.

Chilled fortified wines make suitable apéritifs, especially with soups and first courses. Dry or semidry white wines are generally recommended to accompany white meats, including fish and shellfish. Dry or hearty red wines go well with roast meats and game. Rosé wines complement luncheon meats and some fowl and fish. Dry, white, red, or fortified wines go well with cheeses. Red wine is appropriate with pasta, especially in tomato or meat sauce (Chianti, a dry red wine, is commonly served with spaghetti). Sweet wines should be served with or following desserts.

The most common fault of wine, corked wine, results from a fungus interacting with a chemical used to sanitize corks. Such tainted wine is recognizable by a musty, mildewy aroma and flavor. It has been estimated that 1 in 12 bottles is corked. Plastic corks or screw caps could cure this problem, but have not yet been widely accepted by consumers, although there is a trend to substitute plastic for cork in higher-quality wines. The second most common fault, oxidized wine, often occurs when the bottle is stored upright instead of on its side. The cork dries out and oxygen enters the bottle. Oxidization can also occur from the wine being overexposed to air while being made, aged, bottled, or stored. Air causes wine to taste stale and turn a deep yellow or brownish color. Mild exposure to air can result in wines of low acidity smelling cooked, and those of high acidity smelling burnt. Higher exposure to air results in wines that smell and taste of vinegar, or worse. Sherry and some other wines are deliberately exposed briefly to oxygen to develop their unique character. Madeira is also given a treatment that is very undesirable for other wines: it is baked in its cask for 3–6 months to give it a caramelized flavor. However, "maderized wine" (occasionally "madeirized"), typically produced by severe temperature fluctuations in a short time, is a fault of most wines. This may result from frequently taking the wine in and out of refrigeration, or leaving the wine in a hot car.

Wine should be stored at cool temperatures (10–12 °C or 50–54 °F), with as little temperature fluctuation and as little light as possible (wine bottles are tinted to prevent light from spoiling the wine). Humidity is of little importance (some of the greatest winery cellars are damp and moldy). To prevent dampness from rotting labels, some people spray them with a fixative. Temperature- and humidity-controlled wine cabinets are available for the serious wine collector who lacks an appropriate cellar.

According to the 20-minute rule for chilling wine, whites should be removed from the refrigerator 20 minutes before serving, while reds should be put in 20 minutes before serving. Another rule is that white wines, including Champagne, should not be refrigerated for more than a few hours, as this can dull both the flavor and bouquet. Still another recommendation is that a bottle of champagne should not be completely immersed in ice to chill it; the neck of the bottle should not be chilled or the cork may be difficult to remove.

When passing a food around a table, such as a casserole, to allow seated guests to serve themselves, the dish is passed from the left because most people are right-handed, making it easier for them to manage. However, wine is poured from the right side, where the wine glass is placed (just as dishes are removed by reaching over the right shoulder of each guest).

Wineglasses are of various heights, shapes, and stems, according to the wine served. An all-purpose wineglass holds 192–251 mL (6½–8½ American fluid ounces) of liquid. White wine is customarily served in a long-stemmed glass with a medium bowl. Bordeaux and other red wines are traditionally in a glass with a slightly larger bowl, while Burgundy and other red wines are served in a balloon glass with a large bowl. German wines by custom are served in a small-bowled, long, green-stemmed glass. Brandies are presented in snifters (glasses with short stems and large-bottomed bowls).

The chemical tyramine in aged wine (as well as other aged foods such as cheese, salami, and bacon) can cause migraine headaches in some people. Tyramine is formed when bacteria break down protein, and it causes migraines by constricting blood vessels. Studies have shown that many people susceptible to migraine excrete tyramine relatively slowly, and chemical analysis of a urine sample taken during a migraine attack can be used to determine whether a person is having a tyramine migraine. It has been estimated that perhaps 10% of migraine sufferers get migraines if they eat more than a small amount (6 mg) of tyramine.

Culinary Vocabulary

Note: In the English language, there are more terms descriptive of wine than for any other food. The following represents a selection.

- As noted above, the leaves of grapes are sometimes used as human food. In the eastern Mediterranean, particularly Greece and Turkey, *dolmas* are a delicacy comprising vine leaves stuffed with rice, minced lamb, and seasoning. The word "dolma" is of Turkish origin, derived from *dolmak*, "fill." (While grapes are most often used, dolmas can be made by stuffing any fruit or vegetable, or leaves, with a savory filling, and cooking.)
- The word "raisin" is from the Old French *resin*, from the Latin *racemus*, "bunch of grapes."
- Lexia raisins are raisins with very tiny seeds, like those of figs. These are most often used in fruit cakes to add desired texture. Lexia grapes (also known as Muscat of Alexandria, Muscat Gordo Blanco, and Hanepoot) are an ancient type of grape that makes inferior wine but is used to produce raisins, especially in California.
- Black Corinth grapes are exceptionally small. They are also called Zante currants (Zante is a Greek island where these grapes were cultivated over 2000 years ago) and "champagne grapes," a term that originated in California, and is based on the fact that champagne is often served in glasses garnished with small clusters of these grapes.
- The word "vinegar" comes from the French *vin aigre* (or, more accurately, the Old French *vyn egre*) meaning "sour wine." However, not all vinegar is made from wine, and indeed any food that can be fermented to produce alcohol can also be used to manufacture vinegar. The best vinegars are made from wine or cider. Wine vinegar is made from wine. Balsamic vinegar is made from concentrated grape juice, according to traditional methods; it is usually aged for 4–5 years (occasionally as long as 40 years), producing a dark brown, somewhat syrupy fluid, which is only slightly acidic and has a unique taste. Malt vinegar is made from ale or beer.
- Since Roman times, the word "toast" has been applied to a proposal of health prior to imbibing. This tradition originated because of the Roman practice of adding spiced toast to wine to improve its flavor and absorb sediment.

Muscadine grape (*Vitis rotundifolia*). *Left*, from Bailey (1900–1902); *right*, from Harter (1988).

- The word "brandy" (short for brandywine) is based on the Dutch *brandwijn* (from Middle Dutch *brantwijn*, from *brant*, distilled, + *wijn*, wine). Literally, the Dutch word *bradewijn* means "wine that has been burned," an allusion to distilling the wine over fire. Brandy can be made from fermented grapes or other fruit, such as peaches.
- Information on some prominent names of fortified wines (a topic discussed above) is presented in the following: in the 18th century, it was discovered that when brandy was added during fermentation, a sweet wine of high quality could be produced, and thus was born port wine, named for Oporto, a city of northwest Portugal, where it was processed and from which it was shipped. Another fortified wine, sherry (which ranges from very dry to sweet and from amber to brown) was named after Jerez (Xeres) in southwestern Spain, the town in which most such wine was initially made. The names "brandy" and "armagnac," which define the standards by which brandy is judged, refer to Dutch localities where these wines acquired their initial reputations. ("Cognac" is based on the locale of Charente.) "Malmsey," a sweet fortified wine originally produced in Greece under the name Monembasia, the port from which it was exported, got its name from the Medieval Latin name of the port—Malmasia.
- "Claret" is an English word for Bordeaux wines collectively, the word tracing through the Old French to the Medieval Latin *vinum claratum*, clarified wine.
- The phrase "comet wine" is often applied to wines produced in the year of an important comet, particularly to Portuguese wine bottled in 1811, the year of a great comet.
- Winestone or argol is the hard, crusty deposit that forms in casks of maturing wine, and which is used to refine cream of tartar. It consists mainly of tartar (potassium bitartrate, formed as insoluble crystals from tartaric acid, the principal acid in wine). Cream of tartar, a fine white powder, is employed to give volume and stability to beaten egg whites and to prevent sugar from crystallizing when making frosting and candy.
- *Véronique* (pronounced vay-rho-NEEK) is a French culinary term referring to dishes garnished

with seedless white grapes. A widespread explanation for the name Véronique is that it commemorates a comic opera by that name, written by the French composer André Messager (1853–1929). "Sole Véronique" is a dish of poached sole served with grapes. It was invented by French chef Auguste Escoffier (1846–1935) in London in 1903. (There is also a story that Emile Malley, chef of the Carlton Hotel in London, christened his grape-garnished dish Sole Véronique after the recently born daughter of one of his staff.)

- The French expression *à la Bordelaise* refers to dishes cooked with Bordeaux wine. Bordeaux is a French city that is famous as a wine center, but it is also an important gastronomic center. Some dishes with Bordelaise in the name are not made with Bordeaux wine. Dishes with Bordealise in the name usually have sautéed shallots present.
- The descending tears of wine on the inside of a glass after it has been swirled are called "legs."
- The Pinot Noir is known as the "heartbreak grape" because it is one of the most difficult grapes to grow.
- *Trockenbeerenauslese* (abbreviated as TBA, and often shortened to trock) has been described as the most decadently sweet grade of German and Austrian white wines. The name literally means "dry berry selection," reflecting the practice of letting the grapes stay on the vine, sometimes until December, until they have shriveled up almost like raisins, and then selecting them one by one. *Trock* is easily confused with *trocken* wines, which are extremely dry German white wines.
- A "dry" (as opposed to "sweet") wine is one lacking in sugar. Terms for describing the dryness of champagne (French terms in italics) are as follows: *doux*: extremely sweet; *demi-sec*, dry: very sweet; *sec*: noticeable sweetness; extra dry: slight sweetness; *brut*: no perceptible sweetness; *brut nature*: bone dry. *Brut* (meaning "rough" in French, reflecting the fact that long ago the French preferred sweeter champagne) indicates less than 2% sugar, and is the most widely enjoyed style, even among the French. In California wine parlance, "dry" sometimes means "unfortified."
- Grappa (pronounced GRAH-pah) is an Italian alcoholic spirit, rather like brandy, distilled from the juice extracted from the residue (pomace) left after wine grapes are pressed. It is the same as the English term marc and the French *eau-de-vie de marc*. These beverages are typically slightly sweet, but strong and crude. They have a devoted following.
- Under world trade agreements, a special "*Appellation d'Origine Controlée*" (AOC) is issued for certain products. An excellent AOC white wine is based on the "limited district" of Cadillac in France, and is in demand in North America by wine connoisseurs. However, the French name "Cadillac" was also exported to the New World, most famously by French pioneer Antoine de la Mothe Cadillac (1658–1730), the founder of Detroit, and the man after whom the Cadillac car was named.
- "Zip code wine" is American slang for inexpensive French wine. Rather than being identified with a specific *appellation d'origine*, the producer's address is identified only by a number that is the equivalent of a French mail zip code.
- "Red Ned" is the Australian slang way of referring to cheap, red wine.
- "Cooking wine" is wine to which salt has been added—a concept that apparently traces to the Prohibition era, so that people would use the wine in cooking and not drink it. (The wine would evaporate, leaving its flavor in the food.) No one with any respect for food would purchase this, and regular drinking wines are employed in cooking.
- "Raisin wine" is a sweet wine prepared from dried grapes.
- "Ice wine" (icewine) is produced from grapes that have been allowed to freeze during the winter. The skins shrivel and the juice condenses into a syrup. The frozen bunches of grapes are pressed while still frozen. The concentrated juice produces an extremely flavorful and expensive wine. Ice wine (known as *eiswein*) originated in Germany, but today, Canada is the world's largest producer.
- "Champagne" is sparkling wine made in the traditional fashion in the Champagne region of France. In North America, the word is often used as a generic term to describe all sparkling wines (a practice of course that outrages the French). To avoid confusion, most producers in the United States do not call their sparkling wine "Champagne," even when it is made in the traditional method. Other generic language terms for sparkling wine: *vins mousseux* (the French term for sparkling wines produced outside France's Champagne region), *spumante* (Italian), *sekt* (German), *cava* (Spanish). In North America, sparkling wine, especially Champagne, is often called "bubbly" and "gigglewater."

- "Jewish champagne" is an affectionate term that was sometimes used for seltzer (plain carbonated water) by New York Jews, who frequently drank it to adhere to kosher rules (for example, not drinking milk with meats).
- Wine and cheese are a natural pair of fermented foods, but wine grapes and cheese also combine well in "grape cheese," known in France as *tomme au marc* and *fondue aux raisins*. This is a cheese made with an edible crust of marc (the grape pulp that remains after pressing grapes for wine).
- "Post Grape-Nuts" has neither grapes nor nuts, but is made from wheat and barley. The cereal was created in 1897 by American inventor C.W. Post (1854–1914) who put "grape" in the name because it contained maltose, which Post called "grape sugar," and "Nuts" because of its flavor.
- "Bottle sickness" is a condition of recently bottled wine, which may last for months. Rough handling, filtration, or other processes just before bottling wine results in deterioration that sometimes makes the beverage undrinkable for a period.
- The empty space between the cork and liquid in a bottle (wine or otherwise) is called "ullage."
- The indentation at the bottom of a wine bottle is called a "punt" or a "kick." It strengthens the bottle's structure and traps sediments.
- Wine containers:
    - Split holds 375 mL (12.7 fluid ounces, American measure) or half a standard bottle
    - Bottle holds 750 mL (25 fluid ounces) of wine (also called a "fifth")
    - Magnum holds same as 2 bottles, or 1.5 L (50.7 fluid ounces)
    - Double Magnum equals 4 bottles, or 3 L (101.4 fluid ounces)
    - Jeroboam equals 6 bottles (4.5 L, 156 fluid ounces) for wine or 4 bottles (3 L, 101.4 fluid ounces) for sparkling wine
    - Rehoboam (for champagne) equals 6 bottles, or 4.5 L (156 fluid ounces)
    - Methuselah (for champagne) equals 8 bottles, or 6 L (204.8 fluid ounces)
    - Imperial equals 8 bottles, or 6 L (156 fluid ounces)
    - Salamanzar equals 12 bottles (one case), or 9 L (307.2 fluid ounces)
    - Balthazar equals 16 bottles, or 12 L (415 fluid ounces) [Balthazar is a word that also describes a magnificent feast; for both meanings, it derives from Balthazar, the last king of Babylon, who held a sumptuous banquet for 1000 of his dignitaries, serving the wine in sacred vases that his father, Nebuchadnezzar (see the following), stole from the Temple in Jerusalem. That same night, God punished the transgression.]
    - Nebuchadnezzar holds same as 20 bottles, or 15 L (570 fluid ounces)
- A ton (or tonne, if referring to a metric tonne) is derived from "tun," a large cask used for wine or beer, which in turn gets its name from the French *tonnerre*, thunder, from the deep rumbling sound when the heavy barrel was rolled.

Curiosities of Science and Technology

- The Norse explorer, Leif Ericson, based in Greenland, visited the northeastern coast of what is now Canada about 1000 AD. He was so impressed by the rich growth of native grapes he observed (likely *V. labrusca*) that he dubbed the New World "Wineland the Good." The exact location of "Vinland" as Ericson termed it has been disputed. L'Anse-Aux-Meadows in Newfoundland is the only known Norse settlement in North America, and was occupied for a short time around 1000 AD. However, the nearest wild grapes are found in New Brunswick, several hundred miles to the south. It has been suggested that L'Anse-Aux-Meadows was Ericson's base camp from which he explored nearby areas of North America, which is the presumed "Wineland."
- In past times, the size of ships was intimately related to the size of the largest wine containers. Christopher Columbus's flagship the Santa Maria, in which he sailed to the Americas in 1492, has been reported to be a vessel of "100 ton capacity." The Mayflower, the ship in which the Pilgrim Fathers sailed from England to Massachusetts in 1620, has been reported to be a vessel of "180 ton capacity." These were good-sized ships for their times. Correctly, "ton" in these cases should be called "tun," a different measure (although for practical purposes the actual cargo weight capacity would be about the same). In those days, the size of a merchant vessel was measured in terms of how many "tuns" could be carried safely in the hold. A "tun" was a very large barrel or cask for wine (empty, it would require four men to carry it), capable of carrying a "double hogshead(s)." This was a measure of volume, not of weight, equal to 4 single hogshead(s). Before 1824, a single

hogshead was equal to 63 Imperial wine gallons (75.6 American gallons); today, it is 52.5 Imperial wine gallons (63 American gallons). Modern ship "tonnage" is measured in different ways. A "register ton" is a measure of the internal capacity of a ship, equal to 100 cubic feet. A "displacement ton" is the amount of seawater that the ship displaces, measured in long (metric) tons, and equal to 35 cubic feet. Cargo capacity may also be measured in weight ("deadweight tons") and in volume ("gross tons").

- The Speyer Museum in Germany displays a 1600-year-old bottle of wine.
- In the Middle Ages, wine was used as currency.
- A Virginia law enacted in 1619 required every household in the colony to plant and tend 10 wine grape cuttings. Prizes were offered for the best wines.
- Dom Perignon (Pierre Perignon, 1638–1715), a French Benedictine monk, is alleged to have made the first champagne in 1670. It is said that upon his first taste of champagne he cried, "I am tasting stars." Champagne (and other sparkling wines) likely existed before Dom Perignon's work. His contribution was in inventing the mushroom-shaped cork and wire cage that allowed the effervescent wine to be safely bottled. Previous attempts had all ended with popped corks.
- In old London, customers at a tavern had the right to inspect the wine cellar to verify that the wine hadn't been watered down.
- Through the Middle Ages and down to the 19th century, the idea was widespread that watering seeds with wine instead of plain water would produce stronger, healthier plants.
- The first commercial winery in the United States was established in 1823 in Missouri.
- Dentist Thomas Bramwell Welch (1825–1903) of Vineland, New Jersey, created an unfermented grape sugar beverage in 1869 that he called "Dr. Welch's Unfermented Wine." In effect, he had applied the process that Louis Pasteur invented 4 years earlier to sterilize wine, beer, and milk. Today, Welch Food Inc. of Concord, Massachusetts, annually sells $650 million in juices and other products.
- France is the largest producer of wine. In 1880, the vineyards of France were almost completely destroyed by an aphid called phylloxera [*Phylloxera vitifoliae* (Fitch)], accidentally introduced from eastern North America about 1860. The grape phylloxera produces galls on leaves and roots. The root injury may be very destructive to European grapes, but American varieties are more or less resistant. As New World grapes are essentially immune to root injury, American species have been used as rootstocks since 1880 in Europe, effectively stopping the depredation of phylloxera, although not until 1930 had normal production resumed in Europe. However, wild *V. vinifera* of the Old World has been devastated by this insect. Today, all wines from France are made from fruit of vines grown on American roots. Chile is rather unique in still growing the wine grape on wine grape rootstocks.
- It has been claimed that in Victorian times, estate gardeners never planted grape vines without first burying a dead donkey underneath the plot. While this is doubtful, it is possible that the presence of a large corpse would represent a way of providing minerals for a very long time.
- In Victorian England, a game called Snapdragon was played in which raisins were snatched from a bowl of burning brandy. The game fell out of favor after numerous people were burned.
- Raisin pies were traditionally prepared by Pennsylvania Dutch farm women. These and other types of pastries were stored in cool cellars until required for such occasions as wakes and weddings, and as a result, raisin pies became a traditional "mourning food" served at funerals. Raisin pie became so often associated with funerals in colonial New England and Pennsylvania that an inquiry of whether a raisin pie was needed was a way of asking if someone was seriously ill.
- The Titanic carried 1500 bottles of wine, 2000 wineglasses, and 1500 grape scissors on its fateful voyage of 1912.
- In 1919, the 18th Amendment to the U.S. Constitution was passed, which initiated a period of alcohol prohibition starting in 1920 and lasting until repealed by the 21st Amendment in 1933 (some states had banned all alcoholic beverages as early as 1916; Maine forbade the sale of liquor in 1841). The period of prohibition caused great hardship for US vineyards and the large wine industry. Sacramental wine was allowed for Catholic and Jewish services, and much of New York's grapes were used for the purpose. Moreover, there was another loophole: section 29 of the Volstead Act of 1919, which initiated Prohibition, allowed citizens to make up to 200 gallons a year of "homemade cider" or "nonintoxicating fruit

A giant grape vine, mid 19th century, at Montecito near Santa Barbara, California, with a trunk diameter of 50 cm (about 18 inches) and covering an area of 930 square meters (10 000 square feet), from *Revue de l'Horticulture Belge et Étrangère* (1878, vol. 4).

juice," which was widely understood to be a euphemism for regular alcoholic wine. (Two hundred gallons of home-brewed alcoholic beverage is still the limit in many states.) This peculiarity of the legislation allowed grape growers to sell their fruit to home wine makers, and resulted in a considerable increase of wine grape production in the United States during Prohibition. In 1932, wine with 3.2% alcohol was allowed, and was called "McAdoo wine" after Senator William Gibbs McAdoo of California, who sponsored the bill. During Prohibition, "Grape Brick," made up of dried and pressed wine grape concentrate, was sold, along with an attached packet of yeast, and the warning, "Do not add yeast or fermentation will result." The result of all of these factors was that wine consumption increased considerably during Prohibition, although most of it was not of high quality.

- The oldest wine to have been sold at an auction was a bottle of 1646 Imperial Tokay, which was bought by John A. Chunko of Princeton, New Jersey, and Jay Walker of Ridgefield, Connecticut, for 1250 Swiss Francs (about 725 current US dollars) in Switzerland on 16 November 1984.
- The importance of foods is reflected in the extent to which language terms have been coined, and this is obvious for grapes and wine. Some examples:
  - The expression "sour grapes," refers to the frustration of people who belittle the importance of attainments by others that they themselves wish but have been unable to achieve. The source of the phrase is one of Aesop's fables in which a fox failed to reach a cluster of delicious-looking grapes that dangled from a vine above his head just out of reach of his snapping jaws. The departing fox remarked that they were probably

sour anyway. (The Greek Aesop lived from about 620–564 BC, and is credited with about 250 fables.)

- "Vignette" has several meanings: a short descriptive literary sketch, a brief incident or scene (as in a play or movie), the pictorial part of a postage stamp design as distinguished from the frame and lettering, and a decorative design or illustration placed near or on the title page or at the beginning or end of a chapter. The word originated from the Middle French *vignete*, the diminutive of vine, and originally meant a small design based on a grape vine.
- The use of the term "grapevine" for the informal or secret system that transmits rumors and gossip traces to mid 19th century America, when it was also applied to the rumors themselves.

- In Latin, *uvula* means "little grape." Anatomists adopted this word to refer to the small, teardrop-shaped part of the soft palate that hangs down at the back of the throat.
- Most grapes are reproduced either sexually by seeds, or by vegetative cuttings. In some cultivars, like the ancient Greek variety Black Corinth, which is dried to produce a kind of raisin known as a "currant," the berries arise without fertilization by a process called parthenocarpy. [Cultivars of *Ribes* are often called currant (which see). In the raisin currant, the word is derived by a corruption of Corinth, the Greek port from which this raisin is commonly shipped.] In the Kishmish or Sultanina, the most important raisin variety, seeds abort soon after fertilization.
- Red-wine flavor is particularly due to pigments in the skin of the grapes from which they are made. Some red wines, for example, Cabernets, Merlots, and Chambourcins, are often made from smaller grapes to take advantage of their greater skin to volume ratio.
- According to the *Guinness World Records 2001*, the most expensive commercially available wine is 1787 Chateau d'Yquem Sauternes, costing about $60,000 per bottle.
- According to the *Guinness World Records 2002*, the most expensive glass of wine was the first glass of Beajoulais Nouveau wine released in Beaune, France, bought for $1,382.80 by Robert Denby at an auction in 1983.
- The most expensive bottle of wine, of 1787 Chateau Lafite claret, was sold for £105,000 (150,000 current American dollars) in London on 5 December 1985. The bottle was engraved with the initials of Thomas Jefferson (1743–1826), 3rd president of the United States.
- Hampton Court Palace, built in 1514, and situated along the River Thames in England, is the site of a grape vine planted in 1768, which some have claimed to be the oldest and largest grape vine in the world. In 1807, it produced a record crop of 2245 bunches of grapes, although today it averages 500–700 bunches a year. A very large grape vine (said to be the largest living grape vine in the world) with a stem diameter of 1.5 m (5 feet) is at Graaff-Reinet, South Africa (at Reinet House, a parsonage built in 1811, and now a tourist attraction).
- In an unpruned and untrained plant, *V. vinifera* will spread to amazingly high and distant reaches, and produce startling quantities of fruit. A grapevine in Carpinteria California developed a trunk with a circumference of 2.3 m (7½ feet), and at 53 years of age, bore about 9000 kg (10 tons) of grapes.
- Connoisseurs of the grape have long appreciated that different styles of glass are more appropriate to the physical qualities of particular wines. For example, claret requires a large bowl to contain the volatile aromatic compounds that bring out its bouquet, while champagne should have a thin flute to retain bubbles and keep it from going flat. In experiments at the University of Tennessee at Knoxville, reported in 2002 in *New Scientist* magazine, it was shown that size and shape of wine glasses produced changes in the chemistry of red wine that were sufficiently significant to alert an experienced drinker. Using Merlot wine left for 20 minutes in three different glasses—flute, Martini, and Bordeaux—the scientists found that in the flute and Martini glasses, the red wine retained a high level of gallic acids (which form after pouring because of chemical reactions between tannins from the grapes and oxygen in the air), resulting in the wine tasting tart and sharp. By contrast, in the Bordeaux glass, which allows a larger surface of the wine to be exposed to air, the greater contact with oxygen converted (oxidized) much of the gallic acids into esters, which add aroma and dryness, producing a mellower wine experience.
- The soil of the famed Clos de Vougeot vineyard in the Burgundy district of France is considered so precious that workers are required to scrape it from their shoes before departing.
- Champagne is at its peak between 4 and 10 years of age.

- The wine in an average bottle of Champagne or sparkling wine is under a pressure of 5.5 atmospheres at 10 °C (90 psi at 50 °F), about equal to that of a large, fully inflated tire. Coincidentally, the average bottle of sparkling wine contains 5.5 times its volume in the form of carbon dioxide gas.
- There are approximately 49 million bubbles in an average bottle of sparkling wine. The average diameter of a bubble is 0.5 mm (about a fiftieth of an inch)—much smaller than the large bubbles formed by carbon dioxide that are machine injected into carbonated beverages. The rising bubbles in champagne carry aroma components and contribute to taste and smell, particularly in the air above the glass. Moreover, the tiny bubbles arrive in a steady stream, providing a reasonable period to enjoy the experience.
- In 1999, scientists at the Wujiang Grape Research Institute reported that they had developed a grape variety bearing fruit that weigh an average of 22–25 g (0.8–0.9 ounce) and are the size of table-tennis balls. One monster grape, weighing 36 g (1¼ ounces), was submitted to the Guinness Book of World Records for recognition as the biggest grape ever grown.
- The "French Paradox," high dietary cholesterol and saturated fats, but low death rate from heart disease, has been ascribed to the effect of constant wine consumption by the French. Many scientists attribute this phenomenon to phenolic chemicals found in red wine. Leading cardiologists have endorsed the moderate consumption of wine as a protective measure against heart disease. "Moderate consumption" means a glass or two a day (drinking more will actually increase your risk of heart disease). There is also evidence that wine protects against cancer, Alzheimer's disease, aging, and gastrointestinal infection.
- A small winery in Arkansas sent a bottle of its prized white wine to the University of Arkansas, requesting a chemical analysis. The report came back: "Your horse has diabetes."
- The British Dental Association once warned wine lovers not to swirl their wines in their mouths too enthusiastically because it could rot their teeth. A 52-year-old wine merchant was reported to have lost the enamel from his teeth after tasting an average of 30 wines per day for 40 years (apparently starting at age 12).
- There are various suggestions for removing stubborn red wine stains from fabric: rub with white wine (which will dilute the stain); rub with club soda; add shaving cream and sponge off with water; soak the stain in hot milk brought to the boiling point, then wash; sprinkle with salt and rub with half a lemon. In any event, if the stain can't be treated promptly, soak the article in cold water until it can be treated.
- Wine, especially red wine, oxidizes and becomes unpalatable after being opened, and the deterioration is noticeable after only a few hours. Professional wine bars often use nitrogen displacement dispensers, which force air out of the bottle under the force of pressurized nitrogen gas. The gas does not allow air to contact the wine, which can remain fresh after the bottle has been opened for weeks.
- It has been estimated that 99% of wine is drunk in the year it is released, and that most wine is consumed within 17 minutes of getting it home from the store.
- Fungi are both enemies and friends of wine. Without a special wine yeast (a form of fungus), *Saccharomyces ellipsoideus*, many wines would not ferment properly. "Flor," a special yeast native to Spain's Jerez region and France's Jura district, which appears as a mildew-like growth (the dead yeast cells form a crust over the wine during aging), is an essential addition in the preparation of Spanish sherry. Sauterne wines are allowed to rot by exposure to fungi naturally in the air, in order to develop a sweetness, bouquet, and higher sugar content. However, mildews and black rot, the chief enemies of commercial vineyards today, are controlled by frequent spraying with "Bordeaux mixture," a solution of copper sulphate and hydrated lime.
- Researchers at Glasgow University discovered that after consuming two glasses of wine, members of the opposite sex appeared about 25% more attractive.
- Veterinarians recommend that even small amounts of grapes and raisins should not be fed to dogs. Symptoms produced can include diarrhea, appetite loss, lethargy, abdominal pain, and (most especially) kidney failure. [Reference: McNight, K. 2005. Grape and raisin toxicity in dogs. Vet. Tech. **26(2)**: 1–2.]
- Spain is currently the world's largest wine-growing nation in terms of area (almost 1.2 million ha, or 3 million acres), and global warming is now threatening the $2.4 billion Spanish wine industry. Spain is especially vulnerable to climate change because it is the closest major European wine producer to the equator. One degree of climate change can

Drying raisin grapes in 19th century Australia, from Despeissis (1893).

make wine-growing regions in the Northern Hemisphere similar to regions 200 km (125 miles) further south. Vines can be productive for 80 years, and during this period, temperatures are expected to rise several degrees. Excessive heat and sunlight can increase sugar levels in grapes, resulting in unpalatably high alcohol content. Hot weather can also curb grape acidity and change the flavor. Wine growers in Spain are now developing heat-resistant cultivars, shading and irrigating their vineyards (until 1996 it was illegal to water vineyards in Spain), and moving to mountainside locations.

- In France, a study of grape maturity records was used to reconstruct maximum summer temperatures dating back to 1370. As reflected by especially early grape ripening in the last half century, in recent decades summer temperatures have been about 2 °C higher than in the past. (Reference: Chuine, I., Yiou, P., Viovy, N., Seguin, B., Daux, V., and Ladurie, E. Le Roy. 2004. Historical phenology: grape ripening as a past climate indicator. Nature, **432**: 289–290.)

Key Information Sources

(Note: Thousands of books and tens of thousands of articles have been written about grapes and, more particularly, wine. The following is a selection.)

Adams, L.D., and Novak, B. 1990. Wines of America. 4th edition. McGraw-Hill, New York, NY. 528 pp.

Ahmedullah, M., and Himelrick, D.G. 1990. Grape management. *In* Small fruit crop management. *Edited by* G.J. Galletta and D.G. Himelrick. Prentice Hall, Englewood Cliffs, NJ. pp. 383–471.

Alleweldt, G., Spiegel-Roy, P., and Reisch, B. 1990. Grapes (*Vitis*). *In* Genetic resources of temperate fruit and nut crops. *Edited by* J.N. Moore and J.R. Ballington, Jr. International Society for Horticultural Science, Wageningen, Netherlands. pp. 291–327.

Amerine, M.A. 1956. The maturation of wine grapes. Wines Vine. **27**: 27–55.

Amerine, M.A., and Phaff, H. 1986. A bibliography of publications by the faculty, staff, and students of the University of California, 1876–1980, on grapes, wines, and related subjects. University of California Press, Berkeley, CA. 244 pp.

Bioletti, F.T. 1938. Outline of ampelography for the vinifera grapes in California. Hilgardia, **11**: 227–293.

Brenner, L. 1995. Fear of wine: an introductory guide to the grape. Bantam Books, New York, NY. 310 pp.

Cass, B., and Robinson, J. 2000. The Oxford companion to the wines of North America. Oxford University Press, Oxford, UK. 301 pp.

Clarke, O., and Rand, M. 2001. Oz Clarke's encyclopedia of grapes. Harcourt, New York, NY. 320 pp.

Cox, J. 1999. From vines to wines: the complete guide to growing grapes and making your own wine. Storey Books, Pownal, VT. 235 pp.

Einset, J., and Pratt, C.1975. Grapes. *In* Advances in fruit breeding. *Edited by* J. Janick and J.N. Moore. Purdue University Press, West Lafayette, IN. pp. 130–153.

Foulkes, C. 1994. Larousse encyclopedia of wine. Larousse, New York, NY. 608 pp.

Gohdes, C.L.F. 1982. Scuppernong, North Carolina's grape and its wines. Duke University Press, Durham, NC. 115 pp.

Gray, D.J., and Meredith, C.P. 1992. Grape. *In* Biotechnology of perennial fruit crops. *Edited by* F.A. Hammerschlag and R.E. Litz. pp. 229–262.

Guttadauro, G.J. 1976. A list of references for the history of grapes, wines, and raisins in America. Agricultural History Center, University of California, Davis, CA. 70 pp.

Harris, J.M., Kriedemann, P.E., and Possigham, J.V. 1968. Anatomical aspects of grape berry development. Vitis, **7**: 106–119.

Iverson, J. 1998. Home winemaking, step-by-step: a guide to fermenting wine grapes. 2nd edition. Stonemark Publishing Co., Medford, OR. 172 pp.

Jackson, D., and Schuster, D. 1987. The production of grapes & wine in cool climates. Butterworths of New Zealand, Wellington, New Zealand. 192 pp.

Jancis, J. 1994. The Oxford companion to wine. Oxford University Press, Oxford, UK. 1088 pp.

Jindal, P.C. 1985. Grape. *In* Fruits of India, tropical and subtropical. *Edited by* T.K. Bose. Naya Prokash, Calcutta, India. pp. 219–276.

Jindal, P.C. 1999. Grape. *In* Tropical horticulture, vol. 1. *Edited by* T.K. Bose, S.K. Mitra, A.A. Farooqui, and M.K. Sadhu. Naya Prokash, Calcutta, India. pp. 259–284.

Johnson, H. 1985. The world atlas of wine: a complete guide to the wines and spirits of the world. 3rd edition. Simon and Schuster, New York, NY. 320 pp.

Kasimatis, A.N. 1975. How to produce quality raisins. University of California, Division of Agricultural Sciences, Berkeley, CA. 9 pp.

Kaufman, W.I. 1984. Encyclopedia of American wine, including Mexico and Canada. J.P. Tarcher, Los Angeles, CA. 564 pp.

Kerridge, G., and Antcliff, A.J. 1999. Wine grape varieties. Revised edition. CSIRO Publishing, Collingwood, Victoria, Australia. 205 pp.

Ketsa, S., and Verheij, E.W.M. 1991. *Vitis vinifera* L. *In* Plant resources of South-East Asia, 2, edible fruits and nuts. *Edited by* E.W.M. Verheij and R.E. Coronel. Pudoc, Leiden, Netherlands. pp. 304–310.

Kolpan, S., Smith, B.H., and Weiss, M.A. 1996. Exploring wine: the Culinary Institute of America's complete guide to wines of the world. Van Nostrand Reinhold, New York, NY. 699 pp.

Lavee, S., and Nir, G. 1986. Grape. *In* CRC handbook of fruit set and development. *Edited by* S.P. Monselise. CRC Press, Boca Raton, FL. pp. 167–191.

Liger-Belair, G. 2004. Uncorked: the science of champagne. Princeton University Press, Princeton, NJ. 152 pp.

McGrew, J.R. 1993. Growing wine grapes. G.W. Kent, Inc., Ann Arbor, MI. 91 pp.

Newman, J.L. 2000. Wine. *In* The Cambridge world history of food. *Edited by* K.F. Kiple and K.C. Ornelas. Cambridge University Press, Cambridge, UK. pp. 730–737.

Olmo, H.P. 1995. Grapes. *In* Evolution of crop plants. 2nd edition. *Edited by* J. Smartt and N.W. Simmonds. Longman Scientific & Technical, Burnt Mill, Harlow, Essex, UK. pp. 485–490.

Pearson, R.C., and Goheen, A.C. 1988. Compendium of grape diseases. APS Press, St. Paul, MN. 93 pp.

Pratt, C. 1971. Reproductive anatomy in cultivated grapes. A review. Am. J. Enol. Vitic. **22**: 92–109.

Priew, J. 1999. Wine: from grape to glass. Abbeville Press, New York, NY. 256 pp.

Ramey, B.C. 1977. The great wine grapes and the wines they make. Great Wine Grapes, Inc, Burlingame, CA. 250 pp.

Robinson, J. 1986. Vines, grapes, and wines. Knopf, New York, NY. 280 pp.
Robinson, J. 1996. Jancis Robinson's guide to wine grapes. Oxford University Press, Oxford, UK. 232 pp.
Rombough, L. 2002. The grape grower: a guide to organic viticulture. Chelsea Green Publishing, White River Junction, VT. 289 pp.
Simon, A.L. 1957. The noble grapes and the great wines of France. McGraw-Hill, New York, NY. 180 pp.
Simon, A.L., and Allen, H.W. 1967. Wines of the world. McGraw-Hill, New York, NY. 719 pp.
Stevenson, T. 1997. The new Sotheby's wine encyclopedia. DK Publishing, New York, NY. 600 pp.
Unwin, P.T.H. 1991. Wine and the vine: an historical geography of viticulture and the wine trade. Routledge, New York, NY. 409 pp.
Vine, R.P., Harkness, E.M., and Linton, S.J. 2002. Winemaking: from grape growing to marketplace. 2nd edition. Kluwer Academic/Plenum Publishers, New York, NY. 477 pp.
Wagner, P.M. 1976. Grapes into wine: a guide to winemaking in America. Knopf, New York, NY. 302 pp.
Weaver, R.J. 1976. Grape growing. John Wiley & Sons, New York, NY. 371 pp.
Webb, A.D. 1984. The science of making wine. Am. Sci. **72**: 360–367.

Specialty Cookbooks

Grapes

Barnes, W. 1977. The wine country's great grape cook book. Toll Road Press, Napa Valley, CA. 44 pp.
California Grape Grower. 1925. The use of grapes in the home. Tested recipes for delectable dishes. Dietetic value of the grape. The California Grape Grower, San Francisco, CA. 7 pp.

Raisins

Australian Dried Fruits Association. 1960–1969. Family favourites with currants, Sultanas, and seeded raisins: hints, suggestions, ideas, and dozens of delicious recipes. Australian Dried Fruits Association, Melbourne, Australia. 60 pp.
California Raisin Advisory Board. 1950–1960. Tested raisin formulas. California Raisin Advisory Board, Fresno, CA. 54 pp.
California Raisin Advisory Board. 1967. Bonanza of California raisin originals. California Raisin Advisory Board, Fresno, CA. 26 pp.
California Raisin Advisory Board. 1977. Raisins: everything under the sun. California Raisin Advisory Board, Fresno, CA. 34 pp.
Deacon, P. 1979. The wrinkled grape. Patricia Deacon, Los Angeles, CA. 44 pp. [variously paginated].
Sun Maid Growers of California. 1950s. Downright delicious Sun-Maid raisin recipes. Sun-Maid Raisin Growers of California, Fresno, CA. 32 pp.
Sun Maid Growers of California. 1950s. More downright delicious Sun-Maid raisin recipes. Sun-Maid Raisin Growers of California, Fresno, CA. 24 pp.

Wine

Anderson, S.F., and Anderson, D. 1989. Winemaking: recipes, equipment, and techniques for making wine at home. Harcourt Brace Jovanovich, San Diego, CA. 284 pp.
Ballard, B., and Kittler, P. 1981. Wine in everyday cooking: "cooking with wine for family and friends." Wine Appreciation Guild, San Francisco, CA. 122 pp.
Bowers, W.F., and Bowers, L. 1975. Gourmet cooking with homemade wines. Stackpole Books, Harrisburg, PA. 288 pp.
California Wine Advisory Board. 1978. Gourmet wine cooking, the easy way. Revised edition. Wine Appreciation Guild, San Francisco, CA. 128 pp.
Chase, E. 1960. The pleasures of cooking with wine. Prentice-Hall, Englewood Cliffs, NJ. 243 pp.
Church, R.E. 1976. Entertaining with wine. Rand McNally, Chicago, IL. 174 pp.
Crocker, B. 1989. Betty Crocker's cooking with American wine. Prentice Hall, New York, NY. 112 pp.
Gaulke, J.A. 1972. Cooking with wine. Lane Books, Menlo Park, CA. 79 pp.
Goldstein, J.E., Williams, C., and Goldstein, E. 1999. Food & wine pairing. Time-Life Books, Alexandria, VA. 111 pp.
Hoffman, V., and Hoffman, R. 2003. The great little food with wine cookbook: 76 cooking with wine recipes, pairing food with wine, how and where to buy wine, ordering wine in a restaurant. 2nd edition. Rayve Productions, Windsor, Ca. 128 pp.
Johnson-Bell, L. 1999. Pairing wine and food. Burford Books, Short Hills, NJ. 194 pp.
Logan, A.M. 1972. Wine and wine cooking; entertaining and cooking with American wines. Westover Publishing Co., Richmond, VA. 328 pp.
MacDonald, B. 1976. Wine in cooking and dining. The Culinary Institute, Chicago, IL. 96 pp.
Sarvis, S., and Thompson, R. 1973. American wines and wine cooking. Creative Home Library, Des Moines, IA. 182 pp.
Smith, J. 1986. The Frugal Gourmet cooks with wine. Morrow, New York, NY. 447 pp.
Taylor, G.H. 1963. Treasury of wine and wine cookery. Harper & Row, New York, NY. 278 pp.
Time-Life. 1983. Wine. Time-Life Books, Alexandria, VA. 176 pp.
Werlin, L. 2003. All American cheese and wine book: pairings, profiles, & recipes. Stewart, Tabori & Chang, New York, NY. 335 pp.
Willan, A., and Clay, L. 2001. Cooking with wine. Harry N. Abrams in cooperation with COPIA, New York, NY. 288 pp.

Grapefruit and Pomelo

Family: Rutaceae (rue family)

Names

Scientific names: *Citrus* species

- Grapefruit—*C.* ×*paradisi* Macfad. [probably the hybrid *C. sinensis* (L.) Osbeck × *C. maxima*, i.e., sweet orange × pomelo]
- Pomelo (pummelo)—*C. maxima* (Burm.) Merr. (*C. decumana* L., *C. grandis* Osbeck)

- The grapefruit is so named because John Lunan recorded this name in 1814 in his *Hortus Jamaicensis* "on account of its resemblance in flavor to the grape." An old but apparently erroneous explanation that has often been advanced to explain the name is that the fruit sometimes occurs in huge grape-like clusters (which is incorrect, although sometimes branches can ripen up to 40 fruits, somewhat resembling a cluster).
- The word "pomelo" is of unclear origin, but is related to the Dutch *pampelmoose* (which may have been a name for the pomelo), the French *pamplemousse* (grapefruit), and the French *pomme* (apple).
- Shaddock, an old name for the pomelo, commemorates Captain Shaddock (or Chaddock), a 17th century English ship commander reputed to have brought the seed of this plant from the East Indies to Barbados in 1696.
- "Sweeties" are sweet-tasting grapefruits, the result of a cross between the pomelo and the grapefruit. They look like green grapefruits. "Pomelit" is another sweet-fleshed hybrid of grapefruit and pomelo.
- The ugli (pronounced oo-glee) or uglifruit (or sometimes hoogly) is a hybrid between the grapefruit (or one of its parents, the pomelo) and a tangerine. It resembles a grapefruit in appearance, and also the tangelo (which is a cross between a pomelo and a tangerine). The pulp is orange-like and almost seedless. The uglifruit seems to have been named for its ugliness—it appears semi-deflated (the skin is loose and wrinkled), discolored (the yellowish red skin has a greenish tinge, even when ripe), and altogether rather unattractive, although it combines the succulence of the grapefruit with the sweetness and peelability of the tangerine. The ugli arose in Jamaica, where it was discovered in 1915 or 1916 by F.G. Sharp, owner of Jamaica's Trout Hall estates. Most uglis are still produced in Jamaica, with some cultivation in Brazil. Uglis are generally eaten raw, although in the West Indies, they are frequently baked in hot cinders and consumed with sugar. Canadians import much of the crop. The uglifruit is now marketed under the more politically correct name "Unique Fruit."
- The cami is a hybrid of the tangerine and pomelo (much like the ugli, described above). It has a smooth, orange rind, a knob at the stem end, a sweet–sour flavor, and no seeds.
- In the grapefruit trade, a "seedless grapefruit" is one with no more than five seeds.
- The word "citrus" is based on the citron (*Citrus medica* L.), a plant closely related to the lemon.
- *Paradisi* in the scientific name *C.* × *paradisi* is derived from the Greek *paradeiosos*, a pleasure ground.
- *Maxima* in the scientific name *C. maxima* is Latin for greatest, indicative of the size of the fruit.

Plant Portrait

The grapefruit tree is a subtropical evergreen. In commercial orchards, most trees are 7–10 m (22–30 feet) high. The fruit is more or less globular, 10–15 cm (4–6 inches) wide, pale lemon or sometimes pinkish, with white, pink, or red flesh. Grapefruits weigh 0.45–2.27 kg (1–5 pounds). Although average grapefruit trees produce about 159 kg (350 pounds) of fruit annually, some mature trees can yield up to 680 kg (1500 pounds). While some fruits are seedless or nearly so, there may be up to 90 white seeds.

The grapefruit is quite a recent fruit. It was not described until 1750, from the Barbados, and is believed to have originated in the West Indies. Various conjectures have been advanced to explain the origin of the grapefruit. Seeds of the pomelo or shaddock, a citrus fruit native to Southeast Asia, were transported to the Barbados at the end of the 17th century, and probably hybridized with the orange to produce the grapefruit. A single introduction of grapefruit into Florida from the West Indies in 1823 has been identified as the original source of all known grapefruit varieties that have originated in the United States.

Grapefruit tree (*Citrus ×paradisi*), by B. Flahey.

The grapefruit first became an important crop in the early 1900s, in the southern United States. The United States remains the world's leading producer, with most of the crop grown in Florida, which produces almost 2 million metric tons (over 2 million tons) of grapefruit annually. Other major grapefruit-growing countries are Israel, Cuba, Argentina, and South Africa. The grapefruit is the world's second most important citrus fruit today, after the orange. Major seedy varieties include Duncan, a white-fleshed cultivar, and Foster, a pink-fleshed cultivar. Seedless cultivars include Marsh, a white-fleshed fruit, and Thompson, Burgundy, Redblush, Star Ruby, Flame, and Ray Ruby, which are pink or red cultivars.

The pomelo or shaddock is closely related to the grapefruit, but differs in its loose rind and lower level of juiciness. Its pulp is often rather coarse and dry, although tasty varieties are known. The lemon-yellow fruit has a very thick rind, and is typically pear-shaped (sometimes round), 10–30 cm (4–12 inches) in diameter, and up to 10 kg (22 pounds) in weight. The fragrant rind peels easily, is green, yellow, or pink, and may have a smooth or rough texture. Pomelos are occasionally found in supermarkets, typically looking like huge grapefruits, and these tend to have flesh that is intermediate in taste between a grapefruit and an orange, which can be consumed without sugar. The fruit is the source of bitter naringin (often called "grapefruit extract"), used as a flavor enhancer in sweets and drinks (naringin also has medicinal effects, and is marketed in the medicinal supplement industry). The tree is round-headed, growing up to 5.5 m

(18 feet) in height. This species is probably native to southeastern Asia or the Pacific islands. It is widespread in warm regions as an escape, but has largely been replaced in cultivation by the grapefruit. The pomelo is thought to be the parent of many economically important species of citrus trees, and is presumed to be an ancestor of the grapefruit.

Culinary Portrait

Grapefruit pulp is more or less acidic, sweet, and fragrant. Red grapefruits taste a little sweeter than white grapefruits (see below for additional information). The grapefruit is customarily eaten at breakfast, chilled, cut in half, the sections loosened from the peel and each other by a special curved grapefruit knife (or a special serrated spoon) and the pulp spooned from the half-shell, and sweetened with sugar, honey, or other flavorings. Washing grapefruit before cutting them open is a good idea to get rid of microbes on the skin that may contaminate the flesh. Grapefruit juice is now widely consumed, and occasionally peeled segments of grapefruit are sold in cans. Grapefruit also makes a pleasant marmalade, and the peel may be candied or used as zest (flavoring). Grilled grapefruit sometimes accompanies duck, chicken, pork, or shrimp dishes. When purchasing grapefruits, keep in mind that the biggest are not necessarily the best. Fruit that is compact, firm, springy to the touch, well shaped, and heavy for its size is preferable. Coarse-skinned, puffy fruit is typically not as juicy as is heavy, thin-skinned grapefruit.

Depending on variety, pomelos may be flavorless or tasty, very sweet or very acidic, and seeds may or may not be present. The fruit is often candied or cooked. This fruit is rarely eaten fresh, like grapefruit. Pomelos may be refrigerated for up to a week.

Grapefruits and grapefruit juice can amplify the effect of many medications, and minimize the effect of a few. An enzyme that breaks down medicines is found in the small intestine, and it is suppressed by grapefruit juice, so that more medicine is absorbed into the bloodstream than intended, amplifying the drug's effect. Drugs known to be affected include some taken for anxiety, depression, high blood pressure, HIV/AIDS, cancer, irregular heart rhythms, infections, psychotic problems, erectile dysfunction (including the well-known Viagra), angina, convulsions, gastrointestinal reflux, high cholesterol, and immunosuppressants (particularly Cyclosporin A) taken to stop rejection after organ transplants. Interactions have also been found with some dietary supplements (notably St. John's wort, used to reduce anxiety). When taking medications, the advice of a doctor or pharmacist should be sought to determine if grapefruits and grapefruit juice should be avoided. In the United States, the U.S. Food and Drug Agency requires drugs that could interact with grapefruit juice to carry warnings or precautions. More than three-quarters of medical professionals in the United States in 2002 were warning their patients to avoid drinking grapefruit juice at any time while taking medications. The fear of interactions with drugs has been of great concern to the grapefruit industry, and has been responsible for about a third of Florida's grapefruit growers going out of business in the last decade of the 20th century.

As if the above weren't enough to dissuade people from drinking grapefruit juice, there is some evidence that high consumption (240 mL or 8 American fluid ounces daily) increases the chances of developing kidney stones.

Shaddock (*Citrus maxima*), from Nicholson (1885–1889).

Culinary Vocabulary

- A "Crimea Cooler" is a cocktail prepared with grapefruit juice, vodka, crème de cassis, and ginger ale, and garnished with a sprig of mint.
- A "Salty Dog" is a cocktail prepared from grapefruit juice, vodka, sugar, and salt, with a lime-wedge garnish.
- "Steamboat Gin" is a cocktail prepared with gin, Southern Comfort, grapefruit juice, and lemon juice.
- "Forbidden Fruit" is an American liqueur prepared from pomelo steeped in brandy (see information below for "forbidden fruit").

Curiosities of Science and Technology

- The Frenchman Count Odette Phillipe planted the first grapefruit trees in Florida around Tampa Bay in 1823. Florida now produces more grapefruit than the rest of the world combined.
- It has often been reported that when first encountered in the West Indies, the grapefruit was known as the "forbidden fruit." However, researchers have suggested that the forbidden fruit, alleged to have become extinct over a century ago (it has been rediscovered), was a variety of pomelo, one of the ancestors of the grapefruit.
- During the Great Depression that followed the stock market crash of 1929, grapefruit was not in much demand and was quite unfamiliar to most people. Surplus supplies were often given away free to people on welfare, resulting in a common complaint: no matter how long the fruits were boiled, they remained tough.
- In 1970, consumption of grapefruit was temporarily heightened by a widely promoted "grapefruit diet" plan that claimed to achieve a loss of 4.5 kg (10 pounds) in 10 days.
- Pink grapefruits are pink because they contain more than 25 times as much beta-carotene as white grapefruits. Beta-carotene, the precursor of vitamin A, produces the orange color of the common carrot.
- Salt is said to make grapefruits taste sweeter.
- Most seeds contain just one embryo. Grapefruit seeds often contain several. The extra embryos are

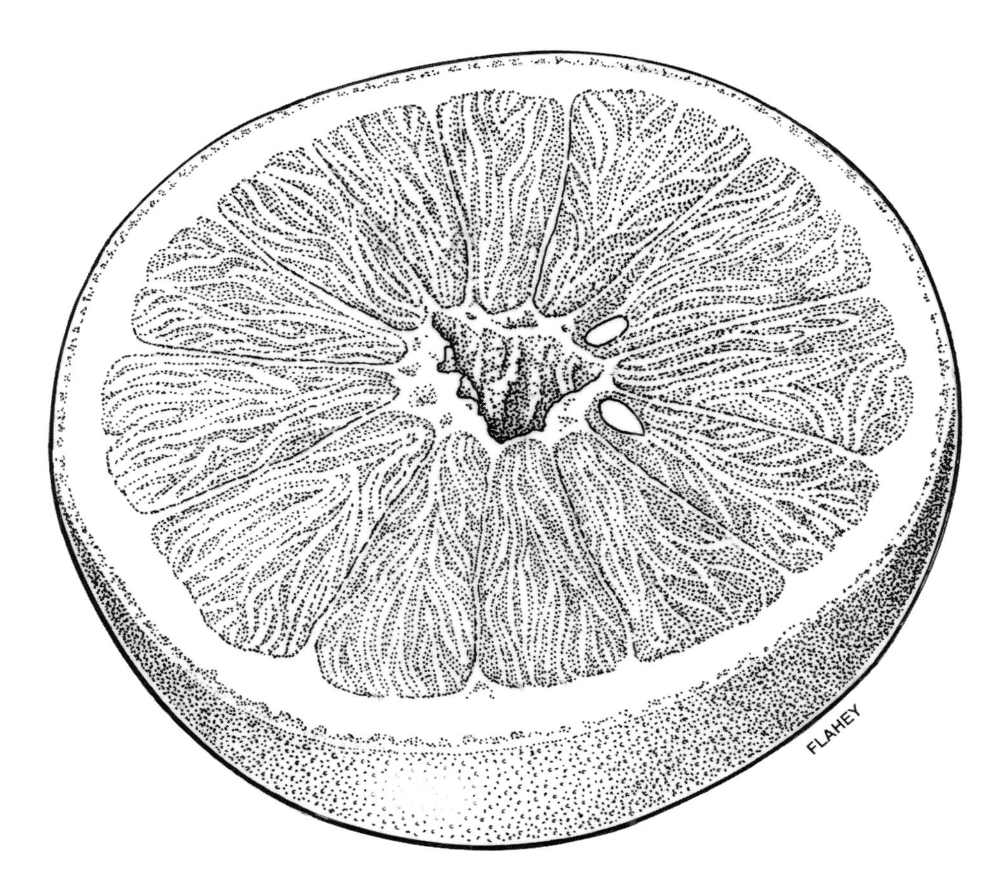

Grapefruit (*Citrus ×paradisi*), by B. Flahey.

derived from maternal tissue, not from sexual fertilization, and so are clones of the maternal plant.

- In old grapefruits or grapefruits that have been left too long on the trees, seeds may germinate within the fruits.

Key Information Sources

Anonymous. 1977. Vitamin C-ooking with oranges, grapefruits and tangerines. Fam. Health, **9**(**2**): 42–44, 46.

Bowman, K.D., and Gmitter, F.G., Jr. 1990. Caribbean forbidden fruit: grapefruit's missing link with the past and bridge to the future? Fruit Var. J. **44**: 41–44.

Brown, M.G., Lee, J.-Y., and Spreen, T.H. 2001. The decline in the domestic market for Florida fresh grapefruit. Institute of Food and Agricultural Sciences, Food and Resource Economics Dept., University of Florida, Gainesville, FL. 6 pp.

Chace, W.G. 1966. Factors affecting the quality of grapefruit exported from Florida. Agricultural Research Service, U.S. Dept. of Agriculture, Washington, DC. 21 pp.

Chomchalow, N. 1984. Genetic wealth of pummelos in Thailand. IBPGR Reg. Comm. South-East Asia Newsl. **8**(**3**): 27–29.

Cohen, A. 1972. Estimation of peel thickness and fruit shape from the specific gravity of grapefruits. Isr. J. Agric. Res. **22**: 161–169.

Corazza-Nunes, M.J., Machaicdo, M.A., Nunes, W.M.C., Cristofani, M., and Targon, M.L.P.N. 2002. Assessment of genetic variability in grapefruits (*Citrus paradisi* Macf.) and pummelos (*C. maxima* (Burm.) Merr.) using RAPD and SSR markers. Euphytica, **126**: 169–176.

Fairchild, G.F., and Tuttell, M. 1984. Florida grapefruit facts and figures. Economic Research Dept., Florida Dept. of Citrus, University of Florida, Gainesville, FL. 51 pp.

Fellers, P.J. 1989. A review of limonin in grapefruit (*Citrus paradisi*) juice, its relationship to flavour and efforts to reduce it. J. Sci. Food Agric. **49**: 389–404.

Kane, G.C., and Lipsky, J.J. 2000. Drug-grapefruit juice interactions. Mayo Clin. Proc. **75**: 933–942.

Kesterson, J.W., and Hendrickson, R. 1953. Naringin, a bitter principle of grapefruit: occurrence, properties and possible utilization. University of Florida Agricultural Experiment Station, Gainesville, FL. 29 pp.

Martin, F.W., and Cooper, W.C. 1977. Cultivation of neglected tropical fruits with promise. Part 3: the pummelo. ARS-S-157, U.S. Dept. of Agriculture, New Orleans, LA. 17 pp.

Morton, J. 1987. Grapefruit. *In* Fruits of warm climates. *Authored by* J.F. Morton. Creative Resource Systems, Winterville, NC. pp. 152–158.

Niyomdham, C. 1991. *Citrus maxima* (Burm.) Merr. *In* Plant resources of South-East Asia, 2, edible fruits and nuts. *Edited* E.W.M. Verheij and R.E. Coronel. Pudoc, Leiden, Netherlands. pp. 128–131.

Robinson, A.J. 1987. Shaddock in Australia. Capricorn Coast Reproductions, Yeppoon, Queensland, Australia. 48 pp.

Sachs, A. 1997. The authoritative guide to grapefruit seed extract. LifeRhythm, Mendocino, CA. 125 pp.

Samson, J.A. 1991. *Citrus × paradisi* Macf. *In* Plant resources of South-East Asia, 2, edible fruits and nuts. *Edited* E.W.M. Verheij and R.E. Coronel. Pudoc, Leiden, Netherlands. pp. 133–135.

Schirra, M. 1993. Behaviour of 'Star Ruby' grapefruits under chilling and non-chilling storage temperature. Postharvest Biol. Technol. **2**: 315–327.

Scora, R.W., and Nicolson, D.H. 1986. The correct name for the shaddock, *Citrus maxima*, not *C. grandis* (Rutaceae). Taxon, **35**: 592–595.

Sinclair, W.B. 1972. The grapefruit: its composition, physiology, and products. Division of Agricultural Sciences, University of California, Berkeley, CA. 660 pp.

Stother, J. 1970. World production and trade in fresh grapefruit. Tropical Products Institute, Ministry of Overseas Development, London, UK. 120 pp.

Ting, S.V. 1983. Citrus fruits. *In* Handbook of tropical foods. *Edited by* H.T. Chan, Jr. Marcel Dekker Inc., New York, NY. pp. 201–254.

Yee, W. 1974. The pummelo in Hawaii. Cooperative Extension Service, University of Hawaii, Honolulu, HI. 8 pp.

Specialty Cookbooks

(Also see Lemon.)

Berry, R.M.F. 1915. Recipes for Florida sealdsweet oranges and grapefruit; how to prepare and serve Florida's luscious citrus fruits and how to use them in cookery and confections, concocting refreshing drinks, etc. Florida Citrus Exchange, Jacksonville, FL. 32 pp.

California Fruit Growers Exchange. 1939. Sunkist fresh grapefruit recipes. California Fruit Growers Exchange, Los Angeles, CA. 18 pp.

Florida Citrus Commission. 1964. 101 ways to enjoy Florida grapefruit. Florida Citrus Commission, Lakeland, FL. 36 pp.

Lane, Mrs. J. 1905. The forbidden-fruit or shaddock, or grape-fruit: how to serve and how to eat it. John Lane, London, UK. 8 pp.

McDaniel. B. 1931. Zestful recipes for every meal: Pure Gold and Silver Seal oranges, lemons, grapefruit. Mutual Orange Distributers, Redlands, CA. 32 pp.

Real Gold Citrus Products. 1950s. Real Gold orange, lemon, grapefruit bases: just dilute and serve! Real Gold Citrus Products, Redlands, CA. 6 pp.

Sunkist Growers, Inc. 1968. Sunkist cook book: lemons, tangerines, citrus treasures of the west, oranges, grapefruit. Sunkist Growers, Inc., Los Angeles, CA. 98 pp.

HYACINTH BEAN

Family: Fabaceae (Leguminosae; pea family)

Names

Scientific name: *Lablab purpureus* (L.) Sweet (*Dolichos lablab* L., *L. niger* Medik., *L. vulgaris* Savi)

- "Hyacinth" in the name "hyacinth bean" is inspired by the many-flowered heads of flowers, somewhat reminiscent of those of the garden hyacinth (*Hyacinthus orientalis* L.).
- Hyacinth bean is also known as bonavist bean, Chinese flowering bean, chink, country bean, dolichos bean, field bean, Egyptian bean, Indian bean, India butter bean, lablab, lablab bean, pharao, pharao bean, poor-man's bean, sem or seem (in India), shink bean, Tonga bean, val, val bean, and wild field bean.
- In China, hyacinth bean is called pig-ears (Mandarin: *bian dou*), for the leaf shape.
- The genus name *Lablab* and "lablab" in the name "lablab bean" are based on the Arabic word for the plant, *lāblāb*. In modern Arabic, the general word for bean is *lubia*.
- *Purpureus* in the scientific name *L. purpureus* is Latin for purple, the usual color of the flowers. The leaves, stems, and pods are also often purple or purplish.

Plant Portrait

The hyacinth bean is a vigorous, herbaceous, frost-tender, twining vine. Varieties that grow as bushes up to 1 m (about 3 feet) tall are also grown. The plant is a short-lived perennial, commonly grown as an annual in the United States and Canada. In frost-free areas, the vines become woody and can grow up to 9 m (30 feet) in length, while in cold areas, the vines remain herbaceous and grow to about 3 m (10 feet). The plants may produce hundreds of spikes of lavender, purple, or white flowers in late summer followed by deep lavender–purple or crimson pods. The flowers are scented and attract butterflies and hummingbirds. The pods are 5–20 cm (2–8 inches) long, flat or inflated, with 3–5 seeds. The seeds are 6–13 mm (1/4–1/2 inch) long, nearly as wide, and flattened, dark brown, black, red, pale tan, or white, and often speckled. The species is native to the Old World tropics. Wild forms are believed to have originated in India, and were introduced into Africa from Southeast Asia during the 8th century. Hyacinth bean has been widely distributed to many tropical and subtropical countries where it has become naturalized. In South and Central America, East and West Indies, Asia, China, and India, the seed and immature pods are used for human food, while the herbage is used as livestock feed. In North America, hyacinth bean is cultivated mainly as an ornamental.

Culinary Portrait

The entire plant is edible, but toxins in the pods and seeds can produce weakness, vomiting, labored breathing, twitching, stupor, and convulsions. The pods, seeds ("beans"), and tubers are edible if

Hyacinth bean (*Lablab purpureus*), from *Revue de l'Horticulture Belge et Étrangère* (1908, vol. 34).

Hyacinth bean (*Lablab purpureus*), from Lamarck and Poiret (1744–1829, plate 610).

thoroughly boiled, changing the water several times. While boiling, the seeds give off a distinct, heavy "beany" smell. The immature pods are consumed like green beans. The seeds can be cooked and eaten directly, used as a source of flour for making bread dough in combination with other flours, processed into bean cakes, or fermented into a kind of tofu. In India, the seeds are dried and split like split peas. The large, starchy tubers can be baked after boiling. The young leaves can be eaten raw in salads, while the older leaves can be cooked and consumed like spinach. The flowers are edible raw. The seeds can be germinated to produce bean sprouts.

Curiosities of Science and Technology

- Thomas Jefferson (1743–1826, 3rd president of the United States) was an extraordinarily competent scientist and student of agriculture. He grew hyacinth bean at his well-known estate, Monticello, in Virginia, and because of this, sometimes in old seed catalogs the hyacinth bean was called the "Thomas Jefferson vine." Today, visitors to Monticello can view the hyacinth bean featured on the arbor in the vegetable garden.
- Hexadecimal is a counting system that uses 16 numerals, starting with zero. Since the standard decimal system provides only 10 different symbols, the letters A through F are used to fill out a set of 16 different numerals. The first 20 numbers in this system are as follows: 0, 1, 2, 3, 4, 5, 6, 7, 8, 9, A, B, C, D, E, F, 10, 11, 12, 13. The system was created in the 1950s for computer applications (e.g., for representing ASCII code). The name of the pea family, "Fabaceae," is the longest English word that can also be a hexadecimal numeral (i.e., consisting only of A, B, C, D, E, and F).

Key Information Sources

Agishi, E.C. 1991. A bibliography of *Lablab purpureus*. ILCA/PSD Working Document A6 ILCA, Addis Ababa (Ethiopia). 76 pp.

Akpapunam, M. 1996. Hyacinth bean (*Lablab purpureus* (L.) Sweet). *In* Food and feed from legumes and oilseeds. *Edited by* E. Nwokolo and J. Smartt. Chapman and Hall, London, UK. pp. 103–108.

Anderson, R.G., Bale, S., and Jia, W. 1996. Hyacinth bean: stems for the cut flower market. *In* Progress in new crops. *Edited by* J. Janick. ASHS Press, Arlington, VA. pp. 540–542.

Chowdhury, A.R., Ali, A., and Quadir, M.A. 1989. Aspects of pollination and floral biology of lablab beans *Lablab purpureus* L. Sweet. J. Jpn. Soc. Hortic. Sci. **58**: 665–672.

Deka, R.K., and Sarkar, C.R. 1990. Nutrient composition and antinutritional factors of *Dolichos lablab* L. seeds. Food Chem. **38**(**4**): 239–246.

Devaraj, V.R., and Manjunath, N.H. 1995. Effect of cooking on proteinase inhibitors of *Dolichos lablab* bean (*Dolichos lablab purpureus* L.). Plant Foods Hum. Nutr. **48**: 107–112.

El-Siddig, O.O.A., El-Tinay, A.H., Abd-Alla, A.W., and Elkhalifa, A.E.O. 2002. Proximate composition, minerals, tannins, in vitro protein digestibility and effect of cooking on protein fractions of hyacinth bean (*Dolichos lablab*). J. Food Sci. Technol. **39**: 111–115.

Fribourg, H.A., Overton, J.R., McNeil, W.W., Culvahouse, E.W., Montgomery, M.J., Smith, M., Carlisle, R.J., and Robinson, N.W. 1984. Evaluations of the potential of hyacinth bean as an annual warm-season forage in the mid-South. Agron. J. **76**: 905–910.

Humphry, M.E., Konduri, V., Lambrides, C.J., Magner, T., McIntyre, C.L., Aitken, E.A.B., and Liu, C.J. 2002. Development of a mungbean (*Vigna radiata*) RFLP linkage map and its comparison with lablab (*Lablab purpureus*) reveals a high level of colinearity between the two genomes. Theor. Appl. Genet. **105**: 160–166.

Kalloo, G., and Pandey, S.C. 1993. Hyacinth bean, *Lablab purpureus* (L.) Sweet. *In* Genetic improvement of vegetable crops. *Edited by* G. Kalloo and B.O. Bergh. Pergamon Press, New York, NY. pp. 387–389.

Konduri, V., Godwin, I.D., and Liu, C.J. 2000. Genetic mapping of the *Lablab purpureus* genome suggests the presence of 'cuckoo' gene(s) in this species. Theor. Appl. Genet. **100**: 866–871.

Latigo, G.V., and Gonzalez, C.L. 1979. *Dolichos lablab* a potential forage legume in south Texas U.S.A. can improve pastures and beautify homes. J. Rio Grande Val. Hortic. Soc. **33**: 121–124.

Liu, C.J. 1996. Genetic diversity and relationships among *Lablab purpureus* genotypes evaluated using RAPD as markers. Euphytica, **90**: 115–119.

Pengelly, B.C., and Maass, B.L. 2001. *Lablab purpureus* (L.) Sweet—diversity, potential use and determination of a core collection of this multi-purpose tropical legume. Genet. Resour. Crop Evol. **48**: 261–272.

Piper, C.V., and Morse, W.J. 1915. The bonavist, lablab, or hyacinth bean. USDA Bull. 318. U.S. Department of Agriculture, Washington, DC. 15 pp.

Schaaffhausen, R. von. 1963. *Dolichos lablab* or hyacinth bean: its uses for feed, food and soil improvement. Econ. Bot. **17**: 146–153.

Shivashankar, G., and Kulkarni, R.S. 1989. *Lablab purpureus* (L.) Sweet. *In* Plant resources of South-East Asia, 1, pulses. *Edited by* L.J.G. van der Maesen and S. Somaatmadja. Pudoc, Leiden, Netherlands. pp. 48–50.

Singh, P., and Pandita, M.L. 1980. Effect of spacing, plants per hill and training on growth and yield of sem (*Dolichos lablab* L.) hyacinth beans. Indian J. Hortic. **37**: 388–391.

Sultana, N., Ozaki, Y., and Okubo, H. 2000. The use of RAPD markers in lablab bean (*Lablab purpureus* (L.) sweet) phylogeny. Bull. Inst. Trop. Agric. Kyushu Univ. **23**: 45–51.

Wood, I.M. 1983. Lablab bean (*Lablab purpureus*) for grain and forage production in the Ord irrigation area. Aust. J. Exp. Agric. Anim. Husb. **23**: 162–171.

KALE AND COLLARD

Family: Brassicaceae (Cruciferae; mustard family)

Names

Scientific names: *Brassica* species and subgroupings (The nomenclature of kales and collards is unsettled, and different authorities treat the groups differently.)

- Kale, collard, borecole—*Brassica oleracea* L. var. *viridis* L. (*B. oleracea* L. var. *acephala* DC.)
- Marrow-stem kale—*B. oleracea* var. *medullosa* Thell.
- Portuguese kale, Portuguese cabbage, seakale cabbage—*B. oleracea* var. *costata* DC.
- Scotch kale, curled kitchen kale—*B. oleracea* var. *sabellica* L.
- Thousand-headed kale—*B. oleracea* var. *fruticosa* Metz.
- Turnip kale—*B. oleracea* var. *gongylodes* L. (This is kohlrabi, not treated in this book.)
- Siberian kale, Hanover salad—*B. napus* L. subsp. *napus* [var. *pabularia* (DC.) Rchb. of some authors]
- Ruvo kale, Italian turnip broccoli—*B. ruvo* L.H. Bailey (often treated as "Ruvo group" of *B. rapa* L.; cf. broccoli raab, discussed in BROCCOLI AND BROCCOLI RAAB)
- Chinese kale, Chinese broccoli—*B. alboglabra* L. [*B. oleracea* L. var. *alboglabra* (L.H. Bailey) Musil]

- "Kale" is a Scottish word representing a variation of *cole*, from the Latin *caulis* (stem, cabbage). In Scotland, kale was used as a potherb in the Middle Ages and was known as "keal." Kale has been so widely used in Scotland that to be asked "to kale" there means to be asked to dine, although not necessarily on kale.
- "Kail" is an alternative British spelling for kale, but is often used to mean cabbage in country districts of the British Isles.
- "Borecole" is from "boer's cole," meaning peasant's cole or cabbage.
- Chinese kale is sometimes known in English by the Chinese names *gai-lohn* and *tsai-shim.*
- "Collard" is an alteration of colewort, which means cole-plant. Collard is usually written in the plural, i.e., collards. In the United States, collards refer to noncrinkled, partly headed varieties. Collards are also called collard greens and collie greens.
- "Cabbage tree" seems to be a phrase applicable to the tree-like forms of kale and collards, but in fact is a name applied to the ti plant (*Cordyline fruticosa* (L.) A. Chev., usually found under the name *C. terminalis* Kunth).
- See CABBAGE for the derivation of the scientific names *Brassica* and *Brassica oleracea.*
- See RAPESEED AND CANOLA for the derivation of *rapa* in the scientific name *B. rapa.*
- *Napus* in the scientific name *B. napus* is based on an old Latin (Roman) name for the turnip (*Brassica rapa* L. subsp. *rapa*).
- *Viridis* in the name *Brassica oleracea* var. *viridis* is Latin for green.
- The "ruvo" in "ruvo kale" and in the scientific name *B. ruvo* is probably based on a region of Italy.
- *Alboglabra* in the scientific name *B. alboglabra* has the meanings white, dead white, pale, and bright.

Plant Portrait

Kales and collards are a collection of similar plants of several species of the genus *Brassica*, which are used as vegetables for humans and also frequently as fodder for livestock. For the most part, kales and collards are nonheading, cabbage-like plants, i.e., they do not produce cabbage-like heads, although the leaves look much like and taste like cabbage leaves. Two types of kales have been developed. The shorter, green or blue–green plants are eaten as human food, whereas the larger or tree-kales have stiff, strong leaves and are used for animal fodder. Collards produce smooth, tender foliage and are used as potherbs. A rosette of leaves developed at the stem apex is cut off and used for cooking. Removing the leaves stimulates additional leaf production. Kales and collards are grown as herbaceous annuals in cold climates (including most of Canada and the northern United States), but are raised as herbaceous biennials in warm temperate areas. They may become woody perennials in the tropics.

The name "kale" is applied to cultivated varieties of several species, including *B. oleracea*, *B. alboglabra*, and *B. rapa*. Cut- or crisp-leaved, green- or purple-leaved varieties are commonly encountered in kitchen gardens and are called kitchen kales. Tree kales produce simple, straight stems that may reach 4 m

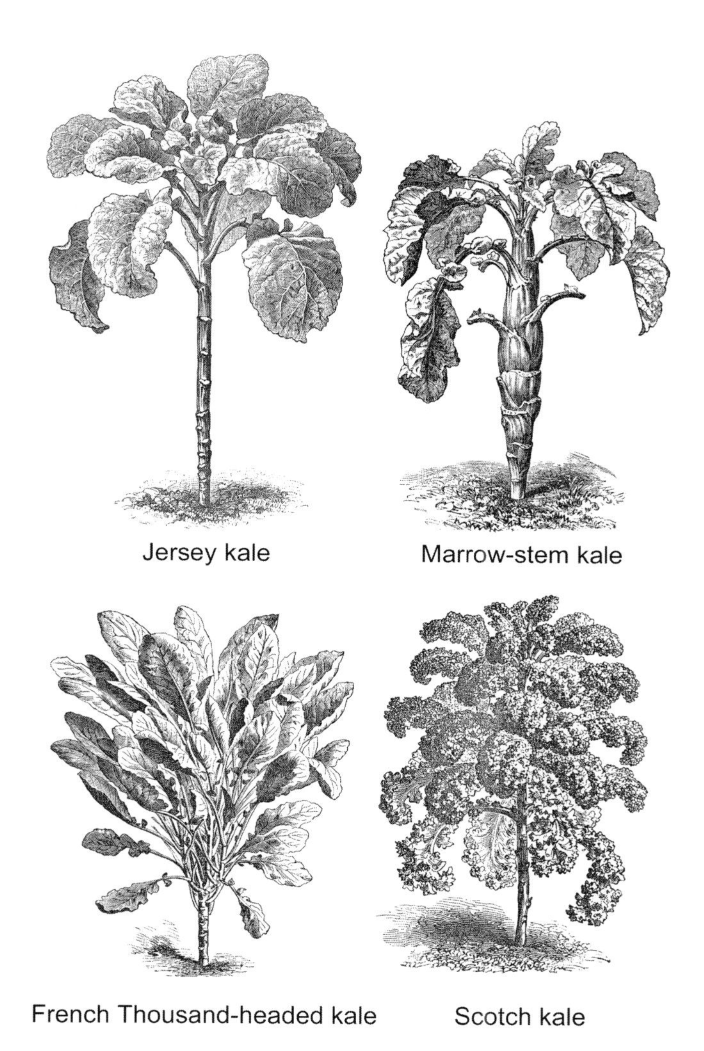

Kale varieties, from Vilmorin-Andrieux (1885), showing Jersey kale (*Brassica oleracea* var. *viridis*), marrow-stem kale (*B. oleracea* var. *medullosa*), French thousand-headed kale (*B. oleracea* var. *fruticosa*), and Scotch kale (*B. oleracea* var. *sabellica*).

(13 feet) in height. Some forms have undivided leaves and others develop fringed or curled foliage. Marrow-stem kales resemble tree kales, but the stem enlarges somewhat, becoming tuberous, with a thick, tender-fleshed trunk. Both are used as animal fodder. Bush-kale, also known as thousand-head kale or borecole, is bush-like and up to 2 m (6½ feet) tall. The Portugese kale is a low, compact form that has large, broad leaves with fleshy stalks and ribs. The fleshy petioles are eaten like celery. Chinese kale is a distinctive annual, grown as a potherb in Asia.

Kales and collards (of *B. oleracea* var. *viridis*) closely resemble the European wild cabbage (*B. oleracea* var. *oleracea*) and were probably the earliest vegetable selected from it, perhaps as early as 4000 years ago. Some consider the kales and collards to be among the more ancient of vegetables, although their time of origin is obscure. During classical Greek times, no distinction was made between kales and collards. The Romans introduced them to Britain and France. The Anglo-Saxons applied the name "cole worts," meaning cabbage plants. In northern Europe today, crinkled or curled collard varieties are popular. These vegetables were introduced to North America by colonists in the 16th century.

Culinary Portrait

Only the tiny, young leaves of kale can be eaten raw. More mature kale leaves are tough and have a pronounced flavor, and therefore must be cooked for use as a vegetable or in soups. Kale may be boiled or steamed (20–30 minutes), or braised in a casserole. Kale leaves are also used in stir-fries, and are employed as wrapping as are cabbage leaves for cabbage rolls. It is said that frost makes the leaves more tender and less pungent. Kale may be stored in a perforated plastic bag in a refrigerator for up to 10 days, but is less bitter when eaten promptly.

Collards or "collard greens" are traditionally grown as a winter source of salad leaves in the southern United States. Collards are best picked when young, fresh, and bright green, and can be frozen. This vegetable is mild (compared to kale), slightly sweet, and cabbage-like in flavor, and can be served with butter, lemon juice, or sauces. The central leaf rib can be discarded, unless the leaves are young and tender. Collards add spice to a salad, but the unusual texture and strong flavor can be overpowering to those not used to it. The flavor can be moderated by blanching the collards in boiling water for several minutes before further cooking. The Southern style of cooking the greens is to boil them with a chunk of bacon or salt pork. Collards are a traditional staple of "soul food," i.e., one of the economical but exceptionally nutritious (and often fattening) dishes that originated with black African slaves of the Old American South. Collards blend well with barley, beans, brown rice, kasha, and potatoes, and are considered an excellent accompaniment of smoked pork. They also combine well with other ingredients in soups, stews, omelets, and quiches. Collards may be stored in a refrigerator for several days, and are best wrapped in a damp paper towel and placed in a perforated plastic bag for this purpose.

Chinese kale has edible leaves and edible floral stems (like broccoli, hence the name Chinese broccoli). It has been called the most delicate-tasting member of the cabbage family.

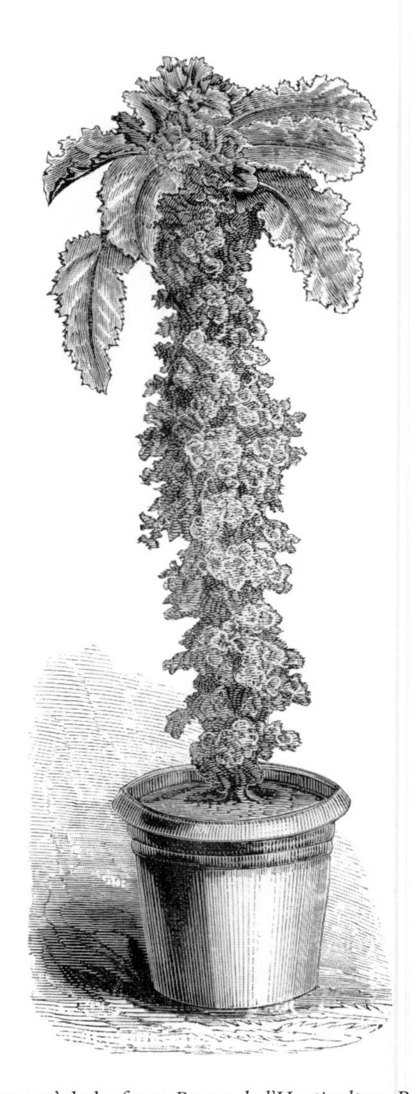

Cottager's kale, from *Revue de l'Horticulture Belge et Étrangère* (1911, vol. 37). This ornamental "kale" is a form of Brussels sprouts (*Brassica oleracea* var. *gemmifera*; see BRUSSELS SPROUTS).

Culinary Vocabulary

- "Potlikker" (a corruption of "pot liquor") is the liquid left over from cooking greens, particularly of vegetables such as collard greens and kale that were once widely consumed among rural people in the American South. Potlikker was once a staple in the southern US, and some people continue to consume it, traditionally by using pieces of corn bread to sop it up.
- "Colcannon" is a traditional Irish dish, typically served on Halloween. Custom calls for it to be made with equal amounts of kale and potatoes, but usually cabbage is substituted for kale (see CABBAGE for more information).
- "Collard pork" and "collard beef" have nothing to do with collards. Collard (or collared) pork is a traditional Irish dish make from a pig's head; Collard (or collared) beef is an English and Irish method of cooking the thin end of a beef flank.
- *Caldo verde* (pronounced KAHL-doh VAIR-deh; literally "green soup"), considered by many to be Portugal's national dish, is based on finely cut kale, as well as sliced potatoes, onions, linguiça sausage, and olive oil.
- One of Ethiopia's national dishes is *yegomen kitfo*, prepared with collard greens and buttermilk curds.

Curiosities of Science and Technology

- For many years, rulers of countries have insisted on agricultural policies that they thought benefited their country. Charlemagne (742–814) was a

Frankish king (the Franks were a Germanic people) who came to be Emperor of the West. The "Capitulare" was an edict from Charlemagne to all those governed within his realm, which included a list of about 90 plants to be grown wherever possible. Kale was among the edible plants that were ordered to be planted.

- Ornamental, highly colorful kales (some of these resemble cabbages and are called "ornamental cabbages") have become extremely popular. These show best during fall and early winter, when it is cool. Thousands have been planted at Disneyland and Disney World. Many restaurants use the colorful leaves as garnishes or as an edible underlay on lavish salad platters. To grow the ornamental kales to perfection, it should be understood that coloration doesn't begin until night temperatures regularly drop below 15.5 °C (60 °F), and full coloration takes 3–4 weeks. The plants need to be almost full grown when coloring time arrives.
- In the Channel Islands (between France and the United Kingdom), the "trunks" of tree kales have been made into walking sticks. A century ago, when the industry was in its heyday, 30 000 walking sticks were exported from Jersey, the largest of the islands. Today, only a few sticks are being made. The plants flourished in the mild winter climate of the Channel Islands, growing as high as 6 m (20 feet), and as they eventually were left with just a crown of leaves at the top, they strongly resembled palm trees. With the appearance of a loose-leaved cabbage atop the tall stems, the plants were sometimes called "giant cabbages." The dried stems were so strong they were used not only for walking sticks but also as rafters.

Key Information Sources

Andersen, C.R. 2002. Collards. Cooperative Extension Service, University of Arkansas, Fayetteville, AR. 2 pp.

Bailey, L.H. 1922, 1930. The cultivated Brassicas. *Brassica alboglabra*. Gentes Herb. **1(2)**: 79–81; **2(5)**: 233–234.

Cartea, M.E., Picoaga, A., Soengas, P., and Ordas,A. 2003. Morphological characterization of kale populations from northwestern Spain. Euphytica, **129**: 25–32.

Chesman, A. 1999. Kale: the cold-weather king. Org. Gard. **46(6)**: 46–49.

Cook, W.P., Griffin, R.P., and Keinath, A.P. 1995. Commercial collard production. Revised edition. Cooperative Extension Service, Clemson University, Clemson, SC. 4 pp.

Dangler, J.M., and Wood, C.W. 1993. Nitrogen rate, cultivar, and within-row spacing affect collard yield and leaf nutrient concentration. HortScience, **28**: 701–703.

Dias, J.S. 1995. The Portuguese tronchuda cabbage and galega kale landraces: a historical review. Genet. Resour. Crop Evol. **42**: 179–194.

Dias, J.S., and Monteiro, A.A. 1994. Taxonomy of Portuguese Tronchuda cabbage and Galega kale landraces using morphological characters, nuclear RFLPs, and isozyme analysis: a review. Euphytica, **79**: 115–126.

Dias, J.S., Monteiro, A.A., and Lima, M.B. 1993. Numerical taxonomy of Portuguese Tronchuda cabbage and Calega kale landraces using morphological characters. Euphytica, **69**: 51–68.

Farnham, M.W. 1996. Genetic variation among and within United States collard cultivars and landraces as determined by randomly amplified polymorphic DNA markers. Am. Soc. Hortic. Sci. **121**: 374–379.

Farnham, M.W., and Garrett, J.T. 1996. Importance of collard and kale genotype for winter production in southeastern United States. HortScience, **31**: 1210–1214.

Hruschka, H.W. 1971. Storage and shelf life of packaged kale. Marketing Res. Rep. No. 923. U.S. Dept. of Agriculture, Washington, DC. 19 pp.

Jarmin, M.L., and Thorton, R.E. 1985. Kale & collard seed production in the Pacific Northwest. Washington State University Cooperative Extension, Oregon State University Extension Service, University of Idaho Cooperative Extension Service, and U.S. Dept. of Agriculture, Corvallis, OR. 3 pp.

Kalloo, G. 1993. Kale, *Brassica oleracea* L. var. *acephala*. *In* Genetic improvement of vegetable crops. *Edited by* G. Kalloo and B.O. Bergh. Pergamon Press, New York, NY. pp. 187–190.

McAvoy, R.J. 1994. Cultural tips for ornamental cabbage and kale. Conn. Greenhouse Newsl. **180**: 13–15.

Miller, J.C. 1934. Collards: a truck crop for Louisiana. Louisiana State University and Agricultural and Mechanical College, Agricultural Experiment Stations, Baton Rouge, LA. 7 pp.

Myers, C. 1991. Collards. Crop Sheet SMC-0014. *In* Specialty and minor crops handbook. *Edited by* C. Myers. The Small Farm Center, Division of Agriculture and Natural Resources, University of California, Oakland, CA. 2 pp.

Myers, C., and Visher, D. 1991. Chinese broccoli, kailan, gai lohn, Chinese kale. Crop Sheet SMC-009. *In* Specialty and minor crops handbook. *Edited by* C. Myers. The Small Farm Center, Division of Agriculture and Natural Resources, University of California, Oakland, CA. 2 pp.

Prendergast, H.D.V., and Rumball, N. 2000. Walking sticks as seed savers—the case of the Jersey kale. Econ. Bot. **54**: 141–143.

Rhee, K.C. 1979. Kale: high-vitamin vegetable for harsh weather. Org. Gard. **26**(7): 60–62.

Sagwansupyakorn, C. 1993. *Brassica oleracea* L. group Chinese kale. *In* Plant resources of South-East Asia, 8, vegetables. *Edited by* J.E. Siemonsma and K. Piluek. Pudoc Scientific Publishers, Wageningen, Netherlands. pp. 115–117.

Sheets, O. 1954. The Nutritive value of collards. Mississippi Agricultural Experiment Station, State College, MI. 48 pp.

Sistrunk, W.A. 1981. Kale greens quality, vitamin retention and nitrate content as affected by preparation, processing, and storage washing, blanching, cooling. J. Food Sci. **45**: 679–681.

Smith, J.W., and Kramer, A. 1972. Palatability and nutritive value of fresh, canned and frozen collard greens. J. Am. Soc. Hortic. Sci. **97**: 161–163.

Valle, C.G. del. 1971. Influence of seeding rate, source and level of nitrogen on collards. J. Am. Soc. Hortic. Sci. **96**: 25–27.

Zeven, A.C., Dehmer, K.J., Gladis, T., Hammer, K., and Lux, H. 1998. Are the duplicates of perennial kale (*Brassica oleracea* L. var. *ramosa* DC.) true duplicates as determined by RAPD analysis? Genet. Resour. Crop Evol. **45**: 105–111.

Specialty Cookbooks

(Also see references for CABBAGE.)

Bryan, B. Collards to cavier [sic]: the Alabama cookbook. Greenberry Publishing Co., Mobile, AL. 342 pp.

Homemakers Club. 1982. New leaves from an old town: collard cookery. Homemakers Club, Ponce de Leon, FL/Circulation Service, Shawnee Mission, KS. [Irregularly paginated, ca. 140 pp.]

LEMON

Family: Rutaceae (rue family)

Names

Scientific name: *Citrus limon* (L.) Burm. f. (*C. limomum* Risso)

- Several cultivars may be of hybrid origin, including Meyer (lemon × mandarin), Perrine (lime × lemon), and Ponderosa (lemon × citron).
- Other lemon-like fruits regarded as lemons include rough lemon (*C. jambhiri* Lush.) and sweet lemon (*C. limetta* Risso).
- The name "lemon," as well as the word *limon* in the scientific name *C. limon*, are ultimately derived from the Arabic, *līmah*, "citrus fruit."
- For an explanation of the term "citrus," see GRAPEFRUIT.

Plant Portrait

The lemon is a small, usually thorny, evergreen, subtropical tree 3–6 m (10–20 feet) tall. It may have originated in northwestern India, and is supposed to have been introduced into southern Italy in 200 AD and to have been cultivated in Iraq and Egypt by 700 AD. Arabs distributed it widely in the Mediterranean region between 1000 and 1150. Today, leading lemon growers and exporters are Italy, Spain, Greece, Turkey, Cyprus, Lebanon, South Africa, and Australia. In the United States, lemons are grown chiefly in California (especially in the southern seacoast areas), Arizona, and to a lesser extent, Florida. The two major varieties of lemons produced commercially in the United States are Lisbon and Eureka. Lemons are generally elongated, with a more or less pronounced characteristic knob or nipple at the tip. Depending on variety, the skin is smooth or rough, thick or thin, the flesh may be sweet or tart, and the fruit may be seeded or seedless. The fruit is high in vitamin content, especially in ascorbic acid (vitamin C), and has long been known to prevent scurvy. Lemon trees can be grown in pots in northern areas, and moved indoors for overwintering. In addition to the culinary uses of lemon oil mentioned below, the oil is widely used to add fragrance to laundry detergents and other cleaning products.

As noted above, several citrus fruits called lemons are hybrids of lemons with other species. Probably, the most popular of these is the Meyer lemon, believed to be a hybrid of lemon and the mandarin orange. This was collected by United States agricultural explorer Frank N. Meyer (1875–1918), who found it growing as an ornamental pot-plant near Peking, China, in 1908. Meyer is a compact variety, sometimes called dwarf lemon. It is relatively cold hardy, suited to containers, and produces relatively sweet lemons.

Culinary Portrait

Lemons generally contain 6–10% citric acid, making the juice too sour to be consumed undiluted. However, they have countless culinary uses, and have been used as a flavoring agent for many centuries. The fruit is in great demand for summer drinks, such as lemonade and punch, as well as other sweet concoctions including ice cream and sherbet. Lemon is a much-used flavoring in baked goods, perhaps most famously lemon meringue pie. Lemon juice is the most widely used souring agent in Western cuisine. It is often an ingredient in sauces and salad dressings, and is an excellent general seasoner for seafood. The fruit is served as a garnish for fish all over the world. Lemon also has numerous culinary uses in addition to acting as a flavorant. Added to rice, it prevents sticking and enhances the white color. Lemon juice prevents fish from sticking to the grill, and reduces frying smells. Onion, fish, and household bleach odors can be removed from hands by rubbing with lemon (but see information below regarding dermatitis).

The outer rind of lemons contains tiny sacs of a pale yellow liquid. This "lemon oil" is extracted and used to flavor carbonated soft drinks, fruit juices, candies, ice creams, and baked goods. "Zest" is obtained by grating the rind, and is used in preparing pastries, sauces, and other cooked dishes. Not only does a medium-sized lemon have about 3 tablespoons (50 mL) of juice, but it can also produce about the same amount of grated peel. Lemon flavoring for sweet dishes can be obtained simply by rubbing a sugar cube against the outside of a lemon. Unfortunately, lemons are sometimes treated with sprays, dyes, or wax, which can make the skin unappetizing.

Lemons are typically picked unripe, and can be stored for 1–4 months while they mature, turning yellow. Once ripe, the fruit deteriorates rapidly unless kept refrigerated. Ripe lemons can be stored in a refrigerator for 2–3 months. However, the fruit should be warmed to room temperature when used as a source of juice, as this is believed to maximize the yield. When purchasing lemons, it may be noted that green-tinted fruit tend to be more acidic, and coarse-skinned fruit tend to have a very thick skin and relatively little flesh.

Culinary Vocabulary

- "Zest" is ground citrus peel, generally prepared by scraping off the peel with a grater. The word "zest" is derived from the French *zeste*, meaning "peel of a citrus fruit."
- "Lemon curd" is a smooth, soft, thick custard made from lemon juice, eggs, sugar, and butter. This British specialty is used as a spread and filler for bread and pastries.
- In Britain, "lemonade" tends to mean a carbonated sweet beverage (particularly Sprite or 7UP), equivalent to "pop" or "soda" in North America. What is called lemonade in North America (i.e., noncarbonated beverages made from lemons or lemon-tasting substitutes) may be called "squash" in Britain (a word that may also refer to other noncarbonated fruit drinks, such as those made from oranges).
- A "lemon spout" is a spout that is screwed into the end of a lemon, and when the fruit is squeezed, seed-free juice comes out.
- The word "punch" refers to a beverage, alcoholic or not, typically with citrus juice, spices, and tea. The word is thought to originate from the Hindi *pāc* (*pauch*), five, similar to the Greek *pente*, five, because the original punch was a drink of India made up of five ingredients: lemon juice, tea, sugar (or honey), water, and arrack (a fermented Eastern beverage made from coconut juice, or from a mash of rice and molasses).
- The French for "lemon sole" is *limande*, and the English name is probably just an alteration of the French name, and has nothing to do with lemons or lemony taste. Moreover, despite its name, the lemon sole does not belong to the sole family. "Lemon sole" is the commercial name of large winter flounders (in the United States, they weigh more than 1.6 kg or 3½ pounds). Lemon sole is called "sole" because this is a much more attractive name than "flounder."
- "Horse's neck" has several culinary meanings: a strip of lemon or orange peel cut from the fruit in a continuous spiral and usually served in a garnish in a cocktail; a tall glass of ginger ale and ice served with lemon peel; a cocktail of ginger ale and bourbon, blended whiskey or gin, with lemon peel; and a drink of moonshine and dry, hard cider. Sometimes, the same beverages are known by the uncomplimentary phrase "horse's ass."
- In the British navy, sailors were issued a daily ration of rum, which was the highlight of life at sea. Since 1740, this was dispensed neat (i.e., without dilution or additions). The first commander to dilute the rum was Admiral Edward Vernon (1684–1757), nicknamed "Old Grog" because of the old cloak of grogram cloth (a coarse loosely woven fabric of silk, sometimes additionally with mohair and (or) wool) that he wore in rough weather. In time, the nickname was transferred to the diluted rum, and when captains found their men too drunk from drinking rum to carry out their duties, they took the opportunity to reduce consumption by further diluting the daily ration with water. When the British Admiralty decreed that a fixed amount of lemon juice should be issued daily to sailors after their fifth or sixth week afloat, it was welcomed by the ships' captains as an ideal opportunity to dilute the rum rations, and so the navy's grog became a mixture of rum, water, and lemon juice (in time, grog came to be diluted to the point that there was only one part of rum to three parts of water). The Royal Navy's rum ration was abolished in 1970, but the word "grog" remains a general term for "alcoholic drink," and "groggy" means drunk.
- *Hamed m'raked* (pronounced hah-mad m'raw-ked) are fermented pickled lemons used in North Africa as condiments.
- "Shandy" (also called shandygaff, panaché, and Alsterwasser) is a beverage prepared with equal parts beer and lemonade (sometimes ginger beer is substituted for lemonade).
- "Vandyked citrus" refers to citrus, especially lemons and oranges, that have been cut across to produce decorative zigzag edges. The triangular edges are supposed to be reminiscent of the beard of Flemish painter Sir Anthony Van Dyke (1599–1641).

Curiosities of Science and Technology

- The Romans are said to have used lemons as mothballs.
- By the year 1299, the Mongolians had invented lemonade (it has also been claimed that lemonade

Lemon (*Citrus limon*), from Köhler (1883–1914).

was invented in 1630 in Paris, when cheap sugar became available there). A lemonade stand has been set up at some time at half of all the houses in the United States and Canada.

- The ladies of the court of Louis XIV (1638–1715) used lemons (probably the so-called "sweet lemons," which are less acidic) to redden their lips.
- By 1795, the importance of eating citrus to prevent scurvy was realized, and lemon or (rarely) lime juice was issued on all British navy ships. Dutch sailors were given sauerkraut. American ships carried cranberries.
- English explorer Captain James Cook (1728–1779) was supplied with a concentrated syrup of lemon juice for his voyages to the Pacific. Unfortunately, this proved ineffective in preventing scurvy among his crew because the preparation had been boiled, deactivating most of the vitamin C. (Cook did dramatically reduce the incidence of scurvy in most of his voyages by providing foods with good vitamin C content.)
- During the California Gold Rush, which started in 1848, miners were willing to pay $1 each for lemons and oranges as a preventative for scurvy.

- Like their fruit, lemon trees are extremely attractive, but have a sour side. The thorns of the lemon tree inflict painful punctures and scratches. Lemon-peel oil may cause contact dermatitis. Parts of the body touched by hands that have peeled lemons may show severe reactions after exposure to the sun. People who suck lemons may suffer irritation and eruptions around the mouth. The wood of lemon trees and its sawdust may induce skin reactions in sensitive woodworkers.
- Oranges do not ripen after they are picked from the tree. By contrast, lemons ripen after they are harvested.
- In India, a 6-year-old tree bore 966 lemons and, at 9 years of age, had produced a total of 3173 fruits.
- Lemon juice is valued in the home as a stain remover, and a slice of lemon dipped in salt can be used to clean copper-bottomed cooking pots, and brass. Lemon rind can be used to clean stainless steel (followed by washing with warm, soapy water).
- Lemon juice has been used for bleaching freckles and is incorporated into some facial cleansing creams. Lemon juice has also been used to bleach fingernails. To whiten socks that have become dingy, add two slices of lemon or 1/2 teaspoon of lemon juice to a pot of recently boiled water and soak the garments in it for 10 minutes, followed by regular washing.
- On a weight basis, lemons contain more sugar than strawberries.
- Whipping cream becomes more stable if a few drops of lemon juice are added before whipping.
- Fruit trees are often grafted onto the roots of related species, usually to dwarf them (fruits from smaller plants are easier to pick) and also often to make the plants hardier. Most Florida orange trees have lemon-tree roots; most California lemon trees have orange-tree roots.
- A 3.9-kg (8.5-pound) record-size commercially grown lemon is credited to C. and D. Knudsen, in Whittier, California, in the 1995 *The Guinness Book of World Records*.
- According to circus lore, one day the operator of a lemonade stand rinsed his red tights in water, and the dye came off. Not wishing to discard his scarce supply of water, he used it to make lemonade. The pink beverage was a big hit, and represented the birth of pink lemonade.
- The sour taste combined with attractive appearance of the lemon has led to its acquiring the meanings "unsatisfactory," "disappointing," and "defective," especially for automobiles that turn out to be mechanical disasters. In this case, a lemon is typically defined as "any recently purchased vehicle that has a substantial problem that is not fixed within a reasonable number of attempts, or has had a certain number of days out of service." Every state in the United States and every province in Canada has some type of "Lemon Law" entitling consumers to certain rights for vehicles, and often for other consumer items.
- In marketplace economics, the phrase "lemons effect" is based on the idea that as with used cars (some of which may be "lemons"), the buyer takes a greater risk than the seller, and so is likely to offer less money in transactions. This concept has been quite influential in recent theories regarding how the balance of information between buyers and sellers influences market values.

Key Information Sources

Ahmed, M., Arpaia, M.L., and Scora, R.W. 2001. Seasonal variation in lemon (*Citrus limon* L. Burm. f) leaf and rind oil composition. J. Essent. Oil Res. **13**: 149–153.

Ayedoun, A.M., Sossou, P.V., Mardarowicz, M., and Leclercq, P.A. 1996. Volatile constituents of the peel and leaf oils of *Citrus limon* L. Burm. f. from Benin. J. Essent. Oil Res. **8**: 441–444.

Ben-Yehoshua, S., Shapiro, B., and Kobiler, I. 1982. New method of degreening lemons by a combined treatment of ethylene-releasing agents and seal-packaging in high-density polyethylene film. J. Am. Soc. Hortic. Sci. **107**: 365–368.

California Fruit Growers Exchange. 1931. The story of California oranges and lemons. California Fruit Growers Exchange, Los Angeles, CA. 30 pp.

Cohen, E., and Schiffmann-Nadel, M. 1978. Storage capability at different temperatures of lemons grown in Israel. Sci. Hortic. **9**: 251–257.

Gulsen, O., and Roose, M.L. 2001*a*. Chloroplast and nuclear genome analysis of the parentage of lemons. J. Am. Soc. Hortic. Sci. **126**: 210–215.

Gulsen, O., and Roose, M.L. 2001*b*. Lemons: diversity and relationships with selected *Citrus* genotypes as measured with nuclear genome markers. J. Am. Soc. Hortic. Sci. **126**: 309–317.

Josan, J.S., Sandhu, A.S., and Kaur, J. 1995. Pericarp anatomy in relation to fruit cracking in lemon (*Citrus limon*). Indian J. Agric. Sci. **65**: 410–413.

Malik, M.N., Scora, R.W., and Soost, R.K. 1974. Studies on the origin of the lemon. Hilgardia, **42**: 361–382.

McDonald, R.E., and Hillebrand, B.M. 1980. Physical and chemical characteristics of lemons from several countries. J. Am. Soc. Hortic. Sci. **105**: 135–141.

McDonald, R.E., Hatton, T.T., and Cubbedge, R.H. 1985. Chilling injury and decay of lemons as affected by ethylene, low temperature, and optimal storage. Hortscience, **20**: 92–93.

Morton, J. 1987. Lemon. *In* Fruits of warm climates. *Authored by* J.F. Morton. Creative Resource Systems, Winterville, NC. pp. 160–168.

Moore, G.A. 2001. Oranges and lemons: clues to the taxonomy of *Citrus* from molecular markers. Trends Genet. **17**: 536–540.

Rock, R.C., and Platt, R.G. 1977. Economic trends in the California lemon industry. Revised edition. Cooperative Extension, Division of Agricultural Sciences, University of California, Berkeley, CA. 21 pp.

Rohidas, S.B., and Chakrawar, V.R. 1982. Studies on floral biology of seedless lemon. Punjab Hortic. J. **22**(**1/2**): 21–28.

Rouse, A.H., and Knorr, L.C. 1969. Maturity changes in pectic substances and citric acid of Florida lemons. Fla. State Hortic. Soc. Proc. **82**: 208–212.

Sauls, J.W., and Jackson, L.K. 1983. Lemons, limes and other acid citrus. Fruit Crops Facts Sheet. University of Florida, Agricultural Extension Service, Gainesville, FL. 4 pp.

Sinclair, W.B. 1984. The biochemistry and physiology of the lemon and other citrus fruits. University of California, Division of Agriculture and Natural Resources, Oakland, CA. 946 pp.

Stephenson, J.W. Lemons. Newsl. Calif. Rare Fruit Grow. **5**(**1**): 8–10. [Lemon varieties.]

Turner, G.W., Berry, A.M., and Gifford, E.M. 1998. Schizogenous secretory cavities of *Citrus limon* (L.) Burm. f. and a reevaluation of the lysigenous gland concept. Int. J. Plant Sci. **159**: 75–88.

Specialty Cookbooks

(Additional references are in GRAPEFRUIT and LIME.)

Bailey, P. 1996. The lemon lovers cookbook. Longstreet Press, Atlanta, GA. 150 pp.

Baker, S., and Sbraga, M. 1995. Lemon tree very healthy cookbook: zestful recipes with just the right twist of lemon. Avery Publishing Group, Garden City Park, NY. 282 pp.

Casson, C., and Lee, S. 1979. The compleat lemon: a cookbook. Holt, Rinehart and Winston, New York, NY. 142 pp.

Chynoweth, K., and Woodson, E. 2003. Lemons: growing, cooking, crafting. Chronicle Books, San Francisco, CA. 120 pp.

Freemark, E. 1993. The lemon cookbook. HPBooks, Los Angeles, CA. 118 pp.

Garamond. 1990. The best of lemons. Garamond, London, UK. 45 pp.

Idone, C. 1993. Lemons. Collins Publishers San Francisco, CA. 95 pp.

JG Press. 1995. The best of lemons. JG Press, North Dighton, MA. 44 pp.

Jones, B. 1993. Lemons. Harlaxton Publishing, Grantham, UK. 64 pp.

Longbotham, L. 2001*a*. Luscious lemon desserts. Chronicle Books, San Francisco, CA. 144 pp.

Longbotham, L. 2001*b*. Lemon zest: more than 175 recipes with a twist. Broadway Books, New York, NY. 256 pp.

Lorenz. 1996. Lemons: a book of recipes. Lorenz, New York, NY. 64 pp.

Ree [no initial]. 1989. Lemon twist: zesty and zingy lemon recipes, garnishes, and menus. Nutrition Unlimited Publications, Arcadia, CA. 204 pp.

Schulte, S., and Scott, L. 1994. Lemons! lemons! lemons! Viking, New York, NY.114 pp.

Siegel, H., and Gillingham, K. 1999. The totally lemons cookbook. Celestial Arts, Berkeley, CA. 96 pp.

Southwater. 2000. Be inspired by lemons. Southwater, London, UK. 64 pp.

Tobias, D., and Merris, M. 1978. The golden lemon: a collection of special recipes. Atheneum, New York, NY. 210 pp.

Ullman, P. 1982. Pass the salt: put flavor into low sodium diets with Sunkist lemons. Consumer Services, Sunkist Growers, Inc., Van Nuys, CA. 15 pp.

LENTIL

Family: Fabaceae (Leguminosae; pea family)

Names

Scientific name: *Lens culinaris* Medik. [*L. esculenta* Moench, *L. orientalis* (Boiss.) Hand.-Mazz.]

- Both the English word "lentil" and the genus name *Lens* trace to the Latin word *lens*, a word that describes an object that is round in outline and convex on both sides. The word was originally applied to very different species with lens-shaped seeds, although many reference works claim that it was first used for lentil.
- Lentils are traditionally consumed during Lent. The similarity in the two words is fortuitous, as they have been derived from different roots. "Lent" comes from the Old English *lengten*, a word allied to the Old High German *lenzin*, meaning spring, alluding to the season of occurrence of Lent.
- *Culinaris* in the scientific name *L. culinaris* is derived from the Latin *culina*, pertaining to a kitchen, namely food, thus signifying something used for food.

Plant portrait

The lentil plant is an annual, upright, branching herb 25–75 cm (10–30 cm) tall. It has small flowers that may be light violet, light blue, white, pink, or a mixture of these colors. The plants bear flat, oblong pods that are short (1–1.5 cm or 2/5–3/5 inch long), with one or two small seeds called lentils. The seeds are clear, light green, deep purple, brown, gray-brown, or red in color, sometimes mottled or speckled. Purple flecking may be so intense in some varieties that the seeds appear black. Hulled lentils (i.e., with the seed coat removed) of commerce are either green or orange to reddish. Green lentils are usually marketed whole, while red lentils are mostly marketed in split form. About 70% of world production is of the red form. Canada and the US produce mainly green lentils, and the rest of the world mainly produces red lentils. In North America, small-seeded lentils are referred to as the "Persian type," and large-seeded lentils as the "Chilean type."

Lentil is a "pulse" or "grain legume" crop, i.e., a member of the pea family, the seeds of which are used as human food. In fact, lentil is the fourth most important pulse crop in the world after common bean (*Phaseolus vulgaris*), pea (*Pisum sativum*), and chickpea (*Cicer arietinum*), all discussed in this book. Although lentils make excellent feed for livestock, they are used mostly as human food. The lentil probably originated somewhere in the Near East. It is one of the most ancient of cultivated plants, and the remains of lentil meals have been found in archeological sites dating to 8000 BC in the Near East. Lentils are said to have been the staple diet of the Egyptian pyramid builders. The crop was brought to North America by early settlers, but was employed mostly in home gardens for use in soups. Commercial cultivation in North America was initiated in 1937 in eastern Washington and northern Idaho, where US production remains centered, with some grown now also in North Dakota and Montana. The United States is the fourth largest producer. Canada, the world's leading exporter of lentils, is the second largest producing country after India, with cultivation mainly in the Prairie Provinces. India, Canada, and Turkey account for nearly 70% of world production. Other major producers are Australia, Ethiopia, Pakistan, Bangladesh, China, Syria, and Spain. About 70% of the world's lentil crop is consumed in the countries where it is produced.

Culinary portrait

The texture and flavor of lentils differ according to variety. The seeds are used in soups, stews, salads, casseroles, curries, snack foods, and vegetarian dishes. They are generally canned or packaged dry for retail sale, sometimes separated into halves like split peas. The cooking time of lentils is shorter than for other pulses. The seeds become mushy if cooked too long. Brown lentils require about 60 minutes of cooking, orange lentils only 20–30 minutes. Brown lentils require 15–20 minutes in a pressure cooker, while orange lentils need as little as 5 minutes. It has been recommended that a little oil be added to lentils in a pressure cooker to prevent the scum formed at the top from blocking the safety valve. Lentil flour is added to cereal flour to make breads, cakes, baby foods, and invalid foods. Lentils are often used as a meat extender or substitute, providing a major source of protein that in rich countries often is based on meat. In some Asian countries, young pods are eaten fresh. In southern Asian countries, split red lentils are used in curries. The seeds are sometimes germinated and consumed as sprouts.

Lentil (*Lens culinaris*), from Thomé (1903–1905, vol. 3).

Lentils are low in fat and sodium and are free of cholesterol. They are a high-quality source of protein and fiber, and an excellent source of the B vitamin folate (an essential nutrient shown to reduce the risk of neural tube defects during pregnancy). Lentil flour is free of gluten and so can be used by people with celiac disease.

Lentils are often part of Old World ethnic holiday meals, as exemplified by the following:

- On New Year's Day, Italians often consume lentils and *cotechino* (pronounced coh-the-KEE-noh; a large, soft sausage made from pork rind and meat, seasoned with nutmeg, cloves, and pepper) as a symbol of money and wealth for the coming year.
- The Persian (Iranian) New Year (variously known as Nouruz, No Rouz, Noruz, Nou-Rooz, and Now Rooz) is celebrated over 13 days, starting with the vernal equinox. In most Iranian homes, a holiday table features Persian New Year symbols, such as goldfish to represent the passage through life's difficult waves, garlic to promote peace, and sprouting lentils to symbolize the life cycle.
- The Parsees (or Parsis) are a Zoroastrian people (Zoroastrianism is a Persian religion founded in the 6th century BC by the prophet Zoroaster) descended from Persian refugees who settled principally in Bombay. The Parsee dish *dhansak* requires a mixture of 3–9 different kinds of lentils. Today, dhansak is served traditionally in central-western India, and in addition to several types of lentils, the combination includes cooked, lamb, dhansak masala (a spicy mixture of toasted coriander seeds and cumin seeds), cinnamon, and cardamom, and typically served with grilled kebabs and tart pickles.

Culinary Vocabulary

- As a result of the biblical tale of Esau selling his birthright to his brother Jacob (see below), the culinary term "Esau" has come to mean lentils. The French expression *potage Esau* means lentil soup.
- "Ezekiel mix" is a mixture of ground lentils and several cereals (usually wheat, barley, spelt, and millet), as specified biblically (Ezekiel 4:9).

Curiosities of Science and Technology

- In ancient Egypt, lentils were sometimes used as money.
- The association between the dead and lentils traces back to the ancient Egyptians, who used lentils as funeral offerings to Horus, god of the sky and light, possibly because the moon (for which Horus was responsible) appeared to die once every month, and the lentil harvest was conceptualized as symbolic of this ending of life. Throughout Europe, people offered lentils (and other legumes, such as peas and beans) to the dead, often placing them in coffins, and serving them at meals commemorating or honoring the dead, a tradition that persists today. All Souls' Day, a Catholic feast day commemorating all the souls who no longer are with the living, is also marked by the serving of legumes, including lentils. In pagan days, many felt that offering the dead food on one day was a way of keeping them from haunting the Earth for the remainder of the year.
- The Greek physician Hippocrates (460?–377 BC), the Father of Medicine, prescribed lentils along with slices of boiled dog as a treatment for liver ailments.
- The famous obelisk that is in front of St. Peter's Basilica at the Vatican was brought to Rome during ancient Roman times in the hold of a ship in which it was buried under tons of lentils.

Lentil (*Lens culinaris*), from Vilmorin-Andrieux (1885).

- In the 6th century, lentils were considered to have the opposite effect of an aphrodisiac, and this was probably the reason why they were included in the meals of monasteries on meatless days.
- Optical lenses were named during the 13th century in Italy for their resemblance in shape to lentils. For more than 300 years, lenses were called "glass lentils." The Italian word for lentils, *lenticchie*, was first used for lenses, but gradually changed to *lente*, and was transformed in English to lens.
- In the 16th and 17th centuries, freckles were called lentils (for their resemblance to small lenses). Today, someone who is heavily freckled may be called "lentiginous" (not to be confused with "litigious," i.e., disputatious or prone to engaging in lawsuits).
- In India, lentil poultices were once used to alleviate the ulcers following smallpox and other slow-healing sores.

Key Information Sources

Adsule, R.N. 1996. Lentil (*Lens culinaris* Medik.). *In* Food and feed from legumes and oilseeds. *Edited by* J. Smartt and E. Nwokolo. Chapman and Hall, London, UK. pp. 109–112.

Adsule, R.N., Kadam, S.S., and Leung, H.K. 1989. Lentil. *In* CRC hand book of world food legumes: nutritional chemistry, processing technology and utilisation, vol. 2. *Edited by* D.K. Salunkehe and S.S. Kadam. CRC Press, Boca Raton, FL. pp. 139–152.

Agrawal, S.C., and Prasad, K.V.V. 1997. Diseases of lentil. Science Publishers, Enfield, NH. 155 pp.

Barulina, E.I. 1930. Lentils of USSR and other countries; a botanico-agronomical monograph. Trudy Prikl. Bot. Genet. Selek. Suppl. **40**: 265–319. [In Russian, English summary.]

Beniwal, S.P.S. 1993. Field guide to lentil diseases and insect pests. International Center for Agricultural Research in the Dry Areas, Aleppo, Syria. 107 pp.

Bhatty, R.S. 1988. Composition and quality of lentil (*Lens culinaris* Medik.): a review. Can. Inst. Food Sci. Technol. J. **21**: 144–160.

Erskine, W. 1985. Selection for pod retention and pod indehiscence in lentils *Lens culinaris*. Euphytica, **34**: 105–112.

Erskine, W., and Saxena, M.C. (*Editors*). 1991. Proceedings of the seminar on lentil in South Asia, 11–15 March 1991, New Delhi, India. International Center for Agricultural Research in the Dry Areas, Aleppo, Syria. 236 pp.

Erskine, W., Tufail, M., Russell, A., Tyagi, M.C., Rahman, M.M., and Saxena, M.C. 1994. Current and future strategies in breeding lentil for resistance to biotic and abiotic stresses. *In* Expanding the production and use of cool season food legumes. *Edited by* F.J. Muehlbauer and W.J. Kaiser. Kluwer Academic Publishers. Dordrecht, Netherlands. pp. 559–571.

Erskine, W, Williams, P.C., and Nakkoul, H. 1985. Genetic and environmental variation in the seed size protein yield and cooking quality of lentils *Lens culinaris*. Field Crops Res. **12**: 153–162.

Ferguson, M.E., Maxted, N., Van Slageren, M., and Robertson, L.D. 2000. A re-assessment of the taxonomy of *Lens* Mill. (Leguminosae, Papilionoideae, Vicieae). Bot. J. Linn. Soc. **133**: 41–59.

Hoffman, T.R. 1991. Observations on international production and trade in red lentil (*Lens culinaris*). IMPACT Center, College of Agriculture and Home Economics, Washington State University, Pullman, WA. 56 pp.

Iliadis, C. 2001. Effects of harvesting procedure, storage time and climatic conditions on cooking time of lentils (*Lens culinaris* Medikus). J. Sci. Food Agric. **81**: 590–593.

Iliadis, C. 2003. Influence of genotype and soil type on cooking time in lentil (*Lens culinaris* Medikus). Int. J. Food Sci. Technol. **38**: 89–93.

Jansen, P.C.M. 1989. *Lens culinaris* Medikus. *In* Plant resources of South-East Asia, 1, pulses. *Edited by* L.J.G. van der Maesen and S. Somaatmadja. Pudoc, Leiden, Netherlands. pp. 51–53.

Ladizinsky, G. 1979. The origin of lentil and its wild gene pool. Euphytica, **28**: 179–187.

Ladizinsky, G. 1993. Lentil domestication: on the quality of evidence and arguments. Econ. Bot. **47**: 60–64.

Ladizinsky, G., and Abbo, S. 1993. Cryptic speciation in *Lens culinaris*. Genet. Resour. Crop Evol. **40**: 1–5.

Ladizinsky, G., Braun, D., Goshen, D., and Muehlbauer, F.J. 1984. The biological species of the genus *Lens*. Bot. Gaz. **145**: 253–261.

Manitoba Agriculture. 1991. Lentil production in Manitoba. Revised edition. Manitoba Agriculture, Winnipeg, MB. 10 pp.

May, D., Lischynski, D., and Leduc, P. 1992. Guidelines for improved storage life of lentils. Saskatchewan. Agriculture Development Fund, SK. 20 pp.

Muehlbauer, F.J., and Ladizinsky, G. 1983. Breeding of the wild genepool of lentil. United States–Israel Binational Agricultural Research and Development Fund, Bet Dagan, Israel. 17 pp.

Muehlbauer, F.J., Cubero, J.I., and Summerfield, R.J. 1985. Lentil. *In* Grain legume crops. *Edited by* R.J. Summerfield and E.H. Roberts. Collins, London, UK. pp. 266–311.

Muehlbauer, F.J., Kaiser, W.J., Clement, S.L., and Summerfield, R.J. 1995. Production and breeding of lentil. Adv. Agron. **54**: 283–332.

Pinkas, R., Zamir, D., and Ladizinsky, G. 1985. Allozyme divergence and evolution in the genus *Lens*. Plant Syst. Evol. **151**: 131–140.

Rajora, O.P., and Mahon, J.D. 1997. Mitochondrial and nuclear DNA variation, genotype fingerprinting and genetic relationships in lentil (*Lens culinaris* Medik.). Can. J. Plant Sci. **77**: 515–521.

Singh, D.K., and Shukla, A. 2000. Classification of germplasm in lentil (*Lens culinaris*, Medik.). Agric. Biol. Res. **16**: 22–29.

Singh, D.P. 1995. Lentil germplasm: evaluation and utilization: a review. Indian J. Genet. Plant Breed. **55**: 173–181.

Skrypetz, S. 2006. Lentils: situation and outlook. Biweekly Bull. (Agric. Agri-Food Can.), **19**(**7**): 1–4.

Summerfield, R.J., Muehlbauer, F.J., and Short, R.W. 1982. Description and culture of lentils. U.S. Dept. of Agriculture, Agricultural Research Service, Washington, DC. 22 pp.

Thomas, G., Solanki, I.S., and Singh, V.P. 1997. Evaluation of selection methods in lentil (*Lens culinaris* Medik.). Crop Improv. **24**: 239–243.

Webb, C., and Hawtin, G.C. (*Editors*). 1981. Lentils. Commonwealth Agricultural Bureaux, Slough, UK. 216 pp.

Zavodna, M., Kraic, J., Paglia, G., Gregova, E., and Morgante, M. 2000. Differentiation between closely related lentil (*Lens culinaris* Medic.) cultivars using DNA markers. Seed Sci. Technol. **28**: 217–219.

Zdunczyk, Z., Godycka, I., Frejnagel, S., Krefft, B., Juskiewicz, J., and Milczak, M. 1993. Nutritional value of lentil seeds (*Lens culinaris*) as compared with beans and peas. Polish J. Food Nutr. Sci. **2**(**3**): 73–81.

Zohary, D. 1973. The wild progenitor and place of origin of the cultivated lentil: *Lens culinaris*. Econ. Bot. **26**: 326–332.

Zohary, D. 1995. Lentil. *In* Evolution of crop plants. 2nd edition. *Edited by* J. Smartt and N.W. Simmonds. Longman Scientific & Technical, Burnt Mill, Harlow, Essex, UK. pp. 271–274.

Specialty Cookbooks

(Also see Pea.)

Day, H. 1970. About rice and lentils. Thorsons, London, UK. 64 pp.

Dixon, P., and Birch, C. 1982. The bean & lentil cookbook: colourful, inexpensive and highly nutritious pulse recipes: includes sweet dishes. Thorsons, Wellingborough, Northamptonshire, UK. 128 pp.

Elliot, R. 1984. Rose Elliot's book of beans and lentils. Fontana Paperbacks, London, UK. 63 pp.

Frets, M. 1990. Lentil & split pea cookbook. Peanut Butter Publishing, Seattle, WA. 193 pp.

Janson, B.L. 1990. Dry peas and lentils: cooking with the "protein twins." Horizon Publishers & Distributors, Bountiful, UT. 218 pp.

Jenner, A. 1984. The Amazing legume: cooking with lentils, dry beans & dry peas. Saskatchewan Pulse Crop Growers' Association, Delmas, Saskatchewan, SK. 134 pp.

Mallos, T. 1985. Wholefood cookery: with beans, peas and lentils. Lansdowne, Sydney, Australia. 128 pp.

Midgley, J. 1992. The goodness of beans, peas, and lentils. Random House, New York, NY. 65 pp.

Ross, T. 2003. Easy beans: fast and delicious bean, pea & lentil recipes. 2nd edition. Big Bean Publishing, Vancouver, BC. 139 pp.

United States Dry Pea and Lentil Council. 2000. The pea & lentil cookbook: from everyday to gourmet. USA Dry Pea and Lentil Council, Moscow, ID. 238 pp.

LETTUCE

Family: Asteraceae (Compositae; sunflower family)

Names

Scientific name: *Lactuca sativa* L.

- The word "lettuce" is derived through the French word for lettuce, *laitue*, from the Latin *lac* (or *lactlac*) for milk, in reference to the milky juice exuding from damaged parts of the plant. The genus name *Lactuca* is also derived from the same Latin word.
- Nonheading, leaf lettuce is sometimes called "cutting lettuce," as it produces a succession of leaves for cutting off.
- The name "iceberg" lettuce came into being in the 1920s when head lettuce, previously called "crisphead," began to be transported in California under mounds of ice to keep the lettuce cool.
- Cos lettuce acquired its name when it was extensively cultivated on the Greek island of Kos, between Greece and Turkey, in the Aegean Sea.
- Romaine lettuce traces its name to a shipment to France of this type of lettuce from the eastern Mediterranean through Rome.
- Bibb lettuce is named after state politician and amateur breeder John B. Bibb (18th century–1884) of Frankfort, Kentucky, who developed the variety, which he called Limestone.
- *Sativa* in the scientific name *L. sativa* is Latin for sown or cultivated.

Plant Portrait

Lettuce, a herbaceous annual, is the Western World's most popular salad plant. The species probably evolved in Asia Minor or the Middle East from wild lettuce, *L. serriola* L. (*L. scariola* L.). Lettuce has been selected for different forms, as noted in the following.

- *Crisphead* is the most commonly grown commercial type of head lettuce in North America. The Crisphead group includes iceberg lettuce.
- *Butterhead* includes the so-called "bibb" lettuces, which have a small soft head and an oily or buttery texture to the inner leaves. Limestone lettuce, a kind of Bibb lettuce, belongs to this category. Boston lettuce is another familiar example.
- *Leaf* types include loose-leaf lettuces, the most common type in home gardens.
- *Cos* or *Romaine* includes cultivars with upright habit, resulting in elongated loose heads with relatively narrow leaves.
- *Stem* lettuce, grown for its edible stem, was originally selected in the Orient. It is often called "celtuce."
- *Latin* lettuce has foliage similar to the Cos class, but the leaves are shorter, leathery, and tough, with a loose, open head.
- *Oilseed* lettuce is a primitive type grown for its large seeds, which are pressed to obtain oil. The ancient Egyptians may have originally cultivated this kind of lettuce as an oil crop.
- *Head* or "cabbage" lettuces were selected relatively recently.

Open, loose types of lettuce with distinct cos-like leaves were used in ancient Egypt, about 4500 BC. Lettuce was popular in the classical Roman and Greek civilizations. It was introduced into China about 600–900 AD, and stem lettuce (celtuce) evolved there. Stem lettuce is still grown commercially, primarily in China. Lettuce use spread throughout the Mediterranean Basin, and by the Middle Ages, Europe was raising cultivars of all the classes of lettuce known today. Today, lettuce is grown to some degree in most countries of the world, although it does best in cool environments and does not tolerate either extreme heat or cold. Lettuce was one of the first vegetables to be introduced to the New World by Europeans. Lettuce has been called "the king of salad vegetables" and "the star of the salad world." In California, lettuce is known as "green gold," reflecting its status as the most important salad vegetable grown in the most important vegetable state. The Salinas–Watsonville district of California, encompassing Salinas Valley, Pajaro Valley, and San Benito Valley, is the largest producer of crisphead lettuce in the world. Along with other vegetables that are raised in this area, this has led to the designation of the region as "the salad bowl of the world."

Culinary Portrait

Most forms of lettuce are used raw in salads and sandwiches, or occasionally heated as in tacos, or braised and creamed. Much of the popularity of lettuce is due to its ability to be trucked long distances at low temperatures while retaining its flavor and crispness.

Varieties of lettuce (*Lactuca sativa*), from Vilmorin-Andrieux (1885).

Stem lettuce is eaten either fresh or cooked. It is cut off down to the leafy portion of the plant, and the outer skin is peeled away to remove the bitter sap. The soft, translucent green core can be eaten fresh, sliced, or diced in a salad. It is also cooked by boiling or frying.

"It's always best to tear lettuce rather than cut it, to prevent discoloration of the cut ends." This rule has been drummed into the minds of countless cooks for centuries. In fact, Harold McGee reported that experiments he conducted (*The Curious Cook*, 1990, Macmillan Publishing, New York) showed that browning of the cut or torn portions was about equal (so long as a sharp knife is employed), and it makes no difference whether lettuce is torn or cut. [Curiously, he found that when he conducted the same exercise for basil leaves, cut pieces quickly developed a black edge, while most of the torn edges stayed green.] McGee found that tearing the tough outer leaves of romaine and iceberg lettuce resulted in detaching large shreds of the cuticle from the underlying tissues, and excessively rapid uptake of oil in salad dressing; so, in this case, cutting was superior. In any event, dressing a salad should be done at the last minute because both vinegar and oil harm the appearance of lettuce. Oil particularly acts rapidly to darken the

Loose-leaved Romaine lettuce (*Lactuca sativa*), from Pailleux and Bois (1892).

color of lettuce and produce a soggy texture. Opaque, water-based mixtures like mayonnaise and cream dressings have the advantages of masking blemishes and slowing the infiltration of oil. Tender lettuce varieties, such as Bibb, benefit from gentle handling, as blemishes can easily be induced.

Culinary Vocabulary

- "Salad" is from the Latin word for salt, *sal*, reflecting the use of little more than salt to dress the first salads made in Rome. "Green salad" is more or less the same as "salad," although some use the former to mean a salad composed only of leafy vegetables, in contrast to a salad with dressing and (or) other additions such as croutons and bacon bits.
- Caesar Cardini (1896–1956), an Italian immigrant, created the now classic Caesar salad in 1924 for the patrons of his Tijuana, Mexico, restaurant chain. The dish became very popular with Hollywood movie people, who often traveled to Tijuana, which is just across the US border. As a result, Caesar salads became popular throughout North America. Indeed, the Caesar salad was voted in the 1930s by the International Society of Epicures in Paris as the "greatest recipe to originate from the Americas in 50 years." The dish was originally made with uncut Romaine lettuce leaves, to be eaten whole using the fingers. Authentic Caesar salads are made also with garlic, olive oil, croutons, Parmesan cheese, Worcestershire sauce, egg yolks, lemon, and anchovies (although Caesar insisted that anchovies should not be included).
- *Stoofsla* (pronounced stoof-slah) is a Dutch dish made by boiling lettuce, topping it with breadcrumbs, and braising the preparation.

Curiosities of Science and Technology

- The ancient Greeks believed that lettuce induced sleep, so they served it at the end of the meal. This belief has been attributed to the white juice of primitive lettuce, which reminded the Greeks of the white juice of the sleep-inducing opium poppy. The Romans continued the custom of serving lettuce at the end of the meal, and during the time of the emperor Augustus (ruling from 27 BC to 14 AD), it was still fashionable to eat salad in this way. However, when Domitian became emperor (ruling from 81 to 96), he insisted that the salad course be served at the beginning of the meal (he believed that it stimulated the appetite). Thus began the debate carried down to our time of whether salad should be served at the beginning or end of a meal!
- Because of the presence of milky latex in the plant, lettuce was believed in classical Roman times to be beneficial to lactating mothers who needed rich milk to feed their babies.

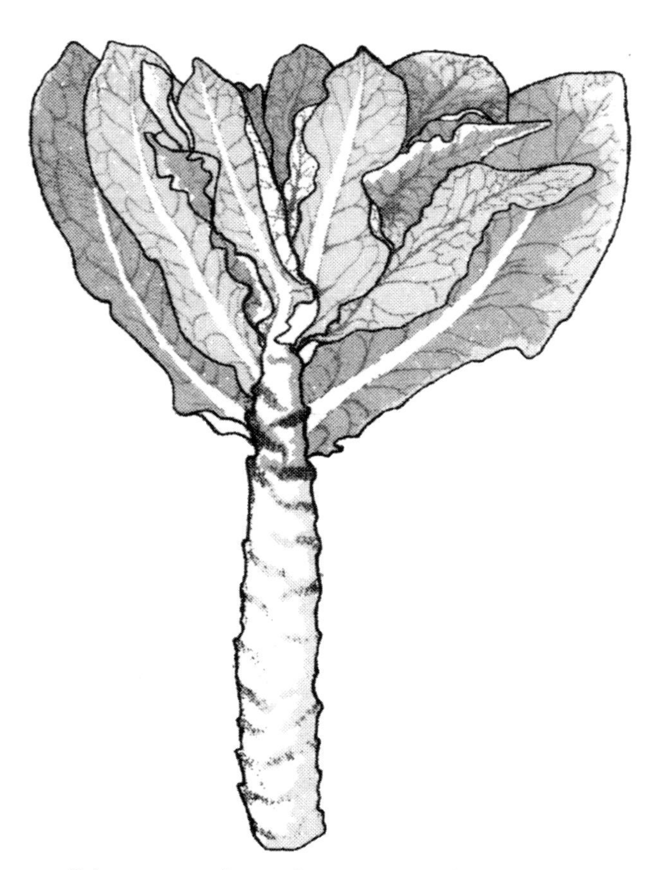

Celtuce or stem lettuce (*Lactuca sativa*), by M. Jomphe.

- Lettuce seeds were apparently first brought to the New World by Columbus in 1494.
- Herbalists in Elizabethan times (i.e., the period of Elizabeth I, 1533–1603) recommended that lettuce be eaten at mealtime and before "indulgence in drink," because, as one of them wrote, "it staieth the vapours that disturb the head and cooleth the hot stomache which some call heart burn." The advice regarding the sobering effect of lettuce probably stemmed from the old Roman belief that lettuce prevents drunkenness.
- The milky juice from lettuce (or wild lettuce, *L. virosa*), known as *lactucarium*, has been considered to be a narcotic since ancient times and has even been used as a substitute for commercial opium. The chemical lactucin in wild lettuce is in fact an analogue of morphine. However, it is probably impossible to consume enough lettuce to become intoxicated!
- Lettuce was the experimental plant used in the 1950s to demonstrate a remarkable physiological process in plants. The seeds of some lettuce cultivars exhibit a kind of dormancy that can be induced or reversed, depending on exposure to different kinds of light. Red light converts a substance called phytochrome to a state that activates the germination process, whereas far-red light or darkness converts phytochrome to another state that inhibits germination.
- Wild lettuce, *L. serriola*, which is closely related to garden lettuce, has leaves that are vertical rather than horizontal. Under the intense rays of the noonday sun, the leaves receive minimum exposure, but when the sun is near the horizon, the leaves receive full insolation—presumably a protective adaptation. The tips of the leaves point north or south, giving rise to the name "compass plant" (a phrase used for several other species that exhibit the same phenomenon).
- Lettuce leaves are 95% water by weight.

Key Information Sources

Aultman, D.A., and Windham, S.L. 1963. Head lettuce production. Mississippi State University, Agricultural Experiment Station, State College, MI. 8 pp.

Davis, R.M. 1997. Compendium of lettuce diseases. American Phytopathological Society, St. Paul, MN. 79 pp.

Friedland, W.H., Barton, A.E., and Thomas, R.J. 1981. Manufacturing green gold: capital, labor, and technology in the lettuce industry. University Press, New York, NY. 159 pp.

Grubben, G.J.H., and Sukprakarn, S. 1993. *Lactuca sativa* L. *In* Plant resources of South-East Asia, 8, vegetables. *Edited by* J.E. Siemonsma and K. Piluek. Pudoc Scientific Publishers, Wageningen, Netherlands. pp. 186–190.

Harlan, J. 1986. Lettuce and the sycomore: sex and romance in Ancient Egypt. Econ. Bot. **40**: 4–15.

Koopman, W.J.M. 1999. Plant systematics as a useful tool for plant breeders: examples from lettuce. *In* Proceedings of the Eucarpia meeting on leafy vegetables genetics and breeding, Olomouc, Czech Republic, 8–11 June 1999. *Edited by* A. Lebeda and E. Kristkova. Palacký University, Olomouc, Czech Republic. pp. 95–105.

Large, J.G. 1972. Glasshouse lettuce. Grower Books, London, UK. 181 pp.

Lindqvist, K. 1960. On the origin of cultivated lettuce. Hereditas, **46**: 319–350.

Lipton, W.J., and Ryder, E.J. 1989. Lettuce. *In* Quality and preservation of vegetables. *Edited by* N.A. Erskin. CRC Press, Inc., Boca Raton, FL. pp. 217–244.

Myers, C. 1991. Celtuce, asparagus lettuce. Crop Sheet SMC-008. *In* Specialty and minor crops handbook. *Edited by* C. Myers. The Small Farm Center, Division of Agriculture and Natural Resources, University of California, Oakland, CA. 2 pp.

Pink, D.A.C., and Keane, E.M. 1993. Lettuce, *Lactuca sativa* L. *In* Genetic improvement of vegetable crops. *Edited by* G.Kalloo and B.O. Bergh. Pergamon Press, New York, NY. pp. 543–571.

Ramadan, A.A.S. 1976. Characteristics of prickly lettuce seed oil in relation to methods of extraction. Die Nahrung, **20**: 579–583.

Rodenburg, C.M., and Basse, H. 1960. Varieties of lettuce; an international monograph. Instituut voor de Veredeling van Tuinbouwgewassen, Wageningen, Netherlands. 228 pp. [In Dutch.]

Ryder, E.J. 1986. Lettuce breeding. *In* Breeding vegetable crops. *Edited by* M.J. Bassett. AVI Publishing Co., Westport, CT. pp. 433–474.

Ryder, E.J. 1999. Genetics in lettuce breeding: past, present and future. *In* Proceedings of the Eucarpia meeting on leafy vegetables genetics and breeding, Olomouc, Czech Republic, 8–11 June 1999. *Edited by* A. Lebeda and E. Kristkova. Palacký University, Olomouc, Czech Republic. pp. 225–231.

Ryder, E.J., and Whitaker, T.W. 1995. Lettuce. *In* Evolution of crop plants. 2nd edition. *Edited by* J. Smartt and N.W. Simmonds. Longman Scientific & Technical, Burnt Mill, Harlow, Essex, UK. pp. 53–56.

Shoeb, Z.E., Osman,, F., El-Kirdassy, Z.H.M., and Eissa, M.H. 1969. Studies and evaluation of Egyptian lettuce seeds *Lactuca scariola* L. Grasas y Aceites, **20**: 125–128.

Thompson, R.C. 1951. Lettuce varieties and cultures. Revised edition. U.S. Dept. of Agriculture, Washington, DC. 41 pp.

Thompson, R.C., Doolittle, S.P., and Henneberry, T.J. 1958. Growing lettuce in greenhouses. U.S. Dept. of Agriculture, Washington, DC. 22 pp.

Tracy, W.B., Jr. 1904. American varieties of lettuce. Bulletin No. 69. United States. Dept. of Agriculture, Bureau of Plant Industry, Washington, DC. 103 pp.

Whitaker, T.W., Ryder, E.J., Rubatsky, V.E., and Pail, P.V. 1974. Lettuce production in the United States. Revised edition. Agricultural Research Service, U.S. Dept. of Agriculture, Washington, DC. 43 pp.

Wien, H.C. 1997. Lettuce. *In* The physiology of vegetable crops. *Edited by* H.C. Wien. CAB International, Oxford, UK. pp. 479–509.

Zohary, D. 1991. The wild genetic resources of cultivated lettuce (*Lactuca sativa* L.). Euphytica, **53**: 31–35.

Specialty Cookbooks

Raymond, D., and Raymond, J. 1979. The Gardens for All book of lettuce & greens. Gardens for All, Burlington, VT. 34 pp.

Schlesinger, C., and Willoughby, J. 1996. Lettuce in your kitchen: where salad gets a whole new spin and dressings do double duty. William Morrow and Co., New York, NY. 264 pp.

United States National Gardening Association. 1987. Book of lettuce & greens. Revised edition. Villard Books, New York, NY. 87 pp.

Lima Bean

Family: Fabaceae (Leguminosae; pea family)

Names

Scientific name: *Phaseolus lunatus* L.

- The lima bean was named for Lima, Peru, where the Conquistadores found it around 1500. Note that the "l" in "lima bean" is not capitalized, by convention, even though the name is based on a city.
- "Madagascar bean" is a name that reflects the growth of the lima bean there shortly after the discovery of the New World. "Burma bean" and "Carolina bean" are other geographically based names.
- "Butter bean" is used for certain varieties of both common bean (which see) and lima bean.
- "Calico bean" is an occasional name for the lima bean. One of the meanings of the word "calico" is blotched or spotted, and as noted below, this is descriptive of the seeds of some varieties.
- See Common Bean for derivation of the genus name *Phaseolus*.
- *Lunatus* in the scientific name *P. lunatus* is from the Latin *luna*, moon, the half moon, shaped like a crescent moon, descriptive of the shape of the pod.

Plant Portrait

The lima bean is a native of tropical America. Lima beans have been found in Peru in archeological sites dated at more than 7000 years ago. Lima bean was not only an important food source but was also incorporated into the art and symbolism of the pre-Columbian culture of Peru. Both climbing forms, 2–4 m (6–13 feet) long, and bush forms, 30–90 cm (1–3 feet) tall, are available today. The species is usually cultivated as an annual, although many varieties have the capacity to grow for more than one season. Some varieties are cultivated as biennials or

Lima bean (*Phaseolus lunatus*). Bush type at left from *Gartenflora* (1899, vol. 48), vine type at right from Vilmorin-Andrieux (1885).

perennials in the tropics. Lima beans are produced in the warmer parts of the world, and have become extremely important in the wet forest regions of tropical Africa. They are extensively grown in tropical America. In the United States, California is the major area of production. The flowers are a combination of greenish or yellow–green and white to yellow–orange, the petals often differently colored. The pods are curved, flat, 5–12 cm (2–5 inches) in length, with 2–4 seeds. The seeds are usually kidney-shaped, compressed, 12–24 mm (1/2–1 inch) long, and ivory white to brown or red. Red–purple and black seeds are sometimes seen, and some varieties have mosaic or spotted patterns. The following types have been recognized:

- Sieva beans (occasionally recognized as a different species, also known as butterbean, a name also used for the common bean, *P. vulgaris*; and also known as civet bean and sewee bean): comparatively thin-podded and with three or four seeds per pod; seeds intermediate in flatness between those of the following groups. These are popular in the southern United States, where they are staple hot-weather vegetables.
- Large lima beans: possessing thicker pods and 4–8 flat seeds per pod. Large lima beans are larger and more vigorous than sieva beans, and they produce large pods with chubby, greenish seeds. They are less resistant to heat than sieva beans, but the larger, thicker pods are easier to shell.
- Potato lima beans: with pods short like sieva beans, and few seeds, but with pods as thick as those of large lima beans, and more spheroidal seeds than those in the above groups.

Lima bean (*Phaseolus lunatus*), from Turner (1892).

Culinary Portrait

Lima bean is much less grown and consumed than common bean (*Phaseolus vulgaris*, which see). Like common bean, it is grown both for the young immature pods and the ripe seeds, which are used as fresh, canned, and frozen vegetables. Young bean pods and young seeds are often eaten raw, but note the warning in the next paragraph. Lima bean seeds (green or dried) are steamed, boiled, and baked. They are served as a side dish and are added to stews, casseroles, and soups. Mature lima bean pods are not eaten. Lima beans have a delicate flavor that is suited to dishes where stronger-tasting beans would mask the subtle taste desired. The beans should not be overcooked, as they tend to become mushy. Recommended cooking times are 15–25 minutes for fresh lima beans in boiling water, and about 1½ hours for dried beans (smaller beans cook faster than larger beans). Considerable foam may be produced during cooking of lima beans, which can be dangerous in a pressure cooker, as the foam could block the safety valve. Fifteen minutes is recommended for large-seeded, dried lima beans in a pressure cooker, while small-seed varieties need only 10 minutes. Sometimes lima beans are germinated to produce "bean sprouts" for consumption.

Mature lima beans contain the cyanogenic glycoside phaseolunatin, and an enzyme that releases hydrocyanic acid from the glycoside. Phaseolunatin is most prevalent in the seeds, and colored beans produce more than white beans. Cooking generally destroys this toxin, and in any case, poisoning is unlikely in humans because of the small quantities ingested. However, there have been occurrences of livestock poisoning when animals have gained access to fields of lima beans and have eaten considerable quantities. Some varieties of lima bean from the Caribbean are especially high in cyanogenic glycosides.

Curiosities of Science and Technology

- In North America, lima beans (as well as common beans) are often cooked with sweet corn to form "succotash," a stew-like mixture concocted originally by the Amerindians (the word is based on the Narraganset Indian, meaning "fragments"). Beans are rich sources of protein, particularly the essential amino acid lysine, but are usually deficient in the sulfur-containing amino acids methionine and cystine. By contrast, cereal grains like corn contain a lower amount of proteins, and these are deficient in lysine but adequate in sulfur-containing amino acids. Together, beans and corn provide a balance of essential protein components, which is especially necessary for survival in diets deficient in meat. Nevertheless, colonists, who learned how to make succotash from the Indians, often added meat in the form of salt pork, bacon, venison, or other game.
- Some lima bean varieties of Asia are quite different from North American cultivars. "Java beans" are somewhat shriveled beans, varying in color from dull purplish red to black. Because they produce large quantities of hydrocyanic acid when ground and moistened, they are poisonous to man and livestock, and suitable only as an agricultural green manure (a crop turned into the soil as fertilizer).
- In 1974, the U.S. Armed Forces conducted a survey to examine food preferences. To check that respondents were paying attention, they invented several imaginary foods, such as "Buttered Ermal" and "Braised Trake." One imaginary food, "Funistrada," was ranked higher than lima beans.
- An article published by Arimura et al. (cited below) reported that lima bean plants under attack from spider mites release chemicals that not only attract the insects' natural predators but also warn nearby plants that there is danger about. Damaged leaves released volatile compounds known as terpenoids. These organic compounds not only made the lima bean leaves more attractive to the carnivorous natural enemies of the mites but also "warned" neighboring leaves with no infestation, prompting them to roll out their own defenses.

Key Information Sources

Allard, R.W. 1953. Production of dry edible lima beans in California. College of Agriculture, University of California, Berkeley, CA. 27 pp.

Arimura, G., Ozawa, R., Nishioka, T., Boland, W., Koch, T., Kuhnemann, F., and Takabayashi, J. 2002. Herbivore-induced volatiles induce the emission of ethylene in neighboring lima bean plants. Plant J. **29**: 87–98.

Arimura, G., Ozawa, R., Shimoda, T., Nishioka, T., Boland, W., and Takabayashi, J. 2000. Herbivory-induced volatiles elicit defence genes in lima bean leaves. Nature, **406**: 512–515.

Baudoin, J.P. 1988. Genetic resources, domestication and evolution of lima bean, *Phaseolus lunatus*. *In* Genetic resources of *Phaseolus* beans: their maintenance, domestication, evolution, and utilization. *Edited by* P. Gepts. Kluwer Academic Publishers, Dordrecht, Netherlands. pp. 393–407.

Baudoin, J.P. 1989. *Phaseolus coccineus* L. *In* Plant resources of South-East Asia, 1, pulses. *Edited by* L.J.G. van der Maesen and S. Somaatmadja. Pudoc, Leiden, Netherlands. pp. 56–60.

Baudoin, J.P. 1993. Lima bean (*Phaseolus lunatus* L.). *In* Genetic improvement of vegetable crops. *Edited by* G. Kalloo and B.O. Bergh. Pergamon Press, Oxford, UK. pp. 391–403.

Cook, W.P. 1985. Commercial lima bean production. Hortic. Leafl. Clemson University Cooperative Extension Service, Clemson, SC. 4 pp.

Crayton, E.F. 1991. Grow your own lima beans. Circ. ANR. Alabama Cooperative Extension Service, Auburn University, Auburn, AL. 8 pp.

Davis, D.W. 1987. Lima bean. *In* Proceedings, grain legumes as alternative crops: a symposium. Univ. Minnesota, July 23–24, 1987. *Edited by* Center for Alternative Crops and Products. University of Minnesota, St. Paul, MN. pp. 103–112.

Degreef, J., Rocha, O.J., Vanderborght, T., and Baudoin, J.P. 2002. Soil seed bank and seed dormancy in wild populations of lima bean (Fabaceae): considerations for in situ and ex situ conservation. Am. J. Bot. **89**: 1644–1650.

Egbe, I.A., and Akinyele, I.O. 1989. Effect of cooking on the antinutritional factors of lima beans (*Phaseolus lunatus*). Food Chem. **35(2)**: 81–87.

Fofana, B., Baudoin, J.P., Vekemans, X., Debouck, D.G., and Du-Jardin, P. 1999. Molecular evidence for an Andean origin and a secondary gene pool for the lima bean (*Phaseolus lunatus* L.) using chloroplast DNA. Theor. Appl. Genet. **98**: 202–212.

Fofana, B., Du Jardin, P., and Baudoin, J.P. 2001. Genetic diversity in the lima bean (*Phaseolus lunatus* L.) as revealed by chloroplast DNA (cpDNA) variations. Genet. Resour. Crop Evol. **48**: 437–445.

Glancey, J.L., Kee, W.E., and Wootten, T.L. 1997. Machine harvesting of lima beans for processing. J. Veg. Crop Prod. **3(1)**: 59–68.

Gutierrez-Salgado, A., Gepts, P., and Debouck, D.G. 1995. Evidence for two gene pools of the lima bean, *Phaseolus lunatus* L., in the Americas. Genet. Resour. Crop Evol. **42**: 15–28.

Holland, A.H. 1953. Production of green lima beans for freezing. Division of Agricultural Sciences, University of California, Berkeley, CA. 22 pp.

Kee, E., Glancey, J.L., and Wootten, T.L. 1997. The lima bean: a vegetable crop for processing. HortTechnology, **7**: 119–128.

Kovach, S., Brown, S., Gazaway, W.S., Patterson, M., and Williams, J.L. 1992. Lima bean production. Circ. ANR. Alabama Cooperative Extension Service, Auburn University, Auburn, AL. 6 pp.

Lyman, J.M., Baudoin, J.P., and Hidalgo, R. 1985. Lima bean (*Phaseolus lunatus* L.). *In* Grain legume crops. *Edited by* R.J. Summerfield and E.H. Roberts. Collins, London, UK. pp. 477–519.

Mackie, W.W. 1943. Origin, dispersal and variability of the Lima bean, *Phaseolus lunatus*. Hilgardia, **15**: 1–29.

Martínez-Castillo, J., Zizumbo-Villarreal, D., Perales-Rivera, H., and Colunga-Garcíamarin, P. 2004. Intraspecific diversity and morpho-phenological variation in *Phaseolus lunatus* L. from the Yucatan Peninsula, Mexico. Econ. Bot. **58**: 354–380.

Nienhuis, J., Tivang, J., Skroch, P., and Santos, J.B. dos. 1995. Genetic relationships among cultivars and landraces of lima bean (*Phaseolus lunatus* L.) as measured by RAPD markers. J. Am. Soc. Hortic. Sci. **120**: 300–306.

Nwokolo, E. 1996. Lima bean (*Phaseolus lunatus* L.). *In* Food and feed from legumes and oilseeds. *Edited by* J. Smartt and E. Nwokolo. Chapman and Hall, London, UK. pp. 144–158.

Specialty Cookbooks

(See BEAN, COMMON.)

LIME

Family: Rutaceae (rue family)

Names

Scientific name: *Citrus aurantifolia* (Christm.) Swingle

- Both the names "lime" and "lemon" are ultimately derived from the Arabic *līmah*, "citrus fruit." The two fruits were often confused in the Middle Ages.
- The name "lime" is also applied to other plants, notably the linden (species of *Tilia*) and sometimes to a species of tupelo or sour gum, known also as the Ogeechee lime (*Nyssa ogeche* W. Bartram ex Marshall). The Ogeechee lime grows in the southeastern United States, and in Georgia, it is used commercially to produce a chutney-like relish.
- The lime has also been called "Adam's apple."
- In the 19th century, lime became an important crop in the Florida Keys. However, the industry was demolished by the infamous hurricane of 1926, and today, limes are grown there merely as a minor resource. The name "Key lime" reflects the former importance of the lime in the Florida Keys.
- The Kaffir lime, *Citrus hystrix* DC., a native of Asia, is cultivated in Southeast Asia, more often for its leaves than for its fruit, although the juice is used as a flavoring. The leaves are employed like bay leaves to flavor soups, curries, sauces, and gravies.
- For an explanation of the genus name *Citrus* and the English word "citrus," see GRAPEFRUIT.
- A*urantifolia* in the scientific name *C. aurantifolia* means "with leaves like those of the orange tree."

Plant Portrait

Lime trees are much-branched, spiny evergreens (spineless in several commercial varieties), 2–4 m (6½–13 feet) high. The lime is native to the Indo-Malayan region. Unknown in Europe before the Crusades, it seems to have been carried to North Africa and the Near East by Arabs and taken by Crusaders from Palestine to Mediterranean Europe. In the mid 13th century, it was cultivated and well known in Italy. The lime was introduced into the Caribbean islands and Mexico by the Spaniards. In contrast to the more subtropical lemon, the lime requires a tropical climate. The plants often flower and bear fruit continually during the year. They are cultivated commercially in India, Egypt, Mexico, the West Indies, tropical America, and throughout the tropics of the Old World. In the United States, the lime is grown to some extent in Florida and in protected valleys in California.

There are two acidic or sour limes in world trade. The Key (Mexican or West Indian) lime has fruit that is nearly round, 2.5–5 cm (1–2 inches) in diameter, with thin rind, acidic, very juicy flavorful pulp, and few to many seeds. The skin is green and glossy when immature, pale yellow when ripe, somewhat rough to very smooth. The pulp is greenish yellow, compartmented into 6–15 segments, which do not readily separate. Tahiti or Persian limes are larger, 5–6.3 cm (2–2½ inches) in diameter, less round (usually oval), with a protrusion or nipple opposite the stem end, greener than the rather yellow Key lime, usually seedless, less acidic, and much more likely to be encountered in supermarkets outside of Florida.

Furanocoumarins are compounds in plants that produce, in some individuals, a condition called photodermatitis—a rash that is developed only after exposure to sunlight. These chemicals are found naturally in the skin of citrus fruits, including limes. Contact with limes has produced several recorded outbreaks of photodermatitis (see Anonymous 1985 and Gross et al. 1987).

Culinary Portrait

The lime is a basic cooking ingredient in much of the world, particularly Latin America, the West Indies, India, Southeast Asia, Africa, and the Pacific Islands. Although limes and lemons are basically interchangeable in recipes, the special bouquet and unique flavor of lime suit it ideally as a garnish and flavoring for fish, meats, ice cream, and desserts, for adding zest to cold drinks, and for making limeade. Lime juice is prized the world over for use in mixed alcoholic drinks like the Margarita. A wedge of lime is a traditional Mexican means of clearing the palate after drinking tequila neat (i.e., undiluted).

Limes are more perishable than lemons, but can be stored in a refrigerator for up to several weeks. They tend to become yellow, losing their desirable acidic taste, when exposed to bright light.

Culinary Vocabulary

- The word "lime" also means "calcium oxide," a building material produced from limestone (calcium carbonate). Surprisingly, limestone is occasionally used for culinary purposes: in Mexican masa harina (tortilla flour), and to enhance release of alkaloids in several narcotic plants (for example, betelnut, *Areca catechu* L.).

- A unique kind of food depending completely on lime juice is the classic Peruvian fresh fish dish "ceviche" (also spelled cebiche, sebiche, and seviche, and pronounced seh-VEE-chee, seh-VEE-cheh, or seh-VEESH), popular in Latin America. Raw fish is marinated overnight in lime juice (occasionally in other citrus juice) and, the next day, seasoned with fresh chiles, cilantro, onions, and tomatoes. The fish proteins denature in an acidic medium, just as they do at high temperatures, so the fish is in a sense "cooked" by the lime juice. This is quite unlike sushi, the Japanese uncooked fish dish.
- "Key Lime Pie," a custard-like pie made with lime juice, condensed milk, and eggs, and served in a pastry shell, is a famous dish of the Florida Keys and southern Florida, particularly since the 1950s. Today, Key Lime Pie is generally not made from Key limes, but largely from the frozen concentrate of the Persian lime.
- The Key lime is so often used in alcoholic beverages that one of its names is "bartender lime."
- The "gimlet" cocktail is made from gin or vodka with lime juice. Rose's Lime Juice is traditionally used. This concoction now contains a variety of ingredients, such as high-fructose corn syrup. The original Rose's Lime juice was made by Lauchlin Rose, a 19th century Scot who patented a process that preserved lime juice without the addition of alcohol. There are two widely circulated stories accounting for the origin of the name "gimlet." Sir T.O. Gimlette, a British navel surgeon during the period 1879–1917, is said to have believed that drinking straight gin was unhealthy and impaired the efficiency of navel officers, so he diluted it with lime juice, inventing the gimlet. A quite different account of the origin of the gimlet begins with the fact that the crew of the Royal Navy mixed their rations of gin and Rose's lime juice to ward off scurvy. A small boring device called a gimlet was used to open kegs of lime juice that were sent to British colonies during the late 18th century. This tool may account for the drink's unique name.
- The "Mai Tai" (Tahitian for "good") cocktail consists of a mixture of lime, rum, and curaçao (a liqueur flavored with the dried peel of the sour orange). The late Victor "Trader Vic" Bergeron of Trader Vic's restaurants is commonly credited with inventing the drink in 1944 in Oakland, California. In recent times, pineapple and other juices are often added to the beverage to increase its tropical flair.
- A "rickey" is basically lime juice and soda water, best known as an alcoholic cocktail (called gin rickey and lime rickey) made usually with gin, sometimes with bourbon, blended whisky, or applejack. The origin of the name is obscure; it has been suggested that it was named after an otherwise unidentified Colonel Rickey. When sugar and lemon juice is added, the cocktail becomes a Tom Collins, often served with a maraschino cherry and served in a Collins glass.
- A "daiquiri" (pronounced dak-ree) is a cocktail made traditionally of rum, line juice (sometimes lemon juice is substituted), and sugar, sometimes with puréed fruit (such as bananas or strawberries) blended into the mix. It is named after a small village on the Cuban coast near Santiago, where the Americans are said to have landed after defeating the Spanish in the 19th century.
- A less known but interestingly named cocktail made with lime juice is the "Suffering Bastard." The version created by 1930s Los Angeles restaurateur Don the Beachcomber (Don Richard Beaumont Gantt) is made with lime juice, curaçao, sugar syrup, orgeat syrup, and rum, served with crushed ice and a garnish of cucumber peel, fresh mint, lime shell, and a fruit stick. (A second kind of Suffering Bastard, said to come from Cairo, Egypt, is made with cognac or brandy, sherry, and ginger ale.)
- A "Shark's Tooth" cocktail is prepared with lime juice, rum, lemon juice, grenadine, rock candy, syrup, and club soda, and is garnished with the shell of a half lime.
- The word "punch" today means a fruit-juice-based beverage, either nonalcoholic or, more often, with any of a variety of liquors or wines. The word comes from the Indian (Hindi) *panch*, five, so named because five ingredients were originally used: lime, sugar, spices, water, and arrack (palm liquor). "Planter's Punch" is a once-popular cocktail served over cracked ice. The following mnemonic indicates its proportions: "one of sour (lime juice), two of sweet (sugar syrup), three of strong (rum), and four of weak (water)."
- The lemon–lime beverage 7UP was created in 1929 (just before the stock market crash) by soda-pop mogul Charles Griggs from Missouri, but did not become popular until 4 years later, when he

renamed it 7UP. The original name was "Bib-Label Lithiated Lemon-Lime Soda." Apparently, the 7UP drink was so named because it contained "seven natural flavors blended into a savory, flavory drink with a real wallop." The "lithiated" in the original name reflected the presence of the antidepressant lithium carbonate, which was removed in the 1940s.

Curiosities of Science and Technology

- Although it is widely believed that limes were the primary source of vitamin C used by the British to prevent sailors from developing scurvy, it was generally the juice of the lemon that was carried aboard English sailing ships of the 18th century for this purpose. In the mid 19th century, lime juice from the West Indies was substituted for juice from Mediterranean lemons, a change that was resented by experienced seamen who realized that lime juice has considerably less vitamin C than lemon juice, and in fact proved to be less effective against scurvy. American sailors (who often carried cranberries to prevent scurvy) commonly taunted their British counterparts by calling them by the pejorative nickname "limeys," an abbreviation of the obsolete phrase "lime-juicers." Limey is sometimes applied to British people generally, especially British immigrants.
- Commodore Matthew C. Perry (1794–1858) led a US expedition to Japan and the China seas during 1852–1854, and this resulted in a number of hitherto unfamiliar foods being introduced to Western culture (see, for example, ADZUKI BEAN). In some regions he visited, cannibalism was still practiced. In Sumatra, he observed that visiting dignitaries were served ears seasoned with lime juice. The dish has not become popular in Western countries.
- In Texas, a necklace of limes was once thought to cure a sore throat.
- Throughout the world, "chew sticks"—tree twigs that are used as a combination toothbrush and toothpick—are employed to clean teeth. Usually, certain tree species are preferred because the texture of the wood and the character of the sap are especially suitable. In past centuries in the West Indies, twigs of the lime tree were commonly used as chew sticks.
- In the 20th century, people in New Caledonia rubbed lime into their hair to repel lice.

Lime (*Citrus aurantifolia*), from Curtis (1787–present, plate 6745).

Key Information Sources

Anonymous. 1971. Limes vary in ascorbic acid content. Agric. Res. (Washington), **20**(**5**): 16.

Anonymous. 1985. Outbreak of phototoxic dermatitis from limes: Maryland. Morb. Mortal. Wkly. Rep. **34**: 462–464.

Castellanos, M., and Hernandez, J. 1982. Relationship between fruit size and its quality in Persian lime (*Citrus aurantifolia*). Cultivos Tropicales (Cuba), **4**: 639–650. [In Spanish, English summary.]

Chamblee, T.S., and Clark, B.C., Jr. 1997. Analysis and chemistry of distilled lime oil (*Citrus aurantifolia* Swingle). J. Essent. Oil Res. **9**: 267–274.

Chavez-Sierra, C., Bosquez-Molina, E., Pelayo-Zaldivar, C., Perez-Flores, L., and Ponce-de-Leon-Garcia, L. 1993. Effect of harvesting season and postharvest treatments on storage life of Mexican limes (*Citrus aurantifolia* Swingle). J. Food Qual. **16**: 339–354.

Chisholm, M.G., Wilson, M.A., and Gaskey, G.M. 2003. Characterization of aroma volatiles in key lime essential oils (*Citrus aurantifolia* Swingle). Flavour Fragrance J. **18**: 106–115.

Dalsgaard, A., Reichert, P., Mortensen, H.F., Sandstrom, A., Kofoed, P.E., Larsen, J.L., and Molbak, K. 1997. Application of lime (*Citrus aurantifolia*) juice to drinking water and food as a cholera-preventive measure. J. Food Protect. **60**: 1329–1333.

Grierson, W., Wardowski, W.F., and Edwards, G.J. 1971. Postharvest rind disorders of "Persian" limes. Proc. Fla. State Hortic. Soc. **84**: 294–298.

Gross, T.P., Ratner, L., De Rodriguez, O., Farrell, K.P., and Israel, E. 1987. An outbreak of phototoxic dermatitis due to limes. Am. J. Epidem. **125**: 509–514.

Hatton, T.T., and Reeder, W.F. 1971. Ascorbic acid concentrations in Florida-grown "Tahiti" (Persian) limes. Proc. Trop. Reg. Am. Soc. Hortic. Sci. **15**: 89–94.

Hatton, T.T., and Reeder, W.F. 1972. Ascorbic acid concentrations in Florida-grown "Tahiti" (Persian) limes. Citrus Ind. **53**(**8**): 7–8.

Hittalamani, S.V., and Rao, M.M. 1977. Comparative study of the dynamics of vegetative growth in 'Kagazi' (*Citrus aurantifolia* Swingle.) and 'Tahiti' (*Citrus latifolia* Tanaka.) limes. Mysore J. Agric. Sci. **11**: 310–314.

Kitat, F.M., El-Azab, E.M., and Wehida, B.M. 1973. Effect of pollination and type of pollen on fruit set, fruit drop and yield in some lime varieties (*Citrus aurantifolia* Swingle). Alexandria J. Agric. Res. **21**: 109–117.

Manley, W.T., and Godwin, M.R. 1960. Retail distribution and merchandising of fresh limes and frozen limeade concentrate. University of Florida Agricultural Experiment Station, Gainesville, FL. 32 pp.

Morton, J. 1987*a*. Mexican lime. *In* Fruits of warm climates. *Authored by* J.F. Morton. Creative Resource Systems, Winterville, NC. pp. 168–172.

Morton, J. 1987*b*. Tahiti lime. *In* Fruits of warm climates. *Authored by* J.F. Morton. Creative Resource Systems, Winterville, NC. pp. 172–175.

Morton, J. 1987*c*. Sweet lime. *In* Fruits of warm climates. *Authored by* J.F. Morton. Creative Resource Systems, Winterville, NC. pp. 175–176.

Morton, J. 1987*d*. Mandarin lime. *In* Fruits of warm climates. *Authored by* J.F. Morton. Creative Resource Systems, Winterville, NC. pp. 178–179.

Nickavar, B., and Mojab, F. 2003. Volatile constituents of the dried fruit of *Citrus aurantifolia* from Iran. J. Med. Aromat. Plant Sci. **25**: 400–401.

Nigg, H.N., Nordby, H.E., Beier, R.C., Dillman, A., Macias, C., and Hansen, R.C. 1993. Phototoxic coumarins in limes. Food Chem. Toxicol. **31**: 331–335.

Oslund, C.R., and Davenport, T.L. 1987. Seasonal enhancement of flower development in 'Tahiti' limes by marcottage. HortScience, **22**: 498–501.

Paredes-Lopez, O., Camargo-Rubio, E., and Gallardo-Navarro, Y. 1974. Use of coatings of candelilla wax for the preservation of limes. J. Sci. Food Agric. **25**: 1207–1210.

Passam, H.C., and Blunden, G. 1982. Experiments on the storage of limes at tropical ambient temperature [*Citrus aurantifolia*]. Trop. Agric. **59**: 20–24.

Pino, J.A., and Rosado, A. 2001. Comparative investigation of the distilled lime oils (*Citrus aurantifolia* Swingle and *Citrus latifolia* Tanaka) from Cuba. J. Essent. Oil Res. **13**: 179–180.

Roy, M., Andrew, C.O., and Spreen, T.H. 1996. Persian limes in North America: an economic analysis of production and marketing channels. Florida Science Source, Lake Alfred, FL. 132 pp.

Sethpakdee, R. 1991. *Citrus aurantifolia* (Christm. & Panzer) Swingle. *In* Plant resources of South-East Asia, 2, edible fruits and nuts. *Edited by* E.W.M. Verheij and R.E. Coronel. Pudoc, Leiden, Netherlands. pp. 126–128.

Spalding, D.H., and Reeder, W.F. 1977. Quality of 'Tahiti' limes stored in a controlled atmosphere or under low pressure. Citrus Ind. **57**(**12**): 5–8, 10.

Spalding, D.H., and Reeder, W.F. 1983. Conditioning 'Tahiti' limes to reduce chilling injury. Proc. Fla. State Hortic. Soc. **96**: 231–232.

Venkateshwarlu, G., and Selvaraj, Y. 2000. Changes in the peel oil composition of Kagzi lime (*Citrus aurantifolia* Swingle) during ripening. J. Essent Oil Res. **12**: 50–52.

Yadav, A.R., Chauhan, A.S., Rekha, M.N., Rao, L.J.M., and Ramteke, R.S. 2004. Flavour quality of dehydrated lime (*Citrus aurantifolia* (Christm.) Swingle). Food Chem. **85**: 59–62.

Specialty Cookbooks

Borden, Inc. 1983. That lively lime twist: recipes with the real difference of Realime lime juice from concentrate. Borden Inc., Borden Consumer Products Division, Columbus, OH. 19 pp.

Donker, R. 2001. The lemon & lime cookbook. Bookmasters, Mansfield, OH. 179 pp.

LaFray. J. 1987. Key lime desserts. Surfside Publications, Surfside, FL. 48 pp.

LaFray, J. 1998. Key lime cookin'. Revised edition. SeaSide Publications, St. Petersburg, FL. 77 pp.

Peterson, L. 2000. A squeeze of lime. Fithian Press, Santa Barbara, CA. 211 pp.

Wallace, I., and Theodorou, S. 2000. The lemon & lime cookbook. Hamlyn, London, UK. 64 pp.

LUPIN

Family: Fabaceae (Leguminosae; pea family)

Names

Scientific names: *Lupinus* species

- White lupin—*L. albus* L
- Blue lupin—*L. angustifolius* L.
- Yellow lupin—*L. luteus* L.
- Andean lupin, pearl lupin—*L. mutabilis* Sweet

- The word "lupin" and the genus name *Lupinus* are based on the Latin *lupus*, wolf. It was thought that lupins took over land, robbing the soil of its fertility, just as wolves robbed shepherds of their livestock. In fact, lupins enrich the soil with nitrogen, which they are able to obtain from the air like other members of the pea family.
- "Lupine" is the standard American spelling; "lupin" is standard elsewhere.
- White lupin (*L. albus*) is also known as field lupin, Egyptian lupin, and wolf bean. *Albus* in the scientific name *L. albus* is Latin for white (for the flowers).
- Blue lupin (*L. angustifolius*) is also known as European blue lupin and narrow-leaved lupin. *Angustifolius* in the scientific name *L. angustifolius* is Latin for narrow-leaved.
- Yellow lupin (*L. luteus*) is also known as European yellow lupin. *Luteus* in the scientific name *L. luteus* is Latin for yellow (for the flowers).
- Andean lupin (*L. mutabilis*) is also known as pearl lupin, South American lupin, and in South America, tarwi. Early English writers named it pearl lupin for the appearance of the seeds, but the International Lupin Association has recommended that the name Andean lupin be employed. *Mutabilis* in the scientific name *L. mutabilis* is Latin for changeable, a reference to the different appearances the plant assumes.

Plant Portrait

The genus *Lupinus* has about 200 species of annual and perennial herbs, and evergreen subshrubs and shrubs. Most grow wild in the New World, with only about a dozen species native to the Old World. Eighty-two species of lupins have been listed as occurring in California alone. The plants are common in the Mediterranean region, North Africa, western North America, and South America, where they are found growing on dry hilly grasslands, coastal sands and cliffs, and along the banks of streams and rivers. Lupins are well-known garden ornamentals, but several species are also employed as cool-season grain legumes or forage crops. All of the agricultural species mentioned here are annuals. The annual lupins are erect herbs, up to 1 m (about a yard) or more in height, with coarse stems, hand-like leaves, and massive spikes of colorful white, yellow, or blue flowers. Except for the Andean lupin, all of the important agricultural species are native to Europe. In the Old World, lupins are consumed primarily in Italy, the Middle East, and North Africa.

For the most part, lupins have traditionally been used for grazing animals, and indeed today their chief agricultural use remains the feeding of livestock, especially cows. Lupins were independently domesticated in both the Old and New Worlds between 3000 and 4000 years ago, and have been employed not only for livestock but also as human food. The classical Egyptians, Greeks, and Romans grew annual lupins (chiefly white lupin) as food for animals and flour for bread. Andean lupin has been cultivated for thousands of years in the Andean Highlands of South America. Yellow lupin and blue lupin have probably also been cultivated for thousands of years. Lupins produce considerable high-protein seed, which has long been a temptation to people as a grain, but certain bitter, poisonous chemicals (primarily alkaloids) make the seeds unpalatable (additionally, other chemicals called tannins and trypsin inhibitors lower the food value of the seeds). People selected less poisonous plants over time, and learned to remove the bitterness of the seeds by soaking, so that they could be eaten, but lupins have remained minor as a source of human food in the Western World.

In 1917, a "Lupin Banquet" was given in Hamburg by the German Botanical Society. At a table covered with a tablecloth of lupin fiber, lupin soup was served, followed by lupin beefsteak roasted in lupin oil and seasoned with lupin extract, then bread containing 20% lupin flour, lupin margarine, and lupin cheese, and finally lupin liqueur and lupin coffee. Lupin soap, lupin-fiber paper, and envelopes made with lupin adhesive were also shown. This generated interest in breeding "sweet" (low-alkaloid) types of lupin, which were developed in the 1920s from the old, bitter types.

Some lupin species used as human food. White lupin from Köhler (1883–1914), yellow lupin from Hallier (1880–1888, vol. 23, plate 2324), blue lupin from Hallier (1880–1888, vol. 23, plate 2323), Andean lupin from Edwards and Lindley (1815–1847, vol. 18).

Lupins are legumes, and share the common trait of the family of "fixing" nitrogen from the atmosphere through the cooperation of certain bacteria housed in nodules on the roots. By this process, inert nitrogen (making up 80% of the atmosphere), which is unavailable to most plants, is turned into a form that all plants can use, and so the requirement for fertilizing the soil is greatly diminished. Because of this ability to improve the soil, lupins are extremely important as "rotation crops," i.e., grown in alternation with other crops (which generally tend to deplete the soil). By the early 1950s, lupins were planted to such an extent in the Coastal Plain of the United States that it was called the "Lupin Belt," but for various reasons the crop declined thereafter.

Andean lupin is of special significance in South America. This is an annual with woody stems and blue, purple, pink, or white flowers. It was domesticated in pre-Incan times, at least 1500 years ago, in the central Andean region. The seeds are still grown for human food in South America, from Ecuador to Chile and northern Argentina. European colonization reduced the importance of the crop. The seeds are astonishingly nutritional, with more than 40% protein, and about 20% oil. However, Andean lupin has a high alkaloid content, producing very bitter, toxic seeds, requiring processing to make them safe and palatable. South American indigenous peoples were and remain expert in doing this, but this has meant that the foods prepared have remained important only locally. In the Andes, lupin dishes include soups, salads, desserts, snacks, and soft drinks (made with papaya juice and Andean lupin flour). Experimentation has shown that up to 15% protein-rich flour can be used in breadmaking. Sweet varieties of the species have very recently been selected, and attempts are underway to develop Andean lupin as a modern crop.

Culinary Portrait

Specialty human-food markets for lupin are developing. Seed preparations are often served plain, or with just a little lemon juice, or consumed as an appetizer or snack, especially in Italy and the Middle East. Sweet lupin seeds (including the hulls) have been shown to be an excellent source of dietary fiber, as an additive to breads and cereals. Lupin flour is been used to enrich soups, pastas, sauces, cereals, cake mixes, breads, and other baked goods. All of the species listed above are being used for these purposes, but white lupin is the most important in most of the world.

The seeds of white and yellow lupins are occasionally roasted, ground, and used as coffee. The seeds of white lupin have been germinated to provide sprouts, sold like alfalfa and mung bean sprouts in Australia.

Curiosities of Science and Technology

- In ancient Roman times, large lupin seeds were used as play money, especially in theatrical productions. This association of lupins and money was well known in past centuries in England, and the lupin flower was sometimes called "penny bean." In one episode of the classic British comedy series *Monty Python's Flying Circus*, a Robin Hood character played by John Cleese robbed people of their lupins, a subtlety that very few of the audience would understand.
- Hundreds of plants were assigned special meanings in the "Victorian Language of Flowers," popular in Victorian times, and many delighted in sending coded messages by this means, especially for romantic purposes. "Lupin" meant "voraciousness" and "imagination."
- In ancient times, "beans" often served as symbols during funerals, and lupins were considered to be a kind of bean. The rapid growth of "beans" (which are seeds) following planting often represented resurrection in the afterlife. Lupins, as well as other beans, were often consumed at funerals. In classical Greece, it was sometimes believed that lupins had magical powers, and could be helpful in contacting the dead. The "Oracle of the Dead" was a shrine on the river Acheron where pilgrims who wanted to contact the spirits of the departed went, first consuming large quantities of lupin seeds to facilitate communication.
- The very bitter Andean lupin (*L. mutabilis*) is used as a protective hedge to prevent damage from grazing animals in South America.
- American Indians used wild American lupins for various purposes. Some of the more interesting uses follow. When sulphur lupins (*L. sulphureus* Dougl. ex Hook.) or manycolored lupins (*L. versicolor* Lindl.) were in flower, the Okanagan and Colville Indians considered it to be a sign that the groundhogs were fat enough to eat. The Kwakiutl rubbed root ash made from seashore lupin (*L. littoralis* Dougl.) into a newborn baby's cradle to make it sleep well. The Cherokee rubbed sundial lupin (*L. perennis* L.) on the hands or body to give a person the power to control horses.
- Seeds of an arctic lupin species (*L. arcticus* S. Wats.) were collected from lemming burrows deeply buried in permanently frozen silt of the central Yukon. The seeds germinated, and it was claimed in an article in the journal *Science* in 1967 that they were at least 10 000 years old, and thus the oldest seeds ever germinated. However, the claim that the seeds were actually so old is difficult to accept. Seeds of Lotus (*Nelumbo nucifera* Gaertn.) are generally thought to hold the record for seed longevity at 1300 years.

Key Information Sources

Bhardwaj, H.L. 2002. Evaluation of lupin as a new food/feed crop in the US Mid-Atlantic region. *In* Trends in new crops and new uses. *Edited by* J. Janick and A. Whipkey. ASHS Press, Alexandria, VA. pp. 115–119.

Bhardwaj, H.L., Hamama, A.A., and Merrick, L.C. 1998. Genotypic and environmental effects on lupin seed composition. Plant Foods Hum. Nutr. **53**: 1–13.

Birk, Y., Dovart, A., Waldman, M., and Uzurean, C. (*Editors*). 1990. Lupin production and bio-processing for feed, food and other by-products. Proceedings of the joint CEC-NCRD Workshop held in Israel, Ginozar Kibbutz, January 1989. Office for Official Publications of the European Communities, Luxembourg. 218 pp.

Commission of the European Communities. 1992. *Lupinus mutabilis*: its adaptation and production under European pedoclimatic conditions: proceedings of a workshop held in Cascais, Portugal, 26 and 27 April 1991. Commission of the European Communities, Luxembourg. 191 pp.

Cowling, W.A., Buirchell, B.J., and Núñez, M.E.T. 1998. Lupin, *Lupinus* L. International Plant Genetic Resources Institute, Rome, Italy. 105 pp.

Gladstone, J.S. 1970. Lupins as crop plants. Field Crop Abstr. **23**: 123–148.

Gladstones, J.S. 1974. Lupins of the Mediterranean region and Africa. West. Aust. Dep. Agric. Tech. Bull. **26**: 1–48.

Gladstones, J.S., Atkins, C.A., and Hamblin, J. 1998. Lupins as crop plants: biology, production, and utilization. CAB International, Wallingford, Oxon, UK. 465 pp.

Gros, R., and Bunting, E.S. 1982. Agricultural and nutritional aspects of lupines: proceedings of the first international lupine workshop, Lima-Cuzco, Peru, 12–21, April 1980. German Agency for Technical Cooperation, Eschborn, Germany. 878 pp.

Haq, N. 1993. Lupins (*Lupinus* species). *In* Pulses and vegetables. *Edited by* J.T. Williams. pp. 103–130.

Henson, P.R., and Stephens, J.L. 1958. Lupines: culture and use. U.S. Dept. of Agriculture, Washington, DC. 12 pp.

Hill, G.D. 1977. The composition and nutritive values of lupin seed. Nutr. Abstr. Rev. **47**: 511–519.

Hill, G.D. 1995. Lupins. *In* Evolution of crop plants. 2nd edition. *Edited by* J. Smartt and N.W. Simmonds. Longman Scientific & Technical, Burnt Mill, Harlow, Essex, UK. pp. 277–282.

Hondelmann, W. 1984. The lupin: ancient and modern crop plant. Theor. Appl. Genet. **68**: 1–9.

International Lupin Association. 1986. Proceedings of the fourth international lupin conference, Aug. 15–22, 1986, Geraldton, Western Australia. Western Australian Department of Agriculture, South Perth, WA, Australia. 350 pp.

International Lupin Association. 1990. Proceedings of the 6th international lupin conference, Nov. 25–30, 1990, Temuco-Pucon, Chile. Asociación Chilena del Lupino, Temuco, Chile. 406 pp.

Kazimierski, T., and Nowacki, E. 1961. Indigenous species of lupins regarded as the initial forms of the cultivated species *Lupinus albus* and *Lupinus mutabilis*. Flora, **151**: 202–209.

Landers, K.F., Sutherland, S., and Sykes, J. 2000. Lupin. 3rd edition. NSW Agriculture, Sydney, NSW, Australia. 44 pp.

Lopez-Bellido, L., and Fuentes, M. 1986. Lupin crop as an alternative source of protein. Adv. Agron. **40**: 239–295.

Martins, J.M.N., and Beirão da Costa, M.L. 1994. Advances in lupin research: proceedings of the 7th international lupin conference, Évora, Portugal, 18–23 April 1993. ISA Press/Instituto Superior de Agronomia, Lisboa, Portugal. 588 pp.

Mask, P.L., Reeves, D.W., Santen, E. van, Mullins, G.L., and Aksland, G.E. 1993. White lupin: an alternate crop for the southern coastal plain. *In* New crops. *Edited by* J. Janick and J.E. Simon. Wiley, New York, NY. pp. 277–278.

Meronuck, R.A., Meredith, H., Kalis-Kuznia, R., and Field, L. 1993. Lupin production and utilization guide. Revised edition. Center for Alternative Plant and Animal Products, University of Minnesota, St. Paul, MN. 27 pp.

Minnesota Extension Service. 1991. Prospects for lupins in North America: proceedings of a symposium held in St. Paul, Minnesota, March 21–22, 1991. Minnesota Extension Service, University of Minnesota, St. Paul, MN. 191 pp.

Pate, J.S., Williams, W., and Farrington, P. 1985. Lupin (*Lupinus* spp). *In* Grain legume crops. *Edited by* R.J. Summerfield and E.H. Roberts. Collins, London, UK. pp. 699–746.

Putnam, D.H. 1993. An interdisciplinary approach to the development of lupin as an alternative crop. *In* New crops. *Edited by* J. Janick and J.E. Simon. Wiley, New York, NY. pp. 266–277.

Santen, E. van. 2000. Lupin, an ancient crop for the new millenium: proceedings of the 9th international lupin conference, Klink/Müritz, Germany, 20–24 June 1999. International Lupin Association, Canterbury, New Zealand. 481 pp.

Todorov, N.A., Pavlov, D.C., and Kostov, K.D. 1996. Lupin (*Lupinus* spp.). *In* Food and feed from legumes and oilseeds. *Edited by* J. Smartt and E. Nwokolo. Chapman and Hall, London, UK. pp. 113–123.

Williams, S.W. 1979. Studies on the development of lupins for oil and protein. Euphytica, **28**: 481–488.

MALANGA (YAUTIA)

Family: Araceae (arum family)

Names

Scientific name: *Xanthosoma sagittifolium* (L.) Schott

- The names "malanga" and "yautia" are encountered in about equal frequency in North America. "Malanga" is an American-Spanish word that probably arose from *kongo*, the plural of *elanga*, water lily. "Yautia" is also American-Spanish, and is based on the same word in the Taino language. The word "malanga" is used in Cuba, "yautia" in Puerto Rico.
- Pronunciations: mah-LAHN-gah or mah-LAHNG-gah; yow-TEEY-ah.
- The species is also known as tannia (tania) and tanier, corruptions of the original Carib name *taia*, the native name in the Antilles. These names are common in the Caribbean.
- The crop is widely known as new cocoyam, especially in West Africa. The name arose by comparison with taro, the older and similar crop (often known as "old cocoyam"). The name "cocoyam" is based on "coco" (from its being planted in coconut groves) + "yam" (for its appearance and use).
- Malanga or yautia is widely confused with the similar taro (which see), and some names are used for both species, and indeed also for yams. In some supermarkets, it may be unclear which species is being sold. In recipes using these names, it may also be unclear, although the vegetables are sufficiently similar that for practical purposes it makes little difference. However, many prefer the taste of malanga over that of taro, and the former stores better than the latter.
- In the West Indies, the small underground storage organs (cormels) are called "nut eddos" (taro is sometimes called eddo).
- In New Guinea, malanga is called kong kong.
- The genus name *Xanthosoma* is based on the Greek *xanthos*, yellow, + *soma*, body, referring to the yellow inner tissue of some species.
- *Sagittifolium* in the scientific name *X. sagittifolium* is Latin for arrow-leaved, which is descriptive of the leaves.

Plant Portrait

The classification of *Xanthosoma* species cultivated particularly for their underground storage organs is unclear, and it may be that several species went into the parentage of modern kinds. Like potatoes, the plant is reproduced vegetatively, and so distinctive types are easily maintained. Also like potatoes, a variety of distinctive kinds are sold in the marketplace. The practice has been to assign most varieties to *X. sagittifolium*, which is by far the most important kind. *Xanthosoma violaceum* Schott (commonly called "*X. nigrum*") is variously known as blue tannia, blue taro, and primrose malanga, and is cultivated in the Caribbean. The interior of its storage organs is distinctly gray–lavender, but whether it really deserves to be called a separate species is unclear.

Malanga is grown for its edible turnip-like or potato-like subterranean storage organs and for the edible young leaves and larger stems that are used as potherbs. The main edible underground part is called a corm (which is stem, not root tissue; these are often called tubers), and the leaves are produced above ground from this. The top of the corm may be at or above ground level. As well, surrounding the "mother corm" are smaller ones (daughter corms or "cormels"). The corms may be 30 cm (1 foot) long, with thick brown or yellow–brown skin, and flesh that is ivory or yellowish, sometimes orange, pink, or reddish. The thin, brownish skin may be smooth, downy, or studded with small protuberances, depending on variety. The leaf blades are large, 30–50 cm (1–2 feet) long, arrow-shaped, and borne on long stalks. Malanga is a herbaceous perennial often grown as an annual. In commercial plantations, harvesting is carried out 10–12 months after planting. If left to flower (which is seldom done), a central compact floral shoot bearing numerous tiny flowers is formed. At the base of this flowering shoot is a white, surrounding, leaf-like structure (technically, a spathe) 12–15 cm (4½–6 inches) long, which only occasionally produces seeds. Malanga is an important food for more than 400 million people worldwide, especially in the tropics and subtropics, particularly in the Caribbean, Central and South America, West Africa, and tropical Asia. In some countries, such as Cameroon, it ranks second only to cassava (which see) as a source of calories in daily diets. Small amounts are grown commercially in

Primrose malanga (*Xanthosoma violaceum*), from Nicholson (1885–1889).

South Florida. The corms and cormels, and the young leaves and shoots, are consumed in a wide range of prepared foods. In the Caribbean and Southeast Asia, young tender leaves are eaten as potherbs. However, limited amounts of leaves are harvested and sold in North America.

Malanga originated in northern South America, and has likely been cultivated in the New World for thousands of years. The crop was spread through tropical America, and when the Europeans arrived, it was known from southern Mexico to Bolivia. From America, the plant reached West Africa, which is now the major producer. There, it has been displacing taro because of its better yield and because it can replace yams for preparing fufu, a very popular food in tropical Africa (the term "fufu" is widely used in West Africa to refer to a sticky dough or porridge prepared from any pounded starchy food). The plant is grown in Africa largely as a subsistence crop.

Culinary Portrait

The starchy corms of malanga are cooked and eaten like yams and potatoes—baked, grilled, fried, mashed, puréed, in soups, and barbecued whole. They can also be deep-fried and served with a sauce. In the West Indies, grated malanga is mixed with vegetables or fish, seasoned with herbs, and a dumpling is prepared. In food combinations, malanga should be added in moderation, as its strong flavor tends to mask that of other foods. The skin is best avoided. Tender corms are washed and peeled before cooking, but some are so hard that they need to be cooked before peeling. Twenty minutes of boiling suffices to prepare the tubers as an accompanying vegetable, either puréed or whole. Cooked malangas may be added to soups or stews at the last minute to prevent overcooking and falling apart. The flavor is strong, reminiscent of hazelnut with a slightly earthy aftertaste.

There is little consumption or consumer familiarity with malanga in North America, outside of ethnic populations. The vegetable is imported from the Caribbean and Latin America, although quality may be poor unless purchased immediately. When buying, it is best to choose very firm tubers with no evidence of soft spots or mold. The juiciness can be determined by pressing the flesh with the fingernail. Malanga can be stored at room temperature for just a few weeks (often for just a few days after purchase). The vegetable spoils rapidly, becoming shriveled and soft, and losing its characteristic hazelnut flavor. It keeps longer if refrigerated at 7 °C (45 °F), but long refrigerated storage at this temperature can produce chilling injury.

As with taro (which see), the corms contain irritating, needle-shaped crystals of calcium oxalate, which cause itchiness and considerable inflammation of tissues. As well, there are toxic chemicals called saponins. Processes such as roasting and cooking are required to neutralize these substances. Once properly prepared, malanga is more digestible than potato. In Cuba, babies and people with mild ulcers are placed on diets of malanga. The leaves are generally consumed as a boiled vegetable, like spinach. In the Caribbean, the leaves are often an ingredient of the popular soup callaloo. Where malanga is grown, it is sometimes used to manufacture alcohol.

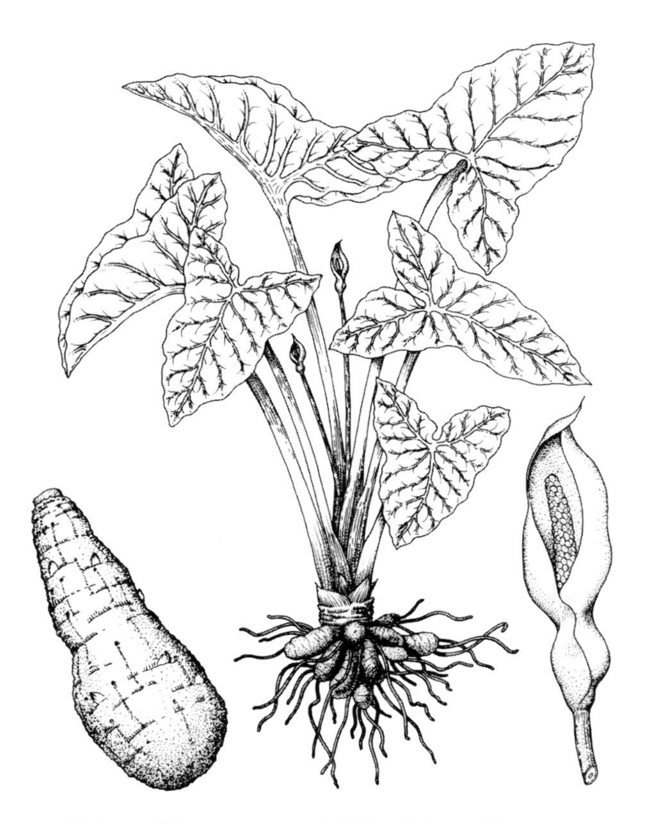

Malanga (*Xanthosoma sagittifolium*), from Giacometti and Léon (1994), reproduced by permission of the Food and Agriculture Organization of the United Nations. Edible storage organ is at bottom left, flowering branch (enveloped in leafy bract) at lower right.

Curiosities of Science and Technology

- Indigenous people in regions of South America where malanga is native sometimes use the large leaves of the plant as a parasol or umbrella.
- The Mexican *Salvia divinorum* Epling & Játiva is hallucinogenic, and was cultivated by Central American Indians for use in religious ceremonies before the conquest of America by Europeans. Its foliage was reportedly kept fresh for a week or longer by wrapping in the large leaves of another Mexican species, *Xanthosoma robustum* Schott, a relative of the edible species of *Xanthosoma*, which is cultivated today as an ornamental.
- The arum family is well known for the presence of needle-like crystals of calcium oxalate, which discourage animals from eating the plants. In malanga, the needles are barbed to cause additional misery.

Key Information Sources

Alamu, S., and McDavid, C.R. 1985. Genetic variability in tannia, *Xanthosoma sagittifolium* (L.) Schott. Trop. Agric. **62**: 30–32.

Cuebas de Escabi, D., and Cedeno-Maldonado, A. 1985. Content and accumulation of oxalic acid in edible species of *Xanthosoma*. J. Agric. Univ. Puerto Rico, **69**: 169–176.

Dizon-Lauzon, R., and Mabesa, L.B. 1988. Utilization of cocoyam *Xanthosoma* sp. flour for food. Ann. Trop. Res. **10**: 16–25.

Giacometti, D.C., and Léon, J. 1994. Tannia, yautia. (*Xanthosoma sagittifolium*). *In* Neglected crops. 1492 from a different perspective. *Edited by* J.E. Hernández-Bermejo and J. Léon. Food and Agriculture Organization of the United Nations, Rome, Italy. pp. 253–258.

Hackett, C. 1984. Tabular descriptions of crops grown in the tropics. 4. Tannia (*Xanthosoma sagittifolium* (L.) Schott). CSIRO Institute of Biological Resources, Division of Water and Land Resources, Canberra, Australia. 51 pp.

Iwuoha, C.I., and Kalu, F.A. 1995. Calcium oxalate and physico-chemical properties of cocoyam (*Colocasia esculenta* and *Xanthosoma sagittifolium*) tuber flours as affected by processing. Food Chem. **54**: 61–66.

Jansen, P.C.M., and Premchand, V. 1996. *Xanthosoma* Schott. *In* Plant resources of South-East Asia, 9, plants yielding non-seed carbohydrates. *Edited by* M. Flach and F. Rumawas. Backhuys Publishers, Leiden, Netherlands. pp. 159–164.

Lemke, D.E., and Schneider, E.L. 1988. *Xanthosoma sagittifolium* (Araceae): new to Texas. Southwest. Nat. **33**: 498–499.

Morton, J.F. 1972. Cocoyams (*Xanthosoma caracu*, *X. atrovirens* and *X. nigrum*), ancient root and leaf vegetables, gaining in economic importance. Proc. Fla. State Hortic. Soc. **85**: 85–94.

Offei, S.K., Asante, I.K., and Danquah, E.Y. 2004. Genetic structure of seventy cocoyam (*Xanthosoma sagittifolium*, Linn, Schott) accessions in Ghana based on RAPD. Hereditas, **140**: 123–128.

Onokpise, O.U., Boya-Meboka, M., and Wutoh, J.T. 1992. Hybridization and fruit formation in macabo cocoyam (*Xanthosoma sagittifolium* (L) Schott). Ann. Appl. Biol. **120**: 527–535.

Onokpise, O.U., Wutoh, J.G., Ndzana, X., Tambong, J.T., Meboka, M.M., Sama, A.E., Nyochembeng, L., Aguegia, A., Nzietchueng, S., Wilson, J.G., and Burns, M. 1999. Evaluation of macabo cocoyam germplasm in Cameroon. *In* Perspectives on new crops and new uses. *Edited by* J. Janick. ASHS Press, Alexandria, VA. pp. 394–396. ["Cocoyam" in this paper refers to malanga.]

Onwuka, N.D., and Eneh, C.O. 1996. The cocoyam, *Xanthosoma sagittifolium*, as a potential raw material source for beer brewing. Plant Foods Hum. Nutrit. **49**: 283–293.

Pereira-Pacheco, F.E., and Molina-Medina, M.R. 1992. Starch extraction from *Xanthosoma sagittifolium*. Trop. Sci. (Lond.), **32**: 203–206.

Plucknett, D.L. 1976. Edible aroids: *Alocasia*, *Colocasia*, *Cyrtosperma*, *Xanthosoma* (Araceae). *In* Evolution of Crop Plants. *Edited by* N.W. Simmonds. Longman, New York, NY. pp. 10–12.

Rodriguez-Sosa, E.J., and Gonzalez, M.A. 1977. Preparation of instant tanier (*Xanthosoma sagittifolium*) flakes. J. Agric. Univ. Puerto Rico, **59**: 26–31.

Sakai, W.S., and Hanson, M. 1974. Mature raphid and raphid idioblast structure in plants of the edible aroid genera *Colocasia*, *Alocasia*, and *Xanthosoma*. Ann. Bot. **38**: 739–748.

Sakai, W.S., Hanson, M., and Jones, R.C. 1972. Raphides with barbs and grooves in *Xanthosoma sagittifolium* (Araceae). Science, **178**: 314–315.

Sefa-Dedeh, S., and Sackey, E.K.A. 2002. Starch structure and some properties of cocoyam (*Xanthosoma sagittifolium* and *Colocasia esculenta*) starch and raphides. Food Chem. **79**: 435–444.

Serviss, B.E., McDaniel, S.T., and Bryson, C.T. 2000. Occurrence, distribution, and ecology of *Alocasia*, *Caladium*, *Colocasia*, and *Xanthosoma* (Araceae) in the southeastern United States. SIDA, **19**: 149–174.

Wilson, J.E 1984. Cocoyam. *In* The physiology of tropical field crops. *Edited by* P.R. Goldsworthy and N.M. Fisher. John Wiley & Sons, Chichester, UK. pp. 589–605. ["Cocoyam" in this paper refers to both malanga and taro.]

MANGO

Family: Anacardiaceae (cashew family)

Names

Scientific name: *Mangifera indica* L.

- The word "mango" may have originated through the Portuguese *manga*, from the Tamil (a South Indian language) word *man cay* (also rendered *mangkay* and *man-gay*), formed from *mān*, mango tree, and *kāy*, fruit.
- The plural of mango is either mangos or mangoes.
- In many languages, the mango is called the "mother of all tropical fruits."
- In the Midwest of the United States, sweet bell peppers are sometimes called mangos. Some produce wholesalers and retailers label boxes "mango fruit" or "mango peppers" to avoid confusion between the two.
- The interestingly named "Pango Mango" is a relatively new, very sweet variety grown in Puerto Rico. It has green skin with a reddish blush, and unlike many other varieties, the skin remains green when the fruit is ripe.
- The genus name *Mangifera* is based on the Latin "carrier of mango."
- *Indica* in the species name means "India," reflective of the home of the mango.

Plant Portrait

The mango is an evergreen (or almost evergreen) tropical tree, 10–40 m (33–128 feet) in height. The tree is long-lived, sometimes surviving to 300 years of age while still fruiting, and typically producing about 100 fruits annually. Mangos are native to southern Asia, especially eastern India and Burma, where they have been cultivated for perhaps the last 4000 years. Leading producers today are India, Thailand, Pakistan, Bangladesh, Brazil, Mexico, the Philippines, Tanzania, the Dominican Republic, and Colombia. Mangos require a frost-free environment, and are grown commercially on a small scale in Florida and California. They are as important in tropical regions as the apple is in temperate regions, and indeed, mangos are sometimes referred to as "the apple of the tropics." In India, which produces about two-thirds of the world's supply, the mango is often called the "king of fruits." The claims are made that on a world-wide basis more mangos are eaten fresh than any other fruit, and that the mango is the most important fruit produced in the Tropics. There are hundreds of varieties. The fruits vary greatly in form, size, color, and quality. Mangos may be round, oval, or somewhat kidney-shaped, and are usually more or less lopsided. They range from 6 to 25 cm (2½–10 inches) in length and from less than 100 g to more than 2 kg (less than 4 ounces to more than 4 pounds). The smooth, leathery skin ranges through tones and mixtures of green, yellow, orange, and red. The peach-like flesh varies from pale yellow to deep orange, and clings to the large, flat, woody stone that is present.

In Hinduism, mangos have acquired considerable symbolic and religious value. Mango leaves and flowers are made into garlands for decorating Hindu temples. Fresh mango leaves are also hung outside the front door during Ponggol (Hindu New Year) and Deepavali (the Festival of Lights, celebrated over 5 days, each day dedicated to a separate philosophy or ideal). Mango leaves are often prominently displayed at weddings in India, and marriages often occur in mango groves, reflecting the association of the tree with marriage, sex, and fertility. The mango is also a symbol of love in India, and a basket of mangos is considered to be a gesture of friendship.

Culinary Portrait

Mangos shipped to northern markets are generally red-skinned and pleasant tasting. The flesh is extremely juicy, with the flavor ranging from very sweet to subacid and tart. The flavor of mango has been described as a delicate blend of peach, pineapple, and apricot. While the best mangos are richly and pleasantly fragrant, some fruits have a turpentine odor and flavor. The most important use of mango is as a dessert fruit when ripe. Mangos are eaten fresh—out of hand, in salads, and added to yogurt, ice cream, and sorbets. They are also cooked, dried, canned, and used in jellies, jams, marmalades, preserves, and curries. The kernel of the stone is frequently cooked and consumed. Unripe mangos are commonly eaten in Asia and the West Indies. In Asia, mangos are often cooked as a vegetable, and served with poultry, pork, or fish. Green, tart mangos are popular in many dishes from India, Thailand, and Malaysia, especially in meats, as proteolytic enzymes in unripe fruit act as a meat tenderizer. Green mangos make excellent chutney and compôte.

The best mangos to purchase yield to gentle pressure, and have no dark spots or blemishes (indicative of very ripe fruit). A ripe mango is not necessarily red, and will have a very fragrant aroma. A sweet, light scent at the stem end usually indicates good flavor, while a slightly sour or alcoholic smell means the mango is overripe and fermenting. If the mangos are still hard and too green, they can be placed in a brown paper bag for a few days. Mangos that are picked too soon will not ripen properly, the skin will often be shriveled, the flesh acidic and unpleasant in taste, and an aroma of turpentine may be present. Ripe mangos can be stored in a plastic bag in the refrigerator for a week or so. Eating varieties are best served chilled. The flesh from a mango may be sliced by reference to the large flat stone. The unpeeled mango can be sliced from stem end to bottom end parallel to the flat stone on either side. This will produce two large slices with most of the flesh. Mangos should be eaten carefully, as the juice leaves indelible stains on clothing.

Poison ivy and poison oak are in the same family as the mango, and the latter also has highly irritant chemicals. A milky sap exudes from the stalk at the base of the mango fruit and becomes pale yellow and translucent when dried. Chemicals in the sap can blister the skin of the average person, so in commercial operations, the sap is washed away before the fruits are marketed. Although not poisonous, the skin is generally not eaten, as it can be irritating to the mouth. As with poison ivy, there is typically a delayed reaction. Hypersensitive persons may react with considerable swelling of the eyelids, the face, and other parts of the body, and may not be able to handle or eat mangos or any food containing mango flesh or juice.

Culinary Vocabulary

- Amchur (amchoor, pronounced AHM-choor) refers to dried, unripe mango, typically in powdered form but also in slices. This is usually available in Indian markets, and is used as a spice to give a tangy, acidic flavor to dishes, mostly Indian cuisines.
- "Mango leather" is a product made in India when there is a glut of mangoes. Juice is squeezed onto trays or plates and dried in the sun. As the juice dries, more is added until a leathery material about 1 cm (1/2 inch) thick has been produced. This is rolled and cut into strips.

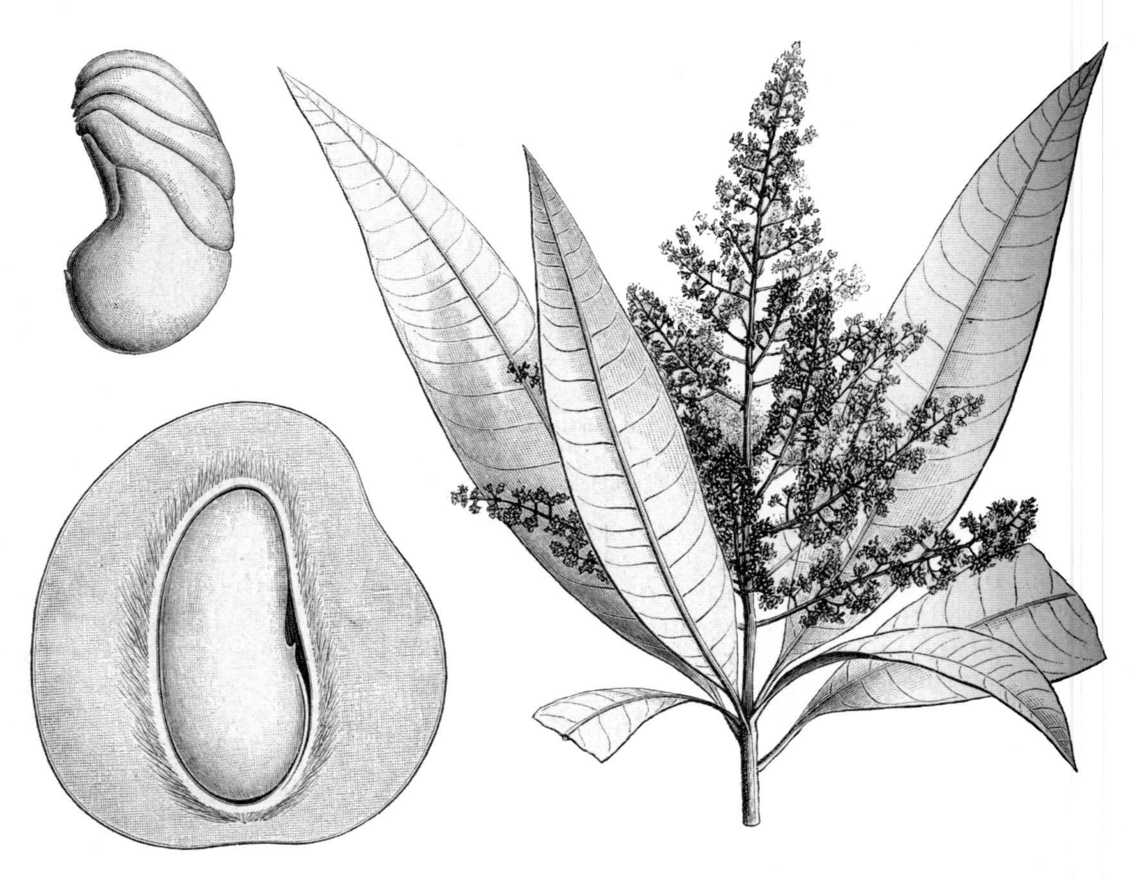

Mango (*Mangifera indica*), from Engler and Prantl (1889–1915). *Lower left*, long section of fruit; *upper left*, seed; *right*, flowering branch.

Mango (*Mangifera indica*), branch with fruit, from Bailey (1900–1902).

Curiosities of Science and Technology

- Many Southeast Asian kings and nobles had their own mango groves. Akbar (1542–1605), the great Mogul emperor of India, ordered the planting of a vast orchard of 100 000 trees. Potentates judged their social standing by the desirability of their mangos, which were sent as gifts. The custom of sending gifts of the choicest mangos remains to this day.
- As noted above, the mango has great symbolic value in Hinduism. In India, Hindus sometimes brush their teeth with mango twigs on holy days (although these are toxic).
- Cows in India were formerly fed mango leaves to obtain a rich yellow dye that is consequently produced in the urine. Since long-term consumption of the leaves may be fatal to the cattle, the practice has been outlawed.
- Mango trees in their prime bear about 500 fruits in good years (typically, the trees bear well every second year). One tree established a record of 29 000 fruits in one season.
- Mango trees with a trunk circumference of 7.6 m (25 feet) have been reported in Brazil.
- Fresh mangos from Hawaii can not be imported into the US mainland, Japan, and various other countries owing to quarantine restrictions related to fruit flies and mango seed weevils.
- Mango wood should never be used in fireplaces or for cooking fuel, or disposed of by burning, as its smoke is highly irritating. Also, poison ivy should never be burned for the same reason.

Key Information Sources

Brekke, J.E. 1975. Mango: processed products. Agricultural Research Service, U.S. Dept. of Agriculture, Western Region, Berkeley, CA. 26 pp.

Caygill, J.C. 1976. The mango (*Mangifera indica* L.): harvesting and subsequent handling and processing: an annotated bibliography. Tropical Products Institute, London, UK. 123 pp.

Cull, B.W. 1991. Mango crop management. Acta Hortic. **291**: 154–173.

De Villiers, E.A. 1998. The cultivation of mangoes. Institute for Tropical and Subtropical Crops, ARC-LNR, Nelspruit, South Africa. 216 pp.

Gangolly, S.R., and Nath, P. 1957. The mango. Indian Council of Agricultural Research, New Delhi, India. 530 pp.

Gruèzo, W.S. 1991. *Mangifera* L. *In* Plant resources of South-East Asia, 2, edible fruits and nuts. *Edited by* E.W.M. Verheij and R.E. Coronel. Pudoc, Leiden, Netherlands. pp. 203–206.

Indian Council of Agricultural Research [C.G. Karup Raghava et al.]. 1967. The mango, a handbook. Indian Council of Agricultural Research, New Delhi, India. 210 pp.

Kostermans, A.J.G.H., and Bompard, J.-M. 1993. The mangoes: their botany, nomenclature, horticulture and utilization. Academic Press, London, UK. 233 pp.

Lakshminarayana, S. 1980. Mango. *In* Tropical and subtropical fruits. *Edited by* S. Nagy and P.E. Shaw. AVI Publishing, Westport, CT. pp. 184–257.

Litz, R.E. (*Editor*). 1997. The mango: botany, production and uses. CAB International, Wallingford, UK. 592 pp.

Majumder, P.K., and Sharma, D.K. 1985. Mango. *In* Fruits of India, tropical and subtropical. *Edited by* T.K. Bose. Naya Prokash, Calcutta, India. pp. 69–123.

Mathews, H., and Litz, R.E. 1992. Mango. *In* Biotechnology of perennial fruit crops. *Edited by* F.A. Hammerschlag and R.E. Litz. CAB International, Wallingford, UK. pp. 433–448.

Mendoza, D.B., and Wills, R.B.H. (*Editors*). 1984. Mango: fruit development, postharvest physiology and marketing in ASEAN. ASEAN Food Handling Bureau, Kuala Lumpur, Malaysia. 111 pp.

Mukerhee, S.K. 1972. Origin of mango (*Mangifera indica*). Econ. Bot. **26**: 260–264.

Mukerhee, S.K. 1985. Systematic and ecogeographic studies on crop genepools 1. *Mangifera* L. IBPGR, Rome, Italy. 86 pp.

Mukerji, S.K. 1949. A monograph on the genus *Mangifera*. Lloyia, **12**: 73–136.

Narain, N., Bora, P.S., Narain, R., and Shaw, P.E. 1998. Mango. *In* Tropical and subtropical fruits. *Edited by* P.E. Shaw, H.T. Chan, Jr., and S. Nagy. Agscience, Auburndale, FL. pp. 1–77.

Pinto, A.C. de Q., Canto Pereira, M.E., and Alves, R.E. (*Editors*). 2004. Proceedings of the seventh international mango symposium, Recife City, Brazil, 22–27 September 2002. International Society for Horticultural Science, Leuven, Belgium. 695 pp.

Sadhu, M.K. 1999. Mango. *In* Tropical horticulture, vol. 1. *Edited by* T.K. Bose, S.K. Mitra, A.A. Farooqui, and M.K. Sadhu. Naya Prokash, Calcutta, India. pp. 170–198.

Schaffer, B., Whiley, A.W., and Crane, J.H. 1994. Mango. *In* Handbook of environmental physiology of fruit crops, Vol. 2. Subtropical and tropical crops. *Edited by* B. Schaffer and P.C. Anderen. CRC Press, Boca Raton, FL. pp. 165–197.

Singh, L.B. 1960. The mango—botany, cultivation and utilization. Interscience Publishers, New York, NY. 438 pp. [Reprinted 1968, Leonard Hill, London, UK.]

Singh, L.B. 1969. Mango (*Mangifera indica*). *In* Outlines of perennial crop breeding in the tropics. *Edited by* F.P. Ferwerda and F. Wit. Misc. Paper 4. Landbouwhogeschool, Wageningen, Netherlands. pp. 309–327.

Singh, L.B. 1976. Mango. *In* Evolution of crop plants. *Edited by* N.W. Simmonds. Longman, London, UK. pp. 3–9.

Singh, R.N. 1978. Mango. Indian Council of Agricultural Research, New Delhi, India. 99 pp.

Soule, M.J. 1950. A bibliography of the mango. Florida Mango Forum, University of Miami, Coral Gables, FL. 89 pp.

Srivastava, R.P. 1998. Mango cultivation. International Book Distributing Co., Lucknow, UP, India. 633 pp.

Stafford, A.E. 1983. Mango. *In* Handbook of tropical foods. *Edited by* H.T. Chan, Jr. Marcel Dekker Inc., New York, NY. pp. 399–432.

Sukonthasing, S., Wongrakpanich, M., and Verheij, E.W.M. 1991. *Mangifera indica* L. *In* Plant resources of South-East Asia, 2, edible fruits and nuts. *Edited by* E.W.M. Verheij and R.E. Coronel. Pudoc, Leiden, Netherlands. pp. 211–216.

Thomson, P.H. 1969. Growing mangos in Southern California. Calif. Rare Fruit Grow. Yearb. **1**: 9–22.

Wainwright, H., and Burbage, M.B. 1989. Physiological disorders in mango (*Mangifera indica* L.) Fruit. J. Hortic. Sci. **64**: 125–135.

Specialty Cookbooks

Harris, M.R., and Smoyer, C.K. 1989. Mangos, mangos, mangos: recipes and art from Hawai'i. M.R. Harris, Honolulu, HI. 231 pp.

Susser, A. 2001. The great mango book. Ten Speed Press, Berkeley, CA. 141 pp.

Thuma, C. 2001. The Mongo mango cookbook and everything you ever wanted to know about mangoes. Pineapple Press, Sarasota, FL. 141 pp.

Maté

Family: Aquifoliaceae (holly family)

Names

Scientific name: *Ilex paraguariensis* A. St-Hil.

- The word maté derives from the Quechua dialect word *matí* (Quechua language is widespread among Indian peoples of Peru, Bolivia, Ecuador, Chile, and Argentina), meaning vessel, widely interpreted to mean the bottle gourd (which see) that is traditionally used as a vessel to prepare and drink the tea. Today, the word maté refers to the species *I. paraguariensis*, the tea made from it, and also to the special cups in which the tea is infused. The cups, made from curiously shaped bottle gourds with small openings cut in the top, are sometimes elaborately decorated with silver.
- Maté is pronounced mah-tay.
- Maté is also known as Brazilian tea, Paraguay tea, and yerba maté.
- The first people to have cultivated this species, rather than simply collecting it from the wild, apparently were the Jesuit missionaries, hence the name "Jesuit's tea." By 1670, the Jesuits had established maté plantations. In time, the settlements of Guarani Indians (of Paraguay) converted to Christianity became economically dependent on maté production. The Jesuits were expelled from the Spanish dominions in 1767, leading to the decline and complete disappearance of the plantations in the settlements of Christianized Indians by around 1820.
- The genus name *Ilex* is based on the Latin name for the evergreen holm oak (*Quercus ilex* L.), which holly (*Ilex aquifolium* L.) was supposed to resemble.
- *Paraguariensis* in the scientific name means "of Paraguayans."

Maté (*Ilex paraguariensis*), from Köhler (1883–1914).

Plant Portrait

Maté is an evergreen tree, typically 4–8 m (13–26 feet) tall, but sometimes growing up to 18 m (59 feet) in height. It is native to Paraguay and adjacent Argentina and Brazil. Archeological evidence indicates that maté tea was widely consumed in pre-Columbian times by Andean Indians and by people in the lowlands of Paraguay. The tea is made from the young leaves and tender shoots of *I. paraguariensis*, and sometimes also closely related species. The highest-quality maté is made only of the leaves of *I. paraguariensis*. Maté is the most popular beverage in southern South America, and the tree is cultivated extensively in Brazil and Paraguay. Over 300 000 tons are produced annually for consumption in Brazil, Argentina, Uruguay, and Paraguay. Paraguayans consider maté to be their "national drink." Exactly the same claim is made by Argentinians and inhabitants of other South American countries. Twenty million people are believed to be regular drinkers of the beverage. The tea is a stimulant, containing considerable caffeine (1–2%), as well as a similar chemical called mateine, claimed to have therapeutic effects.

In addition to the beverage use of maté, it is employed in folk medicines, and marketed in pills and potions in Western countries as a supposedly remarkable tonic useful for alleviating or preventing a wide range of maladies, including stress, decreased libido, sleep difficulties, constipation, high blood pressure, and susceptibility to diseases. Notwithstanding the extent to which these claims are valid, maté is probably at least as healthy a beverage as coffee and tea (which in fact do have health benefits), but because of the high caffeine content, consumption needs to be moderate or sparing by those who are sensitive to this chemical, and especially if pregnant.

Silver drinking vessel and straw with strainer at end used for drinking maté (*Ilex paraguariensis*), from *Hooker's London Journal of Botany* (1842, vol. 1).

Culinary Portrait

Maté tea is made by placing dried leaves in a container of boiled water, reminiscent of regular (Chinese) tea. In southern South America, it is common to see men and women walking around in shopping malls and on beaches carrying maté gourds and thermoses of hot water. Silver vessels in the shape of a gourd are also traditional. Although maté connoisseurs would never think of using glass, pottery mugs, or styrofoam cups, for those who want to experience maté using conventional equipment, the brew can be made using bulk supplies available at specialty stores and a drip coffee maker, or a French press or tea ball. Maté is even available from some distributors as tea bags. The flavor is like that of green tea, somewhat sweet and bitter, with a hint of nuttiness. It is delicious with honey, and can be served hot or cold, with or without lemon, milk, sugar, herbs, and fruit juices. As with coffee, the bitterness of the brew takes getting used to, but the taste grows on you. In South America, liqueurs, ice creams, desserts, and other confections are prepared with maté.

Maté has been termed the "liquid vegetable" of the South American cowboys known as gauchos, whose diet otherwise consists mainly of meat, cheese, and bread. Maté is as constant a companion to the gauchos as is their horses. Maté is generally made with milk rather than water for children. At the other, macho extreme, "real gauchos" do not add sugar under any circumstances.

In South America, preparing and consuming maté may be quite ritualized. Commonly, one person in a group takes the role of the server or cebador(a). The server pours cold water over the loose tea in the hollowed gourd, lets it brew, then sips the concoction through a bombilla—a decorative metal straw with a filter at one end, often made from metals, including gold and silver. The process is repeated, this time pouring hot or cold water over the leaves, passing the gourd to the next person on the server's left (or depending on local tradition, the right), who again finishes the whole gourdful. Making a loud gurgling noise to indicate that one has finished is considered good etiquette. It is considered rude to disinfect the bombilla between passings. A golden rule to be observed is to never stir the bombilla in the gourd, as it is only meant to be sipped from, not touched. The cebador is the only one entitled to move it in case it clogs.

Curiosities of Science and Technology

- Silver-lined maté gourds have been recovered from ancient graves in the Andes region of Argentina. These were meant to allow the dead to prepare maté beverage during their journey to the afterworld.
- Guarani shamans drink maté to help them enter a clairvoyant trance in order to communicate with the spirit world.
- The ritualized custom of the host or server drinking first from the maté gourd is thought to originate from a time when it was important to signify to the guests that the brew has not been poisoned.
- Charles Darwin (1809–1882), the foremost student of biological evolution, learned much about evolution during the voyage of the Beagle, the ship that carried him to South America. He declared maté to be the "perfect stimulant."
- An interesting relative of maté is *Ilex vomitoria* Sol. ex Aiton, variously known as Carolina tea tree, cassina, cassine, Christmas tea-bush, yaupon, and yaupon holly. This is an evergreen shrub, growing from Virginia, south to Florida and Texas. Indians of the southeastern United States used the leaves to produce a bitter stimulant tea, known as cassina, black drink, black tea, and Carolina tea to Europeans. Yaupon is the only plant species native to North America that is known to contain caffeine. While the caffeine content is low (of the order of 0.02% on a dry weight basis), the Indians boiled down preparations to produce a rather potent brew that was consumed in rituals and festivals. During the colonial period in the United States, and especially during the Civil War, cassina was sometimes almost as popular as coffee and tea. *Vomitoria* in the scientific name implies that the plant and its tea are emetic (i.e., they cause vomiting), and this conclusion arose because the Indians habitually vomited after drinking the tea that they made. However, it has been established that the plant is not emetic, and the vomiting seems to have been a cultural practice. Attempts to commercialize cassine as a tea have been unsuccessful.

Fruiting branch of yaupon (cassine, *Ilex vomitoria*), from Sargent (1890–1892, vol. 1, plate 48).

Key Information Sources

Alikaridis, F. 1987. Natural constituents of *Ilex* species. J. Ethnopharm. **20**: 121–144.

Athayde, M.L., Coelho, G.C., and Schenkel, E.P. 2000. Caffeine and theobromine in epicuticular wax of *Ilex paraguariensis* A. St.-Hil. Phytochemistry, **55**: 853–857.

Brotonegoro, S., Campo Gigena, M. del, and Giberti, G.C. 2000. *Ilex paraguariensis* A. St.-Hil. *In* Plant resources of South-East Asia, 16, stimulants. *Edited by* H.A.M. van der Vossen and M. Wessel. Backhuys, Leiden, Netherlands. pp. 83–86.

Clifford, M.N., and Ramirez-Martinez, J.R. 1990. Chlorogenic acids and purine alkaloids contents of maté (*Ilex paraguariensis*) leaf and beverage. Food Chem. **35**: 13–21.

Coelho, G.C., Araujo-Mariath, J.E. de, and Schenkel, E.P. 2002. Populational diversity on leaf morphology of maté (*Ilex paraguariensis* A. St.-Hil., Aquifoliaceae). Brazil. Arch. Biol. Technol. **45**: 47–51.

Eibl, B., Fernandez, R.A., Kozarik, J.C., Lupi, A., Montagnini, F., and Nozzi, D. 2000. Agroforestry systems with *Ilex paraguariensis* (American holly or yerba maté) and native timber trees on small farms in Misiones, Argentina. Agroforestry Syst. **48**: 1–8.

Ferreira, A.G., Kaspary, R., Ferreira, H.B., and Rosa, L.M. 1983. Proportion of sex and pollination in *Ilex paraguariensis* St. Hill. Bras. Flores, **13**(**53**): 29–33. [In Portuguese, English summary.]

Filip, R., Lotito, S.B., Ferraro, G., and Fraga, C.G. 2000. Antioxidant activity of *Ilex paraguariensis* and related species. Nutr. Res. **20**: 1437–1446.

Fiorentino, R., and Dean, G.W. 1974. Rural poverty and government intervention: the case of the Argentine yerba maté industry. Am. J. Agric. Econ. **56**: 751–763.

Fonseca, C.A.S., Otto, S.S., Paumgartten, F.J.R., and Leitao, A.C. 2000. Nontoxic, mutagenic, and clastogenic activities of mate-chimarrao (*Ilex paraguariensis*). J. Environ. Pathol. Toxicol. Oncol. **19**: 333–346.

Gauer, L., and Cavalli-Molina, S. 2000. Genetic variation in natural populations of maté (*Ilex paraguariensis* A. St.-Hil., Aquifoliaceae) using RAPD markers. Heredity, **84**: 647–656.

Giberti, G.C. 1994. Maté (*Ilex paraguariensis*). *In* Neglected crops: 1492 from a different perspective. *Edited by* J.E. Hernández-Bermejo and J. Léon. FAO Plant Prod. Protect. Ser. No. 26. Food and Agriculture Organization of the United Nations, Rome, Italy. pp. 245–252.

Giberti, G.C. 1999. Phytogentic resources relating with the cultivation and exploitation of yerba maté (*Ilex paraguariensis* St. Hil., Aquifoliaceae) in the Cone of South America. Acta Hortic. **500**: 137–144. [In Spanish, English summary.]

Gosmann, G., Guillaume, D., Taketa, A.T.C., and Schenkel, E.P. 1995. Triterpenoid saponins from *Ilex paraguariensis*. J. Nat. Prod. **58**: 438–441.

Grigioni, G., Carduza, F., Irurueta, M., and Pensel, N. 2004. Flavour characteristics of *Ilex paraguariensis* infusion, a typical Argentine product, assessed by sensory evaluation and electronic nose. J. Sci. Food Agric. **84**: 427–432.

Hudson, C.M. 1979. Black drink: a native American tea. University of Georgia Press, Athens, GA. 175 pp. [*Ilex vomitoria.*]

Joyce, T.A. 1921. Yerba maté, the tea of South America. Pan-American Magazine, vol. 33. 23 pp. [cf. Pan American Union 1946, below]

Leitao, A.C., and Braga, R.S. 1994. Mutagenic and genotoxic effects of maté (*Ilex paraguariensis*) in prokaryotic organisms. Braz. J. Med. Biol. Res. **27**: 1517–1525.

Lopez, A. 1974. The economics of yerba maté in seventeenth-century South America. Agric. Hist. **48**: 493–509.

Luxner, L. 1995. Yerba maté comes of age. Tea Coffee Trade J. **167**(**7**): 44, 46, 48–49.

Pan American Union. 1946. Yerba maté: the tea of South America. Bulletin of the Pan American Union, Washington, DC. 23 pp.

Porter, R.H. 1950. Maté, South American or Paraguay tea. Econ. Bot. **4**: 37–51.

Ramallo, L.A., and Albani, O.A. 2004. Prediction and determination of water uptake in packaged yerba maté. Food Sci. Technol. Int. **10**: 35–40.

Reissmann, C.B., Radomski, M.I., and Bianchini de Quadros, R.M. 1999. Chemical composition of *Ilex paraguariensis* St. Hil. under different management conditions in seven localities of Parana State. Brazil. Arch. Biol. Technol. **42**: 187–194.

Santa-Cruz, M.J., Garitta, L.V., and Hough, G. 2003*a*. Sensory descriptive analysis of yerba maté (*Ilex paraguariensis* Saint Hilaire), a South American beverage. Food Sci. Technol. Int. **8**: 25–31.

Santa-Cruz, M.J., Garitta, L.V., and Hough, G. 2003*b*. Relationships of consumer acceptability and sensory attributes of yerba maté (*Ilex paraguariensis* St. Hilaire) using preference mapping. Food Sci. Technol. Int. **9**: 347–352.

Santos, R.P., Mariath, J.E.A., and Hesse, M. 2003. Pollenkitt formation in *Ilex paraguariensis* A. St.Hil. (Aquifoliaceae). Plant Syst. Evol. **237**: 185–198.

Schenkel, E.P., Montanha, J.A., and Gosmann, G. 1996. Triterpene saponins from maté, *Ilex paraguariensis*. *In* Saponins used in food and agriculture. *Edited by* G.R. Waller and K. Yamasaki. Plenum Press, New York, NY. pp. 47–56.

Scherer, R., Urfer, P., Mayol, M.R., Belingheri, L.D., Marx, F., and Janssens, M.J.J. 2002. Inheritance studies of caffeine and theobromine content of maté (*Ilex paraguariensis*) in Misiones, Argentina. Euphytica, **126**: 203–210.

Schmalko, M.E., and Alzamora, S.M. 2001. Color, chlorophyll, caffeine, and water content variation during yerba maté processing. Drying Technol. **19**: 599–610.

Small, E., and Catling, P.M. 2001. Blossoming treasures of biodiversity: 3. Maté (*Ilex paraguariensis*)—better than Viagra, marijuana, and coffee? Biodiversity, **2**(**4**): 26–27.

Tenorio Sanz, M.D. 1991. Mineral elements in maté herb (*Ilex paraguariensis* St. H.) Arch. Latinoam. Nutr. **41**(**3**): 441–454.

Winge, H., Ferreira, A.G., Mariath, J.E. de A., and Tarasconi, L.C. 1995. Erva-mate: biologia e cultura no Cone Sul [Maté: biology and culture in the southern Cone]. Editora da Universidade Federal de Rio Grande do Sul, Porto Alegre, Brazil. 356 pp. [In Portuguese.]

Melon

Family: Cucurbitaceae (gourd family)

Names

Scientific name: *Cucumis melo* L.

- "Melon" is the Late Latin word (shortened from *melopepo*) for this fruit. It is derived from the Greek *melopepon*, from *melon*, apple, + *pepon*, gourd or melon. *Melo* in the scientific name has the same derivation.
- "Cantaloupe" is from the former papal state of Cantalupo or the town of Cantaluppi near Rome, where this variety was cultivated at an early date, apparently from seed received from Asia Minor. Cantaluppi was the summer estate of the popes. The place name "Cantaluppi" has been interpreted as Italian for "singing wolf."
- "Casaba" in the "Casaba melon" is based on the city of Kasaba (now Turgutlu), Turkey, from which the melon was introduced to the United States in the late 19th century.
- The "musk" in "muskmelon" is based on the Persian word for a kind of perfume.
- "Chito" in "Chito melon" is a word of long but obscure standing, probably of geographical significance.
- For information on the Oriental pickling melon (also known as pickling melon and sweet melon), see Cucumber.
- The "Santa Claus melon" or "Christmas melon" looks like a small watermelon, and is often marketed during the Christmas season, hence its names. Its yellowish green flesh is reminiscent of the taste and appearance of honeydew melon.
- "Elijah's melons" refers to certain stones on Mount Carmel, on the Mediterranean coast of Israel. According to legend, the owner of the land refused to supply food for the prophet Elijah, and was punished by having his melons turned into stones.
- The genus name *Cucumis* is based on the ancient Roman name for the cucumber, one of the species in the genus.

Plant Portrait

Cucumis melo is a native of the Old World tropics. Modern melons seem to have been derived from African populations. The melon was apparently unknown to the ancient Egyptians or Greeks, and reached Europe towards the decline of the Roman Empire. Most melon plants are tendril-bearing vines, although some bush-like forms have been bred. The yellow flowers are attractive to insect pollinators, especially honey bees. The fruits are extremely variable in size, shape, color of skin and flesh, texture of surface, and thickness of rind. Melons normally have 10 large longitudinal veins over which there are grooves in the fruit surface of some varieties. The plants are tender, susceptible to frost damage, and so are not adapted to cold climates. California is the largest melon-growing US state. Poor flavor is rather common, reflecting the requirement for careful control of growth conditions. The following groups of melon varieties are found in North America:

- *Muskmelons* (also known as netted and nutmeg muskmelons)—fruits sometimes somewhat smaller than cantaloupes; with a raised network usually lighter than the remainder of the fruit; flesh usually salmon in color, sometimes green; surface netted, with or without grooves or sutures; commonly cultivated in North America, and probably the most important commercial class of melons here. The term "cantaloupe," although a misnomer, has become firmly embedded in American culture to indicate this class of melon.
- *True cantaloupes*—fruit of medium size; surface rough, warty or scaly, not netted; often deeply grooved. Flesh aromatic, usually orange, rarely green; commonly cultivated in Europe, not grown commercially in North America.
- *Winter melons* (also known as Casaba melons and honeydew melons) —surface smooth (honeydew melons) or shallowly corrugated (Casaba melons), but not netted; flesh not strongly scented, white or green; late-ripening; hard-skinned, capable of storage for more than a month; includes the popular Honeydew.
- *Snake melons* (also found under the names Armenian cucumber, banana melon, cucumber melon, Japanese cucumber, and serpent melon) —very elongated melons, 2.5–7.5 cm (1–3 inches) in diameter, 30 cm (1 foot) to more than a meter (more than a yard) in length; smooth-skinned; sometimes coiled; grown as a novelty and sometimes for preserves, and very popular in India, China, and Japan.

- *Chito melon*—This close relative of the muskmelon has fruit 5–8 cm (2–3 inches) long, which looks like an orange or lemon, with a smooth surface, and white or yellow cucumber-like flesh. It is not edible in its natural state, but has been used as an ornament and for making "mango pickles" and chutneys. In North America, it has been marketed under the names mango melon, melon apple, naudin, orange melon, vegetable orange, and vine peach.
- *Melofon*—This recently developed "pickling melon" is similar to pickling cucumbers. It has been bred for its high yield and once-over mechanical harvesting.

Culinary Portrait

Melons are popular appetizers and desserts, the fresh flesh eaten out of the rind after removal of the seeds. The fruit is particularly tasty chilled, and may be served with sifted sugar, grated nutmeg, or ginger. Small cubes are frequently placed in fruit cocktails, and salads. Puréed melon is used to flavor ice cream and sorbets. Melon also blends well with a wide variety of meats, poultry, fish and seafoods in general, and cheeses. An ancient custom practiced by popes was to fill the cavity of a sectioned melon with port wine or some sweet wine. The fruit is also cooked into jam, chutney, and pickles. Melons are not particularly nutritious, containing of the order of 90% water, but

Varieties of melon (*Cucumis melo*), from *Gartenflora* (1898, vol. 47).

white–green-fleshed forms do have good vitamin C content. Deep-yellow varieties have been found to contain more vitamin A than green–white-fleshed varieties. The seeds are sometimes chewed or used as a source of an edible oil (the seeds may contain 45% oil).

Melons are usually harvested before fully ripe, as firmer fruit better withstand the rigors of shipping. However, this makes it difficult to identify a truly tasty melon. Ripeness of melons is very difficult to judge for most consumers, and many have concluded that, of all fruits, judging ripeness is hardest for melons. It has been claimed that ripeness can be detected in muskmelons, if part of the stem is present, by the presence of a thin crack where the stem meets the crown of the melon. If the crack extends all the way around the stem, the melon has been fully ripened on the vine. Indeed, as muskmelon fruits approach sufficient maturity that they should be harvested, a crack develops at the joint attaching the fruit to the stem, and finally, the melon "slips" away from the stem. However, a "slip stage" does not occur in many melons, including Honeydew. Another recommended way of judging whether a melon is ripe is to check if it yields under slight pressure around the stem end. Still another way is to slap a melon in the expectation that when tapped lightly with the palm of the hand a ripe melon will give off a hollow sound. Overly soft melons and those with strong odor or unusual color may be overripe.

Chito melon (*Cucumis melo*), from Morren (1851–1885).

The flesh of a melon is generally obtained after slicing the fruit open. A melon baller can be employed to scoop the flesh out. The seeds should be left in unused portions to keep the pulp from drying out, and the sectioned melon should be wrapped in plastic and stored in a refrigerator.

Some people have difficulty digesting melons. Another problem with melons, especially those with rough and netted skins (so-called cantaloupes), has to do with microbial contamination. Melons are usually grown at ground level, where bacteria are plentiful, and fruit with rough skin is prone to contamination. Moreover, cutting unwashed melons with a knife is likely to spread the bacteria from the rind to the flesh that is consumed. Salmonella has been reported on melon rinds in both the United States and Canada. Cantaloupes should be thoroughly scrubbed before being sliced open.

Culinary Vocabulary

- "Tea melon," also called Chinese pickle, cucumber pickle, pickled cucumber, preserved cucumber, preserved sweet melon, and sweet tea pickle, is a small (5 cm or 2 inches long) cucumber-shaped fruit that is commonly preserved in China for use as a flavoring. It has a sweet, mild flavor and a crisp, crunchy texture. In Asian markets in Western countries, it is sometimes available in cans or jars, preserved in honey and spices, or sometimes in soy sauce. Information does not seem to be available on what kind of melons are used to make tea melons, but probably the snake melon is used.
- "Crystallized winter melon" is winter melon cooked in a sugar syrup, and available as transparent white lumps or slices, covered in sugar. It is consumed as a snack and used in Chinese cuisines to sweeten meat, cakes, and pudding.

Curiosities of Science and Technology

- The names of many popular foods have acquired other meanings in English, especially in vernacular expressions. "Melon" has acquired several slang human anatomical meanings: breast; head or brain; a protruding abdomen. "Melon" also has the slang meaning (especially in the phrase "cut the melon") "large profits to be shared among several people."

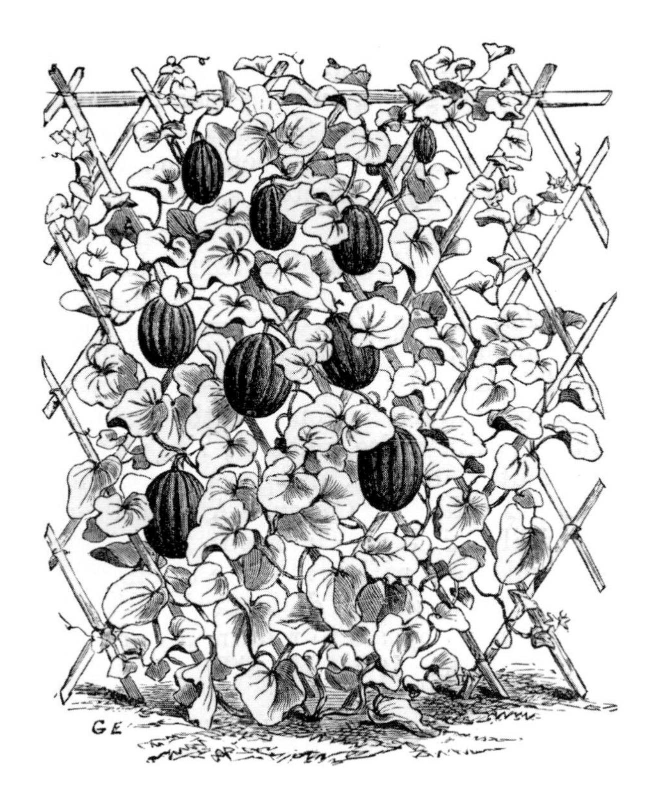

Melon (*Cucumis melo*) growing on a trellis, from Vilmorin-Andrieux (1885).

- Solemn and lemons are anagrams of melons. So is Lemnos, an island of northeastern Greece.
- Several famous people are alleged to have died because of eating melons to excess; these include King Albert II of Germany (about 1298–1358), Pope Paul II (1417–1471), and King Frederick III of Germany (1415–1493) and his son King Maximilian I (1459–1519).
- A 32.8-kg (72.3-pound) cantaloupe was produced by Bill Rogerson of Greenville, North Carolina.
- In the famous shower stabbing scene in Alfred Hitchcock's movie "Psycho," the sound of the knife penetrating flesh was made by stabbing a Casaba melon.

Key Information Sources

(For additional references, see CUCUMBER)

Aguayo, E., Allende, A., and Artes, F. 2003. Keeping quality and safety of minimally fresh processed melon. Eur. Food Res. Technol. **216**: 494–499.

Akashi, Y., Fukuda, N., Wako, T., Masuda, M., and Kato, K. 2002. Genetic variation and phylogenetic relationships in East and South Asian melons, *Cucumis melo* L., based on the analysis of five isozymes. Euphytica, **125**: 385–396.

Davis, G.N., and Whitaker, T.W. 1942. Growing and handling cantaloupes and other melons. University of California, College of Agriculture, Agricultural Experiment Station, Berkeley, CA. 40 pp.

Decker-Walters, D.S., Chung, S.M., Staub, J.E., Quemada, H.D., and Lopez-Sese, A.I. 2002. The origin and genetic affinities of wild populations of melon (*Cucumis melo*, Cucurbitaceae) in North America. Plant Syst. Evol. **233**: 183–197.

Fuller, S., and Hall, C. 1990. Economic trends of the melon industry in Texas and the United States: cantaloupe, honeydew, and watermelon. Texas Agricultural Experiment Station, Texas A&M University System, College Station, TX. 14 pp.

Garcia, S., Lombardero, M., Serra-Baldrich, E., Amat, P., Lluch-Perez, M., and Malet, A. 2004. Occupational protein contact dermatitis due to melon. Allergy, **59**: 558–559.

Goldman, A., and Schrager, V. 2002. Melons: for the passionate grower. Artisan, New York, NY. 176 pp.

Hoberg, E., Ulrich, D., Schulz, H., Tuvia-Alkali, S., and Fallik, E. 2003. Sensory and quality analysis of different melon cultivars after prolonged storage. Nahrung, **47**: 320–324.

Hosoki, T., Ishibashi, A., Kitamura, H., Kai, N., Hamada, M., and Ohta, T. 1990. Classification of Oriental melons based on morphological ecological and physiological differences. J. Jpn. Soc. Hortic. Sci. **58**: 959–970.

Hyams, E. 1952. Melons under cloche. Faber & Faber, London, UK. 80 pp.

Lamikanra, O., Juaraez, B., Watson, M.A., and Richard, O.A. 2003. Effect of cutting and storage on sensory traits of cantaloupe melon cultivars with extended postharvest shelf life. J. Sci. Food Agric. **83**: 702–708.

Lopez-Sese, A.I., Staub, J.E., and Gomez-Guillamon, M.L. 2003. Genetic analysis of Spanish melon (*Cucumis melo* L.) germplasm using a standardized molecular-marker array and geographically diverse reference accessions. Theor. Appl. Genet. **108**: 41–52.

Mallick, M.F., and Masui, M. 1986. Origin, distribution and taxonomy of melons. Sci. Hortic. (Amst.), **28**: 251–261.

McCreight, J.D., Nerson, H., and Grumet, R. 1993. Melon, *Cucumis melo* L. *In* Genetic improvement of vegetable crops. *Edited by* G. Kalloo and B.O. Bergh. Pergamon Press, New York, NY. pp. 267–294.

McGraw, B.D. 1981. Easy gardening—melons. Texas Agricultural Extension Service, Texas A&M University System, College Station, TX. 4 pp.

Mitchell, C.E., and Huelsen, W.A. 1950. Growing melons in Illinois. University of Illinois, College of Agriculture, Extension Service in Agriculture and Home Economics, Urbana, IL. 27 pp.

Mliki, A., Staub, J.E., Zhangyong, S., and Ghorbel, A. 2001. Genetic diversity in melon (*Cucumis melo* L.): an evaluation of African germplasm. Genet. Resour. Crop Evol. **48**: 587–597.

Nerson, H. 2002. Do muskmelons (*Cucumis melo* L.) set parthenocarpic fruits spontaneously? J. Hortic. Sci. Biotech. **77**: 622–628.

Odet, J., and Dumoulin, J. (*Editors*). 1991. Le Melon. Revised edition. Centre technique interprofessionnel des fruits et légumes, Paris, France. 295 pp. [In French.]

Paigalo, K.I. 1930. Critical review of basic literature on the systematics, geography and origin of cultivated and, partially wildly-growing melon. Bull. Appl. Bot. Plant Breed. **23(3)**: 397–442. [Translated from Russian, Al-Ahram Center for Scientific Translations, 1980, 55 pp.]

Paje, M.M., and Vossen, H.A.M. van der. 1993. *Cucumis melo* L. *In* Plant resources of South-East Asia, 8, vegetables. *Edited by* J.E. Siemonsma and K. Piluek. Pudoc Scientific Publishers, Wageningen, Netherlands. pp. 153–157.

Silberstein, L., Kovalski, I., Brotman, Y., Perin, C., Dogimont, C., Pitrat, M., Klingler, J., Thompson, G., Portnoy, V., Katzir, N., and Perl-Treves, R. 2003. Linkage map of *Cucumis melo* including phenotypic traits and sequence-characterized genes. Genome, **46**: 761–773.

Stepansky, A., Kovalski, I., and Perl-Treves, R. 1999. Intraspecific classification of melons (*Cucumis melo* L.) in view of their phenotypic and molecular variation. Plant Syst. Evol. **217**: 313–332.

Specialty Cookbooks

Hazelton, J.W. 1999. Cucumbers, melons & gourds from seed to supper: for good growing, good eating, and good health. Seed to Supper, St. Petersburg, FL. 58 pp.

Raymond, D., and Raymond, J. 1984. The gardens for all book of cucumbers, melons, squash. Gardens for All, Burlington, VT, 36 pp.

Rosen, H., Rosen, R.J., and O'Malley, K.M. 1985. Melon garnishing: with full-color photos and step-by-step instructions. International Culinary Consultants, Elberon, NJ. 94 pp. [cf. Rosen, H., Eidelberg, A., and Rosen, V. 2000. Harvey Rosen presents melon garnishing. International Culinary Consultants, Elberon, NJ. 95 pp.]

Western Growers Association. 1900–1988. Let's cook for a crowd with fresh vegetables and melons. Western Growers Association, Los Angeles, CA. 24 pp.

Western Growers Association. 1968. Western ways with fresh vegetables & melons. Western Growers Association, Los Angeles, CA. 66 pp.

MILLET

Family: Poaceae (Gramineae; grass family)

Names

Scientific names: About 20 species of 10 genera of grasses are called millet. In the widest use of the word "millet," the term includes all true cereals (grasses that produce edible seeds) except those that belong to wheat (species of *Triticum*), barley (*Hordeum*), oat (*Avena*), rye (*Secale cereale*), Triticale (×*Triticosecale*), corn or maize (*Zea*), and rice (*Oryza*). However, some other grass cereals are often not regarded as millets, and the word "millet" is usually not part of their common names. Because of their unique nature, several significant crops often considered to be millets are considered separately (see FONIO, SORGHUM, and TEFF). Other than these species, the following are the most significant millets for human food.

- Pearl millet (African millet, bullrush millet, candle millet, cattail millet, Indian millet, penicillaria)—*Pennisetum glaucum* (L.) R. Br. [*P. americanum* (L.) Leeke, *P. spicatum* (L.) Körn., *P. typhoides* (Burm. f.) Stapf & C.E. Hubb., *P. typhoideum* Rich., *Setaria glauca* (L.) P. Beauv.]
- Proso millet (broomcorn, broomcorn millet, broom millet, common millet, Hershey millet, hog millet, Indian millet, Russian millet)—*Panicum miliaceum* L.
- Foxtail millet (German millet, Hungarian millet, Italian millet, Siberian millet)—*Setaria italica* (L.) Beauv. (*Panicum italicum* L.)
- Finger millet (African millet, birdsfoot millet, coracan, coracana millet, ragi, nagli)—*Eleusine coracana* (L.) Gaertn.
- Barnyard millet (billion-dollar-grass, Japanese millet, Indian barnyard millet, sawa)—*Echinochloa frumentacea* Link [*E. crus-galli* (L.) P. Beauv. var. *frumentacea* (Link) W. Wight, *Panicum frumentaceum* Roxb.]

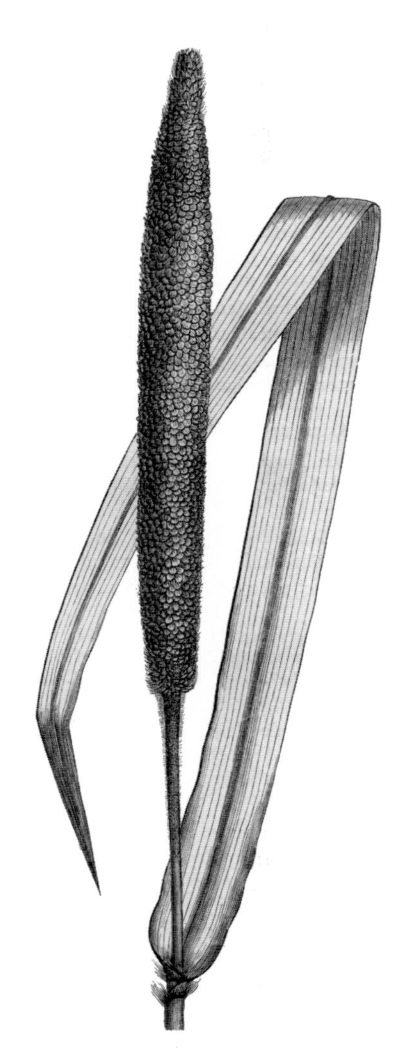

Pearl millet (*Pennisetum glaucum*), from Engler and Prantl (1889–1915).

- The name "millet" is derived from the Latin *milium*, which is similar to the Greek *melinē*, the classical names for millet, based on the word for thousand, referring to the large number of grains of the plant.
- The pearly-white grains of many varieties of pearl millet are responsible for the common name of this species.
- "Proso" in "proso millet" is based on the Russian word *proso*, dark colored, for the seeds of some varieties of the species.
- The seed head of proso millet resembles an old-fashioned broom, hence the alternative names "broomcorn" and "broomcorn millet."
- The seed head of foxtail millet is elongated, rather like the tail of a fox, hence the name.
- The seed head of pearl millet resembles a cattail (genus *Typha*), hence the name "cattail millet." The cattail is known as "bulrush" in Europe, hence the name "bulrush millet." (In North America, the name "bulrush" is also applied to other marsh-dwelling plants, of the genera *Carex* and *Scirpus*.)

- The seed head of finger millet is divided into several segments, reminiscent of fingers.
- "Barnyard millet" is the name of the weedy ancestor [*Echinochola crusgalli* (L.) P. Beauv] of the cultivated form of this species; the wild plant does occupy barnyards, among other habitats.

Plant Portrait

"Millets" are various, usually annual, grass crops whose seeds are harvested as cereal grain (for human consumption), or either the seeds or the entire plants are used as animal feed. Sorghum (which see) is called millet in many parts of Asia and Africa. Millets are some of the oldest of cultivated crops, and have been used since ancient times. They were grown by the lake dwellers of Switzerland in the Stone Age, and were sown by the Chinese in religious ceremonies as early as 2700 BC. Millets have been consumed as cereals and brewed for intoxicating beverages since prehistoric times in Asia, Africa, and Europe. They were widely grown in Europe during the Middle Ages, and were one of the principal foods of the poorer people of Rome and of Europe generally. During the 19th century in Europe, the millets were gradually superseded by wheat, rye, rice, corn, and potatoes. However, the production and consumption of millets have persisted to a greater extent in Eastern Europe and Eurasia. Most millets grown in North America are used to feed livestock and for birdseed, but millets are a staple food for a third of the world's human population. Millets are minor crops except in parts of Asia,

Proso millet (*Panicum miliaceum*). *Left*, from Hallier (1880–1888, vol. 7, plate 568); *right*, from Jumelle (1901).

Africa, China, and Russia. They are generally suited to less fertile soils and poorer growing conditions than most other cereal grains, and indeed, crops such as corn (maize), wheat, and rice do not thrive where millets are grown. Millets are important subsistence food crops in many developing countries.

The well-known Western cereals wheat, barley, and rye, as well as corn, have their grain in "ears," i.e., rather compact fruiting clusters. By comparison, millets often develop their grain in loose branching structures. The grains of millets are also typically smaller than the familiar Western grains, and much rounder.

Pearl millet is the most widely grown of the millets (if one excludes sorghum from the group). This is a tall upright grass, 2–4.6 m (6–15 feet) in height, with coarse stems that grow in dense clumps. The seeds are usually brown, sometimes varying between nearly white and black. The plant looks like corn, although it does not produce cobs. This ancient crop was domesticated sometime before 1000 BC, possibly 5000 years ago, likely in western Africa. It is planted on at least 28 million ha (69 million acres), an area equivalent to the total US corn crop. Pearl millet is grown mainly in Africa and India to produce 10 million tons of grain per year for food to feed at least 500 million people. It is a major food staple in semiarid sub-Saharan Africa, India, and Pakistan. It is also an important summer annual forage crop in the United States, Australia, and South America. In the United States, it is used primarily for birdseed. A high incidence of goiter (enlarged thyroid, visible in the front of the neck) has been reported among people eating very large amounts of pearl millet, and a chemical responsible for this has been identified in the plant.

Proso millet grows up to 1.2 m (4 feet) with stout, erect stems, which may spread at the base. The grains are whitish, straw colored, or reddish brown. In China, records of culture of proso millet extend back to between 2000 BC and 1000 BC. It is the chief cereal in parts of India, Africa, and Russia. Proso millet is also grown in China and western Europe. In the United States, it is cultivated principally in the Dakotas, Colorado, and Nebraska. In the United States and Canada, it is used for feeding poultry and cage birds.

Foxtail millet forms slender, erect, leafy stems varying in height from 30 to 150 cm (1–5 feet). The grains may be white, yellow, brown, red, or black. Foxtail millet probably originated in southern Asia and is the oldest of the cultivated millets. In China, records of culture for foxtail millet extend back more than 5000 years. It is the most important millet in China and Japan, and is also grown in India, Eastern Europe, the United States, and other countries. In North America, it is grown for animal feed. This species comprises 90% of the millets grown in the United States.

Finger millet (*Eleusine coracana*), from Lamarck and Poiret (1744–1829, plate 48).

Finger millet grows 90–120 cm (3–4 feet) tall. It is generally cultivated as a food crop in areas where rice is a crop, in contrast to other millets grown in arid or semiarid conditions. It is raised mainly in India and Africa, where it is an important staple. Like pearl millet, it is nutritious and is used mainly for human food. The seeds can be stored for long periods—10 years or more—so that it is invaluable where drought makes it probable that grain stores will often have to last for more than 1 year. Finger millet is the preferred cereal for brewing beer in Africa.

Barnyard or Japanese millet is an erect grass 30–120 cm (2–4 feet) tall. It resembles the weedy barnyard grass,

Foxtail millet (*Setaria italica*). *Left*, from Metzger (1824); *right*, from *Gartenflora* (1887, vol. 36).

Echinochloa crusgalli, and probably originated from it. It is grown for grain for human food in Asian countries. In the United States, it is raised primarily as a forage.

Culinary Portrait

As food in Old World countries, millet is used as a cereal like rice, prepared as a boiled gruel, or the flour can be incorporated into bread and other baked foods. Because the flour does not contain gluten, millets are unsuitable for preparing leavened breads, although flat breads are widely made in Asia and Africa. Ground millet or millet flour can be added to wheat flour when making breads, cakes, pies, and biscuits. Millets are often strong in flavor, and to those unaccustomed to the taste, they are not liked as a substitute for the main cereals used in Western countries. Millets can be added to egg dishes, meat pies, puddings, and breakfast cereal mixtures. Millet is also sprouted and eaten like alfalfa sprouts. Beer is commonly produced from millets.

Whole-grain millet should be purchased from a store with high turnover of the product. The grain may be stored for several months, if kept in an airtight container, maintained in a cool, dry place. To prepare millets, the grain may first be roasted to bring out the attractive nut-like flavor. The grain can be roasted in a skillet over medium–low heat, stirring constantly. When the millet is golden, 2 cups of cooking liquid may be added for each cup of grain, and the mixture simmered for 30–40 minutes.

Culinary Vocabulary

- "Dolo" is an African beer prepared from fermented millet, and flavored with bitter herbs (but not hops).

Curiosities of Science and Technology

- Foxtail millet was considered to be one of the five sacred crops by the ancient Chinese (the others were soybean, rice, wheat, and barley).

- The Chinese recognized a special constellation of eight stars, one for each of rice, barley, wheat, beans, peas, maize, hemp, and millet. As well, millet had its own constellation of four stars, called T'ien-tsi, which was believed to preside over the grain harvest. If T'ien-tsi shone clearly, it indicated an abundant harvest, but if obscured or invisible, it predicted famine.
- During the Han period in China, millet wine was a more popular beverage than tea.
- The Ainu people of Japan raised bears from the time they were cubs, and then slaughtered them like livestock for meat. To show respect for the spirit of the bear, the skull was hung on a post, and symbolic offerings of millet and other foods were left.
- A 4000-year-old bowl of noodles discovered in western China has been interpreted as the earliest evidence of pasta ever made, and has been used to support the claim that pasta originated in China, not Italy. The noodles had been prepared from a dough of broom corn and foxtail millet, unlike the more common wheat and rice noodles made today. [Reference: Lu, H., Yang, X., Ye, M., Liu, K.B., Xia, Z., Ren, X., Cai, L., Wu, N., and Tung-Sheng Liu, T.-S. 2005. Millet noodles in Late Neolithic China. A remarkable find allows the reconstruction of the earliest recorded preparation of noodles. Nature, **437**(**7061**): 967–968.]

Key Information Sources

(For additional references, see SORGHUM.)

Pearl millet (*Pennisetum glaucum*)

Andrews, D.J., Hanna, W.W., Rajewski, J.F., and Collins, V.P. 1996. Advances in grain pearl millet: utilization and production research. *In* Progress in new crops. *Edited by* J. Janick. ASHS Press, Alexandria, VA. pp. 170–177.

Andrews, D.J., Rajewski, J.F., and Kumar, K.A. 1993. Pearl millet: new feed grain crop. *In* New crops. *Edited by* J. Janick and J.E. Simon. Wiley, New York, NY. pp. 198–208.

Andrews, K.J., and Kumar, K.A. 1992. Pearl millet for food, feed and forage. Adv. Agron. **48**: 89–139.

Brunken, J., Wet, J.M.J. de, and Harlan, J.R. 1977. The morphology and domestication of pearl millet. Econ. Bot. **31**: 163–174.

Brunken, J.N. 1977. A systematic study of *Pennisetum* sect. *Pennisetum* (Gramineae). Am. J. Bot. **64**: 161–176.

Burton, G.W., Wallace, A.T., and Rachie, K.O. 1972. Chemical composition and nutritive value of pearl millet. Crop Sci. **12**: 187–188.

Kumar, K.A., and Andrews, D.J. 1993. Genetics of qualitative traits in pear millet: a review. Crop Sci. **33**: 1–20.

Lopes de Farias Neto, A.L. de. (*Editor*). 1999. Proceedings, international pearl millet workshop. Planaltina, Brazil, June 9–10, 1999. Embrapa Cerrados, Planaltina, Brazil. 218 pp.

Oyen, P.P.A., and Andrews, D.J. 1996. *Pennisetum glaucum* (L.) R.Br. *In* Plant resources of South-East Asia, 10, cereals. *Edited by* C.J.H. Grubben and S. Partohardjono. Backhuys Publishers, Leiden, Netherlands. pp. 119–123.

Pearson, C. 1984. *Pennisetum* millet. *In* The physiology of tropical field crops. *Edited by* P.R. Goldsworthy and N.M. Fisher. John Wiley & Sons, Chichester, UK. pp. 281–304.

Pearson, C.J. (*Editor*). 1985. Pearl millet, special issue. Field Crops Res. **11** (**2**, **3**): 111–290.

Rachie, I.O., and Majmudar, J.V. 1980. Pearl millet. Pennsylvania State University Press, University Park, PA. 305 pp.

Wet, J.M.J. de. 1995. Pearl millet *In* Evolution of crop plants. 2nd edition. *Edited by* J. Smartt and N.W. Simmonds. Longman Scientific & Technical, Burnt Mill, Harlow, Essex, UK. pp. 156–159.

Witcombe, J.R., and Beckerman, S.R. (*Editors*). 1987. Proceedings of the international pearl millet workshop, 7–11 April 1986, ICRISAT Center, India. International Crops Research Institute for the Semi-arid Tropics, Patancheru, Andhra Pradesh, India. 354 pp.

Proso millet (*Panicum miliaceum*)

Bough, M., Colosi, J.C., and Cavers, P.B. 1986. The major weedy biotypes of proso millet (*Panicum miliaceum*) in Canada. Can. J. Bot. **64**: 1188–1198.

Hoek, H.N. van der, and Jansen, P.C.M. 1996. *Panicum miliaceum* L. cv. Group Proso Millet. *In* Plant resources of South-East Asia, 10, cereals. *Edited by* C.J.H. Grubben and S. Partohardjono. Backhuys Publishers, Leiden, Netherlands. pp. 115–119.

Kane, M., and Cavers, P.B. 1992. Patterns of seed weight distribution and germination with time in a weedy biotype of proso millet (*Panicum miliaceum*). Can. J. Bot. **70**: 562–567.

Lloyd, G.S., and Cavers, P.B. 2002. Dormancy, germination patterns and seedling vigour of fresh and aged seeds of different biotypes of proso millet, *Panicum miliaceum* L. Seed Sci. Technol. **30**: 641–649.

Lorenz, K., and Hwang, Y.S. 1986. Lipids in proso millet (*Panicum miliaceum*) flours and brans. Cereal Chem. **63**: 387–390.

Moore, D.R.J., and Cavers, P.B. 1985. A comparison of seedling vigour in crop and weed biotypes of proso millet (*Panicum miliaceum*). Can. J. Bot. **63**: 1659–1663.

Yanez, G.A., Walker, C.E., and Nelson, L.A. 1991. Some chemical and physical properties of proso millet *Panicum miliaceum* starch. J. Cereal Sci. **13**: 299–306.

Foxtail millet (*Setaria italica*)

Prasada Rao, K.E., Wet, J.M.J. de, Brink, D.E., and Mengesha, M.H. 1987. Infraspecific variation and systematics of cultivated *Setaria italica*, foxtail millet (Poaceae). Econ. Bot. **41**: 108–116.

Rahayu, M., and Jansen, P.C.M. 1996. *Setaria italica* (L.) P. Beauvois cv. Group Foxtail Millet. *In* Plant resources of South-East Asia, 10, cereals. *Edited by* C.J.H. Grubben and S. Partohardjono. Backhuys Publishers, Leiden, Netherlands. pp. 127–130.

Rao, K.E.P., Wet, J.M.J. de, Brink, D.E., and Mengesha, M.H. 19987. Infraspecific variation and systematics of cultivated *Setaria italica*, foxtail millet (Poaceae). Econ. Bot. **41**: 108–116.

Wet, J.M.J. de. 1995. Foxtail millet, *Setaria italica*. (Gramineae—Paniceae). *In* Evolution of crop plants. 2nd edition. *Edited by* J. Smartt and N.W. Simmonds. Longman Scientific & Technical, Burnt Mill, Harlow, Essex, UK. pp. 170–172.

Wet, J.M.J. de, Oestry-Stidd, L.L., and Cubero, J.I. 1979. Origins and evolution of foxtail millets (*Setaria italica*). J. Agric. Trad. Bot. Appl. **26**: 53–64.

Finger millet (*Eleusine coracana*)

Hilu, K.W., and Wet., J.M.J. de. 1976. Domestication of *Eleusine coracana*. Econ. Bot. **30**: 199–208.

Hilu, K.W., Wet, J.M.J. de, and Harlan, J.R. 19179. Archaeobotanical studies of *Eleusine coracana* subsp. *coracana* (finger millet). Am. J. Bot. **66**: 330–333.

Jansen, P.C.M, and Ong, H.C. 1996. *Eleusine coracana* (L.) Gaertner cv. Group Finger Millet. *In* Plant resources of South-East Asia, 10, cereals. *Edited by* C.J.H. Grubben and S. Partohardjono. Backhuys Publishers, Leiden, Netherlands. pp. 90–95.

Mehra, K.L. 1963. Differentiation of the cultivated and wild *Eleusine* species. Phyton, **20**: 189–198.

Phillips, S.M. 1972. A survey of the genus *Eleusine* Gaertn. (Gramineae) in Africa. Kew Bull. **27**: 251–270.

Wet, J.M.J. de. 1995. Finger millet. *In* Evolution of crop plants. 2nd edition. *Edited by* J. Smartt and N.W. Simmonds. Longman Scientific & Technical, Burnt Mill, Harlow, Essex, UK. pp. 137–140.

Wet, J.M.J. de, Prasada Rao, K.E., Brink, D.E., and Mengesha, M.H. 1984. Systematics and evolution of *Eleusine coracana* (Gramineae). Am. J. Bot. **71**: 550–557.

Barnyard millet (*Echinochloa frumentacea*)

Lazarides, M. 1994. *Echinochloa*. *In* A revised handbook to the flora of Ceylon. Vol. 8. *Edited by* M.D. Dasssanayake. Amerind Publishing Company, New Delhi, India. pp. 180–189.

Manidool, C. 1992. *Echinochloa colona* (L.) Link, *Echinochloa crus-galli*, (L.) P. Beauv. *In* Plant resources of South-East Asia, 4, forages. *Edited by* L. 't Mannetje and R.M. Jones. Pudoc Scientific Publishers, Wageningen, Netherlands. pp. 125–127.

Partohardjono, S., and Jansen, P.C.M. 1996. *Echinochloa* P. Beauvois. *In* Plant resources of South-East Asia, 10, cereals. *Edited by* C.J.H. Grubben and S. Partohardjono. Backhuys Publishers, Leiden, Netherlands. pp. 87–90.

Wet, J.M.J. de, Prasada Rao, P., Mengesha, M.H., and Brink, D.E. 1983. Domestication of sawa millet (*Echinochloa colona*). Econ. Bot. **37**: 283–291.

Yabuno, T. 1966. Biosystematic study of the genus *Echinochloa*. Jpn. J. Bot. **19**: 277–323.

Yabuno, T. 1987. Japanese barnyard millet (*Echinochloa utilis*, Poacaeae) in Japan. Econ. Bot. **41**: 484–493.

Millets in general

(For additional references, see SORGHUM.)

Acheampong, E., Anishetty, N.M., and Williams, J.T. 1984. A world survey of sorghum and millets germplasm. International Board for Plant Genetic Resources, Rome, Italy. 41 pp.

George Washington University. 1967. The millets; a bibliography of the world literature covering the years 1930–1963. Scarecrow Press, Metuchen, NJ. 154 pp.

McDonough, C.M., Rooney, L.W., and Serna-Saldivar, S.O. 2000. The millets. *In* Handbook of cereal science and technology. 2nd edition. *Edited by* K. Kulp and J.G. Ponte, Jr. Marcel Dekker, New York, NY. pp. 177–201.

Rachie, K.O. 1974. The millets and minor cereals; a bibliography of the world literature on millets, pre-1930 and 1964–69, and of all literature on other minor cereals. Scarecrow Press, Metuchen, NJ. 202 pp.

Rachie, K.O. 1975. The millets—importance, utilization and outlook. International Crops Research Institute for the Semi-Arid Tropics, Hyderabad, India. 63 pp.

Riley, K.W., Gupta, S.S., Seetharam, A., and Mushonga, J.N. (*Editors*). 1994. Advances in small millets. Proceedings of the second international small millets workshop held in Bulawayo, Zimbabwe from April 8–12, 1991. Oxford & IBH Publishing Company, New Delhi, India. 557 pp.

Seetharam, A., Riley, K.W., and Harinarayana, G. (*Editors*). 1990. Small millets in global agriculture. Proceedings of the first international small millets workshop, Bangalare, India, Oct. 29–Nov. 2, 1986. Aspect Publishing, London, UK. 392 pp.

Serna-Saldivar, S.O., McDonough, C.M., and Rooney, L.W. 1990. The millets. *In* Handbook of cereal science technology. *Edited by* K.J. Lorenz and K. Kulp. Markel Dekker, New York, NY. pp. 271–300.

Wet, J.M.J. de. 2000. Millets. *In* The Cambridge world history of food. *Edited by* K.F. Kiple and K.C. Ornelas. Cambridge University Press, Cambridge, UK. pp. 112–121.

Specialty Cookbooks

(See AMARANTH.)

Mung Bean

Family: Fabaceae (Leguminosae; pea family)

Names

Scientific name: *Vigna radiata* (L.) R. Wilczek (*Phaseolus aureus* Roxb., *P. radiatus* L.)

- "Mung" is from ancient languages of India for the plant, the Hindi *mung* and sanskrit *mudga*, related to the Accadian *mangu*.
- "Mung bean" is occasionally spelled as one word, "mungbean."
- Black gram or urd, *Vigna mungo* (L.) Hepper, used for human food as well as forage, is a native crop of India where it is extensively grown. It is also cultivated to some extent in the southern United States, the West Indies, and in various tropical areas. The use of the term *mungo* in the scientific name *V. mungo* has caused considerable confusion with mung bean, *V. radiata*, especially as the two species are closely related and rather similar.
- Cultivated forms often called green gram and gold gram (golden gram) are raised as human food and animal forage, and are important crops in Asia. The "gram" in these names is from the 18th century Portuguese *graõ* (then often spelled "gram"), which in turn came from the Latin *granum*, a seed or grain.
- Mung bean was apparently grown in the United States as early as 1835 under the colorful name of "the Chickasaw pea." Another colorful name is "chop suey bean."
- The genus *Vigna* commemorates the Italian Dominico Vigna (?–1647), professor of botany and director of the botanical garden in Pisa.
- *Radiata* in the scientific name *V. radiata* is from the Latin *radius*, spoke or ray, implying radiant towards every direction, like the spokes in a wheel, a reference to the arrangement of the flowers.

Plant Portrait

Mung bean is a bushy, herbaceous annual plant 30–120 cm (1–4 feet) tall, with long, thin, slightly downy pods 2.5–10 cm (1–4 inches) long, containing 10–20 seeds. This native of India and east Africa is a tropical plant requiring warm conditions. Mung bean was probably domesticated in India, where it has been grown for several thousand years. Areas of cultivation include the Middle East, Pakistan, India, China, southeastern Asia, Africa, Peru, and to a small extent, the United States. Mung bean cultivation in the United States increased greatly during the Second World War when imports from the Orient were cut off. In several Third World countries, mung bean is more extensively used for animal feed than for human food. In the United States, more than 50 000 ha (about 125 000 acres) of mung bean are planted annually of which more than half are plowed under as a green manure (for natural fertilizer). More than 100 cultivars are grown in Asia. Depending on variety, the seeds are green, yellow, brown, olive, or brownish purple, and some are speckled. In the United States, the main cultivars are Golden and Green, both grown for hay and green manure, and the latter also grown for dry beans. Other cultivars grown in North America include Lincoln and Morden. Virtually all domestic US production of mung bean is in Oklahoma. In the United States, 7–9 million kg (15–20 million pounds) of mung beans are consumed annually and nearly 75% of this is imported. Canada imports almost 2 million kg (more than 4 million pounds) annually of mung beans for sprouting, and is currently producing its own cultivars to meet the demand for a domestic supply of mung beans. Cultivars with hard, shining dark green seed coats are preferred for bean sprouts.

It is difficult to establish when the practice of employing mung beans for use of germinated seeds as edible sprouts began. It has been said that germinated mung beans are the oldest of culinary sprouts and have been used for food by the Chinese for nearly 5000 years. However, the ancient use of mung beans for sprouts in either India or China has not been proven. Various legumes were occasionally used as sprouts during past millennia in Asia, including faba bean (*Vicia faba*), pea (*Pisum sativum*), and species of *Phaseolus*; it may well be that mung bean was also used. In any event, mung bean is now the most popular of sprouts.

To produce sprouts, mung bean seeds are soaked for about 12 hours in water and then drained and placed in lightproof bean-sprouting chambers with temperatures above 25 °C (77 °F). For home use, the bean-sprouting chamber can be as simple as a wide-mouth jar or clay pot covered with cheesecloth or plastic screening. The sprouts are rinsed in cool water and drained every few hours and are ready for harvest in 4–6 days. Any loose hulls or unsprouted seeds are removed. The sprouts can be stored in cool temperatures (just above freezing) with high humidity. They are best used fresh. For home storage, the sprouts can be placed in a jar with cold water in a refrigerator for about 1 week, with daily changes of the water.

Culinary Portrait

Mung bean sprouts are commonly sold in supermarkets, specialty food stores, and salad bars and are an important component of Oriental restaurant meals. Demand for this product has been increasing. In Asia, mung beans are commonly sprouted. The sprouts are one of the key ingredients in chop suey, and are also added to other Asian dishes, soups, and salads. The sprouts are sometimes served with sugar as a snack or dessert. The whole seeds are also cooked. Mung beans can be boiled (45–60 minutes; 10 minutes in a pressure cooker for unsoaked beans, 5–7 minutes for soaked beans). They can also be fried and eaten whole, parched and ground into flour, or consumed green as a vegetable. The young pods can be cooked and served like green beans. The mung bean is widely regarded as nonflatulent and easily digested, and cooked mung beans are a good food for small children, invalids, and elderly persons. Mung bean sprouts are high in protein (21%–28%), calcium, phosphorus, and certain vitamins. However, anti-nutritional factors (toxic constituents) are reported to be present in the seeds. Germinating seeds to produce sprouts not only reduces these anti-nutritional factors but also improves content of various nutrients.

Extracts of mung bean are important for preparing various foods, particularly notable of which are noodles; the starch is reported as the best raw material for making transparent starch noodles, known by many names (see below). These slender, gelatinous noodles are widely used throughout China and Southeast Asia. They are almost flavorless, but readily absorb other flavors, and are commonly used in soups, stir-fries, salads, desserts, and beverages. *Tientsin fen pi* is the Asian name for thin round sheets of mung bean starch, which resemble rice paper, but are used to produce noodles. Before cooking, they are soaked in hot water until soft and cut into noodles.

Mung bean (*Vigna radiata*), from Dillenius (1774).

When purchasing sprouts, firm, fat, white shoots are preferable to those that are limp, wilted, long, or slender. Mung bean sprouts may be refrigerated after purchase, preferably at 1–2 °C (34–36 °F), but should be consumed at least within 3 days. Sprouts should not be iced, as they will freeze quickly.

Culinary Vocabulary

- "Silver sprouts" are mung bean sprouts with the roots and seed pods removed. They are usually employed as a garnish.
- During the World Wars, "mung" was US army slang for the much-disliked creamed chipped beef.
- Noodles made of mung bean have numerous Asian names. The English names include bean threads, bean thread vermicelli, cellophane noodles, Chinese vermicelli, crystal noodles, glass noodles, green bean thread noodles, invisible noodles, jelly noodles, mung bean threads, mung pea noodles. peastarch noodles, powdered silk noodles, shining noodles, silver noodles, slippery noodles, translucent noodles, transparent noodles, transparent vermicelli, and vermicelli noodles.

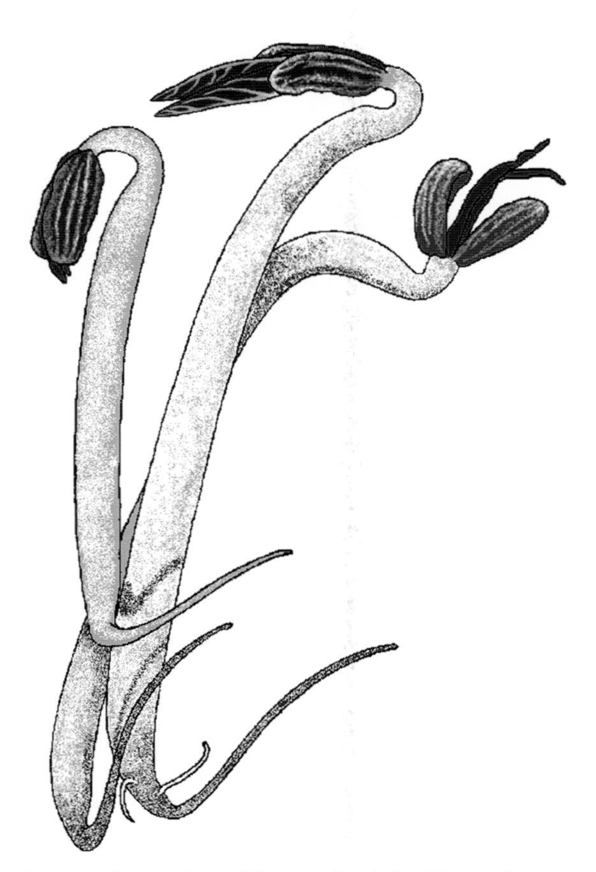

Sprouts of mung bean (*Vigna radiata*), by M. Jomphe.

- *Moong badian* (pronounced moong bah-dee-ahn) is an East Indian dish of fried mung bean dumplings, prepared with puréed yellow mung beans and spinach greens.
- *Payasam* (pronounced pah-yah-sahn) is a southern Indian pudding prepared with yellow mung beans, split peas, and coconut milk.

Curiosities of Science and Technology

- In the martial arts, "hand conditioning" refers to methods used to improve the effectiveness of the hand as a striking weapon, usually by repetitive impact. One of the common methods used in Asia consists of striking a cloth pad filled with mung beans until the beans are ground into flour.

Key Information Sources

Akpapunam, M. 1996. Mung bean (*Vigna radiata* (L.) Wilczek). *In* Food and feed from legumes and oilseeds. *Edited by* J. Smartt and E. Nwokolo. Chapman and Hall, London, UK. pp. 209–215.

Asian Vegetable Research and Development Center. 1988. Diseases and insect pests of mungbean and black gram: a bibliography. Asian Vegetable Research and Development Center, Shanhua, Taiwan. 254 pp.

Bhardwaj, H.L., Rangappa, M., and Hamama, A.A. 1997. Potential of mungbean as a new summer crop in Virginia. Virginia J. Sci. **48**: 243–250.

Buescher, R.W., and Chang, J.S. 1982. Production of mung bean sprouts. Arkansas Farm Res. **31**(**1**): 13.

Cowell, R. (*Editor*). 1978. The 1st international mungbean symposium. Office of Information Services at Asian Vegetable Research & Development Center, Taipei, Taiwan. 262 pp.

Cupka, T.B. 1987. Mung bean. *In* Proceedings, grain legumes as alternative crops: a symposium (Univ. Minnesota, July 23–24, 1987). *Edited by* Center for Alternative Crops and Products. University of Minnesota, St. Paul, MN. pp. 89–96.

DeEll, J.R., Vigneault, C., Favre, F., Rennie, T.J., and Khanizadeh, S. 2000. Vacuum cooling and storage temperature influence the quality of stored mung bean sprouts. HortScience, **35**: 891–893.

Lawn, R.J. 1995. The Asiatic *Vigna* species (*V. radiata, V. mungo, V. angularis, V. umbellata* and *V. aconitifolia*). *In* Evolution of crop plants. 2nd edition. *Edited by* J. Smartt and N.W. Simmonds. Longman Scientific & Technical, Burnt Mill, Harlow, Essex, UK. pp. 321–326.

Lawn, R.J., and Ahn, C.S. 1985. Mung bean (*Vigna radiata* (L.) Wilczek/*Vigna mungo* (L.) Hepper). *In* Grain legume crops. *Edited by* R.J. Summerfield and E.H. Roberts. Collins Publ. & Tech. Books, London, UK. pp. 584–623.

Morton, J.F., Smith, R.E. Poehlman, J.M, and Mayaguez, P.R. 1982. The mungbean: a state of the arts publication. University of Puerto Rico, Dept. of Agronomy and Soils, Mayagüez, Puerto Rico. 136 pp.

Poehlman, J.M. 1991. The mungbean. Westview Press, Boulder, CO. 375 pp.

Seelig, R.A., and Bing, M.C. 1990. Bean sprouts. *In* Encyclopedia of produce. [Loose-leaf collection of individually paginated chapters.] *Edited by* R.A. Seelig and M.C. Bing. United Fresh Fruit and Vegetable Association, Alexandria, VA. 7 pp.

Shanmugasundaram, S., and McLean, B.T. (*Editors*). 1988. Mungbean: proceedings, second international symposium (16–20 Nov. 1987, Bangkok). Asian Vegetable Research and Development Center, Taipei, Taiwan. 730 pp.

Siemonsma, J.S. and Arwooth Na Lampang. 1989. *Vigna radiata* (L.) Wilczek. *In* Plant resources of South-East Asia, 1, pulses. *Edited by* L.J.G. van der Maesen and S. Somaatmadja. Pudoc, Leiden, Netherlands. pp. 71–74.

Smartt, J. 1985. Evolution of grain legumes. Pulses in the genus *Vigna*. Exp. Agric. **21**: 87–100.

Smith, F.W., Imrie, B.C., and Pieters, W.H.J. 1983. Foliar symptoms of nutrient disorders in mung bean (*Vigna radiata*). Commonwealth Scientific and Industrial Research Organization, Melbourne, Australia. 11 pp.

Thavarasook, C. (*Editor*). 1991. Proceedings of the mungbean meeting 90, held in Chiang Mai, Thailand, February 23–24, 1990. Tropical Agriculture Research Center, Bangkok, Thailand. 318 pp.

Tomooka, N., Lairungreang, C., Nakeeraks, P., Egawa, Y., and Thavarasook, C. 1992. Center of genetic diversity and dissemination pathways in mung bean deduced from seed protein electrophoresis. Theor. Appl. Genet. **83**: 289–293.

Yang, C.Y. 1987. Mungbean diseases and their control: a review. Asian Vegetable Research and Development Center, Bangkok, Thailand. 41 pp.

Specialty Cookbooks

(Also see Bean, Common.)

Blanchard, M. 1975. The sprouter's cookbook. Garden Way Publishing, Charlotte, VT. 144 pp.

Braunstein, M.M. 1999. Sprout garden. Revised edition. Book Publishing Company, Summertown, TN. 128 pp.

Larimore, B.B. 1975. Sprouting for all seasons: how and what to sprout, including delicious, easy-to-prepare recipes. Horizon Publishers, Bountiful, UT. 139 pp.

Meyerowitz, S., Parman, M., and Scott, W. 1998. Sprouts. The miracle food: the complete guide to sprouting. Sproutman Publications, Great Barington, MA. 216 pp.

Munroe, E. 1974. Sprouts to grow and eat. S. Greene Press, Brattleboro, VT. 119 pp.

Oliver, M.H. 1975. Add a few sprouts to eat better for less money. Keats Publishing, New Canaan, CT. 126 pp.

Sellmann, P., Sellmann, G., Zweigbergk, K., and Jenkins, P. 1981. The complete sprouting book: a guide to growing and using sprouted seeds. Turnstone Press, Wellingborough, NY. 128 pp.

Whyte, K.C. 1973. The complete sprouting cookbook. Troubador Press, San Francisco, CA. 120 pp.

Wigmore, A. 1986. The sprouting book. Avery Publishing Group, Wayne, NJ. 116 pp.

MUSTARD, MUSTARD GREENS, AND SPINACH MUSTARD

Family: Brassicaceae (Cruciferae; cabbage family)

Names

Scientific names:

- Brown mustard, yellow mustard, mustard greens—*Brassica juncea* (L.) Czern.
- Yellow mustard (commonly called white mustard outside of North America)—*Sinapis alba* L. [*Brassica alba* (L.) Rabenh., *B. hirta* Moench]
- Spinach mustard—*Brassica perviridis* (L.H. Bailey) L.H. Bailey (*Brassica rapa* L. var. *perviridis* L.H. Bailey)

- The word "mustard" is usually thought to come from the Latin *mustum ardens*, meaning burning must, because the classical Romans beat ground seeds with grape must (unfermented grape juice) to develop a burning heat. When mustard seeds are mixed with liquid, a chemical transformation occurs that causes the heat. Another explanation comes from Dijon (famous, of course, for Dijon mustard). In 1382, Philip the Bold, Duke of Burgundy, granted to the town of Dijon, armorial bearings with the motto "Moult me Tarde" (I ardently desire). The arms and motto, engraved on the principal gate, were adopted as a trademark by the mustard merchants, and was shortened to "Moult-tarde," which has quite a different meaning (to burn much) and is presumed to have been transformed into "mustard."
- "Yellow mustard" is a phrase that is commonly encountered, but unfortunately, it can refer to both *Sinapis alba* and *Brassica juncea*.
- The genus name *Sinapis* is based on the Greek *sinapi*, a name used by the ancient Greek scientist Theophrastus (372?–287? BC) for mustard.
- *Alba* in the name *Sinapis alba* is Latin for white, a frequent color of the seeds of yellow (white) mustard.
- The origin of the genus name *Brassica* has been explained in several ways. The most common interpretation is that the name is related to the Celtic *bresic*, variously interpreted as meaning horse, reflecting use as forage for horses, or cabbage, a reference to the removal of leaves from the stem for cattle fodder.
- *Juncea* in the name *Brassica juncea* is Latin for rush-like (although it is arguable that this is an apt description).
- *Perviridis* in the name *B. perviridis* is Latin for very green.

Plant Portrait

Brown mustard, *Brassica juncea*, is an annual herbaceous plant, remarkable for the range of economic forms that have been bred. Numerous vegetable-like characteristics have been selected, mainly in Asia, to produce mimics of kales, cabbages, turnips, kohlrabi, and cauliflower, each with its own distinctive name(s). For example, a remarkable turnip mimic is turnip-rooted mustard (Sichuan large-rooted mustard, *Brassica juncea* (L.) Czern. var. *megarrhiza* N. Tsen & S. N. Lee [*B. juncea* var. *napiformis* (Pailleux & Bois) Kitam.]. (See illustration on page 345.)

A special form of the brown mustard plant—grown for its leaves rather than its seeds—is known as "mustard greens" (also leaf mustard and leaf mustard cabbage; see illustration on page 344). Forms with dissected leaves are known as cut-leaf mustard, dissected-leaf mustard, and ostrich-plume; forms with curled leaves are known as curled mustard and southern curled mustard. Mustard greens is important as a vegetable crop in parts of Asia, eastern Europe, and Africa. It is also used in parts of the United States.

"Spinach mustard" (sometimes known as tendergreen and kabuna), *Brassica perviridis*, is an annual herb (a biennial in mild climates if sown late) that has also been selected for its edible leaves (some forms have been selected for stems and tubers). (See illustration on page 346.) Spinach mustard is used to some extent in Asia and parts of the United States. Some varieties sold in the market are difficult to distinguish from mustard greens, and in any event, the vegetables are used similarly.

This chapter is concerned mostly with the types of *Brassica* selected to produce condiment mustard.

Brassica juncea arose as a hybrid between *B. nigra* and *B. rapa* L. It may have originated in the central Asian Himalayas. Varieties grown for mustard seed are up to 1 m (about 1 yard) or more tall. The reddish-brown-seeded forms are called brown mustard. Yellow-seeded forms arose in the Orient and are known as oriental mustard.

White mustard, *Sinapis alba*, is an annual, herbaceous, yellow-flowered plant, 20–150 cm (8 inches–5 feet) tall (mostly 20–60 cm or 8 inches–2 feet tall as a weed). It is native to the Mediterranean, central Asia, and North Africa. Wild plants of the species often have brown or black seeds unlike the cultivated forms, which only have white or yellow seeds. Unlike the seeds of brown mustard (*B. juncea*), the seeds of *S. alba* have no odor when crushed, and the taste is relatively mild.

In addition to the two primary mustard species described above, two other species deserve mention. *Brassica nigra* (L.) W.D.J. Koch (*Sinapis nigra* L.), black mustard, was once the primary source of mustard seed in the world, and is still often grown. Black mustard seed oil was also used in soap and medicine. This species is an annual Eurasian herb, 1–3 m (3–10 feet) tall, which has become a widespread weed in the United States. *Brassica carinata* A. Braun, known as Abyssinian mustard, is an annual herb up to 120 cm (4.3 feet) tall. It is grown for mustard in India, and the seeds are also used for oil.

Mustard was among the earliest of spices, recorded in about 3000 BC in Sanskrit writings of ancient India. Seeds of the mustards (both of *Brassica* and *Sinapis*) were almost certainly used, and several species were likely domesticated, millennia before classical Hellenistic and Roman times, but there is limited archeological evidence. One rich deposit of seeds of either *Brassica* or *Sinapis* (which are difficult to distinguish in charred archeological remains) has been described from a site dated at 3000 BC in Iraq. Through

Brown mustard (*Brassica juncea*), from Köhler (1883–1914).

history, mustard has been employed to stimulate the appetite, to mask the taste of rotten food, and as a digestive, antiseptic, disinfectant, and laxative.

Mustard did not achieve great popularity in North America until the beginning of the 20th century, when Francis French took the European mustard recipes and toned them down to suit American palates. Today, mustard condiment is one of the best-known and most widely used spices. India is the world's largest producer of mustard seed, with significant production also in Pakistan and Bangladesh. Most Asian mustard seed is crushed for oil. Consumption in Western countries is larger than that of every spice except pepper (which see). Specialty stores and supermarkets stock a variety of jars and fancy pots containing smooth or grainy, wine-soaked or honey-sweet, hot or mild mustards. Mustard use in North America is increasing, reflecting continued popularity in the home, but also in fast-food establishments, where waste is commonly promoted by the use of minipackets. North America, particularly Canada, is the major exporter of mustard seed in the world. The only European countries with significant mustard crops are England and Hungary. In North America, Britain, and most Commonwealth countries, most mustard is sold in both the powdered and paste forms. In Europe, South America, and other parts of the world, the demand is almost exclusively for paste mustards.

Mustard greens (*Brassica juncea*), from Vilmorin-Andrieux (1885).

Culinary Portrait

Mustard is available in many forms. *Powdered mustard*, known as mustard flour or ground mustard, in made by finely grinding the seeds and then removing the hulls by multiple milling, screening, and sifting. The process of preparing powdered mustard was developed in the early 18th century, and became popular because the resulting material kept indefinitely. In North America, powdered mustard is often made from *S. alba*. *English mustard* is made by mixing mustard powder with water and allowing the mixture to stand for 10 minutes to develop flavor. *Prepared mustard*, also called "mustard paste," is a mixture of ground mustard seed, salt, vinegar, and spices. Brown mustard is utilized for its pungency. *Dijon mustard* was originally produced in France, from *B. nigra*, but can be made anywhere in the world. This is prepared with brown mustard seeds, water, white wine, salt, and spices. It is light in color, sharp, salty, not sweet, and with a sharp taste of mustard. Dijon mustard is ground to a very fine paste and contains no tarragon. The light color is the result of the removal of the seed coat. The best-known producer of Dijon mustard is the House of Poupon, famous for its Grey Poupon mustard (see below for additional information). *Bordeaux mustard*, also originating from France, is darker, slightly sweet, milder than Dijon mustard, and is prepared with herbs such as tarragon. *Champsac mustard* is a dark brown, aromatic, smooth blend flavored with fennel seeds. *Champagne mustard* is smooth, pale, mild, and blended with champagne. *Beaujolais mustard* is a blend of coarse-ground seeds and red wine. *Red mustard* is prepared with chili. *Wholegrain mustard* is a hot, crunchy, English preparation of whole mustard seeds. *Honey mustard* is a sweet blend of mustard seeds, honey, raw sugar, vinegar, and spices. French *herb mustard* is usually smooth, mild, and lightly flavored with mixed herbs. *German mustard* is frequently flavored with herbs and spices, and is often somewhat sweet.

Whole seeds of mustard are used in pickling or are boiled with vegetable dishes such as sauerkraut or cabbage. Brown mustard seeds are seldom used as whole seeds in North America but are important in parts of Asia where they are used in curries and pickles. Indian recipes use the whole seeds in "ghee" (clarified butter or vegetable fat), a garnish for many Indian dishes. The treatment destroys the enzyme necessary for the pungent taste, and the seeds become pleasantly nutty, rather like poppy seeds.

The leaves of the "mustard green" type of brown mustard (*B. juncea*), as well as those of spinach mustard (*B. perviridis*), are used as a vegetable, and have a

Turnip-rooted mustard (*Brassica juncea* var. *megarrhiza*), from Pailleux and Bois (1892).

flavor that is more delicate than that of kale, but at the same time more piquant, with a slightly bitter aftertaste. The leaves can be used raw in salads, or cooked. Mustard greens are generally consumed when young, as they become coarse and excessively hot-tasting with age. The vegetable can be cooked like spinach by briefly steaming the leaves and stems. In China, mustard greens are added to soups and salads or stir-fried. Generally, mustard greens go better with highly seasoned rather than delicate dishes, but they are delicious combined with mashed potatoes. Some varieties have swollen stem bases that are boiled and then peeled and sliced. Mustard greens are often available in Chinese markets and some specialty supermarkets. When purchasing, select smaller leaves (or packages with small leaves) that appear to be fresh (not wilted), and without signs of disease, such as spots. Mustard greens can be stored for up to a week in a refrigerator after wrapping in paper towels in a plastic bag. Pickled mustard greens and salted mustard greens (*ham choy*) are also sometimes sold in Chinese markets. The salted form can be stored in a container for up to a week in a refrigerator, and the salt should be rinsed out before consumption. Mustard greens should not be cooked in aluminum or iron pots, as contact with these metals turns the vegetable black. For some people with thyroid conditions, mustard greens can cause the thyroid to increase in size, and should be avoided.

Culinary Vocabulary

- "Made mustard" is the British equivalent of the North American "prepared mustard," i.e., wet mustard kept in a jar, as opposed to dry mustard. The term is archaic.
- *Karashi* is Japanese powdered mustard. It is similar to English mustard but is extremely strong in flavor. Karashi is mixed with water to form a paste, covered, and allowed to stand for 10 minutes to develop flavor.
- "Chow-chow" (chowchow) is an American phrase that came into use in the mid 19th century to designate a spicy relish made up of a potpourri of vegetables, typically pickled in mustard. The preparation originated from a Chinese condiment made up of pieces of orange peel, ginger, and other

constituents, pickled in syrup. The Mandarin Chinese word for the concoction, *cha*, means "mixed," which indicates the character of the product. It is thought that chow-chow was introduced to North America by Chinese laborers working on the railroads of the American West. This vegetable relish was popular when most people lived on farms and could prepare it themselves from the variety of their own produce. Commercial preparations called chow-chow are available, but do not necessarily adhere to the original inclusion of mustard.

- "Mostarda" (pronounced moh-STAR-dah) or *mostarda di frutta* is an Italian relish, chutney, fruit pickle, or fruit preserve, which is often sweet. It is made from candied whole or cut-up fruits (such as pears, cherries, figs, plums, and apricots) preserved in syrup, and includes powdered mustard seed and spices. The Italian word *mostarda* is derived from the French word for mustard, *moustarde*, but mustard is only one of the ingredients (the Italian word for mustard is *senape*; *Mostarda* is the Portuguese word for mustard). Mostarda is one of the standard condiments served with boiled meats in northern Italy. In Italy, it is sold by the barrel, while in other Western countries mostarda is generally marketed as a gourmet item in specialty food stores. Mostarda goes well with cold boiled meat, and to a lesser extent, fowl.
- "Creole mustard," originated from Louisiana's German Creoles and made from vinegar-marinated brown mustard seeds, is a hot, spicy, gourmet mustard flavored with horseradish.
- "Yellow paint," "baby sauce," and "Mississippi mud" are old lunch-counter slang expressions for mustard.

Curiosities of Science and Technology

- In the 6th century BC, Greek scientist and mathematician Pythagoras (who lived about 580–500 BC) used mustard as a remedy for scorpion stings.
- Bornibus, a Burgundian (Burgundy, in eastern France, was once a kingdom), discovered a method of pressing mustard into dry cakes. These cakes were manufactured in Dijon, and in 1634, the vinegar and mustard makers were granted the exclusive right to make mustard in return for wearing "clean and sober clothes."

Spinach mustard (*Brassica perveridis*), by M. Jomphe.

- Frederick the Great (1712–1786, King of Prussia) invented a beverage made with powdered mustard, champagne, and coffee.
- The Reverend Sylvester Graham (1794–1851, of "Graham cracker" fame—see WHEAT) was a health faddist who attracted a huge following in the United States in the early 19th century. He advocated a strict vegetarian diet, drinking no liquids except water, and keeping the windows open at night regardless of the weather. Graham warned that mustard and ketchup cause insanity.
- In Tanganiyaka (now part of Tanzania), dried leaves and flowers of *B. juncea* were smoked like marihuana in order to get in touch with the spirits.
- Grey Poupon mustard was named for Maurice Grey and Auguste Poupon. In 1853, Grey invented a device that made mass production of fine-textured mustard possible. He partnered with mustard-maker Auguste Poupon around 1886, forming the Grey–Poupon firm.
- Mustard, in the form of the R.T. French Company's "cream mustard," first met the hotdog at the 1904 St. Louis World's Fair.
- The poisonous mustard gas of the First World War is made by mixing ethylene with disulfur dichloride, neither of which is a product of mustard seed. The name "mustard gas" was coined by British soldiers, who thought it smelled like English mustard. However, part of the chemical structure of the "tear gas" used in World War I is also part of the structure of the glycoside sinigrin, found in the seeds of black mustard (*B. nigra*).
- Mustard has long been used in mustard plasters for rheumatism, arthritis, chest congestion, aching backs, and sore muscles. Mustard plasters can be made of equal parts of flour and powdered mustard, spread as a moist paste sandwiched inside a doubled piece of cloth, and applied to the body for 10–15 minutes (in no case longer than 20 minutes). The heat that a mustard plaster produces can readily burn skin, and patients who fall asleep are potentially in considerable danger. (If alone, set an alarm clock to avoid waking up like a boiled lobster.) With sensitive skin, four parts of flour can be added to one part of mustard, and olive oil or castor oil can be applied first to the area to be treated. The plaster should not be placed on broken skin. Check occasionally to make sure the skin does not turn raw.
- There are 400 000 brown mustard seeds in a kilogram (185 000 in a pound).
- Some tips for using Colman's dry mustard (from their web site):
  - Make a paste and spread on the back of a loose tile to secure it to the wall. Acts like plaster!
  - Use as fertilizer for better-colored daffodils.
  - Sprinkle dry mustard inside shoes to prevent cold feet and frostbite.
  - To mend leaky car radiators temporarily, pour in contents of 2 oz. tin of mustard while car is running.
  - Smear headlights with a mixture of dry mustard and some water to turn your lights into fog lights.
  - Stuffy nose? Make a paste, spread it generously over a cracker, and take a few bites. This clears it up in seconds!
- There are nearly 1000 varieties of mustard on the market.
- The Mount Horeb Mustard Museum in Wisconsin has a collection of over 3000 jars of specialty or blended mustards of the world.
- American mustards are usually made from yellow-colored mustard seeds, but it is not the mustard seed that makes American mustards yellow. The distinctive yellow is due to the addition of turmeric.
- "Jim Hill mustard" (*Sisymbrium altissimum* L.), more commonly known as tumble mustard (because it produces a tumble weed) and tansy mustard, is a native of Europe that has become an extremely widespread and troublesome weed in North America. "Jim Hill" is James Jerome Hill (1838–1916), US railroad owner, popularly known as "the Empire Builder" for his major role in pushing through the transcontinental line of the Great Northern Railroad, populating the area between St. Paul and Seattle, and working to help settlers improve their land and farming practices. Appropriately, Jim Hill Mustard is especially common along railways. A single plant can produce 1 500 000 seeds, and Indians ground these into a meal used for flavoring soup and making gruel. The plant has been used as greens when the leaves and shoots are young and tender. However, the foliage, when consumed in quantity, has been found to be poisonous to livestock.

Yellow mustard (*Sinapis alba*), from Köhler (1883–1914).

Key Information Sources

Alemaw, G. 1988. Review on breeding of Ethiopian mustard (*Brassica carinata* A. Braun). *In* 7th International Rapeseed Congress. *Edited by* Plant Breeding and Acclimatization Institute. Panstwowe Wydawnictwo Rolnicze i Lesne, Poznan, Poland. pp. 593–597.

Alpine, A. 1978. Untamed mustard. Harrowsmith, **13**: 108, 110, 111.

Anon. 1977. Mustard production. Continental Grain Company, Winnipeg, MB. 8 pp.

Bailey, L.H. 1922. The cultivated Brassicas. Gentes Herb. **1**: 53–108.

Bailey, L.H. 1930. The cultivated Brassicas. Second paper. Gentes Herb. **2**: 211–267.

Bailey, L.H. 1940. Certain noteworthy Brassicas. Gentes Herb. **4**: 318–330.

Chen, S.-R. 1982. The origin and differentiation of mustard varieties in China. Cruciferae Newsl. **7**: 7–10.

Dean, J. 1987. Mustard production in Manitoba. Manit. Agric. Publ. 140-1. 3 pp.

Fenwick, G.R., Heaney, R.K., and Mullin, W.J. 1983. Glucosinolates and their breakdown products in food and food plants. CRC Crit. Rev. Food Sci. Nutr. **18**: 123–201.

Forhan, S.C., and Tisdale, R. 1988. Mustard growers manual. Saskatchewan Agriculture, Saskatoon, SK. 24 pp.

Hemingway, J.S. 1995*a*. Mustards. *In* Evolution of crop plants, 2nd edition. *Edited by* J. Smartt and N.W. Simmonds. Longman Scientific & Technical, Harlow, Essex, UK. pp. 82–86.

Hemingway, J.S. 1995*b*. The mustard species: condiment and food ingredient use and potential as oilseed crops. *In Brassica* oilseeds: production and utilization. *Edited by* D.S. Kimber and D.I. McGregor. CAB International, Wallingford, UK. pp. 373–382.

Love, H.K., Rakow, G., Raney, J.P., and Downey, R.K. 1990. Development of low glucosinolate mustard. Can. J. Plant Sci. **70**: 419–424.

Mackenzie, S.L. 1973. Cultivar differences in proteins of *B. juncea*. J. Am. Oil. Chem. Soc. **50**: 411–414.

Mehra, K.K. 1968. History and ethnobotany of mustard in India. Adv. Front. Plant Sci. **19**: 51–59.

Myers, C. 1991. Specialty mustards. Crop Sheet SMC-032. *In* Specialty and minor crops handbook. *Edited by* C. Myers. The Small Farm Center, Division of Agriculture and Natural Resources, University of California, Oakland, CA. 2 pp.

Opeña, R.T. 1993. *Brassica juncea* (L.) Czernjaew. *In* Plant resources of South-East Asia, 8, vegetables. *Edited by* J.E. Siemonsma and K. Piluek. Pudoc Scientific Publishers, Wageningen, Netherlands. pp. 104–108.

Skrypetz, S. 2007. Mustard seed: situation and outlook. Bi-weekly Bull. (Agric. Agri-Food Can.), **20**(**11**): 1–4.

Too, K.H.C. 1 970. Identification of some common *Brassica* species by their vegetative characters. Malay. Agric. **9**: 53–70.

Toxopeus, H., and Jansen, P.C.M. 1993. *Brassica* L. *In* Plant resources of South-East Asia, 8, vegetables. *Edited by* J.E. Siemonsma and K. Piluek. Pudoc Scientific Publishers, Wageningen, Netherlands. pp. 101–104.

Toxopeus, H., and Lubberts, J.H. 1999. *Sinapis alba* L. *In* Plant resources of South-East Asia, 13, spices. *Edited by* C.C. de Guzman and J.S. Siemonsma. Backhuys, Leiden, Netherlands. pp. 204–207.

Toxopeus, H., and Utomo, I. 1999. *Brassica nigra* (L.) W.D.J. Koch. *In* Plant resources of South-East Asia, 13, spices. *Edited by* C.C. de Guzman and J.S. Siemonsma. Backhuys, Leiden, Netherlands. pp. 85–88.

Vaughan, J.G., and Hemingway, J.S. 1959. The utilization of mustards. Econ. Bot. **13**: 196–204.

Vaughan, J.G., Hemingway, J.S., and Schofield, H.J. 1963. Contribution to a study of variation in *Brassica juncea*. J. Linn. Soc. Bot. **58**: 435–447.

Wei, W.L. 1982. The classification of mustards *Brassica juncea*, varieties, vegetables in China. Yuan I Hsueh Pao Acta Hortic. Sin. (Peking), **9**(**4**): 33–40. [In Chinese, English summary.]

Specialty Cookbooks

(Note: There are dozens of cookbooks dealing with mustards. The following is a selection.)

Alley, L. 1995. Lost arts: a cook's guide to making vinegar, curing olives, crafting fresh goat cheese and simple mustards, baking bread, and growing herbs. Ten Speed Press, Berkeley, CA. 151 pp.

Antol, M.N. 1999. The incredible secrets of mustard: the quintessential guide to the history, lore, varieties, and healthful benefits of mustard. Avery Publishing Group, Garden City Park, NY. 212 pp.

Calvert, R. 1986. The plain & fancy mustard cookbook. East Woods Press, Charlotte, NC. 143 pp.

Costenbader, C.W. 1996. Mustards, ketchups & vinegars: making the most of seasonal abundance. Storey Publishing, Pownal, VT. 96 p.

Creber, A., Williams, C., and Salaverry, P. 1994. Mustards. Weldon Owen, San Francisco, CA. 16 pp.

Doerfer, J. 1984. The pantry gourmet: over 250 recipes for mustards, vinegars, relishes, pâtés, cheeses, breads, preserves, and meats to stock your pantry, freezer, and refrigerator. Rodale Press, Emmaus, PA. 290 pp.

Gordon-Smith, C., and Merrell, J. 1998. Basic flavorings. Mustard. Courage Books, Philadelphia, PA. 64 pp.

Grey Poupon Mustard. 1997. Meals made easy with Grey Poupon mustard. Meredith, Des Moines, IA. 96 pp.

Gunst, K. 1984. Condiments: the art of buying, making, and using mustards, oils, vinegars, chutneys, relishes, sauces, savory jellies, and more. G.P. Putnam's, New York, NY. 258 pp.

Hazen, J. 1993. Mustard: making your own gourmet mustards. Chronicle Books, San Francisco, CA. 71 pp.

Hopley, C. 1991. Making & using mustards. Storey Communications, Pownal, VT. 31 pp.

Jordan, M.A. 1994. The good cook's book of mustard: with more than 100 recipes. Addison-Wesley, Reading, MA. 226 pp.

O'Sullivan, M. 1991. Mustards, pickles and chutneys. Key Porter Books, Toronto, ON. 100 pp.

Roberts-Dominguez, J. 1993. The mustard book. Macmillan Publishers, New York, NY. 163 pp.

Sawyer, H., and Long, C. 2002. Gourmet mustards: the how-tos of making & cooking with mustards. Revised edition. Sibyl Publications, Portland, OR.111 pp.

Stone, S., and Stone, M. 1981. The mustard cookbook. Avon, New York, NY. 234 pp.

OAT

Family: Poaceae (Gramineae; grass family)

Names

Scientific name: *Avena sativa* L.

- The English word "oat" has been said to trace to the Celtic *aten*, from *etan*, to eat.
- Oat is also called common oat(s), oats, and white oat. Some editors prefer the singular (oat) when referring to the plant or the species, and the plural (oats) when referring to the grain.
- The genus name *Avena* is based on the classical Latin word for oats.
- *Sativa* in the scientific name *A. sativa* means sown or cultivated.

Plant Portrait

Oat is one of the principal cereal grains, ranking sixth in world production on a weight basis after wheat, maize, rice, barley, and sorghum, but exceeding sorghum in value. Its origin can be traced back to about 2000 BC in the Middle East, particularly the areas surrounding the Mediterranean Sea. Some of the first evidence of oats was found in ancient Egypt and parts of Switzerland. Oat is now cultivated throughout the temperate zones of the world. It is better adapted to flourish in cold, wet, or warm conditions than most other grains. The world's leading oat producers are Russia, the European Union (especially Germany, Poland, and Finland), Canada, the United States, and Australia. There are several species of oat, but three-quarters of world oat production is of the common oat, *A. sativa*. Less commonly grown is *A byzantina* K. Koch, Mediterranean or red oat (sometimes considered to be a part of *A. sativa*). The oat plant is an annual grass, ranging in height from 60 to 150 cm (2–5 feet), depending on variety and environment. Oat seeds are covered with a thick hull, but hull-less (or "naked") varieties are becoming popular (the hull falls away from the groat at maturity). Most oats are used for livestock feed, with only about 16% of total world production used for human food. Oats are considered to be the premium grain for horses, and prior to the coming of the tractor, great acreages were grown to feed draft horses used for farm power. Most oats are still used to feed livestock, but today primarily dairy cattle and poultry. Oats are also being used for industrial and medicinal extracts. Oat products are now used in the cosmetic industry as talc replacers and in skin care products.

Culinary Portrait

Unlike wheat, the bran (located in the outer layers of the grain) and germ (the embryo) are generally not refined away. Oatmeal, the meal resulting from grinding oats after the husk has been removed, is employed for many dishes, especially porridge, but also in cakes, soups, stews, dumplings, and stuffings. Once considered a poverty food, oats are increasingly being consumed by people, especially in ready-to-eat breakfast cereals, as rolled oats, in cakes, breads, muffins, crackers, granola-type snack bars, and in other pastries. Because oat flour does not contain gluten and will not rise during baking, it is combined with wheat flour (typically, two parts of wheat flour to one part of oat flour) to make bread and other leavened foods. Commercial "oatmeal" breakfast cereals almost always contain sugar, salt, and a variety of additives. Oats are a popular ingredient of infant foods. The grain is particularly important in the traditional foods of Scotland, where it is an ingredient in the famous haggis, in Scottish oatmeal soup, and for coating fish before they are fried. Like the Scots, Austrians also commonly use oats in stews and stuffings. Oats are also used to a minor extent in brewing. Oat cereal is considered to be a healthy, nutritious food for humans, and has been shown to lower cholesterol and to have numerous other dietary advantages. It has high-quality protein and a high concentration of soluble fiber, and is a good source of minerals, essential fatty acids, B vitamins, and vitamin E.

Culinary Vocabulary

- "Oat groat" is the de-husked oat kernel.
- "Oatmeal" is a coarse meal made from ground or rolled oats after the grain has been husked and cleaned ("oat flour" is the same thing, but finely ground).
- "Rolled oats" (also known as "old-fashioned rolled oats") are steamed oat groats rolled into flat cakes (the "rolled" in the phrase comes from the process of crushing the grains with heated steel rollers).
- "Steel-cut oats" (also known as Irish oats and Scotch oats) are toasted oat groats that have been cut (by passing them between steel blades that cut them into slices) into two or three pieces per grain; a longer cooking time is required than for rolled oats. The more finely the grains are cut, the faster they cook.

- "Instant oats" are rolled oats that have been partly cooked and dried before being rolled very thin.
- "Quick-cooking oats" are rolled oats that have been cut into smaller pieces to reduce cooking time.
- As a food, oats are best known in porridge. The word "porridge" is thought to come from *porée*, old French for a sort of vegetable stew, which evolved to potage (a thick soup), and then later to porridge. In Britain, porridge is often called "oatmeal."
- "Stirabout" is an old form of porridge made in the United States, in some places constituting simply a breakfast of oatmeal mush thinned with milk and salt, although the Pennsylvania Dutch of the 19th century made it much more palatable, adding vegetables, saffron, and chicken broth.
- "Haver-cake" is an obsolete northern English word for "oatcake," based on the ancient German word *haver*, "oats." "Haversack" (a knapsack worn over one shoulder) is literally "sack for oats."
- "Bannock" is a traditional British highlands (Scotland and the northern English counties) sweet flatbread made with oats and barley and cooked on a griddle. The unleavened bread was usually made in non-wheat-growing areas, where the word "bannock" became synonymous with bread. In Scotland, bannock was primarily made with barley, while in England, oats predominated. While once considered to be a poor people's food, like the hearth breads typical of many ancient European and Mediterranean cultures, very attractive specialty bannocks resembling fruitcakes and shortbreads are also made. According to superstition, bannock dough must be kneaded from right to left, or bad luck will result. In Scottish wedding tradition, before a bride entered her new home, bannocks would be broken above the head of the couple and pieces of the cake were passed around to everyone.
- "Flummery" (pronounced FLUHM-muh-ree) refers to British, generally oatmeal-based, bland puddings or custards, generally eaten cold. In Britain, flavorings such as cream, raisins, sugar, almonds, or sherry are added. In the United States, flummery refers to a dessert of simmered berries thickened with cornstarch and served with cream. Flummery figuratively means nonsense, blather, or meaningless flattery.
- "Athole brose" (Atholl brose), pronounced AH-thohl broz, is a Scottish concoction of oatmeal, honey, and whiskey (a brose is a porridge-like dish prepared by pouring hot water or milk over oatmeal and adding salt or butter). The "Athole" in the word is said to trace to the 15th century Earl of Athole, who supposedly captured the rebellious Earl of Ross by placing a mixture of honey and whisky in a small well from which Ross used to drink, causing the latter to become drunk and vulnerable.
- "Haggis," a traditional Scottish dish, is made from the lungs, heart, and liver of a sheep, chopped with onions, seasonings, suet, and oatmeal, and then broiled in a bag made from the sheep's stomach. It is traditionally served by the Scottish on the celebration of the birth of Robbie Burns (Jan. 25, 1759), the great Scots poet. The haggis is proudly carried on a platter to the playing of bagpipes, with a bearer of Scotch whisky following the procession. The host of the evening, dressed in a kilt with a sword in hand, slits open the haggis and douses it with whisky to enhance the flavor. This is followed by the reading of the original Robbie Burns poem, *Ode to a Haggis* (available on the Web). The origin of the name "haggis" is unclear. Among the explanations: the word comes from *haggess*, a 17th century name for "magpie"; or from the northern Middle English verb *haggen*, "chop," and French *hachis*, which produced the English word "hash." Facetiously, haggis is named after a mythical Highlands beast with two long legs and two short legs, so it could better scamper around the hillsides.
- "Haggamuggie" is a Shetland Island fish haggis, prepared by stuffing the stomach (*muggie*) of a large fish with oatmeal and chopped fish liver.
- "Crappit heids" is another traditional Scottish dish, as unique as haggis. This delicacy consists of haddock's heads (*heids*) stuffed with a seasoned mixture of oatmeal, suet, and onion, and boiled. *Crappit* means filled or stuffed.

Curiosities of Science and Technology

- Many Scottish cottages once had a "porridge drawer." Porridge was made once a week and then poured into a drawer where it cooled into a slab. Pieces would then be cut off and eaten cold while working out on the land.
- In pre-tractor days, 25% of a farm might be devoted to oats and hay for horsepower.
- A single oat plant can produce more than 100 kernels.
- Ferdinand Schumacher (1822–1908) was a German immigrant to the United States, who

Oat (*Avena sativa*), from Figuier (1867).

became known as "the king of oatmeal" in his day. He is credited with the first advertisement promoting a breakfast cereal—a notice that appeared in 1870 recommending his "Avena oatmeal" for breakfast—which appeared in 1870 in the Akron Beacon of Akron, New York.

- In Ravenna, Ohio, Henry D. Seymour and William Heston established the Quaker Mill Company and in 1877 registered the now famous Quaker trademark that has long been used by the Quaker Oats Company. The symbol of a Quaker man was chosen because of admiration for the purity of living, honesty, and strength of character that was fundamental to the religious movement.
- American cereal magnate Henry Parsons Crowell (1856–1943) purchased the Quaker Oats Company shortly after the brand was registered, and is thought to have been the first to use saturation media advertising to promote a product. In addition to print advertising, he painted signs on barns, ran signs on trains, invented the idea of putting prizes in boxes of a product (he put glasses, bowls, and saucers in his 2-pound boxes of oatmeal), placed a huge sign on the white cliffs of Dover in England, and even made some unsubstantiated medical claims, including "one pound of Quaker Oats makes as much bone and muscle as three pounds of beef." He sold millions of boxes of oatmeal and is regarded as the true "king of oatmeal."
- Cheerios is the brand of breakfast cereal most consumed by children in North America (corn flakes is the overall leading cereal, followed by Cheerios). "Cheerios" was first named "Cheerioats." The Quaker Oats Company sued General Mills, claiming that the name "Cheerioats" infringed on the use of "oats" in its trademarks, and the name was changed to "Cheerios."
- Porridge oats have traveled to the North Pole with American explorer Admiral Richard Byrd (1888–1957) and to the South Pole with Norwegian explorer Ronald Amundsen (1872–1928). The cereal has even traveled in space with US astronauts.
- During the 1980s, oat bran acquired a reputation for adding fiber to the diet and thereby helping to reduce blood cholesterol. Quaker Oats increased its sales of oat bran products from about 454 000 kg/year (500 tons/year) to about 11 000 000 kg/year (12 000 tons/year) by 1989.
- Britons consume 47 million imperial gallons (56.4 American gallons) of porridge every year.
- Oats have contributed words and expressions to the English language. "Avenage" (based on the Latin *avena* for oats) is an old English term referring to oats that were paid to landlords in lieu of rent by the poor. "Oatmeal Monday" was a midterm Monday at Scottish universities when the parents of the poor students were allowed to visit and bring a sack of oatmeal, which was a staple diet for the student for the rest of the term. "To sow one's wild oats" is an expression tracing to the 16th century, which comes from inexperienced farmers who failed to separate away the wild oats in their seeds, and so wasted their resources. Today, it means to indulge in youthful excesses without thinking of the consequences. The expression "feel one's oats," meaning to "feel frisky" or "act self-importantly," traces to America of the 1820s, and originates from the exuberance of horses that have been well fed on oats.
- Many grasses and forages can accumulate toxic levels of nitrates from the soil they are growing in, depending upon fertilization practices. Oats can accumulate excessive amounts of nitrates, especially when heavily fertilized. This can lead to serious poisoning, especially in cattle, and severely affected animals may die. (See PAPAYA for a different example of a problem caused by high nitrates.)

- Oatmeal applied to the skin has been demonstrated to be effective against itching due to allergies. Colloidal oatmeal (i.e., finely ground oatmeal in a suspension solution) is a common ingredient in pet shampoos to treat itching, and is also often employed in medicinal preparations (lotions, soaps, etc.) to treat psoriasis, shingles, acne, and a variety of mild skin irritations (such as sunburn, insect bites, and chapping) in humans. Oatmeal baths are often a simple way to treat itching skin, but oatmeal can be slippery when wet, so caution is advised when getting into or out of a tub.

Key Information Sources

Asgarali, A. 2006. Oats: situation and outlook. Bi-weekly Bull. (Agric. Agri-Food Can.), **19**(**14**): 1–4.

Baum, B.R. 1977*a*. Oats: wild and cultivation. A monograph of the genus *Avena* L. (Poaceae). Agriculture Canada, Ottawa, ON. 463 pp.

Baum, B.R. 1977*b*. Supplement to Oats: wild and cultivated: bibliography. Canada Dept. of Agriculture, Research Branch, Ottawa, ON. 446 pp.

Coffman, F.A. (*Editor*). 1961. Oats and oat improvement. American Society of Agronomy, Madison, WI. 650 pp.

Coffman, F.A., and Stanton, T.R. 1977. Oat history, identification, and classification. Agricultural Research Service, U.S. Dept. of Agriculture, Washington, DC. 356 pp.

Findlay, W.M. 1956. Oats: their cultivation and use from ancient times to the present day. Oliver and Boyd, Edinburgh, UK. 207 pp.

Jones, I.T. 1956. The origin, breeding and selection of oats. Agric. Rev. **2**: 20–28.

Ladizinsky, G., and Zohary, D. 1971. Notes on species delimination, species relationships and polyploidy in *Avena*. Euphytica, **20**: 380–395.

Laszity, R. 1998. Oat grain—a wonderful reservoir of natural nutrients and biologically active substances. Food Rev. Int. **14**: 99–119.

Laverick, R.M. 1997. Winter oats: agronomy review. Semundo Ltd., Cambridge, UK. 72 pp.

Malkki, Y., and Virtanen, E. 2001. Gastrointestinal effects of oat bran and oat gum: a review. Lebensm. Wiss. Technol. **34**: 337–347.

Marquette, A.F. 1967. Brands, trademarks, and good will; the story of the Quaker Oats Company. McGraw-Hill, New York, NY. 274 pp.

Marshall, H.G., and Sorrells, M.E. (*Editors*). 1992. Oat science and technology. American Society of Agronomy, Madison, WI. 846 pp.

McMullen, M.S. 2000. Oats. *In* Handbook of cereal science and technology. 2nd edition. *Edited by* K. Kulp and J.G. Ponte, Jr. Marcel Dekker, New York, NY. pp. 127–148.

Mehra, K.L. 1978. Oats. Indian Council of Agricultural Research, New Delhi, India. 152 pp.

Oakenfull, D. 1988. Oat bran—does oat bran lower plasma cholesterol...and, if so, how? CSIRO Food Res. Quart. (Australia), **48**(**2**): 37–39.

Peterson, D.M., and Murphy, J.P. 2000. Oat. *In* The Cambridge world history of food. *Edited by* K.F. Kiple and K.C. Ornelas. Cambridge University Press, Cambridge, UK. pp. 121–132.

Price-Jones, D. 1976. Wild oats in world agriculture: an interpretative review of world literature. Agricultural Research Council, London, UK. 296 pp.

Thomas, H. 1995. Oats. *In* Evolution of crop plants. 2nd edition. *Edited by* J. Smartt and N.W. Simmonds. Longman Scientific & Technical, Burnt Mill, Harlow, Essex, UK. pp. 132–137.

Valentine, J. 1990. Oats: historical perspectives, present and prospects. J. Roy. Agric. Soc. **151**: 161–176.

Webster, F.H. (*Editor*). 1986. Oats: chemistry and technology. American Association of Cereal Chemists, St. Paul, MN. 433 pp.

Welch, R.W. (*Editor*). 1995. The oat crop: production and utilization. Chapman and Hall, London, UK. 584 pp.

Wood, P.J. (*Editor*). 1993. Oat bran. American Association of Cereal Chemists, St. Paul, MN. 164 pp.

Zohary, D. 1971. Origin of south-west Asiatic cereals. Wheat, barley, oats and rye. *In* Plant life of south-west Asia. *Edited by* P.H. Davis, P.C. Harper, and I.C. Hedge. Royal Botanic Garden, Edinburgh, UK. pp. 235–263.

Specialty Cookbooks

Baggett, N., and Glick, R. 1989. The oat bran baking book: 85 delicious, low-fat, low-cholesterol recipes. Contemporary Books, Chicago, IL. 128 pp.

Cadogan, M., and Bond, S. 1987. The oat cookbook. Optima, London, UK. 127 pp.

Earnest, B., and Schlesinger, S. 1988. The low-cholesterol oat plan: over 300 delicious and innovative recipes for the miracle food. Hearst Books, New York, NY. 352 pp.

Hinman, B. 1989. Oat cuisine. Prima Publishing & Communications, New York, NY. 289 pp.

Marshall, J. 1986. Eats with oats: the new soluble fibre cookbook. Foulsham, London, UK. 128 pp.

Maynard, K., and Maynard, L. 1989. The oat bran cookbook. Rutledge Hill Press, Nashville, TN. 160 pp.

Parkinson, C.M. 1991. Cooking with oats: oat bran, oatmeal, and more. Storey Communications/Garden Way Publishing, Pownal, VT. 32 pp.

Quaker Oats Company. 1989. Quaker oat bran cookbook. Publications International, Lincolnwood, IL. 96 pp.

Quaker Oats Company. 1992. The Quaker Oats treasury of best recipes. Smithmark, New York, NY. 64 pp.

Quaker Oats Company. 1996. Quaker Oats favorite recipe collection. Time-Life Books, Alexandria VA. 96 pp.

Westland, P. 1986. Oat cuisine. Salem House, Manchester, NH. 96 pp.

Wilson, J. 1989. The oat bran way. Berkley Books, New York, NY. 158 pp.

Oil Palm

Family: Arecaceae (Palmae; palm family)

Names

Scientific name: *Elaeis guineensis* Jacq.

- "Oil palm" is so named because of the use of this palm as a source of oil.
- The oil palm is also known as the African oil palm and macaw-fat.
- The genus name *Elaeis* is derived from the Greek *elaia*, olive, indicating the high oil content, like that of the olive.
- *Guineensis* in the scientific name *E. guineensis* is Latin for "of Guinea," a reflection of the species area of origin, Guinea's rain forests in West Africa.

Plant Portrait

The oil palm is a tall tree up to 20 m (66 feet) in height, with an erect trunk usually less than 30 cm (1 foot) in diameter, which bears distinctive rings. The fruits are borne in large clusters of 200–300, each cluster weighing up to 9 kg (20 pounds). The fruit is plum-like, egg-shaped, up to 2.5–5 cm (1–2 inches) long and 2.5 cm (1 inch) or more wide, black when ripe, red at the base, with thick ivory–white flesh. Within the flesh is a "nut" made up of a fibrous covering within which is a kernel. The oil palm is native to the tropical rain forest region of West Africa. It is now widely cultivated throughout the tropics.

Young oil palm trees (less than 10 years of age) grow rapidly, about 50 cm (20 inches) per year. By the age of 20–25 years, the trees have often reached a size of over 12 m (about 40 feet) and grow at a rate of only a few centimeters (1–2 inches) per year. At this height, the economically useful life of the trees is finished, mostly because of the difficulty of harvesting the fruits from tall trees, and they are cut down. However, trees may live for up to 200 years.

The history of the oil palm dates back over 5000 years. While the oil has been used for a very long time for food, medicine, and industrial purposes, it is only over the last century that palm oil has become one of the world's most important vegetable oils. There are many species of palms that can provide oil, but the oil palm is the predominant commercial source. Only soybean is more important than the oil palm for production of vegetable fats and oil, followed respectively by rapeseed, sunflower, peanut, cottonseed, and coconut. In the first decade of the 21st century, palm oil has become almost as important as soybean oil. Because it is a perennial crop that does not have to be replanted every year (the only species in the preceding group grown as a perennial is the coconut, another palm tree), the oil palm is more productive and valuable on a given area of land than other crops, so it is given preference over other possible crops. The oil palm has been described as "the most successful perennial plantation crop in human history," with close to 7 million ha (17.3 million acres) planted in the world. The main producing countries are Malaysia and Indonesia, with lesser amounts in Nigeria, Congo, Sierra Leone, and Ghana.

Two kinds of oil are produced. "Palm oil," the main product, is a reddish orange oil extracted from the pulp of the fruit. This is very high in saturated fat (50% and more), and relatively strong in taste. The oil is used in food products as well as in soap manufacture and in industry. "Palm-kernel oil" is a different

Oil palm (*Elaeis guineensis*), from Köhler (1883–1914).

oil, which is extracted from the nut or kernel. It is nearly colorless or yellowish white, similar to coconut oil, higher in saturated fats but milder tasting than palm oil. Palm-kernel oil is used in food products and in the manufacture of soaps, detergents, and cosmetics. In the early years of the 21st century, about 30 million tons of palm oil are produced annually worldwide, about 10% of it constituting palm-kernel oil.

Culinary Portrait

Palm oil is used in margarine, vegetable shortening, and cooking fat, while palm-kernel oil is used in the manufacture of margarine and other edible fats, ice cream, and mayonnaise. In West Africa, palm oil is used in many prepared foods, including soups and sauces; for frying; as an ingredient in dough made from starchy foods such as cassava, rice, plantains, yams, and beans; and as a condiment or flavoring for bland dishes such as fufu (prepared from cassava). Palm wine (often called "toddy") is made from the sap, which is obtained by tapping the male flowering branches. The central soft center of the top of the stem is called a "cabbage" and is edible.

Palm oil stores indefinitely in a cool place, but solidifies in the container because of the high degree of saturation. The oil can be liquified by placing the container in a pan of warm water for a few minutes. Because so-called "tropical oils," mainly palm oil and coconut oil, are so high in saturated fats, they have been accused of raising cholesterol levels in the body and thereby jeopardizing human health.

Curiosities of Science and Technology

- Oil palms are frequently grown in some of the largest plantations on earth, with plantings often exceeding 20 000 ha (49 000 acres) in size.

Oil palm (*Elaeis guineensis*) in Brazil, from *Flores des Serres* (1845–1880, vol. 14).

Unfortunately, huge areas of rainforest in tropical Asia are destroyed annually in order to plant even more.

- The oil palm is the highest yielding of oil-bearing plants, with an average of between 4 and 10 metric tons of oil per hectare per year (1.8–4.5 tons per acre per year).
- The heraldic shield of Christopher Columbus (1451–1506) displayed palms, symbolic of his tropical voyages.
- The coat of arms of Sierra Leone bears a palm tree, symbolizing the importance of palm oil to the economy of the country.
- "Palm oil" once meant bribes, or money for bribes and related activities. The allusion here is to "lubricating" the palm of the hand with money.

Key Information Sources

Basiron, Y., Jalani, B.S., and Chan, K.W. (*Editors*). 2000. Advances in oil palm research. Malaysian Palm Oil Board, Ministry of Primary Industries, Kuala Lumpur, Malaysia. 2 vols.

Bek-Nielsen, B. 1974. Technical and economic aspects of the oil palm fruit processing industry. United Nations, New York, NY. 40 pp.

Berger, K.G. 1983. Palm oil. *In* Handbook of tropical foods. *Edited by* H.T. Chan, Jr. Marcel Dekker Inc., New York, NY. pp. 433–468.

Berger, K.G. 1986. Food uses of palm oil. Revised edition. Palm Oil Research Institute of Malaysia, Kuala Lumpur, Malaysia. 30 pp.

Berger, K.G., and Martin, S.M. 2000. Palm oil. *In* The Cambridge world history of food. *Edited by* K.F. Kiple and K.C. Ornelas. Cambridge University Press, Cambridge, UK. pp. 397–411.

Corley, R.H.V. 1986. Oil palm. *In* CRC handbook of fruit set and development. *Edited by* S.P. Monselise. CRC Press, Boca Raton, FL. pp. 253–259.

Corley, R.H.V., Hardon, J.J., and Wood, B.J. (*Editors*). 1976. Oil palm research. Elsevier Scientific Publishing Company, Amsterdam, Netherlands. 532 pp.

Farooqi, A.A., and Sreeramu, B.S. 1999. Oil palm. *In* Tropical horticulture, vol. 1. *Edited by* T.K. Bose, S.K. Mitra, A.A. Farooqui, and M.K. Sadhu. Naya Prokash, Calcutta, India. pp. 578–591.

Food and Agriculture Organization of the United Nations. 1977. The oil palm. FAO, Rome, Italy. 40 pp.

Gascon, J.P., Noiret, J.M., and Meunier, J. 1989. Oil palm. *In* Oil crops of the world. *Edited by* G. Röbbelen, R.K. Downey, and A. Ashri. McGraw-Hill, New York, NY. pp. 475–493.

Gunstone, F.D. 1987. Palm oil. Wiley, New York, NY. 100 pp.

Hardon, J.J. 1995. Oil palm. *In* Evolution of crop plants. 2nd edition. *Edited by* J. Smartt and N.W. Simmonds. Longman Scientific & Technical, Burnt Mill, Harlow, Essex, UK. pp. 395–399.

Hardon, J.J., Rajanaidu, N., and Vosen, H.A.M. van der. 2001. *Elaeis guineensis* Jacq. *In* Plant resources of South-East Asia, 14, vegetable oils and fats. *Edited by* H.A.M. van der Vossen and B.E. Umali. Backhuys Publishers, Leiden, Netherlands. pp. 85–93.

Hardon, J.J., Rao, V., and Rajanaidu, N. 1985. A review of oil palm breeding. *In* Progress in plant breeding 1. *Edited by* G.E. Russell. Butterworths, London, UK. pp. 139–163.

Hartley, C.W.S. 1988. The oil palm (*Elaeis guineensis* Jacq.). 3rd edition. Longman, New York, NY. 761 pp.

Kritchevsky, D., Loke, K.H., and Chandra, R.K. (*Editors*). 1992. Health and nutritional aspects of palm oil: selected papers from the Palm Oil Research Institute of Malaysia palm oil international conference, Kuala Lumpur, Malaysia, 12–14 Sept. 1991. Pergamon Press, New York, NY. 232 pp.

Malaysia Institut Penyelidikan Perhutanan. 1994. Utilisation of oil palm tree and other palms, proceedings of the third national seminar, 1994, Malaysia. Forest Research Institute, Kuala Lumpur, Malaysia. 306 pp.

Moll, H.A. 1987. The economics of oil palm. Pudoc, Wageningen, Netherlands. 288 pp.

Ng, S.K. 1972. The oil palm: its culture, manuring, and utilisation. International Potash Institute, Berne, Switzerland. 142 pp.

Onwudike, O.C. 1996. Oil palm (*Elaeis guineensis* Jacq.). *In* Food and feed from legumes and oilseeds. *Edited by* J. Smartt and E. Nwokolo. Chapman and Hall, London, UK. pp. 318–330.

Piggott, C.J. 1990. Growing oil palms, an illustrated guide. The Incorporated Society of Planters, Kuala Lumpur, Malaysia. 152 pp.

Tan, K.S. 1983. The botany of oil palm. The Incorporated Society of Planters, Kuala Lumpur, Malaysia. 32 pp.

Turner, P.D. 1981. Oil palm diseases and disorders. Oxford University Press, Oxford, UK. 280 pp.

Turner, P.D., and Gillbanks, R.A. 1974. Oil palm cultivation and management. The Incorporated Society of Planters, Kuala Lumpur, Malaysia. 672 pp.

Zeven, A.C. 1967. The semi-wild oil palm and its industry in Africa. PUDOC, Wageningen, Netherlands. 180 pp.

Zeven, A.C. 1972. The partial and complete domestication of the oil palm (*Elaeis guineensis*). Econ. Bot. **26**: 247–249.

OLIVE

Family: Oleaceae (olive family)

Names

Scientific name: *Olea europaea* L.

- The English word "olive" and the genus name *Olea* are derived from the Greek *elaia*, olive or olive tree.
- "Russian olive" is *Elaeagnus angustifolia* L. "Autumn olive" is *Elaeagnus umbellata* Thunb. These Old World species (not related to the true olive) have been planted as ornamentals in North America, and sometimes escape to become established as wild-growing plants. Russian olive fruit is occasionally used as a beverage base. The fruit of autumn olive is often used in southeastern Europe like the fruit of the currant (which see), both as a fresh fruit and like raisins.
- "Chinese olive" refers to species of *Canarium* [including Chinese white olive, *C. album* (Lour.) Raeusch.; and Chinese black olive, *C. pimela* Leenh.], with edible fruits (or "nuts").
- *Europaea* in the scientific name *O. europaea* means European.

Plant Portrait

The olive is a principal food plant, as well as a source of lamp oil. The olive tree is an evergreen broad-leaved tree that grows to a height of 7.6–15 m (20–40 feet) and begins to bear fruit between 4 and 8 years of age. Olive trees in well-kept orchards are frequently trained so that they are stout, with massive trunks. Olives are picked before full maturity (when they are green) for use as a food or relish, and when fully mature (and black), they are crushed for salad oil. Black olives are not necessarily mature: methods are available to treat green olives in an aerated lye solution, causing the olives to oxidize and turn an attractive black color. Fresh olives are inedible because they contain an extremely bitter chemical (glycoside) that irritates the digestive tract. The bitterness is removed by chemical treatment and (or) soaking in brine. Typically, treatment with lye or soda is first carried out, followed by pickling, i.e., immersion in a brine or a salt and vinegar mixture, sometimes with spices. Olives for pickling are often twice the size of olives used for oil; pickling olives are up to 2.5 cm (1 inch) in diameter and 3.8 cm (1½ inches) in length.

The olive plant originated in the eastern Mediterranean area, and has been cultivated since ancient times. The trees have been grown in Greece for at least 4000 years, and like wine, were a foundation of ancient Greek civilization. Olives are now cultivated commercially in all countries surrounding the Mediterranean Sea and elsewhere, particularly in South America, Australia, South Africa, China, and in hot interior valleys of California. Spain, Italy, Greece, and Portugal are the largest producers. There are over 600 cultivars, and most countries have their own unique varieties. In California, Manzanillo and Mission are the major varieties, followed by Sevillano, Ascolano, and a few others. Many trees continue to give good yields for hundreds of years.

Olives are frequently grown as ornamental trees in Mediterranean or desert climates, as they have attractive, silver–green foliage, full canopies, and require very little water and maintenance. Unfortunately, the fruits produce black–purple oily stains on sidewalks and cars. Olive pollen is highly allergenic. Some cities such as Tucson, Arizona, have ordinances against planting of cultivars that produce large amounts of pollen. Cultivars are available that produce no pollen, and this is particularly valuable in Arizona, as many people with allergies have moved there.

Culinary Portrait

The pickled olive is well known as an appetizer, garnish, and the classic accompaniment of a martini. Fully grown but still green olives, when pickled, produce a pleasant, tart flavor. Ripe (dark) olives develop 10–30% oil, making them more nutritious than green olives, but also producing a mellower taste after pickling. Olives may be pickled whole, but are often pitted and stuffed with pimento (sweet pepper), anchovy, onion, or other foods and spices. Olives are added to salads, pizza, and numerous meat and poultry dishes. Once opened, a jar of olives should be refrigerated.

The chief importance of the olive is as a source of salad and cooking oil. About 90% of the world's olive production is used to make oil. Olive oil combines well with tomatoes, onions, garlic, basil, and other fresh herbs, and is extensively used in cooked dishes. Although the smoke point is too low to make olive oil

Olive (*Olea europaea*), from Köhler (1883–1914).

ideal for frying, olive oil is extensively used as a cooking oil. It is the primary cooking oil of Greece and much of the Middle East. By general consensus, olive oil is the best tasting of all vegetable oils. "Extra Virgin" olive oil is the juice produced from the first pressing of the olives. It is the most expensive, best for salad oil, and the least desirable for frying. Extra-virgin olive oil ranges in color from champagne through greenish golden to bright green, and generally, the deeper the color, the more intense the olive flavor. Extra-virgin olive oil is not always genuine. "Virgin" olive oil is also a first-press oil, but is somewhat more acidic. "Fino" (Italian for fine) olive oil is a blend of extra-virgin and virgin oils. Products simply labeled "olive oil" (once called "pure olive oil") contain a combination of refined olive oil and virgin or extra-virgin oil. "Extra Light" olive oil is lighter in color and aroma, generally not in calories. Light olive oil has little of the classic olive oil flavor, and is useful for baking and cooking where an olive oil's obvious flavor is undesirable. The filtration process for light olive oil gives it a higher smoke point than regular olive oil, and therefore, it can be used for high-heat frying, whereas regular olive oil is more suited to low- and medium-heat cooking.

Olive oil can be stored in a cool, dark place for up to 6 months (light fades it, heat hastens rancidity). Refrigerated, it lasts up to a year. Chilled olive oil becomes clouded and is too thick to pour, but it will become clear and pour readily when returned to room temperature. The oil easily picks up foreign odors. To prevent this, it should be kept in a tightly closed container.

Olive oil is very high in oleic acid, a monounsaturated fat that is very beneficial for humans. Olive oil seems to reduce the amount of low-density lipoproteins (LDL), which carry cholesterol through the body for cell-building needs, but leave excess cholesterol in arterial walls, endangering health.

Olive tree in the Garden of Gethsemane. This tree stands at the base of the Mount of Olives near Jerusalem and is known as "The Tree of Agony." It is popularly supposed to have witnessed the vigil of Jesus Christ. From *Garden and Forest* (1888, vol. 1).

Olive grove in Italy, from Marilaun (1895).

Culinary Vocabulary

- A "beef olive" is a cookery term that originated in Medieval England, and has nothing to do with olives. A beef olive is made of thick slices of beef rolled with onions, breadcrumbs, and herbs and stewed or braised. The "olive" in a beef olive is based on the Old French *alou*, lark, because of the supposed resemblance of the stuffed roll to the bird, especially headless.
- "Bruschetta" is a slice of bread that is grilled or toasted on both sides and consumed warm with olive oil and often a savory garnish such as garlic. Bruschetta originated in Italy as a way of testing the quality of the season's olive oil. The name arises from the original cooking method, reflecting the Italian verb *bruscare*, "to roast over coals."
- The "muffuletta" is an open sandwich that was created by Italian blue-collar workers in the New Orleans markets in the late 1800s. A mixture of broken green and black olives from the bottom of olive barrels was placed on half a loaf of round Italian bread, commonly called a "muff." Over this were placed slices of ham, salami, and Provolone cheese.
- "St. Lawrence Dressing" is a Canadian salad dressing prepared with olive oil, green olives, orange juice, lemon juice, onions, paprika, Worcestershire sauce, parsley, and mustard.
- A "dirty martini" is a cocktail prepared with gin, dry vermouth, and a little brine from bottled olives, along with an olive as a garnish.

Curiosities of Science and Technology

- The ancient Egyptians are thought to have used olive oil as a lubricant to help move the large blocks of stones they used to construct the pyramids.
- Winners in the Olympian Games were crowned with an olive wreath. Ancient Olympia, the birthplace of the Olympic Games, is overgrown with olive trees, some probably tracing their heritage back to the times when branches from the nearest trees were harvested for making wreaths of victory.
- Olive trees were sacred in ancient Athens, and death was the penalty for cutting one down.
- In ancient Greece, a crown of olive leaves was hung on the door to announce the birth of a boy, and a band of wool if it was a girl.
- The classical Greeks used olive oil as a kind of lubricant in place of soap, covering their bodies and using scraping instruments to remove the oil along with accumulated dirt.
- Some olive trees in the eastern Mediterranean are estimated to be over 2000 years old. In Athens, an olive tree dedicated to the Greek philosopher Plato

(about 427–347 BC) is still alive today, a natural monument around 2500 years old.

- "Olive" has been used as a color term, indicating dull yellowish green, since the 17th century. The color of American army uniforms has been officially described as "olive drab."
- In an effort to cut corners, American Airlines eliminated one olive from each salad served in First Class during 1987. They saved $40,000 that year.
- There are about 800 million olive trees throughout the world. Greece and Italy each have over 100 million olive trees.
- A ton of olives produces about 200 L (53 American gallons) of oil.
- Removing an olive pit before pickling is said to diminish the flavor of the olive.
- Darker olives have higher oil and richer flavor.
- The life of olive oil can be extended by adding a cube of sugar to the bottle.
- When people with glasses fry foods with a cooking oil like olive oil, droplets of oil often accumulate on the glasses—indeed, primarily on the inside of the lenses. Harold McGee reported that experiments he conducted (*The Curious Cook*, 1990, Macmillan Publishing, New York) showed that this occurs because cooks always look down at the stovetop, and the rain of oil droplets produced during frying falls downward onto the inner lenses. A baseball cap was found to do a good job of protecting glasses while frying.

Key Information Sources

Baldoni, L., Guerrero, C., Sossey-Aloui, K., Abbott, A.G., Angiolillo, A., and Lumaret, R. 2002. Phylogenetic relationships among *Olea* species, based on nucleotide variation at a non-coding chloroplast DNA region. Plant Biol. (Stuttgart), **4**: 346–351.

Bartolini, G., Petruccelli, R., and Tindall, H.D. 2002. Classification, origin, diffusion and history of the olive. Food and Agriculture Organization of the United Nations, Rome, Italy. 74 pp.

Bartolini, G., Prevost, G., Messeri, C., and Carignani, G. 1999. Olive cultivar names and synonyms and collections detected in a literature review. Acta Hortic. **474**: 159–162.

Belaj, A., Satovic, Z., Rallo, L., and Trujillo, I. 2002. Genetic diversity and relationships in olive (*Olea europaea* L.) germplasm collections as determined by randomly amplified polymorphic DNA. Theor. Appl. Genet. **105**: 638–644.

Besnard, G., and Berville, A. 2000. Multiple origins for Mediterranean olive (*Olea europaea* L. ssp. *europaea*) based upon mitochondrial DNA polymorphisms. Comptes Rendues Acad. Sci. Paris (Sciences de la Vie, France), **323**: 173–181. [In French, English summary.]

Besnard, G., Khadari, B., Baradat, P., and Berville, A. 2002. *Olea europaea* (Oleaceae) phylogeography based on chloroplast DNA polymorphism. Theor. Appl. Genet. **104**: 1353–1361.

Boscou, D. 1996. Olive oil: chemistry and technology. AOCS Press, Champaign, IL. 161 pp.

Claros, M.G., Crespillo, R., Aguilar, M.L., and Canovas, F.M. 2000. DNA fingerprinting and classification of geographically related genotypes of olive-tree (*Olea europaea* L.). Euphytica, **116**: 131–142.

Fabbri, A., Hormaza, J.I., and Polito, V.S. 1995. Random amplified polymorphic DNA analysis of olive (*Olea europaea* L.) cultivars. Am. Soc. Hortic. Sci. **120**: 538–542.

Ferguson, L., Sibbett, G.S., and Martin, G.C. 1994. Olive production manual. University of California, Division of Agriculture and Natural Resources, Oakland, CA. 156 pp.

Food and Agriculture Organization of the United Nations. 1977. Modern olive production. FAO, Rome, Italy. 251 pp.

Garrido Fernandez, A., Fernandez Diez, M.J., and Adams, M.R. 1997. Table olives, production and processing. Chapman & Hall, London, UK. 495 pp.

International Olive Oil Council. 1997. World encyclopedia of the olive tree. Plaza & Janès, Barcelona, Spain. 479 pp.

King, J.R. 1938. Morphological development of the olive. Hilgardia, **11**: 437–458.

Kiritsakis, A., and Markakis, P. 1987. Olive oil: a review. Adv. Food Res. **31**: 453–482.

Kiritsakis, A.K. 1998. Flavor components of olive oil: a review. J. Am. Oil Chem. Soc. **75**: 673–681.

Lavee, S. 1986. Olive. *In* CRC handbook of fruit set and development. *Edited by* S.P. Monselise. CRC Press, Boca Raton, FL. pp. 261–276.

Martinez Suarez, J.M., and Martínez Moreno, J.M. 1975. Manual of olive-oil technology. Food and Agriculture Organization of the United Nation, Rome, Italy. 164 pp.

Metzidakis, I.T., and Voyiatzis, D.G. (*Editors*). 1999. Proceedings of the third international symposium on olive growing, Chania, Crete, Greece, 12–26 Sept. 1997. International Society for Horticultural Science, Leuven, Belgium. 2 vols.

Moutier, N., and Vossen, H.A.M. van der. 2001. *Olea europaea* L. *In* Plant resources of South-East Asia, 14, vegetable oils and fats. *Edited by* H.A.M. van der Vossen and B.E. Umali. Backhuys Publishers, Leiden, Netherlands. pp. 107–112.

Newmark, H.L. 1997. Squalene, olive oil, and cancer risk: A review and hypothesis. Cancer Epidem. Biomark. Prevent. **6**: 1101–1103.

O'Keefe, S.F. 2000. An overview of oils and fats, with a special emphasis on olive oil. *In* The Cambridge world history of food. *Edited by* K.F. Kiple and K.C. Ornelas. Cambridge University Press, Cambridge, UK. pp. 375–388.

Pallotta, U. 1994. A review of Italian research on the genuineness and quality of extra virgin olive oil. Ital. J. Food Sci. **6**: 259–274.

Pasqualone, A., Caponio, F., and Blanco, A. 2001. Inter-simple sequence repeat DNA markers for identification of drupes from different *Olea europaea* L. cultivars. Eur. Food Res. Technol. **213**: 240–243.

Poli, M. 1979. Literature review on the physiology of alternate bearing in the olive (*Olea europaea* L.). Fruits (Paris), **34**: 687–695. [In French, English summary.]

Rugini, E., and Lavee, S. 1992. Olive. *In* Biotechnology of perennial fruit crops. *Edited by* F.A. Hammerschlag and R.E. Litz. CAB International, Wallingford, UK. pp. 371–382.

Sensi, E., Vignani, R., Scali, M., Masi, E., and Cresti, M.. 2003. DNA fingerprinting and genetic relatedness among cultivated varieties of *Olea europaea* L. estimated by AFLP analysis. Sci. Hortic. **97**: 379–388.

Sibbett, G.S., and Connell, J. 1994. Producing olive oil in California. Cooperative Extension, University of California, Division of Agriculture and Natural Resources, Oakland, CA. 11 pp.

Spennemann, D.H.R., and Allen, L.R. 2000. Feral olives (*Olea europaea*) as future woody weeds in Australia: a review. Aust. J. Exp. Agric. **40**: 889–901.

Tombesi, A. 1994. Olive fruit growth and metabolism. Acta. Hortic. **356**: 225–232.

Vissers, M.N., Zock, P.L., and Katan, M.B. 2004. Bioavailability and antioxidant effects of olive oil phenols in humans: a review. Eur. J. Clin. Nutr. **58**: 955–965.

Vitagliano, C., and Martelli, G.P. (*Editors*). 2001. Proceedings of the 4th international symposium on olive growing, Valenzano, Italy, 25–30 Sept. 2000. International Society for Horticultural Science, Leuven, Belgium. 2 vols.

Zohary, D. 1995. Olive (*Olea europaea*). *In* Evolution of crop plants. 2nd edition. *Edited by* J. Smartt and N.W. Simmonds. Longman Scientific & Technical, Burnt Mill, Harlow, Essex, UK. pp. 379–382.

Specialty Cookbooks

(Note: Over 100 cookbooks on olive oil have been published. The following are examples.)

Antol, M.N. 2004. The sophisticated—olive: the complete guide to olive cuisine. Square One Publishers, Garden City Park, NY. 193 pp.

Aris, P. 2000. Olives. Southwater, London, UK. 64 pp.

California Olive Industry (Association). 1983. The itinerant ripe olive: a guide to the use of California ripe olives. California Olive Industry, Fresno, CA. 31 pp.

California Olive Industry (Association). 1994. International recipes with California ripe olives. California Olive Industry, Fresno, CA. 23 pp.

Dolamore, A. 1989. The essential olive oil companion. Salem House Publishers, Topsfield, MA. 160 pp.

Ferguson, C., and Cassidy, P. 2000. Extra virgin: cooking with olive oil. Ryland Peters & Small, New York, NY. 144 pp.

Flanagan, W., and Hildenbrand, R. 2003. Olives & oils in South Africa. Mbira Press, Hout Bay, South Africa. 87 pp.

Klein, M.B. 1983. The feast of the olive: cooking with olives & olive oil. Aris Books, Berkeley, CA. 223 pp.

Knickerbocker, P. 1997. Olive oil: from tree to table. Chronicle Books, San Francisco, CA. 167 pp.

Krasner, D. 2002. The flavors of olive oil: a tasting guide and cookbook. Simon & Schuster, New York, NY. 226 pp.

Lindsay International. 1985. Simply elegant: entertaining ideas from Lindsay olives. Lindsay International, Lindsay, CA. 96 pp.

Lorenz Inc. 1996. Olives: a book of recipes. Lorenz, New York, NY. 64 pp.

Reichelt, K., and Burr, M. 1997. Extra virgin: an Australian companion to olives and olive oil. Wakefield Press, Kent Town, SA, Australia. 156 pp.

Rogers, F., St. Clair, L., and Cenicola, T. 2002. Olives: cooking with olives and their oils. Ten Speed Press, Berkeley, CA. 128 pp.

Schlesinger, S., and Earnest, B. 1990. The low-cholesterol olive oil cookbook: more than 200 recipes, the most delicious way to eat healthy food. Villard Books, New York. 307 pp.

Seed, D. 1995. Italian cooking with olive oil. W. Morrow, New York, NY. 152 pp.

Spieler, M. 1998. Olives: cooking with the olive and its oil. Chartwell Books, Edison, NJ. 128 pp.

Onion

Family: Liliaceae (lily family; sometimes placed in Alliaceae, the onion family)

Names

Scientific names: *Allium* species

- Common onion—*A. cepa* L.
- Welsh onion, Japanese bunching onion—*A. fistulosum* L.
- Egyptian onion—*A.* ×*proliferum* (Moench) Schrad. ex Willd. (*A. fistulosum* × *A. cepa*)

- The word "onion" has been traced through the French *oignon* to the late Latin *union*, onion, said to be from the Latin *unus*, one, because of the single, perfectly shaped onion bulb, which contrasts with the multitude of cloves in a garlic bulb.
- "Spanish onions" generally refers to large, very mild onions. "Bermuda onions" are also large, mild onions. In the United States, the phrase "Spanish onions" often refers to large, mild, reddish onions.
- The name "Welsh onion" derives from the German *Welsche*, meaning "foreign," or "of southern Europe." This name may have been applied when it was first introduced to Germany in the Middle Ages. It is also possible that it was first used in England, where the Gauls or Welsh were considered foreigners, before Wales was established.
- The Welsh or Japanese bunching onion has also been called cibol and stone leek.
- The Egyptian onion is known by many other names—Catawissa onion, garden rocambole, top onion, topset onion, tree onion, and walking onion. The origin of the name "Egyptian onion" is unclear. It is called "top onion" or "tree onion" because of the bulbils that form on top of stem. It has been called the "garden rocambole" in the United States, but this should not be confused with the rocambole form of garlic. In the United States, the Egyptian onion is sometimes called "walking onion." This refers to the interesting way that the plant propagates vegetatively. Bulbils form on top of a stem, sprout, and when the stem bends over to ground level under the increased weight of the head, the bulbils root, thus taking a "step" away from the base of the parent plant. In Japan, because of its peculiar characteristic of developing bulbils, it is called *kitsune negi*, meaning "foxy" or "mysterious" onion.
- The words "shallots" and "scallions" are explained below, and contrasted with "leeks" and "green onions." Both "scallion" and "shallot" are modifications of the Latin *Ascalonia*, based on Acalon, a port on the coast of Palestine from which onions were exported.
- For information on the genus name *Allium*, see Garlic.
- *Cepa* in the scientific name of the common onion, *A. cepa*, is the classical Latin word for onion.
- *Fistulosum* in the scientific name of the Welsh onion, *A. fistulosum*, means hollow, referring to the hollow, tube-like leaves.
- *Proliferum* in the scientific name of the Egyptian onion, *A.* ×*proliferum*, is Latin for "producing offsets."

Plant Portrait

The common onion is a biennial herbaceous plant grown as an annual. It is unknown in the wild but is presumed to have been selected in the Middle East and parts of Asia. It has a wide variation of bulb characteristics, reflecting extensive selection during domestication. The bulb is a storage organ responsible for vegetative (asexual) reproduction. It consists of condensed stem tissue bearing numerous concentric layers made up of fleshy, juicy, whitish scale leaves with buds in the axils. The bulbs are usually globose to pear-shaped, and are covered with several layers of paper-thin skin. Skin and flesh of the bulbs take many forms, and varieties differ in shape, size, and flavor. The skin may be white, purple, brown, yellow, or red, depending on variety. "Spanish onions" are considered especially mild in flavor. Some types are grown for their edible leaves or for small bulbs that are produced by the flowers. Onions have been cultivated for more than 4000 years. They appear in pyramid wall carvings from the third and fourth dynasties of Egypt (about 2700 BC). The ancient Greeks and Romans used onions. Onions had reached most of Europe by the Middle Ages, and the Spaniards brought them to North America. Today, onions are grown in well over 100 countries.

Onions begin to form bulbs when the day length is of the proper duration for that variety. Different varieties of onions require different day lengths to initiate bulb formation. Accordingly, onion varieties that are grown in the South can usually not be grown in the North and vice versa. Long-day (i.e., long duration of daylight) onions are bred for best performance in the North, and short-day varieties perform best in southern locations. Short-day varieties may develop in the North if the plants are set out very early in the season. However, long-day types may not develop bulbs in the South and so should be avoided there.

Many people confuse shallots, green onions, scallions, and leeks. All but the leek are members of the onion species, *A. cepa*. Green onions have a definite bulb with the typical concentric arrangement that the dry onion has, although they are mostly grown for their edible leaves. Scallions are shoots from the white onion varieties that are pulled *before* the bulb has formed. Leeks (mostly from the species *A. ampeloprasum* L.) are similar in appearance to scallions but have flat leaves, and the white stalk is typically thicker and longer ("shallot" is occasionally used to designate leeks—a confusing and undesirable usage). The shallot is distinguished by its distinctive bulbs, which are made up of a clump of cloves (or small bulbs), reminiscent of garlic. Unlike garlic, the individual bulbs are not closely compacted together by a common membrane. "Potato onions" or "multiplier onions" are a type of shallot, with larger main bulbs and fewer offspring bulbs, which are found in home gardens.

The Welsh or Japanese bunching onion is a perennial herbaceous plant, usually grown as an annual for its edible leaves or white leaf bases, or both. It is unknown in the wild but probably originated in northwestern China from an unknown progenitor. The leaves are circular in cross section, in contrast to the slightly flattened leaves of the common onion. The bases of the leaves overlap to form a neck (pseudostem). However, unlike the common onion *A. cepa*, the Welsh onion develops only a small bulb, with a diameter just slightly larger than that of the neck. The plant flowers earlier in the spring than the common onion, hence, one vernacular name, "spring onion." The Welsh onion has a long history of cultivation. The first written record of it may have been from 3rd century BC in China. However, the written character used was applied to many cultivated members of *Allium*. A definite description of Japanese bunching onion and its culture appeared in 100 BC in China. It was first mentioned in Japanese literature in 720, probably after arriving from China. The Welsh onion appeared in western Europe during or at the end of the Middle Ages and was introduced from there to Russia. Today, some types are widely cultivated from Siberia to tropical Asia. Forms that provide blanched pseudostems are grown in Asia. Other kinds grown only for the green portion of their leaves are often found in North American supermarkets. However, some "bunching onions" are forms of *A. cepa*, common onion. The Welsh or Japanese bunching onion is widely cultivated from Siberia to tropical Asia. In Japan, Korea, China, and Taiwan, it ranks among the top 10 commercial vegetables and is found in markets year-round. The leek-like basal portions are blanched (whitened by protection from the sun), and the green tops are harvested all year.

The Egyptian onion is a perennial herbaceous plant, 0.6–1.0 m (2–3 feet) tall, found only in cultivation. It is a natural hybrid between common onion, *A. cepa*, and Welsh onion, *A. fistulosum*. The place of origin of Egyptian onion is uncertain. It was first noted in Europe in 1587. It is now cultivated in North America, Europe, and Asia. This hybrid plant is sterile, with most or all of the flowers in the inflorescence developing into bulbils or topsets about the size of hazelnuts. Normal flowers, if they develop, are sterile. The bulbils are initially green, and become a brownish red color. These may sprout while still on the mother plant, particularly in the form called "Catawissa onion." The sideshoots (offsets) of Egyptian onion are eaten raw or cooked. It is uncertain if Egyptian onion is grown commercially anywhere in the world. Basically, it is grown as an oddity of home gardens.

Culinary Portrait

Onions are employed to prepare countless dishes, both hot and cold. The bulbs as well as the chopped leaves are used raw or cooked in many vegetable dishes to which they impart an added zest. Onions are boiled, fried, stewed, baked, creamed, roasted, and pickled. They are deep-fried as onion rings. Onions are both a vegetable (i.e., eaten in considerable quantity) and a condiment or spice (i.e., used in small quantities as a flavorant).

Various techniques are used to reduce the irritating, tear-producing vapors of onions. The tears are due to the rupture of the onion's tissues, which release sulfurous molecules that react with air to create the irritant chemical allyl sulfate. Using a sharp knife and keeping the face as far away as possible, or wearing goggles or eyeglasses can all be helpful. Some cooks cut onions under a stream of cold water, or cool onions for an hour in the refrigerator or 15 minutes in the freezer before cutting. Blanching for several minutes in boiling water, followed by soaking in cold water immediately afterwards to stop the cooking process, moderates the sharpness of raw onions, and indeed, onions become milder and sweeter when cooked.

Varieties of common onion (*Allium cepa*), from Vilmorin-Andrieux (1885).

Onion juice is released when onions are cut, and may be absorbed by countertops and wooden cutting surfaces, as well as by hands. Lemon juice or vinegar is sometimes used to remove this juice. Onion bulbs should not be stored in the refrigerator, as their odor tends to be absorbed by other foods (green onions can be kept in the refrigerator for a week). Onions will keep for at least several weeks in a well-ventilated, cool, dry place. Sharp onions tend to keep longer than mild onions.

When purchasing onion bulbs, it is best to select those that are firm, with a dry, smooth outer skin, and no sign of sprouting or mold. However, onions in stores may have been treated by irradiation to prevent sprouting. Dried onion flakes and powder (such as onion salt, which is a mixture of dried powdered onions and salt) are also available for flavoring purposes. Onion powder often has too much salt to be recommended for culinary purposes.

Welsh or Japanese bunching onions are used in soups and stews, meat, poultry, and seafood dishes, stir-fries, and as a cooked vegetable. The green leaves are cooked or consumed raw in soups and salads, and used as a garnish on a variety of dishes including quiches and omelettes. As with chives, it is best to add this type of onion at the end of the cooking cycle. The flavor is milder than that of the common onion, but stronger than chives. The leaves can be used as a substitute for chives, if used in smaller amounts. The white shaft at the base can be used as a substitute for onions. This kind of onion keeps for several days in the vegetable drawer of a refrigerator.

The new green tips of Egyptian onion can be used as a garnish like chives. The white lower stalks can be chopped raw, like common onions, for salads. The young stalks, when thinned from the parent plant, can be consumed, like scallions. The coarser main stalks or fat hollow leaves are sometimes gathered when tender in the spring, slit, and stuffed with cottage cheese. The bulbils (topsets) that form in the inflorescence are cooked or eaten raw. However, they have a strong flavor and should be used sparingly. The bulbils are also pickled, as they have a richer flavor than other onions.

Culinary Vocabulary

- A "boiler onion" is a small, tender, white, mild onion that is usually used in casseroles, stews, or soups, or creamed.
- *À la lyonnaise* (pronounced ah lah lee-oh-NEHZ), named for the cuisine of Lyon, France, is a French method of preparing or garnishing dishes with onions. *Sauce lyonnaise* is made from a demi-glaze flavored with white wine and sautéed onions, and is typically served with poultry and meat.
- *À la boulangère* (pronounced ah lah bu-lahn-jehr, French for "in the style of the baker's wife") is a French method of preparing meats, especially lamb, with the meat roasted on a bed of potatoes and onions.
- "Onion sauce" is an English sauce prepared with onions cooked in butter and cream, and usually served with shoulder of lamb.
- In old American lunch-counter and diner jargon, "take a flower" means "with onions." "Breath" is old lunch-counter slang for onion. "Put out the lights and cry" meant liver and onions. "Two cows, make them cry" meant two hamburgers with onions.
- "Beef Stroganoff" is a dish of strips of beef sautéed in butter typically served with onions or shallots, mushrooms, wine, and sour cream. The Stroganoff responsible for the name is uncertain, either Count Paul Stroganoff, a 19th century Russian diplomat, or another 19th century count, Alexander Grigorievich Stroganov, who lived in the Black Sea port of Odessa and became known for entertainment.
- "Soubise" in English means a white sauce containing onion, while in French, it refers to a preparation of onion, for example, an onion purée. The word commemorates Charles de Rohan, Prince de Soubise (1715–1787), a French general and courtier.

Welsh onion (*Allium fistulosum*), from Vilmorin-Andrieux (1885).

- "Skunk egg" is a cowboy name for an onion.
- The Gibson cocktail is traditionally garnished with a pearl onion (which are produced both from onions and leeks). The cocktail was named after American illustrator Charles Dana Gibson (1867–1944), famous for his drawings of the "Gibson Girl," because (it is claimed) he asked for tiny white onions in his Martini cocktail when a New York restaurant ran out of olives. (Other beverages calling for a cocktail onion are the Gent of the Jury, Patton Martini, Pan Galactic Gargle Blaster #2, and the Yellow Rattler.)
- A "rope of onions" is a customary way of referring to a group of onions (similarly, one refers to a rope or string of pearls).

Curiosities of Science and Technology

- Hammurabi was a King of Babylon who ruled Mesopotamia, and in the year 1750 BC, prepared a legal code consisting of 282 laws inscribed on stone. Hammurabi's Code was the most important of all Mesopotamian contributions to civilization. According to the code, the needy should receive a monthly ration of bread and onions.
- Onions were highly revered by the ancient Egyptians, who used them in religious ceremonies,

occasionally (it has been alleged by ancient writers) even worshipping them as deities, and refusing to eat them because they were divine. Oaths of office were sworn over an onion by the Egyptians, who believed that the layers forming spheres within a sphere symbolized eternity. This reverence for the symbolic onion shape is also reflected in Byzantine architecture, which frequently has onion domes in religious structures.

- The mummy of King Ramses IV, who died in 1160 BC, had small onions in the eye sockets. Onions have also been found in the pelvic region of mummies, in the thorax, and flattened against the ears.
- The classical Greeks used onions to fortify athletes for the Olympic Games. Before competitions, athletes would consume great quantities of onions and drink onion juice, and even rubbed onions on their bodies.
- In the 18th century, the English explorer Captain James Cook (1728–1779) refused to sail to remote areas of the Pacific until each of his crew members had consumed about 14 kg (30 pounds) of onions within 3 days, as a precaution against scurvy. Similarly, during the American Civil War, General Ulysses S. Grant (1822–1885; 18th president of the United States) sent the following wire to the War Department during the summer of 1864: "I will not move my army without onions." The next day, three trainloads of onions were dispatched. Onions were considered essential for controlling dysentery and other ills.
- Pigments obtained from the scales of yellow onions were used in 19th century Germany to color Easter eggs. The skins of two red onions or yellow storage onions are enough to dye a dozen eggs.
- During the American Civil War (1861–1865), doctors in the Union army routinely used onion juice to clean gunshot wounds. (See Garlic for the comparable use of this herb as "Russian penicillin.")
- In 1952, the Lipton company started printing dip recipes on the back of its onion soup packages. The result was a huge surge in popularity of chips and dip as a party food.
- It is claimed that an apple, onion, and potato all have the same taste, the differences in flavor owing simply to their smell, and that one can prove this by asking someone to pinch their nose, take a bite from each, and identify them.
- Turks have the highest per capita yearly consumption of onions: over 36 kg (80 pounds).
- Onions are the third most popular vegetable in the United Sates, after potatoes and lettuce. The average American eats 8.3 kg (18.3 pounds) in a year.
- Why do onions cause tears? The volatile oils that help to give species of *Allium* their distinctive flavors contain a class of organic molecules known as amino acid sulfoxides. Peeling, cutting, or crushing an onion's tissue releases enzymes called allinases, which convert these molecules to sulfenic acids. The sulfenic acids, in turn, spontaneously rearrange to form a volatile gas (syn-propanethial-S-oxide), the chemical that triggers the tears. The gas mixes with the water in the eyes, forming a weak acid that causes the tear ducts to flood the eye to get rid of the irritant. The mechanism is very similar to that in garlic. Rubbing the eyes is a natural reaction, but if onion juices are present on the hands, it will probably worsen the irritation. Some solutions: cut the onion under water or run the tap over it as it is sliced; chill the onion in the refrigerator prior to cutting (this somewhat reduces the release of the gas); wear goggles.
- The tear-producing gases of an onion are concentrated at the root end. Exposure to these gases can be minimized somewhat by cutting the top of an onion off and peeling the skin downwards, and trimming off the root last.
- Placing a slice of cold onion on insect bites has been recommended to stop the pain and swelling.
- Rubbing with lemon juice or salt removes the smell of onions from hand and utensils.
- Water in which onions have been cooked has been used to clean brass-bottomed pots.
- According to the Guinness Book of World Records, the largest onion ever grown weighed 4.9 kg (10 pounds, 14 ounces), and was grown by V. Throup of Silsden, England.
- Some proposed treatments for onion breath: rinse the mouth with half lemon juice, half water; eat an apple; eat parsley; munch on roasted coffee beans; chew a citrus peel; chew an aniseed or dill seed; suck on a piece of cinnamon or a whole clove.
- The Vidalia onion, a yellow onion claimed to be "the world's sweetest onion," is commonly grown in Georgia, particularly around the Vidalia area. The Vidalia Onion harvest is valued at $50 million directly in Georgia, with related downstream marketing activities estimated at $150 million. In 1990, a resolution was passed by the Georgia legislature declaring the Vidalia Onion as Georgia's Official State Vegetable.

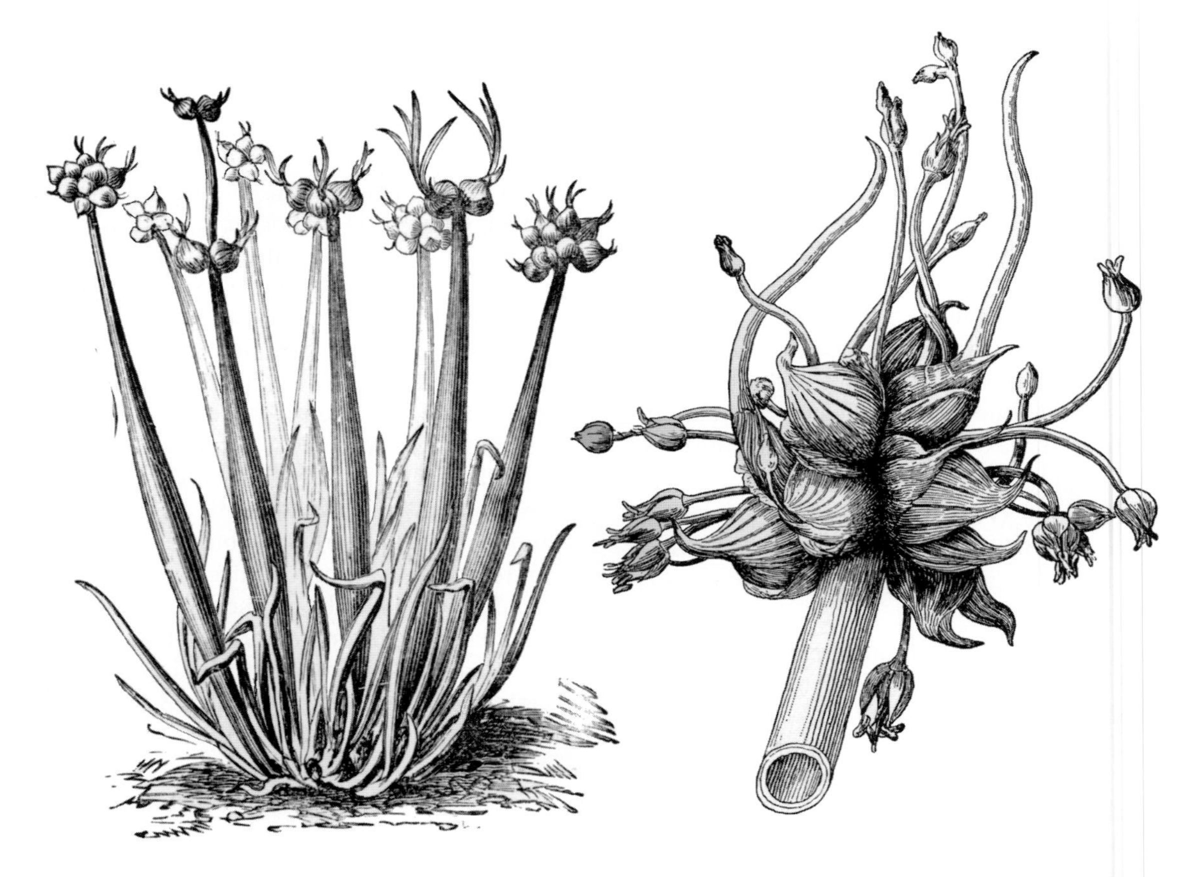

Egyptian onion (*Allium ×proliferum*). *Left*, from Vilmorin-Andrieux (1885); *right*, from Bailey (1900–1902).

- Jainism is a religious sect that originated in India in the 6th century BC. There are more than 2 million Jains, mostly in India. The vegetarian dietary code is demanding—no eating at night (animals must be respected, and tiny flying creatures could be inadvertently eaten), no consumption of parts of plants that are essential to their lives (such as roots), and no eating of any member of the onion family.
- "Onionskin" does not come from onions. It is a thin, strong, translucent paper of very light weight, the name reflecting the thin, translucent, outer wrapping of onions.

Key Information Sources

(See GARLIC for additional references.)

Abdalla, A.A., and Mann, L.K. 1963. Bulb development in the onion (*Allium cepa* L.) and the effect of storage temperature on bulb rest. Hilgardia, **35**: 85–112.

Brewster, J.L. 1977. The physiology of the onion. Hortic. Abst. **47**: 102–112.

Brewster, J.L. 1994. Onions and other vegetable alliums. CAB International, Wallingford, Oxon, UK. 236 pp.

Carson, J.F. 1987. Chemistry and biological properties of onions and garlic. Food. Rev. Int. **3**: 71–103.

Crête, R. 1981. Diseases of onions in Canada. Agriculture Canada, Ottawa, ON. 37 pp.

Currah, L., and Proctor, F.J. 1990. Onions in tropical regions. Bulletin No. 15. National Resources Institute, Chatham Maritime, UK. 232 pp.

Ford-Lloyd, B.V., and Armstrong, S.J. 1993. Welsh onion, *Allium fistulosum* L. *In* Genetic improvement of vegetable crops. *Edited by* G. Kalloo and B.O. Bergh. Pergamon Press, New York, NY. pp. 51–58.

Hanelt, P., Hammer, K., and Knüpffer, H. (*Editors*). 1992. The genus *Allium*—taxonomic problems and genetic resources. Institut Pflanzen. Kulturpflanzenfors, Gatersleben, Germany. 359 pp.

Havey, M.J. 1993. Onion, *Allium cepa* L. *In* Genetic improvement of vegetable crops. *Edited by* G. Kalloo and B.O. Bergh. Pergamon Press, New York, NY. pp. 35–49.

Havey, M.J. 1995. Onion and other cultivated alliums *In* Evolution of crop plants. 2nd edition. *Edited by* J. Smartt and N.W. Simmonds. Longman Scientific & Technical, Burnt Mill, Harlow, Essex, UK. pp. 344–350.

Hecke, E. van. 1977. Contact allergy to onion. Contact Dermatitis, **3**: 167.

Jones, H.A., and Mann, L.K. 1963. Onions and their allies. Botany, cultivation and utilization. Leonard Hill, London, UK. 286 pp.

Lawande, K.E. 2001. Onion. *In* Handbook of herbs and spices. *Edited by* K.V. Peter. Woodhead Publishing, Cambridge, UK. pp. 249–260.

Magruder, R. 1941. Descriptions of types of principal American varieties of onions. U.S. Dept. of Agriculture, Washington, DC. 87 pp.

Mathew, B. 1996. A review of *Allium* sect. *Allium*. Royal Botanic Gardens, Kew, UK. 176 pp.

Matson, W.E., Mansour, N.S., and Richardson, D.G. 1985. Onion storage: guidelines for commercial growers. Oregon State University Extension Service, Corvallis, OR. 15 pp.

Meer, Q.P. van der, and Leong, A.C. 1993. *Allium cepa* L. cv. Group Common Onion. *In* Plant resources of South-East Asia, 8, vegetables. *Edited by* J.E. Siemonsma and K. Piluek. Pudoc Scientific Publishers, Wageningen, Netherlands. pp. 68–713.

Meyer, C., and Meyer, D. 1993. Onions: condiment, nutrient, medicine. Meyerbooks, Glenwood, IL. 105 pp.

Moore, H.E., Jr. 1955. Cultivated Alliums. Baileya, **3**: 156–167.

Organization for Economic Cooperation and Development. 1984. Onions. Revised edition. OECD, Paris, France. 49 pp.

Oyen, L.P.A. 1993. *Allium fistulosum* L. *In* Plant resources of South-East Asia, 8, vegetables. *Edited by* J.E. Siemonsma and K. Piluek. Pudoc Scientific Publishers, Wageningen, Netherlands. pp. 73–77.

Permadi, A.H, and Meer, Q.P. van der. 1993. *Allium cepa* L. group Aggregatum. *In* Plant resources of South-East Asia, 8, vegetables. *Edited by* J.E. Siemonsma and K. Piluek. Pudoc Scientific Publishers, Wageningen, Netherlands. pp. 64–68.

Peterson, J. 2000. The *Allium* species (onions, garlic, leeks, chives, and shallots). *In* The Cambridge world history of food. *Edited by* K.F. Kiple and K.C. Ornelas. Cambridge University Press, Cambridge, UK. pp. 249–271.

Pike, L.M. 1986. Onion breeding. *In* Breeding vegetable crops. *Edited by* M.J. Bassett. AVI Publishing Co., Westport, CT. pp. 357–394.

Rabinowitch, H.D., and Brewster, J.L. (*Editors*). 1990*a*. Onions and allied crops. Vol 1. Botany, physiology, and genetics. CRC Press, Inc., Boca Raton, FL. 273 pp.

Rabinowitch, H.D., and Brewster, J.L. (*Editors*). 1990*b*. Onions and allied crops. Vol. II. Agronomy, biotic interactions, pathology, and crop protection. CRC Press, Inc., Boca Raton, FL. 320 pp.

Rabinowitch, H.D., and Brewster, J.L. (*Editors*). 1990*c*. Onions and allied crops. Vol. III. Biochemistry, food science, and minor crops. CRC Press, Inc., Boca Raton, FL. 288 pp.

Seelig, R.A., and Bing, M.C. 1990. Shallots. *In* Encyclopedia of produce. [Loose-leaf collection of individually paginated chapters.] *Edited by* R.A. Seelig and M.C. Bing. United Fresh Fruit and Vegetable Association, Alexandria, VA. 8 pp.

Takamatsu, E. 1989. Leaf onion, *Allium fistulosum*. *In* Vegetable seed production technology of Japan. Vol. 2. *Edited by* S. Shinohara. Shinohara's Authorized Agricultural Consulting Engineer Office, Tokyo, Japan. pp. 260–287.

Voss, R., and Myers, C. 1991. Japanese bunching onion, Welsh onion, multiplier onion. Crop Sheet SMC-019. *In* Specialty and minor crops handbook. *Edited by* C. Myers. The Small Farm Center, Division of Agriculture and Natural Resources, University of California, Oakland, CA. 2 pp.

Specialty Cookbooks

(Also see GARLIC.)

(Note: There are over 100 cookbooks featuring onions. The following are examples.)

Bailey, L. 1995. Lee Bailey's onions. Clarkson Potter, New York, NY. 80 pp.

Bland Farms, 1996. The Vidalia sweet onion lovers cookbook. Bland Farms, Glennville, GA. 206 pp.

Bothwell, J. 1976. The onion cookbook. Dover Publications, New York, NY. 166 pp.

Cavage, B. 1987. The elegant onion: the art of allium cookery. Storey Communications, Pownal, VT. 156 pp.

Ciletti, B.J. 1998. The onion harvest cookbook. Taunton Press, Newtown, CT. 169 pp.

Cool, J.Z. 1995. Onions. Collins Publishers, San Francisco, CA. 95 pp.

Dille, C., and Belsinger, S. 1996. The onion book: a bounty of culture, cultivation, and cuisine. Interweave Press, Loveland, CO. 96 pp.

Glover, B. 2002. Know your onions. Southwater, London, UK. 128 pp.

Greig, D. 1990. Oh, for an onion! Kangaroo Press, Kenthurst, NSW, Australia. 80 pp.

Griffith, L., and Griffith, F. 1994. Onions, onions, onions. Chapters Publishing, Shelburne, VT. 384 pp.

Mendelsohn, O.A. 1966. A salute to onions; some reflections on cookery and cooks. Hawthorne Books, New York, NY. 190 pp.

Moon, R. 2000. Onions, onions, onions. Firefly Books, Willowdale, ON. 144 pp.

Roberts-Dominguez, J. 1996. The onion book. Doubleday, New York, NY. 339 pp.

Rogers, M.R. 1995. Onions: a celebration of the onion through recipes, lore, and history. Addison-Wesley Publishing Co., Reading, MA. 193 pp.

Wynn, J. 2000. World famous Vidalia sweet onion cookbook, and onions nationwide. Vidalia Cookbooks, Uvalda, GA. 104 pp.

ORANGE

Family: Rutaceae (rue family)

Names

Scientific names: *Citrus* species

- Sweet orange—*Citrus sinensis* (L.) Osbeck (*C. aurantium* L. var. *sinensis* L.)
- Mandarin orange, tangerine, Satsuma orange—*C. reticulata* Blanco (*C. nobilis* Andrews)
- Seville orange—*C. aurantium* L. (*C. bigarradia* Loisel, *C. vulgaris* Risso)
- Bergamot orange—*C. bergamia* Risso & Poit [*C. aurantium* subsp. *bergamia* (Risso & Poit.) Wight & Arn.]

- The word "orange" is derived from the Arabian *naranj*, the Persian *narang*, the Sanskrit *nārangah*, and ultimately, the Tamil *naru*, which means "fragrant." The initial "n" was discarded in French through use of the indefinite article (i.e., *une narange* gave way to the more pronounceable *une orange*), and then the word was taken up in English.
- The word "tangerine" is derived from Tanger or Tangiers, the Moroccan city that was a principal export center of the fruit.
- "Kid-glove orange" is an occasional name for the tangerine, based on the fact that its segments come apart so easily the fruit can be eaten without soiling one's hands.
- The Valencia (sweet) orange is named after the Spanish city Valencia, where it was once cultivated on a large scale. It is considered the best juice variety, and makes up about half of the US crop.
- The name "mandarin orange" was based on the brilliant orange robes once worn by mandarins, i.e., high Chinese government officials.
- "Satsuma," an alternate name for the tangerine or mandarin orange, gets its name from Satsuma, a former province of Kyushu, Japan, where it was first grown.
- "Calamondin orange," "China orange," and "Panama orange" are alternative names of the calamondin [*Citrus madurensis* Lour. or ×*Citrofortunella microcarpa* (Bunge) Wignands], often thought to have arisen as a hybrid of the mandarin orange and the kumquat. Its tangerine-like fruit is eaten raw and made into beverages, marmalades, and jellies, and used as a flavoring, especially in tea. Calamondins are commercially cultivated in the Phillippines, Hawaii, and Florida.
- The "clementine" is a citrus tree that arose from a cross between a tangerine and a Seville orange. The fruit, also known as clementine, is small, spherical, mostly seedless, with an orange–red rind and quite pleasant juicy flesh. The clementine fruit combines the peelability of a tangerine with the bittersweet taste of the Seville orange. This fruit was first cultivated about 1900 near Oran, Algeria, by French missionary priest Père Clément Dozier, who is commemorated in the name "clementine." The name "cleméntine" was first published for the fruit in French about 1902, and first mentioned in English as clementine in 1926.
- For information on the word "citrus," see Grapefruit.
- *Sinensis* in the scientific name *C. sinensis* is Latin for "of China," suggestive of the possible original home of the sweet orange.
- *Reticulata* in the scientific name *C. reticulata* is Latin for reticulate or net-like, a term commonly applied by botanists to leaf venation.
- *Aurantium* in the scientific name *C. aurantium* is Latin for orange colored.
- *Bergamia* in the scientific name *C. bergamia* is a Latinized derivative of the Italian word *bergamotta*, apparently from the Turkish *bey-armuda*, prince's pear. The word is a reference to the pear-like shape of the bergamot orange.

Plant Portrait

The sweet orange tree is evergreen, generally growing to a height of about 8 m (about 25 feet), although very old trees sometimes reach 15 m (about 50 feet). Some trees produce half a tonne of oranges a year for up to 50 years. It is not known where the orange originated, possibly southern China, northeastern India, or southeastern Asia. The orange started to become popular as a fruit in the Mediterranean area in the 15th century, having previously been used by Europeans mainly for medicinal purposes. By the 16th century, the wealthy grew it in private greenhouses called orangeries. After the discovery of the New World, oranges became established from the southern United States to South America. The orange, a subtropical (not a tropical) plant, has become one of the most commonly grown tree fruits in the world. It is an important crop in the Far East, South Africa, Australia, the Mediterranean area, and subtropical areas of South America and the Caribbean. The United States is a leading country in world production, with substantial crops produced in Florida, California, Texas, and Arizona, and much lower production in Louisiana, Mississippi, Alabama, and Georgia. Other major producers are Brazil, Spain, Japan, China, Mexico, Italy, India, Argentina, and Egypt.

There are three main groups of sweet oranges: normal; navel oranges, which develop a distinct navel at the blossom end; and (rarely encountered in North America) blood oranges, with red flesh and juice. There are more than 100 varieties of sweet orange, but only a half-dozen are grown in quantity in the United States. Florida's commercial cultivars are Hamlin, Pineapple, and Valencia, while the principal varieties in California are Washington Navel and Valencia. Some kinds of orange plants produce fruit without the need for pollination and sexual fertilization.

Mandarin oranges or tangerines originated in southeastern Asia. These have fruit segments that readily separate from each other and from the smooth, loose, thin skins, and are widely grown throughout the world in citrus-growing areas. Major producers include Japan, Spain, and Brazil. Some kinds of mandarin orange have seeds, others are seedless. Satsuma oranges are a type of mandarin orange from Japan, and are especially hardy (they can be grown in northern Florida and the coastal regions of the Gulf states).

There are several hybrids of the sweet and mandarin oranges of which the best known is the Temple orange. Tangerine hybrids may be more important economically than pure tangerines. The most important hybrids are tangelos (tangerine × grapefruit), tangors (tangerine × orange), and tangtangelos (tangerine × tangelo). The tangelo is larger than a tangerine, has a pebbly rind ranging from light orange to reddish orange in color, and peels easily. Some tangelo varieties

Orange (*Citrus sinensis*), from Köhler (1883–1914).

have a prominent bump at the stem end. A favorite variety of tangelo is the Minneola, also known as Honeybell. The citrange is a cross between the sweet orange and the trifoliate orange [*Poncirus trifoliata* (L.) Raf., grown as an ornamental and also used as rootstock for citrus trees]; the fruit has a very strong orange, acidic flavor.

The Seville orange originated in Vietnam, but is now widely cultivated. It is also called the bitter or sour orange, and is not suitable for eating raw. It is mainly used for making marmalade. Unlike the sweet orange, the segments of Seville oranges separate, leaving a hollow center.

The bergamot orange is a yellow-skinned, globose or pear-shaped orange with greenish pulp—juicy, but too acidic and bitter tasting to be edible. The tree has been cultivated for centuries in southern Europe, especially in Italy (notably Sicily and Calabria) for oil of bergamot, obtained from the peel. This essential oil is particularly used in perfumery. It is thought that the bergamot orange arose as a cross between the sweet lime (*C. limetta* Risso) and the Seville orange.

Some citrus trees can be grown in pots in northern areas, and moved indoors for overwintering. Satsumi is a compact mandarin variety suited to containers.

Culinary Portrait

Sweet oranges are eaten out of hand and as juice. They are also consumed in fruit salads and used in a wide variety of dishes to impart a sweet, orange taste. Sweet oranges (or just orange juice) are often added to dishes that in the past were made with bitter oranges. They make good additions to sauces, salad dressings, butters, glazes, seafood, poultry, and meat dishes. Blood oranges are frequently employed for decorative culinary purposes.

Mandarin-type oranges are usually eaten on their own as snacks or desserts, but are also employed to prepare fruit salads, sauces, and sweet-and-sour dishes. They may be used for various culinary purposes, as are sweet and bitter oranges. The peel is considered excellent for preparing zest, but is thin, requiring more pressure than used for other kinds of oranges when grating. Tangerine juice is excellent, and mandarin oranges are a special treat when served with ice cream, and topped with chocolate and (or) Grand Marnier liqueur.

The Seville or bitter orange is excellent for cooking, and is ideal for marmalade, jams, and jellies (only bitter oranges have the aromatic, rough, thick rind, suited for the purpose). The zest (ground-up rind) and pulp can be candied. Because oranges are frequently treated with chemicals, they should be washed before using the peel for food. Orange-flower water, often made from bitter orange flowers, is frequently used to add aroma to various dishes, especially desserts. The aromatic oil extracted from Seville orange peel is used in distilling, especially to flavor such liqueurs as Curaçao and Cointreau. Bitter orange sauces are also sometimes prepared. Candied peel is used as an ingredient or decoration for numerous desserts and cakes.

Mandarin orange (*Citrus reticulata*), from Nicholson (1885–1889).

The bergamot fruit is inedible, although the rind is used in confectionery and preserves. The flavor is closer to lemon than to orange. Oil of bergamot is responsible for the distinctive taste of Earl Grey tea. ("Earl Grey tea" is not a unique brand, and some manufacturers use flavorings other than oil of bergamot, particularly lavender.) The oil is also used in confectionery and pastry making, as well as for several nonedible uses. The city of Nancy, France, has specialized in producing bergamot-flavored barley sugar since 1850.

Grapefruit juice is notorious for containing compounds that interfere with the absorption and breakdown of various drugs, making some more potent, others less effective, so that many doctors do not allow their patients taking drugs to also consume grapefruit (see GRAPEFRUIT). Tangelos and Seville oranges are thought to have similar effects.

Culinary Vocabulary

- *Bigerad*, a French word named after the bigarade or Spanish bitter orange, is sometimes encountered as an international culinary term, for example, in *sauce bigerade*, an orange sauce served with duck. The English term is bigarade (pronounced bee-gah-RAHD) sauce, and this is also known as orange sauce. It is made with beef stock, duck drippings, orange juice, lemon juice, blanched orange peel, and occasionally, the liqueur curaçao, all of this sometimes thickened with butter, flour, arrowroot, or duck liver. Although traditionally made with bitter oranges, it is now made with sweet oranges.
- As noted above, the bergamot orange is the source of an oil used to produce the distinctive taste of "Earl Grey tea." This tea commemorates Charles Grey (1764–1845), an English earl who became prime minister and was responsible for abolishing slavery throughout the British Empire. According to tradition, the earl received the recipe for this brew in the 1830s as a gift from a Chinese government official whose life he had saved.
- "Orange flower water" (orange-blossom water) is a distilled product of Seville orange flowers. It is used as a food flavoring, for puddings, cakes, syrups, and confectionery. This is distinct from essential oil, another product of distillation, which is used in perfumery as well as for flavoring.
- The "Maltese orange" is a tart, medium-sized blood orange grown particularly on the Island of Malta. "Sauce maltaise" (Maltese sauce, Maltaise sauce) is hollandaise sauce that has been blended with orange juice and grated orange rind from the Maltese orange. It is used as a topping on cooked vegetables, particularly asparagus and green beans. The French expression *à la maltaise* refers to the use of orange juice (particularly from the Maltese orange) in preparing dishes.
- "Curaçao" (pronounced KYEUR-oh-soh) is a generic term for liqueurs such as Cointreau (the best-known form of curaçao) or Triple sec (French for thrice dry), flavored with the dried peel of bitter or sweet oranges. It is named for the Caribbean island of Curaçao off the coast of Venezuela, where the liqueur may have first been made, or where the bitter oranges first used to make it originated. Curaçao was a Spanish possession from 1527 to 1634 (the word "curaçao" is of Spanish origin), and then became Dutch. In modern times, curaçao is made in different countries, notably Latvia, and while generally colorless or yellow, may also be green, pink, orange, or blue. Curaçao is used in various cocktails, and as a flavoring for cakes, pastries, soufflés, and especially crêpes Suzette.
- The screwdriver cocktail, a mixture of vodka and orange juice, was apparently invented in the 1950s. One widely circulated story concerning its origin is that American oil rig workers in the Middle East used the nearest tool—a screwdriver—to stir the drink, which thereafter became known as a screwdriver. A less-frequently cited claim is that an orange-juice salesman in Bakersfield, California, requested a bartender to serve a mixture of orange juice and vodka to six customers, five of whom liked it, while one said that he would prefer to swallow a screwdriver.
- A "Harvey Wallbanger" is a cocktail of orange juice with vodka or gin, and traditionally with an Italian liqueur such as Galliano floated on top. It apparently was created in the second half of the 20th century in America, and the name is of obscure origin.
- The "sidecar," an obsolescent cocktail made from brandy, lemon juice, and an orange-flavored liqueur, is variously claimed to have originated in Paris during the First World War, or in the United States in the 1920s, and named after the sidecar of a motorcycle in which an officer was said to have been transported to the bar where the drink was created.
- A "Bloody Bronx" is a Bronx cocktail (made with gin, sweet and dry vermouths, and orange juice) made with the juice of a blood orange. The Bronx cocktail was created by bartender Johnnie Solon at the Waldorf-Astoria hotel in New York City at the beginning of the 20th century, the name supposedly referring to the Bronx Zoo.
- A "Hawaiian Orange Blossom" is a cocktail prepared with gin, orange juice, curaçao, and pineapple juice.
- A "Marmalade" is a cocktail prepared with gin, lemon juice, and orange marmalade.
- Danziger Goldwasser [literally "Danzig gold water" (Danzig is the modern port of Gdansk in Poland)]. This is a liqueur, predominantly flavored with orange peel and spices, best known for its tiny bits of gold leaf that float in the bottle. The gold leaf is considered harmless to drink. The beverage has been produced in Europe since the 17th century, and was first concocted for medicinal purposes back when it was thought gold could cure all diseases.

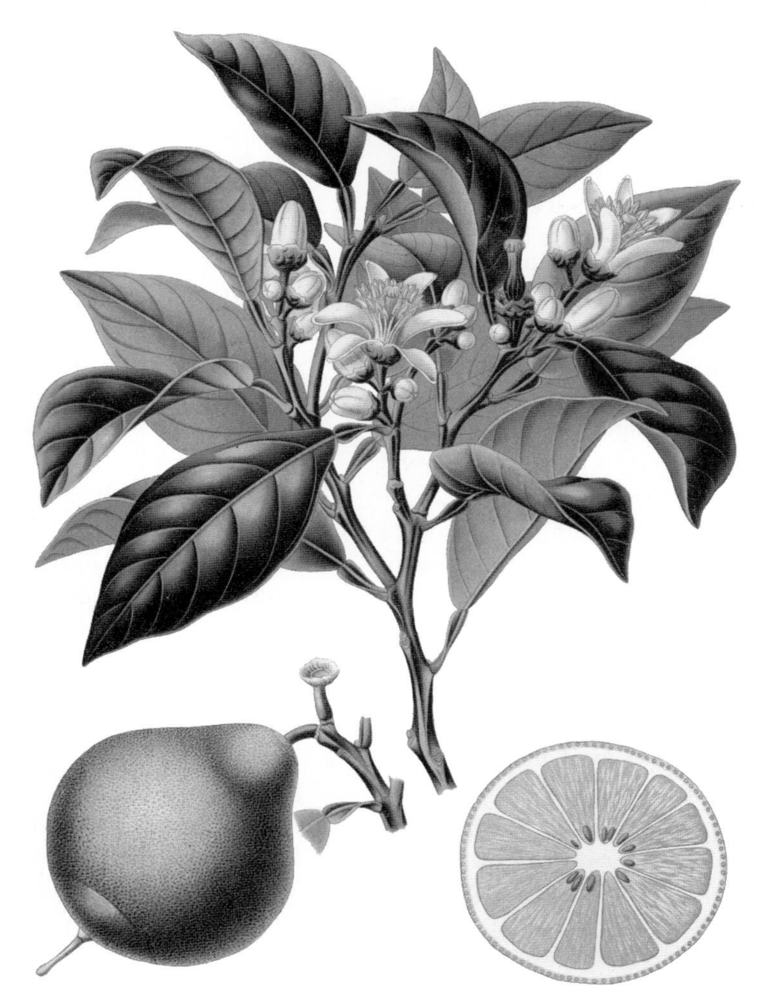

Bergamot orange (*Citrus bergamia*), from Köhler (1883–1914).

Curiosities of Science and Technology

- The Egyptians used orange juice for embalming.
- The ancient Romans used a variety of citrus fruits to keep moths away from their clothes.
- A famous orange tree, known as the Constable, was brought to France in 1421, and produced fruit for 473 years.
- Nostradamus (Michel de Nostrademe, 1503–1566, French physician and astrologer), famous for his predictions, developed cosmetics from both orange peels and orange juice.
- Sir Frances Drake [1540 (or 1543)–1596] was the first Englishman to sail around the world (1577–1580), and also played a leading role in defeating the Great Armada sent by Spain to invade England. Additionally, Drake made three voyages to the New World, plundering Spanish settlements and destroying Spanish ships. In 1586, he sacked the Spanish city of St. Augustine, Florida, and had the orange trees cut down. However, orange trees sprout from the base of the trunks, and in a few years, the orchards recovered.
- During the Elizabethan era (the period of Elizabeth I, 1533–1603), prisoners in the Tower of London were allowed to receive oranges, and perfected the technique of using orange juice as an invisible ink in order to send surreptitious messages.
- French King Louis XIV (1638–1715) was the longest-reigning monarch in European history, and one of the cruelest. He was also one of the most gluttonous people who ever lived. In his palace at Versailles, he had over 1000 orange trees planted in silver tubs. From 1705 to 1709, a million French citizens died of famine while Louis feasted. An autopsy of the king's body when he died revealed that his intestines were twice as long as the average man's, and contained a huge tapeworm.
- King Louis XIV loved perfumes, which in the fashion of the time were used instead of bathing. He

drenched himself more thoroughly with perfumes than any of the other kings of France. As he aged, however, perfume gave him headaches, and finally he could only tolerate water scented with oranges.

- There are over 100 Popsicle flavors, led in popularity by orange, and over 1 billion are sold annually. American Frank Epperson (1894–?), at the age of 11, is credited with inventing the first frozen juice bar when he accidentally left a bottle of soda pop, with a stir stick in it, outdoors in freezing weather. In 1923, he applied for a patent, calling his invention the "Epsicle" (based on Epperson + icicle). Later, the name was changed to "Popsicle" (variously explained as based on soda pop + icicle, lollipop + icicle, or by Eppeson's five children who clamored for their "Pop's icicle"). During the Depression years, Epperson created the twin Popsicle so that the treat could be eaten by two children for only a nickel. Epperson went on to invent the Fudgesicle, Creamsicle, and Dreamsicle.
- Very old, large orange trees in the Mediterranean area may bear 3000–5000 oranges each year.
- Close to 40% of the orange crop in the United States is used in frozen concentrate.
- Navel oranges are named for the resemblance of the blossom end to the human navel. They are actually two fruits in one. Their characteristic "belly button" is another tiny orange growing at the tip of the larger one.
- Orange seeds often have more than one embryo (a phenomenon called "polyembryony"), so planting a single seed often results in more than one plant developing.
- Oranges grown on the south side of the tree tend to be sweeter than those that are grown on the north side.
- The blossom end of an orange is said to be sweeter than the stalk end.
- Hong Kong holds the record for per capita orange consumption—23 kg (50 pounds) per person each year.
- The volatile oils in orange peel can produce dermatitis. People who suck oranges often suffer skin irritation around the mouth.
- Gardenias and orange blossoms produce some of the most attractive odors, but when placed together in the same bouquet, they neutralize each other and there is little smell.
- By coincidence, Disneyland and Walt Disney World amusement parks are in counties with the same name: the former in Orange County, California, the latter in Orange County, Florida.
- Orange juice may taste foul soon after brushing one's teeth because a chemical detergent in toothpaste, sodium lauryl sulfate, reacts with the acids in orange juice to produce an unpleasant, bitter taste.
- A report issued by the British Dental Association in 2002 warned that acidic fruits, such as oranges and tangerines, are best eaten at mealtime, rather than as between-meal snacks. It was stated that dental erosion in adults due to diet is usually a result of excessive consumption of fruits and fruit juices, as the acidity can harm tooth enamel. Eating acidic foods along with nonacidic foods tends to neutralize the effect.
- In 2006, English jam-maker F. Duerr & Son marked its 125th anniversary by producing the world's most expensive marmalade, offered at $10,757 for a 1-kg jar. This was called "Fine Cut Seville Orange Marmalade with Whisky, Champagne and Gold," the cost primarily owing to the inclusion of very expensive alcoholic beverages and flakes of 24-carat gold leaf, all in a custom-made crystal jar.
- Researchers have found that the color of orange juice influences its taste. When subjects were given two cups of the same orange juice, with one cup darkened with food coloring, the members of the sample group thought that the two cups had

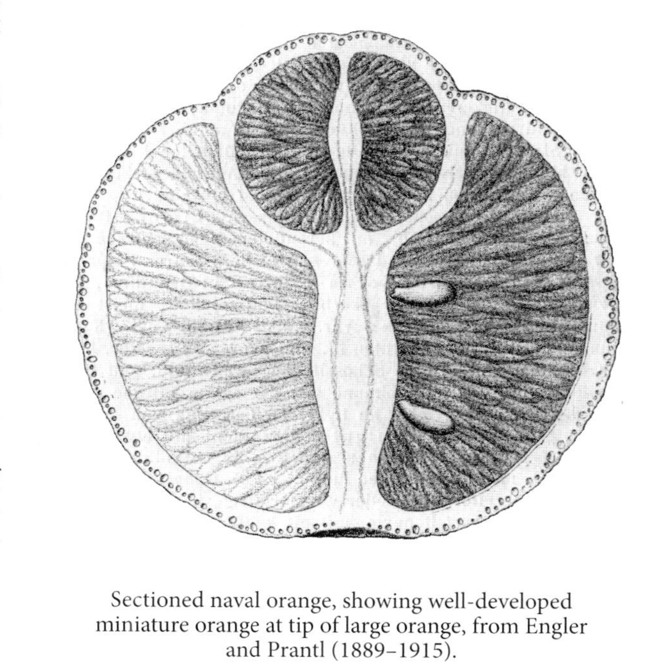

Sectioned naval orange, showing well-developed miniature orange at tip of large orange, from Engler and Prantl (1889–1915).

different tastes. In contrast, given two cups of orange juice of the same color, with one cup sweetened with sugar, the same people failed to perceive taste differences. (Reference: Hoegg, J.A., and Alba, J.W. 2007. Taste perception: more than meets the tongue. J. Consumer Res. **33**: 490–408.)

Key Information Sources

(Additional references are in LEMON.)

Ashari, S., Aspinall, D., and Sedgley, M. 1989. Identification and investigation of relationships of mandarin types using isozymes analysis. Sci. Hortic. **40**: 305–315.

Asjaro. S. 1991. *Citrus reticulata* Blanco. *In* Plant resources of South-East Asia, 2, edible fruits and nuts. *Edited by* E.W.M. Verheij and R.E. Coronel. Pudoc, Leiden, Netherlands. pp. 135–138.

Brown, M.G., and Ying Lee, J.-Y. 1987. World orange juice trends. Florida Dept. of Citrus, Lakeland, FL. 28 pp.

Buxton, B.M. 1993. Comparative costs of producing oranges in selected orange exporting countries. Food and Agriculture Organization, Rome, Italy. 15 pp.

Cooper, W.C. 1990. Odyssey of the orange in China: national history of the citrus fruits in China. William C. Cooper, Winter Park, FL. 122 pp.

Durr, P. 1980. Aroma quality of orange juice. A brief review. Alimenta (Zurich), **19(2)**: 35–36.

Farre, R., Frigola, A., and Roca de Togores, M.C. 1989. Vitamin C content in different varieties of tangerines and oranges and the storage effect. *In* Agriculture, food chemistry and the consumer: proceedings of the 5th European conference on food chemistry, Versailles, September 1989. Vol. 2. *Edited by* Federation of European Chemical Societies. Institut National de la Recherche Agronomique, Paris, France. pp. 503–507.

Fayoux, S.C., Seuvre, A.M., and Voilley, A.J. 1997*a*. Aroma transfers in and through plastic packagings: orange juice and d-limonene. A review. Part I. Orange juice aroma sorption. Pack. Technol. Sci. **10**: 69–82.

Fayoux, S.C., Seuvre, A.M., and Voilley, A.J. 1997*b*. Aroma transfers in and through plastic packagings: orange juice and d-limonene. A review. Part II: overall sorption mechanisms and parameters—a literature survey. Pack. Technol. Sci. **10**: 145–160.

Graumlich, T.R., Marcy, J.E., and Adams, J.P. 1986. Aseptically packaged orange juice concentrate: a review of the influence of processing and packaging conditions on quality. J. Agric. Food Chem. **34**: 402–405.

McPhee, J.A. 1967. Oranges. Noonday Press, New York, NY. 149 pp.

Morton, J. 1987*a*. Sour orange, *Citrus aurantium*. *In* Fruits of warm climates. *Authored by* J.F. Morton. Creative Resource Systems, Winterville, NC. pp. 130–133.

Morton, J. 1987*b*. Orange, *Citrus sinensis*. *In* Fruits of warm climates. *Authored by* J.F. Morton. Creative Resource Systems, Winterville, NC. pp. 134–142.

Morton, J. 1987*c*. Orangelo, orange × grapefruit. *In* Fruits of warm climates. *Authored by* J.F. Morton. Creative Resource Systems, Winterville, NC. p. 160.

Robards, K., and Antolovich, M. 1995. Methods for assessing the authenticity of orange juice. A review. Analyst, **120**: 1–28.

Samson, J.A. 1991. *Citrus sinensis* (L) Osbeck. *In* Plant resources of South-East Asia, 2, edible fruits and nuts. *Edited by* E.W.M. Verheij and R.E. Coronel. Pudoc, Leiden, Netherlands. pp. 138–141.

Seelig, R.A., and Bing, M.C. 1990. Oranges. *In* Encyclopedia of produce. [Loose-leaf collection of individually paginated chapters.] *Edited by* R.A. Seelig and M.C. Bing. United Fresh Fruit and Vegetable Association, Alexandria, VA. 34 pp.

Soost, R.K., and Cameron, J.W. 1975. Citrus. *In* Advances in fruit breeding. *Edited by* J. Janick and J.N. Moore. Purdue University Press, West Lafayette, IN. pp. 507–540.

Stayte, L., and Vaughan, A. 2000. Taking the pith: the impact of the production and consumption of oranges and orange juice on people and the environment. Sustain, London, UK. 19 pp.

United States Dept. of Agriculture. 1978. People on the farm: growing oranges. USDA, Washington, DC. 28 pp.

Specialty Cookbooks

(Additional references are in GRAPEFRUIT and LEMON.)

Anonymous. 1990. The best of oranges. Garamond, London, UK. 45 pp.

California Fruit Growers Exchange, and Sunkist Growers, Inc. 1920. Masterpieces from the chefs of the great hotels of New York featuring Sunkist oranges. California Fruit Growers Exchange. Los Angeles, CA. 29 pp.

Dorman, C. 2003. Oranges & lemons: fabulous recipes with the tangy taste of citrus fruit. Southwater, London, UK. 128 pp.

Food 4 Less Supermarkets, Inc. 1997. Orange Blossom Festival, April 19–20, 1997, the best of the best Orange Blossom Festival recipes and celebrity chef recipes. Food 4 Less Supermarkets, Inc., Riverside, CA. 80 pp.

Jones, B. 1993. Oranges. Harlaxton Publishing, Grantham, UK. 64 pp.

LaFray, J. 1988. Recipes from the orange grove. Surfside Publishing, Tampa, FL. 41 pp.

Lane, J.L. 1909. 365 orange recipes; an orange recipe for every day in the year. G.W. Jacobs & Co., Philadelphia, PA. 158 pp.

Sunkist Orange Blossom Festival. 1996. The best of the best: Orange Blossom Festival recipes, 1995–1996. Sunkist Orange Blossom Festival, Riverside, CA. 80 pp.

Weaver, A.T. 1977. Just oranges: a collection of 785 orange recipes. Weaver, Lake Placid, FL. 246 pp.

Woodward, S. 2001. Oranges & lemons. Cornran Octopus, New York, NY. 144 pp.

PAPAYA

Family: Caricaceae (papaya family; this family is sometimes placed in the Passifloraceae or passion fruit family)

Names

Scientific name: *Carica papaya* L.

- The name "papaya" is derived from the Amerindian (Otomac) *papai*, perhaps ultimately from *ababai*, the Carib Indian name for the fruit.
- The papaya is also known as the pawpaw (paw paw), especially in English-speaking Caribbean islands. This name is best reserved for another species, *Asimina triloba* (L.) Dunal.
- When first encountered by Europeans, the papaya was called "tree melon," a name that still persists today.
- The genus name *Carica* is based on the Greek *karike*, a kind of fig (see FIG for additional information). The name was used for the papaya because of its fig-like leaves.

Plant Portrait

Although often called a tree and resembling a tree palm, the papaya is a gigantic perennial, nonwoody herb. The plants grow 1.8–3 m (6–10 feet) in the first year and eventually reach 6–9 m (20–30 feet) in height, with a hollow stem 30–40 cm (12–16 inches) thick at the base. Male and female flowers are on separate trees except in some cultivated varieties. The papaya has been in cultivation in the New World since prehistoric times, and its ancestral distribution is obscure. Its original range is commonly said to be Central America, but may be northwestern South America. The plant is now very widely cultivated in tropical and subtropical regions of the world. Large commercial production today is carried out in Hawaii, tropical Africa, the Philippines, India, Ceylon, Malaya, and Australia, with smaller-scale cultivation in South Africa, Latin America, and Florida.

The melon-like fruits are pear-shaped ovoid, or spherical, varying from 15 to 50 cm long, and may weigh as much as 9 kg (20 pounds). A "tree" may bear 30–150 fruits per year. The plants are short-lived, and best fruit production is on plants not over 4 years of age. The thin but tough skin is yellow to orange when ripe. The mature flesh varies from pale yellow to orange and red, and is juicy, sweetish, and somewhat like a cantaloupe in flavor, but softer in texture. The hollow inner cavity is lined with many black or dark brown seeds, 5–7 mm (about 1/4 inch) long. Hawaiian varieties have pear-shaped fruit generally weighing about 0.4 kg (1 pound), with yellow skin when ripe, flesh that is bright orange or pinkish, depending on variety, with small black seeds clustered in the center. Mexican papayas are much larger than the Hawaiian types and may weigh up to 4.5 kg (10 pounds), with flesh that may be yellow, orange, or pink, and flavor that is less intense than that of the Hawaiian papaya. The variety Solo, which is about 15 cm (6 inches) long and has yellow–orange flesh, is the most common papaya on the American market. Introduced to Hawaii from Barbados in 1911, it was the first major commercial cultivar. It was named "Solo" because it is just the right size for a one-person snack—as opposed to the more conventional "family size" papaya.

The leaves, trunk, and young fruits exude a white or milky odorless sap when injured. Two protein-digesting enzymes are present in this latex, papain and chymopapain. Papain is widely used as a meat tenderizer, some cooks tenderizing a slab of meat simply by placing it between two slices of papaya. Papain is also used to clear beer, for a variety of medicinal treatments, and for numerous industrial purposes. Papain is harvested by scoring the unripe fruit and letting the latex drip into containers. Papain production is largely in Tanzania and India.

Culinary Portrait

Ripe papayas are most commonly eaten fresh (the thin, smooth skin is inedible). Papaya is an excellent dessert, and is also useful in fruit salads. The flesh can simply be scooped out with a spoon and eaten as is, but may be flavored with sugar, lime or lemon juice, or rum or port. The fruit is used in soft drinks, jam, ice cream, yogurt, pudding, and sorbet. Papayas are also sold dried or canned, and papaya juice is still another product. Papaya purée is used in mixed juices and beverages. The fruit is cooked to make jam, chutney, and ketchup. In Asia, green fruits are served in salads, but are also treated as a vegetable, often cooked. Indeed, green fruit can be used much like winter squash, and can be substituted in many recipes. Papaya can be stuffed with chicken, seafood, or various fruits. Like melon, it goes well with baked ham, smoked salmon, and many other dishes. Papayas vary in taste, and are sometimes bland, sometimes cloyingly sweet, and sometimes musky. Some people don't like the taste of papaya the first time they consume it, expecting it to taste like a muskmelon, which it resembles. Unripe papaya is never eaten raw because of the latex content, but is often boiled as a vegetable. However, it is sometimes desirable to drain off the white, acidic sap. The latex from green papayas has been used in chewing gum.

The seeds look like large peppercorns and are peppery. These are occasionally eaten, and are also ground and made into salad dressing. Leaves are also eaten like spinach in some regions.

The enzyme papain in papaya prevents gelatin from setting and can curdle milk products after a while. When papaya is served in a combination with dairy foods such as ice cream, it should be consumed promptly. When combining papaya with other fruits in fruit salads, it may be desirable to add the papaya shortly before serving to avoid softening of the other fruit by the enzymes present.

Papayas are one of the few fruits that do not turn brown when exposed to air. Therefore, they can be prepared well ahead of serving.

Tips for choosing a papaya: select fruit with a rich-colored skin, which gives slightly to palm pressure. Avoid green, rock-hard, heavily bruised, and oversoft fruits. Slightly green papayas will ripen quickly at room temperature, but completely green fruit were picked too early and will remain flavorless. Refrigerate completely ripe fruit and use as soon as possible. Chilling injury can occur at temperatures below 8 °C (46 °F). The fruit does not freeze well.

Curiosities of Science and Technology

- Columbus recorded in his chronicles how natives of the West Indies often refused to start a meal without papayas. He also noted that natives could consume a large amount of fish and meat without getting indigestion if unripe papaya was eaten after the meal.
- Since 1999, all of the Solo variety papaya grown in Hawaii has been genetically engineered. The plants have a gene inserted to confer resistance to papaya ringspot virus.
- Male papaya plants sometimes completely transform into female plants after being "beheaded" (the tops are often removed when the plants become too tall).

Papaya (*Carica papaya*), from Köhler (1883–1914).

Papaya (*Carica papaya*), from Edwards and Lindley (1815–1847, vol. 6).

- In Queensland, Australia, green papayas are frequently canned. Because of local conditions, the plants may take up high levels of nitrates from the soil, and this causes detinning of ordinary cans. All papayas with over 30 parts per million of nitrates must be packed in cans lacquered on the inside.
- The inner bark of the papaya plant has been used for relief of toothache in Samoa.
- The enzyme papain from papaya is used during surgical procedures to dissolve ruptured spinal discs; it is referred to as "nature's scalpel" because it preferentially degrades dead tissue.
- Papain from papaya is very similar to human stomach pepsin, and is widely used as a meat tenderizer (commonly marketed in North America as Adolph's Meat Tenderizer). Beef cattle are sometimes injected with papain a half-hour before slaughter to tenderize them. Meat treated with papain should not be cooked "rare," as the enzyme will not have been inactivated (and could start to tenderize the consumer, as with the cattle).
- In folk medicine, fresh latex is sometimes smeared on boils, warts, or freckles to remove them (this procedure is not recommended!).
- The babaco, *Carica ×heilbornii* V.M. Badillo var. *pentagona* (Heilborn) V.M. Badillo (*Carica pentagona* Heilborn), also known as highland papaya, is a natural hybrid between two Andean species of *Carica*, which originated in the subtropical Ecuadorean valleys, and is commercially cultivated in Ecuador. The plant is similar to the papaya, but smaller (less than 3 m or 10 feet), producing five-angled fruits reaching 30 cm (12 inches) in length with few or no seeds. The flavor is like pineapple and banana, but not as sweet. Babacos are sterile, so are propagated only by cuttings. The babaco can be grown in tubs, and is more cold resistant than the papaya, but in cold areas must be moved indoors to survive the winter. Cultivated in this manner, they can be grown almost anywhere.

Key Information Sources

Abudu, A.O. 1996, 2000. The pawpaw and its several uses. Dyno-Media, Accra, Ghana. 51 pp.

Agnew, G.W. 1968. Growing quality papaws in Queensland. Queensl. Agric. J. **94**: 24–36.

Ali, Z.M., and Lazan, H. 1998. Papaya. *In* Tropical and subtropical fruits. *Edited by* P.E. Shaw, H.T. Chan, Jr., and S. Nagy. Agscience, Auburndale, FL. pp. 401–445.

Arriola, M.C. de, Calzada, J.F., Menchu, J.F., Rolz, C., Garcia, R., and Cabrera, S. de. 1980. Papaya. *In* Tropical and subtropical fruits: composition, properties and uses. *Edited by* S. Nagy and R.E. Shaw. AVI Publishing Company, Westport, CT. pp. 316–340.

Atkinson, I.C. 1993. Pawpaw growing. 2nd edition. NSW Agriculture & Fisheries, Sydney, Australia. 8 pp.

Becker, S. 1958. The production of papain—an agricultural industry for tropical America. Econ. Bot. **12**: 62–79.

Bhutiani, R.C. 1963. Papaya. Central Food Technological Institute, Mysore, India. 41 pp.

Chan, H.T., and Tang, C.S. 1978. The chemistry and biochemistry of papaya. *In* Tropical foods, vol. 1. *Edited by* G.E. Inglett and G. Charolambous. Academic Press, New York, NY. pp. 33–55.

Chan, H.T., Jr. 1983. Papaya. *In* Handbook of tropical foods. *Edited by* H.T. Chan, Jr. Marcel Dekker Inc., New York, NY. pp. 469–488.

French, C.D. 1972. Papaya: the melon of health. Arco, New York, NY. 96 pp.

Maharaj, R., and Sakat, C.K. 1990. Storability of papayas under refrigerated and controlled atmosphere. Acta Hortic. **269**: 375–385.

Manshardt, R.M. 1992. Papaya. *In* Biotechnology of perennial fruit crops. *Edited by* F.A. Hammerschlag and R.E. Litz. CAB International, Wallingford, UK. pp. 489–511.

Marler, T.E. 1994. Papaya. *In* Handbook of environmental physiology of fruit crops, Vol. 2. Subtropical and tropical crops. *Edited by* B. Schaffer and P.C. Anderen. CRC Press, Boca Raton, FL. pp. 216–224.

Mitra, S.K. 1999. Papaya. *In* Tropical horticulture, vol. 1. *Edited by* T.K. Bose, S.K. Mitra, A.A. Farooqui, and M.K. Sadhu. Naya Prokash, Calcutta, India. pp. 308–318.

Morton, J. 1987. Papaya. *In* Fruits of warm climates. *Authored by* J.F. Morton. Creative Resource Systems, Winterville, NC. pp. 336–346.

Muthukrishnan, C.R., and Irulappan, I. 1985. Papaya. *In* Fruits of India, tropical and subtropical. *Edited by* T.K. Bose. Naya Prokash, Calcutta, India. pp. 320–344.

Nakasone, H.H. 1986. Papaya. *In* CRC handbook of fruit set and development. *Edited by* S.P. Morselise. CRC Press, Boca Raton, FL. pp. 277–301.

Paull, R.E. 1993. Pineapple and papaya. *In* Biochemistry of fruit ripening. *Edited by* G. Seymour, J. Taylor, and G. Tucker. Chapman & Hall, London, UK. pp. 291–323.

Paull, R.E., Nishijima, W., Reyes, M., and Cavaletto, C.G. 1997. A review of post-harvest handling and losses during marketing of papaya (*Carica papaya* L.). Postharvest Biol. Technol. **11**: 165–179.

Sankrat, C.K., and Maharaj, R. 1997. Papaya. *In* Postharvest physiology and storage of tropical and subtropical fruits. *Edited by* S.K. Mitra. CAB International, Wallingford, UK. pp. 167–189.

Seelig, R.A., and Bing, M.C. 1990. Papaya. *In* Encyclopedia of produce. [Loose-leaf collection of individually paginated chapters.] *Edited by* R.A. Seelig and M.C. Bing. United Fresh Fruit and Vegetable Association, Alexandria, VA. 13 pp.

Simonsohn, B. 2000. The healing power of papaya: a holistic handbook on how to avoid acidosis, allergies, and other health disorders. Lotus Light Publications, Twin Lakes, WI. 224 pp.

Storey, W.B. 1969. Papaya (*Carica papaya* L.). *In* Outlines of perennial crop breeding in the tropics. *Edited by* F.P. Ferwerda and F. Wit. Misc. Paper 4. Agricultural University, Wageningen, Netherlands. pp. 389–407.

Storey, W.B. 1976. Papaya. *In* Evolution of crop plants. *Edited by* N.W. Simmonds. Longman, London, UK. pp. 21–24.

Villegas, V.N. 1991. *Carica papaya* L. *In* Plant resources of South-East Asia, 2, edible fruits and nuts. *Edited by* E.W.M. Verheij and R.E. Coronel. Pudoc, Leiden, Netherlands. pp. 108–112.

Yadava, U.L., Burris, J.A., and McCrary, D. 1990. Papaya: a potential annual crop under middle Georgia conditions. *In* Advances in new crops. *Edited by* J. Janick and J.E. Simon. Timber Press, Portland, OR. pp. 364–366.

Specialty Cookbooks

Campbell, L. 1991. Hawaiian papaya, the royal fruit of kings: history, recipes, how to eat them, how to grow at home. Larry Campbell, Makawao, HI. 15 pp.

Hammond, D. 1994. The Australian banana—paw paw recipe book. Southern Holdings, Huonville, Tasmania, Australia. 68 pp.

Hawaii Papaya Administrative Committee. 1991. Perfect papaya recipes from Hawaii's chefs: entries in the first annual Chef's Papaya Recipe Challenge. Hawaii Papaya Administrative Committee, Hilo, HI. 19 pp.

Papaya Administrative Committee, and Pineapple Growers Association of Hawaii. 1994. Chef's papaya & pineapple recipe challenge. 4th edition. Becker Communications, Honolulu, HI. 85 pp. [6th edition, 1996, 40 pp.]

Philippine Council for Agriculture and Resources Research and Development. 1984. The Philippines recommends for papaya. Philippine Council for Agriculture and Resources Research and Development, Los Baños, Laguna, Philippines. 58 pp.

Siewertsen, V. 1986. Papaya (He'i): uses and recipes. Pacific Tropical Botanical Garden, Lawai, Kauai, HI. 42 pp.

Thursby, I.S. 1940. The fruitful papaya: how to serve it. University of Florida. Agricultural Extension Division Bull. 106. University of Florida, Gainesville, FL. 20 pp.

PEA

Family: Fabaceae (Leguminosae; pea family)

Names

Scientific name: *Pisum sativum* L.

- The English word "pea" is derived from the Latin word for pea, *pisum*.
- Other names for the pea include common pea, English pea, field pea, garden pea, and green pea. Edible-podded varieties are called snow pea, sugar pea, and mange-tout—French for "eat everything," and usually in France applied to young French beans, not peas. Despite the "eat everything" name, some cooks cut off the tips of edible-podded peas just before serving them.
- The popular ornamental "sweet pea" (*Lathyrus odoratus* L.) has seeds that are not edible, and indeed are somewhat poisonous. The name "sweet pea" is believed to have first been used by the poet John Keats (1795–1821).
- The genus name *Pisum* is based on the Latin *pisum* and the Greek *pison*, pea.
- *Sativum* in the scientific name *P. sativum* is Latin for cultivated or sown.

Plant Portrait

The common garden pea is most often a climbing or trailing annual vine, growing up to about 2 m (6½ feet) in length. Pea pods are usually 5–15 cm (2–6 inches) long, smooth, straight or slightly curved, either swollen or flat, and contain 2–10 seeds. The seeds are usually round, but some are slightly cubical, and while most peas are green, depending on variety the color may be grayish, whitish, or brownish. Dried peas tend to be yellow or green. Many peas grown in North America are bush-like in growth, and produce a limited number of fruits. Vine-like plants require poles or trellises for support, and continue to produce fruits for a prolonged period. Peas were domesticated very early in history. The plants originated in the Mideast, and were used in Neolithic farming villages in the Near East at least as early as 7000–6000 BC. Peas were known by the classical Greeks and Romans, although neither civilization thought highly of the crop. Whole pea pods were not used in cooking until the Middle Ages. With colonization of North America, the pea became widely distributed in the New World. Today, peas are a major world crop, grown extensively in cool countries, particularly in northern Europe, China, Canada, and the northwestern United States.

Three basic classes of modern cultivated peas have been recognized. "Green peas" have tough pods, with fibrous lining, but the seeds are very sweet. The flowers are usually white, and the seeds are used as human food. Edible-podded peas have pods that are thick, soft, and lack a fibrous lining. The flowers are usually pinkish or purple, and young pods are eaten as human food. "Field peas" have seeds that are used for human food and animal meal, and the plants for fodder and silage. The flowers are usually pink and purple. China and Russia each produce about one-third of the world's supply of dry peas, and Canada about 15%. Green peas and edible-podded peas are used as human food. Green peas are grown commercially primarily for canning or freezing.

Culinary Portrait

Fresh peas are excellent raw, but are also often cooked in boiling water (10–15 minutes), or steamed, or braised. The pea ranks as one of the major cooking vegetables in the Temperate World, along with potatoes, sweet corn, and beans. Peas may be served hot or cold (cold peas are often added to salads), and combine well in soups, stews, and both vegetable and meat dishes. Edible-podded peas are eaten while they are still immature, before much starch has accumulated in the seeds. They can be consumed raw, or steamed, cooked whole, or fried. Raw edible-podded peas are excellent in salads and as appetizers. Edible-podded peas are cooked like green beans, requiring 6–15 minutes when boiled or steamed. If overcooked, they tend to become mushy. They are used increasingly in cooking, and are particularly popular in Oriental cuisine, especially in stir-fries. Field peas are widely used as dry split peas and in the manufacture of soup. Whole dried peas should be soaked for 1–1½ hours before cooking; but halved or split peas do not require pre-soaking. Dried peas should be gently cooked for 1–2 hours, but just until tender, as overcooking turns them mushy and they fall apart. Split peas tend to produce a lot of foam in a pressure cooker, which could block the safety valve, so the use of a pressure cooker is not recommended. Outside of North America, dried peas are typically soaked in water and then used in various cereals, unleavened breads, or rice dishes. Dried pods can be used as "snaps" (i.e., eaten whole) after frying in oil with turmeric and onions, or other condiments. Extracts, including protein, starch, and fiber, have been

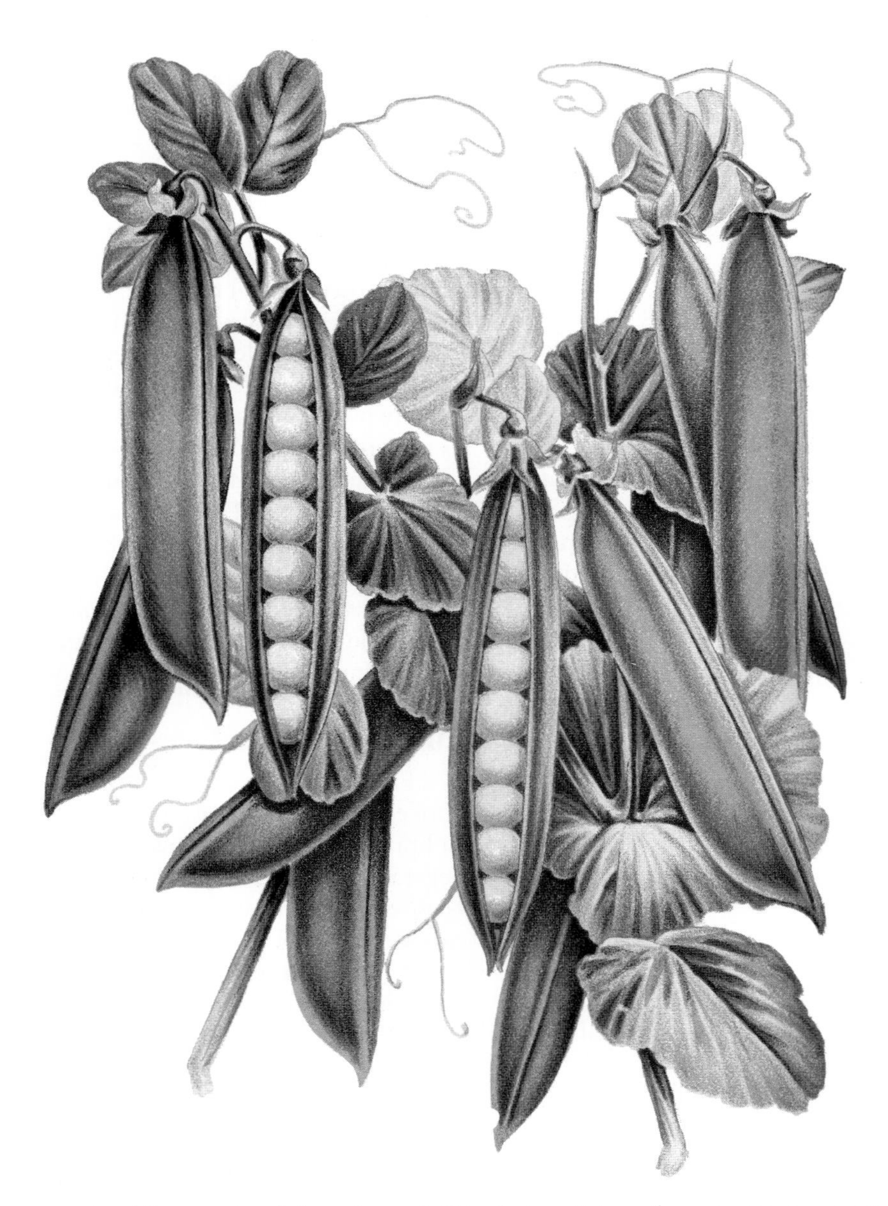

Pea (*Pisum sativum*), from *Revue de l'Horticulture Belge et Étrangère* (1912, vol. 38).

obtained from peas for use in food as well as industrial products. Pea shoots and tendrils are consumed in Hong Kong cuisine (as *dao minu*).

Culinary Vocabulary

- "Petits pois" is an English culinary term for immature garden peas (in French, it means "little peas").
- "Split peas" are peas with the skin removed.
- In the southern United States, the term "pea" usually refers to the black-eyed pea (which see).
- "Pease Pudding" is a British dish of cooked, puréed split peas.
- "Boula" or "boula-boula" (boula boula) is an American soup of turtle and puréed green peas, which originated in the Seychelles. It is flavored with sherry, topped with whipped cream and cheese, and browned. Originally called boula, it is said to have been renamed boula boula (after the silly college song *Boula Boula*) by American President and Mrs. John F. Kennedy who served it at the White House.
- *À la Clamart* (or *à la clamart*; a French expression pronounced ah lah clah-mahr) refers to a French garnish made with whole or puréed green peas. There are many dishes with clamart (or Clamart) in the name, such as Eggs Clamart (poached eggs

served on toast covered with a purée of green peas), Artichokes Clamart (artichokes filled with green peas or pea purée), and Crème Clamart (an elegant cream soup with puréed peas). The reason why Clamart is in the name is obscure; most likely, a small town in France called Clamart was locally famous for its sweet green peas.

- *Saint-Germain* (pronounced san-zhehr-MAHN) is a French term referring to a garnish consisting of green peas or pea purée.

Curiosities of Science and Technology

- Peas are renowned as the subject of studies by the Austrian monk Gregor Mendel (1822–1884). His epochal experiments ushered in the science of genetics. Mendel's studies of heredity in the common garden pea were conducted from 1856 to 1865, in a monastery garden at Brünn (Brno), then in Austria (now the Czech Republic). By examining such contrasting characters as tall versus dwarf size, wrinkled versus smooth seeds, and colors of peas, Mendel recognized that certain characteristics were inherited as units, and in combination, some of these genetic units were expressed (i.e., were dominant) at the expense of others. The pea has become one of the most studied organisms in genetics.
- Peas are such a well-known food that it has influenced the English language. The thick sulfurous fog that regularly choked Londoners of the 19th century and the first half of the 20th century was called "pea soup" or a "pea-souper." The term came to be more widely applied, especially in nautical circles. The small size of the pea is emphasized in the expression "pea-brain" for a dull-witted, stupid person. However, a "pea jacket" has nothing to do with peas, and isn't even pea-green. It is a heavy wool coat, and the name probably comes from the Dutch sailor jacket *pijjekker*, based on *pij*, a coarse cloth, + *jekker*, jacket.
- In the 18th and 19th centuries, a small, rounded, puckered, so-called "bee-stung mouth" was often described as the most important feature of a beautiful woman's face, and it became common practice to promote this appearance by repeating a sequence of words beginning with "p," most popularly "peas, prunes, and prisms." Some women making a "grand entrance" would repeat these words just before making their appearance, and instead of asking their subjects to smile or say "cheese," 19th century photographers often asked them to repeat the "p" words.
- In 1998, a coalition of scientists, religious leaders, health professionals, consumers, and chefs filed a lawsuit to force the U.S. Food and Drug Administration to require safety testing and labeling of genetically engineered foods. The suit was supported by the Alliance for Bio-Integrity and The International Center for Technology Assessment. The plaintiffs created a menu for a "Dinner of Transgenic Foods," all of which were currently marketed or proposed for marketing.

Appetizers

Fingerling Potatoes with Wax Moth Genes (served with sour cream from bovine growth hormone treated cows)

Juice of Tomatoes with Flounder Genes

Entrée

Braised Pork Loin with Human Growth Genes

Boiled New Potatoes with Chicken Genes

Fried Squash with Watermelon and Zucchini Virus Genes

Toasted Cornbread with Firefly Genes

Dessert

Rice Pudding with Pea and Bacteria Genes

Key Information Sources

Adsule, R.N., Lawande, K.M., and Kadam, S.S. 1989. Pea. *In* Handbook of world food legumes: nutritional chemistry, processing technology, and utilization. Vol. II. *Edited by* D.K. Salunke and S.S. Kadam. CRC Press, Inc., Boca Raton, FL. pp. 215–251.

Ali, S.M., Sharma, B., and Ambrose, M.J. 1994. Current status and future strategy in breeding pea to improve resistance to biotic and abiotic stresses. *In* Expanding the production and use of cool season food legumes. *Edited by* F.J. Muehlbauer and W.J. Kaiser. Kluwer Academic Publishers. Dordrecht, Netherlands. pp. 540–558.

Ali-Khan, S.T., and Zimmer, R.C. 1989. Production of field peas in Canada. Revised edition. Agric. Can. Publ. 1710/E. 21 pp.

Ben-Ze'ev, N., and Zohary, D. 1973. Species relationships in the genus *Pisum* L. Isr. J. Bot. **22**: 73–91.

Biddle, A.J. 1986. Seed treatments for peas: a review. Aspects Appl. Biol. **12**: 129–134.

Biddle, A.J., Knott, C.M., and Gent, G.P. 1988. The PGRO pea growing handbook. Processors & Growers Research Org., Peterborough, England. 264 pp.

Casey, R., and Davies, D.R. 1993. Peas: genetics, molecular biology and biotechnology. CAB International, Wallingford, Oxon, UK. 323 pp.

Davies, D.R. 1989. *Pisum sativum* L. *In* Plant resources of South-East Asia, 1, pulses. *Edited by* L.J.G. van der Maesen and S. Somaatmadja. Pudoc, Leiden, Netherlands. pp. 63–64.

Davies, D.R. 1995. Peas. *In* Evolution of crop plants. 2nd edition. *Edited by* J. Smartt and N.W. Simmonds. Longman Scientific & Technical, Burnt Mill, Harlow, Essex, UK. pp. 294–296.

Davies, D.R., Berry, G.J., Heath, M.C., and Dawkins, T.C.K. 1985. Pea (*Pisum sativum*). *In* Grain legume crops. *Edited by* R.J. Summerfield and E.H. Roberts. Collins, London, UK. pp. 147–198.

Great Britain. Ministry of Agriculture, Fisheries and Food. 1969. Peas. 4th edition. Her Majesty's Stationary Office, London, UK. 58 pp.

Gritton, E.T. 1986. Pea breeding. *In* Breeding vegetable crops. *Edited by* M.J. Bassett. AVI Publishing Co., Westport, CT. pp. 283–319.

Hedrick, U.P. 1928. Vegetables of New York. Vol. 1. Peas of New York. J.B. Lyon Company, Albany, NY. 132 pp.

Kalloo, G. 1993. Pea, *Pisum sativum* L. *In* Genetic improvement of vegetable crops. *Edited by* G. Kalloo and B.O. Bergh. Pergamon Press, New York, NY. pp. 409–425.

Lee, C.Y. 1989. Green peas. *In* Quality and preservation of vegetables. *Edited by* N.A. Eskin. CRC Press, Inc., Boca Raton, FL. pp. 159–183.

Matalova, A. 1973. A critical review of different editions of Mendel's *Pisum* paper. Folia Mendeliana, **58**: 243–254.

Muehlbauer, F.J., and McPhee, K.E. 1997. Peas. *In* The physiology of vegetable crops. *Edited by* H.C. Wien. CABI International, Wallingford, Oxon, UK. pp. 429–459.

Selman, J.D. 1978. Review—vitamin C losses from peas during blanching in water. Food Chem. **3**: 189–197.

Shoemaker, D.N., and Delwiche, E.J. 1934. Descriptions of types of principal American varieties of garden peas. U.S. Department of Agriculture, Washington, DC. 39 pp.

Sutcliffe, J.F., and Pate, J.S. (*Editors*). 1977. The physiology of the garden pea. Academic Press, London, UK. 500 pp.

Waines, J.G. 1975. The biosystematics and domestication of peas (*Pisum* L.). Bull. Torrey Bot. Club, **102**: 385–395.

Specialty Cookbooks

(Also see LENTIL.)

Brodie, S. 1978. Cooking with peas, beans and lentils. Rigby, Adelaide, Australia. 64 pp.

Dixon, P. 1980. Pulse cookery: wholesome recipes with peas, beans and lentils, includes sweet dishes. Thorsons, Wellingborough, Northamptonshire, UK. 128 pp.

Green, A. 2000. The bean bible: a legumaniac's guide to lentils, peas, and every edible bean on the planet! Running Press Book Publishers, Philadelphia, PA. 338 pp.

Jenner, A. 1992. The amazing legume: cooking with lentils, dry beans & dry peas. 6th edition. Saskatchewan Pulse Crop Growers' Association, Regina, SK. 134 pp.

Krech, I.M. 1981. Beans & peas. Primavera Books/Crown, New York, NY. 64 pp.

Longnecker, N. 1999. Passion for pulses: a feast of beans, peas and lentils from around the world. Tuart House, Nedlands, WA, Australia. 128 pp.

Pulse Growers Association of Alberta. 1985. The perfect pulses—cooking with Alberta beans, peas, lentils. Pulse Growers Association of Alberta, Lethbridge, AB. 32 pp.

Steffens, G.M. 1985. The National Gardening book of peas & peanuts. Revised edition. National Gardening Association, Burlington, VT 27 pp.

U.S. Dept. of Agriculture. 1957. Dry beans, peas, lentils...: modern cookery. Revised edition. U.S. Dept. of Agriculture, Washington, DC. 24 pp.

Peach and Nectarine

Family: Rosaceae (rose family)

Names

Scientific name: *Prunus persica* (L.) Batsch

- The English word "peach" and *persica* in the scientific name *P. persica* trace to *persicum*, the classical Latin name of the peach (meaning Persian), which reflects the Persian history of the peach.
- Peaches came to Europe and subsequently to the New World via Persia, hence its ancient name "Persian apple."
- The word "nectarine" is based on the Greek *nektar* and Latin *nectar*, the drink of the Greek and Roman gods. "Nectarine" is either a diminutive of "nectar," and is so named for its excellent taste, or an adjectival form meaning "sweet as nectar."
- An "Irish peach" is a potato.
- In Cornwall, England, the peach has been called a "suede apple."
- For information on the genus name *Prunus*, see Plum.

Plant Portrait

Peaches and nectarines are small deciduous trees up to 10 m (33 feet) in height, but generally shorter, with a trunk that sometimes reaches 30 cm (1 foot) in diameter. The trees are short-lived, often bearing fruits as early as the third year, and frequently being removed from orchards before 10 years of age because of declining productivity. Peaches arose from the mountainous areas of Tibet and western China, and were first grown in China almost 4000 years ago. They are now commercially cultivated around the world, mainly in temperate regions. The peach is the world's most popular "stone fruit," i.e., one with a large, stone-like pit, and it is often called "the Queen of Fruits." Leading producers are the United States, Italy, Spain, Greece, and India. Peaches are the third most extensively grown temperate fruit in North America, next to grapes and apples. There are thousands of named fruit cultivars of the peach, in addition to numerous ornamental forms. Peach fruits are fuzzy throughout the growing season, and are usually brushed by machines prior to marketing to remove most of the pubescence. Nectarines have been described as hairless peaches. A misconception that once was fairly common was that nectarines are hybrids of peaches and plums. In fact, a nectarine is merely a minor genetic variation of a peach. The essential difference, that of the fuzz on the fruit, is known to be determined simply by a single gene. Peach pits occasionally give rise to nectarine trees, and vice versa. In practice, nectarines often have fruit that are smaller, firmer, more aromatic, and with a distinct rich flavor. Both peaches and nectarines may be freestone, with a pit that is relatively free of the flesh, or clingstone, with the pit adhering to the flesh. Skin colors include pink-blushed creamy white, greenish white, orange–yellow, and red-blushed yellow, while the flesh varies from greenish white and pinkish white to yellow–gold and red. The flesh near the pit is often stained with red. The skin has a partition line along which it splits easily. Most peaches and nectarines are round, but flat varieties that look like hockey pucks have been known since antiquity and are still marketed today. The Western World prefers round, yellow-fleshed, freestone varieties, although white-fleshed varieties are increasing in popularity.

Culinary Portrait

The peach is largely used as a table fruit (eaten out of hand) or used fresh in fruit salads. Freestone peaches are more popular fresh than the clingstone varieties, but clingstones retain their shape better when canned. Peaches are also dried, canned, frozen, candied, and cooked. They are used in cakes, pies, yogurt, ice cream, and sorbets. Peaches are also often employed as a source of juice for fermentation and distillation, and peach brandy and liqueurs are popular. Peaches can be baked and stuffed, stewed for use in pies and jams, and poached, baked, or grilled. Cooked peaches are an excellent accompaniment for barbecued meats, and are occasionally served along with seafood, poultry, and pork. On New Year's Day in China, peach soup is traditionally consumed. Peaches spoil rapidly, and it is wise to purchase only an amount of ripe, fresh fruit that can be consumed promptly. Underripe peaches can be matured for a few days, and ripening can be hastened by placing the fruit in a paper bag, but loss due to rotting is common. Ripe fruit can be stored for a day or two in a refrigerator. Peaches should not be washed until just before eating or preparation. To preserve the color of peaches once they are cut, the flesh may be sprinkled with citrus juice.

Culinary Vocabulary

- "Peach Melba" (also known by the French *pêches Melba*) and "Melba Toast" are named after renowned Australian opera star Nellie Melba (1861–1931). Helen Porter Mitchell adopted the professional

Peach (*Prunus persica*), from *Garden and Forest* (1888–1895, vol. 5).

name Melba for her birthplace, Melbourne, Australia. Peach Melba is a dessert that was created in the late 1800s by the famous French chef Auguste Escoffier (1846–1935). Escoffier worked for some time at the Savoy Hotel in London, England, where Melba often stayed, and he named dishes in her honor. Two peach halves are poached in syrup and cooled. Each peach half is placed hollow side down on top of a scoop of vanilla ice cream, and topped with Melba sauce (a raspberry sauce) and sometimes with whipped cream and sliced almonds. (Ordering "Peach Melba" from menus often results in a disappointing preparation of inferior ice cream topped by a tinned peach with raspberry jam on top.)

- "Southern Comfort," a US specialty cordial, is a mixture of bourbon whisky, freshly pitted and peeled peaches, and peach liqueur. The name "Southern Comfort" is believed to have been given to the concoction (previously known as "Cuff and Buttons," an allusion to formal white ties and tails) by Louis Herron, a bartender in St. Louis, Missouri, in 1875.
- A "Fuzzy Navel" is a cocktail prepared with peach schnapps and orange juice. A "Fuzzy Pucker" is a cocktail prepared with peach schnapps and grapefruit juice.
- A "Georgia Peach" is a cocktail prepared with vodka, peach brandy, peach preserves, lemon juice, and chopped peaches.
- A "Peach Blow Fizz" is a cocktail prepared with mashed peaches, gin, lemon juice, heavy cream, sugar syrup, and club soda.
- A "Peach Fuzz" is a cocktail prepared with peach brandy, heavy cream, and white crème de cacao, topped with an apple schnapps float.
- A "Peach Tree Street" is a cocktail prepared with peach schnapps, cranberry juice, orange juice, and vodka, and garnished with a peach slice.
- A "Peachy" is an obsolete peach cider, popular in Colonial America.

Curiosities of Science and Technology

- The Ch'in (Qin) Dynasty (221–210 BC) was the first dynasty to unify China. The Qin emperor brutally tried to end all dissent, burning books and burying alive scholars who disagreed with his policies. A much less harmful idea that was initiated during this period was that eating peaches would preserve one's body from deterioration forever. In 1977, the perfectly preserved corpse of the wife of the Marquis of Tai was found in Ch'ang-sha, China, from the 2nd century BC. The tomb contained a bowl of peaches. The belief that peaches have a preservative role survives today in China in the tradition of *shoutao*—the long life peach—actually a steamed roll served on birthdays.
- The Shakers were strict religious Quakers who emigrated from England to the United States in 1774. They enlisted thousands of converts and gathered into 11 communities in Main, New Hampshire, Connecticut, Massachusetts, and New York. Shaker women were first to introduce the use of the peach kernel into North American cookery, and commonly used the peach pits in recipes, obviously not knowing that these contain the poison amygdalin (see ALMOND). The shakers are now extinct, not because of poisoning from peach pits, but because they practiced celibacy (they depended on recruitment of outsiders). The use of peach pits in foods was not common, but recipe books from the 19th century show that during this period pits were occasionally consumed in North America.
- Ann Lee (1736–1784), known as Mother Ann, was the founder of Shaker society in the United States. Mother Ann's Birthday Cake, named for her, is a white cake that is traditionally served on March 1, Lee's birthday. The batter was originally mixed with fresh peach twigs, so that the sap gave the cake a peach flavor, but this is no longer done.
- In England during the time of Queen Victoria (1819–1901, reigning from 1837), high-fashion dinners

were considered incomplete unless a peach was served in a nest of snowy cotton wool. At that time, peaches cost about a guinea each (about $5).

- Basketball is named for the half-bushel peach baskets used as targets by the originator, James A. Naismith (1861–1939), in 1891.
- During World War I, peach pits were commonly collected to make the charcoal used in gas-mask canisters.
- A rather drastic technique sometimes used to produce large peaches and other fruit is to cut the bark off around some of the branches. When girdled, the sugars from photosynthesis in the leaves cannot flow to the trunk and roots, so the fruits receive a large supply of nutrients, and hence grow larger and mature faster. However, the injured branches die, and must be pruned off following fruiting.
- The corporation Crayola has renamed several crayon colors during its history. In 1958, Prussian Blue was renamed Midnight Blue because, according to teachers, children could no longer relate to Prussian history. In 1962, Flesh was renamed Peach to recognize that skin colors differ. In 1999, Indian Red was renamed Chestnut, again in recognition of ethnic sensitivities.
- A record-size peach grown in Britain, weighing 411 g (14½ ounces) and measuring 30.4 cm (12 inches) in circumference, was reported in August 1984 from a 26-year-old tree grown by Mrs. Jean Bird of London.
- In 1932, James Markham obtained the first patent issued for a tree—for a peach tree.
- The peach was adopted as the official fruit of South Carolina in 1984.
- Georgia is widely called "the Peach State," and in 1995 designated the peach as the State Fruit. However, the State Flower of Georgia is the Cherokee rose, its State Tree is the live oak, and South Carolina and California produce more peaches.
- A red cheek on one side of a peach often indicates the side that has been exposed to the sun.
- The peach is extremely closely related to the almond (which see), and is easily hybridized with it. The hybrid has fruits of peach-like appearance, but they are dry.

Nectarine (*Prunus persica*), from *Revue de l'Horticulture Belge et Étrangère* (1908, vol. 34).

Key Information Sources

Childers, N.F. (*Editor*). 1975. The peach—varieties, culture, pest control, storage, marketing: an up-to-date summary of U.S. and Canadian research and world peach situation by over seventy world authorities. 3rd edition. Horticultural Publications, New Brunswick, NJ. 659 pp.

Cooper, J.R. 1955. Factors that influence production, size, and quality of peaches. Agricultural Experiment Station, College of Agriculture and Home Economics, University of Arkansas, Fayetteville, AR. 63 pp.

Coston, D.C., and Ridley, J.D. (*Editors*). 1989? Proceedings 2nd international peach symposium (Clemson University, Clemson, South Carolina, 19–23 June 1988). International Society for Horticultural Science, Wageningen, Netherlands. 362 pp.

Fideghelli, C. (*Editor*). 1985. Proceedings International Conference on Peach Growing (Verona, Italy, July 9–13, 1984). International Society for Horticultural Science, Wageningen, Netherlands. 560 pp.

Fogle, H.W., Keil, H.L., Smith, W.L., Mircetich, S.M., Cochran, L.C., and Baker, H. 1974. Peach production. Agric. Handb. No. 463. U.S. Dept. of Agriculture, Washington, DC. 90 pp.

Hedrick, U.P., Howe, G.H., Taylor, O.M., and Tubergen, C.B. 1917. The peaches of New York. J.B. Lyon Co. (Printer), Albany, NY. 541 pp.

Hesse, C.O. 1975. Peaches. *In* Advances in fruit breeding. *Edited by* J. Janick and J.N. Moore. Purdue University Press, West Lafayette, IN. pp. 285–335.

Hilaire, C., Giauque, P., Saunier, R., Mathieu, V., and Exbrayat, P. 1994. Pêche, les variétés & leur conduite. [Peaches, varieties and management.] Centre technique interprofessionnel des fruits et légumes, Paris, France. 307 pp. [In French, English summary.]

Kemp, W.S. 1973. Peach and nectarine growing. Ministry of Agriculture & Fisheries, Wellington, New Zealand. 51 pp.

LaRue, J.H., and Johnson, R.S. 1989. Peaches, plums and nectarines; growing and handling for fresh market. Cooperative Extension, University of California, Division of Agriculture and Natural Resources, Oakland, CA. 246 pp.

Layne, R.E.C. 1984. Breeding peaches in North America for cold hardiness and perennial canker resistance: review and outlook. Fruit Var. J. **38**: 130–136.

Layne, R.E.C. 1997. Peach and nectarine breeding in Canada: 1911 to 1995. Fruit Var. J. **51**: 218–228.

Monet, R. (*Editor*). 1997. Proceedings 4th International Peach Symposium (Bordeaux, France, 22–26 June, 1997). International Society for Horticultural Science, Leuven, Belgium. 2 vols.

Okie, W.R. 1998. Handbook of peach and nectarine varieties: performance in the southeastern United States and index of names. U.S. Dept. of Agriculture, Agricultural Research Service, Washington, DC. 808 pp.

Organisation for Economic Co-operation and Development. 1969. Peaches. OECD, Paris, France. 79 pp.

Rasmussen, G.D. 1978. How fresh peaches are marketed. U.S. Department of Agriculture, Washington, DC. 27 pp.

Scorza, R., and Okie, W.R. 1990. Peaches (*Prunus*). *In* Genetic resources of temperate fruit and nut crops. *Edited by* J.N. Moore and J.R. Ballington, Jr. International Society for Horticultural Science, Wageningen, Netherlands. pp. 177–231.

Seelig, R.A., and Bing, M.C. 1990. Peaches. *In* Encyclopedia of produce. [Loose-leaf collection of individually paginated chapters.] *Edited by* R.A. Seelig and M.C. Bing. United Fresh Fruit and Vegetable Association, Alexandria, VA. 18 pp.

Weinberger, J.H. 1975. Growing nectarines. U.S. Agricultural Research Service, Washington, DC. 22 pp.

Wolgamot, I.H. 1951. Peaches: facts for consumer education. U.S. Dept. of Agriculture, Bureau of Human Nutrition and Home Economics, Washington, DC. 18 pp.

Zucconi, F. 1986. Peach. *In* CRC handbook of fruit set and development. *Edited by* S.P. Monselise. CRC Press, Boca Raton, FL. pp. 303–321.

Specialty Cookbooks

Bear Wallow Books. 1995. Old southern recipes, with peach recipes, including historical notes. Bear Wallow Books, Indianapolis, IN. 31 pp.

California Canning Peach Growers. 1934. California canned peach recipe book. California Canning Peach Growers, San Francisco, CA. 24 pp.

Cling Peach Advisory Board (San Francisco). 1950–1959. How famous California chefs use canned cling peaches. Cling Peach Advisory Board, San Francisco, CA. 28 pp.

Fawcett, I.N. 1981. Peaches. I.N. Fawcett, Port Moody, BC. 27 pp.

Georgia Agricultural Commodity Commission for Peaches. 1972. Georgia peach recipes. University of Georgia, Athens, GA. 20 pp.

Golden West Publishers. 2003. Peach lovers cookbook. Golden West Publishers, Phoenix, AZ. 93 pp.

Horton, E.M., and Hicks, A. 1983. The peach sampler. At Home Enterprises, Lexington, SC. 195 pp.

McPherson, G. 2000. The passion for peaches cookbook. Quixote Press, Wever, IA. 192 pp.

Pearl City Highlands Elementary School. 1994. A taste of peaches: recipes. Pearl City Highlands Elementary School, Pearl City, HI. 216 pp.

Pearl City Highlands Elementary School. 1996. Another taste of peaches: recipes. Pearl City Highlands Elementary School, Pearl City, HI. 213 pp.

Rhodes, N. 1993. Everything's "peaches." N. Rhodes, Stratford, OK. 72 pp.

Rubin, C., and Rubin, J. 1974. Peach cookbook. Emporium Publications, Newton, MA. 91 pp.

PEANUT

Family: Fabaceae (Leguminosae; pea family)

Names

Scientific name: *Arachis hypogaea* L.

- The word "peanut" is descriptive of the fruit: the pod is like that of the pea, the seeds inside are pea-sized, and the firm, crunchy texture of the seeds is nut-like.
- Botanically, the peanut fruit is not a "nut" (technically, a nut is a hard, indehiscent, one-seeded fruit); a peanut is a legume, i.e., a fruit of the legume family, such as a pea pod.
- There are numerous names for the peanut (so named for the nutty flavor). The names goober, goober pea, groundpea, groundnut, and pindar are frequently used in the southeastern United States, while groundnut (ground nut), earth nut, and peanut are usual in Europe, Africa, and Asia. Most "nuts" are harvested aerially from trees, and such names as earth almond, earthnut, and groundnut indicate that peanuts are obtained from under the ground. Generally, "groundnut" is much more commonly encountered than "peanut" outside of North America. "Ground nut oil" is peanut oil. However, "groundnut" also refers to an edible tuber-bearing species (*Apios americana* Medik.).
- The name "goober" traces to the African (Kongo or Angolese) *nguba*, peanut, a name used by slaves from West Africa who cultivated peanuts in the southern colonies of North America. Goober has acquired the meanings "rustic" or "redneck" in the southern United States. Georgia is sometimes called the "Goober State" because of the importance of peanuts there (but it is more commonly known as "the Peach State").
- The "hog peanut" is *Amphicarpa bracteata* (L.) Fernald, an American species that produces peanut-like fruits.
- *Vigna subterranea* (L.) Verd. [more commonly known as *Voandzeia subterranea* (L.) Thouars ex DC.] is best known as Bambarra (Bambara) groundnut (named for the district of Bambarra in Mali, West Africa), but also as baffin pea, Congo groundnut, Congo goober, earth pea, ground bean, hog peanut, Kaffir pea, Madagascar groundnut, Njugo bean, stone groundnut, voandzou, and underground bean, most names of which could be confused with the names of the peanut. Indeed, the plant is similar to the common peanut, producing pods and edible seeds on or just below the ground. Bambarra groundnut is cultivated throughout its native Africa in large sections of the equatorial region, in Madagascar, Asia, and South America, and occasionally in the warmest parts of the United States, such as Florida.
- Still another food plant in the legume family that produces "nuts" underground is *Macrotyloma geocarpum* (Harms) Maréchal & Baudet (*Kerstingiella geocarpa* Harms), known variously as geocarpa bean, geocarpa groundnut, Kersting's groundnut, and Hausa groundnut. It is grown in the drier parts of West Africa.
- The genus name *Arachis* is based on the Greek, *arachos*, a name used for a plant of the pea family, suggested to be a clover (genus *Trifolium*) or vetch (genus *Vicia*).
- *Hypogaea* in the scientific name is from the Greek *hypo*, beneath, and *gaia*, the earth, a description of the location of the fruits.

Plant Portrait

The common peanut is an annual herb, 15–70 cm (6–28 inches) high, which occurs as erect, bushy forms and as trailing runners. Normally, flowers self-pollinate, and after fertilization, the flower stalk elongates greatly, pushing the developing fruit 2.5–10 cm (1–4 inches) into the soil. The shell of a peanut corresponds to the pod of peas and beans. The common peanut, one of the oldest of domesticated plants, arose in South America, probably in the area from southern Bolivia to northwestern Argentina, likely as a hybrid between two other wild species of the genus *Arachis*. The peanut is one of the oldest of domesticated plants. Archaeological records show it was present in Peru perhaps as long ago as 3000 BC. Today, the peanut is one of the most important oilseed crops and food legumes in the world. It is adapted to areas with hot summers and a long growing season. Peanuts are grown in all tropical and subtropical areas. Leading producers include China, Japan, India, Indonesia, and several African countries, particularly Nigeria and Senegal. In the United States, the peanut grows best in the southeastern region, mainly in Georgia, Texas, Alabama, North Carolina, Oklahoma, Virginia, and Florida.

Peanut (*Arachis hypogaea*), from Köhler (1883–1914).

In the trade, four basic types of peanut are recognized: Virginia—large confectionery or cocktail nuts; Runner—medium-sized nuts sometimes sold as "beer nuts" and increasingly used in peanut butter; Spanish—small red-skinned nuts commonly used in peanut butter, for shelling peanuts, and for oil production; and Valencia—usually with 3–5 nuts per pod, and frequently roasted in the pod and sold as peanuts in the shell. In the early part of the 19th century, commercial peanuts were largely roasted and consumed out of the shell. Peanuts acquired a peculiar attraction at carnivals, zoos, circuses, and sporting event. Later, salted peanuts and peanuts incorporated into confectionery items became widely available.

Culinary Portrait

Peanut oil is the world's second most important vegetable oil, and about two-thirds of the world crop is crushed for oil, used mainly for cooking. The oil has very little taste and is nearly colorless, characteristics considered very desirable. Peanut oil is also used as a salad oil, for producing margarine, for deep-frying, and as shortening in pastry and bread.

In Western countries, the most popular form of consuming peanuts is as peanut butter, a simple product composed of ground, roasted peanuts and salt, with stabilizers sometimes added to prevent oil separation. Dr. John Harvey Kellogg (1852–1943), a physician in St. Louis, and the health-food faddist of breakfast-cereal fame, is commonly credited with inventing peanut butter in 1893. He intended it for use by invalids with poor teeth. There is some dispute about who first created peanut butter in the United States, and a St. Louis physician is often credited with formulating peanut butter as a food for invalids in 1899. Others probably prepared concoctions similar to peanut butter in earlier times; in Africa, India, South America, and Indonesia, types of peanut-butter paste have been made for centuries. "Crunchy" peanut butter has bits of roasted nuts mixed into it. Peanut butter is extremely nutritious, digestible, and cheap, and is one of the best values available for the food dollar. Peanut butter has excellent stability, and has been found to remain edible after 2 years of storage at room temperature, longer when refrigerated. However, eventually peanut butter becomes stale and rancid.

Peanuts are eaten in a wide variety of forms—boiled, fried, steamed, roasted, purêed, and ground, and consumed as a spread, added to soups, soufflés, glazes, ice cream, confections, and numerous other foods. Peanut brittle and peanut butter cookies are particular favorites. Peanut flour and meal are added to a number of food products, especially in the Third World. Green (immature) peanuts, boiled in salted water, are sometimes used as a vegetable in the southern United States. "Dry-roasted" peanuts have been oven-roasted, not deep-fat fried, and tend to have fewer calories than fried peanuts.

It is very important that peanuts not be dried or stored under conditions that promote the growth of molds. A fungus infestation of peanuts produces aflatoxins, which are harmful, dangerous compounds associated with liver cancer. Poor storage conditions in Africa and some Pacific region countries make aflatoxin a significant problem in these areas. In the Orient, people prefer the flavor of crude peanut oil, which is often high in aflatoxins.

Raw peanuts do not keep well, but in Western countries peanuts are generally available roasted. Roasted peanuts can be stored at room temperature for a month or more, but are preferably kept in a refrigerator (up to 9 months if unshelled, 3 months if shelled) or freezer (up to 9 months) to prevent rancidity.

Sensitivities to milk and eggs and some other materials are more prevalent among humans than allergies to peanuts. Nevertheless, peanuts are believed to cause more serious allergic reactions in North America than any other food product. Part of the problem is that peanuts occur in a very wide range of commercial food products. Perhaps 2% of American children have an allergy to proteins in peanuts, although the exact figure has been debated. While many childhood allergies do not persist into adulthood, generally a peanut allergy is permanent. A reaction to peanuts (anaphylaxis) can be caused by only a few milligrams of nuts, and those at risk should wear medic-alert bracelets and carry preloaded epinephrine syringes and antihistamines. Peanut oil (at least the usual grade sold in North America and Europe) contains no detectable protein and is not allergenic to those suffering from peanut allergies. However, in other countries, the oil may have enough protein to cause an allergic reaction. Research published in the *Journal of the American Medical Association* (JAMA, 2001, **285**: 1746–1748) showed that for some nursing mothers who ate 50 g (about half a cup) of dry roasted peanuts, peanut proteins appeared rapidly in their breast milk, often within 1 hour of eating the nuts. This may explain why up to 80% of children with peanut allergies react on what appears to be their first exposure to peanut products. The children must have been previously exposed to develop the allergy. It was recommended that nursing mothers who have someone with a peanut allergy in their immediate family should avoid eating peanuts or peanut products during the times they are nursing their children.

Culinary Vocabulary

- In the United Kingdom, "jelly" is (1) Jello; (2) a seedless preserve made from fruit, sugar, and pectin; and (3) jam. In England, a "peanut butter and jam sandwich" is equivalent to the American "peanut butter and jelly sandwich."
- In 1926, the General Assembly of Virginia passed the following statute: "Genuine Smithfield hams [are those] cut from the carcasses of peanut-fed hogs, raised in the peanut-belt of the State of Virginia or the State of North Carolina, and which are cured, treated, smoked, and processed in the town of Smithfield, in the State of Virginia." Virginia's Smithfield was founded as a town in 1752, and settlers there exported hams to England as early as 1639. Queen Victoria (1819–1901), among many others, was a Smithfield ham fan. The name "Smithfield" traces to London's livestock center, called Smithfield even before the days of Shakespeare.
- A "buckeye" is a candy prepared by dipping small balls of a creamy peanut butter mixture in melted chocolate. It is popular in Ohio, the "Buckeye State."

Curiosities of Science and Technology

- The discovery of an apparent fossilized peanut over 10 000 years old at Chlien-shan-yang in China has cast some doubt on the widely held theory that the peanut arose in South America. The issue of a possible alternate site of origin probably can not be resolved until more evidence is found.
- Peanuts were highly valued by ancient Peruvian Indians, and were used as money, medicine, and status symbols as well as food. The Peruvian tomb of a Moche warrior priest discovered in 290 AD contained gold and silver jewelry shaped like peanuts.
- "Mr. Peanut," the trademark of Planters Nut, now a division of RJR Nabisco, was created by a 13-year-old Virginia schoolboy, Antonio Gentile, in a logo contest sponsored by Planters in 1916. Antonio submitted a drawing of a peanut person with arms and crossed legs, which he labeled "Mr. Peanut," and was awarded the grand prize of $5. Later, a professional illustrator added a top hat, monocle, white gloves, and cane.
- In 1919, the inhabitants of Enterprise, Coffee County, Alabama, erected a monument in appreciation to the boll weevil, which devastated the cotton crop in the early 1900s, compelling the local farmers to turn to the much more profitable cultivation of peanuts, which led to prosperity.
- Dr. George Washington Carver (1864–1943) is of legendary fame for developing new uses for the peanut early in the 19th century. He began his work at Tuskeegee Institute in Alabama in 1903, intending to find a crop that could replace cotton, which was being destroyed by the boll weevil. Carver published a treatise documenting 105 home-kitchen recipes for using peanuts, and is commonly credited with having discovered more than 300 useful derivatives as a result of his research in food chemistry. His peanut-based creations included flour, ink, shaving cream, face cream, soap, shampoo, shoe polish, milk, linoleum, and insulation board. Carver is considered by many to be the father of the peanut industry. He improved the nutrition of poor inhabitants of the southern US, and served as a role model to blacks, at the time subjected to widespread racial prejudice. Carver has been honored in the US by commemorative stamps, a commemorative coin, a national monument, the naming of numerous schools after him, election to many prestigious academic institutions, and by George Washington Carver Recognition Day (Jan. 5).
- George Washington Carver once prepared a meal based only on peanuts for a group of Alabama businessmen in order to demonstrate the versatility of the crop. The menu included peanut soup, peanut "chicken," peanut vegetable, peanut bread, salad with peanut-oil dressing, peanut ice cream, peanut cookies, peanut candy, and peanut "coffee."
- A grandiose attempt to grow peanuts on a huge scale in east Africa was initiated in 1946 by the British government, but had to be abandoned in a few years after over $100 million had been wasted when conditions proved unsuitable for the crop. As a result, the phrase "groundnut scheme" in Britain came to mean an expensive failure or ill-considered enterprise. The fiasco illustrates the need for adequate preliminary study before undertaking large-scale cultivation of new crops.
- As often happens with well-known food plants, the word "peanut" has acquired several meanings in English. "Peanut" is slang for something small or inconsequential, specifically a person who is small in stature or regarded as being insignificant. This is the origin of "Peanuts," the name of Charles Schulz's comic strip. Another informal meaning of peanut is "a very small amount of money; a trifling sum." The phrase "to work for peanuts" reflects the low status in which the peanut was held in North America in early times. It may have originated in the 19th century slave plantations of the southern states. "Peanut Gallery" is an American slang phrase dating to 1888, referring to the balcony section of a theater, and denotes the cheap seats where the lower classes ate peanuts. The term was popularized in the 1950s by the television show Howdy Doody in which the host Buffalo Bob would call the audience of children the peanut gallery. Peanuts have also acquired the meaning of peanut-shaped pieces of polystyrene, used in cushioning items during shipment.
- Peanut oil is used for underwater cooking in submarines because it does not smoke unless heated above 232 °C (450 °F).
- Sylvester, Georgia, claims to be the "Peanut Capital of the world." So does Suffolk, Virginia, and Dothan, Alabama.
- Bar owners offer salted peanuts because it makes their clients thirsty. This happens because of the phenomenon of osmosis—the salty food causes water to flow out of cells of the digestive system in contact with the increased salinity, dehydrating the cells, and producing the sense of thirst.

- The peanut is a member of the legume family, most species of which indirectly obtain nitrogen from the air with the aid of bacteria of such genera as *Rhizobium*. Bacteria live in nodules on the roots and convert nitrogen from the atmosphere into a form that can be used by the plants. Therefore, it is generally unnecessary to apply nitrogen fertilizer, or at least very much nitrogen.
- It has been estimated that the mental energy Shakespeare required to write *Hamlet* (often considered the greatest literary work in the English language) could have been supplied from several peanuts.
- It takes about 550 peanuts to make a 12-ounce (340 g) jar of peanut butter.
- Peanut butter sticks to the roof of your mouth because of hydration of the peanut protein. The high level of protein in the peanut butter draws the moisture away from your mouth, like a sponge soaks up water.
- Arachibutyrophobia (pronounced a-ra-kid-bu-ti-ro-pho-bi-a) has been defined as fear of peanut butter, and more imaginatively as fear of peanut butter sticking to the roof of your mouth.
- Eighty-three percent of Americans purchase peanut butter. Seventy percent of all peanut butter sold is smooth, and the remaining 30% is chunky. Children and women tend to prefer creamy, while men tend to choose chunky.
- Peanut butter and jelly is the most popular sandwich in the United States.
- The world's largest peanut butter and jelly sandwich was created November 6, 1993, in Peanut, Pennsylvania. It measured 12 m (40 feet) in length and contained 68 kg (150 pounds) of peanut butter and 23 kg (50 pounds) of jelly.
- Ninety-six percent of people put the peanut butter on before the jelly when making a peanut butter and jelly sandwich.
- The average American child will have eaten 1500 peanut butter sandwiches by the time of high-school graduation.
- According to a Yale University study, coffee and peanut butter are the two most recognizable food scents.
- According to the United States Food and Drug Administration Guidelines (No. 20), "An average of two rodent hairs per one hundred grams of peanut butter is allowed."
- M&M's® Peanut Chocolate Candies were introduced in 1954 (Chocolate M&M's were first sold in 1941). The tiny white "M" on each candy is put on by a conveyor printing press, but because peanuts vary in size, the press is automatically adjusted to prevent smashing the candies.
- In 2001, food researchers at Oklahoma State University test-marketed sliced peanut butter in Oklahoma, and generated huge media coverage. The sliced peanut butter was packaged like sandwich cheese, and was 90% peanut butter with a filler additive to make the slices hold their form. One newspaper account asked, "Now that we have sliced bread and sliced peanut butter, who is working on sliced jelly?"
- Peanuts are often used by dentists for tests of jaw muscle strength.

Key Information Sources

American Peanut Research and Education Association. 1973. Peanuts—culture and uses [a symposium]. American Peanut Research and Education Association, Stillwater, OK. 684 pp.

Arant, F.S. (*Editor*). 1951. The peanut, the unpredictable legume. National Fertilizer Association, Washington, DC. 333 pp.

Ashley, J.M. 1984. Groundnut. *In* The physiology of tropical field crops. *Edited by* P.R. Goldsworthy and N.M. Fisher. John Wiley & Sons, Chichester, UK. pp. 453–494.

Beattie, J.H., Poos, F.W., and Higgins, B.B. 1954. Growing peanuts. U.S. Dept. of Agriculture, Washington, DC. 54 pp.

Bunting, A.H., Gibbons, R.W., and Wynne, J.C. 1985. Groundnut (*Arachis hypogaea* L.). *In* Grain legume crops. *Edited by* R.J. Summerfield and E.H. Roberts. Collins, London, UK. pp. 747–800.

Dwyer, J.T., and Sandhu, R. 2000. Peanuts. *In* The Cambridge world history of food. *Edited by* K.F. Kiple and K.C. Ornelas. Cambridge University Press, Cambridge, UK. pp. 364–374.

Gibbons, R.W., Bunting, A.H., and Smartt, J. 1972. The classification of varieties of groundnut (*Arachis hypogaea*). Euphytica, **21**: 78–85.

Gregory, W.C., Krapovickas, A., and Gregory, M.P. 1980. Structures, variation, evolution and classification in *Arachis*. *In* Advances in legume science. *Edited by* R.J. Summerfield and A.H. Bunting. Royal Botanic Gardens, Kew, UK. pp. 469–481.

Hermann, F.J. 1954. A synopsis of the genus *Arachis*. U.S. Dept. of Agriculture, Washington, DC. 26 pp.

Johnson, F.R. 1964. The peanut story: a thoroughly researched rags to riches story of the once humble legume which has risen to worldwide importance. Johnson Publishing Co., Murfreesboro, NC. 192 pp.

Maiti, R.K., and Wesche-Ebeling, P. (*Editors*). 2002. The peanut (*Arachis hypogaea*) crop. Science Publishers, Enfield, NH. 376 pp.

McArthur, W.C., Krenz, R.D., and Garst, G.D. 1985. U.S. peanut production practices and costs. U.S. Dept. of Agriculture, Economic Research Service, National Economics Division, Washington, DC. 50 pp.

Mixon, A.C. 1980. Potential for aflatoxin contamination in peanuts (*Arachis hypogaea* L.) before and soon after harvest: a review. J. Environ. Qual. **9**: 344–349.

Moss, J.P., and Feakin, S.D. (*Editors*). 1985. Proceedings of an international workshop on cytogenetics of *Arachis*, ICRISAT Center, Patancheru, India, 31 Oct.–2 Nov. 1983. International Crops Research Institute for the Semi-Arid Tropics, Patancheru, India. 191 pp.

Nwokolo, E. 1996. Peanut (*Arachis hypogaea* L.). *In* Food and feed from legumes and oilseeds. *Edited by* J. Smartt and E. Nwokolo. Chapman and Hall, London, UK. pp. 49–63.

Pattee, H.E., and Young, C.T. (*Editors*). 1982. Peanut science and technology. American Peanut Research and Education Society, Yoakum, TX. 825 pp.

Raman, V.S. 1976. Cytogenetics and breeding in *Arachis*. Today & Tomorrow's Printers & Publishers, New Delhi, India. 84 pp.

Resslar, P.M. 1980. A review of the nomenclature of the genus *Arachis* L. Euphytica, **29**: 813–817.

Shewfelt, A.L., and Young, C.T. 1977. Storage stability of peanut-based foods: a review. J. Food Sci. **42**: 1148–1152.

Shorter, R., and Patanothai, A. 1989. *Arachis hypogaea* L. *In* Plant resources of South-East Asia, 1, pulses. *Edited by* L.J.G. van der Maesen and S. Somaatmadja. Pudoc, Leiden, Netherlands. pp. 35–39.

Singh, A.K. 1995. Grundnut. *In* Evolution of crop plants. 2nd edition. *Edited by* J. Smartt and N.W. Simmonds. Longman Scientific & Technical, Burnt Mill, Harlow, Essex, UK. pp. 246–250.

Young, C.T., and Schadel, W.E. 1990. Microstructure of peanut seed: a review. Food Struct. **9**: 317–328.

Woodroof, J.G. 1983. Peanuts: production, processing, products. 3rd edition. AVI Publishing Co., Westport, CT. 414 pp.

Specialty Cookbooks

(Note: There are over 100 cookbooks dedicated to the peanut (the majority to peanut butter). The following are examples.)

Asher, S. 1977. The great American peanut book. Tempo Books, New York, NY. 190 pp.

Boze, A. 1987. The ultimate peanut butter cookbook. Hillcrest Publishing Co., Los Angeles, CA. 118 pp.

Feinman, J. 1977. Peanuts—in a nutshell! Kensington Publishing Corp., New York, NY. 192 pp.

Frank, D.C. 1976. The peanut cookbook. C.N. Potter, New York, NY. 110 pp.

Hoffman, M. 1996. The peanut butter cookbook. HP Books, New York, NY. 120 pp.

Hubert, M.L. 1984. Peanuts, a southern tradition: a collection of peanut recipes. Georgia Peanut Commission, Tifton, GA. 122 pp.

Kaufman, W.I. 1965. The peanut butter cookbook. Simon and Schuster, New York, NY. 160 pp.

Leahy, L.R., and Maguire, J. 1994. The world's greatest peanut butter cookbook. Villard Books, New York, NY. 117 pp.

National Peanut Council. 1941. The nut that is not a nut: peanuts, their food value and over 100 sure-fire peanut recipes to add zest to your menus. National Peanut Council, Suffolk, VA. 64 pp.

Peanut Museum. 1977. Complete peanut cook-book from the Peanut Museum, Plains, Georgia: over 125 recipes on how to prepare peanuts. Peanut Museum, Plains, GA. 47 pp.

Smith, A.F. 2002. Peanuts: the illustrious history of the goober pea. University of Illinois Press, Urbana, IL. 234 pp.

Zisman, L., and Zisman, H. 1985. The great American peanut butter book: a book of recipes, facts, figures, and fun. St. Martin's Press, New York, NY. 130 pp.

PEAR

Family: Rosaceae (rose family)

Names

Scientific names: *Pyrus* species

- Common pear—*P. communis* L. (a few cultivars such as Kieffer and Le Conte are hybrids between *P. communis* and the Asian pear)
- Asian pear—*P. pyrifolia* (Burm. f.) Nakai (some cultivars are hybrids with *P.* ×*bretschneideri* Rehder, *P. ussuriensis* Maxim., and other species)

- The English word "pear" and the genus name *Pyrus* are derived from the classical Latin word for the pear, *pirus*.
- The Asian pear is also called "pearapple," although not really a hybrid between the apple and the pear. Other names are Chinese pear, Chinese sand pear, Japanese pear, Korean pear, Nashi (Japanese for "pear"), Oriental pear, salad pear, and sand pear.
- The Bartlett Pear, the most popular pear in North America, was discovered in Berkshire, England, in the 17th century by schoolmaster John Stair, who sold cuttings to a horticulturist named Williams, who named it after himself. The Williams pear was brought by the early colonists to Dorchester, Massachusetts, where nurseryman Enoch Bartlett renamed it after himself in 1812. Ever since, the pear has been known as the Bartlett pear in the United States and the Williams pear (or William Bon Chrétien) in England.
- *Communis* in the scientific name *P. communis* is Latin for "common."
- *Pyrifolia* in the scientific name *P. pyrifolia* is Latin for "pear-leaved." The observant reader may ask why anyone would create a name like *Pyrus pyrifolia*, which means "the pear with leaves like a pear tree." Note the "(Burm. f.) Nakai" that follows the name *P. pyrifolia*. These are both the original and the new authors of the scientific name. The first author is "Burm. f.," an abbreviation for "Burman's son" ("f." is the Latin for *filius*, son; the "son" is necessary to avoid confusion with names coined by the father). Burman's son thought that this species was a fig, and his name for the species in Latin was "the fig with leaves like a pear." Later, the second author Nakai transferred the species to the pear genus *Pyrus*, and the technical rules of plant nomenclature require that the descriptive word *pyrifolia* also be transferred, even though the new name created was repetitious.

Plant Portrait

The common pear, a derivative of European and Asian species, was probably cultivated in prehistoric times. Later, the early Greeks and Romans selected a number of varieties. Today, China is the largest pear producer, considerable quantities are produced in Chile, and many countries, notably the United States, Italy, and Russia, produce a large harvest of pears. Pear trees may live 100 or even to 300 years, and if unpruned may reach a height of 15 m (50 feet). Grown on their own roots, pears become very large trees—too large to easily harvest the fruit, so they are commonly dwarfed by growing them on quince rootstocks. Pears range in shape from spherical to bell-shaped. The skin may be various shades of green, brown, yellow, or red, while the flesh is usually white or cream colored. Ripe pears vary in flavor from spicy to sweet and tart. The flesh is usually fine-textured, but is slightly gritty near the center of some varieties, while juiciness and fragrance also differ according to variety. The apple-like core may have up to 10 seeds. There are over 5000 varieties of the common pear. Leading cultivars are Bartlett, D'Anjou, and Bosc, which make up about 90% of US production. In the United States, pears rank second only to apples as the most popular tree fruit. The common pear is hard when picked from the tree, and becomes soft after harvest as the result of the starch converting to sugar. Unlike Asian pears, common pears can not be ripened fully on trees because the interior tissues turn brown and deteriorate. Maturation time depends on variety, some kinds harvested in summer, others in the fall, or even in the winter in some warmer regions. Some varieties change color on ripening, others remain green.

Asian pears are native to eastern Asia, and like common pears, have probably been cultivated for more than 3000 years. They are slightly less cold hardy than the common pear, and are grown mainly in China, Korea, Japan, and Taiwan, with smaller amounts produced in Brazil, New Zealand, and the United States. About 2000 ha (5000 acres) of Asian pears are planted in California, Oregon, and Washington, and Asian pears are also imported to North America. There are hundreds of

Pear tree (*Pyrus communis*), from Loudon (1844, vol. 6).

varieties, and three basic types: (1) round or flat fruit with green-to-yellow skin, (2) round or flat fruit with bronze–russet skin, (3) pear-shaped fruit with green or russet skin. Size ranges from that of small conventional apples to some weighing more than 0.45 kg (1 pound).

Culinary Portrait

Dessert pears are sweet, melting, and juicy at maturity, while pears used mainly for cooking, canning, and preserving are harder. Dessert pears are generally eaten fresh out of hand, but are also consumed in fruit cups. Pears used for eating raw can be cooked, but varieties specialized for cooking do not make suitable dessert pears. Pears are cooked in almost as many ways as are apples, and can be used for making jelly, jam, juice, compôte, vinegar, and liqueurs. The fruit is sometimes cooked in syrup or wine, and goes well with ice cream, chocolate, ginger, apples, and quinces. An Italian tradition is to consume fresh pears with Gorgonzola cheese, and pears also are compatible with Brie, Camembert, Cheddar, goat's cheese, and Roquefort. "Perry" is an alcoholic cider made from pears. In Europe, perry is often made from the snow pear, *P. nivalis* Jacq. Pears are quite perishable. They should be allowed to ripen at room temperature. The fruit is generally ready to eat when the skin yields to slight pressure, especially at the stem end. Pears will keep for a short period in a refrigerator, but they should be stored away from strong-smelling foods to prevent absorption of odors. The fruit discolors when exposed to air; to prevent browning, the fruit may be sprinkled with citrus juice.

Asian pears usually look more like apples than pears, and indeed have hard, crisp, juicy flesh like apples. The fruits are generally crisper and often somewhat grittier than apples, but mildly sweet. Unlike the common pear, which takes a week or so to soften sufficiently to eat after being taken out of cold storage, Asian pears are ready to eat immediately (as noted above, they are picked when ripe), and this feature contributes to their growing popularity. The Asian pear is very firm, even when ripe, but is fragile because of its thin skin, and surface bruises may be present. Although the skin is perfectly edible, the fruit is often peeled in order to better appreciate the subtle flavor of its flesh. Asian pears are often sliced horizontally, showing off the attractive, star-shaped seed core. This fruit is best consumed on its own, as its delicate flavor can be masked when combined with other foods. Nevertheless, Asian pears are added to salads and Chinese-style sautéed dishes. The fruit remains firm when cooked, requiring a longer cooking time than the common pear, and a longer poaching or baking time in the oven. Asian pear blends well with cream cheese and yogurt. The juice is excellent. The pears can be stored in the refrigerator, sometimes for as long as 2 months, depending on variety, preferably individually wrapped in paper towels and placed in a perforated plastic bag.

Culinary Vocabulary

- "Pears Belle Hélène" (often known in France as *poires Hélène*) is a classic French dish with poached pears on a bed of vanilla ice cream topped with chocolate sauce and whipped cream, like a sundae. The dish is named for the heroine of the satirical opera *La Belle Hélène* by the French composer Jacques Offenbach (1819–1890), which was first presented in 1864. Just who gave the name to the dish is not known.
- *Poires Melba* (French for "pears Melba") is like *Pêches Melba* (peach melba), but made with pears instead of peaches (see PEACH).
- *Poires cardinal* is a dessert of poached pears coated with kirsch flavor, sweetened with raspberry sauce, and garnished with chopped almonds.
- *Poires Condé* is a dessert of poached pears arranged on a savarin, served with apricot syrup, and perfumed with kirsch (cherry liqueur). [A savarin is a rich yeast cake named after Anthelme Brillat-Savarin (1755–1826), French politician and food writer.]

Pear (*Pyrus communis*), branch in flower, from Oeder et al. (1761–1883, plate 2118).

- *Poire William* is a French or Swiss clear eau-de-vie prepared from Williams pears, and occasionally marketed with a pear in the bottle (the result of placing a bottle over the budding fruit and letting it grow inside).
- *Poire au vin rouge* is a French dessert of pears poached in red wine.
- "Perry," the alcoholic preparation made from pears, was known as "merry legs" in England because it caused people to get tipsy.

Curiosities of Science and Technology

- Queen Elizabeth I (1533–1603) granted the City of Worcester, England, the right to use the pear as its official insignia for the skill of its citizens in preparing perry.
- Pears ripen faster at room temperature when sealed in a plastic bag with ripe bananas.
- Pear seeds, like those of apples, contain poisonous cyanogenic glucosides, and are toxic if eaten in large quantities.
- The chemical pentyl acetate contributes to the aroma of pears. Pentyl acetate is used in the pear-shaped candy "peardrops," and has become a standard for describing aromas, especially in wines.
- Many cliff-nesting sea birds lay pointed, pear-shaped eggs. Pear-shaped eggs pivot about their axis rather than roll off the cliff, as round eggs would.
- Australian Warren Yeomans harvested a 2.1-kg (4.6-pound) pear, believed to be the world's largest (reported in *The Newsletter of the University of New England*, Volume 40, Number 10, 11 June 1999).

Fruits of pear (*Pyrus communis*), from Anderson (1874). Madam Treyve pear at top, Souvenir du Congress pear at bottom.

- According to an old Korean proverb, "Eating pears cleans the teeth." In fact, pear flesh contains abrasive hard cells (called "stone cells") that will serve like the abrasives added to toothpaste to help clean teeth.
- A pear tree may produce over 50 000 flowers.
- Asian pears are often marketed in plastic mesh "booties" to avoid bruising.

Key Information Sources

Arbury, J., and Pinhey, S. 1997. Pears. Wells & Winter, Maidstone, UK. 103 pp.

Batjer, L.P., and Magness, J.R. 1967. Commercial pear growing. Agricultural Research Service, U.S. Dept. of Agriculture, Washington, DC. 47 pp.

Bell, R.L. 1990. Pears (*Pyrus*). *In* Genetic resources of temperate fruit and nut crops. *Edited by* J.N. Moore and J.R. Ballington, Jr. International Society for Horticultural Science, Wageningen, Netherlands. pp. 657–697.

Bethell, R.S. 1978. Pear pest management. Division of Agricultural Sciences, University of California, Berkeley, CA. 234 pp.

Beutel, J.A. 1990. Asian pears. *In* Advances in new crops. *Edited by* J. Janick and J.E. Simon. Timber Press, Portland, OR. pp. 304–309.

Brooke, J. 1956. Dessert pears. Hart-Davis, London, UK. 106 pp.

Bunyard, G. 1911. Apples & pears. T.C. & E.C. Jack, London, UK. 115 pp.

Carrera, M. (*Editor*). 1989. Fifth international symposium on pear growing, Zaragoza, Spain, 24–27 May 1988. International Society for Horticultural Science, Wageningen, Netherlands. 233 pp.

Chevreau, E., and Skirvin, R.M. 1992. Pear. *In* Biotechnology of perennial fruit crops. *Edited by* F.A. Hammerschlag and R.E. Litz. CAB International, Wallingford, UK. pp. 263–276.

Fidler, J.C. 1973. The biology of apple and pear storage. Commonwealth Agricultural Bureaux, Farnham Royal, East Malling, UK. 235 pp.

Forshey, C.G., Elfving, D.C., and Stebbins, R.L. 1992. Training and pruning apple and pear trees. American Society for Horticultural Science, Alexandria, VA. 166 pp.

Great Britain. Ministry of Agriculture, Fisheries and Food. 1973. Pears. Her Majesty's Stationary Office, London, UK. 71 pp.

Hedrick, U.P., Howe, G.H., Taylor, O.M., Francis, E.H., and Tukey, H.B. 1921. The pears of New York. J.B. Lyon Co., Albany, NY. 636 pp.

Iwahori, S. (*Editor*). 2002. Proceedings of the international symposium on Asian pears commemorating the 100th anniversary of "Nijiseiki Pear," Kurayoshi, Tottori, Japan, 25–29 Aug. 2001. International Society for Horticultural Science, Leuven, Belgium. 2 vols.

Jones, A.L., and Aldwinckle, H.S. 1990. Compendium of apple and pear diseases. APS Press, St. Paul, MN. 100 pp.

Layne, R.E.C., and Quamme, H.A. 1975. Pears. *In* Advances in fruit breeding. *Edited by* J. Janick and J.N. Moore. Purdue University Press, West Lafayette, IN. pp. 38–70.

Lipe, J.A., Lyons, C.G., and Stein, L.A. 1988. Home fruit production, pears. Texas Agricultural Extension Service, Texas A & M University System, College Station, TX. 7 pp.

Meheriuk, M. 1994. Postharvest disorders of apples and pears. Revised edition. Agriculture and Agri-Food Canada, Ottawa, ON. 67 pp.

Oyen, L.P.A. 1991. *Pyrus pyrifolia* (N.L. Burman) Nakai. *In* Plant resources of South-East Asia, 2, edible fruits and nuts. *Edited by* E.W.M. Verheij and R.E. Coronel. Pudoc, Leiden, Netherlands. pp. 272–276.

Patchen, G.O. 1971. Storage for apples and pears. Agricultural Research Service, U.S. Dept. of Agriculture, Washington, DC. 51 pp.

Retamales, J.B. (*Editor*). 1998. Seventh international symposium on pear growing, Talca, Chile, 19–22 Jan. 1997. International Society for Horticultural Science, Leuven, Belgium. 632 pp.

Sugar, D. (*Editor*). 1993. Sixth international symposium on pear growing, Medford, Oregon, 12–14 July 1993. International Society for Horticultural Science, Wageningen, Netherlands. 454 pp.

Van der Zwet, T, and Childers, N.F. 1982. The pear from varieties to marketing. Horticultural Publications, Gainesville, FL. 501 pp.

Williams, M.W. 1978. Pear production. U.S. Dept. of Agriculture, Science and Education Administration, Washington, DC. 53 pp.

Specialty Cookbooks

Baird, E. 1977. Apples, peaches & pears. J. Lorimer, Toronto, ON. 96 pp.

Eckhardt, L.W. 1996. Pears. Chronicle Books, San Francisco, CA. 132 pp.

Flores, B. 2000. The great book of pears. Ten Speed Press, Berkeley, CA.163 pp.

Hazen, J., and Kleinman, K. 1994. Pears. CollinsPublishers, San Francisco, CA. 95 pp.

Nahirney, D., and Jetelina, M. 1995. Okanagan pear delights. Kelowna Museum, Kelowna, BC. 24 pp.

Pacific Coast Canned Pear Service. 1990s. Canned pear microwave recipes. Pacific Coast Canned Pear Service, San Francisco, CA. 6 pp.

U.S. Dept. of Agriculture. 1966. Kieffer pears for home use. Revised edition. Agricultural Research Service, U.S. Dept. of Agriculture, Washington, DC. 6 pp.

PEPPER

Family: Piperaceae (pepper family)

Names

Scientific name: *Piper nigrum* L.

- The word "pepper" is derived from the Sanskrit name *pippali* or *pippalii*, transferred to English from the Greek *péperi* and Latin *piper*, all referring to some type of pepper (not necessarily *Piper nigrum*). The Latin word *piper* is the basis of the genus name *Piper*.
- Black and white pepper from *Piper nigrum* should not be confused with many other spices known as pepper. *Piper nigrum* is the main species of the genus cultivated for pepper, although some other species of the genus are also used as spices, notably the following: *P. longum* L. (long pepper), *P. cubeba* L. f. (cubeb pepper), and *P. guineense* Schumach. & Thonn. (West African, Benin, or bush pepper). Peppers from *Capsicum* are discussed under CHILE PEPPER. Several other unrelated species also have "pepper" in their name. Melegueta (or Guinea) pepper (*Aframomum melegueta* K. Schum.) is a cultivated tropical plant originating from West Africa, used in the spice trade, for example, for flavoring ice cream, candy, and soft drinks. The very similar word "Malagueta" is sometimes applied to certain *Capsicum* peppers, which can be quite confusing. *Xylopia aethiopica* (Dunal) A. Rich., a tropical African tree, is the source of African or Negro pepper, once much used in Europe. Allspice, *Pimenta dioica* (L.) Merr., is sometimes called Jamaican pepper. The genus *Zanthoxylum* (many species of which are called prickly ash) has several species that furnish spices called pepper, and the peppercorns are often called "brown peppercorns," (although the seeds are either brown or black). *Zanthoxylum piperitum* (L.) DC., Japan pepper or Japanese pepper, is a shrub native to and cultivated in Korea, Japan, and northern China. Its bark, unripe fruits, and seeds are used as seasoning, although considered harmful if consumed habitually. *Zanthoxylum simulans* Hance, Szechwan (Sichuan) pepper or Chinese pepper, is a shrub or tree of China. Its fruits produce more pungent whole or ground spice than black pepper. *Zanthoxylum armatum* DC. (*Z. alatum* Roxb.), a shrub or tree native to southern Asia, is another "Chinese pepper" used in China and India as seasoning. *Zanthoxylum rhetsa* DC., another tree of southern Asia, is called Indian pepper, and its seeds are also used as a pepper substitute. *Schinus molle* L., pepper tree or Peruvian pepper, is a native of Peru grown in California, Mexico, Guatemala, and the subtropics of the Old World. The dried, roasted berries are used as a substitute for or adulterant of pepper. The fruit is sweet, but if too much is ingested the result may be intestinal inflammation and hemorrhoids. "Pink peppercorns" are also unrelated to the true pepper plant. They come from the Baies rose plant (*Euonymus phellomanus* Loes.), imported from Madagascar. Also often called pink peppercorn is the seed of *Schinus terebinthifolius* Raddi; this species is also known as Brazilian pepper tree. The berries can be toxic in large quantities and should be used cautiously.
- *Nigrum* in the scientific name *P. nigrum* is Latin for black. The most common preparations of the spice are black, although as explained below, white pepper and green pepper are also marketed.

Plant Portrait

Pepper is a tropical, perennial, woody, climbing vine that often extends 6 m (20 feet), and sometimes more than 9 m (30 feet). It is indigenous to the Malabar coast of South West India. Peppercorns are the fruit of the pepper plant. The pepper vine can

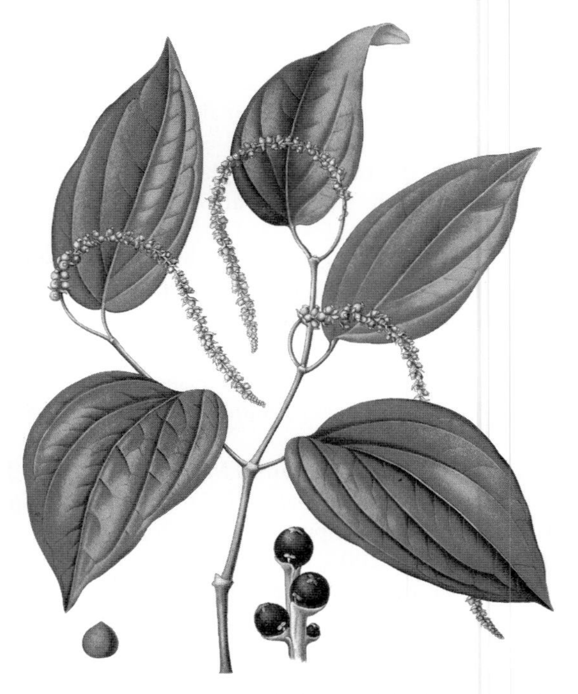

Pepper (*Piper nigrum*), from Köhler (1883–1914).

begin producing peppercorns at the age of 2 years, and continue for as long as 40 years, although in practice, 15 years of production is typical. The pepper berry is about the size of a small pea, bright red when ripe, made up of a single white seed covered with a thin layer of pulp. The berries grow on an elongated flowering branch with 50–60 fruits. This ancient spice has been known in China and India for thousands of years. Pepper was the first Oriental spice to reach Europe, and it became a great luxury and a staple article of trade between India and Europe. The trade in pepper and other spices was the major cause of medieval wars and the main reason for world exploration. Pepper is grown chiefly in southern Asia, and requires a wet, tropical climate. It is produced in India, Indonesia, Malaysia, and Brazil, with lesser amounts in Madagascar, Sri Lanka, and Thailand.

Among the world's finest peppercorns are tellicherry from the Malabar Coast of India and lampong from Sumatra. Black peppercorns are the green unripe berries, which become black and shriveled when dried in the sun. White pepper, which has a milder and less biting flavor than black pepper, is produced by removing the outer skin of the ripening berry and drying the hard cores in the sun. Green peppercorns are the green (immature) berries picked and dried artificially to retain their color. These are milder than black and white peppercorns.

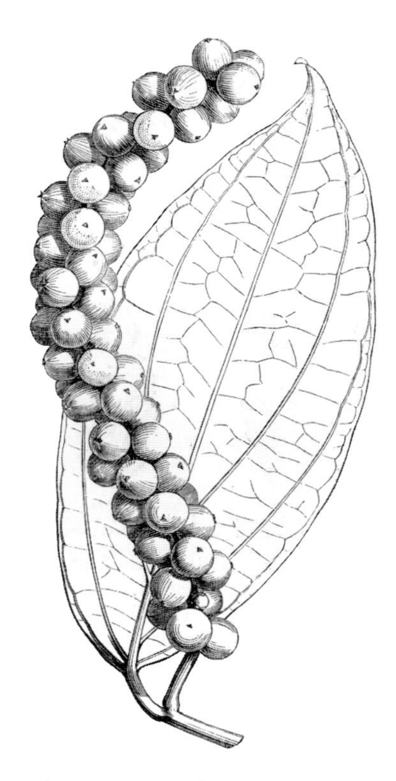

Pepper (*Piper nigrum*), from Maout and Decaisne (1876).

Culinary Portrait

Pepper is one of the most popular spices in the world, and is added to the majority of savory dishes, including vegetables, meats, cold cuts, vinaigrettes, marinades, pâtés, cheeses, soups, stews, and even some desserts. White pepper is often employed on pale-colored dishes such as poultry, fish, and white sauces. Freezing may accentuate the taste of pepper in dishes.

Good-quality black peppercorns will store well for many years, and their flavor is released on grinding. Once ground, the volatile oils soon evaporate, and the pepper loses its flavor. Accordingly, cooks are advised not to use pre-ground powder, but rather to keep a pepper mill for custom grinding. Heat also tends to degrade the taste of pepper, which may become bitter, so addition of this spice should be at the table or near the end of cooking.

The practice of adding salt and pepper to a dish presented at a table, before it is tasted, is a sign of ignorance and an insult to the chef, who has presumably already added the perfect amount of seasoning. Salt and pepper are available at the dining room table because people's taste preferences vary, but salting and peppering before tasting verges on bad manners.

Culinary Vocabulary

- As explained more fully in the chapter on corn, the word "corn" in "peppercorn" has nothing to do with corn (maize), but reflects old usage of "corn" as a small particle. A "peppercorn" is a dried pepper berry.
- "Gray pepper" refers either to black pepper that has been only slightly washed and still has its shell (this is rarely marketed, and is always sold in ground form); or a mixture of black and white pepper.
- *Steak aux trois poivres*, as implied by the French, is steak made with three peppers: black pepper, green pepper, and chile pepper.
- *Sauce poivrade* is any French sauce in which pepper provides the dominant flavor.
- *Garam masala*, literally "hot mixture," is used predominantly in Indian cuisine, and consists of a combination of roasted spices, including black pepper, chili pepper, and coriander, which are ground together. This can often be found in the international section of supermarkets.
- *Pfefferneusse* (German meaning "pepper nut," pronounced FEHF-uhr-noos) is a spicy German cookie made with black pepper, and traditionally eaten at Christmas. Similar cookies, also

named for pepper although other spices are present, are common in Scandinavia: *pepperkaker* (Scandinavia), *pepparnotter* (Norway), and *pebernodder* (Denmark).

- "Pepper pot" (also known as pepper pot and pepper pot stew) refers both to a West Indian stew and a colonial American soup of meat, vegetables, and black peppercorns. The colonial American version is also known as Philadelphia pepperpot (stew or soup), and is said to have arisen during the American Revolution, when a cook at Valley Forge, Pennsylvania, during the grim winter of 1777–1778, was ordered by General Washington to prepare a meal for the troops. The cook, who was from nearby Philadelphia, had nothing but peppercorns and scraps on hand, but it was enough to invent the new dish. Philadelphia pepperpot soup became known as the "soup that won the revolution." In early Philadelphia, during cold winters women sold the dish from carts pushed through narrow streets while crying:

  All hot! All hot!
  Pepper pot! Pepper pot!
  Makes backs strong,
  Makes lives long,
  All hot! Pepper pot!
- "Mike and Ike" and "the twins" is old lunch-counter jargon for salt and pepper shakers. "Sneeze" stood for pepper.

Curiosities of Science and Technology

- About 40 AD, the Greek merchant Hippalus is said to have realized that seasonal monsoons could be used to take sailors back and forth across the ocean from Egypt to the pepper-producing Malabar Coast of India. This led to extensive development of Roman fleets that captured the Indian spice trade from overland routes controlled by Arab traders.
- Pepper was considered one of the five essential luxuries upon which foreign trade with the ancient Roman Empire was based. The others were African ivory, Chinese silk, German amber, and Arabian incense.
- Roman traders established a special spice market in Rome in which the most prominent street was *Via Piperatica*—Pepper Street.
- Barbarians came to sack Rome in 400 AD. Haric (Alaric) the Visigoth is said to have demanded the Romans give him a ransom of 2268 kg (5000 pounds) of gold, 13 608 kg (30 000 pounds) of silver, and 1361 kg (3000 pounds) of pepper for release of the city of Rome in 408 AD. Two years later, he succeeded in getting an annual tribute of 136 kg (300 pounds) of pepper from the city. His assaults on the city continued, and Rome fell on 24 August 410 after the third siege.
- In the Middle Ages, pepper was often more valuable and stable in value than gold, and was frequently used as currency. Wealthy aristocrats kept stores of peppercorns as collateral, and bribes were frequently paid in pepper. Families often provided pepper as their daughter's dowry. The saying that a man "lacked pepper" meant that he was poor.
- In 982 AD in England, during the reign of King Aethelred II, German ships coming up the Thames to trade at London Bridge had to pay a special Christmas and Easter tax in black peppercorns.
- By the end of the 10th century in England, landlords charged a rent tax of as much as 1 pound of pepper a year (which in ancient times was worth its weight in gold). This evolved into the "peppercorn rent"—a token rent of peppercorn in lieu of rent. Prince Charles of England has ritually received peppercorn rent of 1 pound.
- In Elizabethan England (the period of Elizabeth I, 1533–1603), pepper was sold by the individual grain, and guards on the London docks were not allowed to have cuffs on their clothing, and had their pockets sewn up to prevent them from stealing the spice.
- Salem, Massachusetts, was once the "Pepper Capital of the New World," the center of re-exporting of pepper through New England to other countries aboard hundreds of fast schooners built exclusively for the pepper trade. One of Salem's traders, Elias Haskett Derby (1739–1799), became wealthy by importing pepper, and is said to have been America's first millionaire. He used his fortune to endow Yale University.
- The Reverend Sylvester Graham (1794–1851) was an eccentric Presbyterian minister from Connecticut who promoted temperance, vegetarianism, and especially the use of the whole grain of coarsely ground wheat in making flour, hence "graham flour" and "Graham crackers." Graham warned that excessive use of pepper would cause insanity.
- In nature, pepper vines often kill the trees they grow on.
- Pepper is known as the "King of Spices," laying claim to the title of most used spice in the West (but see CHILE PEPPER for the claim of most used condiment in the world). A survey showed that the top five spices

during the winter holidays in the United States are, in order, ground black pepper, ground cinnamon, seasoned salt, garlic powder, and ground nutmeg.

- Official guidelines of the Food and Drug Administration of the United States allow whole pepper to be sold with up to 1% of the volume made up of rodent droppings.
- Tunisians lead nations in pepper consumption with 0.23 kg (1/2 pound) per person per year. On average, Americans use about 0.11 kg (1/4 pound) annually.

Key Information Sources

Central Food Technological Research Institute (India). 1981. Pepper and pepper products: an annotated bibliography, 1970–1980. Central Food Technological Research Institute, National Information Centre for Food Science and Technology, Mysore, India. 53 pp.

Gentry, H.S. 1955. Introducing black-pepper into America. Econ. Bot. **9**: 256–268.

George, P.S., Nair, K.N., and Pushpangadan, K. 1989. The pepper economy of India. Oxford & IBH, New Delhi, India. 88 pp.

Gopalam, A., Zachariah, J., Babu, K.N., and Ramadasan, A. 1990. Effect of different methods of white pepper preparation on the chemical and aroma quality in selected cultivars of *Piper nigrum* L. Indian Perfum. **34**: 152–156.

Govindarajan, V.S. 1977. Pepper—chemistry, technology and quality evaluation. CRC Crit. Rev. Food Sci. **9**: 1–115.

Koizumi, S. 1999. Spicing up India's pepper industry: an economic analysis. Dept. of Agricultural, Resource, and Managerial Economics, Cornell University, Ithaca, NY. 112 pp.

Menon, A.N., Padmakumari, K.P., and Jayalekshmy, A. 2003. Essential oil composition of four major cultivars of black pepper (*Piper nigrum* L.) III. J. Essent. Oil Res. **15**: 155–157.

Pagington, J.S. 1983. A review of oleoresin black pepper and its extraction solvents. Perfum. Flavor. **8(4)**: 29–32, 34, 36.

Peter, K.V., and Kandiannan, K. 1999. Black pepper. *In* Tropical horticulture, vol. 1. *Edited by* T.K. Bose, S.K. Mitra, A.A. Farooqui, and M.K. Sadhu. Naya Prokash, Calcutta, India. pp. 646–656.

Pradeepkumar, T., Karihaloo, J.L., and Archak, S. 2001. Molecular characterization of *Piper nigrum* L. cultivars using RAPD markers. Curr. Sci. (Bangalore), **81**: 246–248.

Rathnawathie, M., and Buckle, K.A. 1984. Effect of berry maturation on some chemical constituents of black, green and white pepper (*Piper nigrum* L.) from three cultivars. J. Food Technol. **19**: 361–367.

Ravindran, P.N. (*Editor*). 2000. Black pepper: *Piper nigrum*. Harwood Academic, Amsterdam, Netherlands. 553 pp.

Ravindran, P.N., and Kallupurackal, J.A. 2001. Black pepper. *In* Handbook of herbs and spices. *Edited by* K.V. Peter. CRC Press, Boca Raton, FL. pp. 62–110.

Russell, G.F., and Else, J. 1973. Volatile compositional differences between cultivars of black pepper (*Piper nigrum*). Ass. Offic. Anal. Chem. J. **56**: 344–351.

Samuel, M.R.A., Gurusinghe, P. de A., Alles, W.S., and Kerinde, S.T.W. 1986. Genetic resources and crop improvement in pepper (*Piper nigrum*). Acta. Hortic. **188**: 117–124.

Spices Export Promotion Council. 1976. Indian pepper: a monograph. Spices Export Promotion Council, Cochin, India. 64 pp.

Spices Export Promotion Council. 1978. Report of the international seminar on pepper, 12th & 13th March, 1976, Cochin. Spices Export Promotion Council, Cochin, India. 110 pp.

Utami, D., and Jansen, P.C.M. 1999. *Piper* L. *In* Plant resources of South-East Asia, 13, spices. *Edited by* C.C. de Guzman and J.S. Siemonsma. Backhuys, Leiden, Netherlands. pp. 183–188.

Waard, P.W.F. de. 1980. Problem areas and prospects of production of pepper (*Piper nigrum* L.): an overview. Department of Agricultural Research, Royal Tropical Institute Amsterdam (Koninklijk Instituut voor de Tropen), Amsterdam, Netherlands. 29 pp.

Waard, P.W.F. de, and Anunciado, I.S. 1999. *Piper nigrum* L. *In* Plant resources of South-East Asia, 13, spices. *Edited by* C.C. de Guzman and J.S. Siemonsma. Backhuys, Leiden, Netherlands. pp. 189–194.

Waard, P.W.F. de, and Zeven, A.C. 1969. Pepper (*Piper nigrum* L.). *In* Outlines of perennial crop breeding in the tropics. *Edited by* F.P. Ferwerda and F. Wit. Misc. Paper 4. Landbouwhogeschool, Wageningen, Netherlands. pp. 409–426.

Ward, J.F. 1960. Black pepper; a review of cultural practices and their application to Jamaica. The Government Printer, Kingston, Jamaica. 15 pp.

United Nations. 1995. Towards a more vibrant pepper economy: studies presented at the international pepper seminar 1994, 17–19 August 1994, Bangkok, Thailand. United Nations, New York, NY. 288 pp.

Zeven, A.C. 1995. Black pepper. *In* Evolution of crop plants. 2nd edition. *Edited by* J. Smartt and N.W. Simmonds. Longman Scientific & Technical, Burnt Mill, Harlow, Essex, UK. pp. 407–409.

Specialty Cookbooks

(Note: Ravindran and Kallupurackal, listed above, present numerous recipes using pepper.)

Cook, S., Slavin, S., and Jones, D. 2003. Salt & pepper: the cookbook. Chronicle Books, San Francisco, CA. 132 pp.

Jemaah Pemasaran Lada. 1994. Sarawak pepper flavours the world. Jemaah Pemasaran Lada Hitam Malaysia (Government of Malaysia) with co-operation from Kuching Hilton. Kuching, Malaysia. 167 pp. (1998 edition, 81 pp.)

Pigeon Pea

Family: Fabaceae (Leguminosae; pea family)

Names

Scientific name: *Cajanus cajan* (L.) Millsp. (*C. indicus* Spreng.)

- The "pigeon pea" (often spelled "pigeonpea") obtained its name in Barbados, where the seeds were once used to feed pigeons. The name pigeon pea was first recorded about 1692 in the West Indies.
- Pigeon pea is also known as Angola pea, Bengal bean, catjang pea, Congo pea, dhal (dahl), goongo pea, goonog, gungo pea (the previous three names are used in Jamaica), gray pea, hoary pea, no-eye pea or no-eyed pea (in contrast with the black-eyed pea), Indian dhal, Puerto Rico pea, Puerto Rico bean, red gram, and yellow gram. Over 350 names have been recorded in various languages.
- Small-seeded forms of the broad bean (which see) are sometimes called pigeon pea.
- The name "dhal" (or "dal") is from the Hindi *dāl*, dahl seed, from Sanskrit *dala*, *dalam*, meaning a piece split off. The allusion is based on the fact that the seeds are split and dried for storage. As well as referring to the plant, the word "dhal" is used to mean dry, split legume seeds (i.e., seeds from species of the pea family) with the seed coat removed before cooking. Dahl also means a thick creamy East Indian stew made with legume seeds. More dhal is made from pigeon peas than from any other legume except chickpea (which see).
- The genus name *Cajanus*, as well as *cajan* in the scientific name *C. cajan*, is based on the Malay *kachang* or *katchang*, meaning a bean or pea of any kind.

Plant Portrait

Pigeon pea is a shrubby perennial that is woody at the base and grows 0.6–4 m (2–13 feet) tall, sometimes reaching 5 m (16 feet). It is frequently cultivated as an annual. The seedpods are up to 8 cm (3 inches) in length and 1.4 cm (1/2 inch) in width, flattish, sometimes hairy, and contain 2–9 seeds. The seeds are round, smooth, 4–9 mm (3/16–3/8 inch) in diameter, green when young, when mature mostly shades of brown and white, but also gray, yellow, purplish black, dark mottled brown, purple–red, and red. The seeds become very hard when mature and dry, but soften and enlarge when soaked. Pigeon pea is one of the world's oldest food crops. It is often said to be native to Africa, but more likely comes from India, and was taken to Africa around 2200 BC. It is claimed that the crop was cultivated in ancient Egypt before 2000 BC, but the evidence is inconclusive. It is now an extremely important crop, cultivated in virtually every tropical country of the world, as well as in the subtropics. Pigeon pea is grown not just for human food but also for livestock, and even to feed silkworms. In addition, there is very extensive usage as a source of twigs for household fuel in countries where wood is in short supply. India is the world's leading producer, and grows over 100 varieties. Considerable amounts are also produced in eastern Africa and the Caribbean, and in Myanmar (Burma), Nepal, and Venezuela. The canning of green pigeon peas is a major industry in Puerto Rico and Trinidad. Small amounts are experimentally grown in the southeastern United States. Although it is very frost sensitive, there is some interest in growing pigeon pea in temperate areas.

Culinary Portrait

Immature (green) seeds of pigeon pea are relatively tasty and tender, and have wider appeal than mature, dried seeds, which are usually yellowish or grayish and have developed a mealy texture and strong flavor. The immature seeds can be eaten raw, but are generally boiled for 75–120 minutes. Some anti-nutritional substances are present, but are destroyed by cooking, which is therefore preferable to raw consumption. Mature, dried seeds require 4–5 hours of cooking, but if soaked overnight, cooking time is reduced to 2–3 hours. The dried seeds are often split and the leathery seedcoat removed, and then the seeds are used in soup or made into a spiced purée. "Five dhal soup" is a traditional dish of India, which includes five types of legume seeds, typically a mixture of pigeon pea, chickpea, mung bean, and yellow and green "split peas" (of various species). "Peas and rice" is perhaps the most famous vegetable dish of the Bahamas, and is usually made with pigeon peas, although black-eyed peas are sometimes substituted. Recipes are available on the Web. Cooked pigeon peas can be substituted in recipes calling for black-eyed peas. Ripe pigeon pea seeds are used as a source of flour, and sometimes the seeds are germinated and used as sprouts. Dried split pigeon peas are often available in Indian markets. Canned, cooked pigeon peas may be available in Latin American markets. Pigeon peas do not freeze well, and frozen green pigeon peas are best avoided.

Culinary Vocabulary

- "Hoppin' John" (Hopping John) is a Caribbean combination of rice and "peas," often black-eyed peas, although more often in the past made with pigeon peas. As noted above, pigeon peas and rice is a favorite dish in the Caribbean. The name Hoppin' John is of uncertain origin, but has been said to have arisen as a corruption of the French name *pois pigeons*, which was perceived as the nearest English equivalent (although it hardly appears similar). Another interpretation holds that the name comes from the custom of inviting guests to eat, with the request "Hop in, John." Still another story suggests the name derives from an old New Year's Day ritual in which children in the house hopped around the table before eating the dish. Hoppin' John is traditionally viewed as a lucky dish, because in the days of slavery in the Caribbean it was considered a fine, even gala, dish that one was lucky to get.

Curiosities of Science and Technology

- Pigeon pea has been used medicinally in several countries. In China, the roots of pigeon pea have had a very strong reputation for countering the effects of all poisons. In the Bahamas, the plant is used in folk medicine to treat poisoning from bad fish. In Columbia, the leaves are used as a remedy for bat bites. In Argentina, a tea made with pigeon pea leaves is prized for treating skin irritations.
- The pigeon pea is used in the Far East to culture scale insects that produce a dark red resin called lac, used to make lacquer and shellac. The most common scale insect used for shellac production is *Kerria lacca* Kerr. (more commonly known as *Laccifer lacca* Kerr.).
- In Madagascar, pigeon peas attract a caterpillar that produces a silk-like cocoon, which is worth more than the peas produced by the plants.
- In India, more than 2 million people make their living growing pigeon peas.
- The dietary combination of pigeon peas and rice in poor areas of the world where meat is in short supply is no accident. Pigeon pea is a "pulse" crop, i.e., a member of the pea family whose seeds are used for human food, while rice is a cereal, i.e., a member of the grass family whose seeds are used for human food. In all great civilizations of the world, a combination of cultivation of pulse crops and cereal crops has been a key to success. This is because protein is essential to human survival. Protein is made up of different building blocks called amino acids. Characteristically, pulse crops are rich in amino acids, particularly lysine, but are usually deficient in the sulfur-containing amino acids methionine and cystine, while cereals contain lower amounts of amino acids, and these are deficient in lysine but adequate in sulfur-containing amino acids. Together, they can furnish perfectly balanced protein nutrition necessary for survival when the supply of meat is insufficient to provide needed protein.

Pigeon pea (*Cajanus cajan*), from Curtis (1787–present, plate 6440).

Key Information Sources

Akinola, J.O.,Whiteman, P.C., and Wallis, E.S. 1975. The agronomy of pigeon pea (*Cajanus cajan*). Commonwealth Agricultural Bureaux, Farnham Royal, UK. 57 pp.

Bisen, S.S., and Sheldrake, R. 1981. The anatomy of the pigeonpea. International Crops Research Institute for the Semi-arid Tropics, Patancheru, Andhra Pradesh, India. 24 pp.

Dahiya, B.S. 1980. An annotated bibliography of pigeon pea 1900–1977. ICRISAT, Patancheru, AP, India. 183 pp.

De, D.N. 1974. Pigeon pea. *In* Evolutionary studies in world crops; diversity and change in the Indian subcontinent. *Edited by* J.B. Hutchinson. University Press, Cambridge, UK. pp. 79–87.

Gooding, H.J. 1962. The agronomic aspects of pigeonpeas. Field Crop Abs. **15**: 1–5.

International Crops Research Institute for the Semi-arid Tropics. 1998. From orphan crop to pacesetter: pigeonpea improvement at ICRISAT. ICRISAT, Patancheru, AP, India. 22 pp.

Maesen, L.J.G. van der. 1983. World distribution of pigeonpea. International Crops Research Institute for the Semi-arid Tropics, Patancheru, AP, India. 40 pp.

Maesen, L.J.G. van der. 1985. *Cajanus* DC. and *Atylosia* W. & A. (Leguminosae). A revision of all taxa closely related to the pigeonpea, with notes on other related genera within the subtribe Cajaninae. Paper 85-4. Agricultural University, Wageningen, Netherlands. 225 pp.

Maesen, L.J.G. van der. 1989. *Cajanus cajan* (L.) Millsp. *In* Plant resources of South-East Asia, 1, pulses. *Edited by* L.J.G. van der Maesen and S. Somaatmadja. Pudoc, Leiden, Netherlands. pp. 39–42.

Maesen, L.J.G. van der. 1995. Pigeonpea. *In* Evolution of crop plants. 2nd edition. *Edited by* J. Smartt and N.W. Simmonds. Longman Scientific & Technical, Burnt Mill, Harlow, Essex, UK. pp. 251–255.

Morton, J.F. 1976. The pigeon pea (*Cajanus cajan* Millsp.), a high-protein, tropical legume. HortScience, **11**: 11–19.

Morton, J.F. 1982. Pigeonpeas (*Cajanus cajan* Millisp. [i.e. Millsp.]): a valuable crop of the tropics. Dept. of Agronomy and Soils, University of Puerto Rico, Mayagüez Campus, Mayagüez, Puerto Rico. 122 pp.

Nene, Y.L., Hall, S.D., and Sheila, V.K. (*Editors*). 1990. The pigeonpea. CAB International, Wallingford, Oxon, UK. 400 pp.

Nwokolo, E. 1996. Pigeon pea (*Cajanus cajan* (L.) Millsp.). *In* Food and feed from legumes and oilseeds. *Edited by* J. Smartt and E. Nwokolo. Chapman and Hall, London, UK. pp. 64–73.

Phatak, S.C., Nadimpalli, R.G., Tiwari, S.C., and Bhardwaj, H.L. 1993. Pigeonpeas: potential new crop for the southeastern United States. *In* New crops. *Edited by* J. Janick and J.E. Simon. Wiley, New York, NY. pp. 597–599.

Reddy, M.V. 1993. Handbook of pigeonpea diseases. International Crops Research Institute for the Semi-Arid Tropics, Patancheru, AP, India. 61 pp.

Salunkhe, D.K., Chavan, J.K., and Kadam, S.S. 1986. Pigeonpea as important food source. CRC Crit. Rev. Food Sci. Nutr. **23**: 103–141.

Sheldrake, A.R 1984. Pigeonpea. *In* The physiology of tropical field crops. *Edited by* P.R. Goldsworthy and N.M. Fisher. John Wiley & Sons, Chichester, UK. pp. 385–417.

Vernon, R.W. 1976. Pigeon pea. *In* Evolution of crop plants. *Edited by* N.W. Simmonds. Longman, New York, NY. pp. 43–54.

Whiteman, P.C., Byth, D.E., and Wallis, E.S. 1985. Pigeonpea. *In* Grain legume crops. *Edited by* R.J. Sumerfield and E.H. Roberts. Collins, London, UK. pp. 658–698.

Specialty Cookbooks

Faris, D.G. 1987. Vegetable pigeonpea: a promising crop for India. International Crops Research Institute for the Semi-arid Tropics, Patancheru, AP, India. 13 pp.

PINEAPPLE

Family: Bromeliaceae (bromeliad or pineapple family)

Names

Scientific name: *Ananas comosus* (L.) Merr.

- The pineapple gets its English name from resemblance to a pine cone (curiously, in old Europe, pine cones were sometimes called "pine apples."). Columbus, the first Westerner to see the pineapple, made this comparison during his second voyage to the New World in 1493. The pineapple is widely called *piña* by Spanish-speaking people, "pine" in Guatemala, and "sweet pine" sometimes in Jamaica.
- The "Pitmaston pine apple" is an apple, not a pineapple. This small, English apple has a conical shape and rich flavor.
- The genus name *Ananas* is derived from the Paraguayan (Guarani Indian) *nana*, meaning excellent fruit. The pineapple is known as *ananas* in Dutch and French, and *nanas* in southern Asia and the East Indies.
- *Comosus* in the scientific name *A. comosus* means "bearing a tuft of leaves," which is descriptive of the crown of leaves on top of a pineapple.

Plant Portrait

The pineapple plant is a biennial or perennial herb, 0.75–1.5 m (2½–5 ft) high with a spread of 0.9–1.2 m (3–4 feet), a very short, stout stem, and a rosette of waxy, strap-like long, stiff, spiny-edged leaves. It is native to southern Brazil and Paraguay where wild relatives occur. Pineapples were spread by the Indians up through South and Central America to the West Indies. In 1493, Columbus found the fruit on the island of Guadeloupe and carried it back to Spain. It was transported around the world on sailing ships, and used for protection against scurvy. The pineapple is now grown throughout tropical regions of the world. Pineapples have been cultivated in home gardens for perhaps 300 years in Hawaii, but in the 19th century, the young entrepreneur James Drummond Dole encouraged the natives to grow the plants as a cash crop. The first commercial plantation was established in Oahu in 1885, and Hawaii produced most of the world's pineapple until the 1960s, when the Philippines became dominant. Southeast Asia, notably the Philippines and Thailand, accounts for most world production, although pineapple is produced in most tropical regions.

The two major commercial varieties in the United States are Cayenne (in Hawaii) and Red Spanish (mainly in Florida and Puerto Rico). The Cayenne is longer and more cylindrical, with golden yellow skin and long, sword-like leaves sprouting from a single tuft. The Red Spanish pineapple is squatter, with a reddish golden brown skin and leaves that radiate from several tufts. Mexico grows the Sugar Loaf, a large, exquisitely flavored pineapple with skin that is still green when ripe. It does not ship well and is rarely imported into North America. Most cultivated pineapples are seedless, and the plant is propagated vegetatively by planting the leafy portion above the fruiting axis, or severed branches (suckers), in soil. Potted pineapple plants in flower are often sold as ornamentals.

Marketed pineapples average 2–4 kg (about 4–9 pounds), but can weigh up to 9 kg (20 pounds). The sweet, juicy, cone-like structure is a multiple fruit

Pineapple (*Ananas comosus*), from Anderson (1874).

composed of many individual berries embedded in a fleshy, edible stem. All varieties have bumpy diamond-patterned skins. Each section or hexagonal marking on the mature, cone-like fruit represents the place where a flower was once attached. The scale-like skin is yellow, green, greenish brown, or reddish brown. The flesh is yellowish, fibrous, sweet, and juicy, and near the base is sweeter, more tender, and more darkly colored. The fruit is picked ripe because the starch will not convert to sugar once it is off the plant. Keeping slightly underripe pineapples at room temperature for several days will reduce their acidity, but will not increase sweetness.

How to grow your own pineapple: Cut the crown (the leafy top) from the fruit from your grocery store. Trim off adhering flesh. Cut the bottom of the crown (its stem) until you see root buds, which are small round structures around the perimeter of the stem base. Remove as little tissue as possible to avoid cutting into young stem. Strip off some of the lower leaves, exposing up to about 2 cm (3/4 inch) of the base of the crown. Place the crown upside down in a dry, shaded place for 5–7 days to allow the cut portions to heal. Plant in a light garden soil with up to 30% well-composted organic matter, or use a commercial potting soil.

Warning: Unripe pineapple is not only inedible but can be poisonous, irritating the throat and acting as a drastic purgative. The flesh of very young (toxic) fruits has been deliberately ingested to produce abortion.

Culinary Portrait

Canned pineapple is the principal product for the fruit, and this may be sliced and cored into rings, crushed, or chopped into bits. There is also a very large market for fresh pineapple, often found in supermarkets with the shell and core removed. Because fresh pineapples do not ripen after they are harvested, unless transported rapidly to market they deteriorate notably in quality. Pineapple is consumed in salads, fruit cocktails, sherbets, ice cream, sauces (especially of the sweet-and-sour variety), pies, cakes (upside-down pineapple cake is a classic dessert in North America), yogurt, ice cream, and a variety of other frozen treats. Pineapple combines well with rice, coleslaw, and cottage cheese. Pineapple juice is also very popular. A small quantity of pineapple is used commercially to make jam. Aside from its use as a dessert fruit, pineapple is employed as a garnish for roast meats, especially ham and pork, and goes well with seafood, chicken, and duck. Frequently, recipes with pineapple have "Hawaii" or "Hawaiian" in their names. American Indians once made wine from pineapples, but little of the fruit is now used for this purpose. However, some brandy and liqueur is manufactured from pineapple.

Pineapples contain the protein-digesting enzyme bromelain, and because of this, they have been used as a meat tenderizer. Fresh and frozen pineapple should not be used in gelatin mixtures because the enzyme prevents setting (canned pineapple does not cause this problem, and agar may be substituted for gelatin). The enzyme also affects the proteins in milk, degrading the taste (which can be demonstrated by adding pineapple juice to milk). Yogurt and ice cream, however, are not affected. Bromelain may cause dermatitis. Workers who cut up pineapples for long periods often have their fingerprints almost completely obliterated from the effects of the enzyme bromelain.

How does one select a good pineapple? This is a difficult exercise, with only fair probability of being certain. Choose fruits that are slightly soft to the touch with a full, strong color and no sign of greening. Thump the fruit with the index finger—a solid fruit should produce a sound similar to the thumping of your wrist. The leaves should be crisp and green with no yellow or brown tips. Avoid fruit with shriveled crowns. Some have recommended checking if a leaf pulls easily from the crown (it should); others have said this is an old wives' tale. Overripe fruit are soft, and there may be dark areas on the skin. One method of judging whether a pineapple is ripe for eating is to smell the base—it should be sweet. Once purchased, pineapple should be consumed promptly. The fruit tends to deteriorate when stored at home, but may be kept for a few days in a refrigerator.

Culinary Vocabulary

- "Chicago" is a slang term meaning pineapple. A "Chicago sundae" is a dessert featuring pineapple. The term originated in Chicago of the 1920s, when gangsters of the city used hand grenades called pineapples as weapons in gang warfare.
- "Hummingbird cake" is a moist layer cake prepared with pineapple and banana, and filled with a frosting of cream cheese.
- A "piña colada" is a cocktail made from pineapple juice, coconut milk, and rum, and often served over a chunk of pineapple. The name is Spanish for "strained pineapple." Ramon Marrero, a bartender at the Caribe Hilton Hotel in San Juan, Puerto Rico, is credited with originating the beverage in 1952. The drink became popular in the 1970s.

- Honolulu bartender Harry K. Yee invented the now-classic Blue Hawaii cocktail in 1957 in Waikiki. The drink is made of rum, vodka, and fresh pineapple juice and blue curaçao (curaçao is a liqueur flavored with the dried peel of the sour orange). The blue liqueur was said to represent Hawaii's ocean.
- A "zombie" is a cocktail made of rum, curaçao, lemon juice, orange juice, pineapple juice, papaya or guava juice, orgeat syrup, grenadine, and Pernod, served with a chunk of pineapple and a sprig of mint. It was created in the late 1930s by Don the Beachcomber (Don Richard Beaumont-Gantt), in his Los Angeles restaurant. According to legend, the customer who first consumed the beverage returned later and remarked that he "felt like the living dead." The cocktail is named zombie (a word derived from the Kongo language term *zumbi*, fetish) for the legendary spirits that reanimate the bodies of dead people in voodoo mythology.
- A "garapina" (presumably based on the Spanish *garapiña*, frozen state) is a drink of Mexico and the Caribbean, made from fermented pineapple juice (which may be flavored with tamarind and oranges, and sweetened with sugar). Garapina also refers to a corn-based fermented beverage, identical to "chicha" (see CORN).
- "Bière douce" (French for "sweet beer") is a Louisiana Creole beer brewed from sugar, rice, water, and pineapple skin.

Pineapple (*Ananas comosus*), from Ives et al. (1831).

- In old lunch-counter slang, a "Hoboken special" was a pineapple soda with chocolate ice cream.

Curiosities of Science and Technology

- In the Caribbean area, pineapple fields were used as a test of manhood for Carib boys, who were expected to run through the sharp-pointed plants grown at the time, while tolerating the resulting wounds.
- Carib Indians placed pineapples or pineapple crowns outside the entrances to their dwellings as symbols of friendship and hospitality. Pineapples were very expensive in Europe, and their presence became associated with good living. The practice of placing pineapple carvings over doorways became common in Spain, England, and later in New England. In New England, pineapples are common on door knockers. Pineapple symbols are still used in hotels and restaurants to signal the presence of hospitality.
- Despite the practice of Carib Indians of using pineapples as a symbol of hospitality, they planted thick hedges of pineapple plants around their villages to keep strangers out.
- The first pineapple grown in France was offered to King Louis XIV (1638–1715). The characteristically greedy and impatient monarch bit into the fruit before anyone had time to peel it, severely cutting his royal lips. Pineapple growing in France ceased until the king's death.
- In the 18th and 19th centuries in Britain, "winter conservatories" (greenhouses) were fashionable among the rich, who often competed to grow the most succulent pineapples.
- In 1935, the Life Savers Corporation added cherry and pineapple flavors to the previous three-flavor combination of lemon, lime, and orange to make the famous five-flavor package, trademarked in 1938. Today, more than 125 million Life Savers are sucked daily, and they have been used for everything from Christmas ornaments to birthday cake candleholders.
- Filippo Tomaso Marinetti (1876–1944) was a well-known Italian futurist (i.e., he predicted future societal and technological trends), many of whose ideas are absurd by modern standards. With the blessing of his friend, the dictator Benito Mussolini, in 1932 he published a book on futurist cooking, *La cucina Futurista*, mostly containing recipes based on mixing unusual combinations of food and flavorings. One of his dishes proposed mixing pineapple, sardines, salami, and nougat in black coffee flavored with eau de Cologne. Another, a purportedly aphrodisiac drink, combined pineapple juice, eggs, cocoa, caviar, almond paste, cayenne pepper, nutmeg, cloves, and Strega (a yellow Italian liqueur flavored with oranges and herbs).
- A pineapple weighing 7.96 kg (17½ pounds) was harvested by Dole Philippines Inc. at South Cotabato, Philippines, in November 1978.
- It requires 18–22 months for a pineapple plant to produce a single 2-kg (4.4-pound) fruit. About a year later, the plant will produce a second fruit, smaller than the first.
- A "pineapple express" is a relatively common weather system in which southwest winds from the Hawaiian tropics bring unseasonably warm winter temperatures to northwestern North America.
- Hummingbirds are the principal pollinators of pineapples. Cultivated pineapples are seedless. To prevent pollination and the development of undesired seeds, importation of the birds has been prohibited in Hawaii.

Key Information Sources

Abd Shukor, A.R., Faridah, A., Abdullah, H., and Chan, Y.K. 1998. Pineapple. *In* Tropical and subtropical fruits. *Edited by* P.E. Shaw, H.T. Chan, Jr., and S. Nagy. Agscience, Auburndale, FL. pp. 137–190.

Bartholomew, D.P., and Kadzimin, S.B. 1977. Pineapple. *In* Ecophysiology of tropical crops. *Edited by* P.T. Alvin and T.T. Kozlowski. Academic Press, New York, NY. pp. 113–156.

Bartholomew, D.P., and Malezieux, E.P. 1994. Pineapple. *In* Handbook of environmental physiology of fruit crops, Vol. 2. Subtropical and tropical crops. *Edited by* B. Schaffer and P.C. Andersen. CRC Press, Boca Raton, FL. pp. 243–291.

Bartholomew, D.P., and Paull, R.E. 1986. Pineapple. *In* CRC handbook of fruit set and development. *Edited by* S.P. Monselise. CRC Press, Boca Raton, FL. pp. 371–385.

Bartholomew, D.P., and Rohrbach, K.G. (*Editors*). 1993. First international pineapple symposium, Honolulu, Hawaii, 2–6 Nov. 1992. International Society for Horticultural Science, Wageningen, Netherlands. 471 pp.

Bartholomew, D.P., Paull, R.E., and Rohrbach, K.G. 2002. The pineapple: botany, production, and uses. CABI Publishing, Oxon, UK. 301 pp.

Broadley, R.H., Wassman, R.C., III, and Sinclair, E. (*Editors*). 1993. Pineapple pests and disorders. DPI Publications, Queensland, Brisbane, Australia. 63 pp.

Chadha, K.L., Reddy, B.M.C., and Shikhamany, S.D. 1998. Pineapple. Directorate of Information and Publications of Agriculture, New Delhi, India. 115 pp.

Collins, J.L. 1948. Pineapples in ancient America. Sci. Mon. **65**: 372–377.

Collins, J.L. 1949. History, taxonomy and culture of the pineapple. Econ. Bot. **3**: 335–359.

Collins, J.L. 1960. The pineapple: botany, cultivation and utilization. InterScience Publishers, New York, NY. 294 pp.

Cunha, G.A.P. da, Cabral, J.R.S., and Souza, L.F. da S. 1999. O abacaxizeiro: cultivo, agroindústria e economia [The pineapple plant: cultivation, agroindustry and economy]. EMBRAPA [Brazilian Corporation for Agricultural Research], EMBRAPA-SCT, SAIN Parque Rural, Brasília, Brazil. 480 pp. [In Portuguese.]

Flath, R.A. 1980. Pineapple. *In* Tropical and subtropical fruits: composition, properties and uses. *Edited by* S. Nagy and P.E. Shaw. AVI Publishing, Westport, CT. pp. 157–183.

Institute for Tropical and Subtropical Crops (South Africa). 1994. The cultivation of pineapple. Institute for Tropical and Subtropical Crops, Agricultural Research Council, Nelspruit, South Africa. 134 pp.

Johnson, M.O. 1935. The pineapple. Paradise of the Pacific press, Honolulu, HI. 306 pp.

Ken, S.K. 1985. Pineapple. *In* Fruits of India, tropical and subtropical. *Edited by* T.K. Bose. Naya Prokash, Calcutta, India. pp. 298–319.

Kramer, J. 1979. The pineapple—top grower's handbook. Prentice-Hall, Englewood Cliffs, NJ. 76 pp.

Leal, F. 1990. On the history, origin and taxonomy of the pineapple. Interciencia, **14**: 235–241.

Leal, F. 1995. Pineapple. *In* Evolution of crop plants. 2nd edition. *Edited by* J. Smartt and N.W. Simmonds. Longman Scientific & Technical, Burnt Mill, Harlow, Essex, UK. pp. 19–22.

Leal, F., and Coppens d'Eeckenbrugge, G. 1996. Pineapple. *In* Fruit breeding, Vol. 1. Trees and tropical fruits. *Edited by* J. Janick and J.N. Moore. John Wiley & Sons, New York, NY. pp. 515–557.

Loison-Cabot, C. 1992. Origin, phylogeny and evolution of pineapple species. Fruits, **47**: 25–32.

Mitra, S.K. 1999. Pineapple. *In* Tropical horticulture, vol. 1. *Edited by* T.K. Bose, S.K. Mitra, A.A. Farooqui, and M.K. Sadhu. Naya Prokash, Calcutta, India. pp. 285–296.

Morton, J. 1987. Pineapple. *In* Fruits of warm climates. *Authored by* J.F. Morton. Creative Resource Systems, Winterville, NC. pp. 18–28.

Okomoto, M.C. 1948. Anatomy and histology of the pineapple inflorescence and fruit. Bot. Gaz. **110**: 217–230.

Platts, P.K. 1957. Pineapple ABC's. Revised edition. Dept. of Agriculture, State of Florida, Tallahassee, FL. 72 pp.

Py, C., Lacoeuilhe, J.J., and Teisson, C. 1987. The pineapple: cultivation and uses. G.P. Maisonneuve & Larose, Paris, France. 568 pp.

Subhadrabandhu, S., and Chairidchai, P. (*Editors*). 2000. Proceedings of the third international pineapple symposium, Pattaya, Thailand, 17–20 Nov. 1998. International Society for Horticultural Science, Leuven, Belgium. 341 pp.

Wee, Y.C., and Charuphant Thongtham, M.L. 1991. *Ananas comosus* (L.) Merr. *In* Plant resources of South-East Asia, 2, edible fruits and nuts. *Edited by* E.W.M. Verheij and R.E. Coronel. Pudoc, Leiden, Netherlands. pp. 66–71.

White, H.A. 1957. James D. Dole: industrial pioneer of the Pacific, founder of Hawaii's pineapple industry. Newcomen Society in North America, New York, NY. 32 pp.

Specialty Cookbooks

Anonymous. 1971. Shelford pineapple party-time. Muller & Retief, Pretoria, South Africa. 63 pp.

Bentley, M.M., Bradley, A., and Splint, S.F. 1927. Hawaiian pineapple as one hundred good cooks serve it: including 100 of the 60,000 recipes contributed by magazine-reading women, in response to an appeal for original, practical ways to serve canned Hawaiian pineapple. Association of Hawaiian Pineapple Canners, San Francisco, CA. 32 pp.

Dobbs, J.C., and Shimabukuro. B. 2002. Hawai'i's favorite pineapple recipes. Mutual Publishing, Honolulu, HI. 146 pp.

Donnelly, P. 1991. The pineapple cookbook. Bess Press, Honolulu, HI. 119 pp.

Hembrow, S. 1977. Pineapple cookbook. Rigby, Adelaide, Australia. 64 pp.

Mullins, J., and Nicholson, M. 1975. Hawaii's king pine: all about pineapples: planting, harvesting, processing, recipes, history: how to prepare fresh pineapple. Jowat, Honolulu, HI. 32 pp.

Pineapple Growers Association. 1960–1969? Make again recipes inspired by canned pineapple. Pineapple Growers Association, San Francisco, CA. 24 pp.

Rosenbaum, H. 1975. How to grapple with the pineapple: from planting pineapple tops to baking upside-down cakes. Hawthorn Books, New York, NY. 129 pp.

South Eastern Michigan Bromeliad Society. 1988. A touch of hospitality: pineapple cuisine. Cookbook Publishers, Inc., Olathe, KS. 188 pp.

Swenson, D. 1984. Tropical treats, the pineapple: preparation and uses. D. Swenson, Molokai, HI. 31 pp.

Weaver, A.T., and [Mrs.] Weaver, A.T. 1976. Just pineapples: a collection of 751 pineapple recipes from Plantation Paradise, the Pineapple Farm. Weaver, Lake Placid, FL. 335 pp.

PISTACHIO

Family: Anacardiaceae (cashew family; also placed in Pistaciaceae, pistachio family)

Names

Scientific name: *Pistacia vera* L.

- The English word "pistachio" and the genus name *Pistacia* are derived from *pistak*, the old Persian name of the nut. "Pistachio" first evolved from the Persian as an Italian word.
- Pistachio is pronounced pih-STASH-ee-oh, pih-STAH-shee-oh, or pih-STAH-see-oh.
- Pistachio is also known as green almond, for the typically green color of the flesh of the nut.
- Kerman, the principal pistachio variety grown in the United States, is named for the famous Persian carpet-making city in Iran near where seed used to establish this variety was collected in 1929.
- *Verum* in the scientific name *P. vera* is Latin for "genuine."

Plant Portrait

The pistachio is a small, deciduous tree, 3–8 m (10–33 feet) tall. Some trees in Syria are known to have a trunk diameter of 1.8 m (6 feet), but such thickness is quite exceptional. The species originated in central Asia, and archaeological evidence in Turkey indicates that pistachio nuts were being used for food as early as 7000 BC. Pistachio trees were cultivated in Mediterranean Europe at the beginning of the Christian era. The trees were not introduced to the United States until 1854. The main producers today are Iran, the United States (in the Sacramento and San Joaquin Valleys in California), Turkey, and Syria. Pistachio trees are either male or female. The females, of course, produce the nuts and so are predominantly grown in orchards, along with relatively few males (as detailed below). The reddish, wrinkled nuts are borne in heavy clusters somewhat like grapes. The nuts are egg-shaped, often 2.5 cm (1 inch) or so long, but there are many varieties, and these differ in size. In California, the main variety grown is Kerman. The external, fleshy hull of the nut loosens at maturity, but must be removed either by hand or mechanically. The nut also has a thin, woody shell, which normally splits along its sutures when ripe. The color of the kernel varies from yellowish through shades of green, which extends throughout the kernel. In general, the deeper the shade of green, the more the nuts are esteemed. The nuts are harvested when the husk or hull covering the shell becomes fairly loose. A single shaking of the tree will bring down the bulk of the matured nuts, which can be caught on a tarp or canvas. A fully mature tree may produce as much as 23 kg (50 pounds) of dry, hulled nuts.

The shells of pistachio nuts are naturally a creamy light beige, but are traditionally colored reddish pink with vegetable dye. This custom may have originated with Joseph A. Zaloom, a Syrian immigrant to the United States who dyed his imported pistachios red to distinguish them from his competitors. An alternative explanation is that the coloring is done to hide blemishes on the shell, which are darker than the natural light beige color of the shell. Today, nuts imported to North America are still mostly dyed, while American-grown pistachios are sold without dye.

Culinary Portrait

Pistachio nuts are considered one of the prime edible nuts, along with almonds, macadamias, and cashews. The flavor of the best selections is mild, although some varieties are resinous in taste. The pistachio is unique in the nut trade owing to its semi-split shell, which enables the processor to roast and salt the kernel without removing the shell and at the same time serves as a convenient form of packaging. Pistachio nuts are sold roasted and salted, shelled or unshelled, and are used for flavoring ice cream, sweets, cakes, breakfast cereals, sauces, stuffings, pâtés, puddings, salads, and meat dishes, especially delicatessen meats. In Indian cuisine, pistachio nuts are frequently puréed and used to flavor rice and vegetables. In Mediterranean and Asian cooking, the nuts are widely used in pastries, and also in poultry and meat dishes. Often the nuts are used in cooking in part because of their decorative green color.

The best nuts to purchase are split open at one end (unopened shells are an indicator of immaturity). The greener the nutmeat, the better the flavor. For cooking, it is preferable to pick pistachios that have not been dyed. The thin, edible brownish layer of skin can be removed from the nutmeats by blanching in boiling water for 2 minutes, draining, and cooling slightly before rubbing off the skins. Pistachios have a limited shelf life, and may be stored in an airtight container in a refrigerator for up to 3 months or (if unshelled) up to 1 year in a freezer. When purchasing shelled nuts, it is preferable to buy vacuum-packed glass jars or cans to ensure freshness.

Pistachio (*Pistacia vera*), from Michaux (1850, vol. 2, plate 103).

Curiosities of Science and Technology

- It has been claimed that a pistachio tree in Iran is 700 years old.
- The creation of pistachio ice cream is credited to James W. Parkinson of Philadelphia, Pennsylvania, around 1940. (Pistachio ice cream has sometimes been faked with other nuts and green coloring.)
- Pistachio nuts did not become popular in the United States until they were sold in the newfangled vending machines that were introduced in the 1930s.
- Iran was the major exporter of pistachio nuts to North America until the Ayatollah Khomeni's "students" held Americans hostage in the US embassy in the late 1970s. As a result, California very rapidly increased production to the point that almost all pistachios consumed in the United States are now produced domestically.
- Male trees hold a rather exalted position in pistachio orchards. They are often planted surrounded by exactly eight female trees laid out in a 3 × 3 square, which means there are three females on

each of the male's four sides. The plants are primarily wind pollinated, and the pattern allows equal mating of the male with his harem. Some males are even more privileged—they are in the center of a 4 × 4 square, bordered by 16 females.

Key Information Sources

(For additional references, see ALMOND.)

Bass, A.W., and Hodge, D. 1986. Pistachio propagation. Dept. of Agriculture, Adelaide, SA, Australia. 23 pp.

Bentley, W.J. 2003. UC IPM pest management guidelines—pistachio. University of California, ANR/ Communication Services, Oakland, CA. Variously paginated.

Caruso, T., Barone, E., and Sottile, F. (*Editors*). 1996. Proceedings of the IX G.R.E.M.P.A. meeting: pistachio; Bronte, Sciacca, Italy, May 20–21, 1993. Renier Tipografia, Salerno, Italy. 140 pp. [In English and French.]

Crane, J.C. 1986. Pistachio. *In* CRC handbook of fruit set and development. *Edited by* S.P. Monselise. CRC Press, Boca Raton, FL. pp. 389–399.

Crane, J.C., and Iwakiri, B.T. 1981. Morphology and reproduction of pistachio. Hortic. Rev. **3**: 376–393.

Crane, J.C., and Maranto, J. 1988. Pistachio production. Division of Agriculture and Natural Resources, University of California, Oakland, CA. 15 pp.

Ferguson, L. 1995. Pistachios in California. Acta Hortic. **419**: 169–173.

Ferguson, L., Beede, R., Teviotdale, B., and Weinberger, G. (*Editors*). 1995. Pistachio production. Center for Fruit and Nut Crop Research and Information, Davis, CA. 160 pp.

Joley, L.E. 1972. Pistachios in Iran and California. CSIRO Plant Intro. Rev. **9**: 12–16.

Kaska, N. (*Editor*). 1995. First international symposium on pistachio nut, 20–24 October 1994, Adana, Turkey. International Society for Horticultural Science, Leiden, Netherlands. 425 pp.

Loukas, M., and Pontikis, C.A. 1979. Pollen isozyme polymorphism in types of *Pistacia vera* and related species as an aid in taxonomy. J. Hortic. Sci. **54**: 95–102.

Maggs, D.H. 1982. An introduction to pistachio growing in Australia. Commonwealth Scientific and Industrial Research Organization, Australia. 36 pp.

Miyamoto, S., and Helmers, S.G. (*Editors*). 1990. Pistachios: a potential crop for far West Texas; proceedings of the first West Texas pistachio conference & workshop, May 31–June 1, 1990, El Paso, Texas. Texas Agricultural Station, Texas Agricultural Extension Service, College Station, TX. 92 pp.

Opiz, K.W. 1975. The pistachio nut. University of California, Division of Agricultural Sciences, Berkeley, CA. 8 pp.

Padulosi, S., Caruso, T., and Barone, E. (*Editors*). 1996. Taxonomy, distribution, conservation and uses of *Pistacia* genetic resources: report of a workshop, 29–30 June 1995, Palermo, Italy. International Plant Genetic Resources Institute, Rome, Italy. 69 pp.

Sanders, S. 1996. Use of pistachios in bakery products. AIB Tech. Bull. **18(11)**: 1–6.

Shrestha, A.B. 1995. Pistachio nut. *In* Evolution of crop plants. 2nd edition. *Edited by* J. Smartt and N.W. Simmonds. Longman Scientific & Technical, Burnt Mill, Harlow, Essex, UK. pp. 14–16.

Western Pistachio Association. 1999. Pistachio '99: challenges in the next millenium [i.e., millennium]. [Conference proceedings, Oct. 5–7, 1999, Hamburg, Germany.] Western Pistachio Association, Washington, DC. 170 pp.

Whitehouse, W.E. 1957. The pistachio nut—a new crop for the Western United States. Econ. Bot. **11**: 281–321.

Zohary, M. 192. A monographical study of the genus *Pistacia*. Palest. J. Bot. Jerus. Ser. **5**: 187–228.

Specialty Cookbooks

California Pistachio Commission. 1970–1979? On the lighter side with California pistachios. California Pistachio Commission, Fresno, CA. 12 pp.

California Pistachio Commission. 1985. California pistachio cuisine. California Pistachio Commission, Fresno, CA. 1 folded sheet.

Carter, P., and Browder, D. 1983. Pistachio panache. Fresno Trading Co., Fresno, CA. Irregularly paginated.

PLUM

Family: Rosaceae (rose family)

Names

Scientific names: *Prunus* species

Main commercial species

- Plum, common plum, European plum, prune—*P. domestica* L.
- Damson plum—*P. insititia* L. [*P. domestica* subsp. *insititia* (L.) C.K. Schneid.]

Minor commercial species

- American plum—*P. americana* Marshall
- Beach plum—*P. maritima* Marshall
- Apricot plum—*P. simonii* Carrière
- Canada plum—*P. nigra* Aiton
- Cherry plum (myrobalan plum) —*P. cerasifera* Ehrh.
- Chicasaw plum—*P. angustifolia* Marshall
- Japanese plum—*P. salicinia* Lindl.
- Munson plum—*P. munsoniana* W. Wight & Hedrick
- Pacific plum—*P. subcordata* Benth.
- Wild-goose plum—*P. hortulana* L.H. Bailey

- The English word "plum" is derived from the Latin *prunus*. The genus name *Prunus* is the classical Latin name of the plum tree, from the Greek *prunos*.
- See APRICOT for the names of apricot-plum hybrids.
- In North America, "prunes" are simply dried plums. However, in Europe, the term is also used to designate a distinct group of plums in which the fruit is usually reddish or blue, elongated, high in sugar content, and firm.
- The "Greengage" plum [so named because it was introduced into England from France in 1725 by botanist Sir William Gage (1656?–1727)] is known in France as the "Reine-Claude," literally "Queen Claude," after Claude de France (1499–1524), wife of King Francis I (1494–1547). She was very fond of the fruit of this plum, and indeed of food in general, and was quite fat. According to legend, the name was said to have been inspired by her ample posterior, reminiscent of the characteristic deep cleft in the plum.
- The expression "plumb crazy" means completely crazy (based on "plumb" in the sense of "completely" vertical). This is frequently found as "plum crazy." In fact, "plumb" was once an accepted spelling of "plum."
- *Domestica* in the scientific name of the common plum, *P. domestica*, is Latin for domesticated, which comes from *domesticum*, belonging to the household.
- *Insititia* in the scientific name of the Damson plum, *P. insititia* "Damson," the popular name for this plum, is a corruption of Damascus, the Syrian city near which this plum is believed to have originated. The alternative name "bullace" is derived from the Latin *bulla*, a large bubble, an allusion to the roundness of these plums ("bullace" is also the name given to a sea fish and a catfish that are consumed in the United States). Historically, attempts were made to use the terms "Damson" and "bullace" to designate different cultivars, but today, these distinctions are not consistently made.

Plum (*Prunus domestica*), from Morren (1851–1885, vol. 2). Cultivars illustrated: de Belle Vu (top two fruits), Victoire Gathoye (middle fruit), Souvenir de Mathilde Gathoye (bottom two fruit).

Plant Portrait

Common plum (*P. domestica*)

The most important of the plums, the common European plum, probably is native to the Caucasus and trans-Caucasus region. The fruit is of various shapes. The skin color may be blue, red, yellow, green, or white. The flesh is typically yellowish. The Reine Claude group of plums is usually placed in *P. domestica*, although it is sometimes assigned to the damson plum. These plums are relatively small, ovoid, usually green or yellow, with firm texture and sweet juicy flesh.

Damson plum (*P. insititia*)

This is native to western Asia and south and southeastern Europe. It is very closely related to the common European plum, and is often considered to be merely a variety of it. The damson plum or bullace is a dwarfish, thorny, hardy tree, with small sweet to sour blue-skinned fruit less than 2.5 cm (1 inch) in diameter. The fruits are generally acidic and tart, making highly esteemed jellies and jams, and are produced commercially for this purpose.

Plums are a very diverse group, with well over a dozen species cultivated for fruit, and thousands of named varieties. Plum species can easily be hybridized, and modern cultivars are often the result of hybridization. Plums were independently domesticated in Europe, Asia, and North America. The common plum originated in Europe, the damson plum in western Asia, the cherry plum in western and central Asia, the Japanese plum and apricot plum in China, and the American plum and other North American plums in North America. The plum, both fresh and in the form of prunes, was one of the staples of prehistoric peoples. Plums were grown by the Romans, and ever since, the fruit has been produced in Europe. Plums were brought to America with the first colonists, but were not nearly as widely planted, as apples, pears, and cherries, and did not achieve notable popularity for some 2 centuries. It is believed that native Indians of North America may have carried out minor selection of native American plums, but only in the last 2 centuries have concerted efforts been made to select varieties of North American species. Plums are a valuable North Temperate zone fruit crop. The two main Eurasian species, the common plum and the damson plum, produce the best fruit, but these require a milder climate than many of the other plum species. Major producers of plums include Russia, China, the United States, and Romania. California produces about 90% of the US crop.

Prunes are simply plums with sufficiently high sugar content that they can be dried successfully without removal of the stone. Fermentation would occur at the pit were it not for the high concentration of sugars. All prune varieties belong to *P. domestica*. Only some plum varieties are suitable for making prunes—generally those with a firm flesh and a high sugar content.

Culinary Portrait

Plums are eaten fresh as dessert fruits, or dried as prunes. They can usually be consumed with their skin, but if peeling is desired, the fruit can be immersed in boiling water for 30 seconds, and then rinsed immediately afterward in cold water. Plums come in both freestone (the pit easily separating from the flesh) and clingstone (the pit clinging to the flesh) varieties, and the former is obviously easier to pit. Plums are made into jelly, jam, juice, liquor, brandy, cognac, and cordials, and are also used in baking and in confections. The fruit is an excellent accompaniment to pork and poultry, and goes well with pies, cakes, muffins, puddings, and ice cream. Plums are commonly used in sweet and sour sauce, and often candied or preserved in vinegar. Overcooking plums turns them into a purée, although stewed plums are popular. When purchasing plums, the presence of a powdery "bloom" on the fruits is an indicator that the fruit has not been handled excessively. Hard, dull-colored fruit are underripe. Fruit that is very soft, bruised, or stained should not be purchased. Plums are moderately perishable. They can be stored at room temperature to ripen, and ripe plums will keep in a refrigerator for several days. Plums and prunes are well known for their laxative effect.

Prunes are usually eaten out of hand or stewed. To stew prunes, add sugar at the end of cooking; otherwise, the sugar will prevent the dried fruit from absorbing enough water. Soaking prunes in water, juice, or alcohol lessens the time required to cook the fruit. Prunes may be used to prepare cakes, muffins, biscuits, puddings, and sauces. Prunes often are served with rabbit, pork, poultry, and wild game, and in Iranian and Near Eastern culinary tradition, prunes are frequently served with lamb. Prunes are sold with or without their pits. The kernels in the pits contain toxic hydrocyanic acid, and should not be consumed. When purchasing prunes, look for those that are plump and fairly soft, without stickiness or mold. Prunes should be stored in an airtight container, avoiding both high and low humidity. They are best kept in a refrigerator.

Culinary Vocabulary

- "Duck sauce" is another name for Chinese "plum sauce" (which may be made from apricots). This name reflects the widespread use to flavor Peking duck in Chinese cuisine.
- A "friar" is a juicy plum suitable for snacks or preserves.
- Concerned about the negative association of prunes as something the elderly take to relieve constipation, the prune industry attempted to market prunes as "plum raisins" in hopes that the new term would appeal more to younger people.
- In 2001, US plum growers won permission from the government to start calling prunes "dried plums," in order to escape the stereotypic association of prunes with laxative use by the elderly. However, "prune juice" remained "prune juice," since it seemed that "dried plum juice" was a contradiction in terms.
- *Slivovitz* (pronounced many ways, often as SLIHV-uh-vihts) is a potent, dry, colorless, slightly bitter, central and eastern European brandy made from plums. The word traces to the Serbo-Croat *sljiva*, plum. It was considered the national drink of the former Yugoslavia, especially of the Bosnian and Serbian republics. (Similar plum brandies of Europe include the German *quetsch*, the French *mirabelle* and *prunelle*, and the Swiss *pfumli*.)
- "Plum pudding" is a steamed or boiled pudding frequently served at holiday times. But plum pudding typically does not contain plums. So why is it called "plum pudding"? Before the 17th century in England, the term "plum" always meant plum. However, by the 17th century, dried plums (i.e., prunes) began to be replaced by raisins in cooking. The dishes once made with plums, despite being now made with raisins, retained the term "plum." Thus, plum pudding, plum cake, plum duff, and others are no longer made with plums. A "sugar plum" is a small candy in a ball or disk, or a sweetmeat, once again not made with plum.
- "Salted plums" are dried, usually heavily salted plums used in Asian (particularly Chinese) cuisine as a condiment (often in place of salt), confection, appetite stimulant, and breath freshener.
- *Ume-boshi* (pronounced OO-meh-boh-shee) are Japanese pickled plums. These are typically pink, with a tart, salty taste.

Cherry plum (myrobalan plum, *Prunus cerasifera*), cultivar Jubilee, from *Revue de l'Horticulture Belge et Étrangère* (1907, vol. 33). This is one of the minor cultivated plum species.

Curiosities of Science and Technology

- The Chinese considered the plum to be one of the "five renowned fruits of antiquity." The others: peach, apricot, jujube, and chestnut.
- England's King Henry VIII (1509–1547) once ordered a 75-kg (165-pound) plum pie that measured 2.7 m (9 feet) in length.
- The Pilgrims had wild plums for dessert during their first Thanksgiving in 1621.
- During the Puritan period in England (16th–17th centuries), plum pudding was outlawed as "sinfully rich."
- Mating is a serious matter with plums, as most varieties are self-sterile, and particular combinations are incompatible. Therefore, a plum grower must know the mating preferences of the plums in his orchard, or they may not produce fruit. European cultivars will not pollinate the Oriental species. Some European cultivars and some others are self-fertile, and still others will set some fruit with their own pollen but not enough to produce a commercial crop. Most Oriental varieties require cross-pollination, either from other Japanese varieties or from American cultivars. Hybrid plums, which are the most important ones grown in the coldest regions of North America, often will

neither pollinate themselves nor any varieties of similar parentage, and are best pollinated by American varieties.

- Species of *Prunus*, in common with many tree species, often have lighter-colored bark in warmer and drier climates. The bark color often becomes lighter towards the south in North America.
- Potatoes should not be planted near plum trees because an aphid that sometimes overwinters in the tree may carry a potato disease.
- As with many fruit trees, a heavy fruit yield one year (called "masting") may cause plums to "take a rest" the following year, and develop a smaller crop.

Key Information Sources

Bailey, L.H. 1892. The cultivated native plums and cherries. Cornell University, Ithaca, NY. 73 pp.

Bartrum, E. 1903. The book of pears and plums. J. Lane, London, UK. 96 pp.

Bernhard, R. (*Editor*). 1990. Fourth international symposium on plum and prune genetics, breeding and pomology, Bordeaux, France, 24–28 July 1989. International Society for Horticultural Science, Wageningen, Netherlands. 333 pp.

Djouvinov, V., Dotchev, D., and Gercheva, P. (*Editors*). 2002. Proceedings of the 7th international symposium on plum and prune genetics, breeding and pomology; Plovdiv, Bulgaria, 20–24 August 2001. International Society for Horticultural Science, Leuven, Belgium. 403 pp.

Fu, Q. 1995. Preliminary research into biochemical classification of plum varieties—gas chromatographic analysis of sugar substances in the fruit of several plum varieties. Acta Hortic. **403**: 211–218.

Gregor, D., Hartmann, W., and Stosser, R. 1994. Cultivar identification in *Prunus domestica* using random amplified polymorphic DNA markers. Acta Hortic. **359**: 33–40.

Gur, A. 1986. Plum. *In* CRC handbook of fruit set and development. *Edited by* S.P. Monselise. CRC Press, Boca Raton, FL. pp. 401–418.

Hedrick, U.P., and Wellington, R. 1911. The plums of New York. J.B. Lyon Co., State Printers, Albany, NY. 616 pp.

Heinkel, R., Hartmann, W., and Stoesser, R. 2000. On the origin of the plum cultivars 'Cacaks Beauty', 'Cacaks Best', 'Cacaks Early' and 'Cacaks Fruitful' as investigated by the inheritance of random amplified polymorphic DNA (RAPD) fragments. Sci. Hortic. **83**: 149–155.

Paunovic, S.A. (*Editor*). 1978. Third symposium on plum genetics, breeding, and pomology, Cacak, Yugoslavia, 16–21 August 1977. International Society for Horticultural Science, Wageningen, Netherlands. 240 pp.

Plich, H. 1999. The effect of storage conditions and date of picking on storability and quality of some plum (*Prunus domestica* L.) fruit cultivars. Acta Hortic. **485**: 301–307.

Ramming, D.W., and Cociu, V. 1990. Plums (*Prunus*). *In* Genetic resources of temperate fruit and nut crops. *Edited by* J.N. Moore and J.R. Ballington, Jr. International Society for Horticultural Science, Wageningen, Netherlands. pp. 235–287.

Ray, G.H. 1978. Plums for the backyard garden: *Prunus salacina* and *Prunus domestica* hybrids, cultivars. Minn. Hortic. (St. Paul), **107(8)**: 224–226.

Reid, W., and Gast, K.L.B. 1993. The potential for domestication and utilization of native plums in Kansas. *In* New crops. *Edited by* J. Janick and J.E. Simon. Wiley, New York, NY. pp. 520–523.

Shimada, T., Hayama, H., Haji, T., Yamaguchi, M., and Yoshida, M. 1999. Genetic diversity of plums characterized by random amplified polymorphic DNA (RAPD) analysis. Euphytica, **109**: 143–147.

Sterling, C. 1953. Developmental anatomy of the fruit of *Prunus domestica* L. Bull. Torrey Bot. Club, **80**: 457–477.

Taylor, H.V. 1949. The plums of England. Lockwood, London, UK. 151 pp.

Waugh, F.A. 1901. Plums and plum culture: a monograph of the plums cultivated and indigenous in North America, with a complete account of their propagation, cultivation and utilization. Orange Judd Co., New York, NY. 371 pp.

Weinberger, J.H. 1975. Plums. *In* Advances in fruit breeding. *Edited by* J. Janick and J.N. Moore. Purdue University Press, West Lafayette, IN. pp. 336–347.

Wight, W.F. 1915. The varieties of plums derived from native American species. U.S. Dept. of Agriculture, Washington, DC. 44 pp.

Specialty Cookbooks

Bestwick, J. 2001. Life's little peaches, pears, plums & prunes cookbook: 101 fruit recipes. Avery Color Studios, Gwinn, MI. 132 pp.

Carver, G.W. 1917. 43 ways to save the wild plum crop. Printed by Tuskegee students from funds supplied by the Ash Fund, Tuskegee, AL. 12 pp.

Eliot, E. 2001. Peaches, pears & plums. World Leisure Marketing, Halifax, NS. 64 pp.

Lawless, D. 1992. Prunes: more than wrinkles. N. Bowyer, Dayton, OR. 53 pp.

Nims, C.C. 2003. Stone fruit: cherries, nectarines, apricots, plums, peaches. WestWinds Press, Portland, OR. 96 pp.

Sunsweet Growers, Inc. 1959. Visions of sugarplums: 29 delightful ways to serve Sunsweet prunes. Sunsweet Growers, Inc., San Jose, CA. 12 pp.

United Prune Growers of California. 1933. Prunes for epicures: 35 intriguing recipes. United Prune Growers of California, San Francisco, CA. 22 pp.

Woodier, O. 2004. Peaches and other juicy fruits: from sweet to savory—150 recipes for peaches, plums, nectarines, and apricots. Storey Publishing, North Adams, MA. 186 pp.

POTATO

Family: Solanaceae (potato family)

Names

Scientific name: *Solanum tuberosum* L.

- The word "potato" is believed to have been derived from the Spanish *patata*, based on the Haitian *batata*, the original West Indian word for sweet potato. When the common potato was introduced to Spain, it was first known as *papas*. In time, the papas were also called patatas, which was translated into English as potato.
- The potato is also called the "white potato," to distinguish it from the sweet potato, the flesh of which is generally yellow to deep orange in varieties consumed in developed countries (although in most developing countries the type produced has white to cream colored flesh). The common potato has also been called "Irish" or "English" potato to differentiate it from the sweet potato.
- "Chinese potato" and "taro potato" are phrases used for taro (which see).
- Potatoes were domesticated in the Andes 10 000 years ago, probably in Peru, and so a name like "Peruvian potato" would be most appropriate. However, the common names that people assign to domesticated plants have often been based not on where they first came from, but rather on where they most recently appear to have come from. Illustrative of this, the potato was called the "Virginian potato" when it first came to England, because Francis Drake in 1586 picked up the vegetable at Cartagena in the Caribbean, but went to Virginia before returning to England, leading to the impression that potatoes came from Virginia. However, in 1718, Irish immigrants to Boston brought the potato back to the Americas, and because it came from Ireland, it acquired the name "Irish potato."
- The word "spud" refers to a spade-like tool. Because spades have been used for hundreds of years to dig potatoes, the term has come to apply also to potatoes.
- The genus name *Solanum* is derived from the Latin name *solanum* for some species of the genus, possibly *S. nigrum* L., the common nightshade, a plant with poisonous potential that is nevertheless used medicinally and for food purposes. It has also been suggested that the genus name is based on the Latin *solamen*, meaning quieting, an allusion to the sedative properties of some of the species.
- *Tuberosum* in the scientific name *S. tuberosum* is Latin for tuberous, descriptive of the potatoes.

Plant Portrait

The potato is the world's most important and best-known vegetable. It is also the world's fourth most important food crop in terms of weight of food produced, after wheat, rice, and corn (maize) (foods that are nutritionally inferior). The potato is superior (per unit area per unit time) to every other food crop in its ability to produce human food. It produces more dry matter and protein per unit area than the major cereal crops, although three times as much raw potato would have to be eaten to achieve equivalent energy to that of cereals.

The potato is a perennial herb, grown as an annual in areas where frost occurs. It grows 0.5–1 m (1.5–3 feet) in height, and has small attractive flowers. The fruits are green or purplish green berries 1.3–2 cm (1/2–3/4 inch) in diameter, but are not produced by many cultivars. The berries are quite poisonous, and indeed, the only edible part of a potato plant is the potatoes. Most potato cultivars are sterile, incapable of producing true seeds. The potato itself is a tuber—an enlarged storage organ produced from an underground stem. Although potatoes can be propagated from seeds, they are normally grown from disease-free certified stock, called "seed potatoes," that are simply potatoes. Seed potato tubers are cut into several pieces and planted. Potatoes come in a rainbow of colors (in addition to black and white), including purple, red, pink, green, gold, and yellow. Although the color of many potatoes is often only skin-deep and the flesh is white or whitish, some potatoes are colorful throughout. These include the cultivars All Blue, All Red, and Yukon Gold. New or early potatoes are dug before full maturity and marketed quickly. They do not stand rough handling as well as the late crop and are more subject to injury. By contrast, potatoes of the late crop are usually more mature when dug and are often stored and shipped to market during the winter and spring.

The potato is thought to have arisen either in a single area of domestication in the central Andes, or also independently in Chile. The earliest known remains of cultivated potato, originating near Lima, Peru, are dated at 7000 BC. The earliest known remains of wild potato, dated at 13 000 BC, come from southern Chile. So-called "moche" ceramic pots, combining potato and human themes, were found through most Andean cultures after the time of Christ. The Spanish conquistadors discovered the potato in 1537 in the Andean village of Sorocota. During the 16th century, Spanish explorers found that potatoes prevented scurvy, and carried them on ships for this purpose. Before the end of that century, the potato had spread through much of western Europe. The potato was introduced by Europeans to North America around 1621 and was also taken to India, China, Japan, and Africa by the end of the 1600s. The fungus *Phytophthora infestans*, which causes late potato blight, was responsible for the Great Irish Potato Famine (1845–1848). The fungus was able to devastate Irish potatoes because the crop was extremely uniform genetically, and vegetative reproduction maintained this uniformity. The potato famine was the worst catastrophe in Europe since the Black Death of 1348. As a result of the famine, about 1 million people died, and 2.5 million emigrated between 1845 and 1860, largely to North America.

Potato (*Solanum tuberosum*), from Engler and Prantl (1889–1915).

There are more than 5000 varieties of potatoes in the world. In the Andes of South America, as many as eight different potato species are cultivated, although in North America only *S. tuberosum* is grown commercially. Potatoes are grown commercially in every continent except Antarctica, with 70% of the world's production in Eastern Europe and Russia.

Chuño (pronounced choon-yo), a Spanish word used in South America, refers to dried vegetables, generally tubers and especially potatoes, preserved in a freeze-dried form in high, cold mountains. In the Andes, potato chuño is used to make soup and sometimes a kind of flour for bread. The word has been anglicized to "chuno" (pronounced choo-noh). Scientists in the US developed much of the "instant potato" industry and acquired many of the relevant patents for freeze-drying potatoes after World War II. However, the Incas prepared freeze-dried dehydrated potatoes (chuño) perhaps 13 000 years ago, so modern scientists can not fairly claim to have exclusively invented instant-potato preparations.

Culinary Portrait

Potatoes contain some indigestible starch, and are cooked before eating to convert this starch to sugar. The vegetable is served separately—baked, boiled, mashed, roasted, sautéed, or braised—or combined with other ingredients in soups, stews, salads, chowder, and dumplings. "Fish and chips" and "meat and potatoes" are long-standing combinations in Great Britain and North America. Small potatoes are typically served whole, with their skins on. Potato skins are an excellent source of vitamin C, and a single small potato boiled in its skin provides half of the adult daily requirement. (The traditional Irish practice of cooking potatoes in their jackets and peeling them at the table takes maximum advantage of the nutritive value of potatoes.) Potato salad, potato casserole, hash browns, and potato pancakes are just some of the large array of popular potato dishes. More exotic uses include crêpes, puff pastries, fritters, pies, and tarts. Increasingly, potatoes are processed into a large variety of foods such as French fries and potato chips, or are dehydrated, flaked, granulated, canned, and frozen. Potatoes are also used as a raw material for production of alcohol, and are the basic source of vodka.

Potato flour (also called potato starch, and often called "farina," although this word is applied to numerous kinds of flour) mixes easily with water, absorbing it quickly. Potato flour is a good binder and thickener and is used commercially as a filler in cooked sausages, where a compact texture is required. Potato flour is also used in some baked goods because it produces a moist crumb.

Culinary quality of potatoes is determined particularly by dry-matter content. Potatoes of high dry-matter content (e.g., Russet) are mealy and preferred for baking and mashing. Such potatoes are called mealy potatoes, baker potatoes, and starchy potatoes. When the dry-matter content is high, however, potatoes may slough apart when boiled (this occurs because the cells of these mealy types tend to separate from each other). Accordingly, potatoes of higher density are more suitable for baking and for mashing than for frying. By contrast, when the dry-matter content in relatively low (in so-called "waxy" types), the tissues remain moist and cohesive, and the tubers are best suited for frying, or for boiling (such potatoes may be called "boiling potatoes") for the purpose of cutting into scallops and chunks for potato salad. Tubers that float in a brine made by dissolving 120 g of table salt in 1 L of water (1 pound/American gallon; alternatively, 1 part salt to 11 parts water) are best for baking and mashing, whereas tubers that sink are suitable for baking or preparing potato chunks.

Some potatoes develop a large, dark-colored region during cooking. This phenomenon, known as "stem-end blackening" because it occurs at the end that was connected to the plant, is caused by a reaction of iron ions in the water and chemicals (phenolic substances) in the potatoes. If it occurs, the problem can be minimized by making the boiling water distinctly acidic. When the potatoes are half done, a teaspoon of cream of tartar can be added for every liter (or American quart) of cooking water.

Much of the nutrients of potatoes are in the skin, and it is a good practice to cook and eat potatoes in their jackets. At the very least, a peeler that removes only a thin layer of skin should be used, and potatoes should be cooked immediately after they have been peeled (leaving potatoes to soak in cold water leaches out soluble nutrients).

Potato plants and potatoes contain toxic chemicals—glycoalkaloids, mainly alpha-solanine and alpha-chaconine, which can irritate the gastrointestinal tract and impair the nervous system. However, the quantities in the tubers are usually harmless unless the latter have been exposed to light or mechanically damaged. Because tubers photosynthesize in the presence of light, causing the potatoes to become green, the presence of green coloration in potatoes is correlated with the amount of toxins. Care should be taken to store potatoes in lightproof containers such as paper bags. Green-colored potatoes should be discarded. Because the toxins concentrate in the skin and eyes of the potatoes, peelings and sprouts destined for compost should be buried to keep them out of sunlight so that animals digging up the remains will not be poisoned.

Another "hazard" associated with the potato is the French fry or "chip"—one of the stars of transnational fast-food chains. By itself, the fried potato chip is not a danger to health, but laden with fat and consumed in the marginal nutritional context of fast food, it has become a major contributor to the plague of obesity that is the most significant food-related threat to health in North America.

Culinary Vocabulary

- Just who invented "French fries" (strips of potato fried in deep fat until brown) and when is debatable, and on that question hinges the question of just how "French" the dish is. Further, the distinction between "fries" and "chips" is problematical. Modern "French fries" are made from deep-fried potatoes, and are further defined by two criteria: they are prepared in long strips or sticks (whether thick, thin, or curled is a matter of preference) and are served hot. By contrast, modern "chips" are also made from fried potatoes, but are prepared in very thin slices (whether flat or corrugated is a matter of preference), which are generally preserved (mostly in sealed bags) and are served at room temperature. Historically, frying of potatoes is an old practice. Fried potatoes were sold on pushcarts by pedlars in Paris, beginning in the mid 19th century (it isn't clear, however, whether potatoes were consistently deep-fried or simply sautéed, i.e., pan-fried). Claims have been made that such potato pushcarts first arose in Belgium. Thomas Jefferson (1743–1826, third American president) has been credited with introducing "French fried potatoes" from France to America when he served them at a White House dinner, but it is not clear that these fried potatoes were French fries in the modern sense; rather, it appears that they were simply potatoes "served in the French style." In France, French fries are called *pommes frites* ("fried potatoes") or *frites* ("fries") for short. Beginning in 1864, French fries were united with fish and sold by pedlars in London as "fish and chips." That is, "potato chips" in England is the equivalent of "French fries" in North America, although "chips" is increasingly becoming an international term equivalent to "French fries." "Chips" in the North American sense might be called "potato crisps" in England. (Sometimes, in England, the term "French fries" is used by fast-food restaurants serving narrow-cut fries prepared in the American style, while reserving the term "chips" for much thicker strips.) "Potato chips" are widely said to

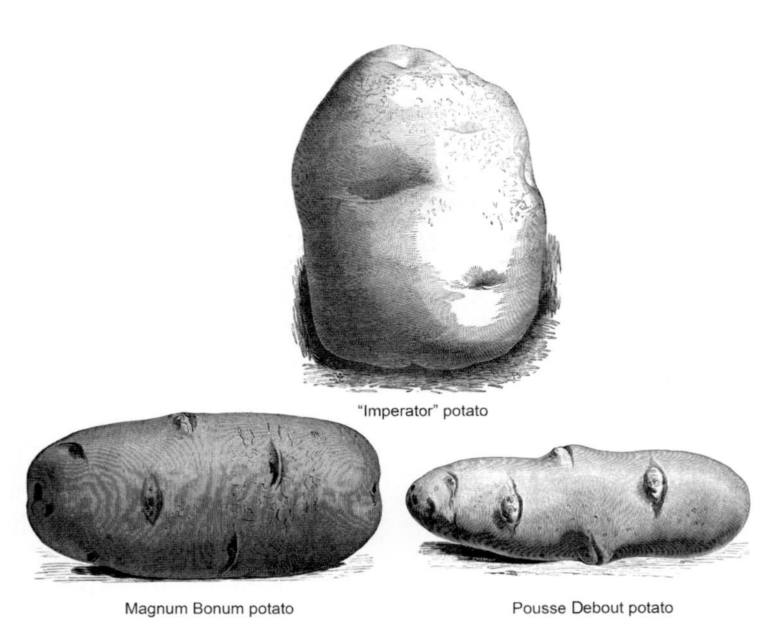

Some varieties of potato (*Solanum tuberosum*). Imperator potato is from *Revue de l'Horticulture Belge et Étrangère* (vol. 33, 1907); the other two varieties are from Vilmorin-Andrieux (1885).

have been invented in 1853 by Chef George Crum at a fashionable resort in Saratoga Springs, New York. According to a widely circulated story, railroad magnate Commodore Cornelius Vanderbilt sent his fried potatoes back to the kitchen, complaining they were too thick. To spite his haughty guest, Chef Crum sliced some potatoes paper thin, fried them in hot oil, and salted them. To everyone's surprise, Vanderbilt loved his "Saratoga Crunch Chips," and thereafter for some time the name "Saratoga chips" became popular. Crum is, accordingly, generally credited with innovating the "potato chip" in the American sense (i.e., sliced as thin as practicable), and the potato chip is therefore an American invention. Interestingly, US-based restaurant chains such as McDonald's and Burger King have become so dominant in many countries that their French fries are often called "American fries." Apart from the question of where frying thin-sliced pieces of potato originated is the issue of where the phrase "French fry" originated. The Oxford English Dictionary, the premier dictionary for judging English word origins, traces "French-fried potatoes" back to 1894, and suggests the phrase is American in origin ("French fries" was first recorded in the 1930s). More generally, "frenching" is a method of cutting potatoes into narrow strips (a "frenched French bean" is one that has been cut lengthwise into very narrow strips), and does not refer to the country of France (leading to the possible conclusion that "French" in "French fries" should not be spelled with a capital "F," as is often done). Some have claimed that "French fries" were invented in Belgium (and therefore should be called Belgian or Belgium fries). One theory holds that during the First World War, American soldiers in Europe encountered elongated fried potato strips being sold by street vendors and in restaurants. The product came to be called "French fries," it is claimed, because the Belgians serving it commonly spoke French. The potato strips were first parfried sufficiently to inactivate their enzymes, and then fried just before serving (this double frying with a pause in the middle has been characterized as "frying in the French manner"). After the war, this technique was adopted in North American restaurants, and today, about one-third of all potatoes are consumed as French fries in North America. To recapitulate: it really isn't known with certainty whether or not "French fries" were first invented in France (or Belgium), whether the "French" refers to "frenching" (i.e., the technique of slicing vegetables into long sticks), or whether somehow Americans just invented the phrase "French fries" because they first associated them with France.

- "Parmentier" is a French term generally used to describe dishes that include potatoes in some form. Antoine Augustin Parmentier (1737–1813) was a French agronomist who popularized the potato in France. His interest in the vegetable started when he was a prisoner of war of the Germans, who forced him to subsist on potatoes. For a time, the potato was known as the parmentière in his honor, and he is commemorated in the names of various culinary dishes featuring potatoes.

- *Pommes* is French for "apples," while *pommes de terre* is French for "potato." However, in French culinary practice, classical potato dishes are usually prefixed by pommes rather than by *pommes de terre*. Dozens of such dishes are recognized, for example,
  - *pommes à l'alsacienne*: new potatoes in butter, with onions, bacon bits, and herbs
  - *pommes à l'anglaise*: steamed or boiled potatoes
  - *pommes à la hongroise*: thin round slices of potato baked with onions sautéed in butter, baked and served with chopped parsley
  - *pommes à la Lyonnaise* (*pommes Lyonnaise*): sliced potatoes in butter with fried onions, sprinkled with chopped parsley
  - *pommes à la menthe*: new potatoes boiled with fresh mint, garnished with a leaf of mint on each potato when served
  - *pommes Anna*: thinly sliced potatoes layered with butter, cooked in a lidded dish until brown and crisp on the outside and soft and buttery on the inside, inverted, and cut into wedges
  - *pommes Annette*: *pommes Anna* made with potatoes cut into slivers
  - *pommes dauphine*: potatoes mashed with butter and egg yolk, blended with choux pastry (cream puff paste), spread into balls or croquettes, breaded, and fried ("Dauphine" in the name of a dish generally refers to any preparation garnished with *pommes dauphine*.)
  - *pommes dauphinoise*: potatoes cut into thick round slices, layered with cream in a gratin dish, rubbed with garlic and butter (or a mixture of eggs, milk, and cream is poured over the potatoes), sprinkled with grated cheese, and baked ("Dauphinois" in the name of a dish refers to preparations of this nature.)
  - *pommes duchesse*: a garnish or dish of mashed potatoes with raw eggs, shaped into pipes or patties and oven-browned ["Duchesse potatoes" (or duchess potatoes) includes a variety of mashed potato dishes that have been formed into croquettes, small cakes, or fancy shapes, and baked or fried until the surface is crisp.]
  - *pommes frites*: French fried potatoes

Statue of Antoine Augustin Parmentier, French expert on potatoes, from *Revue de l'Horticulture Belge et Étrangère* (1888, vol. 14).

  - *pommes soufflées* (soufflé potatoes): thinly sliced potatoes puffed into little pillows by a double-frying process
  - *pommes château* (chateau potatoes): a French dish of potatoes cut into pieces that are 3.8 cm (1½ inch) long, frequently tournéed (cut into a football shape, with seven equal sides and blunt ends), and sautéed in butter until brown.
- *Allumette* (French for "match"; pronounced al-yoo-MEHT) is a culinary expression for foods, especially potatoes, which are cut into a matchstick shape and size. (The word also refers to a thin strip of puff pastry topped with a sweet or savory mixture.)

- "Irish stew" is a stew of mutton, onions, and predominantly, potatoes. The "Irish" in the name reflects the traditional view that potatoes were the staple food of the Irish.
- "Boxty" (pronounced BOX-tee; also "broxty") is a traditional Irish potato dish—a blend of grated raw potatoes and cooked mashed potatoes, with added wheat flour, cooked on a griddle into a pancake. Boxty is celebrated in the old rhyme:

  Boxty on the griddle,
  and boxty in the pan.
  If ye can't make boxty,
  Ye'll never get a man!
- The calorie-packed dish called "poutine" originated from the Bois-Francs area of the province of Quebec in the 1950s. The basic recipe consists of French fries, cheese curds, and gravy. Apparently, poutine was created in the kitchen of a restaurant named *Lutin qui rit* (French for "the laughing elf") located in Warwick, operated by Fernand Lachance. (In France, "poutine" refers to a small or undeveloped fish, usually sardines or anchovies.)
- In English, potatoes baked in their skins are described as jacket potatoes or potatoes in their jackets; this is more charmingly phrased in Italian as *in veste da camera*, "in their nightshirts."
- A *latke* (pronounced LAHT-kuh; a Yiddish word derived from the Russian *latke*, "a pastry"), in Jewish cuisine, is a crisp pancake made from grated potatoes. Typically, other ingredients include eggs, onion, matzo meal, and seasonings. After frying, latkes are served hot, often with applesauce.
- "Stovies" is a Scottish word for potatoes cooked slowly in a very small amount of water.
- Potato salad has long been a popular dish in Germany, and immigrants to North America brought with them the traditional German way of preparing it, with the result that it became known by the phrase "German potato salad." Potato salad in North America is typically made with cooked, sliced, or diced potatoes bound together with mayonnaise to such additions as onions, peppers, eggs, and spices, and usually served cold. German potato salad is similar but is usually served warm or hot, and has a vinegar dressing instead of mayonnaise, and frequently, bacon and (or) bacon fat is added.
- "Straw potatoes" are potatoes sliced into very thin, long sticks that are deep-fried.
- "Delmonico potatoes" are named for the famous 19th century Delmonico Restaurant, the first luxurious restaurant built in New York. The dish was created by the owner–chef, and consists of cooked, creamed, diced (or mashed) potatoes topped with grated cheese and buttered breadcrumbs, and baked until golden brown. Lemon juice, parsley, salt, and pepper are typically sprinkled on top.
- "O'Brien potatoes" is a dish of diced potatoes fried until crisp and brown, along with chopped onions and sweet peppers. The origin of the name is obscure, presumably reflecting the long association of the Irish (represented by the name O'Brien) and potatoes.
- A "potato nail" is a large-headed, sharp-tipped nail that is inserted into a raw potato so that heat is conducted rapidly into the center of the potato and it bakes quickly and uniformly.

Curiosities of Science and Technology

- Toxic levels of glycoalkaloids are present in wild potatoes of the southern United States, Mexico, and South America. Indians of these regions learned to detoxify the tubers and reduce the bitterness due to the glycoalklaoids by boiling the potatoes in a slurry of edible clays. Silicates in the clays are believed to bind with the toxic compounds, thus neutralizing them.
- The Incas of Peru used the duration required to cook potatoes as a unit of time.
- In 1664, a book on potatoes by John Forster was published, and dedicated to King Charles II. Back then, long titles were not unusual. The title: *England's Happiness Increased, Or a sure and Easy Remedy Against all Succeeding Dear Years by a Plantation of the Roots called Potatoes: Whereby (with the Addition of Wheatflower) Excellent Good and Wholesome Bread may be Made Every 8 or 9 Months Together, for Half the Charge as Formerly; Also by the Planting of These Roots Ten Thousand Men in England and Wales Who Know Not How to Live, or What to Do to Get a Maintenance for their Families, may on one Acre of Ground make 30 Pounds per Annum. Invented and Published for the Good of the Poorer Sort.*
- Benjamin Franklin (1706–1790), while ambassador to France, attended a banquet at which the potato was served 20 different ways. Franklin returned to America singing the praises of the potato as the ultimate vegetable. Although the potato was first brought to America in 1621, it did not become popular until Franklin publicized it.
- In the late 1600s, Ireland recognized the food value of potatoes and became the first country in Europe

to plant potatoes as a staple food crop. For about 200 years thereafter, factory workers received 5.4 kg (12 pounds) of potatoes every day. It has been estimated that in the early 1840s, a working man in Ireland ate 5.5–6.3 kg (12–14 pounds) of potatoes daily. A typical Irish family ate more than 113 kg (250 pounds) of potatoes every week. The average American eats only 150 g (5.4 ounces) of potato a day (about one potato daily).

- Several historians have credited the introduction of the potato into Europe with encouraging the rapid expansion of population and the provision of a cheap food source for workers that facilitated the Industrial Revolution in the British Isles.
- Numerous ineffective medicinal uses and beliefs developed in Europe for potatoes, many of them quite silly, and the ideas were often transported to North America. A remedy for rheumatism and eczema was to place a potato in the patient's pocket, especially the first potato of the year. To cure a diseased tooth, a peeled potato was carried in a pocket on the same side as the aching tooth. If an expectant mother ate potatoes, it was feared that the baby would be born with a large head. To alleviate sore muscles and oozing sores, a potato was applied. To get rid of a wart, it was rubbed with a cut potato, and the potato was buried in the ground, with the expectation that as the potato rots, the wart would disappear. In Ireland, the water from boiled potatoes was rubbed on broken bones, aches, and sprains, but such water could not be used for washing, for fear of developing warts. Stones boiled along with potatoes were thought to develop magical curative powers. Delirium was treated by tying raw potatoes behind the ears. In Holland, it was thought that potatoes had to be stolen to be effective as medicinal cures. In Yorkshire, England, they had to be dried in the morning sun and protected from the afternoon sun to be medicinally effective. In Newfoundland, it was thought that a sliced baked potato in a stocking that was tied around the neck would cure a sore throat. In Texas, burns and frostbite were treated by application of a scraped, raw potato.
- In 1806, Napoleon offered 100,000 francs to anyone who could create sugar from a native plant. Russian chemist Konstantin Sigizmundovich (Gottleib Sigismund Constantin) Kirchhof (or more simply, K.S. Kirchhof) later discovered that sulfuric acid added to potato starch would make the conversion. (The potato is not a native plant of Europe. In any event, as Kirchof was a Russian, and so the enemy of France, he was unlikely to collect the prize.)
- The word "pothole," meaning a hole in the road, is derived from the Irish custom of boiling the daily meal of potatoes in a pot. When ready, the pot was set on the ground to cool. Mashing the potatoes in the pot pushed it into the ground, leaving a pothole!
- In the 19th century, hot baked potatoes were sold on London, England, streets from fall through early spring. By 1851, there were over 300 such vendors, selling 10 tons of potatoes each day. In some cases, the hot potatoes seem to have been purchased as hand warmers rather than for food. During the late 18th and early 19th centuries, Americans commonly put hot baked potatoes in their pockets during winter to help keep warm.
- In 1849, 4 bushels of Oregon potatoes were selling for $500 in San Francisco. It has been said that Oregon farmers dug potatoes and struck gold. Indeed, during the Alaskan Klondike gold rush (1897–1898), potatoes were practically worth their weight in gold. Potatoes were so valued for their vitamin C content that miners traded gold for potatoes.
- In Ireland, special "potato rings" were used at table to serve potatoes, and were common until the 20th century. These were large, round, silver rings reminiscent of dish-drying stands, with a napkin placed in the center, and potatoes in the outside ring.
- The U.S. Marine Corps in the 1950s issued the following advice to its military cooks: to keep baked potatoes round and smooth, rub each with a small amount of fat before baking, and cut a small piece off one end to let steam escape during baking. (There is disagreement about whether a potato should be pierced before baking. Piercing a potato, it is claimed, allows the steam to escape, forestalls sogginess, and makes the potato light and fluffy. But others claim that it makes absolutely no difference. There is similar disagreement about whether potatoes should be wrapped in foil before baking, but this does produce softer skin. Some claim that foil-wrapped potatoes are creamier, while others think that they are soggier.)
- John Richard Simplot (1909–2008) of Boise, Idaho, who has been called "America's great potato baron," is credited with inventing frozen French fries (at least, his team of researchers did so in 1953, followed by the first patent). In 1965, he initiated the large-scale commercial consumption of French

fries prepared from frozen material by arranging their sale (in 1966) by the McDonald's fast-food chain. The J.R. Simplot plant in Aberdeen, Idaho, processes about 227 000 kg (500 000 pounds) of potatoes a day (although Simplot is no longer the largest supplier of French fries in North America). Aberdeen is in Bingham County, which grows more potatoes than any other county in the United States. Simplot, an eighth-grade dropout, became a multi-billionaire and one of the richest men in the United States. The license plates of his Lincoln Continental were engraved "Mr. Spud."

- Englishman Eric Jenkins grew a single potato plant that produced 168 kg (370 pounds) of potatoes in 1974.
- A potato weighing 8.3 kg (18 pounds, 4 ounces) grown in England in 1995 was listed as the largest in the Guinness Book of World Records.
- In October 1995, the potato became the first vegetable to be "grown" in space. Four tuber sprouts that sailed off on Space Shuttle Columbia produced marble-sized tubers, the first food ever produced in space. The potato has the distinction of being the first choice of scientists considering what food plants can be grown extraterrestrially to feed space travelers. Its advantages include universal taste appeal, abundance of nutrients, slow release of energy, limited waste, and ease of preparation. The hope is that one day, technology will be perfected to feed astronauts on long space voyages, and eventually feed future space colonies.
- In 1999, Scottish scientists reported that they had developed a potato that will signal farmers when it is thirsty. A gene from the jellyfish, *Aequorea victoria*, which produces a fluorescent protein, was added to the potato. The protein produces a green glow, which becomes more intense if plants need water, and can be detected by a hand-held device placed on a leaf.
- The average American eats 3.6 metric tons (4 tons) of potatoes over a lifetime.
- French fries are ordered more frequently in American restaurants than any other dish.
- Potato chips are the Number 1 selling snack in the United States.
- In promoting new uses for the "Irish spud," Ireland's Horticulture Board suggested that a sliced potato in a muslin bag can help remove bags under the eyes; a dried potato suspended from the neck is helpful for rheumatism; and chopped potatoes in warm water can remove stubborn wine stains at the bottom of a decanter.
- To remove a stain on clothing, let it dry, and rub a fresh potato over the stain for a couple of minutes. Wash the article with laundry detergent and water as normally done. The enzymes in the potato may remove most, if not all, of the stain.
- To remove food stains from hands, slice and rub raw potato on the stains and rinse with water.
- After grating cheese, to clean the grater, rub a raw potato over it before washing. (More generally, experienced chefs, when grating several items, always grate the softest items first, leaving the firmer ones last to keep the grater clean.)
- To remove a broken light bulb from its socket, first disconnect the electricity, then cut a potato in half and press it down over the socket. Slight pressure is needed to cause the potato to move into the inside of the socket and bulb base. While applying pressure on the potato, turn it counterclockwise to unscrew the bulb base. An apple and a green pear work as well. Occasionally, the potato trick doesn't work. Using needle-nose pliers, insert the jaws into the socket, spread the handles apart with pressure, and turn the pliers counterclockwise. Consider just using the pliers in the first place, and eating the potato.
- To keep potatoes from budding, place an apple in the bag with the potatoes.
- Apparently-wild potato populations have been discovered in the mountains of British Columbia. These probably originated from discarded potatoes, which managed to reproduce independent of cultivation, although it has also been suggested that potatoes were introduced to British Columbia by Indian tribes.
- It has been claimed that the first US state to issue a vehicle license plate bearing a slogan was Idaho. The 1928 plate displayed a brown russet potato and the phrase "Idaho Potatoes." The potato appeared again on Idaho plates in 1943 and 1948, and since 1947, all Idaho plates have shown the phrase "Famous Potatoes."
- It has been claimed that of all crops, potato has the largest number of related wild species. The potato genus *Solanum* is huge, containing perhaps 2000 species. Of these, 235 species are in section *Petota* of the genus, and reasonably called "wild potatoes." It is important to identify the wild relatives of crops because these are sources of genes for improving the productivity, vigor, and disease resistance of the cultivated plants.

- Dogs sometimes like to dig up potatoes. One solution is to lay chicken wire flat on the ground over the potato patch. The dogs do not like the feel of wire under their feet and leave the potatoes alone.
- It takes about 4 kg of potatoes to make 1 kg of potato chips.
- About 80% of a potato is water.
- In movies, dried potato flakes are sometimes used to simulate falling snow, and instant mashed potato flakes take the place of fallen snow.
- It has been estimated that less than 1 hectare (2.5 acres) of potatoes can produce enough gasahol to fill up 62 cars.
- Potatoes are grown in all 50 states of the United States, all 10 provinces of Canada, and in about 125 countries.
- In 2003, it was reported that German youths had taken up a dangerous new sport: firing potatoes from homemade cannons fashioned from drainage pipes, and using hair spray ignited with a battery as a propellant. The potatoes leave the barrel of the guns at close to 100 m/s (328 feet/second), and have the capacity to penetrate the wooden wall of a house at 90 m (295 feet). Emergency regulations were enacted to outlaw the practice, which was not previously covered by firearms regulations in Germany.

Key Information Sources

Asiedu, S., Coleman, S.E., Haliburton, T., and Hampson, M.C. 1987. Atlantic Canada potato guide. Atl. Prov. Agric. Serv. Co-or. Comm. Publ. 1300. 47 pp.

Bajaj, Y.P.S. 1987. Potato. Springer-Verlag, Berlin, Germany. 509 pp.

Belknap, W.R., Vayda, M.E., and Park, W.D. (*Editors*). 1994. The molecular and cellular biology of the potato. 2nd edition. CAB International, Wallingford, Oxon, UK. 270 pp.

Beukema, H.P., and van der Zaag, D.E. 1990. Introduction to potato production. Centre for Agricultural Publishing and Documentation, Wageningen, Netherlands. 208 pp.

Bokx, J.A. de. 1972. Viruses of potatoes and seed-potato production. Centre for Agricultural Publishing and Documentation, Wageningen, Netherlands. 233 pp.

Bradshaw, J.E., and Mackay, G.R. (*Editors*). 1994. Potato genetics. CAB International, Oxon, UK. 552 pp.

Burton, W.G. 1989. The potato. Third edition. Longman Inc., New York, NY. 742 pp.

Correll, D.S. (*Editor*). 1962. The potato and its wild relatives; section *Tuberarium* of the genus *Solanum*. Texas Research Foundation, Renner, TX. 606 pp.

Davis, J.W. 1992. Aristocrat in burlap: a history of the potato in Idaho. Revised edition. Idaho Potato Commission, Boise, ID. 203 pp.

Dean, B.B. 1994. Managing the potato production system. Food Products Press, New York, NY. 183 pp.

Donnelly, J.S. 2001. The great Irish potato famine. Sutton Publishing, Phoenix Mill, Gloucestershire, UK. 292 pp.

Douches, D.S., and Jastrzebski, K. 1993. Potato, *Solanum tuberosum* L. *In* Genetic improvement of vegetable crops. *Edited by* G. Kalloo and B.O. Bergh. Pergamon Press, New York, NY. pp. 605–644.

Ewing, E.E. 1997. Potato. *In* The physiology of vegetable crops. *Edited by* H.C. Wien. CABI International, Wallingford, Oxon, UK. pp. 295–344.

Fincher, L.J., and Mountjoy, B.M. 1957. Potatoes: facts for consumer education. U.S. Dept. of Agriculture, Washington, DC. 30 pp.

Food and Agriculture Organization. 1991. Potato production and consumption in developing countries. Plant production and protection paper 110. Food and Agriculture Organization, Rome, Italy. 47 pp.

Grun, P. 1990. The evolution of cultivated potatoes. Econ. Bot. **44**: 39–55.

Hardenburg, E.V. 1949. Potato production. Comstock Publishing Co., Ithaca, NY. 270 pp.

Harris, P.M. (*Editor*). 1992. The potato crop: the scientific basis for improvement. 2nd edition. Chapman & Hall, London, UK. 909 pp.

Hawkes, J.G. 1990. The potato. Evolution, biodiversity and genetic resources. Belhaven Press, London, UK. 259 pp.

Hawkes, J.G., and Hjerting, J.P. 1969. The potatoes of Argentina, Brazil, Paraguay, and Uruguay; a biosystematic study. Clarendon Press, Oxford, UK. 525 pp.

Hawkes, J.G., and Hjerting, J.P. 1989. The potatoes of Bolivia: their breeding value and evolutionary relationships. Oxford University Press, Oxford, UK. 472 pp.

Horton, D.E. 1987. Potatoes: production, marketing, and programs for developing countries. IT Publications, Westview Press, London, UK. 243 pp.

Howard, H.W. 1970. Genetics of the potato: *Solanum tuberosum*. Logos Press, London, UK. 126 pp.

Jellis, G.J., and Richardson, D.E. (*Editors*). 1987. The production of new potato varieties: technological advances. Cambridge University Press, Cambridge, UK. 358 pp.

Kameraz, A.I.A., and Dhote, A.K. 1985. Systematics, breeding and seed production of potatoes. Amerind Publishing Co., New Delhi, India. 213 pp. [Translated from Russian: Kameraz, A.I.A., and Dhote, A.K. 1978. Sistematika, selektsiya i semenovodstvo kartofelya. V.I. Lenin All-Union Academy of Agricultural Sciences, N.I. Vavilov All-Union Scientific Research Institute of Plant Industry, Leningrad, USSR.]

Li, P.H. (*Editor*). 1985. Potato physiology. Academic Press, Orlando, FL. 586 pp.

Lisinska, G., and Leszczynski, W. 1989. Potato science and technology. Elsevier Science Publishing Co., New York, NY. 391 pp.

Messer, E. 2000. Potatoes (white). *In* The Cambridge world history of food. *Edited by* K.F. Kiple and K.C. Ornelas. Cambridge University Press, Cambridge, UK. pp. 187–201.

Ochoa, C.M. 1990. The potatoes of South America. Cambridge University Press, Cambridge, UK. 512 pp.

Oster, M. 1993. The potato garden: a grower's guide. Harmony Books, New York, NY. 128 pp.

Rich, A.E. 1983. Potato diseases. Academic Press, New York, NY. 238 pp.

Salaman, R.N., Burton, W.G., and Hawkes, J.G. 1985. The history and social influence of the potato. Revised edition. Cambridge University Press, Cambridge, UK. 685 pp.

Salunkhe, D.K., Desai, B.B., and Chavan, J.K. 1989. Potatoes. *In* Quality and preservation of vegetables. *Edited by* N.A. Eskin. CRC Press, Inc., Boca Raton, FL. pp. 1–52.

Sieczka, J.B., Thornton, R.E. (*Editors*), and Chase, R.W. 1992. Commercial potato production in North America. Potato Association of America, Orono, ME. 46 pp.

Simmonds, N.W. 1995. Potatoes. *In* Evolution of crop plants. 2nd edition. *Edited by* J. Smartt and N.W. Simmonds. Longman Scientific & Technical, Burnt Mill, Harlow, Essex, UK. pp. 466–471.

Smith, O. 1968. Potatoes: production, storing, processing. Avi Publishing Co., Westport, CT. 642 pp.

Smith, W.L., and Wilson, J.B. 1978. Market diseases of potatoes. Agricultural Research Service, U.S. Dept. of Agriculture, Washington, DC. 99 pp.

Stevenson, W.R. 2001. Compendium of potato diseases. Second edition. American Phytopathological Society, St. Paul, MN. 106 pp.

Spooner, D.M., and Van den Berg, R.G. 1992. An analysis of recent taxonomic concepts in wild potatoes (*Solanum* sect. *Petota*). Genet. Resour. Crop Evol. **39**: 23–37.

Stuart, W., and Davis, K.C. 1937. The potato; its culture, uses, history and classification. 4th edition. J.B. Lippincott, Philadelphia, PA. 508 pp.

Talburt, W.F., and Smith, O. 1987. Potato processing. 4th edition. Van Nostrand Reinhold, New York, NY. 796 pp.

U.S. Dept. of Agriculture. 1994. How to buy potatoes. U.S. Dept. of Agriculture, Washington, DC. 10 pp.

Wagih, M.E., and Wiersema, S.G. 1996. *Solanum tuberosum* L. *In* Plant resources of South-East Asia, 9, plants yielding non-seed carbohydrates. *Edited by* M. Flach and F. Rumawas. Backhuys Publishers, Leiden, Netherlands. pp. 148–154.

Woolfe, J.A. 1987. The potato in the human diet. Cambridge University Press, Cambridge, UK. 231 pp.

Zehnder, G.W. 1994. (*Editor*). Advances in potato pest biology and management. American Phytopathological Society Press, St. Paul, MN. 655 pp.

Zuckerman, l. 1998. The potato: how the humble spud rescued the Western world. Faber and Faber, Boston, MA. 304 pp.

Specialty Cookbooks

(Note: Hundreds of cookbooks are dedicated to the potato. The following are selected examples.)

Bakalar, R. 1969. The complete potato cookbook. Prentice-Hall, Englewood Cliffs, NJ. 312 pp.

Bareham, L. 1991. In praise of the potato: recipes from around the world. Overlook Press, Woodstock, NY. 314 pp.

Barker, A., and Mansfield, S. 1999. Potato: the definitive guide to potatoes and potato cooking. Lorenz, London, UK. 256 pp.

Barker, A., and Mansfield, S. 2000. Discovering potatoes: a cook's guide to over 150 potato varieties and how to use them. Southwater, London, UK. 128 pp.

Davis, M. 1973. The potato book. Morrow, New York, NY. 95 pp.

Fabricant, F. 2001. The great potato book. Ten Speed Press, Berkeley, CA. 148 pp.

Finamore, R., and Stevens, M. 2001. One potato, two potato: 300 recipes from simple to elegant—appetizers, main dishes, side dishes, and more. Houghton Mifflin, Boston, MA. 590 pp.

Jones, J. 1982. Stuffed spuds: 100 meals in a potato. M. Evans, New York, NY. 132 pp.

Klein, V. 1987. The unique potato salad cookbook. Entrepreneurial Workshops Publications, Seattle, WA. 152 pp.

Lauterbach, B. 2002. Potato salad: fifty favorite recipes. Chronicle Books, San Francisco, CA. 108 pp.

Marshall, L. 1992. A passion for potatoes. HarperPerennial, New York, NY. 248 pp.

Reynolds, S. 1997. 50 best mashed potatoes. Broadway Books, New York, NY. 92 pp.

Robyns, G. 1976. The potato cookbook, from thinning to sinning deliciously, from soups to desserts. Stemmer House, Owings Mills, MD. 133 pp.

Scott, M.L., and Scott, J.D. 1991. The incredible potato cookbook. Consumer Reports Books, Yonkers, NY. 211 pp.

Stacey, J. 2000. What's cooking potatoes. Thunder Bay Press, San Diego, CA. 254 pp.

Vezza, D.S. 1993. The perfect potato: over 100 fabulous recipes—from appetizers to desserts—for potato lovers everywhere. Villard Books, New York, NY. 160 pp.

Weinstein, B., and Scarbrough, M. 2003. The ultimate potato book: hundreds of ways to turn America's favorite side dish into a meal. W. Morrow, New York, NY. 262 pp.

Worthington, D.R., and Williams, C. 1993. Potatoes. Time-Life Books, Alexandria, VA 108 pp.

Pumpkin and Squash

Family: Cucurbitaceae (gourd family)

Names

Scientific names: *Cucurbita* species

- Pumpkin, summer squash (less commonly, winter squash and marrow)—*C. pepo* L.
- Winter squash (less commonly, marrow, pumpkin, and turban squash)—*C. maxima* Duchesne ex Lam.
- Winter squash (less commonly, pumpkin—*C. moschata* (Duchesne ex Lam.) Duchesne ex Poir.
- Winter squash, cushaw, cymlin (cymling, cymbling), and pumpkin—*C. argyrosperma* C. Huber (*C. mixta* Pangalo)
- Buffalo gourd—*C. foetidissima* Kunth
- Malabar gourd (also Angora gourd, black-seed squash, black-seeded squash, chilacayote, fig-leaf pumpkin, and Siamese gourd)—*C. ficifolia* Bouché

- The terms "pumpkins" and "squashes" have been applied so inconsistently to different varieties of several species of *Cucurbita* that there really is not any reliable way to distinguish them. This is illustrated by the claim, often made, that "canned pumpkin" is really "canned squash." In reality, the two are the same thing. Confusingly, some species contain some varieties that are called pumpkins and other varieties that are called squashes. Generally, "summer squashes" are the fruits of certain varieties of *C. pepo*, used in the immature state, while "winter squashes" are the fruits of certain varieties of several species of *Cucurbita*, including *C. mixta*, *C. moschata*, and *C. maxima*. In Britain, "pumpkin" is often interpreted in a broader sense than in North America, including many fruits that North Americans would call "squash."
- "Pumpkin" has been attributed to the Old French *pompion*, meaning eaten when "cooked by the sun," or ripe. Perhaps the word is ultimately derived from the Latin *pepo* and the Greek *pepon*—terms for various fruits of the gourd family during classical antiquity.
- "Squash" is derived from the Massachuset Indian (Narraganset and Natick) word *askutasquash*, meaning "eaten raw or uncooked," probably reflecting that some forms of squash were eaten uncooked by the Indians.
- "Courgette" is a variety of vegetable marrow squash that is eaten when immature and small. The name is based on the diminutive of the French *courge*, "gourd." This type of squash generally has white flesh and green skin, although a yellow-skinned variety is known. In North America, this is known as "zucchini," a name brought over from the Old World by Italian immigrants. The name "zucchini" is based on the diminutive of the Italian *zucca*, gourd, which came from the Latin *cucutia*, related to the Latin *cucurbita* (explained as the genus name, below).
- "Gourd" refers to fruits of the gourd family, irrespective of whether they are edible or merely grown for ornament. Gourds of the calabash or gourd tree, *Crescentia cujete* L., belong to another family (the Bignoniaceae).
- "Cushaw," one of the common names of *C. mixta*, is Louisiana French for "big pumpkin."
- "Acorn squash," a winter squash of *C. pepo* so named for its resemblance to a giant acorn, is also known as "Des Moines squash."
- According to James H.J. Gregory of Marblehead, Massachusetts, in a letter published in the December 1847 issue of *Magazine of Horticulture*, Hubbard squash, a variety of *C. maxima*, was acquired by his father from a Mrs. Elizabeth Hubbard, who named the squash for herself.
- The genus name *Cucurbita* is a classical Latin name for some gourd. *Cucurbitare* in old Latin had the meaning "to commit adultery"; it is not clear, however, if this sexual meaning has any relevance to the derivation of the name *Cucurbita*, although melons do interbreed promiscuously.
- *Pepo* in the scientific name *C. pepo* is Latin for a kind of melon; the word is related to the Greek *pepon*, ripe, mellow.
- *Moschata* in the scientific name *C. moschata* is Latin for musky, a reference to the odor of the fruit.
- *Maxima* in the scientific name *C. maxima* is Latin for largest, referring to the size of the fruit of some varieties.
- *Argyrosperma* in the scientific name *C. argyrosperma* is Latin for silver seed.
- *Ficifolia* in the scientific name *C. ficifolia* is Latin for fig-like leaves.

An old pumpkin (*Cucurbita pepo*) cultivar (Etampes), from Vilmorin-Andrieux (1885).

- *Foetidissima* in the scientific name *C. foetidissima* is Latin for very fetid, a reference to the bad smell of the plant.

Plant Portrait

The genus *Cucurbita* is native to southern Mexico and northern Central America. The plants are yellow-flowered, herbaceous, annual or perennial vines or bushes, which climb by tendrils (summer squashes are usually of bush form and lack tendrils). The fruits are remarkably varied in shape, size, and color, and include some of the largest fruits in the plant kingdom. Squashes and pumpkins were unknown in the Old World prior to Columbus' voyage of 1492. The several species now cultivated may have been the earliest domesticated plants of the New World, perhaps dating back 10 000 years, and certainly well known by 5000 BC. Most varieties of *Cucurbita* are extraordinarily productive plants, and home gardeners often find themselves with an embarrassingly large harvest, especially of zucchini. ("Zucchini fairies" are gardeners who dispose of their excess zucchinis by surreptitiously leaving them at neighbors' doorsteps.) Summer squash develop with startling rapidity, and may be harvested in as little as 2 days after the flowers are fertilized. Most gardeners make the error of allowing summer squash to mature excessively. Winter squash can be stored for up to 2 months without refrigeration. Some forms of *Cucurbita* are grown for their production of ornamental, nonedible gourds. The most important cultivated species are discussed in the following.

- *C. pepo* (summer squash and pumpkin; less frequently, marrow and winter squash). This is a very widely grown group of varieties, the fruits varying in size, shape, skin color, flesh color, flesh texture, and other characteristics. Illustrative of the range of variation are small inedible ornamental gourds, and huge jack-o'-lantern pumpkins. The best-known kind is the zucchini squash.
- *C. maxima* (winter squash; less commonly, marrow, pumpkin, and turban squash). This species is highly variable, but not as varied as *C. pepo*. Some of the varieties produce huge fruits. The species is cool tolerant, and so is extensively grown in temperate regions, as well as in India, Ceylon, and the Philippines. Common cultivars include Buttercup and Hubbard. "Banana squashes" are relatively elongated forms. "Turban squashes" have a swollen end reminiscent of a turban on a human head.
- *C. moschata* (winter squash, less commonly, pumpkin). This species is extensively cultivated in the temperate and tropical New World, and is also widely employed in tropical Asia, especially in Japan, where unusual cultivars have been developed. The best-known cultivar is Butternut squash.
- *C. argyrosperma* (winter squash, cushaw, cymlin, pumpkin). For the most part, the flesh of this species tends to be coarse and watery, although the variety Sweet Potato is superior. In the southern United States, cushaws, large pumpkin-like gourds, are often grown and are either baked or fed to livestock.

- *C. ficifolia* (Angora gourd, chilacayot, Malabar gourd, Siamese gourd). The Malabar gourd is native to the highland, tropical regions of Mexico, Central America, and northern parts of South America. It has been cultivated for a long period in the Americas, but there are fewer varieties available than for the above species. This species has watermelon-like fruits of good quality. Its flesh is often impregnated with sugar to produce a candied confection, and the hard rinds are made into containers. The flesh is also fermented to make an alcoholic beverage.
- *C. foetidissima* (buffalo gourd). This wild perennial of the western United States and Mexico produces very large fleshy storage organs, which may reach weight of 50 kg (110 pounds). All parts of the plant are bitter and obnoxious in taste and smell, and Indian groups commonly treated young fruits by various cooking methods to render them palatable. Attempts have been made to domesticate the plant in the southern United States as a source of seed oil, seed meal, and perhaps root starch. This and some other species of *Cucurbita* are potentially useful crops of hot, arid or semiarid lands.

Culinary Portrait

The flesh of pumpkins and squashes is consumed immature as a fresh vegetable, or used in cooking, and the mature flesh is used in cooking and in canning, especially for pie stock. Pumpkin pie was served at the Pilgrims' second Thanksgiving in 1623, and has remained a traditional Thanksgiving dessert in the United States ever since. In North America, the top Thanksgiving Day pie preference is still pumpkin, followed by apple. Traditional pumpkin dishes in New England include pumpkin pudding, pancakes, bread, butter, dried chips, and even beer (pumpkin beer was popular among the Pilgrims). Mature pumpkins and winter squashes are roasted (especially with butter and brown sugar), stewed, steamed, boiled as vegetables, added to soups, and candied.

Pumpkin seed is the source of a high-quality vegetable oil, mostly produced in Europe. Mutant varieties of pumpkin with hull-less (hulless) seeds arose in Europe centuries ago. The most popular is the Steiermark mutation, which has been recognized as *Cucurbita pepo* var. *styriaca* Grebenšč, named for the Styria region of Austria, where it is the basis of the popular Steiermark pumpkin seed oil. This oil is used primarily for salads, and is not suitable as a cooking oil. When toasted (the usual form in which it is available), Steiermark pumpkin seed oil has a strong flavor (which takes getting used to), which is useful for increasing the pumpkin taste of pumpkin dishes. The Steiermark mutant arose as a trailing plant (like most pumpkin plants), but is also available as a bush form (*C. pepo* var. *oleifera* Pietsch). This latter name is used more broadly to include var. *styriaca*, i.e., all hull-less central European oilseed cultivars, and also sometimes to include edible-seed varieties developed in the Americas. Some soft-seeded types of pumpkin or squash seeds are sold as a confection. Roasted, salted seeds are a popular snack in Latin America. The seeds have a nutty flavor, and are consumed worldwide—raw, boiled, or roasted, usually with the coat removed. Lady Godiva, a variety of *C. pepo* with seeds lacking a seed coat, is favored for snacks in the United States. In Mexico, pumpkin seeds prepared as food are known as *pepipitas* (pronounced puh-PEE-tahs).

Aboriginal Americans dating back to Aztec times have eaten the relatively large male flowers of squash or pumpkin species. The large blossoms can be used to add flavor and color to stews, soups, and salads, and may be stuffed, fried, or battered. In Mexico and Asia, young leaves and shoots are used as potherbs.

Of the various species, *C. maxima* and *C. moschata* produce the strongest-tasting flesh, which is mildly sweet, with a somewhat musky flavor. As a result, these species are favored for canning. The fruit of *C. maxima* is the richest in vitamins and finest in texture, and so is preferred for mashing into baby food.

Summer squashes are picked while very young, typically 2–7 days after flowering. Their skin and seeds are tender and edible. These squashes are quite perishable, and do not store well. The best known is the zucchini or courgette, which is at its most flavorful when measuring between 15 and 20 cm (6–8 inches) in length, the flavor and quality declining as the fruit becomes larger. Zucchini flowers are often stuffed or deep fried in batter.

Winter squashes are harvested when completely ripe. Like melons, they have a hollow central cavity containing hard, fully developed seeds. The seeds are edible, but the thick, hard shells of winter squashes are not. The rind of some varieties is so hard that a mallet is required to strike a knife hard enough to penetrate the gourd. Depending on variety, winter squashes can be stored for 1–6 months.

Pumpkin and squash can be boiled, steamed, baked, microwaved, and cooked in a pressure cooker. Boiling tends to make squash watery and less tasty. Steaming is highly recommended.

Culinary Vocabulary

- "Marrow," in Great Britain, denotes fruits of varieties of *C. pepo* and *C. maxima* used in the mature state. The word "marrow" arose because it appeared to some that the soft flesh of these squashes was like relatively soft bone marrow. In Great Britain, "marrow squash" is often used to mean summer squash in general.
- "Squash" has an unrelated meaning in England: a drink prepared by squeezing the juice from the pulp of citrus fruit. Usually, soda water is added.
- *Pepitas* (pronounced peh-PEE-tah) are hulled, roasted pumpkin seeds, consumed in Mexican cuisine as snacks. When the seeds are ground to a powder and used as a thickener, the preparation is known as *pepitoria* (pronounced peh-PEE-toh-ree-ah).
- *Cappellacci con la zucca* (Italian for "little hats with pumpkin") are made with fresh egg pasta cut into circles to which small amounts of freshly cooked, puréed pumpkin is added, seasoned with a little nutmeg, formed into small "hats," cooked like ravioli, and served with melted butter and Parmesan cheese.

Curiosities of Science and Technology

- The advanced pre-Columbian civilizations of the Americas were dependent on a combination of corn, beans, and squash, a mixture that produces excellent nutrition. This famous combination was referred to by several Indian groups as "the three sisters."
- Pumpkin seeds were used as a cure for tapeworms and roundworms by American Indian medicine men, and European settlers adopted the practice.
- The Pilgrims used pumpkins for many different purposes. They dried pumpkins and made the shells into bowls for eating as well as jars for storage. They brewed beer from persimmons, hops, maple sugar, and pumpkins.
- In the colonial United States, pumpkins were used as an ingredient for the crust of pies, not the filling.
- The expression "pumpkinhead," first applied to New Englanders, originated from the practice of using pumpkin half-shells as guides for haircuts in colonial Connecticut to ensure a round, uniform style.

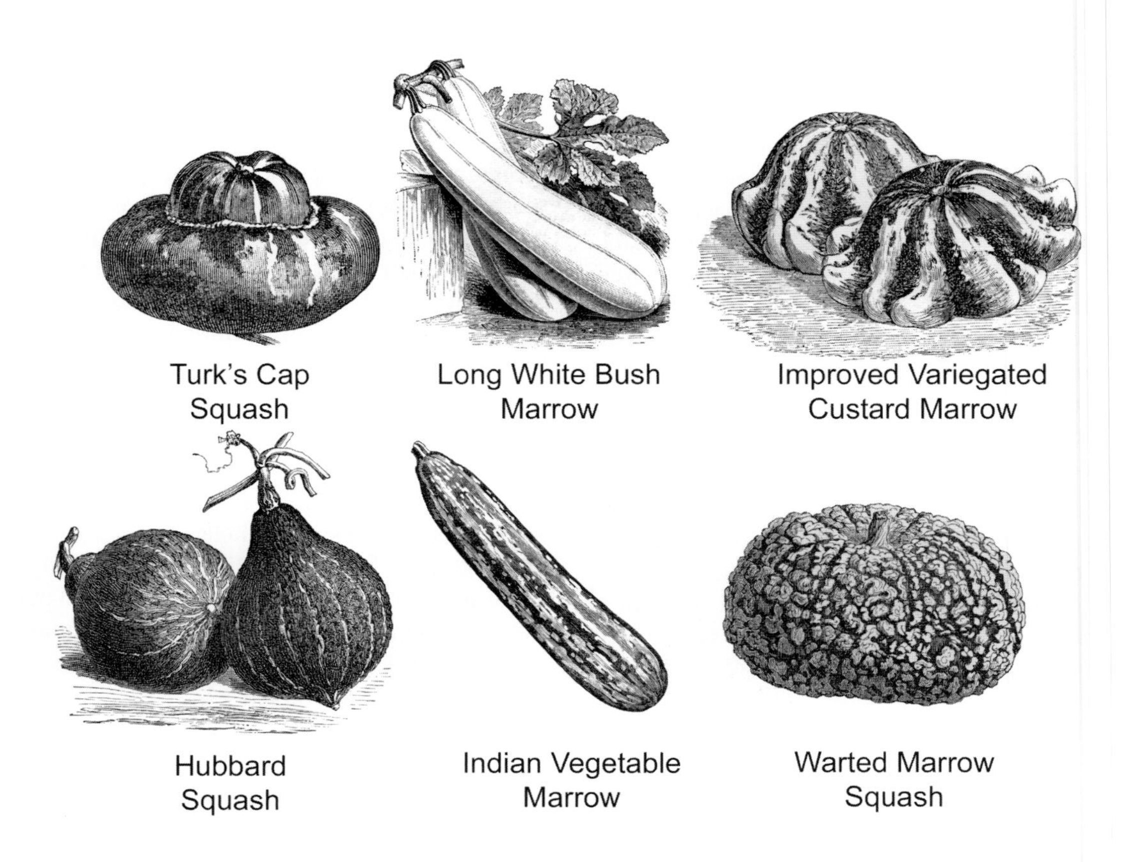

Varieties of squash (*Cucurbita* species), from Vilmorin-Andrieux (1885).

- Pumpkin seed oil, which is greenish, strongly stains clothing. In the words of Gernot Katzer, author of a very large website on spice plants, "Pumpkin seed oil stains have terminated the career of many garments, particularly shirts and ties."
- Pumpkins are 90% water.
- Joe Jutras from North Scituate, Rhode Island, grew a world-record (766 kg or 1689 pound) pumpkin in 2007.
- Pumpkins have been observed to develop roots whose total length reached more than 24 km (15 miles).
- A widespread fallacy is that squashes and pumpkins will cross-pollinate with other genera of the gourd family, including watermelon and muskmelon, causing the fruit of one to take on the taste attributes of the other.
- The flowers of pumpkin and squash species close about dusk and open about sunrise. Squash and gourd bees, the natural pollinators, commonly enter the closing flowers at dusk to pass the night protected, and leave at dawn when the flowers open.
- Botanically speaking, pumpkins and squashes are a type of fruit technically classified as a "berry," but are sold as "vegetables."
- Halloween dates back at least 3000 years in Europe to the Celtic harvest celebration of Samhain (pronounced "sow-ain," meaning "summer's end"). Since pumpkins had not yet been discovered by the Old World, jack-o'-lanterns were carved from turnips or gourds. Burning lumps of coal were used inside as a source of light, later to be replaced by candles. When European settlers arrived in America, particularly the Irish, they found the native pumpkin to be larger, easier to carve, and the perfect choice for jack-o'-lanterns.
- Ninety percent of pumpkins offered in stores are sold for Halloween jack-o'-lanterns.
- To delay shriveling of a carved Halloween jack-o'-lantern, rub vegetable oil or petroleum jelly onto the fresh-cut edges, or soak the pumpkin in water, submerging it completely, but not longer than 8 hours (which could cause premature cracking). Spraying the inside and outside surfaces with an antiseptic spray to kill bacteria may also make the jack-o'-lantern last longer.

Malabar gourd (*Cucurbita ficifolia*), from Vilmorin (1906).

- The racket ball game of squash, which has nothing to do with the vegetable squash, was invented in Harrow school in Britain around 1830, at which time the similar game of rackets ball was being played. The pupils discovered that a punctured rackets ball, which "squashed" on impact with the wall, produced a game with a greater variety of shots and a more interesting game. In 1864, the first squash courts were constructed at the school and squash was officially founded as a sport in its own right.
- A "pumpkin moon" may be observed when the moon acquires an orange–red glow due to atmospheric conditions and the refraction of light. This often happens during a total eclipse of the moon, when the moon is covered by the earth's shadow cast by the sun. Despite the name "total eclipse," the moon does not become entirely blotted out, but just how dark it gets and whether or not it develops an orange–red coloration depends on how deep it travels into the earth's shadow.

Key Information Sources

Andres, T.C. 1990. Biosystematics, theories on the origin, and breeding potential of *Cucurbita ficifolia*. *In* Biology and utilization of the Cucurbitaceae. *Edited by* D.M. Bates, R.W. Robinson, and C. Jeffrey. Comstock Publishing Associates, Syracuse, New York, NY. pp. 102–119.

Bailey, L.H. 1902. A medley of pumpkins. Memoirs Hortic. Soc. New York, **1**: 117–124.

Bailey, L.H. 1943. The domesticated cucurbitas. Gentes Herb. **2**: 62–115.

Bemis, W.P., Berry, J.W., Weber, C.W., and Whitaker, T.W. 1978. The buffalo gourd: a new potential horticultural crop. HortScience, **13**: 235–240.

Bird, R. 2003. Growing squashes and pumpkins: a directory of varieties and how to cultivate them successfully. Lorenz, London, UK. 64 pp.

Cutler, H.C., and Whitaker, T.W. 1956. *Cucurbita mixta* Pang. Its classification and relationships. Bull. Torrey Bot. Club, **83**: 253–260.

Cutler, H.C., and Whitaker, T.W. 1961. History and distribution of the cultivated cucurbits in the Americas. Am. Antiquity, **26**: 469–485.

Decker, D.S. 1988. Origin(s), evolution, and systematics of *Cucurbita pepo* (Cucurbitaceae). Econ. Bot. **42**: 4–15.

Decker-Walters, D.S. 1990. Evidence for multiple domestications of *Cucurbita pepo*. *In* Biology and utilization of the Cucurbitaceae. *Edited by* D.M. Bates, R.W. Robinson, and C. Jeffrey. Cornell University Press, Ithaca, NY. pp. 96–101.

Decker-Walters, D.S., and Walters, T.W. 2000. Squash. *In* The Cambridge world history of food. *Edited by* K.F. Kiple and K.C. Ornelas. Cambridge University Press, Cambridge, UK. pp. 335–351.

Decker-Walters, D.S., Walters, T.W., Posluszny, U., and Kevan, P.G. 1990. Genealogy and gene flow among annual domesticated species of *Cucurbita*. Can. J. Bot. **68**: 782–789.

DeVeaux, J.S., and Schultz, E.B. 1985. Development of buffalo gourd (*Cucurbita foetidissima*) as a semiaridland starch and oil crop. Econ. Bot. **39**: 454–472.

Goldman, A. 2004. The compleat squash: a passionate grower's guide to pumpkins, squash, and gourds. Artisan, New York, NY. 224 pp.

Hurd, P.D., Linsley, E.G., and Michelbacher, A.D. 1974. Ecology of the squash and gourd bee, *Peponapis pruinosa*, on cultivated cucurbits in California (Hymenoptera: Apoidea). Smithsonian Institution Press, Washington, DC. 17 pp.

Langevin, D. 2003. How-to-grow world class giant pumpkins (version III). Annedawn Publishing, Norton, MA. 192 pp.

Lazos, E.S. 1986. Nutritional fatty acid, and oil characteristics of pumpkin and melon seeds. J. Food Sci. **51**: 1382–1383.

Loy, J.B. 1990. Hull-less seeded pumpkins: a new edible snackseed crop. *In* Advances in new crops. *Edited by* J. Janick and J.E. Simon. Timber Press, Portland, OR. pp. 403–407.

Loy, J.B., and Broderick, C.E. 1990. Growth, assimilate partitioning, and productivity of bush and vine cultivars of *Cucurbita maxima*. *In* Biology and utilization of the Cucurbitaceae. *Edited by* D.M. Bates, R.W. Robinson, and C. Jeffrey. Cornell University Press, Ithaca, NY. pp. 436–447.

Merrick, L.C. 1990. Systematics and evolution of a domesticated squash, *Cucurbita argyrosperma*, and its wild and weedy relatives. *In* Biology and utilization of the Cucurbitaceae. *Edited by* D.M. Bates, R.W. Robinson, and C. Jeffrey. Cornell University Press, Ithaca, NY. pp. 77–95.

Merrick, L.C. 1995. Squashes, pumpkins and gourds. *In* Evolution of crop plants. 2nd edition. *Edited by* J. Smartt and N.W. Simmonds. Longman Scientific & Technical, Burnt Mill, Harlow, Essex, UK. pp. 97–105.

Mutschler, M.A., and Pearson, O.H. 1987. The origin, inheritance, and instability of butternut squash (*Cucurbita moschata* Duchesne). HortScience, **22**: 535–539.

Paris, H.S. 1989. Historical records, origins, and development of the edible cultivar groups of *Cucurbita pepo* (Cucurbitaceae). Econ. Bot. **43**: 423–443.

Provvidenti, R. 1990. Viral diseases and genetic sources of resistance in *Cucurbita* species. *In* Biology and utilization of the Cucurbitaceae. *Edited by* D.M. Bates, R.W. Robinson, and C. Jeffrey. Cornell University Press, Ithaca, NY. pp. 427–435.

Roxas, V.P. 1993. *Cucurbita ficifolia* Bouché. *In* Plant resources of South-East Asia, 8, vegetables. *Edited by* J.E. Siemonsma and K. Piluek. Pudoc Scientific Publishers, Wageningen, Netherlands. pp. 165–167.

Saade, L, and Hernández, S.M. 1994. Cucurbits. (*Cucurbita* spp.). *In* Neglected crops. 1492 from a different perspective. *Edited by* J.E. Hernández-Bermejo and J. Léon. Food and Agriculture Organization of the United Nations, Rome, Italy. pp. 63–77.

U.S. Agricultural Research Service. 1969. Growing pumpkins and squashes. Revised edition. U.S. Agricultural Research Service, Washington, DC. 21 pp.

Whitaker, T.W. 1971. Squash and gourd bees (*Peponapis xenoglossa*) and the origin of the cultivated *Cucurbita*. Evolution, **25**: 218–234.

Whitaker, T.W., and Bemis, W.P. 1975. Origin and evolution of the cultivated *Cucurbita*. Bull. Torrey Bot. Club, **102**: 362–368.

Whitaker, T.W., and Bohn, G.W. 1950. The taxonomy, genetics, production and uses of the cultivated species of *Cucurbita*. Econ. Bot. **4**: 52–81.

Whitaker, T.W., and Robinson, R.W. 1986. Squash breeding. *In* Breeding vegetable crops. *Edited by* M.J. Bassett. AVI Publishing Co., Westport, CT. pp. 209–242.

Widjaja, E.A., and Sukprakarn, S. 1993. *Cucurbita* L. *In* Plant resources of South-East Asia, 8, vegetables. *Edited by* J.E. Siemonsma and K. Piluek. Pudoc Scientific Publishers, Wageningen, Netherlands. pp. 160–165.

Wien, H.C. 1997. The cucurbits: cucumber, melon, squash and pumpkin. *In* The physiology of vegetable crops. *Edited by* H.C. Wien. CABI International, Wallingford, Oxon, UK. pp. 345–386.

Wilson, H.D. 1990. Gene flow in squash species. Bioscience, **40**: 449–455.

Specialty Cookbooks

(Note: There are dozens of cookbooks dedicated to pumpkin and (or) squash. The following are selected examples.)

Bard, R., and Kellogg, C. 1985. Zucchini and all that squash. W.H. Publishing, Tacoma, WA. 103 pp.

Boisset, C. 1997. Pumpkins & squashes. The Reader's Digest Association, New York, NY. 120 pp.

Brabb, E., Krupski, H., and Paddock, B.T. 1996. Pumpkin companion. Brick Tower Press, New York, NY. 94 pp.

Damerow, G. 1997. The perfect pumpkin. Storey Publishing, Pownal, VT. 219 pp.

Dandar, H.O., and Dandar, E.B. 1977. The complete pumpkin cook book. Sterling Specialities Cook Books, Penndel, PA. 108 pp.

Gabbert, B.B. 1995. Pumpkin lovers cook book. Golden West Publishers, Phoenix, AZ. 126 pp.

Gilberg, R.L. 1982. Pumpkin cookbook. Gilmar Press, Newcastle, CA. 120 pp.

Hill, N. 1996. The pumpkin cookbook. Hamlyn, London, UK. 64 pp.

Kowbel, A.K. 1987. Fabulous cooking with zucchini, pumpkin and squash. Fabulous Cooking, Saskatoon, SK. 112 pp.

Libby, McNeill & Libby. (Corporate author.) 1984. The Great pumpkin cookbook: a harvest of Libby's favorite recipes. Rutledge Books, New York, NY. 128 pp.

MacCallum, A.C. 1986. Pumpkin, pumpkin!: lore, history, outlandish facts, and good eating. Heather Foundation, San Pedro, CA. 203 pp.

McMonigal, V. 1994. The perfect pumpkin and other squash. Bay Books, Pymble, NSW, Australia. 96 pp.

Moon, R. 1998. The American harvest cookbook: cooking with squash, zucchini, pumpkins, and more. Chartwell Books, Edison, NJ. 128 pp.

Nizzoli, A. 1998. The squash: history, folklore, ancient recipes, celebrated in recipes from ancient times to the present. Könemann, Köln, Germany. 141 pp.

Petersen-Schepelern, E. 2000. Pumpkin, butternut & squash: 30 sweet and savory recipes. Ryland Peters & Small, London, UK. 62 pp.

Ralston, N.C., Jordan, M., and Chesman, A. 2002. The classic zucchini cookbook: 225 recipes for all kinds of squash. Storey Books, North Adams, MA. 311 pp.

Tarr, Y.Y. 1978. The squash cookbook. Random House, New York, NY. 223 pp.

QUINOA

Family: Chenopodiaceae (goosefoot family)

Names

Scientific name: *Chenopodium quinoa* Willd.

- The word "quinoa" is based on the American Spanish *quínoa*, which is from the Quechua *kinua* or *kinoa*. In the Quechua language of the Incas, quinoa is the *chisiya mama* or "mother grain."
- Quinoa is sometimes spelled *quinua*, although *quinoa* is essentially the standardized spelling today.
- Quinoa is pronounced KEEN-wah (although quin-OH-a is an accepted alternative pronunciation in some dictionaries).
- Because of its importance in the highlands of South America, quinoa has been termed "vegetable caviar" and "Inca rice." Less complimentary, quinoa has been called "petty rice."
- The genus name *Chenopodium* is based on the Greek *chen*, goose, and *pous*, foot, an allusion to the shape of the leaves.

Plant Portrait

Quinoa is a "pseudocereal," i.e., it is not a true cereal (a member of the grass family with many small seeds that are collected as grain), but nevertheless similarly yields large quantities of edible small seeds. (Amaranth is another pseudocereal discussed in this book.) The plant is an annual, reaching a height of 0.5–2 m (1½–6½ feet), very rarely 3 m (10 feet). Quinoa has a thick, erect, woody stalk that may be branched or unbranched, and alternate, wide leaves that resemble the foot of a goose. The large seedheads, which make up nearly half the plant, vary dramatically in color, often displaying a rainbow of red or wine red, purple, green, rose, lavender, orange, black, and yellow. The seed is small, 1–2.6 mm (about 0.03–0.10 inch) in diameter, circular and flattened on two surfaces, resembling an aspirin tablet. Seeds with their coats left on can be black, purple, brown, red, pink, orange, yellow, or white.

Quinoa has been cultivated continuously in the Andean Highlands of South America since 3000 BC. In 1532, the Spanish explorer Francisco Pizarro and his army of 158 men reached the Andes, and proceeded to destroy the quinoa fields, kill the god-king, and force the Inca culture into submission. Under Pizarro's rule, the Indians were forbidden to practice their ceremonial rituals that centered on quinoa, which were replaced with Catholicism and potatoes, wheat, and barley. Despite having their hands cut off or even being killed for growing the grain, the Incas managed to maintain their interest in quinoa in hidden locations. The descendants of the Inca Empire, 8–10 million Quechua and Aymara Indians, still use quinoa as an important component of their diet. Although a staple for thousands of years, since the European conquest of South America quinoa has been considered a food of low social prestige. However, this prejudice is slowly changing. Attention to the crop has revived since about 1975, with increasing areas planted in South America, as well as in the United States, Canada, and Europe, initially as a health food but also as a source of materials for industrial uses. Industrial interest is in the small starch granules, which have properties different from those of cereal starches. There is also exploration of the saponin chemicals of the seedcoat, which have medicinal properties.

The plant is adapted to cold, dry climates, and to highland areas. Quinoa is cultivated in South America from Columbia to Chile and Argentina, with major production areas in Columbia, Ecuador, Peru, and Bolivia. The crop has been grown successfully in areas with an environment similar to that of the Bolivian Altiplano, where quinoa originated. Areas of cultivation include high elevations in Colorado, and also in Oregon, Washington, western Canada, Sweden, and Great Britain. At other locations, particularly at lower elevations, plants have not developed seeds.

Culinary Portrait

Quinoa grain is now a widely sold product in health food outlets and gourmet food shops in Western Europe and North America, and is increasingly found in many well-stocked groceries. It is sold in bulk bins as well as in packages, generally at high prices. Its nutty flavor has been compared to hazelnut, corn, squash, and couscous, and its slightly crunchy texture to North American wild rice and caviar. Quinoa can be substituted for most cereals, particularly rice. It is cooked like rice (2–4 parts liquid to 1 part quinoa) after removal of the seedcoat, but requires half the time (about 15 minutes). Quinoa expands 3–5 times its original volume. To toast quinoa, the grain can be sautéed in a frying pan for 10 minutes and then boiled in a double quantity of water for 10–15 minutes. Quinoa can also be curried, served as a side dish or a meat substitute, and added to salads, soups, pasta, breakfast porridges, and puddings. Quinoa flour can be used in cookies, muffins, cakes, and other pastries, but is unsuitable for yeast breads. The grain is also used in puffed breakfast cereals and in extruded foods (it blends well with corn and with oats). The flour works well as a starch extender in combination with wheat flour or cornmeal for breads, biscuits, and processed food. The grain can also be used to produce alcohol. In South America, the alcoholic beverage *chicha* is often prepared using quinoa. Most quinoa sold in the United States is in the form of whole grain that is cooked separately like rice or in combination dishes such as pilaf. Quinoa should be stored in an airtight container, preferably in a refrigerator or freezer if kept for more than a month. The leaves are occasionally eaten as a leafy vegetable, like spinach, although the seeds are the chief product of culinary interest, and fresh leaves are very rarely available in North America.

The seedcoat of quinoa contains a toxic chemical (saponin) that must be removed prior to consumption. The saponin, in the form of a bitter, soapy resin, forms a lather on contact with water, and needs to be washed off before the seeds are eaten. The Indians of the Andes traditionally accomplished this by hand-scrubbing in alkaline water. Today, detoxification is carried out by washing, or by simply removing the seedcoats. Some "sweet" varieties are now available with very low amounts of saponin in the seeds. Processing the grain to remove the seedcoat or at least the saponins produces white or whitish grain. Varieties with larger seeds are most likely to be found in stores, and the seed is usually white. Because some of the bitter saponin covering on the grain can still be present, even though most quinoa is washed before being sold, it is advisable to thoroughly rinse the dry seeds until the water runs clear, with no soapiness or bitterness.

Quinoa is high in protein, and the protein is of high quality (i.e., it contains an excellent balance of most amino acids). Lysine, an essential amino acid deficient in most grain crops, exceeds the concentrations necessary for proper amino acid nutrition in humans. In addition, the sulfur-containing amino acids cystine and methionine are found in concentrations that are unusually high compared to other food crops.

Curiosities of Science and Technology

- The armies of the Incas of South America frequently marched for days, sustaining themselves by eating a mixture of quinoa and fat known as "war balls."
- When the Incas thrived in Bolivia, relay teams of barefoot runners would carry news from one region to another, often covering 240 km (150 miles) in a 24-hour period at an elevation over 3600 m (12 000 feet) above sea level, an altitude where oxygen is considerably reduced. The stamina required has been attributed in part to the runners chewing coca leaves (from *Erythroxylum* species, especially *E. coca* Lam.) and ash from the quinoa plant, holding the material in their cheeks. The combination is said to increase the body's supply of oxygen because quinoa ash releases alkaloids from the coca.

Quinoa (*Chenopodium quinoa*), from Curtis (1787–present, plate 3641).

- Because there is little level ground in Bolivia, the Huarpa Indians developed terrace farming to cultivate quinoa, and many of the terraces have survived to the present. Sometimes, the terraces proceed great distances up the side of mountains, like a giant staircase, the steps often less than 3.7 m (12 feet) wide, and the top of the upper terrace sometimes more than 600 m (2000 feet) higher than the lowest terrace.
- In past centuries in Peru, it was widely thought that useful plants were animated by a divine being who caused their growth. According to the particular plant, these divine beings were called, for example, Maize-mother (Zara-mama), the Quinoa-mother (Quinoa-mama), the Coca-mother (Coca-mama), and the Potato-mother (Axo-mama). Figures of these divine mothers were made, depending on the plant, of ears of maize or leaves of the quinoa or coca plants, dressed in women's clothes, and worshipped.
- The Spanish invasion of South America brought many new crops back to the outside world that have become staples, including corn (maize), tomatoes, peanuts, and potatoes. The simple fact that the Spanish conquistadors ignored quinoa because they thought it was a cheap plant, fit only for peasants, denied almost all of the world the beneficial properties of this extraordinary food plant for hundreds of years, to say nothing of retarding centuries of probable progress on developing superior varieties. The story of quinoa is by no means unique—there are dozen of other crops that the Incas developed that have been ignored and have come to be called "the lost crops of the Incas." Famine relief experts believe that quinoa, with its high nutritional content and ability to grow in cool, harsh environments, could play a major role in relieving world hunger, so that the failure hundreds of years ago by the Spanish to recognize its potential is indeed tragic. The need to revisit the lost crops of the Incas is well known to the scientific community, but unfortunately, agricultural research seems inevitably concentrated on the priorities of the day.
- Because of its excellent nutritional characteristics, quinoa has been listed as a potential crop for NASA's Controlled Ecological Life Support System (CELSS). The CELSS concept utilizes plants to remove carbon dioxide from the atmosphere and generate food, oxygen, and water for the crew of long-term human space missions.
- A single plant of quinoa produces so many seeds that those from just 10 plants are sufficient to seed a hectare (the seeds of 4 plants are sufficient for an acre).
- Quinoa is gluten free, therefore likely to be suitable for those prone to food allergies induced by grains from the grass family such as wheat, rye, and barley.
- Seeds of wild plants are often poisonous and (or) bitter as protection to keep from being consumed by animals, so that they can survive and produce new plants. The bitter-tasting saponins in the seedcoats of quinoa protect the seeds from birds and insects.
- In 1998, Colorado State University (CSU) surrendered (i.e., chose not to renew) US Patent 5 304 718 on Apelawo, a cultivar of quinoa. The event was celebrated as a victory for Andean quinoa farmers. A very public anti-patent campaign had been waged for 14 months by a coalition of church agencies, including Agricultural Missions of the United States and Canadian Lutheran World Relief, and a variety of other socially minded organizations that inevitably made the situation embarrassing for the university, and for the well-motivated scientists who were conducting their activities completely within currently accepted norms. The Apelawo patent was named after a village on Lake Titicaca where the CSU scientists first picked up seed samples. The patent covered a method of hybridizing quinoa that also involved 43 other traditional Andean quinoa varieties named after villages from Ecuador to Chile. If the patent had been enforced, it was argued, Andean exports to the growing quinoa markets in North America and Europe would have been threatened. The subject of intellectual property rights unilaterally claimed by those in Western nations based on material originating from developing countries (so-called "biopiracy") is one that is extremely controversial, and thousands of similar situations have produced considerable friction. It needs to be stated that there is often right on both sides of this issue about which United Nations agencies are attempting to draft agreements.

Key Information Sources

Becker, R., and Hanners, G.D. 1991. Composition and nutritional evaluation of quinoa whole grain flour and mill fractions. Lebensm.-Wiss. u.-Technol. **23**: 441–444.

Cusack, D. 1984. Quinoa: grain of the Incas. The Ecologist, **14**: 21–31.

Fleming, J.E., and Galwey, N.W. 1995. Quinoa (*Chenopodium quinoa*). *In* Cereals and pseudocereals. *Edited by* J.T. Williams. Chapman & Hall, London, UK. pp. 3–83.

Galwey, N.W. 1989. Exploited crops. Quinoa. Biologist, **36**: 267–274.

Galway, N.W. 1993. The potential of quinoa as a multipurpose crop for agricultural diversification: a review. Ind. Crops Prod. **1**: 101–106.

Galwey, N.W. 1995. Quinoa and relatives. *In* Evolution of crop plants. 2nd edition. *Edited by* J. Smartt and N.W. Simmonds. Longman Scientific & Technical, Burnt Mill, Harlow, Essex, UK. pp. 41–46.

Galwey, N.W., Leakey, C.L.A., Price, K.R., and Fenwick G.R. 1990. Chemical composition and nutritional characteristics of quinoa (*Chenopodium quinoa* Willd.). Food Sci. Nutr. **42F**: 245–261.

Jacobsen, S.E., Mujica, A., and Ortiz, R. 2003. The global potential for quinoa and other Andean crops. Food Rev. Int. **19**: 139–148.

Johnson, D.L. and Ward, S.M. 1993. Quinoa. *In* New crops. *Edited by* J. Janick and J.E. Simon. Wiley, New York, NY. pp. 219–221.

Koziol, M.J. 1992. Chemical composition and nutritional evaluation of quinoa (*Chenopodium quinoa* Willd.). J. Food Compos. Anal. **5**: 36–68.

Koziol, M.J. 1993. Quinoa: a potential new oil crop. *In* New crops. *Edited by* J. Janick and J.E. Simon. Wiley, New York, NY. pp. 328–336.

Mastebroek, H.D., Soest, L.J.M. van, and Siemonsma, J.S. 1996. *Chenopodium* L. (grain chenopod). *In* Plant resources of South-East Asia, 10, cereals. *Edited by* C.J.H. Grubben and S. Partohardjono. Backhuys Publishers, Leiden, Netherlands. pp. 79–83.

Mujica, A., and Jacobsen, S.-E. 2002. Genetic resources and breeding of the Andean grain crop quinoa (*Chenopodium quinoa* Willd.). PGR Newsl. **130**: 54–61.

National Research Council. 1989. Quinoa. *In* Lost crops of the Incas. Report of an ad hoc panel of the Advisory Committee on Technology Innovation, Board on Science and Technology for International Development, National Research Council. National Academy Press, Washington, DC. pp. 148–161.

Rafats, J. 1986. Quinoa (*Chenopodium quinoa*), high fiber, high protein grain, 1970–86: 43 citations. U.S. Dept. of Agriculture, National Agricultural Library, Beltsville, MD. 6 pp.

Risi, J., and Galwey, N.W. 1984. The *Chenopodium* grains of the Andes: Incas crops for modern agriculture. Adv. Appl. Biol. **10**: 145–216.

Risi, J., and Galwey, N.W. 1989. Chenopodium grains of the Andes: a crop for temperate latitudes, *In* New crops for food and industry. *Edited by* G.E. Wickens, N. Haq, and P. Day. Chapman and Hall, London, UK. pp. 222–234.

Schlick, G., and Bubenheim, D.L. 1996. Quinoa: candidate crop for NASA's Controlled Ecological Life Support Systems. *In* Progress in new crops. *Edited by* J. Janick. ASHS Press, Arlington, VA. pp. 632–640.

Simmonds, N.W. 1965. The grain chenopods of the tropical American highlands. Econ. Bot. **19**: 223–235.

Wilson, H.D. 1980. Artificial hybridisation among species of *Chenopodium* sect. *Chenopodium*. Syst. Bot. **5**: 253–263.

Wilson, H.D. 1988*a*. Quinua biosystematics I: domesticated populations. Econ. Bot. **42**: 461–477.

Wilson, H.D. 1988*b*. Quinua biosystematics II: free-living populations. Econ. Bot. **42**: 478–494.

Wilson, H.D. 1990. Quinua and relatives (*Chenopodium* sect. *Chenopodium* subsect. *Cellulata*). Econ. Bot. **44**(3 Suppl.): 92–110.

Specialty Cookbooks

(For additional cookbooks, see AMARANTH and TEFF.)

Kijac, M.B. 1991. The art of cooking with quinoa. PM Publishing, Deerfield, IL. 35 pp.

RAPESEED AND CANOLA

Family: Brassicaceae (Cruciferae; mustard family)

Names

Scientific names: *Brassica* species

- Turnip rape, Polish rape, biennial turnip rape, winter turnip rape, Canola—*B. rapa* L. subsp. *oleifera* (DC.) Metzg. [*B. campestris* subsp. *rapifera* (Metzg.) Sinsk. of most North American authors]. Other subspecies are grown, especially in Asia: (*a*) subsp. *dichotoma* (Roxb.) Hanelt [*B. rapa* var. *dichotoma* (Roxb.) Kitam., *B. campestris* subsp. *dichototoma* (Roxb.) Olsson of some North American authors, B. *campestris* var. *dichotoma* Roxb.) G. Watt], known as brown sarson, canola, Indian rape, spring turnip rape, toria; and (*b*) subsp. *trilocularis* (Roxb.) Hanelt [*B. campestris* var. *sarson* Prain, *B. campestris* subsp. *trilocularis* (Roxb.) Olsson of some North American authors, *B. rapa* subsp. *sarson* (Prain) Denford, *B. rapa* subsp. *sarson* (Prain) Denford var. *trilocularis* (Roxb.) Kitam., *B. trilocularis* (Roxb.) Hook. f. & Thomson], known as yellow sarson and Indian colza.
- Argentine rape, Swede rape, Swedish rape, winter rape, Canola—*Brassica napus* L. subsp. *napus* [*B. napus* var. *napus* of most North American authors, subsp. *oleifera* Metzg. of some authors, and still other nomenclature by others; *B. campestris* subsp. *napus* (L.) Hook. f. & T. Anderson]

- The name "rape" is derived from the Latin *rapa*, an old Roman name referring to the rape plant. *Rapa* in the scientific name *B. rapa* has the same base. Rapeseed or rape is not a single species. In addition to the species highlighted in this chapter, several other species of *Brassica* and other members of the crucifer family are grown as "rape(seed)" in Asia, notably the mustards (see MUSTARD) and rocket (*Eruca sativa* Mill.). The word "rape" applied to plants has no connection with the word "rape" applied to criminal sexual violations. However, the use of the substitute name "Canola" was in part intended to escape from the negative image of the word "rape."
- The word "Canola" is a trade-marked acronym for "Canada Oil Low Acid." It refers to types of rapeseed varieties selected in Canada for edible oil and animal feed, with low amounts of toxic substances (notably erucic acid) that are present in the varieties used for industrial purposes. Canola includes varieties of both *B. napus* and *B. rapa*, and recently, a new Canola crop derived from the mustard species *B. juncea* has been made available in Canada [see MUSTARD; also see Burton et al. (1994), cited below]. This is more heat and drought tolerant than the traditional Canola species, but produces oil indistinguishable from *B. rapa* and *B. napus*.
- Although Canola is correctly spelled with an initial capital *C*, the word is often spelled canola, and used in a generic sense to refer to cultivars that can produce oil suitable for human consumption.
- See CABBAGE for the origin of the genus name *Brassica*.

Brassica rapa

- The "turnip" in "turnip rape" reflects the fact that the turnip is in the same species.
- Turnip rape is also called Polish (type) rape and British rape.
- *Oleifera* in the scientific name *B. rapa* subsp. *oleifera* is Latin, meaning producing or containing oil.

Brassica napus

- This species is called "Argentine (type) rape" in the New World (it was introduced to Canada from Argentina in 1942); and "Swede rape," "Swedish rape," and "continental rape" in Europe.
- *Napus* in the scientific name *B. napus* is based on an old Latin (Roman) name for the turnip.

Plant Portrait

Brassica napus is an erect plant with branching stems reaching up to about a meter (about a yard) in height, and is somewhat taller than the similar *B. rapa*. The fruit capsules of these two rape species are brownish, 5–10 cm (2–4 inches) long, with blackish or reddish brown seeds. The seeds are the edible part of rape plants. Both *B. rapa* and *B. napus* are grown as annuals ("summer" or "spring" types, planted in the spring and harvested in late summer or fall) and biennials ("winter" types, planted in the fall, and harvested in the early summer of the next year).

The recorded history of *B. rapa* is considerably longer than that of *B. napus*. Wild forms of *B. napus* subsp. *oleifera* are common weeds in both Europe and

Swedish rape (*Brassica napus* subsp. *napus*), from Köhler (1883–1914).

North America. The wild forms closely resemble the cultivated varieties, but usually have black, not red-brown or yellow seeds. Two centers of origin have been proposed. The European forms are biennials and had their origins in the eastern Mediterranean region. Annual forms come from eastern Afghanistan and adjoining Pakistan, this region considered to be another major center. The time and place of domestication is unknown, but it occurred in the preclassical era, as there are ancient Arab and Hebrew names for the crop. Sarson, an annual form of rapeseed, has been used in ancient India since about 2000 BC. Oilseed rapes were not cultivated by the Greeks or Romans. "Rape" was known in Europe at least since the Middle ages, but *B. rapa* and *B. napus* have been reliably distinguished only since the early 19th century, so that the early history of these oilseeds is obscure. Both species were apparently grown in 16th century Germany, as seeds of both have been found in grist mills of old German settlements. Initially, rapeseed oil in Europe was used mostly for illumination, but later it also was used in foods and as a cooking oil. With the development of steam power, it was found that rapeseed oil would cling to water- and steam-washed metal surfaces better than any other lubricant, and the oil became widely used for this purpose. North and South America adopted oilseed rape as a crop prior to and during the Second World War. In Canada, the commercial growing of rapeseed started in 1942. Initially, *B. napus* was grown in western Canada as a war measure to supply oil for lubrication of marine engines. Shortly after, *B rapa* was grown. Until the early 1950s, interest in rape as a North American crop was erratic. The first edible rapeseed oil was extracted in 1956–1957, and marked the beginning of a western-Canadian-based industry. Expansion was so spectacular during the 1960s that rape, known as Canola, has been called a "Cinderella crop."

Rapeseed has become the world's second most important edible vegetable oil, and is one of the very few that can be consistently and successfully raised in the north temperate climate. Rape is of considerable significance in Europe, Chile, China, India, Pakistan, Canada, and the United States. It is Canada's most important oilseed, and is the most widely used cooking oil in Canada, Japan, and some other countries. Oilseeds like rape are crushed to produce oil and meal. The oil is refined to manufacture a wide array of edible products, as well as soaps and detergents. The meal contains about 40% high-quality protein, used as a supplement for livestock feeds. Rapeseed oil contains erucic fatty acid. This is nutritionally undesirable, and breeding in Canada led to the creation of "Canola" cultivars with low amounts of this

Canola growing on the Canadian prairies, from Agriculture and Agri-Food Canada photograph collection.

component. Rapeseed meal also contains anti-nutritional chemicals called glucosinolates. Glucosinolates are desirable in some cabbage and mustard crops, because they result in flavor or pungency, but they can harm animals that eat the meal. Breeding in Canada has resulted in low-glucosinolate cultivars (also called "Canola").

Culinary Portrait

Rapeseed oil is used in vegetable cooking oils, salad dressings, margarines, processed cheese, "non-dairy" products, chips, fried foods, cookies, and pastries. The very low content of saturated fat (the lowest of any major cooking oil) has resulted in Canola being viewed as the most healthy of cooking oils.

Culinary Vocabulary

- The term "colza" has been used to refer to refined rapeseed oil.
- The phrase "lear oil" (based on "Low Erucic Acid Rapeseed") has been used to refer to Canola oil.

Curiosities of Science and Technology

- In the mid 1990s, a campaign of vilification against Canola oil was circulated widely on the Internet. Quite unfounded claims were made that Canola is toxic, often based on industrial varieties of rapeseed, which as outlined above do have some potentially anti-nutritional substances. In fact, Canola oil is among the healthiest of vegetable oils.
- The first genetically engineered variety of rapeseed was Calgene's "high-laurate rapeseed," a specialty edible type of rape first grown commercially in Georgia as a winter crop in the 1995–1996 growing season. Today, 80% of the acreage of Canola grown in North America is based on genetically engineered varieties, mostly altered to be resistant to herbicides.
- A fuel (rape methyl ester) derived from oilseed rape (*B. rapa* subsp. *oleifera*) is increasing in popularity in Europe for buses, taxis, and tractors. Although slightly more expensive than diesel, it has been reported to produce almost no sulfur dioxide and less carbon dioxide.

Key Information Sources

Adolphe, D. 1979. Canola, Canada's rapeseed crop. Rapeseed Association of Canada, Winnipeg, MB. 52 pp.

Ahmad, H., Hasnain, S., and Khan, A. 2002. Evolution of genomes and genome relationship among the rapeseed and mustard. Int. J. Biotech. (Pak.), **1**: 78–87.

Appelqvist, L.-Å., and Ohlson, R. 1972. Rapeseed. Cultivation, composition, processing and utilization. Elsevier, Amsterdam, Netherlands. 393 pp.

Beckie, H.J., Warwick, S.I., Nair, H., and Seguin-Swartz, G. 2003. Gene flow in commercial fields of herbicide-resistant canola (*Brassica napus*). Ecol. Appl. **13**: 1276–1294.

Bell, J.M. 1984. Nutrients and toxicants in rapeseed meal: a review. J. Anim. Sci. **58**: 996–1010.

Bell, J.M. 1993. Factors affecting the nutritional value of canola meal: a review. Can. J. Anim. Sci. **73**: 679–697.

Bhardwaj, H.L., Hamama, A.A., and Rangappa, M. 2003. Characterization of nutritional quality of canola greens. HortScience, **38**: 1156–1158.

Burton, W.A., Ripley, V.L., Potts, D.A., and Salisbury, P.A. 2004. Assessment of genetic diversity in selected breeding lines and cultivars of canola quality *Brassica juncea* and their implications for canola breeding. Euphytica, **136**: 181–192.

Busch, L., Gunter, V., Mentele, T., Tachikawa, M., and Tanaka, K. 1994. Socializing nature: technoscience and the transformation of rapeseed into canola. Crop Sci. **34**: 607–614.

Casey, M.F., and Cooke, P.W. (*Editors*). 1992. Canola cache: the farmer's handbook for growing canola. Kondinin Group, Mt. Lawley, WA, Australia. 135 pp.

Downey, R.K., and Röbbelen, G. 1989. *Brassica* species. *In* Oil crops of the world. *Edited by* G. Robbelen, R.K. Downey, and A. Ashri. McGraw-Hill, Toronto, ON. pp. 339–362.

Downey, R.K., Klassen, A.J., and McAnsh, J. 1974. Rapeseed: Canada's Cinderella crop. 3rd edition. Rapeseed Association of Canada, Vancouver, BC. 52 pp.

Friesen, L.F., Nelson, A.G., and Van Acker, R.C. 2003. Evidence of contamination of pedigreed canola (*Brassica napus*) seedlots in western Canada with genetically engineered herbicide resistance traits. Agron. J. **95**: 1342–1347.

Gomez-Campo, C. (*Editor*). 1999. Biology of *Brassica* coenospecies. Elsevier Scientific Publishing, Amsterdam, Netherlands. 489 pp.

Gulden, R.H., Shirtliffe, S.J., and Thomas, A.G. 2003. Secondary seed dormancy prolongs persistence of volunteer canola in western Canada. Weed Sci. **51**: 904–913.

Gunstone, F.D. (*Editor*). 2004. Rapeseed and canola oil: production, processing, properties and uses. CRC Press, Boca Raton, FL. 222 pp.

Jansman, A.J.M., Hill, G.D., Huisman, J., and Poel, A.F.B. van der (*Editors*). 1998. Recent advances of research in antinutritional factors in legume seeds and rapeseed: proceedings of the third international workshop on antinutritional factors in legume seeds and rapeseed. Wageningen Pers, Wageningen, Netherlands. 476 pp.

Kimber, D.S., and McGregor, D.I. (*Editors*). 1995. Brassica oilseeds. CABI Publishing, Wallingford, UK. 394 pp.

Kramer, J.K.G., Sauer, F.D., and Pidgen, W.J. (*Editors*). 1983. High and low erucic acid rapeseed oils: production, usage, chemistry, and toxicological evaluation. Academic Press, New York, NY. 582 pp.

Mailer, R.J., Scarth, R., and Fristensky, B. 1993. Discrimination among cultivars of rapeseed (*Brassica napus* L.) using DNA polymorphisms amplified from arbitrary primers. Theor. Appl. Genet. **87**: 697–704.

Morgan, B. 2006. Canola: situation and outlook. Bi-weekly Bull. (Agric. Agri-Food Can.), **19(17)**: 1–4.

Naczk, M., Amarowicz, R., Sullivan, A., and Shahidi, F. 1998. Current research developments on polyphenolics of rapeseed/canola: a review. Food Chem. **62**: 489–502.

Niewiadomski, H. 1990. Rapeseed: chemistry and technology. Elsevier, Amsterdam, Netherlands, and PWN–Polish Scientific Publishers, Warszawa, Poland. 433 pp.

Nydahl, M., Gustafsson, I.B., Ohrvall, M., and Vessby, B. 1995. Similar effects of rapeseed oil (canola oil) and olive oil in a lipid-lowering diet for patients with hyperlipoproteinemia. J. Am. Coll. Nutr. **14**: 643–651.

Palmer, J.D., Shields, C.R., Cohen, D.B., and Orton, T.J. 1983. Chloroplast DNA evolution and the origin of amphidiploid *Brassica* species cabbage, turnip, mustard, rapeseed. Theor. Appl. Genet. **65**: 181–189.

Phillips, W.B., and Khachatourians, G.G. (*Editors*). 2001. The biotechnology revolution in global agriculture: innovation, invention, and investment in the canola industry. CABI Publishing, Wallingford, Oxon, UK. 360 pp.

Raymer, P.L. 2002. Canola: an emerging oilseed crop. *In* Trends in new crops and new uses. *Edited by* J. Janick and A. Whipkey. ASHS Press, Alexandria, VA. pp. 122–126.

Shahidi, F. (*Editor*). 1990. Canola and rapeseed: production, chemistry, nutrition, and processing technology. Van Nostrand Reinhold, New York, NY. 355 pp.

Singh, D. 1958. Rape & mustard. Indian Central Oilseeds Committee, Hyderabad, India. 105 pp.

Stringam, G.R., Ripley, V.L., Love, H.K., and Mitchell, A. 2003. Transgenic herbicide tolerant canola: the Canadian experience. Crop Sci. **43**: 1590–1593.

Szabo, T.I. 1985. Variability of flower, nectar, pollen and seed production in some Canadian canola (rapeseed) varieties. Am. Bee J. **125**: 351–354.

Toxopeus, H. 2001. *Brassica* L. (oilseed crops). *In* Plant resources of South-East Asia, 14, vegetable oils and fats. *Edited by* H.A.M. van der Vossen and B.E. Umali. Backhuys Publishers, Leiden, Netherlands. pp. 65–70.

Specialty Cookbooks

Agriculture Canada and Alberta Dept. of Agriculture. 1973. Favorite recipes with rapeseed oil. Agriculture Canada, Ottawa, ON. 17 pp.

Canola Council of Canada. 1990s–1999? Cooking with canola oil: a guide to using canola oil and a collection of canola oil recipes. Canola Council of Canada, Winnipeg, MB. 32 pp.

Gore, M. 1996. Light and easy cooking with canola. Custom Book Company, Beecroft, NSW, Australia. 57 pp.

Long, D. 2000. Canola cooks! A culinary celebration of canola! Canola Info, Lloydminster, SK. 100 pp.

RICE

Family: Poaceae (Gramineae; grass family)

Names

Scientific names: *Oryza* species

- Asian rice—*O. sativa* L.
- Red rice, African rice—*O. glaberrima* Steud.

- The English word "rice" is derived from the Greek word *oryza*, through the French *ris* and the Old Italian *riso*. An unidentified Asian word for rice is believed to be the ultimate root. The genus name *Oryza* is the Latin form of the Greek word.
- Red rice is called by this name because the bran layer is red; it is also called African rice because it is commonly grown in West Africa.
- "Tuscarora rice" is an old name for ground corn produced by the first patented invention in the American colonies (see CORN).
- Rice should not be confused with "wild rice" (*Zizania* species).
- *Sativa* in the scientific name *O. sativa* is Latin for sown or cultivated.
- *Glaberrima* in the scientific name *O. glaberrima* is Latin for almost hairless or smooth. *Oryza glaberrima* has glabrous (hairless) flower clusters (spikelets) and leaf blades, while *O. sativa* cultivars are mostly pubescent, although most cultivars in the United States are glabrous.

Plant Portrait

Asian rice is an annual native grass of the tropics and subtropics of Southeast Asia, and grows best in a hot, moist climate. The plants are typically 60–180 cm (2–6 feet) tall. Domestication of rice occurred perhaps as far back as 10 000 years, in the river valleys of South and Southeast Asia and China. Rice is now commercially cultivated in 112 countries, representing all continents of the world except Antarctica.

There are more than 15 000 different varieties, organized into three basic groupings: (1) Indica—usually tropical and long grain, the most widely consumed type; (2) Japonica—also known as Sinica, usually temperate climate and shorter grained, the type used to make sweet "rice pudding" along with milk, eggs, and sugar, because it cooks quickly and amalgamates more easily than the long-grained variety; and (3) Javanica—also known as bulu, tropical to sub-tropical, medium to long grain. Indica rice is cultivated in Indochina, Thailand, India (hence the name Indica), Pakistan, Brazil, and the southern United States. Japonica is mostly grown in colder-climate countries, including Japan (hence the name Japonica), Korea, northern China, and California. Javanica is grown particularly in Indonesia (the name Javanica is based on the Indonesian island of Java). With the creation of more and more varieties, the distinctions of these three types of rice has become increasingly blurred.

"Upland rice" is grown without submersion, usually on terraced hillsides; "lowland rice," the predominant type, is grown in beds that are flooded during much of the growing season (with a water level of 10–15 cm or 4–6 inches), with the water drained off as the plants approach maturity. In Asia, the monsoon (a season of torrential rains followed by a dry spell) naturally provides the necessary cycle of wetness and dryness.

More than 90% of the world's rice production is in Asia. China, India, Indonesia, Bangladesh, and Thailand are the largest producers. While rice is responsible for 80–90% of the dietary intake of Asians, at the latitude north of Beijing (Peking), called the "Rice Line," wheat replaces rice as the staple food. India tends to be divided into two food areas: wheat and meat in the north, rice and vegetables in the tropical south. Italy is the leading rice-producing nation in Europe. In the United States, rice is grown in Arkansas, California, Louisiana, Mississippi, Missouri, and Texas (but not in Georgia and South Carolina, where American rice production was once centered). The United States, curiously, is the world's largest rice exporter. Rice, which is widely considered to be the world's most important food crop, is the primary staple of more than two billion people in Asia and hundreds of millions of people in Africa and Latin America. Throughout history, failures of the rice crop have caused widespread famine and death; conversely, in human history, there is a close relationship between expansion of rice cultivation and rapid rises in population growth.

Red or African rice is thought to have been domesticated as early as 3500 years ago in the delta of Niger from which it was introduced into other parts of West Africa, where diverse cultivars were selected.

The reddish color of the grain has been said to be one of the reasons why this rice is much less popular than Asian rice, the white color of the latter being generally preferred. Asian rice is more easily produced, harvested, and milled. African rice continues to be produced mostly by small-scale farmers, mainly for local consumption, often in remote areas of Africa. It is cultivated principally in Nigeria, Mali, Niger's inland delta, and to a lesser extent on the hills near the Ghana–Togo border and in Sierra Leone. African rice is important in a number of West African rituals.

Amylose is a kind of starch that is present in the rice grain. "Nonwaxy" rice is rich in amylose, and "waxy rice" is low in this material. Waxy rices absorb less water while cooking, and are sticky after cooking. Waxy rices are said to be "sticky" or "glutinous," although the stickiness has nothing to do with gluten (see WHEAT and RYE), which is not present in rice. Examples of nonwaxy rice are basmati, Carolina Gold, and Texmati. Examples of waxy rices with low amounts of amylose include Thai jasmine, Italian, Spanish, and some Japanese rices. Examples of waxy rices with very low amylose content include Thai and Chinese sticky rice (known also as sweet rice). As noted above, rice varieties have traditionally been classified as indica, japonica, and javanica. High- and intermediate-amylose rices (i.e., nonwaxy types) are usually indicas (i.e., long grain). Low-amylose rices (i.e., waxy types) tend to be japonicas (Thai jasmine is an exception). Javanica tends to be intermediate in amylose content.

Because rice flour is nearly pure starch and free from allergens, it is the main component of infant formulas and face powders. Because of the low fiber content, rice powder is useful for polishing camera lenses and expensive jewelry.

Culinary Portrait

As noted above, one of the ways rice is classified is by length of grain, and a second way of classifying rice is by stickiness or "waxiness." These features, as well as tradition in various countries, determine to a considerable extent the method of preparation and usage. Rice is the main accompaniment to much Chinese food, Indian curries, Turkish kebabs, and many other Oriental dishes. Beans and rice is a popular dish in many countries. Long-grained rice is mainly boiled and drained, and used as a side dish for meat, fish, or eggs, served as a vegetable in place of potato, or used in salads and soups. In Thailand, the taste is for long-grained rice, cooked firmly, each grain separate, and eaten with fingertips of the right hand or with spoons and forks. The Spanish rice dish paella is usually made with short-grain rice, cooked with seafood and chicken, flavored with saffron, peppers, onion, and garlic ("paella" is the Spanish word for "frying pan" and also the name of the dish cooked in it). "Spanish rice" is a medium-grain rice grown in Valencia, Spain, considered best for paella. In China, a small- or medium-grain rice is preferred, and steamed rice is the usual way it is prepared, the grains not cooked dry but left fluffy and clinging so they can be easily picked up with chopsticks. Korean rice is similarly served. In Japan, rice is also preferred small- or medium-grained, cooked fluffy, and slightly clinging, so that it can be consumed with chopsticks, as in China. Small-grained rice is also preferred by Indonesians, but they like the grains separated, very firm, and not clinging. Arborio rice, from Italy, has short, shiny, pearl-smooth grains that are considered excellent for making risotto because it can absorb large quantities of liquid without becoming excessively mushy.

Glutinous (i.e., low-amylose or waxy) types of rice are used in various dishes, cakes, soups, pastries, breakfast foods, and pastes. Glutinous types are also used in Asia for special purposes as sweetmeats. This kind of rice is excellent for those who like to cook rice ahead of time, since it does not harden on standing. Such rices absorb less water than intermediate- or high-amylose varieties, and have a clingy texture after cooking. Rices with very low amylose content absorb so little water during cooking that they are frequently steam-cooked. They are often shiny after cooking. High- and intermediate-amylose rices (i.e., nonwaxy rices) absorb more water than glutinous rices, tend to expand more when cooked, are firmer, whiter, duller (less shiny), and often drier or fluffier than the glutinous types.

Basmati rice is a fragrant rice (the Hindi *bāsmatī* means "fragrant") from India and Pakistan, which is also grown in the United States. It has long tender grains and a distinctively nutty taste, and is considered the rice of choice for curries and pilafs. Basmati rice is an example of a "perfumed rice" (jasmine rice is another example).

All rices seem to improve when allowed to stand or rest for 10–15 minutes after cooking. It appears that this allows the starches present to firm up, much like the newly softened starches in baked bread must set for a time before the bread can be easily cut.

Puffed rice, which was made popular by Quaker Cereals at the 1904 World's Fair, is actually shot from gun-barrel devices, at least in commercial production. The grains are superheated, and with a sudden

Rice (*Oryza sativa*), from Köhler (1883–1914).

release in pressure, they expand (i.e., puff) to several times the original size. This method is also applied to wheat. In India, puffed rice is popular and is prepared simply by heating polished rice under pressure in a metal container. "Instant rice" is precooked rice that has been dried (Minute Rice is a familiar brand). The cracked grains allow water to enter so that the rice cooks almost instantly. The taste is inferior, cost is higher, and nutritional value is lowered, but food preparation is at least fast and the cooked rice is dry and fluffy in appearance. "Seasoned rice" is typically made from precooked or parboiled rice (explained below), together with seasonings and food additives. Rice cakes were a trendy dessert for some time in North America, desired for their limited content of calories, although some people equate them with "cardboard cakes." Rice flour contains no gluten, and is therefore unsuitable by itself for making bread, but it is excellent for puddings, pastries, and cakes, and can be substituted for wheat flour, especially where the latter causes an allergic reaction. "Ground rice" is similar to rice flour, but the texture is slightly coarser. Rice flour is used by the food industry to prepare cakes, noodles, and other pasta products, baked goods, puddings, snack foods, infant formulas, and fermented preparations, including beer, wine, and vinegar.

Rice with the inedible hulls removed but the bran left on is brown rice (also known as whole rice, cargo rice, and rough rice), whereas rice that is further milled to remove the bran is white ("polished") rice, which may be further milled to produce "pearled" rice. "Black rices," of which there are many varieties, have a black-colored bran layer (the grain is white underneath). Some rices even have red-colored bran. Brown rice has a nutty flavor that is stronger than that of white rice. Most people (including those in Asia, where rice consumption is centered) prefer white rice to brown, although the latter is more nutritious (often white rice is enriched with iron, niacin, and thiamine in order to restore some of the nutritional value). Brown rice is difficult to digest because of the high fiber content, and tends to become rancid if stored for a long period (the oil-rich germ is the part that goes rancid). Since brown rice contains bran and germ, it is rich in vitamin B_1. A disease called beriberi, characterized by inflammation of the peripheral nervous system and a loss of muscle tone in the arms and legs, is caused by a dietary deficiency of vitamin B_1. This disease reaches epidemic proportions in Asia and Africa among people who subsist on a diet of white rice. Brown rice generally takes three times as long to cook as white rice, although faster-cooking types are marketed (instant brown rice requires a cooking time of only 5 minutes, with a 5-minute rest period, compared to the normal cooking time of 45 minutes for brown rice). Brown rice also uses twice as much water. The color is not as attractive in many dishes, and brown rice usually contains some green grains (a result of uneven maturation of grain of a given plant), a phenomenon that also occurs with white rice, but is much less noticeable in the latter because of the greater degree of milling away of the surface layers. Brown rice is available in many food stores in both short- and long-grain varieties, and is preferred by some health-conscious people, who appreciate its nutritional superiority over white rice. Brown rice does not keep as well as white rice, and is preferably stored in a refrigerator. "Semi-milled rice" is a compromise between white and brown rices. It has flecks of bran still attached to the grain, and may still have some germ also attached. It tastes better than brown rice, which can be quite chewy, is more nutritious than white rice, and requires only a little more cooking time than white rice. Semi-milled rices are produced occasionally in Asia and Europe, and are becoming available in North America.

"Converted" (parboiled) rice is steamed and dried before the hull and bran are removed, resulting in a fluffy rice with grains that do not stick together. (Converted, rather than parboiled, is the preferred term in the United States.) Another advantage of parboiling rice is that the keeping qualities are improved. Parboiling actually can increase the required cooking time, but commercial parboiled rice is often precooked to some extent, so that cooking time can be relatively short. The parboiling process moves some of the water-soluble vitamins and minerals of the germ and outer layers to the interior, so that after polishing, the remaining rice is more nutritious than normal polished rice, although there is some loss of nutrients (which may be replaced in commercial preparations). Parboiled rice absorbs water less easily than regular rice, and so is less suitable for preparing rice dishes like paella and risotto where absorption of flavors is important. Parboiled brown rice is prepared like parboiled white rice, but the bran and germ are not milled away. The required cooking time for converted brown rice may be significantly shorter than that for normal brown rice (25 minutes vs. 45 minutes).

Rice grain, malt, and water are used to make rice wine or sake, much consumed in Japan. Sake is clear or pale amber in color, with a mild, sweet flavor, and an alcoholic content of 14–18%. Sake is traditionally taken along with meals, sipped out of tiny porcelain cups. In Japanese restaurants, sake is customarily presented hot (sometimes warm or cold), in a small ceramic pitcher, and patrons are provided with small ceramic glasses (by Japanese custom, no one should pour sake for himself or herself).

African rice is prepared in the same ways as Asian rice, but in Africa, the grain is also traditionally fermented to produce regionally popular beers. African rice is often made into flour, which is used to prepare a variety of baked goods.

Traditionally, rice was picked over by consumers to remove debris, but commercial rice has generally been well cleaned. However, when buying specialty rices in bulk, it may be wise to closely inspect the rice before cooking it.

Another traditional process by consumers has been to wash rice in several changes of cold water, in order to remove dirt and dust from the threshing process. Rinsing (i.e., "washing") rice before cooking is still widely recommended, but should not be necessary for standard brands of rice sold in Western supermarkets. When rice is rinsed before cooking, it results in the premature release of starch, which can prevent the grains from thickening. In the past, washing rice grains was recommended because millers often coated rice with glucose to make it look better; this resulted in nutrients, especially B vitamins, being rinsed away (coating with glucose and talc or with magnesium silicate is still sometimes encountered). In the United States, most polished rice is enriched with a dusting of iron, niacin, and thiamine, and the American Rice Council has advised that such enriched rice should not be washed before use, as the added mineral and vitamins would be reduced. In Asia, rinsing rice before cooking is widely practiced—deliberately—not only to clean the rice but also to remove some surface starch to reduce the stickiness of the grains. Washing does have the advantage of removing milling and polishing dust that is still present, which tend to cook to a glue, sticking grains together. In any event, the best advice is to follow the manufacturer's instructions. Some authorities advocate soaking rice in water for a period, while others insist this spoils the flavor. Some cooks recommend against washing parboiled, long-grain white, and instant rice, but do recommend washing medium- and short-grain rice to prevent sticking. Soaking is not generally advised, except when making pilaf in the classical manner. When using basmati (and other perfumed rices), the rice should be soaked in cold water before cooking; it should be stirred and the water replaced when it turns milky, this process repeated four or five times until the water is clear (at which stage all of the loose starch or other powder or coating that might be present has been removed).

Rice can be cooked in several ways, and while some cooking methods produce good results for many varieties, some varieties are best prepared in a particular way. In addition to whether rice should be rinsed or soaked, and whether this should be done with cold water or boiling water, there are several other points over which rice culinary authorities vehemently disagree. In Asia, salt is often not added to the water in the belief that it ruins the rice, but elsewhere, rice is generally cooked with salt. Some claim that lifting the lid on a pot of cooking rice to check whether the rice is done will ruin it; but rice is often cooked in an open pot. It is sometimes said that rice must be added to boiling water, although most are content to start rice cooking in cold water, and not everyone agrees that once the water starts boiling the heat should be lowered.

- Boiling

This is often called the "absorption method," since a given amount of rice is cooked in a given amount of water, generally resulting in all of the water being absorbed. (However, many rices cook very well, like pasta, in a higher amount of water than can be absorbed. In this technique, the rice should be drained immediately and thoroughly in a sieve after cooking, and rinsed with tepid water to stop further cooking. This technique is not recommended for lower-amylose rices.) A heavy pot or saucepan with a tight-fitting lid is recommended. A heavy bottom is best, thick enough to hold and distribute the heat to prevent burning or sticking. The heat should be lowered once the water begins boiling to prevent burning the rice. The rice should be no more than 5 cm (2 inches) deep (if deeper, there will be hard, undercooked grains on top and (or) some mushy grains near the bottom, owing to uneven cooking; parboiled rice, however, is more tolerant). Alternatively, a rice cooker can be used. Rice cookers are electrically heated containers with markings on the inside, showing how much water and rice to add to produce a given volume of rice. They have automatic timers to regulate the heating cycle. Rice cookers are widely used in Asia.

There are numerous methods of boiling rice, which can be carried out in water, milk, stock, or juice. Using 1 part rice and 2 parts liquid, the mixture can be brought to a boil, the heat lowered, the pot covered, and the rice simmered gently until the liquid is absorbed. Alternatively, the liquid can be brought to a boil first before adding the rice. Or, the rice can be cooked without a cover, drained, and excess moisture can be removed by placing in an oven at 149 °C (300 °F) for 7–15 minutes.

Cooking time depends on type of rice and personal preferences, and recommendations accompanying the package of rice should also be considered. Some general recommendations: brown, 40–45 minutes; parboiled, 25 minutes; white, 14 minutes; instant, 5 minutes.

A good recommendation is to use a large pan or pot to allow rice to expand and lie shallowly; otherwise, the lower rice becomes compacted, cooks faster, and becomes soggy. Firmer rice can be produced by reducing the quantity of water and not overcooking; conversely, softer rice requires more water. If rice is to be reheated, reduce the original cooking time. Stirring rice during cooking tends to prevent it from sticking.

Often, rice is left on low heat after cooking to "steam" or "fluff." To achieve maximum fluffing, advanced cooks may remove rice immediately after cooking, sieve the water away, rinse with hot or cold water, and reheat gently, either with steam, in a low-temperature oven, or in butter or oil in a pan.

- Steaming

The rice is placed on the top of a double boiler, over boiling water, covered, and simmered over moderate to high heat. The rice is not immersed in the water, but is cooked by the steam. Steaming is a preferred method for sticky rice and for some low-amylase rices.

- Cooking in fat

This is the traditional way of preparing rice pilaf, risotto, paella, Greek-style rice, and Creole rice. The rice is cooked in a little fat or butter for several minutes while stirring, then twice its volume of liquid is added, the pan is covered, and the mixture simmered until the liquid is absorbed. Such cooked rice remains firm and does not stick. Frying is an excellent method of using up leftover cooked rice.

- Microwaving

Microwaving is actually a blend of boiling and steaming. Instructions provided with the microwave should be followed.

- Pressure-cooking

A pressure cooker is a way of cooking rice at high altitudes. Because of the lower atmospheric pressure, boiling would require a longer period, risking turning the rice into mush.

Some rice equivalents:
1 cup precooked rice yields 2 cups cooked
1 cup uncooked long grain rice yields 3 cups cooked
1 cup parboiled or converted rice yields 4 cups cooked
1 cup brown rice yields 4 cups cooked

Leftover rice can be refrigerated in a sealed container for up to 3 days, but is best eaten within a day. Sticky rices harden so much that they are difficult to revive. Nonsticky rices may cake or produce lumps, but this can be broken up with a utensil, and the rice can be reheated in a pot with a little added water on a conventional stove, or in a sealed container in a microwave oven.

Like other starchy foods, letting uneaten rice harden on dishes or in pots before they are washed in a dishwasher is undesirable. Unless rinsed off, the hot water makes the starches harden and stick.

Rice has a relatively long storage life, but should be kept in a dark, cool, place, away from high humidity. In hot, humid weather, it is preferable to buy rice in smaller quantities if it will not be used up in a few months. Insect infestations are a potential problem for rice stored in bags. Because of the higher oil content, brown rice will become rancid if stored too long in a warm place.

When the expression "rice paper" is used, it generally refers to paper made from the white stem pith (soft central tissue) of the rice-paper plant, *Tetrapanax papyrifer* (Hook.) K. Koch ("*T. papyriferus*" of many authors), a small Asiatic tree or shrub. In Japanese homes, translucent sliding panels of rice paper mounted on wooden frames are used as partitions and doors. However, paper (the kind you write on) is also made to a small extent from the straw of the rice plant. Moreover, there is an edible kind of rice paper (also called "wafer paper"), which is made from rice flour and water, and looks like translucent tortillas. Edible rice paper is quite uncommon, and is most obtainable from large Asian markets, where it will be found in packages of dry material. Do not confuse packages of rice paper with spring roll wrappers, which are meant to be deep-fried. Some sophisticated and adventurous cooks use edible dyes on edible rice paper to write edible messages suitable for festive occasions.

Culinary Vocabulary

- Types of rice (note that the international community defines long grain, medium grain, and short grain by the ratio of length to width of grain, not on absolute length):
  - Long grain: rice typically 6 mm (1/4 inch) or more long, with a length that is more than three times as long as wide; when cooked, the grains are firm, fluffy, and easily separated, producing a light, fluffy rice; an all-purpose rice, particularly suitable for pilafs and side dishes consumed with curries, chicken, meat dishes, and stews. Well-known examples of long-grain rice include basmati, Thai jasmine, and Carolina Gold.
  - Medium grain: rice typically 5–6 mm (1/5–1/4 inch) long, which is shorter than long grain, plump but not round (from 2½ to 3 times as long as wide); less starchy than short grain, when cooked it produces relatively moist, tender grains that start to stick together as the grains cool. Medium-grain rice is often sold as "short grain."
  - Short grain (also known as round grain): rice typically 4–5 mm (less than 1/5 inch) long, with fat, almost-round grain (less than twice as long as wide) that is softer than medium- and long-grain rice, and has a high starch content; when cooked, the grains are tender, moist, and sticky, and tend to stick together; good for sushi and rice pudding, favored for the steamed rice eaten with Japanese and Chinese food (the stickiness of the rice makes it easy to use chopsticks), and also much used for risotto and croquettes. Short-grain rice requires a little less water than long-grain rice to cook, and unlike most kinds of rice, tends to lose flavor eventually on standing. Informally, "short-grain rice" often refers both to medium-grain and short-grain rice collectively.
  - Glutinous rice, also known as mochi rice, pearl rice, sticky rice, and sweet rice, is a short-grained rice with a sweet, sticky consistency when cooked, and is popular in Asian cooking for making sweets and snacks.

- "Waxy rice" (i.e., glutinous or low-amylose rice) is called waxy because the uncooked grains look opaque and solid white. By contrast, uncooked grains of nonwaxy (high-amylose) rice look translucent.
- Sake (also spelled saki and saké, and pronounced SAH-kee) is the traditional national alcoholic beverage of Japan. This word for Japanese wine made from fermented rice is said to be based on *sasake*, bamboo smell, tracing to an old Chinese legend. A flock of sparrows is said to have picked some grains of rice to store for the winter in a piece of bamboo, but the bamboo was flooded and the rice fermented, producing rice wine.
- Fermented rice wine in China is called *samshu*, from *sha*, three, and shao, *fire* (i.e., thrice-fired or distilled). The Chinese ideograph for this recalls the sparrow and flood legend, as it is a combination of the ideographs for bird and water. Shaoxing wine of China is also similar to sake.
- Although commonly called "rice wine," the process of manufacturing sake from fermented rice is more reminiscent of brewing than of wine-making (i.e., sake is fermented from a grain, not from a fruit, making it a beer, not a wine). Because sake was the only alcoholic beverage in Japan for many years, "sake" acquired the meaning "alcohol" in general.
- In addition to sake, a dry "wine" that is used for both drinking and cooking, the Japanese also produce another type of rice wine, mirin (pronounced mee-REEN), which is sweet, and used just for cooking. Also known as "sweet sake" and "sweet rice wine," this is relatively low-alcohol (some varieties contain about 8%), syrupy, thin, and golden-colored, and is used to add sweetness and flavor to sauces, glazes, and a variety of Japanese dishes. It has been claimed that half of all Japanese dishes are flavored with mirin.
- A "saketini" is a cocktail made with gin and sake, and (like a martini) garnished with an olive or lemon twist.
- "Mochi" (pronounced MOH-chee) refers to a short-grain, sweet, gelatinous rice with a high starch content used in Japan to make confections, and also to traditional, chewy, Japanese cakes made with this rice. Prepared mochi is sold in Japan as small balls and as sheets, and while eaten year round, it is a traditional New Year's food.
- "Nuka" (pronounced noo-KAH) is a Japanese rice bran powder mostly used as a pickling agent. It is generally available in Japanese food stores. In Japan, it is also wrapped (e.g., placed in a cotton bag) and used as a dishwashing tool, and is also employed to polish woodwork.
- A *dosa* is a thick pancake prepared in southern India from rice flour. It is often served at breakfast and as a snack, as well as at main meals.
- *Chicha* is a traditional Andean drink made from fermented maize or rice. To stimulate the fermentation process in some rural parts of Ecuador, chicha makers chew the ingredients and spit them back in the pot to brew. When prepared in this unsanitary manner, this brew has been found to transmit hepatitis B.
- *Rijsttafel* is a popular dish that is considered to be the national dish of Indonesia (Rijsttafel literally means "rice table" in Dutch, reflecting the fact that most of the islands of Indonesia were formerly under the control of the Dutch East India Company). Traditional etiquette requires that rijsttafel be served by at least 20 waiters. Rijsttafel refers more to the way the dish is served, rather than the actual ingredients. Robert Ripley (1893–1949) of *Ripley's Believe It or Not!* was fascinated with food habits in exotic places, and in 1932, he filmed a rijsttafel meal that was served by 22 waiters in Java (an island of modern Indonesia). Each waiter carried a separate dish. The first waiter served boiled rice. The last waiter served *kroepoek*, crinkly rice flour biscuits as large as plates.
- The well-known Northern Italian dish "risotto" is made with simmered rice and vegetables, sometimes meat, and usually finished off with grated cheese. Risotto refers not only to this, but also to the associated general cooking method for grains, i.e., the grains are lightly sautéed in butter, a liquid is gradually added, and the mixture is simmered with constant stirring until the still-firm grains merge with the cooked liquid. Pilaf (pilaff, pilav, pilaw, a word of Turkish origin) is a dish in which rice (or sometimes some other grain) is cooked in stock, generally with various ingredients such as vegetables, meat, or fish. Pilaf and risotto are essentially the same thing, pilaf originating from the Middle East, while risotto is Italian. Pilau (pronounced pi-loe) or purloo (pronounced puhr-loo) refers to a rice dish from the American south, in which long-grain rice is cooked in an aromatic broth until almost dry, and meat, fish or shellfish is usually added, along with garnish.
- *Arroz con pollo* (pronounced ah-ROHS con POH-yoh), literally "rice with chicken," is a Spanish and

Mexican dish also made with tomatoes, green peppers, seasonings, and sometimes saffron.

- In Japan, *okazu* means "anything that can be eaten with rice." The word for rice, *goban*, also is the word for a whole meal. Breakfast is *asa goban*, "morning rice"; dinner is *hiru goban*, "afternoon rice"; and supper is *yoru goban*, "evening rice." This linguistic dominance of meals reflects the historical importance of rice in Asia, where rice customarily has provided at least 80% of food calories, and is the only food staple of the majority of people.
- In Thailand, dishes are almost always served with rice. In fact, the generic term for all other foods is "not rice" (or "with the rice"), and "come and eat rice" is the invitation to a meal.
- "Ashley bread" is a southern United States batter bread of South Carolina origin, made from rice flour. It has been suggested that the name commemorates Anthony Ashley Cooper (1621–1683, first Earl of Shaftesbury), an early landowner in the American colony of Carolina.
- "Dirty rice" is a Louisiana Cajun dish that combines cooked rice, chicken gizzards or livers, onions, and bacon fat. The name is based on the meat giving the rice a "dirty" unappetizing appearance.
- "Sizzling rice soup" is a Chinese broth combining chicken or pork, and sometimes shrimp, with various vegetables, served over deep-fried rice cakes in soup bowls. The rice sizzles and pops when the broth is ladled over the cakes.
- *Riz à l'impératrice* (French for "rice as the empress likes it") is a very rich rice pudding prepared with vanilla custard, whipped cream, and candied fruit (often soaked in kirsch). The phrase *à l'imperatrice* is employed to describe various rich sweet or savory dishes.
- A "ricer" (also known as a food ricer and a potato ricer) is an old-fashioned tool used to cut a cooked food, such as potato or carrot, into rice-like pieces. Food is placed in a cylindrical basket with perforated holes, and a lever is used to push a metal disk against the food, pushing it through the perforations. The invention of the food processor made the ricer unnecessary.
- The classic rice dish paella, described above, is often finished with a "socarrat"—a caramelized crust of rice at the bottom of the pan.
- "Kobe beef" is an extremely expensive (often more than $100/pound) meat from cattle raised in Kobe, Japan. The cattle are massaged with sake and fed large amounts of beer, resulting in exceptionally tender, finely marbled, and tasty meat. The techniques have been adopted by some beef-cattle ranchers in the United States.
- "Bash ful" is an interesting name for a Bangladeshi parboiled rice that is sold in the United States.
- "Oryziverous" is a pretentious word meaning "pertaining to eating rice."

Curiosities of Science and Technology

- Ice cream was invented in China around 2000 BC when a milk-and-rice mixture was packed in snow.
- Built in 495, China's Shaolin Temple in the province of Henan is famous as an important Buddhist shrine and as the ancient center of development of kung-fu. At the beginning of the 7th century, a tiny army of 13 Shaolin monks were reputed to have saved the future Tang Dynasty emperor Li Shimin by defeating an entire division of the ruling Sui Dynasty's army. In the martial arts, "hand conditioning" refers to methods used to improve the effectiveness of the hand as a striking weapon, usually by repetitive impact. According to the "iron palm" techniques developed by the monks of the Shaolin Temple, the hands should be toughened by repeatedly spearing them into pails of abrasive materials, gradually increasing the hardness of the contents of the containers. In one suggested sequence, the trainees first use a pail of rice, next beans, then sand, and finally ball bearings.
- The tradition of throwing rice at newlyweds traces to the ancient use of rice in Asia as a symbol of fertility in religious ceremonies. Rice was also considered to be a symbol of health and prosperity, and a means of appeasing evil spirits so they would not bother the wedding couple.
- Willem van Ruysbroeck ("William of Rubrouck," ca. 1215–ca. 1295) was a French Franciscan friar and traveler, who was sent by King Louis IX of France on a mission to the Mongol Empire in 1253, and prepared an account of the trip that is considered to be the best of any medieval Christian voyager. He reported that the Mongol prince Mangu Khan had a silver fountain with four spouts, respectively, dispensing kumiss (a fermented liquor made from mare's milk), wine, mead, and rice wine. [Alcoholic beverages made with animal constituents are not usual today, but in Asia, records show "wines" were made of deer, tiger bones, tortoises, snakes, dogs, and mutton. Eighteenth century English and American cookbooks had a recipe for making

"cock ale" using 10 imperial (12½ American) gallons of ale, one large and elderly cock, raisins, mace, and cloves.]

- In feudal Japan, salaries were often paid in rice.
- Women of Honshu Japan, while harvesting rice, wore stylized masks centuries ago to make them unattractive to male supervisors. Such masks are still sometimes worn today by rice harvesters, but as protection against insects and sunburn.
- In the late 19th century, over 25% of the sailors in the Japanese Navy developed beriberi—the nutritional disease resulting from insufficient quantities of the vitamin thiamine (additional information is provided above). It was later found that a diet of brown rice—as opposed to white rice—prevented the disease. Beriberi had become more common because of the introduction of improved polishing techniques that removed the brown outer layers of the rice grain in which thiamine (vitamin B_1) occurs.
- Carolina rice was so important in the old American south that it was known as "Carolina gold" and was actually used as currency. South Carolina has the nickname the "Rice State." Although there is no longer any significant commercial rice production in South Carolina, some companies still market good-quality long-grain American-grown rice under the name "Carolina rice."
- At the conclusion of the American Revolution, the British removed all the rice out of the former Carolina colonies, including seed rice. Thomas Jefferson (1743–1826, 3rd American president) smuggled rice that would grow in dry fields out of Italy, and introduced it to the Carolinas in 1787, restarting the American rice industry. Jefferson's action was risky, as exporting the rice at the time carried the death penalty.
- In some Asian cultures, rice becomes more valuable as it ages. This is often the case for basmati rice.
- Rice is so important in China, where people consume about 0.45 kg (1 pound) every day, that the word for "agriculture" is the same word as for "rice." A common greeting equivalent to "How do you do?" is "Have you had your rice yet?"
- Traditional table etiquette in China calls for placing one large spoonful and one small spoonful of rice in one's bowl. The large one is for yourself and the small one is for the gods.
- In a 1951 survey of food preferences in the US armed services, rice pudding was the least-favored dessert.
- The U.S. Marine Corps in the 1950s issued the following advice to its military cooks: to make rice whiter and fluffier, add one teaspoon of lemon juice to each quart of water while cooking.
- Fifty percent of the world's rice is eaten within 13 km (8 miles) of where it was grown.
- More than a billion people are involved in the growing of rice.
- Snap! Crackle! Pop! is a registered trade mark of Rice Krispies, known for the curious sounds produced when milk is added. The puffed rice is filled with tiny air bubbles of different sizes, and when liquid is added, it is unevenly absorbed. This results in uneven swelling of the kernel's starch structure, producing a breaking action, which is accompanied by the well-known noises.

Key Information Sources

Bajaj, Y.P.S. (*Editor*). Rice. 1991. Biotechnology in agriculture & forestry, 14. Springer-Verlag, Berlin, Germany. 645 pp.

Chang, T.T. 1976*a*. The origin, evolution, cultivation, dissemination, and diversification of Asian and African rices. Euphytica, **25**: 425–441.

Chang, T.T. 1976*b*. Manual on genetic conservation of rice germplasm for evaluation and utilization. International Rice Research Institute, Los Baños, Philippines. 77 pp.

Chang, T.T. 1985. Crop history and genetic conservation: rice—a case study. Iowa State J. Res. **59**: 425–455.

Chang, T.T. 1995. Rice. *In* Evolution of crop plants. 2nd edition. *Edited by* J. Smartt and N.W. Simmonds. Longman Scientific & Technical, Burnt Mill, Harlow, Essex, UK. pp. 147–155.

Chang, T.T. 2000. Rice. *In* The Cambridge world history of food. *Edited by* K.F. Kiple and K.C. Ornelas. Cambridge University Press, Cambridge, UK. pp. 132–149.

Chang, T.T., and Harahap, Z. 1996. *Oryza* L. *In* Plant resources of South-East Asia, 10, cereals. *Edited by* C.J.H. Grubben and S. Partohardjono. Backhuys Publishers, Leiden, Netherlands. pp. 102–106.

De Datta, S.K. 1981. Principles and practices of rice production. John Wiley, New York, NY. 618 pp.

Duistermaat, H. 1987. A revision of *Oryza* (Gramineae) in Malaysia and Australia. Blumea, **32**: 157–193.

Food and Agriculture Organization of the United Nations. 1966. Rice: grain of life. FAO, Rome, Italy. 93 pp.

Food and Agriculture Organization of the United Nations. 1970*a*. Wet paddy or swamp rice. FAO, Rome, Italy. 40 pp.

Food and Agriculture Organization of the United Nations. 1970*b*. Upland rice. FAO, Rome, Italy. 30 pp.

Food and Agriculture Organization of the United Nations. 1974. Rice terminology. FAO, Rome, Italy. 82 pp.

Gariboldi. F. 1984. Rice parboiling. Food and Agriculture Organization of the United Nations, Rome, Italy. 73 pp.

Grist, D.H. 1986. Rice. 6th edition. Longmans, London, UK. 599 pp.

Hettiarachchy, N.S., Ju, Z.Y., Siebenmorgen, T., and Sharp, R.N. 2000. Rice: production, processing, and utilization. *In* Handbook of cereal science and technology. 2nd edition. *Edited by* K. Kulp and J.G. Ponte, Jr. Marcel Dekker, New York, NY. pp. 203–221.

Holder, S.H., and Grant, W.R. 1979. U.S. rice industry. U.S. Dept. of Agriculture, Economics, Statistics, and Cooperatives Service, Washington, DC. 141 pp.

International Rice Research Institute. 1970. Rice production manual. 3rd edition. International Rice Research Institute and College of Agriculture, University of the Philippines, Laguna, Philippines. 382 pp.

International Rice Research Institute. 1991. Rice germplasm: collecting, preservation and use. International Rice Research Institute, Los Baños, Philippines. 173 pp.

Juliano, B.O. (*Editor*). 1985. Rice: chemistry and technology. 2nd edition. American Association of Cereal Chemists, St. Paul, MN. 774 pp.

Juliano, B.O. 1993. Rice in human nutrition. Food and Agriculture Organization of the United Nations, Rome, Italy. 162 pp.

Khush, G.S., and Tonniessen, G.H. 1991. Rice biotechnology. C.A.B. International, Wallingford, Oxon, UK. 320 pp.

Kumar, T.T. 1988. History of rice in India: mythology, culture, and agriculture. Gian Publishing House, Delhi, India. 242 pp.

Luh, B.S. 1991. Rice. 2nd edition. Van Nostrand Reinhold, New York, NY. 2 vols.

Matsubayashi, M. (*Editor*). 1967. Theory and practice of growing rice. Fuji Publishing Co., Tokyo, Japan. 527 pp.

Mossman, A.P. 1983. Rice in the tropics. *In* Handbook of tropical foods. *Edited by* H.T. Chan, Jr. Marcel Dekker Inc., New York, NY. pp. 489–536.

Small, E. 2008. Rice: a valuable but fragile servant of humanity. Biodiversity, **9**(**1, 2**): 35, 130.

United States Agricultural Research Service. 1973. Rice in the United States: varieties and production. Revised edition. U.S. Department of Agriculture, Washington, DC. 154 pp.

Vaughan, D.A. 1994. The wild species of rice: a genetic resources handbook. International Rice Research Institute, Los Baños, Philippines. 137 pp.

Vaughan, D.A., and Chang, T.T. 1992. In-situ conservation of rice genetic resources. Econ. Bot. **46**: 368–383.

Vergara, B.S., and De Datta, S.K. 1996. *Oryza sativa* L. *In* Plant resources of South-East Asia, 10, cereals. *Edited by* C.J.H. Grubben and S. Partohardjono. Backhuys Publishers, Leiden, Netherlands. pp. 106–115.

Yoshida, S. 1981. Fundamentals of rice crop science. International Rice Research Institute, Los Baños, Philippines. 269 pp.

Specialty Cookbooks

(Note: There are thousands of books dedicated to the cooking of rice. The following are examples.)

Agranoff, R.B. 1997. Risottos, paellas, and other rice specialties. Bristol Publishing Enterprises, San Leandro, CA. 171 pp.

Alford, J., and Duguid, N. 1998. Seductions of rice: a cookbook. Artisan, New York, NY. 453 pp.

Casas, P. 1999. Paella! Spectacular rice dishes from Spain. Henry Holt, New York, NY. 226 pp.

Eve, P. 1973. Cooking with rice. Elm Tree Books, London, UK. 182 pp.

Fretz, S. 1995. Pilaf, risotto, and other ways with rice. Little, Brown, Boston, MA. 248 pp.

Gin, M., and Ng, W. 1975. Ricecraft: a gathering of rice cookery, culture & customs. Yerba Buena Press. San Francisco, CA. 113 pp.

Harris, V. 1998. Risotto! risotto! 80 recipes and all the know-how you need to make Italy's famous rice dish. Cassell, London, UK. 160 pp.

Hensperger, B., and Kaufmann, J. 2002. The ultimate rice cooker cookbook: 250 no-fail recipes for pilafs, risotto, polenta, chilis, soups, porridges, puddings, and more, from start to finish in your rice cooker. Harvard Common Press, Boston, MA. 368 pp.

Owen, S. 1993. The rice book. St. Martin's Press, New York, NY. 402 pp.

Robertson, R. 2000. Rice & spice: 100 vegetarian one-dish dinners made with the world's most versatile grain. Harvard Common Press, Boston, MA. 150 pp.

Scott, M.L., and Scott, J.D. 1985. Rice. Times Books, New York, NY. 362 pp.

Seed, D. 2000. The top one hundred Italian rice dishes: including over 50 risotto recipes. Ten Speed Press, Berkeley, CA. 137 pp.

Simmons, M. 2003. The amazing world of rice: with 150 recipes for pilafs, paellas, puddings, and more. William Morrow, New York, NY. 274 pp.

Teubner, C., Witzigmann, E., and Khoo, T. 1999. The rice bible. Viking Studio, New York, NY. 240 pp.

Uncle Ben's, Inc. 1977. Uncle Ben's rice cookery: how to prepare 231 delicious recipes with nourishing, economical rice: main dishes with meats, poultry, or seafood; side dishes; vegetable with rice; quick skillet dishes; desserts; traditional favorites for entertaining. H.P. Books, Tucson, AZ. 176 pp.

RYE AND TRITICALE

Family: Poaceae (Gramineae; grass family)

Names

Scientific names:

- Rye—*Secale cereale* L.
- Triticale—×*Triticosecale* Wittm. [*Triticum* spp. (wheat species) × *Secale cereale* L.]

- The English word "rye" traces to the Old English *ryge*, and seems similar to the Old High German *rocko* and Lithuanian *rugys*.
- The genus name *Secale* is based on the Latin name for some cereal, perhaps rye. One interpretation is that the name was originally based on *seco*, to cut, in reference to the cereal that was cut during the harvest.
- *Cereale* in the scientific name *S. cereale* is the Latin word for grain, which in turn is related to Ceres, the Roman goddess of vegetation.
- Triticale (from the generic names *TRITIcum* and *SeCALE*) and the similarly composed genus name ×*Triticosecale* are names constructed to reflect the parental genera of this hybrid group. Pronunciation: triht-ih-KAY-lee.
- Triticale has sometimes been called ryewheat (representing the parents).

Plant Portrait

Rye

Rye is an annual grass, although perennial varieties have been developed. The plant is generally 1–1.5 m (3–5 feet) tall, occasionally reaching 2.4 m (8 feet), and is slightly bushy at the base. Rye grains resemble those of wheat, but are longer and less plump, varying in color from yellowish brown to greenish gray. The species is believed to have originated in southwestern Asia, and was first cultivated rather late in human history in comparison to most major cereal grains, perhaps as recently as 2000–3000 years ago. The ancient Greeks and Romans consumed rye, but it was not highly regarded, and in most of Europe, except for Scandinavia and eastern Europe, rye never became very popular as human food. Like wheat, there are winter varieties (planted in the fall) and spring varieties (planted in the spring). Rye is now widely cultivated in the temperate regions of the world, with the majority of the world's supply produced in northeastern Europe. Compared to most of the other cereals, rye is more tolerant of poor soil and cold temperatures. Major producers are Russia, Germany, Czechoslovakia, Poland, Ukraine, Belarus, the United States, Canada, France, Hungary, Spain, the Netherlands, Argentina, China, and Turkey. Ninety percent of world production comes from Europe, more than half from Russia. Japan is the world's largest rye importer, accounting for about a third of world imports. Use of rye as human food has declined in recent decades, and most rye that is grown today is used as pasture, hay, or as a cover crop (to prevent soil erosion) or as a green manure crop (ploughed under to increase soil fertility).

Triticale

Triticale is an artificial cereal crop genus, the first created by crossing two different genera, wheat (*Triticum*) and rye (*Secale*). It combines the quality of wheat and the hardiness of rye, and is grown particularly where resistance to drought is important. Different cultivars of triticale have been developed, and the plant has become important in many countries. Triticale is mainly used as a feed grain for livestock, but the quantity used for human consumption is increasing.

Culinary Portrait

Rye

Like wheat and barley, the hull is removed from the rye grain before human consumption, and the grains are employed whole or cracked, or are ground into flakes or flour. Rye flakes can be used like rolled oats. As with the other cereal grains, the more of the grain's germ and bran that are retained in the flour, the greater is the nutritional value, although at the cost of taste and digestibility. The whiter the refined flour, the more it has been stripped of nutrients. One of the nutritional benefits of rye is the presence of fibrous complex carbohydrates called pentosans that may reduce certain types of cancer and heart disease.

Rye is second only to wheat for flour production, despite the production of wheat exceeding that of rye by at least a factor of 10. Rye flour does not develop true gluten, but it has proteins that give it the capacity for making a nutritious leavened bread. Rye gluten is less elastic and retains less gas than wheat, so the resulting bread rises less and is denser and more compact than that of wheat. Rye bread keeps longer and does not dry out as quickly as wheat bread. Relatively moist breads, such as rye bread, also freeze better than drier, white breads. Rye bread is usually darker than wheat bread, and is often called "black bread," especially the kind called pumpernickel. Pumpernickel is a dark, coarse, moist, slow-baked unleavened bread, said to have originated in the 15th century in Westphalia, Germany. The firm texture of pumpernickel makes it ideal for canapés and open sandwiches, and the aromatic, acid aroma seems to go well with a hearty cheese or sweet butter. Store-bought pumpernickel in North America varies in texture and quality, and the best is usually found in German and Jewish neighborhood bakeries. Authentic pumpernickel is made with coarsely ground, or at least minimally refined rye flour, and the taste is slightly sourish. Baked goods made with rye flour have a distinctive strong flavor, which is often not appreciated by those brought up on a diet of white bread. Rye is usually mixed with 25–50% wheat flour for breadmaking, and most so-called rye bread sold in North America is a mixture of rye and wheat flour. Genuine pumpernickel is quite dark in color. However, domestic pumpernickel often has so much wheat flour that it is not much darker than ordinary rye bread. To make such breads appear to be genuine pumpernickel, caramel coloring is often added to achieve a really dark color.

In addition to bread, rye is also used to make rusks, muffins, and other baked goods. Rye can also be simply cooked and eaten like the other cereal grains, particularly rice. Whole grains can be soaked overnight in 2–3 cups of water per cup of grain, then boiled until tender. Rye flakes can be cooked to produce hot breakfast cereals, or added to cereal mixes.

Substantial quantities of rye are made into distilled alcoholic beverages, particularly whisky and gin, but also beer and vodka. Rye was the predominant type of whisky first made in America. Many of the early immigrants who made their homes in Maryland and Pennsylvania were German and Scots-Irish farmers who used their surplus grain to make distilled spirits, but were unfamiliar with the use of corn for the purpose. George Washington (1732–1799), first president of the United States, was a notable rye whisky producer of his time.

Triticale

In many modern varieties of triticale, the nutritional qualities of the grain are more or less comparable to those of wheat, although often triticale cultivars have superior protein (amino acid) balance. Triticale meals and flours are currently marketed (although not nearly to the same extent as wheat flour), and are used commercially to make specialty bread, biscuits, cookies, shortbreads, and confectionery items. Triticale is available in some health-food stores and supermarkets as "whole berry" (whole kernel), flakes, and flour. Home cooks use it in a variety of cooked dishes, including cereals, casseroles, and pilafs. Because it is low in gluten, bread made from triticale alone is quite heavy. Accordingly, triticale flour is blended with wheat flour (often half and half) or with other cereals to achieve desired market characteristics. There is a need for breeders to increase the palatability of triticale, which in general is not yet comparable to wheat.

Culinary Vocabulary

- It was once frequent practice to sow rye and wheat together in a field, so that the grains were harvested and mixed together to produce a cereal mixture called "maslin" (variously spelled). The word "maslin" (thought to trace back to the Latin *mixtus*, mixed) refers to any grain mixture, but normally means a combination (the plants, grain,

Rye (*Secale cereale*), from Baillon (1876–1892, vol. 4).

flour, or meal) of rye and wheat. Maslin, which was the most common flour in Europe from the 1300s to the 1600s, also refers to the bread made from such flour.

- The rye-based bread "pumpernickel" is most commonly interpreted as based on the German *pumpern*, to fart, + *Nickel*, devil, demon, or goblin, based on the amusing idea that this dark bread has the power to produce demonic outbursts of flatulence. At least two other interpretations have been advanced: that the word combines "pumper," the sound of a person falling, and *Nickel*, as explained above; and that the word is a corruption of the French *pain pour Nicole*, bread for Nicole, tracing to the fondness of Napoleon Bonaparte's horse Nicole for black bread.
- In England, pumpernickel is sometimes called "brown George."
- Schwartzbröt, the German word for "black bread," is a very dark, Russian-style, black rye bread. It has been described as the blackest of the black breads. In Northern and Central Europe, the word *schwartzbröt* is used as a generic term for "very black pumpernickel." The flavor is often sweet because of the addition of molasses.

Rye (*Secale cereale*), infected with ergot (note inflated diseased structures replacing some of the much smaller grains), from Bulliard (1784)

- "Boston brown bread" is a bread made with molasses and rye flour (or sometimes Graham flour, i.e., coarsely ground flour made from whole wheat). Still available today, it was once served by the Puritans on the Sabbath along with Boston baked beans.
- "Rye" is a popular term for ordinary blended whisky in the northeastern United States, despite the absence or limited content of genuine rye whisky. "Blended whisky," which is lighter in body and flavor than straight whisky, usually has no more than 20% straight whisky (of any grain), the rest made up of neutral spirits. The use of the term "rye" as a generic term for "blended whisky" in the northeastern United States may have originated from historical times, when rye was indeed the chief grain used to manufacture whisky. "Straight rye whisky" or "straight rye" in the United States must be distilled from a mash consisting of at least 51% rye grain, while Bourbon whisky must be made with at least 51% corn. The majority of Bourbons also use rye. Other requirements for the legal use of the term "straight rye" include distillation at a proof no higher than 160, maturation for at least 2 years in charred, white oak barrels, and dilution at bottling time to a proof no lower than 80. By comparison with Irish or Scotch whisky, genuine rye whisky is sweeter and fuller bodied. ["Whisky" and whiskey" are alternative spellings; however, "Scotch whisky," which is distilled from barley, is traditionally spelled without the "e" (i.e., not "whiskey"). "Canadian whisky" is also traditionally spelled the same way, while "whiskey" is traditional in "Irish whiskey" and in America.]
- "Rock and rye" is a commercial liqueur prepared from rye whisky and rock-candy syrup, with the addition of neutral spirits and fruit flavoring, usually lemon, orange, or cherry. The beverage can be prepared at home by combining rock-candy syrup, rye whisky, and a slice of lemon. (Rock candy, a hard candy, is clarified, concentrated sugar syrup that has been crystallized around a small wooden stick or a piece of string.)
- "Whisky down" is old lunch-counter jargon for rye toast.

Curiosities of Science and Technology

- Ergot is a fungus disease of the fruiting head of rye, which makes the grain poisonous. Such grains must be removed before rye grain can be safely consumed. Ergot was responsible for several epidemics in medieval times, including the widespread disease called St. Anthony's Fire. In 944 AD in southern France, 40 000 people died of ergotism. Symptoms often include gangrene, and people would lose parts of their extremities, such as toes, fingers, ear lobes, or in more serious cases, arms and legs. The tails, ears, and hooves of farm animals that grazed on affected plants were also often lost. Hallucinations were common in humans, not surprising, since ergot is a natural source of the psychedelic drug LSD, and indeed is the original source from which LSD was first isolated. The disease was called "Fire" because it caused burning sensations at the extremities from gangrenous infection, and was named after St. Anthony because, beginning in 1039, the hospitals set up to treat the disease in Europe were dedicated to Saint Anthony. Outbreaks occurred from time to time until 1816. Today, pharmaceutical companies deliberately infect fields of rye with the ergot fungus in order to harvest it for pharmaceutical products, notably Ergotamine, which is taken for various causes of headaches, including migraines, and Ergonovine, used in obstetrics, to control postpartum hemorrhage.

- A study (published in *Science*, April 1976) suggested that the Salem, Massachusetts, witch trials in 1692 resulted from hallucinations of important community members who were exposed to contaminated rye. Many of the women who believed they were bewitched lived among or near fields of rye. Nineteen witches were hanged, and one was pressed to death in Salem.
- An old European custom was to leave some sheaves of rye in rye fields at harvest time, in the belief that the devils who were resident in the fields would quarrel over them and leave the farmers alone. In Germany, a "rye wolf" (*Roggenwolf*) was said to inhabit rye fields, and to take up residence in the last sheaf harvested. Sometimes, the last sheaf was molded into the shape of a wolf, and in some areas, it was tradition for the man harvesting the last sheaf to howl and pretend to bite other harvesters, like a wolf. The presence of threatening spirits in rye has been attributed to the association of the crop with ergotism—the awful effects seeming to provide proof of evil. The purple–brown ergot-infected ears of rye were often referred to as rye-wolves, and interpreted as the children of the rye devils inhabiting the plants or fields.
- According to the *Guinness World Records 2001*, a single rye plant was shown to have produced 623 km (387 miles) of roots in 51 L (1.8 cubic feet) of earth.
- In 2002, Thomas Hofmann and colleagues presented evidence in the *Journal of Agricultural and Food Chemistry* that bread crust is the healthiest part of bread, because the crust is a rich source of cancer-fighting antioxidant chemicals. Generally, dark-colored breads, such as pumpernickel made from rye, and whole-wheat bread, provided more of these antioxidants than light-colored breads.

Key Information Sources

Rye

Anonymous (*Editor*). 1986. Proceedings, EUCARPIA meeting of the cereal section on rye: breeding for quality, breeding for resistance, hybrid breeding, population breeding, genetics and biotechnology (June 11–13th 1985, Svalöv, Sweden). The Cereal Dept., Svalöv AB, Svalöv, Sweden. 2 vols.

Anonymous. (*Editor*). 1996. International symposium on rye breeding & genetics, University of Hohenheim, Stuttgart, Germany, 27–29 June 1996. Erhältlich durch Saatgut-Treuhandverwaltung, Bonn, Germany. 321 pp.

Asgarali, A. 2006. Rye: situation and outlook. Bi-weekly Bull. (Agric. Agri-Food Can.), **19(8)**: 1–4.

Behre, K.-E. 1992. The history of rye cultivation in Europe. Veg. Hist. Archaeobot. **1**: 141–156.

Briggle, L.W. 1959. Growing rye. U.S. Dept. of Agriculture, Washington, DC. 16 pp.

Bushuk, W. 1976. Rye: production, chemistry and technology. American Association of Cereal Chemists Inc., St. Paul, MN. 181 pp.

Darwinkel, A. 1996. *Secale cereale*. L. *In* Plant resources of South-East Asia, 10, cereals. *Edited by* C.J.H. Grubben and S. Partohardjono. Backhuys Publishers, Leiden, Netherlands. pp. 123–127.

Deodikar, G.B. 1963. Rye: *Secale cereale* Linn. Indian Council of Agricultural Research, New Delhi, India. 152 pp.

Evans, G.M. 1995. Rye. *In* Evolution of crop plants. 2nd edition. *Edited by* J. Smartt and N.W. Simmonds. Longman Scientific & Technical, Burnt Mill, Harlow, Essex, UK. pp. 166–170.

Helbaeck, H. 1971. The origin and migration of rye, *Secale cereale* L.; a palaeo-ethnobotanical study. *In* Plant life of South-West Asia. *Edited by* P.H. Davis, P.C. Harper, and I.G. Hedge. Royal Botanic Garden, Edinburgh, UK. pp. 265–280.

Küster, H. 2000. Rye. *In* The Cambridge world history of food. *Edited by* K.F. Kiple and K.C. Ornelas. Cambridge University Press, Cambridge, UK. pp. 149–152.

Langille, J.E., and MacLeod, J.A. 1976. Growing fall rye for grain in the Atlantic provinces. Agriculture Canada, Ottawa, ON. 12 pp.

Lorenz, K. 2000. Rye. *In* Handbook of cereal science and technology. 2nd edition. *Edited by* K. Kulp and J.G. Ponte, Jr. Marcel Dekker, New York, NY. pp. 223–256.

McLelland, M.B. 1999. Fall rye production. Revised edition. Alberta Agriculture, Food and Rural Development, Edmonton, AB. 13 pp.

McLeod, G. 1992. Fall rye reference manual. Agriculture Canada and Prairie Pools Inc., Swift Current, SK. 46 pp.

Murray, J. 1998. Classic bourbon, Tennessee & rye whiskey. Prion Books, London, UK. 272 pp.

Nuttonson, M.Y. 1958. Rye-climate relationships and the use of phenology in ascertaining the thermal and photo-thermal requirements of rye. American Institute of Crop Ecology, Washington, DC. 219 pp.

Poutanen, K., and Autio, K. (*Editors*). 1995. International rye symposium: technology and products. (Helsinki, Finland, 7–8 December 1995). Valtion teknillinen tutkimuskeskus, Espoo, Finland. 221 pp.

Stutz, H.C. 1972. The origin of cultivated rye. Am. J. Bot. **59**: 59–70.

Triticale

Brown, W.L. 1989. Triticale: a promising addition to the world's cereal grains: report. National Academy Press, Washington, DC. 105 pp.

Bushuk, W., and Larter, E.N. 1980. Triticale: production, chemistry, and technology. *In* Advances in cereal science and technology, vol. 3. *Edited by* Y. Pomeranz. American Association of Cereal Chemists, St. Paul, MN. pp. 115–157.

Darvey, N.L., and Naeem, H. 2000. Triticale: production and utilization. *In* Handbook of cereal science and technology. 2nd edition. *Edited by* K. Kulp and J.G. Ponte, Jr. Marcel Dekker, New York, NY. pp. 257–274.

Forsberg, R.A. 1985. Triticale: proceedings of a symposium. Crop Science Society of America and American Society of Agronomy, Madison, WI. 82 pp.

Guedes-Pinto, H., Darvey, N.L., and Carnide, V.P. (*Editors*). 1996. Proceedings 3rd international triticale symposium, triticale: today and tomorrow, June 1994, Lisbon, Portugal. Kluwer Academic Publishers, London, UK. 898 pp.

Gupta, P.K., and Priyadarshan, P.M. 1982. Triticale: present status and future prospects. Adv. Genet. **21**: 255–345.

Hörlein, A.J., and Valentine, J. 1995. Triticale (× *Triticosecale*). *In* Cereals and pseudocereals. *Edited by* J.T. Williams. Chapman & Hall, London, UK. pp. 187–221.

Hulse, J.H., and Laing, E.M. 1974. Nutritive value of triticale protein (and the proteins of wheat and rye). International Development Research Centre, Ottawa, ON. 183 pp.

Larter, E.N. 1995. Triticale. *In* Evolution of crop plants. 2nd edition. *Edited by* J. Smartt and N.W. Simmonds. Longman Scientific & Technical, Burnt Mill, Harlow, Essex, UK. pp. 181–183.

Pena, R.J., and Amaya, A. 1992. Milling and breadmaking properties of wheat-triticale grain blends. J. Sci. Food Agric. **60**: 483–487.

Tsen, C.C. (*Editor*). 1974. Triticale: first man-made cereal. American Association of Cereal Chemists, St. Paul, MN. 291 pp.

Varughese, G., Pfeiffer, W.H., and Pena, R.J. 1996*a*. Triticale: a successful alternative crop (part 1). Cereal Foods World, **41**: 474–482.

Varughese, G., Pfeiffer, W.H., and Pena, R.J. 1996*b*. Triticale: a successful alternative crop (part 2). Cereal Foods World, **41**: 636–645.

Specialty Cookbooks

(Additional references are in AMARANTH and BARLEY.)

Rye

Collister, L. 2001. Bread: from sourdough to rye. Ryland Peters & Small, New York, NY. 144 pp.

General Mills. 1948. The Gold Medal rye dictionary. General Mills, Inc. Minneapolis, MI. 99 pp.

Greenstein, G. 1993. Secrets of a Jewish baker: authentic Jewish rye and other breads. Crossing Press, Freedom, CA. 368 pp.

Gross, S.A. 1974. Sourdough rye and other good breads. 2nd edition. Kitchen Harvest Press, St. Charles, IL. 15 pp.

Peavey Company. 1900–1999? The high altitude Hungarian whole grain bake book: whole wheat and rye flour recipes for breads, rolls, cakes, cookies, pastries and desserts. Peavey Company, Minneapolis, MI. 60 pp.

Triticale

Cooper, K.V. 1985. The Australian triticale cookery book. Savvas Publishing, Adelaide, Australia. 79 pp.

Rosen, J., and Deethardt, D. 1977. Bounty from the bin: a collection of recipes introducing new uses for whole wheat and triticale cereal grains. Agricultural Experiment Station, South Dakota State University, Brookings, SD. 12 pp.

TritiRich Products Limited. 1979. Triticale cook book. TritiRich Products Limited, Winnipeg, MB. 139 pp.

TritiRich Products Limited. 1982. Triticale new harvest recipes: a collection of recipes from our triticale kitchens. TritiRich Products Limited, Winnipeg, MB. 163 pp.

SAFFLOWER

Family: Asteraceae (Compositae; sunflower family)

Names

Scientific name: *Carthamus tinctorius* L.

- The name "safflower" is derived through the words of various languages for the saffron (*Crocus sativus* L.), the source of a very expensive yellowish spice obtained from the flowers.
- Safflower is sometimes called bastard saffron, false saffron, and saffron thistle, because the floral parts resemble those of saffron in color, and have often been used to adulterate this very much more expensive spice. In Latin American shops, safflower flowers are sold as "Mexican saffron" (*azafrán*).
- The genus name *Carthamus* is the Latinized form of the Hebrew word *kartami*, which means to dye, a word borrowed from the Arabic *quartum* or *gurtum*, which also refers to dye.
- *Tinctorius* in the scientific name is related to the Latin *tinctor*, dyer, a person who dyes.

Plant Portrait

The safflower is an annual, thistle-like herb, growing 0.4–2 m (1¼–6½ feet) in height. The species is thought to have originated in southern Asia, and has been cultivated in China, India, Persia, and Egypt for thousands of years, and more recently in Europe and elsewhere. Over half of the world's safflower is produced in India, and most of the remainder comes from the United States, Mexico, Ethiopia, Argentina, and Australia. China produces much of its safflower for use in traditional medicines. In North America, most safflower is grown in several western states and Canadian prairie provinces. California produces about half of the safflower in the United States, while North Dakota and Montana grow most of the remaining American crop. The flowers are usually yellow or orange, although some varieties have red or white flowers. The flowers have widely been used as a source of red and yellow dyes for coloring textiles, but the dye use of saffron has largely been replaced by synthetic aniline dyes. Rarely, safflower dye is still sometimes used in rouge. Safflower oil does not yellow with age, and so is used in preparing industrial products such as varnish, paint, and linoleum. The type of safflower oil most useful for this purpose is high in the fatty acid linoleic acid, not the type widely used in salad oils, which is high in the fatty acid oleic acid. Safflower is now cultivated mainly for the edible oil in the seed.

Culinary Portrait

Safflower oil has a very high percentage of unsaturated fatty acids and a very low percentage of saturated fatty acids, making it a particularly high-quality, edible, vegetable seed oil, considered extremely desirable in the diet. Safflower oil lowers blood cholesterol levels and is used to treat heart diseases. The oil is light in color, and is used in salad and cooking oils and margarines. Safflower is commonly used as a substitute for saffron, at least to mimic the latter's color. However, in India and Afghanistan, "saffron rice" is commonly made with safflower, which gives it an orange color rather than the gold color of saffron. The seeds can be eaten roasted or fried and used in chutney.

Curiosities of Science and Technology

- Safflower has been collected from 4000-year-old Egyptian tombs. It was used to dye the bandages of Egyptian mummies.
- Safflower was imported from China to Japan in the 5th century and has been cultivated there ever since. In the late 8th century, a red dye produced from the safflower plant became a color forbidden to most people. Only women of high rank and those favored by the Empress could wear it.
- The safflower is the official flower of Yamagata prefecture (Japan has 47 administrative units called prefectures, roughly equivalent to "counties").
- The taproot of safflower can penetrate the soil to depths of 2.4–3 m (8–10 feet), and as a result, safflower is more tolerant of drought than many other crops.
- Safflower is often used in birdfeed. The relatively high oil content makes the seeds a high-energy food source ideally suited for larger birds. Safflower is said to be the food of choice when attracting cardinals. Striped sunflower seeds, rather than unstriped seeds, are widely preferred as a confection, but striped safflower seeds are also available, which are higher in oil and protein content than

Safflower (*Carthamus tinctorius*), from Hallier (1880–1888, vol. 30, plate 3154).

unstriped safflower seeds. Curiously, the birdfood industry prefers safflower seeds that are white or unstriped, even though the striped types are more nutritious.

- As of 2009, SemBioSys Genetics Inc., a biotechnology company based in Calgary, Alberta, continued to pursue development of commercial production of insulin from genetically modified safflower. Such "bio-farming" (or "bio-pharming") pharmacological ventures utilize plants with genes inserted into their DNA so that they produce desired proteins for therapeutic use. To achieve commercial success, it is necessary that the cost of using plants to produce insulin be less than current methods (utilization of bacteria, yeast, or extraction from animals). Because of concern that crossbreeding with cultivated safflower used for food production could occur, thus spreading the insulin-producing gene, it is also necessary to isolate the modified plants.

Key Information Sources

Ashri, A. (*Editor*). 1985. Sesame and safflower: status and potentials; proceedings of expert consultation. Food and Agriculture Organization of the United Nations, Rome, Italy. 223 pp.

Ashri, A., Knowles, P.F., Urie, A.L., Zimmer, D.E., Cahaner, A., and Marani, A. 1977. Evaluation of the germ plasm collection of safflower, *Carthamus tinctorius*, 3: oil content and iodine value and their associations with other characters. Econ. Bot. **31**: 38–46.

Baltensperger, D.D., Frickel, G., Lyon, D., Krall, J., and Nightingale, T. 2002. Safflower management and adaptation for the high plains. *In* Trends in new crops and new uses. *Edited by* J. Janick and A. Whipkey. ASHS Press, Alexandria, VA. pp. 183–186.

Bassiri, A. 1977. Identification and polymorphism of cultivars and wild ecotypes of safflower based on isozyme patterns. Euphytica, **26**: 709–719.

Beech, D.F. 1969. Safflower. Field Crop Abstr. **22**: 107–109.

Beech, D.F. 1979. Safflower cultivation, breeding, diseases. Aust. Field Crops **2**: 161–182.

Chavan, V.M. 1961. Niger and safflower. Indian Central Oilseeds Committee, Hyderabad, India. 150 pp.

Dajue, L., and Mündel, H. 1996. Safflower. *Carthamus tinctorius* L. Promoting the conservation and use of underutilized and neglected crops. 7. Institute of Plant Genetics and Crop Plant Research, Gatersleben and International Plant Genetic Resources Institute, Rome, Italy. 83 pp.

Hanelt, P. 1963. Monographische Ubersicht der Gattung *Carthamus* (Compositae). Feddes Repert. **67**: 41–180. [In German.]

Herdrich, N. 2001. Safflower production tips. Washington State University Cooperative Extension, Pullman, WA. 16 pp.

Johnson, A., and Marter, A. 1993. Safflower products: utilization and markets. Natural Resources Institute, Chatham Maritime, UK. 26 pp.

Johnson, R.C., Bergman, J.W., and Flynn, C.R. 1999. Oil and meal characteristics of core and non-core safflower accessions from the USDA collection. Genet. Resour. Crop Evol. **46**: 611–618.

Knowles, P.F. 1955. Safflower—production, processing and utilization. Econ. Bot. **9**: 273–299.

Knowles, P.F. 1958. Safflower. Adv. Agron. **10**: 289–323.

Knowles, P.F. 1969*a*. Modification of quantity and quality of safflower oil through plant breeding. J. Am. Oil. Chem. Soc. **46**: 130–132.

Knowles, P.F. 1969*b*. Centers of plant diversity and conservation of crop germplasm: safflower. Econ. Bot. **23**: 324–329.

Knowles, P.F. 1989. Safflower. *In* Oil crops of the world. *Edited by* G. Robbelen, R.K. Downey, and A. Ashri. McGraw-Hill, Toronto, ON. pp. 363–374.

Knowles, P.F., and Ashri, A. 1995. Safflower. *In* Evolution of crop plants. 2nd edition. *Edited by* J. Smartt and N.W. Simmonds. Longman Scientific & Technical, Burnt Mill, Harlow, Essex, UK. pp. 47–50.

Knowles, P.F., and Miller, M.D. 1965. Safflower. Division of Agricultural Sciences, University of California, Berkeley, CA. 51 pp.

Kumar, H., Tsuchiya, T., and Gupta, P.K. 1991. Cytogenetics of safflower. *In* Chromosome engineering in plants: genetics, breeding, evolution. Part B. Amsterdam (Netherlands). *Edited by* P.K. Gupta and T. Tsuchiya. Elsevier Science Publishers, Amsterdam, Netherlands. pp. 251–277.

Kupsow, A.J. 1932. The geographical variability of the species *Carthamus tinctorius*. Bull. Appl. Bot. Genet. Plant Breed. **9**: 99–181.

Larson, N.G. 1962. Safflower, 1900–1960: a list of selected references. National Agricultural Library, U.S. Dept. of Agriculture, Washington, DC. 31 pp.

Li, D., and Han, Y. 1994. (*Editors*). Proceedings, third international safflower conference, Beijing, China, June 14–18, 1993. Beijing Botanical Garden, Institute of Botany, Chinese Academy of Sciences, Beijing, China. 906 pp.

Madan-Kumar, S.S. 1977. Annotated bibliography on safflower (*Carthamus tinctorius* L.). University of Agricultural Sciences, University Library, Bangalore, India. 146 pp.

Mündel, H.H. 1992. Safflower production on the Canadian Prairies. Research Station, Agriculture Canada, Lethbridge, AB. 29 pp.

Oyen, L.P.A., and Umali, B.E. 2001. *Carthamus tinctorius* L. *In* Plant resources of South-East Asia, 14, vegetable oils and fats. *Edited by* H.A.M. van der Vossen and B.E. Umali. Backhuys Publishers, Leiden, Netherlands. pp. 70–76.

Pavlov, D.C., and Todorov, N.A. 1996. Safflower (*Carthamus tinctorius* L.). *In* Food and feed from legumes and oilseeds. *Edited by* J. Smartt and E. Nwokolo. Chapman and Hall, London, UK. pp. 245–257.

Peterson, W.F. 1965. Safflower culture in the West-Central Plains. Agricultural Research Service, U.S. Dept. of Agriculture, Washington, DC. 22 pp.

Ranga-Rao, V., and Ramachandram, M. 1991. Proceedings 2nd international safflower conference, Hyderabad, India, Jan. 9–13, 1989. Indian Society of Oilseeds Research, Indian Council of Agricultural Research, Directorate of Oilseeds Research, Hyderabad, India. 419 pp.

Sakakibara, J., and Okuyama, H. 1997. Antioxidative compounds isolated from safflower (*Carthamus tinctorius* L.) oil cake. Chem. Pharm. Bull. **45**: 1910–1914.

Smith, J.R. 1996. Safflower. AOCS Press, Champaign, IL. 606 pp.

Stern, W.R., and Beech, D.F. 1965. The growth of safflower (*C. tinctorius* L.) in a low level environment. Aust. J. Agric. Res. **16**: 801–816.

Ta-Chüeh, L., and Mündel, H.H. 1996. Safflower, *Carthamus tinctorius* L. International Plant Genetic Resources Institute, Rome, Italy. 83 pp.

U.S. Dept. of Agriculture. 1966. Growing safflower, an oilseed crop. Revised edition. U.S. Dept. of Agriculture, Washington, DC. 13 pp.

Vema, M., Shukla, Y.N., Ram, M., Jain, S.P., and Kumar, S. 1997. Chemistry and biology of the oil and dye crop *Carthamus tinctorius*: a review. J. Med. Aromatic Plant Sci. **19**: 734–744.

Weiss, E.A. 1971. Castor, sesame and safflower. Barnes & Noble, New York, NY. 901 pp.

Specialty Cookbook

Betty Crocker and General Mills Inc. 1966. Golden good cooking with Betty Crocker's Saff-o-life safflower oil: recipes for special diets using oil. General Mills Inc., Minneapolis, MN. 21 pp.

SEAWEEDS

Names

Red Algal Group (Rhodophyta)

About 200 species of red algae are harvested, and of these, about 90 are used directly as food, especially the following.

- Nori or laver—*Porphyra* species, especially *P. yezoensis* Ueda
- Irish moss—*Chondrus crispus* Stackhouse
- Dulse—*Palmaria palmata* (L.) O.Kuntze [*Rhodymenia palmata* (L.) Greville]

- The name "laver" has been used for several edible seaweeds of the genera *Porphyra* (of the red algae) and *Ulva* (of the green algae). Below, purple laver of the red algae is mentioned. "Nori" is the term used in Japan (and now elsewhere), while laver is a term of English origin.
- Irish moss is found on North Atlantic shorelines, particularly on those of Ireland, which is responsible for the "Irish" in its common name. It is a seaweed, i.e., an alga, not a "moss."
- Irish moss is also known as carrageen (or carageen) moss, a name that is popularly interpreted as based on a place in Ireland that carries its namesake, Carrageen (Carragheen) Head in County Donegal. The town has been claimed to have been an important shipping point for Irish moss. However, there apparently is no such place (Mitchell, M.E., and Guiry, M.D. 1983. Carageen: a local habitation or a name? J. Ethnopharmacol. **9**: 347–351.) Carrageen moss is known as carraig'n in Gaelic. The name "carageen moss" is also applied to *Gigartina stellata* (Stackh.) Batters, which is used similarly.
- Other names for Irish moss are Iberian moss, pearl moss, and sea moss.
- The name "dulse" may have come from Ireland, and likely is based on the Gaelic word *duileasq*.
- In the 19th century, the term "dulse" was applied to algae that are now known as separate genera, *Dilsea* and *Iridaea*.
- Dulse is also known as dillisk, and is often marketed as "sea parsley."

Brown Algal Group (Phaeophyta)

About 70 species of brown algae are harvested of which the most important are the "kelp" species.

- Kelp—numerous edible forms belonging to several families of the order Laminariales (orders are groups of families). *Laminaria* and *Undaria* are the two most economically important edible kelp genera. The most important Asian species of *Laminaria* is *L. japonica* Aresch., and other extensively cultivated species include *A. angustata* Kjell., *L. ochotensis* Miyabe, and *L. religiosa* Miyabe. The Asian *Laminaria* species that are consumed in Asia do not occur outside of Asia. In North America, other species of *Laminaria* are substituted, particularly *L. saccharina* (L.) Lamouroux, *L. bongardiana* Postels et Ruprecht (*L. groenlandica* Rosenvinge), and *L. pyrifera* (L.) Lamouroux, but the volume harvested is not comparable to the Asian species. The most important edible *Undaria* is *U. pinnatifida* (Harv.) Sur. The edible *Undaria* species harvested in Asia are also restricted to Asian waters. Young *Alaria* plants have been substituted in North America, and *A. esculenta* (L.) Grev. has been experimentally cultivated in Great Britain.

- The name "kelp" arose when seaweeds in 17th century France, particularly brown algae, and especially species of *Laminaria*, were burned to obtain sodium (soda) and potassium (potash). The ash (soda ash), which was used in the production of soap, glass, and potash, was called kelp (a word of unknown origin), and the name was gradually transferred to the brown algae.
- Kelp is used both as singular and plural. "Kelps" is also encountered as the plural of kelp.
- Kelp are eaten primarily in Asia. Edible *Laminaria* species are called kombu in Japan and *haidai* in China. Edible *Undaria* species are called wakame in Japan and *qundai-cai* in China. The name "wakame" usually refers to *Undaria* but (especially in North America) is also used for *Alaria*.
- In Japan, culinary puns based on food names are given considerable significance. *Kombu* refers not

only to kelp but also is close to the word *yorokobu*, "happiness," and accordingly, *kombu* is often served during celebratory or ceremonious meals. *Kombu* also means a pregnant woman, and so kelp is often present at weddings, symbolizing fertility.

Green Algal Group (Chlorophyta)

- Sea lettuce—*Ulva lactuca* L.

- The name "sea lettuce" is descriptive of the appearance of these species, which are reminiscent of leaf lettuce (not head-type lettuce).

Plant Portrait

General Information

Seaweeds are not really "weeds." They are "macroalgae" (literally large algae), a heterogenous assemblage currently placed in three groups of species, called the red, brown, and green algae, although it is now known that this basis of division is not satisfactory and rearrangements will be necessary after additional studies. Estimates of the number of species worldwide vary, from 5000 or 6000 species to twice this number. Only the most important commercial seaweeds used for food are discussed here, but numerous others are also harvested and eaten. In fact, while some species are much more edible than most, there does not seem to be a poisonous seaweed (although there are poisonous algae). Most seaweeds, true to their name, live in the sea (or at least in brackish water). Red and brown algae are almost exclusively marine, while green algae are also common in freshwater and in terrestrial situations. In the sea, the green seaweeds tend to inhabit the shallowest zones along the shore, the browns are usually found in deeper waters, and the reds tend to be in the deepest waters. Some of the green algae are the ancestors of the flowering plants, but are of limited commercial value. Just like flowering plants, many seaweeds are highly edible, and indeed are sometimes collectively called "sea vegetables." They are harvested from the wild for food and other products, but today, most edible seaweed is cultivated. The earliest archaeological evidence for the consumption of algae dates to 2500 BC, along the coast of Peru. In Asia, edible seaweeds were consumed for food beginning at least 2500 years ago in China. Europeans have significantly collected seaweeds for food only for about 500 years. Some species have been commercially cultured for over 300 years. Seaweeds are eaten directly in large quantities only in the Far East. However, most people in Western countries ingest red or brown algae products in such everyday foods as chocolate milk, candy, ice cream, and salad dressing.

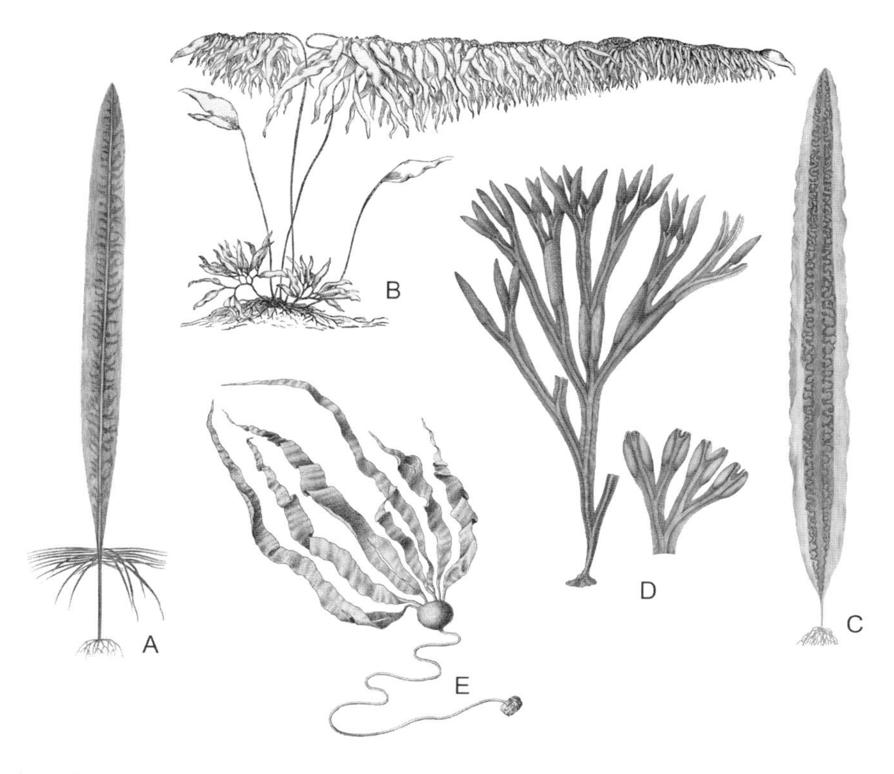

Brown seaweeds. A, "wakame," "edible kelp" (*Alaria esculentus*), from Oeder et al. (1761–1883, plate 417). B, giant kelp (*Macrocystis pyrifera*), from Engler and Prantl (1897, vol. 1(2)). C, sugar kelp (*Laminaria saccharina*), from Oeder et al. (1761–1883, plate 416). D, bladderwrack (*Fucus vesiculosus*), from Oeder et al. (1761–1883, plate 1127). E, bull kelp (*Nereocystis lutkeana*), from Engler and Prantl (1897, vol. 1(2)).

Uncontrolled harvesting has led to reductions of wild supplies of some edible seaweeds, but seaweed farms are well established in parts of the world, reducing the pressure on natural stands. "Polyculture" (combining fish farming and seaweed culture) is a clever way of using seaweeds to metabolize by-products of the fish culture. China is a major seaweed producer, growing over 2.5 million tonnes (2.8 million tons) of *Laminaria japonica* annually. Japan's primary aquaculture seaweed is nori, with an estimated value of $1.5 billion annually. By contrast, the edible seaweed market in the United States is currently valued at only about $30 million annually. North America's development of a cultivated seaweed industry lags far behind that of several countries with low labor costs and a warm climate that allows year-round cultivation. There has been some cultivation of *Laminaria saccharina* and *L. bongardiana* (*L. groenlandica*) on the Pacific coast, for the Asian and health food markets.

The majority of cultivated seaweed is not used directly for food, but rather for extraction of substances—principally agar-agar, algins, and carregeenins, which are used in food and other products. Agar is a major constituent of the cell wall of some red algae. It is used extensively in microbiology as a culture medium. In modern molecular biology and genetic engineering, agar gels are used extensively in electrophoresis and chromatography, two of the fundamental analytical techniques. Carrageenin (carragheenin) is similar to agar, and also comes from red algae (it was once obtained mostly from Irish moss). Carageenins are used in stabilizing and gelling foods such as ice cream and puddings, and also in cosmetics, pharmaceuticals, and industrial products such as paints and inks. Algins are a major constituent of all brown algae. Most commercial algin come from the giant kelp, *Macrocystis* and *Nerocystis*. Algin from kelp and other algae is used in the manufacture of more than 300 commercial products, and is particularly valued for its ability to suspend agents in food, cosmetics, and a variety of commercial liquid mixtures. Illustrating food applications, algin stops ice crystal formation in ice cream, acts as an emulsifier in salad dressings, sherbets, and cheeses, clarifies and stabilizes the foam of beer, is a filler in candy bars and a thickener in gravies and puddings. Nonedible products using alginates include plastics, paints, adhesives, and rubber tires. Algin can be used as a fiber in fire-resistant clothes and in audio speakers.

Since algae are quite different from the higher (flowering) plants, a special terminology has developed for them. Some of the terms applicable to seaweeds are as follows:

Stipe: a stem-like portion of the plant.
Holdfast: a structure that anchors the plant to the ground or to a rock, the counterpart of a root of flowering land plants.
Frond: a flattened expansion, the counterpart of a leaf of flowering land plants.
Intertidal zone: the area between the high tide and low tide.

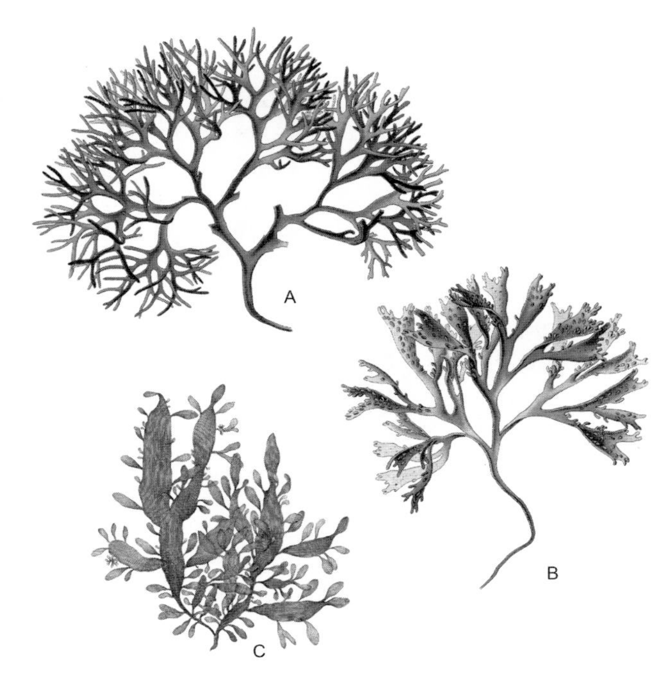

Red seaweeds. A, Irish moss (*Chondrus crispus*), from Köhler (1883–1914). B, carrageen moss (*Gigartina mamillosa*), from Köhler (1883–1914). C, Dulse (*Rhodymenia palmata*), from Oeder et al. (1761–1883, plate 1128).

Laver or Nori

Nori (the Japanese name) or purple laver (the name in the West), a member of the red algae, is the most widely consumed seaweed in the world. It is also the first seaweed to be cultivated, and has been grown in Tokyo Bay for 300 years. The plants grow in thin sheets on rocky shores, attached by a holdfast. Color differs, depending on the species and the amount of exposure to sun, but most are a reddish hue. Although a member of the red seaweed family, it becomes black or green when dry. The species are typically found from mid to low tide in the intertidal zone. The Japanese are the largest growers of nori (in Japan, nori is known as the "Imperial Japanese seaweed"), followed by the Chinese. Nori is also produced in the Marinan Islands, Saipan, Guam, and the Philippines. In North America, nori is commonly found in health food stores.

Kelp

Kelp species constitute a very large and important group of edible seaweeds. Many are huge, tough, leathery plants. The genus *Macrocystis* has species that are among the most impressive of the kelp, sometimes developing weights of over 200 kg (247 pounds). The longest known *Macrocystis* measured 47 m (154 feet). Wakame (*Undaria pinnatifida*) is native to the coasts of China, Japan, and Korea. It has a short stipe and divided blades, and is extensively used in Asian cooking. *Alaria esculenta*, sometimes called edible kelp, has long fronds up to 3.5 m (12 feet) and grows on northern Atlantic shores. Edible species of *Alaria* are also found in the Pacific Ocean. Most of the kelp are found in the Pacific Ocean, with a restricted number in the Atlantic. Kelp forests constitute important marine ecosystems in coastal areas of much of the world, and support important commercial and recreational fisheries.

Another brown algal group, the order Fucales, also contains algae sometimes marketed medicinally as kelp (and also often as "fucus"). The species usually encountered in commerce are *Fucus vesiculosus* L. (bladderwrack), frequent on the rocky coasts of the Atlantic and Pacific oceans, and *Ascophyllum nodosum* (L.) Le Jolis (knotted wrack) of the coasts of the North Sea, the western part of the Baltic Sea, and the east coast of Canada. The pharmacological properties of these are comparable to those of the true kelp, but they are inferior for culinary purposes.

Kelp are fast growing despite their often cool environment, and provide food and habitat for numerous species of marine organisms. Sea urchins are very significant herbivores of kelp, but their numbers are controlled through predation by lobsters, crabs, and sea otters. Widespread destruction of kelp beds in shallow water on both the East and West Coasts of Canada has been attributed to the unchecked overfishing of lobsters and crabs in the East, and the early extirpation of sea otters in the West to supply fur for the European market during the 1700s. Sea otters were recently reintroduced from Alaska to northwestern Vancouver Island and are extending their range along the west coast of North America. The reintroduction has led to a decline in sea urchin populations and a resulting revival in barrier kelp, which was once so extensive that it provided a safe navigation route of sheltered water between the offshore beds and the land.

Seaweed products were used in traditional Chinese herbal medicine as early as the 16th century, and the kelp have also been employed in western folk medicine. Until very recently, however, there has been limited evidence of pharmacological value, aside from the effects of compensating for iodine deficiency. Kelp tablets and powders have become popular herbal preparations in North America, and claims have been made that kelp products are useful in treating a variety of ailments. The therapeutic properties of kelp have been attributed particularly to content of trace minerals, especially iodine, which is typically 20 000 times as concentrated in the seaweed by comparison with its aquatic habitat. However, the level of iodine has been found to vary considerably among algal species and even within a species, so that commercially available products may deliver different dosages. The very high iodine content of brown algae led to their use in goiter medicines, but the variability of concentrations and the varying absorption conditions for bound and unbound iodine in the plant has made such algal therapy obsolete. Required daily intake of iodine in adults is only 150 µg, and the thyroid gland normally does not make use of excess iodine, and indeed, very large doses of iodine can induce or intensify hyperthyroidism. Sixty percent of all edible salt in the world is now iodized, reducing iodine deficiency disorders, the world's leading cause of preventable mental retardation until recently. Before 1990, approximately 40 million children were born each year at risk of mental impairment due to iodine deficiency in their mothers' diets. By 1997, that figure was close to 28 million. Kelp are currently marketed as a weight-reducing agent, supposed to be a result of increased production of thyroid hormones that increase metabolism and so remove deposited fats. While beneficial effects on obesity have been claimed, the medical community considers such therapy as potentially dangerous and highly inadvisable.

Although kelp are thought to inhibit heavy metal absorption in humans, kelp growing in polluted waters may accumulate very high levels of heavy metals such as strontium and cadmium. Toxic levels of lead can develop. Most dangerously, kelp accumulates arsenic, although in nonpolluted waters this should not pose a risk. Manufacturers of kelp products should of course ensure that toxic levels of any of these are not present.

Irish Moss

Irish moss grows 15–25 cm (5–10 inches) in length and is attached underwater to a rock by its holdfast. The stipe is compressed at the base, opening up gradually to a flattened frond which forks repeatedly. This seaweed is a dark, reddish brown, but ranges through purplish red to brown, green, yellow, or white, depending on the exposure to sunlight. Irish moss grows on rocky shores in pools in the lower intertidal and shallow subtidal zones of the Atlantic Ocean. It is very common around the North American and European coastline. Irish moss is over 60% carrageenin, which is extracted and used as a gel in industry and pharmacy and as a thickener in soups and dairy products.

Dulse

Dulse is a red seaweed that grows attached to rocks in the North Atlantic and Northwest Pacific, particularly on the coasts of Europe and North America. It grows from the mid-tide portion of the intertidal zone (the area between the high tide and low tide) to deep water. Fronds may vary from rose to reddish purple, and range from 20 to 40 cm (8–16 inches). It is harvested from June through September. This seaweed is commonly used in Ireland and Atlantic Canada as food and medicine. Fresh dulse is sometimes available in the produce section of markets, and dried dulse is found in many health food stores.

Sea Lettuce

Sea lettuce (*Ulva*) species, the largest of the green seaweeds, grow in bright green, thin, crinkled lettuce-like sheets or ribbons up to 90 cm (3 feet) long. Sea lettuce is attached by its holdfast to exposed rocks or detached in tide pools. It thrives in moderate pollution, and should not be collected from polluted locations.

Culinary Portrait

"Phycophagy" is not as bad as it sounds: it means eating of algae, and humans consume a wide variety, mostly seaweeds. Seaweeds are often available from health-food stores, and sometimes from specialty markets. Recipes are available on the Web (and in the cookbooks cited at the end of this chapter). Algae are consumed as food in a variety of ways, depending on the dish: hot or cold, raw in salads, pickled or fermented in relish, cooked as vegetables, and mixed into soups, stews, and sauces. Algae can be served as appetizers, garnishes, desserts, teas, side dishes, and main dishes. Cooking may be by boiling, roasting, steaming, frying, or sautéing. As noted above, algal

Marine landscape, from Harter (1988).

products like agar, alginates, and carrageenans are extracted from some seaweeds and employed in a wide variety of foods. In powdered form, agar is the "gelatin" preferred by vegetarians, because it is a vegetable product. Ordinary gelatin is a protein (not a carbohydrate, like agar) derived from animal bones. In addition to numerous commercial culinary usages, the home cook may use agar as a thickener for jellies, fruit juices, and puréed fruit.

Seaweed requires washing before consumption, as sand, shells, and small animals are often present. Dried seaweed usually must be soaked, sometimes for as long as an hour, before being eaten or cooked.

Carbohydrates are a major component of seaweeds, typically about 50% of the plant, and these are often difficult to digest, although people who habitually consume seaweeds experience reduced problems compared to those who only occasionally eat them.

Laver or Nori

These seaweeds are eaten fresh as a vegetable, sautéed with other vegetables, or sun-dried. They can also be dry roasted and crumbled, and used as a condiment in soups, sauces, omelets, grains, popcorn, and salads, or sprinkled on baked potatoes. Laver bread is a traditional Welsh delicacy, made by soaking, mixing with oats, and frying. In Asia, cultivated nori are processed into sheets for sushi wrap (wrapped around raw seafood and rice to hold the two together). In Asia, a tea is prepared from nori, and this seaweed is often cooked with fish, tofu, vegetables, rice, and pasta. Roasted nori is said to have a pleasant, sardine-like taste.

Some Oriental swifts (or swiftlets) use a salivary secretion to line their nests or to cement collected material together to build their nests; these nests are used to make bird's nest soup in Chinese cuisine. Laver is a preferred food of many of these birds.

Kelp

Depending on species, kelp may be cooked, sliced, and added to soups, beans, or stews. Some species may be used for pan-fried crispy chips and dry roasted flakes or they may be pickled in vinegar. In Japan, powdered kelp is used to make a tea called *kombuchu*. Kelp have large amounts of sodium, and so should be avoided in salt-restricted diets.

Kelp carbohydrates are not normally digested by humans, and three-quarters of the dry bulk of the most edible brown seaweeds is indigestible. Nevertheless, kelp are used directly as vegetables and condiments, principally in China and Japan. Like other seaweeds, kelp are also harvested as dried fodder for terrestrial livestock in coastal areas (despite the limited digestibility), and sometimes grown as forage for cultivated aquatic animals, such as abalone.

Herring often deposit their eggs on kelp (especially *Macrocystis*), and on the West Coast of North America, kelp blades coated with a thick layer of eggs has long been a favorite food of Indigenous People. The combination of the eggs ("roe") on the raw kelp is also greatly valued as a delicacy in the sushi bars of Japan, and has been produced in British Columbia by herring spawning in penned kelp enclosures. The industry is largely managed by native people, and the harvest is valued at over $20 million annually.

Irish Moss

During the great famine, which devastated mid 19th century Ireland, many Irish families in coastal localities avoided the ravages of starvation by eating Irish moss. It was often boiled and made into a highly nutritious jelly, used in soups, stews, and puddings, or simply boiled with milk to form a highly nutritious food supplement. Today, Irish moss remains a popular food, used most frequently to prepare blancmange, a delicious vanilla-flavored pudding.

Dulse

Dulse has been used as a food in Scotland and Ireland since the 10th century. Fresh dulse is rubbery but quite edible raw, and can be eaten directly off the rocks. Sun-dried dulse is eaten as is (preferably after soaking), or is ground to flakes or a powder. It can also be quickly pan fried into tasty chips; baked in an oven and covered with cheese and salsa; or microwaved briefly for a crispy treat. Dulse is also used in soups, chowders, sandwiches, and salads, and added to bread and pizza dough.

Sea Lettuce

This seaweed has a taste like sorrel, and is eaten fresh as a side dish or in salads. The leaf shape lends itself to wrapping around a fillet of fish that can be gently steamed for a delicate, tender dish. Dried sea lettuce is first soaked in water for 3 or 4 minutes, then consumed raw or cooked for 5 minutes. Sea lettuce can be cooked in various dishes, used in soups, and deep-fried for a crispy treat.

Culinary Vocabulary

- The algal species "dulse" should not be confused with "dulce." The latter term (Spanish for "sweet," and pronounced DOOL'th-eh or DOOL-say) refers to a sweetening agent added to some sherries when they are bottled, or to an intensely sweet Spanish confection prepared with sugar and cream.

- "Grass jelly" is a black jelly made with seaweed and cornstarch, which is available in cans in Asian grocery stores. It is diced and used in sweet drinks and fruit salads in much of Asia.
- "Seaweed noodles" (also called Chinese seaweed noodles and seaweed threads, and known in Chinese as *yang fun*) are thread-like, transparent, gelatinous noodles made from seaweed. They are used in Asian cuisine as a food extender.
- Agar (agar-agar) is also known as "Japanese gelatin."
- "Agar agar noodles" (agar-agar noodles) are strips of agar-agar gelatin, often served cold in a salad. They are used in Asian, especially Chinese cuisines.
- "Hair seaweed" (also known as hair vegetable, black moss, and cow hair) are strands of a dark-purple seaweed, *Bangia fusco-purpurea*, which belongs to the red algal group. This seaweed resembles tangled hair, hence the name. When cooked in water, it absorbs the flavor of other foods. It can be added to soups and salads, or made into rolls and fried in oil until crisp. It is an ingredient of "Lohan's Delight," a Buddhist vegetarian dish traditionally prepared with dried oysters during the first 2 weeks of the Chinese New Year. It seems particularly desired because its Chinese name, *faht choy*, sounds similar to the Chinese New Year greeting, *Gung hei fat choy* ("Good luck!").

Curiosities of Science and Technology

- The Ainu in northern Japan collected kelp for food as early as the 8th century, and during a later period, only the privileged classes of Japan were allowed to eat this sea vegetable.
- In 701, the emperor of Japan established the Law of Taiho, which listed seaweeds among the marine products paid to the court as a tribute.
- British Columbia coastal Indians stretched kelp stipes to make fishing lines and used the hollow bulbs and stipe bases as bottles.
- Indians of Sitka, Alaska, once used the bull kelp (*Nereocystis*) as a cure for headaches. A hollow float of the seaweed would be placed on a hot stone until steam was generated and passed through the attached hollow stem into the ear of the patient, supposedly curing the headache.
- The bladders of bladderwrack (*Fucus vesiculosus*) have been cut off, filled with water, and compressed to squirt water in old times as water pistols are used today.
- In recent times, native peoples of the Andean highlands reserved the right to come down to the coast in order to gather and dry seaweed, and bring it back to the mountains. There, algae are so greatly valued that they are used as money. Hypothyroidism, due to lack of iodine, is endemic in the Andes, and can be prevented by consumption of seaweeds, which are rich in the element.
- The dried stipes of the kelp genus *Laminaria* expand 3–5 times their original circumference on wetting. These dried stipes have been used to produce noninstrument mechanical dilation of the cervical canal during birth and gynecological treatment.
- Japanese stores sell boxes of kelp that are given as gifts the same way as chocolates are in the West.
- When young abalones feed on red seaweed, their shells turn red.
- Sea otters anchor themselves with kelp to keep from drifting far out to sea during the night.
- Pacific giant kelp can grow up to 46 cm (18 inches) a day.
- Seaweeds are 75–90% water. In this respect, they are comparable to herbaceous land plants. The adult human body is 50–65% water, while a child's body is approximately 75% water.
- Flowering land plants as well as nonflowering related groups (such as ferns, which produce spores) possess special stem tissues that allow for transportation of water and nutrients from one part of the plant to another (much as higher animals have blood vessels). Kelps are the only group of algae that seem to have developed a specialized internal transportation tissue, which is comparable to the conductive tissue called phloem in land plants that transports sugars from the leaves to other parts of the plant. The need for such a system is obvious in giant kelps, which are like huge trees, with the need to transport sugars manufactured by photosynthesis in the leaf-like fronds located in sun-lit areas to the lowest parts of the plant, which are often too shaded to make sufficient sugar to survive. The phenomenon of completely unrelated species coming up with similar means of solving a common problem is called convergent evolution.

- Monosodium glutamate (MSG) is a widely used, tasteless food additive, sold as a white powder, which enhances the flavor of some foods, particularly soups, meat, fowl, fish, and vegetables. Monosodium glutamate was first isolated in 1908 in Japan from *Laminaria japonica*, which in fact had been used by the Japanese for centuries, as a seaweed broth, to bring out the flavor of food. Today, MSG is made by a fermenting process using starch, sugar beets, sugar cane, or molasses. Chemically, MSG is a sodium salt of glutamate, and only in the form free of the sodium is glutamate effective. Glutamate, an amino acid, occurs naturally in human bodies, and is also naturally present and enhances the taste of protein-containing foods such as cheese, milk, meat, peas, and mushrooms. Indeed, even if MSG is not added to food, "glutamates" may be present in the form of "hydrolyzed proteins." In 1968, an allergic reaction to MSG termed "Chinese Restaurant Syndrome" was found to be due to the heavy use of MSG in some Chinese restaurants. Even small amounts of MSG will produce allergies in susceptible individuals. Symptoms include the following: burning sensation in the back of the neck, forearms, and chest; numbness in the back of the neck, radiating to the arms and back; tingling, warmth and weakness in the face, temples, upper back, neck, and arms; facial pressure or tightness; chest pain; headache; nausea; rapid heartbeat; bronchospasm (difficulty breathing), particularly in MSG-intolerant people with asthma; drowsiness; and weakness. For most people, it appears that MSG is safe "when eaten at customary levels."
- The Sargasso Sea is a large area in the Atlantic Ocean between the West Indies and Africa (between 20°–35° north latitude and 30°–70° west longitude), which is relatively little affected by ocean currents. It gets its name from the abundance of species of *Sargassum*, seaweeds that belong to the brown algal group. When Christopher Columbus reached the Sargasso Sea and observed all the seaweed, he thought land was nearby, and even fathomed the sea, only to find no measurable bottom. His sailors feared that their boats would be permanently entangled in the seaweed. In complete contrast to extremely strong ocean currents around the Sargasso Sea, it is extremely calm. Anything that drifts onto any of the surrounding currents eventually ends up in the seaweed of the Sargasso Sea, and this led to the myth of the "Bermuda Triangle," a phrase first used in an article written by Vincent H. Gaddis for *Argosy* magazine in 1964. The corners of the Bermuda Triangle are Florida, the islands of Bermuda, and Puerto Rico, an area which is at the heart of the Sargasso Sea. In the article, Gaddis claimed that in this strange sea, ships and planes had disappeared without explanation. However, he was not the first to make such claims (either for ships or aircraft). The Sargasso Sea has long had a reputation for being an inescapable trap for ships, and indeed, legends of a "sea of lost ships" in the area considerably predate the concept of the Bermuda Triangle. Derelict, deserted, shipshape vessels were often found in the Sargasso Sea. Early paintings showed sailing vessels being devoured by the seaweed. Although exaggerated, the Sargasso Sea was dreaded by the sailors with some justification, because its calm waters could be deadly. Spanish sailors often found themselves becalmed for weeks, and were frequently forced to jettison their war horses in order to conserve water. Hence, the areas known as the "Horse Latitudes" traverse the Sargasso Sea. Modern ships have no reason to fear the region, since they are not dependent on the wind or sea currents. Examination of the frequency of accident records in modern times has shown that the deadliness of the Bermuda Triangle is a myth.

Key Information Sources

Aaronson, S. 1986. A role for algae as human food in antiquity. Food Foodways, **1**: 311–315.

Aaronson, S. 2000. Algae. *In* The Cambridge world history of food. *Edited by* K.F. Kiple and K.C. Ornelas. Cambridge University Press, Cambridge, UK. pp. 231–249.

Bardach, J.E., Ryther, J.H., and McLarney, W.O. 1972. Aquaculture, the farming and husbandry of freshwater and marine organisms. John Wiley & Sons, New York, NY. 868 pp.

Bird, C.J., and van der Meer, J.P. 1993. Systematics of economically important marine algae: a Canadian perspective. Can. J. Bot. **71**: 361–369.

Bird, K.T., and Benson, P.H. (*Editors*). 1987. Seaweed cultivation for renewable resources. Developments in aquaculture and fisheries science 16. Elsevier, Amsterdam, Netherlands. 381 pp.

Bold, H.C., and Wynne, M.J. 1978. Introduction to the algae: structure and reproduction. Prentice-Hall, Inc., Englewood Cliffs, NJ. 706 pp.

Booth, E. 1979. The history of the seaweed industry. Part 4. A miscellany of industries. Chem. Ind. **1979**: 378–383.

Chapman, V.J. 1970. Seaweeds and their uses. 2nd edition. Methuen & Co., Ltd., London, UK. 304 pp.

Critchley, A.T., and Ohno, M. (*Editors*). 1998. Seaweed resources of the world. Japan International Cooperation Agency, Nagai, Yokosuka, Japan. 431 pp.

Fritsch, F.E. 1935. The structure and reproduction of the algae. Vol. 1. Cambridge University Press, London, UK. 791 pp.

Fritsch, F.E. 1955. The structure and reproduction of the algae. Vol. 2. Cambridge University Press, London, UK. 939 pp.

Hoppe, H.A., Levring, T., and Tanaka, Y. (*Editors*). 1979. Marine algae in pharmaceutical science. Water de Gruyter, Berlin, Germany. 807 pp.

Lembi, C.A., and Waaland, J.R. (*Editors*). 1988. Algae and human affairs. Cambridge University Press, Cambridge, UK. 590 pp.

Levring, T., Hoppe, H.A., and Schmid, O.J. 1969. Marine algae. A survey of research and utilization. Botanica Marina Handbooks 1. Cram, de Gruyter & Co., Hamburg, Germany. 421 pp.

Lobban, C.S., and Harrison, P.J. 1994. Seaweed ecology and physiology. Cambridge University Press, Cambridge, UK. 366 pp.

Lobban, C.S., and Wynne, M.J. 1981. The biology of seaweeds. University of California Press, Berkeley, CA. 786 pp.

Lüning, K. 1990. Seaweeds. Their environment and ecophysiology. John Wiley & Sons, New York, NY. 529 pp.

Margulis, L., and Schwarz, K.V. 1982. Five kingdoms. An illustrated guide to the phyla of life on earth. W.H. Freeman and Company, New York, NY. 338 pp.

McHugh, D.G. 1996. Seaweed production and markets. FAO/GLOBEFISH Research Programme. Vol. 48. Food and Agriculture Organization of the United Nations, Rome, Italy. 73 pp.

McHugh, D.J. (*Editor*). 1987. Production and utilization of products from commercial seaweeds. FAO Fisheries Technical Paper 288. Food and Agriculture Organization of the United Nations, Rome, Italy. 189 pp.

Michanek, G. 1975. Seaweed resources of the oceans. FAO Fish. Tech. Pap. **138**: 1–137.

Morgan, K.C., Wright, J.L.C., and Simpson, F.J. 1980. Review of chemical constituents of the red alga, *Palmaria palmata* (dulse). Econ. Bot. **34**: 27–50.

Newton, L. 1951. Seaweed utilisation. Sampson Low, London, UK. 188 pp.

Ohno, M., and Critchley, A.T. 1993. Seaweed cultivation and marine ranching. Japanese International Cooperation Agency, Yokosuka, Japan. 151 pp.

Prud'homme van Reine, W.F., and Trono, G.C., Jr. (*Editors*). 2002. Plant resources of South-East Asia, 15, cryptogams: algae. Prosea Foundation, Bogor, Indonesia. 318 pp.

Round, F.E. 1981. The ecology of algae. Cambridge University Press, Cambridge, UK. 653 pp.

Shelef, G., Soeder, C.J., and Balaban, M. 1980. Algae biomass: production and use. Elsevier, Amsterdam, Netherlands. 852 pp.

Taylor, W.R. 1960. Marine algae of the eastern tropical and subtropical coasts of the Americas. The University of Michigan Press, Ann Arbor, MI. 870 pp.

Tiffany, L.H. 1958. Algae, the grass of many waters. C.C. Thomas, Springfield, IL. 199 pp.

Tseng, C.K. 1983. Common seaweeds of China. Science Press, Beijing, China. 316 pp.

van den Hoek, C., Mann, D.G., and Jahns, H.M. 1995. Algae. An introduction to phycology. Cambridge University Press, Cambridge, UK. 623 pp.

Venkataraman, G.S. 1969. The cultivation of algae. Indian council of Agricultural Research, New Delhi, India. 319 pp.

Xia, B., and Abbott, I.A. 1987. Edible seaweeds of China and their place in the Chinese diet. Econ. Bot. **41**: 341–353.

Zanefeld, J.S. 1959. The utilization of marine algae in tropical South and East Asia. Econ. Bot. **13**: 89–131.

Specialty Cookbooks

Arasaki, S., and Arasaki, T. 1981. Low calorie, high nutrition. Vegetables from the sea to help you look and feel better. Japan Publications, Tokyo, Japan. 196 pp.

Bradford, P., and Bradford, M. 1985. Cooking with sea vegetables: a collection of naturally delicious dishes using to the full the bountiful harvest of the oceans. Healing Arts Press, Rochester, VT. 144 pp.

Ellis, L. 1999. Seaweed: a cook's guide: tempting recipes for seaweed and sea vegetables. Fisher Books, Tucson, AZ. 100 pp.

Fortner, H.J. 1978. The limu eater: a cookbook of Hawaiian seaweed. University of Hawaii Sea Grant College Program, Honolulu, HI. 107 pp.

Gusman, J. 2003. Vegetables from the sea: everyday cooking with sea greens. William Morrow, New York, NY. 131 pp.

Joie, Inc. 1984. Seaweed and vegetables. Joie, Inc., Tokyo, Japan. 100 pp.

Madlener, J.C. 1977. The seavegetable book. Crown Publishers, Toronto, ON. 288 pp.

McConnaughey, E. 1985. Sea vegetables: harvesting guide & cookbook. Naturegraph Publishers, Happy Camp, CA. 239 pp.

Rhoads, S.A., and Zunic, P. 1978. Cooking with sea vegetables. Autumn Press, Brookline, MA. 136 pp.

Surey-Gent, S., and Morris, G. 2000. Seaweed. Whittet, Stowmarket, UK. 160 pp.

Williams, B. 1973. Seaweed cookery. Williams, Corning, CA. 21 pp.

SESAME

Family: Pedaliaceae (pedalium family)

Names

Scientific name: *Sesamum indicum* L. (*Sesamum orientale* L.)

- The name sesame and the genus name *Sesamum* trace to ancient names for the plant, particularly the Greek *seésamon*, which in turn likely originates from Afro-Asiatic languages (such as the Arabic *saasim*).
- Sesame is sometimes known as sum-sum, the Hebrew for sesame seed. In Israel, the popular children's program Sesame Street is called Rehov Sum Sum.
- *Indicum* in the scientific name *S. indicum* means Indian. As noted below, the species may have originated in India.

Plant Portrait

Sesame is an annual (occasionally perennial) bushy herb, 0.5–2.5 m (2–9 feet) tall, depending on variety and conditions (0.6–1.2 m or 2–4 feet is typical). Its seed capsules are 2.5–7.5 cm (1–3 inches) long, and contain 50–100 or more small, oval, flat, edible seeds from which oil is obtained. The seed is covered by a fibrous hull, which according to variety, may range in color from yellowish white through red and brown to black. Most of the seed that comes to North America is light tan in hull color. The hulled seed in all varieties is pearly white with a glossy finish.

Sesame is one of the oldest oilseed plants, and was a highly prized oil crop of ancient Babylon and Assyria. It is among the top 10 oil seed crops in the world, and there are many named varieties. The species probably originated in Africa (or possibly India), and is now extensively grown in subtropical and tropical regions of the world, including China, India, Burma, Turkey, Nigeria, Uganda, Ethiopia, El Salvador, Nicaragua, Sudan, Mexico, and Guatemala. There is also some cultivation in the United States, and plants escaped from cultivation now grow wild in the southern United States. The seeds contain 50–60% oil, which has excellent stability, keeping for several years before going rancid, owing to the presence of natural antioxidants. In addition to the culinary uses noted below, sesame oil is used in paints, soaps, lubricants, insect sprays, and pharmaceutical and skin care products.

Culinary Portrait

Sesame oil, also known as gingelly (often misspelled as gingerly) and teel oil, is bland in taste and almost odorless, and is used in margarine, salad oils, and cooking oils. The low cholesterol content and high proportion of polyunsaturated fats make it a healthy dietary choice. Sesame seed is 25% protein, and the protein is rich in methionine and tryptophan, two amino acids that are lacking in many other sources of vegetable protein. The hulled seeds may be eaten directly, but are typically sprinkled on buns, breads, pastries, crackers, and confections, and indeed, almost all sesame consumed in North America is used as a spice for food products such as hamburger buns and other bakery goods. The seeds are also added to stews for the sweet nutty taste they develop upon cooking. The seeds are ground into a nutritious paste called tahini, which is a popular spread in Middle Eastern countries, used to flavor main dishes, sauces, and desserts. "Sesame butter" is a thicker version of tahini, resembling peanut butter. The Turkish/Jewish/Arabic confection halvah (halva, halavah) is largely made of

Sesame (*Sesamum indicum*), from Köhler (1883–1914).

crushed and sweetened sesame seeds, almonds, and honey. Sesame flour is also made from the seeds, and used in baking, usually in combination with other flours. In much of Asian cooking, sesame oil is used as a condiment and seasoning, as well as for cooking. For example, the Lebanese dish hummus is typically made by combining chickpeas and sesame oil.

Culinary Vocabulary

- "Tahini," as noted above, is ground sesame seed. The word is derived from the Arabic for "to grind."
- Slaves brought sesame to America in the 17th and 18th centuries and used the seeds in puddings, stews, and candy. In some parts of the southern US, sesame is known as "benne," which was its name in the African (Bantu) dialect. The slaves considered benne seeds to be a symbol of good luck. "Benne wafers" are thin, crisp cookies made traditionally in the Old South with brown sugar, pecans, and sesame seed. "Benne cookies" and "benne cakes" are sometimes termed "good luck cookies" in South Carolina.
- "Sesame balls" are a deep-fried Chinese dessert or pastry prepared with rice, sweet bean paste, and sesame seeds.
- "Sesame chile oil" is sesame oil that has been flavored with chile pepper. It is used as a condiment in Asian cuisines.

Curiosities of Science and Technology

- Sesame may be the world's oldest condiment. The Chinese used it 5000 years ago.
- Sesame was used by the Egyptians as a medicine as early as 1500 BC.
- Italian explorer Marco Polo (1254?–1324?) sampled some sesame oil in Abyssinia (Ethiopia) and pronounced it to be the best oil he had ever tasted.
- For centuries, sesame oil has been burned to make soot for the finest Chinese ink blocks for printing.
- "Open Sesame" was the magical password that opened the entrance to the cave in *Ali Baba and the Forty Thieves*. The hero of the story, Ali Baba, observed a band of robbers enter a cave by means of the password, and after they left, he was able to steal the thieves' treasure. This saying is related to the fact that ripe sesame seed pods often open with a sharp pop at the slightest touch, scattering the seeds. Because of this, considerable hand labor is needed to prevent loss of the seeds. However, nonscattering varieties are now available, making mechanized harvesting of the crop possible. [An alternative, less plausible explanation has been advanced to explain the expression "Open Sesame." High-quality sesame oil was used to oil locks and make them work well, so sesame could have been thought to act magically in opening doors.]
- In 1975, McDonald's launched its famous "Big Mac" slogan: "Two all-beef patties, special sauce, lettuce, cheese, pickles, onions, and a sesame seed bun." There are approximately 178 sesame seeds on the average Big Mac hamburger bun.
- Sesamoid bones were so named because of their apparent resemblance in shape to sesame seeds. These are generally small, oval, nodular masses of bone or cartilage in a tendon, especially at a joint. The largest sesamoid bone in the human body is the patella (kneecap bone). Unlike most bones of the body, sesamoid bones develop independently of other bones. Sesamoid bones are thought to provide protection from excessive friction or to change the direction of the pull of tendons.

Key Information Sources

(For additional references, see SAFFLOWER.)

Ahuja, D.B., and Bakhetia, D.R.C. 1995. Bioecology and management of insect pests of sesame: a review. J. Insect Sci. **8**: 1–19.

Ashri, A. (*Editor*). 1981. Sesame, status and improvement: proceedings of expert consultation, Rome, Italy, 8–12 December 1980. Food and Agriculture Organization of the United Nations, Rome, Italy. 198 pp.

Ashri, A. 1998. Sesame breeding. Plant Breed. Rev. **16**: 179–228.

Bedigian, D. 2000. Sesame. *In* The Cambridge world history of food. *Edited by* K.F. Kiple and K.C. Ornelas. Cambridge University Press, Cambridge, UK. pp. 411–421.

Bedigian, D. 2003. Evolution of sesame revisited: domestication, diversity and prospects. Genet. Resour. Crop Evol. **50**: 779–787.

Bedigian, D. 2004. History and lore of sesame in southwest Asia. Econ. Bot. **58**: 329–353.

Bedigian, D., and Harlan, J.R. 1983. Nuba agriculture and ethnobotany with particular reference to *Sesamum indicum* and *Sorghum bicolor*. Econ. Bot. **37**: 384–395.

Bedigian, D., and Harlan, J.R. 1986. Evidence for cultivation of sesame in the ancient world. Econ. Bot. **40**: 137–154.

Bedigian, D., Seigler, D.S., and Harlan, J.R. 1985. Sesamin, sesamolin and the origin of sesame. Biochem. Syst. Ecol. **13**: 133–139.

Bedigian, D., Smyth, C.A., and Harlan, J.R. 1986. Patterns of morphological variation in sesame. Econ. Bot. **40**: 353–365.

Beroza, M., and Kinman, M.L. 1955. Sesamin, sesamolin, and sesamol content of the oil of sesame seed as affected by strain, location grown, ageing, and frost damage. J. Am. Oil Chem. Soc. **32**: 348–350.

Bhat, K.V., Babrekar, P.P., and Lakhanpual, S. 1999. Study of genetic diversity in Indian and exotic sesame (*Sesamum indicum* L.) germplasm using random amplified polymorphic DNA (RAPD) markers. Euphytica, **110**: 21–33.

Bisht, I.S., Mahajan, R.K., Loknathan, T.R., Gautam, P.L., Mathur, P.N., and Hodgkin, T. 1999. Assessment of genetic diversity, stratification of germplasm accessions in diversity groups and sampling strategies for establishing a core collection of Indian sesame (*Sesamum indicum* L.). Plant Genet. Resour. Newsl. **119**(Suppl.): 35–46.

Brar, G.S., and Ahuja, K.L. 1979. Sesame: its culture, genetics, breeding and biochemistry. *In* Annual Review Plant Science. *Edited by* C.P. Malik. Kalyani Publishers, New Delhi, India. pp. 245–313.

Ikeda, S. 2001. Dietary sesame seeds elevate α-γ tocopherol concentrations. J. Nutr. **131**: 2892–2897.

John, C.M., Narayana, G.V., and Seshadri, C.R. 1950. The wild gingelly of Malabar. Madras Agric. J. **37**: 47–50.

Johnson, L.A., Suleiman, T.M., and Lusas, E.W. 1979. Sesame protein: a review and prospectus. J. Am. Oil Chem. Soc. **56**: 463–468.

Joshi, A.B. 1961. *Sesamum*. Indian Central Oilseeds Committee, Hyderabad, India. 109 pp.

Kamen, B. 1987. Sesame: the superfood seed and how it can add vitality to your life. Keats Publishing, New Canaan, CT. 25 pp.

Langham, D.R., and Wiemers, T. 2002. Progress in mechanizing sesame in the US through breeding. *In* Trends in new crops and new uses. *Edited by* J. Janick and A. Whipkey. ASHS Press, Alexandria, VA. pp. 157–173.

Maiti, S., Raoof, M.A., Sastry, K.S., and Yadava, T.P. 1985. A review of *Sesamum* diseases in India. Trop. Pest Manage. (Lond.), **31**: 317–323.

Manning, S. 1991. The genera of Pedaliaceae in the southeastern United States. J. Arnold Arbor. Suppl. Ser. **1**: 313–347.

Morris, J.B. 2002. Food, industrial, nutraceutical, and pharmaceutical uses of sesame genetic resources. *In* Trends in new crops and new uses. *Edited by* J. Janick and A. Whipkey. ASHS Press, Alexandria, VA. pp. 153–156.

Namiki, M. 1995. The chemistry and physiological functions of sesame. Food Rev. Int. **11**: 281–329.

Nanthakumar, G.K., Singh, N., and Vaidyanathan, P. 2000. Relationships between cultivated sesame (*Sesamum* sp.) and the wild relatives based on morphological characters, isozymes and RAPD markers. J. Genet. Breed. **54**: 5–12.

Nayer, N.M. 1995. Sesame. *In* Evolution of crop plants. 2nd edition. *Edited by* J. Smartt and N.W. Simmonds. Longman Scientific & Technical, Burnt Mill, Harlow, Essex, UK. pp. 404–406.

Nayar, N.M., and Mehra, K.L. 1970. Sesame: its uses; botany, cytogenetics, and origin. Econ. Bot. **24**: 20–31.

Nicolson, D.H., and Wiersema, J.H. 2004. Proposal to conserve *Sesamum indicum* against *Sesamum orientale* (Pedaliaceae). Taxon, **53**: 210–211.

Ram, R., Catlin, D., Romero, J., and Cowley, C. 1990. Sesame: new approaches for crop improvement. *In* Advances in new crops. *Edited by* J. Janick and J.E. Simon. Timber Press, Portland, OR. pp. 225–228.

Weiss, E.A. 1971. Sesame. *In* Castor, sesame and safflower. *Authored by* E.A. Weiss. Barnes & Noble, New York, NY. Longmans, London, UK. pp. 311–525.

Weiss, E.A., and Cruz, Q.D. de la. 2001. *Sesamum orientale* L. *In* Plant resources of South-East Asia, 14, vegetable oils and fats. *Edited by* H.A.M. van der Vossen and B.E. Umali. Backhuys Publishers, Leiden, Netherlands. pp. 123–128.

Specialty Cookbooks

Andersen, J., and Andersen, S. 1983. Juel Andersen's sesame primer: a compendium of sesame seed cookery. Creative Arts Communications, Berkeley, CA. 63 pp.

Clemson Agricultural College. 1956. Recipes and commercial uses for South Carolina sesame. South Carolina Agricultural Experiment Station of Clemson Agricultural College, South Carolina, Clemson, SC. 16 pp.

Miller, D., and Shane, S. 1983. The magic of tahini: open sesame. Hecuba's Daughters, Inc., Bearsville, NY. 89 pp.

SHEA NUT

Family: Sapotaceae (sapote family)

Names

Scientific name: *Vitellaria paradoxa* C.F. Gaertn. [*Butryospermum paradoxum* (C. F. Gaertn.) Hepper, *B. parkii* (G. Don) Kotschy, *B. niloticum* Kotschy]

- "Shea" is based on the Bambara word *si.* (Bambara or Bambarra refers to a people of West and Central Africa, as well as their language.)
- The shea nut is also known as the karite nut, a name based on the Dioula language word for "life." *Karité* is the French word for shea nut.
- Other names for the shea nut are butterseed, shea buttertree, and shea tree.
- The genus name *Vitellaria* is based on the Latin *vitellus*, yolk, + *aria*, like, a reference to the valuable kernel within the fruit.
- *Paradoxa* in the scientific name *V. paradoxa* is Latin for strange or contrary to expectation, a reference to the ability of the tree to grow in extremely dry regions where few trees survive.

Plant Portrait

The shea nut is an economically important tropical African nut tree, which grows up to 20 m (66 feet) in height. It is distributed across Central Africa, from West Africa to East Africa. The nuts are produced primarily in West and Central Africa in the semiarid Sahel, referred to by traders as the "Shea Belt." The main producing countries are Burkina Faso, Mali, Cote d'Ivoire, Benin, Togo, Ghana, Nigeria, Guinea, and Central African Republic. Wild trees are used as the source of the nuts. The trees can live to an age of over 300 years, starting to produce nuts at about 25 years, and reaching full production at 45–50 years. The fruits have been compared to small avocados. They are ellipsoid, greenish, up to 6.5 cm (2½ inches) long and 4.5 cm (1¾ inches) in diameter, with a sweet pulp surrounding a seed. Occasionally, up to three seeds occur in a fruit. The seed is shining, dark brown, up to 5 cm (2 inches) long and 3.5 cm (1⅜ inches) in diameter.

The seeds contain 50% or more fat, and the oily kernels are used to produce an edible, solid, whitish or yellowish vegetable "butter," as well as a liquid oil. In Africa, shea butter is used in the preparation of cooking fat, illuminant oil, candles, medicinal ointment, hair dressing, and soap. A large volume of shea nuts is exported from West Africa, mainly to Holland and Belgium. Shea butter is used extensively in the cosmetic and pharmaceutical industries, particularly as a base in products for the treatment of dry hair and skin, burns, and multiple skin ailments. Shea butter acts as a moisturizer, softener, and skin protector, and is becoming increasingly popular as an ingredient in cosmetics and soaps, especially in France and the United States. The primary market niche today of shea butter is as a substitute for cocoa butter in the chocolate and confectionery industry. However, many countries, including the United States, forbid the manufacture of "CBEs" (cocoa butter equivalents), so the main importers tend to be in Europe, with minor amounts also shipped to Japan.

Culinary Portrait

In Africa, shea nut is used mostly as a cooking fat, although the flesh of the fruit is also sometimes consumed. In Europe, the fat is used for frying and cooking, and in the manufacture of margarine. The seed fat is also used in Europe and in Japan to prepare pastry, because of its high dough pliability, and in confectionery as a substitute for cocoa butter, which it resembles closely in physical and chemical properties.

Curiosities of Science and Technology

- The funeral beds of kings in West Africa were sometimes carved out of the "noble wood" of an old shea tree.
- Shea butter has the distinction in West Africa of being used for certain rites of passage for both men and women: as a healing agent for accelerating the cicatrization (scarring) of the umbilical cord, and after a circumcision.
- Caterpillars of *Cirina butryospermi* consume the leaves of shea nut trees, causing defoliation. In parts of Africa, these insects are considered a delicacy, and they have long been an article of food in Nigeria under the name *monemone.*
- The black bark of the shea tree is very toxic, and an infusion (boiled tea) of the bark has been used as a poison. Nevertheless, it has been employed in a medicinal bath in West Africa, sometimes even for babies and during childbirth.

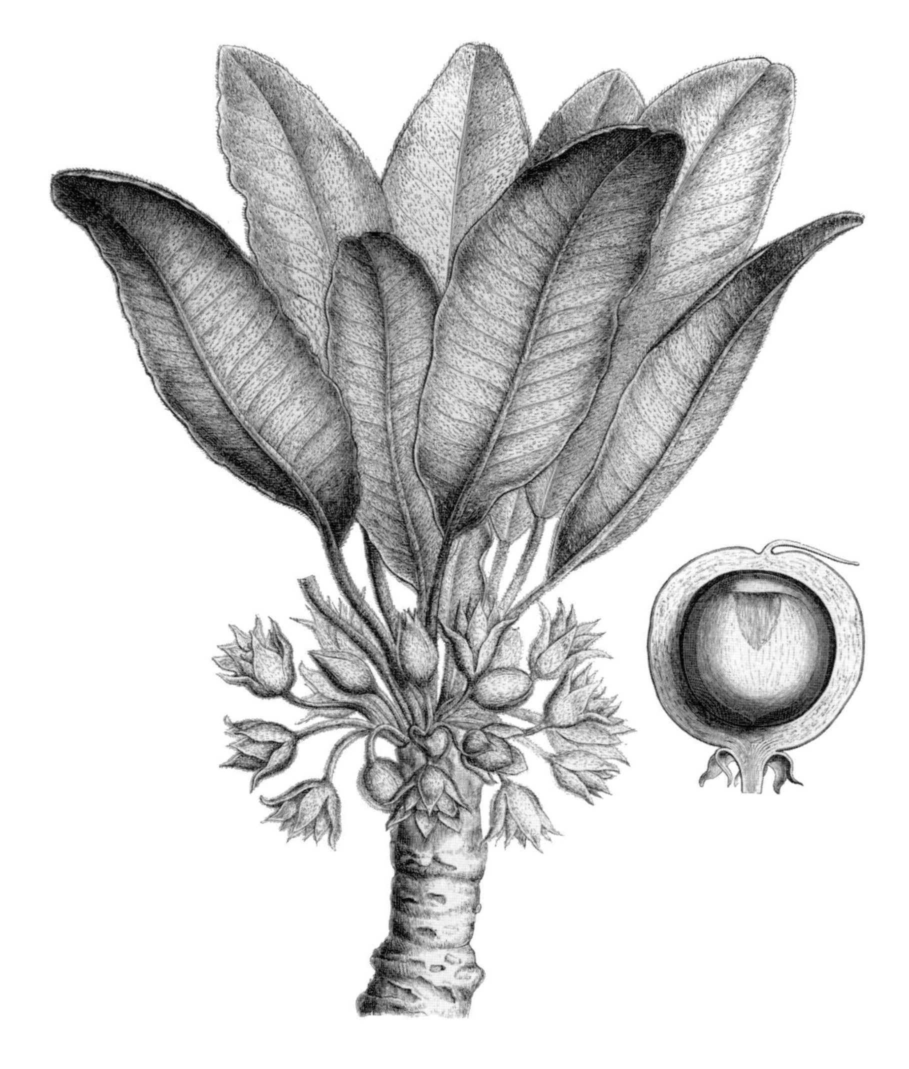

Shea nut (*Vitellaria paradoxa*), from Engler and Prantl (1889–1915).

- In West Africa, shea butter is sometimes called "women's gold," since the laborious harvesting and processing of shea is primarily an activity of rural women. Shea butter is one of the few economic assets controlled by women in Sahelian Africa (the semiarid regions of Sub-Saharan Africa). With the support of Third World development groups specializing in women's issues, there has been some success in elevating this less-than-subsistence, informal-sector activity into a formalized and systematized cottage industry capable of generating increased incomes and new skills and opportunities for women. Although shea butter is now mostly extracted chemically by Western companies, some firms use only traditionally produced material in their products in order to support the African women.
- Desertification (the transformation of a region into a desert) is a major problem in Sub-Saharan Africa. The problem is caused in part by cutting down the relatively few scattered indigenous trees for construction and firewood, since these are well adapted to using minimal water, while binding the soil against wind erosion. The shea tree is a principal asset in the fight against desertification, because people in Africa understand that the tree is a critical source of food and income, and generally conserve the tree, even in the middle of cultivated fields. Helpful to the conservation of the tree is the belief that it has mystical and beneficial properties.

Key Information Sources

Adu-Ampomah, Y., Amponsah, J.D., and Yidana, J.A. 1995. Collecting germplasm of sheanut (*Vitellaria paradoxa*) in Ghana. Plant Genet. Resour. Newsl. **102**: 37–38.

Bayala, J., Mando, A., Ouedraogo, S.J., and Teklehaimanot, Z. 2003. Managing *Parkia biglobosa* and *Vitellaria paradoxa* prunings for crop production and improved soil properties in the sub-Sudanian Zone of Burkina Faso. Arid Land Res. Manage. **17**: 283–296.

Bouvet, J.M., Fontaine, C., Sanou, H., and Cardi, C. 2004. An analysis of the pattern of genetic variation in *Vitellaria paradoxa* using RAPD markers. Agrofor. Syst. **60**: 61–69.

Bruinsma, D. 1998. Shea nut processing—possibilities and problems in the choice of technology for women. Food Chain, **22**: 3–6.

Hall, J.B. 1996. *Vitellaria paradoxa*: a monograph. School of Agricultural and Forest Sciences, University of Wales, Bangor, UK. 105 pp.

Hyman, E. 1991. A comparison of labor-saving technologies for processing shea nut butter in Mali. World Dev. **19**: 1247–1268.

Kapseu, C., and Kayem, J. (*Editors*). 2000. International workshop on drying and improvement of shea and canarium, Ngaoundéré, Cameroon, 1–3 December 1999. Presses universitaires de Yaoundé, Yaoundé, Cameroon. 463 pp. [In French.]

Kater, L.J.M., Kante, S., and Budelman, A. 1992. Karite (*Vitellaria paradoxa*) and nere (*Parkia biglobosa*) associated with crops in South Mali. Agrofor. Syst. (Neth.), **18**: 89–105.

Kelley, B.A., Hardy, O.J., and Bouvet, J.M. 2004. Temporal and spatial genetic structure in *Vitellaria paradoxa* (shea tree) in an agroforestry system in southern Mali. Mol. Ecol. **13**: 1231–1240.

Kelly, B.A., Bouvet, J.M., and Picard, N. 2004. Size class distribution and spatial pattern of *Vitellaria paradoxa* in relation to farmers' practices in Mali. Agrofor. Syst. **60**: 3–11.

Kershaw, S.J., and Hardwick, E. 1981. Heterogeneity within commercial contract analysis samples of shea-nut kernels. J. Am. Oil Chem. Soc. **58**: 706–710.

Lovett, P.N., and Haq, N. 2000*a*. Evidence for anthropic selection of the sheanut tree (*Vitellaria paradoxa*). Agrofor. Syst. **48**: 273–288.

Lovett, P.N., and Haq, N. 2000*b*. Diversity of the sheanut tree (*Vitellaria paradoxa* C.F. Gaertn.) in Ghana. Genet. Resour. Crop Evol. **47**: 293–304.

Maranz, S., and Wiesman, Z. 2003. Evidence for indigenous selection and distribution of the shea tree, *Vitellaria paradoxa*, and its potential significance to prevailing parkland savanna tree patterns in sub-Saharan Africa north of the equator. J. Biogeog. **30**: 1505–1516.

Maranz, S., Kpikpi, W., Wiesman, Z., Saint Sauveur, A. de, and Chapagain, B. 2004. Nutritional values and indigenous preferences for shea fruits (*Vitellaria paradoxa* C.F. Gaertn. F.) in African agroforestry parklands. Econ. Bot. **58**: 588–600.

Maranz, S., Wiesman, Z., and Garti, N. 2003. Phenolic constituents of shea (*Vitellaria paradoxa*) kernels. J. Agric. Food Chem. **51**: 6268–6273.

Maranz, S., Wiesman, Z., Bisgaard, J., and Bianchi, G. 2004. Germplasm resources of *Vitellaria paradoxa* based on variations in fat composition across the species distribution range. Agrofor. Syst. **60**: 71–76.

McAllan, A., Aebischer, D. (*Editor*), and Tomlinson, H. (*Editor*). 1996. *Parkia biglobosa*, the dawadawa tree (néré), and *Vitellaria paradoxa*, the shea butter tree (karité): a handbook for extension workers. School of Agricultural and Forest Sciences, University of Wales, Bangor, UK. 30 pp.

Odebiyi, J.A., Bada, S.O., Omoloye, A.A., Awodoyin, R.O., and Oni, P.I. 2004. Vertebrate and insect pests and hemi-parasitic plants of *Parkia biglobosa* and *Vitellaria paradoxa* in Nigeria. Agrofor. Syst. **60**: 51–59.

Okullo, J.B.L., Hall, J.B., and Obua, J. 2004. Leafing, flowering and fruiting of *Vitellaria paradoxa* subsp. *nilotica* in savanna parklands in Uganda. Agrofor. Syst. **60**: 77–91.

Olaniyan, A.M., and Oje, K. 2002. Some aspects of the mechanical properties of shea nut. Biosyst. Eng. **81**: 413–420.

Teklehaimanot, Z. 2004. Exploiting the potential of indigenous agroforestry trees: *Parkia biglobosa* and *Vitellaria paradoxa* in sub-Saharan Africa. Agrofor. Syst. **61/62**: 207–220.

Specialty Cookbooks

Lamien, N., Sidibe, A., and Bayala, J. 1996. The joy of cooking: recipes for the success of shea tree. Agroforestry Today (Oct.–Dec.), International Council for Research in Agroforestry (ICRAF), Nairobi, Kenya. pp. 10–11.

SORGHUM

Family: Poaceae (Gramineae; grass family)

Names

Scientific name: *Sorghum bicolor* (L.) Moench (*S. vulgare* Pers.)

- The common name sorghum and the genus name *Sorghum* are based on an East Indian name for the plant. This has been interpreted as originating from the Latin *surgo*, arise, so named because of the height of the plant. An alternative speculation is that the name arose from an early Romance language phrase, *syricum gramen*, "Syrian grass."
- Other names for sorghum are broomcorn, durra, Guinea-corn, kaffir-corn, milo, and shattercane.
- *Bicolor* in the scientific name *S. bicolor* is from the Latin *bis*, two, and *color*, color, referring to the green leaves that have white midribs.

Plant Portrait

Sorghum is a corn-like grass, generally tall (sometimes surpassing 4 m or 13 feet), although dwarf forms about 1 m (about a yard) high are available. Sorghum is usually annual or weakly perennial, although perennial forms that live for years by reproducing vigorously from underground stems (rhizomes) are known. Each plant typically has many stalks.

Sorghum culture goes back to antiquity. It is believed to be an ancient African crop, perhaps domesticated as long as 7000 years ago. The savanna zone south of the Sahara has been advanced as a likely site of selection. Possibly, cultivated sorghums were also developed independently in India and China. The main center of distribution of cultivated sorghums is in Africa. The crop has been grown in Ethiopia for more than 5000 years. Sorghum has not been found either in the ancient tombs of Egypt or in early farming sites in the Near East. Probably well before the beginning of the Christian Era, domesticated sorghum was spread to India. Sorghum is thought to have reached China at least by the beginning of the Christian Era. Grain sorghum was first brought to America with the slave trade from West Africa. Forage sorghums were introduced to the United States about 1850.

Today, sorghums are widely distributed throughout the tropics, subtropics, and warm temperate areas of the world. Sorghum is grown mostly in parts of the world south of 40° north latitude. About one-third of world production occurs in Asia, one-third in the United States, and one-quarter in Africa. About 50 million ha (124 million acres) of sorghum are grown annually in the world. On a world basis, somewhat more grain is used to feed humans than animals, although almost all sorghum raised in North America is used for livestock. The sorghum grain belt of the United States includes the Great Plains from its southern boundary in Texas through southern Nebraska. There is also considerable production in other parts of Texas, in Arizona, California, and several southern states. Attempts to grow sorghum in Canada have had limited success, as it is not suitable for northern climates. However, sorghum is rarely surpassed by other crops in ability to grow vigorously during hot conditions, and to survive dry weather. It is far more productive than corn in this respect.

Sorghum (*Sorghum bicolor*), from *Flores des Serres* (1845–1880, vol. 10).

Sorghum is grown for five principal uses: grain for humans; grain for wild and pet birds and for livestock; entire plants for animal food (forage and fodder); syrup or sugar; and industrial use of the stems and fibers. The human food uses are emphasized below.

Sorghum is one of the world's most important food crops, ranking only behind wheat, rice, corn, barley, and potato, and perhaps on a par with soybean and oats. Sorghum grain (including grain for livestock) ranks about fifth in world acreage, after wheat, rice, corn, and barley. In the United States, Australia, and other western countries that grow sorghum, most of the grain harvested is used as livestock feed, but in the Orient and Africa, most is used as food. Grain sorghum is a staple cereal in hot dry tropics, the threshed grain ground into a wholesome flour. Sorghum is a major source of nutrition for people in Africa, China, India, and other countries of the Near East and Middle East, and has also become important in Central America. In much of Africa, where sorghum is the most important native cereal, it is the main food crop of subsistence agriculture.

Grain sorghums are enormously varied. Kafir sorghums from South Africa have thick, juicy stems, large leaves, and seeds that may be white, pink, or red. Milo sorghums from East Africa have stems that are less juicy than in Kafir, leaf blades that are wavy with a yellow midrib, and seeds that are pale pink to cream. Feterita sorghums came from Sudan, and have seeds that are chalky white. Durra sorghums, from the Mediterranean area, the Near East, and Middle East, have flattened seeds. Sballu sorghums from India have seeds that are pearly white. Koaliang sorghums from China, Manchuria, and Japan have seeds that are brown and bitter in taste. These and other groups have been hybridized, producing a very wide range of varieties.

Some quantities of grain sorghums go into industrial use, both for food and nonfood purposes. Starch is manufactured in the same way that cornstarch is produced from corn. The starch is then used in foods, for adhesives, and for various other industrial products. Alcohol can also be produced from sorghum.

Some sorghums produce considerable sucrose (sometimes as much as 10% of the stalks), and are used for syrup production, and also as forage. These are called "sorgos" in North America, and are also known as saccharine sorghum, sweet sorghum, and sugar sorghum. Most sorgos have smaller kernels than found in grain-type sorghums, but have thick, juicy stems. The development of a syrup industry for sorghum took place mostly in the United States.

Fruiting head of sorghum (*Sorghum bicolor*), from Metzger (1824).

Major production is in areas adapted to growing of cotton, and sorgos are extensively cultivated in the United States in the Great Plains and Gulf states. Sorghum syrup is popular in the nearly two dozen states where it is produced in appreciable quantities. Notable areas of production are Mississippi, Alabama, Iowa, Tennessee, and Kentucky.

Sorghums may contain toxic quantities of some chemicals, including dhurrin. The amount of dhurrin depends on variety and environmental conditions, and may be present both in the green parts of the plant, and in the grain. This can be converted to hydrocyanic (prussic) acid, which is quite poisonous. Drought is known to increase hydrocyanic acid content notably. Prussic acid level falls rapidly in properly dried sorghum, so poisoning of livestock and humans can be avoided. Dark brown and red seed coats have high concentrations of tannin, unlike white and yellow seeds. Tannins are bitter, but not only do they decrease palatability, they decrease protein digestibility by binding with seed proteins and digestive enzymes, so that white and yellow seeds are better for food products. However, birds love sorghum seeds, are a major pest problem, and prefer the lighter-colored seeds.

Culinary Portrait

For food use, sorghum grain may be roughly ground and made into bread-like preparations, used after grinding and stewing as a mush or porridge, or made into flour for mixing with wheat flour for breads. Sorghum flour does not contain gluten, and so is not suitable alone for preparing leavened breads, although it is widely used to prepare flat breads. Varieties with considerable starch have properties similar to tapioca. The grain is also a principal ingredient of African beers, particularly red-grained, bitter-tasting types.

Sweet sorghum syrup was once the sweetener of choice among rural cooks in North America, when sugar was somewhat rare and expensive. While sorghum syrup is not competitive with that of other sugar syrup sources today, sorghum syrup has a distinctive taste and is considered a table delicacy. It is often available in health and specialty food stores. In the southern United States, sorghum is used in a variety of dishes, including sorghum cake (made with sour milk, spices, and baked in a loaf pan), and it is used as a syrup or spread for biscuits, pancakes, waffles, corn cakes, and other breakfast foods. Sorghum syrup spoils quickly, like molasses, and should be stored in the refrigerator for no longer than 2 months after the container is opened.

Culinary Vocabulary

- "Charlie Taylor" was a mixture of bacon grease and syrup, usually sorghum syrup, which was used by Western Americans as a substitute for butter.
- Just as "maple sugar boils," which combine preparation of maple sugar with a party atmosphere, are traditional in areas where sugar maples grow, so "sorghum boils" were once common events in the American South, joining preparation of sorghum syrup with a social event.

Curiosities of Science and Technology

- The economic classes in ancient Egypt were distinguishable on the basis of what kind of bread they ate. The rich ate bread made of wheat. The common people ate bread made of barley. And the very poor ate bread made of sorghum.
- Some varieties of sorghum, known as broomcorn, develop flowering heads that look like and are used as brooms after the seeds are removed. European broomcorn seems to have first become known in Italy in the 17th century, although the Chinese had developed a type of broom sorghum earlier. There has been considerable production of brooms using broom sorghum in the United States. However, such natural brooms are being replaced by plastic ones.
- American statesman and philosopher Benjamin Franklin (1706–1790) is credited with the introduction of broomcorn into the United States from England in 1725. It is reported that he planted a seed from an imported broom, thereby growing the first broomcorn in North America.
- Hybrids have been made between sorghum and sugar cane (*Saccharum officinarum*), and also between sorghum and corn (*Zea mays*). All three species are in the grass family, so that they are at least distantly related, but crosses between such very different species is extraordinarily unusual in the plant world.

Key Information Sources

Bennett, W.F., Tucker, B., and Maunder, A.B. 1990. Modern grain sorghum production. Iowa State University Press, Ames, IA. 169 pp.

Clayton, W.D., and Renvoize, S.A. 1982. *Sorghum. In* Flora of tropical East Africa. Gramineae (part 3). *Edited by* R.M. Polhill. A.A. Balkema, Rotterdam, Netherlands. pp. 726–731.

Dendy, D.A.V. 1995. Sorghum and millets: chemistry and technology. American Association of Cereal Chemists, St. Paul, MI. 406 pp.

Doggett, H. 1988. Sorghum. 2nd edition. Longman Scientific & Technical, London, UK. 512 pp.

Doggett, H., and Prasada Rao, K.E. 1995. Sorghum. *In* Evolution of crop plants. 2nd edition. *Edited by* J. Smartt and N.W. Simmonds. Longman Scientific & Technical, Burnt Mill, Harlow, Essex, UK. pp. 172–180.

Food and Agriculture Organization of the United Nations. 1995. Sorghum and millets in human nutrition. Food and Agriculture Organization of the United Nations, Rome, Italy, and Lanham, MD. 184 pp.

Frederiksen, R.A., and Odvody, G.N. 2000. Compendium of sorghum diseases. 2nd edition. American Phytopathological Society Press, St. Paul, MN. 78 pp.

Freeman, K. 1986. Sweet sorghum culture and sirup production. U.S. Dept. of Agriculture, Washington, DC. 55 pp.

George Washington University. 1967. Sorghum: a bibliography of world literature covering the years, 1930–1963. Scarecrow Press, Metuchen, NJ. 301 pp.

Gomez, M.I., House, L.R., Rooney, L.W., and Dendy, D.A.V. (*Editors*). 1992. Utilization of sorghum and millets. International Crops Research Institute for the Semi-Arid Tropics, Patancheru, AP, India. 244 pp.

Harlan, J.R., and Wet, J.M.J. de. 1972. A simplified classification of cultivated sorghum. Crop Sci. **12**: 172–176.

House, L.R. 1985. A guide to sorghum breeding. 2nd edition. International Crops Research Institute for the Semi-Arid Tropics, Patancheru, AP, India. 206 pp.

Hulse, J.H., Laing, E.M., and Pearson, O.E. 1980. Sorghum and the millets: their composition and nutritive value. Academic Press, London, UK. 997 pp.

Leslie, J.F. 2002. Sorghum and millets diseases. Iowa State Press, Ames, IA. 504 pp.

Maii, R.K. 1996. Sorghum science. Science Publishing, Lebanon, NH. 352 pp.

Milliano, W.A.J. de, Frederiksen, R.A., and Bengston, G.D. 1992. Sorghum and millets diseases: a second world review. International Crops Research Institute for the Semi-Arid Tropics, Patancheru, AP, India. 370 pp.

Peacock, J.M., and Wilson, G.L. 1984. Sorghum. *In* The physiology of tropical field crops. *Edited by* P.R. Goldsworthy and N.M. Fisher. John Wiley & Sons, Chichester, UK. pp. 249–279.

Rains, G.C., Cundiff, J.S., and Welbaum, G.E. 1993. Sweet sorghum for a piedmont ethanol industry. *In* New crops. *Edited by* J. Janick and J.E. Simon. Wiley, New York, NY. pp. 394–399.

Rockefeller Foundation, and George Washington University. 1973. Sorghum: a bibliography of the world literature, 1964–1969. Scarecrow Press, Metuchen, NJ. 393 pp.

Rooney, L.W., and Serna-Saldivar, S.O. 2000. Sorghum. *In* Handbook of cereal science and technology. 2nd edition. *Edited by* K. Kulp and J.G. Ponte, Jr. Marcel Dekker, New York, NY. pp. 149–175.

Rooney, L.W., Khan, M.N., and Earp, C.F. 1980. The technology of sorghum products. *In* Cereals for food and beverages: recent progress in cereal chemistry. *Edited by* G.E. Inglett and L. Munch. Academic Press, New York, NY. pp. 513–554.

Rooney, L.W., Kirleis, A.W., and Murty, D.S. 1986. Traditional foods from sorghum: their production, evaluation and nutritional value. *In* Advances in Cereal Science and Technology, Vol. 8. *Edited by* Y. Pomeranz. American Association of Cereal Chemists, St. Paul, MN. pp. 317–353.

Smith, C.W., and Frederiksen, R.A. 2000. Sorghum: origin, history, technology, and production . Wiley, New York, NY. 824 pp.

Stenhouse, J.W., and Tippayaruk, J.L. 1996. *Sorghum bicolor* (L.) Moench. *In* Plant resources of South-East Asia, 10, cereals. *Edited by* C.J.H. Grubben and S. Partohardjono. Backhuys Publishers, Leiden, Netherlands. pp. 130–136.

Wall, J.S., and Ross, W.M. (*Editors*). 1970. Sorghum production and utilization. AVI Publishing Co., Westport, CT. 702 pp.

Wet, J.M.J. de. 1978. Systematics and evolution of *Sorghum* sect. *Sorghum* (Gramineae). Am. J. Bot. **645**: 477–484.

Wet, J.M.J. de. 2000. Sorghum. *In* The Cambridge world history of food. *Edited by* K.F. Kiple and K.C. Ornelas. Cambridge University Press, Cambridge, UK. pp. 152–158.

Wet, J.M.J. de, and Harlan, J.R. 1972. The origin and domestication of *Sorghum bicolor*. Econ. Bot. **25**: 128–135.

Wet, J.M.J. de, and Huckabay, J.P. 1967. The origin of *Sorghum bicolor*: distribution and domestication. Evolution, **21**: 787–802.

Specialty Cookbooks

Bear Wallow Books. 1991. Old-fashioned honey, maple syrup, sorghum recipes. Bear Wallow Books, Indianapolis, IN. 31 pp.

Governor's Council on Agriculture. 1981. Honey & sorghum: the flavor of Kentucky. Governor's Council on Agriculture, Frankfort, KY. 12 pp.

Maxwell, M. 1972. Sorghum cook book. Nutri-Books, Denver, CO. 64 pp.

Mitchell, P. 1989. Sweet 'n' slow: apple butter, cane molasses, and sorghum syrup recipes. Revised edition. Patricia B. Mitchell Foodways Publications, Chatham, VA. 37 pp.

Mitchell, P. 1991. Just naturally sweet: recipes utilizing honey, molasses, sorghum, and maple syrup, no refined sugar. Revised edition. Patricia B. Mitchell Foodways Publications, Chatham, VA. 36 pp.

National Sweet Sorghum Producers and Processors Association. 1991. Sorghum treasures: a compilation of recipes—old and new. National Sweet Sorghum Producers and Processors Association/Jumbo Jack's Cookbooks, Audubon, IA. 160 pp.

Orsburn, L., and Brumley, K. 1993. Grandpa's sorghum molasses cookbook. Grandpa Leroy, Wewoka, OK. 99 pp.

Vander Hart, P., Van Klompenburg, C., and Hoksbergen, K. 1992. Sweet memories: sorghum making at Lower Grove Park. Lower Grove Books, Pella, IA. 44 pp.

SOYBEAN

Family: Fabaceae (Leguminosae; pea family)

Names

Scientific name: *Glycine max* (L.) Merr. [*Dolichos soja* L., *Glycine gracilis* Skvortzov, *G. hispida* (Moench) Maxim., *Phaseolus max* L. *Soja hispida* Moench, *S. max* (L.) Piper]

- The English "soy" is based on the Dutch *soja*, from the Japanese *shoyu*, and Old Chinese *sou* or *shi-yow*, "salt bean + oil."
- Soybean is also known as soya bean and soja bean. Obsolete names include Chinese pea, Japanese pea (Japan pea), and Manchurian bean.
- The Chinese word for soybean, *ta-too*, means "greater bean."
- The genus name *Glycine* is based on the Greek *glykys*, sweet, and is probably an allusion to the sweetness of the fruits of some of the species.
- The origin of *max* in the scientific name *G. max* is unclear. One explanation is that it is an old name taken up by the Swedish botanist Linnaeus when he first created a Latin name for the soybean. It has been claimed that the word *max* is the result of a Portuguese transcription of the Persian name of the plant. It has also been suggested that *max* is related to *maxima*, Latin for extremely large, referring to the large nodules on the root system.

Plant Portrait

Soybean is an annual, herbaceous plant, erect and bushy (sometimes vine-like), from 30 cm (1 foot) to as much as 2 m (6½ feet) tall. The pods are 25–75 mm (1–3 inches) long, and contain 1–5 seeds, typically two or three, and both the young pods and mature seeds are used for food.

Soybean is said to have been described as such as long ago as 2800 BC by legendary Chinese emperor Shen-Nung. It was considered to be one of the "five sacred grains" (or "five essential grains of life") considered indispensable for the existence of ancient Chinese civilization (the others were the cereals rice, wheat, barley, and millet). The plant was domesticated in China, possibly starting during the Shang Dynasty (about 1700–1100 BC) or earlier. The first specific reference to consumption of immature soybean pods as a vegetable comes from the 2nd century BC. Soybean was cultivated early in the histories of Korea, Manchuria, and Japan and was introduced to Europe in the late 1700s and to the United States in 1765. However, the potential of soybean in the Western World did not become realized until the early 20th century. The phenomenal expansion of soybean cultivation, beginning in the 1930s in much of the Western World, earned it the label "Cinderella Crop." In the United States, soybeans became the nation's Cinderella crop earlier—in the 1920s. It is now widely cultivated in tropical and temperate areas.

Soybean has become the world's most important oilseed and grain legume crop. It is the world's leading oilseed [other important oilseeds are cottonseed, peanut, sunflower, canola/rapeseed, and oil palm (including both copra and palm kernel), all of which are discussed in this book]. The world's leading producer is the United States, with large amounts also produced in China, Brazil, and Argentina. One-third of the soy produced by the United States is exported, notably to Japan. Of the soybean that stays in the United States, more than 90% is fed to agricultural animals. Soybeans are grown for their seeds, which are rich in protein (average 40%) and oil (average 21%). In many industrialized countries, most of the crop is marketed for processing, which involves hulling the seeds and extracting the oil. Most of the residual meal is used in livestock feeds.

Thousands of cultivars of the soybean are available (a recent estimate claims there are more than 20 000), many specifically selected for producing forage, industrial oil, vegetable beans, and sprouts. Cultivars suitable for use as a vegetable are different from those used as forage or for obtaining oil. Seed colors of different varieties range from yellow to gray and black. However, in Western commerce, yellow seeds are usually demanded because use of brown or black seeds results in undesirable coloration. Black seeds are usually employed to make soya sauce.

Culinary Portrait

The edible oil is used in cooking and in the production of salad dressings, margarines, whipped toppings, coffee whiteners, icings, ice creams, confections, shortenings, frozen desserts, and soups. Soya flour is low in starch and lacks gluten, and accordingly, it cannot be used alone in making bread. It is best employed as a protein supplement for foods, so

Soybean (*Glycine max*), from Turner (1891).

soybean protein in various forms is added to flour or processed into protein concentrates and textured and modified proteins. As is well known, combining seeds of cereals (like wheat) and legumes (like soybean) often produces an excellent balance of essential amino acids for human nutrition.

Soybeans are probably used in more ways for food than any other single vegetable species. In the Far East, the soybean is consumed in many ways. Sprouts are popular, although not as common as mung bean sprouts. Soybeans are used as an immature green vegetable, often termed "vegetable soybean," although edamame is the preferable name (from the Japanese *eda mame*, fresh soybean). The immature pods or shelled seeds are cooked with various kinds of meats, rice dishes, stews, and cereal. The seeds are employed to make "tofu" (bean curd; this has been described as "soft, vegetable cheese"), bean flour (known as *kinako* in Japan), soybean milk, okara (a by-product of the fabrication of soy milk, this has a fine, crumbly texture like that of freshly grated coconut), soya sauce, and various forms of fermented soybean [known in Japan as *miso* and *natto*, and in Indonesia and China as *tempe*]. The sauces produced by fermentation constitute the basic flavoring agents of Oriental foods. To produce soybean milk, used in parts of the East much like cow's milk in the West, soybeans are ground and added to water. To form tofu, soybean milk is allowed to curdle. Tofu, a kind of boneless meat substitute used in the Orient, is now available in Western supermarkets. In southern Asia and developing African and Latin American countries, the crop is increasingly viewed as an important infant food. This multipurpose crop has even been used as a coffee substitute.

Soy sauce, a salty sauce for flavoring meat, fish, vegetables, marinades, and other sauces, almost makes it unnecessary to use ordinary table salt with Japanese and Chinese food. The saltiness comes from doubly fermenting soybeans along with bruised, roasted wheat in a solution of salty water. Soy sauce is mostly dark in Western nations ("dark soy" or "table soy" has caramel added; "heavy soy" or "black soy" has molasses added); however, light versions are also available in the East. Soy sauce is almost invariably found on the tables of Chinese restaurants in Western nations. However, in China, soy sauce is not used as a table condiment, as such behavior is considered to be an insult to the chef.

Soybeans should be thoroughly cooked so that the anti-nutrients (toxins) present are deactivated (the numerous methods of fermentation by which soybeans are prepared in Asia have the effect of neutralizing the anti-nutrients). Dried soybeans are first soaked, and then cooked for at least 3 hours (up to 9 hours are required for some varieties). The beans are cooked until they can be easily mashed with a fork. Ground soybean has had the outer shell removed from the seeds, which are then ground into fine granules. This cooks much more quickly than whole soybean (about 30 minutes, using 4 cups of water for each cup of ground soybean).

Culinary Vocabulary

- As noted above, most soybean in North America is produced for seed oil, industrial purposes, and animal feed, but some is produced for direct use as a vegetable. Vegetable soybean is called "edamame" (the Japanese name for vegetable soybean), pronounced ay-duh-MAH-may. Edamame is becoming the standard name, replacing the bewildering variety of other names that have been used, which include beer bean, edible soybean, edible green soybean, fresh green soybean, garden soybean, garden-type soybean, green soybean, green vegetable soybean, immature soybean, large-seeded soybean, vegetable soybean, and vegetable-type soybean. The use of the word "green" is confusing because mature green soybean seeds are also called green soybeans.
- The Chinese figuratively call tofu "meat without bones." However, the original Chinese *dòufu*, from which the Japanese "tofu" arose, means "rotten beans."
- "Soybean pudding," "tofu pudding," and "soft soybean curds" are soft curds prepared from soy milk, and usually sold in an unpressed or lightly pressed form.
- "Cotton tofu" has a firm texture and low moisture content. Its irregular surface pattern is caused by the weave of the cotton fabric used to press it.
- "Silk tofu" has a soft texture and high moisture content. Its smooth surface is due to a silk fabric used to press it, not leaving a pattern.
- "Tofutti" is a frozen mixture of tofu products, sweeteners, and flavoring agents made up to resemble ice cream in flavor and texture.
- "Tofurkey" (based on *tofu + turkey*) is a vegetarian dish prepared to mimic the texture and taste of turkey. (As one critic remarked, "tofurkey is a vegetable-based imitation turkey product that tastes just like turkey without all the flavor.")
- Dried sticks made of soybean curd are generally simply called "bean curd" or "dried (bean curd) sticks," but are sometimes called "bamboo" (and "second bamboo") because of the resemblance to bamboo in texture. These are found in Chinese and Japanese markets, and are soaked to make them pliable enough to cook.
- "Soy jam" is a paste-like sediment of soy sauce, sold in cans in Chinese food stores. It is more viscous than soy sauce, and may or may not be better tasting, as some versions are less salty, other much saltier.
- "Tofu noodles" [also known as beancurd noodles (bean curd noodles), soybean curd noodles, soy noodles, soy vermicelli, and tofu shreds) are chewy noodles made from tofu. They look like a pack of rubber bands, and are usually served in salads, soups, and stir-fries. They are often sold in the refrigerated or frozen foods section of Asian markets.
- "Miso" (pronounced ME-so) is a thick paste made from fermenting soybeans and either rice or barley; the mixture is inoculated with yeast and allowed to ferment for various periods, ranging from 3 weeks to 3 years. Miso is heavily salted, and occasionally sweetened, and there are specialty misos with additional ingredients, such as sake, nuts, spices,

and seaweed. Miso is used as a seasoning and soup base, and in Asian markets is available in refrigerated tubs, jars, and plastic bags. Shiromiso or white miso is made with rice. Akamiso or red miso (which may be quite brown) is made with barley, and has a much stronger flavor. Aside from the salt, miso is extremely nutritious, and in Asia, it is added to almost every kind of dish. Miso soup is often consumed for breakfast in Japan, and is served as a first course in Japanese restaurants.

- "Natto" (pronounced na-TOH and NAH-toh) is light brown, steamed, fermented, and mashed soybean used to season Japanese cuisine. It is often combined with other ingredients, such as soy sauce, mustard, and chives. Traditionally, natto is wrapped in straw to allow it to breathe, but is also sometimes available in refrigerated packages. Natto has a strong, cheesy odor and a slippery, gelatinous, stringy texture, both of which are offensive to many, but natto is very nutritious and easily digested.
- *Kinako* (pronounced KEE-nah-koh), a Japanese word for a strong-flavored soy flour, is high in protein and often available in natural-food stores. It spoils within a few weeks at room temperature, and is best stored in a freezer.
- "Tempe" (tempeh, pronounced TEHM-pay) is a fermented bean curd cake, important in Indonesia, Japan, and China. It looks like Rice Crispy confections. Although usually made from soybean, some forms are prepared from other beans, grains, and occasionally even coconut. Tempe is widely available in the refrigerators of Asian markets and health-food stores, and can be stored in the home refrigerator for up to 3 weeks, or for 6 months when frozen. The taste (yeasty and nutty) and texture vary considerably, depending on the stage of fermentation. Popular in Asian and vegetarian cuisines, tempe is high in protein and so nutritious that is has been characterized as a "nearly perfect food."
- "Hoisin sauce" (also called haisein sauce and Peking sauce) is a sweet, thick, mahogany-colored sauce made from soybeans and other ingredients. It is available in cans or jars in Chinese food stores, and is used as a seasoning or condiment. *Chee how* is a spicy version of hoisin sauce.
- "Tamari" (pronounced tuh-MAH-ree) is a Japanese sauce traditionally made from fermented soybean (but more recently tending to combine soy and wheat). It is usually aged, and thicker, darker, and mellower than most soy sauces.
- *Ketjap manis* (*ketjap benteng*) is a syrupy, sweet, Indonesian soy sauce, which is much thicker, darker, and sweeter than Chinese and Japanese soy sauces. *Ketjap* is an Indonesian word, meaning a flavoring sauce for food, and is the root of the English words ketchup and catsup. (For additional information, see TOMATO, CULINARY VOCABULARY.)
- "Yellow bean sauce," prepared from salted, preserved yellow soybeans, and sometimes flavored with chile pepper, is commonly encountered in Chinese and Thai cuisines.

Curiosities of Science and Technology

- The archaic Chinese ideogram *su* was used to denote the soybean. The ideogram has a row of little marks at the base of the character, which are believed to represent the roots of the plant. The roots are highlighted because, like other plants of the legume family, they have nodules, which house bacteria that take nitrogen out of the air and make it available to the plant, thereby increasing the fertility of the soil.
- Soybeans were used in ancient Chinese folk medicine as a specific remedy for proper functioning of the bowels, heart, kidney, liver, and stomach.
- During the American Civil War, soybeans were frequently substituted for coffee beans, especially by Union army soldiers.
- France was the first country in Europe to plant soybeans and experiment with their cultivation. The first organization in the Western World to promote the use of soybean was the Society for Acclimatization, centered in Étampes, France. In 1911, the group served an "all-soya meal" to demonstrate the versatility of the soybean.
- During World War I, British doctor Charles Fearn was asked by President Wilson to assist the US war effort by developing soybean products. His Soy-O-Pancake mix, sold in health-food stores, was one of the first soy products to enter the US market. During the Second World War, soy flour was used extensively by the allied armies in field camps, and soybean was commonly called the "Wonder Bean" and the "Miracle Bean."
- The U.S. Armed Forces serves over 45 million kg (100 million pounds) of soy protein each year.
- Dr. John Harvey Kellogg (1852–1943), famous for his breakfast foods, produced the first soymilk and meat substitutes made from soy in the United

States. By the 1920s, he was marketing America's first foods made from soy.

- During World War II, the basic survival ration carried by Japanese solders was a bag of soybean flour.
- Henry Ford (1863–1947) was obsessed with soybeans. He once wore a suit and tie made from soy-based material, and had a 16-course meal prepared entirely from soybeans. Henry Ford's laboratories discovered many industrial uses for the soybean. The oil, combined with other components, could be used to mould parts of automobile bodies. By 1935, a full bushel of soybeans went into the manufacture of each Ford automobile.
- With the possible exception of corn, soybean surpasses all other crops in the diversity of its industrial usage. In addition to edible products, there are thousands of manufactured goods, including soaps, resins, lubricants, rubber substitutes, paints, varnishes, enamels, inks, stains, adhesives, sealing and caulking compounds, linoleum, oilcloth, explosives, diesel fuel oil, and various industrial oils.
- Soybeans have health-promoting constituents. They are used in food for diabetics because of the low starch content. Soybean oil, high in unsaturated fatty acid, is recommended to combat hypercholesteremia (dangerous levels of cholesterol deposition). Soybean contains high amounts of lecithin, reported to be a potent vasodepressor; that is, it dilates blood vessels.
- Soybeans yield more usable protein per acre than does any other common food crop. Whole soybeans contain 40% protein. Soy flour contains approximately 2.5 times the amount of protein as meat. Soybean provides a far less expensive protein than animal foods. It has been calculated that 0.4 ha (1 acre) of soybeans could keep a moderately active man alive for 2200 days, but that same area would produce only enough beef to keep him alive for 75 days. It has been said that if everyone ate soybeans instead of meat, the world could support a population of 16 billion people.

Key Information Sources

Beachat, L.R. 1994. Fermented soybean foods. Food Technol. **38**: 64–70.

Beckman, C. 2005. Vegetable oils: competition in a changing market. Bi-weekly Bull. (Agric. Agri-Food Can.), **18(11)**: 1–6.

Berk, Z. 1992. Technology of production of edible flours and protein products from soybeans. Food and Agriculture Organization of the United Nations, Rome, Italy. 178 pp.

Brar, G.S., and Carter, T.E., Jr. 1993. Soybean, *Glycine max* (L.) Merrill. *In* Genetic improvement of vegetable crops. *Edited by* G. Kalloo and B.O. Bergh. Pergamon Press, New York, NY. pp. 427–463.

Caldwell, B.E., and Howell, R.W. (*Editors*). 1973. Soybeans: improvement, production, and uses. American Society of Agronomy, Madison, WI. 681 pp.

Carter, T.E., Jr., and Shanmugasundaram, S. 1993. Vegetable soybean (*Glycine*). *In* Pulses and vegetables. *Edited by* J.T. Williams. Chapman & Hall, New York, NY. pp. 219–239.

Centro Nacional de Pesquisa de Soja (Brazil), and Food and Agriculture Organization of the United Nations. 1994. Tropical soybean: improvement and production. FAO, Rome, Italy. 254 pp.

Erdman, J.W., and Fordyce, E.J. 1989. Soy products and the human diet. Am. J. Clin. Nutr. **49**: 725–737.

Erickson, D.R. (*Editor*). 1995. Practical handbook of soybean processing and utilization. AOCS Press, Champaign, IL. 584 pp.

Hartman, G.L., Sinclair, J.B., and Rupe, J.C. 1999. Compendium of soybean diseases. 4th edition. APS Press, St. Paul, MN. 100 pp. + plates.

Heatherly, L.G., and Hodges, H.F. 1999. Soybean production in the midsouth. CRC Press, Boca Raton, FL. 394 pp.

Hermann, F.J. 1962. A revision of the genus *Glycine* and its immediate allies. U.S. Dept. of Agriculture, Washington, DC. 82 pp.

Higley, L.G., and Boethel, D.J. 1994. Handbook of soybean insect pests. Entomological Society of America, Lanham, MD. 136 pp.

Hinson, K., Hartwig, E.E., and Minor, H.C. 1982. Soybean production in the tropics. Revised edition. Food and Agriculture Organization of the United Nations, Rome, Italy. 222 pp.

Holt, S. 1998. The soy revolution: the food of the next millennium. M. Evans, New York, NY. 214 pp.

Hume, D.J., Shanmugasundaram, S., and Beversdorf, W.D. 1985. Soyabean (*Glycine max* (L.) Merrill). *In* Grain legume crops. *Edited by* R.J. Summerfield and E.H. Roberts. Collins Professional and Technical Books, London, UK. pp. 391–432.

Hymowitz, T. 1990. Soybeans: the success story. *In* Advances in new crops. *Edited by* J. Janick and J.E. Simon. Timber Press, Portland, OR. pp. 159–163.

Hymowitz, T. 1995. Soybean. *In* Evolution of crop plants. 2nd edition. *Edited by* J. Smartt and N.W. Simmonds. Longman Scientific & Technical, Burnt Mill, Harlow, Essex, UK. pp. 261–266.

Hymowitz, T., and Harlan, J.R. 1983. Introduction of soybean to North America by Samuel Bowen in 1765. Econ. Bot. **37**: 371–379.

Hymowitz, T., and Newell, C.A. 1981. Taxonomy of the genus *Glycine*: domestication and uses of soybeans. Econ. Bot. **35**: 272–288.

Liu, K. 1997. Soybeans: chemistry, technology, and utilization. Chapman & Hall, New York, NY. 532 pp.

Newell, C.A., Delannay, X., and Edge, M.E. 1987. Interspecific hybrids between the soybean and wild perennial relatives. J. Hered. **78**: 301–306.

Norman, A.G. 1978. Soybean physiology, agronomy, and utilization. Academic Press, New York, NY. 249 pp.

Nwokolo, E. 1996. Soybean (*Glycine max* (L.) Merr.). *In* Food and feed from legumes and oilseeds. *Edited by* J. Smartt and E. Nwokolo. Chapman and Hall, London, UK. pp. 90–102.

Scott, W.O., and Aldrich, S.R. 1983. Modern soybean production. 2nd edition. S & A Publications, Champaign, IL. 230 pp.

Shanmugasundaram, S. (*Editor*). 1991. Vegetable soybean: research needs for production and quality improvement: proceedings, workshop held at Kenting, Taiwan, 29 April–2 May 1991. AVRDC, Taipei, Taiwan. 151 pp.

Shanmugasundaram, S., and Sumarano [no initial]. 1989. *Glycine max* (L.) Merr. *In* Plant resources of South-East Asia, 1, pulses. *Edited by* L.J.G. van der Maesen and S. Somaatmadja. Pudoc, Leiden, Netherlands. pp. 43–47.

Shurtleff, W., and Aoyagi, A. 1979. The book of tofu. Revised edition. Ballantine Books, New York, NY. 431 pp.

Shurtleff, W., and Aoyagi, A. 1991. Bibliography of soy sprouts: 655 references from 3rd century A.D. to 1991, extensively annotated. Soyfoods Center, Lafayette, CA. 185 pp.

Singh, S.R., Rachie, K.O., and Dashiell, K.E. (*Editors*). 1987. Soybeans for the tropics: research, production, and utilization. Chichester, West Sussex, UK. 230 pp.

Smith, A.K., and Circle, S.J. (*Editors*). 1978. Soybeans: chemistry and technology. Revised edition. Avi, Westport, CT. 470 pp.

Snyder, H.E., and Kwon, T.W. 1987. Soybean utilization. Van Nostrand Reinhold, New York, NY. 346 pp.

Sorosiak, T. 2000. Soybean. *In* The Cambridge world history of food. *Edited by* K.F. Kiple and K.C. Ornelas. Cambridge University Press, Cambridge, UK. pp. 422–427.

Verma, D.P.S., and Shoemaker, R.C. 1996. Soybean: genetics, molecular biology, and biotechnology. CAB International, Wallingford, UK. 270 pp.

Wilcox, J.R. (*Editor*). 1987. Soybeans: improvement, production and uses. 2nd edition. American Society of Agronomy, Madison, WI. 888 pp.

Specialty Cookbooks

(Note: There are hundreds of books dedicated to soybean cooking. The following are examples.)

Evans, K.L., and Rankin, C. 2000. Giant book of tofu cooking. Sterling Publishing, New York, NY. 256 pp.

Greenberg, P., and Hartung, H.N. 1988. The whole soy cookbook: 175 delicious, nutritious, easy-to-prepare recipes featuring tofu, tempeh, and various forms of nature's healthiest bean. Three Rivers Press, New York, NY. 221 pp.

Gundy Jones, D. van, and Lager, M.M. 1963. The soybean cookbook. Revised edition. Arco Publishing Co. New York, NY. 239 pp.

Hayter, J. 2000. The tofu cookbook: 50 classic and new recipes using bean curd. Apple, London, UK. 80 pp.

Honda, K., and Nagai, K. 1997. Tofu & soybean cooking: the Japanese healthy way. Graph-sha Ltd., Tokyo, Japan. 64 pp.

Jacobi, D. 2001. Amazing soy: a complete guide to buying and cooking this nutritional powerhouse with 240 recipes. Morrow, New York, NY. 364 pp.

Kale, F.S. 1937. Soya bean: its value in dietetics, cultivation, and uses, with 300 recipes. 2nd edition. Baroda State Press, Baroda, India. 375 pp.

Mallard, G. 1976. Soy bean magic: delicious recipes with soy beans, flour & grits. Hancock House, Seattle, WA. 80 pp.

McBride, R.W. 1991. Cooking with tofu: for those who hate tofu but don't know any better. 2nd edition. Gylantic Publishing Co., Littleton, CO. 62 pp.

O'Brien, J. 1983. The magic of tofu and other soybean products. Thorsons Publishers, New York, NY. 128 pp.

Sass, L.J. 1998. The new soy cookbook: tempting recipes for tofu, tempeh, soybeans, and soymilk. Chronicle Books, San Francisco, CA. 120 pp.

Treloar, B., and Inge, K. 2002. Healthy soy: cooking with soybeans for health and vitality. Periplus, Boston, MA. 111 pp.

Watanabe, T., and Kishi, A. 1984. The book of soybeans: nature's miracle protein. Japan Publications, Tokyo, Japan. 191 pp.

Winter, R. 1996. Supersoy: the miracle bean: includes a cookbook of 50 soy recipes. Crown Trade Paperbacks, New York, NY. 191 pp.

Yew, B. 2002. Everything bean curd! Times Editions, Singapore, Singapore. 200 pp.

SPINACH

Family: Chenopodiaceae (goosefoot family)

Names

Scientific name: *Spinacia oleracea* L.

- The English word "spinach" is derived through a series of languages: Old French *espinache*, Medieval Latin *spinachium*, Arabic *'isfanah*, and ultimately from the Persian *espenaj*, or *espenakh*.
- Other plants cultivated for edible spinach-like leaves and called spinach include New Zealand spinach (*Tetragonia tetragonioides* (Pall.) Kuntze), Malabar spinach (*Basella alba* L), and water spinach (*Ipomoea aquatica* Forsk.). These are relatively minor crops.
- The genus name *Spinacia* is probably based on the Latin *spinus*, spine, an allusion to the spiny fruit.
- *Oleracea* in the scientific name *S. oleracea* is the Latin for vegetable garden, or kitchen garden. Many vegetables and herbs have this word in their scientific names.

Plant Portrait

Spinach is an annual herb cultivated throughout the temperate parts of the world. Wild types are not known, but spinach is thought to have arisen from wild plants in Asia. This vegetable may have first been used in what is now Afghanistan and Tajikistan. It was apparently unknown to the ancient Greeks and Romans. Chinese writings of 647 AD described spinach as originating from Nepal. It spread to the rest of Europe and was well known in England and France by 1500. By 1806, at least three cultivars were being grown in North America. Different leaf types include crushed, crumpled, savoy (wrinkled), semisavoy, and smooth. Spinach intended for processing is usually smooth, or sometimes semisavoy. The savoy cultivars are preferred for the fresh market. Spinach is a cool-season plant that grows best when the temperature is moderate. Abundant crops of leaves are produced only in early spring or autumn (or in winter in southern North America). Spinach is grown in most temperate regions, especially in the United States, the Netherlands, and Scandinavia.

Culinary Portrait

Spinach somehow has acquired a reputation for being the archetypical bad-tasting vegetable, which although "it's good for you," and although it is the tonic gulped by the canful by muscular, squinting Popeye the sailor man to acquire miraculous strength, it remains the most universally disliked vegetable by children in the Western World. So bad is spinach's reputation that *Webster's Third New International Dictionary* gives as meanings for spinach, in addition to its being a potherb, "something repellant, obnoxious, or nonexistent...something spurious or unwanted...an untidy overgrowth...an inessential, irrelevant, or inharmonious excrescence." In the United States in the 1930s and 1940s, "spinach" was a colloquial term for "nonsense." Properly prepared and served, however, spinach is a wonderful complement to many foods.

Varieties of spinach (*Spinacia oleracea*), from Vilmorin-Andrieux (1885).

Flowering stem of spinach (*Spinacia oleracea*), from Lamarck and Poiret (1744–1829, plate 814).

Spinach is eaten by itself as a separate dish, or incorporated into a very wide variety of culinary preparations, more commonly cooked, but sometimes also raw in salads and sandwiches. It is a filling for soufflés, stuffings, and pizzas, a component of pancakes and croquettes, a classic accompaniment of veal, fish, and poultry, and part of numerous ethnic dishes. Spinach goes well with milk (which offsets the acidity) and eggs, and is often combined with them in omelets, quiches, and other dishes. "Green pastas" are various forms of pasta that include spinach. Spinach is popular intermixed with lettuce and other vegetables in mixed salads, and many restaurants feature fresh spinach salad, typically combined with bacon bits, raw mushrooms, and chopped hard-boiled eggs.

Because of the folds in its leaves and because it is a low-lying plant, fresh spinach from the garden requires thorough washing to remove all grit. The same is true of spinach sold in stores that has not been prewashed. The leaves may be swished in a sink of cold water and drained, and the process repeated until all the sand is gone. Washing spinach is best done just before using it, and it should not be allowed to soak in water because this softens the leaves. When cooked alone, only a small amount of liquid is added, as the leaves release considerable fluid naturally. Cooking reduces the volume. The water remaining on the leaves after spinach has been washed and briefly drained usually suffices for cooking, which only requires 1–3 minutes in a covered saucepan over high heat. Overcooked spinach tends to become brown. In simmered dishes, spinach is best added at the end of the cooking cycle. Pressure cookers should be avoided, as they tend to overcook spinach. Steaming tends to result in bitterness.

Spinach is valued as a vegetable for the high vitamin and iron content of its leaves. However, the high oxalic content is a health concern. One of the difficulties with high-oxalate foods is interference with calcium absorption. This problem can be overcome by serving spinach with a calcium-rich dressing of milk, cream, yogurt, cheese, or egg. Glass and stainless steel cookware and utensils will prevent oxidation (discoloration) of spinach, while untreated aluminum and cast-iron pots and pans should not be used for this acidic vegetable.

Culinary Vocabulary

- The French expression *à la florentine* refers to a French method of food preparation associated with the cuisine of Florence, Italy, in which spinach is used. Dishes are often prepared on a bed of spinach on which the principal ingredient is laid and covered with sauce and sometimes cheese; for example, Eggs Florentine is poached eggs on a bed of spinach, covered with a Mornay sauce. Crème

Florentine is cream of spinach soup. This style traces to Catherine de' Medici (1519–1589), who left Florence in 1533 to become queen of France. Catherine brought her skilled Italian chefs and contributed to the improvement of French cooking. She was so fond of spinach that it was said she had to have it with every meal, and the royal cooks learned to adapt it to numerous dishes.

- "Long-life spinach noodles" are a traditional food served during the Chinese New Year. Made with spinach, flour, and eggs, they symbolize, as the name implies, longevity.
- "Callaloo greens" refer to the edible leaves of a variety of food plants popular in the Caribbean area, including malanga, amaranth, and taro (all treated in this book). In the Caribbean area, callaloo greens are sometimes called "spinach."
- "Bongo bongo soup" is made with oysters and spinach. It was created by Victor Jules "Trader Vic" Bergeron (1902–1984), owner of Trader Vic restaurants in San Francisco and elsewhere. Bergeron said that he first tasted the soup during World War II in New Zealand, where it was made with clams instead of oysters.
- "Verdi sauce" or "Verdi mayonnaise" is a French mayonnaise sauce prepared with chopped spinach, gherkins, chives, and sour cream.

Curiosities of Science and Technology

- Because of its iron content, spinach is valued as part of a diet to treat anemia. During the First World War, spinach juice was added to wine given to French soldiers weakened by hemorrhage. But perhaps reflecting the general dislike of its taste, only 2% of the beverage was spinach juice.
- Although spinach contains iron, it is not in a readily assimilable form, and there are better food choices for those desiring especially iron-rich foods. The iron absorbed from a small serving of beef equals the iron from a big serving of spinach. This is because some vegetables, including spinach, contain iron inhibitors that interfere with the absorption of iron. The iron from animal sources (meat, fish, fowl, eggs) is "heme" or "hemo" iron, while that from plant sources is "non-heme" [referring simply to the fact that animals have blood (i.e., "heme") while plants do not]. Hemo iron food, especially red meat, is not only rich in iron, but that iron is in a form that is easily absorbed by the body, and even promotes iron absorption from vegetables eaten at the same time. This is not to deny that plant foods are also useful sources of iron. Spinach is an excellent vegetable source of iron, but so are beans, whole grains, and dried fruits. The average woman needs 18 mg of iron a day. The average man needs 10 mg, and infants and children need 10–15 mg. Iron deficiency is commonly associated with anemia, characterized by low levels of hemoglobin in the blood, causing oxygen starvation in tissues. Women of childbearing age are commonly iron deficient, because they lose blood (and iron) during their monthly period, and often do not eat enough to get the iron required. Certain nutrients, such as vitamin C, enhance the absorption of iron, so eating high-iron foods with, for example, a glass of orange juice, increases the amount of iron the body can use. Conversely, some foods, like tea, inhibit the absorption of iron.
- Like many other leafy, green vegetables, spinach tends to accumulate nitrogen in the nitrate form (NO_3), particularly when nitrate fertilizer is added to increase green color and succulence. The result can be a poisoning in the digestive tract of mammals, called methemoglobinemia. This can be serious for ruminant (cud-chewing) animals such as cows, but also for human infants who consume large amounts of high-nitrate leafy vegetables. The disease has been dubbed "blue baby" (blue baby syndrome is a condition generally associated with low-oxygen availability in babies; in fact, ingestion of high levels of nitrates can destroy haemoglobin in humans, the molecule responsible for carrying oxygen). Different varieties, soils, and methods of growing spinach and other plants have been found to influence nitrate accumulation in leafy vegetables, and growers are well aware of the problem.
- Spinach is one of the very few cultivated crops that usually (but not always) have separate male and female plants. Some others are hemp (*Cannabis sativa* L.), hop (*Humulus lupulus* L.), and date palm (see DATE). As is often the case with such plants, the males flower earlier than the females and die shortly afterwards. (In the case of spinach, both male and female plants are harvested before they flower.)
- The Green Giant corporation's entire spinach crop was wiped out by the 1989 San Francisco earthquake.
- Researchers at the University of South Florida reported in 2001 that rats fed spinach became smarter. The scientists attributed this to the large amount of antioxidants in spinach, which can block the effects of free radicals that may deteriorate

mental abilities. It was suggested that the lifelong accumulation of free radicals in the brain is linked to mental decline in old age, and may be a factor in Alzheimer's and Parkinson's. Perhaps eating spinach and other foods rich in antioxidants, such as blueberries, can stave off age-related declines of intellect.

Key Information Sources

Anonymous. 1992. Grow spinach to the finach. Org. Gard. **39**(7): 26–30.

Bhattacharjee, S., Dasgupta, P., Paul, A.R., Ghosal, S., Padhi, K.K., and Pandey, L.P. 1998. Mineral element composition of spinach. J. Sci. Food Agric. **77**: 456–458.

Breimer, T. 1982. Environmental factors and cultural measures affecting the nitrate content in spinach. M. Nijhoff/Dr. W. Junk Publishers, The Hague, Netherlands. 102 pp.

Castenmiller, J.J.M., Poll, C.J. van de, West, C.E., Brouwer, I.A., Thomas, C.M.G., and Dusseldorp, M. van. 2000. Bioavailability of folate from processed spinach in humans: effect of food matrix and interaction with carotenoids. Ann. Nutr. Metab. **44**: 163–169.

Djurovka, M., Lazic, B., and Markovic, V. 1988. General characters of spinach varieties suitable for industrial processing. Acta Hortic. **220**: 159–164.

Elia, A., Santamaria, P., and Serio, F. 1998. Nitrogen nutrition, yield and quality of spinach. J. Sci. Food Agric. **76**: 341–346.

Grevsen, K., and Kaack, K. 1996. Quality attributes and morphological characteristics of spinach (*Spinacia oleracea* L.) cultivars for industrial processing. J. Veg. Crop Prod. **2**(**2**): 15–29.

Howard, L.R., Pandjaitan, N., Morelock, T., and Gil, M.I. 2002. Antioxidant capacity and phenolic content of spinach as affected by genetics and growing season. J. Agric. Food Chem. **50**: 5891–5896.

Huyskes, J.A. 1971. The importance of photoperiodic response for the breeding of glasshouse spinach. Euphytica, **20**: 371–379.

Labib, A.A.S., Abd El Latife, S.A., and Omran, H. 1997. Quality indices of Jew's mallow and spinach during frozen storage. Plant Foods Hum. Nutr. **50**: 333–347.

Magruder, R. 1938. Descriptions of types of principal American varieties of spinach. U.S. Dept. of Agriculture, Washington, DC. 60 pp. + plates.

Okutani, I., and Sugiyama, N. 1994. Relationship between oxalate concentration and leaf position in various spinach cultivars. HortScience, **29**: 1019–1021.

Pandey, S.C., and Kalloo, G. 1993. Spinach, *Spinacia oleracea* L. *In* Genetic improvement of vegetable crops. *Edited by* G. Kalloo and B.O. Bergh. Pergamon Press, New York, NY. pp. 325–336.

Parlevliet, J.E. 1967. The influence of external factors on the growth and development of spinach cultivars (*Spinacia oleracea* L.). H. Veenman & Zonen, Wageningen, Netherlands. 80 pp.

Parlevliet, J.E. 1968. Breeding for earliness in spinach (*Spinacia oleracea* L.) as based on environmental and genetic factors. Euphytica, **17**: 21–27.

Pleasant, B. 1996. Grow a cool crop of spinach. Org. Gard. **43**(7): 46–49.

Sistrunk, W.A., Mahon, M.K., and Freeman, D.W. 1977. Relationship of processing methodology to quality attributes and nutritional value of canned spinach. HortScience, **12**: 59–60.

Sneep, J. 1982. The domestication of spinach and the breeding history of its varieties. Euphytica 1982. Suppl. 2. 27 pp.

Vossen, H.A.M. van der. 1993. *Spinacia oleracea* L. *In* Plant resources of South-East Asia, 8, vegetables. *Edited by* J.E. Siemonsma and K. Piluek. Pudoc Scientific Publishers, Wageningen, Netherlands. pp. 266–268.

Zornoza, P., and Gonzalez, M. 1998. Varietal specificity in growth, nitrogen uptake, and distribution under contrasting forms of nitrogen supply in spinach. J. Plant Nutr. **21**: 837–847.

Specialty Cookbooks

Feldt, L.D. 2003. Spinach and beyond: loving life and dark green leafy vegetables: a cookbook. Moon Field Press, Ann Arbor, MI. 152 pp.

Grigson, J. 1976. Cooking spinach. Abson Books, Bristol, UK. 35 pp.

Henry, N. 1983. Turning over a new leaf: 126 outstanding spinach recipes. Nancy Henry, Guilford, IN. 85 pp.

Miller, E. 1936. Spinach and other greens. Utah Cooperative Extension Service, New Circular Series. No. 84. Logan, UT. 4 pp.

Olivier, B.L. 1999. I love spinach. Green Toque, Lafayette, LA. 286 pp.

STRAWBERRY

Family: Rosaceae (rose family)

Names

Scientific name: *Fragaria* ×*ananassa* Duchesne

- The word "strawberry," derived from the Anglo-Saxon *streoberie*, is peculiar to the English language, and does not seem to have originated from other languages. It was used as early as 1000 AD in England. The name has a variety of possible origins. Straw was commonly used to mulch the plants during the winter, and as weed and soil control to keep the berries cleaner. In London, England, children used to collect the berries, string them on pieces of straw, and then sell them at the markets as "straws of berries." The runners that the plants produce are said to be strewn or dispersed around the plant, and indeed in some literature, the fruit is called strewberry, so the "straw" in strawberry may be a corruption of "strew."
- In 1712, Captain François Amédée Frézier, a French explorer spy observing Spanish fortifications on the west coast of South America, brought back to France five plants of the Chilean strawberry, destined to become one of the parents of the modern strawberry. The French word for strawberry, *fraise*, commemorates the Captain.
- The ancient Romans were very fond of *F. vesca* L., the wild strawberry native to the Alps. The Romans called it *fragaria* (which was adopted as the genus name), based on the Latin word *fragrans*, fragrant, in tribute to its intense aroma.
- The "strawberry tree" is *Arbutus unedo* L., a shrub or small tree native to Eurasia and Africa, and cultivated in warmer regions of the world, including the southern United States. Its strawberry-like fruit is used in preserves and alcoholic beverages.
- *Ananassa* in the scientific name *F.* ×*ananassa* is Latin for pineapple. The first modern strawberry was known as the pineapple or pine strawberry, owing to its distinctive flavor.

Plant Portrait

The strawberry is a low-growing perennial herb, with prostrate stems or runners (see Glossary) that spread out from the base of the plant, take root, and grow into new plants. Wild strawberries were no doubt collected in prehistoric times, and both the classical Greeks and Romans ate them. Wild strawberries are quite small, but often have a delightful taste, and to this day continue to be collected, and sometimes offered in elite restaurants at very expensive prices. The Romans cultivated the wild European strawberry (*F. vesca*), which also occurs in North America. It has very small berries, and is now raised to a minor extent in home gardens, mostly as the "alpine strawberry." The modern cultivated strawberry is a hybrid of two American species, the Virginia strawberry (*F. virginiana* Duchesne) of eastern North America and the Chilean strawberry [*F. chiloensis* (L.) Duchesne], which grows from Chile to the mountains of western North America. The Chilean strawberry was probably cultivated in South America long before the Spanish arrived, while North American Indians regularly ate the wild Virginian strawberry. The cross of the two strawberry species to produce the modern strawberry was first made in France about 1750. Most of today's hundreds of varieties were subsequently selected from the original hybrid. Major producers

Strawberry (*Fragaria* ×*ananassa*), variety Keen's Seedling, from Anderson (1874).

Strawberry (*Fragaria ×ananassa*), variety Veitch's Prolific, from *The Gardeners' Chronicle* (July 30, 1898, p. 378).

include the United States, Poland, Japan, Spain, Italy, Russia, and Korea, but many other countries grow strawberries. The fruit is a fleshy body with numerous seeds sunken in pits on the surface. (Botanically, a strawberry is made up of a swelling of the end of the floral stalk, and what appear to be seeds on the surface are actually tiny true fruits.) Strawberries are 1.3–3.8 cm (1/2–1½ inches) in diameter, generally conic, some nearly round. The strawberry is the most popular small fruit grown in the home garden, and often everbearing ("day-neutral") types that produce fruit throughout the season are grown, rather than the spring-producing ("summer") commercial types. The everbearing types (which tend to have several fruiting periods during the summer) usually need to be replaced in the garden every 2 or 3 years. The summer-bearing plants tend to have bigger berries, and require replacement every 5–8 years.

Culinary Portrait

Strawberries are often consumed fresh—whole, sliced, or crushed—often after being preserved frozen. Fresh strawberries are excellent with yogurt, ice cream, cream, or sprinkled with liqueur, and they may be added to fruit salads, omelets, and other dishes. The fruit is commonly placed on appetizer and cheese platters. Different traditions of consuming the fruit have developed in various countries: in Venice—with a squeeze of lemon and sugar; in France—splashed with red wine vinegar; in Greece—half-dipped in chocolate fondant; in Belgium—with a lemon wedge and fresh ground pepper. Strawberries are also widely used in cooked preparations, including jams, preserves, confectionery, flavoring syrups, sorbets, custards, and pastries.

Strawberries stop ripening when picked, and should be purchased with this in mind. The best fruits to pick are bright red, with fresh-looking green caps. Berries with white shoulders were picked underripe and their flavor is undeveloped. Strawberries should not be purchased if there is any sign of mold in the container. Fresh strawberries should be consumed quickly, at least within 3 days. To store, damaged fruit should first be discarded, then the remainder placed in a large container with a dry paper towel at the bottom. The berries should be separated by layers of paper towels, and should not be washed until just before use. The berries are usually washed under running cold water, but should not be left sitting in water, as this will deteriorate the color and flavor.

When consumed in large quantities, strawberries may act as a laxative. Curiously, strawberry leaf tea is sometimes used to relieve diarrhea, while strawberry plant root is used to increase urination. Some people develop a temporary skin rash after eating strawberries.

Culinary Vocabulary

- "Arizona strawberries," "prairie strawberries," and "Mexican strawberries" are not strawberries. They are joke names created by American cowboys for beans (which were often red, like strawberries).

Similarly, "Boston strawberries" is an obsolete 19th century term for "Boston baked beans."

- "In the hay" is old lunch-counter jargon for a strawberry milkshake (a pun on hay as straw).
- "Bagatelle" (also known by the French *le fraisier*) is a French strawberry cake.
- *Schweinerei* (pronounced SHVINE-eh-ree) is an American Jewish dish prepared with cottage cheese, strawberry preserves, and chopped cold vegetables.
- "Chambraise" is a light French vermouth flavored with wild alpine strawberry juice.
- "Strawberries Romanoff" (sometimes rendered Romanov) was created by renowned French cook Marie Antoine Carême (1784–1833) for Czar Alexander I (1777–1825) of Russia (the dish has no connection to the late Mike Romanoff of Hollywood). The original recipe called for crushing very ripe strawberries, adding a good red Port wine, stirring, and allowing the concoction to stay in a cool spot overnight, and the next day sieving the mixture and pouring it over fresh strawberries. More recently, the dish is often prepared by soaking strawberries in orange-flavored liqueur (or orange juice and a liqueur) and topping it with whipped cream.

Wild European (alpine) strawberry (*Fragaria vesca*), from Oeder et al. (1761–1883, plate 2235).

Curiosities of Science and Technology

- Medieval stonemasons carved strawberry designs, symbolizing perfection and righteousness, on altars and around the tops of pillars in churches and cathedrals. Consistent with this symbolism, strawberries were served at important occasions of state and festivals to ensure peace and prosperity.
- Cardinal Wolsey (1475?–1530) introduced the English to wild strawberries and cream, before modern strawberries were invented.
- Madame Tallien, a prominent figure at the French court of Emperor Napoleon, was famous for bathing in the juice of 10 kg (22 pounds) of fresh strawberries.
- Several hundred years ago, strawberry leaves symbolized the rank of a Duke in England, which indicates how highly the berries were regarded.
- Strawberries were once considered to be a medicinal cure for almost everything; it was even thought that a lotion made of their roots could fasten loose teeth by strengthening the gums. Strawberry was used for the treatment of gout by the 18th century botanist Linnaeus.
- American Indians used to crush strawberries in a mortar and mix them with meal to make a strawberry bread. After trying this bread, the colonists developed their own version: strawberry shortcake.
- In 1780, the first strawberry hybrid variety, Hudson, was developed in the United States. (In 1838, Charles M. Hovey introduced a strawberry grown from seed produced by hybridization. The Hovey variety of strawberry has been said to be the first fruit cultivar that originated through breeding on the North American continent, but is obviously predated by Hudson.)
- The most popular ice cream toppings in the United States, in decreasing order, are hot fudge, chocolate fudge, caramel, butterscotch, and strawberry.
- Juan Metzger, a former executive of the Dannon Company located in the Bronx, is credited with first adding fruit to the bottom of containers of yogurt. The first flavor was strawberry.
- A "strawberry moon" is a traditional American name for a full moon in June.
- In baseball sports medicine, a "strawberry" refers to the bruised thigh suffered by sliding base runners.
- Strawberries are grown in every state in the United States and every province in Canada.

- Botanically speaking, a strawberry is not a true berry (berries have seeds on the inside, not the outside), nor is the fleshy "fruit" a true fruit (true fruits arise from flowers; most of the strawberry is actually the enlarged swollen top of the stalk that held up many flowers), nor are the little "seeds" on the surface true "seeds" (they are actually true fruits, which admittedly do contain seeds).
- Strawberries are 90% water by weight.
- The Nebraska Supreme Court ruled (Baker's vs. Dept. of Agriculture, 1996) that strawberries can be sold in stores either by weight (e.g., pounds) or by volume (e.g., pints), but not both by weight and by volume in the same store at the same time, although they could sell by weight one day and by volume the next day.
- Worldwide sales of Jell-O are over 1 000 000 packages a day, or 13 boxes per second. The most popular flavor is strawberry.
- Wild strawberry species are very widely distributed in the temperate world, in part because it is impossible for animals (or humans) to eat the fruit without eating the seeds. The seeds pass through the digestive system to be deposited unharmed with a useful application of natural fertilizer.
- A museum built in Belgium was dedicated to strawberries.
- Oxnard, California, proclaims that it is the "Strawberry Capital of the World. Wepion, Belgium, makes the same claim. Plant City, Florida, calls itself "The Winter Strawberry Capital of the World." Portage La Prairie, Manitoba, says it is the "Strawberry Capital of Canada."

Key Information Sources

Anderson, 1966, 1969. The strawberry: a world bibliography, 1920–1966. Scarecrow Press, Metuchen, NJ. 731 pp.

Avigdori-Avidov, H. 1986. Strawberry. *In* CRC handbook of fruit set and development. *Edited by* S.P. Monselise. CRC Press, Boca Raton, FL. pp. 419–448.

Bertelsen, D.R. 1995. The U.S. strawberry industry. U.S. Dept. of Agriculture, Economic Research Service, Washington, DC. 34 pp.

Bringhurst, R.S. 1990. Cytogenetics and evolution in American *Fragaria*. Hortic. Sci. **25**: 879–881.

Childers, N.F. (*Editor*). 1981. The strawberry: varieties, culture, pests and control, storage, marketing: proceedings and added information from the national strawberry conference ["Strawberry challenges in the 1980s"], Saint Louis, Missouri, 1980. Horticultural Publications, Gainesville, FL. 514 pp.

Craig, D.L. 1976. Strawberry culture in Eastern Canada. Canada Dept. of Agriculture, Ottawa, ON. 44 pp.

Dale, A., and Luby, J.L. (*Editors*). 1991. The strawberry into the 21st century: proceedings of the third North American strawberry conference, Houston, Texas, 14–16 February 1990. Timber Press, Portland, OR. 288 pp.

Darrow, G.M. 1966. The strawberry: history, breeding, and physiology. Holt, Rinehart and Winston, New York, NY. 447 pp.

Fletcher, S.W. 1917. The strawberry in North America; history, origin, botany, and breeding. The Macmillan Company, New York, NY. 234 pp.

Galletta, G.J. 1990. Strawberry management. *In* Small fruit crop management. *Edited by* G.J. Galletta and D.G. Himelrick. Prentice Hall, Englewood Cliffs, NJ. pp. 83–156.

Galletta, G.J., Maas, J.L., and Rosati, P. (*Editors*). 1989. Proceedings of the international strawberry symposium, Cesena, Italy, May 22–27, 1988. International Society for Horticultural Science, Wageningen, Netherlands. 2 vols.

Hancock, J. 1999. Strawberries. CABI Publishing, Oxon, UK. 237 pp.

Hancock, J.F., Maas, J.L., Shanks, C.H., Breen, P.J., and Luby, J.J. 1990. Strawberries (*Fragaria*). *In* Genetic resources of temperate fruit and nut crops. *Edited by* J.N. Moore and J.R. Ballington, Jr. International Society for Horticultural Science, Wageningen, Netherlands. pp. 491–546.

Hietaranta, T. (*Editor*). 2002. Proceedings of the fourth international strawberry symposium, Tampere, Finland. International Society for Horticultural Science, Leuven, Belgium. 2 vols.

Hyams, E. 1962. Strawberry growing complete: a system of procuring fruit throughout the year. Revised edition. Faber & Faber, London, UK. 159 pp.

Jones, J.K. 1966. Evolution and breeding potential in strawberries. Sci. Hortic. **18**: 121–130.

Jones, J.K. 1995. Strawberry. *In* Evolution of crop plants. 2nd edition. *Edited by* J. Smartt and N.W. Simmonds. Longman Scientific & Technical, Burnt Mill, Harlow, Essex, UK. pp. 412–418.

Maas, J.L. 1998. Compendium of strawberry diseases. 2nd edition. APS Press, St. Paul, MN. 98 pp.

Maas, J.L., and Galletta, G.J. (*Editors*). 1993. Proceedings of the 2nd international strawberry symposium, Beltsville, Maryland, 13–18 September 1992. International Society for Horticultural Science, Wageningen, Netherlands. 520 pp.

Organisation for Economic Co-operation and Development. 1979. Strawberries. Revised edition. O.E.C.D., Paris, France. 27 pp.

Ourecky, D.K. 1976. The strawberry grower's handbook, including special PYO supplement. American Fruit Grower, Willoughby, OH. 98 pp.

Plakidas, A.G. 1964. Strawberry diseases. Louisiana State University Press, Baton Rouge, LA. 195 pp.

Riotte, L. 1998. Grow the best strawberries. Revised edition. Storey Books, Pownal, VT. 32 pp.

Scheer, H.A.T. van der, Filip, L., and Dijkstra, J. (*Editors*). 1997. Proceedings third international strawberry symposium, Veldhoven, Netherlands, 29 April–4 May 1996. International Society for Horticultural Science, Leiden, Netherlands. 2 vols.

Scott, D.H., and Lawrence, F.J. 1975. Strawberries. *In* Advances in fruit breeding. *Edited by* J. Janick and J.N. Moore. Purdue University Press, West Lafayette, IN. pp. 71–97.

Scott, D.H., Lawrence, F.J., and Draper, A.D. 1979. Strawberry varieties in the United States. Revised edition. U.S. Dept. of Agriculture, Washington, DC. 26 pp.

Smith, C.R., and Childers, N.F. 1963. The strawberry: varieties, culture, marketing, pest control. Rutgers—The State University, New Brunswick, NJ. 184 pp.

Staudt, G. 1989. The species of *Fragaria*, their taxonomy and geographical distribution. Acta Hortic. **265**: 23–33.

Staudt, G. 1999. Systematics and geographic distribution of the American strawberry species: taxonomic studies in the genus *Fragaria* (Rosaceae: Potentilleae). University of California Press, Berkeley, CA. 162 pp.

Sukumalanandana, C., and Verheij, E.W.M. 1991. *Fragaria × ananassa* (Duchesne) Guédès. *In* Plant resources of South-East Asia, 2, edible fruits and nuts. *Edited by* E.W.M. Verheij and R.E. Coronel. Pudoc, Leiden, Netherlands. pp. 171–175.

Ulrich, A., Mostafa, M.A.E., Allen, W.W., and Davis, P.A. 1980. Strawberry deficiency symptoms: a visual and plant analysis guide to fertilization. University of California, Division of Agricultural Sciences, Berkeley, CA. 58 pp.

United States Dept. of Agriculture. 1975. Strawberry insects: how to control them. Revised edition. U.S. Dept. of Agriculture, Washington, DC. 17 pp.

University of California. 1994. Integrated pest management for strawberries. Integrated Pest Management Program, University of California, Division of Agriculture and Natural Resources, Oakland, CA. 142 pp.

Waldo, G.F., Bringhurst, R.S., and Voth, V. 1971. Commercial strawberry growing in the Pacific Coast states. Revised edition. U.S. Dept. of Agriculture, Washington, DC. 22 pp.

Wilhelm, S. 1974. The garden strawberry: a study of its origin. Am. Sci. **62**: 264–271.

Wilhelm, S., and Sagen, J.E. 1974. A history of the strawberry, from ancient gardens to modern markets. University of California, Division of Agricultural Sciences, Berkeley, CA. 298 pp.

Wright, D.M. 1973. Gardening with strawberries. Drake Publishers, New York, NY. 207 pp.

Specialty Cookbooks

(Note: There are over 100 books dedicated to strawberry cuisine. The following are examples.)

Allardice, P. 1993. Strawberries. Chronicle Books, San Francisco, CA. 71 pp.

Anonymous. 1974. National strawberry festival cookbook. West Graf, Manistee, MI. 125 pp.

Buszek, B.R. 1984. The strawberry connection: strawberry cookery with flavour, fact, and folklore, from memories, libraries, and kitchens of old and new friends and strangers. Nimbus Publishing, Halifax, NS. 212 pp.

Clark, M., and Clark, G. 1994. Cooking with strawberries. Bearly Cooking, La Crescenta, CA. 112 pp.

Elliot, E., and Beveridge, J. 1998. Strawberries. Formac Publishing, Halifax, NS. 64 pp.

North American Strawberry Growers Association. 1997. Strawberry eats & treats: the guide to enjoying strawberries. Amherst Press, Amherst, WI. 108 pp.

Pitzer, S. 1985. Simply strawberries. Garden Way Publishing, Pownal, VT. 123 pp.

Shirkey, C. 1992. Strawberries: capture the essence. New Win Publishing, Clinton, NJ. 174 pp.

Siegel, H., and Gillingham, K. 1999. The totally strawberries cookbook. Celestial Arts, Berkeley, CA. 95 pp.

Waldron, M. 1977. Strawberries. 101 Productions, San Francisco, CA. 96 pp.

Whiteaker. S. 1985. The compleat strawberry. Crown Publishers, New York, NY. 128 pp.

York, L. 2001. Simple strawberry sensations! Simple Sensations Marketing, Tampa, FL. 76 pp.

SUGAR CANE

Family: Poaceae (Gramineae; grass family)

Names

Scientific name: *Saccharum officinarum* L. (Modern cultivars are the products of hybridization with other species of *Saccharum*.)

- The English word "sugar" traces to ancient languages: Old Italian *zucchero*, Greek *sachar*, Arabic *sukkar*, Persian *shakar*, and Sanskrit *sarkara*. The Sanskrit word means "grain."
- Sugar cane is often spelled as one word, sugarcane.
- There are other sweet compounds called "sugar," notably dextrose (grape or corn sugar), fructose (levulose), lactose (milk sugar), and maltose (malt sugar).
- The genus name *Saccharum* is based on the Greek word for sugar, *sakcharon*, which is based on an Asiatic word, apparently the Malay *singkara*.
- *Officinarum* in the scientific name *S. officinarum* is Latin, meaning "of the shops," i.e., referring to its sale in past times in medicinal shops, reflecting the fact that sugar was (and continues to be in some countries) widely used to treat diseases.

Plant Portrait

Sugar cane is a tall (3–5 m or 6–15 feet, sometimes 7 m or 23 feet) perennial grass with arching stems and long, broad leaves. The canes (stems) are solid, yellowish green, and 2–5 cm (3/4–2 inches) thick. The pith (soft central tissue) of the stems contains 12–15% sugar. The native land of sugar cane is uncertain, but the plant is thought to have originated from wild species in Southeast Asia and neighboring islands. Sugar cane was known in China and India 2500 years ago, reached the Mediterranean countries in the 8th century, and the Americas in early colonial times. As for many other extremely labor-intensive, major crops, for many centuries Europeans and European colonists employed slaves imported from Africa to grow and process the crop, especially in the Caribbean area, usually under appalling conditions. Today, sugar cane is grown in every tropical and semitropical country. Brazil and India are the largest producers, and other major producers are Pakistan, China, Thailand, Mexico, and Cuba. In the United States, sugar cane is cultivated from Florida to Texas (Louisiana is sometimes called the "Sugar State" because of its large sugar cane crop). The canes become tough and turn pale yellow when ready for cutting, 12–20 months from when they were planted. They are cut close to the ground, as the root end of the cane is richest in sugar. The rhizomes (underground stems) continue to produce harvestable crops for at least 3 years, sometimes more than 8 years. Sugar cane is pressed to produce a syrup that is used to make sugar. The cane yields about 10% of its weight in raw sugar, and this is converted to refined sugar with 96% efficiency. Cane syrup, molasses, wax, rum, and alcohol are other products of sugar cane. In Brazil, more than half of the sugar crop is turned into ethanol used as fuel for cars, and the country leads the world in running vehicles on ethanol. Refuse cane, called bagasse, is used in the manufacture of paper, cardboard, and fuel. White sugar is made by removing all the molasses during preparation, while some is left in brown sugar. Icing sugar is very fine-textured, and is used particularly for decorating cakes. Table sugar (sucrose) is also produced from sugar beet (see Beet), which accounts for about 40% of the world's sugar intake. Sugar is cheap, easily transported, easily stored, and relatively imperishable, characteristics that have contributed to its importance as a food.

Culinary Portrait

Sugar is the world's most widely used sweetener, and probably the cheapest form of food produced. It represents between 15 and 30% of the total calorific intake of the world, and 24% of the calories consumed by the average American. Huge amounts of sugar are used in the manufacture of alcoholic beverages, soft drinks, candy (the word has been traced to the Arabic *quand*, cane sugar), and prepared cereals. A cup of sugar-coated cereal contains almost 2 tablespoons of sugar. A can of cola has over 3 tablespoons. The canning industry adds sugar to canned fruits, jams, marmalades, jellies, and evaporated milk. Meat packers use sugar to cure ham, bacon, and other meat products. Sugar is present in condiments, canned soups, vegetables, and hundreds of other food products. Fresh cane stems are often chewed for their sweet taste. Sugar sweetens without changing the flavor of food and drink, unlike such traditional sweeteners as fruit syrups, honey, and maple syrup.

Sugar has a number of additional, very significant properties in prepared foods. It is a preservative agent in many foods. For example, in jams and jellies, sugar binds water, thereby inhibiting the growth of microorganisms. Sugar helps to maintain texture in fruit, and fruit frozen after being coated with sugar, or frozen in 40% sugar solution, tends to remain intact. Sugar promotes elasticity of gluten, and cakes rise better; the protein of eggs in cakes and custards similarly remains more elastic. Sugar combines with pectin in fruit, when fruit acids are present, to form jelly. Sugar adds bulk and texture to cakes and ice cream. Sugar makes meat more tender. Saltpeter, which is used to cure ham and bacon, has less of a hardening effect in the presence of sugar.

Air hardens brown sugar (when the moisture in the sugar evaporates), a problem that has frustrated many people. To prevent this, sugar experts recommend purchasing only as much brown sugar as can be used up in a short time. Also, brown sugar should be removed from the package and stored in an airtight, large-mouth jar (jars with rubber rings used for preserves are excellent for this purpose). A small piece of apple placed in a jar of brown sugar is commonly used to prevent caking, but too large a piece can encourage mold. Special sponges that one adds to the jar, which should be resoaked every few months, are also available. To keep brown sugar soft, some people store it in a refrigerator (not recommended by some, but recommended by others), or preferably in a freezer (which requires several hours of thawing for use, and in humid areas, could result in ice crystals forming, which could cake the sugar on thawing). A standard oven or a microwave oven can be used to soften hardened brown sugar (instructions are available on the internet; this information is not provided here, as there is danger that unless done properly, and unless used promptly, the sugar simply rehardens). A small amount of dampened paper towel, or a wedge of fresh apple, can be temporarily added to a container of hardened brown sugar, but softening can require several days, and requires monitoring every few days to ensure that too much moisture is not being added.

Molasses is a thick syrup produced as the uncrystallized residue from sugar refining. The color ranges from bright amber to dark brown, the lighter color corresponding to finer grades. Molasses is used in breads, cookies, and pastries, in various confectionery, and for the manufacture of rum.

Rum is a liquor distilled mainly from fermented molasses and other products of sugar refining, but also sometimes from sugar cane juice. Rum is generally manufactured wherever sugar cane is grown. The natural color is clear, but often caramel is added to produce a darker color, ranging from amber to deep

Sugar cane (*Saccharum officinarum*), from Köhler (1883–1914).

mahogany. The Daiquiri, a combination of rum, sugar, and lime juice, is the best-known rum cocktail, while rum and cola is a classic mixed rum drink.

Refined sugar has been claimed to be associated with health problems such as tooth decay, obesity, diabetes, cardiovascular disease, and behavioral changes in children (although it is often unclear whether sugar consumption alone can be blamed). Aside from simply supplying calories, sugar has very little nutritional value, and is often said to contain "empty calories." Dentists have recommended brushing teeth after eating sweet substances that stick to tooth enamel. It has been widely recommended that sugar consumption be reduced, which is not easy, since human taste buds crave sweetness. Among the techniques that have been proposed to reduce sugar consumption:

- Gradually reduce the addition of sugar to foods, and therefore become habituated to lower sweetness levels. In most prepared desserts that require more that 2/3 cup (175 mL) of sugar, the amount required can be halved.
- Instead of adding sugar, substitute spices such as cinnamon, ginger, and nutmeg.
- Use fruit instead of sugar to sweeten dishes.

Culinary Vocabulary

- "Invert sugar" is sugar that has been has been heated as a syrup with an acid (such as lemon juice) or an enzyme, breaking ("inverting") the original sucrose molecule into two simpler sugars, glucose and fructose. Sold in liquid form, invert sugar blends easily into breads, candies, and syrups.
- "Castor sugar" (caster sugar, superfine sugar) is a British phrase for an exceptionally fine-grained white sugar, suitable for use in a castor (also called a muffineer), a small pot with a perforated lid used for sprinkling or "casting" sugar on food. This finely granulated sugar is used in beverages and frosting because it dissolves quickly. Castor sugar may be called "verifine" or "superfine" in North America. It is finer than granulated table sugar, but not as fine as confectioner's sugar.
- "Confectioner's sugar" (also known as powdered sugar) is the most powdery grade of sugar. Cornstarch is often added. This very finely powdered sugar is used in baking and candy making, where its ability to soak up liquids is useful.
- "Bar sugar" is a very fine sugar that dissolves quickly, and is used mostly in cocktails.
- The British expression "demerera sugar" refers to a light brown, granulated sugar. The British "soft brown sugar" equals the American "light brown sugar." The British "icing sugar" is the same as the American "powdered sugar."
- "Blown sugar" is a concoction of sugar (sucrose, glucose), tartaric acid, and coloring that has been shaped into decorative objects using an air pump.
- "Crystal sugar" (also known as sanding sugar and pearl sugar) is refined sugar processed into large grains, used for decorating baked goods.
- "Butterscotch" is a flavoring, confection, or candy made principally from brown sugar and butter (or substitutes), the name probably referring to its origin in Scotland.
- "Caramel" (pronounced KAIR-ah-mehl or KAR-ah-mehl), derived from Medieval Latin meaning "honey-cane," has two modern meanings. First, it refers to sugar that has been cooked ("caramelized") until it melts and becomes a thick, dark liquid used to flavor various culinary preparations, and second, to a firm, chewy candy made with sugar and other ingredients.
- "Treacle" is sugar syrup. The word "treacle" is much more common in British English than in American, and in Britain it designates either molasses or sweet syrup made from molasses and corn syrup. However, up to the 19th century, the meaning of treacle was quite different: it referred to antidotes for poison, and traces to the classical Greek *thēriakē*, antidote. The "treacle well" of Lewis Carroll's *Alice in Wonderland* is related to so-called treacle wells in Britain, with water said to be curative (not sweet). The application of "treacle" to sugar syrup seems to date from the 17th century, probably from the practice of sugaring bitter medicinal pills.
- "Golden syrup" (sometimes called treacle in Britain) is a sweet, amber-colored syrup made from slightly caramelized refined sugar. It tastes of butterscotch, and is very popular in Britain, where it is used much like corn syrup and maple syrup are in North America.
- "Cane syrup" (cane ribbon syrup) is a syrup prepared from cane sugar. Popular in the southern United States, it is clearer than molasses and has a much lighter consistency.
- Rum originated in the West Indies after Columbus brought sugar cane cuttings there and the planters found that the liquor could be easily produced from the products of sugar production. The first crude rum that was produced was called "kill devil," and in England it became known as the

"comfortable waters of Barbados." The next name to become popular was "rumbullion," a Devonshire word meaning "great tumult," and this was shortened to "rum."

- "Dunder" is the leftover part after juice has been expressed from sugar cane. It is used in making heavier rums.
- "Sugar basin" is the traditional British equivalent to the North American "sugar bowl." However, "sugar bowl" is increasingly being used in Britain.
- The North American "cube sugar" is translated as "lump sugar" in Britain.
- "Spun sugar" is a sort of candy in the form of fine, golden threads, which are used to decorate desserts and pastries. It is prepared by cooking sugar in water with cream of tartar to about 153 °C (310 °F), and dipping a fork or whisk into the mix and drawing out hair-like strands (which can be used directly on dishes, or put on waxed paper for later use).
- "Rock candy" is a hard candy prepared by crystallizing concentrated sugar syrup around a small wooden stick or string. Alternatively, the sugar can simply be left to evaporate slowly, for up to a week.
- "Rock sugar" is sugar that has been cooked until it begins to color, producing amber-colored crystals. Not as sweet as regular granulated sugar, it is used to sweeten some Chinese teas and meat glazes.
- "Rock and rye" is a liqueur based on American rye whiskey, flavored with lemon or orange, with a chunk of rock candy in the bottom of each bottle. More generally, any blended whiskey base and any fruit can be used, with rock-candy syrup substituted for the hard rock candy.
- A "bite and stir box" is a box that was used by the New York Dutch. It had two compartments, one holding sugar lumps to bite and chew, the other holding granulated sugar to stir into tea or coffee.

Curiosities of Science and Technology

- In 14th century England, a kilogram (about 2 pounds) of sugar was of equal value to a pig, or the wages of a carpenter for 10 days.
- By the beginning of the 17th century, the British demand for sugar had become huge. *The English Housewife*, a cookbook published in Shakespeare's day, called for sugar in almost every dish—salads, omelets, fritters, pancakes, broth, boiled meat, stewed fish, roast meats, meat pies, and all manner of desserts. Often sugar was even added to wines to make them sweeter. The proverbial English bad teeth have been attributed to this centuries-old love of sugar.
- Sugar was once used to starch petticoats.
- Many of the foods sold in Victorian England (the time of Queen Victoria, 1819–1901) had poisons added to improve their appearance. Venetian lead was added to sugar.
- Molasses was so important in colonial America that the founders of the colony of Georgia promised every new resident, including children, that if they survived a year they would receive 64 quarts (73 L) of molasses as a reward.
- Some historians have argued that the British tax on America's tea was not the cause of the American Revolution (1765–1783), but rather that it was the Molasses Act of 1733, which imposed a heavy tax on sugar and molasses coming from any source except the British sugar islands in the Caribbean. Just before the War of Independence, rum (made from molasses) may have been the most important alcoholic beverage, with colonists downing an average of 3 imperial (3¾ American) gallons per person (including children) per year.
- Cheap rum was very common in the late 17th and early 18th centuries, and drinking often produced unbearable cramps, which came to be called "dry bellyache." The symptom was in fact the result of lead poisoning, lead leaching into the rum from lead fittings in the stills. The same problem was to affect American hooch (illegal whisky, named after the Hoochinoo Chinook Indians of Alaska who made a distilled drink and sold it to US soldiers occupying the new Alaskan territory in the late 19th century), as illicit stills were often made of old automobile parts that leaked deadly amounts of lead.
- Although dentists regularly advise children to avoid sugar, sugar was first added to chewing gum by a dentist—William Semple in 1869.
- In 1872, English sugar merchant Henry Tate (1819–1899) patented a method of cutting sugar into small cubes. Part of his resulting fortune was spent on a superb collection of modern paintings, which he contributed to the nation of England, along with the Tate Gallery in London (which opened in 1897) to house the collection.
- Cotton candy is made of sugar and coloring, spun at very high speed in a large tub. The sugar is first melted, and the liquid is spun through tiny holes, where it cools and forms cottony threads, which are collected and served on a paper cone. Cotton

candy was first prepared by American snack vendor Thomas Patton and introduced in 1900 at the Ringling Brothers and Barnum and Bailey Circus. Patton acquired a patent for the cotton- candy machine in 1900. (The unsupported claim is made that Josef Delarose Lascaux, a dentist from New Orleans, invented cotton candy and the cotton-candy machine.)

- On January 15, 1919, a large 15 m (50 feet) high storage tank in Boston burst and sent a tidal wave of over 7.6 million L (2 million gallons) of molasses traveling at over 48 km/hour (30 miles per hour). Buildings were crushed, 21 people died, over 150 were injured, and damage was in the millions of dollars.
- Experienced cooks avoid making meringue on a humid day because the sugar absorbs moisture, which tends to create meringues that are too soft and gooey, or ones that bead.
- When whipping cream, the best time to add sugar is when the cream is mostly whipped, as the cream will whip to a larger volume. Adding the sugar at the beginning results in a smaller volume.
- Sugar tends to absorb liquid in the body, explaining why people become thirsty after eating sweet foods. The same phenomenon occurs with salt.
- Stunt men and women use bottles and plate-glass windows made of sugar when shattering of glass is required.
- The average American consumes 3 metric tons (3.5 tons) of sugar over a lifetime.
- A teaspoon of sugar can be added to a glass of water, completely filled to the brim, without causing a drop of water to spill.
- "Sure" and "sugar" are almost the only words in the English language that are spelt "su" and pronounced "sh."
- Some moths and butterflies can detect sweetness when the sugar ratio is only 1–300 000. By comparison, cats, lions, and tigers are never able to detect sweetness.
- Some humans can detect sweetness in a solution of 1 part sugar to 200 parts water. By comparison, saltiness is detectable at 1 part in 400, sourness at 1 part in 130 000, and bitterness at 1 part in 200 000. The greater sensitivity to sour and bitter food is likely adaptive, warning humans that food is unsafe. In nature, sweet-tasting foods are usually safe to eat, while plant poisons almost always taste bitter. A day-old baby can detect a sweet taste, but does not respond to saltiness until 6 weeks of age.

Key Information Sources

Alexander, A.G. 1973. Sugarcane physiology. A comprehensive study of the *Saccharum* source-to-sink system. Elsevier Scientific Publishing Co., Amsterdam, Netherlands. 752 pp.

Artschwager, E., and Brandes, E.W. (*Editors*). 1958. Sugarcane (*Saccharum officinarum* L.): origin, classification, characteristics, and descriptions of representative clones. Agricultural Research Service, Crops Research Division, Washington, DC. 307 pp.

Baikow, V.E. 1982. Manufacture and refining of raw cane sugar. 2nd edition. Elsevier, Amsterdam, Netherlands. 588 pp.

Bakker, H. 1999. Sugar cane cultivation and management. Kluwer Academic/Plenum Publishers, New York, NY. 679 pp.

Barnes, A.C. 1974. The sugar cane. 2nd edition. L. Hill, London, UK. 572 pp.

Birch, G.G., and Parker, K.J. 1979. Sugar: science and technology. Applied Science Publishers, London, UK. 475 pp.

Blackburn, F. 1984. Sugar cane. Longman, London, UK. 414 pp.

Blume, H. 1985. Geography of sugar cane. Verlag Dr. Albert Bartens, Berlin, Germany. 371 pp.

Chen, J.C.P. 1985. Meade-Chen cane sugar handbook. John Wiley & Sons, Chichester, UK. 1134 pp.

Chen, J.C.P., and Chou, C.C. 1993. Cane sugar handbook: a manual for cane sugar manufacturers and their chemists. 12th edition. Wiley, New York, NY. 1090 pp.

Clarke, M.A., and Godshall, M.A. (*Editors*). 1988. Chemistry and processing of sugarbeet and sugarcane: proceedings of the symposium on the chemistry and processing of sugarbeet, Denver, Colorado, April 6, 1987, and the symposium on the chemistry and processing of sugarcane, New Orleans, Louisiana, September 3–4, 1987. Elsevier, Amsterdam, Netherlands. 406 pp.

Clements, H.F. 1980. Sugar cane crop logging and crop control, principles and practices. Pitman Publishing Ltd., London, UK. 520 pp.

Dillewijn, C. van. 1952. Botany of sugarcane. Chronica Botanica Co., Waltham, MA. 371 pp.

Edgerton, C.W. 1959. Sugarcane and its diseases. 2nd edition. State University Press, Baton Rouge, LA. 301 pp.

Fauconnier, R. 1993. Sugar cane. Macmillan, London, UK. 140 pp.

Galloway, J.H. 1989. The sugar cane industry: an historical geography from its origins to 1914. Cambridge University Press, Cambridge, UK. 266 pp.

Galloway, J.H. 2000. Sugar. *In* The Cambridge world history of food. *Edited by* K.F. Kiple and K.C. Ornelas. Cambridge University Press, Cambridge, UK. pp. 437–449.

Heinz, D.J. (*Editor*). 1987. Sugarcane improvement through breeding. Elsevier, Amsterdam, Netherlands. 607 pp.

Hugot, E., and Jenkins, G.H. 1986. Handbook of cane sugar engineering. 3rd edition. Elsevier, Amsterdam, Netherlands. 1166 pp.
Humbert, R.P. 1968. The growing of sugar cane. Revised edition. Elsevier Publishing Co., Amsterdam, Netherlands. 779 pp.
Hunsigi, G. 1993. Production of sugarcane: theory and practice. Springer-Verlag, Berlin, Germany. 245 pp.
Husz, G.S. 1972. Sugar cane, cultivation and fertilization. Ruhr-Stickstoff A.G., Bochum, Germany. 116 pp.
Jenkins, G.H. 1966. Introduction to cane sugar technology. Elsevier Publishing Co., Amsterdam, Netherlands. 478 pp.
Koibe, H. 1988. Sugar-cane diseases: a guide for field identification. Food and Agriculture Organization of the United Nations, Rome, Italy. 127 pp.
Kuntohartono, T., and Thijsse, J.P. 1996. *Saccharum officnarum* L. *In* Plant resources of South-East Asia, 9, plants yielding non-seed carbohydrates. *Edited by* M. Flach and F. Rumawas. Backhuys Publishers, Leiden, Netherlands. pp. 143–148.
Naik, G.R. 2001. Sugarcane biotechnology. Science Publishers, Enfield, NH. 165 pp.
Paturau, J.M. 1982. By-products of the cane sugar industry, an introduction to their industrial utilization. Elsevier, Amsterdam, Netherlands. 274 pp.
Rao, G.P., et al. (*Editors*). 1999–? Sugarcane pathology. Science Publishers, Enfield, NH. 3 vols.
Roach, B.T. 1995. Sugar canes. *In* Evolution of crop plants. 2nd edition. *Edited by* J. Smartt and N.W. Simmonds. Longman Scientific & Technical, Burnt Mill, Harlow, Essex, UK. pp. 160–166.
Smith, D. 1978. Cane sugar world. Palmer Publications, New York, NY. 240 pp.
Stevenson, G.C. 1965. Genetics and breeding of sugar cane. Longmans, London, UK. 284 pp.
Williams, J.R. (*Editor*). 1969. Pests of sugar cane. Elsevier, Amsterdam, Netherlands. 568 pp.

Specialty Cookbooks

Sugar

(Note: There are hundreds of cookbooks dedicated to sugarless or sugar-reduced cuisine. As is evident below, most cookbooks advocating the use of sugar have been sponsored by sugar companies.)

American Sugar Refining Company. 1950s. Festive Domino sugar recipes. American Sugar Refining Company, New York, NY. 10 pp.
American Sugar Refining Company. 1962. Sugar spoon recipes from the Domino sugar bowl kitchen. American Sugar Refining Company, New York, NY. 51 pp.
Bilheux, R., Escoffier, A., and Michalet, P. 2000. Petits fours, chocolate, frozen desserts, and sugar work. 2nd edition. John Wiley & Sons, Inc., New York, NY. 240 pp.
British Columbia Sugar Refining Company. 1973. Recipes for young adults: copies of recipes for young adults available from the B.C. Sugar Refining Co. Ltd. British Columbia Sugar Refining Company, Vancouver, BC. 104 pp.
Cooper, J. 1983. Sweet talk [recipes from the Domino sugar chef]. Amstar Corp., American Sugar Division [place of publication not stated]. 94 pp.
Hertzler, A.A. 1993. Sugar. Revised edition. Virginia Cooperative Extension, Blacksburg, VA. 4 pp.
Irish Sugar Company. 1999. Sure-set jam sugar: information and recipes. Irish Sugar Company, Carlow, Ireland. 34 pp.
Jacobson, C. 1997. Sugar pouring, pulling and blowing. J.B. Fairfax, Rushcutter Bay, NSW, Australia. 48 pp.
Lodge, N., Elwood, J., Becker, J., Grey, M., and Tann, G. 1986. The art of sugarcraft: sugar flowers. Merehurst Press, London, UK. 127 pp.
Roth, J. 1969. How to use sugar to lose weight. Award Books, New York, NY. 157 pp.
Sugar Industry Information Service (Australia). 1987. Sugar in Australia. Sugar Industry Information Service, Everton Park, Queensland, Australia. 10 pp.
Towers, J.C. 1993. Imperial Sugar 150th anniversary cookbook: a collection of 150 all-time favorite recipes from Imperial Sugar bags, cartons, and cookbooks. Imperial Sugar Co., Sugar Land, TX. 160 pp.
Turner, I., and Williams, R.E. 1960. Favorite sugar recipes from Louisiana plantations. Louisiana Agricultural Extension, Baton Rouge, LA. 22 pp.

Molasses

American Molasses Company. 1930. One hundred and seventy practical and tasty recipes using Grandma's Old-fashioned Molasses. American Molasses Company, New York, NY. 31 pp.
American Molasses Company. 1962. Molasses classics for modern cooks. American Molasses Co., New York, NY. 35 pp.
Crosby's Molasses Co. Ltd. 2003. Crosby's: Newfoundland's winning molasses recipes. Crosby's Molasses Co. Ltd., St. John's, NL. 51 pp.
Franks, T. 1977. The molasses book. Gerald Hodges [Printer], Booneville, MS. 36 pp.
Hanssen, M. 1980. Country kitchen recipes with molasses. Tharsons, Wellingborough, UK. 32 pp.
Jordan, R.W. 1934. Old-fashioned molasses goodies. Penick & Ford, New Orleans, LA. 48 pp.
Nickell, E.B. 1981. My favorite molasses recipes. E.B. Nickell, West Liberty, KY. 114 + 60 pp.
Penick & Ford [Company]. 1948. Brer Rabbit's New Orleans molasses recipes: for delicious desserts, candies, beverages and other tasty foods. Penick & Ford, New Orleans, LA. 48 pp.

SUNFLOWER

Family: Asteraceae (Compositae; sunflower family)

Names

Scientific name: *Helianthus annuus* L.

- The name "sunflower" is an allusion to the similarity of the flower to the sun's shape, having a circular head around which there are emanating rays (petals).
- The Spanish name for the sunflower, *girasol*, and the French name, *tournesol*, mean "turn with the sun." While many green plants, particularly the sunflower, grow towards light sources, and during early growth bend towards the east in the morning and towards the west in the evening, once the sunflower opens its flowers this "phototropism" no longer occurs. Nevertheless, the belief is extremely widespread that the sunflower head actually follows the sun.
- The genus name *Helianthus* is from the Greek *helios*, the sun, and *anthos*, a flower, named for the same reason as explained above.
- *Annuus* in the scientific name is Latin for annual.

Plant Portrait

The sunflower is a fast-growing, annual herb, generally ranging from 0.6 to 4.2 m (2–14 feet) in height. "Dwarf" types are considered to be no higher than 1.4 m (4.6 feet) tall. The taproot sometimes penetrates to a depth of 3 m (10 feet). The sunflower is native to western North America, and until recently, it was widely accepted that domesticated varieties arose from wild sunflowers around 1000 BC within the area occupied by the United States. However, domesticated sunflower seeds from 5000 BC were recently found in Mexico, and it has been suggested that cultivated varieties originated in ancient times in Mexico, not the United States. The first Europeans in the New World observed sunflower cultivated in many places from southern Canada to Mexico. Spaniards brought the sunflower to Europe in 1510, but it did not develop into a significant European crop until the late 1800s, when Russians selected varieties with very high oil content. In the 1940s, Americans brought back to North America Russian sunflower varieties with seed oil contents of nearly 50%—twice as high as existing American varieties. Over 2000 varieties of sunflowers have been recognized. Confectionery type seeds are large, with striped hulls that can be separated from the seeds relatively easily. The hulls represent about 15% of the weight of the seeds, and they are used for livestock bedding, and occasionally as a source of fiber for cattle feed. The main areas of production are Russia, Argentina, Eastern and Western Europe, China, and the United States. Most North American sunflower seed is produced in North and South Dakota and Minnesota. In Canada, sunflower is a minor crop in southern Manitoba and southeastern Saskatchewan. Sunflower is cultivated mostly for its seeds (technically fruits: see Glossary), which yield one of the world's most important source of edible oil (after soybean, palm, and canola/rapeseed). About 95% of world production is the oilseed type of plant, and only about 5% is the confectionery type. In addition to its culinary uses noted below, sunflower oil is employed for producing lubricants, soaps, paints, and varnishes. As with most other oilseeds, the press-cake left after expressing the oil is used as food for livestock. Both wild and domesticated birds love sunflower seeds.

Culinary Portrait

Sunflower oil is used for cooking and in margarine, shortening, and salad dressings. It is generally considered to be a premium oil because of its light

Sunflower (*Helianthus annuus*), from *Revue de l'Horticulture Belge et Étrangère* (1914, vol. 40).

color, high level of unsaturated fatty acids, bland flavor, and high smoke point (a measure of ability to withstand high cooking temperatures). Sunflower kernels are eaten by humans as a snack food, raw or roasted and salted, or made into flour for bread, pancakes, cookies, and cakes. The seeds can be added to many dishes, including salads, stuffings, sauces, and yogurt, to add crunchiness and a nutty taste. Although sunflower seeds are nutritionally excellent, commercially sold snack seeds are often roasted in saturated oil and bathed in salt and additives. Sunflowers are easily grown around the home, and the seeds harvested. Shelling the seeds, however, is more difficult. A seed mill is best, but an electric mixer can be used (pour a small amount of seeds into the bowl, turn the mixer on for a few seconds, then separate the seeds from the shell by putting them in water—the light shells will float to the surface and can be skimmed away; strain the seeds and dry them). Fresh sunflower seeds can be purchased (shelled or unshelled) and cooked with little effort. They can be roasted in a frying pan, without oil, over medium heat, stirring constantly; or baked for 10 minutes in an oven at 93 °C (200 °F), stirring occasionally. A small amount of oil may be added afterwards in order that salt will stick to the seeds. Sunflower seeds often turn an olive-green when cooked, because of a reaction that occurs under alkaline pH conditions, between the protein and chlorogenic acid that is present (this is harmless). One of the first uses of sunflowers when brought to Europe from the New World in the 16th century seems to have been consumption of the leaf stalks and young flowers as vegetable delicacies. Indeed, the flower buds can be eaten like artichokes. Sunflower seeds can be stored in a cool, dry place, but once shelled, they should be stored in a refrigerator, or frozen, to prevent them from becoming rancid.

Curiosities of Science and Technology

- American Indians used sunflower as a hunting calendar. When the plants were tall and in bloom, the buffalo were fat and the meat good.
- A number of Indian tribes of the southwestern United States, including the Hopi, extracted yellow and purple dyes from sunflowers, which were used to color basketry and decorate bodies.
- Stems of sunflowers were used by Native Americans to construct hoods over baking stoves.
- The Holy Orthodox Church of Russia imposed very strict dietary regulations during the 40 days of Lent and during Advent, and in the early 19th century nearly all foods rich in oil were prohibited. But the sunflower was not on the list, perhaps because the church authorities had not yet become familiar with it. As a result, the Russian people eagerly took to consuming sunflower seeds and oil, and by mid-century Russia had become the world's leader in sunflower production.

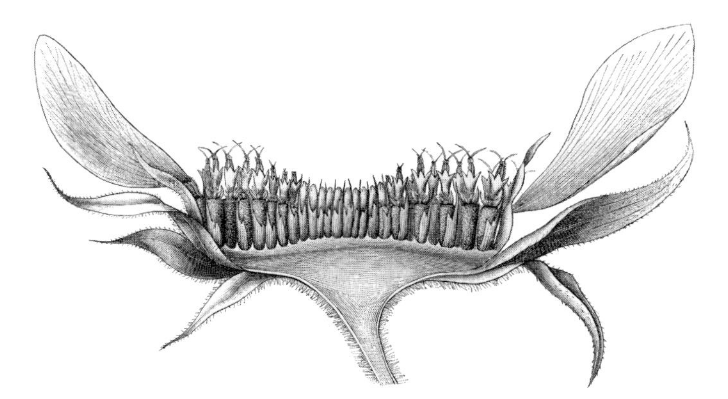

Section of flowering head of sunflower (*Helianthus annuus*), from Engler and Prantl (1889–1915).

- During the reign of the Czars in Russia, sunflower seeds were a staple food for many. Soldiers in the field were sometimes issued 1 kg (2.2 pounds) of sunflower seeds as "iron rations," and at times had little else to eat.
- In Victorian times, the sunflower became a symbol for the Aesthetic Movement, a reaction against the Industrial Age. The flower's image was carved into chair backs, glazed onto vases, and emblazoned on iron railings.
- Sunflower stems were used to fill life jackets before the advent of modern flotation materials.
- Sunflower seeds in a bird feeder attract a greater variety of wild birds than any other bird food.
- Species of the sunflower family are popular with butterflies because the wide flower head makes a good landing platform, and the numerous individual flowers make for a high probability of finding nectar. Monarch butterflies are commonly seen sucking nectar from sunflowers during their fall migration.
- During their peak growing period, sunflowers can grow as much as 30 cm (12 inches) per day.
- According to the *Guinness World Records 2001*, in 1986 M. Heijms of Oirschot, Netherlands, grew a sunflower that was 7.7 m (25 feet 5 inches) tall.
- According to the *Guinness World Records 2002*, the most heads on one sunflower was produced in 2000 by Rose Marie Roberts of Waterford, Michigan, who grew a sunflower with 129 heads.
- The shortest mature sunflower was just over 5 cm (2 inches) tall, and was grown in Oregon using a bonsai technique.

- A record diameter sunflower head was grown in Maple Ridge, British Columbia, and measured 82 cm (32¼ inches).
- "Sunflowers instead of missiles in the soil would ensure peace for future generations." This was a statement by U.S. Secretary of Defense William J. Perry on June 4, 1996, while celebrating Ukraine officially giving up its nuclear weapons. Perry joined Russian and Ukrainian defense ministers in a ceremony planting sunflowers on a former missile base in the Ukraine. The symbolism of using sunflowers arose from their use to soak up irradiation-contaminated soil resulting from the Chernobyl nuclear plant explosion.
- The apparent "flower" of species in the composite family is actually a head of miniature flowers. In many of the "radiate" species, including the sunflower, there are two kinds of flowers: the inner ones with quite inconspicuous petals and the outer ones with a large strap-shaped petal facing outward. Reminiscent of the rays of the sun, the large petals facing outward are called "ray petals." In *Helianthus*, the ray flowers are always sterile, while the inner flowers produce the seeds. The sunflower "flower" (i.e., head) has more flowers than any other genus in the composite family. There are often between 1000 and 8000 individual flowers present in the flowering head.
- Some specialists contend that the largest family of flowering plants is the sunflower family, with about 24 000 species. The sunflower family is rivaled in size only by the orchid family (Orchidaceae), with at least 20 000 species (some specialists think this is the largest family), and the legume family (Leguminosae or Fabaceae), with about 18 000 species.

Sunflower (*Helianthus annuus*) from Hallier (1880–1888, vol. 29, plate 2997).

- Living things often grow in mathematical patterns, and the best-known such pattern is the Fibonacci series. Fibonacci numbers are a sequence of numbers with several fascinating properties. The most evident property is that each term is the sum of the two previous terms: 0 +1 = 1, 1 + 1 = 2, 1 + 2 = 3, 2 + 3 = 5, 3 + 5 = 8, etc., leading to: 0, 1, 1, 2, 3, 5, 8, 13, 21, 34, 55, 89, 144, 233, 377, 610, 987, 1597... Another property has to do with the "golden ratio" or "divine ratio," which is approximately 0.62 (for example, a picture is said to look best if its height is 0.62 times its width). The sequence of ratios between every consecutive pair of Fibonacci numbers is alternatively greater than, or less than, the golden ratio, and gets closer and closer to it (for example, 1/1 = 1.00, 1/2 = 0.500, 2/3 = 0.667, 3/5= 0.6000, 5/8 = 0.625, 8/13 = 0.615). Fibonacci numbers were first discussed by an Italian mathematician, Leonardo of Pisa (1175–1250), whose nickname was Fibonacci. Fibonacci numbers are most apparent in plants in *phyllotaxis*, the spiral patterned arrangement of organs on plants, such as the leaves on stems. Demonstrate this for yourself with the stem of any herbaceous plant. Start at the bottom with a given leaf, move up the stalk, counting the leaves until you reach a leaf that is directly above the first one (do not count the first one), and you will have a Fibonacci number. In addition, the number of times you have circled the stalk will be another Fibonacci number. Leaf arrangements in most plant species demonstrate a ratio of Fibonacci numbers, the one representing the number of units (leaves in this case) before one gets to the same position, and the other the number of spirals to get to the same position. The arrangements of seeds in sunflower heads spectacularly demonstrate Fibonacci ratios, but these are more difficult to count than leaves on a stem. A sunflower head may have Fibonacci ratios of 21/34, 34/55, 55/89, or 89/144. In a good specimen of a sunflower, observe two sets of spirals superimposed or intertwined, one spiral turning clockwise and the other counterclockwise, with each seed filling a dual role by belonging to both spirals.
- Most ornamental sunflower varieties available to home gardeners today are hybrids. Those who find that they have grown a particularly huge plant or one with a huge seed head are often tempted to

collect the seeds to grow next year. However, this is generally unwarranted with hybrid plants, as the seeds will not produce plants with the same characteristics as the parental plant.

- Mature sunflower heads in the Northern Hemisphere generally face east, probably (it is often said) because this prevents overheating of the seeds by excessive exposure to the hot sun. (As noted earlier, only young plants bend towards the sun as it progresses across the sky.) Breeders have created some plants with heads that face downward, so that birds find it harder to reach the seeds.

Key Information Sources

Berglund, D.R. 1994. Sunflower production. North Dakota State University Extension Service, Fargo, ND. 98 pp.

Blamey, F.P.C., Edwards, D.G., and Asher, C.J. 1987. Nutritional disorders of sunflower. Dept. of Agriculture, University of Queensland, St. Lucia, Queensland, Australia. 72 pp.

Carter, J.F. (*Editor*). 1978. Sunflower science and technology. American Society of Agronomy, Madison, WI. 505 pp.

Cobia, D.W., and Zimmer, D.E. 1978. Sunflower: production and marketing. North Dakota State University of Agriculture and Applied Science, Fargo, ND. 73 pp.

Dedio, W. 1980. Sunflower seed crops. Agriculture Canada, Ottawa, ON. 31 pp.

Fick, G.N. 1989. Sunflower. *In* Oil crops of the world. *Edited by* G. Röbbelen, R.K. Downey, and A. Ashri. McGraw-Hill, New York, NY. pp. 301–318.

Heiser, C.B. 1951. The sunflower among the North American Indians. Proc. Am. Philosoph. Soc. **98**: 432–448.

Heiser, C.B. 1976. The sunflower. University of Oklahoma Press, Norman, OK. 198 pp.

Heiser, C.B., Jr. 1995. Sunflowers. *In* Evolution of crop plants. 2nd edition. *Edited by* J. Smartt and N.W. Simmonds. Longman Scientific & Technical, Burnt Mill, Harlow, Essex, UK. pp. 51–53.

Heiser, C.B., Jr. 2000. Sunflower. *In* The Cambridge world history of food. *Edited by* K.F. Kiple and K.C. Ornelas. Cambridge University Press, Cambridge, UK. pp. 427–430.

Heiser, C.B., Smith, D.M., Clevenger, S.B., and Martin, W.C., Jr. 1969. The North American sunflowers (*Helianthus*). Mem. Torrey Bot. Club, **22**: 1–218.

Kansas State University. 1993. Sunflower production handbook. Kansas State University, Cooperative Extension Service, Manhattan, KS. 24 pp.

Kolte, S.J. 1985. Diseases of annual edible oilseed crops. Sunflower, safflower and nigerseed diseases. Vol. 3. CRC Press, Boca Raton, FL. pp. 97–136.

McCormick, I., Davison, C.W., and Hoskin, R.L. 1992. The U.S. sunflower industry. U.S. Dept. of Agriculture, Economic Research Service, Washington, DC. 58 pp.

McMullen, M.P. 1985. Sunflower: production and pest management. North Dakota State University, Fargo, ND. 76 pp.

Nwokolo, E. 1996. Sunflower (*Helianthus annuus* L.). *In* Food and feed from legumes and oilseeds. *Edited by* J. Smartt and E. Nwokolo. Chapman and Hall, London, UK. pp. 259–269.

Poncavage, J. 1996. Totally sunflowers. Celestial Arts, Berkeley, CA. 94 pp.

Pustovoit, V.S. 1976. Selection, seed culture and some agrotechnical problems of sunflower. Indian National Scientific Documentation Centre, New Delhi, India. 461 pp.

Riesberg, L., and Seiler, G. 1990. Molecular evidence and the origin and development of the domesticated sunflower (*Helianthus annuus*, Asteraceae). Econ. Bot. **44** (Suppl. 3): 79–91.

Rogers, C.E. 1980. Natural enemies of insect pests of sunflower: a world view. Texas Agricultural Experiment Station, Texas A & M University System, College Station, TX. 30 pp.

Rogers, C.E., Thompson, T.E., and Seiler, G.J. 1982. Sunflower species of the United States. National Sunflower Association, Bismarck, ND. 75 pp.

Schilling, E.E., and Heiser, C.B. 1981. Infrageneric classification of *Helianthus*. Taxon, **30**: 293–403.

Schneiter, A.A., Seiler, G.J., Miller, J.F., Charlet, L.D., and Bartels, J.M. (*Editors*). 1997. Sunflower technology and production. Agronomy Series No. 35. American Society of Agronomy, Madison, WI. 834 pp.

Seiler, G. 1992. Utilization of wild sunflower species as sources of genetic diversity for improvement of cultivated sunflower. Field Crops Res. **30**: 195–230.

Skoric, D. 1992. Achievements and future direction of sunflower breeding. Field Crops Res. **30**: 231–270.

Vossen, H.A.M. van der, and Duriyaprapan, S. 2001. *Helianthus annuus* L. *In* Plant resources of South-East Asia, 14, vegetable oils and fats. *Edited by* H.A.M. van der Vossen and B.E. Umali. Backhuys Publishers, Leiden, Netherlands. pp. 101–107.

Specialty Cookbooks

Flores, B. 1997. The great sunflower book: a guidebook with recipes. Ten Speed Press, Berkeley, CA. 130 pp.

Halperin, A. 1970. Cooking and baking for low fat diets: a selection of recipes compiled particularly for sufferers from coronary thrombosis and kindred ailments, using ingredients free from animal fats and the whites of eggs only and sunflower or corn oil. Bailey Bros. & Swinfen, Folkestone, UK. 113 pp.

Millette, R.A. 1973. Seeds from the sunflower. Cooperative Extension Service, North Dakota State University, Fargo, ND. 4 pp.

Peel, L. 1997. The ultimate sunflower book. HarperCollins Publishers, New York, NY. 110 pp.

Sitton, D.M. 1995. Sunflowers: growing, crafting, and cooking with the sunniest of plants. Gibbs Smith, Salt Lake City, UT. 54 pp.

SWEET POTATO

Family: Convolvulaceae (morning-glory family)

Names

Scientific name: *Ipomoea batatas* (L.) Lam.

- When the sweet potato was first observed by Europeans on the islands off the coast of Yucatan and Honduras, it was called *batatas* by the natives. Note this word in the scientific name, *I. batatas*. The name *batatas* evolved into *patata* in Spanish and eventually "potato" in English. The word "potato" was transferred to the common potato (see POTATO). The sweet potato has sometimes been called "long potato" in the United States.
- According to Seelig and Bing (1990), "Since the sweetpotato is not a potato, it is desirable to spell it as one word to differentiate it. The United States Department of Agriculture has adopted the single-word form."
- Confusion between sweet potatoes and yams in the United States began in colonial times when African slaves thought the sweet potatoes they were fed were like their familiar African yams called *nyami*. The African name was eventually shortened to yam. Since then, sweet potatoes have been a mainstay of African-American cooking.
- Moist fleshed varieties of the sweet potato are often but inappropriately called "yam," a term better reserved for species of the genus *Dioscorea*, some of which produce edible tubers similar to the sweet potato.
- Sweet potato should not be confused with wild sweet-potato vine [*Ipomoea pandurata* (L.) G. Mey.], a cultivated ornamental vine with a large, inedible tuberous root.
- The genus name *Ipomoea* is formed from the Greek *ips*, worm, vine-like plant, + *homoios*, similar to. Many of the species are twining vines, hence the phrasing "like a vine" or "like a worm."

Plant Portrait

Sweet potato (which is not related to the common or Irish potato) is a herbaceous vine or bushy perennial, cultivated for its edible roots. Wild forms of sweet potato have been reported from coastal Ecuador, Columbia, and Mexico. While it is clear that the species is native to the Americas, there is debate as to where it was domesticated. Sweet potatoes have been dated from Peru at 10 000 BC, although this vegetable may have been introduced from farther north. When Europeans first arrived in the New World, sweet potato was already widely used, and several cultivated forms were known. Living plants reached Spain about 1550. Some evidence exists that sweet potato was dispersed westward across the Pacific, perhaps by the Polynesians, before European exploration. It has been identified archaeologically in New Zealand from the 14th century. Also, the first Spaniards to reach the Solomon Islands, the Marianas, and the Philippines reported that it was cultivated there. Sweet potato was a major food source during the American Civil War. Today, nearly all sweet potatoes are grown in developing Asian countries, about 80% in China alone. Other major growers are Indonesia, Vietnam, Uganda, Japan, and India. The sweet potato is a major food crop in all tropical and subtropical countries. It is far more important in subtropical and tropical areas than the Irish potato because it thrives in a hot, moist climate, while the latter requires a cool climate.

Sweet potato (*Ipomoea batatas*), from Nicholson (1885–1889).

Sweet potato (*Ipomoea batatas*), variety Rose de Malaga, from Vilmorin-Andrieux (1885).

The edible portion of a sweet potato is a fleshy storage root, often called a tuber (see Glossary). Sweet potatoes do not have eyes (which are embryonic shoots that in "seed potatoes" grow into new plants), but they can develop adventitious buds on vines or cuttings, and these are useful for vegetative reproduction. There are hundreds of cultivars. Two important kinds of sweet potatoes are grown commercially. In most developing countries, where sweet potato is consumed, the type produced has white to cream colored, dry-textured flesh and a bland, nonsweet flavor. In developed countries, the flesh color is generally yellow or deep orange, the texture is moist, and the sugar content is high. The yellow or orange flesh color is directly related to beta-carotene, a precursor of vitamin A.

Culinary Portrait

The enlarged storage roots of sweet potato are commonly used as a vegetable—boiled, baked, fried, and glazed. Sweet potatoes are used to make pies as frequently as pumpkin, and are also made into biscuits, cakes, bread, pudding, marmalade, cookies, and muffins. This vegetable can replace winter squash in most recipes, and seems to go particularly well with pork, ham, and poultry. The roots are canned, frozen, or dehydrated, and are also used as a source of starch or alcohol, and as animal fodder. In North America, sweet potato purchases increase dramatically for the Thanksgiving and Christmas seasons, reflecting the popularity of the vegetable at these times. Candied sweet potatoes, prepared with orange peel and juice, is a classic Thanksgiving dish.

Hints: Select sound, firm roots. Handle carefully to prevent bruising. Store in a dry, unrefrigerated bin kept at 13–16 °C (55–60 °F). Do not refrigerate, because temperatures below 13 °C (55 °F) will chill this tropical vegetable, giving it a hard core and an undesirable taste when cooked. Sweet potatoes are sometimes dyed or given a waxy coating, and therefore some people prefer to remove the peel.

Culinary Vocabulary

- In Cuba, white sweet potato, known as *boniato*, is fried and sprinkled lightly with salt and sugar. In addition to its white rather than yellow or orange flesh, Cuban white sweet potatoes tend to be irregular in shape, and skin color (reddish to cream colored) and are drier and not as sweet as other varieties of sweet potato. Nevertheless, the word *boniato* has become slang for "sweet and harmless."

Curiosities of Science and Technology

- In the 19th century, noted African-American scientist George Washington Carver [1864–1943, much better known for developing new uses for the peanut (which see)] devised over a hundred ways of using the sweet potato, and indeed is said to have had 500 ways to prepare them.

- Gerry Thomas, an executive of C.A. Swanson and Sons of Omaha, Nebraska, invented the frozen "TV dinner" in the early 1950s. The first TV dinner meal consisted of turkey, cornbread dressing and gravy, buttered peas, and sweet potatoes. Five million were sold in the first year of marketing (1954). The first Swanson package was placed in an exhibit at the Smithsonian museum in Washington, DC.
- Roots of sweet potatoes produce chemicals that inhibit growth of other plants. Field studies have shown that the sweet potato cultivar Regal greatly reduced the growth of a common weed, yellow nut sedge (*Cyperus esculentus* L.). Sweet potato residues in soil also inhibited growth of several plants, including sweet potato cuttings.
- German researchers have inserted an antisense gene into sweet potatoes that blocks the production of an enzyme that converts sucrose to starch in the roots. The result is that the roots remain very sweet.
- The deeper the color of the sweet potato, the higher the content of vitamin A.
- To grow an attractive sweet potato vine as a house plant, place a sweet potato in a jar of water with an opening that will support it, with its narrow end down. Put the jar in a warm, dark place and keep the jar filled with water. New roots should grow. When the stem starts to develop, in about 10 days, transfer the jar to a sunny window.

Key Information Sources

Austin, D.F. 1978. The *Ipomoea batatas* complex —1. Taxonomy. Bull Torrey Bot. Club, **105**: 114–129.

Bohak, J.R., Dukes, P.D., and Austin, D.F. 1995. Sweet potato. *In* Evolution of crop plants. 2nd edition. *Edited by* J. Smartt and N.W. Simmonds. Longman Scientific & Technical, Burnt Mill, Harlow, Essex, UK. pp. 57–62.

Bouwkamp, J.C. (*Editor*). 1985. Sweet potato products: a natural resource for the tropics. CRC Press, Inc., Boca Raton, FL. 271 pp.

Clark, C.A., and Moyer, J.W. 1988. Compendium of sweet potato diseases. American Phytopathological Society, St. Paul, MN. 74 pp.

Cooley, J.S. 1961. The sweet potato—its origin and primitive storage practices. Econ. Bot. **5**: 378–386.

Edmond, J.B., and Ammerman, G.R. 1971. Sweet potatoes: production, processing, marketing. Avi Publishing Co., Westport, CT. 334 pp.

Hahn, S.K., and Hozyo, Y. 1984. Sweet potato. *In* The physiology of tropical field crops. *Edited by* P.R. Goldsworthy and N.M. Fisher. John Wiley & Sons, Chichester, UK. pp. 551–567.

Hall, M.R., and Phatak, S.C. 1993. Sweet potato, *Ipomoea batatas* (L.) Lam. *In* Genetic improvement of vegetable crops. *Edited by* G. Kalloo and B.O. Bergh. Pergamon Press, New York, NY. pp. 693–708.

Hill, W.A., Bonsi, C.K., and Loretan, P.A. (*Editors*). 1993. Sweet potato technology for the 21st century. Tuskegee University, Tuskegee, Al. 607 pp.

House, H.D. 1908. The North American species of the genus *Ipomoea*. Ann. N. Y. Acad. Sci. **18**: 181–263.

International Potato Center. 1988. Exploration, maintenance, and utilization of sweet potato genetic resources: report of the first sweet potato planning conference, 1987. International Potato Center, Lima, Peru. 369 pp.

International Potato Center, Indian Council of Agricultural Research, and Central Tuber Crops Research Institute (India). 1988. Improvement of sweet potato (*Ipomoea batatas*) in Asia: report of the "Workshop on Sweet Potato Improvement in Asia," held at ICAR, Trivandrum, India, October 24–28, 1988. International Potato Center, Lima, Peru. 253 pp.

Jansson, R.K., and Raman, K.V. 1991. Sweet potato pest management: a global perspective. Westview Press, Boulder, CO. 458 pp.

Jones, A., Dukes, P.D., and Schalk, J.M. 1986. Sweet potato breeding. *In* Breeding vegetable crops. *Edited by* M.J. Bassett. AVI Publishing Co., Westport, CT. pp. 1–35.

Martin, F.W., and Jones, A. 1972. The species of *Ipomoea* closely related to the sweet potato. Econ. Bot. **26**: 201–215.

Martin, F.W., and Jones, A. 1986. Breeding sweet potato. *In* Plant breeding reviews, vol. 4. *Edited by* J. Janick. Avi Publishing Co., Westport, CT. pp. 313–345.

McDonald, J.A., and Austin, D.F. 1990. Changes and additions in *Ipomoea* section *Batatas* (Convolulaceae). Brittonia, **42**: 116–120.

Nishiyama, I. 1971. Evolution and domestication of the sweet potato. Bot. Mag. (Tokyo), **84**: 377–387.

O'Brien, P.J. 1972. The sweet potato: its origin and dispersal. Am. Anthropol. **74**: 342–365.

O'Brien, P.J. 2000. Sweet potatoes and yams. *In* The Cambridge world history of food. *Edited by* K.F. Kiple and K.C. Ornelas. Cambridge University Press, Cambridge, UK. pp. 207–218.

Price, R.H. 1896. Sweet potato culture for profit. A full account of the origin, history and botanical characteristics of the sweet potato. Texas Farm and Ranch, Dallas, TX. 107 pp.

Purcell, A.E., Walter, W.M., Jr., and Wilson, L.G. 1989. Sweet potatoes. *In* Quality and preservation of vegetables. *Edited by* N.A. Eskin. CRC Press, Inc., Boca Raton, FL. pp. 285–304.

Roberts, R.E. 1979. Keys to profitable Texas sweet potato production, storage and marketing. Texas Agricultural Extension Service, Texas A & M University System, College Station, TX. 7 pp.

Seelig, R.A., and Bing, M.C. 1990. Sweet potatoes. *In* Encyclopedia of produce. [Loose-leaf collection of individually paginated chapters.] *Edited by* R.A. Seelig and M.C. Bing. United Fresh Fruit and Vegetable Association, Alexandria, VA. 18 pp.

Takagi, H., Kuo, C.G., and Sakamoto, S. 1996. *Ipomoea batatas* (L.) Lamk. *In* Plant resources of South-East Asia, 9, plants yielding non-seed carbohydrates. *Edited by* M. Flach and F. Rumawas. Backhuys Publishers, Leiden, Netherlands. pp. 102–107.

Tsuno, Y. 1970. Sweet potato: nutrient physiology and cultivation. International Potash Institute, Berne, Switzerland. 73 pp.

Vilareal, R.L., and Griggs, T.D. (*Editors*). 1982. Sweet potato. Proceedings of the first international symposium. Asian Vegetable Research and Development Center, Shanhua, Taiwan. 481 pp.

Woolfe, J.A. 1992. Sweet potato: an untapped food resource. Cambridge University Press, Cambridge, UK. 643 pp.

Yen, D.E. 1974. The sweet potato and Oceania. Bishop Museum Press, Honolulu, HI. 389 pp.

Specialty Cookbooks

Key, G.M., and Yarborough, R.H. 1971. Sweet potato recipes by homemakers of Virginia's Eastern Shore. Virginia Sweet Potato Commission, Richmond, VA. 100 pp.

Louisiana Sweet Potato Commission. 1957. 75 easy yam recipes with a romantic past, from the magic Evangeline land of Louisiana. Louisiana Sweet Potato Commission, Opelousas, LA. 36 pp.

Lund, D.R. 2000. 150 ways to enjoy potatoes, including recipes, preparation tips, soups and sweet potato recipes. Adventure Publications, Cambridge, MN. 103 pp.

Martin, F.W., Ruberté, R.M., and Herrera, J.L. 1989. The sweet potato cookbook. Educational Concerns for Hunger Organization, North Fort Myers, FL. 86 pp.

Na Lima Kokua, and Pacific Tropical Botanical Garden. 1983. Sweet potato = 'Uala:' uses and recipes. Na Lima Kokua, Lawai, HI. 24 pp.

Sweet Potato Council of the U.S. 1900–1982? Cooking with sweet potatoes. Sweet Potato Council of the U.S., Inc., College Park, MD. 28 pp.

Sweet Potato Festival Committee. 1970–1979? A book of favorite sweet potato recipes. Sweet Potato Festival Committee, Vardaman, MS. 103 pp.

Talmadge, L.N. 1998. The sweet potato cookbook. Cumberland House, Nashville, TN. 228 pp.

Texas Department of Agriculture. 1980. Texas sweet potato recipes from the Texas Department of Agriculture. Texas Department of Agriculture, Austin, TX. 4 pp.

Thompson, V., and Varner, J. 2001. Pass the sweet potatoes, please! Writers Club Press, San Jose, CA. 94 pp.

Wall, V.W. 1993. Cooking with nature's pre-sweeten vegetable: *Ipomoea batatas* (sweet potatoes): a collection of our favorite recipes using the sweet potatoes which have high energy value and contain vitamins A and C. V.W. Wall and VATCO Gifts, Memphis, TN. 142 pp.

Taro

Family: Araceae (arum family)

Names

Scientific name: *Colocasia esculenta* (L.) Schott (*Arum esculentum* L., *C. antiquorum* Schott)

- The Malay name *tales* is the base of the word "taro."
- Taro is also known as Caribbean cabbage, Chinese potato, coc, coco, cocoyam, dalo, dasheen, eddo(e), elephant ear, elephant's ear, elephant's ear plant, gabic, kalo, madumbe, malanga (but see MALANGA), and taro potato. There are hundreds of local names in various languages.
- The name "dasheen" may have come from the French *de Chine* ("from China"), although this is probably not the place of origin of the crop but instead where its use was first documented.
- "Eddo" often designates a variety of taro with small purplish tubers. The word is of Niger-Congo origin, akin to *fante edwo*, yam.
- The name "elephant ear" (also elephant's ear and elephant's ears) arose because of the similarity of the huge leaves to the ears of elephants. Several plants have this name, particularly *Xanthosoma sagittifolium* (L.) Schott, another tropical member of the arum family that produces edible tubers. It has also been termed tannia, kongkong, and American taro. This is used as a starchy vegetable (see MALANGA).
- "Cocoyam" is a name of West African origin used to designate taro. The name was first recorded in 1922, and may have been based on the appearance of the large tubers, which look like yams with the fibrous outer coating of a coconut. In the Caribbean, the abbreviation "coco" is often used. "Cocoyam" could be confused with "new cocoyam," *Xanthosoma sagittifolium* mentioned above ("old cocoyam" is *Colocasia esculenta*).
- The word "taro" appears in the names of several other tropical vegetables of the arum family. Blue taro is *Xanthosoma violaceum* Schott. Giant taro is *Alocasia macrorrhizos* (L.) G. Don, and Chinese taro is *A. cucullata* (Lour.) G. Don.
- The genus name *Colocasia* is from the Arabic word for the plant, *kolkas* or *kulcas*.
- *Esculenta* in the scientific name *C. esculenta* is Latin for edible.

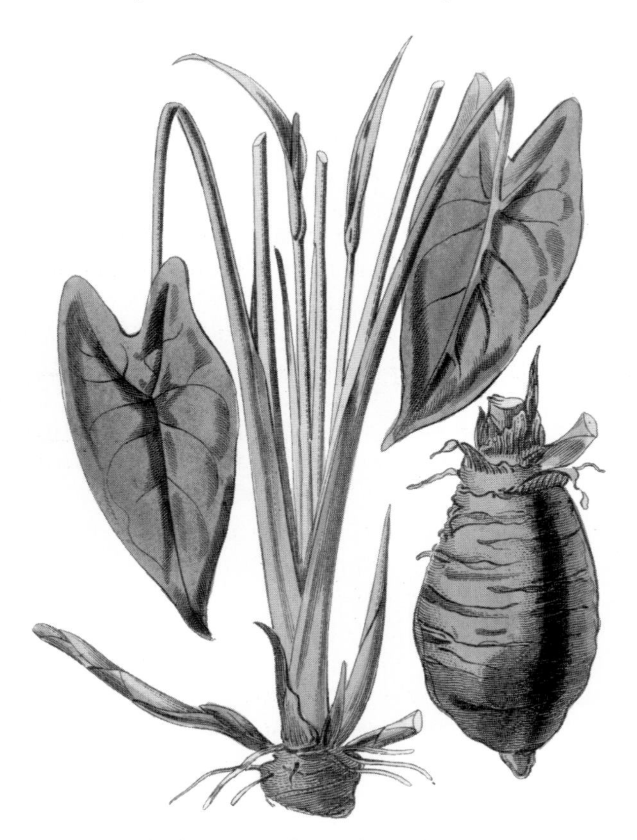

Taro (*Colocasia esculenta*), from Rhind (1855).

Plant Portrait

Taro is a large perennial herb, growing up to 2.4 m (8 feet) in both height and spread. It has large, broad, heart-shaped to arrow-shaped, dark green leaf blades as long as a meter (about 1 yard) in length, with leaf stalks that are thick, succulent, often purplish, and about a meter long. Unlike most leaves, the leaf stalk attaches near the center of the leaf blade, not at its basal edge. Taro has a large, starchy, edible "corm" (an enlarged, underground, modified stem) that is generally called a root or tuber. The corm is similar in consistency to a potato, but a little sweeter (taro is sometimes called "the potato of the tropics"), and is typically top-shaped, brown on the outside, with rough ridges, lumps, and spindly roots. Internally, it is gray, beige, white, cream, pink, or purple, depending on the variety, and sometimes the flesh has pink or brown veins. The corms vary in weight from 1 to 3.5 kg (about 2–8 pounds). On its sides, the corm bears one or more small secondary tubers called "cormels" or "eddoes," which reproduce the plant vegetatively.

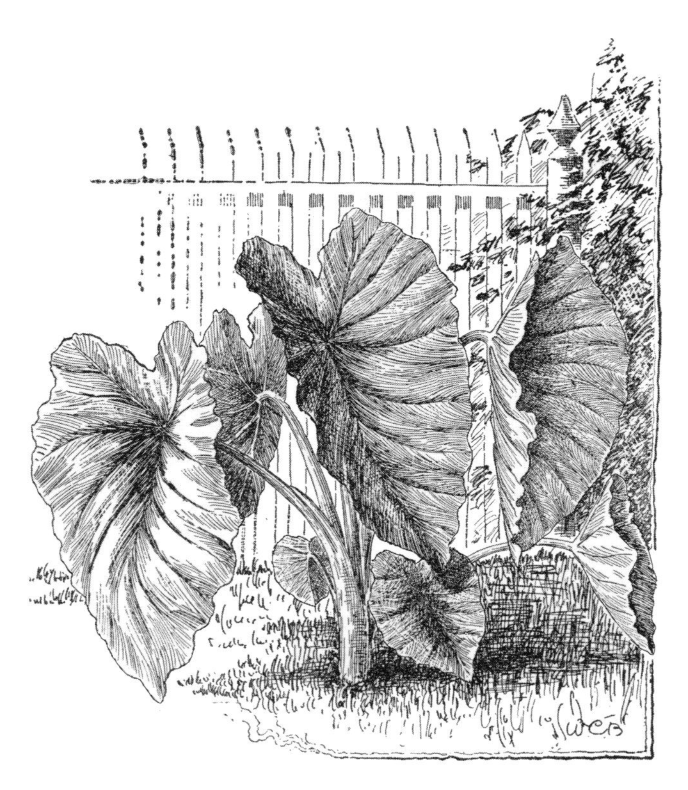

Taro (*Colocasia esculenta*), from Bailey (1900–1902). Note cultivation on dry land.

Taro is a member of the arum family, which includes caladium, dieffenbachia, and philodendron, and like them, it produces large spikes bearing tiny flowers. Hundreds of varieties were selected in Hawaii of which many no longer exist. Some varieties are best grown in water, others in upland culture. Wetland taro is grown so that the root is almost always covered by moving water. Dryland taro is grown in regions without much running water, the crop relying on rainfall for growth.

Taro is one of the oldest of cultivated crops. It was domesticated between 2000 and 5000 BC. It originated from southern Asia, perhaps India, and was once an important food source in the Nile Valley, as early as 500 BC. Taro was cultivated in China 2000 years ago. It came to the South Pacific from Asia, and was one of the plants that the first Polynesian voyagers brought to Hawaii about 1500 years ago. English navigator and explorer Captain James Cook (1728–1779) arrived in the Hawaiian Islands in 1778, to find an estimated 300 000 people living chiefly on taro, sweet potato, fish, seaweed, and a few green vegetables and fruits. They used no grain or animal milk in their diet, and animal proteins were a rarity. Yet the good physique and excellent teeth of the Polynesian people testified to an adequate diet. In past times in Hawaii, taro was also used medicinally to soothe an upset stomach and as a poultice for wounds, infections, and insect stings. Taro is usually cultivated in a complex terraced system of ponds, and in the past, taro fields were extensive on all the major islands of Hawaii. Today, because of the scarcity of water and agricultural land, only a few such areas remain. Poi made with taro, which was once the most important staple in the Hawaiian diet, is now relatively infrequently eaten. Other foods, primarily rice and white bread, have become major dietary sources of carbohydrates in Hawaii. However, taro remains a staple in Asia, the Caribbean, West Africa, and South America, and is grown and eaten in over 60 countries. Recently, taro was introduced by the U.S. Department of Agriculture to the southern United States as a supplement to potatoes.

Taro known as "elephant ear" is commonly grown as an ornamental in hot climates, in or near water gardens. It can also be grown as a potted plant or as an annual in temperate climates. Its huge, heart-shaped, glossy leaves are very attractive.

Culinary Portrait

Taro is best known in its cooked form as poi, a sticky, purple–gray paste, the traditional staple food of native Hawaiians (it has been called "the soul food" of Hawaii). Poi is made by cooking peeled taro root (often with the addition of breadfruit, sweet potato, and banana) until it is soft enough to mash with water in a bowl. Cooked taro is very firm and requires strength to mash. In past times, a stone and a pounding board would be used to mash it. Poi is traditionally eaten by dipping one or two fingers in the mixture and putting it in the mouth, without the use of utensils. The thickest poi is called "one-finger poi" (since only one finger is required to lift it to the mouth); a thinner form is called "two-finger poi," and a still thinner form is called "three-finger poi." Traditionally, Hawaiians often let poi stand for a few days until it fermented and turned sour. The longer the fermentation, the sourer it becomes; "one-day" poi is sweeter than "three-day poi." Although poi is considered delicious by many, it is not to the taste of everyone. Its paste-like texture has led to its description by some as tasting like library paste.

Taro is often available in Asian food stores and large supermarkets, and the adventurous cook may wish to experiment with it. Taro should be peeled and boiled or steamed like a potato, to produce a tender vegetable. It can also be used in stews and stir-fries, or braised and combined with meat dishes. Soups and stews are thickened by the addition of taro, which like potato, absorbs the flavor of foods with which it is combined. Taro is preferably served hot, and the texture of the flesh changes when it is cooled.

Taro (*Colocasia esculenta*), from Baillon (1876–1892, vol. 2). Note aquatic cultivation.

While taro made from the corms is the main food, young, unopened leaves are also employed as greens. Young taro leaves are used as a principal vegetable throughout Melanesia and Polynesia. The leaves can be cooked and eaten plain, or employed to make traditional Hawaiian foods like "laulau" (pork and fish wrapped in taro leaf and steamed), or a thick, green soup called "lu'au." Taro leaves are often used in Asian markets as a wrapper for steamed food.

Needle-like crystals of calcium oxalate are abundant in the roots of taro, and are found in varying amounts in all parts of the plant. Farmers contacting the plants experience a stinging sensation on the skin from the crystals, but many seem to gradually lose sensitivity to the irritant. The crystals can cause a burning, stinging, itching sensation in the mouth and throat, and a stomach ache, when eaten. Cooking is essential to destroy these crystals. Anyone handling fresh taro roots should wear gloves. There are varieties of taro that are nearly free of calcium oxalate. Despite this problem, taro is "hypoallergenic" (much less likely to cause allergies) and can be processed into a white flour for people who are allergic to other starches, for example, in cereals. Taro flour and other products have been used extensively for infant formulae in the United States and have formed an important constituent of proprietary canned baby foods.

Culinary Vocabulary

- A "luau" is a Hawaiian feast. The word originally referred to only the leaves of the taro plant, eaten as a vegetable. It came to refer to dishes prepared with the leaves, and finally to the feasts at which the dishes were served.
- The Caribbean term "callaloo" refers to a variety of edible greens, and a soup made from them. The principal green used is the leaves of taro.
- "Taro pudding," known as *kulolo* in Hawaii where it is a specialty often served at a luau, is prepared with taro, brown sugar, honey, and coconut milk.

Curiosities of Science and Technology

- Early Hawaiians are said to have eaten up to 7 kg (15 pounds) of poi daily.
- The state seal of Hawaii shows a number of objects, including eight taro leaves.
- In 1910, the U.S. Department of Agriculture introduced taro into Florida as a substitute for potatoes, which did not grow well in the moist soils of the Southeast. As late as 1982, experiments to grow taro for food or fuel were being conducted in the Florida Everglades. Unfortunately, taro became a weed of moist areas, and programs are now underway to control it.
- In old Hawaii, sea voyagers often took dried poi with them, in the same way that hikers and backpackers carry dried food. In the most updated version of "instant poi," Soviet astronauts were reported to have eaten dehydrated taro in space, adding water to the packets.

Key Information Sources

Allen, O.N., and Allen, E.K. 1933. The manufacture of poi from taro in Hawaii: with special emphasis upon its fermentation. University of Hawaii, Honolulu, HI. 32 pp.

Begley, B.W. 1979. Taro in Hawaii. Oriental Publishing Co., Honolulu, HI. 28 pp.

Bird, B.K., and Rotar, P.P. 1982. Bibliography of taro and other edible aroids (supplement, 1977–1982) with author, KWIC, and source indexes. Hawaii Institute of Tropical Agriculture and Human Resources, College of Tropical Agriculture and Human Resources, University of Hawaii at Manoa, Honolulu, HI. 123 pp.

Ferentinos, L., and Vargo, A. (*Editors*). 1993. Taro production systems in Micronesia, Hawai'i and American Samoa. College of Tropical Agriculture and Human Resources, University of Hawaii, Honolulu, HI. 86 pp.

Gunawardhana, S.A. 1982. A review paper on aroids: taro (*Colocasia esculenta* (L) Schott); Kongkong taro (*Xanthosoma sagittifolium* Schott). Keravat, Papua New Guinea. 23 pp.

Hill, A.F. 1939. The nomenclature of the taro and its varieties. Bot. Mus. Leafl. Harv. Univ. **7**: 113–118.

Hollyer, J.R., and Sato, D.M. (*Editors*). 1990. Proceedings of taking taro into the 1990's, a taro conference, Hilo, Hawaii, August 17, 1989. Hawaii Institute of Tropical Agriculture and Human Resources, College of Tropical Agriculture and Human Resources, University of Hawaii, Honolulu, HI. 96 pp.

Ivancic, A., and Lebot, V. 2000. The genetics and breeding of taro. CIRAD, Montpellier, France. 194 pp.

Jackson, G.V.H. 1980. Diseases and pests of taro. South Pacific Commission, Noumea, New Caledonia. 51 pp.

Joubert, F.J. 1997. Taro (*Colocasia esculenta*), an under exploited tuber crop in South Africa. J. S. Afr. Soc. Hortic. Sci. **7**: 39–43.

Lambert, M. (*Editor*). 1982. Taro cultivation in the South Pacific. South Pacific Commission, Noumea, New Caledonia. 146 pp.

Manrique, L.A. 1995. Taro, production principles and practices. Manrique International Agrotech, Honolulu, HI. 215 pp.

Manrique, L.A. 1996. Glossary of taro terms. Manrique International Agrotech, Honolulu, HI. 61 pp.

Okonkwo, C.A.C. 1993. Taro, *Colocasia* spp. *In* Genetic improvement of vegetable crops. *Edited by* G. Kalloo and B.O. Bergh. Pergamon Press, New York, NY. pp. 709–715.

Ooka, J.J. 1994. Taro diseases: a guide for field identification. Hawaii Institute of Tropical Agriculture and Human Resources, College of Tropical Agriculture and Human Resources, University of Hawaii, Honolulu, HI. 13 pp.

Plucknett, D.L. 1976. Edible aroids: *Alocasia*, *Colocasia*, *Cyrtosperma*, *Xanthosoma* (Araceae). *In* Evolution of crop plants. *Edited by* N.W. Simmonds. Longman, London, UK. pp. 10–12.

Plucknett, D.L., de la Pena, R.S., and Obrero, F. 1970. Taro (*Colocasia esculenta*). Field Crop Abstr. **23**: 413–426.

Pollock, N.J. 2000. Taro. *In* The Cambridge world history of food. *Edited by* K.F. Kiple and K.C. Ornelas. Cambridge University Press, Cambridge, UK. pp. 218–229.

Rotar, P.P., Plucknett, D.L., and Bird, B.K. 1978. Bibliography of taro and edible aroids. Dept. of Agronomy and Soil Science, College of Tropical Agriculture and Human Resources, University of Hawaii at Manoa, Honolulu, HI. 245 pp.

Secretariat of the Pacific Community. 1998. Proceedings, taro genetic resources conservation and utilisation workshop, 1998, Auckland, N.Z. Secretariat of the Pacific Community, Suva, Fiji. 20 pp.

Singh, U. (*Editor*). 1992. Proceedings of the workshop on taro and tanier modeling, held at the University of Manoa, Honolulu, Hawaii, August 8–14, 1991. University of Hawaii, College of Tropical Agriculture & Human Resources, Honolulu, HI. 92 pp.

Spier, R.F.G. 1951. Some notes on the origin of taro. Southwest. J. Anthropol. **7**: 69–76.

Sreekumari, M.T. 1997. Cytological studies in taro—a review. J. Root Crops (India), **23**: 1–7.

Taha, R.M. 2003. *Colocasia esculenta* (L.) Schott. *In* Plant resources of South-East Asia, 12(3), medicinal and poisonous plants 3. *Edited by* R.H.M.J. Lemmens and N. Bunyapraphatsara. Backhuys Publishers, Leiden, Netherlands. pp. 130–131.

Thompson, S.A. 2000. *Colocasia*. *In* Flora of North America north of Mexico. Vol. 22. *Edited by* Flora of North America Editorial Committee. Oxford University Press, New York, NY. pp. 137–139.

Wang, J.K., and Higa, S. (*Editors*). 1983. Taro, a review of *Colocasia esculenta* and its potentials. University of Hawaii Press, Honolulu, HI. 400 pp.

Whitney, L.D., Bowers, F.A.I., and Takahashi, M. 1939. Taro varieties in Hawaii. Hawaii Agricultural Experiment Station, Honolulu, HI. 86 pp.

Wilson, J.E. 1984. Cocoyam. *In* The physiology of tropical field crops. *Edited by* P.R. Goldsworthy and N.M. Fisher. John Wiley and Sons, Chichester, UK. pp. 589–605.

Wilson, J.E., and Siemonsma, J.S. 1996. *Colocasia esculenta* (L.) Schott. *In* Plant resources of South-East Asia, 9, plants yielding non-seed carbohydrates. *Edited by* M. Flach and F. Rumawas. Backhuys Publishers, Leiden, Netherlands. pp. 69–72.

Specialty Cookbooks

(Also see Breadfruit.)

Governor's Agriculture Coordinating Committee (Hawaii). 1980–1991? Taro and poi. Market Development Branch, Division of Marketing and Consumer Services, Hawaii Dept. of Agriculture, Honolulu, HI. 24 pp.

Hollyer, J. 1997. Taro: mauka to makai: a taro production and business guide for Hawai'i growers. College of Tropical Agriculture & Human Resources, University of Hawaii at Manoa, Honolulu, HI. 108 pp.

Na Lima Kokua. 1977. Taro = kalo: uses and recipes. Na Lima Kokua, Pacific Tropical Botanical Garden, Lawai, HI. 20 pp.

Pereira, A.S. 1983. Cooking with taro and poi. A.S. Pereira, Honolulu, HI. 56 pp.

TEA

Family: Theaceae (tea family)

Names

Scientific name: *Camellia sinensis* (L.) Kuntze (*C. thea* Link, *C. theifera* Griff., *Thea bohea* L., *T. viridis* L.)

- The word "tea" comes from early Chinese dialect words for the beverage, including *Tchai*, *Cha*, and *Tay*. In China, tea is usually pronounced ch'a, but in the southeast of the country, it is pronounced as t'e in some dialects. Collectively, variations of these words are said to be the most universal term in all languages. Most tea arrived in Europe via ships from ports in the southeast of China, so the most common word used in Western languages is a variation of t'e.
- Chai or cha is the word for tea in many parts of the world (both the plant and the beverage). Cha is used in most of China, Japan, and Portugal. Chai is used in Russia, and similar words are found in Turkey, South Asia, and Arab countries. In North America, chai has come to mean an Indian-style preparation of tea, that is, sweet, with spices and milk.
- "Tea" has acquired the general meaning of a hot beverage prepared by soaking a herb in warm or hot water. Technically, in this broad sense, there are two kinds of preparations: infusions (made by pouring hot water over herbal material) and decoctions (prepared by boiling herbal material in water). Often the word "tisane" is used to designate a "herbal tea," indicating one that is produced from the leaves or other parts of plants other than the tea plant.
- Several *Melaleuca* species (Myrtaceae family) of Australia are known as "tea tree" (or teatree). [The essential oil extracted from the leaves of tea trees (known as tea tree oil) is employed externally for medicinal and cosmetic purposes. Unfortunately, some people are misled by the term "tea tree oil" into thinking that it can be used internally (the oil is somewhat toxic). Tea tree oil can be confused with "tea oil," a sweet seasoning and cooking oil extracted from pressed seeds of the tea plant discussed in this chapter (*Camellia sinensis*) or from the "tea oil plant" (*Camellia oleifera* C. Abel).]
- The genus *Camellia* is named for Georg Josef Kammel (1661–1706), a Jesuit pharmacist born in Brno, Moravia (now in the Czech Republic). He carried out botanical research starting in 1688 in Luzon, Philippines, and published an account of Philippine plants in 1704 under his Latinized name, Camellus. It is said that he returned from the Philippines with potted shrubs that he presented to the queen of Spain, Maria Theresa, who picked the white flowers and used them effectively to cheer up her husband, Ferdinand, with the result that camellias became widely planted in the royal greenhouses.
- *Sinensis* in the scientific name *Camellia sinensis* means "of China."

Plant Portrait

The tea plant is a small evergreen tree that grows as high as 16 m (52 feet) tall, but is usually pruned back to a low shrub in cultivation so that the leaves can be easily picked. It is native to the Assam–Burma–Yunnan triangle, and has been planted widely in tropical and subtropical areas. Tea has been drunk in China for over 1000 years (claims of 3000 years require verification), but for less than 400 years in most Western nations. The most important

Tea (*Camellia sinensis*), from Woodville (1810).

Sorting tea leaves (mid 17th century), from Works of Linnaeus (1746, endleaf).

tea-growing countries are China, Taiwan, Japan, India, Sri Lanka, and Indonesia. Tea is also grown in two dozen other countries, notably Iran, Georgia, Cameroon, Kenya, Brazil, and Argentina. China is the leading producer, but India, Ceylon, and East African countries are the leading exporters. Assam in northern India is the world's largest black-tea-producing region. Darjeeling (which translates as "land of the thunderbolt") is a tea-growing region of India, said to produce the "champagne of teas." Keemum tea, from China's Quimien area, is considered the champagne of Chinese black teas. Tea is the most popular beverage in the world, next to water (the same claim is often made for coffee). Caffeine is present at an average level of 3%, making tea a stimulant beverage.

In the early 17th century, Europeans, beginning with the Dutch, began trading commodities with the Chinese for their tea. The Chinese did their best to keep Europeans ignorant of how tea was prepared in order to maintain their monopoly. Tea seeds provided to the Europeans were boiled so that they could not be germinated. So successful were the Chinese in keeping the Europeans from knowing their secrets that it was only in the 19th century, after 250 years of the tea trade, the Europeans learned that black and green tea came from the same plant. In the early 19th century, a botanist, J.C. Gordon, managed to collect 80 000 seeds representing diverse varieties of tea, and when they were grown in India, the British discovered that all varieties of tea came from the same species. The British sent gunboats to China, and in what has been called the Opium Wars (1840–1842), overwhelmed the Chinese defences. China was forced by The Treaty of Nanking in 1842 to accept the opium poppy, to open the country to trade, and to cede Hong Kong to the British. Robert Fortune, an English agent, entered China in 1848 and for 3 years roamed about the country's tea regions disguised as a Chinese merchant, unlocking many of China's remaining tea secrets.

In 1767, the British passed the Townshend Acts, which levied a three-pence-per-pound tax on tea imported to the colonies. In 1773, Britain's Tea Act gave the British East India Company a near monopoly on the selling of tea. When 50 or 60 "Sons of Liberty," disguised as Mohawks, protested the British tax on tea by boarding three English ships and dumping chests of the popular drink into Boston Harbor on December 16, 1773, and (less well known) repeated the performance on March 7, 1774, they cost the British around $3 million in modern money. The "Boston Tea Party" was not the only American protest over the British tea tax. The "Chestertown Tea Party" is held every Memorial Day weekend (since 1973) in Chestertown, Maryland, to commemorate an event that took place on May 1774 (5 months after the Boston Tea Party), which was much like the uprising in Boston. Citizens of Chestertown boarded the British ship Geddes, and dumped its cargo of tea into the Chester River. The citizens of Chestertown also wrote a set of resolves that prohibited the buying, selling, or drinking of tea.

Leaf buds and young leaves are used in making tea. For ancient Chinese emperors, only the top bud was collected. Today, the highest quality, most expensive tea is made from the first buds and the smallest leaves, ideally a bud and the youngest two leaves. The remaining leaves can be machine harvested, and usually become part of lower-grade teas, as explained below.

Green tea leaves are steamed and dried, while black tea leaves are withered, rolled, fermented or chemically processed, and dried. Essentially, black teas are green teas that have been fermented (i.e., oxidized). About 98% of tea traded on the world market is black tea. Oolong tea (Oolong means "black dragon"), a favorite in North America, is semi-fermented. Some of the best Oolong teas are grown on the steep slopes of the mountains of China, and because the leaves often can only be picked by trained monkeys, the teas are called "monkey teas." "Yellow tea" or Mandarin tea is a very fine tea. White teas, the rarest of teas, are made up only of buds plucked from a very rare tea plant variety, and are prepared without fermentation. The tea is very light, pale golden in color, and has a lingering taste.

Most tea consumed in North America, and indeed in the West in general, is low-grade black, while green teas are consumed mostly in Japan, China, and some of the Arab countries. In Muslim North Africa, green tea is common, because in the early days of tea introduction there, religious and political leaders misinterpreted the word "fermentation," thought that black tea was an alcoholic form of green tea, and so opted for the latter. Green teas come mainly from China, India, and Sri Lanka. Chinese teas are often given descriptive names, such as Water Fairy, Dragon's Well, Bright Virtue, and Iron Goddess of Mercy. Tea is sometimes allowed to absorb the scent from various flowers, such as jasmine and mango, although more often flavoring is achieved with essential oils. For example, Earl Grey is flavored with a citrus oil (see ORANGE). Lapsang souchong is a curious case: this large-leaved black tea is scented with pinewood smoke.

There are thousands of tea varieties, with the highest diversity in China (it has been said that there are more kinds of tea in China than there are wines in France). Teas are named according to where they originate, color, and the combinations of tea in mixtures.

Teas are partly classified based on the fineness of the leaf. Grades of black tea mainly conform to standards that were created by the British tea industry in 19th century India and Ceylon (Sri Lanka). For black tea, the leaf bud is termed "pekoe tip"; the youngest fully opened leaf is "orange pekoe"; and the next categories recognized by increasing size of leaf are "pekoe," "pekoe souchong," and "souchong." Similarly, for green tea, the grades are "twanky" (which just uses the bud), then "gunpowder" (made from tight, small balls of younger leaves), "imperial" (larger and looser balls), "young hyson," and "hyson" (long, loose leaves of mixed ages; hyson is a Chinese term meaning "flourishing spring"). Probably, most consumers think that "orange" in "orange pekoe" means orange-flavored. Etymologically, the word orange as it relates to tea traces to the ancient Chinese use of a certain size of tea leaf that was scented with orange blossoms (the Chinese word *pekoe* is based on the Mandarin *peh*, white, + *ho*, down, a reference to the delicate down covering the lower surface of the emerging leaf). To reiterate the point, orange pekoe tea is neither orange in color or taste nor made with oranges—it is simply a black tea of a certain particle size. Small pieces of leaves, the debris from the processing of whole leaves, are called "dust," or "fannings" if the particles are very small, and are mostly used in tea bags and brick tea.

Culinary Portrait

Tea is occasionally used as a flavorant in recipes for various food preparations (e.g., sherbets and pastries can be prepared with tea; dried fruit take on a pleasant taste when soaked in tea; green tea is used to flavor soba noodles). However, tea is mainly used to make tea. The highly ritualized Japanese and Chinese tea ceremonies, and the English Tea Party, are reflective of the central role tea plays in different societies.

As a counterpart to Irish coffee, other liquors (such as rum, brandy, bourbon, and whiskey) can be added to tea.

Tea tips

- Tea varies greatly in quality, and the best teas are more costly. Tea bags can be as good as tea leaves, but since some large companies hide inferior ingredients in these, it is especially important to buy a reliable brand.
- To keep tea fresh, store it in a dark, airtight container. Tea bags should be transferred from paper and cellophane packaging to a tightly sealed container. Storing tea in the refrigerator is unnecessary and could result in unwanted condensation.
- Ideally, teas should be used within a year of production.
- Dry tea leaves absorb moisture, smells, and tastes easily, and accordingly spices, perfumes, and even other strong-smelling teas should be kept separate.
- The ideal teapot is made of plain glazed earthenware or porcelain. Never use teapots made of aluminum, unlined copper, or enamel that is chipped. Glass and stainless steel are acceptable, but ceramic pots have superior heat retention, and

if kept clean, are less likely to introduce unwanted tastes. Metal pots draw essential heat away from the tea. Infusers and tea balls for tea leaves are best made of high-quality stainless steel.

- The same pot is best not used for perfumed teas and teas of pronounced flavor, as the latter tend to leave flavor traces that can prevent full appreciation of the subtle taste of light teas.
- Teapots do not require regular cleaning after every use like other kitchen utensils, but should be kept clean to avoid the bitterness of tannin residues. Many authorities recommend only regular rinsing with boiling water. Teapots should not be cleaned with soap or detergent. Should a thorough cleaning become necessary, put 2 tablespoons of bicarbonate of soda into the teapot, fill the pot with hot water and let stand for 3 hours before rinsing well with plenty of cold water. To remove lime deposits from teapots, fill the kettle with equal parts of vinegar and water, bring to a boil, and let stand overnight.
- The ideal source of water for tea is probably a pure mountain spring, but tap water is often adequate. Use good-tasting, fresh water; if necessary, use bottled spring water or charcoal-filtered tap water. Never use softened water. Distilled water makes flat tea and should be avoided. Never reboil water, as boiled water contains less oxygen. Chlorine and fluoride in tap water are said to harm the flavor of tea. High mineral content in water brings out the richness and sweetness of green tea. Ideally, tea water should have an alkaline pH around 7.9.
- Add cold water that has been allowed to run from the tap for a minute to the kettle. Such water is fully oxygenated and makes better tea.
- Warm the teapot with hot water; tea brews better in hotter water, and stays hotter longer. Some traditionalists "hot the pot" by filling it with boiling water and letting it sit for 3–5 minutes.
- "Bring the pot to the kettle, not the kettle to the pot," to prevent cooling of the water.
- Place a carefully measured amount of tea in the empty teapot—one tea bag or one heaping teaspoon of tea leaves for each (8 ounce) cup of tea is usually recommended. An extra tea bag or spoonful of leaves is often added to the pot for stronger tea, or just because experience indicates this makes for better-tasting tea; however, an extra teaspoon or teabag "for the pot" is only appropriate when making six or more cups of tea. While these ratios are useful for groups of people, experiment to establish your personal preference for how strong the tea should be. The most experienced tea drinkers in the world, the Chinese, generally prefer less strong tea.
- If a tea ball or similar infuser is used, do not fill it more than half full. The leaves will double in size as they absorb water, and room is required for the water to flow through them.
- While allowing the tea to come to a rolling boil is generally recommended for black teas, greener teas are more delicate and are better with a lower temperature. Green teas are ruined by boiling water; the temperature is best around 77–85 °C (170–185 °F). Oolongs made with underboiled water are more fragrant.
- As soon as the water is ready, pour it over the tea. Leaving the water boiling drives away oxygen and weakens flavor.
- The smaller the size of the leaf particles, the less time is required. Some general guidelines: tea bags and small leaf teas: 3 minutes; Darjeeling: 3–5 minutes; white tea: 30 seconds to 2 minutes; green tea: 1–3 minutes; green oolong tea: 2–3 minutes; black oolong tea: up to 5 minutes; all other loose and black teas: 5–6 minutes. Stir the tea no more than once while it is brewing (some authorities recommend not stirring).
- If possible, wrap the teapot with a tea towel or tea cozy while steeping to retain the heat.
- After the steeping period, promptly remove tea bags or tea leaves without squeezing, which can press out some of the bitter tasting polyphenols that remain.
- Pour the tea into cups. Use a clean mesh strainer if tea leaves have been employed.
- If you prefer milk, add it first as it dissolves better in hotter liquid. Never add cream, as it has too much fat for the delicate flavor of tea. Adding sugar, honey, lemon, etc., is fine, but putting lemon and milk together will curdle the milk. Green teas, green Oolong, and scented teas are best appreciated with nothing added that could mask their delicate flavors.
- Relax and enjoy!

Iced Tea

Iced tea is an American invention, traditionally prepared by steeping hot tea using 50% more tea than normally used, and pouring the tea over ice in tall glasses. Refrigerated tea tend to cloud, which does not affect taste but is less attractive. A clouded pitcher can be cleared by adding a little boiling water and stirring.

(Ceylon, a Sri Lankan black pekoe tea, is considered ideal for serving iced because it does not produce a cloudy mix when cold.) "Thai tea" is a Thai-American strong black tea prepared with cream and sugar and served chilled over ice. To make perfect iced tea (4 servings):

- Place six tea bags in a 1-L (1 American quart) pitcher. If desired, add 45 mL (3 tablespoons) granulated sugar and 2–4 sprigs of fresh mint.
- Pour 300 mL (1¼ cups) of freshly boiled water over tea bags.
- Steep 5 minutes and remove tea bags, and if used, mint sprigs. Fill pitcher with fresh cold water.
- Place four ice cubes in each water glass, and pour tea over ice.
- Add more sugar to taste. Garnish with lemon slices and (or) mint leaves.

Chemicals in tea (condensed catechin tannins) have been linked to high rates of esophageal cancer in some areas where tea is heavily consumed. (Tannins are also undesirable in that they hinder the absorption of iron from foods.) Adding milk binds the tannins, preventing harmful effects. However, the relation of tea and cancer is mixed, some reports suggesting both black and green tea reduce some cancers. In particular, green tea is not fermented, so many of its active constituents remain unaltered. Chemicals called polyphenols are believed to be responsible for most of green tea's roles in promoting good health, including protection against cardiovascular disease, some cancers, and even dental plaque. Because several cupfuls daily seem necessary to produce significant benefits, decaffeinated tea is helpful in reducing insomnia, anxiety, and other symptoms that may be caused by the herb's caffeine content.

Culinary Vocabulary

- A "shotty tea" sounds bad, but is one that has been well manufactured, usually by traditional labor-intensive methods.
- Tipping to promote good service was born in the tea gardens of England. A small wooden box inscribed with TIPS for "To Insure Prompt Service" was placed on tables to receive coins.
- "Rosie Lee" (also "Rosie" and "Rosy Lee") is British rhyming slang for "tea," and appears to have originated in the early 20th century. "Snow broth" is old British slang for "cold tea" (also, very cold liquor; "snow-broth," with a hyphen, has nonculinary meanings). "Mash" is English slang for "to brew a cup of tea."
- The British expression "high tea" refers to food served late in the afternoon, usually including a cooked item such as eggs or sausages. High tea is a substantial meal served at about tea time. This should not be confused with "tea" or "afternoon tea," which is merely a cup of tea with a cake or very small sandwich.
- "Morning tea" in England may refer to tea served to guests in their hotel room before going to the dining room for breakfast. "Breakfast" is the name of a popular tea made in Scotland for over a century, and requesting "breakfast tea" may result in confusion.
- In Australian slang, a container for boiling tea is called a billy.
- In New Zealand slang, "tea" means dinner or the evening meal.
- "Cuppa" is British and Australian slang for a cup of tea.
- "Instant tea" is a fine powder prepared, much like instant coffee, by dehydrating a brew of the beverage.
- "Cambric tea," an American slang term, refers to a hot drink of milk, water, sugar, and if desired, a small amount of tea. It was a favorite of children and the elderly in the late 19th and early 20th centuries. The name is based on cambric (linen or cotton) fabric, which is white and thin like the tea (cambric fabric takes its name from the French textile manufacturing town of Cambrai). Cambric tea is also known as "hot water tea." "Coffee milk" (see COFFEE) is the coffee equivalent to cambric tea. Cambric is pronounced KAYM-brihk.
- "Hot spot" is old lunch-counter jargon for a glass of tea, while "cold spot" meant a glass of iced tea. "Spot with a twist" indicated a cup of tea with lemon. The term "PT" referred to a pot of tea. "Boiled leaves" was another way of indicating tea. A "muffin fight" was a tea party.
- "Agony of the leaves" is an expression describing the relaxation of curled tea leaves during steeping. (Gunpowder tea, for example, is a green tea that has been rolled into pellets, which unfurl during brewing in hot water.)
- A "tea egg" is prepared by thoroughly cracking the shell of a hard-cooked egg but leaving the shell on while it simmers in strong tea (and salt) to create a marbled effect on the surface. It is an appetizer in Chinese cuisine. The egg tastes mildly of tea, and tends to be tough.

- *Genmai cha* (pronounced GEHN-ma-ee CHA) is an unusual tea—a combination of green tea and toasted rice kernels—which is often available in Japanese food stores and is frequently served with sushi and other Japanese dishes.
- "Tsiology" is either (1) the study of tea or (2) a dissertation on tea.
- "Long Island Iced Tea" does not contain tea. It is a cocktail prepared with tequila, vodka, gin, rum, and a cola-flavored soda.
- A "mote spoon" is a utensil with a long sharp-pointed handle at one end (used to push tea leaves out of a teapot spout) and a pierced spoon at the other end (used as a tea strainer).

Curiosities of Science and Technology

- In the 18th and 19th centuries in England, during the days when there was little control of what additives were placed in food (Britain's first Adulteration of Foods Act was passed in 1860), a number of despicable practices were common. "Tea" was often made from other commonly available plants, such as ash, elder, and sloe. Tea leaves that had been used to brew tea were bought from hotels and restaurants, and were colored and flavored with toxic substances such as black lead, Prussian blue, copper carbonate, and lead chromate, and sold as new tea. "Gunpowder tea," a fine variety of green Chinese tea, was sometimes actually contaminated with gunpowder.
- Tea was first introduced in the American colonies, with the instructions that the leaves should be boiled. Many served the tea leaves with sugar or syrup after throwing away the water in which they had been boiled.
- After the American Revolution, Britain continued to dominate the tea trade. However, in 1859, Americans George Huntington Hartford and George Gilman established a very large tea trading company, the Great Atlantic and Pacific Tea Company, and underpriced the British competition. The company went on to develop into the well-known supermarket chain "A & P."
- At the St. Louis World's Fair in 1904, an enterprising British salesman, Richard Blechynden, had a tea concession. On a very hot day, none of the fairgoers were interested in drinking hot tea. Blechynden served the tea cold, and thus has generally been credited with inventing iced tea. In fact, iced tea was available earlier, although Blechynden did bring iced tea to the world's attention.
- The tea bag was invented in 1908 by tea importer Thomas Sullivan, who developed the practice of sending small samples of tea in silk bags to retailers. The retailers mistakenly thought Sullivan had intended them to infuse the whole bag.
- Solid blocks of tea were used as money in Siberia until the 19th century.
- The English, Irish, Norwegians, and Danish have, at one time or another, all been credited with the invention of the cosy (cozy). According to the Irish interpretation, in the 1600s, Irish farmers typically consumed a large pot of tea with dinner each evening. One farmer's hat fell off, covering the tea pot. Later, when he went to pour more tea and removed the hat, he discovered the tea was still very warm. His wife decided to make a cover for their tea pot and called it a "cosy."
- In the first quarter of the 20th century, tea cigarettes were smoked in England, particularly by women.
- In Tibet, horses and mules have been provided with large vessels of tea to increase their capacity to work.
- In Mongolia, distance is sometimes measured in number of cups of tea, with 3 cups equal to 8 km (5 miles).
- American Classic Tea, the official White House tea since 1987, is the only tea grown in the United States. It is produced on the subtropical sea island of Wadmalaw, 32 km (20 miles) south of Charleston, South Carolina.
- In Egypt, a host may pour tea into the glass of a guest until it overflows into its saucer, the generous overflow indicating respect to the guest.
- Different estimates have compared the amount of beverage that can be made from a given weight of coffee and tea. A kilogram of tea can yield up to 660 cups of beverage (a pound up to 300), although 440 cups per kilogram (200 per pound) is a more reasonable average estimate. A kilogram of coffee typically produces only 66–88 cups (a pound, 30–40 cups). Approximately, five times as much tea can be made from a given weight of tea compared to the amount of coffee that can be made from the equivalent weight, helping to explain the relative cheapness of tea as a beverage.
- A folk remedy for hemorrhoids is to apply a warm, wet tea bag on the affected area. Tannic acid in tea is known for its astringent (skin-tightening) and anti-inflammatory properties, which could be helpful for hemorrhoids.

Chinese coolies carrying 136-kg (300-pound) loads of tea in the early 19th century, from Ward (1911). Note use of pole to support heavy load while at rest.

- Tea stains can be removed by rubbing vigorously with a paste of baking soda and water.
- In 1978, the United Nations Food and Agricultural Organization estimated that "all the tea in China" amounted to approximately 356 000 metric tons (about 400 000 North American tons).
- Nearly 75% of the tea used in the Western World comes in tea bags.
- One American teaspoon has a volume of 5 mL or 5 cc (or, more precisely, "4.929 mL U.S. Standard"), which is about 0.17 fluid ounce. One tablespoon equals 3 teaspoons. An Australian teaspoon also has a volume of 5 mL, but an Australian tablespoon equals 4 teaspoons, i.e., 20 mL. One United Kingdom teaspoon equals 1.2 American teaspoons equals 6.16 mL (1 United Kingdom tablespoon equals 1.2 American teaspoons equals 18.48 mL). One metric teaspoon equals 5 mL (1 metric tablespoon equal 15 mL).
- Ireland is the leading tea-drinking nation, consuming an average of 1390 cups per person each year. Britain ranks second, with an average of 1113 cups. The average in the United States is only 150 cups per year.
- The following traditional folk remedy procedure has been recommended as a means of deodorizing smelly feet. Boil 3 or 4 tea bags in a quart of water for 10 minutes. Add cold water to reduce the temperature sufficiently to be able to put your feet into the liquid. Soak your feet in the warm tea for 20–30 minutes. Repeat twice a day until the odor is under control. Then soak twice a week to keep odors away. Tannic acid in tea has germicidal properties, and so could destroy odor-causing bacteria.

Key Information Sources

Anderson, J.L. 1991. An introduction to Japanese tea ritual. State University of New York Press, Albany, NY. 348 pp.

Balentine, D.A., and Paetau-Robinson, I. 2000. Teas as a source of dietary antioxidants with a potential role in prevention of chronic diseases. *In* Herbs, botanicals & teas. *Edited by* G Mazza and B.D. Oomah. Technomic, Lancaster, PA. pp. 265–287.

Castile, R. 1971. The way of tea. Weatherhill, New York, NY. 329 pp.

Chow, K.B., and Kramer, I. 1990. All the tea in China. China Books and Periodicals, San Francisco, CA. 187 pp.

Dusinberre, D. 1992. The book of tea. Flammarion, Paris, France. 256 pp.

Dutta, A.K. 1999. Tea. *In* Tropical horticulture, vol. 1. *Edited by* T.K. Bose, S.K. Mitra, A.A. Farooqui, and M.K. Sadhu. Naya Prokash, Calcutta, India. pp. 440–473.

Eden, T. 1976. Tea. 3rd edition. Longmans, London, UK. 205 pp.

Ellis, R.T. 1995. Tea. *In* Evolution of crop plants. 2nd edition. *Edited by* J. Smartt and N.W. Simmonds. Longman Scientific & Technical, Burnt Mill, Harlow, Essex, UK. pp. 22–27.

Evans, J.C. 1992. Tea in China: the history of China's national drink. Greenwood Press, New York, NY. 169 pp.

Faulkner, R. 2003. Tea: east & west. V & A, London, UK. 128 pp.

Forrest, D.M. 1973. Tea for the British: the social and economic history of a famous trade. Chatto and Windus, London, UK. 320 pp.

Forrest, D.M. 1985. The world tea trade: a survey of the production, distribution, and consumption of tea. Woodhead-Faulkner, Cambridge, UK. 243 pp.

Goodwin, J. 1991. A time for tea: travels through China and India in search of tea. Knopf, New York, NY. 287 pp.

Harler, C.R. 1963. Tea manufacture. Oxford University Press, London, UK. 126 pp.

Harler, C.R. 1964. The culture and marketing of tea. Revised edition. Oxford University Press, London, UK. 262 pp.

Harler, C.R. 1966. Tea growing. Oxford University Press, London, UK. 162 pp.

Katiyar, S.K., and Mukhtar, H. 1996. Tea consumption and cancer. World Rev. Nutr. Diet. **79**: 154–184.

Labaree, B.W. 1964. The Boston tea party. Oxford University Press, New York, NY. 347 pp.

Macfarlane, A., and Macfarlane, I. 2004. The empire of tea: the remarkable history of the plant that took over the world. Overlook Press, Woodstock, NY. 308 pp.

Moxham, R. 2003. Tea: addiction, exploitation, and empire. Carroll & Graf, New York, NY. 271 pp.

Pettigrew, J. 1997. The tea companion. Apple, London, UK. 192 pp.

Pettigrew, J. 2001. A social history of tea. National Trust, London, UK. 192 pp.

Pintauro, N. 1977. Tea and soluble tea products manufacture. Noyes Data Corp., Park Ridge, NJ. 261 pp.

Pratt, J.N. 1982. New tea lover's treasury. Publishing Technology Associates, San Francisco, CA. 210 pp.

Schoorel, A.F., and van der Vossen, H.A.M. 2000. *Camellia sinensis* (L.) Kuntze. *In* Plant resources of South-East Asia, 16, stimulants. *Edited by* H.A.M. van der Vossen and M. Wessel. Backhuys, Leiden, Netherlands. pp. 55–63.

Scott, J.M. 1965. The great tea venture. Dutton, New York, NY. 203 pp.

Sealy, J. 1958. A revision of the genus *Camellia*. Royal Horticultural Society, London, UK. 239 pp.

Ukers, W.H. 1936. The romance of tea; an outline history of tea and tea-drinking through sixteen hundred years. A.A. Knopf, New York, NY. 276 pp.

Visser, T. 1969. Tea, *Camellia sinensis*. *In* Outlines of perennial crop breeding in the tropics. *Edited by* F.P. Ferwerda and F. Wit. Misc. Paper 4. Wageningen Agricultural University, Wageningen, Netherlands. pp. 459–493.

Walsh, W.I. 1986. The rise and decline of the Great Atlantic & Pacific Tea Company. L. Stuart, Secaucus, NJ. 254 pp.

Weisburger, J.H., and Comer, J. 2000. Tea. *In* The Cambridge world history of food. *Edited by* K.F. Kiple and K.C. Ornelas. Cambridge University Press, Cambridge, UK. pp. 712–720.

Werkhoven, J. 1974. Tea processing. Food and Agriculture Organization of the United Nations, Rome, Italy. 196 pp.

Wight, W. 1951. Tea classification revised. Curr. Sci. (India), **8**(**31**): 289–299.

Willson, K.C., and Clifford, M.N. 1992. Tea: cultivation to consumption. Chapman & Hall, London, UK. 769 pp.

Yamamoto, T. 1997. Chemistry and applications of green tea. CRC Press, Boca Raton, FL. 160 pp.

Zhen, Y., Chen. Z., Cheng, S., and Chen, M. [*Editors*]. 2002. Tea: bioactivity and therapeutic potential. Taylor & Francis, London, UK. 267 pp.

Specialty Cookbooks

(Note: Hundreds of books on preparing and using tea are available in English. The following are among the most popular.)

Campbell, D. 1995. The tea book. Pelican Publishing Co., Gretna, LA. 223 pp.

Compestine, Y.C. 2000. Cooking with green tea. Avery, New York, NY. 194 pp.

Foley, T., and Calvert, C. 1987. Having tea: recipes & table settings. C.N. Potter, New York, NY. 87 pp.

Gustafson, H. 2001. The green tea user's manual. Clarkson Potter, New York, NY. 111 pp.

Israel, A., and Mitchell, P. 1987. Taking tea: the essential guide to brewing, serving, and entertaining with teas from around the world. Weidenfeld & Nicolson, New York, NY. 144 pp.

Kaufman, W.I. 1966. The tea cookbook. Doubleday, Garden City, NY. 188 pp.

Perry, S., and Miksch, A. 2001. The new tea book: a guide to black, green, herbal, and chai teas. Revised edition. Chronicle Books, San Francisco, CA. 120 pp.

Pettigrew, J. 2001. Tea-time recipes. Revised edition. National Trust, London, UK. 152 pp.

Pratt, J.N., and Rosen, D. 1996. The tea lover's companion: the ultimate connoisseur's guide to buying, brewing, and enjoying tea. Carol Publishing Group, Secaucus, NJ. 185 pp.

Pruess, J., and Harney, J. 2001. Eat tea: a new approach to flavoring contemporary and traditional dishes. Lyons Press, Guilford, CT. 116 pp.

Rasmussen, W., and Rhinehart, R. 1999. Tea basics: a quick and easy guide. John Wiley, New York, NY. 181 pp.

Resnick, J. 1997. Loving tea. Berkley Books, New York, NY. 212 pp.

Siegel, J., Siegel, M., and Simon, J. 1996. Cooking with tea: cookbook. Park Lane Press, New York, NY. 128 pp.

Slavin, S., Petzke, K., and Berry, L. 1998. Tea: essence of the leaf. Chronicle Books, San Francisco, CA. 96 pp.

Tekulsky, M. 1995. Making your own gourmet tea drinks: black teas, green teas, scented teas, herb teas, iced teas, and more! Crown Publishers, New York, NY. 95 pp.

Thompson, F. 2002. Iced tea: 50 recipes for refreshing tisanes, infusions, coolers, and spiked teas. Harvard Common Press, Boston, MA. 96 pp.

Tokunaga, M. 2004. New tastes in green tea: a novel flavor for familiar drinks, dishes, and desserts. Kodansha International, Tokyo, Japan. 127 pp.

Waller, K. 1999. The pleasures of tea. Hearst Books, New York, NY. 128 pp.

Ward, M. 1996. The top 100 international tea recipes: how to prepare, serve and experience great cups of tasty, healthy tea and tea desserts. Lifetime Books, Hollywood, FL. 216 pp.

Wemischner, R., and Rosen, D. 2000. Cooking with tea: techniques and recipes for appetizers, entrées, desserts, and more. Periplus Editions, Boston, MA. 146 pp.

TEFF

Family: Poaceae (Gramineae; grass family)

Names

Scientific name: *Eragrostis tef* (Zuccagni) Trotter [*E. abyssinica* (Jacq.) Link]

- Why "teff" is so named is uncertain, despite linguistic analysis. One interpretation is that the name is based on the word for the grain in Amharic (the official, Semitic language of Ethiopia, which has about 80 languages). The Amharic word *teffa* means lost or disappeared, which probably is related to the very small size of the grain, which is easily lost if dropped. The Amharic *tef meret* denotes unused, uncropped, usually degraded land, and since teff is adapted to such land, this may explain why it is so named. The word teff has also been said to come from the Arabic *tahf* used in South Arabia. Still another explanation is that the word is a Cushite name of the crop (Cushite is one of the languages of Ethiopia).
- Teff is alternatively spelled tef, and sometimes t'ef.
- Teff is also known as annual bunch grass, lovegrass, teff grass, and warm season annual bunch grass.
- The genus name *Eragrostis* is based on the Greek *Eros*, the god of love, and *agrostis*, a grass, which once referred to some Old World grass species long known as love grass, perhaps because of its beauty. Species of *Eragrostis* are commonly known as love grass.

Plant Portrait

Teff is an annual grass growing 25–135 cm (10 inches to over 4 feet) in height. The plant is a fine-stemmed "bunch grass," i.e., it grows in tufts because it develops "tillers" (branches produced from the base of the plant). The grain color ranges from pale white to ivory white and from very light tan to deep brown to reddish brown purple. Teff seed has been described as "smaller than a pinhead," and is indeed very small, ranging from 1 to 1.7 mm (0.04–0.07 inch) in length and 0.6 to 1 mm (0.02–0.07 inch) in diameter. One grain of wheat typically weighs as much as 150 grains of teff. Teff is believed to be native to northeastern Africa and southwestern Arabia, and was domesticated in Ethiopia, starting as long ago as 4000 BC. Today, it remains primarily a cereal crop of Ethiopia, where it is the most important food plant of the country. However, teff is expensive in Ethiopia, as it requires labor-intensive harvesting and processing techniques, and produces especially low yields, and so is purchased mostly by those who are relatively wealthy. Several dozen cultivated varieties are grown in Ethiopia. Teff is relatively unknown as a food crop elsewhere. In Ethiopia, teff is also raised for livestock forage, or the straw remaining after the grains are threshed away is used as fodder. The straw is also utilized to reinforce mud or plaster used in the construction of buildings. Outside of Ethiopia, teff is grown in Yemen, Kenya, Malawi, South Africa, India, Australia, and to a small extent in Canada and the United States. Outside of Eurasia, it has been cultivated mostly for feeding livestock. Teff is also grown for grain in Australia and a few western states of the United States to supply a small specialty health-food market.

Culinary Portrait

Teff has a mild, slightly molasses-like sweetness. In nutritional value, it is similar to the traditional cereals, but unlike wheat, it contains very little gluten, necessary to make a raised or leavened loaf of bread. People with allergies to gluten are being attracted to teff. However, the cereal is little known as a food outside of Ethiopia, where the grain is ground into flour, fermented for several days, and made into *injera* (also spelled aenjera, enjera, and ingera), a sour-dough bread that has been described as a soft, spongy, porous, thin, giant (typically, 60 cm or 2 feet in diameter), bubbly, moist, chewy, almost elastic pancake, with a sour taste. Although described as "flat bread," in fact it is intermediate between a flat bread and a raised one. Injira is an Ethiopian staple, said to comprise 92% of the rural diet. Ethiopians use pieces of injera to scoop up spicy stews. Today, injera is sometimes served in ethnic or specialty restaurants in major international cities. (Injera can be made from other grains, especially buckwheat, but tends to harden after a day, whereas teff-made injera keeps its soft spongy texture for 3 days). In Ethiopia, teff flour is often mixed with other cereal flours, although the flavor and quality of injera made from mixtures is considered less tasty. The longer the fermentation process, the sourer the product. If the dough is fermented for only a short period of time, injera has a

Teff (*Eragrostis tef*), from Jacquin (1781–1786, plate 17). Copy of illustration and the right to reproduce it courtesy of The LuEsther T. Mertz Library of The New York Botanical Garden, Bronx, New York.

tasty sweet flavor. Ethiopians also consume teff as porridge, and use it as an ingredient of home-brewed alcoholic drinks.

Teff grain and flour have found a niche in the health-food market, and are often available in specialty and natural food stores in Western countries. White grains produce a blander flavor, and are more expensive than the darker-colored (reddish, brown, or black) grains that produce a richer, more robust flavor preferred by teff aficionados. However, injera made with white grain is said to have a longer shelf life. Uncooked teff can be added to most kinds of baked goods, including breads, biscuits, cookies, cakes, stir-fry dishes, casseroles, soups, stews, and puddings. Cooked teff is gelatinous and adds body to puddings and icebox pies. Teff is a good thickener for soups, stews, and gravies. When teff is used as a substitute for other grains or nuts, because of its small size and high density, less is required than the amount of the grain substituted. One-half cup of teff can be used to replace 1 cup of sesame seeds. To prepare teff porridge, add 3 cups of water or milk for each cup of grain and simmer for about 15 minutes. Numerous teff recipes are available on the Web. Teff grain or flour should be stored in a cool, dry area in a sealed glass or plastic container, because air, moisture, and sunlight can cause the oils to become rancid.

Teff is a special "millet," i.e., a minor true cereal (see MILLET for additional information), which has some nutritional advantages over other millets. It has a high iron content, nearly double that of other food grains, and a calcium content almost 20 times more than other grains. The high level of these two nutrients provide a favorable nutritional value for teff compared to other millets.

Curiosities of Science and Technology

- Seeds of *Eragrostis* were found in a brick of the Dassur Egyptian pyramid built in 3359 BC. These have been identified as teff by some, and as other species of the genus by others.
- Teff grain may either be white or a very deep reddish brown. In the past in Ethiopia, the upper class consumed the white grain, while the dark grain was reserved for soldiers and servants. At one

time, white-grained varieties were grown only in the king's fields, and never sold in open markets. This is reminiscent of how white bread was long a status symbol in Western countries.

- Injera bread made from teff is used as an edible plate or tablecloth in Ethiopia, often covering the table.
- In the central Sahara region, indigenous peoples have a long-standing relationship with *Eragrostis* species and ants. The ants collect the small seeds and bring them to their nests, where some of the seeds germinate and grow into plants. People recognize the plants and know that they might find ant nests nearby from which they can pillage a store of seeds.
- Teff is considered an outstanding forage to feed livestock in South Africa. During the Boer Wars (the South African Wars of 1880–1881 and 1899–1902, which were fought between the British and the descendants of the Dutch settlers called Boers), teff was extensively used to feed the horses and oxen used in combat by both sides.
- Teff grass is highly recommended for the feeding of the white rhinoceros (*Ceratotherium simum*) in captivity.
- Tef is the name of a Turkish tambourine, consisting of a metal or wooden hoop over which a skin is stretched.
- There are two, widespread classes of plants that differ in the way they carry out photosynthesis. Most familiar temperate zone plants are called C3 plants because the first stable compound formed when carbon dioxide is processed is a three-carbon compound, i.e., C3. The C4 plants are so named because the first organic compound incorporating CO_2 is a four-carbon compound. Many tropical plants have the C4 type of photosynthesis, which allows for increased growth if temperature and light intensity are sufficiently high. Teff is a C4 plan, but is intermediate between tropical and temperate species in its physiology, adapted to warm but not the very hot climates of many C4 plants.
- Teff is considered to be an excellent "rescue crop," "catch crop," and "nurse crop." A "rescue crop" is one that is needed to replace another crop in mid-season because the original crop did not develop. Teff produces a good hay or pasture crop when late season plantings are required owing to a crop failure. A "catch crop" is a more general category, referring to any quick-growing crop that is sown to make use of temporary idleness of the soil or to compensate for the failure of a main crop. Teff holds down soil and prevents erosion until perennial plants develop sufficiently to assume these tasks. Teff is also an excellent "nurse crop," i.e., one that is planted along with another to facilitate its growth. As a nurse crop, teff is often planted together with perennial grasses. During the first season, teff grows well and prevents weeds from establishing but allows the perennial grasses to develop. The next year, the perennial grasses take over. Because of teff's abilities to grow in circumstances too difficult for the survival of most crops, it has been called "a reliable cereal for an unreliable climate."
- Ethiopians commonly mix legume seeds such as fenugreek, lentils, peas, and faba beans into their injera teff bread. Beans are rich sources of protein, particularly the essential amino acid lysine, but are usually deficient in the sulfur-containing amino acids methionine and cystine. By contrast, cereal grains like teff contain lower amounts of proteins, and these are deficient in lysine but adequate in sulfur-containing amino acids. Together, teff and beans provide a balance of essential protein components. In Ethiopia, it is said that one such pancake provides enough protein to sustain life without another protein source, such as meat. It has been estimated that teff provides two-thirds of the protein requirements of the people of Ethiopia.

Key Information Sources

Assefa, K., Ketema, S., Tefera, H., Nguyen, H.T., Blum, A., Ayele, M., Bai, G., Simane, B., and Kefyalew, T. 1999. Diversity among germplasm lines of the Ethiopian cereal tef. Euphytica, **106**: 87–97.

Assefa, K., Tefera, H., Merker, A., Kefyalew, T., and Hundera, F. 2001. Quantitative trait diversity in tef [*Eragrostis tef* (Zucc.) Trotter] germplasm from Central and Northern Ethiopia. Genet. Resour. Crop Evol. **48**: 53–61.

Bai, G., Ayele, M., Tefera, H., and Nguyen, H.T. 2000. Genetic diversity in tef and its relatives as revealed by random amplified polymorphic DNAs. Euphytica, **112**: 15–22.

Bekele, E., and Lester, R.N. 1981. Biochemical assessment of the relationship of *Eragrostis tef* (Zucc) Trotter with some wild *Eragrostis* species (Gramineae). Ann. Bot. **48**: 717–725.

Bekele, E., Fido, R.J., Tatham, A.S., and Shewry, P.R. 1995. Heterogeneity and polymorphism of seed proteins in tef (*Eragrostis tef*). Hereditas, **122**: 67–72.

Boe, A., Sommerfeldt, J., Wynia, R., and Thiex, N. 1986. A preliminary evaluation of the forage potential of teff. Proc. S. D. Acad. Sci. **65**: 75–82.

Bultosa, G., Hall, A.N., and Taylor, J.R.N. 2002. Physico-chemical characterization of grain tef (*Eragrostis tef* (Zucc.) Trotter) starch. Starch, **54**: 461–468.

Costanza, S.H., DeWet, J.M.J., and Harlan, J.R. 1979. Literature review and numerical taxonomy of *Eragrostis tef* (t'ef). Econ. Bot. **33**: 413–424.

Ebba, T. 1969. T'ef (*Eragrostis tef*). The cultivation, usage, and some of the known diseases and insect pests. Exp. Stn. Bull. 60. Part 1. Haile Selassie University, College of Agriculture, Dire Dawa, Ethiopia. 56 pp.

Ebba, T. 1975. T'ef (*Eragrostis tef*). Cultivars: morphology and classification. Exp. Stn. Bull. 69. Part II. Addis Ababa University, College of Agriculture, Dire Dawa, Ethiopia. 73 pp.

Endeshaw, B. 1995. Variations in basic amino acids including lysine and total protein in *Eragrostis tef* (zucc.) Trotter. SINET (Ethiop. J. Sci.), **18**: 175–194.

Hundera, F. 1998. Variations of morpho-agronomic characters and grain chemical composition of released varieties of tef (*Eragrostis tef* (Zucc.) Trotter). J. Genet. Breed. **52**: 307–311.

Ingram, A.L., and Doyle, J.J. 2003. The origin and evolution of *Eragrostis tef* (Poaceae) and related polyploids: Evidence from nuclear waxy and plastid rps16. Am. J. Bot. **90**: 116–122.

Jones, B.M.G., Ponti, J., Tavassoli, A., and Dixon, P.A. 1978. Relationships of the Ethiopian cereal t'ef [*Eragrostis tef* (Zucc.) Trotter]: evidence from morphology and chromosome number. Ann. Bot. **42**: 1369–1373.

Kebede, H., Johnson, R.C., and Ferris, D.M. 1989. Photosynthetic response of *Eragrostis tef* to temperature. Physiol. Plant. **77**: 262–266.

Kefyalew, T., Tefera, H., Assefa, K., and Ayele, M. 2000. Phenotypic diversity for qualitative and phenologic characters in germplasm collections of tef (*Eragrostis tef*). Genet. Resour. Crop Evol. **47**: 73–80.

Ketema, S. 1993. Tef (*Eragrostis tef*): breeding, agronomy, genetic resources, utilization, and role in Ethiopian agriculture. Institute of Agricultural Research, Addis Abeba, Ethiopia. 102 pp.

Ketema, S. 1997. Tef. *Eragrostis tef* (Zucc.) Trotter. Promoting the conservation and use of underutilized and neglected crops. 12. Institute of Plant Genetics and Crop Plant Research, International Plant Genetic Resources Institute, Rome, Italy. 50 pp.

Mengesha, M.H. 1966. Chemical composition of teff (*Eragrostis tef*) compared with that of wheat, barley and grain sorghum. Econ. Bot. **20**: 268–273.

Mengesha, M.H., Pickett, R.C., and Davis, R.L. 1965. Genetic variability and interrelationship of characters in teff, *Eragrostis tef* (Zucc.) Trotter. Crop Sci. **5**: 155–157.

National Academy of Sciences. 1996. Teff. *In* Lost Crops of Africa. Vol. 1. Grains. National Academy of Sciences, Washington DC. pp. 215–235.

Parker, M.L., Umeta, M., and Faulks, R.M. 1989. The contribution of flour components to the structure of injera, an Ethiopian fermented bread made from tef (*Eragrostis tef*). J. Cereal Sci. **10**: 93–104.

Robinson, R.G. 1986. Amaranth, quinoa, ragi, tef, and niger: tiny seeds of ancient and modern interest. Agricultural Experiment Station, University of Minnesota, St. Paul, MN. 23 pp.

Seyfu, K. 1991. Germplasm evaluation and breeding work on teff (*Eragrostis tef*) in Ethiopia. *In* Plant genetic resources of Ethiopia. *Edited by* J.M.M. Engels, J.G. Hawkes, and W. Melaku. Cambridge University Press, Cambridge, UK. pp. 323–328.

Seyfu, K. 1993*a*. Phenotypic variations in tef (*Eragrostis tef*) germplasm: morphological and dynamic traits: a catalogue. Institute of Agricultural Research, Addis Abeba, Ethiopia. 96 pp.

Seyfu, K. 1993*b*. Tef (*Eragrostis tef*): breeding, genetic resources, agronomy, utilization and role in Ethiopian agriculture. Institute of Agricultural Research, Addis Abeba, Ethiopia. 102 pp.

Shiferaw, B., and Baker, D.A. 1996. An evaluation of drought screening techniques for *Eragrostis tef*. Trop. Sci. **36**: 74–85.

Stallknecht, G.F., Gilbertson, K.M., and Eckhoff, J.L. 1993. Teff: food crop for humans and animals. *In* New Crops. *Edited by* J. Janick and J.E. Simon. Wiley, New York. pp. 211–218.

Stewart, R.B., and Getachew, S. 1962. Investigations of the nature of injera. Econ. Bot. **16**: 127–130.

Tadesse, D. 1993. Study of genetic variation of landraces of teff (*Eragrostis tef* (Zucc.) Trotter) in Ethiopia. Genet. Resour. Crop Evol. **40**: 101–104.

Tefera, H., Ketema, S., and Tesemma, T. 1990. Variability, heritability and genetic advance in tef (*Eragrostis tef* (Zucc.) Trotter) cultivars. Trop. Agric. **67**: 317–320.

Teferra, T., Tefera, H., Simane, B., and Tuinstra, M. 2000. The influence of drought stress on yield of tef (*Eragrostis tef*). Trop Sci. **40**: 40–45.

Umeta, M., and Faulks, R.M. 1988. The effect of fermentation on the carbohydrates in tef (*Eragrostis tef*). Food Chem. **27**: 181–189.

Zerihun, T. 1996. The agro-ecology and production technology of tef (*Eragrostis tef*). *In* Research achievements and technology transfer attempts: vignettes from Shewa. *Edited by* D. Abera and B. Seboka. IAR, Addis Abeba, Ethiopia. pp. 2–19.

Specialty Cookbooks

Jones, M.H. 1990. Superfoods allergy recipes: amaranth, buckwheat, quinoa, spelt, teff. Nutri Books, Denver, CO. 36 pp.

Y.W.C.A. (Addis Ababa). 1970. Around the injera basket. Central Printing Press, Addis Ababa, Ethiopia, 104 pp.

TEQUILA AND MEZCAL (*AGAVE* SPECIES)

Family: Agavaceae (agave family)

Names

Scientific names: the most important culinary *Agave* species are the following.

Species Used to Make Tequila

- Blue agave, Weber blue agave, azul agave, tequila—*A. tequilana* F.A.C. Weber

Species Used to Make Mezcal

- Century plant, maguey, American aloe—*A. americana* L.
- Pulque agave—*A. salmiana* Otto ex Salm-Dyck
- Maguey lechugilla—*A. angustifolia* Haw.
- + others

- "Agave" is the common name for species of the genus *Agave*. The English name "agave" as well as the genus name *Agave* is based on the Greek word *agavos*, meaning admirable or illustrious, an apt description of the magnificent appearance of many of the species.
- The name "century plant" is often applied to agave species, based on the idea that the plants live for a century, produce a spectacular flowering stem, and subsequently die. The popular belief that agaves rarely flower probably arose from observation that many cultivated agaves (particularly those grown for fiber) rarely flower. In fact, wild agaves seldom take more than 40 years to flower, typically living for 12–15 years, and not all species die after fruits are produced.
- There are several explanations of the origin of the word "tequila" (pronunciation: teh-KEE-luh). It has been claimed that it comes from the Nahuatl words *tequitl* (work, duty, job, or task) and *tlan* (place). Another explanation holds that the word is a corruption of the name of a volcano, *tetilla*, which looks like a small woman's breast. And still another claim is that the word is a corruption of the name of the native tribes *Ticuilas* or *Tiquilos*. The name "tequila" has also been said to come from some ancient Indian word, variously interpreted as meaning "the place of harvesting plants," "the place of wild herbs," "the place where they cut," "the place of work," "the place of tricks," and "the rock that cuts." A simple explanation is that the word "tequila" comes from the Mexican city of Tequila.
- "Maguey" (Spanish, of Cariban origin) means any of various American plants of the genus *Agave*. It can be used as a synonym for mezcal, but may also mean any of various plants of the related genus *Furcraea*, from which fiber is also obtained.
- "Mescal" or "mezcal" was once a generic term used by native Mexicans for agaves, and occasionally, the word is still used to refer to *Agave* species used to prepare the corresponding alcoholic beverage. The word (pronunciation: mess-KAHL) is American-Spanish, from the Nahuatl *mexcalli* (*metl*, maguey plant, + perhaps *ixca, xca*, to bake), meaning mezcal liquor.
- A "mescal button" is peyote, the fresh or dried button-like tubercles of peyote cactus [*Lophophora williamsii* (Lem. ex Salm-Dyck) J.M. Coult.]. In the 1950s, tequila became popular with trendy Californians because of the erroneous rumor that it contained the psychedelic chemical mescaline (based on the correct idea that tequila is a type of mezcal, and the incorrect idea that mescaline, which is found in the peyote cactus, and mezcal are the same thing). The spelling "mescal" has become less popular than "mezcal," because of the potential but incorrect association with mescaline and the drug culture.

Plant Portrait

Agaves are centered in Mexico, some species extending north to the southern United States, while others grow as far south as northern South America. There are as many as 300 species of *Agave*, ranging in size from less than 10 cm (a few inches) in height to massive plants sometimes exceeding 12 m (39 feet) and weighing more than a ton. Agaves produce a basal rosette of sword-shaped, rigid, usually succulent leaves on a short stem. The largest plants have leaves that are sometimes longer than 3 m (about 10 feet). Agaves are perennial, or reproduce new plants annually at the base of the mother plant. After building up energy reserves for years, then rapidly producing a flowering stalk in a few months, many of the species die. The enormous amounts of energy and sap used in a short time to develop the flowering stalk seem to fatally exhaust such species.

Agaves have provided the inhabitants of Mexico with food and drink since ancient times. Remnants of agave have been found in 9000-year-old mummified human feces found in caves in Tamaulipas and Tehuacán, Mexico, demonstrating that the plants have been eaten almost as long as humans have occupied North America. Flower buds, flowering branches, leaves, and stems of certain species were roasted and consumed in the past.

Before the Spanish conquest of Mexico, agave was a critical agricultural crop. It has been said that the early Aztec, Maya, and other Indians of Mexico made such extensive use of agaves that these plants were outranked in value only by corn (maize) and potatoes. Farming of agaves was highly advanced in Mexico, while further north, in what is now the southwestern United States, planting and processing agaves was on a much lesser scale. Hernando Cortez (1485–1547) and his Spanish army arrived from Cuba in 1519, seeking gold, other minerals, and slaves. In conquering Mexico, they destroyed the most advanced civilization of the Americas, and set back the development of agave cultivation. Nevertheless, agave continued to be important to the native inhabitants. By the early 20th century, a small elite of Mexican landlords controlled the production of agave alcoholic beverages, and wielded enormous power and wealth while the peasants lived in poverty. This situation was one of the causes of the Mexican Revolution of 1910. The huge haciendas of the agave barons can still be seen in and around Mexico City.

Agaves build up a store of starch in their stems in preparation for the development of a flowering stalk. Roasting transforms the starch into sweet-tasting sugars. Indians in Mexico traditionally prepared the condensed stem that bears the leaves by cutting it off at its base, and then cutting off the leaves at their bases, leaving an odd-looking plant "heart" called a *cabeza* or *piña*. This was roasted in a pit, and eaten like artichoke: the leaves were scraped against the teeth, pulling off the edible portion, leaving the fibers behind. By comparison with an artichoke, however, agave hearts are often huge—sometimes over 68 kg (150 pounds). Indeed, a cabeza has been described as an artichoke with a thyroid problem!

Agaves of the Mexican uplands, from Marilaun (1895).

Some agave species used to produce alcoholic beverages. *Left*, *A. tequilana* (by B. Brookes); *center*, *A. americana* (from Harter 1988); *right*, *A. salmiana* (by B. Brookes).

In addition to providing food, wild agaves have long been used by native peoples as a source of fiber. Ropes, bags, mats, baskets, sandals, clothes, and paper were all made out of agave fiber at least 5000 years ago. Today, several agave species are cultivated in plantations as a fiber crop, especially sisal (*A. sisalana* Perrine) and henequen (*A. fourcroydes* Lem.).

Culinary Portrait

There are four basic categories of agave beverages, discussed below.

Honey water

The most innocuous (i.e., nonalcoholic) drink from agaves is *aguamiel* or "honey water." At least 10 species of agave are used to obtain honey water for fermentation, including *A. americana*, *A. salmiana*, and *A. angustifolia*. The central leaves of large plants are removed, leaving a hemispherical cavity at the center of the plant, which is scraped to encourage exudation of honey water, that accumulates in the cavity. The sweet juice is removed twice daily, sometimes for as long as 6 months, and some plants can produce up to 1200 L (317 American gallons). In some arid regions, honey water is the only source of uncontaminated water. While honey water is used locally in Mexico, a pasteurized agave juice called "miel de maguey," is now exported to the United States.

Pulque

"Pulque" (pronunciation: POOL-keh) is a thick fermented alcoholic beverage that has been made in Mexico for as long as 2000 years from various species of agave. The word is from the American Spanish, based on the Nahuatl *poliuhqui* or *puliuhqui*, decomposed, spoiled, lost, perhaps a reference to the short shelf life. Honey water is mostly used as the starting material. Pulque is usually drunk like beer, typically containing 3–4% alcohol. This beverage is rather nutritious, and has sometimes been used to prepare bread. The taste has been compared to sour milk, and is traditionally improved by adding fruits, vegetables, nuts, grains, herbs, spices, eggs, or cream. In Mexico, one can sometimes purchase pulque flavored with fruit extracts, but this beverage is now generally confined to rural areas, and most young Mexicans have never heard of it. Generally, pulque must be consumed quickly, as it is not pasteurized, and so does not enter the commercial market. However, an agave beer called Tequiza was launched in the United States by Anheuser-Busch in 1999. So-called "pulque" offered to tourists in Mexico often is quite palatable, but bears little resemblance to the genuine article.

Mezcal

The Spanish introduced distillation into Mexico in the 16th century, and this led to the creation of mezcal in the modern sense, i.e., a distilled, alcoholic beverage based on agave. Agave cabezas are the starting material. While wild species are often used, cultivated plants are manipulated to produce especially large cabezas. This is done by removing the early flower stalk, which forces the plant to store its nutrients in the stem, which becomes large, weighing 27–55 kg (70–150 pounds). The cabezas are harvested, roasted, mashed, fermented, and the resulting liquid is distilled. Cooking of the cabezas converts the stored polysaccharides (inulins) to fructose and glucose, most of which is converted to alcohol by fermentation. The quality of the mezcal depends on whether the process is carried out as a cottage industry or in modern factories. The best mezcal is distilled twice, and aged for up to 4 years. Mezcal has a distinct, smoky flavor that is quite different from tequila, described below. Mezcal is commonly sweetened and flavored with fruits, herbs, and nuts to make it more palatable. Mezcal exported to the United States often has been subdued by these and other methods.

Mexican law allows use of the name mezcal only for agave alcoholic beverages made in six counties around the city of Oaxaca (southeast of Mexico City) in the mountainous southern part of Mexico, as well as in the states of Guerrero, Durango, San Luis Potosi, and Zacatecas. Most production is in the Oaxaca region, where there are about 500 distillers, mostly small and run by Zapotec Indians using traditional methods. *Agave angustifolia* and *A. salmiana* are favored for mezcal production, although 10 other species have been collected from the wild for the purpose. It has been estimated that some 50 000 cultivated plants are harvested monthly in Oaxaca for production of mezcal.

In the northern state of Sonora, a trendy new type of mezcal called bacanora, made from wild *A. angustifolia*, has become the status drink of the region. It was illegal until 1992, previously only available from bootleggers, and is still rare. About 500 families are currently employed to make this beverage using wild *A. angustifolia* from the central Sonoran foothills.

Tequila

Tequila is a special kind of mezcal, doubly distilled and usually aged for a lengthy period, often more than 4 years. Most factories are located in the western Mexican state of Jalisco, in the region of the town of Tequila (population 18 000), west–northwest of Mexico City. *Agave tequilana* is used for tequila production, and by Mexican law the only variety that can be used is Azul (Spanish for blue), so named for the bluish tint of the leaves. About 55 000 ha (136 000 acres) in Mexico, mostly in Jalisco, are dedicated to the cultivation of Azul. The first bottled tequila was produced in the late 18th century. There are now about 600 brands of tequila, which typically have 40–45% alcohol by volume (and are 80–90 proof). Exports of tequila from Mexico earn the country about $200 million (by contrast, exports of mezcal are only worth about $6 million annually). By Mexican law, at least 60% (until 1995 only 51%) agave juice must be used to produce tequila (the rest may be cane and sorghum sugar), but 100% agave tequila will be labeled as such (and will be expensive—$30–$50 or more in a collector's bottle). Tequila has become Mexico's most widely recognized export product. Mexico's best customers for tequila exports are (in decreasing order) the United States, Canada, France, and Japan. Under world trade agreements, a special "Appellation d'Origine Controlée" (AOC) is issued for certain products. In 1996, Mexico received final AOC recognition that the word "tequila" can only be used to designate Mexican tequila.

Four AOC categories of tequila are recognized by Mexican law:

- White or Silver (*Tequila Blanco* or *Plata*): fresh, clear tequila from the still, proof adjusted with demineralized water. The quality varies from smooth to fiery.
- Gold (Tequila Joven Abocado): Silver Tequila with the addition of coloring and flavorings to mellow the flavor. Uncommon in Mexico, but a popular North American choice for shooters, margaritas, and other mixed drinks.
- Rested (*Tequila Reposado*): aged for 2–12 months in oak tanks or barrels; flavoring and coloring agents, and demineralized water may be added. Popular in Mexico, the better quality Rested tequila is ideal for sipping and for margaritas and other mixed drinks.
- Aged (Añejo): Rested Tequila aged for at least 1 year in government-sealed oak barrels. The best Añejo Tequila subtly blends the taste of oak and tequila; lesser quality Añejo is dominated by the taste imparted from the wood or by flavorings. The best quality is most suitably drunk straight, not in mixed drinks.

Tequila, in common with cognac and other fine spirits, does not improve with age in the bottle.

According to *The Art of Bartending* by Mark Barrett, "The traditional way to do a shot of tequila is as follows: lick the back of your hand between the thumb and forefinger; sprinkle salt on the moist area and lick the salt; immediately swallow the shot of tequila; bang the shot glass down on the bar; pick up a wedge of lemon or lime and suck on it; smack your lips and say 'Ahhh.'"

Tequila is most famous as a key ingredient in the margarita, which generally contains about two parts of tequila, one part of lemon or lime juice, and 1/5 to 1/2 part of the orange liqueur Triple Sec. Margaritas are blended with crushed ice and served in a glass of which the rim has been moistened with a citrus rind and then dipped in salt.

Culinary Vocabulary

- The ritual Mexican toast for mezcal is "arriba [above], abajo [below], al centro [the center], para dentro [within]."
- A "caballito" (literally "little horse," also called a tequillita) is the traditional tall drinking/shot glass, with a flat bottom and wider mouth, used for tequila. It is also the name of a tequila-based cocktail.
- "Denomination of Origin (DOA)" refers to law governing legal use of a term with respect to geographical origin. In the wine and spirits industry, there are only four drinks recognized with DOA: sherry, cognac, champagne and tequila.
- NOM stands for "Normas Official Mexicana de calidad," the Mexican government system of production standards. Every tequila distillery gets a NOM identification number, and this should be present on the label to ensure the tequila meets the minimum standards. The NOM number does not guarantee quality, only that the distiller is appropriately licensed.
- A *Perla o concha* (pearl or conch) is a bubble that remains on the surface of the tequila after serving it or stirring it, and is said to denote a fine tequila. To check the quality of a bottle of tequila, close it tightly, hold it upside down, then turn it right side up: the bubbles should appear and continue to float.
- *Pulquerías* are Mexican bars that dispense pulque, the lowest grade of agave alcoholic beverage. They are rapidly disappearing from Mexico, as beer and other alcoholic drinks become popular.
- *Calichal* (pronounced de-lee-CHAL) is a Mexican beverage prepared with one part beer and four parts pulque.
- A "Tequila Maria" is a tequila cocktail with tomato juice, grated horseradish, Worcestershire sauce, Tabasco sauce, celery salt or celery seed, white pepper, tarragon, oregano, dill, and lime juice.
- A "Tequila Sunrise" is a cocktail made with tequila, orange juice, and grenadine (optionally with lime juice and (or) a lime slice garnish). It may have been invented at the Agua Caliente racetrack bar in Mexico during Prohibition. Supposedly, the name reflects the practice of visitors from California of adding tequila to their morning orange juice as a pick-me-up at sunrise. The "Tequila Sunset" substitutes the grenadine with pineapple juice.
- A "Brave Bull" is a cocktail prepared with white tequila and Kahlua, and garnished with a twist of lemon. A "Mexican Bull Shot" combines beef consommé, tequila, lime or lemon juice, Worcestershire sauce, and celery salt or celery seed, with a lime wedge.
- A "Sangrita" (pronounced sahn-gah-REE-tah) is a tequila (or gin) cocktail, with numerous additives: tomato juice, orange juice, lime juice, minced onion, Tabasco sauce, Worcestershire sauce, white pepper, and celery salt.
- A "Sneaky Pete" cocktail is made with tequila, white crème de menthe, pineapple juice, and lime or lemon juice, with a lime slice garnish.
- A "Vampiros" is a Mexican cocktail, made simply by preparing a Bloody Mary (see TOMATO) with tequila instead of vodka.
- "Mexican Coffee" is a cocktail prepared with tequila; Kahlua or sugar syrup and strong hot black coffee; and a garnish of whipped cream.
- *Salsa borracha* (pronounced boh-rah-tchah) is a Mexican salsa prepared with poblano chile pepper, orange juice, onion, and tequila.

Curiosities of Science and Technology

- Some gender roles in past times seem to have evolved for no particular reason. For example, in the southwestern United States, among the Pima Indians, men harvested the agave plants and women gathered wood, while among the Havasupai the roles were exactly reversed.
- As noted above, "mescal" is an ancient Indian name for certain agaves. A subtribe of the Apaches used agaves in so many ways that they became known as the Mescalero Apaches.
- In 1830, the Mexican Congress ordered that no legal document should be written on any other material than the paper made from the national plant, the maguey (agave).

- Chewed pieces of agave were once used as wads to tamp down gunpowder in muzzle-loader rifles.
- Some native peoples of Mexico removed the spine at the tip of an agave leaf while leaving a string of vascular tissue attached to the spine, thus cleverly preparing a "needle and thread."
- The cuticle and adjacent tissue of the inner leaves of some agaves is peeled off and used in Mexico as a translucent, edible, food wrap that imparts a subtle flavor. The delicacy "mixiote," prepared for special occasions such as Christmas and Easter, frequently uses agave wrap. Unfortunately, there is considerable poaching of agave cuticles, especially from large cultivated plants. Bandits often simply use the terminal leaf spine as their cutting tool. To prevent such theft, agave farmers often mutilate the leaves of their own plants to make the cuticle unusable.
- It has been estimated that there are about 100 million *Agave tequilana* plants that are being cultivated for tequila in the Mexican state of Jalisco.
- The larger *A. tequilana* farms in Mexico hire security firms to protect their plants from agave rustlers.
- In past times, some Indian groups in the Americas planted larger agaves close together as fences around their homes, since the stiff, saber-like leaves are more threatening than barbed wire. Today, such armed fences constructed of agave plants are common in Mexico.
- In Mexico, tequila is considered to have many virtues— useful as a powerful aphrodisiac, aftershave lotion, insect repellant, lighter fuel, tonic, a remedy for indigestion, and a laxative.
- Agaves are not grown for their flowering stalks, but these too are employed to make useful products. They have been used like bamboo poles to support plants, to make fences, and as fishing poles. The stalks are even being used today in California to make musical instruments and surf boards.
- Agaves have a type of photosynthesis called crassulacean acid metabolism (CAM), which adapts them to arid and semiarid environments. The CAM plants fix atmospheric carbon dioxide mostly at night when temperatures are lower than during the daytime, thereby decreasing the loss of water by transpiration that occurs whenever plants open their pores (stomata) to take in carbon dioxide.
- Pollination of agaves is by bats, various birds, and insects. Long-tongued, nectar-feeding bats are known to be the chief pollinators of many agave species, and are in serious decline owing to loss of habitat. The greater long-nosed bat (*Leptonycteris nivalis*) is a principal pollinator of agaves. It was declared endangered by the United States in 1988 and by Mexico in 1991.
- The national flower of Antigua and Barbuda (in the British West Indies of the Caribbean, the largest of the British Leeward Islands) is an agave, the West Indian dagger log, *A. karatto* Miller. In the past, fishing rafts were made from the flower's stem (the so-called "log"), and fishing bait was made from the white interior pulp of the leaves.
- "Worms" (actually moth larvae) are associated with species of agave. The larvae or grubs of brownish moths (*Aegiale hesperarius*, known in Mexico as *gusanos blancos*) bore into the heart of the stem of some agave species. These boring insects are about 3 cm (1¼ inches) long, are considered a great delicacy in Mexico when fried, and are sold in many Mexican markets. Small red worms (moth larvae of *Comadia redtenbachi*, known in Mexico as *gusanos rojos*) live in the roots of some agave species. It seems that in past times, an agave worm was used in the way miners once used canaries to check for toxic mine gases: a worm was put into a bottle of mezcal, and if it was still wriggling when it reached the bottom of the bottle, the mezcal was considered safe to drink. Sometimes a mixture of powdered agave worm, chili, and salt is supplied in a bag attached to the bottle, to be used as a condiment! About 1950, a merchant began deliberately introducing an agave worm in his bottles of mezcal to take advantage of the old mystique of mezcal having agave worms, and others followed his lead. However, this marketing ploy can not be used for true tequila bottled in Mexico, since it would be contrary to Mexican laws. Nevertheless, some tequila that is exported to the United States in bulk and is not subject to Mexican tequila laws has the worm introduced.
- The agave cactus (*Leuchtenbergia principis* Hook.) is a true cactus that resembles agaves. Contrary to popular opinion, agaves are not cactus plants, although like cacti they are succulent plants that dwell in hot, dry climates.

Key Information Sources

Bahre, C.J., and Bradbury, D.E. 1980. Manufacture of mescal in Sonora, Mexico. Econ. Bot. **34**: 391–400.

Barrios, V.B. de. 1971. Guide to tequila, mezcal and pulque. Mexics, Mexico City, Mexico. 65 pp.

Benn, S.M., and Peppard, T.L. 1996. Characterization of tequila flavor by instrumental and sensory analysis. J. Agric. Food. Chem. **44**: 557–566.

Breitung, A.J. 1968. The agaves. Abbey Garden Press, Reseda, CA. 107 pp.

Cedeno, C.M. 1995. Tequila production. Crit. Rev. Biotech. **15**: 1–11.

Cutler, L. 2000. The tequila lover's guide to Mexico and mezcal. 2nd ed. Wine Patrol Press, Vineburg, CA. 259 pp.

Diódoro, G.S. 1993. Los agaves en México. Universidad Autónoma Chapingo, Mexico. 252 pp. [In Spanish.]

Emmons, B. 2003. The book of tequila: a complete guide. 2nd edition. Open Court Publishing, Chicago, IL. 312 pp.

Gentry, H.S. 1982. Agaves of continental North America. University of Arizona Press, Tucson, AZ. 670 pp.

Gorman, M., and Alba, F.P. de. 1976. Tequila book. Regnery, Chicago, IL.184 pp.

Iniguez-Covarrubias, G., Diaz-Teres, R., Sanjuan-Duenas, R., Anzaldo-Hernandez, J., and Rowel-Roger, M. 2001. Utilization of by-products from the tequila industry. Part 2: potential value of *Agave tequilana* Weber Azul leaves. Bioresour. Technol. **77**: 101–108.

Iniguez-Covarrubias, G., Lange, S.E., and Rowell, R.M. 2001. Utilization of by-products from the tequila industry. Part 1: agave bagasse as a raw material for animal feeding and fiberboard production. Bioresour. Technol. **77**: 25–32.

Irish, M., and Irish, G. 2000. Agaves, yuccas, and related plants: a gardener's guide. Timber Press, Portland, OR. 312 pp.

Martinez-Morales, R., and Meyer, S.E. 1985. A demographic study of maguey verde (*Agave salmiana* ssp. *crassispina*) under conditions of intensive utilization. Desert Plants, **7**: 61–64, 101–102.

Nobel, P.S. 1990. Environmental influences on carbon dioxide uptake by agaves, CAM plants with high productivities. Econ. Bot. **44**: 488–502.

Nobel, P.S. 1994. Remarkable agaves and cacti. Oxford University Press, Oxford, UK. 166 pp.

Nobel, P.S., and Valenzuela, A.G. 1987. Environmental responses and productivity of the CAM plant, *Agave tequilana*. Agric. For. Meteor. **39**: 319–394.

Nobel, P.S., Castañeda, M., North, G., Pimienta-Barrios, E., and Ruiz, A. 1998. Temperature influences on leaf CO_2 exchange, cell viability and cultivation range for *Agave tequilana*. J. Arid Environ. **39**: 1–9.

Reveal, J.L., and Hodgson, W.C. 2002. *Agave*. *In* Flora of North America north of Mexico. Vol. 26. *Edited by* Flora of North America Editorial Committee. Oxford University Press, New York, NY. pp. 442–461.

Salinas, M.L., Ogura, T., and Soffchi, L. 2001. Irritant contact dermatitis caused by needle-like calcium oxalate crystals, raphides, in *Agave tequilana* among workers in tequila distilleries and agave plantations. Contact Dermatitis, **44**: 94–96.

Small, E., and Catling, P.M. 2002. Blossoming treasures of biodiversity: 6. Agave—are we running out of tequila? Biodiversity, **3**(**3**): 47–48.

Tello-Balderas, J.J., and García-Moya, E. 1985. The mezcal industry in the Altiplano Potosino-Zacatecano of north-central Mexico. Desert Plants, **7**: 81–87.

Valenzuela-Zapata, A.G. 1985. The tequila industry in Jalisco, Mexico. Desert Plants, **7**: 65–70.

Valenzuela-Zapata, A.G., and Diego-R., G.E. Fertilization of (*Agave tequilana* Weber) of Tequila, Jalisco, Mexico. Terra (Mexico), **13**: 81–95. [In Spanish.]

Valenzuela-Zapata, A.G., and Nabhan, G.P. 2004. Tequila: a natural and cultural history. University of Arizona Press, Tucson, AZ. 160 pp.

Vega, K.G, Chavira, M.G., Vega, O.M. de la, Simpson, J., and Vandemark, G. 2001. Analysis of genetic diversity in *Agave tequilana* var. Azul using RAPD markers. Euphytica, **119**: 335–341.

Villalvazo-Rosales, A.S. 1986. Mezcal (*Agave tequilana* Weber) cultivation in Tequila, Jalisco (Mexico) region. Universidad Autonoma Chapingo, Chapingo, Mexico. 91 pp. [In Spanish.]

Specialty Cookbooks

(Note: Cookbooks on tequila often have information other than recipes; Hutson, in particular, is an excellent source of general information; conversely, some of the above monographs contain recipes.)

Collins, W. 2000. Little book of tequila cocktails. Hamlyn, London, UK. 96 pp.

Hutson, L. 1995. Tequila! Cooking with the spirit of Mexico. Ten Speed Press, Berkeley, CA. 158 pp.

Jones, S. 1979. Cooking with Stan Jones' tequila sauza. BarGuide Enterprises, Los Angeles, CA. 192 pp.

Nusom, L. 1993. The tequila cook book. Golden West Publishers, Phoenix, AZ. 126 pp.

TOMATO

Family: Solanaceae (potato family)

Names

Scientific name: *Lycopersicon* species

- Tomato—*L. esculentum* Mill. (*Solanum lycopersicum* L.)
- Currant tomato—*L. pimpinellifolium* (L.) Mill. (*Solanum pinpinellifolium* L.)

  [Note: Some modern authorities have concluded that the genus name *Solanum* should be used for tomato species, because phylogenetic studies indicate that *Lycopersicon* is part of *Solanum*. Alternatively, *Solanum* may deserve to be split into several groupings, one of which is *Lycopersicon*, in which case the genus name *Lycopersicon* should be retained. The delimitation of the very large genus *Solanum* is presently too poorly understood to resolve this taxonomic issue.]

- The word "tomato" is derived from the Spanish *tomate*, which in turn is based on the South American Indian (Nahuatl) word for the plant, *tomatl*.
- The name of the currant tomato is based on the size and appearance of the fruits, which are barely larger than those of the currant (which see).
- Tomatoes were first described in Europe in 1554, in Italy, where they were called "pomi d'oro" (gold apples), indicating that some early introductions were yellow-fruited. Later, as tomatoes became more widely appreciated, they were known as "love apples" from the early French name *pomme d'amour*. However, it has been argued that the French originated from the Italian *pomi dei Moro*, Moor's apples (for the Moors of Spain; in 16th century Europe, Spaniards were commonly called Moors), and that the French translation as love apple was a misunderstanding (which led to the idea that tomatoes had aphrodisiac properties).
- "Mexican green tomatoes" are tomatillos (*Physalis philadelphica* Lam.).
- "Ugly tomatoes" is a phrase that arose in the early 21st century in Florida. The expression traces to a beefstake-style cultivar of the tomato, named Ugly Ripe, which is derived from an old French heirloom tomato variety called Marmande. Like most heirloom varieties that were popular in old times, the taste is much superior to currently popular varieties, but there are marketing disadvantages. The Ugly Ripe tomato is indeed ugly (at maturity it becomes "cat faced," i.e., misshapen by cracks at the stem end), but it is delicious. Since the introduction of the Ugly Ripe tomato to stores in 1999, the term "ugly tomatoes" has been applied by some to all heirloom tomatoes.
- In addition to tomatoes of the genus *Lycopersicon*, the word "tomato" is associated with two other genera that produce edible fruits. The tree tomato [*Cyphomandra betacea* (Cav.) Sendtn.] produces tomato-like fruit and is grown in semitropical areas. Husk tomatoes or strawberry tomatoes (*Physalis* species) produce small, tomato-like fruit enclosed in a papery bract, and the fruit is generally used in cooking.
- The genus name of the tomato, *Lycopersicon*, comes from the Greek *lykos*, wolf, and *persikon*, peach—figuratively, "like a wolf in sheep's (peach's)

Tomato (*Lycopersicon esculentum*), from Vilmorin-Andrieux (1885).

clothing"— which reflects the old belief that the tomato appears tasty but is poisonous. One of the tomatoes early European names was "Devil's wolf apple."

- *Esculentum* in the scientific name *L. esculentum* is Latin for edible.

Plant Portrait

The tomato is a tender, tropical, perennial herb that is grown as an annual in temperate parts of the world. Small-fruited forms known as cherry tomato [variety *cerasiforme* (Dunal) A. Gray] are widely regarded as the progenitor of the large-fruited common tomato (variety *esculentum*). Wild cherry tomatoes are found in Bolivia, Peru, and Columbia in South America, most Central American states, and parts of Mexico. Mexico is often thought to be the major area of domestication, at least of the familiar large-fruited varieties of commerce. It has also been suggested that some tomatoes may have originated in the eastern Andes. The tomato was already used by the Aztecs when Europeans first arrived in the Americas. The Spanish are generally thought to have introduced tomatoes to Europe in post-Columbian times (although some literature is suggestive of the presence of tomatoes in the Old World prior to 1492).

Because the tomato is a member of the "nightshade family" (Solanaceae, here referred to as the potato family), it was considered poisonous in many countries. The history of tomatoes in the United States is exemplary of the difficulties encountered. The Pilgrims considered growing tomatoes to be an abomination, and those caught with tomatoes were displayed in the public square, where they were ridiculed. An Italian painter brought tomato seeds to Salem, Massachusetts, in 1802, but found it difficult to interest anyone in even tasting the fruits. Thomas Jefferson (1743–1826), third president of the United States, in 1871 is said to have become the first American to grow tomatoes. He tried to interest his countrymen in eating tomatoes, but as in other areas where this vegetable was unfamiliar, it was avoided. According to a frequently repeated story, in a dramatic demonstration that tomatoes were not poisonous, as was widely thought, Colonel Robert Gibbon Johnson ate a basketful on the courthouse steps in Salem, New Jersey, at noon on September 26, 1820. In fact, Robert G. Johnson was a prominent leader in Salem and New Jersey during the first half of the 19th century, and promoted agricultural innovation and the introduction of new crops. However, there is no historical evidence proving that Johnson actually carried out the courthouse step event, which seems to have been made up in the 20th century. In 1860, the first of the "women's magazines," *Godey's Lady's Book*, which was published in Philadelphia, recommended that tomatoes "should always be cooked for 3 hours." This extraordinary long time reflected lingering suspicion of the time that there were poisonous constituents present, and that these could be deactivated by sufficient cooking.

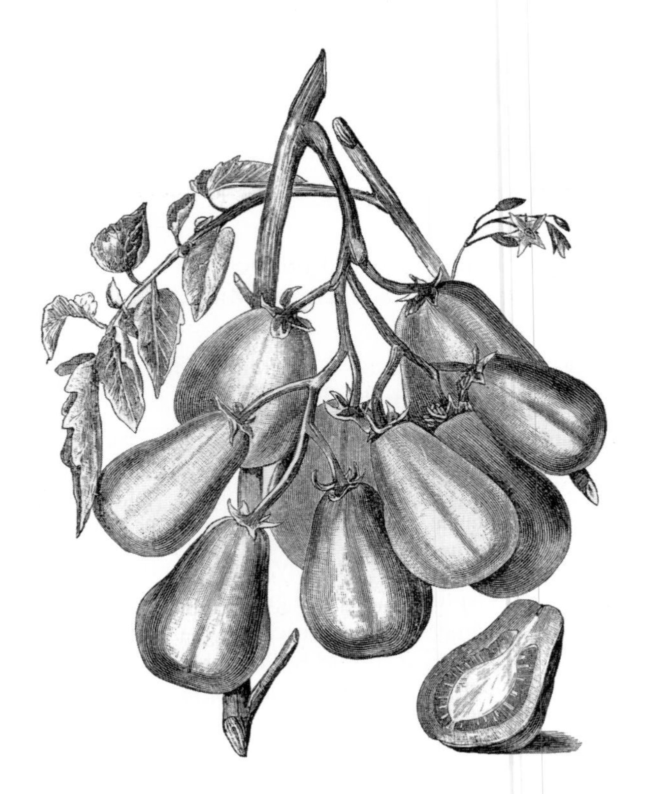

Italian Marvel cultivar of tomato (*Lycopersicon esculentum*), from *Gartenflora* (1886, vol. 35).

Gradually, tomatoes were accepted as a vegetable. By the 1860s, attempts were made to improve the tomato and develop new cultivars. There are now more than 4000 varieties of tomatoes, including one with square fruit, selected by American plant breeders in 1984 for convenient packing. Depending on variety, the color of mature fruits may be green, red, pink, orange, or yellow. Shape also varies greatly, depending on variety, and size ranges from fruit the size of cherries to large ones weighing a kilogram (over 2 pounds). "Determinate" plants are bush-like in growth, with a terminal inflorescence (flowering stem) that produces a limited number of fruits. "Indeterminate" plants are more vine-like, lack a terminal inflorescence, require staking, and continue to produce tomatoes for a prolonged period.

The currant tomato is a wild perennial herbaceous vine from Peru. It produces miniature tomatoes (about 1 cm or 3/8 inch in diameter), which make excellent pickles. The plant is also useful as a ground cover on poor soils in warm climates.

Culinary Portrait

The tomato is the second most-consumed vegetable worldwide, next to the potato. Ripe tomato fruits can be used raw in salads or cooked as an ingredient of stews and soups, and of meat, fish, soufflés, omelets, and pasta dishes. Tomatoes are a major component of sauces and condiments used in cooking in many parts of the world, exemplified by Italian pastas and Mexican salsa. This vegetable can be stewed, pan-fried, stuffed, grilled, simmered in soups and sauces, and turned into excellent juice. Mature green tomatoes can be fried or added to stir-fries, stews, and even mince meat for pies and tarts. Tomatoes are processed into many kinds of pastes, purées, chutneys, sauces, relishes, and other condiments. Among the memorable convenience foods associated with the tomato are pizza and tomato ketchup. Tomato varieties are often specialized for particular culinary uses. The cherry tomato, for example, is used in salads and as appetizers. Plum tomatoes (also called Italian or Roma tomatoes) are egg- or plum-shaped and are ideal for sauces, stews, and other cooked dishes. Tomatoes are best not cooked in aluminum pots, as their acid corrodes the metal, contaminating food and producing an off-taste.

Tomatoes should be kept at room temperature, as they lose flavor and change in texture if stored in the refrigerator. But the advice to "never refrigerate tomatoes" is not always easy to apply. Stores often refrigerate tomatoes to prolong their shelf life. This is a terrible thing to do to unripe tomatoes, because cold temperatures stop the ripening process, and the fruits never fully recover when returned to room temperature. Because tomatoes purchased from supermarkets have frequently been refrigerated, they tend not to keep as well unrefrigerated as do garden tomatoes picked when mature or nearly mature. In any event, once tomatoes are mature, if not eaten or refrigerated, they simply become overripe. Ripe tomatoes keep reasonably in a refrigerator for 2 or 3 days.

In the tomato industry, the expression "vine ripe" rather dishonestly refers to green tomatoes that have developed yellowish or reddish patches of color on the skin, an indication that if the fruit is picked it will ripen off the vine. However, only when a tomato is left to actually ripen fully on the vine will it develop its best flavor.

To ripen tomatoes more quickly, place them in a paper bag along with an apple or banana. Tomatoes, like many other fruits, have a natural ripening hormone called ethylene. Placing unripe tomatoes close to ripe fruit (including other tomatoes) speeds ripening because ripe fruit gives off a greater amount of ethylene gas than unripe fruit. Although it has long been common practice to ripen tomatoes on a sunny windowsill, picked tomatoes should never be ripened in direct sunlight.

Tomato paste is made by cooking tomatoes, with or without seasonings, until a thick paste is formed. It is useful for giving body to sauces. Tomato paste should be cooked slowly over low heat. To preserve tomato paste left in a can after it has been opened, a small amount of vegetable oil can be added. This will float on top, keeping the air away. To use, tip the can to expose fresh tomato paste, and spoon out. Alternatively, seal the rim with plastic wrap and freeze the tomato paste in its can. The bottom of the can, while still frozen, can be opened with a can opener, and in a few minutes, the bottom can be used to force the frozen paste over the top in the quantity desired.

The foliage and immature green fruits of tomatoes contain the toxic alkaloid tomatine, and poisoning of livestock has been reported. However, the tomatine content in colored fruits and in almost-mature green fruits is minimal or nonexistent.

Culinary Vocabulary

- The nickname of Queen Mary I of England and Ireland (Mary Tudor, 1516–1558) was "Bloody Mary" because of her notorious, violent persecution of Protestants. The Bloody Mary is generally said to have been named after her. This is a vodka and tomato juice drink—also often with lemon juice, Worcestershire sauce, and a dash of Tabasco sauce—usually served with a slice of lime or with a large stalk of celery. The concoction is believed to have been created by Fernand "Pete" Petiot, an American bartender at Harry's New York Bar in Paris, France, in 1921 (the owner, Harry MacElhone, is also often credited with preparing the first Bloody Mary). However, it was first called the "Bucket of Blood," so named by American entertainer Roy Barten, after a nightclub in Chicago. In 1933, Petiot was brought to New York to work the King Cole Bar at the St. Regis Hotel, at which time the beverage first became popular, under the name "Red Snapper." The name "Bloody

Mary" seems to have started in the United States during the Second World War.

- "Aurore sauce" is bechamel sauce with enough tomato purée added to tint it pink. (Bechamel is a basic French white sauce made by stirring milk into a cooked mixture of butter and flour.)
- A "Bloody Maria" is a variation of the Bloody Mary, made with tequila instead of vodka. A "Bloody Bull" is vodka-based, and made with beef bouillon (or broth or consommé) and tomato juice. A "Bloody Caesar" combines tomato and clam juice. A "Bloody Mary Quite Contrary" uses Japanese saké as the source of alcohol. A "Virgin Mary" (also called a "Bloody Shame") is a Bloody Mary without alcohol.
- The word "ketchup" (= catsup, catchup) has been said to be derived from the Malay *kechap*, via the Amoy Chinese dialect *kōetsipa* or *kētsipa*, combining *kōe*, minced sea-food, and *tsiap*, sauce. This spiced fish sauce was much more like modern Worcestershire sauce than today's tomato-based ketchup. (For additional information, see SOYBEAN, NAMES.) Early concoctions that appeared in the late 17th century under the name catchup and finally ketchup in England were sauces made of anchovies, walnuts, mushrooms, and kidney beans. Henry J. Heinz is believed to have not started making tomato ketchup until 1876. The spelling "ketchup" is established in Britain and Canada, while "catsup" is quite common in much of the United States.
- In old lunch-counter and diner jargon, a "full house" was a grilled cheese, bacon, and tomato sandwich. "Hemorrhage" referred to ketchup. "Paint it red" meant put ketchup on it ("paint a bow-wow red": put ketchup on a hot dog). A "lighthouse" was a bottle of ketchup. A "splash of red noise" was a bowl of tomato soup.
- "Chowder" (based on the French *chaudière*, kettle or cauldron) was a hearty soup made by early Quebecers from salt pork and fish. Maine acquired the basic recipe, using water, clams, salt pork, and potatoes. Massachusetts added milk. But when Manhattan and Connecticut added tomatoes, a famous food controversy, still unsettled, began regarding whether chowder should be made with tomatoes. "Manhattan-style" clam chowder is made with a tomato base, and is preferred by many North Americans. In Rhode Island, legislators passed a law specifying that the state's "official shore dinner" include tomato-clam chowder. However, in 1939, Assemblyman Seeder introduced into the Maine Legislature a bill to make adding tomatoes illegal in clam chowder.
- "Red sauce," at least in North America, refers to tomato-based Italian-style sauces, sometimes with meat, sometimes clam-flavored.
- *Strattù* (pronounced strah-TOO) is an Italian sauce prepared from sun-dried tomatoes.
- *Alla bolognese* (pronounced ahl-lah boh-loh-nay-see) is an Italian expression referring to food garnished with a tomato and cheese sauce.
- *Cioppino* is a seafood/tomato sauce stew now encountered across North America, but particularly popular in the Italian fishing community in San Francisco. The stew is made in San Francisco with whatever is caught on a particular day, combined with a variety of vegetables. The name has been explained, tongue in cheek, to have arisen during the early part of the 20th century when Italian fishermen returning to Fisherman's Wharf in San Francisco were asked to chip in a few small fish or other seafood to a communal stew, and the request to "Chip in! Chip in!" was changed to "cioppino" in Italian.
- The French expression *à la niçoise* (pronounced ah lah nee-SWAHZ) refers to a method of food preparation associated with the cuisine of Nice, France, in which the dishes are almost always charactered by the use of tomatoes, and often also garlic, black olives, green beans, and anchovies.
- "Mystery Cake" is a layer cake prepared with canned, condensed tomato soup, spices, and raisins.

Curiosities of Science and Technology

- Sir Walter Raleigh (1552?–1618) presented a gift of a tomato plant to Elizabeth I (1533–1603), starting a fad in England of growing the species as an ornamental curiosity. (Unfortunately, Raleigh also introduced tobacco to England.)
- Ketchup was once used as a patent medicine in the United States. In 1837, Dr. Archibald Miles of Cincinnati marketed "Dr. Miles's Compound Extract of Tomato," to be used as a remedy for indigestion, diarrhea, liver disease, and as a cholera preventive. Soon, "Dr. Phelp's Compound Tomato Pills" appeared, promising to cure "gravel" (kidney stones), colic, influenza, fevers, nervous diseases, acid stomach, and many other conditions. Drs. Miles and Phelps engaged in a heated war of

Currant tomato (*Lycopersicon pimpinellifolium*), from Vilmorin-Andrieux (1885).

words, even accusing their competitor's product of not containing tomatoes.

- In 1865, James H.W. Huckins received a US patent for making "Improved Tomato Soup," and became the first known soup canner. After he began advertising his canned tomato soup in 1876, he became extremely successful, and other canners entered the field, including the Campbell Soup Company. Campbell's, the largest soup manufacturer in North America, sells 300 million cans of its tomato soup annually in the United States. Chicken soup is Campbell's best-selling soup, although if all the varieties of soups with tomatoes in them are considered, tomato soups are number one in the soup world.
- In 1887, J.T. Dorrance developed a unique line of condensed soups for the Campbell company, which used the red and white school colors of Cornell University in Ithaca, New York, to produce a distinctive can. The label for Campbell's tomato soup is still red and white.
- The first Harley Davidson motorcycle, built in 1903, used a tomato can for a carburetor.
- In 1991, sales of tomato-based salsa in the United States exceeded those of ketchup for the first time.
- Heinz catsup has been reported to travel at a blistering 40 km/year (25 miles/year) as it leaves the bottle.
- The H.J. Heinz Company introduced "Blastin' Green" ketchup in 2000 for the child market, selling more than 10 million bottles of green ketchup that year. In 2001, purple ketchup was introduced, made with red and blue food coloring, but tasting the same as traditional red ketchup.
- For a crispier pizza crust, it has been recommended that the cheese be placed on the pie first, followed by the tomato sauce.
- Tomato plants may easily be grafted to potato plants, and such graft hybrids are termed pomatoes or topatoes. This clever combination can

produce edible underground storage organs and edible fruits simultaneously.

- Tomatoes were once grafted experimentally to rootstocks of jimson weed (*Datura stramonium* L.) in an attempt to combat root-infecting nematode worms. However, toxic alkaloids generated by the root system of the jimson weed turned up in the fruits, producing very poisonous tomatoes!
- In 1962, a wild Peruvian tomato was collected, the genes of which have been estimated to have a value for the improvement of tomatoes of more than $20 million annually.
- What is a vegetable? In 1887, this question became the subject of a famous legal battle, when a merchant launched an attempt to evade paying duties on tomatoes imported into New York form the West Indies. The tax had been collected under a tariff act that levied duties on vegetables. In 1893, his argument that tomatoes were exempt because they really were fruits was carried to the Supreme Court of the United States. The Supreme Court ruled as follows:

  > "Botanically speaking, tomatoes are the fruit of a vine, just as are cucumbers, squashes, beans and peas. But in the common language of the people, whether sellers or consumers of provisions, all these are vegetables, which are grown in kitchen gardens, and which whether eaten cooked or raw, are, like potatoes, carrots, parsnips, turnips, beets, cauliflower, cabbage, celery and lettuce, usually served at dinner in, with, or after the soup, fish or meats, which constitute the principal part of the repast, and not, like fruits generally, as dessert."
- By contrast, the European Union Jam Directive, written in the 1980s to describe the constituents that can be labeled as jam or marmalade, stated that " tomatoes, the edible part of rhubarb stalks, carrots, sweet potatoes, cucumbers, pumpkins, melons and water melons are considered to be fruit."
- A Harvard University study of 47 000 middle-age male health professionals found that men who eat 10 or more servings of tomato-based foods weekly cut their prostate cancer risk by 45%. Tomatoes are rich in the oxidant lycopene, which may be the health-promoting factor.
- The first genetically engineered crop was developed in 1982 at Washington University in St. Louis, Missouri. The first genetically engineered crop plant approved for commercial marketing (in 1994) was the Flavr-Savr tomato—designed to slow fruit ripening and increase shop life. This variety produces fruit that remains fresh for 2 weeks after harvesting, nearly twice as long as conventional tomatoes. The genetic engineering was achieved by antisense RNA technology that suppresses the production of a softening enzyme, polygalacturonase, which attacks pectin in the cell walls of ripening fruits, rendering them soft. Unfortunately for the Flavr Savr, its taste wasn't good enough to make it a commercial success.
- A woman in New Jersey grew a 2.8-kg (6.16-pound) tomato in 1997, setting a record.
- A wild form of *Lycopersicon esculentum*, *L. esculentum* var. *minor* Rick [*L. cheesmanii* Riley var. *minor* (Hook.) Mill.], grows on the coasts of the Galápagos Islands, where the fruits are eaten by the Galápagos tortoises. The seeds germinate well only if they pass through the digestive tracts of the tortoises, which takes from 1 to 3 weeks.
- Tomatoes are about 94% water.
- A full-grown tomato plant produces as many as 25 000 seeds.
- So-called "black tomatoes" (usually various shades of brown or purple) are available from garden catalogs.
- The tomato is the most popular garden plant, and is found in over 90% of home gardens.
- Tomato juice is commonly recommended to neutralize the effects of skunk odor on sprayed dogs and cats. This has been said to be a myth, and in any event is not as effective as commercial skunk-odor removers and some home-made alternatives. One recommendation: mix 1 L (approximately 1 American quart) of 3% hydrogen peroxide, 50 mL (1/4 cup) of sodium bicarbonate or baking soda, and a little (5 mL or 1/4 teaspoon) dishwashing detergent. Wash the areas that have been sprayed, taking care to avoid the pet's eyes, ears, and mouth. Afterwards, wash your pet with pet shampoo to get rid of the chemicals, and rinse with water.
- The jelly-like substance around the seeds contains the highest concentration of vitamin C in tomatoes.
- A tomato horn worm can eat an entire tomato plant by itself in one day.
- A "pica" is an eating disorder marked by persistent cravings to eat non-nutritive material, or food items in abnormally high amounts. The materials eaten, for example, may be ice, detergent, starch, clay, or dirt. The specific pica in which people often

crave hugely excessive amounts tomatoes has been dubbed tomatophagia. (Reference: Marinella, MA. 1999. Tomatophagia and iron-deficiency anemia. N. Engl. J. Med. **341**: 60–61). Picas, including tomatophagia, are often correlated with iron-deficiency anemia.

Key Information Sources

American Horticultural Society. 1982. Tomatoes. American Horticultural Society, Mount Vernon, VA. 144 pp.

Atherton, J.G., and Rudich, J. (*Editors*). 1986. The tomato crop. A scientific basis for improvement. Chapman and Hall, New York, N.Y. 661 pp.

Bennett, J. 1997. The tomato handbook. Firefly Books, Willowdale, ON. 95 pp.

Blancard, D. 1995. A colour atlas of tomato diseases: observation, identification and control. Halsted Press, New York, NY. 212 pp.

Cutler, K.D. 1997. Tantalizing tomatoes: smart tips & tasty picks for gardeners everywhere. Brooklyn Botanic Garden, Brooklyn, NY. 111 pp.

Davies, J.N., and Hobson, G.E. 1981. The constituents of tomato fruit—the influence of environment, nutrition, and genotype. CRC Crit. Rev. Food Sci. Nutr. **15**: 205–280.

Doty, W.L., Sinnes, A.C., and Lammers, S.M. 1981. All about tomatoes. Revised edition. Ortho Books, San Francisco, CA. 96 pp.

DuBose, F. 1985. The total tomato: America's backyard experts reveal the pleasures of growing tomatoes at home. Harper & Row, New York, NY. 206 pp.

Esquinas-Alcazar, J.T. 1981. Genetic resources of tomatoes and wild relatives: a global report. International Board for Plant Genetic Resources, Rome, Italy. 65 pp.

Foster, C.O. 1975. Terrific tomatoes: all about how to grow and enjoy them. Rodale Press, Emmaus, PA. 262 pp.

Frenkel, C., and Jen, J.J. 1989. Tomatoes. *In* Quality and preservation of vegetables. *Edited by* N.A. Eskin. CRC Press, Inc., Boca Raton, Fl. pp. 53–73.

Gould, W.A. 1983. Tomato production, processing and quality evaluation. AVI Publishing, Westport, CT. 445 pp.

Grewe, R. 1988. The arrival of the tomato in Spain and Italy: early recipes. J. Gastronomy, **3**: 67–81.

Hayward, H.E. 1938. *Lycopersicon esculentum*. *In* The structure of economic plants. *Edited by* H.E. Hayward. Macmillan, New York, NY. pp. 550–579.

Hendrickson, R. 1977. The great American tomato book: the one complete guide to growing and using tomatoes everywhere. Doubleday, Garden City, NY. 226 pp.

Hobson, G.E., and Davies, J.N. 1971. The tomato. *In* The biochemistry of fruits and their products. Vol. 2. *Edited by* A.C. Hulme. Academic Press, London, UK. pp. 437–482.

Hogenboom, N.G. 1979. Incompatibility and incongruity in *Lycopersicon*. *In* The biology and taxonomy of the Solanaceae. *Edited by* J.G. Hawkes, R.N. Lester, and A.D. Skelding. Academic Press, London, UK. pp. 435–444.

Jarvis, W.R., and McKeen, C.D. 1991. Tomato diseases. Revised edition. Agriculture Canada, Ottawa, ON. 70 pp.

Jenkins, J.A. 1948. The origin of the cultivated tomato. Econ. Bot. **2**: 379–392.

Jones, J.B. 1991. Compendium of tomato diseases. APS Press, St. Paul, MN. 73 pp.

Jones, J.B. 1999. Tomato plant culture: in the field, greenhouse, and home garden. CRC Press, Boca Raton, FL. 199 pp.

Kalloo, G. 1991. Genetic improvement of tomato. Springer-Verlag, Berlin, Germany. 358 pp.

Kalloo, G. 1993. Tomato, *Lycopersicon esculentum* Miller. *In* Genetic improvement of vegetable crops. *Edited by* G. Kalloo and B.O. Bergh. Pergamon Press, New York, NY. pp. 645–666.

Kinet, J.M., and Peet, M.M. 1997. Tomato. *In* The physiology of vegetable crops. *Edited by* H.C. Wien. CABI International, Wallingford, Oxon, UK. pp. 207–258.

Kramer, M.G., and Redenbaugh, K. 1994. Commercialization of a tomato with an antisense polygalacturonase gene: the Flavr Savr™ tomato story. Euphytica, **79**: 293–297.

Long, J. 2000. Tomatoes. *In* The Cambridge world history of food. *Edited by* K.F. Kiple and K.C. Ornelas. Cambridge University Press, Cambridge, UK. pp. 351–358.

Male, C.J. 1999. 100 heirloom tomatoes for the American garden. Workman, New York, NY. 246 pp.

Martineau, B. 2001. First fruit: the creation of the Flavr Savr tomato and the birth of genetically engineered food. McGraw-Hill, New York, NY. 269 pp.

McCue, G.A. 1952. The history of the use of the tomato: an annotated bibliography. Ann. Mo. Bot. Gard. **39**: 289–348.

Nevins, D.J., and Jones, R.A. (*Editors*). 1987. Tomato biotechnology. Alan R. Liss, Inc., New York, NY. 339 pp.

Opeña, R.T., and Vossen, H.A.M. van der. 1993. *Lycopersicon esculentum* Miller. *In* Plant resources of South-East Asia, 8, vegetables. *Edited by* J.E. Siemonsma and K. Piluek. Pudoc Scientific Publishers, Wageningen, Netherlands. pp. 199–205.

Organisation for Economic Co-operation and Development. 1988. Tomatoes. 2nd edition. O.E.C.D., Washington, DC. 66 pp.

Papadopoulos, A.P. 1991. Growing greenhouse tomatoes in soil and in soilless media. Agriculture Canada, Ottawa, ON. 79 pp.

Rick, C.M. 1978. The tomato. Sci. Am. **239(8)**: 66–76.

Rick, C.M. 1995. Tomato. *In* Evolution of crop plants. 2nd edition. *Edited by* J. Smartt and N.W. Simmonds. Longman Scientific & Technical, Burnt Mill, Harlow, Essex, UK. pp. 452–457.

Rick, C.M., and Bowman, R.I. 1961. Galapagos tomatoes and tortoises. Evolution, **15**: 407–417.

Rick, C.M., and Holle, M. 1990. Andean *Lycopersicon esculentum* var. *cerasiforme*: genetic variation and its evolutionary significance. Econ. Bot. **44**: 69–78.

Smith, A.F. 1994. The tomato in America: early history, culture, and cookery. University of South Carolina Press, Columbia, SC. 224 pp.

Smith, A.F. 2000. Souper tomatoes: the story of America's favorite food. Rutgers University Press, New Brunswick, NJ. 236 pp.

Spooner, D.M., Peralta, I.E., and Knapp, S. 2005. Comparison of AFLPs with other markers for phylogenetic inference in wild tomatoes [*Solanum* L. section *Lycopersicon* (Mill.) Wettst.]. Taxon, **54**: 43–61.

Tigchelaar, E.C. 1986. Tomato breeding. *In* Breeding vegetable crops. *Edited by* M.J. Bassett. AVI Publishing Co., Westport, CT. pp. 135–171.

Varga, A., and Bruinsma, J. 1986. Tomato. *In* CRC handbook of fruit set and development. *Edited by* S.P. Monselise. CRC Press, Boca Raton, FL. pp. 461–481.

Villareal, R.L. 1980. Tomatoes in the tropics. IADS Development-oriented literature series. Westview Press, Boulder, CO. 174 pp.

Warnock, S.J. 1988. A review of taxonomy and phylogeny of the genus *Lycopersicon*. HortScience, **23**: 669–673.

Wittwer, S.H., and Honma, S. 1969. Greenhouse tomatoes; guidelines for successful production. Michigan State University Press, East Lansing, MI. 95 pp.

Yoder, J.I. (*Editor*). 1993. Molecular biology of tomato: fundamental advances and crop improvement. Proceedings, symposium, August 17–19, 1992, University of California, Davis. Technomic Publishing Co., Lancaster, PA. 314 pp.

Specialty Cookbooks

(Note: There are hundreds of books dedicated to the culinary aspects of tomatoes. The following are examples.)

Bailey, L. 1992. Lee Bailey's tomatoes. Clarkson Potter, New York, NY. 80 pp.

Ballantyne, J. 1982. Garden Way's red & green tomato cookbook. Garden Way Publishing, Charlotte, VT. 158 pp.

Buff, S. 1999. The great tomato book. Burford Books, Short Hills, NJ. 166 pp.

Davis-Hollander, L. 2004. The tomato festival cookbook: 150 tempting recipes for your garden's lush, vine-ripened, sun-warmed, fat, juicy, ready-to-burst heirloom tomatoes. Storey Publishing, North Adams, MA. 310 pp.

Della Croce, J. 1996. Salse di pomodoro: making the great tomato sauces of Italy. Chronicle Books, San Francisco, CA. 132 pp.

Doty, W.L., and Sinnes, A.C. 1982. Tomatoes. American Horticultural Society, Mount Vernon, VA. 144 pp.

Dribin, L., and Marina, D. 1990. Cooking with sun-dried tomatoes. Fisher Books, Tucson, AZ. 160 pp.

Garden Way Publishing. 1991. Tomatoes! 365 healthy recipes for year-round enjoyment. Storey Communications, Pownal, VT. 284 pp.

Hériteau, J. 1975. Tomato gardening & cooking. Grosset & Dunlap, New York, NY. 79 pp.

Hoffman, M. 1994. The complete tomato cookbook. HPBooks, New York, NY. 127 pp.

Hunt-Wesson Foods. 1976. Hunt's complete tomato sauce cookbook. Hunt-Wesson Foods, Fullerton, CA. 160 pp.

Ibsen, G., and Nielsen, J. 1999. The great tomato book. Ten Speed Press, Berkeley, CA. 150 pp.

Jordan, M.A. 1995. The good cook's book of tomatoes: with more than 200 recipes. Addison-Wesley Publishing Co., Reading, MA. 310 pp.

Lundy, R., Stehling, J., and Ciletti, B.J. 2004. In praise of tomatoes: tasty recipes, garden secrets, legends & lore. Lark Books, New York, NY. 176 pp.

Nimtz, S., and Cousineau, R. 1994. Tomato imperative! From fried green tomatoes to summer's ripe bounty. Little, Brown, Boston, MA. 253 pp.

Tarr, Y.Y. 1977. The tomato book. Vintage Books, New York, NY. 220 pp.

VANILLA

Family: Orchidaceae (orchid family)

Names

Scientific name: *Vanilla planifolia* Jacks. [*V. fragrans* (Salisb.) Ames]

- Hernan Cortés (1485–1547), the Spanish conqueror of Mexico, brought vanilla pods back to Spain. The Spaniards called the pod *vainilla*, a diminutive of *vaina*, meaning sheath, vagina, pod, an allusion to the sheath-like shape of the fruit. This led to the English word "vanilla" and the genus name *Vanilla*.
- Vanilla has also been called Bourbon vanilla, because the French successfully established plantations in the Bourbon Islands, now known as Réunion and the Comoros (additional information below).
- "Tahitian vanilla" is *V. tahitensis* J.W. Moore (thought to be a hybrid of *V. planifolia* and *V. pompona* Schiede), introduced to Tahiti from South America. It has a more floral or anise aroma than regular vanilla, said to go especially well with fruit and savory dishes.
- "West Indian vanilla," *V. pompona*, has shorter, thicker, banana-shaped pods and a less pleasant aroma compared to regular vanilla.
- The word *planifolia* in the scientific name *V. planifolia* is Latin for flat leaves, which is descriptive of the striking, flat shape of the leaves.

Plant Portrait

The vanilla plant is a leafy, climbing orchid native to hot, wet, tropical Central America and the West Indies. The plant has roots in the air (like many other orchids), as well as in the ground. The aerial roots of the vanilla vine hold it to the trees upon which it climbs. The vanilla vine can reach lengths of 30 m (100 feet). The plant is grown for its long, slender, black seed pods which are the source of vanilla. The Totonacs, Indian inhabitants of the eastern coastal area of Mexico for hundreds of years, are believed to have been the first to use vanilla as a flavoring, possibly over 1000 years ago. When the Aztecs defeated the Totonaco Indians, one of the most important tributes they demanded was the fruit of the vanilla vine. Bernal Díaz, an officer in the army of Cortes, is thought to have been the first European introduced to vanilla. The Aztec Indians in Mexico used vanilla pods to flavor their chocolate drink, and the custom of combining chocolate and vanilla was brought back to Europe by the Spaniards. Hugh Morgan, apothecary to Queen Elizabeth I, in 1602 was the first European to suggest that vanilla could be used as a flavoring by itself. Vanilla was brought to the United States in the late 1700s by Thomas Jefferson (1743–1826), third president. Vanilla is now cultivated throughout the Tropics, principally in Madagascar (the largest producer), the Comoros, Tahiti, Reunion, and Mexico. World production is about 8000 metric tonnes (8800 tons) of pods.

Vanilla pods are commonly called "beans" in American usage, owing to their resemblance to green string beans. The word "bean" is not strictly a technical botanical term, and refers to both seeds and immature fruits of the legume family (such as the string bean), as well as of similar nonlegumes. The

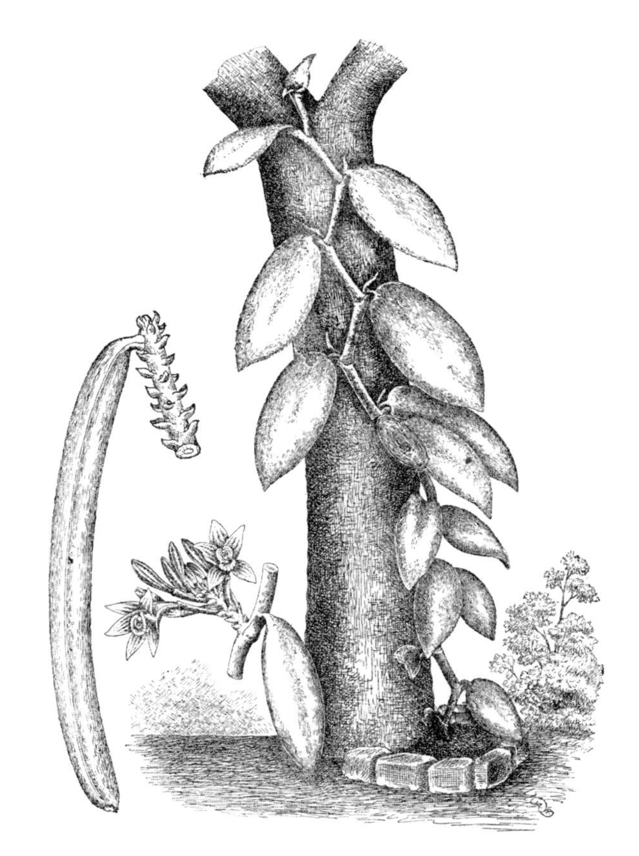

Vanilla (*Vanilla planifolia*), from Bailey (1900–1902).

British "vanilla pod" is typically expressed as "vanilla bean" in the United States.

The flowers are hand pollinated during the morning of the first day the flower is open, for unless pollination occurs, the flower drops from the vine the next day and a pod will not be produced. The pods are ready for harvesting 6–9 months after pollination. They grow in bunches of 6–10, somewhat resemble bananas, and are 13–25 cm (5–10 inches) long, yellow green, with tiny black seeds. The pods are watery and tasteless. The pleasant aroma and taste of vanilla must be brought out by curing, which takes 3–6 months. These processes are responsible for the high cost of vanilla, which is the second costliest spice, after saffron. Vanillin is the primary flavor constituent of vanilla beans. Vanillin has been synthesized, and many contend that the synthetic substitute, which is much cheaper, is actually more flavorful than true vanilla. Much of the world's vanilla is now synthetic, prepared from eugenol, the chief component of clove oil.

Culinary Portrait

Vanilla is one of the world's most important flavoring materials. It is the most popular flavor in the baking and chocolate industries in North America. Plain old vanilla is the favorite ice cream flavor, accounting for 29% of all sales. Vanilla is also used as a flavoring of beverages, confections, cakes, custards, yogurt, compôtes, puddings, and candy. In very small amounts, vanilla is sometimes added to savory foods, such as fish soup, oysters, and poultry. Vanilla is also employed in a variety of beverages, such as wines, distilled alcohols, punches, and hot chocolate.

Bottled pure vanilla extract contains alcohol, which is not permitted in some religions. Powdered vanilla (vanillin) is made with artificial ingredients that give vanilla-like aroma, but the taste is not as true. However, it is alcohol free. A better choice is bottled vanilla that substitutes glycerine for alcohol, although it is more difficult to locate a store that sells it. Liquid vanilla can lose much of its flavor during cooking, and is best added at the end of the cooking cycle (although this is not always possible).

Vanilla is also available as pods and seeds, which are less likely to be fake. The pods are best stored in an airtight container. Vanilla seeds can be chopped or powdered in a blender, and used to flavor syrups, milk, and fruits. The whole pods can simply be split lengthwise, steeped in cold liquid and heated before the liquid is used for flavoring, then the seeds can be removed, rinsed, and used again up to four times. "Vanilla sugar" can be prepared by placing a whole pod or part of a pod in a container of sugar.

Vanilla (*Vanilla planifolia*), from Köhler (1883–1914).

Culinary Vocabulary

- Natural vanilla is referred to as natural vanilla extract, pure vanilla extract, and vanilla extract. "Vanilla essence" generally means a highly concentrated, expensive, pure vanilla extract, but in Britain, it means imitation vanilla.
- Artificial vanilla is known by many names: artificial vanilla extract, artificial vanilla flavor, artificial vanilla flavoring, imitation vanilla extract, imitation vanilla flavor, imitation vanilla flavoring, vanilla essence, and vanilla flavoring.
- "Vanilla powder" is an alcohol-free powder made from vanilla beans that have been dried and pulverized. Depending on brand, sugar may be added. It is used to make cooked desserts such as custard.
- Vanilla sometimes means "bland" when applied to food, not necessarily that vanilla is actually in the food. For example, "vanilla wonton soup" means ordinary wonton soup, as opposed to hot-and-sour wonton soup.
- "Black and white" is a term with several culinary usages, sometimes referring to black coffee with a container of cream on the side, but generally to soda fountain confections mixing chocolate and vanilla flavors.
- Reunion (the island in the West Indian Ocean) was formerly known as Bourbon Monarchy, but the French, who had every reason to hate the name (the Bourbons ruled France tyrannically during the 17th and 18th centuries), rechristened the island; vanilla beans from the area, notwithstanding, are still called Bourbon beans, and (as noted above) vanilla used to be called Bourbon vanilla.
- The French expression *à la mode*, has the general meaning "in the style of," but also is used (particularly in the United States) specifically for a piece of pie served with a generous scoop of ice cream. It has been claimed that vanilla ice cream and apple pie are the only acceptable bases for using the expression.

Curiosities of Science and Technology

- Vanilla was once believed to be a tonic for the brain.
- A tiny bee, the melipone, found only in Mexico, is responsible for the natural pollination of the vanilla flower. Vanilla could only be produced in Mexico until 1836, when Charles Moran of Liege, Belgium, discovered how to hand pollinate the vanilla flower, and ended the Mexican monopoly of the vanilla bean. Not surprisingly, the procedure generally requires the small hands and patience of women. An average worker pollinates 1000–2000 flowers a day.
- People who harvest vanilla pods frequently develop dermatitis, particularly on their hands, as a result of contacting the pods.
- In Madagascar, the largest producer of vanilla, extraordinary measures have been taken to protect the crop from theft. Some plantation owners brand each pod with a tattoo while it is still attached to the vine.
- There are as many as 25 000 species of orchids, and the orchid family is perhaps the largest of all of the more than 300 plant families. Nevertheless, vanilla is the only member of the orchid family that has commercial significance as food, although there are many orchids that are grown as ornamentals.
- The war between vanilla and chocolate has reached a new intensity with the marketing of Crave Control™, a British-invented patch worn on the skin, reminiscent of the nicotine patch intended to help people break the tobacco habit. This new type of patch gives off a vanilla smell that allegedly drastically cuts the craving for chocolate.
- Baskin-Robbins, a major producer and franchiser of ice cream, claimed to have developed 548 flavors as of 1995, with vanilla as the all-time favorite.
- Regulations concerning food labeling can be confusing. In the United States, if a label states "vanilla ice cream," the product may only be made with pure vanilla extract and (or) vanilla beans. If the label says "vanilla-flavored ice cream," the material may contain up to 42% artificial flavoring.

Key Information Sources

Abeysinghe, A. 1974. Vanilla: cultivation, processing, and marketing. Planning Division, Ministry of Plantation Industries, Colombo, Sri Lanka. 8 pp.

Ackerman, J.D. 2002. *Vanilla. In* Flora of North America north of Mexico, Vol. 26. *Edited by* Flora of North America Editorial Committee. Oxford University Press, New York, NY. pp. 507–510.

Anand, A., and Smith, A.E. 1986. The market for vanilla. Tropical Development and Research Institute, London, UK. 33 pp.

Arditti, J. 1971. Vanilla: an historical vignette. Am. Orchid Soc. Bull. **40**: 610–613.

Baruah, A., and Saikia, N. 2002. Vegetative anatomy of the orchid *Vanilla planifolia* Andr. J. Econ. Taxon. Bot. **26**: 161–165.

Besse, P., Silva, D. da, Bory, S., Grisoni, M., Le Bellec, F., and Duval, M.F. 2004. RAPD genetic diversity in cultivated vanilla: *Vanilla planifolia*, and relationships with *V. tahitensis* and *V. pompona*. Plant Sci. **167**(issue 2): 379–385.

Bouriquet, G. 1954. Le vanillier et la vanille dans le monde. [The vanilla plant and products in the world.] Editions Lechavalier, Paris, France. 739 pp. [In French.]

Childers, N.F., and Cibes, H.R. 1948. Vanilla culture in Puerto Rico. U.S. Dept. of Agriculture, Washington, DC. 94 pp.

Correll, D.S. 1953. Vanilla: its botany, history, cultivation and economic import. Econ. Bot. **7**: 291–358.

Dignum, M.J.W., Kerler, J., and Verpoorte, R. 2001. Vanilla production: technological, chemical, and biosynthetic aspects. Food Rev. Int. **17**: 199–219.

Ecott, T. 2004. Vanilla: travels in search of the ice cream orchid. Grove Press, New York, NY. 278 pp.

Feldkamp, C.L. 1945. Vanilla: culture, processing and economics: a list of references. U.S. Dept. of Agriculture Library, Washington, DC. 28 pp.

Fouche, J.G., and Coumans, M. 1992. Vanilla: pollination. Am. Orchid Soc. Bull. **61**: 1118–1122.

Fouche, J.G., and Jouve, L. 1999. *Vanilla planifolia*: history, botany and culture in Reunion island. Agronomie (France), **19**: 689–703.

Krishnakumar, V., and Potty, S.N. 2003. Evaluation of performance of vanilla (*Vanilla planifolia*) under different nutritional management practices. J. Med. Arom. Plant Sci. **25**: 46–49.

Marles, R.J., Compadre, C.M., and Farnsworth, N.R. 1987. Coumarin in vanilla extracts: its detection and significance. Econ. Bot. **41**: 41–47.

Martin, G.E., Ethridge, M.W., and Kaiser, F.E. 1977. Determining the authenticity of vanilla extracts. J. Food Sci. **46**: 1580–1583, 1586.

Nauman, C.E. 1991. Vanilla: the fragrant, flavorful orchid. Fairchild Trop. Gard Bull. **46**(**2**): 10–14.

Ngere, O. 2001. Vanilla. NARI, Papua New Guinea. 42 pp.

Nielsen, L.R., and Siegismund, H.R. 1999. Interspecific differentiation and hybridization in *Vanilla* species (Orchidaceae). Heredity, **83**: 560–567.

Peter, K.V., and Kandiannan, K. 1999. Vanilla. *In* Tropical horticulture, vol. 1. *Edited by* T.K. Bose, S.K. Mitra, A.A. Farooqui, and M.K. Sadhu. Naya Prokash, Calcutta, India. pp. 706–711.

Rain, P. 2004. Vanilla: the cultural history of the world's most popular flavor and fragrance. Jeremy P. Tarcher/Penguin, New York, NY. 371 pp.

Ramachandra-Rao, S., and Ravishankar, G.A. 2000. Vanilla flavour: production by conventional and biotechnological routes. J. Sci. Food Agric. **80**: 289–304.

Ranadive, A.S. 1994. Vanilla cultivation, curing, chemistry, technology and commercial products. *In* Spices, herbs and edible fungi. Developments in Food Science 34. *Edited by* G. Charalambous. Elsevier Science Publishers, Amsterdam, Netherlands. pp. 517–577.

Riley, K.A., and Kleyn, D.H. 1989. Fundamental principles of vanilla/vanilla extract processing and methods of detecting adulteration in vanilla extracts. Food Technol. **43**: 64, 66, 68, 70, 75, 77.

Rosenbaum, E.W. 1977. Vanilla extract and synthetic vanillin. *In* Elements of food technology. *Edited by* N.W. Desrosier. Avi Publishing Co., Westport, CT. pp. 683–689.

Stern, W.L., and Judd, W.S. 1999. Comparative vegetative anatomy and systematics of *Vanilla* (Orchidaceae). Bot. J. Linn. Soc. **131**: 353–382.

Straver, J.T.G. 1999. *Vanilla planifolia* H.C. Andrews. *In* Plant resources of South-East Asia, 13, spices. *Edited by* C.C. de Guzman and J.S. Siemonsma. Backhuys, Leiden, Netherlands. pp. 228–223.

Summers, M.K. 1962. The vanilla story. Marion Kay Products, Brownstown, IN. 48 pp.

Theodose, R. 1973. Traditional methods of vanilla preparation and their improvement. Trop. Sci. **15**: 47–57.

Vinning, G. 1992. Vanilla: prospects and prospectives. 1992. J. South Pacific Agric. (Samoa), **1**(**1**): 15–23.

Specialty Cookbooks

Benning, L.E. 1992. Make mine vanilla. Simon & Schuster, New York, NY. 192 pp.

Hazen, J. 1995. Vanilla. Chronicle Books, San Francisco, CA. 72 pp.

Herbst, S.T. 1996. The food lover's guide to chocolate and vanilla. Morrow, New York, NY. 118 pp.

Lewis, B.L. 1951. The story of pure vanilla, with recipes. Vanilla Bean Association of America, [New York?]. 32 pp.
[= Flavoring Extract Manufacturers' Association of the United States, Vanilla Bean Association of America, Vanilla Growers of Madagascar, and Bernard L. Lewis, Inc. 1964. Make mine real vanilla: recipes and information on the world's most popular, natural flavor. Published cooperatively by Flavoring Extract Manufacturers Association, Vanilla Bean Association of America and Vanilla Growers of Madagascar. New York, NY. 32 pp.
= Vanilla Information Bureau, 1972. Vanilla desserts cookbook: a collection of new and old fashioned recipes made with real vanilla. Vanilla Information Bureau, New York, NY. 32 pp.]

Rain, P. 1986. Vanilla cookbook. Celestial Arts, Berkeley, CA. 124 pp.

Rain, P. 2002. The vanilla chef. Vanilla Queen Press, Santa Cruz, CA. 190 pp.

Watkins Incorporated. 1994. Vanilla: the Watkins kitchen collection. Watkins Incorporated, Winona, MN. 111 pp.

WALNUT

Family: Juglandaceae (walnut family)

Names

Scientific names: *Juglans* species

- Black walnut—*J. nigra* L.
- Persian or English walnut—*J. regia* L
- Japanese walnut—*J. ailantifolia* Carrière (*J. sieboldiana* Maxim.)
- Heartnut—*J. ailantifolia* var. *cordiformis* (Makino) Rehder

- The word "walnut" has been said to come from the Old English *wealhhnutu*, literally "foreign nut" or "Gaul nut" (the Gauls or Welsh were considered foreigners) because the walnut was imported into England during the old Roman period at which time it was indeed considered foreign. In Old German, *welsche Nuß* means Welsch nut or nut of southern Europe, and this use may simply have been transferred to English. Another interpretation is that Germanic peoples considered the walnut of the Romans and Gauls to be "foreign," and so the Old German name, rather than the Old English name, was the first to capture the idea of "foreign nut." Still another interpretation is that the name walnut comes from the drying of the nuts on top of English garden walls.
- One Chinese name for walnut means "a toad," presumably based on the bumpy surface of the shell.
- The genus name *Juglans* is Latin for "nut of Jupiter," the classical king of the gods, an indication that the ancient Romans considered the nuts extremely tasty.
- The black walnut *J. nigra* is so named for the dark color of the bark. *Nigra* in the scientific name is Latin for black.
- The English walnut, *J. regia*, is called "English" by inhabitants of England, and subsequently in North America because the species was imported from England. The species is called the "Chinese walnut" in China, the "Italian walnut" in Italy, and the "French walnut" in France. The "Persian walnut" is a much more appropriate name, since it actually seems to have come originally from Persia. *Regia* in the scientific name is Latin for "royal," reflecting the belief of the ancient Greeks and Romans that the nut was the choice of the gods.
- "White walnut" is an occasional name for butternut (*Juglans cinerea* L), and very uncommonly for shagbark hickory (*Carya ovata* (Mill.) K. Koch).
- Mediterranean varieties of the English walnut were brought to California in the 18th century by Spanish missionaries, and became known as "mission walnuts."
- Presbyterian minister Paul C. Crath of Toronto conducted missionary work in western Ukraine (then under Polish occupation) from 1924 to 1936. During this time, he became well known for introducing hardy strains of Persian walnut from the Carpathian Mountains in Poland. His discovery was featured in many newspaper and magazine articles with headlines such as "Greatest tree find of the century." Cold-hardy strains of Persian walnuts in North America generally became referred to as Carpathian walnuts, or simply "Carpathians."

Persian or English walnut (*Juglans regia*), from Figuier (1867).

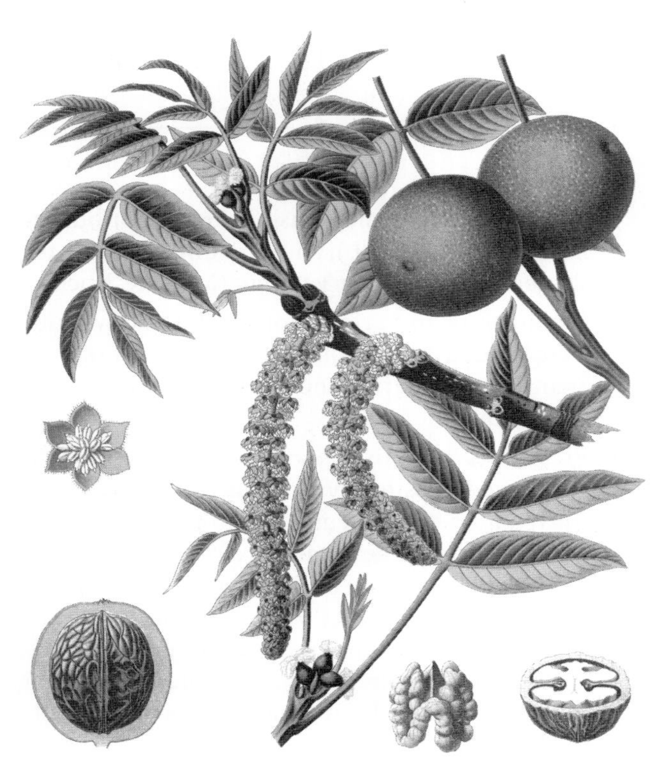

Persian or English walnut (*Juglans regia*), from Köhler (1883–1914).

- A variety of the Persian walnut called the "Titmouse walnut" has been described from France, so named because the shell is so thin that birds, especially the titmouse, can break it and eat the kernel.
- The Persian walnut variety Bijou produces large nuts 4–5 cm (about 2 inches) in diameter. As the name of the cultivar indicates (bijou = jewel), the shells were once in demand as jewel boxes. The shells were polished, fitted with a metal hinge and clasp, and lined with velvet. They were also used to hold gloves as presents, and made up as cradles for miniature dolls. Reminiscent of these practices, the Chinese employed walnut shells as snuff-boxes. The Chinese also used intricately carved walnut shells to transport musically trained, singing crickets, believed to be good omens.
- American horticulturist Luther Burbank (1849–1926) was a genuine but erratic genius (often called "the plant wizard") when it came to creating amazing new hybrid varieties of plants. In 1891, he apparently crossed the black walnut (*J. nigra*) and Hind's black walnut [*J. hindsii* (Jeps.) Rehd.] to produce the "Royal Walnut"—a huge tree that received sensationalistic press coverage. The name royal walnut is actually a translation of the Latin name of the English walnut, *Juglans regia*.
- *Juglans ailantifolia* is called the Japanese walnut, because it comes from Japan. The heartnut variety is so named because its nut and kernel are both heart-shaped. The Japanese walnut is also sometimes called Siebold walnut, after Philipp Franz von Siebold (1796–1866), a German physician and botanist who studied the natural history of Japan. *Ailantifolia* in the scientific name means "leaves like *Ailanthus*" (which is the tree-of-heaven, a small ornamental tree sometimes grown in North America, which has very large compound leaves—of course like walnuts, only bigger).

Plant Portrait

There are about 20 species of *Juglans*, all of which produce edible nuts. The most often encountered commercial food walnuts are mentioned above. There are many cultivars of these species, as well as varieties that are of hybrid origin. Walnuts are deciduous trees. The Persian and black walnuts are particularly attractive valuable landscape as well as nut trees, with tall straight trunks. The black walnut and Persian walnut are the tallest, and the longest-lived of the species of *Juglans*, sometimes reaching 50 m (164 feet) in height, 2 m (about 6 feet) in diameter, and more than 250 years of age. The black walnut is a native of eastern North America. The Persian walnut, a native of southeastern Europe and western Asia, is the most widely cultivated of the walnuts and is the chief walnut of commerce. Its shells have been found in the Neolithic Era Swiss Lake dwellings, dated at about 7000 BC. These were likely just gathered from wild trees, not cultivated. Persian walnuts were cultivated by the early Chinese, Greeks, and Romans. The United States, China, Iran, and Turkey are leading producers, and substantial quantities are also grown in Europe, notably in France, Romania, and the Balkan countries. Superior varieties of the Persian walnut were imported from Chile into California in the 18th century, and California subsequently developed into the major walnut-raising area of the New World, a status it still enjoys. The Japanese walnut, a native of Japan, grows as high as 18 m (60 feet). The heartnut variety of this species (var. *cordiformis*) has a nut that is somewhat heart-shaped, flattened, and thin-shelled.

Culinary Portrait

Walnuts, primarily from the Persian walnut, are one of the most widely consumed of nuts, and among dessert nuts only the almond and the Brazil nut exceed it in quantities traded in the world's markets. The nuts are eaten directly, used in stuffings, sandwiches, sauces, pastries, and desserts, and as a flavoring for ice cream, syrup, and other food preparations. Maple-walnut ice cream and walnut fudge are favorite treats. Green Persian walnuts when immature (about the size of large grapes) are pickled to prepare a relish, which is popular in Britain and is sometimes available in food specialty stores in North America. Extracts from walnut shells are used to flavor liqueurs, ratafias, and wines.

The nuts of the black walnut are very strong in flavor and are enclosed in a very hard shell that is difficult to open (for most people, the easiest way is to drive over the nuts with a car). Unlike many other nuts, black walnuts retain their flavor well when cooked. The slightly bitter, distinctive taste of black walnuts is strongly desired by many gourmets. Although not done commercially, black walnut trees can be tapped in the spring to obtain syrup.

Walnuts sold in vacuum-packed jars or cans are usually very fresh in taste. The freshness of shelled walnuts sold in bulk is uncertain, and buying a small sample to taste is probably the best assurance of quality. The kernels should be crunchy, not soft, shriveled, or rancid in taste. Unshelled walnuts should be free of cracks and holes, and should appear to be heavy and full, but once again their quality can not be judged with certainty until they are tasted. A paint smell from cracked or shelled walnuts is an indication that the product is too old. Shelled nuts should be brittle and snap easily, and should not be soft and pliable. Once purchased, walnuts should be stored in a sealed container. Away from heat and humidity and in a cool area, they should remain fresh for several months. The nuts should not be shelled until they are to be eaten. Shelled walnuts can be stored in a refrigerator for 6 months, and in a freezer for up to a year. Unless they are in a tightly sealed container, they can absorb odors from other foods.

Walnuts are rich in heart-healthy omega-3 fatty acids, which are believed to reduce the risk of heart disease by making the blood less sticky (i.e., platelets tend not to clump and form dangerous clots).

Many people prefer the flavor and crispness of fresh walnuts, and harvest them directly from trees for consumption. If this is done it should be noted that the pellicle or kernel skin of fresh walnuts is often very bitter, and should be removed. As the nuts age, this bitterness disappears. It should also be kept in mind that walnuts leave a brown stain on bare hands that some have described as "impossible to remove."

Culinary Vocabulary

- "Nougat," the French confection, literally means "made from walnuts," and indeed the walnut was originally used in its preparation.
- The Greek stew *stefado*, made with beef, veal, or rabbit, is traditionally topped with feta cheese and walnuts.
- The culinary expression "butter the size of a walnut" is a traditional measure of butter volume, about

Black walnut (*Juglans nigra*), from *Gardeners' Chronicle* (Oct. 26, 1901, p. 303).

equal to 2 tablespoons (approximately 58 mL), which is about half the size of the volume indicated by the expression "butter the size of an egg."

Curiosities of Science and Technology

- Petrified remains of whole walnuts and kernels in the Naples Museum in Italy are among the items found in the ruins of Pompeii and Herculaneum, destroyed by the Mount Vesuvius eruption in 79 AD. Walnuts were left at a table, uneaten by priests whose meal was terminally interrupted.
- In primitive times, medicinal practice was more often than not based on the "Doctrine of Signatures"—the idea that plants with characteristics reminiscent of human organs could cure diseases of those organs. The ancient Greeks interpreted the walnut's shell as a human skull (their word *caryon* for the walnut meant head) and the nut as a human brain (the joined halves of a kernel of a walnut certainly looks like the brain). Accordingly, they believed that walnuts cured headaches. Others have thought that low intellect and madness could be cured by eating walnuts.
- A paste of ground Persian walnuts mixed with white lead was a recommended cure for baldness in China. In North America, burnt kernels of black walnut in red wine was supposed to stop hair from falling out. (Again, note the association of the walnut with the human head.)
- The nuts and leaves of the black walnut are thought to repel insects to some degree, and have been used, both by Native Americans and colonists, as an insect repellent. Walnut leaves were sometimes rubbed on the faces of cattle and horses to repel flies. It was once customary to plant a walnut in the area of a stable, in the expectation that the smell from the very aromatic leaves would keep flies away from the animals.
- The "mound builders" were peoples who built mounds in a large area from the Great Lakes to the Gulf of Mexico and from the Mississippi River to the Appalachian Mountains. The greatest concentrations of mounds are found in the Mississippi and Ohio valleys. It is thought that they were the ancestors of Native Americans, and were sedentary farmers who lived in permanent villages. The mounds vary considerably in shape

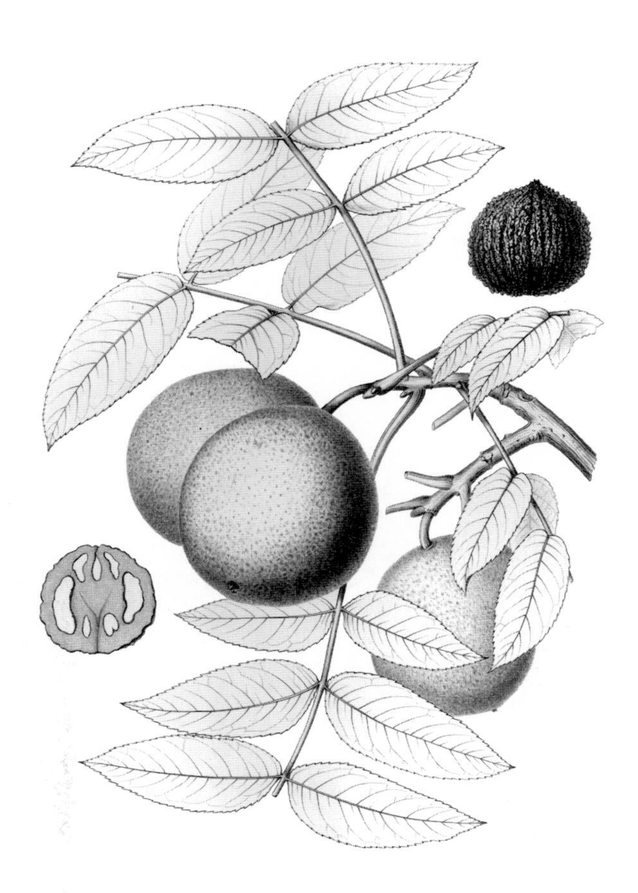

Black walnut (*Juglans nigra*), from Sargent (1890–1892, vol. 7, plate 234).

and size (from less than 0.4 ha or 1 acre to more than 40 ha or 100 acres). The mounds were used chiefly as burial places but also as foundations for buildings. Mounds also vary in age, some dating back to the 6th century, while others were built since Europeans colonized North America. Black walnuts shells carved in the form of birds and pierced to served as earrings were discovered in the mounds of Ohio and Indiana.

- In the 17th century in parts of Europe, a young man was not allowed to marry until he had planted a sufficient number of walnut trees that his prospective father-in-law and the community at large were convinced of his worth.
- During the 17th and 18th centuries, English walnut was the European wood of choice for fine furniture. Old walnut furniture, for example, a Queen Anne walnut bureau, are valued antiques.
- Walnut wood to this day is considered excellent for gunstocks. "Shouldering walnut" once meant enlisting in the military. In 1806, 12 000 walnut trees were reportedly needed annually in France for manufacturing muskets. Kaiser Wilhelm II (1859–1941) is reported to have purchased large quantities of black walnut wood as a stockpile for gunstocks in anticipation of World War I.
- The Gardener's Chronicle of London of 1852 described a 7.7 m (25 feet) wide table made out of a single plank of walnut, which was used for a banquet given by the Emperor Frederick III in Lorraine. The same source mentioned a 1000-year-old tree in the Crimea that yielded 80 000–100 000 nuts annually, and was shared by five Tartar families. Walnut trees this large or old are improbable. More reliable is a report of a walnut from Norfolk, England, which had a trunk circumference near ground level of almost 10 m (32 feet), which was 27 m (90 feet) tall, spread its branches over 110 m (120 yards), and produced 54 000 nuts in one season.
- The shells of walnuts were a valuable war material. In the First World War, the shells were converted to high-quality charcoal that was used extensively in gas mask filters. In the Second World War, ground nutshells proved excellent as an abrasive for cleaning aircraft pistons and cylinder heads.
- Toyo Tire & Rubber Co. Ltd. has manufactured winter tires (notably its Observe tires) with ground walnut shells in the tread compound to increase grip on ice. The shells act as sandpaper grit to improve traction.
- At one time, airplane propellers were made of walnut wood.
- Actors once used yellow dye from walnut fruits and a dark brown dye from the tree's roots to stain their skins.
- The flesh of black walnuts has a food value equal to the best beef. During the 15th century, the aboriginal city of Cahokia Mound on the Mississippi River was larger than London, England, at the same time. Its large population was sustained by the protein value of black walnuts.
- A black walnut tree in Tennessee produced more than 6000 nuts in one season.
- A single black walnut tree on the Lower Missouri was said to have furnished 200 fence posts.

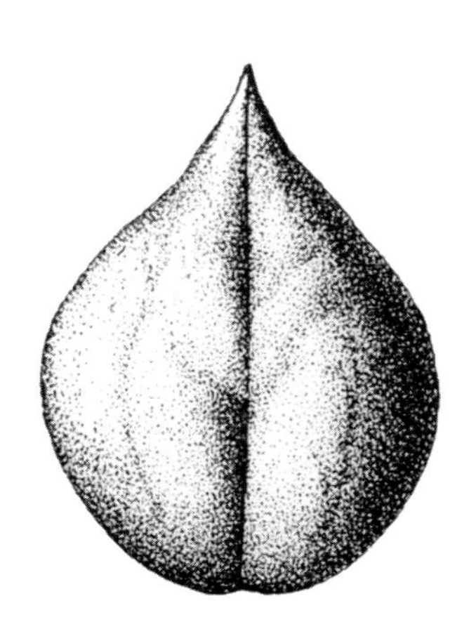
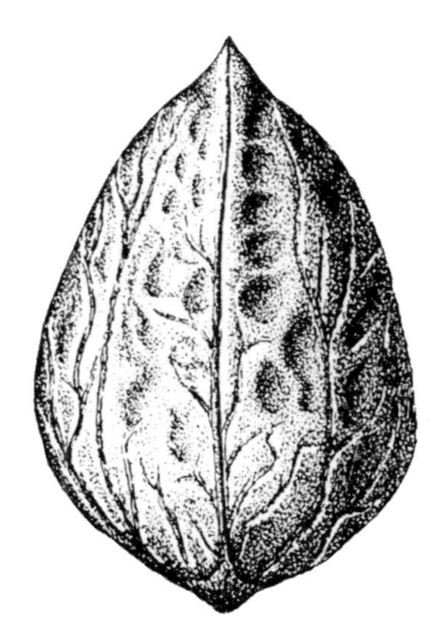

Right, Japanese walnut (*Juglans ailantifolia*); *left*, heartnut (*J. ailantifolia* var. *cordiformis*), from United States Department of Agriculture, Division of Pomology (1896, plate 7).

- Black walnut is a bad neighbor for other plants. The toxic substance juglone is present in the roots, leaves, and seed husks, and this has been reported to kill several species of pines when they grow nearby. Such crops as alfalfa, potatoes, and tomatoes are easily injured by juglone, and fruit trees such as apple may not bear fruit if planted too close to walnut trees. Members of the heath (Ericaceae) family, such as rhododendrons, azaleas, blueberries, and mountain laurel, are particularly susceptible. The toxic zone from a mature tree extends on average over a 15–18 m (50–60 feet) radius from the trunk, but can be up to 24 m (80 feet) away. Persian walnut also exhibits similar toxicity, and this was recorded by the Roman scholar Pliny (23–79).
- Horses hauling logs or lumber at sawmills or bedded on sawdust or wood shavings from black walnut sometimes suffered from lameness owing to an inflammation of the hoof (laminitis). Still an occasional problem today, this is believed to be caused by juglone and other chemicals present in large amounts in the heartwood.
- Walnut lumber is highly valued. Walnut veneers may be sliced less than 1 mm (1/25th inch) in thickness after the wood as been softened by hot water and steam. A veneer company paid $39,000 for a black walnut tree growing near Johnson City, Iowa. There have been cases of highly organized "rustling" of mature trees from their owners' properties.
- A black walnut is 90% shell.
- The shell of the black walnut has intricate internal cavities. It has been suggested that these are an adaptation to confuse rodents trying to locate the edible nut inside the shell.
- Walnut oil resembles linseed oil in its drying properties, and has been used for centuries in the making of fine artists' paints.
- Walnut bark is used as toothpaste in Pakistan.

Key Information Sources

Aletà, N., Girona, J., and Tasias, J. (*Editors*). 1993. Proceedings international walnut meeting, IRTA–Generalitat de Catalunya, Terragona, Spain, 21–25 October, 1991. International Society for Horticultural Science, Wageningen, Netherlands. 315 pp.

Allen, W.F. 1912. English walnuts; what you need to know about planting, cultivating and harvesting this most delicious of nuts. W.F. Allen, Lawrenceville, NJ. 29 pp.

Beineke, W.F. 1983. The genetic improvement of black walnut for timber production. Plant Breed. Rev. **1**: 236–266.

Brooks, M. 1951. Effect of black walnut trees and their products on other vegetation. Agricultural Experiment Station, College of Agriculture, Forestry and Home Economics, West Virginia University, Morgantown, WV. 31 pp.

Chenoweth, B. 1995. Black walnut: the history, use, and unrealized potential of a unique American renewable natural resource. Sagamore Publishing, Champaign, IL. 334 pp.

Davis, E.F. 1928. The toxic principle of *Juglans nigra* as identified with synthetic juglone, and its toxic effects on tomato and alfalfa plants. Am. J. Bot. **15**: 620.

Dillow, M.K., and Funk, D.T. 1975. Bibliography of walnut, supplement no. 2. North Central Forest Experiment Station, Forest Service, U.S. Dept. of Agriculture, St. Paul, MN. 26 pp.

Dillow, M.K., Hawker, N.L., and Funk, D.T. 1971. Annotated bibliography of walnut, supplement no. 1. North Central Forest Experiment Station, Forest Service, U.S. Dept. of Agriculture, St. Paul, MN. 23 pp.

Forde, H.I. 1975. Walnuts. *In* Advances in fruit breeding. *Edited by* J. Janick and J.N. Moore. Purdue University Press, West Lafayette, IN. pp. 439–455.

Funk, D.T. 1966. Annotated bibliography of walnut and related species. North Central Forest Experiment Station, Forest Service, U.S. Dept. of Agriculture, St. Paul, MN. 48 pp.

Funk, D.T. 1970. Genetics of black walnut. U.S. Dept. of Agriculture, Washington, DC. 13 pp.

Gergely, I., and Szentiványi, P. (*Editors*). 1990. First international symposium on walnut production, Budapest, Hungary, 25–29 September 1989. International Society for Horticultural Science, Wageningen, Netherlands. 369 pp.

Germain, E. (*Editor*). 2001. Proceedings of the fourth international walnut symposium, Bordeaux, France, 13–16 September, 1999. International Society for Horticultural Science, Leuven, Belgium. 616 pp.

Gomes Pereira, J.A., Martins, J.M.S., and Pinto de Abreu, C. (*Editors*). 1997. Third international walnut congress, Alcobaça, Portugal, 13–16 June, 1995. International Society for Horticultural Science, Leuven, Belgium. 445 pp.

Kessler, K.J., Jr., and Weber, B.C. (*Coordinators*). 1979. Walnut insects and diseases: workshop proceedings, June 13–14, 1978, Carbondale, Illinois. North Central Forest Experiment Station, Forest Service, U.S. Dept. of Agriculture, St. Paul, MN. 100 pp.

McGranahan, G., and Leslie C. 1990. Walnuts (*Juglans*). *In* Genetic resources of temperate fruit and nut crops. *Edited by* J.N. Moore and J.R. Ballington, Jr. International Society for Horticultural Science, Wageningen, Netherlands. pp. 907–949.

North Central Forest Experiment Station. 1966. Black walnut culture; papers presented... August 2 and 3, 1966. North Central Forest Experiment Station, St. Paul, MN. 94 pp.

North Central Forest Experiment Station. 1973. Black walnut as a crop. U.S. Dept. Of Agriculture, Forest Service general technical report NC 4. North Central Forest Experiment Station, St. Paul, MN. 114 pp.

North Central Forest Experiment Station. 1982. Black walnut for the future. U.S. Dept. of Agriculture, Forest Service, North Central Forest Experiment Station, St. Paul, MN. 151 pp.

North Central Forest Experiment Station. 1985. How to start walnut trees from seed. U.S. Forest Service, North Central Forest Experiment Station, St. Paul, MN. 6 pp.

Ramos, D.E. 1985. Walnut orchard management. Cooperative Extension, University of California, Division of Agriculture and Natural Resources, Oakland, CA. 178 pp.

Ramos, D.E. 1997. Walnut production manual. University of California, Division of Agriculture and Natural Resources, Oakland, CA. 320 pp.

Rink, G. 1985. Black walnut. Revised edition. Forest Service, U.S. Dept. of Agriculture, Washington, DC. 7 pp.

Sambeek, J.W. van. (*Editor*). 1997. Knowledge for the future of black walnut: proceedings of the fifth black walnut symposium, Hammons Conference Center, Springfield, Missouri and Sho-Neff Plantation, Stockton, Missouri, July 28–31, 1996. North Central Forest Experiment Station, U.S. Dept. Of Agriculture Forest Service, St. Paul, MN. 256 pp.

Schlesinger, R.C., and Funk, D.T. 1977. Manager's handbook for black walnut. U.S. Dept. of Agriculture, Forest Service, North Central Forest Experiment Station, St. Paul, MN. 22 pp.

Thompson, B.S. 1976. Black walnut for profit. 2nd edition. Timber Press, Forest Grove, OR. 285 pp.

Thomsen, M.E., Davis, E.G., and Rush, B.R. 2000. Black walnut induced laminitis. Vet. Hum. Toxicol. **42**: 8–11.

Whittemore, A.T., and Stone, D.E. 1997. *Juglans*. *In* Flora of North America north of Mexico, Vol. 3. *Edited by* Flora of North America Editorial Committee. Oxford University Press, New York, NY. pp. 425–428.

University of California Integrated Pest Management Program. 1993. Integrated pest management for walnuts. 2nd edition. University of California, Statewide Integrated Pest Management Project, Division of Agriculture and Natural Resources, Oakland, CA. 96 pp.

U.S. Dept. of Agriculture. 1978. Growing black walnuts for home use. Revised edition. U.S. Dept. of Agriculture, Science and Education Administration, Washington, DC. 9 pp.

Williams, R.D. 1974. Planting methods & treatments for black walnut seedlings. North Central Forest Experiment Station, Forest Service, U.S. Dept. of Agriculture, St. Paul, MN. 6 pp.

Specialty Cookbooks

Bough, C. 1993. Cooking with black walnuts: recipes collected by Carole Bough. Missouri Dandy Pantry, Stockton, MO. 151 pp.

California Walnut Growers Association. 1930s. 100 delicious walnut recipes. Diamond Walnut Growers and California Walnut Growers Association, Los Angeles, CA. 31 pp.

California Walnut Growers Association. 1936. To win new cooking fame just add walnuts. California Walnut Growers Association, Los Angeles, CA. 31 pp.

California Walnut Growers Association. 1950–1959? Around the kitchen clock with walnuts. California Walnut Growers Association, Los Angeles, CA. 39 pp.

Carroll, J.P. 1993. California walnuts 7 day meal plan. Walnut Marketing Board, Sacramento, CA. Unpaginated.

Cunningham, M. 1984. California walnuts: talk of the town: a collection of delicious recipes, menus and tips featuring America's favorite nut. Walnut Marketing Board, Sacramento, CA. 35 pp.

Diamond Walnut Growers. 1950–1959? The new walnut cook book. Diamond Walnut Growers, Stockton, CA. 28 pp.

Diamond Walnut Growers. 1978. Diamond walnut recipe favorites. Diamond Walnut Growers, Stockton, CA. 112 pp.

Diamond Walnut Growers. 1990. The Diamond walnut collection. Diamond Walnut Growers, Stockton, CA. 80 pp.

Ryan, F. 1993. The walnut diet and fitness plan. Swift, Bromsgrove, Worcestershire, UK. 62 pp.

Toussaint, J.-L. 1998. The walnut cookbook. Ten Speed Press, Berkeley, CA. 206 pp.

Turner, M. 1978. The Napa Valley walnut cookbook: from soup to nuts. M. Turner, Rutherford, CA. 56 pp.

Walnut Marketing Board. 1993–2002? Cooking with taste: expanding your taste with California walnuts. Walnut Marketing Board, San Francisco, CA. 8 pp.

West Virginia Black Walnut Festival, Inc. 1980. Black walnut festival baking competition: award winning recipes. West Virginia Black Walnut Festival, Inc., Spencer, WV. 130 pp.

WATERMELON

Family: Cucurbitaceae (gourd family)

Names

Scientific name: *Citrullus lanatus* (Thunb.) Matsum. & Nakai

- The word "watermelon" was first recorded in 1615, and is an apt description because the fruit contains 82% water by volume.
- The citron watermelon, described below, is also known as preserving melon (since it is principally eaten as a preserve) and stock watermelon (because it is fed to livestock).
- The genus *Citrullus* is based on the new Latin (not the classical Latin of Rome, but the Latin of Middle Ages Europe) *citrull*, in turn derived from Old French *citrulle*, a cucurbit (i.e., member of the gourd family).
- *Lanatus* in the scientific name *C. lanatus* is Latin for woolly, descriptive of the foliage.

Plant Portrait

Watermelons originated in the Kalahari Desert of southern Africa, where native peoples relied on them as a source of water. The great missionary–explorer David Livingstone (1813–1873) provided critical evidence indicating that the watermelon is African in origin. He observed wild watermelon plants growing over large area of central Africa. The plant is a ground-hugging annual vine. Wild watermelons growing throughout Africa are not much larger than apples. The fruit of cultivated varieties may be spherical, egg-shaped, or pear-shaped, generally greenish skinned, often striped or spotted. Size varies greatly, some fruits only a kilogram (2.2 pounds) in weight, some over 45 kg (100 pounds). The rind is thick but fragile. The flesh is generally reddish, although some are orange-, pinkish-, or yellow-fleshed, and sometimes white. The pulp is seedless in some varieties, but usually contains numerous smooth seeds that may be black, brown, white, green, yellow, or red. Watermelons are a very old crop, grown by the ancient Egyptians 5000 years ago. By the 1600s, watermelon had made its way to Great Britain, Spain, China, and beyond. It arrived in North America courtesy of European colonists and African slaves. The plant is cultivated in tropical, subtropical, and temperate regions, and generally thrives in hot areas. In commercial cultivation, the average yield is about one large watermelon for each vine. The main producers are China, Turkey, Russia, the United States, and Italy. The main US producers are California, Indiana, and Texas. There are hundreds of varieties. Watermelons are classified into four general categories:

- Picnic: 6.8–20.4 kg (15–45 pounds), red or yellow flesh, round or oblong, light to dark-green rind, with or without stripes.
- Ice-box: 2.26–6.8 kg (5–15 pounds), red or yellow flesh, round, dark or light-green rind.
- Seedless: 4.5–11.3 kg (10–25 pounds), red or yellow flesh, oval to round, light-green rind with dark-green stripes.
- Yellow flesh: 4.5–13.6 kg (10–30 pounds), oblong to long, light-green rind with mottled stripes, yellow to bright-orange flesh.

The citron or citron watermelon is a special type of watermelon, reminiscent of the wild African watermelon, and it should not be confused with the citron of the citrus family (*Citrus medica* L.), the peel of which is often candied and preserved. Citron watermelons resemble small watermelons, and are usually oval, up to 15 cm (6 inches) long, light green with darker green stripes, and smooth-surfaced. The flesh is white, and too tough to consume in the fresh state (the fruits are so resilient that they can usually be bounced on a floor with little chance of bursting). The rinds of all watermelons, especially those with a white rind, have traditionally been made into sweet pickles, but the citron watermelon is especially suited to the purpose.

Culinary Portrait

Watermelon is a widely appreciated, thirst-quenching dessert fruit, usually eaten out of hand. Watermelons are sweet because they contain (typically) about 8% natural sugar. The pulp is over 90% water, making it extremely refreshing. After the seeds are removed, the pulp can be puréed to make sorbet or jam. The juice is bottled commercially, and made into syrup and wine. Unripe watermelons are sometimes stuffed and cooked as a vegetable (like summer squash). The seeds are dried, roasted, salted, and consumed, especially in Asia. Occasionally, the seeds are ground up to prepare a sort of flour used to make bread or cereal. An edible oil is sometimes produced from the seeds, and used for cooking.

Black-seeded watermelon

Red-seeded watermelon

Varieties of watermelon (*Citrullus lanatus*), from Vilmorin-Andrieux (1885).

The fruit rind of citron watermelons is used in preserves, jellies, and to make pickles (marinades) or conserves. Preserves are prepared by boiling the rind for 20 minutes in a sugar and water solution that is strongly flavored, for example, with lemon and ginger. Like the true citron (the citrus fruit), the rind is sometimes candied.

Unlike some fruits, watermelons do not become sweeter after they have been picked.

How do you pick a good watermelon? On the vine, the stalk of the watermelon is usually green and fleshy before maturity, and dead and dried up when the melon is ripe. But in a supermarket, aside from tasting it is very difficult to tell when the fruit is ripe.

- Appearance: good-quality watermelons are firm, evenly shaped, and heavy for their size; the "ground spot" (where the melon rested on the ground) turns yellow on ripening, and watermelons are often whitish or pale green on the ground side, but the rind generally should have a healthy sheen. If the ground spot is not apparent, it may mean the fruit was picked prematurely.
- Sound tests: Thump the melon with the knuckles; a deep, dead, muffled "hollow thunk" indicates a ripe melon, while a sharp, metallic, ringing sound indicates that the melon is not ripe. Equally, the fruit should have a deep-pitched tone when slapped with an open palm. When using the open palm test, avoid melons with a high-pitched tone or a dead, thudding sound. A high-pitched tone indicates green or underripe product. A dull sound or dead thud indicates an overripe melon. Slap a number of melons one after the other and you will be able to hear the differences in tone. Small melons are the hardest to test for ripeness, since most will sound green even if they are ripe.

Once harvested, watermelons are best stored in a refrigerator, as the flesh becomes dry and fibrous at room temperature. Also, watermelons taste best when chilled. Cut portions should be covered with plastic wrap to prevent drying out and absorption of odors from other foods.

Culinary Vocabulary

- "August ham" is an old Southern United States name for watermelon, the phrase reflecting the time of availability and the large size.

Citron watermelon (*Citrullus lanatus*), from Bailey (1900–1902).

Field of watermelons (*Citrullus lanatus*), from Henderson (1890).

Curiosities of Science and Technology

- Abraham Lincoln (1809–1865, 16th president of the United States) was asked to participate in a ceremony on Aug. 27, 1853, that gave Lincoln, Illinois, its name. He christened the new town by taking a ripe melon from a wagon, breaking it open, and squeezing the juice on the ground.
- During the American Civil War (1861–1865), the Confederate Army boiled down watermelons to produce sugar and molasses.
- The watermelon was one of the first crops in which disease resistance was transferred from an inedible wild plant to a fruit crop. W.A. Orton in the 1900s transferred Fusarium wilt resistance from wild watermelon to the cultivated type.
- Hope, Arkansas, birthplace of former president Bill Clinton, claims it is the "watermelon capital of the world," based on a world-record melon grown there more than 60 years ago. Cordele, Georgia, also calls itself the Watermelon Capital of the World. It has been said that there are way too many "Watermelon Capitals" in the United States, and that the claims are hazy, conflicting, and hard to care about.
- It has been suggested that the watermelon should be declared to be the "national fruit" of Mexico, as the green skin, white seeds, and rind of some varieties, and red flesh reflect the colors of the Mexican flag. However, the cactus pear [*Opuntia ficus-indica* (L.) Mill.], which produces an edible fruit, has been adopted as a formal symbol of Mexico.
- *The Guinness Book of World Records 1996* lists a watermelon weighing 118.8 kg (262 pounds), grown by B. Carson of Arrington, Tennessee, in 1990.
- It has been claimed that the world's largest "man-made watermelon" is a water tower built in 1999 in Luling, Texas, painted to look like a watermelon. It has a diameter of 17 m (56 feet), and a weight of 1906 metric tons (2101 tons) when full.
- The average American eats 7.3 kg (16 pounds) of watermelon in a year.
- Some South American indigenous people wear watermelon rinds around their foreheads to relieve the pain of migraine headaches.
- In June 2001, about 400 "square" (cubic) watermelons were produced in Zentsuji on the western Japanese island of Shikoku, by growing the fruit in glass moulds. While their shape was ideal for storage in home refrigerators, their high cost (about $85) made them a luxury novelty.
- A standard commercial watermelon may contain as many as 1000 seeds.
- Since seeds cannot be saved from seedless watermelons, how are they obtained? Seedless watermelons are produced by hybridizing normal watermelon plants (with two sets of chromosomes) with plants that have had their chromosome number doubled so that they have four sets of chromosomes (accomplished by exposing seedlings to the chemical colchicine). The resulting seeds are triploid (i.e., they have three sets of chromosomes), and while capable of producing plants with fruit, they are

sterile and cannot produce new seeds. Seeds for growing seedless watermelons are available from many major seed companies. Unlike watermelons, which require fertilization before the fruit develops, some fruit crops have varieties that can develop without fertilization (e.g., banana, fig, melon, pineapple), a phenomenon called parthenocarpy.

Key Information Sources

Allred, A.J., and Lucier, G. 1990. The U.S. watermelon industry. U.S. Dept. of Agriculture, Economic Research Service, Commodity Economics Division, Washington, DC. 73 pp.

Andrus, C.F., Seshadri, V.S., and Grimball, P.C. 1971. Production of seedless watermelons. U.S. Agricultural Research Service, Washington, DC. 12 pp.

Beattie, J.H., and Doolittle, S.P. 1951. Watermelons. Revised edition. U.S. Dept. of Agriculture, Washington, DC. 30 pp.

Brewer, J.W. 1974. Watermelon pollination by honey bees. Colorado State University Experiment Station, Fort Collins, CO. 5 pp.

Bright, L. 1978. Producing giant watermelons—an Arkansas tradition. Etter Printing Co., Hope, AR. 58 pp.

Bush, A. 1978. Citron melon *Citrullus lanatus* var. *citroides* for cash and condiment. Econ. Bot. **32**: 182–184.

Doolittle, S.P. 1962. Commercial watermelon growing. Agricultural Research Service, U.S. Dept. of Agriculture, Washington, DC. 31 pp.

Elmstrom, G.W., and Maynard, D.N. 1995. Growing seedless watermelons. Revised edition. Cooperative Extension Service, Institute of Food and Agricultural Sciences, University of Florida, Gainesville, FL. 4 pp.

Fehér, T. 1993. Watermelon, *Citrullus lanatus* (Thunb.) Matsum. & Nakai. *In* Genetic improvement of vegetable crops. *Edited by* G. Kalloo and B.O. Bergh. Pergamon Press, New York, NY. pp. 295–311.

Fulks, B.K., Scheerens, J.C., and Bemis, W.P. 1979. Natural hybridization of two *Citrullus* species: wild watermelon, *Citrullus colocynthis* and the domesticated watermelon, *Citrullus lanatus*. J. Hered. **70**: 214–215.

Fursa, T.B. 1981. Intraspecific classification of water-melon under cultivation. Kulturpflanze, **29**: 297–300.

Lee, G.S. 1989. Growing giant pumpkins & watermelons. Cooperative Extension Service, University of Arkansas, Fayetteville, AR. 26 pp.

Lee, S.J., Shin, J.S., Park, K.W., and Hong, Y.P. 1996. Detection of genetic diversity using RAPD-PCR and sugar analysis in watermelon (*Citrullus lanatus* (Thunb.) Mansf.) germplasm. Theor. Appl. Genet. **92**: 719–725.

Levi, A., Thomas, C.E., Keinath, A.P., and Wehner, T.C. 2001. Genetic diversity among watermelon (*Citrullus lanatus* and *Citrullus colocynthis*) accessions. Genet. Resour. Crop Evol. **6**: 559–566.

Longbrake, T., Parsons, J., and Roberts, R.E. 1980. Keys to profitable watermelon production. Texas Agricultural Extension Service, Texas A & M University System, College Station, TX. 7 pp.

Maynard, D.N. 1992. Watermelon production guide for Florida. Florida Cooperative Extension Service, Institute of Food and Agricultural Sciences, University of Florida, Gainesville, FL. 65 pp.

Maynard, D.N. 2001. Watermelons: characteristics, production, and marketing. ASHS Press, Alexandria, VA. 227 pp.

Mohr, H.C. 1986. Watermelon breeding. *In* Breeding vegetable crops. *Edited by* M.J. Bassett. AVI Publishing Company, Westport, CT. pp. 37–66.

Orton, W.A., and Meier, F.C. 1922. Diseases of watermelons. U.S. Dept. of Agriculture, Washington, DC. 31 pp.

Paje, M.M., and Vossen, H.A.M. van der. 1993. *Citrullus lanatus* (Thunberg) Matsum. & Nakai. *In* Plant resources of South-East Asia, 8, vegetables. *Edited by* J.E. Siemonsma and K. Piluek. Pudoc Scientific Publishers, Wageningen, Netherlands. pp. 144–148.

Parris, G.K. 1952. Diseases of watermelons. University of Florida Agricultural Experiment Station, Gainesville, FL. 48 pp.

U.S. Dept. of Agriculture. 1966. Watermelons for the garden. Revised edition. U.S. Dept. of Agriculture, Washington, DC. 8 pp.

Zamir, D., Navot, N., and Rudich, J. 1984. Enzyme polymorphism in *Citrullus lanatus* and *Citrullus colocynthis* in Israel and Sinai. Plant Syst. Evol. **146**: 163–170.

Zhang, J. 1996. Breeding and production of watermelon for edible seed in China. Cucurbit Genet. Coop. Rep. **19**: 66–67.

Specialty Cookbooks

Brattleboro Memorial Hospital Auxiliary. 1978. Watermelon and other good things to eat. Brattleboro Memorial Hospital Auxiliary, Brattleboro, VT. 85 pp.

Ficklen, E. 1984. Watermelon. American Folklife Center, The Library of Congress, Washington, DC. 64 pp.

Lynch, L.T. 1991. Mastering the art of carving watermelon centerpieces. Dunlap International Press, Pompano Beach, FL. 109 pp.

Massie, L.B., and Massie, P. 1990. Walnut pickles and watermelon cake: a century of Michigan cooking. Great Lakes Books edition. Great Lakes Books, Wayne State University Press, Detroit, MI. 353 pp.

Matson, H. 1992. Delicious watermelon recipes. Kansas State Board of Agriculture, Marketing Division, Topeka, KS. 1 sheet.

Swansboro United Methodist Women. 1993. Collard greens, watermelons and "Miss" Charlotte's pie: a cookbook. Swansboro United Methodist Women, Swansboro, NC. 381 pp.

Texas Dept. of Agriculture. 1900–1985? Texas watermelon. Texas Dept. of Agriculture, Austin, TX. 8 pp.

WHEAT

Family: Poaceae (Gramineae; grass family)

Names

Scientific names: *Triticum* species

Some of the most important wheat species are listed below.

- Common wheat (bread wheat)—*T. aestivum* L. subsp. *aestivum* (*T. sativum* Lam., *T. vulgare* Vill.) [Kernels may be red or white, hard or soft. This is the source of most of the wheat varieties cultivated in North America, and represents 90% of the world's wheat production.]
- Club wheat—*T. aestivum* subsp. *compactum* (Host) MacKey (*T. compactum* Host) [The acreage is limited.]
- Durum wheat (macaroni wheat)—*T. turgidum* L. subsp. *durum* (Desf.) Husn. (*T. durum* Desf.) [Includes most spring wheats with white or red kernels, used mainly for the manufacture of semolina, which is made into macaroni, spaghetti, and related products. The acreage is small compared to common wheat, but this is the second most important species in North America, after common wheat.]
- Einkorn—*T. monococcum* L. [A rarely grown primitive kind of wheat tracing to prehistoric times, with both winter and spring forms, and pale red seeds.]
- Emmer—*T. turgidum* subsp. *dicoccon* (Schrank) Thell. [*T. dicoccon* (Schrank) Schübl.] [An ancient cultivated cereal, grown either as a winter or spring wheat, has red or white kernels, rarely cultivated in North America.]
- Polish wheat (rivet wheat, cone wheat)—*T. turgidum* subsp. *polonicum* (L.) Thell. (*T. polonicum* L.) [Spring wheats with tall stems, hard kernels, grown extensively in Mediterranean countries, but not in North America.]
- Poulard wheat—*T. turgidum* L. [Related to durum wheat, cultivated extensively in Mediterranean countries, but not in North America. Often called "kamut," the ancient Egyptian word for wheat.]
- Spelt—*T. aestivum* L. subsp. *spelta* (L.) Thell. (*T. spelta* L.) [Winter or spring wheats with red kernels, rarely cultivated in North America, once cultivated extensively in Germany, Switzerland, and France, the hulls are difficult to remove but, once hulled, spelt is a good replacement for rice.]

- The English word "wheat" may derive from the Old High German word *weizzi*, meaning wheat, which in turn seems based on *hwiz* or *wiz*, white, because of the white flour made from wheat.
- "Corn" means wheat in England (where corn is called "maize"). In the Bible, corn means grain.
- The genus *Triticum* is based on the same word denoting wheat in classical Latin. It has been suggested that the word was coined in classical times from the Latin *tritum*, rubbed, an allusion to its being originally rubbed down (threshed) to obtain the grain.
- Einkorn (*T. monococcum*) is a German name meaning "one grain." This is also called German wheat, and is indeed still occasionally cultivated in Germany, as well as in other parts of Europe.
- *Aestivum* in the scientific name *T. aestivum* is Latin for "of summer," reflecting the annual nature of the species, which lives for just one summer.
- *Durum* in the scientific name *T. turgidum* L. subsp. *durum* is Latin for hard, reflecting the fact that this kind of wheat produces hard grains.
- *Turgidum* in the scientific name *T. turgidum* is Latin for swollen or inflated, probably originally intended as a description of the grain.
- *Monococcum* in the scientific name *T. monococcum* is Latin for one-seeded, descriptive of the fruiting head.

Plant Portrait

Wheat is the most important food grain of the temperate zones, and indeed rivals rice as the most important food plant in the world. Wheat has been consumed by mankind since the beginning of agriculture. Wild grasses related to modern cultivated wheats were harvested dating back at least 20 000 years, and in time, certain types were cultivated and selected. Common wheat, the predominant modern cultivated wheat, was selected probably somewhere in the Near East, likely in the Fertile Crescent, located in the upper reaches of the Tigris-Euphrates drainage basin. Wheat spread throughout Europe not later than the Stone Age. Today, the most significant wheat-producing nations are Russia, China, India, Pakistan, France, the United States, Canada, Argentina, and Australia. The "Durum Triangle" is a region in north-central North Dakota, which produces most of the durum wheat grown in the United States. The location is superbly adapted to production of durum wheat, and is one of the most intensely farmed areas in the world.

Modern domesticated wheats have been so altered by human selection during the last 10 000 years that they are unable to thrive in the wild. It has been pointed out that not only has the course of wheat evolution been fundamentally altered by humans, but the reverse has also occurred—i.e., human evolution has been altered by wheat. (The biological phenomenon of unrelated species cooperating in ways that promote their survival is called "mutualism.") Wheat, which today feeds a third of human beings, fundamentally altered human physical and social development. The great civilizations of Mesopotamia, Egypt, Greece, and Rome came into existence because wheat (and other cereals) allowed the development of large concentrations of people who could live on stored grains, whereas formerly seasonal availability of foods made it necessary for humans to exist in small bands of nomadic hunter–gatherers, whose physical characteristics were necessarily more robust than those of modern humans. The large concentrations of people that developed because of the cereals led to specialized crafts, extraordinary technology, the arts, and religion. Different cereals played the role of wheat in other parts of the world—rice in Asia, sorghum in Africa, and corn in the Americas.

Wheat plants are annual grasses, with stems growing from 30 to 120 cm (1–4 feet) or more in height, depending on kind and growing conditions. Color of the kernel varies from dark red through light brown, and commercial wheats are accordingly classified as red, white, cream, or yellow (and amber in durum wheat). The wheat kernel is made up of three main parts: (1) an outer covering, the bran—separated from the flour during most milling processes, and constituting about 12% of the kernel weight; (2) a starchy portion (endosperm), making up about 85% of the kernel, and most of white flour; (3) the germ (the embryo), which makes up only about 2.5% of the kernel, and like the bran, is separated out in most milling processes. Wheat is grown for its seeds (grain; see Glossary), used whole or ground. The coarsest degree

Common wheat (*Triticum aestivum* subsp. *aestivum*), from Baillon (1876–1892).

of milling is called "whole wheat," "whole meal," or "Graham flour." It is often stoneground, and is brown ("wholemeal" is the common term in Britain, while "whole wheat" and "Graham" are more often used in the United States). It contains the germ, endosperm, and bran. Although considered especially healthy, whole wheat has limited shelf life because of the presence of oils in the germ, which become rancid, and the bran may cause intestinal problems in some individuals. At the other extreme, fine white flour (often whitened with the addition of bleaching chemicals) contains little or no germ and bran, and relatively limited vitamin content (but is generally enriched with thiamine, nicotinic acid, iron, riboflavin, calcium, and phosphorus, the vitamins and elements removed during milling).

In addition to its culinary uses, wheat is used to manufacture industrial alcohol, and a variety of industrial products. Wheat is also cultivated as a livestock feed.

The two main classes of wheat are winter wheat and spring wheat. Winter wheat is grown in temperate regions, and is sown in the fall. Spring wheat is sown in the spring, and is grown in areas where winters are extremely cold, as in Canada and the northern United States. There are both "hard" and "soft" varieties of both winter and spring wheat (the different culinary usages of hard and soft wheat are discussed below).

Durum wheat production and consumption is well established in the hot dry regions around the Mediterranean, such as North Africa, southern Europe (especially Italy, Spain, France, and Greece), Turkey, and Syria. Other areas of production include Kazakhstan, India, Australia, and Mexico. In North America, western North Dakota and southern Saskatchewan are the major growing regions, with small amounts produced under irrigation in the Arizona and California deserts.

Culinary Portrait

Wheat flour is the world's chief breadmaking flour. Indeed, "flour" means wheat flour, unless qualified as coming from some other sources, such as corn flour or pea flour. The grain is also used to make alcoholic beverages. Wheat is classified into hard and soft varieties. Extra gluten (made up of proteins) in hard wheat makes bread dough sticky, so that it rises better during baking, and consequently, hard wheat is used chiefly for making bread. By contrast, the "soft" wheats (mostly cultivars of common wheat) tend to be rich in starch and low in gluten, lack stickiness, and are used mainly for cakes, cookies, biscuits, and pastry. Hard and soft wheat are often mixed together and sold as "all-purpose flour." "Prepared" or "self-raising" flour contains baking powder. Macaroni wheats (mostly durum wheat varieties) are more suitable for the manufacture of spaghetti and pasta in general than for bread. Wheat is the only grain with sufficient gluten content to make a raised or leavened loaf of bread, although proteins in rye serve essentially the same purpose, so that both wheat and rye can be used for leavened bread. The gluten formed by wheat proteins when combined with water results in a sticky adhesive mass that can be rolled and pummeled while holding together and expanding to contain rather than release the gases produced by yeast. Wheat gluten is used in the manufacture of the flavor enhancer monosodium glutamate. "Seitan" is another interesting product made from wheat gluten, which in many respects is like tofu produced from soybean. Seitan, which can be obtained from natural health food stores, is an excellent mimic of meat, with similar taste, texture, and cooking properties. It may be added to soups, stews, sauces, stuffings, sandwiches, and similar dishes.

In Europe and North America, pasta products such as spaghetti and macaroni are generally produced exclusively from durum wheat. Common wheat can be used to make pasta, but discriminating pasta consumers prefer pasta made from 100% durum wheat. In North Africa, durum is preferred for the production of couscous, a staple food in the region. Couscous is small balls of semolina steamed and prepared in a similar manner to rice. Pasta made from durum semolina maintains a desirable firm texture during cooking, and has a natural amber color that is associated with good quality. Pasta made from common wheat, even if made from high protein hard red spring wheat, tends to absorb more water in cooking and produce a softer, stickier product, and is white unless artificial color is added. In Africa, traditional breads are made from durum wheat.

The coarsest wheat grains are used for pilafs (i.e., they are lightly sautéed in hot fat and then a hot liquid stock is added, and the mixture is simmered without stirring until the grains absorb the liquid). Medium grains are employed for cereals, and the finest grain for various forms of tabbouli, a cold grain salad (described below). Farina or "cream of wheat" has had most of the bran and germ removed, and is usually encountered in the form of a breakfast cereal. Cracked wheat, rolled wheat, and wheat flakes are other forms of wheat commonly used in baked products.

Celiac disease is a condition where proteins consumed from the closely related cereals wheat, rye, barley, and triticale damage the small intestines. Celiac disease affects one in 250 people, usually beginning in infancy. Common symptoms include constipation, diarrhea, bloating, gas, and burping, but there are many other problems that may develop. Treatment involves removing gluten and its related proteins from the diet. Rice, corn, flax, quinoa, tapioca, amaranth, potato, nuts, and beans are considered safe. Pure oats are also safe, at least for most people with celiac disease, but are difficult to obtain commercially without wheat and other gluten sources. Specialty stores sell gluten-free cereals, bread, and pastries.

Some of the rarer cultivated wheats, including einkorn, emmer, and spelt, produce food products that are relatively hypoallergenic in comparison to those made from the common bread wheats. Some individuals suffer allergic reactions to common bread wheat products, including wheat gluten, but their reactions are often absent when consuming spelt products.

Culinary Vocabulary

- By definition, cereals are grasses yielding seed crops. More broadly, prepared foodstuffs from grains (particularly from cereal grasses, but possibly from other seed crops) are also known as cereals. The word "cereal" comes from the Latin *cerealis*, meaning of Ceres. Ceres was the Greek goddess of sowing, reaping, harvesting, and of agriculture in general She was depicted as holding a torch, ears of corn, wheat, and poppies, and it was said she drove a chariot drawn by winged dragons.
- "Kibbled wheat" is coarsely ground wheat, used in meusli (musli, a variously concocted breakfast cereal) and as a sprinkling on top of bread and rolls.
- Gluten in contact with water forms a sticky mass, and indeed the word "gluten" is based on the Latin *glu*, glue. Gluten, formed by the proteins in wheat, is responsible for the texture, appearance, and volume of dough. Gluten gives dough the ability to rise by allowing it to retain gas released by a leavening agent such as baking powder or yeast.
- Where did the expression "baker's dozen" (meaning 13) come from? In 1266, the English parliament passed a law that strictly regulated bread weight. Because it was difficult to make loaves of uniform weight, bakers added a thirteenth loaf to every dozen they sold to ensure they wouldn't be prosecuted.
- The Reverend Sylvester Graham (1794–1851) was an eccentric Presbyterian minister from Connecticut who promoted temperance, vegetarianism, and especially the use of the whole grain of coarsely ground wheat in making flour, hence "Graham flour" and "Graham crackers." He was correct in recognizing the merits of high fiber instead of refined white flour, particularly the importance of retaining the bran. During the 19th century, vegetarianism was named Grahamism after him. (Note that the American expression "Graham crackers" is equivalent to the British "digestive biscuits.")
- "Shredded Wheat" was patented in 1893 by Henry D. Perky of Colorado and William Ford of Watertown, New York. When the original patent expired, the name "shredded wheat" became generic, and could be used by anyone, as ruled by a Federal U.S. court on 14 November 1938 (Kellogg Co. vs. National Biscuit Co., 305 U.S. 111).
- In old lunch-counter slang, "baled hay" meant shredded wheat.
- "Frumenty" (also furmenty) is an old English form of whole-wheat porridge flavored with milk, sugar, salt, eggs (or egg yolks), and sometimes saffron.
- Pablum (in ancient Rome, *pabulum* simply meant food) was invented in 1930 by three Toronto doctors at the Hospital for Sick Children, and patented in the early 1930s by Mead Johnson & Company of Evansville, Indiana. The baby cereal, a mixture of wheat, dehydrated alfalfa leaf, and powdered beef bone, was an instant hit with parents, although it is no longer sold today. The name for this bland, mushy cereal has since come to mean bland and mushy.
- Couscous and bulgar (bulgur, bulghur, bulghur wheat, bulghur cracked wheat) are products of durum wheat. Neither word is found in the average dictionary. Couscous (pronounced KOOS-koos) is the name for both the coarsely ground semolina pasta that is a staple side dish in North African cuisine, and also for the spicy meat and vegetable stew often served with it (which is considered the national dish of Algeria, Morocco, and Tunisia). Couscous has become increasingly popular in recent years as a gourmet item. "Instant couscous" is precooked, requiring only a short time to heat. Bulgar (bulgur, bulghur; pronounced BUHL-guhr, and tracing to the Turkish *burghal*), aside from denoting a Bulgarian, is a quick-cooking form of whole wheat, which when cooked has a nutty flavor and chewy texture. It is made by soaking and

Marquis wheat, the most important variety in Canada in the early 20th century. From the Agriculture and Agri-Food Canada photograph collection.

cooking the whole-wheat kernel, drying it and then removing part of the bran and cracking the remaining kernel into small pieces. Its uses are numerous, from salads to soups and from breads to desserts. Bulgar is widely used in vegetarian cooking, and is found particularly in the Lebanese tabbouli salad, described below. Bulgar absorbs twice its volume in water and can be used in place of rice in any recipe.

- Tabbouli (pronounced tah-BOO-lee, and variously spelled, for example, tabbouleh, tabouli, tabbouley, tabouly) is a fine grind of bulgar. The word also refers to a Middle Eastern salad made with bulgar, parsley, and olive oil, and often mixed with mint, onion, tomato, and lemon juice.
- Semolina is a kind of coarse wheat flour made from durum or other hard wheat, and used for such preparations as macaroni, spaghetti, and vermicelli. (Semolina may mean any very coarse flour, such as rice semolina or corn semolina, but unqualified, the word always means semolina from wheat.) The North American hot cereal "Cream of Wheat," which is made from very fine semolina, may be called "semolina" in Britain.
- "Grano" (the Italian word for wheat) refers to polished durum wheat kernels that have been polished (i.e., the bran, or at least some of it has been removed). This is traditionally cooked like rice to produce a nutty-tasting side dish with al dente texture (i.e., firm but not hard).
- *Fu* (foo) is a preparation of gluten flour from wheat. It is high in protein and low in starch, and is popular in Asian cuisine. In the West, it is available from Asian grocery stores, usually in dried form. In Japan, fresh fu is used in many dishes, including those cooked at the table, such as sukiyaki. Dried fu is used like croutons, or is softened in warm water for 5 minutes, causing it to puff up like spongy bread, and then the water is pressed out and the fu is added to soups and noodle dishes. The Chinese counterpart is *seitan* (Chinese style wheat gluten).

Curiosities of Science and Technology

- Egyptian pharaohs were buried with wheat to help nourish them on their voyage into the afterlife. "Mummy wheat" refers to wheat that allegedly originated from the ancient tombs of Egyptian mummies. In the early 19th century, following an expedition by Napoleon's army to Egypt, there were reports in Europe that wheat grains from tombs up to 6000 years old were found to be extraordinarily productive. The seeds were said to produce as many as seven fat ears of regular wheat, and sold for as much as $10 apiece at today's prices. In fact, attempts to resuscitate wheat grains from ancient tombs have not been successful, and it has been demonstrated that although Egyptian tombs are excellent for keeping seeds alive, wheat seeds can not live for more than a century under such conditions.
- In ancient Greece, grain was often used as money. In 625 BC, metal coins were introduced, stamped with a likeness of an ear of wheat. The association of wheat and money has carried down to modern times. In most of the European languages, money is colloquially called wheat (e.g., in French, *blé*). A similar linguistic parallelism in American English is the use of "dough" to mean money.
- Wheat is traditionally a symbol of fruitfulness. Before rice became the standard cereal ceremoniously showered on a bride and groom, wheat was often used.

- During the Middle Ages, some convicts were put into a prison cell with 2 bushels of wheat. When they had eaten all of the wheat, they were released.
- An 1899 recipe for making your own face powder: sift a quarter pound of wheat starch through a fine sieve or a piece of lace; add 8 drops of rose oil, 30 drops of lemon oil, and 15 drops of bergamot oil, mix thoroughly.
- Shredded wheat was the first ready-to-eat commercial breakfast cereal. It was introduced in Denver in 1893 by Henry D. Perky (died 1908). However, the first modern breakfast cereal was probably developed in 1863 by Dr. James C. Jackson (1811–1895; well known as a slavery abolitionist) of Danville, New York. Jackson prepared a heavy dough out of Graham flour and water, baked it slowly until the loaves were very dry, broke these up into small chunks, baked them again, and finally ground the brittle chunks into smaller pieces. He called this preparation "Granula," which closely resembles Post's Grape Nuts, and it was served for breakfast after soaking overnight in milk. John Harvey Kellogg (see the following item) in 1877 created a breakfast food made of wheat, oats, and cornmeal, and took up the name Granula for it, but when he was sued by Dr. Jackson for infringing on the use of his word, Kellogg changed the name to "Granola." Kellogg also had contact with Perkey, the inventor of shredded wheat, and tried to form a partnership with him to make the product, but Perkey decided not to cooperate with Kellogg. Rebuffed, Kellogg decided to make his own cereal, a decision which eventually led in 1906 to his creation of corn flakes flavored with barley malt.
- John Harvey Kellogg (1852–1943), a physician in St. Louis, and the health-food faddist of breakfast-cereal fame, introduced his first precooked, flaked cereal, made with wheat, in 1894. His invention was inspired by a patient who asked for payment of $10 to replace her false teeth after she consumed a hard bread formula that he had provided. (Kellogg went on to produce about 60 more cereals. For additional food introductions by him, see Peanut and Soybean.)
- Charles W. Post (1854–1914) was an American inventor and salesman who was treated for chronic illnesses by Dr. Kellogg in his sanitarium in Battle Creek Michigan, starting in 1891. After 10 months without progress, Post left, and was "cured" by a local Christian Science advocate who told him to eat what he pleased. In 1895, Post marketed Postum, a coffee substitute made of wheat, bran, and molasses, which was quite similar to the caramel coffee served at Kellogg's sanitarium. Post then created an immensely successful version of Jackson's wheat-based cereal, Granula (see above), which he called "Elijah's Manna." When it was condemned in pulpits across American and England for exploiting the Bible, it was renamed in 1908 as "Post Toasties," the same name under which it is marketed today by General Foods.
- Cream of Wheat boxes carry the image of a kindly black chef holding a steaming bowl of cereal. Rastus, the Cream of Wheat chef, was first drawn about 1890 by Emery Mapes, one of the owners of the Diamond Milling Company of North Dakota. This illustration was used until the 1920s, when a Chicago waiter was paid $5 to pose in a chef's hat and jacket. The face of that unknown man has been featured on Cream of Wheat boxes with only slight modifications until today.
- Filippo Tomaso Marinetti (1876–1944) was a well-known Italian futurist (i.e., he predicted future societal and technological trends), many of whose ideas are absurd by modern standards. With the blessing of his friend, the dictator Benito Mussolini, and his ruling Fascists, he attacked pasta as a food for dull, unimaginative people, promoting locally grown rice over the expensive imported wheat needed for the traditional staple of Italian cooking. While Marinetti succeeded in lowering the esteem with which pasta was held by some Italians, this was only temporary.
- In the early 1950s, there was a radio show for children, transformed into a TV show from 1953 to 1957, called *Sergeant Preston of the Yukon* (this evolved from *The Challenge of the Yukon*, lasting from 1947 to 1953, which in turn came from earlier radio versions dating back to 1938). The leading characters were Sergeant Preston of the Royal Canadian Northwest Mounted Police, his horse Rex, and his dog Yukon King. The series was sponsored by Quaker cereals, and to promote two of their products associated with the show, Quaker Puffed Rice and Quaker Puffed Wheat, the company carried out one of the most famous advertising initiatives in history. Quaker Oats paid $1,000 for 7.7 ha (19 acres) of Yukon land near Dawson, divided the land into square inches, and printed 21 million "deeds" to the parcels. In January 1955, the deeds were packaged in 21 million boxes of Quaker Puffed Wheat and Puffed Rice. Consumers bought up all 21 million

boxes within 2 weeks. The Klondike Big Inch Land Company, the dummy corporation the cereal company had set up in Illinois to own and distribute the land, went out of business, and the land that was the site of the millions of parcels was repossessed by the Yukon government in 1965 for nonpayment of an outstanding $37.20 tax bill. Some memorabilia experts claim the old deeds are now worth as much as $100.

- One bushel of wheat contains approximately 1 million individual kernels.
- A modern combine can harvest 1000 bushels of wheat per hour.
- One bushel of wheat yields enough flour for 73 one-pound (0.45 kg) loaves of white bread.
- The wheat that produces a loaf of bread may have required 2 tons of water to grow.
- Norman Borlaug was awarded the Nobel Peace Prize in 1970. As the "Father of the Green Revolution," he developed high-yielding dwarf strains of wheat while working at the Rockefeller-financed CIMMYT Agricultural Station in Mexico City. Use of such seed has allowed tropical countries to double their wheat productivity. Along with improvements in rice productivity at a similar center in the Philippines and other crops at yet more agricultural stations, the "Green Revolution" came into being, contributing to much higher production of food in such countries as India, Pakistan, and Mexico. Unfortunately, the high-yielding varieties require high input of fertilizers, contributing to environmental degradation and higher use of fossil fuels. And, unfortunately, increase in human population has continued, wiping out gains in reducing world hunger.
- Wheat is harvested somewhere in the world every month of the year.
- Puffing cereals is rather like popping corn, and has been known for more than a century. Cereals made from puffed wheat and rice are now breakfast staples worldwide. The puffing process involves placing the grain in a sealed chamber and heating until the pressure rises. Then, the chamber or "puffing gun" is suddenly opened. With the pressure relieved, the water vapor in the grain expands rapidly, blowing up the grain to many times the original size (for wheat, 816 times; for rice, 6–8 times). Finally, the puffed grains are toasted and dried until crisp.
- Wheat prices in 17th century England have been shown to have been influenced by sunspot numbers (L. Pustilnik and Y. Din, New Scientist, 20 Dec. 2003). Low sunspot activity corresponded to peaks in the price of wheat, indicating a lower crop yield, possibly because sunspots change level of cloud cover and accordingly crop productivity.

Key Information Sources

Aykroyd, W.R., and Doughty, J. 1970. Wheat in human nutrition. Food and Agriculture Organization of the United Nations, Rome, Italy. 163 pp.

Cook, R.J., and Veseth, R.J. 1991. Wheat health management. APS Press, St. Paul, MN, 152 pp.

Damania, A.B. (*Editor*). 1993. Biodiversity and wheat improvement. Wiley, New York, NY. 434 pp.

Evans, L.T., Peacock, W.J., and Frankel, O.H. (*Editors*). 1981. Wheat science, today and tomorrow. Cambridge University Press, Cambridge, UK. 290 pp.

Fabriani, G., and Lintas, C. (*Editors*). 1988. Durum wheat: chemistry and technology. American Association of Cereal Chemists, St. Paul, MN. 332 pp.

Feldman, M., and Sears, E.R. 1981. The wild gene resources of wheat. Sci. Am. **244**(**1**): 102–112.

Feldman, M, Lupton, F.G.H., and Miller, T.E. 1995. Wheats. *In* Evolution of crop plants. 2nd edition. *Edited by* J. Smartt and N.W. Simmonds. Longman Scientific & Technical, Burnt Mill, Harlow, Essex, UK. pp. 184–192.

George Washington University. 1971. Bibliography of wheat. Scarecrow Press, Metuchen, NJ. 3 vols.

Ginkel, M. van, and Villareal, R.L. 1996. *Triticum* L. *In* Plant resources of South-East Asia, 10, cereals. *Edited by* C.J.H. Grubben and S. Partohardjono. Backhuys Publishers, Leiden, Netherlands. pp. 137–143.

Gooding, M.J., and Davies, W.P. 1997. Wheat production and utilization: systems, quality, and the environment. CAB International, Wallingford, Oxon, UK, and New York, NY. 355 pp.

Hanson, H.E., Borlaug, N.E., and Anderson, R.G. 1982. Wheat in the Third World. Westview Press, Boulder, CO. 174 pp.

Heyne, E.G. (*Editor*). 1987. Wheat and wheat improvement. 2nd edition. American Society of Agronomy, Madison, WI. 765 pp.

Inglett, G.E. (*Editor*). 1974. Wheat, production and utilization. Avi Publishing Co., Westport, CT. 500 pp.

International Maize and Wheat Improvement Center (CIMMYT). 1985. Wheat for more tropical environments. CIMMYT, Mexico D.F., Mexico. 354 pp.

Kent, N.L. 1975. Technology of cereals with special reference to wheat. 2nd edition. Pergamon Press, Oxford, UK. 306 pp.

Klatt, A.R. (*Editor*). 1988. Wheat production constraints in tropical environments. International Maize and Wheat Improvement Center (CIMMYT), Mexico D.F., Mexico. 410 pp.

Knight, J.W. 1965. The chemistry of wheat starch and gluten, and their conversion products. L. Hill, London, UK. 156 pp.

Knott, D.R. 1989. The wheat rusts: breeding for resistance. Springer-Verlag, Berlin, Germany. 201 pp.

Line, R.F. 2002. Stripe rust of wheat and barley in North America: a retrospective historical review. Annu. Rev. Phytopathol. **40**: 75–118.

Lupton, F.G.H. (*Editor*). 1987. Wheat breeding: its scientific basis. Chapman and Hall, New York, NY. 566 pp.

McCorriston, J. 2000. Wheat. *In* The Cambridge world history of food. *Edited by* K.F. Kiple and K.C. Ornelas. Cambridge University Press, Cambridge, UK. pp. 158–174.

Peterson, P.D. 2001. Stem rust of wheat: from ancient enemy to modern foe. APS Press, St. Paul, MN. 156 pp.

Peterson, R.F. 1965. Wheat: botany, cultivation and utilization. L. Hill Books, Interscience Publishers, London, UK. 422 pp.

Pomeranz Y. (*Editor*). 1988. Wheat: chemistry and technology. 3rd edition. American Association of Cereal Chemists, St. Paul, MN. 2 vols.

Pomeranz Y. (*Editor*). 1989. Wheat is unique: structure, composition, processing, end-use properties, and products. American Association of Cereal Chemists, St. Paul, MN. 715 pp.

Posner, E.S. 2000. Wheat. *In* Handbook of cereal science and technology. 2nd edition. *Edited by* K. Kulp and J.G. Ponte, Jr. Marcel Dekker, New York, NY. pp. 1–29.

Royo, C., Nachit, M., and Fonzo, N. di. (*Editors*). 2005. Durum wheat breeding. Food Products Press, Binghamton, NY. ca. 299 pp.

Satorre, E.H., and Slafer, G.A. 1999. Wheat: ecology and physiology of yield determination. Food Products Press, New York, NY. 503 pp.

Saunders, D.A. (*Editor*). 1991. Wheat for the non-traditional warm areas. International Maize and Wheat Improvement Center (CIMMYT), Mexico D.F., Mexico. 549 pp.

Saunders, D.A., and Hetel, G.P. (*Editors*). 1994. Wheat in heat-stressed environments: irrigated, dry areas and rice-what farming systems. International Maize and Wheat Improvement Center (CIMMYT), Mexico D.F., Mexico. 402 pp.

Schofield, J.D. (*Editor*). 2000. Wheat structure, biochemistry and functionality. Royal Society of Chemistry, Cambridge, UK. 379 pp.

Shewry, P.R., and Tatham, A.S. (*Editors*). 2000. Wheat gluten. Royal Society of Chemistry, Cambridge, UK. 548 pp.

Stallknecht, G.F., Gilbertson, K.M., and Ranney, J.E. 1996. Alternative wheat cereals as food grains: einkorn, emmer, spelt, kamut, and triticale. *In* Progress in new crops. *Edited by* J. Janick. ASHS Press, Alexandria, VA. pp. 156–170.

Wiese, M.V. 1987. Compendium of wheat diseases. 2nd edition. American Phytopathological Society, St. Paul, MN. 112 pp.

Zohary, D., Harlan, J.R., and Vardi, A. 1969. The wild diploid progenitors of wheat and their breeding values. Euphytica, **18**: 58–65.

Specialty Cookbooks

Anonymous. 1974. Naturally good wheat germ cookbook. Avon, New York, NY. 144 pp.

Barger, C.D. 1979. Food from the kernel: over 380 original 100% whole wheat recipes. The Kernel, Inc., Downey, ID. 125 pp.

Beck, H.B., and Beck, A.R. 1986. Whole wheat cookery: treasures from the wheat bin. The Wheat Bin, Halstead, KS. 165 pp.

Bedrosian, S.K. 1980. Heritage wheat: the complete bulgur cookbook. Heritage Kitchens, Montvale, NJ. 108 pp.

Jacob, B., and Jacob, L. 1994. Cooking with seitan: the complete vegetarian "wheat-meat" cookbook. Avery Publishing Group, Garden City Park, NY. 185 pp.

Jewell, D.R., and Jewell, C.T. 1989. The oat and wheat bran health plan. Bantam Books, New York, NY. 279 pp.

Kinderlehrer, J. 1977. The art of cooking with love and wheat germ (and other natural foods). Rodale Press, Emmaus, PA. 355 pp.

Moulton, L. 1997. The amazing wheat book. LM Publications, Provo, UT. 222 pp.

Partain, K. 1983. Honey delights: cooking with honey and whole wheat flour: over 250 sugarless recipes. San Diego Publishing Co., San Diego, CA. 149 pp.

Rosenvall, V.G., Miller, M.H., and Flack, D.D. 1975. The classic wheat for man cookbook: more than 300 delicious and healthful ways to use stoneground whole wheat flour. Revised edition. Woodbridge Press Publishing Co., Santa Barbara, CA. 229 pp.

Shandler, N., and Shandler, M. 1980. How to make all the "meat" you eat out of wheat: international gluten wheat "meat" cookbook. Rawson, Wade Publishers, New York, NY. 241 pp.

Skidmore, L. 2003. Wheat dishes de Savor: food with whole grain wheat that tastes great. Marquise Publishing, Solihull, UK. 80 pp.

Tyler, L.D. 1974. The magic of wheat cookery: the beginning of a new adventure in wheat cooking. Magic Mill, Salt Lake City, UT. 146 pp.

YAM

Family: Dioscoreaceae (yam family)

Names

Scientific names: *Dioscorea* species

The Most Important Species of Yam Used as Sources of Food

English names	Scientific name(s)	Origin	Principal cultivation areas
Yellow Guinea yam, yellow yam, atoto yam (attoto yam), Lagos yam, Negro yam, twelve-months yam	*D. ×cayenensis* Lam.	West Africa	*D. rotundata* and *D. × cayensis* are the most commonly cultivated yams in Africa
White Guinea yam	*D. rotundata* Poir.	West Africa	*D. rotundata* and *D. × cayensis* are the most commonly cultivated yams in Africa
White yam, water yam, Guyana arrowroot, ten-months yam, winged yam, greater yam	*D. alata* L.	Tropical Asia	Old World tropics; this species and *D. esculenta* are the most commonly cultivated yams in Asia
Asiatic yam, Chinese yam, lesser yam	*D. esculenta* (Lour.) Burkill	Indian subcontinent, Indo-China, Malesia	Tropics, especially the subcontinent of India, southern Vietnam, and South Pacific islands; this species and *D. alata* are the most commonly cultivated yams in Asia
Chinese yam, Chinese potato, cinnamon vine	*D. batatas* Decne. ("*D. opposita*" of various authors)	China, eastern Asia (has become a weed in temperate regions of the world)	Cultivated in temperate regions of the world, particularly East Asia
Cush-cush, cush-cush yam, yampee	*D. trifida* L. f.	South America	South America and elsewhere; the most commonly cultivated yam in the New World
Air potato, aerial yam, bitter yam, cheeky yam, air potato yam, potato yam, devil's potato	*D. bulbifera* L.	Africa, China, Indian subcontinent, Indo-China, Malesia, Australia	Old World tropics; not commonly cultivated
Hottentot bread, tortoise plant, turtle plant, elephant's foot	*D. elephantipes* (L'Hér.) Engl.	South Africa	Occasionally wild plants are a source of food; grown basically as an ornamental

- The word "yam" is from the Portuguese *inhame* or obsolete Spanish i*gname*, *iñame*, both from Portuguese and English Creole *nyam*, to eat, of West African origin (from the Mande *niam*, or the Temne *enyame*).
- The true "yams" are species of the genus *Dioscorea* (whether edible or not), but in North America, the word "yam" is sometimes applied loosely to many tropical root crops, most especially the sweet potato (which see). Indeed, very frequently in the southern United States the word "yam" means orange-fleshed sweet potatoes. Moreover, canned sweet potatoes are frequently labeled yams. Yams and sweet potatoes are not related, although they look similar and are used in much the same way.
- The names "tortoise plant" and "turtleback plant" for *D. elephantipes* point out that the woody, above-ground part of its huge tuber resembles the shell of

a tortoise or turtle. "Elephant foot" is the way some Africans perceived the aboveground tuber. During severe drought conditions, the gigantic tubers were cooked and eaten as famine food by Hottentots, hence the name "Hottentot bread."

- The name cinnamon vine for *D. batatas*, one of the more popular edible species and the one best suited to cultivation in North America, is due to the cinnamon-scented small flowers.
- The alternative name "Chinese potato" for the cinnamon vine is also used to refer to several other edible species. Rarely, the name is used for jicama (*Pachyrhizus erosus* (L.) Urb.), Chinese arrowhead (*Sagittaria sagittifolia* L.), and taro (which see). For the most part, the phrase "Chinese potato" is used simply to mean regular potatoes with a Chinese connection (e.g., a Chinese culinary dish made with potatoes).
- The name "yam bean" is applied to several edible species of the pea family, most notably *Pachyrhizus erosus* (L.) Urb. (which is also known as jicama and Mexican turnip). This is a crisp, sweet turnip-shaped root vegetable used raw in salads or cooked in stews.
- "Devil's tongue yam" and "elephant yam" are names of *Amorphophallus konjac* C. Koch (*A. rivieri* Durieu, more often known as "konjac"). This native of southern China and Vietnam is often grown as an ornamental, and its edible roots are used as a vegetable and a source of extractives made into noodles and other products. It is not a true "yam."
- The genus name *Dioscorea* commemorates the ancient Greek physician and botanist Dioscorides (about 40–90). His classic work *De materia medica* was for 1500 years the authority of facts for botany and medicine. It discussed over 600 plants and a number of animal and mineral substances of medical use.

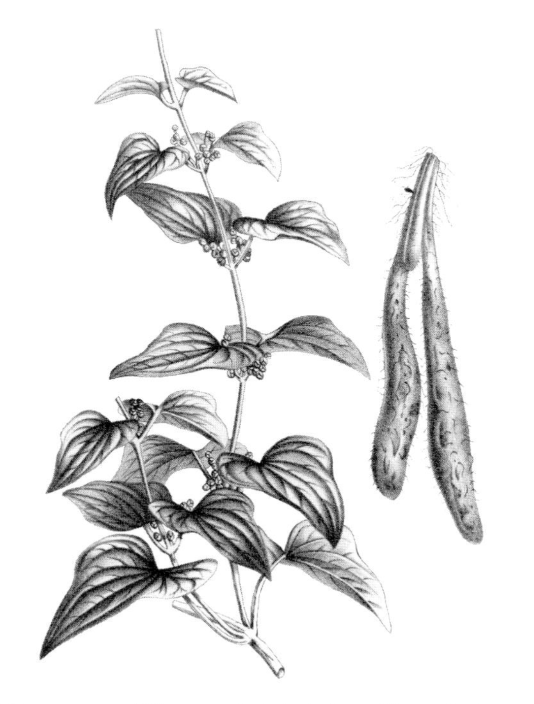

Chinese yam (*Dioscorea batatas*), from *Flores des Serres* (1845–1880).

Plant Portrait

About 150 of the 600 or more species of *Dioscorea* are grown in various countries. The most important food species are noted above. Yams may have been cultivated as early as 8000 BC in Asia. They are now produced mostly in tropical areas of the world, but also in northern China and Japan. Domestication of yams seems to have occurred independently on different continents. All of the important yams are of Old World origin except the cush-cush yam, which is a native of South America and the West Indies. Except for aerial potatoes mentioned below, the edible part of the yam plant is the underground storage organ. Just as with potatoes, these are not "roots," but "tubers" (swollen underground stems). The tubers of some species are used for feeding livestock, but it is as human food that yams are most significant. The important edible yams are perennial herbaceous vines that are grown as annuals on a trellis. Yam tubers of the common edible species can range in size from that of a small potato to giants over 2 m (6½ feet) long and 54 kg (120 pounds) in weight. Some yams have been known to exceed 3 m (about 10 feet) in length, and extend as far as 2 m (6 feet) deep in the ground, making them difficult to dig up.

Yams are one of the most important food crops in the world. They are the third most important tropical "root" crop after cassava and sweet potato. Yams are the second most important root/tuber crop in Africa, with production reaching just under one third the level of cassava. Nigeria accounts for about 70% of the world's production. More than 95% of the current global area under yam cultivation is in sub-Saharan Africa, particularly in West Africa. Yams are also popular in the West Indies and parts of Asia, as well as South and Central America. In addition to being cultivated throughout the tropics, yams are produced in parts of the subtropics and temperate zones, notably in northern China and Japan. There is some cultivation of yams for food in the southern United States.

White yam (*Dioscorea alata*), from Rhind (1855).

The white yam, yellow Guinea yam, white Guinea yam, and Asiatic yam (one of the yams called "Chinese yam") and cush-cush are the principal cultivated tropical yams. The above table points out where they are most commonly cultivated. These are not suitable for commercial cultivation in temperate North America.

The "Chinese yam" *D. batatas*, unlike most edible species of yams that are tropical, is from the mountains of northern Japan and can tolerate temperatures down to about −20 °C (−4 °F), so it can be grown outdoors in much of North America. This vigorous twining climber can grow like runner beans up a frame or on bamboo sticks. The edible root can be up to 1 m (about a yard) long and weigh 2 kg (4½ pounds) or more if grown in deep soil. The root is club-shaped, about as wide as an adult's finger at the top, thickening to the size of an arm at the base. Harvesting requires digging deep. Because yams produce a very vertical root, it is possible to grow a number of plants very close to each other, perhaps at 20–30 cm (8–12 inches) spacing. Like the air potato described next, this species also bears edible aerial tubers in the leaf axils, although these are much smaller than those of the air potato.

The air potato is a peculiar yam species that produces edible tubers on the aboveground stem rather than underground. The large, potato-like tubers that form in the leaf axils can weigh more than a kilogram (2¼ pounds). This native of South Asia is cultivated to a limited extent in the United States. In Florida, escaped plants have become weeds, capable of growing over 21 m (70 feet) in length, and sometimes overtopping and shading out tall trees. Uncultivated wild or weedy plants tend to be bitter, compared to cultivated varieties.

Elephant's foot was once cooked and eaten as a famine food in Africa, and is sometimes cultivated in greenhouses as a curiosity. Its huge underground tuber weighs up to 318 kg (700 pounds).

There is little consumption of yams in the United States and Canada, except by ethnic populations. Yams are imported to some extent from the Caribbean area and from Latin America, and can sometimes be purchased in supermarkets. However, the quality of imported yams is often poor.

Culinary Portrait

Yams vary widely in taste and appearance. The taste ranges from sweet to bitter to tasteless. Depending on the variety, a yam's flesh may be various shades of off-white, yellow, purple, or pink, and the skin from off-white to dark brown. The texture of this vegetable varies from moist and tender to coarse, dry, and mealy.

Yams are consumed as cooked, starchy vegetables. They are peeled and served in various ways, especially baked, boiled, grated, and mashed, sometimes by roasting or frying as chips or fritters, and often in soups. Yams are also grated and fried, or grated and steamed for bread and cakes. Because of their bland flavor they are rarely eaten alone, but are usually mixed with strongly flavored, salty, fatty, or spicy foods. In West Africa, yams are often pounded into a thick paste after boiling, and eaten with soup. Yams are also processed into flour that is used in the preparation of the paste. Although yams are most often boiled and then mashed into a sticky paste or dough in Africa, in Western cultures they can be fried, roasted, or baked in the manner of potatoes. In Japan, yams are battered and deep-fried.

In West Africa, yams are typically prepared as *fufu* (the term "fufu" is widely used in West Africa to refer to a sticky dough or porridge prepared from any pounded starchy food). This is a communal, laborious ritual carried out by women. The yams are washed, peeled, cut up, and boiled until soft. Then two or three women pound the cooked yams in a pestle with thick sticks until the preparation has the consistency of baker's dough. The noise the fufu pounders make is one of the most instantly recognisable sounds. Like pâtes, fufu is eaten with sauces. Peanut (known in Africa as groundnut), goat, and palm nut are popularly served with fufu.

Most yams contain a bad-tasting chemical (diosgenin, discussed below) that is neutralized in cooking. Some kinds of yams are poisonous (some are extremely poisonous), but normally, these varieties are not eaten or sold as food. Yams should not be eaten raw, as they may contain calcium oxalate crystals just underneath the skin. The crystals are eliminated by peeling and cooking. Yams mostly contribute just starch to human nutrition. Potatoes are much more nutritious, and can be substituted with little difference in flavor or texture.

Fresh white yams are often available in Latin American and Caribbean markets. They should be firm, unblemished, with no signs of wrinkling, and are best when they weigh less than 0.45 kg (1 pound). Canned white yams may be available in African and Latin American markets. Yams store very well in a cool dry place, and can be kept for up to a week in a refrigerator.

Asiatic yam (*Dioscorea esculenta*), from Paxton (1834–1849, vol. 3).

Curiosities of Science and Technology

- In some cultures, only men are allowed to grow yams. In some of the Melanesian islands (northeast of Australia), when the yam vines are being trained, it was the practice for the men to sleep near the gardens and never approach their wives; should they enter the garden after breaking this rule of continence, it was thought that the fruits of the garden would be spoiled.
- In West Africa and New Guinea, yams are so important that they are the subject of religious cults, with their own priesthoods.
- In New Guinea and Melanesia, special ceremonial yams weighing over 54 kg (120 pounds) are raised to reflect the grower's status in the community. The yams are used for gifts and ritualized exchanges. A yam festival is held at harvest, and the tubers are covered by elaborate woven masks.
- Some yam species produce vines that twine clockwise, while other species wind counterclockwise. For example, *D. alata*, *D. rotundata*, and *D. ×cayenensis* twine right, and *D. esculenta* and *D. trifida* twine left. ("Right-hand twining" occurs when the thumb of the right hand is held up and the fingers are in the direction of vine growth; left-hand twining is vice versa.)
- Why do yam species often produce gigantic storage organs? Like desert cacti with enlarged stems, the swollen tubers are a way of storing moisture for extended periods of drought, as well as for storing carbohydrates.
- In 1940, steroids were discovered in yams that proved useful for the manufacture of cortisone and sexual hormones. As a result, the cost of these hormones dropped from $80 to $2 per gram. The chemist Russell Marker discovered how to manufacture progesterone from the yam steroids. He was unable to receive support to further his work in the United States, and moved to Mexico City, forming a joint venture named Syntex, which eventually manufactured testosterone and 19-norprogesterone, a chemical analog of progesterone that was so effective at inhibiting ovulation that it became "The Pill." Today, about 50 species of *Dioscorea* are used as sources of diosgenin, the chemical that is transformed by drug companies into medical steroids.

Key Information Sources

Akoroda, M.O. 1993. Yams, *Dioscorea* spp. *In* Genetic improvement of vegetable crops. *Edited by* G. Kalloo and B.O. Bergh. Pergamon Press, New York, NY. pp. 717–733.

Ayensu, E.S., and Coursey, D.G. 1972. Guinea yams, the botany, ethnobotany, use and possible future of yams in West Africa. Econ. Bot. **26**: 301–318.

Burkill, I.H. 1960. The organography and the evolution of the Dioscoreaceae, the family of yams. J. Linn. Soc. Bot. **56**: 319–412.

Coursey, D.G. 1967. Yams. An account of the nature, origins, cultivation and utilisation of the useful members of the Dioscoreaceae. Longmans, London, UK. 230 pp.

Coursey, D.G. 1975. The origins and domestication of yams in Africa. *In* Gastronomy: the anthropology of food and food habits. *Edited by* M.L. Arnott. Mouton, The Hague. pp. 187–212.

Coursey, D.G. 1983. Yams. *In* Handbook of tropical foods. *Edited by* H.T. Chan, Jr. Marcel Dekker Inc., New York, NY. pp. 555–601.

Degras, L., and Coste, R. 1993. The yam: a tropical root crop. Macmillan, Wageningen, Netherlands. 408 pp.

Hahn, S.K. 1995. Yams, *Dioscorea* spp. (Dioscoreaceae). *In* Evolution of crop plants. 2nd edition. *Edited by* J. Smartt and N.W. Simmonds. Longman Scientific & Technical, Burnt Mill, Harlow, Essex, UK. pp. 112–120.

Hamon, P., and Bakary, T. 1990. The classification of the cultivated yams (*Dioscorea cayensis–rotundata* complex) of West Africa. Euphytica, **47**: 179–187.

Jos, J.S. , and Vehkateswaralu, T. 1978. Twining in relation to distribution among Asian yams. J. Root Crops, **4**: 63–64.

King, G., and Hackett, C. 1986*a*. Tabular descriptions of crops grown in the tropics. 13. Greater yam (*Dioscorea alata* L.). Technical Memorandum 86(4). Commonwealth Scientific and Industrial Research Organization, Canberra, Australia. 53 pp.

King, G., and Hackett, C. 1986*b*. Tabular descriptions of crops grown in the tropics. 14. Lesser yam (*Dioscorea esculenta* L.). Technical Memorandum 86(15). Commonwealth Scientific and Industrial Research Organization, Canberra, Australia. 52 pp.

Lawani, S.M., and Odubanjo, M.O. 1976. A bibliography of yams and the genus *Dioscorea*. International Institute of Tropical Agriculture, Ibadan, Nigeria. 192 leaves.

Martin, F.W. 1972. Yam production methods. Agricultural Research Service, U.S. Dept. of Agriculture, Washington, DC. 17 pp.

Martin, F.W. 1974*a*. Tropical yams and their potential. Part 1. *Dioscorea esculenta*. Agricultural Handbook 457. U.S. Dept. of Agriculture, Washington, DC. 18 pp.

Martin, F.W. 1974*b*. Tropical yams and their potential. Part 2. *Dioscorea bulbifera*. Agricultural Handbook 466. U.S. Dept. of Agriculture, Washington, DC. 20 pp.

Martin, F.W. 1976. Tropical yams and their potential. Part 3. *Dioscorea alata*. Agricultural Handbook 495. U.S. Dept. of Agriculture, Washington, DC. 40 pp.

Martin, F.W., and Degras, L. 1978*a*. Tropical yams and their potential. Part 5. *Dioscorea trifida*. Agricultural Handbook 522. U.S. Dept. of Agriculture, Washington, DC. 26 pp.

Martin, F.W., and Degras, L. 1978*b*. Tropical yams and their potential. Part 6. Minor cultivated *Dioscorea species*. Agricultural Handbook 538. U.S. Dept. of Agriculture, Washington, DC. 23 pp.

Martin, F.W., and Rhodes, A.M. 1977. Intra-specific classification of *Dioscorea alata*. Tropical Agriculture (Trinidad), **54**: 1–13.

Martin, F.W., and Sadik, S. 1977. Tropical yams and their potential. Part 4, *Dioscorea rotundata* and *Dioscorea cayenensis*. Agricultural Handbook 502. U.S. Dept. of Agriculture, Washington, DC. 36 pp.

Miège, J., and Lyonga, S.N. (*Editors*). 1982. Yams. Papers presented at the first international seminar on yams held in Cameroon, 1978. Clarendon Press, Oxford, UK. 411 pp.

Onwueme, I.C. 1978. The tropical tuber crops: yams, cassava, sweet potato, and cocoyams. Wiley, New York, NY. 234 pp.

Onwueme, I.C. 1984. Yam. *In* The physiology of tropical field crops. *Edited by* P.R. Goldsworthy and N.M. Fisher. John Wiley & Sons, Chichester, UK. pp. 569–588.

Onwueme, I.C. 1996*a*. *Dioscorea* L. *In* Plant resources of South-East Asia, 9, plants yielding non-seed carbohydrates. *Edited by* M. Flach and F. Rumawas. Backhuys Publishers, Leiden, Netherlands. pp. 85–90.

Onwueme, I.C. 1996*b*. *Dioscorea esculenta* (Lour.) Burkill. *In* Plant resources of South-East Asia, 9, plants yielding non-seed carbohydrates. *Edited by* M. Flach and F. Rumawas. Backhuys Publishers, Leiden, Netherlands. pp. 93–95.

Onwueme, I.C., and Ganga, Z.N. 1996. *Dioscorea alata* L. *In* Plant resources of South-East Asia, 9, plants yielding non-seed carbohydrates. *Edited by* M. Flach and F. Rumawas. Backhuys Publishers, Leiden, Netherlands. pp. 90–93.

Osagie, A.U. 1992. The yam tuber in storage. Postharvest Research Unit, Dept. of Biochemistry, University of Benin, Benin City, Nigeria. 247 pp.

Osuji, G. (*Editor*). 1985. Advances in yam research: the biochemistry and technology of the yam tuber. Papers presented at a symposium held at the Anambra State University of Technology, Enugu, May 1983. Biochemical Society of Nigeria in collaboration with Anambra State University of Technology, Enugu, Nigeria. 362 pp.

Raz, L. 2002. Dioscoreaceae R. Brown, yam family. *In* Flora of North America north of Mexico. Vol. 26. *Edited by* Flora of North America Editorial Committee. Oxford University Press, New York, NY. pp. 479–485.

Udoessien, E.I., and Ifon, E.T. 1992. Chemical evaluation of some antinutritional constituents in four species of yam. Trop. Sci. (U.K.), **32**: 115–119.

Specialty Cookbooks

Jacob, J., and Ashkenazi, M. 2007. The world cookbook for students. Greenwood Publishing Group, Westport, CT. 5 vols.

Appendix 1. Technical Aspects of Plant Names

> "I wonder whether mankind could not get along without all these names, which keep increasing every day."
>
> —Herman Melville (1819–1891), US author

Names of plants are indispensable for information retrieval and communication. Of course, plants receive different names in different languages, but this is simply a matter of translation. Much more serious problems of communication have been created when a plant acquires more than one name in the same language, and when different plants get exactly the same name. This is a critical problem for food plants, many having numerous names, and many known by exactly the same name. As explained below, there are three systems in use concerned with plant names, and it is important to understand the applications and limitations of these systems, and how they relate to each other.

Common Names

The familiar words in English used to refer to plants are called "common," "colloquial," and "vernacular" names. These are often highly misleading, since some names are used in different places or by different people to mean different plants.

Plants often have many common names. Wintergreen (*Gaultheria procumbens*) has also been called aromatic wintergreen, boxberry, Canadian tea, checkerberry, chickenberry, chinks, cowberry, creeping wintergreen, creeping wintergreen checkerberry, deerberry, drunkards red pollom, eastern teaberry, gaultheria, groundberry, ground holly, ground tea, grouseberry, hillberry, ivyberry, ivory plum, mountain tea, nannyberry, partridgeberry, rapper dandies, spiceberry, redberry tea, spring wintergreen, spicy wintergreen, teaberry, spicy wintergreen checkerberry, wax cluster, and winter berry. Country borage (*Plectranthus amboinicus*), facetiously called "the herb of a hundred names," is also known as allspice, bread-and-butter plant, broadleaf thyme, coleus thyme, cowsleaf, cowslip, Cuban oregano, East Indian thyme, five seasons herb, fivespice, French thyme, Indian borage, Indian mint, Mexican mint, Mexican thyme, mint Swedish ivy, mother of herb, oregano, Puerto Rico oregano, sage, soup mint, Spanish thyme, stinging thyme, sweet-scented coleus, sugánda, tropical oregano, wild oregano, and wild thyme (all names that could be confused with the names of unrelated herbs).

Many plants are known by different common names as a consequence of people in different regions adopting different names for the same plant. The North American "eggplant" is equivalent to the European "aubergine." Belgian endive, French endive, and Brussels endive are exactly the same plant (*Cichorium endivia*). *Juglans regia* is known as "Chinese walnut" in China, "Italian walnut" in Italy, and "French walnut" in France, although it originally came from Persia, and is also known as the "Persian walnut."

Often plants are deliberately given a new, more attractive name for marketing purposes. The fruit of the prickly pear (*Opuntia ficus-indica*) is now usually sold under the less prickly name "cactus pear." The "monkey peach" has become the kiwi. The Jerusalem artichoke is sold as "sunchoke." The "hog peanut" (*Apios americana*) has become the groundnut.

The same name is often used for different plants. "Groundnut" refers to the peanut (*Arachis hypogaea*) in much of the world, especially Africa, but to *Apios americana* in North America. "Corn" refers to *Zea mays* in North America, but to various cereals, especially wheat and barley, in England (corn is most often called "maize" internationally). In North temperate areas, "sorrel" refers to culinary herbs of the genus *Rumex*, but in the tropics, sorrel means the roselle plant (*Hibiscus sabdariffa*). In a supermarket, "anise" might refer on the one hand to the culinary herb species *Pimpinella anisum*, or to the quite different vegetable species *Foeniculum vulgare* subsp. *azoricum*. "Pawpaw" might mean the tropical fruit *Carica papaya*, or the temperate region fruit *Asimina triloba*. In the southern United States, "pea" likely refers to *Vigna unguiculata* subsp. *unguiculata* (the black-eyed pea), and not to the conventional garden pea (*Pisum sativum*). In the Midwest of the United States, sweet bell peppers (*Capsicum annuum*) are sometimes called "mangos," a name usually used for the tropical fruit *Mangifera indica*. Several species are referred to equally as "pumpkin" and "squash." "Citron" refers to the fruit of a citrus tree (*Citrus medica*), but also to a kind of watermelon.

Linguistic confusion for plant names is sometimes very dangerous. For example, the name "Japanese star anise" is used both for an edible spice plant (*Illicium verum*) and a widely cultivated poisonous plant (*I. anisatum*), and this has sometimes led to marketing of the poisonous plant as the edible one. Ginkgo nuts are often sold as "silver almonds," and people who think they really are some kind of almond may eat poisonous amounts. The word "fiddlehead" is commonly used to refer to a quite edible species (*Matteuccia struthiopteris*) and to the cancer-causing *Pteridium aquilinum*, also collected for food. The ambiguity of common names is why "scientific names" are so important.

There have been attempts to have just one agreed-on common name for economically important plant species. Groups dedicated to particularly significant classes of plants, such as the European Plant Protection Organization and the Weed Science Society of America, both concerned with weeds, have undertaken attempts to produce reference lists of standardized English names for weeds. Examples of publications with recommended standardized English names include Beetle (1970, for American horticultural plants), Wiersema and Léon (1999, for numerous economic plants), McGuffin (2000, for culinary and medicinal herbs in American commerce), and Darbyshire et al. (2000, for Canadian weeds). However, at this time, there is no universal agreement on standard English common names that should be used by everyone.

There also have been attempts to enforce rules governing the spelling of English common names, particularly with regard to compound names (names with several words), and whether the words making up a name should be separated, joined with hyphens, or simply run together (e.g., sugar beet, sugar-beet, or sugarbeet). General guidelines for spelling common names of plants are found in Hamilton (1969) and Kartesz and Thieret (1991). However, many of the names in works that follow such guidelines simply do not reflect the way the names are most often spelled. Usage in large dictionaries is generally the best guideline to preferred spelling, but numerous names are simply not included.

Scientific Names

This book presents the complete, correct scientific names for all of the plants discussed, which is usually a species name (a combination of two words) followed by the abbreviated name of the author who coined the name [Brummitt and Powell (1992) provide standardized author abbreviations; this is available online at http://www.ipni.org/index.html]. Scientific names are very important, because historically many names have been used for some species, and to avoid confusion, only one name should be used. The rules for choosing that one correct name are complex, governed by an internationally agreed upon *Botanical Code of Nomenclature*, which is revised and updated periodically. The latest version available in 2009 is McNeill et al. (2006). For additional discussion of scientific names of plants, see Gledhill (2002).

The most comprehensive general source for finding scientific names is *The International Plant Names Index*, available online at http://www.ipni.org/index.html. This is certainly the best database of scientific names, but many of the names listed require checking, and sometimes conflicting information is provided. *GRIN Taxonomy* (http://www.ars-grin.gov/cgi-bin/npgs/html/index.pl) is an extremely useful source of information for scientific names of economic plants [this is the online version of Wiersema and Léon (1999), cited elsewhere here]. *The International Seed Testing Association List of Stabilized Plant Names* (http://www.ars-grin.gov/~sbmljw/istaintrod.html) is another useful source of information for scientific names of numerous economically important plants.

Occasionally, a widely used scientific name is changed when it is discovered that another name is correct. This causes confusion because the older scientific name continues to be encountered in the literature produced up to that point, and people are often reluctant to adopt the correct but unfamiliar name for some time. In such cases, commonly used older scientific names are also given in this book. Occasionally, experts disagree on the classification of a species, and there are competing scientific names. When this is the case for the plants discussed in this book, a choice of one name was made, and the competing name presented as a synonym.

Occasionally, scientific names of animals are given in this book. The scientific nomenclatural code for animals differs somewhat from that for plants (for example, "varieties" of animals are not recognized as a formal category).

Cultivar Names

In addition to the above botanical code, which deals with the naming of both wild and cultivated plants (including fungi), there is a second code specifically devoted to the provision of rules for coining the names for "cultivated varieties" or "cultivars" of plants. This is known as *The International Code of Nomenclature for Cultivated Plants* (Brickell et al. 2004).

Readers should be aware that the *International Code of Botanical Nomenclature* uses the term "variety" in a much more comprehensive sense by comparison with the way the same term is used by *The International Code of Nomenclature for Cultivated Plants.*

The *International Code of Botanical Nomenclature* emphasizes Latin words in botanical names. Latin words are typographically distinguished, usually in italics, and varieties are always indicated by the abbreviation var. (which actually stands for "varietas"). For example, *Brassica oleracea* var. *italica* is the scientific name for broccoli. Such usage indicates a relatively large grouping of plants that specialists in plant classification have decided merits recognition. Another example is celeriac, *Apium graveolens* var. *rapaceum*, which is distinguished from its relative celery, *Apium graveolens* var. *dulce*.

By contrast, plant names governed by *The International Code of Nomenclature for Cultivated Plants* are at least partly non-Latin, usually recognizable by the presence of at least one non-italicized word in the name (see examples below). In this system, the term "variety" is more correctly called "cultivar," short for "cultivated variety." A cultivar is a named selection of a cultivated plant, i.e., a variation that has been perpetuated by people.

The 2004 code of nomenclature for cultivated plants (which supersedes all previous codes) provides that cultivar names must be constructed by combining two elements. The first element is any one of the following: (*a*) the scientific name of a plant; (*b*) the genus name of a plant (permissibly followed by the name of a group of cultivars); (*c*) an unambiguous common name. The second element is a term or phrase in the common language (i.e., not in Latin), and enclosed in single quotation marks. The following examples illustrate the three accepted ways of constructing a complete cultivar name.

(*a*) *Planta alba* 'Magnificent Red'

(*b*) *Planta* 'Magnificent Red' or *Planta* Group Northern 'Magnificent Red'

(*c*) Strawberry 'Magnificent Red' or 'Magnificent Red' strawberry

Before the 1995 code of nomenclature for cultivated plants made the practice unacceptable, an alternative format was to use the abbreviation "cv." (abbreviation for "cultivated variety") in lieu of single quotes, in the fashion *Planta alba* cv. Magnificent Red, and older literature frequently uses this format.

Officially constructed cultivar names are long, and in practice in popular writing, such as found in magazines and newspapers, the non-Latin portion of the name is generally used as the name of the cultivar, when it is clear what the species is. For example, in a discussion of spearmint, a writer might refer simply to Kentucky Colonel or 'Kentucky Colonel'.[1] A discussion of celeriac might offer as examples of cultivars: Cascade, Monarch, Mars, and Giant Prague (instead of 'Cascade' celeriac, 'Monarch' celeriac, 'Mars' celeriac, and 'Giant Prague' celeriac). For brevity, the English names alone are usually used in this book, and without single quotations (although both practices are officially forbidden).

Illustrative Example

The following complex name is given simply to illustrate the various conventions that botanists use in naming plants, using both codes of nomenclature referred to above. Normally, names are not this difficult. Items 1–12 are covered by the *International Code of Botanical Nomenclature*. Item 13 is covered by *The International Code of Nomenclature for Cultivated Plants*.

Planta	×	*alba*	A. Smith	ex	B. Jones	subsp.	*alba*	var.	*grandifolia*	(L.)	L. f.	'Magnificent Red'
1	2	3	4	5	6	7	8	9	10	11	12	13

[1]Notice the format "'Kentucky Colonel'." The typographical convention followed by most printers of placing a final double quotation mark (") *after* a period (or any other punctuation mark) is reversed in the case of the single quotation marks used for cultivars.

1. The first word of a scientific name indicates the genus, *Planta* in this case. A genus may be composed of one or many species. The genus name may commemorate an individual, but frequently is based on an old plant name, often from classical Latin or Greek.

2. Sometimes one sees an × (the mathematical multiplication symbol, not the letter x) in a scientific name. This indicates that the name designates a plant of hybrid origin.

3. The names of species are "binomials," i.e., a combination of two words, in this example *Planta alba*. Often it is said that the second word in the name is the "species name," but this is based on ignorance (the "species name" is always a combination of two words). Technically, the second word is called a "species epithet" or "specific epithet," and is generally an adjective in Latinized form. Latin adjectives are of different types, but must agree with the gender of the genus name. The people who coin names try to select a specific epithet that is descriptive of some characteristic of the species. In this case, the specific epithet *alba*, which is Latin for white, may indicate that the leaves are whitish. Sometimes, the specific epithet is inaccurate as a description, but for naming purposes this does not matter.

4–6. In a complete name citation, the author(s) of a name, abbreviated as much as possible without possibility of confusion with another author of a plant name, follow(s) the name. Sometimes, as in this example, the word "ex" is present, which indicates that an author following the ex adopted the specific epithet from a name used (but not validly published) by the person preceding the ex. In this case, A. Smith first used the word *alba* in some way relating to the present name, and it was taken up by B. Jones when he coined the name *Planta ×alba*. For technical reasons relating to the code of nomenclature, Smith has not been given credit as the person who named the species, but at least it is indicated that his name was the inspiration that led Jones to adopt the name.

7. The abbreviation subsp. (or ssp.) means "subspecies," which is a subdivision of a species.

8. The word *alba* here is the "subspecific epithet," a word descriptive of the subspecies, just as the specific epithet is descriptive of the species. However, a special case is illustrated here, indicated by two things: the same word *alba* is used both for the specific epithet and for the subspecific epithet, and there is no citation of author(s) following the subspecific epithet. By convention, whenever a botanist splits a species into subspecies or lower groupings, one of the groupings must be given the same epithet as the species, and no authors for this are recognized.

9. One very rarely sees a name in which both subspecies and varieties are given, as in this example. Normally, if a plant species is divided into smaller groupings, either subspecies or varieties are recognized, not both. In this case, a subspecies has been divided into varieties.

10. The word *grandifolia* is a varietal epithet, descriptive of the variety. In this example, it is Latin for "large-leaved." Although the majority of descriptive terms actually do more or less accurately indicate some characteristic of the species, words that are quite inaccurate are permitted (*Planta chinensis* need not come from China).

11. Whenever one sees an author's name in brackets in a scientific name, it indicates that this author was the first to use the epithet in connection with the species, but did so in a different way, for example, assigning the species to a different genus, or using the epithet at a different rank (e.g., subspecies instead of variety). In this case, L. stands for Linnaeus, the "Father of Biological Nomenclature," who recognized more species names than any other individual in history.

12. In this example L. f. is the author who took the epithet first used by the author in brackets (explained in 11) and was the first to use it in the present name. "L. f." stands for "Linnaeus fils," i.e., "son of Linnaeus."

13. 'Magnificent Red' in Roman (non-italicized) type and in single quotation marks is the portion of the name that is added to the scientific name (in italics) to produce the full cultivar name of the plant.

Literature cited

Beetle, A.A. [for American Joint Committee on Horticultural Nomenclature]. 1970. Recommended plant names. Agricultural Experiment Station, University of Wyoming, Laramie, WY. 124 pp.

Brickell, C.D. (*Chairman*), Baum, B.R., Hetterscheid, W.L.A., Leslie, A.C., McNeill, J., Trehane, P., Vrugtman, F., and Wiersema, J.H. (*Editors*). 2004. International code of nomenclature for cultivated plants incorporating the rules and recommendations for naming plants in cultivation. 7th edition. International Society for Horticultural Science, Leuven, Belgium. 123 pp.

Brummitt, R.K., and Powell, C.E. 1992. Authors of plant names: a list of authors of scientific names of plants, with recommended standard forms of their names, including abbreviations. Royal Botanic Gardens, Kew, UK. 732 pp.

Darbyshire, S.J., Favreau, M., and Murray, M. 2000. Common and scientific names of weeds in Canada. Agriculture and Agri-Food Canada, Ottawa, ON. 132 pp.

Gledhill, D. 2002. The names of plants. 3rd edition. Cambridge University Press, Cambridge, UK. 326 pp.

Hamilton, R.A. 1969. The spelling of English common names. *In* Common and botanical names of weeds in Canada. *Edited by* Canada Weed Committee. Canada Dept. of Agriculture, Ottawa, ON. p. vi.

Kartesz, J.T., and Thieret, J.W. 1991. Common names for vascular plants: guidelines for use and application. Sida, **14**: 421–434.

McGuffin, M. 2000. Herbs of commerce. American Herbal Products Association, Austin, TX. 421 pp.

McNeill, J., Barrie, F.R., Burdet, H.M., Demoulin, V., Hawksworth, D.L., Marhold, K., Nicolson, D.H., Prado, J., Silva, J.P.C., Skog, J.E., Wiersema, J.H., and Turland. N.J. 2006. International code of botanical nomenclature (Vienna Code). Regnum Vegetabile 146. A.R.G. Gantner, Verlag KG, Koenigstein, Germany. 568 pp.

Wiersema, J.H., and León, B. 1999. World economic plants. A standard reference. CRC Press, New York, NY. 749 pp.

Appendix 2. Sources of Illustrations Presented in this Book

Note: Some of the figures were drawn by artists associated with the Ottawa center of Agriculture and Agri-Food Canada, as indicated in Acknowledgments. Sources of several photographs are given in their captions. For a few illustrations, a full citation is given in the caption, and is not repeated here. Permission to reproduce some copyrighted illustrations and to employ provided copies of some non-copyrighted illustrations is indicated in the captions for figures, in several cases using required language. While the originals of most of the figures corresponding with the illustrations presented in this book are copyright-free, the present versions have been electronically enhanced and (or) altered in significant respects, and so constitute new works. In many cases, the original works list the illustrations under different names; in all cases, the correct names by today's standards are given.

Anderson, J. 1874. The new practical gardener and modern horticulturist. William Mackenzie, London, UK. 988 pp.

Bailey, L.H. (*Editor*). 1900–1902. Cyclopedia of American horticulture. Virtue & Company, Toronto, ON. 4 vols.

Bailey, L.H. 1910. Manual of gardening: a practical guide to the making of home grounds and the growing of flowers, fruits, and vegetables for home use. Macmillan, New York, NY. 539 pp.

Bailey, L.H. (*Editor*). 1916. The standard cyclopedia of horticulture. Macmillan, Toronto, ON. 3 vols.

Baillon, M.H. 1876–1892. Dictionnaire de Botanique. Librarie Hachette, Paris, France. 4 vols.

Britton, N.L., and Brown, A. 1896–1898. An illustrated flora of the Northern United States, Canada and the British possessions from Newfoundland to the parallel of the southern boundary of Virginia and from the Atlantic westward to the 102d meridian. Charles Scribner's Sons, New York, NY. 3 vols.

Bulliard, P. 1784. Histoire des plantes vénéneuses et suspectes de la France. Bulliard, Paris, France. 2 vols.

Chevalier, A., and Perrot, É. 1911. Les végétaux utiles de l'Afrique tropicale française. Fasc. 6: Les kolatiers & les noix de kola. A. Challamel, Paris, France. 483 pp. + 16 plates.

Corel Corporation. 1998. Corel gallery 1,000,000. One million images. (Software package of distributable clip art.)

Curtis, W. (*Editor*). 1787–present. The Botanical Magazine. London, UK. [Various publishers, beginning with Curtis; this is the longest-lived and most important English-language horticultural serial publication.]

Despeissis, J.A. 1893. The vineyard and the cellar. Agric. Gaz. N. S. W. (Aust.), **4**: 33–36 + plate.

Dewey, L.H. 1897. Wild garlic, *Allium vineale* L. United States Department of Agriculture, Division of Botany. Circular No. 9. 8 pp.

Dillenius, J.C. 1774. Hortus elthamensis. Cornelius Haak, Leiden, Netherlands. 2 vols.

Duthie, J.F. 1893. Field and garden crops of the north-western provinces and Oudh, with illustrations. Part 3. Department of Land Records and Agriculture, N.W. Provinces and Oudh, India. 65 pp.

Edwards, S.T., and Lindley, J. 1815–1847. Edward's botanical register. James Ridgway, London, UK. 33 vols.

Engler, H.G.A., and Prantl, K.A.E. (*Editors and contributors*). 1889–1915. Die natürlichen Pflanzenfamilien, ed. 1. Wilhelm Engelmann, Leipzig, Germany. Numerous volumes, irregularly numbered.

Figuier, L. 1867. The vegetable world: being a history of plants, with their botanical descriptions and peculiar properties. Chapman, London, UK. 576 pp.

Flore des serres et des jardins de l'Europe: annales générales d'horticulture. 1845–1880. L. van Houtte, Gand, Belgium. 23 vols.

Foord, J. 1906. Decorative plant & flower studies. B.T. Batsford, London, UK. Unpaginated.

Garden and Forest. A journal of horticulture, landscape art, and forestry. Vol. 1, 1888–Vol. 10, 1897. Published by Garden and Forest Publishing Co., New York, NY.

Gardeners' Chronicle (The). 1874–1956. Serial journal published in London, UK. (Illustrations taken for the years 1898 and 1901.)

Gartenflora: Zeitschrift für Garten-und Blumenkunde. 1852–1938. Serial journal published in Berlin, Germany.

Giacometti, D.C., and Léon, J. 1994. Tannia, yautia. (*Xanthosoma sagittifolium*). *In* Neglected crops. 1492 from a different perspective. *Edited by* J.E. Hernández-Bermejo and J. Léon. Food and Agriculture Organization of the United Nations, Rome, Italy. pp. 253–258.

Hallier, E.H. 1880–1888. Flora von Deutschland, edition 5 of D.F.L. von Schlechtendal et al. F.E. Köhler, Gera-Untermhaus, Germany. 39 vols.

Hare, H.A., Caspari, C., Jr., and Rusby, H.H. 1908. The national standard dispensatory. 2nd edition. Lea & Febiger, New York, NY. 2011 pp.

Harter, J. (*Editor*). 1988. The plant kingdom compendium. A definitive volume of more than 2,400 copyright-free engravings. Bonanza Books, New York, NY. 374 pp.

Henderson, P. 1890. Henderson's handbook of plants and general horticulture. Peter Henderson & Company, Jersey City, NJ. 528 pp.

Hitchcock, A.S., and Chase, A. 1951. Manual of the grasses of the United States, 2nd edition. United States Department of Agriculture Miscellaneous Publication 200. 1051 pp. (Public domain.)

Hooker, W.J. (*Editor*). Hooker's London Journal of Botany. Vol. 1 (1842), Vol. 2 (1850).

Humboldt, A. de, and Bonpland, A. 1807. Plantae aequinoctiales, Part 5 [of 17]. Lutetiae, Paris, France.

Ives, E., Tully, W., and Leveonworth, M.C. 1831. Catalogue of the phenogamous plants and the ferns growing without cultivation, within five miles of Yale College, Ct. Hezekiah Howe, New Haven, CT. Irregularly paginated.

Jackson, J.R. 1890. Commercial botany of the nineteenth century. Cassell & Company, London, UK. 168 pp.

Jacquin, N.J. 1781–1786. Icones plantarum rariorum. Vindobonae, Wien, Austria. Vol. 1 (Folio 4), plate 17. (A book of 200 plates.)

Jumelle, H. 1901. Les cultures coloniales, plates industrielles & médicinales. J.B. Baillière et Fils, Paris, France. 360 pp.

Köhler, H.A. 1883–1914. Köhler's Medizinal Pflanzen, 4 vols. Verlag von F.E. Köhler, Germany. (All illustrations used are from vols. 1–3, published 1883–1898.)

Lamarck, J.B.A.P.M. de, and Poiret, J.L.M. 1744–1829. Encyclopédie méthodique. Botanique. Chez Panckoucke, Paris, France. 22 vols.

Linnaeus, C., Works of, 1746. Kongl. Swenska Wetenskaps Academiens Handlingar. Lorentz Ludwig Grefink, Stockholm, Sweden. Vol. 7.

Loudon, J.C. 1844. The trees and shrubs of Britain. Longman, Brown, Green, and Longmans, London, UK. 8 vols.

Maout, E. le, and Decaisne, J. 1876. A general system of botany. Longmans, Green, and Co., London, UK. 1066 pp.

Marilaun, A.K. von. 1895. The natural history of plants. (Translated from German by F.W. Oliver.) Henry Holt, New York, NY. 2 vols. (each with 2 parts).

Martius, C.F.P. von. 1831. Die Pflanzen und Thiere des tropischen America. Vol. 2. Verfasser, Munich, Germany.

Metzger, J. 1824. Europaeische Cerealen. C.F. Winter, Heidelberg, Germany. 74 pp. + plates.

Michaux, F.A. 1850. The North America sylva; or a description of the forest trees of the United States, Canada, and Nova Scotia. G.P. Putnam, New York, NY. 3 vols. (Translated from French.)

Morren, C. (*Editor*). 1851–1885. La Belgique Horticole, Journal des jardins, des serres, et des vergers. Liège, Belgium. 35 vols.

Nicholson, G. (*Editor*). 1885–1889. The illustrated dictionary of gardening, a practical and scientific encyclopedia of horticulture for gardeners and botanists. L. Upcott Gill, London, UK. 4 vols.

Oeder, G.C., Müller, O.F., and Vahl, M. (*Editors*). 1761–1883. Flora Danica Copenhagen, Denmark. 17 vols.

Paillieux, A., and Bois, D. 1892. Le potager d'un curieux: histoire, culture & usages de 200 plantes comestibles peu connues ou inconnues. 2e édition. Librairie agricole de la maison rustique, Paris, France. 588 pp.

Pallas, P.S. 1784–1831. Flora Rossica. Imperiali J.J. Weitbrecht, Petropoli (St. Petersburg), Russia. 2 vols.

Paxton, J. (*Editor*). 1834–1849. Paxton's magazine of botany, and register of flowering plants. Orr and Smith, London, UK. 16 vols.

Prain, D. (*Editor*). 1916. Hooker's icones plantarum. Fourth Series, Part 3. Dulau & Co., London, UK.

Revue de L'Horticulture Belge et Étrangère. Vol. 1, 1875–Vol. 40, 1914.

Rhind, W. 1855. A history of the vegetable kingdom. Blackie and Son, London, UK. 720 pp.

Sargent, C.S. 1890–1892. The silva of North America. Houghton, Mifflin and Company, Boston, MA. 14 vols.

Strasburger, E. (rewritten by H. Fitting, H. Sierp, R. Harder, and G. Karsten). 1930. Strasburger's textbook of botany. MacMillan, London, UK. 818 pp.

Thomé, O.W. 1903–1905. Prof. Dr. Thomé's Flora von Deutschland, Österreich und der Schweiz. H.V. Verlag, Berlin-Lichterfelde, Germany. [First 4 vols., reissued with amended text.]

Turner, F. 1891. New commercial crops for New South Wales. The cultivation of the soy bean (*Soja hispida* Moench.). Agric. Gaz. N. S. W. (Aust.), **2**: 648–650 + plate.

Turner, F. 1892. New commercial crops for New South Wales. The cultivation of the dwarf or bush lima bean (*Phaseolus lunatus* Linn). Agric. Gaz. N. S. W. (Aust.), **3**: 644–647 + plate.

United States Department of Agriculture, Division of Pomology. 1896. Nut culture in the United States. Government Printing Office, Washington, DC. 144 pp.

Vilmorin, P.L. de. 1906. Hortus Vilmorinianus.Verrières-le-Buisson, Paris, France. 371 pp.

Vilmorin-Andrieux, M.M. 1885. The vegetable garden. John Murray, London, UK. 620 pp.

Ward, A. 1911. The grocer's encyclopedia. James Kempster Printing Company, New York, NY. 748 pp.

Woodville, W. 1810. Medical botany, 2nd edition. William Phillips, London, UK. 4 vols.

Appendix 3. General Literature Sources of Information

English Common Name and Culinary Word Origins

Ayto, J. 1994. A gourmet's guide. Food and drink from A to Z. Oxford University Press, Oxford, UK. 387 pp.

Ayto, J. 1998. The Oxford dictionary of slang. Oxford University Press, Oxford, UK. 474 pp.

Ayto, J., and Simpson, J. 1992. The Oxford dictionary of modern slang. Oxford University Press, Oxford, UK. 299 pp.

Barnette, M. 1992. A garden of words. Times Books, New York, NY. 188 pp.

Barnette, M. 1997. Ladyfingers & nun's tummies. A light-hearted look at how foods got their names. Times Books (Random House, Inc.), New York, NY. 213 pp.

Gove, P.B., and the Merriam-Webster Editorial Staff. 1981. Webster's third new international dictionary. Merriam-Webster, Inc., Springfield, MA. 2662 pp.

Jacobs, J. 1995. The eaten world. The language of food, the food in our language. Carol Publishing Group, New York, NY. 254 pp.

Kays, S.J., and Silva Dias, J.C. 1995. Common names of commercially cultivated vegetables of the world in 15 languages. Econ. Bot. **49**: 115–152.

Morton, M. 2004. Cupboard love. A dictionary of culinary curiosities. 2nd revised edition. Insomniac Press, Toronto, ON. 336 pp. [Information provided is not always accurate.]

Oxford University Press. 1989. Oxford English dictionary. 2nd edition. Oxford University Press, Oxford, UK. 20 vols. [Available as personal computer version, and online, with several updated versions or versions with added material made available since 1989. Widely acknowledged to be the most authoritative dictionary ever compiled.]

Prior, R.C.A. 1879. On the popular names of British plants. 3rd edition. Frederic Norgate, London, UK. 294 pp.

Scientific Plant Name Origins

Brown, W.B. 1956. Composition of scientific words (revised). Published by the author. Printed by Reese Press, Baltimore, MD. 882 pp. [Probably the best basic source of information on Latin and Greek words used in biological names.]

Durant, M. 1976. Who named the daisy? Who named the rose? A roving dictionary of North American wildflowers. Dodd, Mead & Company, New York, NY. 214 pp. [An easy-to-read, popular treatment of many plant names.]

Fernald, M.L. 1950. Gray's manual of botany. 8th ed. Dioscorides Press, Portland, OR. 1632 pp. [Provides information on scientific names of plants of northeastern United States and adjacent Canada.]

Hitchcock, C.L., Cronquist, A., Ownbey, M., and Thompson, J.W. 1955–1969. Vascular plants of the Pacific Northwest. University of Washington Press, Seattle, WA. 5 vols. [Provides information on genus names of plants of Washington, and parts of Oregon, Idaho, Montana, and British Columbia.]

Huxley, A., Griffiths, M., and Levy, M. (*Editors*). 1992. The new Royal Horticultural Society dictionary of gardening. 4 vols. Stockton Press, New York, NY. [Provides information on the origin of genus names of cultivated plants of the world.]

Lewis, C.T., and Short, C. 1879 (reprinted 1958). A Latin dictionary. Clarendon Press, Oxford, UK. 2019 pp. [A monumental dictionary of Latin words and their English translations, especially useful for trying to ascertain the meaning of Latin words when other sources of information prove unhelpful.]

Paxton, J. 1868. Paxton's botanical dictionary. Revision by S. Hereman. Bradbury, Evans, & Co., London, UK. 623 pp. [Obsolete, but occasionally provides information apparently overlooked by more recent reference works.]

Quattrocchi, U. 1999. CRC world dictionary of plant names: common names, scientific names, eponyms, synonyms, and etymology. CRC Press, Boca Raton, FL. 4 vols. [The most comprehensive and authoritative source of information on the origin of genus names of plants.]

Stearn, W.T. 1992*a*. Stearn's dictionary of plant names for gardeners. Cassell, London, UK. 363 pp. [Provides information on Latin names of garden plants.]

Stearn, W.T. 1992*b*. Botanical Latin. 4th edition. David & Charles, Newton Abbot, Devon, UK. 546 pp. [Provides information on the use of Latin in botany, as well as a substantial vocabulary of words commonly used in Latin names.]

Correct Scientific Plant Names

Mabberley D.J. 1997. The plant-book: a portable dictionary of the vascular plants, 2nd edition. Cambridge University Press, Cambridge, UK. 706 pp.

Wiersema, J.H., and Léon, B. 1999. World economic plants. A standard reference. CRC Press, Boca Raton, FL. 749 pp. [An extremely useful guide to correct Latin plant names of economically useful plants. Also lists many common names.]

Food History

Amory, C., Beebe, L.M., Bullock, H.C.D., McCully, H., and the editors of American Heritage. 1964. The American Heritage cookbook and illustrated history of American eating & drinking. American Heritage Publishing Co./Simon and Schuster, Inc., New York, NY. 590 pp.

Kiple, K.F., and Ornelas, K.C. (*Editors*). 2000. The Cambridge world history of food. Cambridge University Press, Cambridge, UK. 2 vols. [Botanical information provided is occasionally flawed.]

Moerman, D.E. 1998. Native American ethnobotany. Timber Press, Portland, OR. 927 pp.

Pillsbury, R. 1998. No foreign food. The American diet in time and place. Westview Press, Boulder, CO. 262 pp.

Prance, G.T., and Nesbitt, M. (*Editors*). 2005. The cultural history of plants. Routledge, New York, NY. 452 pp.

Root, W. 1980. Food. An authoritative and visual history and dictionary of the foods of the world. Simon and Schuster, New York, NY. 602 pp.

Root, W., and Rochemont, R. de. 1976. Eating in America. A history. William Morrow and Company, New York, NY. 512 pp.

Smith, B.D. 1998. The emergence of agriculture. New edition. Scientific American Library, W. H. Freeman & Co., New York, NY. 231 pp.

Tannahill, R. 1973. Food in history. Eyre Methuen, London, UK. 448 pp.

Toussaint-Samat, M. 1992. A history of food. Blackwell Publishers, Cambridge, MA. 801 pp. [Translated by A. Bell, from the French: Histoire naturelle et morale de la nourriture, 1987, Bordas, Paris, France.]

Trager, J. 1970. The enriched, fortified, concentrated, country-fresh, lip-smacking, finger-licking, international, unexpurgated food book. Grossman Publishers, New York, NY. 578 pp.

Wason, B. 1962. Cooks, gluttons & gourmets. A history of cookery. Doubleday & Company, Garden City, NY. 381 pp.

Food Encyclopedia

Anonymous. 1996. The visual food encyclopedia. Éditions Québec/Amérique, Montreal, Canada. 685 pp.

Bader, M.H. 1993. 4001 food facts and chef's secrets. Mylin Enterprises, Auburn, CA. 282 pp.

Bartlett, J. 1996. The cook's dictionary and culinary reference. A comprehensive, definitive guide to cooking and food. Contemporary Books, Chicago, IL. 488 pp.

Chalmers, I. 1994. The great food almanac. Collins, San Francisco, CA. 368 pp.

Chan, H.T., Jr. (*Editor*). 1983. Handbook of tropical foods. M. Dekker, New York, NY. 639 pp.

Claiborne, C. 1985. Craig Claiborne's The New York Times food encyclopedia. Times Books (Random House Inc.), New York, NY. 496 pp.

Coyle, L.P. 1982. The world encyclopedia of food. Facts on File, Inc., New York, NY. 790 pp.

Davidson, A. 1999. The Oxford companion to food. Oxford University Press, Oxford, UK. 892 pp.

Fitzgibbon, T. 1976. The food of the Western World. Hutchinson & Co., London, UK. 529 pp.

Furia, T.E., and Bellanca, N. 1975. Fenaroli's handbook of flavor ingredients. 2nd edition. CRC Press, Cleveland, OH. 2 vols.

Herbst, S.T. 1990. Food lover's companion. Comprehensive definitions of over 3,000 food, wine and culinary terms. Barron's, Hauppauge, NY. 582 pp.

Labensky, S., Ingram, G.G., and Labensky, S.R. 2001. Webster's new world dictionary of culinary arts. 2nd edition. Prentice Hall, Upper Saddle River, NJ. 522 pp. [Possibly the largest collection of culinary terms ever assembled.]

Mariani, J.F. 1983. The dictionary of American food and drink. Ticknor & Fields, New York, NY. 477 pp.

McCully, H. 1967. Nobody ever tells you these things about food and drink. Holt, Rinehart and Winston, New York, NY. 308 pp.

McGee, H. 1984. On food and cooking. The science and lore of the kitchen. Scribner, New York, NY. 684 pp.

McGee, H. 1990. The curious cook. Macmillan, New York, NY. 339 pp.

Stobart, T. 1981. The cook's encyclopedia. Ingredients and processes. Harper & Row, New York, NY. 547 pp.

Voorhees, D. 1995. Why does popcorn pop? And 201 other fascinating facts about food. Citadel Press/Carol Publishing Group, New York, NY. 250 pp.

Plant Encyclopedia

Bailey, L.H. [*Editor and contributor*]. 1909. Cyclopedia of American horticulture. 4 vols. Macmillan, New York.

Bailey, L.H., and Bailey, E.Z. 1976. Hortus third. Revised by staff of L.H. Bailey Hortorium. MacMillan Publishing Co., New York, NY. 1,290 pp.

Bois, D. 1927. Les plantes alimentaires chez tous les peuples et à travers les âges. Paul Lechavalier, Paris, France. 3 vol.

Brouk, B. 1975. Plants consumed by man. Academic Press, New York, NY. 479 pp.
Brücher, H. 1989. Useful plants of neotropical origin and their wild relatives. Springer-Verlag, New York, NY. 296 pp.
Facciola, S. 1998. Cornucopia II. A source book of edible plants. Kampong Publications, Vista, CA. 713 pp.
Hackett, C., and Carolane, J. 1982. Edible horticultural crops. A compendium of information on fruit, vegetable, spice and nut species. Academic Press, New York, NY. 673 pp.
Hanelt, P. (*Editor*). 2001. Mansfeld's encyclopedia of agricultural and horticultural crops. Springer-Verlag, Heidelberg, Germany. 6 vols.
Hedrick, U. P. (*Editor*). 1972. Sturtevant's edible plants of the world. Dover Publications, New York, N.Y. 686 pp. [Reprint of original edition published in 1919.]
Hernández Bermejo, J.E., and Léon, J. (*Editors*). 1994. Neglected crops: 1492 from a different perspective. FAO Plant Production and Protection Series No. 26. Food and Agriculture Organization of the United Nations, Rome, Italy. 341 pp.
Huxley et al. 1992. Listed in SCIENTIFIC PLANT NAME ORIGINS.
Jaques, H.E. 1958. How to know the economic plants. Wm. C. Brown, Dubuque, IA. 173 pp.
Kuhnlein, H.V., and Turner, N.J. 1991. Traditional plant foods of Canadian indigenous peoples—nutrition, botany and use. Gordon and Breach Science Publishers, Philadelphia, PA. 633 pp.
Kunkel, G. 1984. Plants for human consumption: an annotated checklist of the edible phanerogams and ferns. Koeltz Scientific Books, Koenigstein, Germany. 393 pp.
Levetin, E., and McMahon, K. 2007. Plants and society. 5th edition. McGraw Hill, New York, NY. 544 pp.
Martin, F.W. (*Editor*). 1984. CRC handbook of tropical food crops. CRC Press, Boca Raton, FL. 296 pp.
Morton, J.F. 1971. Exotic plants. Golden Press, New York, NY. 160 pp.
Roecklein, J.C., and Leung, P. 1987. A profile of economic plants. Transaction Books, New Brunswick, NJ. 623 pp.
Schery, R.W. 1972. Plants for man. 2nd edition. Prentice-Hall Inc., Englewood Cliffs, NJ. 657 pp.
Seelig, R.A., and Bing, M.C. 1990. Encyclopedia of produce. United Fresh Fruit and Vegetable Association, Alexandria, VA. Irregularly paginated.
Simmonds, N.W. (*Editor*). 1976. Evolution of crop plants. Longman, London, UK. 339 pp. [Includes a number of minor crops not in the 2nd edition, mentioned below.]
Simpson, B.B., and Ogorzaly, M.C. 2001. Economic botany: plants in our world. 3rd edition. McGraw-Hill, New York, NY. 529 pp.
Smartt, J., and Simmonds, N.W. (*Editors*). 1995. Evolution of crops plants. 2nd edition. Longman Scientific & Technical, Burnt Mill, Harlow, Essex, UK. 531 pp.
Tanaka, T. 1976. Tanaka's cyclopedia of edible plants of the world. Keigaku Publishing Co., Tokyo, Japan. 924 pp.
Uphof, J.C. Th. 1968. Dictionary of economic plants. 2nd edition. Verlag Von J. Cramer, Lehre, Germany. 591 pp.
Vaughan, J.G., Geissler, C.A., and Nicholson, B.E. 1997. The new Oxford book of food plants. Oxford University Press, Oxford, UK. 239 pp.
Westphal, E. (*General Editor*) and others. 1989–2001. Plant resources of South-East Asia. Backhuys, Leiden, Netherlands. 16 volumes (some published as more than one book).
Zeven, A.C., and de Wet, J.M.J. 1982. Dictionary of cultivated plants and their regions of diversity. Pudoc, Wageningen, Netherlands. 263 pp.
Zohary, D., and Hopf, M. 1993. Domestication of plants in the Old World. 2nd edition. Oxford University Press, New York, NY. 278 pp.

Fruit

Bose, T.K. (*Editor*). 1985. Fruits of India: tropical and subtropical. Naya Prokash, Calcutta, India. 637 pp.
Morton, J.F. 1987. Fruits of warm climates. Creative Resource Systems, Winterville, NC. 506 pp.
Nagy, S., Shaw, P.E., and Wardowski, W.F. (*Editors*). 1990. Fruits of tropical and subtropical origin. Composition, properties and uses. Florida Science Source, Inc., Lake Alfred, FL. 391 pp.
Nakasone, H.Y., and Paul, R.E. 1998. Tropical fruits. CABI International, Wallingford, Oxon, UK. 445 pp.
Sampson, J.A. 1986. Tropical fruits. 2nd edition. Longman Scientific & Technical, Burnt Mill, Harlow, Essex, UK. 336 pp.
Van Wyk, B.-E. 2005. Food Plants of the World. Timber Press, Portland, OR. 480 pp.

Vegetables

Buishand, T., Houwing, H.P., and Jansen, K. 1986. The complete book of vegetables. Gallery Books, W.H. Smith Publishers, New York, NY. 180 pp.
Halpin, A.M. (*Editor*). 1978. Unusual vegetables. Rodale Press, Emmaus, PA. 443 pp.
Harrington, G. 1978. Grow your own Chinese vegetables. MacMillan Publishing Co., New York, NY. 268 pp.
Herklots, G.A. 1972. Vegetables in south-east Asia. London George Allen & Unwin Ltd., London, UK. 525 pp.
Kay, D.E. 1973. Root crops. Tropical Products Institute, London, UK. 245 pp.
Laws, B., and Green, H. 2004. Spade, skirret and parsnip: the curious history of vegetables. Sutton Publishing, Stroud, UK. 216 pp.

Nonnecke, I.L. 1989. Vegetable production. Van Nostrand Reinhold, New York, NY. 657 pp.
Organ, J. 1960. Rare vegetables for garden and table. Faber and Faber Limited, London, UK. 184 pp.
Rubatzky, V.E., and Yamaguchi, M. 1996. World vegetables. Principles, production and nutritive values. 2nd edition. Chapman and Hall, New York, NY. 843 pp.
Rupp, R. 1987. Blue corn & square tomatoes: unusual facts about common vegetables. Storey Communications, Pownal, VT. 222 pp.
Stephens, J.M. 1982. Know your minor vegetables. Revised edition. Florida Cooperative Extension Service, Institute of Food and Agricultural Sciences, University of Florida, Gainesville, FL. 93 pp.
Splittstoesser, W.E. 1990. Vegetable growing handbook. 3rd edition. AVI Book, Van Nostrand Reinhold, New York, NY. 355 pp.
Tindall, H.D. 1983. Vegetables in the tropics. AVI Publishing Co., Westport, CT. 533 pp.
United States Agricultural Research Service. 1977–1981. Vegetables for the hot, humid tropics. United States Agricultural Research Service, New Orleans, LA and Mayagüez, Puerto Rico. 8 vols.

Herbs and Spices

Crockett, J.U., and Tanner, O. 1977. Herbs. Time-Life Books, Alexandria, VA. 160 pp.
Farrell, K.T. 1990. Spices, condiments, and seasonings. 2nd edition. AVI/Van Nostrand Reinhold, New York, NY. 414 pp.
Grieve, M. 1931. A modern herbal. Reprinted 1978. Penguin Books, New York, NY. 912 pp. [Obsolete but nevertheless often a source of information not available elsewhere.]
Parry, J.W. 1945. The spice handbook. Chemical Publishing, Brooklyn, NY. 254 pp.
Peter, K.V. (*Editor*). 2001. Handbook of herbs and spices. Woodhead Publishing, Cambridge, UK. 319 pp.
Peter, K.V. (*Editor*). 2004. Handbook of herbs and spices. Vol. 2. Woodhead Publishing, Cambridge, UK. 360 pp.
Purseglove, J.W., Brown, E.G., Green, C.L., Robbins, S.R. 1981. Spices. Longman Inc., New York, NY. 2 vols.
Rosengarten, F., Jr. 1969. The book of spices. Livingston Publishing, Philadelphia, PA. 489 pp.
Rosengarten, F., Jr. 1973. The book of spices. Jove Publications Inc., New York, NY. 475 pp. [Revised, abridged edition of Rosengarten 1969.]
Simon, J.E., Chadwick, A.F., and Craker, L.E. 1984. Herbs: an annotated bibliography, 1971–1980. The Shoe String Press, Inc., Hamden, CT. 770 pp.
Small, E. 2006. Culinary herbs, 2nd edition. NRC Press, Ottawa, ON. 1036 pp.
Staples, G.W., and Kristiansen, M.S. 1999. Ethnic culinary herbs. A guide to identification and cultivation in Hawai'i. University of Hawai'i Press, Honolulu. 123 pp.
Stobart, T. 1977. Herbs, spices and flavourings. Penguin Books, New York, NY. 320 pp.

Nuts

Riotte, L. 1975. Nuts for the food gardener. Growing quick, nutritious crops anywhere. Garden Way Publishing, Charlotte, VT. 179 pp.
Rosengarten, F., Jr. 1984. The book of edible nuts. Walker and Company, New York, NY. 384 pp.

Monographs on special groups

Duke, J.A. 1981. Handbook of legumes of world economic importance. Plenum Press, New York, NY. 345 pp. [Reviews of the world's most important species of the legume family (Leguminosae or Fabaceae), many of which are food plants. Written at a technical level for professionals.]
Heiser, C.B. 1979. The gourd book. University of Oklahoma Press, Norman, OK. 248 pp.
Summerfield, R.J. (*Editor*) 1988. World crops: cool season food legumes. Kluwer Academic Publishers, Norwell, MA. 1179 pp.
Whitaker, T.W., and Davis, G.N. 1962. Cucurbits. Interscience Publishers Inc., New York, NY. 249 pp.

Cookbooks for Unusual Food Plants

(Note: There are thousands of books available on the preparation of ethnic and unusual foods. The following is a selection.)

Algar, A.E. 1985. The complete book of Turkish cookery. Kegan Paul International, London, UK. 335 pp.
Arndt, A. 1999. Seasoning savvy. How to cook with herbs, spices, and other flavorings. Haworth Press, New York, NY. 265 pp.
Barash, C.W. 1993. Edible flowers—from garden to palate. Fulcrum Publishing, Golden, CO. 250 pp.
Belsinger, S. 1991. Flowers in the kitchen. Interweave Press, Loveland, CO. 128 pp.
Brennan, J. 1981. The original Thai cookbook. Richard Marek Publishers, New York, NY. 318 pp.
Chantiles, V.L. 1984. The New York ethnic food market guide & cookbook. Dodd, Mead & Co., New York, NY. 370 pp.
Claiborne, C., and Lee, V. 1972. The Chinese cookbook. J.B. Lippincott Co., New York, NY. 451 pp.
Clifton, C. 1984. Edible flowers. McGraw-Hill, New York, NY 94 pp.
Corey, H. 1962. The art of Syrian cookery. Doubleday & Company, Garden City, NY. 186 pp.

Creasy, R. 1982. The complete book of edible landscaping. Sierra Club Books, San Francisco, CA. 379 pp.
Crowhurst, A. 1973. The flower cookbook. Lancer Books, New York, NY. 198 pp.
Dahlen, J., and Phillipps, K. 1983. A popular guide to Chinese vegetables. Crown Publishers Inc., New York, NY. 113 pp.
Day, A., and Stuckey, L. 1964. The spice cookbook. David White Co., New York, NY. 623 pp.
Farah, M. 1979. Lebanese cooking. 4th edition. Lebanese Cuisine, Portland, OR. 159 pp.
Gilbertie, S. 1988. Kitchen herbs. The art and enjoyment of growing herbs and cooking with them. Bantam Books, Toronto, ON. 253 pp.
Hansen, B. 1980. Mexican cookery. H.P. Books, Tucson, AZ. 192 pp.
Hawkes, A.D. 1984. A world of vegetable cookery. Simon and Schuster, New York, NY. 283 pp.
Holt, G. 1989. Recipes from a French herb garden. Stoddard, Toronto, ON. 160 pp.
Hughes, P. 1982. Pueblo Indian cookbook. Museum of New Mexico Press, Sante Fe, NM. 64 pp.
Jaffrey, M. 1975. An invitation to Indian cooking. Vintage Books/Random House, New York, NY. 302 pp.
Jaffrey, M. 1981. Madhur Jaffrey's world-of-the-East vegetarian cooking. Alfred A. Knopf, Inc. New York, NY. 460 pp.
Kennedy, D. 1978. Recipes from the regional cooks of Mexico. Harper & Row, New York, NY. 289 pp.
Kennedy, D. 1989*a*. The art of Mexican cooking. Bantam Books, New York, NY. 526 pp.
Kennedy, D. 1989*b*. The cuisines of Mexico. Harper & Row, New York, NY. 391 pp.
Leonard, J.N. 1971. Latin American cooking. Time-Life Books, New York, NY. 206 pp.
Mallos, T. 1979. The complete Middle East Cookbook. McGraw-Hill, New York, NY. 400 pp.
McCormick [& Company, Inc.]. 1979. Spices of the world cookbook. McGraw-Hill, New York, NY. 431 pp.
Morash, M. 1982. The Victory Garden cookbook. Alfred A. Knopf, New York, NY. 374 pp.
Ngo, B., Zimmerman, G. 1986. The classic cuisine of Vietnam. New American Library, New York, NY. 250 pp.
Owen, M. 1978. A cook's guide to growing herbs, greens, and aromatics. Alfred A. Knopf, New York, NY. 263 pp.
Payne, S., and Payne, W.J.A. 1979. Cooking with exotic fruit. Batsford, London, UK. 144 pp.
Richardson, J. 1990. Worldwide selection of exotic fruits and vegetables. Les Éditions Héritage Inc., Saint-Lambert, QC. 256 pp.
Schneider, E. 1986. Uncommon fruits and vegetables. A commonsense guide. Harper & Row Pub., New York, NY. 547 pp.
Schneider, E. 2001. Vegetables from amaranth to zucchini: the essential reference. William Morrow/HarperCollins, New York, NY. 777 pp.
Sing, P. 1981. Traditional recipes of Laos. Prospect Books, London, UK. 318 pp.
Solomon, C. 1992. The complete Asian cookbook. Raincoast Books, Vancouver, BC. 511 pp.
Thompson, T. 1986. Cajun-Creole cooking. HP Books, Tucson, AZ. 176 pp.
Tsuji, S. 1980. Japanese cooking. Kodansha International/USA Ltd., New York, NY. 517 pp.
Welanetz, D. von, and Welanetz, P. von. 1982. The von Welanetz guide to ethnic ingredients. J.P. Tarcher, Inc., Los Angeles, CA. 731 pp.
Wolfe, L. 1972. Recipes: the cooking of the Caribbean Islands. Revised edition. Time-Life Books, New York, NY. 144 pp.

Edible Wild Plants

Brackett, B., and Lash, M. 1975. The wild gourmet. D.R. Godine, Boston, MA. 160 pp.
Couplan, F. 1998. The encyclopedia of edible plants of North America. Keats Publishing, New Canaan, CT. 584 pp.
Duke, J.A. 1992. Handbook of edible weeds. CRC Press, Boca Raton, FL. 246 pp.
Fernald, M.L., and Kinsey, A.C. (*Revised by* R.C. Rollins). 1958. Edible wild plants of eastern North America. Harper & Row, New York, NY. 452 pp.
Fleurbec (le groupe). 1981. Plantes sauvages comestibles. Le Groupe Fleurbec, Saint-Cuthbert, QC. 167 pp.
Gaertner, E.E. 1995. Reap without sowing. Wild food from nature's cornucopia. The General Store Publishing House, Burnstown, ON. 182 pp.
Gibbons, E. 1962. Stalking the wild asparagus. David McKay Company, Inc., New York, NY. 303 pp.
Gibbons, E. 1966. Stalking the healthful herbs. David McKay, New York, N.Y. 303 pp.
Kavasch, B. 1979. Native harvests—recipes and botanicals of the American Indian. Random House, New York, NY. 202 pp.
Kavena, J.T. 1980. Hopi cookery. The University of Arizona Press, Tucson, AZ. 115 pp.
Kindscher, K. 1987. Edible wild plants of the prairie. An ethnobotanical guide. University Press of Kansas, Lawrence, KS. 276 pp. [1992 edition: 340 pp.]
Kirk, D.R. 1975. Wild edible plants of the western United States. Naturegraph, Healdsburg, CA. 307 pp. + unpaginated pages.
Launert, E., 1981. The Hamlyn guide to edible and medicinal plants of Britain and Northern Europe. Hamlyn, London, UK. 288 pp.
Medsger, O.P. 1974. Edible wild plants. Collier MacMillan, New York, NY. 323 pp. [Reprint of 1939 edition.]

Michael, P. 1980. All good things around us. A cookbook and guide to wild plants and herbs. Holt, Rinehart and Winston, New York, NY. 240 pp.

Niethammer, C. 1974. American Indian food and lore. Macmillan, New York. 191 pp.

Szczawinski, A.F., and Turner, N.J. 1978. Edible garden weeds of Canada. National Museums of Canada, Ottawa, ON. 184 pp.

Szczawinski, A.F., and Turner, N.J. 1980. Wild green vegetables of Canada. National Museums of Canada, Ottawa, ON. 179 pp.

Thompson, B. 1978. Syrup trees. Walnut Press, Fountain Hills, AZ. 164 pp.

Turner, N.J., and Szczawinski, A.F. 1978. Wild coffee and tea substitutes of Canada. National Museums of Canada, Ottawa, ON. 111 pp.

Nutritional Periodicals

Annual review of nutrition [Published since 1981 by Annual Reviews Inc., Palo Alto, CA.]

Current topics in nutrition and disease [Published since 1977 by A. Liss, New York, NY.]

Nutrition reviews [Published since 1942 by Nutrition Foundation, Washington, DC.]

World review of nutrition and dietetics [Published since 1959 by S. Karger, New York, NY.]

Food Plant Periodicals

Economic Botany [Published since 1947 for the Society for Economic Botany by the New York Botanical Garden Press, Bronx, NY. A gold mine of authoritative information.]

Books on Healthy Eating

(Note: There are thousands of books available on the topic of nutritious foods. The following is a selection.)

Clark, N. 2003. Nancy Clark's sports nutrition guidebook. 3rd edition. Human Kinetics, Champaign, IL. 406 pp.

Duyff, R.L. 1998. The American Dietetic Association's complete food & nutrition guide. Chronimed Publishing, Minneapolis, MN. 637 pp.

Ely, L. 2001. Healthy foods: an irreverent guide to understanding nutrition and feeding your family well. Champion Press, Vancouver, WA. 192 pp.

Hausman, P., and Hurley, J.B. 1989. The healing foods: the ultimate authority on the curative power of nutrition. Rodale Press, Emmaus, PA. 461 pp.

Hudnall, M. 1999. Vitamins, minerals, and dietary supplements. Chronimed Publishing, Minneapolis, MN. 111 pp.

Margen, S. 1992. The wellness encyclopedia of food and nutrition: how to buy, store, and prepare every fresh food. Rebus, New York, NY. 512 pp.

Mayo Clinic, University of California, Los Angeles, and Dole Food Company. 2002. Encyclopedia of foods: a guide to healthy nutrition. Academic Press, San Diego, CA. 516 pp.

Medical Economics Books. 1995. The PDR family guide to nutrition and health. Medical Economics, Montvale, NJ. 782 pp.

O'Neil, C., and Webb, D. 2004. The dish on eating healthy and being fabulous. Atria Books, New York, NY. 371 pp.

Ronzio, R.A. 1997. The encyclopedia of nutrition & good health. Facts on File, New York, NY. 486 pp.

Rosenfeld, I. 1995. Doctor, what should I eat? Nutrition prescriptions for ailments in which diet can really make a difference. Random House, New York, NY. 425 pp.

Sadler, M.J., Strain, J.J., and Caballero, B. (*Editors*). 1999. Encyclopedia of human nutrition. Academic Press, San Diego, CA. 3 vols.

Schwartz, R. 2003. The enlightened eater's whole foods guide: harvest the power of phyto foods. Viking Canada, Toronto, ON. 290 pp.

Tribole, E. 2004. Eating on the run. 3rd edition. Human Kinetics, Champaign, IL. 216 pp.

Weil, A. 2000. Eating well for optimum health: the essential guide to food, diet, and nutrition. Knopf, New York, NY. 307 pp.

GLOSSARY

Note: Many words that are widely used and well understood in common English have specific, restricted meanings in technical botanical terminology (e.g., fruit, nut, berry, pod, bulb, root, variety). In general, in this book, such words are used in their common, widely understood senses. Where there is a need for the reader to understand the technical use of terms, they are explained in the text. As reviewers of the preliminary manuscript for this book identified a need to also have some technical meanings available, these are included in this glossary. However, most of this glossary is based on words or phrases that are relatively common in botany and (or) agriculture. As with the vocabulary in the text, this glossary attempts to be comprehensible without burdening the reader with additional technical words. Sufficient information is provided that the reader will appreciate the context of how the word is used in the book.

Accadian An ancient language of a people (called Accadians) thought to have lived in Babylonia. Although obsolete for more than 3700 years, some of the Latin and occasionally also vernacular names used today for plants trace ultimately back to this language.

Acidic Having a pH level of less than 7. Cf. Basic, pH.

Adventitious In botany, refers to organs (such as buds, roots, and shoots) produced in an unusual location. For example, adventitious roots arise from locations other than from the primary root and its branches (adventitious roots regularly arise from the stems of corn and vanilla).

Alkaloid One of a large, heterogeneous group of nitrogen-containing compounds found naturally in plants. These are usually alkaline (they have a basic reaction), and react with acids to form salts. They are usually very bitter, and although often poisonous, may have pharmacological value. Their names generally end in "ine." Examples: caffeine, morphine, nicotine, quinine. The term "alkaloid" is also applied to similar synthetic substances such as procaine.

Allergies, food Inborn or acquired intolerances of certain foods, which usually can not be overcome by becoming habituated to them. Symptoms may include itching, swelling, muscle spasm, and lung and throat tightening as in asthma. The most commonly reported allergenic plant foods are legumes (such as soybeans), tree nuts, and wheat, and the substances responsible are usually proteins. Peanut (see the chapter on it) is the most common cause of life-threatening reactions. Also see Anaphylaxis, Dermatitis.

Amino acids Chemically defined as molecules containing both amine and carboxyl functional groups. Amino acids serve various functions in metabolism, the most important of which is as the building blocks of proteins, which are chains of any of 20 particular amino acids. Proteins are essential agents for promoting metabolic processes, as well as basic construction materials for many components of the body. Eight "essential amino acids" cannot be synthesized from other compounds by humans in amounts needed for normal growth, and must be obtained from food. Four others are semiessential for children because they can only partly synthesize the amounts required. Amino acids, unlike fat and starch, are not stored in the body, and need to be consumed regularly. Animal foods usually supply all of the needed amino acids, but given plant species usually are deficient in one or more essential amino acids. A varied diet is the best guarantee of obtaining not only adequate and balanced levels of amino acids, but also of other constituents that promote health. Also see Protein and the chapter HUMAN NUTRITION IN RELATION TO PLANT FOODS.

Amygdalin A bitter substance present in the kernels of some members of the rose family, including apple, peach, apricot, plum, and especially bitter almond. Under certain conditions, amygdalin is converted into toxic prussic acid (hydrogen cyanide), which can be dangerously poisonous (see the chapter ALMOND). Peach pits, although mildly toxic, were once occasionally consumed in North America (see the chapter PEACH AND NECTARINE). Amygdalin, largely from apricot pits, came to be called the drug Laetrile or "Vitamin B17" (although not a vitamin), and elicited great interest as a cancer tumor destroyer during the 1970s; however, it is as an example of a fraudulent cancer cure (see the chapter APRICOT).

Anaphylaxis An acute allergic reaction that may occur after ingestion, skin contact, or other kind of exposure to an allergen. If very serious or life threatening, it is known as anaphylactic shock. Of all foods, allergies to peanuts are the most frequent cause of anaphylaxis in North America (see the chapter PEANUT for recommendations).

Angiosperms (Angiospermae) Any of hundreds of thousands of mostly terrestrial plants that produce flowers. Cf. Gymnosperm.

Annual A plant that completes its life cycle in one season. Many crops that in their natural habitat grow as perennials or biennials are raised as annuals. This is the case for most domesticated grains, crops like tomato that are insufficiently cold hardy for the local climate, and biennials like carrot that lose much of their food value if allowed to flower and fruit in the second year. True annuals often live for more than 1 year if prevented from setting seed. "Winter annuals" are sown (or naturally self-sow) and germinate in late fall, pass the winter as small plants, and complete their growth the next season.

Anthocyanins Purple, red, or blue water-soluble pigments, belonging to a class of chemicals called flavonoids, found in most tissues of higher plants. They are well known for providing color to some vegetables (e.g., beet, purple eggplant, red cabbage), and fruits (e.g., black currant, blueberry, cherry, cranberry, purple grape, raspberry). Anthocyanins are antioxidants (which see), and so extremely desirable in foods. Accordingly, fruits and vegetables with colorful skins and pulp are considered nutritious. In fruits, the color attracts animals that eat the fruits and disperse the seeds. In flowers, the color attracts pollinators. It has been hypothesized that in leaves and stems, anthocyanins act as a kind of sunscreen.

Antioxidant A substance that prevents or delays deterioration caused by the action of oxygen in the air. Chemically, oxidation consists of an increase of positive charges on an atom or the loss of negative charges. Most biological oxidations are accomplished by the removal of a pair of hydrogen atoms (dehydrogenation) from a molecule. At the cellular level, oxidative reactions produce energy; although crucial for life, they can also be damaging. Along with other oxidizing agents, free radicals (see Free Radical) can damage membranes and other cell components and interfere with regulatory systems. Antioxidants such as vitamins C and E are able to counteract the damage from oxidation from oxygen-based free radicals. Beta carotene (the precursor of vitamin A) is an antioxidant said to protect cells against oxidation damage that can lead to cancer. Some plant foods, such as spinach and blueberries, are rich in antioxidants.

Appellation d'Origine Controlée (AOC) This French phrase translates as "controlled term of origin" and is the French certification granted to geographical areas of France specializing in wines, cheeses, butters, and other agricultural products. Many countries have similar systems modeled on the French AOC classification. Under world trade agreements, a special AOC designation is issued for certain products. For example, in 1996, Mexico received final AOC recognition that the word "tequila" can only be used to designate Mexican tequila. Compare Normas Official Mexicana de Calidad (NOM) and Denomination of Origin (DOA).

Aromatherapy The practice of utilizing the presumed medicinal properties of essential oils extracted from plants. Treatments may be administered through inhalation, external application (e.g., bath, massage, compress, or topically) or (untypically) by ingestion. The essential oils of many spices and edible fruits and herbs are often employed. Users should be aware that often there is a potential for toxicity. Also see Essential Oil.

Axil The junction of a leaf with the stem.

Basic (Alkaline) Having a pH level of more than 7. Cf. Acidic, pH.

Bean An informal term applied most often to the edible seeds (and sometimes fruits) of several species of the legume family (Fabaceae), but also to a variety of seed-like products from species in other plant families, such as coffee beans and cocoa beans, and pod-like products from other plant families, such as vanilla beans (called vanilla pods in Britain). Although plant seeds are usually small, the edible seeds called beans are relatively large. "Green beans" refers to the edible pods of legumes. Cf. Pulse.

Berry See Fruit.

Biennial A flowering plant that completes its life cycle in 2 years. Short-lived perennials sometimes mimic biennials, but only true biennials always die in the second year (however, plant breeders have produced annual cultivars of several biennials). In biennials, vegetative growth occurs in the first year, and the plant becomes dormant, typically passing the winter as a rosette (which see) low to the ground. In the spring of the second year, a flowering stalk is produced (see Bolting), fruits and seeds are developed, and the plant dies. Biennials are much less common than annuals and perennials (which see). Biennials cultivated for edible leaves or roots are grown as annuals, and include beet, Brussels sprout, cabbage, carrot, celery, parsley, and Swiss chard. Biennials cultivated for edible fruits ("seeds") are grown for 2 years, and include celery and parsley.

Bolting Rapid development of a flowering stalk of a herbaceous plant with consequent production of fruits and seeds, and typically the senescence or death of the plant. Bolting may be part of the natural life cycle (mostly of annual and biennial plants), or may be induced by environmental stresses such as drought or heat, which in the case of crops is usually undesirable, as it terminates growth and lowers the quality of edible vegetative parts.

Bran The hard outer layer of grain (i.e., of grass fruit). In milling cereals, bran is usually removed. Its high oil content makes grain susceptible to becoming rancid, and many people find removing bran from cereals makes them more palatable and digestible. However, bran is usually rich in a variety of nutrients, dietary fiber, and antioxidants, and is often collected as a by-product. Bran is often employed to enrich bread, muffins, and breakfast cereals.

Brandy A general term for distilled wine, usually 40–60% ethyl alcohol by volume. Unless otherwise specified, brandy is made from wine grapes and is generally colored with caramel to imitate the effect of aging in wooden casks. However, many fruits, such as peaches, are used to make brandy, in which case the beverage is usually not aged or colored. Brandy is usually consumed as an after-dinner beverage.

Bulb A term often used informally in a very general sense to refer to plants that have underground, fleshy storage structures (whether they produce true bulbs, corms, tubers, tuberous roots, or rhizomes). True bulbs are underground (more or less) storage organs from which the plant grows anew in the next season. Bulbs are formed from the plant's stem and leaves. Most of a bulb is made up of layers of fleshy scales, which are modified leaves, and in the center, there is a bud for the next year's flower stalk. The fleshy scales provide food for the new, growing plant. At the bottom of the bulb, there is a thin, flat disc from which roots grow. Typically, new bulbs are formed around the base of the old bulb. See the chapters on GARLIC and ONION for information on the structure of the bulbs of these plants.

C3 plant The most widespread of three classes of flowering plants that differ in the way they carry out photosynthesis (cf. C4 plant and Crassulacean Acid Metabolism). Over 90% of flowering plants are C3 plants. The first stable compound formed when carbon dioxide is processed by C3 fixation is a three-carbon compound, i.e., "C3." The mode of photosynthesis in C3 plants seems adapted to conditions where the intensity of sunlight is moderate, temperatures are also moderate, and ground water is in good supply.

C4 plant One of three classes of flowering plants that differ in the way they carry out photosynthesis (cf. C3 plant and Crassulacean Acid Metabolism). About 1% of the world's flowering plants are C4 plants. The first stable compound formed when carbon dioxide is processed is a four-carbon compound, i.e., "C4." The mode of photosynthesis in C4 plants seems adapted to conditions of drought and high-temperature limitation, and sometimes also limited availability of nitrogen—conditions often found in the tropics, and indeed many C4 plants grow in tropical areas. C4 plants include amaranth, corn (maize), finger millet, sorghum, and sugarcane. Some C4 plants that originated in hot climates, such as corn, grow well in temperate regions, provided that high temperatures and light intensity are available. Teff (see the chapter on this) is a C4 plan, but is intermediate between tropical and temperate species in its physiology, adapted to warm but not the very hot climates of many C4 plants.

Carbohydrates One of the three major classes of macronutrients (substances eaten regularly in fairly large quantities) by animals; the others are fats and proteins, which see. Carbohydrates are organic molecules in which carbon, hydrogen, and oxygen bond to each other in the ratio one carbon atom to two hydrogen atoms to one oxygen atom. They are the main energy source for the human body, i.e., they are broken down to release energy. Most carbohydrates are manufactured by plants during the process of photosynthesis. The most common carbohydrate is cellulose, the main structural material of plants, which cannot be digested by humans, although cellulose fiber is essential in the human diet to maintain the health of the digestive tract (see Fiber). Cellulose can be digested by bacteria in the digestive tract of cows, and the resulting meat and milk are human foods. All carbohydrates are composed of units of sugar (which see). Two of the most common sugars are glucose and fructose. Glucose is the main form of sugar stored in the human body for energy, while fructose is the main sugar found in most fruits. Table sugar, which occurs for example in sugar cane, is sucrose. Starch, a carbohydrate where the sugar is based on glucose, is widespread in plants, which use it for food storage. Edible plant carbohydrates are mostly transformed in humans into a simpler carbohydrate called glycogen.

Carnivorous Referring to organisms that mainly or exclusively eat meat. Fungi that capture microscopic animals are often called carnivorous. Plants that capture and digest insects are called carnivorous plants, although in reality they synthesize most of their food. Cf. Herbivorous, Omnivorous.

Carotenoids A class of fat-soluble pigments found principally in plants. Carotenoids range in color from pale yellow through bright orange to deep red. In plants, carotenoids absorb light for use in photosynthesis. Animals can not synthesize carotenoids. Human health is enhanced by consuming plant parts rich in carotenoids. For example, beta-carotene from carrots is a precursor of vitamin A, which is essential for good vision. Carotenoids also act as antioxidants (see Antioxidant).

Caryopsis A fruit of members of the grass family (Poaceae), consisting of a single seed with the thin seed coat fused with the fruit wall. Also known as a true grain. In a caryopsis, most of the fruit is the contained seed. The seed components rather than the fruit wall provide nutrition to humans, unlike the case with many fleshy fruits, so it is easy to think of grains as "seeds," and in common language, cereal grains are often called seeds. See Fruit, Bran, Endosperm, Germ.

Catch crop A quick-growing crop sown to make use of temporary idleness of the soil or to compensate for the failure of a main crop. Catch crops are grown simultaneously with, or between, successive plantings of a main crop. For example, quick-growing radishes can be raised between rows of slower-growing vegetables, and harvested before the latter mature. Catch crops can often be planted between the spring harvest and fall planting of other crops.

Celiac disease A condition where the consumption of glutens and related proteins from the closely related cereals wheat, rye, barley, and triticale results in defective absorption of nutrients and damage to the small intestines. Celiac disease affects one in 250 people, usually beginning in infancy. Common symptoms include malnutrition, constipation, diarrhea, bloating, gas, and burping, but there are many other problems that may develop. Treatment involves removing gluten (which see) and its related proteins from the diet. Rice, corn, flax, quinoa, tapioca, amaranth, potato, nuts, and beans are considered safe. Pure oats are also safe, but are difficult to obtain commercially without wheat and other gluten sources. Specialty stores sell gluten-free cereals, bread, and pastries.

Cereal A species of the grass family (Poaceae) that yields abundant seeds that can be eaten by humans. In common language, "cereal" also means a foodstuff prepared from grains, particularly from cereal grasses, but possibly from other seed crops. The principal cereals are rice, wheat, corn, barley, oat, and sorghum. The many species called millet are also cereals, and there are many minor cereals, such as fonio and teff. Compare Grain, Pseudocereal.

Cholesterol The most common steroid in the body, produced in the liver and carried in the bloodstream by lipoproteins. Low-density lipoprotein cholesterol ("bad" cholesterol) is associated with an increased risk of coronary artery disease. High-density lipoprotein cholesterol is "good" cholesterol. Cholesterol, a fat-like substance, is found in all food from animal sources, and is an essential component of body cells and a precursor of bile acids and certain hormones. Negligible amounts of cholesterol are found in plants and fungi. Abnormally high cholesterol levels (a condition called hypercholesterolemia) are associated with cardiovascular disease. Beef, dairy products, and poultry are major dietary sources of cholesterol. For additional information, see the chapter HUMAN NUTRITION IN RELATION TO PLANT FOODS.

Cinderella crop Like the fairy tale character, a crop that arises from obscurity to acquire tremendous status and importance over a short period. For example, soybean in the US and Canola (rapeseed) in Canada, both became extremely important in the 20th century.

Citrus Any of several trees or their fruits, which are characteristically pulpy, thick-rinded, and tartly acidic. The principal citrus fruits are orange, tangerine, lemon, lime, and grapefruit.

Clingstone Referring to a stone fruit with flesh that adheres to the surface of the pit, sometimes making it difficult to remove. A semi-clingstone fruit is intermediate between clingstone and freestone (which see).

Clone In biology, clones are identical copies, whether of DNA fragments (produced by molecular cloning), cells (from cell cloning), or organisms (produced by asexual reproduction). Asexual reproduction usually produces identical individuals, but mutation and other genetic processes can generate genetic differences. Cultivars of almond, apple, banana, peach, potato, grape, and many other crops are often clones.

Common name The name for groups of organisms (usually species) used in common language, i.e., English in this book. See Appendix 1.

Conifer A group of about 700 cone-bearing, woody, terrestrial seed plants, with vascular (water- and nutrient-transporting) tissue. Examples are cedar, Douglas fir, cypress, fir, juniper, larch, pine, redwood, spruce, and yew. The wood of coniferous trees is known as softwood (because it is relatively soft), and is widely employed for timber and paper production. Conifers are of minor use for food (e.g., pine nuts). Most conifers are evergreen (which see) and all are gymnosperms (which see).

Corm A compressed underground storage organ formed from the plant stem from which the plant develops anew in the next growing season. Unlike bulbs, corms are a mass of solid tissue, without separate fleshy scales, although there are usually dry papery, protective, scale-like leaves on the outside. Buds form at the top of the stem, and roots on the underside. Compare Corm, Rhizome, and Tuber. Taro and Malanga are examples of plants with corms.

Cover crop A crop grown to protect soil beneath, prior to replanting. Cover crops are useful to prevent (1) erosion by wind or rain, (2) leaching of nutrients, (3) compaction of the soil, (4) weeds. Cover crops can also be turned-in as a green manure (which see).

Crassulacean acid metabolism (CAM photosynthesis) One of three mechanisms of photosynthesis carried out by different flowering plants (cf. C_3 plant and C_4 plant). This mechanism occurs in about 7% of flowering plants, and is adaptive to arid and semi-arid environments. CAM plants fix atmospheric carbon dioxide mostly at night when temperatures are lower than during the daytime, thereby decreasing the loss of water by transpiration that occurs whenever plants open their pores (stomata) to take in carbon dioxide. For an example, see the chapter TEQUILA AND MEZCAL.

Culinology A term coined by Winston Riley, former president and a founder of the Research Chefs Association (RCA, based in Atlanta, Georgia), to describe a fusion of two disciplines—culinary art and food technology. In practice, the term has been applied since 2004 to preprepared food for a mass market. For a critique, see the chapter HUMAN NUTRITION IN RELATION TO PLANT FOODS. Cf. Slow Food.

Cultivar A kind of cultivated plant that is (1) selected for desirable characteristics that are retained when propagated and (2) given a unique, non-Latinized name. Cf. Variety. Also see Appendix 1.

Cyanogenic Capable of producing cyanide. See Cyanogenic Glycosides.

Cyanogenic glycosides Compounds made up of cyanide bound to a sugar molecule. These serve to protect numerous plants, including such food plants as bitter almond, cassava, lima bean, and sorghum, and are present in quite small amounts in mangos and apple seeds. When fresh material is chewed, enzymes in the animal combine with the cyanogenic glycosides to release hydrogen cyanide, a respiratory poison. When present in food plants, people have learned how to detoxify preparations before consumption.

Deciduous Falling off at maturity, in reference to plant parts. The term is mostly used to refer to plants with perennial aboveground stems from which the leaves fall off, by means of a special zone of weakness in the petiole (leaf stalk), at the end of the growing season. By contrast, evergreen plants have leaves most of which remain functional on the plant for more than 1 year. Cf. Evergreen.

Decoction A method of extracting compounds of interest from plant material by preparing a solution, i.e., soaking the material in a liquid. Some "teas" are sometimes considered to be decoctions. However, decoctions are often distinguished from teas, infusions, and tisanes in that decoctions are usually boiled during preparations. Also, preparing a decoction often first involves mashing the material to facilitate extraction.

Dehiscent The spontaneous opening at maturity of a plant structure, such as a fruit or anther to release the contents upon maturity.

Denomination of Origin (DOA) This refers to law governing legal use of a term with respect to geographical origin. While countries have their own laws governing such names, international trade agreements have increasingly resulted in widespread agreements requiring respect for certain names. In the wine and spirits industry, there are only four drinks recognized with DOA: sherry, cognac, champagne, and tequila. Compare Appellation d'Origine Controlée (AOC) and Normas Official Mexicana de Calidad (NOM).

Dermatitis A general term meaning inflammation of the skin. Skin rash is a very common reaction to both handling and eating plants, and individuals differ greatly in their susceptibility to particular species. Also see Anaphylaxis, Raphide, and Photophytodermatitis.

Endosperm A tissue within the seeds of most flowering plants (notably not in orchid seeds), that surrounds and provides nutrition for the germinating embryo. This is usually starch, but may also contain oils and protein. In some species, the endosperm is absorbed during the development of the embryo, so that none is left in mature seeds (e.g., common bean), but in others, particularly in cereals, there is considerable endosperm present, and it is a very important human food.

Essential oil The "essential" in essential oil implies that the oil contains essential aroma and flavor. Essential oils do not have specific chemical properties in common. They are complex mixtures of volatile (i.e., quickly evaporating) constituents, and so are often called volatile oils (they are also called ethereal oils). By contrast, fixed (non-evaporating oils) are common in plants, often in much larger amounts (particularly in so-called oilseeds). About 2000 species of flowering plants are known to contain essential oil. Orange, lemon, mint, cedarwood, eucalyptus, and clove are among the most popular of essential oils. Essential oils are employed in perfumes, cosmetics, medicines, for scenting cleaning products, and for flavoring food and drink. They are also used in aromatherapy, which see.

Estrogen (oestrogen) The female, steroid, sex hormones. In humans, estrogen is formed particularly in the ovary, but has various functions in both sexes. It is responsible for the development of the female secondary sex characteristics, and during the menstrual cycle, it acts on the female genitalia to produce an environment suitable for the fertilization, implantation, and nutrition of the early embryo. Estrogen is used in oral contraceptives and in the relief of the discomforts of menopause. Estrogen deficiency in women can lead to osteoporosis. Some plants (e.g., wild yams) are sources of precursors of estrogen, or have estrogenic effects (e.g., soybean, flaxseed, some fruits and vegetables).

Evergreen A plant with leaves that remain functional for more than 1 year. Cf. Deciduous.

Fast food Mass-marketed, preprepared foods that can be heated and served quickly but which by and large sacrifice nutrition and taste to achieve cheapness, convenience, and speed. Fast foods are usually too fat, too salty, too sweet, and too monotonous. For additional information, see the chapter HUMAN NUTRITION IN RELATION TO PLANT FOODS. Cf. Culinology, Slow Food.

Fats One of the three major classes of macronutrients (substances eaten regularly in fairly large quantities) by animals; the others are carbohydrates and proteins, which see. Fats are a group of compounds that serve to store energy for the body. Fats also are converted to structural components of cell membranes, and are starting materials for the production of steroids and hormones with various metabolic functions. Less important, body fat serves as insulation for heat and as a cushion protecting delicate internal organs from physical damage. Chemically, fats are molecules made up of a combination of one molecule of glycerol and three fatty acids (which see). A part of fatty acid molecules is made up of chains of linked carbon atoms. If the carbon atoms in these chains are linked to as many hydrogen atoms as possible, the fat is said to be "saturated"; if not, it is "unsaturated." Saturated fats tend to be solid at room temperature, like butter and lard. Saturated fatty acids are abundant in red meat, lard, butter, hard cheeses, and some vegetable oils (particularly tropical oils such as palm, coconut, and cocoa butter, and partially hydrogenated oils). Eating saturated fats is relatively undesirable, as it can lead to arteriosclerosis (hardening of the arteries) and heart disease (see Hypercholesteremia). By contrast, unsaturated fats are usually liquid at room temperature (in which case they are called oils). Unsaturated fats can be either "monounsaturated" (with relatively few hydrogen atoms linked to the carbon chain as described above) or "polyunsaturated" (with relatively more hydrogen atoms linked to the carbon chain). Polyunsaturated fats are found in high concentrations in sunflower, corn, soybean, and flaxseed oils, and also in foods such as walnuts, flaxseeds, and fish. Polyunsaturated fats are healthier than saturated fats, but the healthiest fats for humans are the monounsaturated fats, found in sources such as canola, olive, and peanut oils; avocados; nuts such as almonds, hazelnuts, and pecans; and seeds such as pumpkin and sesame seeds.

Fats, saturated See Fats.

Fats, unsaturated See Fats.

Fatty Acids A fatty acid is a carboxylic acid (an organic acid usually represented by the formula -COOH) attached to a long unbranched chain of carbon atoms, which is either saturated or unsaturated (see Fats). Natural fatty acids range in length of chain from 4 to 28 carbon atoms. The human body can produce all but two of the fatty acids, linoleic acid and alpha-linolenic acid, which are widely distributed in plant oils. Because they must be supplied in food, these fatty acids are called essential fatty acids. Essential fatty acids are used mostly to produce hormone-like substances that regulate many body functions, such as blood pressure, immune response, and response to injury and infections. Fish oils contain long-chain "omega-3 fatty acids," eicosapentaenoic acid and docosahexaenoic acid, and while these are not essential fatty acids, their presence in the human diet is known to promote health. Also see the chapter Human Nutrition in Relation to Plant Foods.

Favism A genetically based disease in which the enzyme glucose-6-phosphate dehydrogenase can not be synthesized, resulting in hemolytic anemia (destruction of red blood cells). About 200 million people suffer from favism, often in regions where malaria is present, suggesting that the condition is associated with some resistance to this disease. Favism is aggravated by eating broad beans.

Fiber In nutrition science, a kind of desirable but indigestible carbohydrate. Fiber is thought to reduce the risk of heart disease, diabetes, diverticular disease, constipation, and other conditions. Good sources include whole fruits and vegetables, whole-grain breads and breakfast cereals, and all types of beans.

Flat bread A simple, compact bread, as found in pizza, made from flattened dough. Flatbreads are often unleavened, i.e., made without yeast so they will not rise.

Flatulence Excessive intestinal gas, a concept difficult to quantify. The average person passes about 2 L of gas each day through burping and farting.

Flowering plants Plants that have flowers. This includes most terrestrial plants on earth, but excludes algae, mosses, ferns, gymnosperms, and some other groups. Most human food and most food that is fed to livestock comes from flowering plants.

Fodder Dried plant material used to feed animals. Often fodder and forage are used as synonyms. Cf. Forage.

Forage Living plants used as food by grazing animals, typically in pastures or wild grasslands. Cf. Fodder.

Free radical Compounds or atoms with an unpaired electron in the outer electron shell, an unstable situation because the compounds or atoms attempt to gain another electron from surrounding materials. Free radicals are essential to the body, used, for example, to destroy invading bacteria and viruses. However, excessive free radicals cause aging and degenerative diseases such as cancer, tumors, cardiovascular diseases, Alzheimers, Parkinsons, and arthritis. Environmental factors such as pollution, radiation, and cigarette smoke also produce free radicals. Antioxidants such as vitamins C and E neutralize free radicals, thereby helping to prevent cell and tissue damage that could lead to disease. See Antioxidant.

Freestone Referring to a stone fruit in which the pit is easily removed from the flesh. Cf. Clingstone. Many peaches sold for consumption as fresh fruit are freestone fruits, since consumers tend to prefer them, while canned peaches tend to be clingstone, because they hold together well in liquid. Numerous other stone fruits come in freestone and clingstone varieties, including plums, apricots, cherries, and nectarines.

Fruit In common language, "fruit" refers to edible, generally fleshy, sweet plant products like apples, oranges, and watermelons that are purchased in the fruit sections of stores and that are typically consumed fresh as desserts or snacks. In technical botanical terminology, fruits are the mature seed-bearing containers produced by plants, and this includes a much wider range of plant structures (such as "grains" as well as many so-called "nuts" and "vegetables") than covered in the first definition, and also excludes a few objects (such as rhubarb, which might be considered to be a fruit in a supermarket, but botanically is a petiole, i.e., a leaf stalk). There is not universal agreement on how the different kinds of fruits should be technically defined, and different authorities assign some fruits to different groups. Key considerations are whether or not the fruit is dry or fleshy; whether it originated only from the ovary or ovaries of a single flower (the part of the flower analogous to the uterus of an animal, i.e., the container(s) in which embryos develop), or in combination with other parts of the plant, or indeed from non-floral parts of the plant; and whether multiple flowers are involved. The term "false fruit" ("pseudocarp," "accessory fruit") is sometimes applied to a fruit like the fig (a multiple-accessory fruit; see below) or to a plant structure that resembles a fruit but is not derived from a flower or flowers. In some fruits, a significant amount of other floral parts makes up the fruit, in which case the fruit is called an accessory fruit. The following classification of fruits is restricted to those for which edible examples are discussed in this book.

A. Dry-walled fruits (originating from one or more ovaries in a single flower).

- Achene. A single-seeded dry indehiscent fruit in which the seedcoat is not part of the fruit coat; e.g., sunflower.
- Capsule. A dry fruit that splits open to release the seeds. This is the most common fruit type; e.g., Brazil nut, cotton.
- Caryopsis. A dry indehiscent fruit, like an achene, but with the seedcoat fused with the fruit coat, occurring in members of the grass family (Poaceae); e.g., barley, corn, fonio, millet, oat, rice, rye, sorghum, teff, triticale, wheat.
- Legume. A dry dehiscent pod that splits on two sides, all in members of the pea family (Fabaceae); e.g., adzuki bean, black-eyed pea, broad bean, chickpea, common bean, hyacinth bean, lentil, lima bean, lupin, mungbean, pea, pigeon pea, peanut, soybean.
- Nut. A large single hardened achene; e.g., filbert, walnut.
- Schizocarp. A dry fruit that splits at maturity into two or more closed, one-seeded parts; e.g., anise, carrot.
- Silique. A dry dehiscent fruit that is long and thin and splits down two long sides, with a papery membrane ("septum") between the halves; e.g., mustard, rapeseed.

B. Fleshy-walled fruits (originating from one or more ovaries in a single flower; in some of these, a significant amount of other floral parts than the ovary contributes to the fruit wall).

- Berry. A fleshy fruit formed from a single flower, usually containing a number of seeds; e.g., avocado, banana, chile pepper, chocolate, currant, eggplant, gooseberry, grape, tomato, pomegranate.
- Drupe or stone fruit. A fleshy fruit formed from a single flower, containing a hard stone which contains one seed; e.g., almond, apricot, cherry, coconut, olive, plum, peach.
- Hesperidium. A berry with a tough, aromatic rind, containing several seeds; includes most citrus fruit; e.g., lemon, lime, grapefruit, orange, tangerine.
- Pepo. A berry with hardened, leathery skin, containing several seeds; all members of the gourd family (Cucurbitaceae); e.g., cucumber, pumpkin and squash, watermelon.
- Pome. A fleshy fruit with a thin skin, formed from a single flower, but developing from another part of the plant than the ovary. The seeds are in central chambers in the center of the fruit. Also called accessory fruits; e.g., apple, pear.

C. Aggregate fruits. Many fruits from different ovaries, all from a single flower; e.g., strawberry, which is not a true berry.

D. Multiple fruits. Many fruits from different flowers, packed or fused closely together; e.g., fig, mulberry, pineapple.

Fungi A kingdom of species that obtains its food energy from living or dead plants and animals, and possesses cell walls made of a material called chitin. Most of the species grow as multicellular filaments called hyphae (collectively a mycelium), and reproduction is commonly by spores. Although traditionally associated with "plants," fungi are quite distinctive. Fungi are significant for human food in the form of mushrooms and as promoters of fermentation of various food products, such as wine, beer, and soy sauce, and in imparting flavors to some cheeses such as Stilton and Roquefort. However, several species of contribute to diseases of crops and livestock. Although only about 70 000 species have been described, it has been estimated that there may be about 1.5 million species.

Germ The term used for the embryo of cereals, which occupies a relatively small part of the grain, and is often milled away to improve keeping qualities and taste. See Whole Grain.

Germplasm Refers to a collection of desired genetic resources for an organism (species or variety). For plants, germplasm is commonly stored as a seed collection, or for trees and plants that are propagated vegetatively, in a nursery. Sometimes germplasm is stored as tissues.

Gluten A mixture of water-soluble proteins (gliaden and glutenin) in the grains of wheat, rye, barley, and some other cereals. In contact with water, gluten forms a sticky mass. It is responsible for the texture, appearance, and volume of dough, giving it the ability to rise by retaining gas released by a leavening agent such as baking powder or yeast. Although important nutritionally, gluten is responsible for celiac disease (which see), that results from an inappropriate immune system response.

Goiter Enlargement of the thyroid gland (usually visible as a swelling in the front portion of the neck). This can be associated with normal, elevated (hyperthyroidism) or decreased (hypothyroidism) thyroid hormone levels in the blood. Both deficiency and excessive intake of iodine can lead to inadequate production of thyroid hormone from the thyroid gland (hypothyroidism). Graves' disease, the most common cause of hyperthyroidism, is the result of too much thyroid hormone. The seaweeds discussed in this book have been used in the past to alleviate goiter caused by insufficient availability of iodine.

Grain The fruit ("seed") of cereal species of the grass family (Poaceae). Sometimes the word "grain" is used to also include edible seeds used like cereal grains but harvested from species not in the grass family. See Pseudocereal.

Grain legume See Pulse.

Green manure A crop that is turned into the soil to add nutrients and organic matter. Plants in the pea family (Fabaceae) are often used as green manure because they add nitrogen to the soil (see Nitrogen Fixation).

Ground cover In agriculture, ground cover refers to plants growing over an area that serve to maintain or improve the soil. Cf. Green Manure. In horticulture (especially in gardening), ground cover often refers to ornamental plants (often low-growing or prostrate) that are used to cover the ground instead of establishing a lawn. In ecology, the ground cover is the layer of vegetation below the shrub layer, often measured by the areal percentage of all plant material covering a region.

Gymnosperm A species of a group of less than 1000 seed-bearing, terrestrial plants with the ovules (essentially analogous to the unfertilized eggs of animals) on the edge or blade of bracts (leaf-like structures), which are usually arranged in cone-like structures. The other major group of seed-bearing plants, the angiosperms (flowering plants), have their ovules enclosed inside a structure. Most gymnosperms are conifers (which see). Gymnosperms are of minor significance as food sources for humans.

Herb A traditional botanical definition of a herb is that it is any nonwoody plant. This gives rise to the adjective "herbaceous," meaning soft-stemmed. However, more broadly a herb is any plant valued for its medicinal properties, flavor, or scent. In Britain in recent years, herb is pronounced "herb," i.e., the "h" is pronounced, and grammatically one writes "a herb." In most of North America, herb is increasingly pronounced "erb," and grammatically one writes "an herb."

Herbivorous Referring to organisms that mainly or exclusively eat plants. When humans choose to avoid meats and mainly or exclusively eat plants, they are called vegetarians (see Vegetarian). Cf. Carnivorous, Omnivorous.

Heterosis (hybrid vigor) A term used in genetics and selective breeding to describe superior characteristics (e.g., size, growth rate, resistance to disease) in hybrids compared to their parents. While heterosis is often developed in hybrids, frequently hybrids are inferior in many respects to their parents. Heterosis is the opposite of inbreeding depression, which produces poorer offspring, and occurs with inbreeding or breeding between very closely related individuals.

Hormone A chemical released from a gland into the bloodstream that affects organs or tissues elsewhere (than the gland) in the body.

Horticulture A term that always refers to the cultivation of crops. It may refer to (1) small-scale growing of crops, (2) the early historical stages of crop cultivation in agriculture, or (3) the growing or ornamental crops. Usages (1) and (3) are most common.

Hydrogenation See Trans Fat.

Hypercholesteremia (hypercholesterolemia, hypercholesterolaemia) The presence of abnormally high, dangerous concentrations of cholesterol in the bloodstream; this can lead to heart disease, hardening of the arteries, heart attacks, and strokes; combated (in part) by a diet high in unsaturated fatty acids. Also see Fats.

Inbreeding Reproduction involving breeding between close (i.e., genetically very similar) relatives or within an individual (selfing). If recessive, deleterious traits are present, over time the result is inbreeding depression. However, many plant species are naturally adapted to selfing, and plant breeders often employ inbreeding to "fix" desired characteristics within a cultivar. Cf. Heterosis.

Inflorescence A stalk or stem, often branched, bearing several to many flowers.

Infusion A solution made by soaking plant material in water or oil to produce a consumable liquid, usually employed for culinary or medicinal purposes after straining away the plant material. Sometimes infusions are called teas. An infusion is usually made by pouring boiling water over the material and letting it steep, and is usually stronger than a tea. Sometimes preparations made by adding concentrated extracts to water are called infusions, but this is not the usual meaning. Also see Decoction, Tea, Tincture.

Internode The part of a stem between nodes. See Node.

Invertebrates Animals lacking a vertebral column, including 97% of all animal species. Cf. Vertebrates.

Kosher Descriptive of foods satisfying the prescribed standards of Jewish dietary laws.

Lacto-ovo vegetarian A person who eats plant-based foods, dairy products (like milk and cheese), and eggs, but avoids all other animal foods. Cf. Vegan.

Land race (landrace; traditional variety) A local variety of a crop or livestock, typically uniquely adapted to local environmental conditions, developed and maintained by farmers in a region. Land races have not been developed by modern plant breeders and are not usually marketed by seed companies. Many of the world's farmers grow land races, although improved varieties are increasingly replacing them. Land races often possess unique genes, which need to be preserved as germ plasm (which see).

Laxative A substance that promotes evacuation of the bowel. The terms "purgative," "cathartic," and "laxative" are often used to indicate progressively gentler action. Laxatives usually cause a more or less normal evacuation of the bowel, usually without griping or irritation.

Legumes Members of the bean family (Fabaceae). Also, their type of dry fruit that splits along two opposite sides at maturity. See Fruit.

Lipid Any of a heterogenous group of fats and fat-like substances characterized by being water-insoluble, extractable by nonpolar (or fat) solvents such as alcohol, ether, chloroform, benzene, etc. Lipids include fatty acids, neutral fats, waxes, and steroids. Lipids are easily stored in the body and serve many functions. For example, they are a source of fuel, and an important constituent of cell structure.

Liqueurs Sweet alcoholic beverages, often flavored with plant materials, and sometimes cream. Liqueurs are usually not aged for long periods. They are difficult to distinguish from cordials, spirits, and sometimes also flavored wines. Liqueurs are usually flavored with herbs, while "cordials" are generally prepared with fruit pulp or juices. Some flavored spirits are much like liqueurs, but are generally not prepared by infusion. Liqueurs are often served with or following dessert, and are also used in cooking.

Malt Germinated grain, usually barley. In the malting process, the germinated grains are dried quickly. Malted grain is used to make beer, whisky, and malt vinegar, and extracts are used in various prepared foods.

Masting A reproductive behavior widely observed in trees, whereby fruit and seed production is very high in some years, between which seed production is relatively low. This phenomenon is widely believed to be a strategy to satiate herbivores in years of high production, leaving many seeds to germinate and produce new plants.

Millet A minor true cereal. "Millets" are various, usually annual, grass crops, the fruits ("seeds") of which are harvested as cereal grain (for humans consumption), or either the seeds or the entire plants are used as animal feed. About 20 species of 10 genera of grasses are called millet. In the widest use of the word "millet," the term includes all true cereals (grasses that produce edible seeds) except those that belong to wheat (species of *Triticum*), barley (*Hordeum*), oat (*Avena*), rye (*Secale*), Triticale (×*Triticosecale*), corn or maize (*Zea*), and rice (*Oryza*). However, some other grass cereals are often not regarded as millets, and the word "millet" is usually not part of their common names (e.g., Fonio, Sorghum, Teff). See the chapter MILLET for the most significant millet species used for human food.

Mineral In geology, a mineral is a naturally occurring substance formed through geological processes that has a characteristic chemical composition, a highly ordered atomic structure, and specific physical properties. However, in biology, a mineral is simply a chemical element needed for growth. Plants require more than 15 mineral elements to complete their life cycles. Those needed only in trace amounts (of which there are at least seven) are called micronutrients. For mineral requirements of humans, see the section Minerals in the chapter Human Nutrition in Relation to Plant Foods.

Monoculture A single crop grown more or less continuously over a wide area (e.g., wheat fields, apple orchards, grape vineyards). The advantages are mostly based on large-scale production. At present, it seems impossible to satisfy the world's demand for food without monocultures, and most monocultures are of crops for which there is a very large demand. Unfortunately, monocultures substantially exclude natural species (i.e., they reduce biodiversity), and the crops are especially susceptible to insects and diseases, often requiring large chemical inputs. Cf. Polyculture.

Monounsaturated fats See Fats.

Mycorrhiza (plural mycorrhizae) Nonparasitic fungi that form symbiotic associations with the roots of higher plants, in which the fungal hyphae are located either on the surface or inside the cells of the roots.

The fungi facilitate the uptake of nutrients, and the plants provide food in the form of carbohydrates for the fungi.

Nitrates Compounds composed of nitrogen and oxygen. Nitrates are abundant in food because plants take up nitrogen from the soil. However, some plants grown under conditions of stress can accumulate potentially toxic levels. Many grasses and forages can accumulate toxic levels of nitrates from the soil they are growing in, depending upon fertilization practices, but this primarily represents a danger to livestock. Like many other leafy, green vegetables, spinach tends to accumulate nitrates, particularly when nitrate fertilizer is added to increase green color and succulence. The result can be a poisoning in the digestive tract of mammals, called methemoglobinemia. This can be serious for human infants who consume large amounts of high-nitrate leafy vegetables. Different varieties, soils, and methods of growing spinach and other plants have been found to influence nitrate accumulation in leafy vegetables, and growers are well aware of the problem.

Nitrogen fixation The conversion of inert nitrogen (N_2) into nitrogenous compounds such as ammonia, nitrate, and nitrogen dioxide. This is carried out naturally by certain bacteria and other microorganisms. Most nitrogen-fixing plants are in the legume family (Fabaceae), and the nitrogen-fixing bacteria occur in nodules on the roots. A few non-leguminous plants can also fix nitrogen, by association with certain microorganisms. Nitrogen can also be artificially fixed for use in fertilizer, explosives, and other products, and indeed, such artificial nitrogen fixation now accounts for most nitrogen used by humans.

Node The part of a stem where one or more leaves are attached. Cf. Internode.

Nodule Pertaining to legumes (Fabaceae), spherical growths on the roots that are inhabited by bacteria that fix nitrogen.

Normas Official Mexicana de calidad (NOM) The Mexican government system of production standards for tequila distilleries that issues a NOM identification number, which should be present on the label to ensure the tequila meets minimum standards. See the chapter TEQUILA AND MEZCAL. Cf. Denomination of Origin (DOA) and Appellation d'Origine Controlée (AOC).

Nurse crop A crop planted with another to facilitate the latter's growth. Often, annual crops such as oats are nurse crops for establishing perennial crops, and often leguminous plants such as alfalfa and clover serve as nurse crops for establishing perennial grasses. Nurse crops can reduce the density of weeds, prevent erosion, and prevent excessive sunlight from reaching tender seedlings.

Nut In technical botanical terminology, a nut is a kind of fruit (see Fruit). In common language, nuts are hard-textured, usually oily, kernel-like edible plant materials housed in a frequently hard shell. Generally, nuts in the common sense are seeds or fruits. Nuts are usually also nuts in the technical sense (i.e., a simple, dry fruit with one (rarely two) seeds, the seeds not attached or fused with the stony or woody fruit wall; e.g., walnut, filbert), but sometimes not (e.g., peanuts, which are a type of legume fruit, are sold as nuts). The edible reproductive structures of some gymnosperms are often called nuts (e.g., ginkgo nuts, pine nuts), although they are not true nuts.

Omega-3 fatty acids Polyunsaturated fatty acids (which see) that are essential (necessary in the diet), available particularly from fish, flaxseeds, walnuts, and oils such as flaxseed, Canola, and soybean. For additional information, see the chapter HUMAN NUTRITION IN RELATION TO PLANT FOODS.

Omnivorous Referring to organisms that naturally eat a combination of meat and plants. Cf. carnivorous, omnivorous. Humans are naturally omnivorous (see the chapter Human Nutrition in Relation to Plant Foods).

Osteoporosis A bone disease that leads to increased risk of fractures, particularly prevalent in women after menopause. Exercise and a diet with adequate calcium and vitamin D are important to prevention and (or) treatment.

Oxalates Chemicals that provide an acidic, tangy taste to many vegetables. However, excessive oxalates in some plant foods (such as rhubarb and spinach) can bind calcium, leading to a medical condition called hypocalcemia in susceptible individuals. People with kidney disorders (especially kidney stones), gout, rheumatoid arthritis, and other conditions are often advised to avoid foods high in oxalic acid or oxalates.

Pepo The fruit of members of the Cucurbitaceae (squash family). See Fruit.

Perennial A perennial plant or perennial is capable of living in nature for more than 2 years. Some people artificially restrict the term "perennial" to herbaceous plants, excluding shrubs and trees. In warm climates, perennials grow continuously, but in regions where there are winters or seasonal drought, growth is limited to the hospitable season. The aboveground portions of herbaceous perennials typically die at the beginning of the winter in cold climates, and are regenerated from underground parts of the plant in the spring. Most perennials flower and fruit over many years, but a few do so only once and then die (see the chapter TEQUILA AND MEZCAL). In tropical regions, perennial crops are much more commonly cultivated than in temperate regions, because they are more efficient at taking advantage of the year-round availability of sunlight.

pH A scale of 1–14, where 7 is neutral, less than 7 is acidic, and more than 7 is basic. The pH of the soil in which plants grow is always of concern, since plant species are usually adapted to growing well in a particular pH range. The pH of food preparations is also of concern, often bearing on taste and food value. See Acidic, Basic.

Photophytodermatitis A rash produced only after exposure to sunlight in some individuals after consuming some food or touching a plant. Several species (including angelica, buckwheat, lemon, and parsnip) can produce this condition in susceptible people. Cf. Dermatitis.

Photosynthesis Biochemical mechanisms that "fix" carbon dioxide (bind the gaseous molecules to dissolved compounds inside the plant) for sugar production. The process captures energy from the sun and uses that energy to synthesize sugar from water and carbon dioxide, which can then be converted by the plant into cellulose, starch, and other organic substances. See C3 Plant, C4 Plant, Crassulacean Acid Metabolism.

Phytochemical Plant-derived chemical compounds, many of which are used as drugs. "Phytonutrients" are nutritional phytochemicals from edible plants.

Pod A dry-walled fruit with a definite line or seam. Often pods are defined as the dehiscent fruit of the family Fabaceae, which at maturity split along two seams to release the enclosed seeds. More generally, pod is an informal term referring to a wide variety of more or less similar fruits.

Polyculture The raising of several crops in the same space, either simultaneously or sequentially. Polyculture may also involve both animals and plants, for example, combining fish farming and seaweed culture. Cf. Monoculture.

Pome Fruit of members of the Maloideae (the apple subfamily of the Rosaceae or rose family). See Fruit.

Protein Any of a group of high molecular weight, complex, organic compounds that contain carbon, hydrogen, oxygen, nitrogen, and usually sulphur. These are the principal constituents of the protoplasm of all cells, consisting mostly of combinations of 20 different linked amino acids. Proteins serve as enzymes, structural elements, hormone, etc. Also see Amino Acids. Proteins are one of the three major classes of macronutrients (substances eaten regularly in fairly large quantities) by animals; the others are carbohydrates and fats, which see.

Pseudocereal A species that yields abundant grain, but does not belong to the grass family; e.g., amaranth, buckwheat, quinoa.

Pulse (grain legume) A crop of the pea family, the seeds of which are collected, dried, and used as human food. A few people enlarge the term to include crops like soybean and peanuts that are used mainly for oil extraction, but are also dried and subsequently eaten. Still others use the term to denote any edible seed or plant of the legume family. Of course, a crop like pea can furnish both dried and fresh seeds. Legume species, the seeds or pods of which are consumed fresh, are classed as "vegetables" although the edible parts are from fruits.

Rancidity The degraded state of fats, oils, and other lipids that have become oxidized. Antioxidants are often added to fat-containing foods in order to retard rancidity. The rate of rancidification can be decreased by storing fats and oils in a cool, dark place with little exposure to oxygen. See Fats.

Raphide (plural raphides) A needle-shaped crystal, usually of calcium oxalate, formed as a metabolic byproduct in some plant cells. Raphides occur in many plant families and in many parts of plants, but they are of particular significance in certain food plants because they can be damaging if consumed, and sometimes cause dermatitis if plants are merely handled. In taro, for example, raphides are abundant in the roots, and in varying amounts in all parts of the plant. Farmers contacting taro experience a stinging sensation on the skin, and eating the crystals causes a burning, stinging, itching sensation in the mouth and throat, and a stomach ache. Cooking is essential to destroy these crystals (although there are varieties of taro that are nearly free of calcium oxalate). Similarly, yams should not be eaten raw, as they may contain raphides just underneath the skin, but these can be eliminated by peeling and cooking.

Ratafia A liqueur or cordial flavored with fruits or seeds such as peach, cherry kernels, or bitter almonds, or other plant materials. Cf. Liqueurs.

Rescue crops Crops used to replace other crops in midseason because the original crops failed.

Resin A hydrocarbon secretion of many plants, particularly coniferous trees. Resins are used in varnishes, adhesives, incense, perfume, and many other products, a few of which are used for flavoring.

Rhizome A swollen stem bearing leaves and roots, which grows horizontally on or just below the soil surface. Rhizomes often serve the plant as food storage organs, as well as a means of spreading and reproducing vegetatively. Rhizomes usually have scaly reduced leaves along their surface, and often resting buds in the axils. A rhizome may be propagated by division after which the resting buds often grow and produce leaves for a new plant. Rhizomes are often mislabeled as "roots." Sugar cane is an example of a plant with rhizomes. A rhizome is similar to a runner, which see for distinction.

Root Plant organs that usually are under the surface of the soil, serving to absorb water and inorganic nutrients and to anchor the plant to the ground. In some plants, roots grow above the ground, serving, for example, to acquire air for the plant or to anchor vines to trees. Roots often also function as food-storage organs.

Rootstock A plant or the basal part of a plant on which some desirable crop (e.g., grapes, roses) is often grafted. Rootstocks are commonly used to grow cultivars of fruiting plants. The rootstock supporting the cultivar has been selected for its ability to support the cultivar, while the cultivar has been selected for its desirable fruit or ornamental characteristics.

Rosette A plant growing close to the ground, the aboveground portion consisting of a circle of leaves growing closely together and radiating like spokes from the short central stem. Rosettes are most often found in biennial plants. See Biennial.

Rotation crops Crops grown in alternation with one or more other crops. Generally, the more desirable crops tend to deplete the soil, while the least desirable ones from a harvest perspective benefit the soil. Crop rotation tends to reduce the buildup of pathogens and pests that often occurs when one species is continuously cultivated.

Runner (stolon) Horizontally growing shoots with quite elongated internodes. They grow either underground or aboveground and sometimes bear leaves and roots, and sometimes new shoots that grow upright. They serve to spread and propagate the plant vegetatively. Strawberry is an example. See Rhizome, which is very similar, but usually has short internodes and rarely produces leaves. Some authorities restrict the term "rhizome" to underground laterally spreading stems, and the term "runner" to aboveground spreading stems.

Saturated fats See Fats.

Schizocarp A type of fruit that develops from multiloculate (i.e., several-chambered) ovaries that separate when ripe into achenes (see Fruit). The separated achenes are called mericarps. In the carrot family (Apiaceae), the mericarps are sometimes called cremocarps. In commonplace language, these usually small, hard fruits are called "seeds."

Scientific name The Latin name assigned to organisms. In the case of plants, this is sometimes called the "botanical name." See Appendix 1.

Scurvy A disease characterized by spongy gums and loose teeth, resulting from a lack of vitamin C. Scurvy was once common among sailors at sea longer than perishable fruits and vegetables could be stored, and by soldiers and explorers similarly isolated for extended periods.

Seaweeds Relatively large ("macroscopic") marine algae, often found near seashores. The three main groups are the red algae, green algae, and brown algae, and sometimes tuft-forming bluegreen algae are also called seaweeds. Many seaweeds are edible or provide edible extracts, as discussed in the chapter on seaweeds.

Seed A plant embryo, often accompanied by nutritive tissues, enclosed in an outer protective coat. Seeds serve to propagate the species. In many species, in addition to the outer protective coat ("seed coat"), there are dry coats derived from the fruit, in which case the entire structure is technically a "fruit," although commonly called a seed. See Fruit.

Shattering A spontaneous release or scattering of the fruits or seeds of plants at maturity. In crop plants, especially cereals, selection has resulted in the fruits or seeds remaining on the plants so that they can be harvested.

Silage Anaerobically fermented, high-moisture fodder for feeding cud-chewing animals like cattle and sheep, usually made from grass crops such as maize and sorghum, using the entire plant or mainly the stalks. Silage can be made from many other field crops.

Slow food A phrase coined in an attempt to counter the negative aspects associated with the development of "fast food." Slow foods are frequently slowly prepared, but are often advantageous in maximizing taste and nutrition. Slow foods use traditional or pure ingredients while avoiding artificial chemical preservatives and taste enhancers. Slow foods are often ethnic meals, and are typically served in a calm, relaxed, family or group atmosphere. For additional information, see the chapter HUMAN NUTRITION IN RELATION TO PLANT FOODS. Cf. Culinology, Fast Food.

Sprouts Sprouted (i.e., germinated) seeds can be highly nutritious because they contain stored food for the young plants into which they will develop. Sprouting is generally carried out in the dark, which increases tenderness. Although many seeds contain toxins, often these are transformed when germination occurs into more edible and palatable substances. Sprouts are a valuable food because of their generally high content of vitamins, mineral, carbohydrates, and proteins, and most sprouted seeds are traditionally consumed raw. For legumes, some of which have toxins, it is sometimes preferable to serve them boiled, steamed, stir-fried, or added to soups. The pea family (Fabaceae) furnishes the most popular of sprouts, mung bean (which see), as well as adzuki bean, alfalfa, broad bean, lentil, pea, soya bean, and others. The mustard family (Brassicaceae) supplies cabbage and mustard sprouts, as well as radish and cress. Several cereals of the grass family (Poaceae) make good sprouts, including barley, rice, rye, triticale, and wheat. Occasionally, commercial sprouts are contaminated with bacteria, and it is essential that production and marketing be conducted with a very high degree of care.

Steroid A group of lipids (which see) characterized by a carbon skeleton with four fused rings. Hundreds of steroids are found in plants, animals, and fungi. They influence numerous processes in the body, for example, acting as hormones in reproductive development.

Stolon See Runner.

Stone fruit A type of fruit with a large, stone-like pit. See Fruit.

Sugars Generally edible, crystalline, carbohydrate substances. Glucose (sometimes called dextrose) is one of the most important molecules serving as an energy source for plants and animals. It is found in the sap of plants (it is a primary product of photosynthesis), and in the human bloodstream where it is called "blood sugar." Corn syrup is mostly glucose. Other important sugars include sucrose (table sugar), lactose, and fructose. Cf. Carbohydrates.

Symbiosis A mutualistic interaction between two species that is beneficial to both.

Tannin A general term for a heterogenous group of compounds; a subclass of chemicals called phenolics. These astringent chemicals are widely distributed in plants, where they act as defensive compounds, counteracting bacteria and fungi by complexing with their proteins. Tannins are responsible for the astringency of certain foods, such as tea. The sensation of astringency is the result of the "tanning" of proteins on the mucous membranes of the mouth, and in the saliva, resulting in reduction of lubrication and a contraction of the surface tissues. Many tannins are considered to be anti-nutrients, because they reduce the amount of protein that is digested, and may also hinder the absorption of iron from foods. However, some tannins, including perhaps some of those in tea, may have healthful functions. Tannins are used in tanning, dyeing, and preparation of ink.

Tea In a strict sense, indeed in a legal sense in many countries, tea is an infusion made by steeping processed leaves, buds, or twigs of the tea bush (*Camellia sinensis*) in hot water for several minutes. "Herbal tea" (= tisane) is an infusion of any other plant similarly made for drinking a beverage with a desirable flavor. However, "flavored tea" made by adding herbs or oils to conventional tea (*C. sinensis*) is always considered to be tea. Sometimes teas are distinguished from decoctions by using as a criterion boiling: teas are not boiled (although boiling water may be poured over material), while decoctions are boiled preparations. Also see Decoction.

Tendril A specialized stem, leaf, or petiole with a thread-like shape that is employed by climbing plants for support and attachment, generally by twining around objects contacted.

Tillers Branches or suckers produced from the base of a plant.

Tinctures In medicine, alcoholic solutions of compounds extracted from plants, or a solution of a nonvolatile substance such as iodine or mercurochrome. Cf. Decoction.

Tisane A herbal tea, usually not as strong as an infusion, often made with flowers. Compare Decoction, Infusion, Tea, Tincture.

Tonic An agent or preparation that stimulates the restoration of physical or mental tone. Tonics improve or promote general health in a slow, continuous manner.

Topical Applied to a certain area of the skin and usually affecting only that area.

Trace element In biochemistry, a chemical element that is needed in minute quantities for the proper growth, development, and physiology of an organism. Also referred to as a micronutrient.

Trans fat (transfat) Fat made by heating liquid vegetable oils in the presence of hydrogen gas, a process called hydrogenation. This makes the oils more stable and resistant to spoilage, and converts them into solids. The oils also withstand repeated heating without breaking down, and are in this respect ideal for frying fast foods. Trans fats have been widely used in commercially prepared baked goods, margarine, processed foods, and fast foods. However, trans fats are very deleterious for human health, raising bad cholesterol and lowering good cholesterol (which see), and increasingly they are being eliminated from the Western diet.

Tuber In a restricted sense, the term "tuber" refers to underground (sometimes aboveground), fleshy storage organs that are formed from stem tissue, and so can easily develop new shoots (e.g., potato, which can produce new shoots from "eyes" (buds) on the surface; many yam species similarly produce stem tubers). However, similar structures formed from roots are also called tubers (e,g., sweet potato), but these typically do not produce new shoots, serving exclusively for food storage. Similarly, "tuberous roots" are simply elongated storage roots that are relatively enlarged with storage tissue but do not form swollen compact structures like tubers. See Yam for information on the air potato, which produces peculiar tubers on the aboveground stems. Corms (which see) are often (erroneously) called tubers.

Unleavened bread See Flat Bread.

Unsaturated fats See Fats.

Variety A category denoting a group of organisms within a species. In common language, a variety refers to a distinctive kind of organism that is recognizable as different from the remaining individuals of the species. Secondly, a variety may be a cultivar (which see). Thirdly, a variety (technically varietas) is a formal scientific category, requiring a Latin name. See Appendix 1 for detailed information.

Vegan A person who adheres to the philosophy of eating foods only of vegetable origin. Strict vegans exclude all animal products from their diet, including eggs and milk. Cf. Vegetarian, Lacto-ovo Vegetarian.

Vegetable A general term sometimes employed to denote plants in general (as opposed to animals and minerals), but usually used to mean edible plant products with limited or no sweetness (in contrast with "fruits") that are consumed in relatively large amounts (in contrast with herbs and spices used in small amounts for flavor). Some vegetables are also used as spices (onions, garlic, and pungent peppers are used in very large amounts in some cultures, but only as flavoring material in other circumstances). Technically, vegetables can be seedlings (e.g., bean sprouts), fruits (e.g., bean pods, corn, cucumber, eggplant, pea pods, pumpkin, squash, sweet pepper, tomato), seeds (e.g., beans, peas), leaves (e.g., celery, lettuce, kale and collard greens, spinach), stems (e.g., asparagus), condensed shoots (stem + leaves: onions, garlic), roots (e.g., parsnip, beet, radish, turnip), stems (e.g., potato), or flower buds (e.g., broccoli, cauliflower) in the restricted technical meanings of these terms. Edible mushrooms (which are fungi) and edible seaweeds (which are algae) are considered to be vegetables.

Vegetarian A person who consumes a diet predominantly or exclusively of plant foods. Cf. Vegan, Lacto-ovo Vegetarian.

Vertebrates Animals with backbones, including fish, reptiles, amphibians, birds, and mammals. Cf. Invertebrates.

Vitamin Carbon-containing compounds essential in small quantities for normal functioning of the body. For vitamin requirements of humans, see the section Vitamins in the chapter HUMAN NUTRITION IN RELATION TO PLANT FOODS.

Whole grain Cereal grains in which the bran and germ as well as the endosperm are retained, in contrast to refined grains, which retain only the endosperm. Whole-meal products are prepared using whole-grain flour. Whole-grain products commonly include oatmeal, popcorn, brown rice, whole-wheat flour, and whole-wheat bread. Recognizing whole-grain products can be difficult, depending on local regulations. Many breads are misleadingly colored brown (often with molasses) to make them look like whole-grain bread. The higher oil content (because the germ is retained) of whole grains makes them more expensive than refined grains, because they are susceptible to rancidification, and will not store as well. Whole grains are nutritionally superior to refined grains, richer in dietary fiber, antioxidants, protein, minerals, and vitamins. However, manufacturers often add vitamins and mineral to refined grain products.

Wine An alcoholic beverage made from the fermentation of grape juice. Other fruits can also be fermented, in which case the wine is usually named after the fruit from which they are produced (e.g., elderberry wine). So-called barley wine and rice wine (e.g., sake), are made from starch-based materials and resemble beer and spirits more than wine.

Wine, fortified Wine to which additional alcohol has been added, most commonly brandy. Popular fortified wines include Sherry, Port, Madeira, and Vermouth. Fortified wines are called dessert wines in the US and liqueur wines in Europe.

Index 1: Complete Index of Common Names[1]

[1]Names of well-known cultivated varieties and of classes and kinds of cultivars are included. A few non-English common names are also included.

Index 2: Complete Index of Scientific Names

Index 3: Culinary Terms and Dishes[1]

[1] Many of the common names that are applied to plants (e.g., chile, chocolate, cinnamon, cocoa, coconut, coffee, citrus) that are listed in Appendix 1 also serve as the words for edible extracts or are the basis of culinary dishes. If a culinary word of interest is not listed here, check Index 1.